North Devon

Mid-Devon

East Devon

Teignbridge

Torbay

Plymouth

South Hams

West Devon

Torridge

Exeter

0   10   20km

64 (Lundy, inset)

---

25 Goodleigh
26 Pilton West
27 Ashford
28 Heanton Punchardon
29 Fremington
30 Barnstaple
31 Instow
32 Westleigh
33 Horwood, Lovacott, Newton Tracey
34 Tawstock
35 Bishop's Tawton
36 Landkey
37 Swimbridge
38 East & West Buckland
39 Filleigh
40 North Molton
41 Twitchen
42 Molland
43 West Anstey
44 East Anstey
45 Bishop's Nympton
46 Queen's Nympton
47 George Nympton

48 South Molton
49 Chittlehampton
50 Atherington
51 Satterleigh & Warkleigh
52 Chittlehamholt
53 Burrington
54 King's Nympton
55 Mariansleigh
56 Romansleigh
57 Meshaw
58 Rose Ash
59 Knowstone
60 Rackenford
61 Witheridge
62 East Worlington
63 Chulmleigh

Torridge
64 Lundy
65 Hartland
66 Welcombe
67 Clovelly
68 Woolfardisworthy
69 Parkham

70 Alwington
71 Abbotsham
72 Northam
73 Bideford
74 Littleham
75 Landcross
76 Weare Giffard
77 Huntshaw
78 Alverdiscott
79 Yarnscombe
80 High Bickington
81 Roborough
82 St Giles in the Wood
83 Great Torrington
84 Monkleigh
85 Frithelstock
86 Buckland Brewer
87 Bulkworthy
88 East Putford
89 West Putford
90 Abbots Bickington
91 Sutcombe
92 Bradworthy
93 Pancrasweek

94 Bridgerule
95 Pyworthy
96 Holsworthy
97 Holsworthy Hamlets
98 Cookbury
99 Thornbury
100 Milton Damerel
101 Newton St Petrock
102 Shebbear
103 Langtree
104 Peters Marland
105 Little Torrington
106 Merton
107 Beaford
108 Ashreigney
109 Winkleigh
110 Dowland
111 Dolton
112 Huish
113 Petrockstowe
114 Buckland Filleigh
115 Sheepwash
116 Black Torrington
117 Bradford

118 Ashwater
119 Hollacombe
120 Clawton
121 Tetcott
122 Luffincott
123 Northcott
124 St Giles on the Heath
125 Virginstow
126 Broadwoodwidger
127 Halwill

West Devon
128 Meeth
129 Iddesleigh
130 Broadwoodkelly
131 Monkokehampton
132 Hatherleigh
133 Highampton
134 Northlew
135 Inwardleigh
136 Exbourne
137 Jacobstowe
138 Sampford Courtenay
139 Bondleigh
140 North Tawton
141 Spreyton
142 South Tawton
143 Sticklepath
144 Belstone
145 Okehampton
146 Okehampton Hamlets
147 Beaworthy
148 Bratton Clovelly

149 Germansweek
150 Thrushelton
151 Bridestowe
152 Sourton
153 Common to Bridestowe & Sourton
154 Lydford
155 Lewtrenchard
156 Coryton
157 Marystow
158 Stowford
159 Lifton
160 Kelly
161 Bradstone
162 Dunterton
163 Milton Abbot
164 Brentor
165 Mary Tavy
166 Peter Tavy
167 Tavistock
168 Lamerton
169 Sydenham Damerel
170 Gulworthy
171 Whitchurch
172 Sampford Spiney
173 Horrabridge
174 Walkhampton
175 Bere Ferrers
176 Buckland Monachorum
177 Meavy
178 Sheepstor
179 Dartmoor Forest
180 Chagford
181 Gidleigh
182 Throwleigh
183 Drewsteignton

Mid-Devon
184 Wembworthy
185 Brushford
186 Eggesford
187 Chawleigh
188 Coldridge
189 Nymet Rowland
190 Lapford
191 Morchard Bishop
192 Kennerleigh

193 Woolfardisworthy
194 Washford Pyne
195 Thelbridge
196 Puddington
197 Poughill
198 Cruwys Morchard
199 Templeton
200 Tiverton
201 Loxbeare
202 Washfield
203 Stoodleigh
204 Oakford
205 Bampton
206 Morebath
207 Clayhanger
208 Hockworthy
209 Huntsham
210 Uplowman
211 Sampford Peverell
212 Holcombe Rogus
213 Burlescombe
214 Culmstock
215 Hemyock
216 Clayhidon
217 Uffculme
218 Kentisbeare
219 Cullompton
220 Willand
221 Halberton
222 Butterleigh
223 Bradninch
224 Silverton
225 Bickleigh
226 Cadeleigh
227 Cheriton Fitzpaine
228 Stockleigh English
229 Sandford
230 Copplestone
231 Down St Mary
232 Zeal Monachorum
233 Bow
234 Clannaborough
235 Colebrooke
236 Hittisleigh
237 Cheriton Bishop
238 Crediton Hamlets
239 Crediton
240 Upton Hellions
241 Newton St Cyres
242 Shobrooke
243 Stockleigh Pomeroy
244 Cadbury
245 Thorverton

Exeter

Teignbridge
315 Tedburn St Mary
316 Whitestone
317 Holcombe Burnell
318 Dunsford
319 Moretonhampstead
320 Bridford
321 Doddiscombsleigh
322 Dunchideock
323 Ide
324 Shillingford St George
325 Exminster
326 Powderham
327 Kenn
328 Ashton
329 Trusham
330 Christow
331 Hennock
332 Bovey Tracey
333 Lustleigh
334 North Bovey
335 Manaton

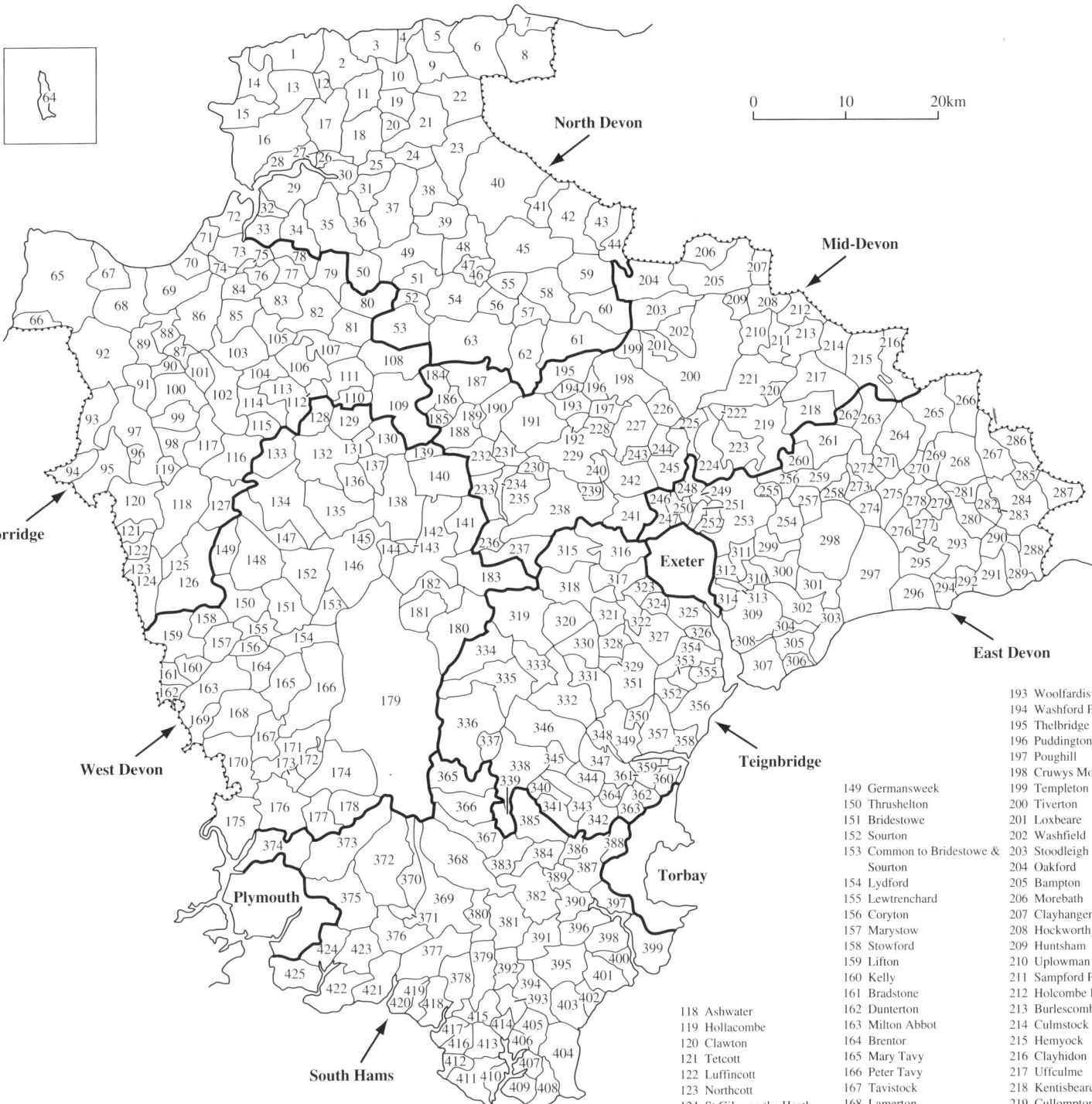

246
247
248
249
250
251
252
253 Broadclyst
254 Whimple
255 Clyst St Lawrence
256 Clyst Hydon
257 Talaton
258 Feniton
259 Payhembury
260 Plymtree
261 Broadhembury
262 Sheldon
263 Dunkeswell
264 Luppitt
265 Upottery
266 Yarcombe
267 Membury
268 Stockland
269 Cotleigh
270 Monkton
271 Combe Raleigh
272 Awliscombe
273 Buckerell
274 Gittisham
275 Honiton
276 Farway
277 Northleigh
278 Offwell
279 Widworthy
280 Shute
281 Dalwood
282 Kilmington
283 Common to Axminster & Kilmington
284 Axminster
285 All Saints
286 Chardstock
287 Hawkchurch
288 Uplyme
289 Combpyne Rousdon
290 Musbury
291 Axmouth
292 Seaton
293 Colyton
294 Beer
295 Southleigh
296 Branscombe
297 Sidmouth
298 Ottery St Mary
299 Rockbeare
300 Aylesbeare
301 Newton Poppleford & Harpford
302 Colaton Raleigh
303 Otterton
304 Bicton
305 East Budleigh
306 Budleigh Salterton
307 Exmouth
308 Lympstone
309 Woodbury
310 Farringdon
311 Clyst Honiton
312 Sowton
313 Clyst St Mary
314 Clyst St George

343 Torbryan
344 Ogwell
345 Bickington
346 Ilsington
347 Newton Abbot
348 Teigngrace
349 Kingsteignton
350 Ideford
351 Chudleigh
352 Ashcombe
353 Mamhead
354 Kenton
355 Starcross
356 Dawlish
357 Bishopsteignton
358 Teignmouth
359 Shaldon
360 Stokeinteignhead
361 Haccombe with Combe
362 Coffinswell
363 Kingskerswell
364 Abbotskerswell

Torbay

South Hams
365 Holne
366 West Buckfastleigh
367 Dean Prior
368 South Brent
369 Ugborough
370 Harford
371 Ivybridge
372 Cornwood
373 Shaugh Prior
374 Bickleigh
375 Sparkwell
376 Ermington
377 Modbury
378 Aveton Gifford
379 Loddiswell
380 North Huish
381 Diptford
382 Harberton
383 Rattery
384 Dartington
385 Staverton
386 Littlehempston
387 Berry Pomeroy
388 Marldon
389 Totnes
390 Ashprington
391 Halwell & Moreleigh
392 Woodleigh
393 Buckland-tout-Saints
394 East Allington
395 Blackawton
396 Cornworthy
397 Stoke Gabriel
398 Dittisham
399 Kingswear
400 Dartmouth
401 Stoke Fleming
402 Strete
403 Slapton
404 Stokenham
405 Frogmore & Sherford
406 Charleton
407 South Pool
408 Chivelstone
409 East Portlemouth
410 Salcombe
411 Malborough
412 South Huish
413 West Alvington
414 Kingsbridge
415 Churchstow
416 South Milton
417 Thurlestone
418 Bigbury
419 Ringmore
420 Kingston
421 Holbeton
422 Newton & Noss
423 Yealmpton
424 Brixton
425 Wembury

Plymouth

HISTORICAL
ATLAS
OF SOUTH-WEST ENGLAND

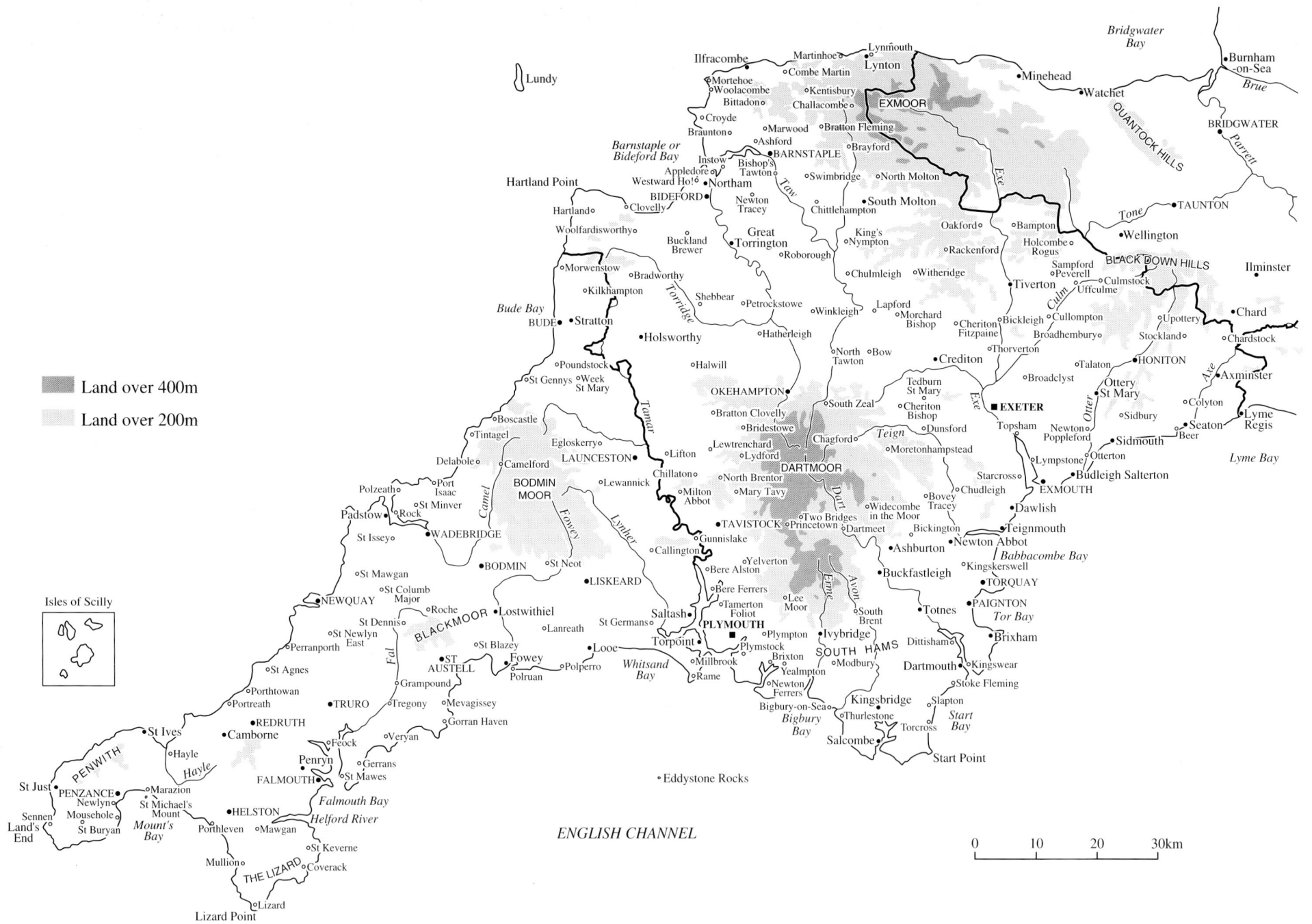

Land over 400m

Land over 200m

Isles of Scilly

Lundy

Ilfracombe
Martinhoe
Lynmouth
Minehead
Watchet
Mortehoe
Woolacombe
Combe Martin
Lynton
Kentisbury
Bittadon
Challacombe
EXMOOR
QUANTOCK HILLS
BRIDGWATER
Burnham-on-Sea
Croyde
Marwood
Bratton Fleming
Ashford
Brayford
Braunton
North Molton
Bridgwater Bay
Brue

Barnstaple or Bideford Bay
Instow
Bishop's Tawton
BARNSTAPLE
Swimbridge
Exe
Appledore
Westward Ho!
Northam
Newton Tracey
Chittlehampton
South Molton
Tone
TAUNTON
Hartland Point
BIDEFORD
Clovelly
Great Torrington
King's Nympton
Oakford
Bampton
Wellington
Hartland
Woolfardisworthy
Buckland Brewer
Roborough
Rackenford
Holcombe Rogus
BLACK DOWN HILLS
Ilminster

Morwenstow
Bradworthy
Shebbear
Petrockstowe
Chulmleigh
Witheridge
Sampford Peverell
Culmstock
Uffculme
Kilkhampton
Torridge
Tiverton
Culm
Bickleigh
Cullompton
Upottery
Chard
Bude Bay
BUDE
Stratton
Holsworthy
Hatherleigh
Winkleigh
Lapford
Morchand Bishop
Cheriton Fitzpaine
Broadhembury
Stockland
Chardstock

Poundstock
Halwill
North Tawton
Bow
Crediton
Thorverton
Ave
HONITON
Axminster
St Gennys
Week St Mary
OKEHAMPTON
South Zeal
Tedburn St Mary
Broadclyst
Talaton
Ottery St Mary
Colyton
Lyme Regis
Boscastle
Bratton Clovelly
Cheriton Bishop
EXETER
Newton Poppleford
Sidbury
Seaton
Beer
Tintagel
Bridestowe
Chagford
Dunsford
Topsham
Sidmouth
Lyme Bay
Delabole
Egloskerry
Lewtrenchard
Lydford
Teign
Moretonhampstead
Starcross
Lympstone
Budleigh Salterton
Camelford
LAUNCESTON
Lifton
Dartmoor
Otterton
Bovey Tracey
Chudleigh
EXMOUTH
Polzeath
Port Isaac
Lewannick
Chillaton
North Brentor
Mary Tavy
Dart
Widecombe in the Moor
Bickington
Dawlish
St Minver
BODMIN MOOR
Milton Abbot
Two Bridges
Princetown
Dartmeet
Newton Abbot
Teignmouth
Padstow
Rock
Camel
Fowey
Lynher
TAVISTOCK
Ashburton
Babbacombe Bay
St Issey
WADEBRIDGE
Gunnislake
Yelverton
Erne
Avon
Buckfastleigh
Kingskerswell
Callington
Bere Alston
South Brent
Totnes
TORQUAY
St Mawgan
BODMIN
St Neot
LISKEARD
Bere Ferrers
Lee Moor
PAIGNTON
Tor Bay
St Columb Major
Roche
Tamerton Foliot
Plympton
Ivybridge
Dittisham
Brixham
NEWQUAY
BLACKMOOR
Lostwithiel
Lanreath
St Germans
Saltash
PLYMOUTH
SOUTH HAMS
Modbury
Dartmouth
St Dennis
St Blazey
Fowey
Looe
Torpoint
Plymstock
Brixton
Yealmpton
Kingswear
Perranporth
St Newlyn East
Fowey
Polruan
Polperro
Whitsand Bay
Millbrook
Rame
Newton Ferrers
Stoke Fleming
St Agnes
AUSTELL
Grampound
Mevagissey
Kingsbridge
Slapton
Start Bay
Porthtowan
Portreath
Tregony
Gorran Haven
Bigbury-on-Sea
Thurlestone
Torcross
REDRUTH
TRURO
Veryan
Bigbury Bay
Salcombe
Start Point
St Ives
Camborne
Fal
Feock
Gerrans
Eddystone Rocks
Hayle
Penryn
St Mawes
PENWITH
PENZANCE
FALMOUTH
Falmouth Bay
St Just
Newlyn
Marazion
Helford River
Sennen
Mousehole
St Michael's Mount
HELSTON
Land's End
St Buryan
Mount's Bay
Porthleven
Mawgan
ENGLISH CHANNEL
Mullion
THE LIZARD
St Keverne
Coverack
Lizard
Lizard Point

St Austell
Holsworthy
ST

0   10   20   30km

HISTORICAL
ATLAS
OF SOUTH-WEST ENGLAND

EDITED BY
ROGER KAIN AND WILLIAM RAVENHILL

CARTOGRAPHY
HELEN JONES

FOREWORD
HRH THE PRINCE OF WALES

UNIVERSITY
of
EXETER
PRESS

SPONSORS

BARING FOUNDATION
THE BRITISH ACADEMY
MR AND MRS CLAUDE PIKE
TOWRY LAW CHARITABLE TRUST
THE ESTATE OF DR RALEGH RADFORD
UNIVERSITY OF EXETER DEPARTMENT OF GEOGRAPHY
UNIVERSITY OF EXETER CENTRE FOR SOUTH-WESTERN HISTORICAL STUDIES

THE EDITORS THANK THE FOLLOWING INDIVIDUALS AND INSTITUTIONS

Editorial Assistant
TODD GRAY

Photography
BARRY PHILLIPS and ANDREW TEED

Archives and Libraries
DEVON COUNTY RECORD OFFICE
CORNWALL COUNTY RECORD OFFICE
WESTCOUNTRY STUDIES LIBRARY EXETER
DEVON AND EXETER INSTITUTION LIBRARY

First published in 1999 by University of Exeter Press
Reed Hall, Streatham Drive, Exeter, Devon EX4 4QR, UK
www.ex.ac.uk/uep/

British Library Cataloguing in Publication Data
A catalogue record of this book is available from the British Library

ISBN 0 85989 434 7

Designed and typeset in Palatino by Topics—The Creative Partnership, Exeter
Printed and bound in Singapore by Tien Wah Press (PTE) Ltd

Front cover and front of slipcase
Satellite map of South-West England reproduced by permission of M-SAT Ltd

Back of slipcase
The Saxton Atlas, Devon map (C7C1) and Cornwall map (C.7.cPG NO. 1)
reproduced by permission of the British Library

For many people, including myself, the South West is the quintessence of rural England, synonymous with holidays and an idyllic countryside and coastline. It is all too easy to forget that this enchanting part of our country also has an important and fascinating economic, social and political history. This book reminds us of those histories, and records for posterity how Devon and Cornwall have become what they are today. All lovers of the South West will, I am sure, be grateful to the University of Exeter for producing this unique atlas.

The team of geographers and historians brought together by the Editors has recorded many unfamiliar treasures from the history of the region, making the Atlas both an anthology to enjoy as well as a work to put on every shelf of reference books in the two counties and beyond. The scale of the task undertaken by the Editors must, at times, have seemed overwhelming and the book is a credit to them and to their scholarship. The University can take pride in this bequest to the next millennium.

Charles

CONTENTS

CHAPTERS, MAPS AND FIGURES

THEMES IN THE HISTORY OF SOUTH-WEST ENGLAND SINCE THE NORMAN CONQUEST

POPULATION

THE CONTRIBUTORS

DR ANDREW ALEXANDER, School of Management Studies for the Service Sector, University of Surrey

✝ MR JAMES BARBER

DR JONATHAN BARRY, Department of History, University of Exeter

PROFESSOR MARK BLACKSELL, Department of Geographical Sciences, University of Plymouth

DR MARK BRAYSHAY, Department of Geographical Sciences, University of Plymouth

MR LEONARD BURGE

PROFESSOR ROGER BURT, Department of History, University of Exeter

DR CHRISTOPHER J. CASELDINE, Department of Geography, University of Exeter

VERONICA CHESHER, Department of Continuing and Adult Education, University of Exeter

DR BRUCE I. COLEMAN, Department of History, University of Exeter

DR GRACE DAVIE, Department of Sociology, University of Exeter

DR MICHAEL DUFFY, Department of History, University of Exeter

DR STEPHEN FISHER, Department of History, University of Exeter

DR HAROLD S.A. FOX, Department of English Local History, University of Leicester

CYNTHIA GASKELL BROWN, Mount Edgcumbe House, Cornwall

DR PETER GAUNT, Department of History, University College Chester

DR SANDY GERRARD, English Heritage

ANDREW W. GILG, Department of Geography, University of Exeter

DR TODD GRAY, Department of History, University of Exeter

PROFESSOR JUSTIN GREENWOOD, School of Public Administration and Law, Robert Gordon University, Aberdeen

FRANCES GRIFFITH, Devon County Council

✝ DR FREDERICK L. HARRIS

MICHAEL A. HAVINDEN, Department of History, University of Exeter

DR DEREK HEARL, Department of Sociology, University of Exeter

CHRISTOPHER HENDERSON, Exeter Archaeology

DR ROBERT A. HIGHAM, Department of Archaeology, University of Exeter

PROFESSOR CHRISTOPHER HOLDSWORTH, Department of History, University of Exeter

DR DELLA HOOKE, Institute for Advanced Research in the Humanities and Social Sciences, University of Birmingham

DR PETER HOWARD, Department of Art and Design, University of Plymouth

PROFESSOR ROGER J.P. KAIN, Department of Geography, University of Exeter

DR JOHN KANEFSKY, Department of History, University of Exeter

DR VALERIE A. MAXFIELD, Department of Archaeology, University of Exeter

IAN MAXTED, Exeter Central Library

PROFESSOR MALYN NEWITT, Department of Portuguese and Brazilian Studies, King's College London

DR RICHARD R. OLIVER, Department of Geography, University of Exeter

PROFESSOR NICHOLAS ORME, Department of History, University of Exeter

DR OLIVER PADEL, Department of Anglo-Saxon, Norse and Celtic, University of Cambridge

DR PHILIP PAYTON, Institute of Cornish Studies, University of Exeter

PROFESSOR SUSAN PEARCE, Department of Museum Studies, University of Leicester

DR MARTIN PURVIS, School of Geography, University of Leeds

HENRIETTA QUINNELL, Department of Continuing and Adult Education, University of Exeter

✝ PROFESSOR WILLIAM L.D. RAVENHILL

DR ALISON ROBERTS, Ashmolean Museum, Oxford

PROFESSOR MICHAEL RUSH, Department of Politics, University of Exeter

DR ROGER R. SELLMAN

DR GARETH SHAW, Department of Geography, University of Exeter

DR TERRY R. SLATER, School of Geography and Environmental Science, University of Birmingham

ROBIN STANES, Department of Continuing and Adult Education, University of Exeter

JEFFREY STANYER, Department of Politics, University of Exeter

DR DAVID STARKEY, Department of History, University of Hull

PROFESSOR ALLAN STRAW, Department of Geography, University of Exeter

PROFESSOR CHARLES THOMAS, Institute of Cornish Studies, University of Exeter

DR FRANK THORN

PROFESSOR MALCOLM TODD, Trevelyan College, Durham

PROFESSOR ALLAN WILLIAMS, Department of Geography, University of Exeter

DR SARAH A.H. WILMOT, Darwin Correspondence Project, Cambridge University Library

HISTORICAL ATLAS
OF
SOUTH-WEST ENGLAND

INTRODUCTION

ROGER J.P. KAIN

All historical atlases help lay to rest that most unhelpful (and politically incorrect) of sayings: 'geography is about maps, history is about chaps'. This *Historical Atlas of South-West England* is all about writing history through the medium of maps. The South West has a long and rich history, which we have tried to reveal in a fresh light by the interdisciplinary approaches of geography and history, and with maps.

In 1954 south-west England was characterized by Arthur Davies, then professor of geography at Exeter University, as 'a peninsula surrounded on three sides by the sea'. His tautological description still brings a smile to some faces, but it does serve to stress that there are few places in south-west England where human activities are unaffected by the interplay of land and sea. The South West is also a peripheral region of Britain, but it is one that is far from remote and backward in cartographical terms—it has a rich legacy of early maps. John Norden's manuscript maps of Cornwall (*c.*1597) occupy a special place in the history of English county cartography (**Maps 1 and 2**), Joel Gascoyne's map of Cornwall (1699) was the first county map to portray parish boundaries (**Map 3**), Benjamin Donn's map of Devon (1765) was the first recipient of the (Royal) Society of Arts prize for a truly accurate map of an English county (**Map 4**), while the Ordnance Survey conducted some experimental mapping around Plymouth from which its nineteenth- and twentieth-century national topographical map series were to grow. Now south-west England is the canvas for the first historical atlas of a major region of England.

The *Historical Atlas of South-West England* project has been nurtured by the University of Exeter Centre for South-Western Historical Studies. It grew out of conversations in 1986 between the editors and Michael Havinden and Bob Higham (founder chairman and director respectively of the Centre) and Malcolm Todd, then professor of archaeology at Exeter University. The Centre was founded on a belief in the intellectual value of the regional scale for historical research, both as a context for local studies and also for rooting national history in regional experiences. A project to compile a historical atlas of the region embodied these ideals.

South-west England is defined for the purposes of the Atlas as the counties of Cornwall, Devon, a small part of west Somerset (to include all of Exmoor), and the Isles of Scilly. Many maps in the Atlas confirm that there is more to the South West than the simple sum of its two principal county components with their different cultural-political histories. Nor is the River Tamar which divides Devon and Cornwall the only powerful sub-regional boundary. There are socio-economic distinctions between the north and south of the peninsula as a whole, between urban and rural, and especially between coastal and inland areas that are as remarkable as some of the contrasts between Devon and Cornwall. And if one looks down on to the landscape of the South West from the viewpoint of an earth-orbiting satellite (see front cover), there does seem to be as much that unites as divides south-west England. In short, the Atlas argues for the coherence of the South West while also analysing and revealing its spatial differentiation.

The temporal coverage extends from prehistory through the whole of the historic period up to the 1990s. The Atlas opens with chapters which review, first, the physical context of human activity, and second, the influence of geology on the style and form of the built environment. The pre-Conquest content of the Atlas which follows is organised chronologically, but thereafter our reconstruction of human occupation in the South West is structured thematically: administrative divisions, politics and military affairs, religion, education, language, dissemination of knowledge, agriculture, mining and manufacturing, inland transport, maritime activities, towns, and service industries (tourism and retailing). Finally, for reasons of their particular importance, the two cities of Exeter and Plymouth are accorded separate treatment as well as being included in the thematic chapters. Unfortunately a few notable gaps remain within this broad agenda: the maritime history of Cornwall in the later seventeenth and eighteenth centuries, for example, has yet to find its researcher, so we have no maps.

Maps are the rationale of any historical atlas. An overwhelming majority of those in the *Historical Atlas of South-West England* are purpose-drawn; a few are photographic reproductions of contemporary maps. The endpapers are maps of the present-day parishes of Devon and Cornwall. These maps, together with the general map of south-west England placed as the frontispiece, are intended to serve as general reference maps. The diagrams which support the maps in some chapters are mostly graphs which present data for which temporal change rather than spatial variation is a key element. Finally, many chapters contain additional

Map 1
John Norden's general map
of Cornwall, *c*.1597.

SOURCE: reproduced by
permission of the Master and
Fellows of Trinity College,
Cambridge.

Map 2
John Norden's map of Kerrier hundred, Cornwall, *c.*1597.

SOURCE: reproduced by permission of the Master and Fellows of Trinity College, Cambridge.

illustrations by way of embellishment. These are almost all pictures, drawings and photographs contemporary with the time period reviewed in the particular chapter.

A fundamental concept underpinning the Atlas is the idea that the spatial distribution of features plotted out on maps can aid our understanding and help explanation—as demonstrated by the analysis of the distribution of archaeological finds in relation to components of the physical environment. For some purposes, a map is an indispensable method of communication; for the analysis of town plans, for example, maps are both the source of data and the only practicable way of presenting the results. Maps, though, can be slippery media. In the history of mining in Devon and Cornwall, for instance, they impart a spurious impression of a decline in activity in the nineteenth century as company amalgamations left fewer 'mines' to map, though overall output actually increased dramatically.

The chapter texts are not intended merely to duplicate in words what is shown graphically on the maps and diagrams, though most chapter texts do draw attention to salient elements of the mapped distributions, especially where patterns are complex.

The texts also provide some socio-economic and political context for the events or phenomena which are mapped and introduce other, non-map material. They provide a brief discussion of source material for maps where this is especially problematic—as in compiling maps of population in the pre-nineteenth-century period. Chapter texts also reveal the mapping methodology and data assumptions where these are not clear from the maps themselves. Authors provide some short lists of further reading for those who wish to take a particular topic further. Detailed notes and references for each chapter are gathered at the end of the book.

The *Historical Atlas of South-West England* is a truly collaborative venture. About half of the contributors are drawn from within the University of Exeter and the remainder are researchers at other universities who specialize on south-western topics. A majority of the contributors are geographers, archaeologists and historians, but there are also valuable contributions from political scientists, sociologists, educationalists and the region's museum, library and archive services. All authors have given their time generously, especially those who

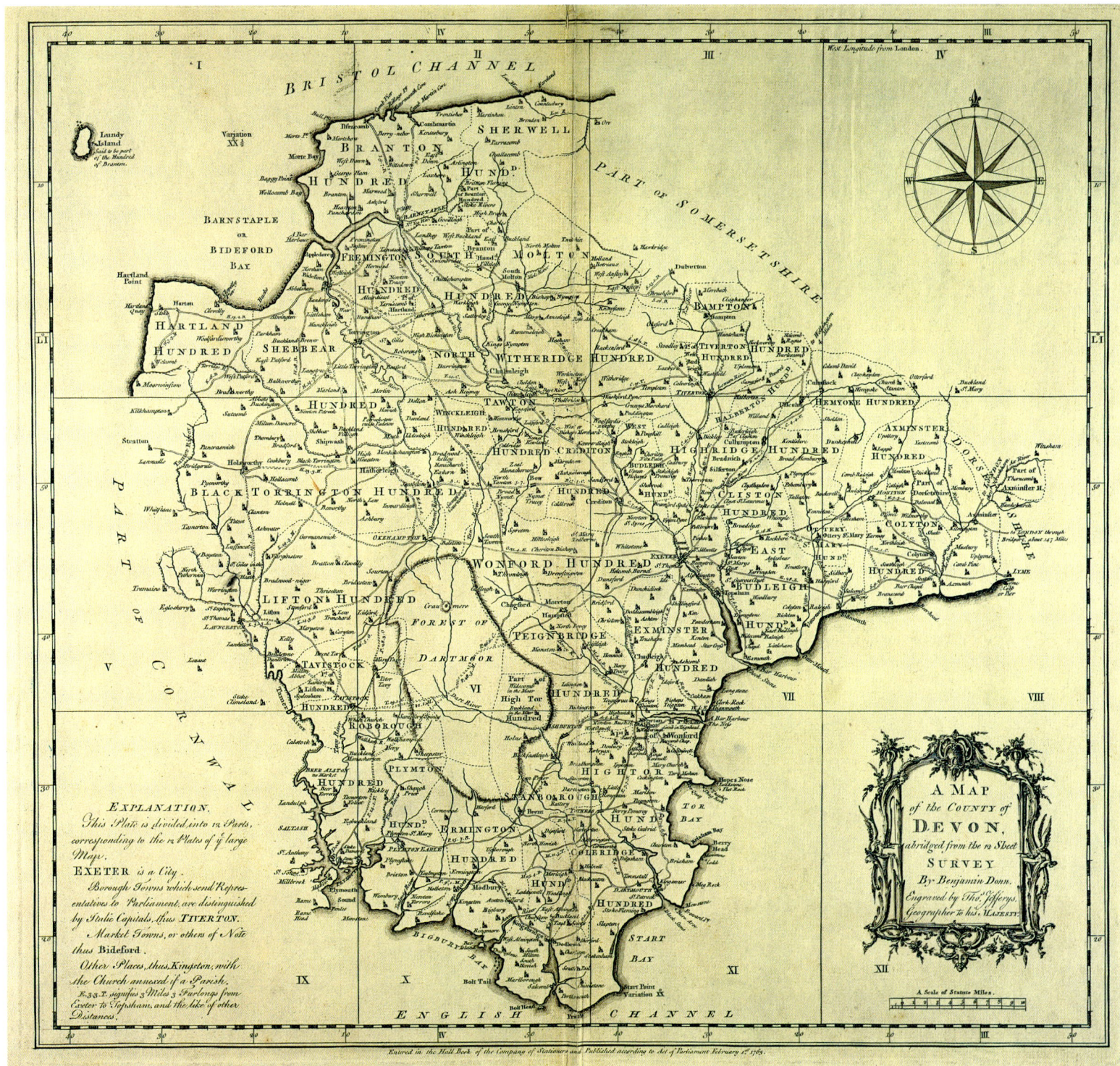

Map 4
A Map of the County of Devon abridged from the 12 Sheet Survey by Benjamin Donn, 1765.

SOURCE: reproduced by permission of the Department of Geography, University of Exeter.

have contributed several chapters. I am grateful to them all. Each author was given the opportunity to frame his or her own contribution within broad parameters set by the editors. In the terminology of today, the Atlas is very much a 'bottom-up' venture. The editorial hand has in most chapters been applied with restraint. It was our great good fortune to recruit a team of contributors who are both acknowledged specialists and who were prepared to be patient and stay with what by any definition has been a complex and long-term project.

The Atlas did not only need a band of willing contributors but also required financial backing, particularly to pay for some nine years of cartography. The editors' special gratitude goes to the British Academy, the Baring Foundation and the Department of Geography, University of Exeter for grants which underpinned the cartographical effort. More recently, generous donations from the University of Exeter Development Appeal, Mr and Mrs Claude Pike, the Estate of Ralegh Radford and the Towry Law Charitable Trust have enabled us to introduce some colour printing, and have made it possible for the University of Exeter Press to offer the Atlas for sale at a price which should be within the reach of most interested individuals. A further donation from Towry Law Charitable Trust has gifted a copy of the Atlas into the library of every secondary school in Devon, Cornwall and the Isles of Scilly. I am especially grateful for all this financial support.

The *Historical Atlas of South-West England* has been with me for twelve years. For the most part it has been deeply satisfying, not least for the contact it has brought with so many first-rate collaborators. And there was the pleasure of witnessing its gradual coming into being as Helen Jones worked on the maps. Helen has a real gift for visualization and also possesses the skills with both mapping pen and computer to put that vision into effect. Her commitment and dedication to achieving the best possible result has been unwavering throughout. That she kept her cheerful sense of humour during the many months of painstaking checking, editing and rechecking the maps was little short of miraculous. An atlas stands or falls on the quality of its cartography. I know that the contributors to our Atlas are as one in sharing my admiration of Helen's work.

In recent months, the Atlas has benefited from the professional skills of a number of others whom I would like to acknowledge by name. Barry Phillips and Andrew Teed took most of the studio and location photographs, Jane Raistrick copy-edited a complicated text with great care, Andy Jones supervised the typesetting and design on behalf of University of Exeter Press where Anna Henderson was a reassuring presence as she directed the production schedule. Both Simon Baker, University of Exeter Publisher, and David Rogers, Director of Development at the University, kept faith with the Atlas, which was essential for its completion. I thank HRH Prince Charles for his gracious foreword.

Todd Gray, already a contributor, latterly took on much more on behalf of the Atlas. From the summer of 1997 to spring 1998 Todd was employed as editorial assistant and researcher and his energy gave much-needed impetus to the final stages of the project as he galvanised us all to action and brought the home straight within reach. Much of his work lies unseen in the project's correspondence files but the product of one part of his work is highly visible. His broad knowledge of the history of south-west England and his familiarity with its libraries and archives brought to light most of the non-map illustrations for the Atlas which do so much more than just enliven the pages.

Undoubted joys there have been. One of the great sadnesses is that six people associated with the project did not live long enough to see it published. Margaret Hewitt and Jeff Porter died before they could complete their work; James Barber and Fred Harris have their contributions published posthumously. Rodney Fry, technical superintendent of the Geography Department until his untimely death in April this year, offered much wise advice at the inception of the project and was a loyal friend and supporter throughout. They are all missed—but none more so than Bill Ravenhill. Bill was part of the Atlas from the very beginning. He and I established its outline together and recruited its contributors together. We fund-raised together and worked on its contents together right up to his death on 9 October 1995. Much has been written about Bill's contribution to the archaeology, historical geography and the history of cartography of south-west England and this is not the place to say more, other than to note that it was he who brought to public notice the unique and nationally significant south-west England cartographical contributions of John Norden, Joel Gascoyne and Benjamin Donn which illustrate this introduction. With the publication of this Atlas, Bill Ravenhill has been associated with another cartographical 'first' for the South West. This *Historical Atlas of South-West England* invites the whole region to look back to its past as it also looks to its future on this, the eve of a new millennium.

Exeter
21 June 1999

THE PLATES

SEDIMENTARY ROCKS

g go	Oligocene Palaeogene	TERTIARY
ku	Upper	CRETACEOUS
kl	Lower	
jl	Lower	JURASSIC
t		TRIASSIC
pu	Upper	PERMIAN
pl	Lower	
cw	Westphalian } Upper	CARBONIFEROUS
cn	Namurian	
cl	Dinantian } Lower	
dc		DEVONIAN-CARBONIFEROUS
du	Upper	DEVONIAN
dm	Middle	
dl	Lower	
e	Undivided	LOWER PALAEOZOIC

METAMORPHIC ROCKS

B	Undivided	BASEMENT

IGNEOUS ROCKS
Extrusive

c	CARBONIFEROUS AND PERMIAN
d	DEVONIAN

Intrusive

D	Acid
G	Basic
U	Ultrabasic

— Intrusive dyke, undifferentiated

—— Geological boundary

Fault at surface } Crossmark on
Fault at depth } downthrow side

Thrust, slide or shear zone

0 10 20 30km

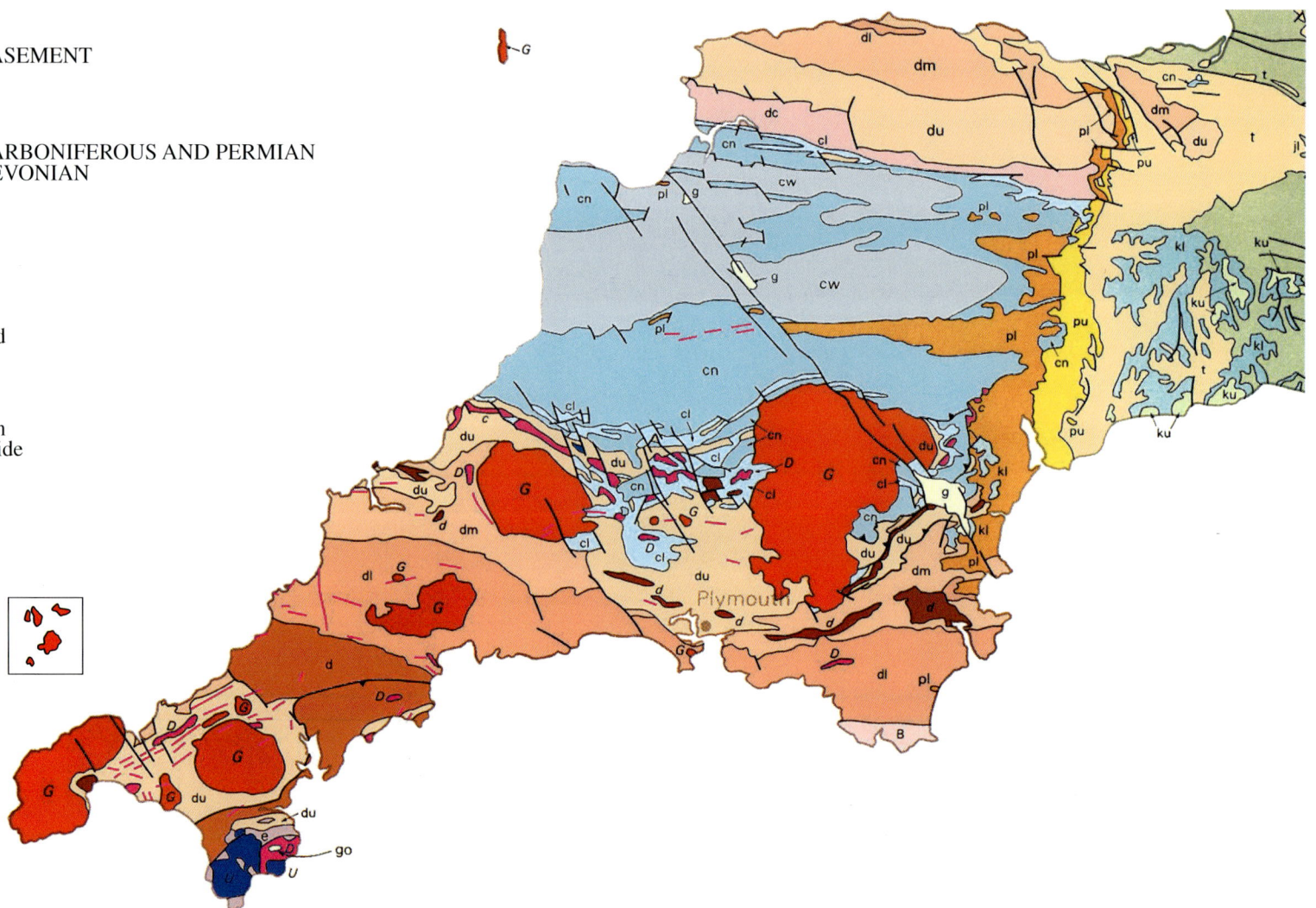

Map 1.1
Geology.

(See Chapter 1, page 26.)

SOURCE: reproduced by permission of *British Geological Survey* (BD/IPR/12–22), © NERC, all rights reserved.

SEDIMENTARY ROCKS

gonp	Pliocene-Oligocene	NEOGENE AND PALAEOGENE
pt	Undivided	TRIASSIC AND PERMIAN
dc	Undivided	CARBONIFEROUS AND DEVONIAN

d	Undivided
du	Upper
dm	Middle
dl	Lower

DEVONIAN

METAMORPHIC ROCKS (of possible Lower Palaeozoic age)

ZTQ	Treleague Quartzite
ZMS	Lizard Mica Schists (equivalent to Old Lizard Head Series)
ZHS	Lizard Hornblende Schists (including Traboe Schists and Granulites and Landewednack Schists)
ZKG	Lizard Gneisses (including Kennack Gneiss and Man o'War Gneiss)

Lizard Complex

IGNEOUS ROCKS
Extrusive

| (lava symbol) | Basic Lava | In Devonian |

Intrusive

G	Granite
F	Elvan (mainly quartz porphyry and microgranite)
D	Dolerite and Greenstone
E	Gabbro and Troctolite
U	Serpentine and Lamprophyre (L)

| ʚ | Basement, undivided (offshore area only). Probably mainly metamorphic rock related to the Lizard Complex |

LITHOLOGICAL ORNAMENT

Quartzite
Pelite
Lava

———— Geological boundary

—⊥— Fault or thrust; crossmark on downthrow side where known

++++ Outer margin of metamorphic aureole

• ₂₀ Offshore core with registered IGS number

↗+60 Mean direction of stable natural remanent magnetization

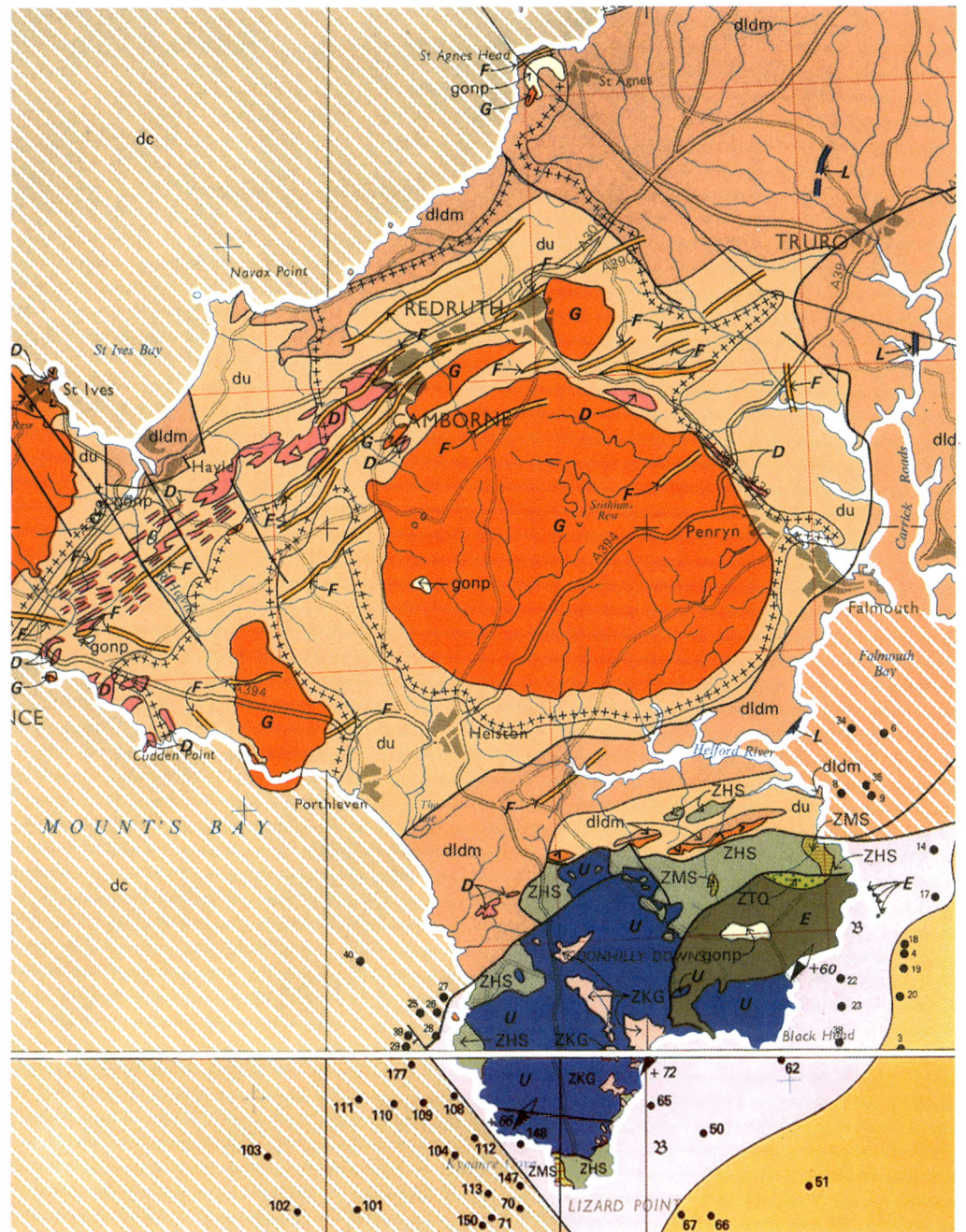

Built-up area

0   5   10km

Map 1.2
Geology of the Lizard and west Cornwall.

(See Chapter 1, page 26.)

SEDIMENTARY ROCKS

gego	Oligocene-Eocene	
ge	Eocene	} PALAEOGENE
gp	Palaeocene	

ke	Cenomanian	Upper, ku }
ka	Albian	Lower, kl } CRETACEOUS

pt		Undivided } TRIASSIC and PERMIA
pu		Upper }
pl		Lower } PERMIAN

cw		Westphalian
cn		Namurian } CARBONIFEROUS
cl		Lower, Undivided
du		Upper
dm		Middle } DEVONIAN
dsde	Emsian to Siegenian }	
dgds	Siegenian to Gedinnian }	Lower

IGNEOUS ROCKS
Extrusive

Lava (Basalt)   In Permian

Tuff   In Carboniferous

Basic (Spilitic) lava }
Basic (Spilitic) tuff }   In Devonian

Intrusive

D   Dolerite

G   Granite

LITHOLOGICAL ORNAMENT

} Sandstone

Gravel, conglomerate and breccia
Limestone
Lava
Tuff

—— Geological boundary

—— Fault at surface

⇌ Transcurrent fault, with sense of relative movement

⊤ ⊤ Thrust

◇—◇ Anticlinal axis

×—× Synclinal axis

++++ Limit of the metamorphic aureole of the Dartmoor Granite

↑² Apparent dip in degrees, along line of geophysical traverse

•₁₂ Offshore core with registered IGS number

⬟-25 Mean direction of stable natural remanent magnetization

■ Built-up area

0   5   10km

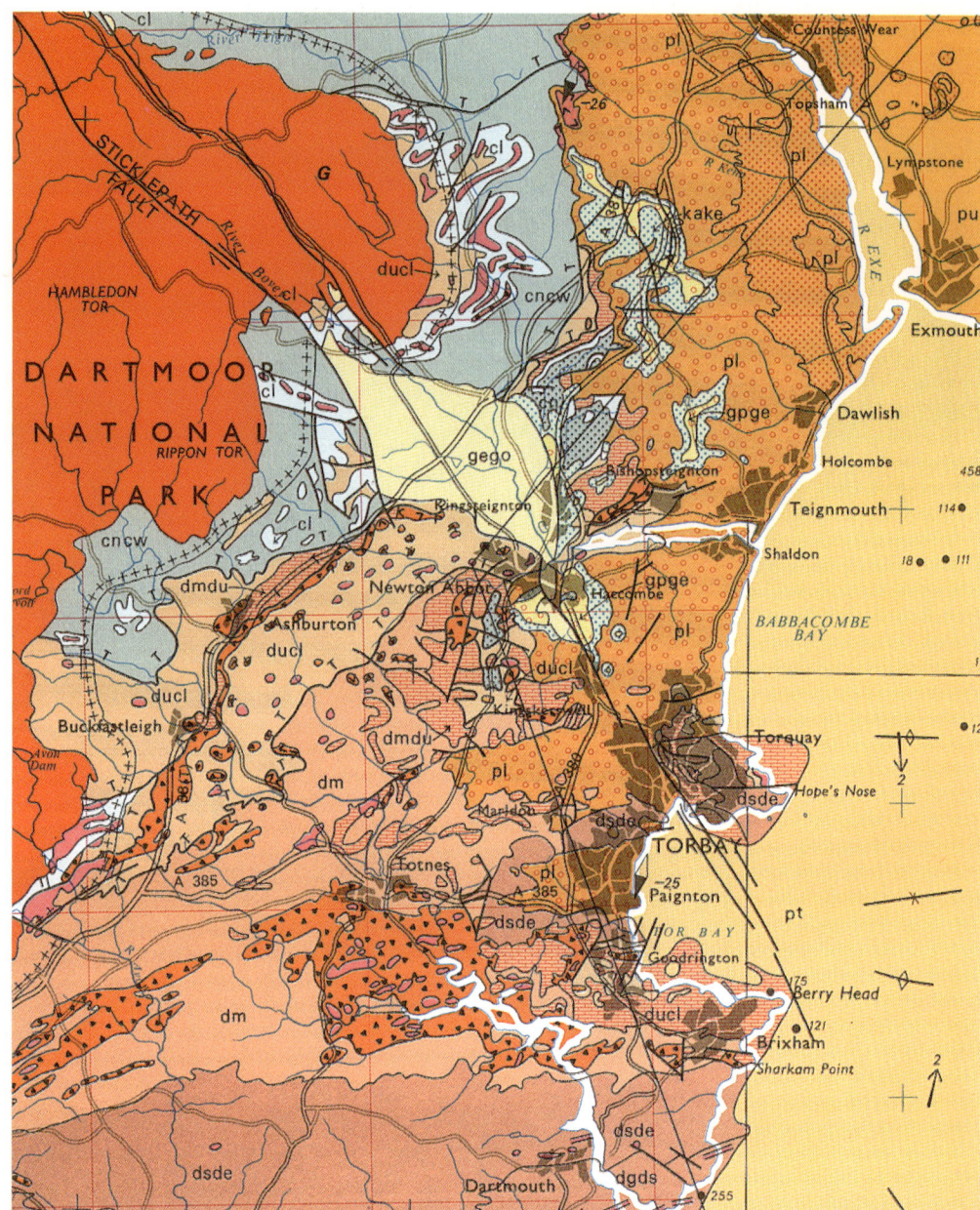

Map 1.3
Geology of south Devon.

(See Chapter 1, pages 26 and 28.)

Fig. 2.1 (above left)
St Just in Penwith, Cornwall: typical granite detail in a seventeenth-century house near the church—ornamental kneelers, roof coping, chimney detail.

Fig. 2.2 (below left)
Week St Mary, Cornwall: walls of local rubble masonry are dominated by the great granite ashlar chimney in this early-sixteenth-century college building.

Fig. 2.3 (above right)
Sanders Farmhouse, Lettaford, North Bovey, Devon: very large blocks of squared granite are used for the wall face at the shippon end of this Dartmoor longhouse.

Fig. 2.4 (centre right)
Old Post Office, Tintagel, Cornwall: a medieval house with a heavy rag roof and rubble walling of local slate stone.

Fig. 2.5 (below right)
Totnes parish church, Devon: pale Cretaceous limestone-dressed masonry contrasts with the stronger colours and rougher textures of local sandstone and volcanic stone walling.

(See Chapter 2, pages 36–38.)

Fig. 2.6 (above left) Old Post Office, Tintagel, Cornwall: a slab of local greenstone is used for the window opening.

Fig. 2.7 (above right) Near Ashbury, Devon: cob wall and sandstone masonry contrasted in a ruined barn.

Fig. 2.8 (below right) Trewerry Mill, St Newlyn East, Cornwall: fine elvan detail and ashlar masonry in a modest mill building dated to 1630.

Fig. 2.9 (below left) Gribbleford Bridge, Northlew, Devon: a cottage built of the distinctive volcanic stone from a quarry close to the bridge.

(See Chapter 2, pages 38–40.)

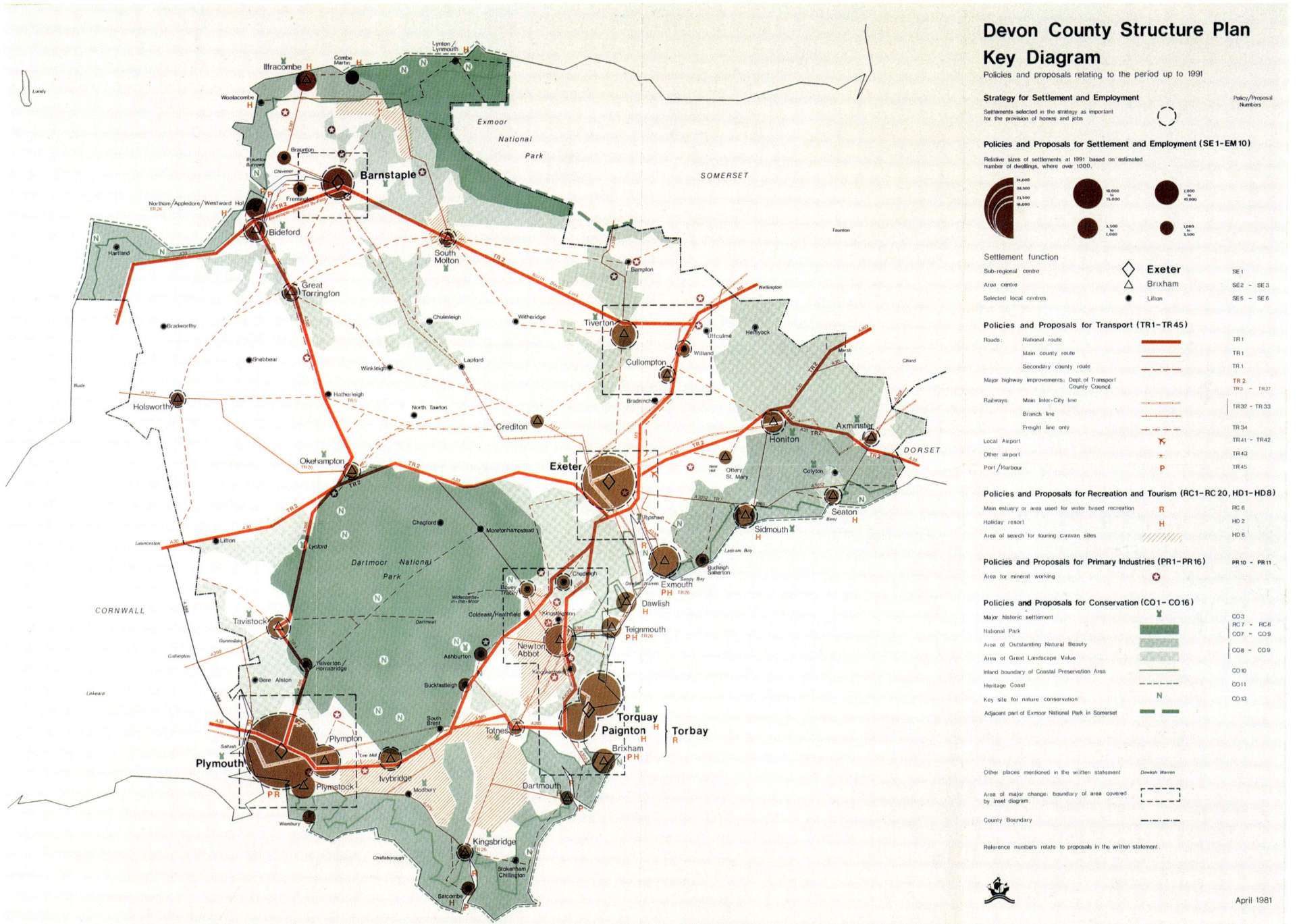

Devon County Structure Plan Key Diagram

Policies and proposals relating to the period up to 1991

Strategy for Settlement and Employment

Settlements selected in the strategy as important for the provision of homes and jobs

Policies and Proposals for Settlement and Employment (SE1–EM10)

Relative sizes of settlements at 1991 based on estimated number of dwellings, where over 1000.

Settlement function

Sub-regional centre	◇	**Exeter**	SE1
Area centre	△	Brixham	SE2 – SE3
Selected local centres	●	Lifton	SE5 – SE6

Policies and Proposals for Transport (TR1–TR45)

Routes:	National route		TR1
	Main county route		TR1
	Secondary county route		TR1
Major highway improvements: Dept. of Transport County Council			TR2 / TR3 – TR27
Railways:	Main Inter-City line		TR32 – TR33
	Branch line		
	Freight line only		TR34
Local Airport			TR41 – TR42
Other airport			TR43
Port / Harbour	P		TR45

Policies and Proposals for Recreation and Tourism (RC1–RC20, HD1–HD8)

Main estuary or area used for water based recreation	R	RC6
Holiday resort	H	HD2
Area of search for touring caravan sites		HD6

Policies and Proposals for Primary Industries (PR1–PR16)

| Area for mineral working | ⊙ | PR10 – PR11 |

Policies and Proposals for Conservation (CO1–CO16)

Major historic settlement		CO3 RC7 – RC8 CO7 – CO9
National Park		
Area of Outstanding Natural Beauty		CO8 – CO9
Area of Great Landscape Value		
Inland boundary of Coastal Preservation Area		CO10
Heritage Coast		CO11
Key site for nature conservation	N	CO13
Adjacent part of Exmoor National Park in Somerset		

Other places mentioned in the written statement

Area of major change: boundary of area covered by inset diagram

County Boundary

Reference numbers relate to proposals in the written statement.

April 1981

Map 55.6
Devon County Council,
Structure Plan Key Diagram,
1981.

(See Chapter 55, page 441.)

Cornwall County Structure Plan

KEY DIAGRAM

COUNTY WIDE POLICIES

EMPLOYMENT, INDUSTRY AND COMMERCE	–	7ABCDEFGHJK
HOUSING	–	8ABCDEFGHJKLMNPQRS
TRANSPORTATION	–	9BCDEGHJKLM
TOWNS, VILLAGES AND SHOPPING	–	10ABCDEFG
EDUCATION, HEALTH AND SOCIAL SERVICES	–	11AB
COAST AND COUNTRYSIDE	–	13ABEFHJKLNSTUVWXYZ
HOLIDAY INDUSTRY	–	14BCDEFGHJKMN
RECREATION	–	15ABC
MINING AND QUARRYING	–	16ABCC1G
REFUSE DISPOSAL	–	17A

TOURISM RESTRAINT AREAS

1. NORTH OF BUDE
2. PENTIRE POINT TO MILLOOK
3. NEWQUAY TO PORTHCOTHAN
4. ST. AGNES
5. GODREVY POINT TO PORTREATH
6. ST. IVES
7. PENWITH COAST AND MOORS
8. LIZARD PENINSULA AND HELFORD RIVER
9. FALMOUTH AREA
10. ROSELAND AND MEVAGISSEY TO PORTHPEAN
11. POLKERRIS TO THE TAMAR

DO NOT SCALE
CORNWALL COUNTY COUNCIL

POLICY NUMBERS

NATIONAL ROUTES		
COUNTY ROUTES		9A
MAJOR URBAN ROAD SCHEMES		9F
RAILWAY WITH PASSENGER SERVICE		9L
AIRPORT WITH SCHEDULED PASSENGER SERVICE		
HELIPORT		
FERRY ON COUNTY ROUTE		
ISLES OF SCILLY SHIPPING SERVICE		
MAIN PORTS FOR CHINA CLAY		
AREA OF OUTSTANDING NATURAL BEAUTY		13C, 13V, 13X, 13Y, 14D, 16A
PROPOSED ADDITIONS TO AREA OF OUTSTANDING NATURAL BEAUTY AND OTHER AREAS TO BE INVESTIGATED FOR DESIGNATION		13C, 13V, 13X, 13Y, 14D, 16A
AREAS OF GREAT LANDSCAPE VALUE		13D, 13X, 14D, 16A
AREAS OF GREAT SCIENTIFIC VALUE		13G, 13X, 14D, 16A
AREAS OF GREAT HISTORIC VALUE		13M, 13X, 14D, 16A
HISTORIC SETTLEMENTS		10E
NATIONAL NATURE RESERVE		13E
HERITAGE COAST		13P, 13O, 13R
ST. AUSTELL CHINA CLAY AREA		16D, 16E, 16F
TOURISM RESTRAINT AREAS		14A, 14M
DISTRICT BOUNDARIES		
STRUCTURE PLAN AREAS		
SUB-DIVISIONS OF STRUCTURE PLAN AREAS		
STRUCTURE PLAN AREA POLICY NUMBERS		
COUNTY BOUNDARY		

Morwenstow
Remote parts of Bude
Bude
BUDE 12.21A 12.21B
Holsworthy
DEVON
Remote parts of Bude
CAMELFORD 12.20A 12.20B 12.20C
Camelford
Remote parts of Launceston
LAUNCESTON 12.22A 12.22B
Launceston
Pentire Point
St. Minver
Padstow
PADSTOW 12.19A
WADEBRIDGE 12.18A 12.18B
Wadebridge
BODMIN MOOR SOUTH 12.24
CALLINGTON 12.27A 12.27B 12.27C
Tavistock
Gunnislake
GUNNISLAKE 12.27B 12.27C
BODMIN 12.17A 12.17B
Bodmin
Callington
Watergate Bay
St Columb Major
LISKEARD 12.23A 12.23B 12.23C
SALTASH 12.26A 12.26B
Saltash
Newquay
CENTRAL RESTORMEL 12.13A 12.13B 12.13C
LOSTWITHIEL 12.16A 12.16B
Lostwithiel
Liskeard
S.W. Caradon
LOOE 12.25A
TORPOINT 12.28A 12.28B
Plymouth
Perranporth
NEWQUAY 12.12A 12.12B 12.12C
St Austell
FOWEY 12.15A
Looe
Torpoint
PERRANPORTH & ST. AGNES 12.11A
ST. AUSTELL 12.14A 12.14B 12.14C
Fowey
RAME 12.29A
TRURO 12.9A 12.9C
St Austell Bay
Dodman Point
Truro
Mevagissey
Camborne
Redruth
ROSELAND 12.10A
CAMBORNE & REDRUTH 12.5A 12.5B 12.5C 12.5D
St Ives
Hayle
HAYLE 12.4A 12.4B
Falmouth Estuary
Penryn
PENZANCE 12.3A
HELSTON 12.6A 12.6B
Falmouth
St Just
ST JUST 12.1A
Penzance
FALMOUTH & PENRYN 12.8A 12.8D 12.8E
Helford River
THE LIZARD
Lizard

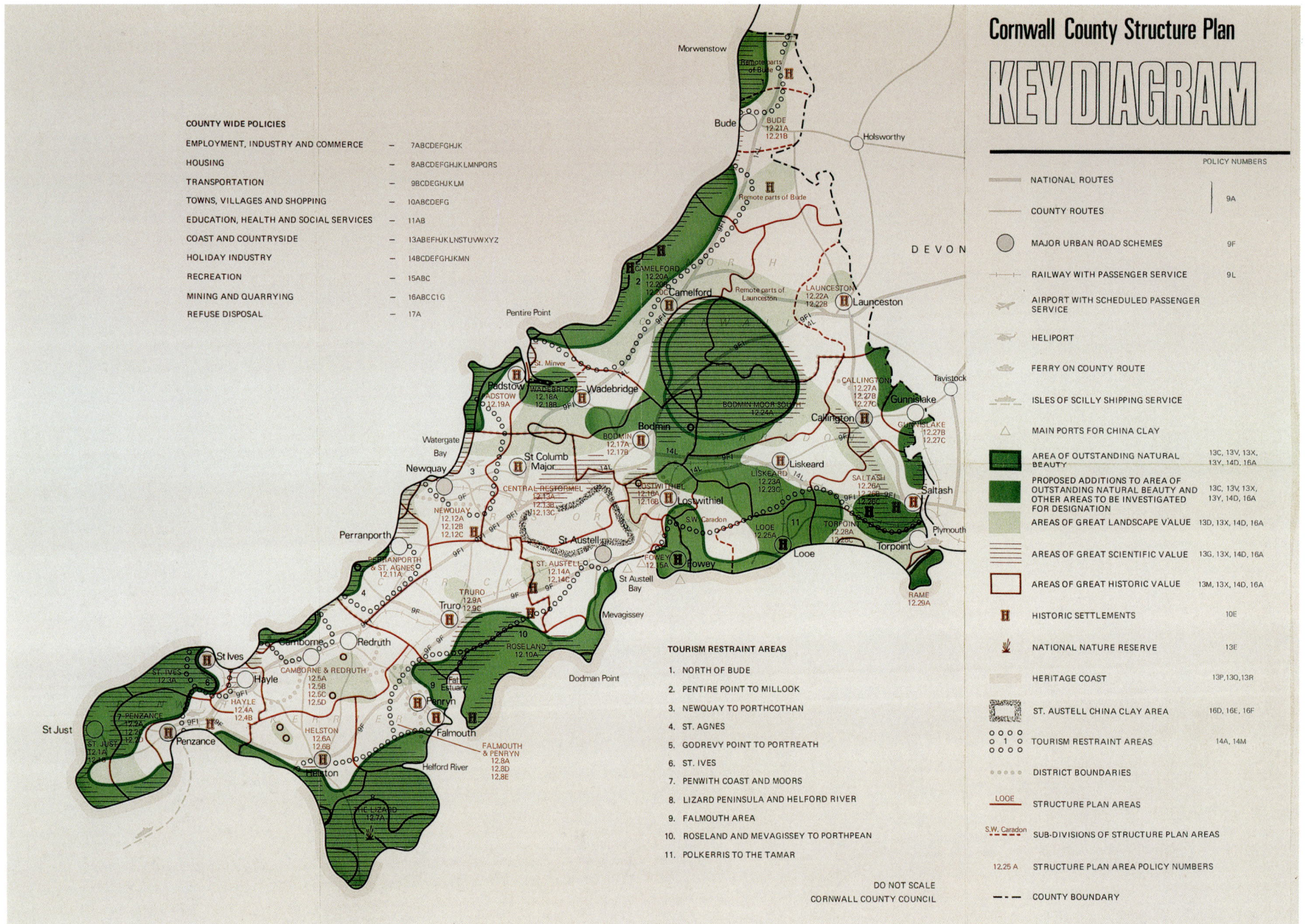

Map 55.11
Cornwall County Council,
Structure Plan Key Diagram,
1981.

(See Chapter 55, pages 443–46.)

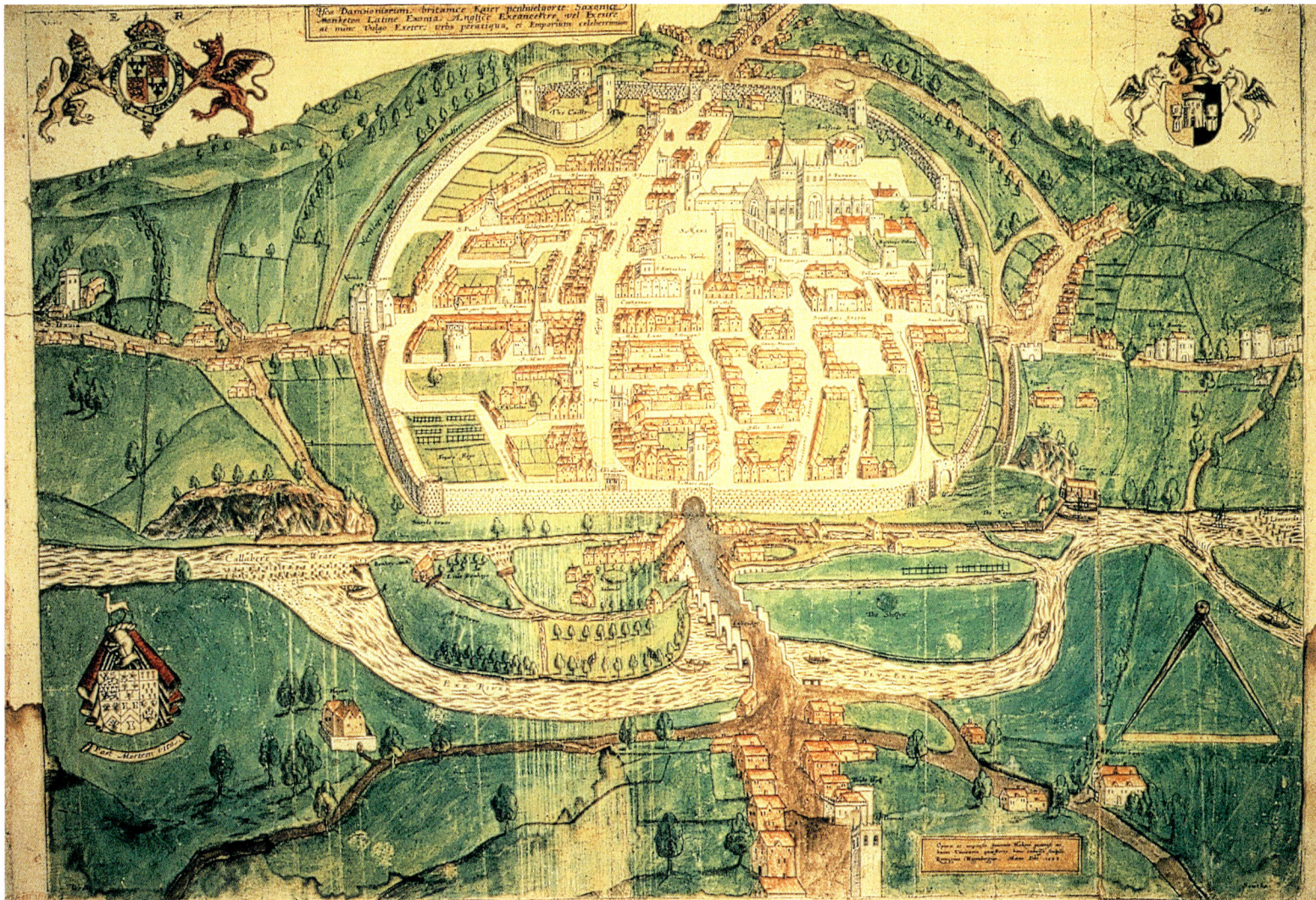

Map 61.12
John Hooker, *Exeter, c.*1587.

(See Chapter 61, pages 492–93.)

SOURCE: photograph
reproduced by permission of
the British Library Board
(BL Maps c.5.a.3).

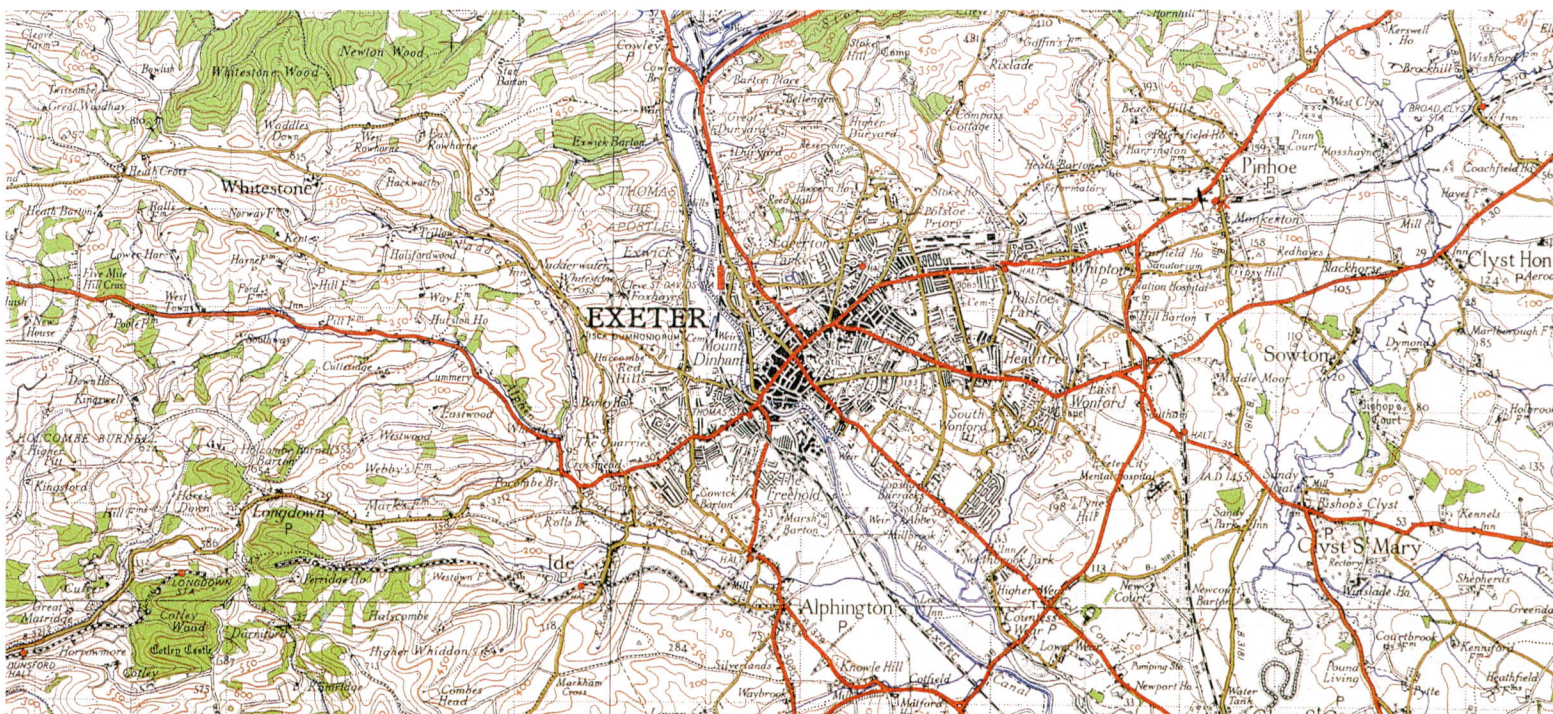

Map 64.1
Exeter, 1946: Ordnance Survey
one-inch to one mile New
Popular map.

(See Chapter 64, page 512.)

SOURCE: photograph
reproduced by permission
of University of Exeter,
Department of Geography.

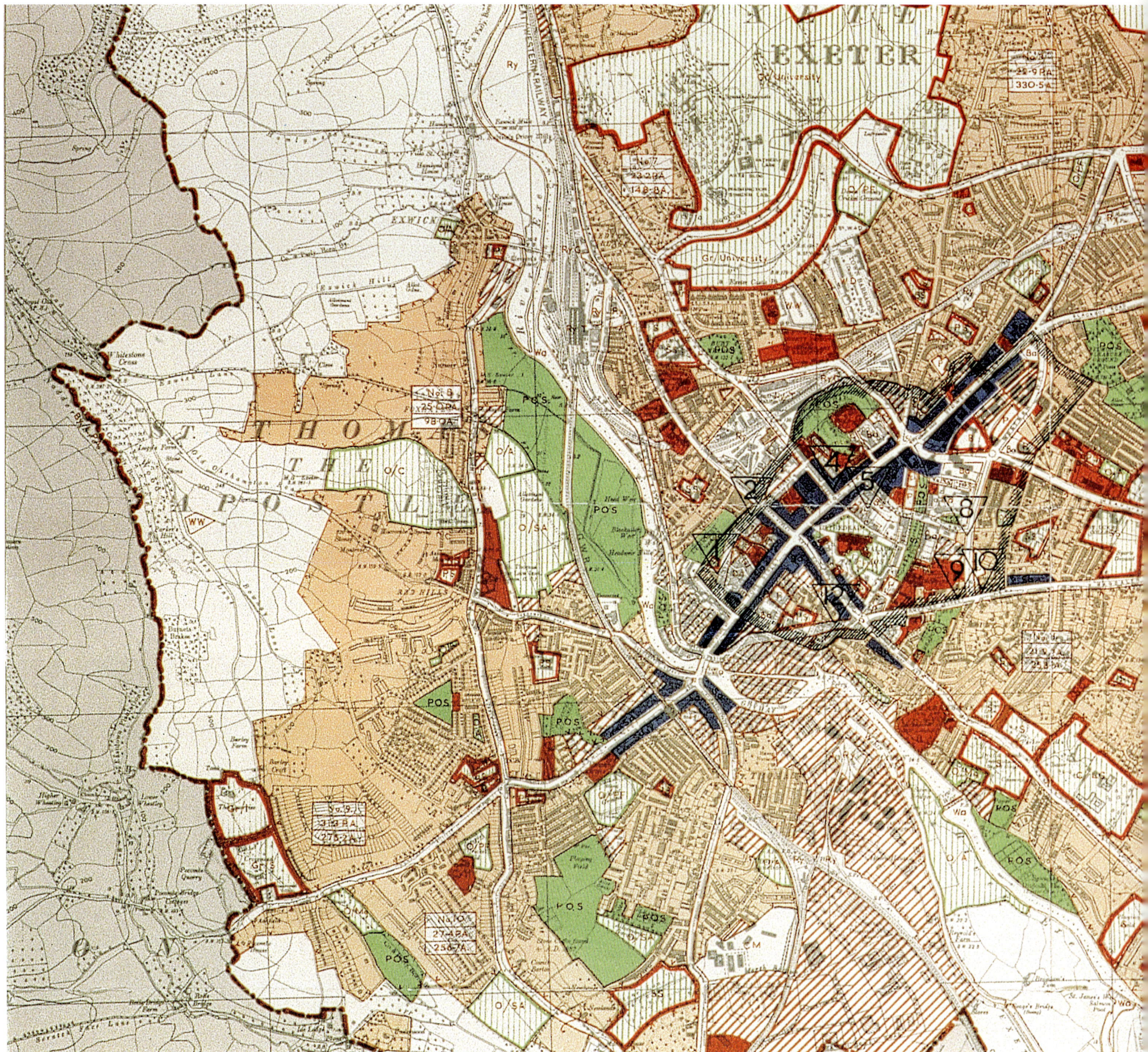

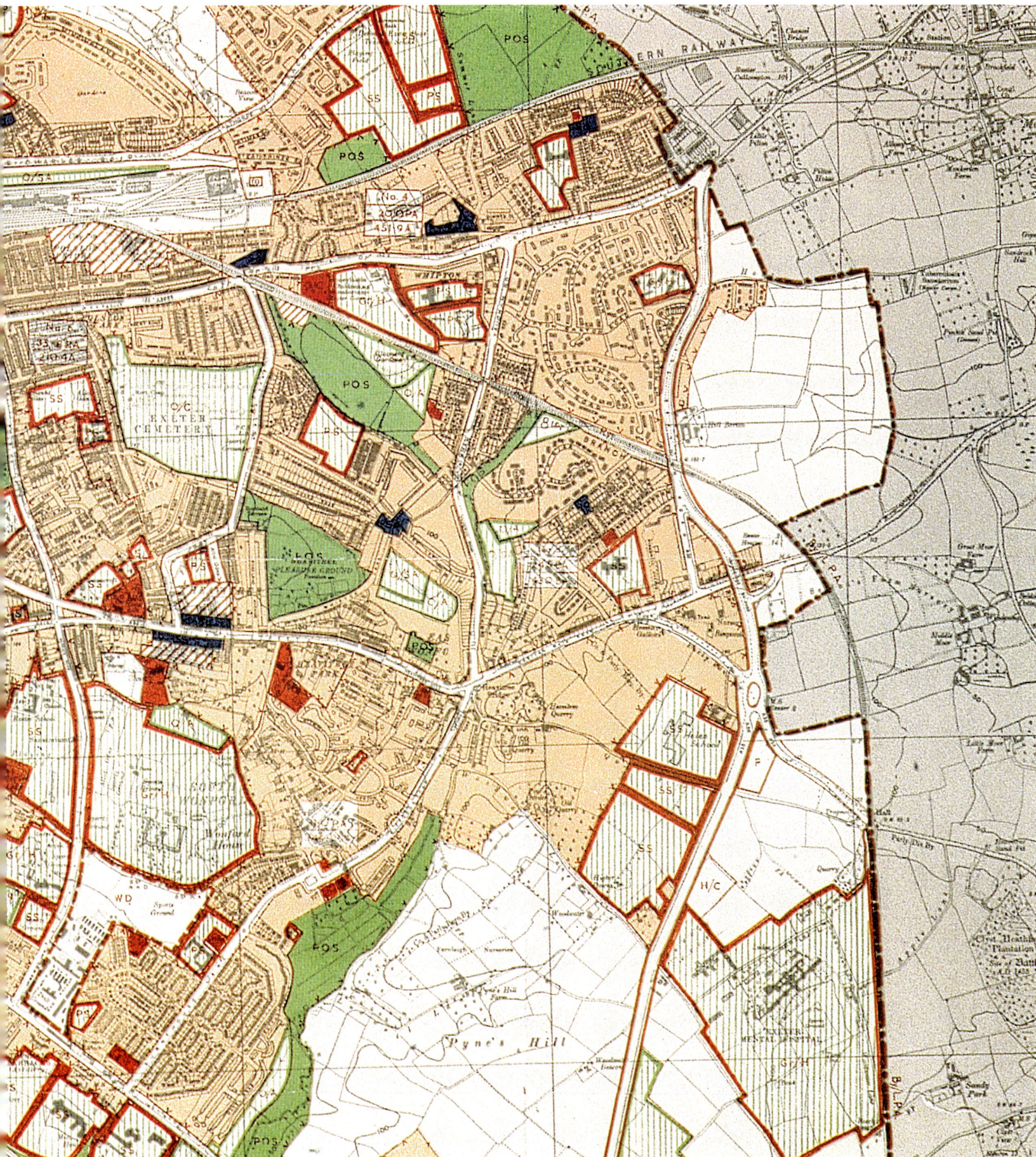

Map 64.2
Exeter Development Plan, 1963.

(See Chapter 64, pages 512–13.)

Map 64.3
Exeter Local Plan, 1984.

(See Chapter 64, pages 512–13.)

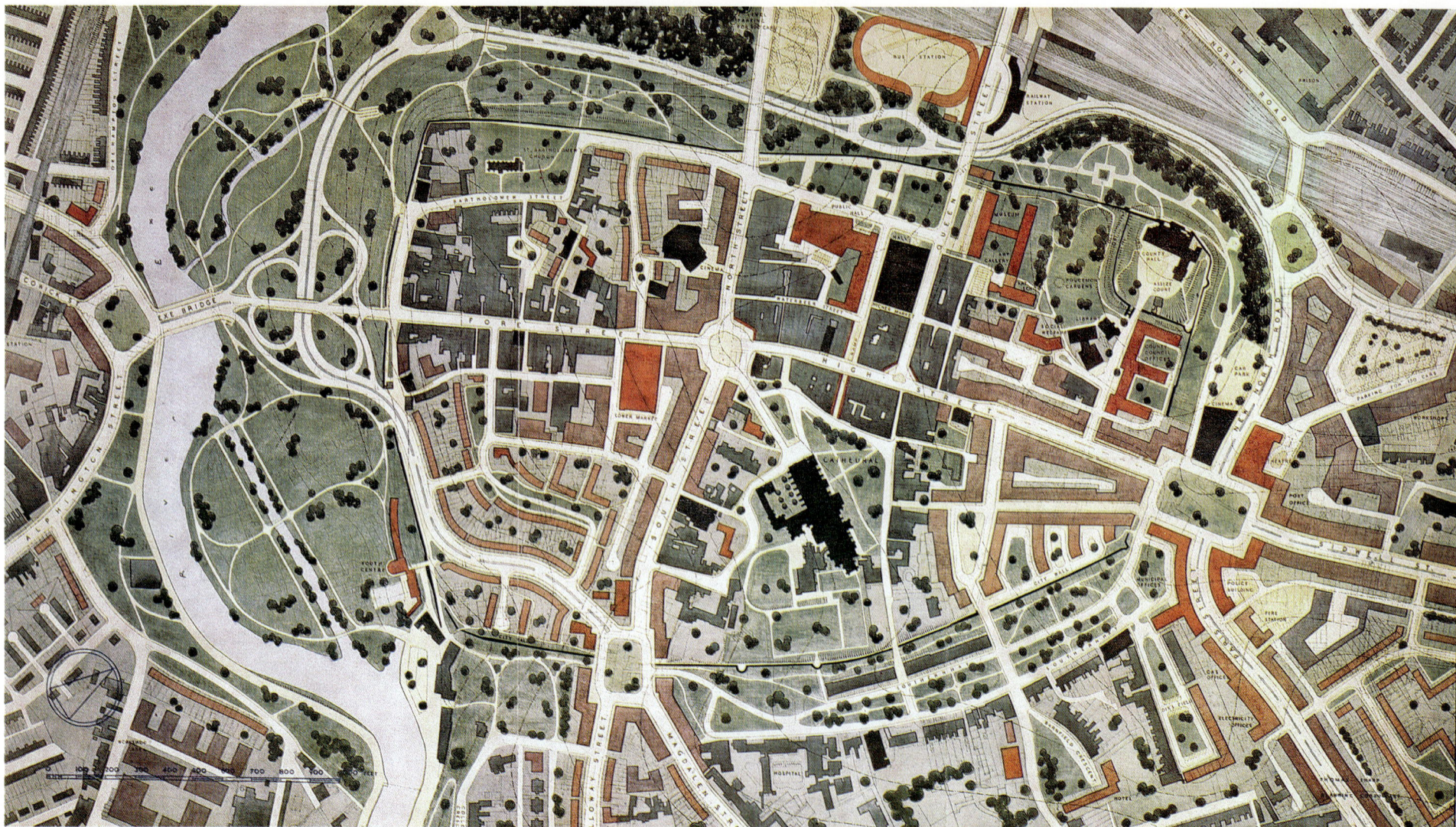

Map 64.4
Thomas Sharp, Proposed
Central Redevelopment Plan,
1946.

(See Chapter 64, page 513.)

SOURCE: *Exeter Phoenix*
(London, 1946).

Fig. 64.1 (above left) Debenham's modern movement style department store built in the late 1960s punctuates the eastern end of the rebuilt High Street.

Fig. 64.2 (below left) Exeter's inner bypass road looking west towards the River Exe from a viewpoint close to the location of the city's former South Gate. Part of the Roman city wall and a recent pedestrian footbridge giving access along the top of the Roman wall to a car park can also be seen.

Fig. 64.3 (below right) The city's Higher Market of 1840 has been restored as the main entrance to the Guildhall shopping centre.

Fig. 64.4 (above right) Pre- and post-1970 approaches to new building in what is now the Central Conservation Area: C&A and Marks & Spencer stores, respectively.

Fig. 64.5 (centre) A new 'regency' terrace of office accommodation contrasts with the modernist Civic Centre.

(See Chapter 64, page 513.)

Map 65.14
The Plan for Plymouth, 1943. The plan devised by Professor Patrick Abercrombie and James Paton Watson for the rebuilding of Plymouth after the Second World War is regarded as one of the great mile-stones in the history of British town planning. Abercrombie's vision (shown here) for the redevelopment of the bomb-damaged central area represented a startling departure from the notion of restoring the pre-war townscape. Though modified somewhat in its practical realization, Plymouth's city centre clearly bears Abercrombie's distinctive signature.

(See Chapter 65, page 536.)

SOURCE: author's collection.

Fig. 65.4
Plymouth Sound from above Stoke Church, *c*.1670–1690; watercolour by W. Du Busc. This view pre-dates the start of the building of the Royal Dockyard *c*.1690 and the enlargement of Stoke Damerel Church which commenced in 1715, but post-dates the construction of the Royal Citadel. Plymouth, with St Andrew's Church, is depicted on the extreme left. Stonehouse appears below Stoke Church, with the upper creek yet to be drained and reclaimed. On the far right, Mount Edgcumbe House is shown set in its extensive, wooded park.

(See Chapter 65.)

SOURCE: Plymouth City Museum and Art Gallery.

Fig. 65.5
The Eddystone Lighthouse, Plymouth, 1864; oil on canvas by William Gibbons. Two earlier lighthouses had been destroyed respectively by storm and fire. John Smeaton's replacement, commenced in 1756 and completed in 1759, lasted until 1877 when, because its foundations were being eroded, it was dismantled and replaced by the Eddystone's fourth tower, which still survives to warn shipping of the treacherous waters around the reef, 14 miles south of Plymouth. Smeaton's remarkable tower, formed of great interlocking blocks of granite and Portland limestone, was re-erected on Plymouth Hoe in 1882 where it has become a landscape icon, symbolizing the city.

(See Chapter 65, page 529.)

Fig. 65.9
Aerial view of modern Plymouth looking west. The modern yacht marina at Queen Anne's Battery occupies the foreground. The original medieval core of Plymouth is discernible in the small area between St Andrew's Church down to Sutton Pool. Similarly, the Tudor extensions of the town, which occupied the space on the southern side between the Parade inlet and the Hoe, can be identified. The modern, post-war city centre can be seen in the middle distance, dominated by the Guildhall tower (left of St Andrew's Church) and the City Council Office tower block beyond. At the top left of the picture, Millbay Docks are visible and, beyond that, Stonehouse and the Royal William Victualling Yard can be seen. The extreme (oldest) southern portion of Devonport Dockyard is also just visible on the distant horizon. Bretonside, Royal Parade and Union Street form the lengthy, east–west artery running across the right of the picture and thereby linking Plymouth, Stonehouse and Devonport by road.

(See Chapter 65, page 540.)

Source: Western Morning News Company Ltd.

ENVIRONMENTAL SETTING

C.J. CASELDINE

In introducing the environmental setting for a range of human activities from the economy of the Palaeolithic to tourism in the twentieth century, it is only possible to draw at most a skeletal picture with the small number of maps presented in this chapter. This text highlights those aspects of the environment which have had a significant impact upon the human settlement and economy of the South West. Without being too deterministic, it is often possible to establish a strong link between the exploitative potential of the environment and the way human communities have viewed and capitalized upon available resources.

As an introductory generalization it is tempting to follow the summary penned by Malcolm Todd in his introduction to *The South West to AD 1000*: 'The South West is different, in its geology, its landscapes, its climate, and in the tenor of its life. There is nothing quite like high Dartmoor, or the salty moors of West Penwith, or the warm red land of Devon, in any other part of Britain. Just as sea and land are mingled together, so there is an un-English promiscuity of upland and lowland defying the ruling of Sir Cyril Fox that this is part of his Highland Zone.'[1] While it may not be true that there are no parallels to be found in the specific elements of the physical landscape of the South West, it is their combination in a peninsula only 200km long by 120km wide which has given south-west England its most distinctive characteristics.

It is necessary, however, to keep in mind that although much of the essential physical form of the landscape may have changed little over time, it is the perception of the resources offered by the land and the abilities of human groups to utilize them that have changed, and continue to change. The attraction of the extensive coastline has varied, for example, from its value as a food resource base in the Mesolithic, through the historical period when it offered both safe anchorages and dangerous reefs, to become a key element of the late-twentieth-century leisure and recreation economy providing as it does an important focus for tourists. The Bronze Age communities which farmed and settled parts of Dartmoor were unaware of the future value of the kaolin which lay immediately below the surface, and which, when extracted, was able to play a part in allowing archaeological excavation to understand more clearly the nature of human activity in the second millennium BC.

Although in terms of land form there has been relatively little impact from human activity, except in limited areas of, for example, quarrying and mining or sea reclamation, there have been major alterations in virtually all of the critical physical resources, particularly the soils and vegetation of the region. When viewing the landscape of the South West in its widest sense, therefore, it is important not to see it through the eyes of the present or even recent centuries, a common pitfall when considering more 'natural' environments such as the upland and moorland areas:

> My dear Holmes,
>
> My recent letters and telegrams have kept you pretty well up to date as to all that has occurred in this God-forsaken corner of the world. When you are out upon its [Dartmoor] bosom … you are conscious everywhere of the homes and the work of prehistoric people. The strange thing is that they should have lived so thickly on what must always have been the most un-fruitful of soil.[2]

Andrew Fleming is quick to use the above as representing a misleading picture of the past when introducing the environmental context within which the Dartmoor reaves (boundaries of Bronze Age territorial units) were constructed in the second millennium BC, but its true significance is much wider.[3] Although upland Dartmoor provides a prime example of a radically altered biotic environment, the same may well be true for almost the whole of the South West, with areas now being farmed which previously had little apparent agricultural potential. The whole field of environmental change in south-west England touches an area which presents enormous problems and challenges, and which represents considerable deficiencies in our knowledge, especially for lowland areas.[4] Thus it is inevitable that for much of the prehistoric (and indeed the historic) period, reconstruction of the former physical environment relies heavily on extrapolation and analogy, a procedure fraught with pitfalls in an area as singular as that outlined by Malcolm Todd.

GEOLOGY

For **Maps 1.1–1.3** see pages 8–10 in the colour section.

The geology of the South West provides the dominant influence on the present physical landscape, and from the Devonian onwards (i.e. from almost 400 million years ago) all eras are represented somewhere in the region (**Map 1.1**). Earlier strata, probably from the Pre-Cambrian (older than 500 million years), are represented by the igneous and metamorphic rocks of the Lizard peninsula (**Map 1.2**) with their highly distinctive flora, and offshore by the Eddystone reef on which the famous lighthouse stands (see chapter 65). Fragments of pre-Devonian rocks in the form of quartzites and limestones occur as part of later breccias and conglomerates, showing how earlier strata were removed by subsequent surface erosion and then redeposited during later geological time periods, as can be seen happening in river valleys and along coasts today.

Most of the South West is known geologically as the Cornubian massif, comprising rocks principally of Devonian and Carboniferous age (395–280 million years ago). Apart from the early Devonian hornblende schists of Start Point, like the Lizard a resistant outlier on the current southern shoreline of Devon and Cornwall, the Devonian is characterized by the Old Red Sandstones formed under a tropical desert climate when Devon lay at the southern margin of a large continent. Deposits of this time were not all sandstone; they include shales, submarine volcanic rocks and even limestones, formed as coral reefs. It is the latter which outcrop in the Torbay area and which today provide evidence, through much later deposits in caves subsequently formed in the limestone, for some of the earliest human occupation in the region. Deposition into a principally marine environment continued into the Carboniferous, which is represented by shales and sandstones known as the Culm Measures. It has been argued that the word *culm* derives from *col*

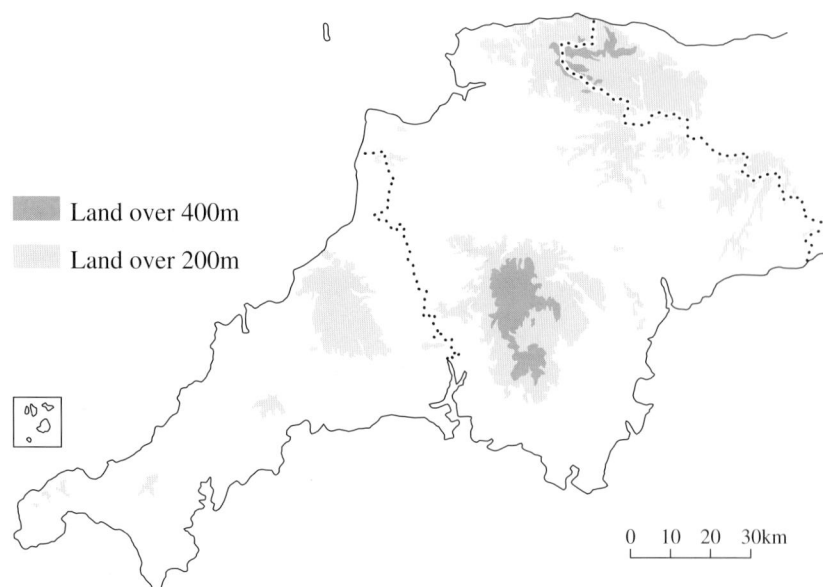

Land over 400m
Land over 200m

0 10 20 30km

Map 1.4
Land physiography.

or coal, found as a rather poor form in limited outcrops, but there are no real coal deposits in the area and the major geological resource of this period is the range of minerals found in association with the great granitic batholiths which were intruded around 290 million years ago (**Maps 1.2, 1.3 and 1.4**). The intrusion of these huge areas of igneous rocks caused the surrounding country rocks to be altered by metamorphic activity, giving rise to minerals such as tin, copper, lead, zinc, silver, arsenic, antimony, iron, manganese, tungsten, cobalt, nickel, uranium, baryte and fluorspar, which have each been exploited as technologies and the demands of economic activity have changed over the last few millennia (**Map 1.5**). The principal batholiths which have now been exposed by erosion of the surrounding rocks are those of Dartmoor and Bodmin Moor, but these same granites comprise most of West Penwith, the Scilly Isles and Carnmenellis, as well as the area around St Austell where the emphasis is on the extraction of kaolin, a very fine-grained 'china clay' produced by hydrothermal alteration of granite minerals during the final stages of cooling.

Apart from modern signs of mineral extraction, the landscape of south-west England also bears the signs of earlier mining processes. Tin was often mined along the lodes or veins in which it occurred, and is also found reworked in alluvial deposits (see chapters 41 and 43). Extraction of these sources has created the very uneven ground surface of many Dartmoor valleys, as can be seen to the south-east of the main road crossing Dartmoor near the Warren House Inn (SX 674 809). Where deeper mining using shafts has taken place to exploit lodes which often follow the joint patterns in the granite, as for example in the St Ives district of Cornwall, then the surface expression is less obvious except for spoil and the now-derelict mining buildings. Minerals, notably lead and iron ores, are also found in north Devon, where they occur within both Carboniferous and Devonian rocks. They may have formed earlier than the granite-related deposits but their formation is still not fully understood. Also in north Devon there is a further granite outcrop which forms the island of Lundy. This is, however, much later than the majority of granites in the South West and represents the southernmost expression of the Tertiary intrusive rocks extensive in Northern Ireland (e.g. Antrim) and western Scotland (e.g. Arran and Staffa), and was formed only 50 million years ago.

Characteristic of the eastern part of the peninsula are the younger Permian rocks, including the New Red Sandstones which 'lend warmth and charm to the local scenery' and provide the outcrops for the majority of the cliffs that form the coastline from Torbay to Seaton.[5] As well as sandstones, the Permian strata (280–230 million years ago) include breccias, marls, conglomerates and mudstones, and give rise to soils with a greater potential than those derived from earlier formations, especially showing contrasts where they outcrop adjacent to the more acid and wetter soils of the Culm Measures. Durrance and Laming evoke a picture of Devon in the Permian as looking 'similar to the southern fringes of the Sahara Desert today … though the desert then would have

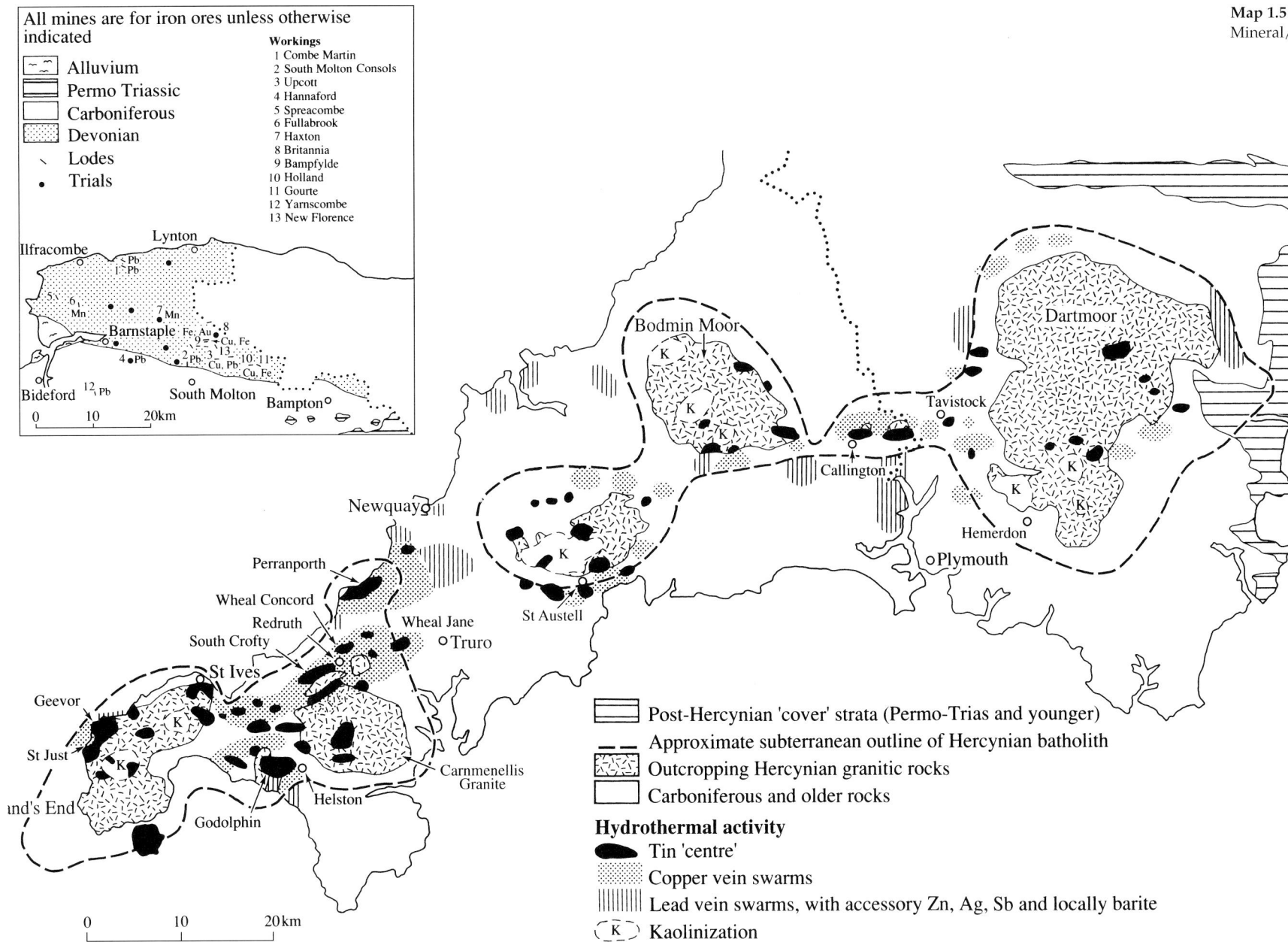

Map 1.5
Mineral/extractive deposits.

All mines are for iron ores unless otherwise indicated

Workings
1 Combe Martin
2 South Molton Consols
3 Upcott
4 Hannaford
5 Spreacombe
6 Fullabrook
7 Haxton
8 Britannia
9 Bampfylde
10 Holland
11 Gourte
12 Yarnscombe
13 New Florence

Alluvium
Permo Triassic
Carboniferous
Devonian
Lodes
Trials

Post-Hercynian 'cover' strata (Permo-Trias and younger)
Approximate subterranean outline of Hercynian batholith
Outcropping Hercynian granitic rocks
Carboniferous and older rocks

Hydrothermal activity
Tin 'centre'
Copper vein swarms
Lead vein swarms, with accessory Zn, Ag, Sb and locally barite
Kaolinization

been much more lifeless'.[6] In the subsequent Triassic (from 230 million years ago) little changed except for the deposition of further conglomerates from impermanent streams flowing across the desert that is now the English Channel. Further east still, the coastline east of Seaton is dominated by the complex of Jurassic and Cretaceous strata. The limited exposures of Jurassic rocks (from around 195–180 million years ago) are known world-wide, through their extensive fossil record. This includes the blue lias, shales-with-beef and black ven marls with their fossil ammonites which can appear as almost solid pavements. The juxtaposition of shales and marls with the overlying Cretaceous rocks creates not only some of the most beautiful and interesting coastal scenery, but also leads to extremely active slope movements today and rapid rates of cliff recession.

From 105 million years ago almost the whole of south-west England was covered by the sea, leading to the deposition of the Gault and Upper Greensand which cap many of the eastern hills providing a very distinctive landscape. With deepening water, limestones and eventually chalk were deposited. Much of this Cretaceous cover which must have overlain much of the South West has now been removed by erosion, but the chalk remains around Beer and in east Devon. Recent studies of the offshore geology of both the north and south coasts of the peninsula have revealed extensive tracts of both Cretaceous and Permo-Triassic cover, such that during periods of extremely low sea-level, as during Quaternary glacial maxima (see below), a very different geology from that of current mainland Devon and Cornwall would have been available to human groups. This also helps, in

For **Maps 1.1–1.3** see pages 8–10 in the colour section.

part, to explain the origins of a lot of the beach material found at present along the coastline, especially in the south. Outcrops of Tertiary strata are relatively rare in the South West. Land uplift caused the extensive erosion of the Cretaceous cover, producing the flint gravels which cover the Haldon Hills and which are found in small outcrops near Bideford and Sidmouth. The most important Tertiary deposits are those of the Bovey Basin in Devon (**Map 1.3**). Here subsidence allowed the deposition of over 1,200m of sediments. These are best known for the ball-clay, a form of kaolin, which is still extensively quarried, but the sequence represents a range of depositional environments including lignites comparable to the brown coals of northern Germany and which contain tropical plant species including *Sequoia* and *Cinnamomum* (Cinnamon). Isolated Tertiary sediments including clays are also found elsewhere, as at Petrockstow (with over 600m of sediments) and St Erth, the latter probably from the Pliocene, the final stage of the Tertiary.

QUATERNARY CLIMATE AND SEA-LEVEL CHANGES

By the end of the Tertiary, approximately 2.4 million years ago, the broad structure of the South West was in place and much of the surface geology must have been very similar to that which we see today. Over the most recent geological period, the Quaternary, this surface was subjected to a sequence of rapidly changing climates (rapid in a geological sense), with alternating cycles of climates as warm or slightly warmer than today separated by much longer cooler phases with, at times, extremely severe cold climates when ice sheets extended south over the British Isles, though these probably never reached the

South West. At these times the surface was exposed to periglacial activity with deeply frozen ground and the shattering of rock outcrops, creating the cliff exposures around Start Point and Prawle for example, and contributing to the development of the tors on Dartmoor when surrounding deeply weathered granite was stripped away leaving the core stones to form the tors. The rounded slopes of Bodmin Moor and the granite outcrops to the west also owe much to modification by periglacial processes during the Quaternary period. Apart from the modification of slopes through the erosion of rock debris by freezing and thawing, and the transportation of this dissociated material by solifluction, the slow downslope movement of the sediment mass created head deposits which now blanket many surfaces, sometimes as very thin cappings but occasionally, as along the Devon coast between Prawle and Salcombe, as thick aprons now exposed in cliff sections. On the uplands this same process occurred and is testified more noticeably by spreads of boulders coming away from source rocks as in the clitter fields which surround some tors. These spreads can sometimes be found as alignments downslope known as stone stripes. In the east of the region, river terrace sequences were formed by the reworking of Cretaceous gravels, principally under cold-climate conditions; there is an impressive suite of terraces in the Axe valley. These gravels are now quarried and this activity has uncovered Palaeolithic stone tools made from flint and chert, and at least 250,000 years old. Neither Devon nor Cornwall was covered by loess deposits as was Brittany to the south. Loess is a fine-grained, wind-blown deposit usually derived from the margins of ice masses or from an exposed continental shelf, and whilst not forming discrete units in south-west England, traces of it can be found in the uppermost cold-stage deposits which show enhanced proportions of silt-sized particles and produce soils with a distinctive silt component.

During the height of the glacial phases, the last of which occurred about 20,000 years ago, when 5 per cent of the world's water was tied up in the major ice sheets, sea-level around the South West was probably 120m lower than today. The whole of the peninsula was part of the continental land mass of Eurasia, and could be seen as the very western tip of a huge continent stretching from Siberia in the east, rather than as the tip of an oceanic island. Indeed for much of the Quaternary, when sea-levels, except for short interglacial phases, were lower than at present, Britain was not an island, and movement by plants and animals to and from what is now mainland Europe would have been facilitated. As the ice sheets of the last cold stage, the Devensian, melted and world sea-level rose, the coastline approached that of the present, until by about 6,500 years ago Britain was once more an island. The changing nature of the coastline over this recent geological time, the Holocene, can be estimated from **Map 1.6** on which 10m and 40m submarine contours are plotted. By the end of the last cold stage at 10,000 years ago, sea-level was at -40m, and rapidly rose to -20m at around 8,000 years ago. Relatively stable sea-levels close to the

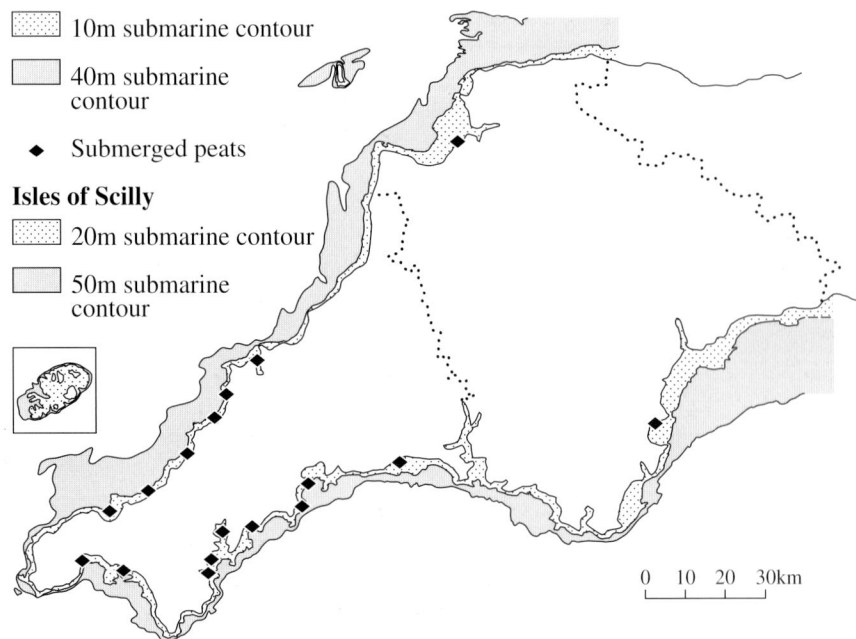

10m submarine contour

40m submarine contour

◆ Submerged peats

Isles of Scilly

20m submarine contour

50m submarine contour

0 10 20 30km

Map 1.6
Marine physiography.

present altitude were probably attained 3,000 years ago. Earlier lower Holocene sea-levels can also be inferred from the extensive offshore submerged forests described at many locations around the South West (**Map 1.6**). As the sea encroached on the land and water tables rose, woodland was engulfed in peat and submerged, only to be exposed at extremely low tides or when covering sand and shingle is removed during storms as at Hallsands in south Devon. At Westward Ho! on the north Devon coast, the submerged peats have also revealed traces of early human activity, in this instance from the Mesolithic period. The rising sea-levels of the current interglacial have also been responsible for the drowned valleys or rias which are found along much of the south coast, as, for example, in the Fal estuary. Traces of earlier Quaternary higher sea-levels, formed in previous interglacials which were thought to have been warmer allowing world sea-level to rise to higher altitudes, can be traced at a number of locations along both the north and south coasts, as at Baggy Point and Berry Head.

During the current interglacial there has been relatively little alteration to the land form of the South West. Once current sea-level was reached, coastal modification by erosion and deposition continued with occasional exceptional forms such as the bars across Slapton Ley and Looe Pool. Soundings in the lower reaches of the main river valleys—Exe, Teign, Dart, Erme, Tamar and Taw–Torridge—have shown the existence of buried valleys, cut during cold stages when sea-level was much lower and then filled in with fluvial sediments to their current altitude. In many cases the final infill was accelerated only in later prehistoric and historical times as ploughing accelerated erosion in their catchments. Overall slope modification has been relatively slight since the end of the last cold stage 10,000 years ago, such that if the vegetation cover could be stripped off, the configuration of the landscape would look remarkably similar to the way it did then.

SOILS

Soils are intimately related to the underlying geology, as well as being influenced by other factors such as climate, aspect, altitude, slope position, and, also of great importance, human modification. Any cursory view of the pattern of soils in the South West illustrates the link with geology and highlights the enormous complexity of soils over small areas (**Maps 1.7 and 1.8**). These patterns were just as important in the past as now, but what the maps fail to show is the way in which almost all soils have been influenced by human activity. Under the woodland cover that extended over almost all the land surface following the end of the last cold stage, soils would have been more nutrient rich and capable of sustaining a more diverse and stable flora, even on acid parent materials such as the extensive areas of granite. With deforestation for agriculture, a process which has been going on for over 5,000 years, soils have been much changed. Studies of soils buried under late Bronze Age monuments on granite on Bodmin Moor have shown the persistence of acid brown earths with an earthworm population until the first millennium BC. These buried soils contrast sharply with the acid, peaty podsols of the area at present, except where soils have been reclaimed by considerable input of capital in the form of fertilizer and labour. The areas of Dartmoor now covered by blanket peats were once woodland with a reasonably freely drained soil. The transition to peat began as early as about 8,000 years ago through the activities of early Mesolithic communities using fire as part of a strategy to assist in the hunting of animals on the high moor. An exception to the changing nature of soils over time may well be the gleys, soils which are prone to waterlogging from either surface water or groundwater. Unless there have been significant changes in local hydrology, which may have affected surface water gleys, it is likely that these soils were always prone to high water levels and not easily managed for agriculture.

It has long been thought that the areas of lowland, especially in Devon, were only deforested very late, perhaps only in the last 2,000–3,000 years, but recent archaeological finds and a limited amount of palaeoenvironmental evidence now point to human interference in these areas over a longer timescale, and hence the soils we see today must owe a greater debt to human modification than is traditionally acknowledged. The precise details of the transformation of the South West from a wooded landscape to the present largely open landscape are still sketchy; much more work is needed, if suitable deposits can be found. In broad terms, the natural woodland was largely dominated by oak and hazel, but with increasing amounts of birch on higher ground and a more diverse woodland in the lowlands with elm, and probably lime, on richer soils, and extensive ash woodland on the areas of base-rich parent material as in parts of south Devon. How these woodlands were cleared and modified is very difficult to demonstrate on the basis of the limited palaeoenvironmental data sources available in Devon and Cornwall. What can be said with confidence is that for the most part the major influence for change was human, with patterns being determined by need, will and capability rather than natural factors. Climatic limitations were only felt in marginal environments such as the extending moorland fringe where, with increasing deforestation and exposure, peat was able to spread. Even here on Dartmoor the final development of peaty soils at the margins of cultivation, as can be seen by the distribution of former field systems and settlements, only really accelerated in the later first millennium BC. A similar picture appears to be emerging from Bodmin Moor.

In assessing the status of soil at present it is of particular value to use the concept of land capability (**Map 1.9**). This grades land on a scale of 1 to 7 from land with no limitations for agricultural activity to that with extreme limitations. This map shows that the South West is dominated by grade 3 land, the main limitation being climate, with pockets of grade 2 in the east and in parts of Cornwall. Around the moors limitations increase as might be expected. Poorer grade 4 land is also a feature of the Culm Measures in mid-Devon and the greensand and gault clays to the east.

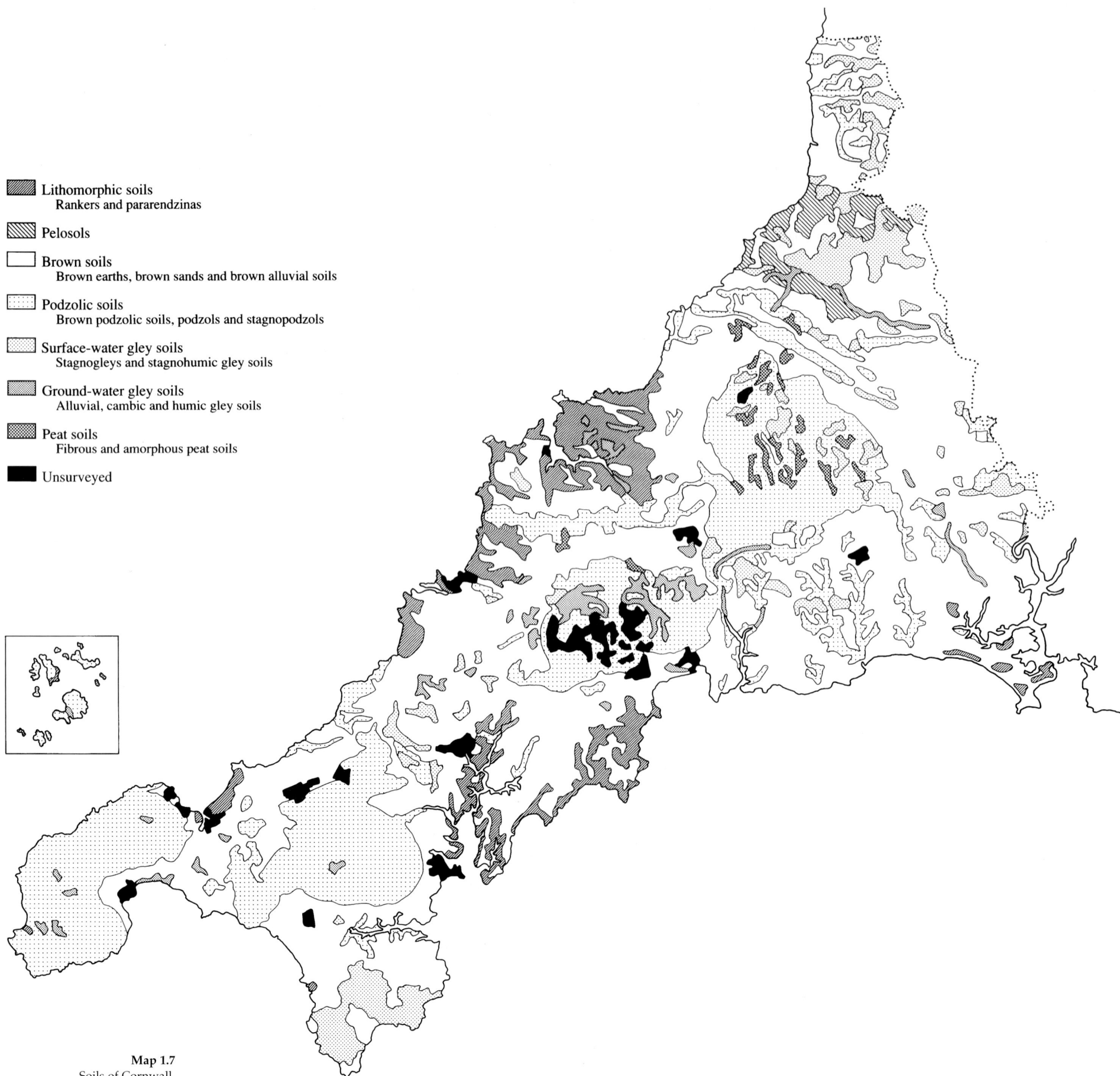

Lithomorphic soils
Rankers and pararendzinas

Pelosols

Brown soils
Brown earths, brown sands and brown alluvial soils

Podzolic soils
Brown podzolic soils, podzols and stagnopodzols

Surface-water gley soils
Stagnogleys and stagnohumic gley soils

Ground-water gley soils
Alluvial, cambic and humic gley soils

Peat soils
Fibrous and amorphous peat soils

Unsurveyed

Map 1.7
Soils of Cornwall.

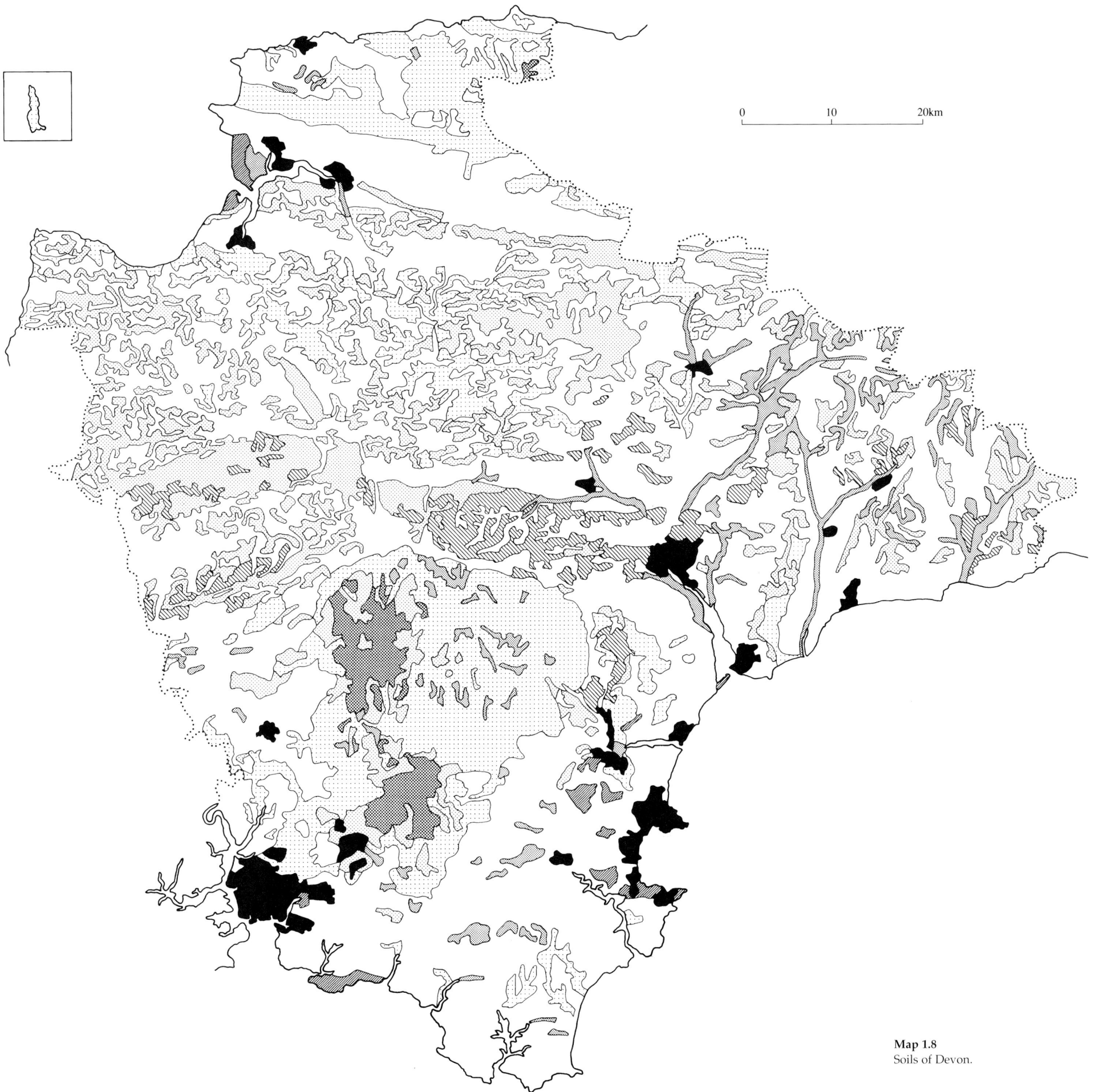

Map 1.8
Soils of Devon.

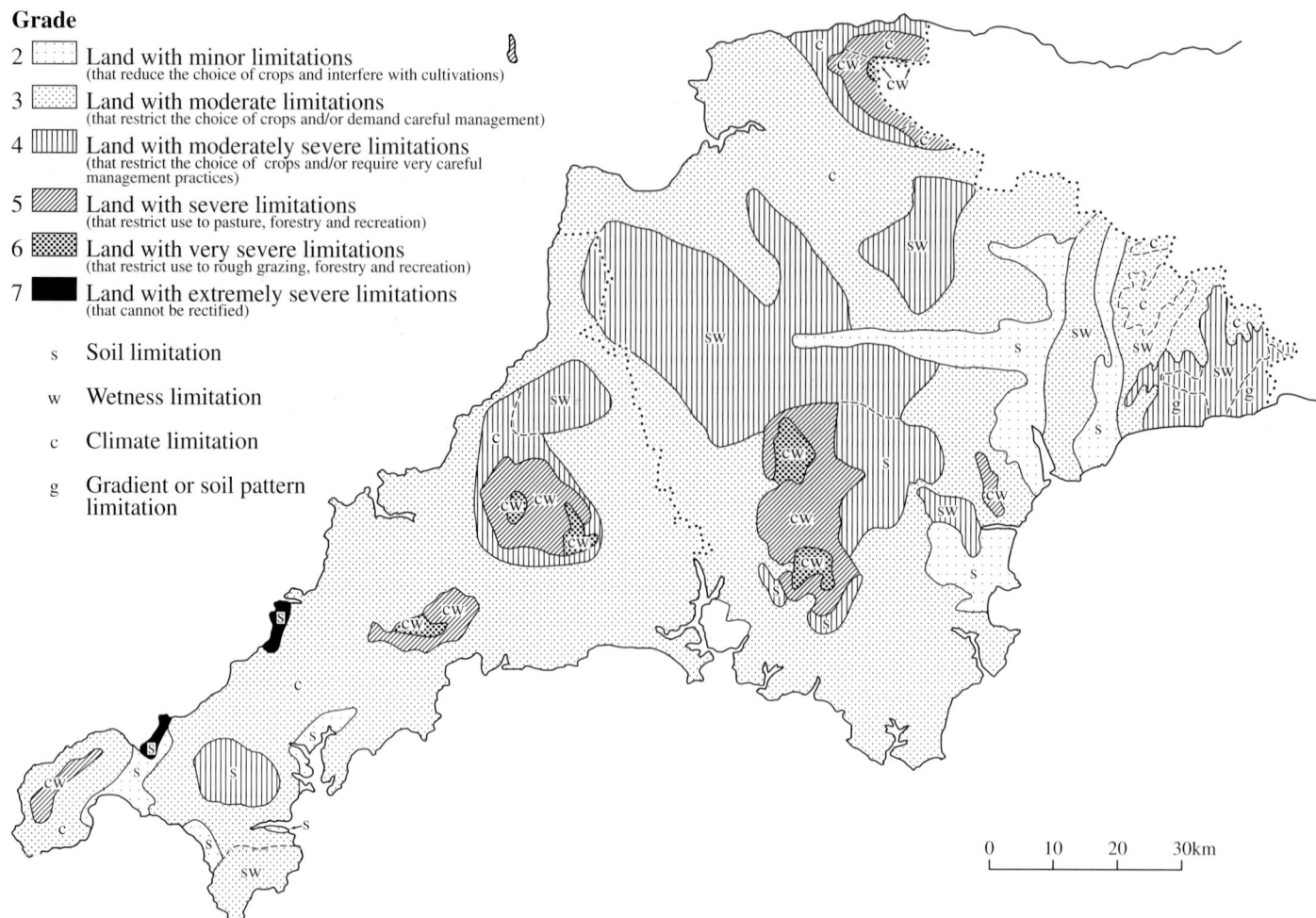

Map 1.9
Land capability.

Grade

2 ⬚ Land with minor limitations
(that reduce the choice of crops and interfere with cultivations)

3 ⬚ Land with moderate limitations
(that restrict the choice of crops and/or demand careful management)

4 ⬚ Land with moderately severe limitations
(that restrict the choice of crops and/or require very careful management practices)

5 ⬚ Land with severe limitations
(that restrict use to pasture, forestry and recreation)

6 ⬚ Land with very severe limitations
(that restrict use to rough grazing, forestry and recreation)

7 ⬛ Land with extremely severe limitations
(that cannot be rectified)

s Soil limitation

w Wetness limitation

c Climate limitation

g Gradient or soil pattern limitation

0 10 20 30km

CLIMATIC CHARACTERISTICS

Examination of the limitations defined on **Map 1.9** reveals the problem of the climate of the South West. High rainfall and steep rainfall gradients across the area are a feature of the region, with variations of over 1,200mm in annual rainfall between Torbay and Dartmoor, a distance of only 20km (**Map 1.10**). West of the Exe valley the majority of the area has rainfall in excess of 1,000mm per year. In contrast, the relative warmth and mild nature of the climate leads to a long growing season, peaking at over 325 days along the coasts of both northern and southern Cornwall (**Map 1.11**). Climate has not, of course, been constant over the last 10,000 years as the current landscape has evolved. At the end of the last cold stage, January temperatures would have registered a mean around -20°C, with mean July temperatures probably around 8°C. At the onset of the present interglacial, although summer temperatures quickly reached values similar to or slightly higher than today, there was still a much more marked contrast between winter and summer, a contrast which was not finally reduced until the separation from the Continent. Over the last 5,000 years changes have been less perceptible and would only really have had any significant effect on human communities in extremely

marginal locations. Even then it is questionable whether these effects would have been recognizable as trends at more than a decadal timescale. Our understanding of climate and its impact is influenced greatly by our perception of its impact on our lives. Thus in 1968 in the British Regional Geology memoir it is affirmed that 'in terms of water resources there is no water supply problem in south-west England'.[7] This preceded the drought of 1976, which, although an exceptional climatic event based on recent experience, may be more common in the future. The problems were also exacerbated by the ever-increasing demand for water as a result of rising numbers of tourists in the South West and greater use of water on a *per capita* basis.

Looking to the future, the main influence on the environment of the South West, apart from changing socio-economic demands, new policies on agriculture and the ever-increasing demands on a fragile resource for recreational needs, will undoubtedly be global warming. If predictions are fulfilled then the South West will experience less change than many other areas of the British Isles—an increase of 1.5°C in winter and only around 1°C in summer by 2050—but as world sea-level rises the coastal fringes of the peninsula will begin to notice change, perhaps with an increase of over 10cm by the same time. In a globally warmed world, even if

Map 1.10
Average annual rainfall,
1941–1970.

-1000- Rainfall in mm

Land over 400m

Land over 200m

0   10   20   30km

Map 1.11
Average length of the growing
season, 1941–1970.

−275− Growing season in days

Land over 400m

Land over 200m

0   10   20   30km

temperatures have not risen significantly in the South West relative to elsewhere, tourists will still probably expect a shift towards a more Mediterranean regime, and increase their demands on what will remain a small but incredibly diverse landscape.

FURTHER READING

The single simplest overall view of the geology of south-west England remains E.A. Edmonds, M.C. McKeown and M. Williams, *British Regional Geology: South West England*, 4th edn (London, 1985). The Geologists' Association publishes local guides of which only A. Hall, *West Cornwall* (Bath, 1994) is available for the region. More detailed geological coverage is available for both Devon and Cornwall in E.M. Durrance and D.J.C. Laming (eds), *The Geology of Devon* (Exeter, 1982) and E.B. Selwood (ed.), *The Geology of Cornwall* (Exeter, 1998). For soils there is a thorough and detailed account in the local Soil Survey Memoir: D.C. Findlay, G.J.N. Colborne, D.W. Cope, T.R. Harrod, D.V. Hogan and S.J. Staines, *Soils and their Use in South West England*, Soil Survey of England and Wales Bulletin No. 14 (Harpenden, 1984). An outline of the archaeology of the South West with useful environmental comments may be found in M. Todd, *The South West to AD 1000* (London, 1987).

TRADITIONAL BUILDING MATERIALS
AND THEIR
INFLUENCE on VERNACULAR
STYLES

VERONICA CHESHER

The subterranean geography … of Devonshire must be peculiarly interesting to minds in the least disposed to philosophical reflection … (Revd Richard Polwhele)[1]

As Dr Polwhele inferred, the geology of Devon and in fact of the whole of the south-west peninsula is extremely complex (see chapter 1). This is of interest not only to philosophical minds but to anyone concerned with local buildings, building materials and their history. Geologically there are two factors of prime importance in the story of building in Devon and Cornwall. One is the great range of different types of available stone (**Map 2.1**). Foremost among these are the sedimentary rocks, themselves in great variety both visually and in what they offer to builders, as they range widely in composition, texture and colour and include contrasting types of both sandstone and limestone. Second, there are the metamorphosed sedimentary rocks, notably slates, again in great variety of type and colour, and third, the igneous rocks of which granite is dominant in Devon and Cornwall. The region is the only area in England that has used granite extensively in local building and to many people it is the rugged solidity of this stone which gives building in the South West a special character. Another factor of particular interest in historic building especially in Devon (and Cornwall to a lesser extent) is that builders often turned to an entirely different material—cob. This material, literally home-made on site from clay, straw, gravel and water, was used to produce buildings in such large numbers and with such characteristic outlines that they form a dominant element in the Devon scene.

Builders in the South West were thus presented with a wide range of potential materials. Three prime factors affected their choice for particular buildings: availability, durability and workability. First, availability was a crucial factor at the level of vernacular building in past times: i.e. for the smaller to middling houses and associated buildings where cost and easy access were paramount considerations. The expense of bringing materials from a distance was much greater proportionally than either the cost of 'winning' them or the actual building operation. It has been estimated that in medieval building operations the cost of carriage began to exceed the cost of the stone itself once the distance exceeded 12 miles. Sixteenth- and seventeenth-century Devon building accounts, such as for work on the city walls of Exeter and the building of the

Orphans Aid orphanage in Plymouth, show transport as a considerable item.[2] The building of Exeter Cathedral involved considerable carriage of stone, but although the stone was clearly chosen for its quality, it is noticeable that each type came from quarries accessible by sea—Caen, Salcombe Regis, Beer and Swanage, from whence it could be carried up the Exe esturary. At a local building level, the importance of the availability factor is clearly shown in the *Memoirs* of the British Geological Survey relating to Devon and Cornwall, which contain many references to small quarries serving very localized areas. In some cases these provided stone for a very small group of buildings or even a single house in close proximity to the quarry.[3]

A second factor affecting a builder's choice of material is durability: is the material weatherproof, hard-wearing and reliable for building? The importance of this as an imperative is more debatable. Clearly some stones were recognized for their durability, particularly in standing up to the demands of west-country weather. Richard Carew in his *Survey of Cornwall*, written in the later sixteenth century, commented on the severity of the maritime storms whereby 'even the hard stones and iron bars of the windows do fret to be so continually grated', and he commends granite for its ability 'to withstand the fretting weather'.[4] Certainly granite was used widely, well beyond its own area, for exterior door and window frames, and, as Carew pointed out, it 'countervaileth its great hardness in working with the profit of long endurance'.[5] On the other hand, there are plenty of examples in both counties of poor building stone being used because it was easily accessible. On the Lizard peninsula, for example, serpentine was used in building despite being notoriously unreliable in the severe weather encountered there. In some medieval churches in south Devon local limestone was used, although it tended to let in damp: to counter this, walls were rough-cast. Some kind of rendering, rough-casting or stucco was a solution where poor local stone was used. The evidence both of surviving buildings and in documents suggests that a very large proportion of stone exterior walls were protected by rendering even during the Middle Ages.[6]

One of the great attractions of the calcareous sandstone of the Salcombe Regis area is that it is both durable and easily worked. How important was this third factor of 'workability' to west-country builders in the past? Where granite was used—and it was used extensively—workability was obviously a lesser factor than its durability (as Carew remarked). Many other obdurate stones were

For **Figs 2.1–2.9** see pages 11–12 in the colour section.

Map 2.1
Predominant historic building stones.

This generalized map of stone types should be read in conjunction with **Map 1.1.** on page 8 in the colour section.

Legend:
- Sandstone
- Sandstone with some slate and shale
- Red Permian, sandstone and breccia
- Calcareous sandstone
- Slate
- Schist
- Devonian limestone
- Cretaceous limestone
- Flint/chert
- Granite
- Elvan
- V Volcanic
- Lizard Complex (see Map 1.2)

used in building in the South West, including the hard sandstone of parts of north-west Devon, some of the greenstones in Cornish coastal areas and the Lizard gabbro. At the other end of the spectrum, soft rocks presented different problems in workability: some breccia crumbled when attempts were made to use it for dressed stone work. And it was in dressing stone for detail in buildings that masons had the narrowest limits in terms of choice, which is why Beer stone and even Ham stone from across the border in Somerset were used in far away Devon churches: the most desirable qualities were for stone which dressed easily when first quarried but hardened with exposure. The popularity of granite was a case of durability outweighing difficulties of dressing. Over the centuries masons in the granite areas of Devon and Cornwall developed their own techniques for working granite. However, even granite varies in hardness and buildings can often be seen—for example in the Chagford and in the St Austell areas—where softer, less durable but more easily worked varieties were used.

Some of this variety of building materials can be seen in the set of colour photographs which illustrate this chapter and which are located on **Map 2.2**. In the sections which follow, particular building stones and materials of the region are discussed briefly.

GRANITE (Figs 2.1, 2.2, 2.3)

The great granite deposits of Cornwall and Devon have played a tremendous role in traditional building in the region, creating a special relationship between buildings and landscape in areas such as Dartmoor, Bodmin Moor and West Penwith. Each granite area has its distinctive types: some are coarser and more difficult to work; some emerge as large blocks (granite to the north east of St Austell is notable for this). Finer stone tends to lie within coarser outer layers and soft granite can occur on the edge of deposits or close to the surface. Although granite is usually thought of as a drab grey stone, its varying elements can produce warm creams and buffs and occasionally even pink shades. Around Chagford a relatively soft buff-coloured granite was sometimes used for garden and outbuilding walls, being more easily worked if less durable; a similar stone was used in West Penwith buildings. Near Burrator Reservoir on the fringe of west Dartmoor a rusty pink granite with dark spots introduces an exotic element in local masonry. All granite buildings give an impression of solidity and strength which is particularly in tune with the rugged moors on which many are sited.[7] Until the nineteenth century, granite used for building in the region was

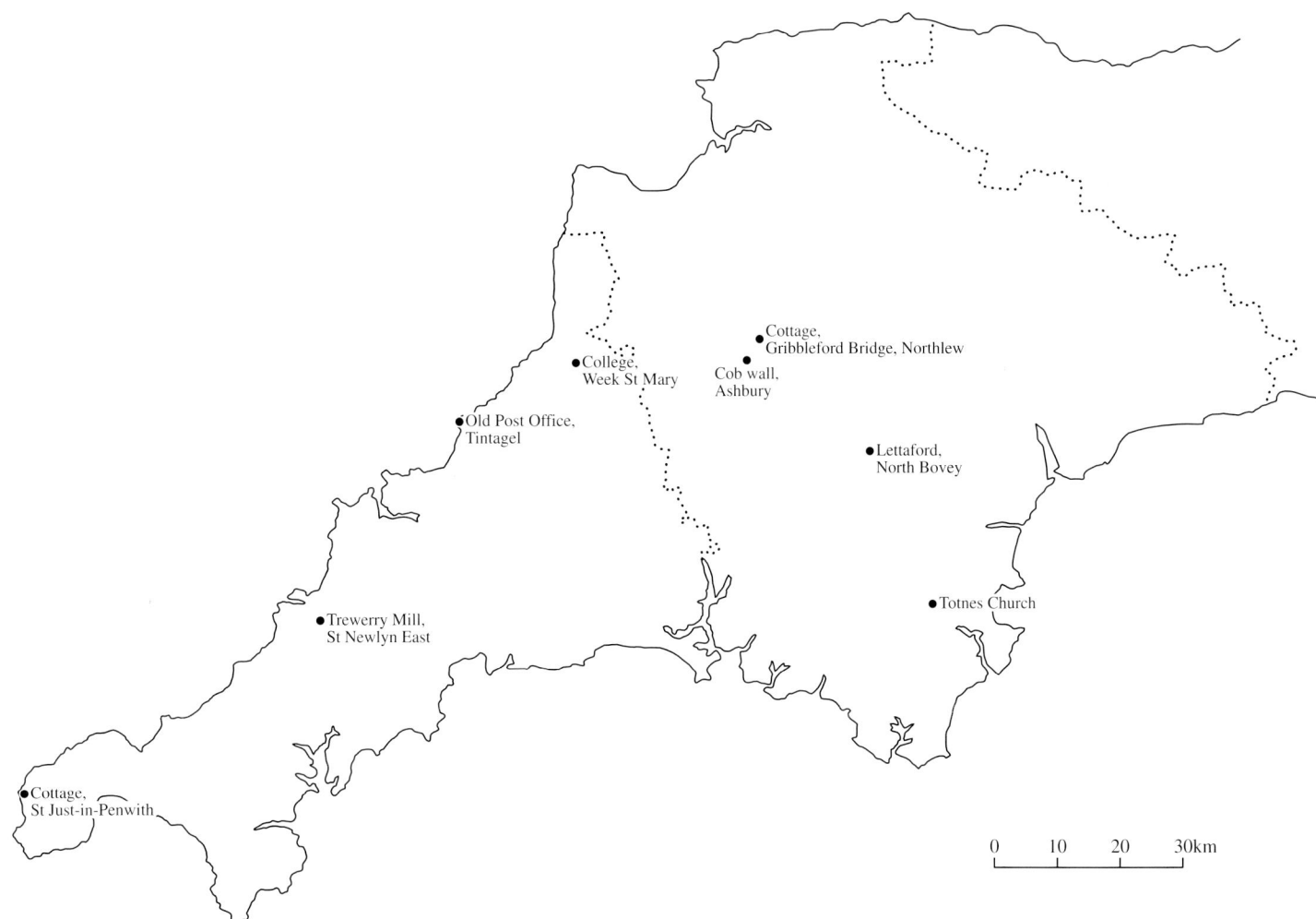

Map 2.2
Location of buildings
photographed.

See pages 11–12.

● Cottage,
Gribbleford Bridge, Northlew

● College,
Week St Mary

Cob wall,
Ashbury

● Old Post Office,
Tintagel

● Lettaford,
North Bovey

● Trewerry Mill,
St Newlyn East

● Totnes Church

● Cottage,
St Just-in-Penwith

0   10   20   30km

taken directly from the surface of those moors where it lay in abundance: for this reason it was known as 'moorstone'. The term 'granite' is a relatively recent one, emerging as the accepted geological name for the stone at about the same time that quarrying techniques were developed to exploit it and which created a major west-country industry in the nineteenth century.

The use of granite was by no means confined to vernacular building in the region. At high levels of wealth and status it was used to produce impressive mansions. An early seventeenth-century example is the Robartes family's great house at Lanhydrock, faced with granite ashlar. In the eighteenth century granite emerged more or less successfully as a 'polite' building material, rugged but tamed, in the spate of mansion-building resulting from new wealth and the new classical style of architecture.

SLATE (**Fig. 2.4**)

Slate is the most widely found building stone in Cornwall and Devon and one of the most frequently used in past times. The essential character of slate is its fissility, the result of regional metamorphism, which allows slate-stone to be split or cleaved repeatedly. The quality of cleavage, essential to the role of slate as a building stone, varies with the composition of the rock.

The slate-stone areas in Devon occur across the county (**Map 2.1**) in both the Devonian and Carboniferous formations, the latter in the northern part of the county and the Devonian slates to the west and south of the central granite mass. In Cornwall, where Devonian rocks underlie much of the county, slate is even more widespread.[8] However, its varying qualities as a building stone, depending on the composition of the original sediments and their subsequent metamorphism, make for noticeable contrasts in slate building in both counties. In Cornwall, for example, there is a notable contrast between the areas on either side of the Camel estuary and along the north coast, where coarser slate-stones in the south of the area contrast with the smooth, closely textured, highly fissile roofing slates produced from the Upper Devonian deposits around Tintagel and Delabole. Delabole slates were highly prized from early medieval times for their lightness, durability and ability to throw off rain. Documentary and archaeological evidence shows that Devon and Cornish 'blue' roofing slates were transported by sea to areas as far as Southampton and its hinterland, and also to Bristol and

For **Figs 2.1–2.9** see pages 11–12 in the colour section.

South Wales. According to Richard Carew, Cornish roofing slates were also exported to Brittany and the Netherlands.[9]

As a building stone slate is much more workable than granite, both in the quarry and on site. Even when machine sawing and planing largely superseded hand tools in the nineteenth century, the hand-splitting of slate continued. Being easy to cleave into relatively narrow blocks it is easy to handle and lay, usually in some form of coursing, and in the highest-quality slate walling very fine jointing can be achieved. The fissile qualities of slate make it a particularly versatile building material with many uses, in addition to providing good roofing and walling. Slate provided one of the best domestic paving materials in the past: well laid flags lasted for generations and were being laid in houses in the region, particularly in entrance passages, kitchens and dairies, well into the nineteenth century. In dairies, slabs were used for shelving and tables. Slate was also an alternative for meat-curing troughs and water cisterns; the slabs were fitted together with watertight grooves that were a tribute to the maker's skill. The same expertise produced slate slab leats; one good example is the leat which carries water several hundred yards along the top of a wall to a water wheel in Lemail farmyard, Egloshayle.

A characteristic use of slate in west-country building in which the skills of roofing and walling literally overlapped was in slate-hanging. In a technique resembling roofing, courses of overlapping slates provided a weatherproof exterior on a wall. Slate-hanging grew in popularity from the later sixteenth century; an early dated example is a slate-hung house by the church steps at St Columb Major, of 1638. Here fish-scales provide a more decorative alternative to rectangular slates, as they do in numerous slate-hung houses in the region. Nineteenth-century builders used slate-hanging as a more utilitarian technique which could often cover up much jerry-building and insubstantial walling—a further example of the very varied role of slate in west-country buildings.[10]

LIMESTONES (**Fig. 2.5**)

Limestones are widespread throughout Devon but deposits in Cornwall are smaller and more localized. The varieties of limestone in the South West show dramatic contrasts in appearance and texture as building stones. In the far south-eastern corner of Devon there is the shelly Cretaceous limestone of the Middle Chalk formation known as Beer stone from its very localized place of origin. It is soft enough to be sawn on extraction but, like some other limestones, hardens on exposure. It consists almost entirely of coarse shelly fragments and has a uniform texture and close grain. For all these reasons it has been valued as a high-quality freestone from medieval times when it was used for churches and other prestigious buildings far beyond its area of origin; these included the King's Chapel at Westminster and old London Bridge. Since it could be extracted in blocks weighing as much as 6–8 tons, contemporary documents refer to the 'great stones of Beer'.[11] Elsewhere in Devon and Cornwall limestones are of a very different character. The much

older Devonian limestones of south Devon occur primarily in the Ashburton–Chudleigh–Newton Abbot area, in the coastal belt of Torbay, and at Plymouth. They are hard, rugged and difficult to work. However, they have an attractive colour range combining soft greys and pinks with darker veining and when polished can be transformed into a decorative stone with a resemblance to marble—hence Torquay, Chudleigh or Ashburton 'marble'. Thin lenticular deposits of other limestones occur elsewhere in Devon, particularly within the deep water Lower Carboniferous deposits which lie in an arc around the granite of north Dartmoor. Limestone was never of importance as a local building stone in Cornwall. However, it was an essential constituent of good-quality mortars, renders and plasters in both Cornwall and Devon.

SANDSTONE (**Fig. 2.6**)

Sandstone occurs in a variety of forms over much of Devon and in some areas of Cornwall (**Map 2.1**). Sandstones were laid down in varying continental and marine environments, which explains their great variety in type, colour and building qualities. In the far east of Devon, the Salcombe Regis and Branscombe area produces calcareous sandstone from the Upper Greensand: it is creamy-grey in colour and has some similarity to the much better known Cretaceous limestone found at Beer, not far away. In the Middle Ages Salcombe Regis sandstone was in demand as a high-quality building stone and was used not only in a number of local churches but well beyond the area for prestige buildings such as Exeter Cathedral.[12] In the areas of north and west Devon where slates and sandstones commonly occur in the same quarries, builders usually preferred the sandstones, the slate being of low quality and less durable. Many farmhouses and farm buildings and a number of churches feature local sandstone in a variety of shades from the predominant greys north of South Molton to the creams and browns farther west.

The really dramatic impact of sandstone as a building material is provided by the continental hot desert deposits of the New Red Sandstone formed in the Permo-Jurassic age. The impact is heightened where this sandstone associates with red breccia conglomerates (e.g. Heavitree stone) and the pinks and purples of Exeter Volcanic Trap (see below). Freestones from long-established quarries in the Whipton and Broadclyst districts give many buildings of the Exeter area their distinctive character. Towards Ottery St Mary and westward to Crediton the colours of these sandstones become more subtle. Medieval builders used their soft purples and greyish pinks to great effect in exterior walling and interior detail, as for example at Crediton.

FLINT AND CHERT

Flint and the related stone, chert, are more generally associated with building in the neighbouring counties of Dorset, Somerset and

Wiltshire than in Devon or Cornwall. However, there are extensive deposits of chert and flint in the Upper Greensand and Chalk formations on the Devon–Dorset border and these extend over a wide area of east Devon from the Dorset border to the coast around Sidmouth. When 'in lump' (rough-hewn) both are very workaday materials but their decorative possibilities when knapped were widely exploited by Devon builders of the area, particularly in houses of the sixteenth and seventeenth centuries. One of the most dramatic displays of the material is in the seventeenth-century courtyard of Cadhay near Ottery St Mary, where it is combined with sandstone in striking chequered and zig-zag patterns.

Chert and flint are both difficult building materials. The irregularly rounded shapes of flint nodules make coursing and tying-in difficult, if not impossible, without a great deal of manipulation. A large proportion of strong lime mortar is needed, which tends to restrict the height of walling raised at a time and calls for extended drying-out periods.

COB (Fig. 2.7)

Cob must be the building material most associated with Devon in the minds of many visitors taking home memories of picturesque 'cob and thatch' villages. Houses and cottages of cob predominate in many areas of the county from the north and parts of the west, through mid-Devon and towards the southern coast (**Map 2.3**).[13] Cornwall is not thought of as cob country but the material was used in a number of areas including north-west of Launceston, around Padstow, on the Roseland peninsula, and in the area around Truro and the Fal and Helford estuaries. Cob also occurs on the fringe of the rugged granite terrain of West Penwith.

Many west-country cob houses date back to the seventeenth century and some survive from the Middle Ages. The cob of Devon and Cornwall is a conglomerate material made up of a carefully balanced mixture of clay, straw and an aggregate such as small stones, 'shillet' or gravel, and then mixed with water. The straw ingredient was vital as a binder and to minimize the effects of shrinkage as the mixture dried. Sometimes dung is found in old cob walling but there is uncertainty as to its role; the fact that the laborious mixing of cob was often performed by cows trampling it in an enclosure may be the chief reason for its presence. Suitable clay for cob can be very localized, sometimes existing in one area of a parish and not in another. Cob walls reflect the colour of local soil and stone types; there are the browns, yellows and buffs of north and west Devon and the pinks and plums of New Red Sandstone territory of parts of mid-Devon and the south and south-east. Most house walls of cob were rendered, again in shades which reflected local subsoils.

The great number of cob-built houses and farm buildings which survive in the South West, together with the even greater number that must have disappeared, emphasize the importance of cob as a vernacular material. It was locally available and could be used by local labour in a process which involved traditional 'know-how' rather than formal craft skills. Cob house-building continued into the nineteenth century and a few quite substantial cob town houses survive in Devon. Interest in cob revived after the First World War, when it was promoted as a building material which combined economy and practicality.[14] However, nothing in Devon and Cornwall resulted comparable in scale to Dorset cob council house building between the wars.

BRICK

Brick was not a vernacular building material in Cornwall or in most areas of Devon. It was not widely produced or generally available until relatively late. In Cornwall it was viewed with suspicion and considered not reliable in Cornish weather conditions. Richard Carew voiced these suspicions and Thomas Tonkin reiterated them in his notes on Carew a hundred years later.[15] One reason why brick became accepted in the region was that bricks came into west-country ports from the Low Countries. It was traditionally supposed that they originally came as ballast but records show that brick was imported into Exeter as a product in its own right as early as the fifteenth century.[16] Whether as import or ballast, bricks used for building appeared increasingly in coastal towns in Devon and to a lesser extent in Cornwall from the turn of the sixteenth and seventeenth centuries. By the late seventeenth and early eighteenth centuries, more of the gentry

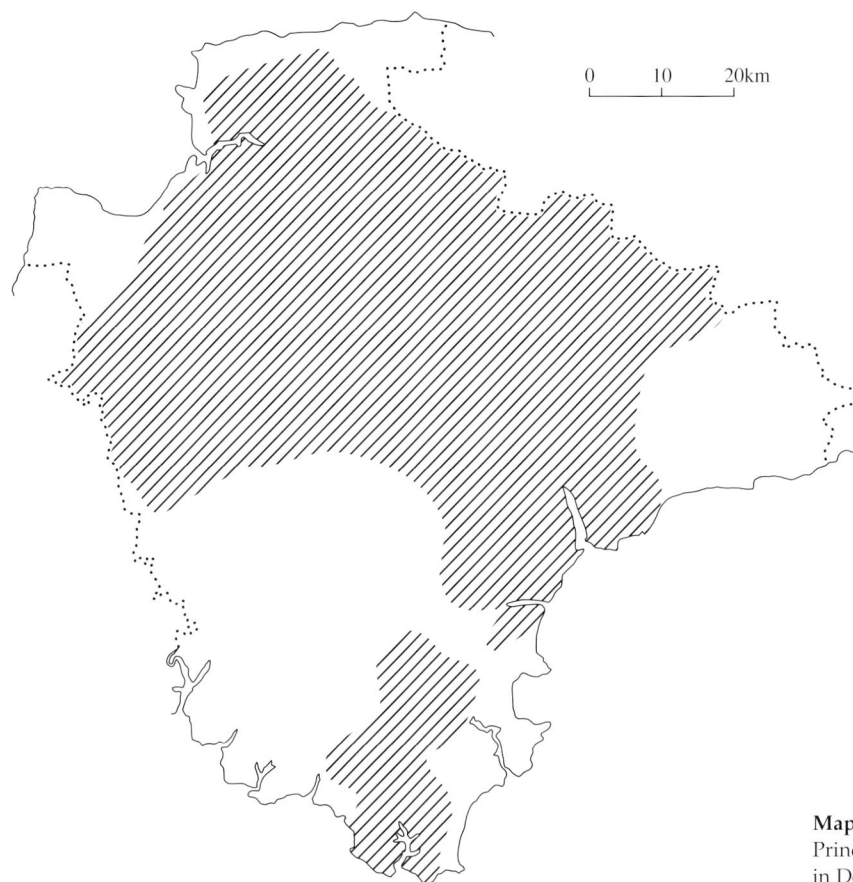

Map 2.3
Principal areas of cob building in Devon.

For **Figs 2.1–2.9** see pages 11–12 in the colour section.

were responding to new architectural styles by erecting classical mansions in brick. The adoption of brick as a building material in the South West was accompanied by experimentation in local brick-making and the gradual spread of brickworks over the two counties. Many of these were small concerns using newly discovered local clay deposits. The older *Geological Memoirs* for Devon and Cornwall mention many now forgotten brickworks, but their bricks, in a variety of colours from creams and yellows to shades of red, still give local character to towns such as Ilfracombe, Tiverton, Newton Abbot and the mining towns of west Cornwall. In the far South West, the great Cornish mine engine-house chimneys are perhaps some of the most impressive monuments to the mastery of a material that was, by the nineteenth century, no longer unfamiliar and mistrusted.

MINOR BUILDING STONES OF CORNWALL AND DEVON

In addition to the deposits of better-known rock types in Devon and Cornwall described above, there are some more localized and lesser-known building stones. Some were used only because there was nothing better in the area; others made first-class materials and added a special quality to the buildings in which they were used.

Elvan or quartz porphyry is one of the latter and is among the most attractive of the higher-quality building stones of the South West (**Fig. 2.8**). It is an igneous rock occurring in extremely localized pockets or 'dykes' within the geological shadow of the granite masses. With some similarity to granite but finer grained and more easily worked, its colour ranges from buff to cooler cream and greys. In Devon the principal occurrence is on Roborough Down and in adjacent parishes on the western fringe of Dartmoor where it was known and worked from the Middle Ages. South of St Austell there are a number of deposits as far as the coast at Pentewan, where the stone given that name was highly regarded from Norman times to the eighteenth century. Elvan is a stone which deserves a higher profile in discussions of the architectural history of the South West.

Some of the other less common west-country rocks occur in such limited areas that their idiosyncratic colour and texture give a particularly local character to the buildings in which they were used. At Dartmouth there is a distinctive greenish hue to the local dolerite used in the castle and nearby buildings. Up the river at Totnes, a reddish-brown volcanic tuff from a very localized source occurs in buildings in and around the town. Another volcanic stone, specific to the Tavistock area, is Hurdwick, named from a quarry in which it occurs a mile to the north of the town. Because it cut easily into relatively small blocks it was popular with the builders of Tavistock Abbey on whose land the quarry lay. The duke of Bedford used Hurdwick stone from his quarries in the extensive Victorian rebuilding of Tavistock, most notably the Town Hall and market area and adjoining streets. In poor light it is a rather drab stone but evening sunlight can give it a blue-green luminosity.

The Lizard peninsula's outstandingly complex geology (see **Map 1.2**) yields a kaleidoscope of rocks of varying suitability for building. Horneblende to the west, serpentine in the centre and south, and gabbro to the south-east were all tried. Sometimes they mingle indiscriminately in buildings, suggesting a strong element of experimentation, desperation or resignation in local builders. Hornblende is a difficult stone to work and gabbro is a coarse igneous rock which can only be worked into rough lumps for uneven random walling. Serpentine has a reputation for not being rainproof.

Volcanic activity occurred sporadically in the South West from early Devonian to Permian times and is responsible for most of the more exotic rocks of the region (**Fig. 2.9**). One of these is Exeter Trap, a continental basaltic rock originating from cooling lava flows in Permian times. It is centred in Exeter where a principal source was in the quarry near the city centre at Northernhay, but it also occurs well to the north-east of the city in the Thorverton and Killerton areas, and south to Dunchideock and west to beyond Crediton. Trap is a hard but workable stone with colours varying from brown to pinkish-grey, often with distinctive pale veining. Its qualities were appreciated from early times; Exeter Trap occurs in the Roman city walls and the Norman castle of Rougemont, close to the quarry.

Another material of the same geological era is the red breccia found over a wide area around Exeter. Like the red sandstones, it derives from desert conditions which produced its mixture of coarse sand and angular fragments with strong staining from iron oxides. Its soft, crumbly texture makes it unreliable but it was popular as a cheap and easily worked material. For centuries it was produced in large quantities, most notably in Heavitree quarry, until overtaken in the nineteenth century by another economy material: brick. Before that period red sandstone, breccia and Trap were used in buildings all over south and central Devon, mingling and toning with the red earth of the area to provide a striking display of the relationship between geology, landscape and local buildings.

FURTHER READING

Peter Beacham (ed.), *Devon Building* (Exeter, 1990) has a useful chapter on building materials. Bridget Cherry and Nikolaus Pevsner (eds), *The Buildings of England: Devon* (London, 1989) has a very full introduction including a section on Devon building materials by Alec Clifton Taylor. Nikolaus Pevsner and Enid Radcliffe (eds), *The Buildings of England: Cornwall* (London, 1970) has discussion on Cornish building materials by Alec Clifton Taylor. See also V.M. and F.J. Chesher, *The Cornishman's House* (Truro, 1968); Alec Clifton Taylor, *The Pattern of English Building* (London, 1980), and E.A Edmonds, M.C. McKeown and M. Williams, *British Regional Geology: South-West England* (London, 1985).

SOUTH-WEST
ENGLAND
BEFORE THE
NORMAN CONQUEST

PALAEOLITHIC

THE EARLIEST HUMAN OCCUPATION

ALLAN STRAW

Map **3.1** records findspots of artefacts referable to the Lower, Middle and Early Upper Palaeolithic cultural stages. The vast majority of the artefacts are in secondary context; only in Kents Cavern in Torquay and at Broom in the Axe valley does material in near-primary context survive. The map has been compiled from reports for Devon and Cornwall of the Southern Rivers Palaeolithic Project by Wessex Archaeology, from publications by Roe and by Campbell, and from information provided by Dr Alison Roberts, author of the next chapter on the Mesolithic.[1] At most locations a single artefact only has been recovered. The most prolific area is the Axe valley in east Devon, and Kents Cavern provides the best record of the three stages together under reasonable stratigraphic control. **Fig. 3.1** relates deposits to cultural stages and oxygen isotope ratio stages.

valley.[2] Caves, like Kents Cavern, apart from attraction as shelters or dens, also acted as natural traps for sediments, including at times plant and animal fossils and artefacts. Flattish interfluve areas suffered relatively little loss by erosion and artefacts found on them have probably not travelled far from their point of abandonment.

The most recent period of excessive slope modification and valley-floor aggradation was the Late Devensian stage (20,000–10,000 years ago), for Devon and Cornwall a period mostly of periglacial frost-action and high snowfall. The earlier part of this stage was too cold for human occupation, and this permits a discussion of the grossly disturbed and displaced evidence for the Lower, Middle and Early Upper Palaeolithic stages separately from the less disturbed succession from the Late Upper Palaeolithic to the present day.

SPATIAL DISTRIBUTION

The distribution of findspots is the consequence of the interaction of three factors. It is primarily a record of finds and at once an uneven and small sample of the material potentially to be discovered, it is constrained by the original extent and density of human occupation, and it is a legacy of the displacement by geomorphological processes subsequent to occupation.

Most findspots lie eastward of Dartmoor and no finds to date have been made across north and central Devon or in the Tamar basin. Early man certainly reached Cornwall on one or more occasions during the Quaternary and there are no physiographic or biogeographic reasons why early man should not have traversed central Devon and Exmoor to reach the north coast. Future discoveries might well be made but it can be observed that the Tamar basin retains few deposits of terrace gravels, while those in the Exe basin, although far more extensive, are not commercially attractive and remain unquarried.

The findspots can be placed into three geomorphologically determined groups: on interfluves, on valley floors (often in river terrace gravels) and in caves. Unless early man deliberately avoided valley sides, this grouping is most probably a consequence of erosion and transport processes that carried soils, rock-waste and any contained artefacts down toward valley floors, as in the Axe

CHRONOLOGICAL DISTRIBUTION

In the last few decades studies of ocean-floor sediments have provided the most complete record of Quaternary climatic changes to date.[3] Based on variations in oxygen isotope ratios (OI) in fossilised shells of minute marine organisms, a sequence of warm (odd-number) and cold (even-number) stages has been established for most of the past one million years against which British terrestrial evidence can be evaluated and correlated.[4] However, the only correlations accepted at all widely in Britain are that the Anglian glaciation (the most extensive) occurred in OI Stage 12, the Late Devensian glaciation (the most recent) in OI Stage 2, and the Ipswichian interglacial in OI Stage 5e (see **Fig. 3.1**). Much evidence for other interglacials and glaciations exists but allocation to stages remains controversial.

The record of early man in Britain has also been subject to revision. The typology of handaxes, for example, is no longer an infallible indication of age; rather the taphonomy of artefacts must be carefully assessed and, where possible, a variety of independent dating techniques must be employed.[5] For the South West this is only possible for the artefacts in the Axe terrace gravels and for those at some of the cave sites.

THE LOWER PALAEOLITHIC

Most locations on the map are *surface finds* which, being in secondary context, provide no information of the times when they were made, used or lost, A single artefact might date from anywhere within the 400,000 years of the British Lower Palaeolithic.

The oldest artefacts from Kents Cavern, Torquay, including at least fourteen Acheulean handaxes, were recovered by William Pengelly during his 1865–80 excavations.[6] They occurred randomly, in secondary context, within the breccia (**Fig. 3.1**) which has every appearance of an extraneously derived flow-earth that penetrated several openings (now blocked) in the rear of the cave system.[7] The breccia is composed of slate, siltstone and sandstone fragments in a clayey-silt matrix derived mostly from higher ground outside the cave system.[8] While it is possible that the artefacts had previously been abandoned inside the cave, the strongest likelihood is that they are a small proportion of a relatively large assemblage that had accumulated near, and probably in, the cave mouths that existed at that time.

The artefacts are indubitably older than the Crystalline Stalagmite (**Fig. 3.1**) that overlies the breccia. Samples from the basal parts of this speleothem have yielded Uranium-series and Electron Spin Resonance dates which indicate that it began to form at least as early as OI Stage 9 and probably in OI Stage 11.[9] The breccia, older and of periglacial character, is surely of Anglian date (OI Stage 12), so that the artefacts may have originally been made and abandoned either during an early Anglian interstadial or in the preceding stage. Certain derived elements of the fossil fauna of the breccia have a Cromerian affinity, which supports a view that the artefacts are as old as those discovered elsewhere in Britain at Boxgrove, Weston-sub-Mendip and High Lodge.[10]

The Axe valley, in particular the Broom locality, has yielded thousands of Acheulean handaxes, some flakes, and debitage from the fluvial gravel of the Axe terrace formation.[11] At Broom the deposits comprise some 2m of moderately stratified clay, sand and gravel sediments separating over 5m of lower gravels from up to 15m of upper gravels. Most of the unrolled handaxes have been found in association with these middle beds (probably river flood-plain sediments) in near-primary context.[12] Other artefacts, most rolled and randomly distributed, have been recovered from the upper gravel.[13] This is a cold-climate aggradation revealing braided-stream structures. The lower gravel is also probably periglacial, but pollen extracted from the finer middle beds

Map 3.1
Findspots of Palaeolithic artefacts.

SOURCES: Wessex Archaeology, 'The South West and South of the Thames', *Report No. 2*, 1992–3, Southern Rivers Palaeolithic Project (Trust for Wessex Archaeology Ltd and English Heritage, 1993); D.A. Roe, 'A gazetteer of British Lower and Middle Palaeolithic sites', *Research Report of the Council for British Archaeology* (1968), 8, 42–8, 256–61; D.A. Roe, *The Lower and Middle Palaeolithic Periods in Britain* (London, 1981); J.B. Campbell, *The Upper Palaeolithic of Britain: A Study of Man and Nature in the Late Ice Age*, 2 vols (Oxford, 1977).

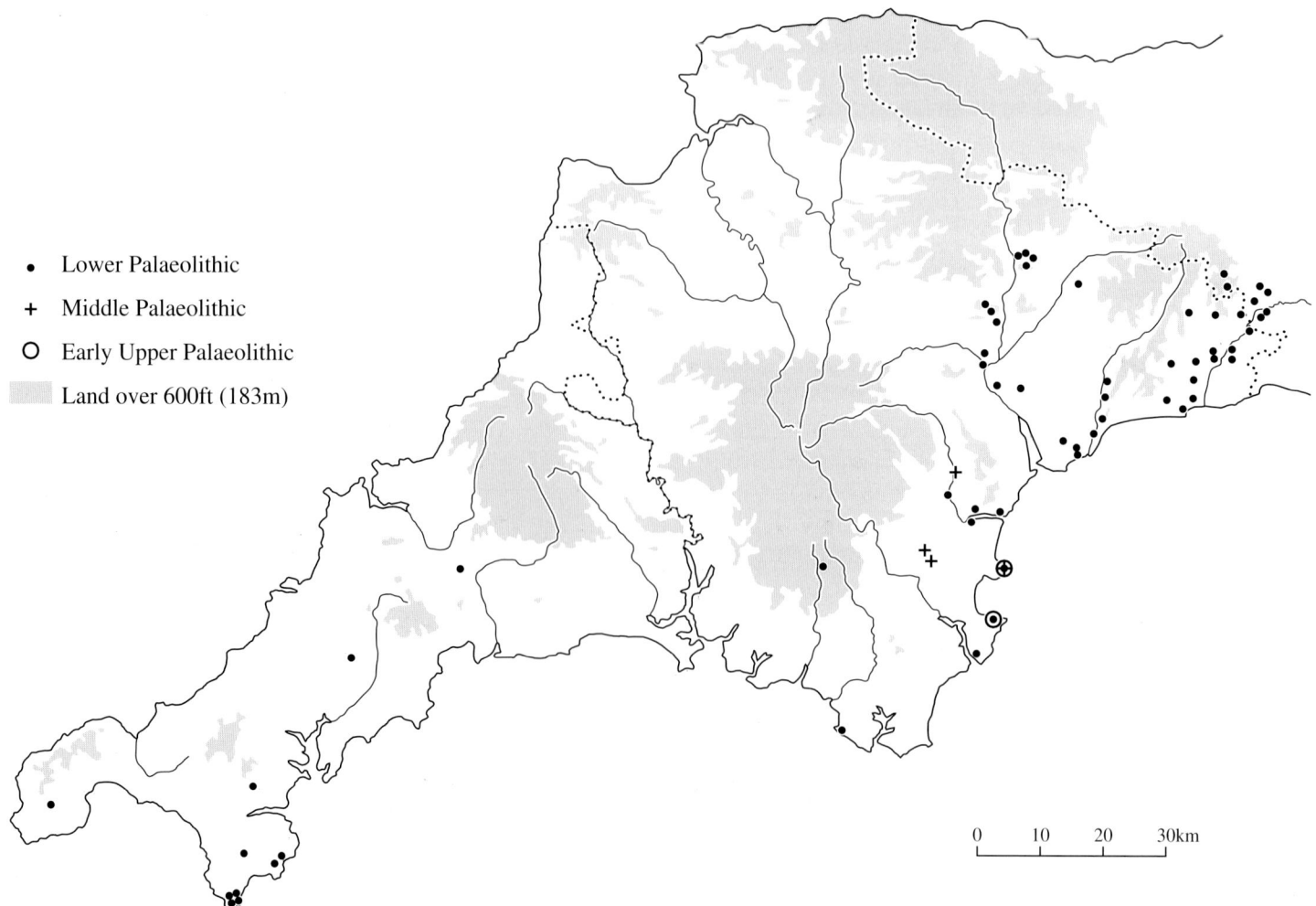

- • Lower Palaeolithic
- + Middle Palaeolithic
- ○ Early Upper Palaeolithic
- ▨ Land over 600ft (183m)

0   10   20   30km

Deposits	Thickness (m)		Cultural stages	Postulated oxygen isotope ratio stages
Black Mould	0 - 0.3			1
Granular Stalagmite	0 - 1.6	Includes material from Crystalline Stalagmite and Breccia	Late Upper Palaeolithic	
Cave Earth - Stony Loamy	0 - 10		Early Upper and Middle Palaeolithic	4 - 2
Fluvial silts, sands and fine gravels	0 - 1			
MAJOR EPISODE OF DISLOCATION AND EROSION				
Crystalline Stalagmite	0 - 4			11 - 5e
Breccia	0 - 3	Includes pieces of older speleothem	Lower Palaeolithic	12
Fluvial silts, sands and fine gravels	0 - 1			

Fig. 3.1
Relationships between Palaeolithic deposits, cultural stages and oxygen isotope ratio stages.

indicate their deposition during a more temperate phase.[14] The axes may have been made by Acheulean hunters on a former flood-plain of the River Axe from materials rich in chert, the original source of which is the Upper Greensand that crops out around the Axe valley.

The Axe terrace today is elevated and fragmented along both sides of the valley. Its surface from Chard Junction to Kilmington declines at 1:600, compared with the present flood-plain gradient of 1:250. Below Kilmington to Seaton, the surface slope is steeper at 1:300 and the deposits, over 25m thick, rest on a near-plane surface of Mercia Mudstones up to 10m above the present flood-plain. By comparison with similar valley accumulations in south Devon, it is most probable that younger gravels of Devensian age (OI Stages 3, 4) lie beneath the flood-plain alluvium and pass upstream at least as far as Chard Junction Dairy, where they are at least 6m thick.[15] The upper gravel at Broom might then be ascribed to one of the colder phases of the Wolstonian (OI Stages 6–10), and the finer sediments with handaxes to a Wolstonian interstadial (OI Stage 9) or to the Hoxnian interglacial (OI Stage 11).[16] During construction of the Axminster bypass in 1990 at least 5m of bedded sand and cherty gravel were exposed at 68m OD with the base some 18m higher than the surface of the Axe terrace. This vestige of a higher terrace might represent the Anglian cold phase (OI Stage 12). However, the latter (incorporating the most extensive British glaciation) might have been responsible for the upper gravel of the Broom sequence (the major deposit in the valley), and if so the artefacts might then be as old as those in Kents Cavern and elsewhere.

Clearly a precise age for the Axe valley artefacts remains elusive, but they constitute a remarkably rich assemblage associated closely with fluvial deposits (the middle beds) for which an early Wolstonian interstadial (OI Stage 9) is preferred. On this basis, the peoples responsible for the Kents Cavern artefacts had left Devon some 150,000 years before the arrival of those who occupied the Axe valley. Since they left, presumably because of climatic deterioration, there is no firm evidence for early man in the South West for another 200,000 years.

THE MIDDLE PALAEOLITHIC

When humans did reappear in Britain, probably after 66,000 years ago, they brought Middle Palaeolithic cultures, but the evidence is sparse compared with that found on the Continent.[17] Artefacts are generally referred to the 'Mousterian of Acheulean Tradition', and it is likely that they were associated with *Homo sapiens neanderthalensis*.[18] Only four findspots are recorded in the South West, all caves.[19] In the Teign valley, Cow Cave has yielded several flakes. Three Holes Cave and Tor Point Cave in the Torbryan group have each produced a single artefact.

The richest site, certainly in Devon and perhaps for Britain, is Kents Cavern, where some material is believed to occur in primary context.[20] Artefacts and debitage lie within lower parts of the Loamy Cave Earth (**Fig. 3.1**), not in the much older breccia as erroneously reported by Rosenfeld.[21] Roe acknowledges forty-five Mousterian artefacts: thirty-three surviving from Pengelly's haul of over 1,000 pieces (including Early Upper Palaeolithic) from the Loamy Cave Earth and twelve discovered by Ogilvie between 1926 and 1941.[22] The assemblage includes handaxes, side-scrapers, saws and awls. The Cave Earth consists of clasts of sandstone and particularly limestone in a loose silty-clay matrix, and its fossil fauna is predominantly of cold character with mammoth, reindeer, woolly rhinoceros and brown bear, but dominated by hyena and horse.[23] Substantial amounts of Cave Earth remain in the Wolf's Cave so that the possibility of obtaining better stratigraphical control on artefact distribution exists. However, the original incremental accumulation of the Cave Earth by mass movement and wash processes, and its disturbance by hyenas and rodents in the past, will continue to cause problems.

THE EARLY UPPER PALAEOLITHIC

In Europe the Middle Palaeolithic was replaced relatively suddenly by the Upper Palaeolithic around 40,000 years ago with the immigration of wholly modern humans (*Homo sapiens sapiens*).[24]

Kents Cavern was occupied by Early Upper Palaeolithic peoples on several occasions and large numbers of artefacts have been recovered by Pengelly and later workers. The artefacts are associated stratigraphically in higher parts of the Loamy Cave Earth with a rich mammal fauna, dominated by horse and brown bear with fewer hyena and no mammoth, largely a consequence of hunted and scavenged animals being taken into the cave by both humans and hyenas.

Radio-carbon dates broadly delimit the age range of the artefacts between 38,000 years and 27,500 years ago.[25] A piece of human jaw with three teeth discovered in the Cave Earth by Ogilvie in 1927 has recently been dated at almost 31,000 years, the oldest known date on a modern human fossil in Europe.[26] In stark contrast to the wealth of material in Kents Cavern, only two other

sites in Devon have yielded Early Upper Palaeolithic material. A single 'leaf-point' artefact, the typical element, has been found at Bench Cavern and another at Windmill Hill Cave, both located in Brixham.[27]

CONCLUSION

The meagre occurrences of the Middle and Early Upper Palaeolithic within 20km of Tor Bay, like the Lower Palaeolithic finds, cannot be truly representative of either the original range of the people in the South West or survival of their artefacts. Further discoveries will be made, but the profound reworking of soils and regolith, especially on sloping ground, and the transport and rearrangement of valley-floor sediments by rivers during the Late Devensian cold stage make it highly unlikely that more material in primary context will be found outside caves. Thus **Map 3.1** remains, no more and no less than is claimed, a record of findspots, but Devon undoubtedly contains two important sites: Kents Cavern, which provides some evidence for the earliest occurrence of early man in Britain and contains a highly valuable record of Middle and Early Upper Palaeolithic occupation, and the Axe valley with its prodigious evidence for Lower Palaeolithic occupation of an open valley floor.

FURTHER READING

R. Leakey, *The Making of Mankind* (London, 1981) is a thoughtful and personal explanation and enquiry into human origins by a member of a family long involved in the archaeology and anthropology of early man in Africa, leading into a balanced discussion of world-wide human evolution, spread and social development, and drawing on material gathered for a BBC TV film series. J.J. Wymer, *The Palaeolithic Age* (London, 1982) is a comprehensive and scholarly global review of the evolution, distribution, artefacts and ways of life of people of the Old Stone Age. R. Leakey and R. Lewin, *Origins Reconsidered* (London, 1992) is a reflective and evocative account of fossil-hunting in Africa, and of current thinking about human evolution, especially the meaning and mystery of 'human'. R. Lewin, *The Origins of Modern Humans* (New York, 1993) is an authoritative, though somewhat journalistic account and discussion of the ancestry, distribution and distinctiveness world-wide of our own species, *Homo sapiens sapiens*.

LATE UPPER PALAEOLITHIC AND MESOLITHIC HUNTING-GATHERING COMMUNITIES 13,000–5,500 BP

ALISON ROBERTS

The period covered by **Map 4.1** includes the end of the last ice age (Pleistocene period) and the beginning of the following warm interglacial period in which we live today (Holocene period). All of the humans who lived in Britain and Europe during this period had a nomadic hunting and gathering lifestyle.

THE LATE UPPER PALAEOLITHIC: *c.*13,000–10,000 BP

Britain was abandoned by humans during the coldest part of the last ice age (Last Glacial Maximum, *c.*18,000 BP), but was recolonized during the subsequent warm period known as the Lateglacial Interstadial. Exactly when the humans returned is still uncertain, but it appears to have taken place by *c.*12,600 BP.[1] There are three separate archaeological traditions known during this period in Britain: Creswellian, Final Upper Palaeolithic, and 'Long Blade' industries. All of these industries have close parallels with contemporary archaeological traditions on the European mainland. This situation is not surprising as throughout the period sea-levels were lower than today, and Britain was connected to the rest of Europe by extensive areas of dry land which now lie underneath parts of the North Sea and English Channel. There would have been no obvious physical barrier to present-day Britain for mobile hunting-gathering groups, and it is likely that the same peoples were occupying the entire north-west European territory including Britain.

The earliest of the British Late Upper Palaeolithic assemblages is the 'Creswellian', which is assumed to be a regional variant of the European Late Magdalenian, a period perhaps best known for its art.[2] Creswellian sites tend to be very small and located in limestone caves such as those in the Creswell Crags (Derbyshire) from which the assemblage type takes its name. **Map 4.1** shows that they are so far known in the South West only from a few caves in south Devon. There is no known cave art of this period in Britain, and very few examples of engraved bone. Creswellian assemblages are characterized by distinctive angle-backed blades which are presumed to be spear-tips, although the edges of some examples also seem to have been used, perhaps as knives (**Fig. 4.1A**). Creswellian sites usually contain only a limited range of tools, and few contain evidence for

flint-working at the locale. Three Holes Cave (Devon) is typical of British Creswellian sites and has been interpreted as a short-term hunting camp where horse was the main prey animal.[3] Kent's Cavern (Devon) and Gough's Cave (Somerset) are unusual Creswellian sites and both contained larger assemblages and a wider range of artefact types. They seem to represent longer-term occupations, and may have served as 'base camps' for hunting groups exploring a region. The latter also contains human skeletal remains which seem to show cut marks from stone tools, perhaps as part of a burial ritual.[4] A high-quality flint is used at all of the known Creswellian sites, and it is probable that the material found at the Devon sites was brought in from a considerable distance. The only geological source of high-quality flint in the South West is at Beer in east Devon, and this does not appear to have been used during this period. Creswellian assemblages seem to occur *c.*12,600–12,000 BP, a time during which the landscape in Britain would have been relatively open.

British Final Upper Palaeolithic industries are seen as being comparable with the north European *Federmesser* (Penknife point) tradition, and are characterized by similar curve-backed blades and points to those found on the Continent (**Fig. 41B**).[5] Sites and findspots of Final Upper Palaeolithic type are far more frequent than Creswellian ones and are found in both cave and open air localities. In the South West, they are found in Cornwall as well as Devon, and one 'Penknife point' has even been found on the Isles of Scilly (**Fig. 4.1C**).[6] Unlike in the Creswellian period, locally available lithic raw materials appear to have been used for tool-making at these sites. The preference for local lithic materials may indicate changing patterns of mobility by hunter-gatherers in a landscape now dominated by birch woodland. These industries seem to have been prevalent throughout the twelfth millennium BP, a period which saw temperatures steadily declining until the warm interstadial ended about 11,000 BP.

During the following millennia temperatures became sufficiently cold to allow the regrowth of glaciers in northern Europe. Much of the region seems again to have been abandoned by humans as polar conditions returned, and Britain does not appear to have been occupied again until *c.*10,200 BP. The sparse archaeology of this period includes 'Long Blade' lithic assemblages characterized by blades *c.*12cm and over in length. These have been compared to the

Map 4.1
Finds of tools dated to the
Late Upper Palaeolithic
and Mesolithic.

SOURCE: the author
acknowledges the
assistance of Peter
Berridge.

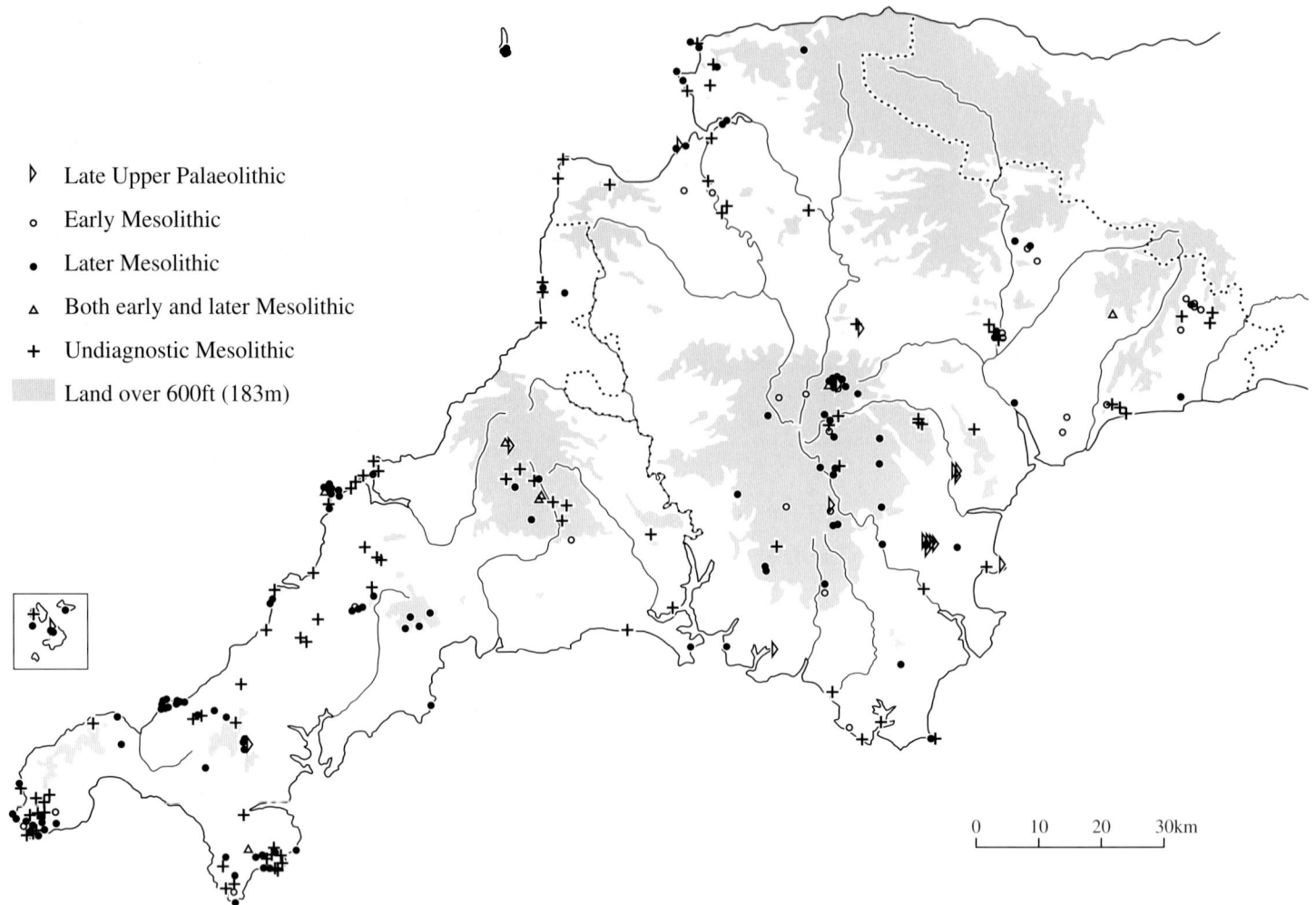

▷ Late Upper Palaeolithic

○ Early Mesolithic

● Later Mesolithic

△ Both early and later Mesolithic

+ Undiagnostic Mesolithic

░ Land over 600ft (183m)

0   10   20   30km

Ahrensburgian 'tanged point' industries of northern Europe: the Reindeer hunter tradition which evidences the first known use of the bow as a hunting weapon in Europe.[7] No sites of this type are known as yet from west of the Upper Thames region. However, potentially of the same age are stray finds of small tanged points including one from Doniford Cliff (Somerset). No such finds are yet known from Devon or Cornwall, but may very well be found in the future.

This final cold phase of the last ice age ended abruptly c.10,000 BP, which is the date traditionally taken as marking the start of the Holocene. At this time there was a rapid climatic warming across Britain and western Europe, and temperatures seem to have risen to as high or slightly higher than those of the present day. It is in this warm, increasingly forested environment that the first Mesolithic industries are found.

THE MESOLITHIC: c.9500–5500 BP

It is customary to present dates for Mesolithic periods in years BP (Before Present) which are age estimates based on radiocarbon

analysis. Such age estimates need adjustment (calibration) to convert them to calendar years BC.

As originally defined, the Mesolithic (Middle Stone Age) is the period in between the Palaeolithic (Old Stone Age/mobile hunter-gatherers) and the Neolithic (New Stone Age/sedentary farmers). As such, the period has often been treated as a transitional phase between these two contrasting lifestyles. The Mesolithic is perhaps better regarded as being the period of early Holocene hunter-gatherers in northern Europe.

In Britain, the period has been divided into two parts: the Early, c.9600–8500 BP, and Later Mesolithic, c.8500–5500 BP.[8] The chronological boundary is based on the date formerly accepted for the opening of English Channel and the separation of Britain from the European continent. Although there is now evidence that the Channel did not open completely until much later,[9] on the basis of other criteria a start date for the Later Mesolithic of c.8500 BP still seems a reasonable estimate.

The diagnostic tool types of Mesolithic industries are transversely sharpened flint axeheads and small microlith points (**Fig. 4.1D–G**). Both of these tools are interpreted as being innovations related to survival in a forested environment. The axes are presumed to be specialist tree-felling and/or woodworking tools. Complete examples are rare in south-west Britain, but their use is shown by broken axe fragments and resharpening debris. These large tools do not seem to have been made of locally occurring flint, and must have been imported from some distance.[10] As such they are likely to have been valued items which were carefully maintained and preserved. The only exception to this pattern is the large number of axes made of greensand chert which have been found in the Yarty valley of east Devon close to a source of large nodules of this raw material.[11]

Microliths are assumed to have been the tips and barbs of arrows, at least during the Early Mesolithic period. There may have been alternate uses for some of the microlith types during the Later Mesolithic.[12] Typological variation in microlith form and the composition of microlith assemblages are the major means by which different Mesolithic industries and assemblage types are defined. Although the variations within these assemblages are not yet fully understood, the groupings do seem to have a chronological and geographic coherence and it is likely that they represent, at least in part, the design preferences of particular hunting-gathering peoples.

Three different stylistic groupings have been so far recognized for the Early Mesolithic, only two of which are well-represented in the South West.[13] Here, Early Mesolithic sites and findspots are concentrated in the flint- and chert-rich river gravels of east Devon and along the present-day coastline. Isolated finds from inland locations, as well as from Dartmoor and Bodmin Moor, however, suggest that humans were utilizing some of the hinterland, if only on a sporadic basis for hunting trips or as a means to travel between more favoured locations. As yet there are no diagnostic artefacts of this period known from the Scillies.

There is considerably more variation in microlith assemblages during the Later Mesolithic throughout the country. This, combined with a lack of reliable associated radiocarbon dates, results in the fact that the period is understood poorly on a national level. The first attempt to define the period in the South West was in 1979, when Roger Jacobi suggested that there was a distinct Later Mesolithic south-western territory characterized by assemblages dominated by small convex-backed and lanceolate microliths, as opposed to the small scalene triangle, oblique point and straight-backed microlith assemblages common elsewhere in

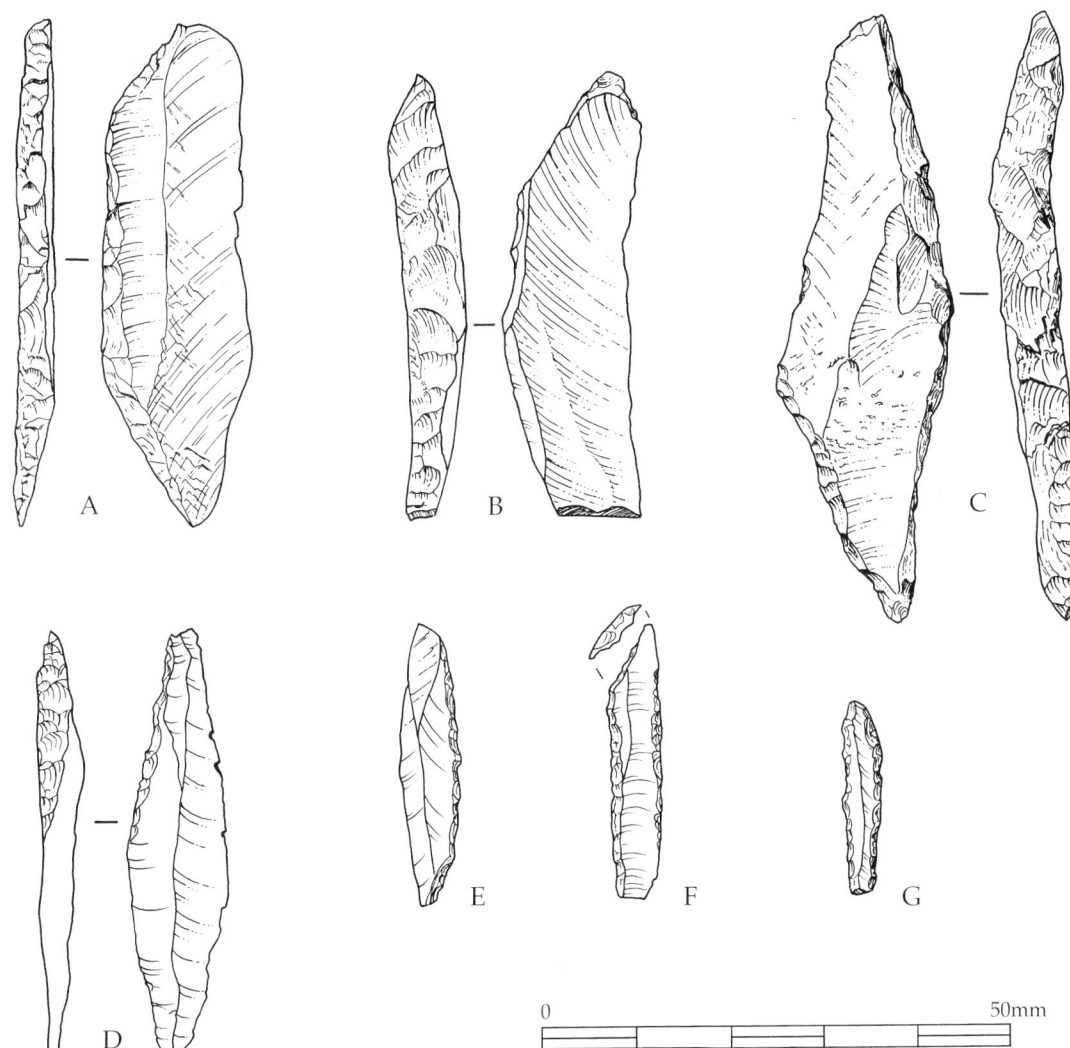

Fig. 4.1
(**A**) Creswellian angle-backed blade (Cheddar Point), Three Holes Cave, Devon;
(**B**) Final Upper Palaeolithic curve-backed point, Pixie's Hole, Devon;
(**C**) Final Upper Palaeolithic 'Penknife Point', St Mary's, Isles of Scilly;
(**D**) Early Mesolithic oblique point, Trevose Head, Cornwall;
(**E–G**) Later Mesolithic curve-backed point, micro-scalene triangle, rod, all from Trevose Head, Cornwall.

SOURCE: original drawings for **A** and **B** by Karen Hughes and **C–G** by Jeff Wallis.

Britain. The idea persists despite the fact that Jacobi now regards Devon and Cornwall as part of a much broader 'Southern English' grouping.[14] In fact, a number of distinct microlith assemblage types exist in the region, and it can be suggested that there is a chronological succession of Later Mesolithic microlith assemblages in the South West: i.e. assemblages dominated by curved-backed forms and small oblique points were replaced by those dominated by small scalene triangles and straight-backed pieces some time in the early ninth millennium BP; these, in turn, were replaced by those dominated by micro-scalene triangles and rods in the early seventh millennium BP (**Fig. 4.1E–G**).[15] However, to what degree the use of these assemblages overlapped is still uncertain.

As **Map 4.1** shows, Later Mesolithic activity can be identified throughout the region including the Scillies. Several clusters of Later Mesolithic sites on headlands along the coast suggest that these were favoured locations. Many of these headlands were located next to estuaries: habitats rich in food resources as well as pebbles of flint and chert. No such clusters are known elsewhere on the present coast of the peninsula, presumably due to the submergence of the Mesolithic coast by the rising sea-level. The granite uplands of Dartmoor and Bodmin Moor dominate the landscape of the region, and, not surprisingly, also seem to have been focuses for human activity during this period. Indeed, there is evidence that humans deliberately used fire to maintain open areas of heathland on Dartmoor during this period, in order to attract game.[16] The rivers and streams which drain the moors would have provided the easiest transport routes to and across them during this time of dense and increasing forest growth. The South West was apparently covered by deciduous woodland throughout the Mesolithic, and this reached its maximum extent during the final part of the period (see chapter 1). That rivers served as Mesolithic travel routes is supported by evidence from inland lowland Later Mesolithic findspots, most of which are located in the valleys of the major river systems and their tributaries. For example, the assemblage from Three Holes Cave, located on a tributary of the River Dart, contains clear evidence of contact with the coast as well as having very close comparisons to flint assemblages from sites farther up the river valley on Dartmoor.[17] In short, the hunter-gatherers who occupied the South West during this period were apparently using the landscape to a much greater extent and in a more complex manner than during the earlier periods.

FURTHER READING

See end of chapter 9.

NEOLITHIC SETTLEMENT, LAND USE AND RESOURCES

FRANCES GRIFFITH AND HENRIETTA QUINNELL

The Neolithic, covering the later fifth, the fourth and third millennia BC, is the period during which, traditionally, agriculture has been regarded as becoming established in Britain, accompanied by a distinctive range of artefacts such as ground stone axes, leaf-shaped flint arrowheads and the first pottery.[1] Such artefacts are still considered innovations, but the importance of agriculture is being increasingly questioned.[2] The evidence supports the presence of domesticated cattle, sheep, goats and pigs, for which substantial woodland clearance would be appropriate, but cereals may have been less important than gathered wild plants.[3] During the Neolithic the first monumental constructions, with varying links to burial rituals, occur. Such monument construction forms a sequence extending into the Bronze Age (see chapter 6).

Map 5.1 and the larger-scale map for the Isles of Scilly (Map 5.2) attempt to show an overall picture of Neolithic activity and of disturbance of woodland cover. There is little definite evidence for decrease in tree pollen in the South West.[4] Concentrations of lithic finds (flint and chert artefacts and waste from their manufacture) may be presumed to indicate intensive, if short, periods of activity.[5] 'Concentrations' represent finds of twenty or more pieces at a location, and in many areas, such as the lower Exe valley, these occur sufficiently frequently to be presented as continuous given the small scale of Map 5.1. Scattered evidence for Neolithic activity has been derived from areas with sparser lithics than 'concentrations' and with round barrow groups. Round barrow construction nationally began during the later fourth and third millennia BC, and continued throughout the second, peaking during its earlier part. Examination of soils, and the pollen they contain, beneath almost all excavated round barrows in the South West, indicates a period of woodland clearance before construction,[6] much of which, given the range of barrow dates, may relate to activity during the third if not the fourth millennium BC. The data, put together, show extensive areas over which human activity may have affected the landscape, even if only for a short while, during the Neolithic. The information derived from lithic finds is closely related to areas of fieldwork carried out during the last century and can confidently be expected to be extended with future work. Those areas shown blank represent terrain where little fieldwork has been carried out or where it is not possible, especially in built-up areas. It is important to stress that no area has yet been demonstrated to be devoid of activity in this period.

A few excavated sites, usually on hilltops, show evidence of concentrated Neolithic activity. The perimeters of some are surrounded by the pattern of interrupted ditches distinctive of causewayed enclosures: Hembury (Fig. 5.1), High Peak, Membury and Raddon.[7] Causewayed enclosures, found across southern Britain, probably served a variety of purposes ranging from ceremonial meeting places to defence;[8] not enough is known about south-western sites for their particular function to be identified. Recent evaluation at Haldon indicates that causewayed ditches may be present, but excavation of the hilltop at Hazard Hill did not identify any perimeter feature.[9] In Cornwall two excavated sites with Neolithic artefacts, Carn Brea and Helman Tor, have stone-built enclosing walls and may represent an upland variation of causewayed enclosures.[10] These two sites are sometimes referred to as 'tor forts'; at least six others, described in this way after surveys, are omitted from Map 5.1 as their dates are conjectural.[11] All these excavated sites have pottery, and sometimes radio-carbon dates, appropriate for the fourth to early third millennium BC, the Earlier Neolithic. This handmade pottery consists of round-based vessels, mainly bowls (Fig. 5.2A), with little decoration. Other sites with pottery of this date are of indeterminate function and vary in quality and amount of available data and in topography: Seaton, Nymet Barton, Torbyran, Bulleigh Meadow and Gwithian. The sites mapped represent only a very small proportion of surviving evidence for Early Neolithic activity. Hembury, High Peak and Raddon were found because later earthworks were visible. Some of the numerous, largely unexcavated hillforts (see chapter 7) must mask other Neolithic sites. Haldon, Hazard Hill and Membury were located by excavation in areas with concentrations of lithic finds, which Map 5.1 shows to be numerous. The concentration of Earlier Neolithic pottery in the Isles of Scilly (Map 5.2) is due to regular programmes of monitoring of development work and coastal erosion since the mid-1980s, while that in the Lizard is the result of systematic field survey; neither of these Cornish areas should be interpreted as having had exceptional levels of activity in the Neolithic.[12]

Map 5.1
Neolithic settlement, land use and resources.

SOURCES: authors and data from Cornwall Sites and Monuments Record, Cornwall County Council, and Devon Sites and Monuments Register, Devon County Council.

- ● Occupation site/causewayed enclosure (excavated)
- ○ Findspot of Neolithic pottery
- GROUP I Suggested source of igneous rock used for axes
- ///// Concentration of lithic finds
- ▓ Scattered evidence for Neolithic activity
- ‑‑‑‑ Schematic indication of position and density of beach flint

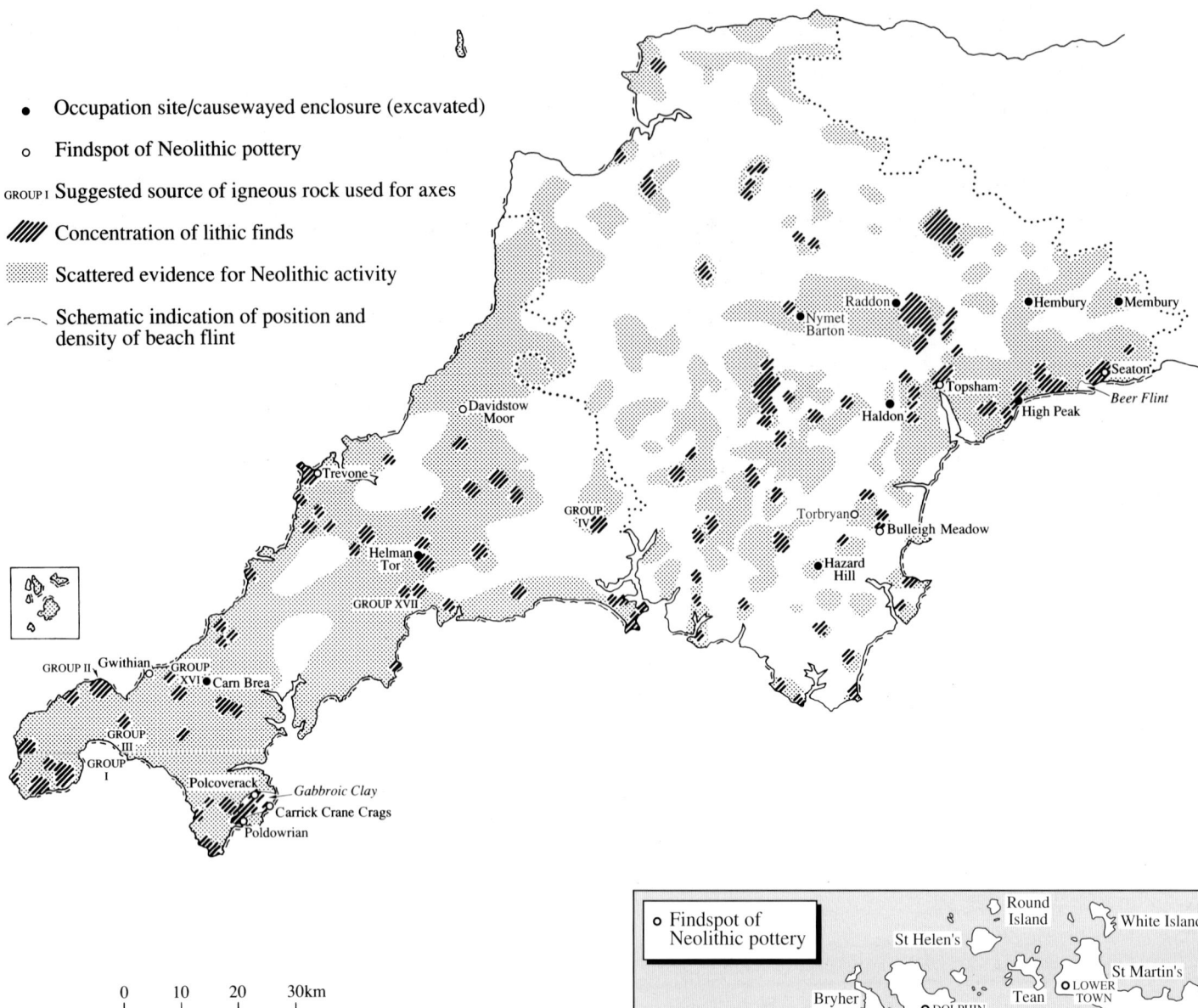

Raddon
Nymet Barton
Hembury
Membury
Topsham
Seaton
Davidstow Moor
Haldon
High Peak
Beer Flint
Trevone
GROUP IV
Torbryan
Bulleigh Meadow
Helman Tor
GROUP XVII
Hazard Hill
GROUP II Gwithian GROUP XVI ● Carn Brea
GROUP III
GROUP I
Polcoverack
Gabbroic Clay
Carrick Crane Crags
Poldowrian

0   10   20   30km

Map 5.2
Neolithic pottery in the Isles of Scilly.

SOURCES: Isles of Scilly Coastal Erosion Project, 1989–93, *The Pottery and Other Significant Artefacts from Sites with Recorded Stratigraphy*, archive report prepared by H. Quinnell for Cornwall Archaeological Unit; data used with their permission.

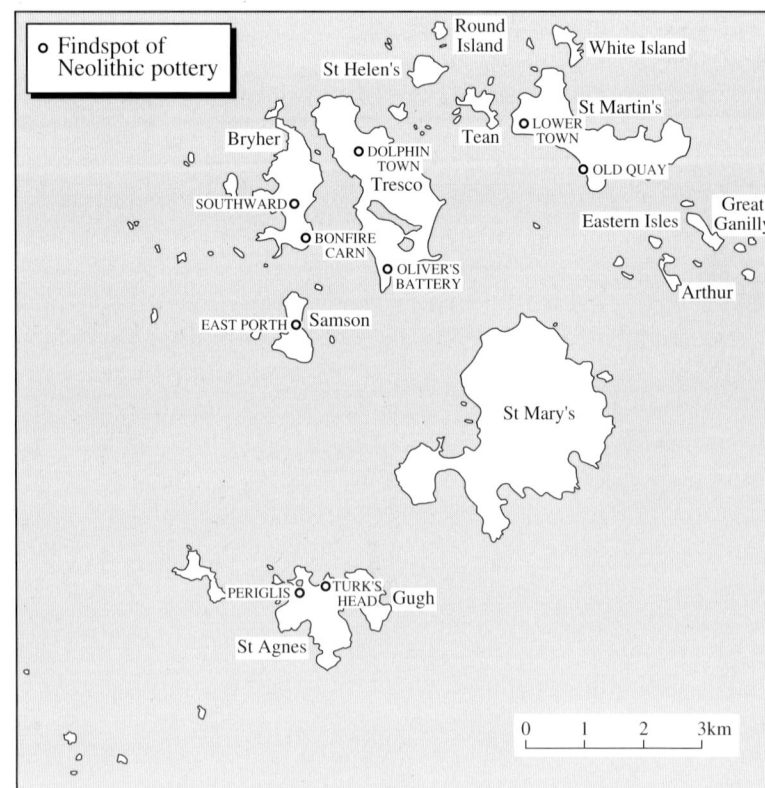

○ Findspot of Neolithic pottery

Round Island
White Island
St Helen's
St Martin's
Bryher
Tean
LOWER TOWN
DOLPHIN TOWN
Tresco
OLD QUAY
SOUTHWARD
Great Ganilly
Eastern Isles
BONFIRE CARN
OLIVER'S BATTERY
Arthur
EAST PORTH ○ Samson
St Mary's
PERIGLIS
TURK'S HEAD Gugh
St Agnes

0   1   2   3km

A few sites named on **Map 5.1** have pottery of the Later Neolithic (**Fig. 5.2B**), dating to the third millennium BC.[13] Such pottery is more highly decorated than the Earlier styles and may have flat bases. Many of the flint scatters contain diagnostic artefacts such as transverse arrowheads of Later Neolithic type. In other parts of Britain recorded activity appears on more varied geology and topography in the Later Neolithic as compared to the Earlier.[14] Much of this is in areas of intensive subsequent farming and settlement which can restrict fieldwork. No conclusions can be drawn from present data regarding the extent of Later as opposed to Earlier Neolithic activity in the South West as only small samples of both have been definitely identified.

Much of the pottery is identified as 'gabbroic', thought to be made from clays weathered out from the distinctive gabbroic rocks of the Lizard, although the actual source and the mechanisms of distribution have been the subject of much dispute.[15] All pottery from Carn Brea, 25km from the suggested source, is gabbroic,[16] but the proportion of gabbroic to other fabrics decreases to the east. Recent work in Devon indicates that fabrics are much more varied than previously supposed, but that at sites such as Haldon and Hazard Hill about 10 per cent of the assemblages are gabbroic.[17]

There is good evidence that Neolithic communities used materials obtained by exchange networks from sources at varying distances for both pottery and lithics. Good quality flint in the South West is only found at Beer in Devon. This is visually distinctive and was widely used both in Cornwall and Devon.[18] Haldon was a source of lower grade flint but otherwise flint was mostly obtained from beaches. The extent to which such flint occurs around the coast of the South West is not generally recognized; **Map 5.1** presents the results of observations by one of the authors (HQ) over twenty-five years.

The use of suitable igneous rocks to make axes in the Neolithic has long been recognized. Many axes from south-west England are of rocks which can be matched to sources in the region, although the debris from manufacture at 'axe factories' has not been found here as it has been, for example, in Wales. Igneous rock types regularly used for artefacts are conventionally grouped by Roman numerals, e.g. Group XVI in the Camborne area.[19] Recently the detailed basis for much of this identification has been questioned,[20] but general movement eastwards of Cornish-rock axes may be accepted with reasonable certainty.

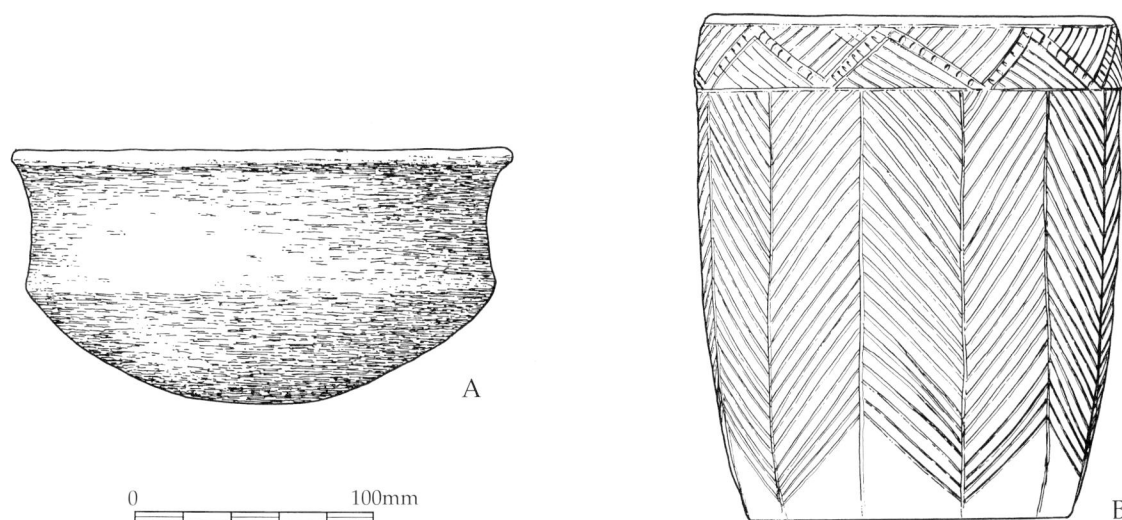

Fig. 5.1
Hembury, east Devon. The visible ramparts are those of the second phase of the Iron Age hillfort but this strong end-spur was first defended in the Neolithic period and later was also the site of a Roman fort.

SOURCE: photograph by Frances Griffith, reproduced by permission of Devon County Council (copyright reserved).

A

0   100mm

B

Fig. 5.2
(**A**) Earlier Neolithic bowl from Carn Brea; (**B**) Later Neolithic Grooved Ware, Trevone.

SOURCES: Drawings by S. Brouillard after R.J. Mercer, 'Excavations at Carn Brea, Illogan, Cornwall', *Cornish Archaeology*, 20 (1981), 167; D.G. Buckley, 'The excavation of two slate cairns at Trevone, Padstow, 1972', *Cornish Archaeology*, 11 (1972), 15.

In conclusion, **Maps 5.1 and 5.2** present a view of the landscape of the South West during the fourth and third millennia BC based on current knowledge. They must be used with the clear understanding that this view co-ordinates the results of at least a century of study, during which there has been great variety both in the intensity and quality of research spread across the two counties. It must also be recognized that the period is one in which all aspects of data, especially definition of chronological horizons, function of sites, bases of subsistence and modes of distribution, are subject to ongoing review.

FURTHER READING

See end of chapter 9.

BARROWS AND CEREMONIAL SITES IN THE NEOLITHIC AND EARLIER BRONZE AGE

FRANCES GRIFFITH AND HENRIETTA QUINNELL

Maps 6.1–6.5 present a range of sites with ceremonial, burial and ritual affinities which belong within a broad time band covering the late fifth to the second millennia BC. This time band can be divided for convenience into Earlier Neolithic (late fifth and fourth millennia BC), Later Neolithic (third millennium BC), Early Bronze Age (late third to early second millennium BC) and Middle Bronze Age (later second millennium BC). These periods are based on refinements of the nineteenth-century Three Age System and are typified by a succession of artefact styles. They can provide a convenient way of referring to bands of time, but their dating, distinctive features, and, for the Neolithic, even their names vary with different researchers. In parts of the country, construction of some types of site appears to be restricted to particular periods, but in general the use of period names in discussing prehistoric ceremonial monuments may generate an artificial separation of categories in a continuum, whereas it is clear that sites continued to have significance long after their creation. Dates of monument construction may differ across the British Isles. In the South West few types have been firmly dated, and here as elsewhere new research is rapidly changing perceptions. We take a broad chronological approach which in our judgement is most appropriate to current knowledge.

For cartographic presentation sites have been categorized into classes convenient for description and discussion. The categories apply to the most morphologically distinctive monument types among a broad spectrum of merging variations. Although sites are classified by their surface appearance, excavation, both in the South West and elsewhere, demonstrates that numerous sites are multi-period, and frequently of different form at successive stages. Round barrows may mask stone constructions which should be categorized as cairns. Wood was frequently used, especially in early phases of a construction sequence, as in the post rings subsequently sealed beneath round barrows. 'Monuments', as we now see them, are the final product of a sequence of activity. Monument construction also varied with regard to the materials available and to human decisions relating to the values and beliefs of the societies concerned, decisions which we cannot now reconstruct.

Sequences of ceremonial monuments were built or altered over a long period of time, with earlier constructions maintaining meaning and influencing later phases. Such activities result in the patterning now described as 'ceremonial landscapes'.[1] Our perceptions of these have been gradually evolving since the recognition of round barrow groups as cemeteries by early antiquarians. The general impression that ceremonial monuments tend to occur in elevated terrain can now be seen to be the product of the comparative lack of agricultural disturbance, and so of monument destruction, in particular on much of the moorland of south-west England. The recognition, following systematic programmes of aerial reconnaissance from the 1980s in Devon and the 1990s in Cornwall, of sites in forms such as ring ditches and oblong ditches, identifiable from their sub-surface surviving remains, begins to lessen the focus on higher ground. But aerial reconnaissance can only locate sites with detectable elements such as ditches or differentially coloured soil. The remains of monuments made, for example, of timber are still under normal conditions almost undetectable. Modern research is demonstrating that 'ceremonial landscapes' encompass larger areas and a wider variety of terrains than previously supposed, and the reality of apparent gaps between or within such areas cannot be established. An additional complexity not addressed by the maps is the fact that some natural features, for example tors, springs and indeed trees, appear to have had ceremonial significance.[2] This can be demonstrated when a tor such as Corn Ridge is surrounded by a stone ring cairn,[3] but can only otherwise be surmised.

Questions of topography and the durability of construction materials are factors additional to the general problems of destruction through clearance, agriculture and all types of development and to those of differential fieldwork. Bodmin Moor was the focus of intensive archaeological fieldwork, supported by professional surveying, during the 1980s, while Dartmoor has been covered by two separate exercises based on sketch transcriptions of air photographs.[4] Leslie Grinsell covered the whole of Devon, but not Cornwall, in his series of comprehensive barrow surveys.[5] The maps present current knowledge of surviving data;[6] we may never be able to establish how far they truly reflect the extent of ceremonial monuments during the fifth to second millennium BC.

Long mounds, sparsely scattered, such as the Woolley Barrow, Morwenstow, Uplowman, Tiverton, or the Bodmin Moor group located during the recent survey programme, may relate to the ditched earthen long barrows concentrated on the

Map 6.1
'Ceremonial sites' in
the Neolithic and
Earlier Bronze Age.

SOURCES: authors and
Devon Sites and
Monuments Register,
Devon County Council.

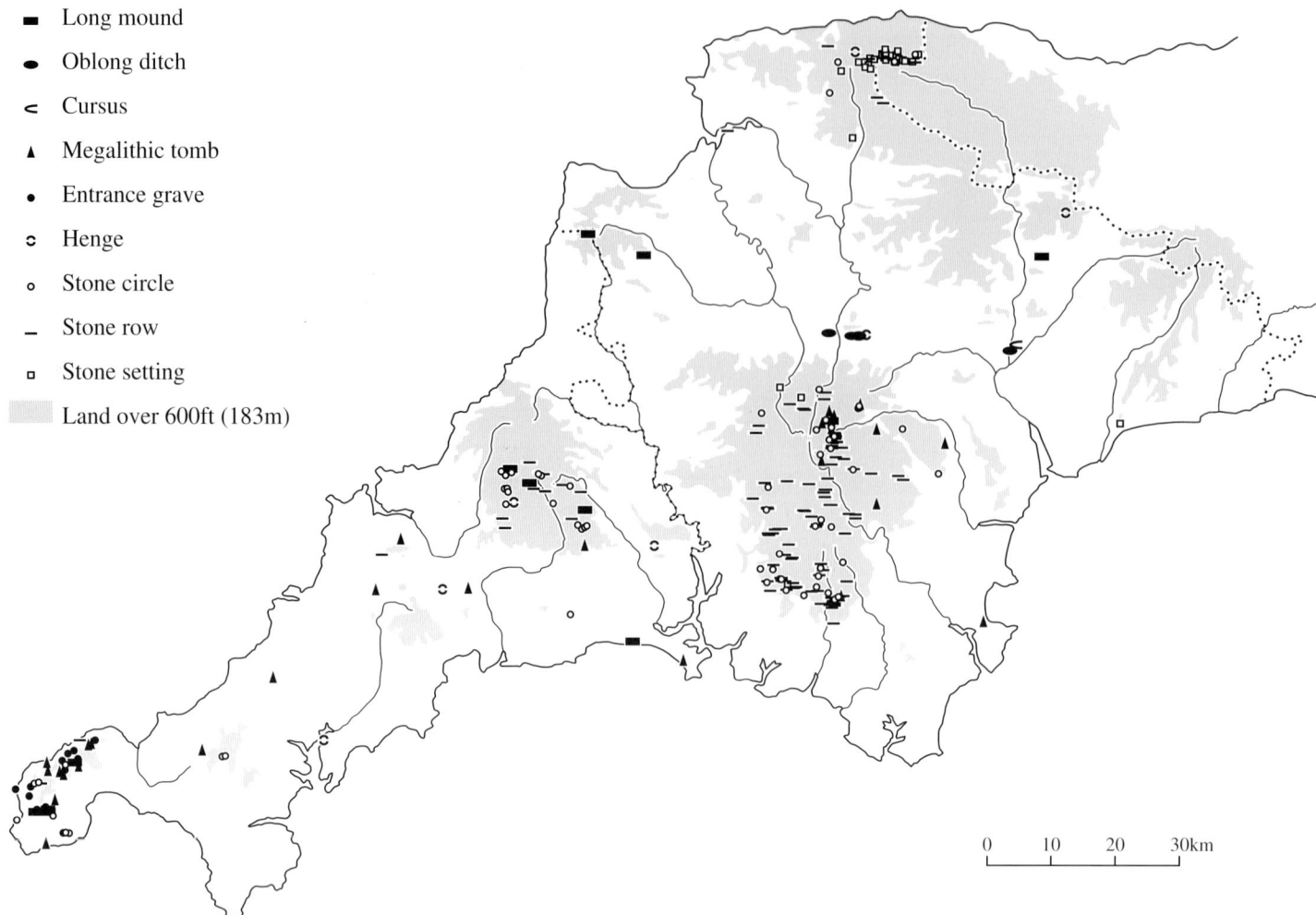

- ■ Long mound
- ● Oblong ditch
- ⊂ Cursus
- ▲ Megalithic tomb
- • Entrance grave
- ⊃ Henge
- ○ Stone circle
- ― Stone row
- ▫ Stone setting
- ▨ Land over 600ft (183m)

0   10   20   30km

Map 6.2
Entrance graves in
the Isles of Scilly.

SOURCE: *Isles of Scilly
Management Plan*, 1988;
data reproduced by
permission of J. Ratcliffe,
Cornwall Archaeological Unit
and Cornwall County Council.

chalk of southern Britain.[7] Excavations in this area have provided a good range of dates within the fourth millennium BC and demonstrate regular use of wood for chambers, but no excavation in the South West has provided data for structural detail or chronology. Some in upland areas, as at Corringdon Ball on Dartmoor, can be seen to have included megalithic stone structures. Some megalithic tombs in Brittany and Ireland with long mounds belong to the third rather than the fifth/fourth millennia BC.[8]

Oblong ditches recorded so far at North Tawton and Nether Exe in Devon survive mainly as sub-surface features.[9] They may originally have contained mounds, in which case they would represent eroded, ditched, long mounds or barrows. These may be related to the ditched oval barrows of southern Britain, such as that at Dorchester, Oxfordshire, which date from the end of the earthen long barrow sequence in the late fourth millennium BC.[10]

The *cursus* (**Fig. 6.1**), only so far recognized at Nether Exe, in proximity to an oblong ditch,[11] has been identified elsewhere in Britain. A cursus consists of an elongated ditched enclosure, possibly a development from the longer ditched earthen long barrows and bank barrows. Dates extend through the fourth into the third millennium BC.[12] Some are of great length, the Dorset cursus running for 10km, but only one end of the Nether Exe cursus is known.

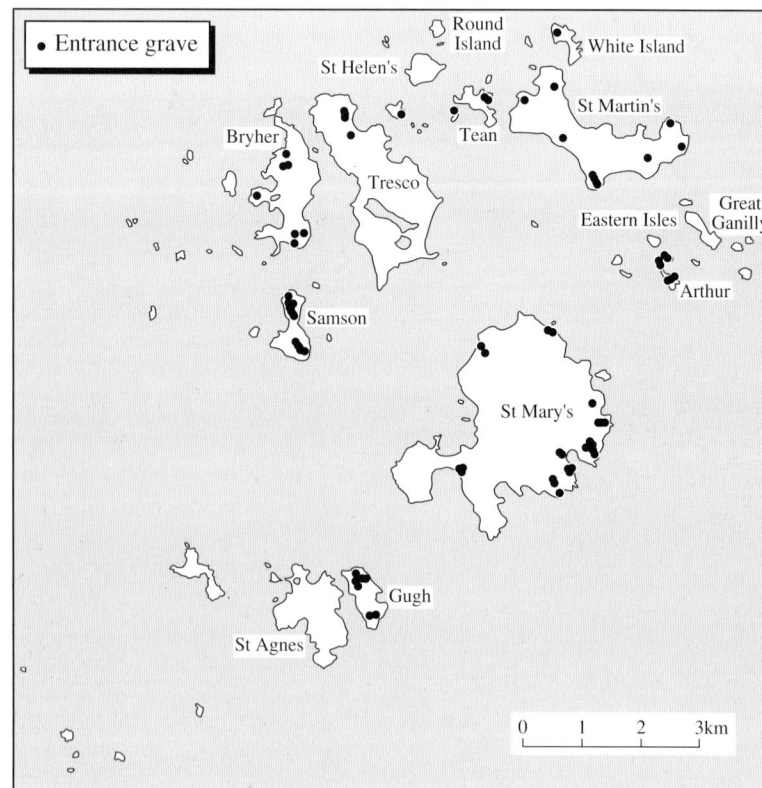

• Entrance grave

Round Island
White Island
St Helen's
St Martin's
Bryher
Tean
Tresco
Great Ganilly
Eastern Isles
Arthur
Samson
St Mary's
Gugh
St Agnes

0   1   2   3km

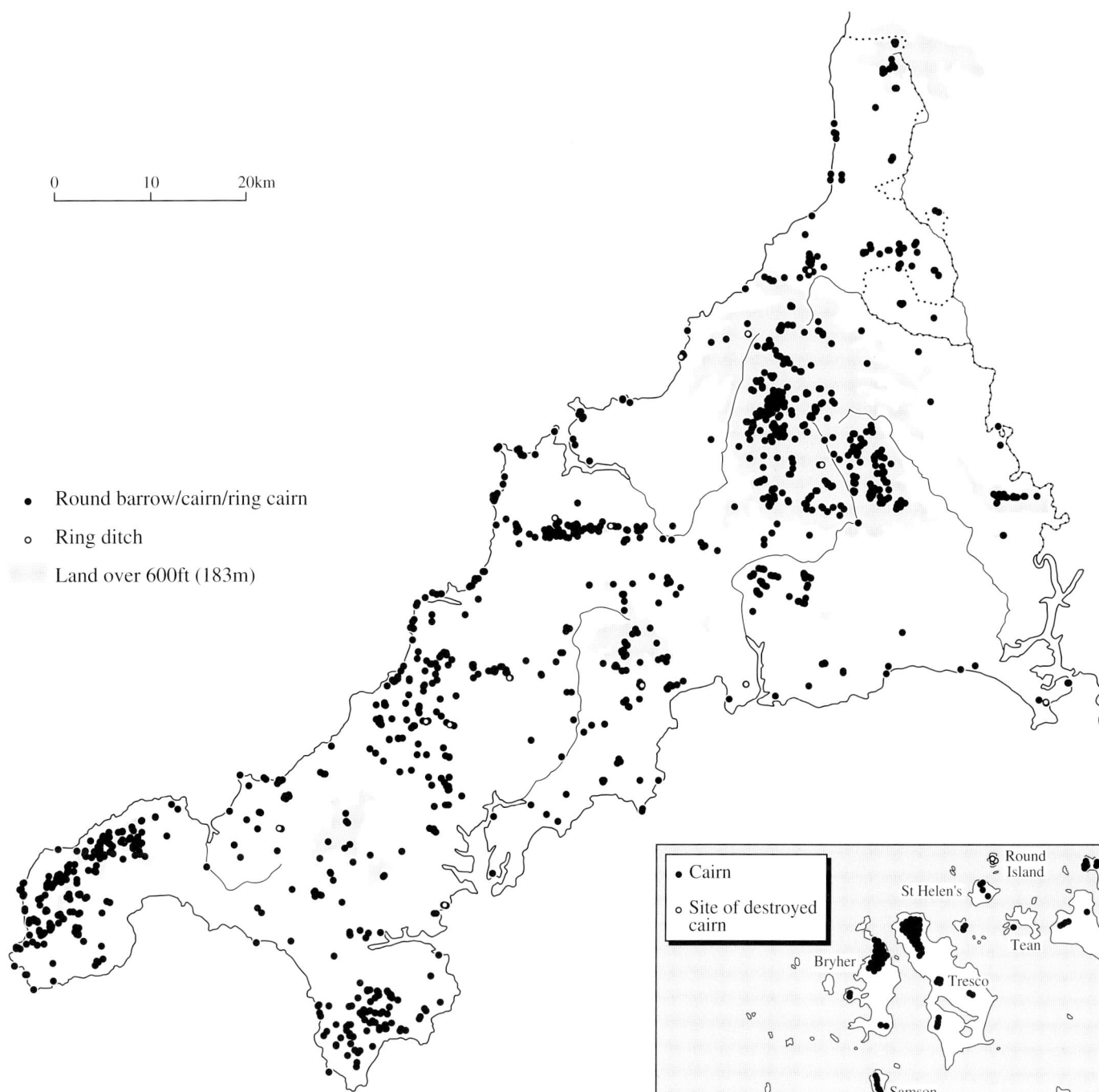

Map 6.3
Cornwall: barrows, cairns,
ring cairns and ring ditches,
*c.*2500–1000 BC.

SOURCE: Cornwall and Scilly
Sites and Monuments Record,
Cornwall Archaeological Unit
and Cornwall County Council.

- Round barrow/cairn/ring cairn
- Ring ditch

Land over 600ft (183m)

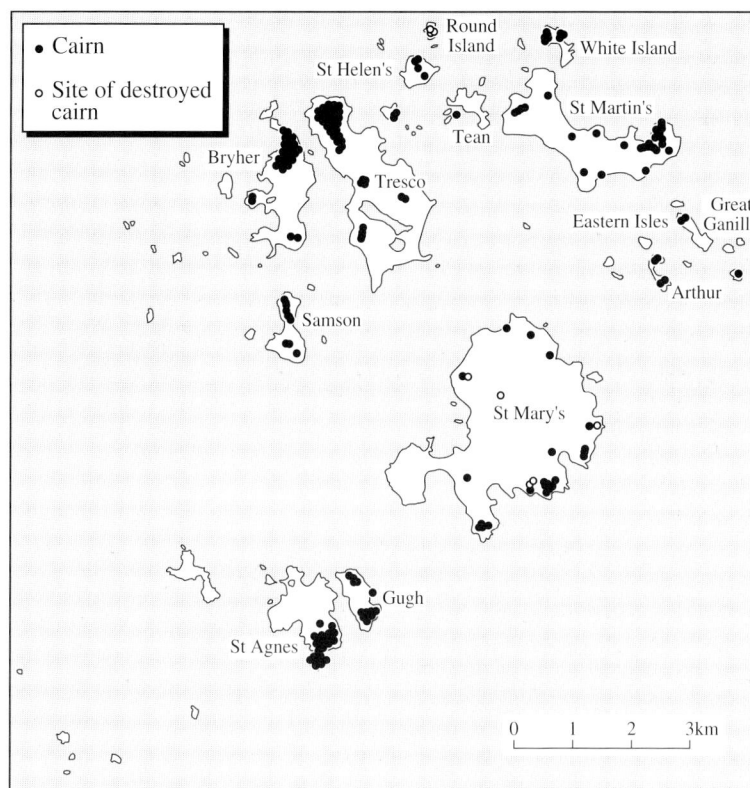

Map 6.4
Cairns in the Isles of Scilly.

SOURCE: as **Map 6.2**.

Megalithic tombs are depicted on the maps where there is evidence for a stone chamber; these may be embedded in mounds, usually largely of stone, which have often attracted attention as quarries and therefore have been frequently damaged or denuded. There are two concentrations: one across the north edge of Dartmoor and the other in West Penwith.[13] Megalithic tombs were built with a wide variety of plans across northern Europe and parts of Britain.[14] Variants range, in Britain, across the fourth and third millennia BC. Funerary elements, as with earthen long barrows, tend to involve the remains, often partial, of numerous individuals. Only a few, small-scale excavations have taken place, of which one resulted in the interpretation of

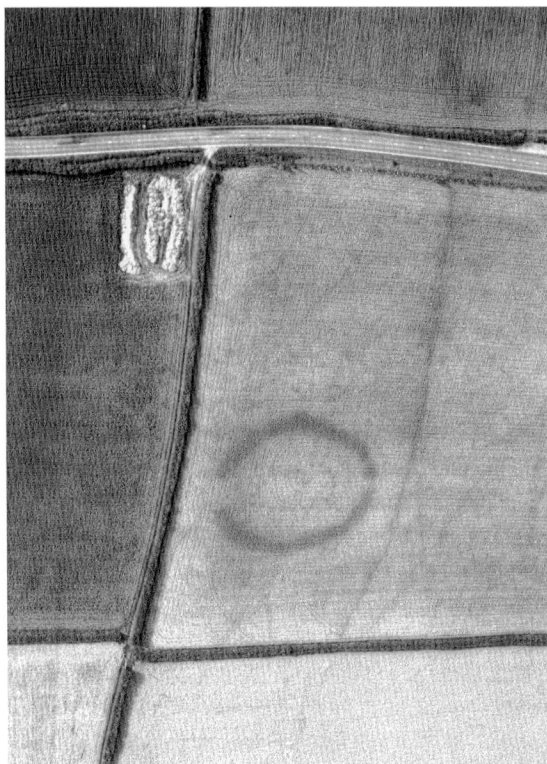

Broadsands, Paignton, as a passage grave, a chamber approached by a passage within a round mound.[15] It is possible that some other large round cairns on Dartmoor, Exmoor and Bodmin Moor also mask internal round chambers. In the South West one or two portal dolmens (megalithic tombs in which the side stones extend forward beyond the chamber to form an entrance area) have also been recognized, as at Zennor.

A special form of megalith, the *entrance grave*, with a simple passage and chamber within a revetted drum-like mound and multiple cremation burials, is concentrated in West Penwith and, more particularly, Scilly (**Map 6.2**). Excavations have not produced radio-carbon dates but have demonstrated the regular occurrence, most notably at Bant's Carn, of pottery well-dated on Scillonian settlement sites to the second millennium BC; the few finds of Neolithic pottery are from older excavations.[16] Entrance graves appear to demonstrate the continuance of the use of one localized form of megalith into the second millennium.

Henges, ditched enclosures with external banks, have a broad distribution in the British Isles, and a range of associations with post rings, stone circles and burials, usually cremation, which support their general interpretation as ceremonial centres. Most dates for construction belong in the third millennium;[17] small excavations of a few sites in the South West—at the Stripple Stones and at Castilly—have provided nothing with regard to function or dating. The only certain Devon henge, at Bow (**Fig. 6.2**), was located by aerial reconnaissance during the 1980s and forms part of a spread of ceremonial sites across the upper Taw valley.[18]

Stone circles are found concentrated in groups on the moors of south-west England.[19] Considerable excavation, usually linked to programmes of re-erection, as at the Hurlers, is remarkable for the lack of information produced. Unlike in some other areas of the country, south-western stone circles do not appear to contain burials (unless inhumations have decayed in acidic soils). In other parts of the country, stone circles can be demonstrated to be additions to henge monuments; however the Stripple Stones is the only henge containing a stone circle in the South West. Dates are generally difficult to establish but a range through the third millennium BC and into the second seems probable for these monuments nationally.

Stone rows (**Fig. 6.3**) have, since the antiquarian work of the nineteenth century, been seen as a monument type almost unique to Dartmoor, but recent fieldwork has established a good grouping on Bodmin Moor; a small number have also been known for a long while on Exmoor. Stone rows, generally shorter than those in the South West, occur infrequently elsewhere in the British Isles, but provide no help with chronology.[20] The only excavation in south-west England (at Cholwichtown on Dartmoor) produced neither dating, artefacts, nor evidence of function.[21] The role of the monuments remains enigmatic, although their association with a range of cairns and ring cairns (see below) provides some broad indication of chronology. In their extended linear form they may seem analogous with cursuses, although the rows never form enclosures. Occasional finds elsewhere indicating wooden post rows associated with ceremonial monuments remind us that wooden equivalents may have been present in the landscape. A date range similar to that of stone circles seems appropriate from the data currently available.

Menhirs or standing stones occur intermittently across the South West with little direct evidence for date or function. Some are isolated, but when associated with stone rows or circles they, and their vanished timber equivalents, may represent further components of 'ritual landscapes'. They are not shown on the maps because in many instances, especially with smaller stones, it is not always possible to ascertain whether they may have been created for more recent utilitarian functions, such as cattle scratching stones.

On Exmoor small *stone settings* with rectilinear shapes are a distinctive local feature which have been the subject of recent intensive survey.[22] There have been no excavations and no artefactual or radio-carbon dates have been obtained, but the use of orthostats relates the settings to stone rows and provides the only tenuous indication of chronology.

Round barrows and their *cairn* equivalents in piled stone are numerous in the region. There is great variety in size and form;

Map 6.5
Devon: barrows, cairns,
ring cairns and ring ditches,
*c.*2500–1000 BC.

SOURCE: Devon Sites and
Monuments Register,
Devon County Council.

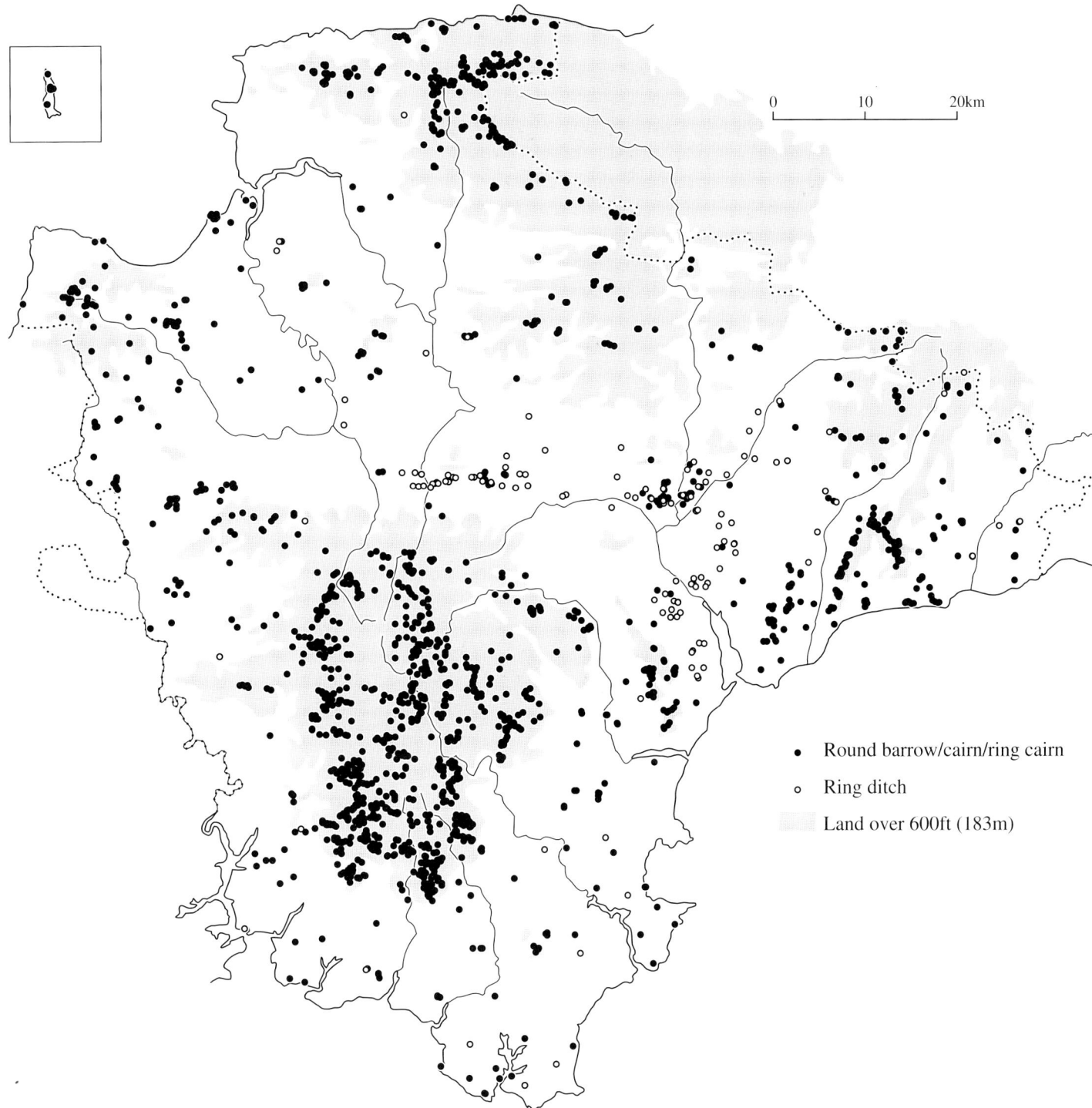

0   10   20km

• Round barrow/cairn/ring cairn

○ Ring ditch

░ Land over 600ft (183m)

sites may range from 3m to 30m in diameter and from 0.1m to 5m in height. Mounds may be rounded or flat-topped, and incorporate a bank, with or without a ditch; excavations frequently reveal a sequence of construction, the present-day shape representing the final phase.[23] The form(s) chosen reflect materials available in the local environment (**Fig. 6.4**), so that stone was used to visual effect in areas such as the granite moors, whereas ditches are more frequent on softer, easier bedrocks. However, turf was extensively used both on granite and elsewhere in the region. During the 1970s *ring cairns* were recognized as variants distinguished by the pre-eminence of a visually significant annular element. Recent studies demonstrate their variety, and the difficulties of categorization.[24] As ring cairns cannot be clearly divided from barrows/cairns, and indeed a constructional sequence may include both, they are not distinguished from barrows and cairns on the maps. In south-west England, excavated round barrows and cairns do not always have an obvious funerary function. However, in almost all cases evidence of ritual or ceremonial activity can be seen.

Round barrows in Britain sometimes date as early as the fourth millennium BC,[25] but no such early dates are recorded from the South West. These monuments increase with the spread of Beaker pottery during the later third millennium BC; Fernworthy is a south-west England example. About twenty-five barrows and related monuments in south-west England now have radio-

Fig. 6.3
Stone row at Drizzlecombe, Dartmoor.

SOURCE: photograph by Henrietta Quinnell.

carbon dates, some of them multiples.[26] These dates concentrate in the period c.2200–1500 BC, with scattered examples earlier in the third millennium, and a few extending the range to c.1000 BC. The main dates are more or less coincident with those of Early Bronze Age metal artefacts.

Ring ditches are sites on which the only surviving feature is a ditch, generally located by aerial reconnaissance. Such sites were recognized as aerial survey developed from the 1920s in areas such as the Thames valley. In some cases ring ditches originally surrounded low mounds which, due to the intensive arable use of the types of land they occupy, have been totally eroded.[27] The comparatively recent programme of reconnaissance in the South West has shown ring ditches to be present in lower-lying areas, often, as in the upper Taw valley, where no monuments in the barrow class had been thought to have existed.[28] A few ring ditches in Devon have now been excavated, and have produced evidence of mortuary practices and ceramic forms comparable with those found in round barrows.[29]

The range of forms in ceremonial monuments is distinctively regional, reflecting both the materials available and the choices made about their use. Earlier ideas confining the South West to a role peripheral to Wessex, in which south-western complexities of barrow form were seen as reflections of a 'Wessex Culture', are now considered unsustainable. The relationship of human burials to ceremonial monuments of different types is complex. Barrows can no longer be considered as simple burial monuments but as structures which frequently incorporate human remains. Conversely other sites such as henges may include burials among other deposits and activities of a ritual nature. Current evidence suggests that, from around 4000 BC, landscapes opened up by woodland clearance were increasingly marked out by ceremonial monuments. The development of ceremonial monuments may have peaked in the centuries around 2000 BC but declined during the second millennium, around the time when communities constructed habitations, fields and enclosures as life became more focused on settled arable farming.

Maps 6.1–6.5 make evident the much greater monument numbers for the later third and the second millennia, several thousands for Devon and Cornwall, as opposed to the few dozen for the earlier period back to c.4000 BC. Far more excavation has been carried out on the numerous round barrows and related monuments in the later part of the ceremonial sequence than on the small numbers which may be presumed earlier, such as long barrows and megalithic tombs. Some of the metalwork discussed in chapter 8 comes from round barrows, which are also the main source of pottery types traditionally assigned to the Early Bronze Age: collared urns, food vessels, and cord-impressed Trevisker pottery.[30]

Round barrows frequently occur in 'cemeteries', which survive on higher land. Linear cemeteries, such as that on the Taphouse Ridge in east Cornwall, may reflect distinctive local topography, but concentrations with more complicated patterning, such as Denzell Downs, north Cornwall, or Farway,

Fig. 6.4
A complex Cornish barrow under excavation at Colliford reservoir, Bodmin Moor, showing central cairn, turf mound and outer stone ring.

SOURCE: photograph by Chris Guy (copyright reserved).

east Devon, are more usual. Upland or marginal areas will tend to preserve monuments better; indeed sites close together may protect one another. The recognition of monument clusters was reinforced by studies from the late nineteenth century onward of stone rows and circles often found in close conjunction with cairns or ring cairns. The plotting of monuments at Merrivale in the 1890s was an early demonstration of a ceremonial landscape.[31]

Aerial reconnaissance has demonstrated that monument clusters in the South West include a wider range of forms than previously recognized and that they exist in low-lying as well as upland areas. The North Tawton/Bow concentration, with henge, oblong and ring ditches, and that at Nether Exe

with cursus, oblong ditch and ring ditches, are both ceremonial landscapes on lower-lying land similar to those now being recognized over most of Britain where there has been adequate aerial reconnaissance. Our maps demonstrate the increasing recognition that, in the South West as elsewhere, ceremonial monuments and the landscapes they mark out occupied a variety of topographical positions.

FURTHER READING

See end of chapter 9.

SETTLEMENT
*c.*2500 BC TO *c.*AD 600

FRANCES GRIFFITH AND HENRIETTA QUINNELL

The term 'settlement' means substantially different things in different periods, and is used by archaeologists to describe a range of sites from those recognized primarily by concentrations of artefacts in ploughsoil to highly visible constructions such as hillforts.

Maps 7.1, 7.2 and 7.4 cover a wider timespan than most others in this volume, and overlap periods covered by other maps, notably those giving information on the Neolithic and on prehistoric ceremonial monuments, prehistoric metalwork, as well as Roman military occupation and earlier medieval settlement.

The reason that this wide time frame is used derives from the nature of the evidence itself. Although excavation will usually provide an indication of date for the occupation of a site, the great majority of settlement sites within this date range have never been subjected to excavation, and their external characteristics do not permit their attribution to a particular period with any degree of certainty. The sites shown here range from south-west England's first town—Exeter, founded in the Roman period in the first to second centuries AD—to hillforts and settlement enclosures, and sites known only as artefact scatters. It is important, therefore, to use the maps with an understanding of the limitations of what they can show. Substantial stone-built enclosures surviving on the uplands of the South West will obviously be more strongly represented as a percentage of the original distribution of such sites than will lowland settlements, where structures will not have been made of stone and where, if there was any enclosure at all, it was either of impermanent material or has been erased by subsequent cultivation. Further, the uplands preserve for us evidence of the fields and grazing territories associated with the habitation sites; in the lowlands we can glimpse these only infrequently.

In the last fifteen years aerial reconnaissance, and increasingly geophysical survey, have had a major impact on our perception of the density of some forms of prehistoric settlement in the lowlands, particularly when enclosed; for some other forms of settlement we remain largely dependent on discoveries during ground disturbance.[1] Equally, though the 'settlement' sites shown here predominantly date from the second millennium BC and later, it will be clear from examination of other maps relating to the prehistoric period in this volume that earlier activity was taking place over much of the area, despite the fact that the actual settlement sites currently escape present-day archaeologists. It is also true that some periods will be under-represented in the archaeological record because clear dating evidence is hard to find: for example, in the post-Roman period the local manufacture of pottery, except in west Cornwall, virtually ceased. Sites can, therefore, only normally be ascribed to this period if they were sufficiently well-connected to contain imported pottery (usually from the Mediterranean area), or if other forms of dating such as tree-ring or radio-carbon dating are available.[2]

Our understanding of the archaeology of these periods, both in the uplands and the lowlands, has changed substantially in the course of the last twenty-five years. On the uplands, particularly Dartmoor and Bodmin Moor, systematic and detailed field survey has permitted a much closer appreciation of the prehistoric archaeology and its articulation, while in the lowlands a combination of rescue archaeological work, aerial reconnaissance and intensified fieldwork has transformed the picture.[3] Nevertheless, though it is now possible to dismiss some of the more sweeping generalizations of the past about settlement in the South West in the prehistoric period—for example, the assertion that the uplands were densely occupied while the lowlands were covered with impenetrable forest until the Saxon period—archaeologists are still not able to suggest with any confidence the rate at which the landscape was cleared, nor how this process varied through time in different areas.[4] The palaeoenvironmental work that will provide the evidence for the vegetation cover in earlier periods is still lacking for much of the region (especially off the moorlands), as is a sufficient number of well-dated excavated settlement sites. Such work is also necessary to explore the farming and other land-use practices of the time, as well as the environmental conditions that prevailed. Nevertheless, the state of archaeological knowledge is now considerably better than previously, and current research continues to refine the picture.

HILLFORTS

Hillforts—defended areas, usually topographically determined, surrounded by strong bank and ditch circuits—represent some of the most substantial prehistoric

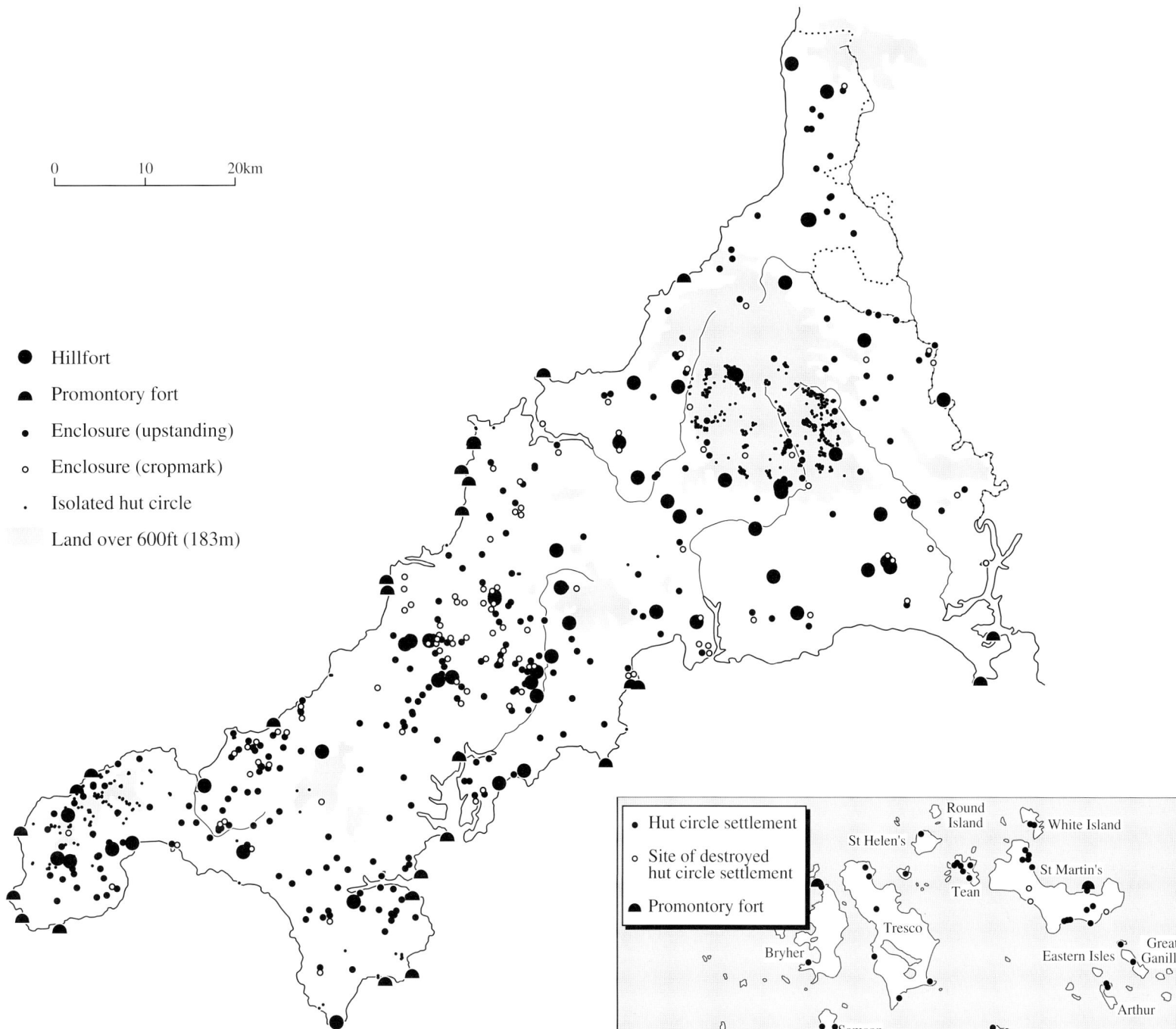

Map 7.1
Settlement in Cornwall from
c.2500 BC to AD 600.

SOURCE: Cornwall and Scilly
Sites and Monuments Record,
Cornwall Archaeological Unit
and Cornwall County Council.

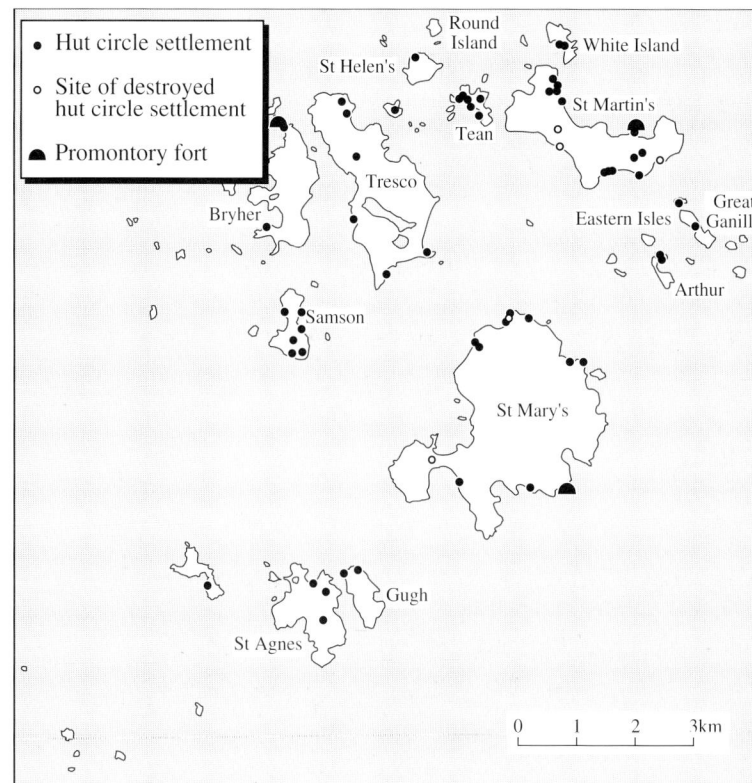

0   10   20km

- ● Hillfort
- ◖ Promontory fort
- • Enclosure (upstanding)
- ○ Enclosure (cropmark)
- . Isolated hut circle
- ▬ Land over 600ft (183m)

Map 7.2
Settlement in the Isles of Scilly.

Hut circles on the Isles of Scilly
usually consist of clusters
built up against one another,
best shown by the excavated
settlement on Nornour. Many
of these, now situated on
the coastline and subject to
erosion, have produced
pottery of the second
millennium BC, and the
majority of surviving hut
circle groups appear to be of
this date, A few sites, such as
Nornour itself, continued in
occupation into the Roman
period. A rise in sea level,
mostly in the post-Roman
period, produced the present
pattern of islands and
flooded much land likely to
have been occupied; this adds
yet another factor to those
discussed in the text which
affect the interpretation of
early settlement patterns.

SOURCE: Isles of Scilly
Management Plan, 1988;
data reproduced by permission
of J. Ratcliffe, Cornwall
Archaeological Unit and
Cornwall County Council.

- • Hut circle settlement
- ○ Site of destroyed
  hut circle settlement
- ◖ Promontory fort

Round Island
St Helen's
White Island
St Martin's
Tean
Tresco
Great Ganilly
Bryher
Eastern Isles
Arthur
Samson
St Mary's
Gugh
St Agnes

0   1   2   3km

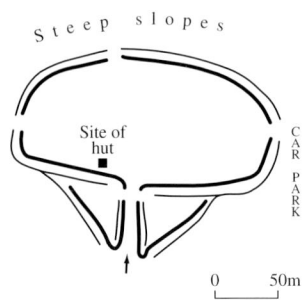

'monuments' in Britain. None in south-west England has quite the massive quality of a site such as Maiden Castle in Dorset, but in the east of the area hillforts such as Hembury (**Fig. 5.1** in chapter 5), with hilltop siting and closely spaced ramparts, more closely resemble the hillforts of central and eastern England than the majority of sites conventionally called hillforts further west.[5] The hillforts of the South West may have more in common with those of Wales than with those of central-southern England. South-western hillforts very often have more widely spaced ramparts, or 'multiple enclosures', as at Clovelly Dykes or Warbstow Bury (**Fig. 7.1**), and this feature has been suggested, by Aileen Fox and others, to indicate a particular concentration upon a stock-rearing economy in the South West.[6] Others have a single rampart, or a combination of a single enclosing rampart and a cross-ridge bank on the most accessible side for greater defence. Such sites as Warbstow Bury may be situated on slopes rather than hilltops, and their difference from 'enclosures' (see below) may be one of scale alone. The traditional Cornish word 'round', most commonly used to describe enclosures, discussed below, also sometimes refers to sites coming within the hillfort class, particularly multiple enclosure forts.

Hillforts are conventionally dated to the first millennium BC, or to the Iron Age. However, there is increasing evidence that hilltop defended enclosures may have much earlier origins. The Neolithic hilltop sites at Carn Brea and Helman Tor have been discussed in chapter 5. A pre-Iron Age origin is doubtless also true for others: at Raddon a Neolithic enclosure has been found in excavation, succeeded by two phases of first millennium BC defences, on the hilltop. Similar evidence comes from many other parts of the country. Equally, it has been demonstrated in other areas that almost all hillforts appear to have fallen out of regular use before the end of the first millennium BC. In view of the generally very limited nature of excavation even on well-known and highly visible sites, the term 'hillfort' has been used on **Maps 7.1, 7.2 and 7.4** to denote a substantial defensible enclosed site, whether now seen in earthwork or cropmark form, of probable later prehistoric date.

While most hillforts appear to have fallen into disuse before the coming of the Romans to south-west England, there is evidence from a range of hillforts and promontory forts (see below) of their reoccupation in the late and post-Roman period. Cadbury Castle, in Somerset, is one of the best documented examples of this, but post-Roman reuse has been recognized at Chûn Castle, in west Cornwall, and at High Peak, in east Devon.

No hillfort in south-west England has so far been excavated on a sufficiently large scale to provide a clear indication of the extent of permanent occupation, as compared with its possible role as a place of refuge, or as a 'central place'—perhaps serving a wide community who gathered there at particular times, but itself having little or no permanent habitation. A high proportion of those English and Welsh hillforts that have been subjected to extensive excavation have produced finds indicating processing of raw materials by specialists—the working of iron, glass making etc. There are also indications, in the form of four-post structures believed to represent granaries, that hillforts and other enclosures played a role in providing a defended place for storing foodstuffs. This is borne out by more extensive excavations in other areas, where regularly laid out lines of four-post structures have been found—for example at Danebury, Hampshire, or at the Breiddin in north Wales. Limited work at Killibury, near Wadebridge, and at Castle Dore, near Fowey, both in Cornwall, suggested dense occupation within the inner enclosure dating to the fourth to first centuries BC, whereas the small trenches cut within Blackbury hillfort, east Devon (**Map 7.3**), produced only slight structural traces.[7] This site, however, offered both evidence of industrial activity, in the form of iron slag, and also structural evidence indicative of a serious defensive function. The entrance is defended by a complex series of earthworks, and in addition no fewer than 1,271 slingstones were found, predominantly around the entrance.

Where any significant degree of excavation at a hillfort has been carried out, evidence of change and development over the lifetime of the site's use is found. At Hembury, in east Devon, the earlier Iron Age rampart was found to have been completely replaced with one of more complex construction. Such detail is normally only resolved by detailed excavation, as are questions of the comparative dating of hillforts. It will be clear from the above that some hillforts in the South West have early origins, but the sequence of construction and questions of the contemporaneity of occupation of different sites cannot yet be answered.

PROMONTORY FORTS

The fortified coastal sites known as promontory forts occupy highly defensible cliff top sites, usually on a spur projecting into the sea that can be protected on the landward side by one or more banks and ditches across the narrowest point of the neck leading out to it, as at Bolt Tail (**Fig. 7.2**). Such sites are a characteristic feature of the western seaboard of Britain and the coasts of Ireland and Brittany. The bleak and exposed situation of these, also known as 'cliff castles', has led to the natural suggestion that they were only used in emergency, as a place of refuge. However, this is not borne out by excavation, either in the South West or elsewhere, which tends to demonstrate the presence of houses and finds comparable with those of contemporary hillforts. The date range of those promontory forts so far excavated lies in the later first millennium BC, perhaps continuing a little into the first century AD.

ENCLOSURES

Even hillforts have clearer dates than the sites known collectively as 'enclosures'. This term covers a range of sites from those only 20m across to major, lightly defended settlement sites. In Cornwall, the term 'round' is often used for these sites. Some have strong surviving banks or walls surrounding them, while others may never have had strong 'defences', but may have been constructed for keeping stock, or indeed simply to define the boundary of the curtilage of a hamlet, house or farm. However, excavation also shows that some sites, indistinguishable on field evidence from settlement enclosures, have strongly non-domestic characteristics. The 'enclosure' may be marking off an area for ceremonial or special use, and its perimeter may have formed a symbolic rather than a physical barrier.

On the whole, we can only identify enclosures that have either an extant earth or stone bank or else a perimeter ditch or ditches that we can observe in cropmark form (**Fig. 7.3**) or in excavation. These two types are differentiated on the maps, not because they were different in purpose—each enclosure was built with the locally available materials—but to demonstrate the effects that construction material and subsequent land use have

upon recognition patterns. The sites shown as 'cropmark enclosures' on the maps have only been recognized in the last thirty—mainly the last fifteen—years; previously the then known settlement pattern was that represented by the stone-built or surviving earthwork sites. When only these sites were known, it is unsurprising that a very different view of the settlement pattern was obtained.

Variation in building materials can also confuse our perception of the degree of protection required: Grimspound, one of the more substantial Dartmoor enclosures, gives no indication of a particularly defensive objective, but being made of granite boulders it presents a more impressive aspect today than many earthwork enclosures that may originally have been more strongly defensive in design. Further, the existence of many other forms of enclosure that leave no lasting traces (for example by the use of dead thorn bushes) are known from many other parts of the world. Enclosures so far excavated in south-west England have provided dates ranging between the Neolithic and the middle of the medieval period. Their rather simple forms—in general rectilinear or curvilinear, sometimes with more than one perimeter circuit—mean that without excavation it is generally not possible to assign a date. Equally, the length of occupation that is known from some excavations is considerable; the round at Trethurgy, near St Austell, for example, was occupied between the second and sixth centuries AD.

Limited excavations on upland sites have provided information on both date and function: for example the enclosure on Shaugh Moor, excavated before its burial under a china clay waste heap, provided dates in the later second millennium BC. The

Fig. 7.3
Enclosures seen in cropmark form, Tregony, south Cornwall. Two contrasting bivallate enclosures intersect, which indicates that they were not used at the same time.

SOURCE: photograph by Stephen Hartgroves, Cornwall Archaeological Unit, reproduced by permission of Cornwall County Council (copyright reserved).

Map 7.4
Settlement in Devon from
*c.*2500 BC to AD 600.

SOURCE: Devon Sites and
Monuments Register,
Devon County Council.

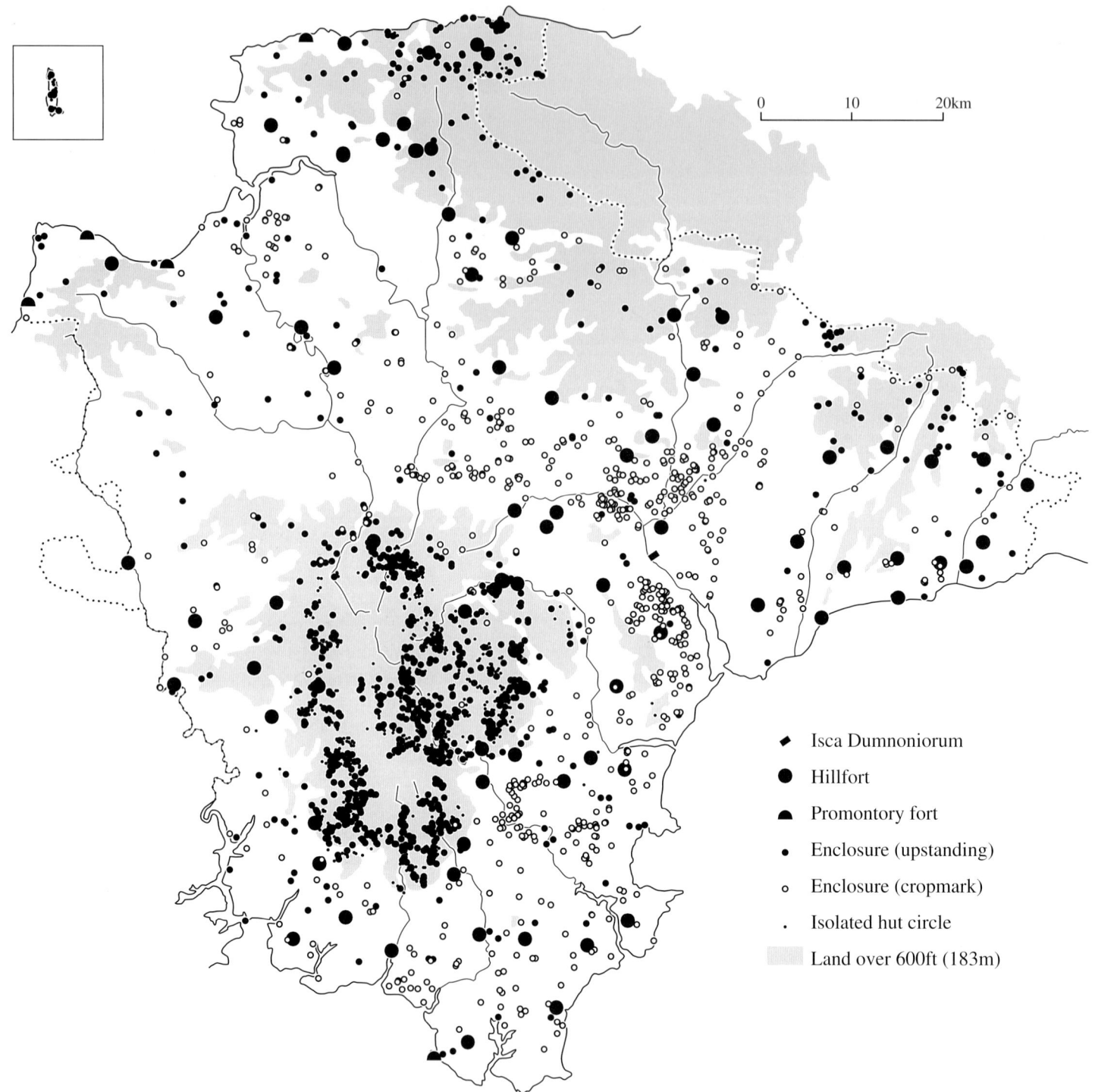

0   10   20km

◢ Isca Dumnoniorum

● Hillfort

▲ Promontory fort

• Enclosure (upstanding)

○ Enclosure (cropmark)

· Isolated hut circle

▨ Land over 600ft (183m)

excavators suggested a period of over 1,000 years for the lifetime of the settlement, though it was not clear whether it was occupied continuously. Interestingly, the site also produced evidence that some of the houses predated the construction of the enclosure wall—the settlement had an earlier, unenclosed phase. The excavations also challenged conventional ideas of the purpose of enclosed sites in other ways: the enclosure wall was continuous and had no entrance. Presumably the enclosure was entered by a stile—in other words, it appears that it may have been intended to keep animals out, rather than in.[8]

Off the high ground, recent excavations have shown the extent of the date range of enclosed sites. Excavations in advance of the A30 road near Exeter in 1997 have investigated some previously known and also newly discovered enclosure sites of dates which range from the Middle Bronze Age to the Roman period. At all of these, traces of structures, round houses, post settings etc. were found and excavated. Such features are too slight normally to give rise to cropmarks, and, therefore, can only be seen in excavation or geophysical survey. Excavations on the line of the same road in the west of Cornwall produced enclosures which provided dates of occupation from the second half of the first millennium BC to the fourth century AD. At Penhale, the excavation found evidence of repeated complex refurbishment of the enclosure line—a feature that could not have been recognized without excavation. The date range can, however, extend further. Excavations in west Devon in advance of the construction of Roadford Reservoir showed that enclosures indistinguishable in outward form from some of prehistoric age were still being constructed to surround medieval farmsteads in the twelfth or thirteenth centuries AD.[9] While the recognition of such sites in the landscape is a valuable indication of past activity, their precise dates of occupation can usually only be established by excavation.

Roman villas in south-west England are discussed in chapters 9, 10 and 11, but it is worth noting that two of the few Roman villas so far recognized in the area (Holcombe and Magor) are known to have been constructed within pre-existing enclosures. It is possible that in the South West, unlike for example in the Cotswolds, 'villas' and Romanized lifestyles may represent a continuum of settlement with other enclosed farms. Such sites in Devon and Cornwall are not structures that would have been thought of as 'villas' by a Roman: they are really small farms using Roman building types, rather than the grandiose establishments known further east.

UNENCLOSED SETTLEMENT

Predominant on the maps are 'hut circles', but, as discussed above, this is largely a product of their survival. When settlements take the form of stone structures, such as the round houses or 'hut circles' of the uplands, where subsequent land-use pressures have been slight, the present pattern may not be much different from that at the time of their occupation, though the landscape in which they sat would have looked very different. However, in recent years our picture of the occupation even of these areas has been substantially modified by the discovery of wooden houses beneath or among stone round houses and broadly contemporary with them, at both Holne Moor and Shaugh Moor on Dartmoor.[10] It is likely that there are many other undiscovered examples in those areas that we see as 'spaces' between the enclosed and unenclosed settlements of the uplands. The open settlement preceding construction of the enclosure at Shaugh Moor has been mentioned above: such a transition has many parallels elsewhere. At present the majority of the round house settlements of the uplands are believed to date from broadly the second millennium

BC, with a falling off of settled exploitation of these areas about the end of the second millennium and the beginning of the first millennium BC. However, recent work at Gold Park, near Grimspound on Dartmoor, has identified settlement in a modified form of round house in an exposed position, datable to the late Iron Age; this serves to demonstrate the uncertainty of current knowledge.[11]

Maps 7.1, 7.2 and 7.4 show that the vast majority of unenclosed settlements so far recognized are those surviving as stone structures in the uplands. In the lowland areas, where building stone is absent and the lack of enclosing ditches means that even cropmarks are unable to reveal site locations, the upsurge of routine rescue archaeological work in the last twenty years has helped to increase our knowledge of site location. Work at Trethellan Farm near Newquay is typical; this is a site discovered during routine observation of building operations and proved to be an unenclosed settlement of some complexity, datable to the fifteenth to thirteenth centuries BC.[12] No hint of the presence of the settlement had previously been known. Much slighter traces of settlement occupation can also be recognized in careful observation of construction work, for example in association with laying pipelines. Rescue archaeological observations of this sort have been carried out more intensively in the last twenty years than ever before, and this work is gradually expanding our understanding of the settlement of the lowland areas now so heavily modified by millennia of cultivation. Of potentially even greater value is the recent emergence of geophysical survey equipment capable of examining extensive areas swiftly. This has already demonstrated its value for reconnaissance in advance of major development schemes, but there will in the future be potential for this to be used as a primary survey technique, particularly if it can be linked to current developments in airborne multispectral scanning.[13]

Away from the domestic settlements themselves, we know varying amounts about the agricultural and industrial practices of this wide period. In the uplands, where land-use pressures have historically been quite light, we can still see the fields that were in use by the second millennium BC. Those on Dartmoor and Bodmin Moor have been extensively studied over the last twenty years, and evidence has been mapped both from air photographs and by ground survey.[14] In other cases excavation shows the picture to be more complicated: the upstanding field systems known as the Dartmoor 'reaves' (**Fig. 7.4**) have been shown in excavation to have earlier, non-surviving predecessors, in the form of fences and lynchets, demonstrating that cultivation was taking place prior to the layout of the major reave systems.

In the lowlands, surviving evidence for early farming is harder to find. In exceptional cases, evidence of early ploughing may be preserved under a later feature, as at Gwithian, while on some very shallow soils the lynchets of later prehistoric fields have survived because they have not been heavily ploughed more recently.[15] In general, however, the farming systems of this period have to be reconstructed with evidence from excavation, and

Fig. 7.4
Dartmoor reaves.

These are extensive systems of parallel banks which can stretch across country for several miles. Excavation shows that the forms that we see now date to the Middle Bronze Age, but there is evidence that elements of the system had forerunners in the form of earlier field boundaries or fences. Houses, enclosures and small fields are sometimes incorporated within the reave system and there is evidence of cultivation within the fields, in a climate that was probably somewhat better than today. In other areas of Dartmoor, settlements, enclosures and houses lie outside any visible layout of reaves and may well be earlier.

SOURCE: photograph reproduced by permission of the Royal Commission on the Historic Monuments of England (copyright reserved).

much work remains to be done to reconstruct how farming, both pastoral and agricultural, developed over the period of time reviewed in this chapter.

The exploitation of the geological resources of the region was also a significant factor affecting settlement patterns during this broad period of time. In the earlier prehistoric period, flint was an important raw material and its exploitation is discussed in chapter 5. Somewhat later, tin deposits were exploited (see next chapter). The availability of materials such as clay suitable for ceramics and other raw materials will also have affected choices in settlement location. By the later prehistoric or Roman period iron was being extracted quite widely over the region: recent work has demonstrated extraction from both pits and open lodes, and primary iron smelting.[16] This activity will in turn have produced significant impacts on the landscape. Processing of metal ores at that time demanded large quantities of fuel, mainly charcoal. Either broad areas of woodland were cleared for this or,

more probably, substantial tracts of woodland were already being sustainably managed for production of wood for charcoal in the same way as we know they were managed for timber and wood production in, for example, the Somerset Levels. The production of pottery is known to have been widespread.[17] Many domestic utensils would, however, in these early times have been made from wood or bone: ubiquitous raw materials which survive only infrequently for archaeological discovery. The exploitation of resources in this period is thus a further factor which must be taken into account in any consideration of the influences affecting the settlement pattern indicated on **Maps 7.1, 7.2 and 7.4.**

FURTHER READING

See end of chapter 9.

BRONZE AGE METALWORK

SUSAN M. PEARCE

The people of the Bronze Age, who lived very roughly between 2500 and 600 BC, have left us their metalwork as one of their most distinctive classes of evidence.[1] The study of Bronze Age metal depends to a great extent on the creation of typological sequences, crucial to which are, first, the multiple finds known as hoards, and second, an understanding of the broad and detailed contexts from which the metalwork comes. It has proved possible to show only a small part of this information on the maps, but it has all been taken into account in the following discussion, which concentrates upon the role of metalwork in the social development of communities in south-west England.[2]

Much argument has raged over the sources of the copper and (especially) the tin used in the manufacture of British bronze, usually in the proportion of 90 per cent copper and 10 per cent tin.[3] There is tin ore in Cornwall and Devon, and in Brittany, north-west Spain and parts of central Europe: copper occurs in all these places, and in others also, including Ireland. The evidence for the use of south-western tin ore on a large scale in the Early and Middle Bronze Age is difficult to interpret, and, however strange and frustrating this seems, a considerable quantity of the tin in use during this period in the British Isles and the Continent must be supposed to have come from continental sources.

The period from roughly c.2500 BC to 1600 BC spans what is conventionally called the Early Bronze Age, when metal objects are few but evidently socially significant. Two interlocking traditions seem to be important: an axe-making tradition with Irish roots, and a dagger tradition rooted in Beaker practices of eastern Britain and continental Europe (**Map 8.1**).

Relatively little of the earliest metalwork, made of copper, has been found in the South West. An important early bronze find comes from Harlyn Bay, which gives its name to the succeeding phase. Here two great gold collars, Irish in character, known as lunulae were found with an early flat axe. The next phase is called the Trenovissick after the pair of elaborately decorated later flat axes found there which date to roughly 2000–1700 BC. Material belonging to the corresponding dagger tradition, known as Wessex I, is virtually absent in Devon and Cornwall, although the knife daggers are difficult to date, and a few low flanged axes probably belong here. The succeeding Plymstock/Wessex II period

(c.1700–1600 BC) is well represented by gold objects like those from barrows at Rillaton and Hameldown and by the hoard from Plymstock which had high flanged axes, a spearhead, contemporary daggers and a tool.

During this whole phase we can perceive a series of ritual monuments like henges, or stone alignments, which seem to have acted as focal points for all aspects of social life. Such organization implies leaders, and leaders need to maintain their position by visible prestige, which means among other things the control of impressive objects. This is likely to have been the role played by the metalwork: it is hard to see the magnificent lunulae as anything but sacred regalia worn by the highest of the high at ceremonies for which the communities' monument was the stage. The rich male barrow burials characteristic of Wessex II suggest a shift in power towards groups that had succeeded in establishing themselves within the old communities and were developing exchange routes for precious goods from Brittany and beyond. Behind their power probably stands increased agricultural prosperity. This seems to have included a much greater exploitation of the higher moorland, to which the land organization of fields and grazing lands represented by the Dartmoor reave system bears witness.

The Plymstock find, with its axes, tool, daggers and single spearhead, stands on the frontier of the next big development, called the Middle Bronze Age, or the Taunton phase, after an important Somerset find. The Plymstock material is contemporary with the metal found in the Wessex II phase barrows, but the character of the find, which was not with a burial, links it with the bronze finds of the next period, between roughly 1600–1200 BC. Taunton period bronzework had several phases, but is characterized by four broad types: axes, often of the clumsy types called palstaves; rapiers, short double-sided thrusting weapons; spearheads; and ornaments, especially heavy armrings, neckrings and pins like those which have been found in Cornwall at Helston and Tredarvah (**Map 8.2**).

Taunton phase metalwork is often found in what are conventionally called 'hoards', that is as several pieces together. Analysis shows that these multiple finds usually belong to one of two types. Some are hoards properly speaking, in that they include a small group of pieces deliberately buried together, like those

Map 8.1
Metalwork during the
copper-using Harlyn,
Trenovissick/Wessex I and
Plymstock/Wessex II phases of
the Early Bronze Age
(c.2500–1600 BC).

SOURCE: S. Pearce,
*The Bronze Age Metalwork of
South Western Britain*
(Oxford, 1983).

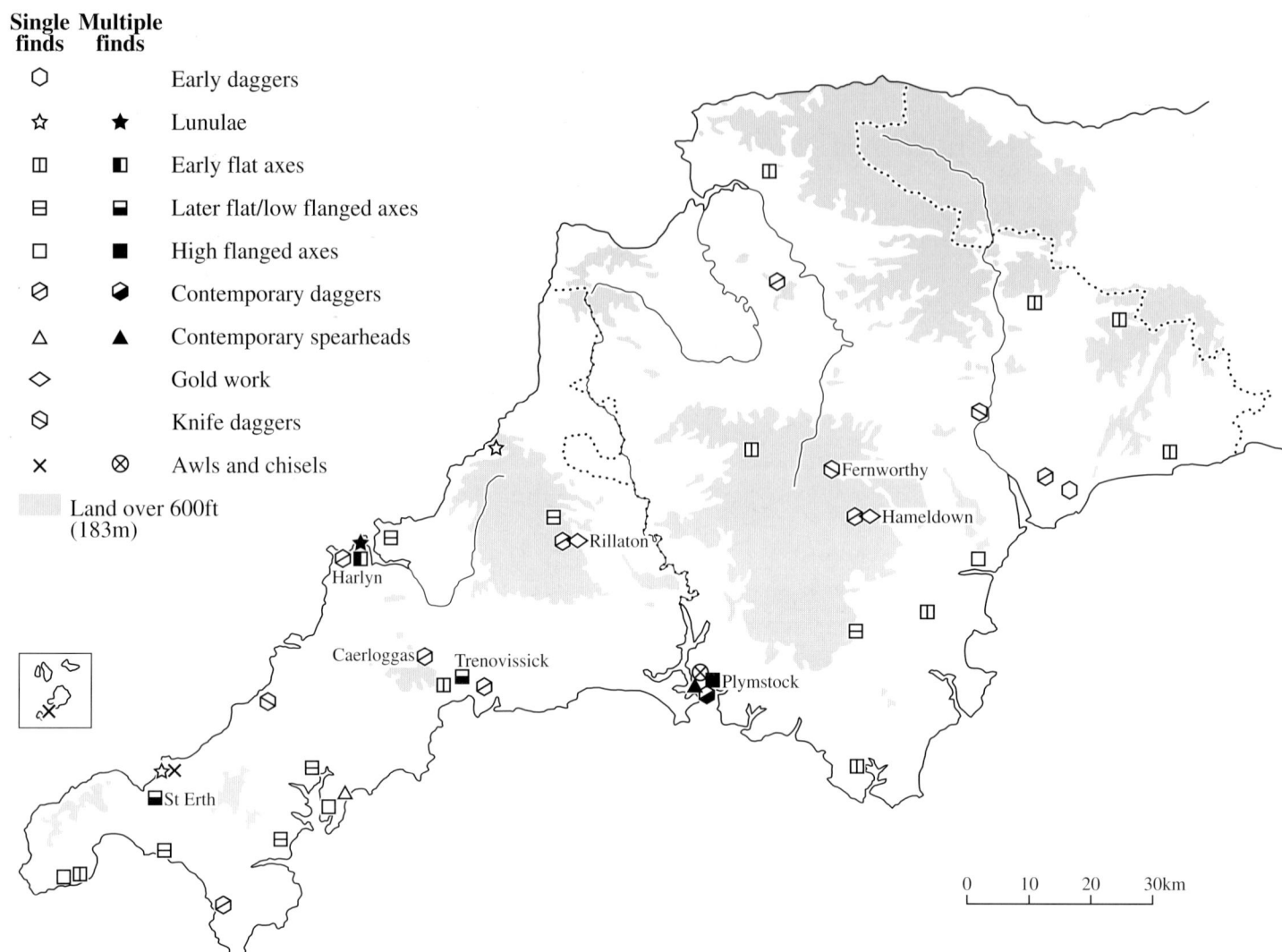

Single finds	Multiple finds	
⬡		Early daggers
☆	★	Lunulae
⊞	◧	Early flat axes
⊟	◨	Later flat/low flanged axes
☐	■	High flanged axes
⬡	⬢	Contemporary daggers
△	▲	Contemporary spearheads
◇		Gold work
⬡		Knife daggers
✕	⊗	Awls and chisels

Land over 600ft
(183m)

from New Quarry, Truro, and also the Week find which had two axes buried in a field bank. Others are scattered finds in which bronzes are mixed with household rubbish and often, significantly, the remains of bronze smithing operations: the Tredarvah find was of this type. These scattered finds suggest that a 'local bronze smith' model may best account for how the bronzes were produced. The true hoards may have been offerings to the gods or acted as collections of bride wealth or something similar. The need to accumulate bronze is likely to have supported the position of leading men, whose superiority may have been bound up with the control of land and of the metal supply which, as large finds like Lovehayne suggest, may have sometimes arrived from France in the form of axes for re-casting. Their status symbols were rapiers,

sometimes themselves gathered into hoards like that from Talaton: two rapier moulds were found at Chudleigh.

Around 1200 BC worsening weather conditions, and perhaps over-exploitation of land, caused social problems which we see as the disruption of the old bronzework patterns. The metalwork is an outward expression of the new conditions, and the period is referred to as the Worth (c.1200–1000 BC) and the Dainton (c.1000–900 BC) phases after two important finds. Modified palstaves continue, but rapiers and bronze ornaments disappear. In their place come slashing swords, more elaborate and efficient spearheads, gold ornaments like those from Towednack, and early efforts at the manufacture of the sheet bronze vessels called buckets and cauldrons. These pieces are sure signs of social

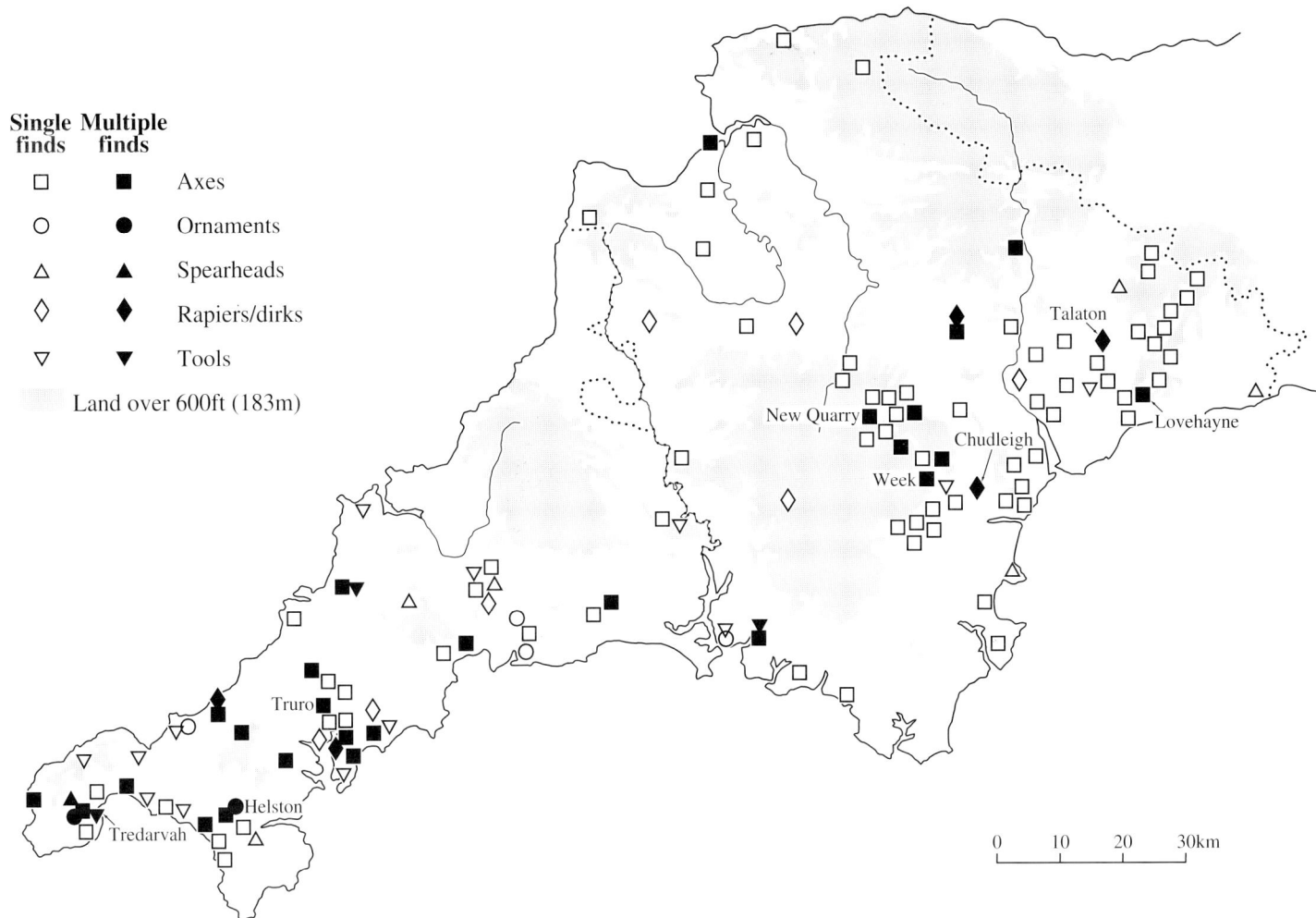

Map 8.2
Metalwork during the
Chagford/Taunton phases
of the Middle Bronze Age
(c.1600–1200 BC).

SOURCE: as **Map 8.1.**

division and the emergence of a fighting elite. Warriors need warrior rituals, and perhaps this is what the hoards of weapons, like that found at Worth, represent. Towards the end of the period special barbed spearheads were produced, apparently specifically as offerings to the gods: one group of these was found at Bloody Pool, South Brent (**Map 8.3**).

The early part of the conventional Late Bronze Age, known as the Ewart Park phase in Britain generally, and the Stogursey after an important Somerset find in the South West, is characterized by a great proliferation of numbers and types, and by significant advances in technique. A range of important south-western multiple finds, like those from Kent's Cavern, Kenijack (St Just), Lelant and Carn Brea, characteristically

contain a large range of mixed types, including socketed axes, swords, socketed tools, socketed spears and casting material including bun-shaped copper ingots, waste and scrap (**Map 8.4**). The bivalve stone axe moulds from Helsbury show how most of the heavy pieces were made, although swords may well have been cast in clay moulds.

The copper ingots seem to be part of an exchange network which ran into western France and down the Atlantic coast, and possibly into the western Mediterranean. The sources of the tin are still elusive, but if the south-western tin ore deposits were exploited on any scale at any time in the Bronze Age, then this seems to be the most likely period. Gold ornaments are also a feature of contemporary metal, usually in the South West taking the form of

Map 8.3
Metalwork during the
Worth/Dainton phases of
the Middle Bronze Age
(*c*.1200–900 BC).

SOURCE: as **Map 8.1**.

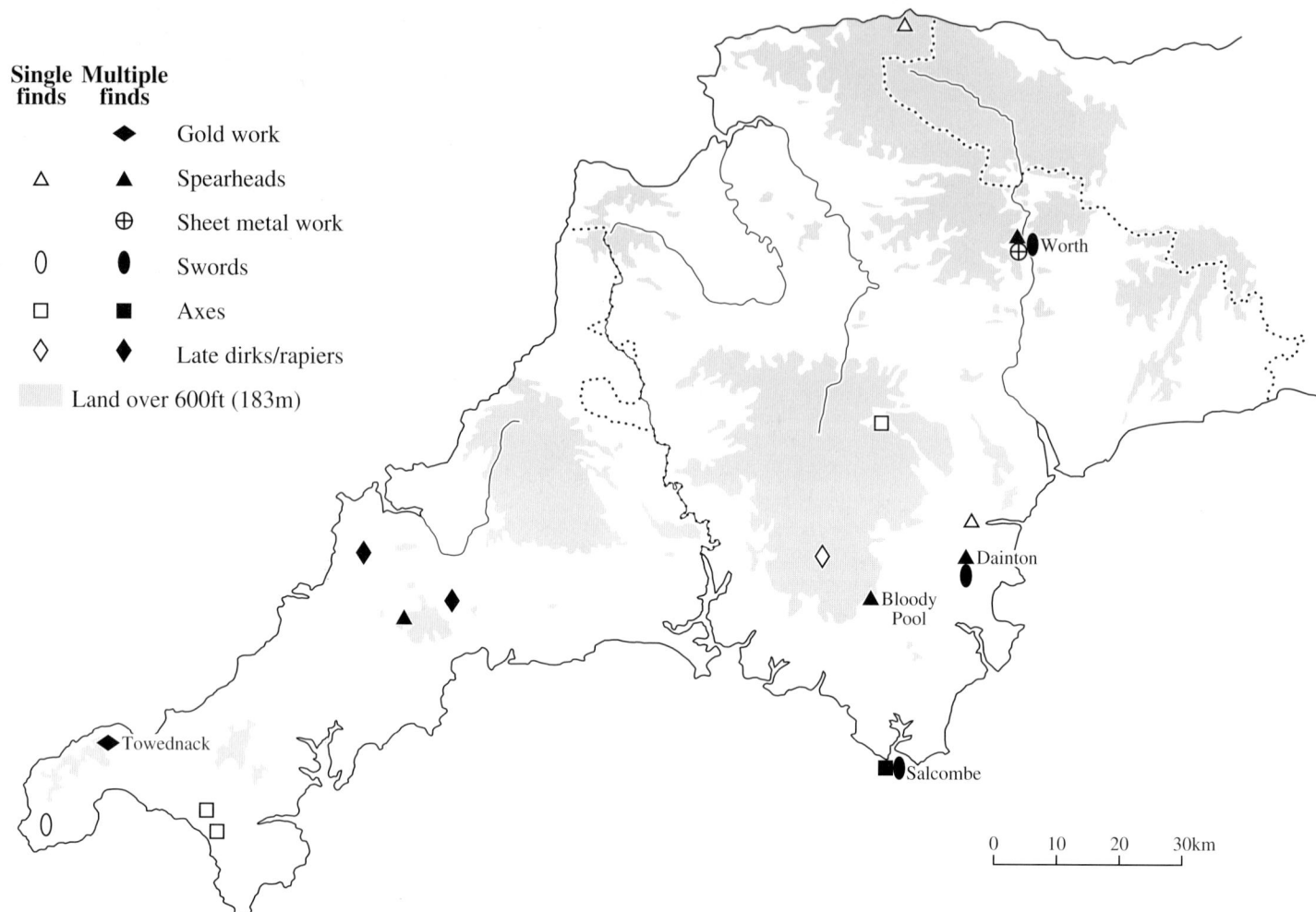

armrings. Armring hoards are known from Morvah, and from two recent finds, one at Rosemorran near Penzance and the other from Colaton Raleigh in east Devon. The gold is conventionally thought to have come from Ireland, but some of it may be south-western.

Around 700 BC a new range of metalwork types appeared in southern Britain, which included long, narrow Gündlingen swords, ornamental harness fittings, chunky neckrings and armrings, elaborate razors and the late type of socketed axes known as Sompting axes: all this is known as the Sompting or Hallstatt C material. This material is very rare in the South West, but contemporary with it is another group, the socketed 'American' axes from Brittany and related types. These sometimes have a very high tin content and seem, in effect, to be

tin ingots, while the others may have acted as supply bronze. Axes of this broad type are known from Carn Brea and Mount Batten, where they were found with a wide range of material. This period is called the Mount Batten phase in south-west England.

Some of the Stogursey–Mount Batten finds come from 'high' sites which were to become full-blown hillforts in the Iron Age, like Kenijack Castle and Carn Brea in Cornwall and Woodbury and Trendlebere in Devon. There are signs that bronze was being worked on these hilltops. Sometimes too, a bronze comes from earthworks which seem to have been in the process of developing into substantial land boundaries. This suggests a warrior elite in a position to dominate tracts of country, manage the cross-Channel metal supply from trading points like Mount's Bay and Mount

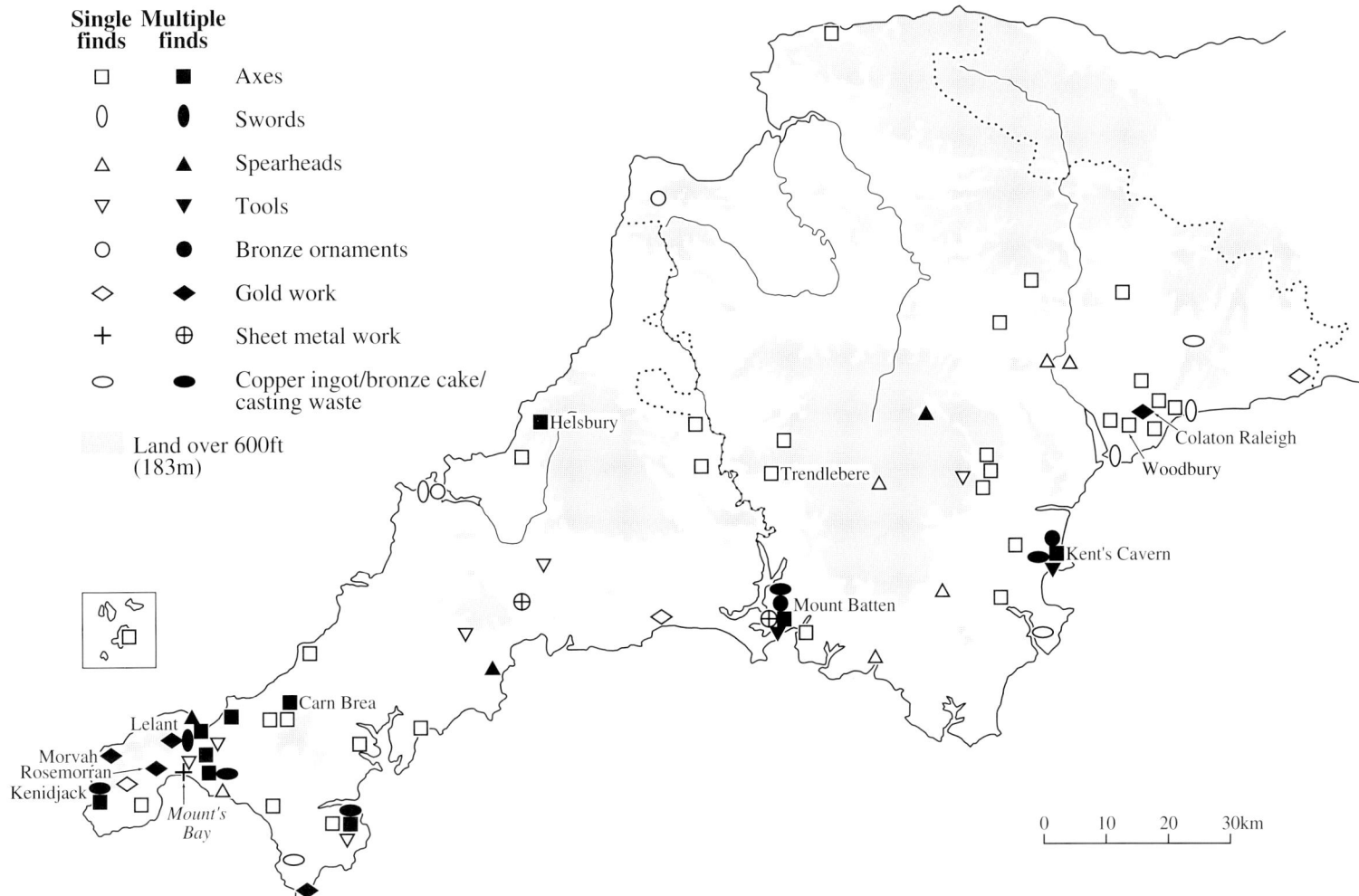

Map 8.4
Metalwork during the
Stogursey/Mount Batten
phases of the Late Bronze Age
(*c*.900–600 BC).

SOURCE: as **Map 8.1.**

Batten, and control the production and distribution of weapons for themselves and axes and tools for their dependent peasants and craftsmen.

Bronze production for all except some ornamental work seems to have come to an abrupt end as knowledge of iron-working spread from about 600 BC. Nevertheless, Bronze Age developments in the South West, as in the rest of southern Britain, were crucial. By the end of the Bronze Age, Britain had become the hierarchical peasant-based and generally sedentary society it was to remain until the Industrial Revolution. Analysis of the metalwork finds recorded on the maps in this chapter helps us to understand the social and economic processes that brought about this fundamental change.

FURTHER READING

R. Bradley, *The Passage of Arms: An Archaeological Analysis of Prehistoric Hoards and Votive Deposits* (Cambridge, 1990); R.J. Taylor, *Hoards of the Bronze Age in Southern Britain*, British Archaeological Reports, British Series No. 228 (Oxford, 1993); J. Coles and A.F. Harding, *The Bronze Age in Europe: An Introduction to the Prehistory of Europe, c.2000–700 BC* (London, 1979); I.M. Stead, *The Salisbury Hoard* (Stroud, 1998).

IRON AGE to ROMAN BUILDINGS, STRUCTURES, AND COIN and OTHER FINDSPOTS

FRANCES GRIFFITH AND HENRIETTA QUINNELL

Map 9.1 illustrates the scattered but extensive use of south-west England in the Roman period.[1] **Map 9.1** and the text of this chapter should be read in conjunction with the maps and text of chapter 7, which discusses the background to the many unexcavated and undated occupation sites, especially enclosures, which may have been in use during the Roman period. Roman period settlement in Devon and Cornwall was largely a continuance of traditions dating well back into the prehistoric period and continuing into the post-Roman centuries. There are insufficient finds from the Iron Age for useful cartographic presentation but their distribution extends from the Dorset border to West Penwith. Iron Age coins and metalwork are not common and the authenticity of coins, particularly hoards, is often dubious, with probably only four hoards of genuine Iron Age date.[2]

Two groups of sites span the Iron Age and Roman periods: inhumation cemeteries (not plotted on **Map 9.1**) and fogous. Fogous, sometimes termed souterrains, are trench-built underground chambers, mostly constructed in the Later Iron Age but which remained in use throughout the Roman period. There is a considerable literature on fogous, the function of which has been variously interpreted as being for storage, for refuge, or for religious purposes.[3] Similar monuments are found in parts of Ireland and Scotland, dating in general later than those in Cornwall but with similar problems regarding function. Fogous are presented for clarity in West Penwith and the Lizard on an inset.[4] The few cemeteries recorded are all in Cornwall and the Isles of Scilly except Stamford Hill, above Plymouth Sound. Three sites—Harlyn Bay, Trethellan and Trelan Bahow—belong entirely within the Iron Age, but about six were in use from the Later Iron Age until the Roman period.[5]

It has long been recognized that structures built in Roman style were not common in the South West. Romanized buildings on **Map 9.1** include both villas and simpler edifices whose presence has been demonstrated by excavation, aerial reconnaissance or recovery of building materials. Three close to the Dorset border—Membury, Holcombe and Seaton—have been known since the nineteenth century and preserve the best evidence for elaborate Roman features such as mosaic floors and baths.[6] More recently, excavations in advance of developments in the Topsham area have revealed a variety of buildings in a complex along the Exe, while work at Otterton, following chance surface finds,

showed a group of simple stone buildings.[7] Crediton, largely unexcavated as yet, is the only site so far located by aerial reconnaissance.[8] Several excavations conducted on medieval or later sites in Devon have produced building materials, in particular tile, which indicate the presence of Romanized buildings scattered as far as Exmouth, Barnstaple and Totnes.[9] Not enough is known about the context of any sites on **Map 9.1** for their status and function to be decided. The known distribution of Romanized buildings has been gradually expanded from Exeter and Devon east of the Exe to the north and south of the county. No buildings or building material of Roman age are recorded west of Dartmoor, except at Magor in the far south-west, found and excavated in 1931.[10] It should be noted that both the buildings at Magor and Holcombe are on sites with enclosures; that at Holcombe has been demonstrated by excavation to precede the stone building.

Stone-walled courtyard houses, often surviving as visible monuments, have long been recognized as a distinctive structural group; their construction has now been clearly placed in the Roman period, with some possible occupation as late as the sixth century.[11] Their distribution, despite extensive fieldwork over the whole region, is still restricted to West Penwith and the Isles of Scilly.[12]

The continued building and occupation into the Roman period of enclosed settlements or rounds demonstrated by excavation has already been emphasized in the discussion of **Maps 7.1, 7.2 and 7.4** in chapter 7. In Cornwall, and possibly in Devon, round houses appear to have been replaced by those of elongated oval shape.[13] Pottery and other Roman period finds from excavated rounds are represented by the 'material' symbols on **Map 9.1**. Known unenclosed occupation sites are very rare because of the difficulties in their recognition. Most of those known tend to be in coastal locations, where they are revealed by erosion, such as the salting at Trebarveth on the Lizard.[14]

The 'material findspots' include both pottery scatters noted in fieldwork or as chance finds and all other significant artefacts or artefact groups. Pottery scatters in the Lizard reflect the input of an intensive programme of field survey c.1980.[15] Most pottery from Cornwall is gabbroic, probably made from Lizard clays, where production is presumed to have been centred, although no actual potteries have yet been located.[16] Other sources of ceramics, especially for Devon, are South Devon ware, probably from the Dart valley, and South-East Dorset

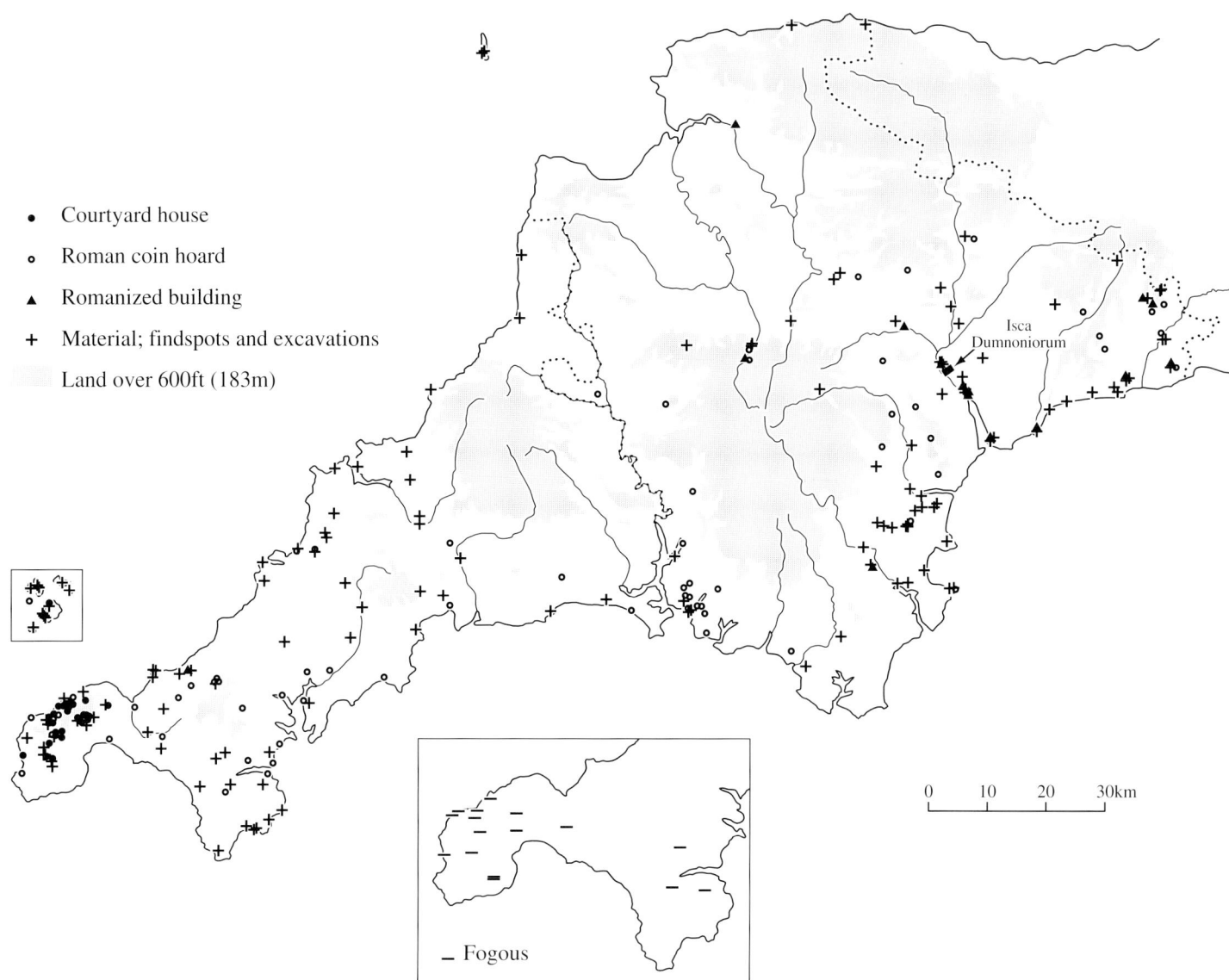

Map 9.1
Structures assigned to the Iron Age and Roman periods and finds of Roman period material.

SOURCES: authors and Devon Sites and Monuments Register, Devon County Council.

- Courtyard house
- Roman coin hoard
- Romanized building
- Material; findspots and excavations
- Land over 600ft (183m)

Isca Dumnoniorum

0   10   20   30km

— Fogous

Black-burnished ware from the Poole area.[17] Additionally, artefacts include such diverse items as pewter, tin ingots, the altar from Hugh Town, Isles of Scilly (**Fig. 9.1**), and the bronze centaur statuette from Sidmouth beach.[18] Finds in the Plymouth area, together with coin hoards, suggest a possible harbour.[19] Three sites may be religious centres, in the absence of 'Romano-Celtic' temples common in other parts of southern Britain. Two are deep shafts, at Cadbury in mid-Devon and Bosence in west Cornwall, where, respectively, deposits included bronze bracelets and a pewter dish with an inscription to the god Mars; these may reflect a widespread indigenous tradition of northern Europe.[20] The third is a stone structure on Nornour, Isles of Scilly, the focus for deposition of numerous coins, brooches and pipeclay figurines.[21]

Coin hoards are a clearly definable class of datable artefacts, although their significance is uncertain as coins are not common finds even on excavated sites in the area and the regular establishment of a monetary economy in the South West has been questioned.[22] The pattern of distribution is certainly affected by factors such as antiquarian activity in West Penwith since the eighteenth century, and disturbance caused by more recent tin extraction in Cornwall. The broad scatter across south and east Devon reflects the comparative amount of archaeological work carried out, until recently, in the south as opposed to the north of the county. Coin hoards in general occur in areas with other finds from the period. The majority, especially those of large size, belong to the third and fourth centuries.[23] Single coin finds are not shown because there is no reliable data base and many are of dubious provenance.

Two further features of **Map 9.1** deserve particular comment. West Penwith is characterized by a clustering of symbols. This is because so many courtyard houses survive as prominent monuments and have attracted investigation since the nineteenth century. By contrast there is a marked absence of Roman period

Fig. 9.1
Roman altar from St Mary's,
Isles of Scilly.

SOURCE: drawing reproduced
by permission of Paul Ashbee
(based on records held by
the Society of Antiquaries
of London).

finds from Bodmin Moor and Dartmoor. This may reflect real absence of occupation: the frequent, if early, excavations on Dartmoor have investigated numerous hut circles, and peat-cutting on both moors has produced no Roman material. However, these moorland lacunae need not reflect real lack of activity, because settlement may have had foci away from the visible stone-walled sites. Moreover, the moorland peat is not a good preservative for finds of any kind except lithics (see **Maps 7.1, 7.2 and 7.4** in chapter 7).

In conclusion, the evidence of Roman material from findspots and excavations and coin hoards indicates extensive use of the countryside in the Roman period and thus the existence of a sizeable population in south-west England at this time. Both these facts have gone largely unrecognized until recently because of the concentration of attention on Roman type structures.

FURTHER READING FOR CHAPTERS 4–9

The following list of books and journals introduces the now extensive archaeological literature on south-west England. More specialized books and articles are listed in the notes to each chapter.

The annual journals of the Cornwall and of the Devon Archaeological Societies, *Cornish Archaeology* (from 1962) and *Proceedings of the Devon Archaeological and Exploration Society* (from 1929), provide a corpus of background material. *Cornish Archaeology: Silver Jubilee Volume 1961–1986*, 25 (1986) presents a review of the county's archaeology period by period from the Mesolithic onward. *The Archaeology of Dartmoor: Perspectives from the 1990s*, the Devon Archaeological Society's *Proceedings*, 52 (1994), contains papers both on Dartmoor and the relationship of its archaeology to that of Devon as a whole.

A. Fox, *South West England* (London, 1964) is the classic text; its wide range of illustrations is especially useful. S.M. Pearce, *The Archaeology of South West Britain* (London, 1981) and M. Todd, *The South West to AD 1000* (London, 1987) provide more recent coverage and references.

English Heritage with Batsford have produced a popular and readable series which provides background for the archaeology of the South West: N. Barton, *Stone Age Britain* (London, 1997); M. Parker-Pearson, *Bronze Age Britain* (London, 1993), which also covers much of the Neolithic; B. Cunliffe, *Iron Age Britain* (London, 1995); M. Millett, *Roman Britain* (London, 1995). S. Gerrard, *Dartmoor* (London, 1997) and A. Fleming, *The Dartmoor Reaves* (London, 1988), concentrate on Dartmoor archaeology.

A selection of in-depth studies: P. Ashbee, *Ancient Scilly* (Newton Abbot, 1974); N. Johnson and P. Rose, *Bodmin Moor: An Archaeological Survey, Vol. 1, The Human Landscape to c.1800* (London, 1994); J. Butler, *Dartmoor Atlas of Antiquities*, 5 vols (Exeter, 1991–7).

THE
ROMAN ARMY

VALERIE A. MAXFIELD

When the third edition of the Ordnance Survey map of Roman Britain was published in 1956, it identified no Roman military sites in Cornwall, and just three in Devon—the fortlets at Martinhoe, Old Burrow and Stoke Hill. Twenty-two years later, the revised edition (published in 1978) added to this total one fort in Cornwall, and one fort, one fortress and two temporary camps in Devon. Aerial reconnaissance and ground survey have since added substantially to this total, and doubtless will continue to do so. The distribution of sites on **Map 10.1** indicates a Roman military presence over much of Devon, though no further sites have been as yet identified in Cornwall.

The sparse surviving literary account of the Roman conquest of southern Britain (see chapter 11), combined with archaeological evidence from neighbouring Dorset, suggests that campaigns had been launched westwards very soon after the initial invasion in AD 43. Exactly when the invaders first entered Devon and Cornwall remains unclear: their earliest bases will have been the temporary encampments which they put up each night when on the march. Occupied for such short periods and by troops using robust metal mess-tins rather than breakable pottery, these temporary camps are notoriously difficult to date. The two known in Devon, at North Tawton and Alverdiscott (**Map 10.1**), are totally undated.[1] Their major interest is in indicating likely lines of advance— around the north side of Dartmoor and up over the Taw–Torridge watershed towards the north coast.

The more permanent military bases, established in the wake of the initial invasion, belong, where dated, within the range *c.*AD 50–80. The earliest site appears to be Hembury, where there is a brief Roman occupation, starting in the late 40s, within the area of a disused hillfort.[2] For such a hilltop site, atypical of Roman fort siting (which in general opted for accessibility and mobility rather than defensibility), to represent the earliest strands of military settlement makes good sense, for such sitings tend characteristically to represent the earliest phase of occupation within the areas to which they belonged.

It is singularly unlikely that all of the fort sites shown were occupied at the same time; had they been they would together have accommodated nearly a quarter of the Roman army estimated to be in Britain at the time. At least two phases may be detected in the occupation of several sites. The juxtaposition of two separate bases at North Tawton, one of at least 5ha (and conceivably as large as 20ha), the other of about 3ha, indicates two separate periods of use of that site, while excavation at Tiverton shows that the fort, while remaining on the same site and unchanged in size, underwent a total rebuild and redesign of its gateway (**Fig. 10.1**).[3] Surface observation at Cullompton suggests that here too the fort was redesigned during its lifetime, while at Exeter a major garrison reduction is suggested by the reduction in size of the legionary bathhouse as well as by adjustments to other internal buildings.[4]

The location of these various military bases will have been influenced by a number of related factors. A ready water supply was one of the basic needs, and forts were commonly sited on rivers; Exeter, in common with the vast majority of legionary bases, was accessible by water as well as by land, greatly facilitating the bringing in of supplies. Forts were sited along the major communication routes for ease of access, so that the soldiers who occupied them could be deployed wherever they might be needed, which, during the latter part of the period of their use, will have been in Wales and the north, the scenes of the major campaigning in Britain in the last quarter of the first century.

The forts were essentially winter bases, occupied by the troops when they were not out on campaign during the summer months. The splitting up of the army into relatively small groups (most of the forts in Devon and Cornwall can have accommodated no more than about 500 soldiers) is perhaps in part a response to the need to police the local population—to protect friends and keep a wary eye on actual or potential enemies. In part, too, the fragmentation of the army may have been a way of dividing up the not inconsiderable burden of provisioning the troops, both through the winter and in readiness for the forthcoming summer campaigns, and lessening the land pollution inevitably caused by the imposition of large numbers of men. For some of the sites a more specific role may be detected. The two north Devon fortlets are clearly sited with a view to maintaining a watch over the Bristol Channel, northwards towards the land of the hostile Silures people in South Wales. The fortlet at Stoke Hill, above Exeter, and its probable companion site across the river at Ide, provide excellent look-out points, enjoying as they do wide views over the Exe estuary, the fortress site and its approaches. The fort at Nanstallon, near Bodmin, is situated

Map 10.1
Roman army bases and camps.

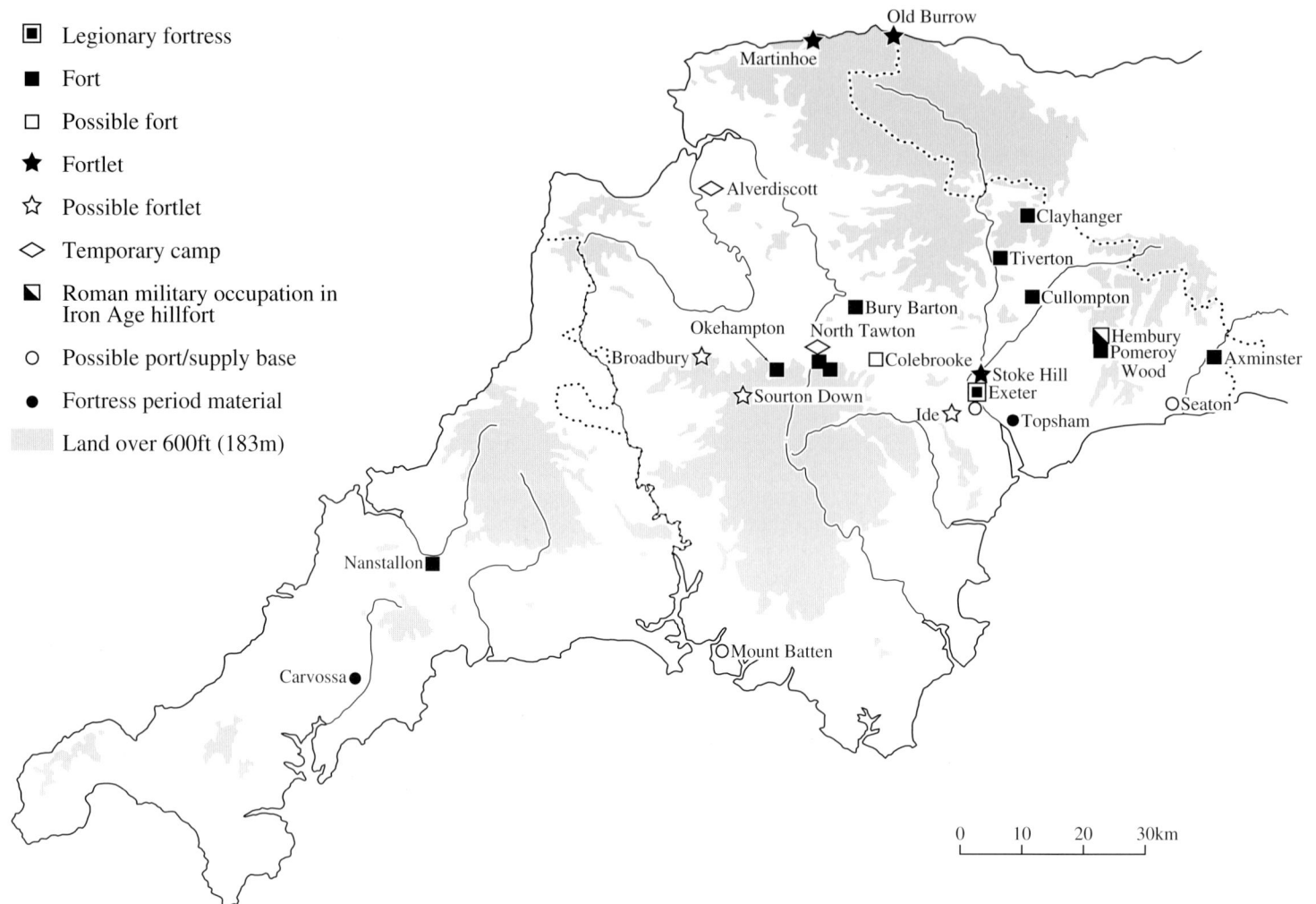

Legionary fortress

Fort

Possible fort

Fortlet

Possible fortlet

Temporary camp

Roman military occupation in
Iron Age hillfort

Possible port/supply base

Fortress period material

Land over 600ft (183m)

Old Burrow

Martinhoe

Alverdiscott

Clayhanger

Tiverton

Cullompton

Bury Barton

Okehampton   North Tawton

Hembury

Broadbury

Pomeroy
Wood

Colebrooke

Axminster

Stoke Hill

Sourton Down

Exeter

Ide

Seaton

Topsham

Nanstallon

Mount Batten

Carvossa

0   10   20   30km

in a region rich in mineral deposits: the soldiers who occupied it could have been charged with supervision of the exploitation of these resources—a role commonly assumed by the army.[5]

Of coastal or riverine supply bases there is little positive evidence. Topsham is commonly quoted as the port of Exeter in the Roman period, but no military structures are known there and it is more than possible that the Exe was navigable as far as Exeter at this time, thus eliminating the need for a port down-river.[6] Military-type timber buildings and a ditched compound outside the fortress may relate to just such an establishment.[7] A tile from Seaton, stamped 'LEG II AVG', may indicate the existence of a coastal base here in the third century (the date to which the tile belongs), while it is difficult to believe that the army did not make use of the site of the existing entrepôt at Mount Batten, by Plymouth.[8] As yet the only evidence that they might have done so is a small quantity of first-century AD material among the finds from the site.[9]

FURTHER READING

I.A. Richmond, revised by M. Todd, *Roman Britain*, 3rd edn (London, 1993) includes a discussion of the conquest of southern Britain in chapter 2. V.A. Maxfield, 'The army and the land in the Roman South-West', in R.A. Higham (ed.), *Security and Defence in South West England before 1800* (Exeter, 1987), 1–25, is an account of the Roman conquest of the South West, the nature and location of the Roman military sites and the relationship between the army and the local population. V.A. Maxfield, 'Devon and the end of the Fosse frontier', *Proceedings of the Devon Archaeological Society*, 44 (1986), 1–8, is a discussion of the Fosse Way and its interpretation as a frontier or a line of communication, with particular reference to its south-western terminus in Devon.

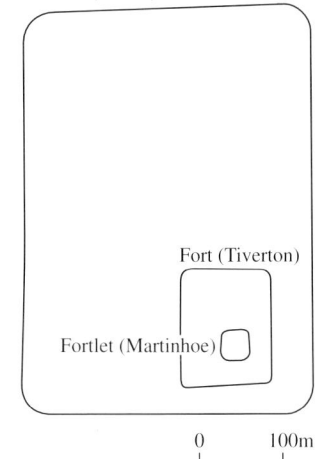

Map 10.2
Exeter fortress, Tiverton fort, and Martinhoe fortlet.

A *fortress* was a base for a Roman legion of approximately 5,000 soldiers. The only example in Devon and Cornwall is Exeter, just under 17ha in area, base for *legio II Augusta*; a *fort* was to house auxiliary (non-citizen) troops, organized into cavalry and infantry regiments nominally five hundred and a thousand strong, and examples in the South West range in size from North Tawton (3ha) to Nanstallon (0.88ha); a *fortlet* provided accommodation for a detachment of soldiers outposted from their regular base for policing and surveillance duties, and sizes vary according to the task in hand, but examples such as Martinhoe, with an internal area of 0.06ha, provided space for about eighty soldiers (*a century*).

Fig. 10.1
Reconstruction of the west gateway of Tiverton fort in period.

SOURCE: drawing by B. Chandler in *Proceedings of the Devon Archaeological Society*, 49 (1991), Fig. 12.

CLASSICAL SOURCES
FOR
ROMAN PLACE-NAMES

MALCOLM TODD

Four surviving documents provide information on the Roman names of sites in south-western Britain (**Map 11.1**).[1] These are the *Geography* of Ptolemy, compiled between AD 140 and 150 but relying heavily upon earlier sources, the *Antonine Itinerary*, a compilation of the early third century detailing routes followed by the emperor Caracalla, the *Peutinger Table*, a map which originated in the third century, was revised in the fifth and which is largely extant in a thirteenth century form, and the *Ravenna Cosmography*, a compilation by an anonymous cleric of Ravenna, shortly after AD 700, of lists of places in the Roman world. The last-named work provides the largest number of place-names for the south-west peninsula, but its value is severely restricted by obvious corruption of the text and by uncertainty about the principles underlining its arrangement.[2] The Ravenna text is nevertheless the fullest surviving source on the place-names of south-west England and its evidence must be seriously considered, imperfect (and often infuriating) as it is. It is unfortunate that the south-western area of Britain which would have been depicted on the Peutinger Table was lost before the existing copies were made. Its information might have enabled closer identification of at least some of the places listed in the Ravenna Cosmography.

Ptolemy's *Geography* gives us the ancient names for three great headlands in the South West: Antivestaeum or Belerium (Land's End), Damnonium (= Dumnonium) or Ocrinum (The Lizard) and the promontory of Hercules (probably Hartland Point).[3] Three rivers were also known to Ptolemy: Isca (Exe), Tamarus (Tamar) and Cenio. The last-named presents a problem. It is identified with the Kenwyn by Rivet and Smith but without obvious reason.[4] It may be the stream noted in the Ravenna Cosmography as the Conio(s) discussed later in this chapter. Only one place is named by Ptolemy, the legionary base at Exeter (Isca), an obvious reflection of the scattered tribal geography of the Dumnonii.

The Antonine Itinerary and the Peutinger Table provide little evidence for the nomenclature and siting of places in the South West. Both name Exeter (Isca Dumnoniorum) and Moridunum, a place 15 Roman miles to the east, but nothing else. The location of Moridunum has provoked as much debate as any other single name in the itineraries. Seaton (of which Moridunum has been seen as a doublet: sea-fort), Sidford, Hembury, Gittisham and Axminster have all had their proponents, but the matter remains unresolved.[5] The least satisfactory arguments are those which rest merely on the form of modern names. From this point of view, Seaton has nothing to recommend it. Almost equally unsatisfactory are deductions which involve arbitrary and unjustified emendations of the recorded distance between Isca Dumnoniorum and Moridunum. This is given as 15 miles by both the Antonine Itinerary and the Peutinger Map. To ignore or override this fact is a serious error. An interval of 15 miles east of Exeter brings us to the Otter valley west of Honiton and the southern fringes of the Blackdown Hills. A crossing-place of the Otter at or near Fenny Bridges, where the Roman road passes the river, is very close to 15 miles from Exeter and the site is well suited to a roadside station. Also 15 miles from Exeter is the nearby hillfort of Hembury, where excavation has revealed early Roman military occupation within the prehistoric defences.[6] The name Moridunum, whatever its meaning, could have been originally applied to the fort and later transferred to the roadside site only 2½ miles away.[7] The Ravenna Cosmography also mentions Moridunum, possibly twice, but provides no further clue as to its position.

The list provided by the Ravenna Cosmography begins in south-western Britain and appears to run from west to east, the first group of names leading to Exeter.[8] The list runs as follows:

> Giano; Eltabo; Elconio; Nemetotatio; Tamaris; Purocoronavis; Pilais; Vernilis; Arduaravenatone; Devionisso; Statio Deventiasteno; Duriarno; Uxelis; Vertevia; Melamoni; Scadum Namorum.

Some of the textual errors are obvious and easily corrected. The final 'i' of Melamoni should clearly begin the next name in the list, which thus reads Isca Dumnamorum (= Dumnoniorum). An 's' has fallen out of Nemetostatio and Purocoronavis is an evident corruption of some such name as Durocornovium or Durocornoviorum. Arduaravenatone is probably two names: Ardua and Ravenato.[9] The two names beginning in El- are fairly certainly rivers: Fl(umen) Tabus (or better Tavus) and Fl(umen) Conio(s). Giano is presumably Gianum (= Elanum?) in its nominative form.[10] There may be dittography in Devionisso and Deventiasteno, while there is evident corruption in Pilais and Melamon. All this may seem to offer unpromising fare, but a few points can be established with some probability. It is certain that this group of names leads to Exeter and thus it is a reasonable presumption that these places lie to the west of that city. It is natural to look for any points that can be fixed with greater or lesser probability. Tamaris clearly denotes the river Tamar (Tamara?) or a site on the river, perhaps at a major crossing, which has taken its name. Nemetostatio equally clearly derives its name from the *nemeton* (sacred grove) in north-central Devon which is still commemorated by several villages and hamlets.[11] Apart from Fl. Tavus (the river Taw?) and Fl. Conio(s), Uxelis and Vertevia are possible river-names or place-names derived from rivers. None of the other names in the list has any obvious resonance, except that Durocornoviorum, if correctly reconstructed, seems to record a fort (*duro-*) of the Cornovii, the sept or sub-tribe whose name survives in that of Cornwall.

Given the prevailing un-Roman character of the Dumnonian civitas, it is remarkable that so many names are listed for the region in this document. A possible explanation is that a number of these are the names of natural features, notably rivers, rather than settlements. There is a further possibility. It is difficult to believe that all the recorded names lay on a single east–west route through the peninsula. More probably our source has recorded at least two series of names, associated with routes running to the north and to the south of Dartmoor. If there is any consistency in the record at all, the first five or six names could belong to the northern route, which linked the Taw (Tavus), perhaps the Torridge (Conio), Nemetostatio (the *nemeton* in north Devon) and the Tamar. The place named 'fort of the Cornovii' (Durocornoviorum) would make most sense if it lay on the borders of that sept, possibly in the Tamar valley. The middle group of names, in which deformations are unfortunately most evident, is virtually opaque, though Ardua would be a colourable name for a site on the inhospitable western side of Dartmoor. The *statio* (official post) which appears in this group of names might also lie on or close to Dartmoor, possibly having a connection with official Roman interests in the varied mineral deposits on the south-western moor. The four names Duriarno, Uxelis, Vertevia and Melamon might then lie to the south of Dartmoor, between the Tamar and the Exe. Duriarno is 'fort on the river Arnos', while Uxelis and perhaps Vertevia as well should be river-names. If these are *all* river-names, or names derived from rivers, the four obvious candidates in south Devon are the Erme, Avon, Dart and Teign. Some such structure to the Ravenna list does not seem wholly fanciful, while possessing the merit of according well with the known character of Dumnonian land.

The following group of six names was seen by Richmond and Crawford as belonging to places lying to the north-east of Exeter: Termonin; Mestevia; Milidunum; Apaunaris; Masona; Alovergium.[12] The suggestion of Richmond and Crawford is no

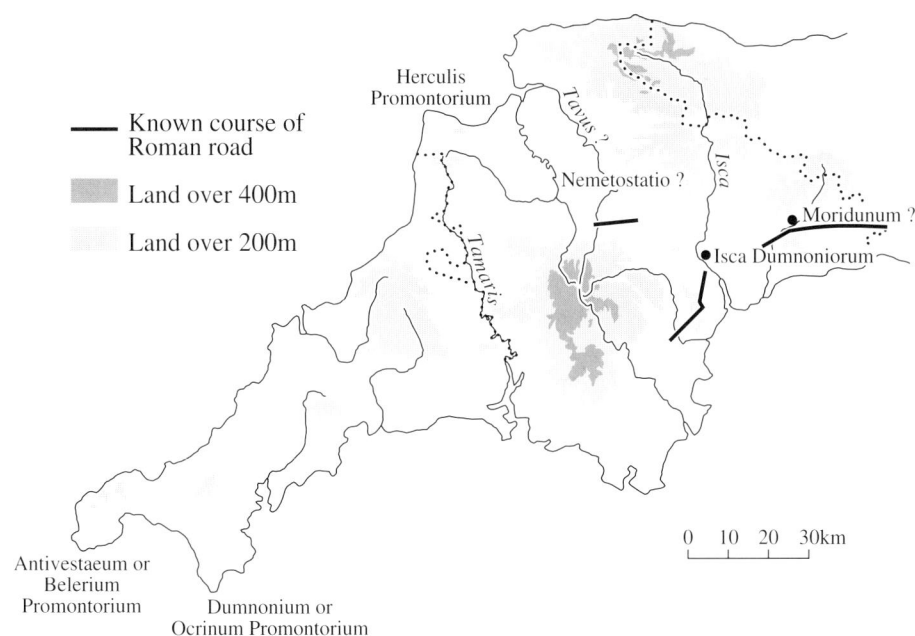

Map 11.1
Roman place-names.

more than a guess: there is no certain evidence as to where this group lay. Milidunum may be a corruption of Moridunum, Mestevia of Vertevia, but neither is certain. More plausibly, Apaunaria is a deformation of Ad Tamarim, 'at the Tamar', in which case the names may mark another route across the peninsula.[13] Probably Termonin does not contain the element 'terminus' and thus does not mark the end of a route or some other boundary. The usual formula would be Fines or Ad Fines. Dillemann links Alovergium with Ptolemy's *Ovepyibly T*, one of the two seas west of Britain, an ingenious but not wholly convincing proposal. These six names thus present many difficulties. Perhaps the only gleam of light is the name of the Tamar behind Apaunaris.

FURTHER READING

L. Dillemann, 'Observations on Chapter V, 31, Britannia in the Ravenna Cosmography', *Archaeologia*, 106 (1979), 61–73; I.A. Richmond and O.G.S. Crawford, 'The British section of the Ravenna Cosmography', *Archaeologia*, 93 (1949), 1–50; A.L.F. Rivet, 'The British section of the Antonine Itinerary', *Britannia*, 1 (1970), 34–82; A.L.F. Rivet, 'Some aspects of Ptolemy's Geography of Britain', in R. Chevallier (ed.), *Littérature gréco-romaine et géographie historique. Mélanges offerts a Roger Dion* (Paris, 1974), 55–82; A.L.F. Rivet and C. Smith, *The Place-Names of Roman Britain* (London, 1979); M. Todd, *The South West to AD 1000* (London, 1987).

EARLY CHRISTIAN DUMNONIA

A.C. THOMAS

Apart from the incidence of native names on formal Roman inscriptions, the oldest monuments displaying the British (Celtic) language—the speech that became Cumbric, Welsh, Cornish and Breton—are the lettered memorial stones of northern and western Britain, dating between the fifth and eighth centuries AD. They are also among the earliest evidence for (post-Roman) Christianity, and have a human interest in that they name individuals. There are small groups in south-west Scotland and the Isle of Man, and in north and north-west Wales; a much larger (one hundred-plus) spread across South Wales; and fifty-plus in Devon and Cornwall. Their importance to history and archaeology, as well as to language and epigraphy, is once again being realized and reflected through increased modern study.

The distributional emphasis, principally west of *Isca Dumnoniorum*, Roman Exeter, stands well apart from any continuing aspect of the Roman period (**Map 12.1**). It is entirely post-Roman. A line down the middle of the peninsula might show rather more symbols north than south of it; any refinements would confirm that most of these inscribed memorials do indeed betray new influences from the *north* (in this case, south and west Wales, with direct landfalls at, for example, the Camel estuary).[1] Only one memorial—(479) CUNAIDE, Hayle, later fifth century[2]—with eleven horizontal lines in Roman capitals is genuinely exotic. The wording suggests contemporary Gaulish Christian practice and the stone may be a spin-off from maritime commercial contact.

Throughout the peninsula these inscriptions seem to commemorate the death and burial of (at least notional) Christians, mostly male (including a few likely priests) and probably from the upper reaches of society. The messages are formulaic. Using capital letters, then progressively mixed with 'lower-case' (e.g. half-uncial) letters borrowed from book-hands, most say little beyond 'Of A, the son of B' or 'Here lies (HIC IACET, IACIT) A, son of B'. Six of them additionally show the same names (rarely, words) in ogam (or ogham), a twenty-symbol stroke script invented in Ireland in the third and fourth centuries (**Map 12.2**). We might see this as implying arrival by sea, at first in North Cornwall, of immigrant families including Christians whose memorial customs still embraced the use of ogam. 'Bilinguals', texts using both Roman letters and ogam

symbols, denote not 'Irish invasions', but persons of mixed Irish-British origin. **Map 12.2** suggests secondary diffusion around Bodmin Moor and *via* the Tamar into Devon's South Hams. Such incomers, in their hundreds rather than thousands, and arriving perhaps over the course of a century, were coming across from south-west Wales since shortly before 500.

Refinement of this historical model comes from individual names (**Map 12.3**). This combination of Irish and continuing-Roman names and 'scholastic' perpetuation of ogam can only be paralleled in the Pembroke–Carmarthen region where it dates from the later fifth century. At Worthyvale, north Cornwall, we find (470) LATINI IC IACIT FILIUS MACARI (Latinus, Macarius) with ogam *LATINI*. 'Hic iacet' (here lies …) is a Christian phrase spreading, 460 onwards, from Atlantic Gaul. Other stones have Christian features. At Lewannick near Launceston a priest's (?) memorial of *c*.500 reads (466) INGENUI MEMORIA, with ogam *IGENAWI MEMOR*: the *memoria* 'Christian grave-monument' of Ingenuus and its Primitive Irish equivalent.

The thesis of origins in south-west and south-central Wales, not just of erecting and inscribing such monuments—mostly lettered vertically down a slab or pillar—but also of Christianity as a still-minority faith in a landowning class, is partly borne out when we note that individual *names* found in Devon and Cornwall are matched with records of the same in Wales (**Map 12.4**). Repetitions are too many to be ascribed to chance. With (472) ULCAGNI FILI SEVERI at St Breoke near Wadebridge, and (373) SEVERINI FILI SEVERI at Newchurch, Carmarthenshire, we may have sixth-century Britons of mixed descent (*Ulcagnas* is Primitive Irish), one brother remaining at the family estate, another emigrating to Dumnonia.

Attempts to date individual stones are fraught with problems. **Map 12.5** is in this sense hypothetical, using thirds of centuries only as indicators of probability. Criteria include the nature of the cut (chiselled, pocked) letters, or epigraphy; the degree of linguistic change shown, allowing for conservatism; the context, if known; and some minimal 'art' as the simplest Christian symbols. Again the few supposedly oldest memorials suggest the pattern of an initial north-coast introduction.

from *c*.700 could introduce minster churches to Somerset and Devon, had no clear ecclesiastical foci. Field archaeology points to a growth in enclosed consecrated Christian cemeteries (many to become medieval churchyards); but **Map 12.8** antedates the rise, by the ninth century, of such monastic centres as St Buryan, Probus, Padstow and Bodmin. The significance of the inscribed memorial stones is that often they constitute the *only* early evidence.

FURTHER READING

The principal corpus of Irish and British memorial stones, from which any numbers (above) are cited, is R.A.S. Macalister, *Corpus Inscriptionum Insularum Celticarum*, 2 vols (Dublin 1945, 1949). The post-Macalister discoveries from Devon and Cornwall are

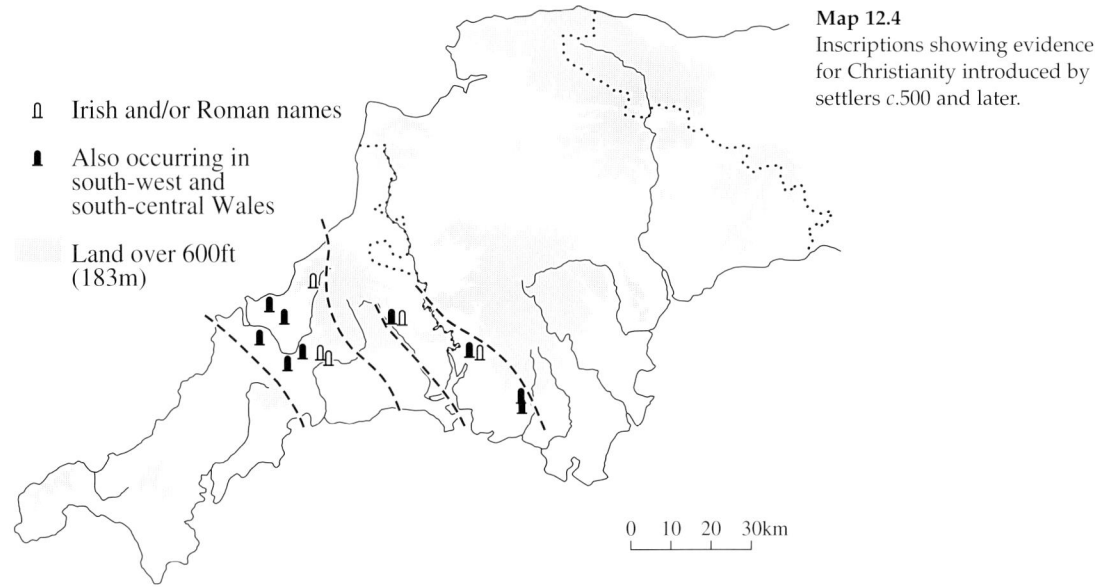

Map 12.4
Inscriptions showing evidence for Christianity introduced by settlers *c*.500 and later.

Legend:
- Irish and/or Roman names
- Also occurring in south-west and south-central Wales
- Land over 600ft (183m)

Legend:
- *c*.450 - 470
- *c*.470 - 500
- *c*.500 - 530
- *c*.530 - 570
- *c*.570 - 600
- Land over 600ft (183m)

Map 12.5
Dates of intrusive inscriptions by thirds of centuries.

Map 12.6
Intrusive memorials
and native inscriptions
with British names.

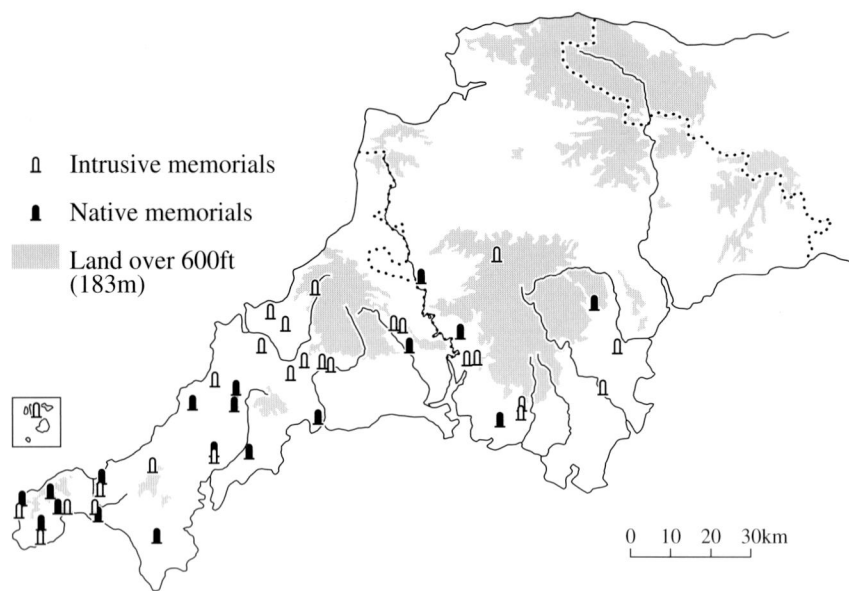

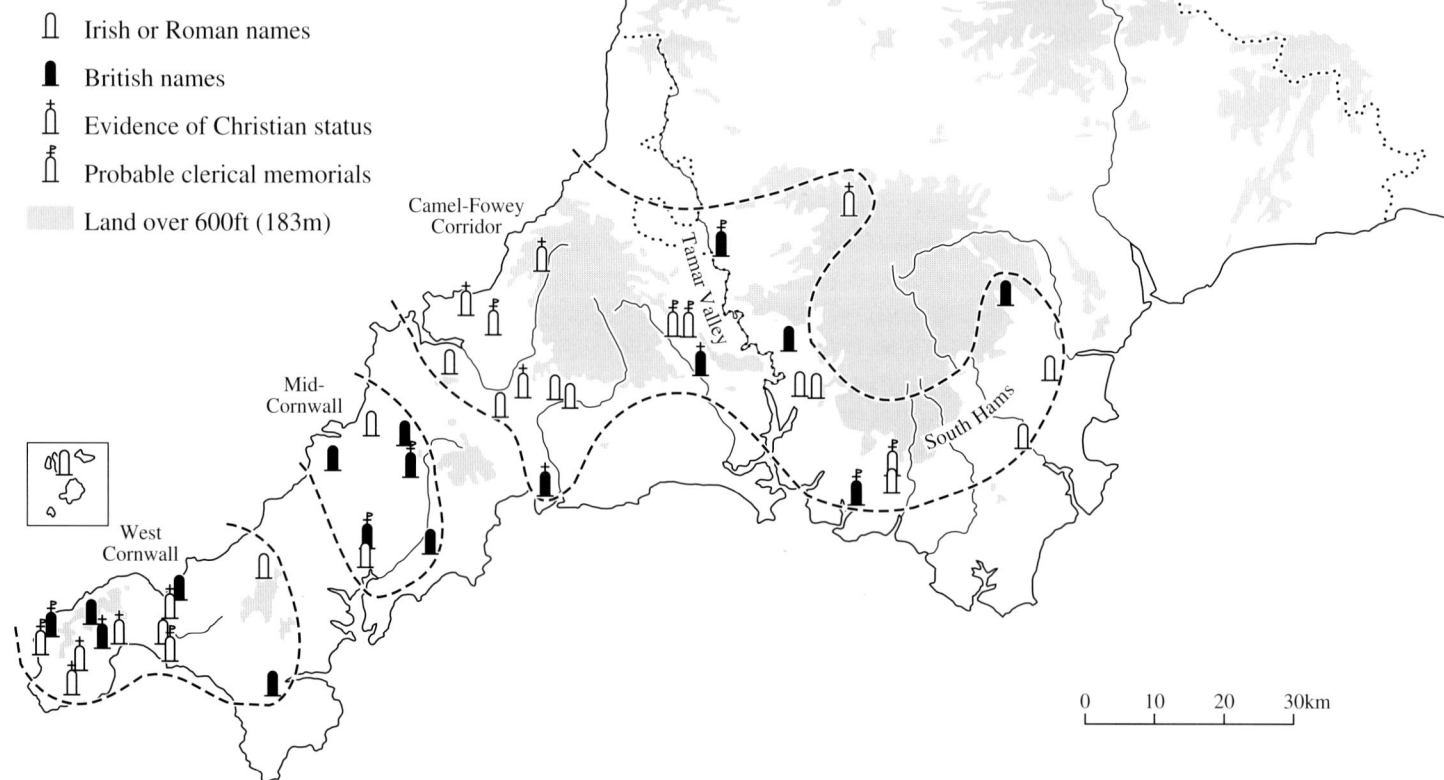

𝄑 Intrusive memorials

♟ Native memorials

▨ Land over 600ft
(183m)

0   10   20   30km

𝄑 Irish or Roman names

♟ British names

𝄑 Evidence of Christian status

𝄑 Probable clerical memorials

▨ Land over 600ft (183m)

Camel-Fowey
Corridor

Tamar Valley

South Hams

Mid-
Cornwall

West
Cornwall

0   10   20   30km

Map 12.7
Correlation of inscriptions
with evidence of
established Christianity.

Map 12.8
Cornwall: early Christian sites
and later hundred boundaries.

SOURCE: B.L. Olson,
Early Monasteries in Cornwall
(Woodbridge, 1989).

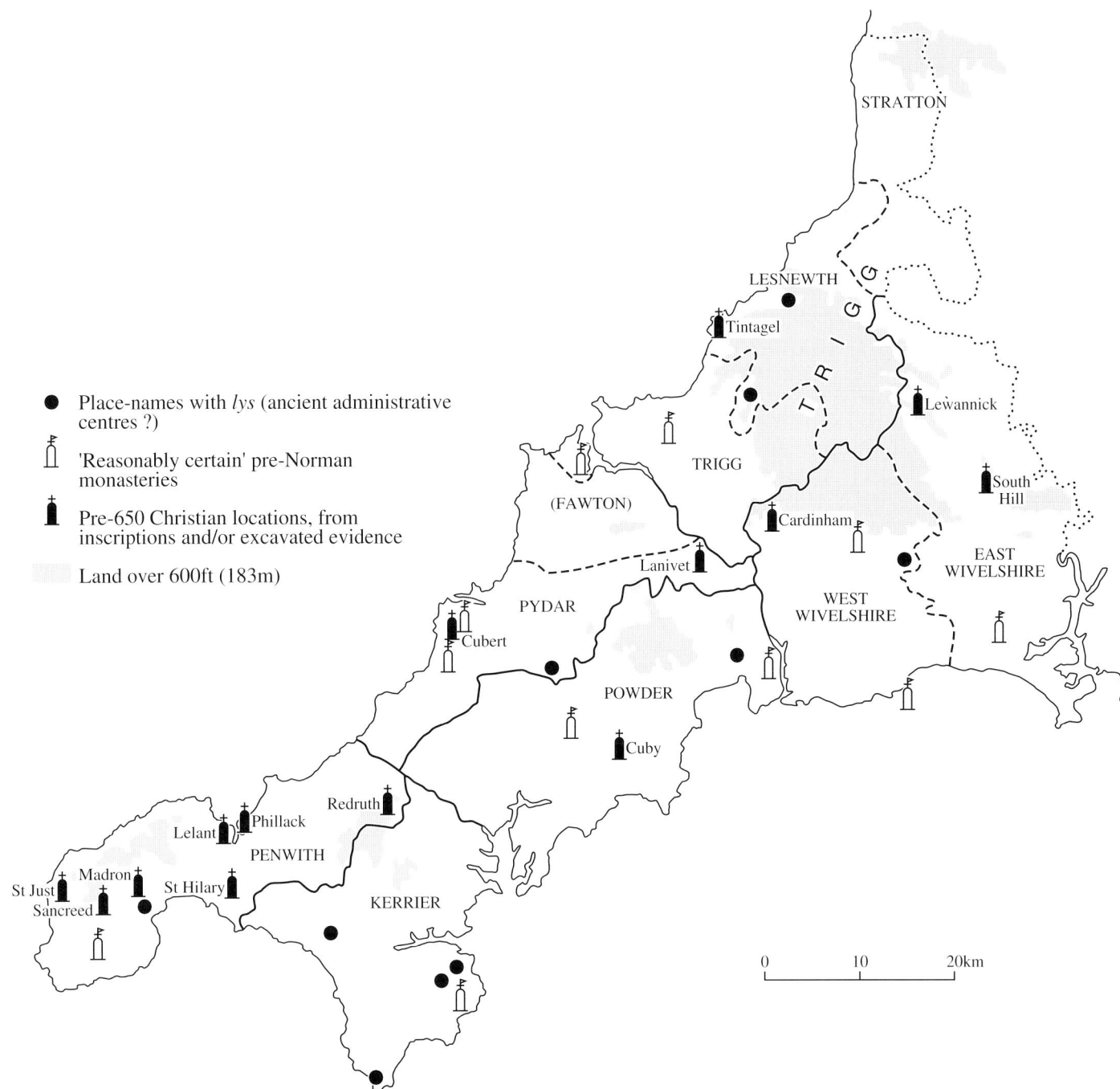

STRATTON

LESNEWTH

Tintagel

TRIGG

Lewannick

● Place-names with *lys* (ancient administrative
centres ?)

∐ 'Reasonably certain' pre-Norman
monasteries

Pre-650 Christian locations, from
inscriptions and/or excavated evidence

Land over 600ft (183m)

(FAWTON)

TRIGG

South
Hill

Cardinham

Lanivet

EAST
WIVELSHIRE

WEST
WIVELSHIRE

PYDAR

Cubert

POWDER

Cuby

Redruth

Phillack

Lelant

PENWITH

Madron

St Just

St Hilary

Sancreed

KERRIER

0   10   20km

included (with numbers) and discussed with all such inscriptions in a wider context in Charles Thomas, *And Shall These Mute Stones Speak? Post-Roman Inscriptions in Western Britain* (Cardiff, 1994). Elizabeth Okasha's *Corpus of Early Christian Inscribed Stones of South-West Britain* (Leicester, 1993) forms a catalogue with full bibliography for each stone, but avoids the problem of closer dating, and many of the readings of inscriptions seem unusually timid or inadequate.

The best background to Cornwall's earliest Christianity is Lynette Olson's *Early Monasteries in Cornwall* (Woodbridge, 1989), and for the contemporary setting of Devon as well, see Susan M. Pearce, *The Kingdom of Dumnonia: Studies in History and Tradition in South-Western Britain, A.D. 350–1150* (Padstow, 1978).

PLACE-NAMES

O.J. PADEL

As Anglo-Saxon settlers occupied Devon and Cornwall, from the seventh century onwards, they replaced the native Celtic place-names with English ones of their own creation. The extent to which they renamed existing settlements or founded new ones of their own is largely an unanswerable question, as is the greater one of to what extent the Celtic population remained and was assimilated into the incoming English population, or moved away or was killed. The place-names provide part of the evidence used in assessing these questions, although they cannot provide definite answers.

THE PLACE-NAME ELEMENT *TRE*—'HAMLET, FARMSTEAD, ESTATE'

The evidence of **Map 13.1** is the most clear-cut, and appears to provide very definite evidence.[1] The Cornish place-name element *tre* or *tref*, 'hamlet, farmstead, estate', is numerous over almost the whole of Cornwall but is virtually absent from Devon. The exceptions are themselves instructive. In Cornwall, the blank spaces are primarily the moorland areas, notably the high ground of Bodmin Moor, the Carnmenellis granite uplands north-east of Helston, and the higher parts of the Land's End peninsula. There are also two other areas lacking names in *tre*, both on the eastern border with Devon, one being the northernmost part of the county (north of the River Ottery) and the other a part of south-east Cornwall around Callington, between the Rivers Lynher and Tamar. As will be seen below, English place-names indicate that these areas were the parts of Cornwall heavily settled by the Anglo-Saxons at an early period. Apart from these areas, it can be seen that the county boundary, the River Tamar, forms a true linguistic boundary, with the Cornish-language place-names stopping short at it.

The implication is either that the *tre*-names were formed at a date later than the Anglo-Saxon settlement of Devon, or that existing *tre*-names, which the Anglo-Saxons may have found when they arrived in Devon, were almost entirely dropped by them. There is some evidence in favour of both interpretations. There are various cogent reasons for considering that the main period of the formation of place-names in *tre* in Cornwall was the fifth to eleventh centuries. This period includes, and extends either side of, the period when the Anglo-Saxons must have arrived in Devon (probably the seventh to eighth centuries). Place-names in *tre* were only very rarely coined after the eleventh century, as far as can be established.[2] This means that the British inhabitants of Devon should have been coining such names before the arrival of the Anglo-Saxons, and that the few which are found there can be seen as survivors of an older, and perhaps denser, distribution, not as freak intruders from across the western border. However, it also means that such place-names were probably still being formed *after* the English take-over had reached its fullest extent, and that the linguistic boundary shown on **Map 13.1** represents in some sense a national boundary which lasted for some time, after the main English expansion had ceased. The great contrast in numbers between the two counties (three such names in Devon, compared with some 1,300 in Cornwall) must represent a durable difference in language and culture. In addition to the three marked on the map in Devon, there are a few more place-names in Devon (perhaps three or four) starting with *Tr-*, which are not readily explicable as English place-names, yet which cannot obviously be explained as Celtic names either. Trusham, near Exeter, is one such. If further work were able to show that these were actually *tre*-names, the map of Devon would be altered slightly, although the great contrast with Cornwall would still remain.

However, another reason for the contrast between Devon and Cornwall might be the conservative effect of written documents. The later the Anglo-Saxons arrived in an area, the more likely it was that the native Celtic place-names might exist in written form, at least the major ones such as are represented by those in *tre*. This would have acted to preserve the names, in contrast with those areas further east where the Anglo-Saxons had arrived earlier. In Devon, it is doubtful whether any but a very few of the native Celtic place-names would have existed in written form, and the same probably applies to those two areas of east Cornwall where the place-names suggest considerable and early (probably eighth-century) settlement by the Anglo-Saxons. However, by the time that the English were in secure domination of the rest of Cornwall, in the later ninth or the tenth century, there might well have been native written documents concerning some of the more important Cornish estates.

The power of the written word in conserving names is great. The spelling *Woolfardisworthy*, for the name pronounced 'Woolsery', is out of date by at least

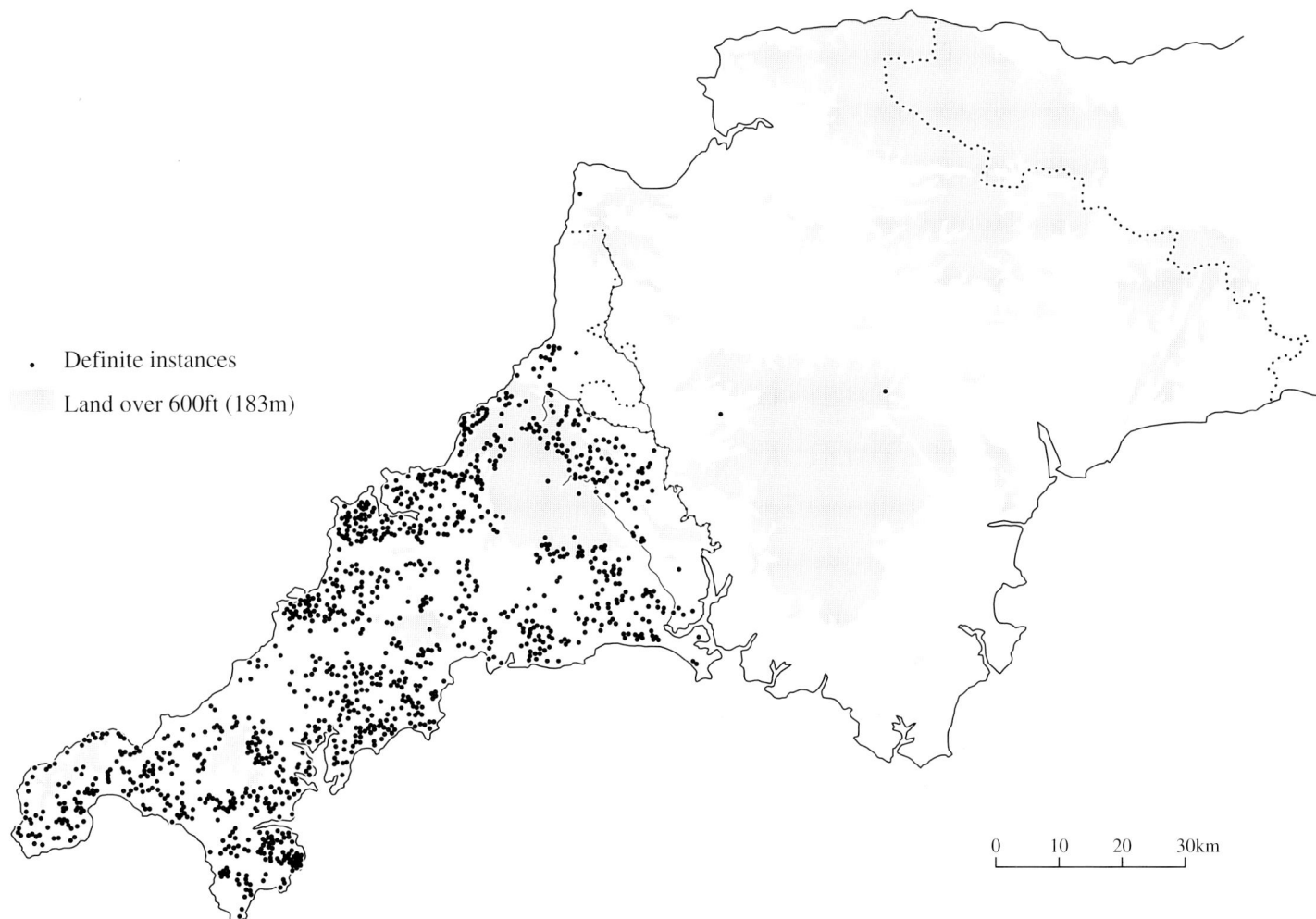

Map 13.1
Cornish *tre*, 'hamlet,
farmstead', as generic (first)
element in place-names.

. Definite instances

Land over 600ft (183m)

0   10   20   30km

three centuries (the spelling *Woolsry* is found in the seventeenth century); while the name of the Cornish farm spelt *Treviades* (in Constantine parish) has been pronounced 'Trevizzies' for seven or eight centuries.[3] The spellings are retained from an earlier period, while the pronunciation moves on. Nowadays, with widespread literacy and mobility, many people first learn their local place-names in written form, through maps and sign-posts, rather than in spoken form. This makes the power of the written word even greater, and people are liable to pronounce a name such as *Woolfardisworthy* as it is written, instead of using the authentic local pronunciation. The extent to which this conservative force could work across a change of language, such as that from Cornish to English, is less clear. However, there are, just occasionally, hints in the earliest-recorded Cornish place-names that a native written source might lie behind an early form found in an Anglo-Saxon or Norman document. This factor could have worked to preserve some native place-names in Cornwall from replacement by English ones, in contrast with Devon. If that were so, then the great contrast between the two counties might, to some extent, be merely a chronological distinction: the

distribution of Cornish place-names would show, in part, the area where the native Cornish remained dominant long enough to put some of their place-names into writing, and so ensure their preservation after the English take-over. However, if that were all, then we would not expect the boundary to form such a clear line: such a marked boundary must indicate a political and/or linguistic distinction at some period, probably at about the eighth to the tenth or eleventh centuries.

It is interesting that there is an Anglo-Saxon charter concerning the estate represented by one of the three *tre*-names in Devon; yet the charter does not mention the place-name in that form. Treable, easily the most easterly of the *tre*-names (SX 719 928), is first recorded in 1242, as *Tryfebel*, and it consists of Cornish *tre(f)* plus a personal name *Ebel*, 'Ebel's estate or farmstead'. But it is the same place as that granted in an Anglo-Saxon charter of 976, where the estate is called *Hyples eald land*, 'Hyple's old estate'.[4] The name *Hyple* is likely to be an English rendering of the Old Cornish personal name *Ebel*; and it may well be that *eald land*, 'old estate', is a loose translation of *tre(f)*. If so, this might have provided an interesting case where a Celtic place-name in Devon could be seen

Map 13.2
Old Cornish *bod*, 'dwelling',
as generic (first) element
in place-names.

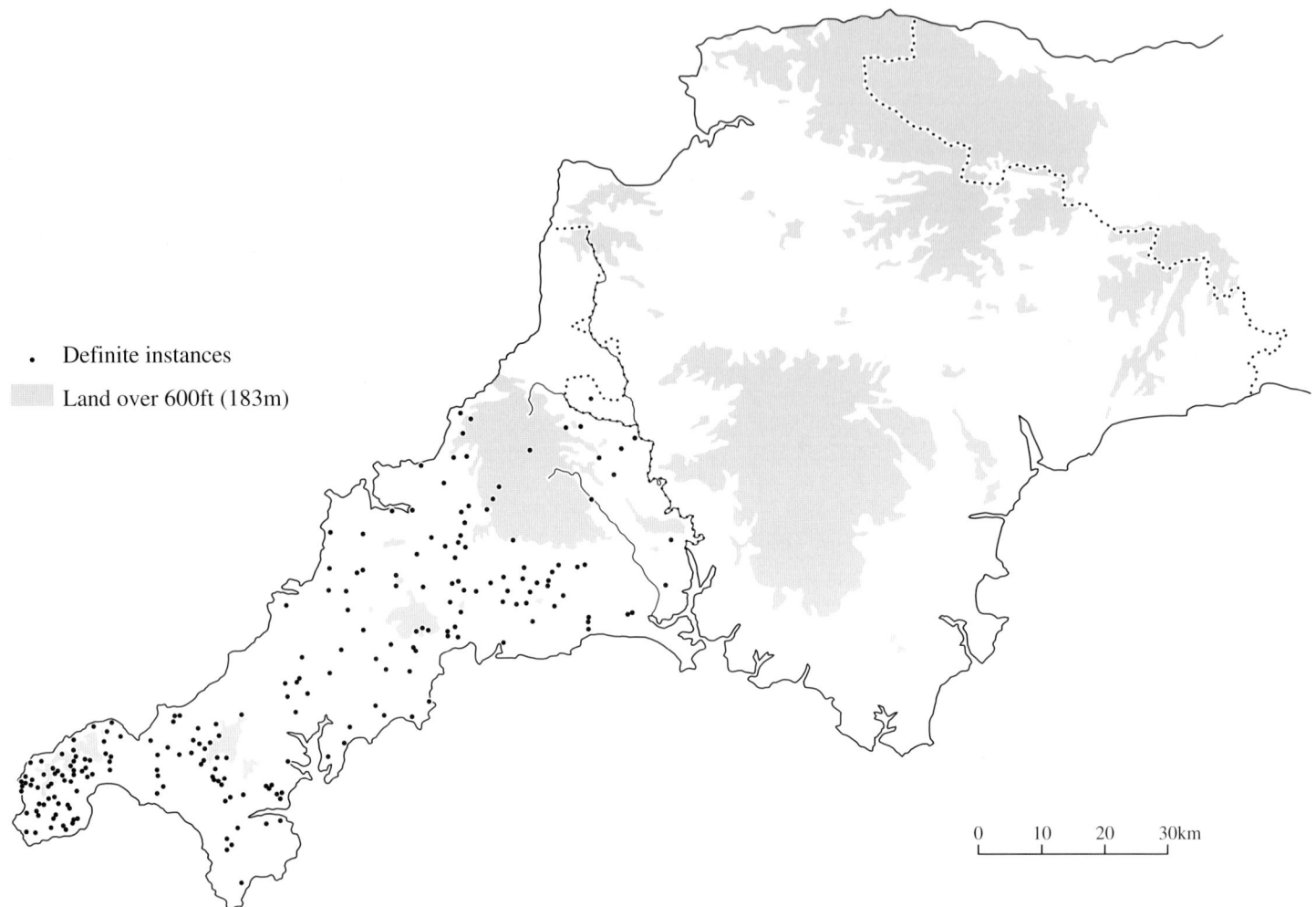

Definite instances

Land over 600ft (183m)

0   10   20   30km

in the process of disappearing; but in fact the name did survive, as shown by the next reference in 1242 (the place is not named in Domesday Book), and by the modern name. In any case it is instructive to see the Anglo-Saxon treatment of one Celtic place-name which probably survived because it was that of an estate, even though it does not appear in the earliest surviving document referring to the estate.

THE PLACE-NAME ELEMENT *BOD*—'DWELLING'

Map 13.2 indicates the distribution of another Cornish habitative place-name element, *bod*, 'dwelling'. This is less widespread than *tre*, with some 240 examples throughout Cornwall, as compared with about 1,300 instances of *tre*; and its distribution within Cornwall is more problematical. But at the border with Devon it gives the same message, just as clearly. Place-names containing it are found right up the River Tamar, but not beyond it.[5] Like *tre*, *bod* is largely absent from Bodmin Moor (Bodmin itself lies south-west of the moor) and also from the far north-east of the county, the

ancient hundred of Stratton. As with *tre*, too, it seems probable that names containing *bod* were formed in the period between the fifth and eleventh centuries, though there is not enough information to refine the dates more closely than that. It is noticeable that *bod* is present in two areas where *tre* is lacking, giving it in some respects a distribution complementary to that of *tre* within the county: these areas are the moorlands of Carnmenellis and of the Land's End peninsula. It seems that *bod* is more willing than *tre* to move into the less desirable land, which might be compatible with the idea that a place named with *bod* was a lower-status habitation than one named with *tre*. This does not, however, explain the remarkably dense distribution of *bod* in the Land's End peninsula. This distribution was already peculiar in the fourteenth century, by which time most of the *bod*-names in the area are already attested: indeed, a substantial proportion are found in the thirteenth century, and if the late copy of King Æthelstan's charter to the church of St Buryan represents a genuine grant of the earlier tenth century, then already at that early date we have five *bod*-names attested, as compared with only two *tre*-names in the same document. It would seem that the

Map 13.3
Old English *tun*, 'farmstead,
settlement', as generic (second)
element in place-names.

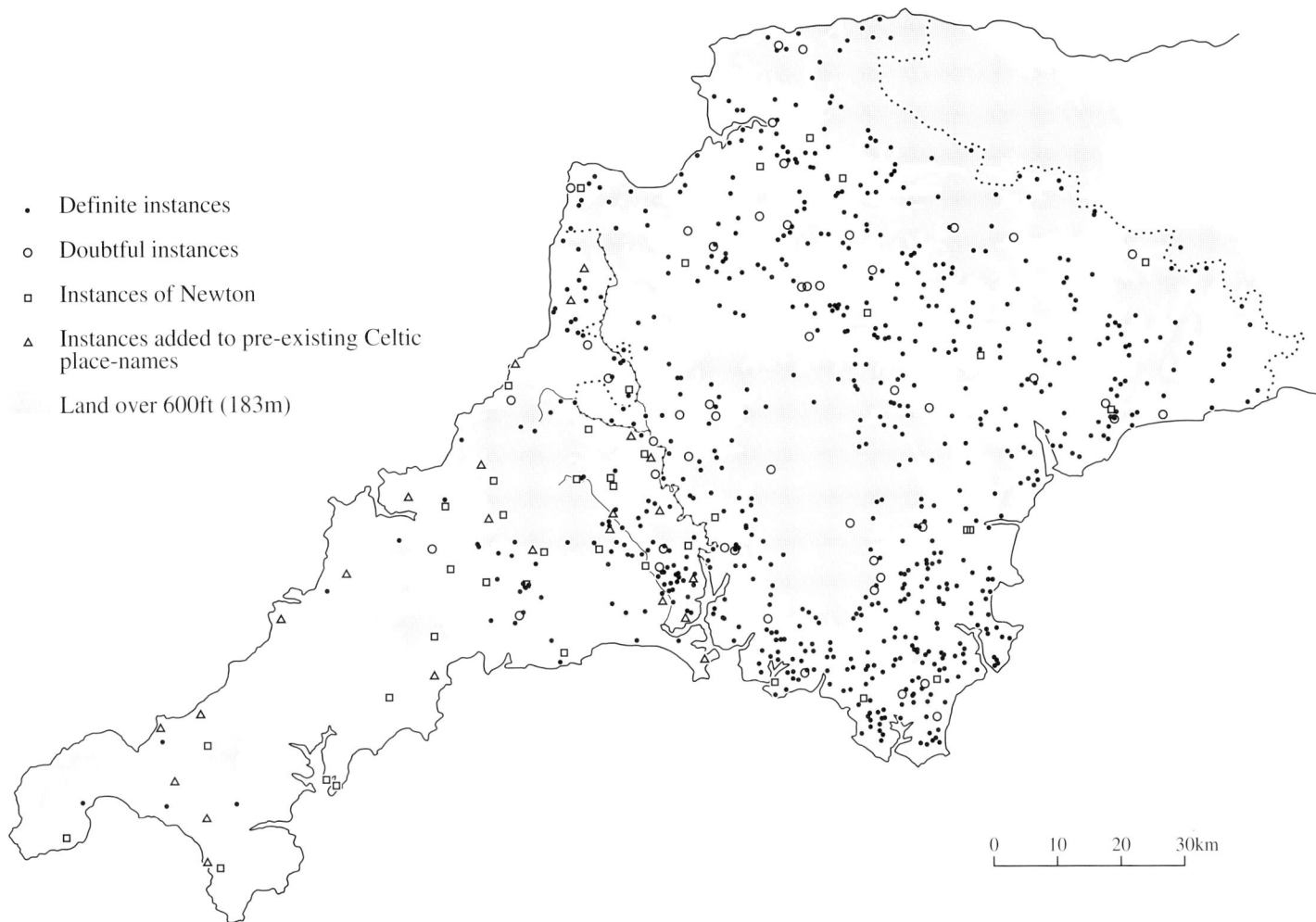

- Definite instances
- Doubtful instances
- Instances of Newton
- Instances added to pre-existing Celtic place-names

Land over 600ft (183m)

0   10   20   30km

distinctiveness of the Land's End peninsula, in terms of its nomenclature, had already come into being at this early date.

OLD ENGLISH PLACE-NAME ELEMENTS

Maps 13.3, 13.4 and 13.5, which indicate the distribution of three Old English place-name elements, ought to show a reverse or complementary picture to that seen in **Maps 13.1 and 13.2**. To some extent they do so. Map 13.3 plots the distribution of the element *tun*, 'farmstead, estate', probably the closest in meaning to Old Cornish *tre*. Cornwall largely lacks such names, while in Devon, by contrast, names in *tun* occur in almost all parts, and particularly thickly in the south. However, two parts of Cornwall do show names in *tun* in significant numbers, and they are precisely those areas where *tre* is largely absent, namely the southern part of the land between the Rivers Lynher and Tamar (the area around Callington), and the north-eastern extremity of the county. The greater density of *tun* in the southern of these two areas probably reflects the greater density of *tun* in southern, as opposed to northern, Devon. Both **Maps 13.1 and 13.3** thus show in different ways the English replacement of Celtic names in all of Devon and in two areas of east Cornwall; this must to some extent reflect the settlement of Anglo-Saxon incomers.

When were these English names given? It seems to be harder to put a date to Old English place-name elements than it is to Old Cornish ones. However, two facts suggest that *tun* had largely dropped out of use by the time of the Norman Conquest. One is the relative scarcity of the element further west in Cornwall, and the other that it rarely occurs with Middle English personal names. There is an interesting case of a place called Alverton, near Penzance, which was held in 1066 by a tenant called *Aluuard*, and was evidently named from him, at about that time, as 'Aluuard's *tun*'. It is notable that this is also the furthest west of the *tun*-names in Cornwall. It may be that most of the thin scatter of such names in the western half of the county was given in the later tenth or the eleventh century, when the English had consolidated their hold on the county, and were perhaps settling in the western half in small numbers, and either creating or renaming settlements accordingly.

Map 13.4
Old English *worðig*,
'enclosure, farm', as generic
(second) element in
place-names.

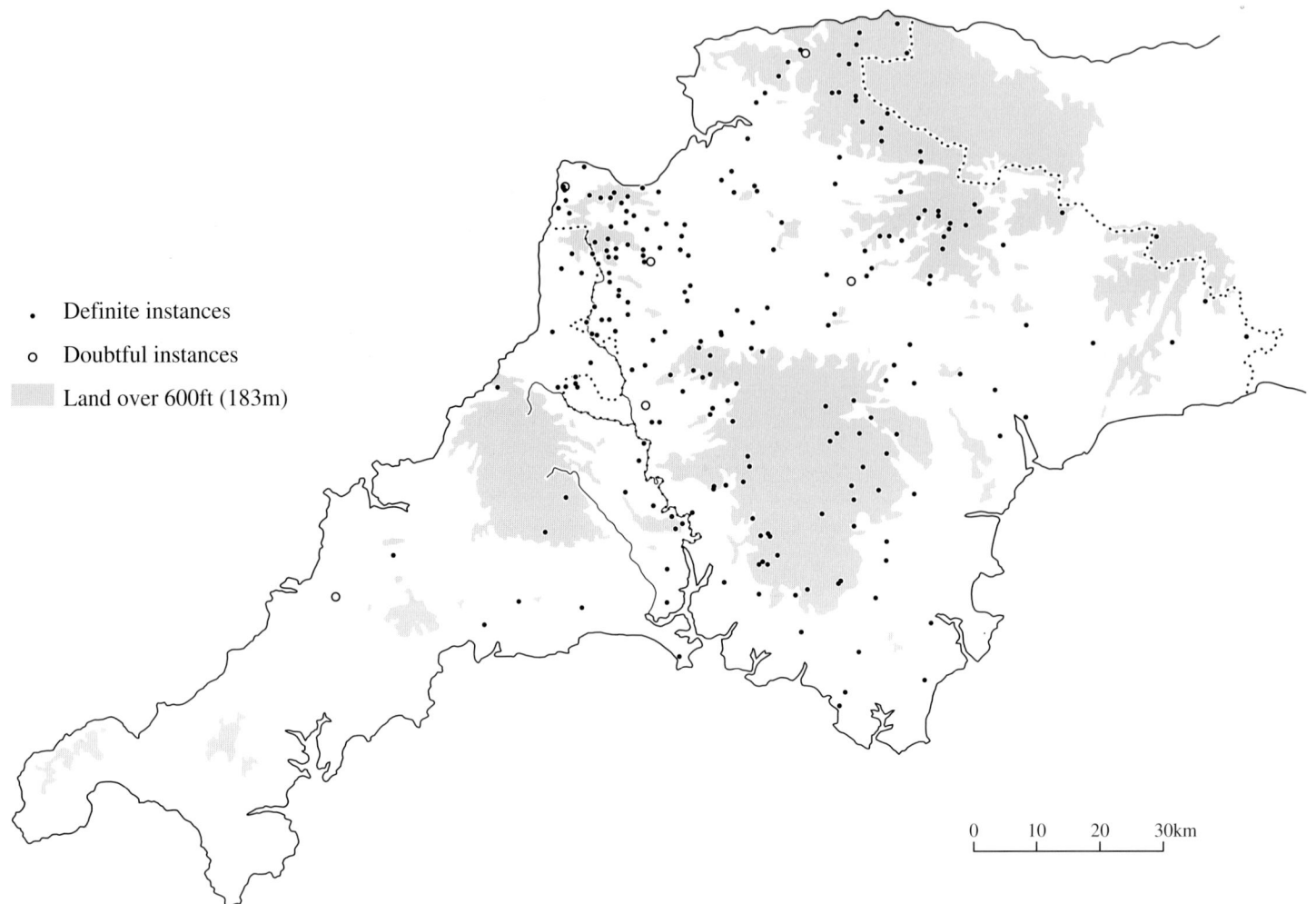

- Definite instances
- Doubtful instances
- Land over 600ft (183m)

0   10   20   30km

Although *tun* is largely absent from most of Cornwall, there are three exceptions to this. One is the scatter of names in central Cornwall, just west of the upland of Bodmin Moor. It seems possible that this is related to the town of Bodmin (the only place in Cornwall described as a town in 1086), which was also the centre of the powerful monastery of St Petrock. The Cornish towns are generally thought of as centres of English influence in the county. It may be that these *tun*-names resulted from English settlements emanating from the town, perhaps in the eleventh or twelfth centuries if *tun* continued in use that late. Second, examples of *Newton* penetrate into the county in proportionately greater numbers than the ordinary *tun*-names; they are presumably due to that name continuing to be bestowed at a date after ordinary *tun*-names had largely ceased to be given. It seems likely that *Newton* in central and western Cornwall may well date from the Middle English period, say the twelfth to fifteenth centuries. Third, there are the cases of *tun* added to an existing Celtic place-name. These represent a rather different phenomenon, the English administrative take-over of an existing Celtic estate. A good example is seen in Helston, in west

Cornwall, which in the Cornish language retained its name *Hellas* (Cornish *hen-lys*, 'old court') as late as the fourteenth or fifteenth century, but in English was already known as *Henlistone*, with the addition of Old English *tun*, at the time of its first mention in 1086. These names are often those of royal or other important manors in the Middle Ages, and it seems that *tun* had a special meaning in these names (often used interchangeably with Old English *lond*, 'land', one name referring to the manorial centre, and the other to the whole area of the estate). The usage is well-attested in Devon (in purely English names) as well as Cornwall, and since it is a special administrative usage, its frequency in west Cornwall is less significant than ordinary instances of *tun*, such as Alverton.

The uneven distribution of *tun* in Devon, with a greater density in the north than in the south, is not explained; but it does to some extent provide an answer, however incomplete, to a problem posed by **Maps 13.4 and 13.5**. These show the distribution in Devon and Cornwall of two other Old English elements, *worðig*, 'enclosure, farm' (usually found as *-worthy* in place-names), and *cot*, 'cottage, dwelling'. Both of these have a

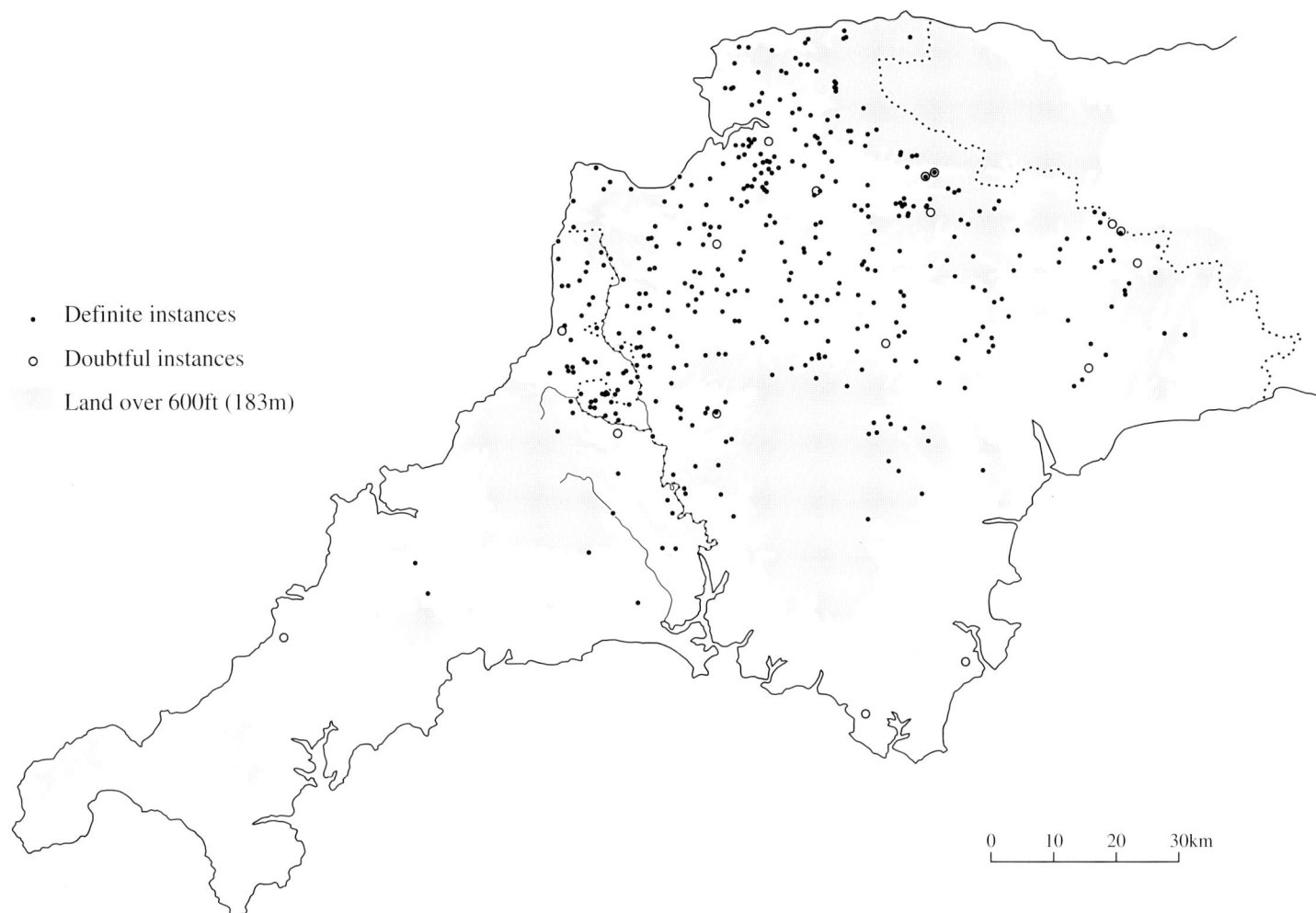

Map 13.5
Old English *cot*, 'cottage',
as generic (second) element
in place-names.

- Definite instances
- Doubtful instances
- Land over 600ft (183m)

0   10   20   30km

predominantly northern distribution, that of *cot* being particularly marked. It will be seen that both elements also penetrate less far into Cornwall than does *tun*, though when their relative frequency is taken into account, *worðig* at least seems to show about the same degree of penetration as *tun*.

In both cases, the border is the most interesting aspect of the map. Both show, more markedly than *tun*, and almost as clearly as did *tre*, a sharp cut-off, not coinciding exactly with the county boundary on the River Tamar, but rather with the River Ottery. This is noteworthy, because until 1974 the River Ottery was, in fact, the county boundary for part of its length. Tenurial reasons in the later Middle Ages were suggested by Finberg for this anomaly, but it may be that the boundary retained in some sense a memory of the fact that the part of Cornwall north of the River Ottery belongs, in terms of its place-names, much more with Devon than with Cornwall.[6] However, this leaves unexplained why the remainder of the area, further north than these two intrusive Devonian parishes, was allotted to Cornwall. In terms of its place-names, the whole of the north Cornish hundred of Stratton ought to have been in Devon rather than Cornwall. It is

very likely that the reason why it was in fact placed in Cornwall was that it had anciently been so, in pre-English times. It was certainly reckoned in Cornwall in the late ninth century, at the time of King Alfred's will, and the names of the regions indicate that this reflected an older, Celtic arrangement. If so, then the use of the River Tamar as the county boundary, if this was done in the tenth century as later sources suggest, would have been a denial of the then reality, as an act of generosity to native Cornish sentiment and ancient tradition.

The Devon place-names in *worðig* have been the subject of a recent discussion.[7] More than the *tun*-names, they occasionally contain a Middle English personal name as their first element. In addition, **Map 13.4** shows that *worðig* more than *tun* appears on the higher, less attractive ground. The element penetrates further on to Dartmoor than does *tun* (though still only on the edges), and there are a few examples of its penetration, in the same manner, on to the higher edges of Bodmin Moor, in east Cornwall, as well. These two facts together suggest that some of the *worðig*-names may belong to the expansion of farming which belongs to the eleventh to thirteenth centuries.[8] If so, then these names would represent rather low-status sites, farms which were finding an existence at the limits of cultivable existence during a colonizing expansion in the post-Conquest period. They are perhaps the equivalent of the *bod*-names in Cornwall. However, if *worðig* was in use at a later date than *tun*, it is curious that it does not penetrate as far into Cornwall as that element does. The answer may lie in the relative status of the settlements represented by the two elements, or of the dates when the names were being formed. Perhaps, if *tun*-names often represent relatively high-status settlements, they were more liable to supplant older Celtic names than the lower-status *worðig*; alternatively, if *worðig* was partly (or mainly?) in use in the period after the eleventh century, it may have failed to supplant older names because by then more of the names had become fixed through the use of written documents. What remains unexplained is the very different distribution of the element within Devon, with a far greater frequency in the north of the county than in the south. In this respect it is complementary to the map of *tun* (**Map 13.3**), which shows a higher density in the south. The reason for the contrast is not known.

The element *cot* (**Map 13.5**) shows the same discrepancy as *worðig* in Devon, though it is even more marked, with a total absence of the element in south Devon, apart from a pair of doubtful examples. No reason for this marked pattern within Devon is known. A *cot* was originally, presumably, a rather low-status site; but already by the time of Domesday Book, there were examples in Devon which had grown and attained the status of manors. Like *worðig*, *cot* fails to penetrate far into Cornwall, with the River Ottery forming a marked boundary. Curiously, it penetrates further south in Cornwall than it does in Devon, mainly in the area around Callington, accompanying the other English names there. However, these instances of *cot* in southern and western Cornwall consist predominantly of the compound *ceald-cot*, 'cold cottage', usually appearing in modern form as Challacott, Chilcott or the like. As with Newton, it seems likely that this compound (perhaps a derogatory term for a bleak hut or shelter, or for a very poor dwelling) continued to be used as a place-name after the element *cot* had ceased to be used to form other place-names; hence its deeper penetration into the county.

CONCLUSION

The five elements studied here represent only a small part of the evidence to be gleaned from the study of place-names and what this can tell us about the early history of our area. In particular, it should be emphasized that recent work on English place-names has tended to play down the role of habitative terms (such as all of those studied here) in the Anglo-Saxon settlement of England. Natural-feature names (such as those whose generic term is some word such as *ford*, *wudu* 'wood', or some word for 'hill') are likely to have played just as major a part as habitative names did in the naming of habitations during the settlement period. However, such elements are less easily quantified and plotted. One reason for this is the problem of knowing at what time a natural-feature name was transferred from its natural feature and became the name of a settlement, whereas one can be sure that a place-name with a habitative term as its generic element has always denoted a settlement.

FURTHER READING

S.M. Pearce, *The Kingdom of Dumnonia: Studies in History and Tradition in South-Western Britain, A.D. 350–1150* (Padstow, 1978) is a general survey of the region in the period, dated but a useful introduction. W.G. Hoskins, *The Westward Expansion of Wessex* (Leicester, 1960) is a dated but useful discussion of the Anglo-Saxon take-over of Devon and Cornwall. Lynette Olson, *Early Monasteries in Cornwall* (Woodbridge, 1989) is a thorough assessment of one important dimension of pre-Norman Cornwall.

J.E.B. Gover, A. Mawer and F.M. Stenton, *The Place Names of Devon*, 2 vols, English Place-Name Society 8–9 (Cambridge, 1931–2) is a full treatment, thorough though now old, of the place-names of the county, with early spellings and explanations, and a historical introduction. O.J. Padel, *Cornish Place-Name Elements*, English Place-Name Society 56–57 (Nottingham, 1985) is a dictionary of the Cornish words used to form place-names, with discussion of meaning and examples of place-names containing them. O.J. Padel, *A Popular Dictionary of Cornish Place-Names* (Penzance, 1988) contains 600 of the main place-names of the county, with early forms and explanations.

SAXON CONQUEST
AND SETTLEMENT

DELLA HOOKE

Conquest

I n the sixth century King Constantine of Dumnonia drew upon himself the wrath of Gildas, a British writer, for murdering two royal youths 'at the holy altar', a record which immediately evokes both the Christianity that was already accepted in the south-west peninsula and the waywardness of its rulers.[1] Devon was, however, soon to be brought under English domination. Cornwall remained distinctly different from Devon throughout the centuries preceding the Norman Conquest, the contrast being displayed in numerous facets of its administrative, ecclesiastical and linguistic characteristics (see chapter 13.

King Gerent of Cornwall was already under attack from Ine of Wessex in AD 710. By this date the British had lost control over the region around Exeter and a minster had been established there by 690 (probably by 670). The campaigns by Ine completed the conquest of Devon but the English were defeated and held in check at a battle which took place in 721 or 722 at *Hehil*, a location not securely identified but perhaps at Hele in Jacobstow.[2] The conquest of the remainder of the peninsula is unrecorded. Here Cornish kings may have maintained a number of *lys* or courtly centres and coastal trading stations such as, perhaps, Tintagel on Cornwall's wilder northern coast.[3] Place-names incorporating the *lys* term include Liskeard, 'the court of Kerwyd', and Helston (in Trigg), 'the ancient court' (as opposed to Lesnewth near by, 'the new court').[4]

Continued British resistance was sufficient to invoke reprisal: '*hit gelamp þæt westwealas ónhofon hi ongean ecgbriht cyng*': 'It happened that the West Welsh [the Cornish] rose against King Egbert'.[5] In 815, 'King Egbert raided in Cornwall from east to west'.[6]

The British of Cornwall continued to raid and in 838, in their last recorded battle, even allied unsuccessfully with the Vikings against Egbert. The English king, as a result, became masterful enough to annex large personal domains, also granting generously to the Church 'a tenth part of the land'. He granted estates as far west as Pawton, near Padstow, to the Church of Sherborne (S 1296). The last recorded king of Cornwall is Dumngarth, who reigned c.870. It was only during the reign of Æthelstan, in the tenth century, that Cornwall was finally subdued and at last given recognition, and the Tamar fixed as the boundary between Cornwall and Devon. From this time Cornwall was administered as an English shire, but cultural assimilation was a much slower process, as the language and place-names of Cornwall clearly indicate.

The Vikings had continued to attack throughout the ninth century and *The Anglo-Saxon Chronicle* records numerous battles at which they were vanquished, usually by 'the men of Somerset … the men of Dorset' and 'the men of Devon', although not before they had caused 'much harm'. In one raid, in AD 878, 840 Vikings are said to have been slain and 'the banner which they called the Raven' was taken. A number of sites were fortified against them in the ninth century— Exeter is first described as a *fæsten*, 'stronghold', in the *Chronicle* when occupied by the Danes in 877, but as a *burh* in 893, probably fortified by Alfred between 880 and 892. The walls of a *burh* protected the inhabitants against attackers but also provided a refuge for the folk of the surrounding countryside. Other sites that were to be fortified at this time included Lydford, Halwell and Pilton. Others, like Barnstaple and Totnes, were probably fortified by Edward the Elder in the first decade of the tenth century.[7] The terror caused by the Viking attacks is indicated under the entry for a later raid of 997:

> Here in this year the raiding-army travelled round Devonshire into the mouth of the Severn, and there raided, both in Cornwall and in Wales and in Devon; then they went up at Watchet, and wrought great harm there by burning and by slaughtering of men, and after that turned back round Penwith Tail [i.e. Land's End] to the south side, and then turned into the mouth of the Tamar, and then went up until they came to Lydford, and burned and killed everything that they met, and burned down Ordwulf's monastery at Tavistock, and brought indescribable war-booty with them to the ships.

The *Chronicle* further admits to the defeat of 'an immense army from the Devon people and Somerseters' at Pinhoe in 1001, where the Vikings 'made great slaughter there, and then rode over the countryside; and each succeeding occasion was always worse than the last; and they brought much war-booty with them to the ships … In every way it was a heavy time, because they never left off

Map 14.1
Minsters and monasteries:
pre-Conquest foundations.

SOURCES: B.L. Olson,
*Early Monasteries in
Cornwall* (Woodbridge, 1989);
B.L. Olson and O.J. Padel,
'A tenth-century list of Cornish
parochial saints',
*Cambridge Medieval
Celtic Studies*, 12 (1986).

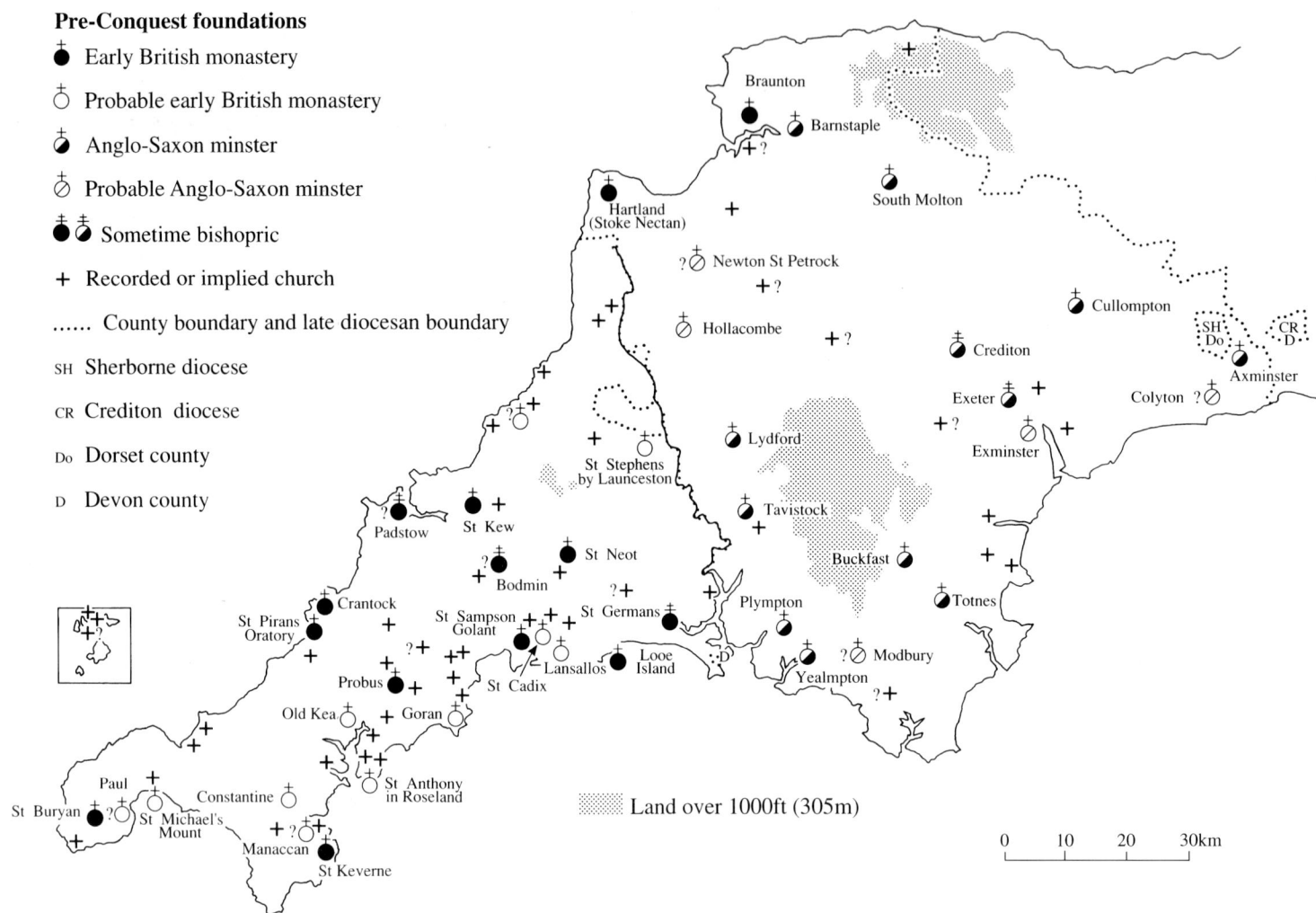

Pre-Conquest foundations

● Early British monastery

○ Probable early British monastery

◗ Anglo-Saxon minster

◎ Probable Anglo-Saxon minster

●◗ Sometime bishopric

+ Recorded or implied church

...... County boundary and late diocesan boundary

SH Sherborne diocese

CR Crediton diocese

Do Dorset county

D Devon county

▨ Land over 1000ft (305m)

their evil.' In 1003 the Vikings were able to sack Exeter because of the treachery of its French reeve, burning the minster and destroying its documents. Having offered allegiance to Harold Godwin, the south-west peninsula was to become involved in more unrest at the time of the Norman Conquest, Exeter being besieged for eighteen days in 1068 before it fell.[8] Exeter at this time was a thriving centre, possibly the seventh city in England according to the number of coins minted.[9] Much of its prosperity may have been based upon tin-working on Dartmoor and it was heavily involved in trade with the Continent.[10]

CONQUEST AND THE CHURCH

The relationship between the 'Celtic' church and the institutions introduced by the conquerors was not easy. The former had been established here since at least the fifth century AD. Monasteries were an early feature: one in Cornwall is noted by Gildas in the sixth century.[11] Memorial stones, which are particularly numerous in Cornwall, also throw light on early Christianity, betraying early links with South Wales, as does the earliest form of the *Life of St Samson*.[12] Early burial grounds and literary traditions are also informative. In the late seventh or early eighth century, Aldhelm, abbot of Malmesbury, warned King Gerent and all *sacerdotes* of Dumnonia of the worthlessness of monastic or eremetic contemplation outside the Catholic Church, presumably referring to British monasticism.[13]

Map 14.1 shows those early monasteries which can be authenticated, but knowledge of the provision of pastoral care in the Celtic church is derived from later sources. From these it is possible to deduce that bishops were associated with the site of *Dinuurrin* (?Bodmin) in the ninth century and with *Lanalet* (St

Land over 500ft (152m)

Detached portions of parishes

B   Bow

DR   Drewsteignton

H   Hittisleigh

CR   Crediton

BS   Brampford Speke

beonnan ford

Kennerleigh

ðeflisc

Morchard Bishop

nimed

scipbroc

crydian

CR

Sandford

Chaffcombe

swincumb

B

Colebrooke

cridian brycge

Crediton

herpað

sulhford

Brampford Speke

Newton St Cyres

healre dunæ

Upton Pyne

tettanburnan

C

wealdan cumbes ford

BS

DR

nimed

Hittisleigh

Cheriton Bishop

lyllan broc

doddan hrycg

H

grenan dune

herpað

exan

River Exe

DR

francancumb

Drewsteignton

wonbroc

paðford

teng

River Teign

●   Crediton minster

——   Anglo-Saxon routeway

– – –   Approximate area of Crediton Hundred (excludes vill of Morchard Bishop)

– · – · –   Additional area of S255

·········   Nineteenth-century parish boundary

herpað   Selected charter landmark

◆   Domesday manor

✚   Sixteenth-century church

✛   Medieval chapel

0   1   2   3km

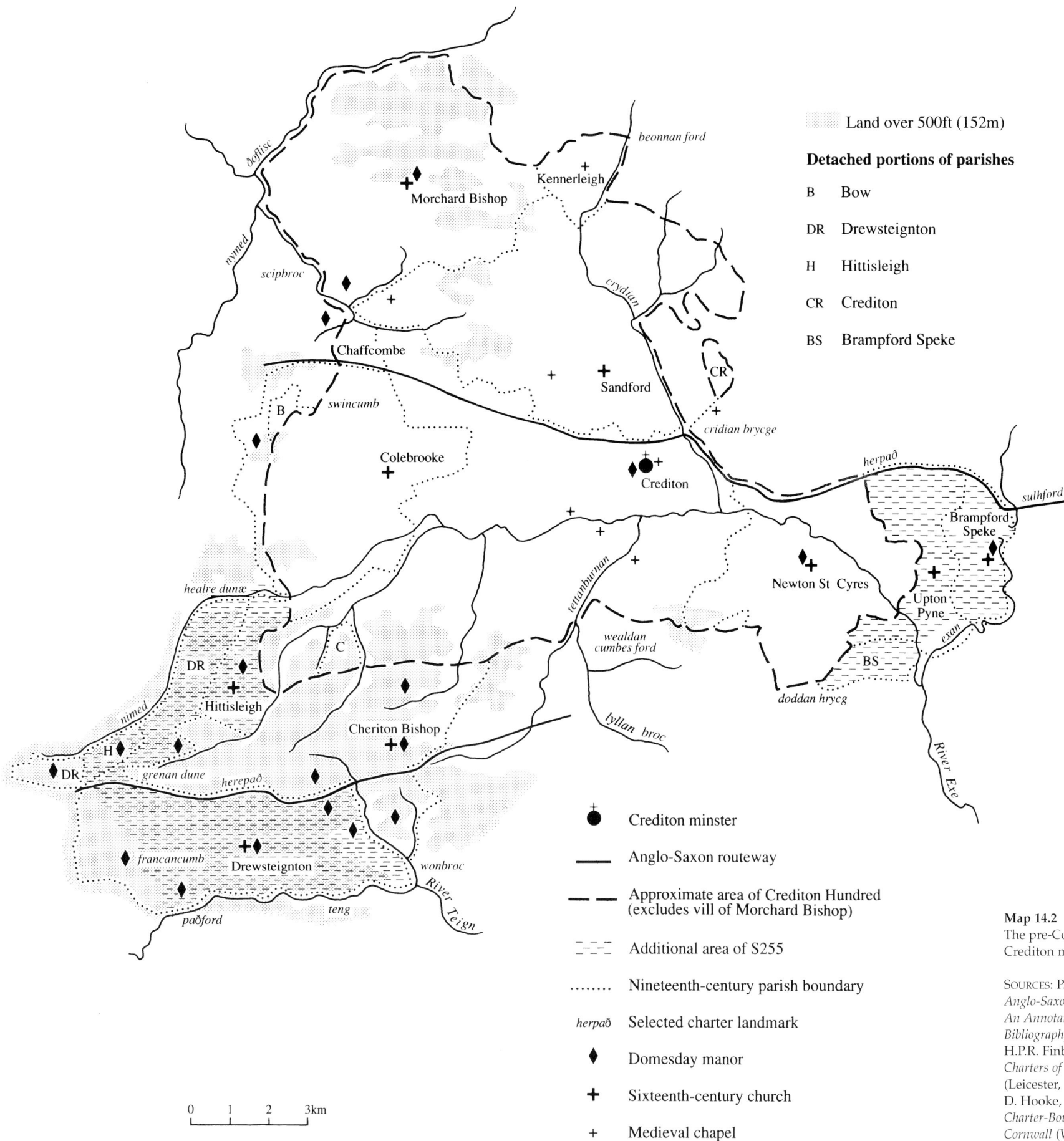

Map 14.2
The pre-Conquest estates of Crediton minster.

SOURCES: P.H. Sawyer,
*Anglo-Saxon Charters:
An Annotated List and
Bibliography* (London, 1968);
H.P.R. Finberg, *The Early
Charters of Devon and Cornwall*
(Leicester, 1953);
D. Hooke, *The Pre-Conquest
Charter-Bounds of Devon and
Cornwall* (Woodbridge, 1994).

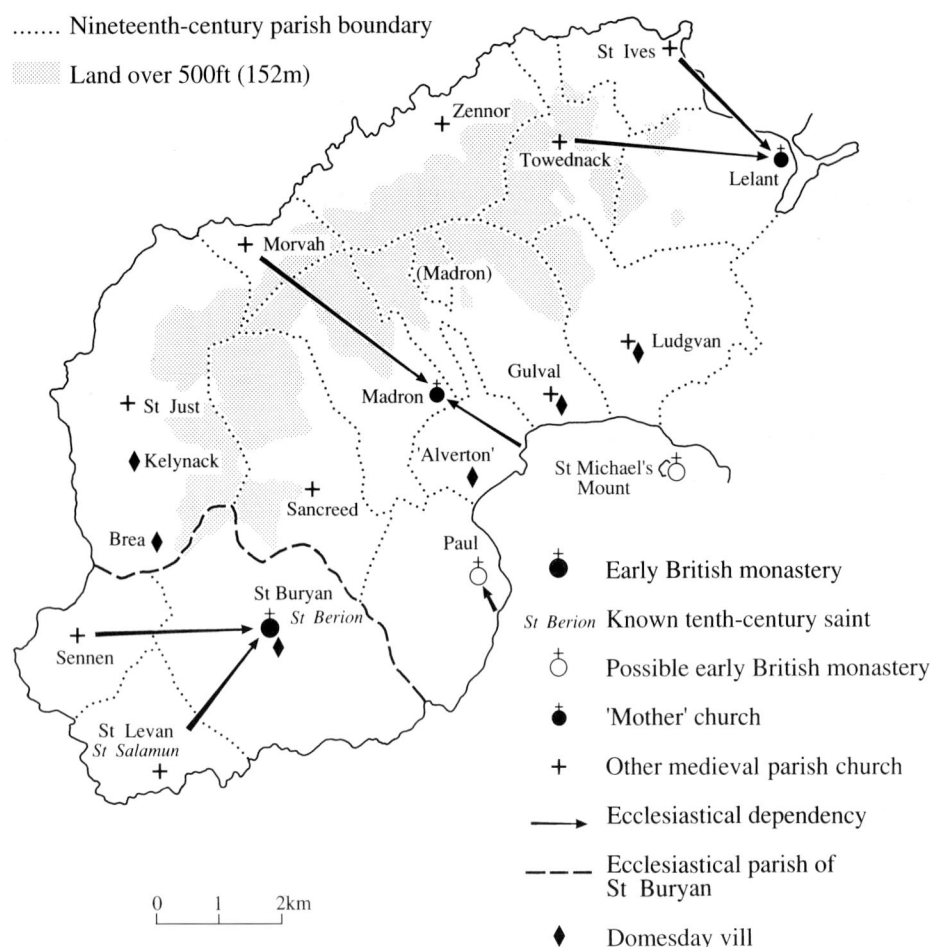

.... Nineteenth-century parish boundary

Land over 500ft (152m)

Early British monastery

St Berion Known tenth-century saint

Possible early British monastery

'Mother' church

Other medieval parish church

Ecclesiastical dependency

Ecclesiastical parish of St Buryan

Domesday vill

Map 14.3
Churches and chapelries in West Penwith.

SOURCES: C. Thomas, 'Christians, chapels, churches and charters—or, proto-parochial provisions for the pious in a peninsula (Land's End)', *Landscape History*, 11 (1989); parish boundaries from the tithe surveys.

Germans) in the tenth century, but several sources suggest the earlier pre-eminence of the house of St Petroc at Padstow. Several pilgrim monks were associated with the peninsula and their missionary zeal led them to carry their faith to the continental heathen. St Samson, a monk from South Wales, founded a monastery in Cornwall, probably at Golant, before crossing to Brittany, and later St Boniface, born at Crediton, preached to the Saxons and others living east of the Rhine.

The Celtic church was soon in retreat in the face of English expansion, for secular authority made use of the English church to confirm and sustain its political control. Anglo-Saxon kings had already been converted to Christianity before the time of this westward expansion. While Devon was soon drawn into its sphere, the minster having been established at Exeter by AD 690 and probably by AD 670, the church in Cornwall was less readily assimilated. Egbert's victory over the Cornish accelerated ecclesiastical control. Initially the whole of the peninsula formed part of the diocese of Sherborne, but Æthelstan subdivided this large territory, at first creating a new diocese of Crediton for Devon and Cornwall in the early tenth century (S 421), apparently placing Cornwall under the authority of the people of Devon because 'they

had formerly been disobedient' (S 1296). Shortly before 936 Cornish individuality was acknowledged in the recognition of the independence of the see of St Germans (S 880), given full episcopal rights in 994. Before the Norman Conquest the two sees were again to be united and in 1950 the seat of the bishopric was transferred from Crediton to the prosperous town of Exeter (S 1021), where the minster had probably been newly endowed by Æthelstan in the tenth century. The early minster here appears to have lain beneath the later church of St Mary Major, the present cathedral dating only from the twelfth century.[14]

Pastoral care in the English church was mainly provided by groups of priests based in minsters. A number of minsters are likely to have begun as British monasteries, including the majority of those reorganized in Cornwall, but others were also to be established. It was in AD 739, for instance, that King Æthelheard gave Forthhere, the bishop of Sherborne, land at *Cridie*, Crediton, to found (? or re-found) a minster in central Devon (**Map 14.1**) which would care for a *parochia* in the surrounding district (S 255). Crediton had been the birthplace of St Boniface, a missionary monk who had carried the faith to the Saxons east of the Rhine. Unfortunately, no early *parochia* here can be reconstructed with certainty, but that of Crediton was probably at least partly represented by Crediton hundred. In the eleventh century, Crediton church claimed that it had been granted twenty hides at its foundation and produced a boundary clause which encompassed an even greater area (S 255) (**Map 14.2**).[15] If the church had ever possessed rights over all the land within these bounds, many of the small manors in the south had been lost to individual thegns before the Norman Conquest and other lands in the east had been taken away by the king. New minsters are also likely to have been established in at least some of the tenth-century *burhs*, as at Barnstaple and Totnes.[16]

The *parochiae* of the minsters were to become fragmented as additional churches were established within their territories, often by manorial lords, and the more distant estates were often the first to be lost. This process continued long after the Norman Conquest, the boundaries of the ecclesiastical parishes surviving as administrative units to the present day. Some of the early churches are noted in Domesday Book, others suggested by place-names, but the record is obviously incomplete (**Map 14.1**). A hierarchy of mother and daughter churches (the former including the early minsters) can sometimes be recognized in later documents. In addition to new foundations, many outlying shrines and pre-existing churches had been drawn into the pattern of ecclesiastical provision, especially in Cornwall. The number of parishes which bear the name of a Cornish saint known from a tenth-century list (which may indicate early church foundations) reveals the strength of the Celtic church in this county. This and other sources suggest the early formation of parishes by the Cornish church.[17] **Map 14.3** shows the pattern of church dependencies in the West Penwith peninsula of Cornwall. The dedication of St Levan church to St Salamun, a known saint recorded in the tenth century, suggests that fragmentation of the *parochia* of St Buryan had already begun by that date. In the Lizard peninsula a district known as Meneage, its name derived from *manahoc*, 'the monkish land', appears to have been centred upon an early monastery of St Keverne, but was being broken up by secular land grants in the late tenth century.[18] The holy wells of Cornwall hold a special place in the ecclesiastical history of the county, a not insignificant number associated with churches (some 30–35 of 210 parish churches), but few can be ascribed to a precise date of origin and most of the stone crosses which abound are likely to be of post-Conquest origin.[19]

GRANTS OF LAND AND PRIVILEGES

The pre-Conquest charters of Devon and Cornwall record grants of land and privileges to both laymen and the Church, but the latter was particularly anxious to have documentary proof of its estate ownership. The authenticity of many such charters is open to suspicion. After the destruction of the Exeter library in 1003 by the Danes, new documents were produced there which were clumsy reconstructions of the lost originals.[20] The charter evidence shows grants being made to the churches of Glastonbury (mostly in Devon), of Sherborne (the estate at Maker in Cornwall claimed as a gift from the Cornish king Gerent *c*.705) and of Crediton, although these churches claim to have acquired many estates for which no documents survive (**Map 14.4**).[21] Tenth-century charters grant land to Milton Abbey in Dorset (S 391), to St Petroc's (S 388) and to the church of St Buryan (S 450), both in Cornwall (in the latter case, perhaps confirming land already held by the church); an eleventh-century lost charter recorded the grant of a Devon manor to Buckfast Abbey (F 57); shortly before the Norman Conquest an estate at Ottery St Mary, Devon, was granted to St Mary's, Rouen (S 1033). Other documents which have survived include the rules of an Exeter guild of *c*.950 (F 33), copies of the foundation charter of Tavistock Abbey of 981 (S 838), and the will of Ælfwold, bishop of Crediton (S 1492), made between 1008 and 1012. A charter erecting St Germans into a diocesan see dates from 994 (S 880), while the foundation charter of the see of Exeter (S 1021) dates from 1050. There are also records of exchanges of land

Map 14.4
Pre-Conquest charters.

SOURCES: as **Map 14.2**.

Map 14.5
King Æthelwulf's estate in the
South Hams AD 847.

Sources: as **Map 14.2** and
D. Hooke, 'Studies on Devon
charter boundaries',
*Transactions of the
Devonshire Association*,
122 (1990).

denewaldes stan

ðone torr

merce cumb

Wastor

weg

ðæs wælles heafod

OCHENEBERIE

ðære flodan ?

broc

afene

alda suin haga

ænne
beorg

SURLEI

sueordleage wælle
wulfwælles heafod

hreodpol

LEGE

weal weg

BICHEBERIE

ða lytlan burg
ða burg

CUMBE

ðone stan

secgwælles heafod

NOTONE

ðone herepað

REIMORE

ðone healdan weg
odencolc

dic ?

ðæs wælles
heafod

græwan stane

wealdenesford

ALVINTONE

hunburge
fleot

smalan cumbes
heafde

TORLESTAN

MIDELTONE

smalan cumb

SOUTH HAMS S298

ULSISTONE

BACHEDONE

HEWIS

ALWINESTONE

EDDETONE

WALEMENTONE

COLETONE

BADESTANE

BOLTESBERIE

PORLEMUTE

BOTEBERIE

SURE

—·— Boundary of S298, *on homme*

NOTONE Domesday manor

·········· Nineteenth-century parish boundary

——— Anglo-Saxon routeway

0   1   2   3km

(e.g. S 791), of a mortgage of land by the bishop of Crediton and several documents relating to the manumission of slaves. One unattached boundary clause (S 1547) describes the boundary of an extensive land unit around Ashburton on the fringes of Dartmoor (**Map 14.4**). Such grants are late in date and are derived mainly from the English tradition, although it seems that British documentation existed earlier, as suggested by documents referring to personal manumissions[22] and to a recently discovered tenth-century Cornish charter of Lanlawren in Lanteglos.[23] The lateness of the date of the surviving authentic charters for Cornwall reflects the lengthy period in which Cornwall remained beyond the control of the English.

A number of the charters referring to land grants are accompanied by boundary clauses which enable the estates to be identified with some precision (**Map 14.4**), although the charter boundaries of the South West still present many problems.[24] Land rights were ultimately vested in the crown, but the king could also hold land in private possession. In a ninth-century charter (S 298) concerning land in Devon, King Æthelwulf grants twenty *manentes* in the South Hams to himself, thus freeing it from many dues and services and also enabling him to grant or bequeath it freely. The extent of this estate has often been over-estimated.[25] The bounds show that it covered most of the two peninsulas between the River Erme and the Kingsbridge estuary, a fertile region largely underlain by Devonian sandstones with an above-average density of population and plough-teams in 1086 (**Map 14.5**).[26]

In Cornwall, pre-existing land-units were incorporated into the English administrative system and the later Cornish land grants often concern portions of parishes. In St Keverne (**Map 14.6**) these are small *tref* holdings which may once have been part of the large ecclesiastical estate, noted above, which in the tenth century was being granted into secular ownership. In time, the Anglo-Saxon estates of Devon and the proto-parishes of Cornwall were to form the basis for the ecclesiastical parish divisions of the English church, a network complete by medieval times (**Map 14.7**). The pattern was tidied up in the nineteenth century as detached holdings and extra-parochial areas were amalgamated into the nearest parish, but already by late Anglo-Saxon times detached parcels can be recognized in Devon, often patches of woodland or pasture. The estate at Topsham, for instance (**Map 14.8**), in one of the eleventh-century Exeter charters (S 433), possessed additional land at *Æschyrste* (from the name likely to have been woodland or wood-pasture). Later boundaries show several parishes holding detached blocks of woodland on the watershed ridge to the west of Exeter, while Kenton possessed an outlier on Dawlish Warren which was probably rough grazing land on the salt marsh.

THE EARLY MEDIEVAL LANDSCAPE

In common with many other Anglo-Saxon charters concerning estates elsewhere in England, accompanying boundary clauses

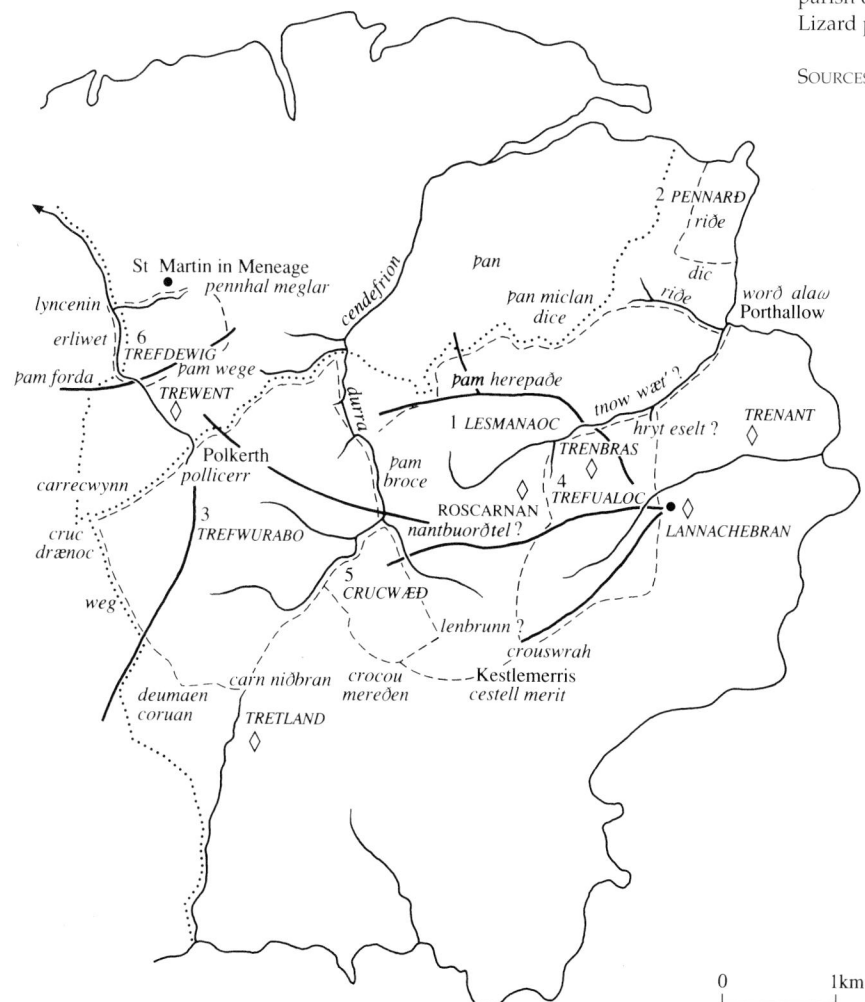

Map 14.6
Tenth-century estates in the parish of St Keverne on the Lizard peninsula.

SOURCES: as **Map. 14.2.**

---- Approximate boundary of pre-Conquest land unit

1 Lesneage   4 Trevallack
2 Pennare   5 Grugwith
3 Traboe   6 Trethewey

◊ TRENANT Domesday manor

······· Nineteenth-century parish boundary

——— Anglo-Saxon routeway

are full of local topographical detail.[27] However, the long supremacy of the British language in Cornwall led to the use of many Cornish terms in the charters of that county, words occasionally mangled out of all recognition by later English scribes. In a peninsula, the sea naturally finds mention and in Cornwall the bounds of Lesneage (S 755) begin at a *porth*, Cornish for 'cove, harbour' (Porthallow), those of Lamorran (S 770, **Map 14.9**) at *penpoll lannmoren*, 'Lamorran creek's head' (Cornish *pen pol*), while in Devon those of the South Hams (S 298) end at *hunburgefleot*, a creek of Kingsbridge estuary (Old English *fleot*, 'an

Detached portions of nineteenth-century parishes

EX   Exminster
KN   Kenton
WD   Woodbury
CL   Clyst Lawrence
FA   Farringdon
CR   Colyton Raleigh

Map 14.7
Parishes and Domesday
manors in the Exe basin.

SOURCES: Domesday Book
and tithe surveys.

⊚   Iron Age hillfort

RR   Roman road

———   Anglo-Saxon routeway

—·—·—   Charter boundary

·········   Nineteenth-century parish boundary

TEIGNE   Eleventh-century or earlier estate name
(after 'Place-names of Devon')

halsforda   Charter place-name

Clyst   Early river name

+   Church

0   1   2   3km

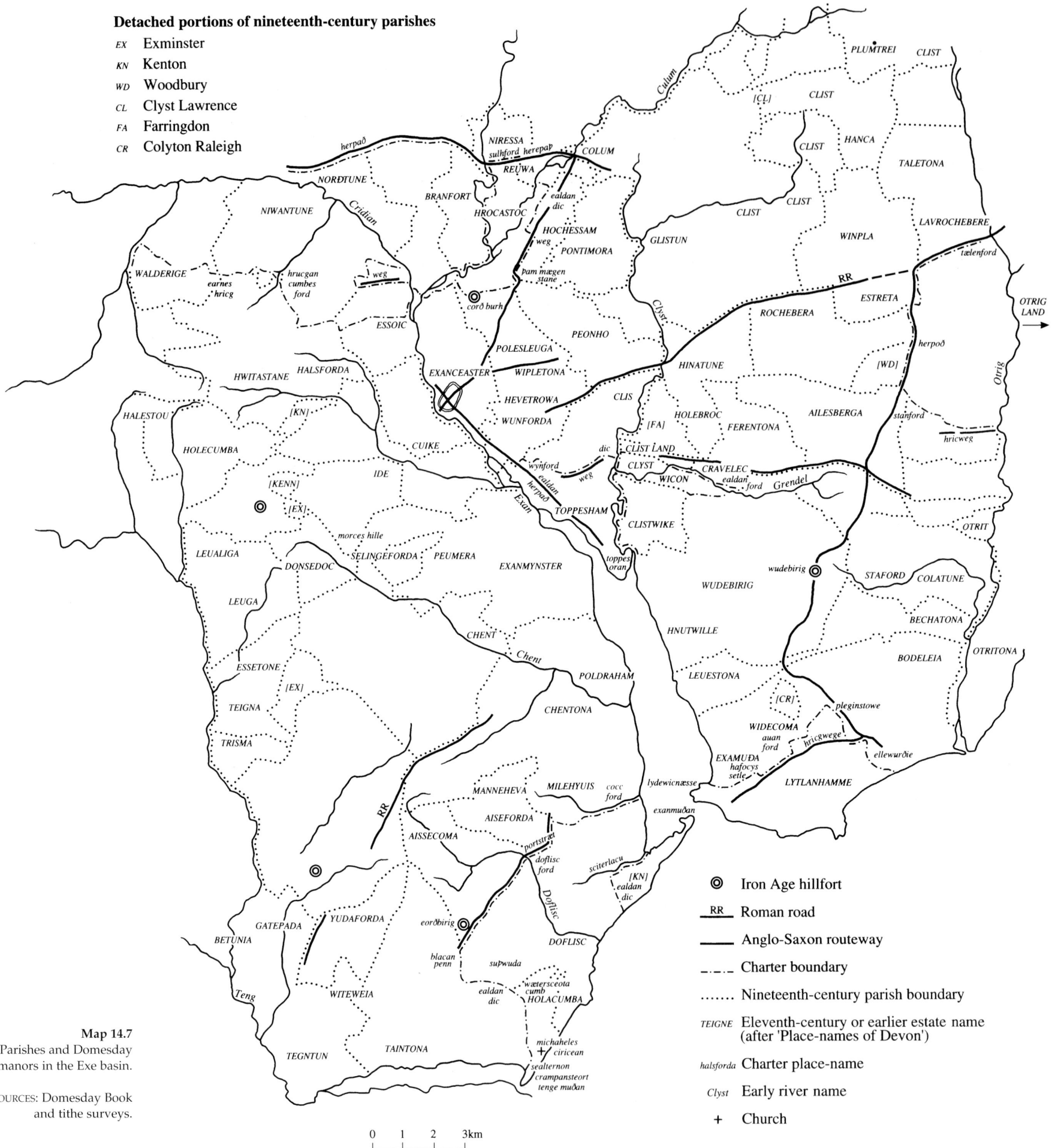

Map labels (estate names, place-names, rivers):

PLUMTREI, CLIST, CLIST, HANCA, TALETONA, [CL], LAVROCHEBERE, NIRESSA, herpað, sulhford, herepaþ, COLUM, NORÐTUNE, REUWA, ealdan dic, WINPLA, tælenford, NIWANTUNE, BRANFORT, HROCASTOC, HOCHESSAM, weg, PONTIMORA, GLISTUN, CLIST, ESTRETA, WALDERIGE, earnes hricg, hrucgan cumbes ford, weg, pam mægen stane, ROCHEBERA, RR, herpoð, OTRIG LAND, corð burh, ESSOIC, PEONHO, POLESLEUGA, Clyst, HINATUNE, [WD], HWITASTANE, HALSFORDA, EXANCEASTER, WIPLETONA, CLIS, HOLEBROC, FERENTONA, AILESBERGA, stanford, hricweg, HALESTOU, [KN], HEVETROWA, WUNFORDA, [FA], CLIST LAND, CUIKE, dic, CLYST, CRAVELEC, HOLECUMBA, IDE, wynford, ealdan, weg, WICON, ealdan ford, Grendel, OTRIT, [KENN], ⊚, [EX], morces hille, herpoð, TOPPESHAM, CLISTWIKE, toppes oran, LEUALIGA, DONSEDOC, SELINGEFORDA, PEUMERA, EXANMYNSTER, wudebirig ⊚, STAFORD, COLATUNE, LEUGA, WUDEBIRIG, BECHATONA, CHENT, HNUTWILLE, BODELEIA, OTRITONA, Chent, POLDRAHAM, LEUESTONA, ESSETONE, [EX], CHENTONA, [CR], pleginstowe, TEIGNA, WIDECOMA, auan ford, hricgwege, TRISMA, MANNEHEVA, MILEHYUIS, cocc ford, lydewicnæsse, ellewurðie, EXAMUÐA, hafocys setle, LYTLANHAMME, AISEFORDA, exanmuðan, RR, AISSECOMA, portstræt, doflisc ford, sciterlacu, [KN], ealdan dic, BETUNIA, GATEPADA, YUDAFORDA, eorðbirig ⊚, DOFLISC, blacan penn, supwuda, wæterscéota cumb, HOLACUMBA, Teng, WITEWEIA, ealdan dic, michaheles +ciricean, sealternon, crampansteort, tenge muðan, TEGNTUN, TAINTONA

estuary, inlet, arm of the sea'). *Fleote* was also the early name for Seaton (S 910), referring to the estuary of the River Axe which forms the eastern boundary of the estate.

In both counties, streams and rivers frequently served as boundaries, and references to natural hills and valleys reflect the broken nature of the terrain. Valleys in Devon are often described as 'coombs' (Old English *cumb*), denoting the small, enclosed valleys of much of this county (a Cornish word for valley which occurs in charters is *nans*). References to tors (Old English *torr*) reflect the prominence of these granite, or similar, rock intrusions in the landscape, although in Cornwall these are referred to by the Cornish word *carn*, 'rock-pile', such as *carn winnioc*, 'the white

tor', on the north-eastern boundary of Trenowth. A Cornish word for 'hill', which also survives in some Devon place-names, is *pen*, 'head, top, end'. Hackpen Hill is noted on the boundary of Culmstock (S 386) and another hill on the boundary of Ipplepen is also referred to as *pen* (S 601), although in Cornwall the term is more often used to refer to the head of a marsh.

References to archaeological features abound, although crosses and stones are more numerous in Cornwall. These may have served as markers in an open moorland landscape, whereas individual trees are more commonly referred to on boundaries in the gentler regions of Devon. The charter boundaries within St Keverne parish, for example, refer to barrows and stones, one of which may have been the cist known as 'The Three Brothers of Grugwith', and to a cross, *crouswrah* (Cornish *crous* with *gruah*, 'witch', hence 'the witch's cross'), which gave its name to Crousa on the high moorland around Goonhilly (S 755, S 832, S 1027). In Probus parish, two rounds (circular enclosures of probable Roman date), *cayr uureh*, 'the witch's round', and *caer lydan*, 'the wide round' (Cornish **ker*), occur on the boundaries of Tregellas in the north of the parish, although their earthworks have now been ploughed away (S 770). There are other references to stone rows, to hillforts and earthworks. Another Christian feature (Old English *cristel-mæl*, 'a crucifix') is noted on the boundary of the manor of Tywarnhayle in Perranzabuloe (S 684) and from the boundary clause can be identified as a cross or crucifix near the site of the remains of the early church of St Piran on Penhale Sands, perhaps the cross which survives amidst the dunes today.

The long-settled nature of much of the region is illustrated by references to the roads and ways which linked the scattered settlements, fords across streams being especially numerous. A *herepath*, a 'highway', may, for instance, be traced across the Crediton estate to Creedy Bridge, at one point on the Sandford boundary overlooked by a hill described as a *weard setl*, 'a look-out' or 'watch place'. To the east of Creedy Bridge, a branch ran

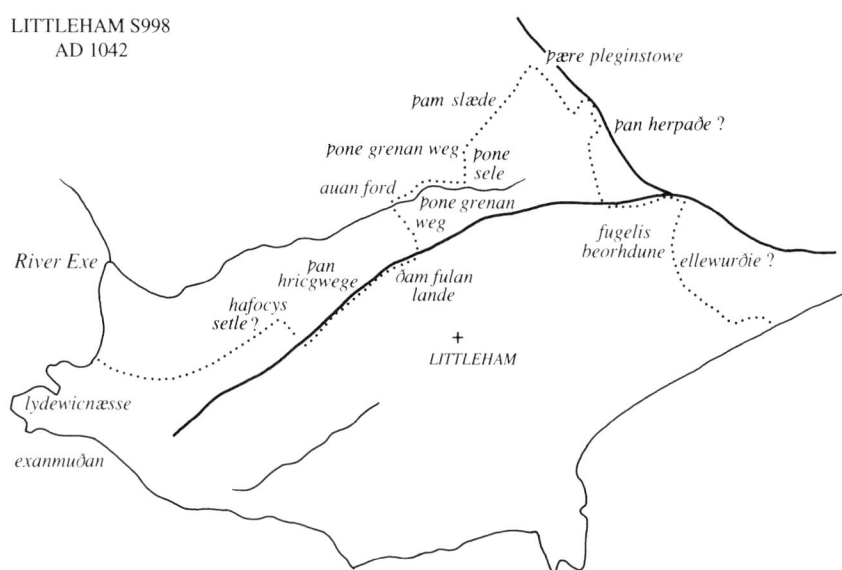

TOPSHAM S433
? AD 937

Map 14.8
Pre-Conquest charter boundaries, Topsham, Devon.

SOURCES: as **Map 14.2.**

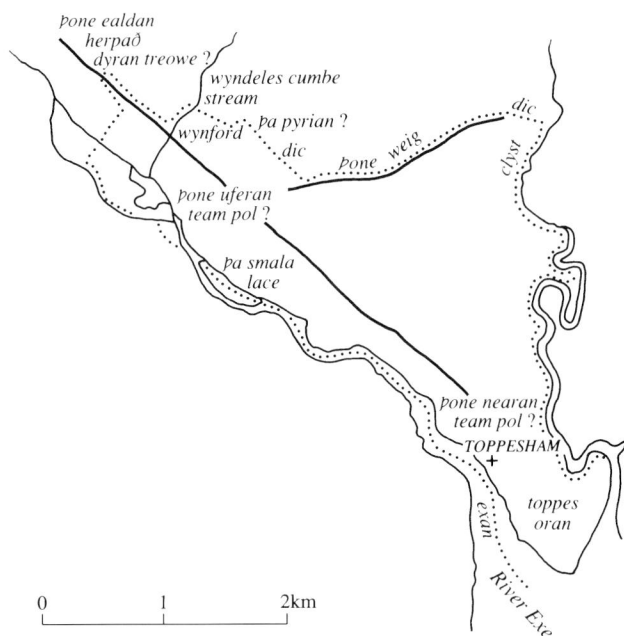

LAMORRAN S770
AD 969

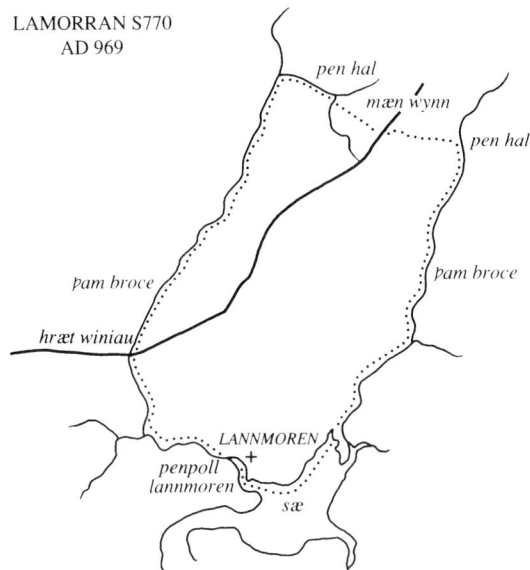

LITTLEHAM S998
AD 1042

Map 14.9 (left)
Pre-Conquest charter boundaries, Lamorran, Cornwall.

SOURCES: as **Map 14.2.**

Map 14.10 (right)
Pre-Conquest charter boundaries, Littleham, Devon.

SOURCES: as **Map 14.2.**

south-eastwards and followed the southern boundary of Shobrooke before crossing the Exe at *sulhford*, at the north-western corner of Stoke Canon, and then the Culm at the 'long ford', the latter probably at Columbjohn (**Map 14.7**). One can similarly trace a number of the routeways which ran seawards across the South Hams, often following the ridges (**Map 14.5**).

The charters refer to many settlement nuclei, often naming those unrecorded in Domesday Book, as in the parish of St Keverne on the Lizard peninsula (**Map 14.6**). As in all boundary clauses, references to settlements are few, largely because these were usually well within the boundaries, but in Devon a number of *worthig* features named were probably farmsteads. These include 'Burgheard's farmstead' near the eastern boundary of Culmstock, and *ellewurthie* on the coastal downs above Budleigh Salterton, a clue to the dispersed nature of settlement in much of the South West.

Many boundaries seem to have been demarcated by dykes, possibly the raised banks of earth and stones characteristic of much of the region today. 'The great dyke' of Lesneage still marks part of the northern parish boundary of St Keverne; another runs along the northern boundary of Culmstock in Devon. Other references to woods, meadows and ploughland help to present a picture of the ancient, developed landscape more fully described in the Domesday survey. The charter of Ayshford and Boehill notes 'many hills which one may plough', which may be a reference to land that could be used intermittently as outfield.[28] An eleventh-century grant of Trerice in St Dennis adds that 'the enclosures [farmsteads] and the barley land, and the mill and the out-leap are common' (S 1019), a sure reference to common ownership of community resources in Cornwall at that date.

Some boundary clauses provide glimpses of peasant life: the coombs where sheep or pigs were pastured and pits where wolves were captured; the king's mill on the creek at Tinnell in Cornwall; the salterns on the coast at Dawlish where brine was evaporated; or the pinfold in Ottery St Mary where straying stock would have been impounded. At Littleham (S 998, **Map 14.10**), on the downland above the coast, was the earliest recorded *pleginstowe*, 'play-place', a name which survives in later field-names throughout the peninsula, and presumably was a place where traditional games were held. These and many other references provide the earliest documented descriptions of the landscape of the south-west peninsula, a fragmentary picture yet a familiar landscape to those who know it today.

FURTHER READING

H.P.R. Finberg, *The Early Charters of Devon and Cornwall*, Department of English Local History Occasional Paper 2, University of Leicester (Leicester, 1953); W.G. Hoskins, *The Westward Expansion of Wessex*, Department of English Local History Occasional Paper 13, University of Leicester (Leicester 1960), including 'Supplement to the Early Charters of Devon and Cornwall' by H.P.R. Finberg; Jeremy Haslam, 'The towns of Devon', in J. Haslam (ed.), *Anglo-Saxon Towns in Southern England* (Chichester, 1984), 249–83; B. Lynette Olson, *Early Monasteries in Cornwall* (Woodbridge, 1989); Susan Pearce, *The Kingdom of Dumnonia: Studies in History and Tradition in South-Western Britain, A.D. 350–1150* (Padstow, 1978); A. Preston-James and P. Rose, 'Medieval Cornwall', *Cornish Archaeology*, 25 (1986), 135–85; Malcolm Todd, *The South West to AD 1000* (London, 1987); D. Hooke, *The Pre-Conquest Charter-Bounds of Devon and Cornwall* (Woodbridge, 1994).

THEMES
IN THE
HISTORY OF
SOUTH-WEST ENGLAND
SINCE THE
NORMAN CONQUEST

POPULATION

POPULATION DISTRIBUTION
FROM THE
DOMESDAY BOOK OF 1086

WILLIAM RAVENHILL

Reference is made later in chapter 35 to the fact that the information in the Domesday Book is replete with difficulties and uncertainties; interpretation of the seemingly rich amount of data it contains has to be qualified with reservations. All these caveats are equally applicable to the enumeration of the population by Domesday commissioners and scribes. The mass of the population was recorded in the three main categories of bordars, villeins and serfs, but in addition there were burgesses and a miscellaneous group. Priority in the enumeration is given to listing systematically the working rural population but there are many puzzling omissions even among these. There are twenty-six entries in Cornwall and no fewer than thirty-eight in Devon with no mention of inhabitants, even though ploughlands and in some cases even ploughteams are recorded. A particularly intriguing example is the fairly full entry for St Stephens by Launceston where, among other items, three demesne teams and six teams with 'the villeins' have been written down but neither villeins nor any other inhabitants are actually enumerated. Villeins constituted the most numerous group in the population of Devon, amounting to 49 per cent of the total, while in Cornwall bordars were significantly the largest element, they being the sole recorded inhabitants on some holdings.

An urban element in the population is suggested for five places in Devon and one in Cornwall (**Map 15.1**). Exeter was clearly the largest urban centre, followed by Totnes, Lydford, Barnstaple and Okehampton where the small number of only four burgesses suggests the recent making of a new borough in connection with the building of a castle and the establishment of a market. Such detail for Okehampton throws into relief how imperfect the Domesday record is, for one can infer the recent building of castles at Exeter, Lydford and Barnstaple from the references to houses wasted at each of these places. Bodmin is the only place in Cornwall where the Domesday folios hint, albeit indirectly, at the existence of urban life, although there were two castles and four markets elsewhere in the county.

The miscellaneous group in Devon was a most varied one. Noted particularly for their large numbers were the swineherds, the figure of 370 being greater than that for any other county. In addition there were many coscets, coliberts and cottars (all types of villein, the freemen), salt-workers, and a small number of bee-keepers, iron-workers, fishermen and smiths. Cornwall is strangely different in that serfs feature prominently and contribute 21 per cent to the total population, a figure surpassed in only one other county. In the miscellaneous group also only two categories are recorded. These are the forty-nine coliberts on two manors only and the forty *cervisarii* on the royal manor of Helston; this word has been variously translated but is usually considered to refer to tenants who paid their dues in ale. The absence of craftsmen is a surprising feature, and equally noteworthy is the fact that not one fisherman nor one tin-worker is to be found among the population recorded for the county, even though both activities are known to have been present at this time.

That which is problematic and partial with regard to the actual data cannot be remedied by the application of cartographic techniques, which also have their own discrete deficiencies. Foremost among these is the fact that the population data can only be assigned to the identified named manors. It is now well accepted that the majority of the place-names recorded are those of the demesne manors, the associated farms not being individually inscribed. For example, it is not possible that Pawton, the largest of the Domesday manors in Cornwall, with a recorded population of eighty-six and the only place mentioned in the hundred of that name, was the sole inhabited settlement, particularly when there is no evidence to suggest that there has ever been anything more than an episcopal barton on that site. A similar notable example from Devon is another episcopal manor, that of Crediton, where the bishop held a vast area extending over that fertile zone of red soils. It is again not credible that a manor which had 264 villeins out of a total recorded population of 407 consisted of the one settlement. Actually in the landscape there would have been the demesne and a village at Crediton itself and, separated from the *caput*, a considerable number of hamlets and isolated farms. An arrangement over the two counties such as this imposes severe limitations on the interpretation of **Map 15.1** since, by the absence of symbols, as around Pawton, the map suggests unreal areas of sparse population. Thus while there is abundant information to be mapped, the maps need to be interpreted with considerable caution.

The above caveats notwithstanding, some useful pointers as to the relative distribution of population can be discerned when the apparent voids around the major settlements are assigned their due significance. Areas with relatively light

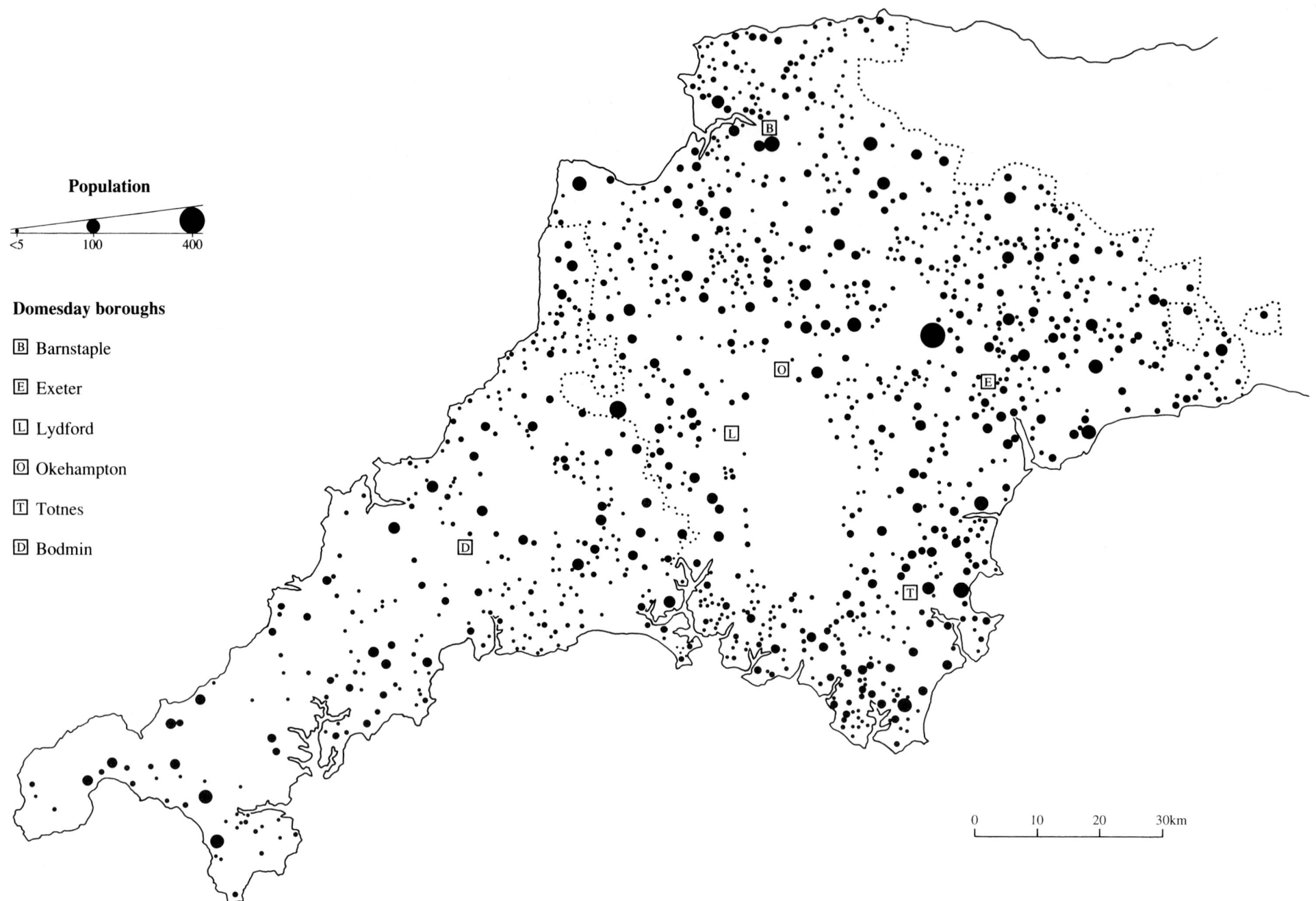

Population

<5   100   400

Domesday boroughs

B Barnstaple

E Exeter

L Lydford

O Okehampton

T Totnes

D Bodmin

0   10   20   30km

Map 15.1
Population distributions at the time of the Norman Conquest.

SOURCE: H.C. Darby and R. Welldon Finn (eds), *The Domesday Geography of South-West England* (Cambridge, 1967).

population or none at all coincide with the high granite moorlands, particularly the two large masses of Dartmoor and Bodmin Moor. Much of the former lies above the 400m contour and in some places heights exceed 600m. The large modern parish of Lydford contains no Domesday place-name except that of the borough of Lydford itself and so appears convincingly on the map as an empty space. A comparable tell-tale absence of symbols highlights four of the five granite masses of Cornwall. Bodmin Moor, the largest and highest, has a good deal of land over 250m above sea-level, with Brown Willy, its highest point, exceeding 420m. Further west lie the Hensbarrow Downs, most of which lies

above 150m. West of Truro is the third granite mass, again with extensive tracts above 125m rising to 250m. Further west still is the granite upland of the Land's End peninsula with much of its area over 125m and some parts reaching over 250m. Although these last three masses are generally lower than the moors or Bodmin and Dartmoor, their extreme exposure makes for inhospitable environments. There is a fifth granite mass which has been dissected by erosion and drowned by submergence to produce the numerous Isles of Scilly, but no mention is made of this archipelago in Domesday Book. Relatively intractable upland for farming is not entirely confined to the actual granite masses.

The high ground continues to the north of Bodmin Moor on to the Davidstow Moor, and to the north of Hensbarrow also on to the St Breock Downs. Both of these experience a similar degree of exposure to the granite and appear conspicuous on the map by an absence of symbols. Again the reason for this is in part due to the convention of allocating population to settlements. Not all these areas of highland were entirely without value, as witnessed by the large amount of pasture which was credited to many of the less elevated farms. Strangely one other extensive area of highland in the north-western part of Exmoor in Devon is not so conspicuously lacking in population, even though there are considerable tracts of moorland above 300m. Domesday manors with population were recorded here at the surprising altitudes of 350m. Also outstanding are the tongues of high-level plateaux in east Devon developed mainly on Upper Greensand and in places with cappings of clay-with-flints.

Areas with the highest recorded Domesday populations are not so clearly differentiated as those considered above. One that can be located extends from the Dart estuary through the Torbay coastal belt northwards into the valleys of the Rivers Exe and Culm. Most of this countryside lies below the 125m contour line and enjoys much protection from being in the lee of Dartmoor. Similar high populations occur in the Taw and Torridge lowlands and their coastal tract, all of which are comparatively fertile and well-drained. No such prominent areas with sizeable populations existed in Cornwall, the most densely peopled being found in a strip adjacent to the county boundary with Devon. Overall Cornwall, according to the record, was sparsely settled, but when it is compared with Devon and indeed with other counties to the east it is difficult not to sense that the friction of distance was taking its toll on the conscientiousness and ardour of the commissioners, for the entries exhibit a lack of attention to detail. When Domesday Book was in the making, Cornwall must have appeared a remote and strange land, 'not only the ends of the earth, but the very end of the ends thereof'.

FURTHER READING

William Ravenhill, 'The geography of Exeter Domesday: Cornwall and Devon', in Christopher Holdsworth (ed.), *Domesday Essays*, Exeter Studies in History No. 14 (Exeter, 1986), 29–49; I.S. Maxwell, *The Domesday Settlements of Cornwall* (London, 1986); H.C. Darby and R. Welldon Finn (eds), *The Domesday Geography of South-West England* (Cambridge, 1967).

POPULATION DISTRIBUTION AND GROWTH IN THE EARLY MODERN PERIOD

JONATHAN BARRY

SOURCES

Before the census of 1801, population totals on a region-wide basis are hard to establish. Two main sources of information survive: those generated by the state and those produced by the ecclesiastical authorities. Unfortunately for the historian, from 1377 until the late seventeenth century governments eschewed the levying of poll taxes, and documentation of those for the 1660s and 1690s does not survive for much of our region. In the sixteenth century most taxation took the form of 'subsidies', which targeted those most able to pay and which, from the 1540s onwards, bore increasingly little relation to changing regional or individual wealth, let alone population.[1] The most comprehensive tax with extensive coverage is the hearth tax of the 1660s and 1670s. However, these records are only complete for a minority of hundreds and there are problems with the listing of 'exemptions' (those too poor to pay), and with the relationship between 'hearths' and 'households' and the multiplier which should be used to produce population figures.[2] These sources have thus been used mainly to check the plausibility of other data in this chapter as discussed below.

The most useful state-generated sources are, in fact, non-taxation ones. During Elizabeth I's reign repeated efforts were made to create an efficient militia, and this generated relatively complete lists of 'able-bodied men' able to serve, as well as of those prosperous enough to provide military equipment. Such lists have survived for 1569 for both Cornwall and Devon.[3] The Cornish list has been mapped (erroneously dated 1567) by Professor Pounds.[4] The Devon return, unlike the Cornish one, is incomplete in two respects. Twenty-five parishes are omitted entirely, while in several hundreds (Lifton, Roborough and Plympton) it is made explicit that tinners subject to the stannary courts are omitted.[5] As a consequence it is not possible to map the county-wide distribution of population with any accuracy, although the problems are limited to western, central and southern Devon. No attempt has been made to map the returns, therefore, but they are used in the following discussion to suggest changes in the century before 1660 and to test the reliability of the figures used for the seventeenth century. Needless to say, there are problems in converting numbers of 'able-bodied men' into total population figures, not only in allowing for women and children but also in estimating how many men were considered 'unable', as this is known to have varied from place to place.[6]

The least problematic source is the listing of adult males provided by the Protestation Oath returns of 1641–2.[7] This was a state-administered expression of loyalty to the Protestant establishment, acceptable to all but a few Roman Catholics (whose tiny numbers in our region are clear from evidence presented in chapter 28). All adult males were expected to take the oath so, once again, a multiplier is necessary to convert this into a number of inhabitants and the standard figure of three has been applied (a doubling to allow for women and a multiplier of 1.5 to allow for children). Some returns are missing, notably for large parts of eastern Devon (Axminster, East Budleigh and Colyton hundreds).

Of ecclesiastically generated sources, the most extensive are parish registers, and any attempt at a comprehensive study of the overall population history of this period would need to be based on the record which these registers contain of baptisms, marriages and burials.[8] However, their sheer scale renders a regional analysis problematic, while it has always been notoriously difficult to translate register figures into static population totals for any one time. For much of the period reviewed in this chapter, parish registers are not extant, and by the eighteenth century, when more are available, there are substantial problems of under-registration connected with nonconformity and declining Anglican control.[9] There are also major variations, both over time and between types of location (e.g. rural and urban), concerning the level of population implied by the various types of 'vital event'. Burials in particular could run at very different levels, but so could rates of marriage and of fertility. Without establishing these trends in some detail it is very hard to apply register figures to regional totals, let alone use them to trace varied population totals within our region.

In view of these problems, historians seeking to establish overall population levels have relied more on the various ecclesiastical surveys which sought to elicit, usually from parish incumbents, the numbers of their flock. Here Devon and Cornwall (the old diocese of Exeter) is deficient, for no detailed returns survive for inquiries made concerning the number of communicants in both

1563 and 1603, though we do have the grand total for 1603, discussed below.[10] The first usable survey is that of 1676, the so-called Compton Census, conducted in order to establish the number of dissenters, both Roman Catholic and Protestant, in comparison to the number of 'conformists'.[11] The religious ambiguities this involved are discussed in chapter 28, but for population purposes the main problem lies in the ambiguity of the question regarding 'conformists': some incumbents may have returned all parishioners, others all those eligible to take communion, others those actually doing so, or perhaps only the males. The figures in each case thus have to be evaluated before they can be multiplied to reflect the number of inhabitants. The default assumption in standard accounts, and that which is used in this analysis unless evidence suggested otherwise, is that the return is of those of communicable age, including (for this diocese at least) the dissenters then numbered separately, and that this should be multiplied by 1.5 to allow for children.[12] However, as we shall see, in this diocese there may be many cases where this assumption does not hold.

COMPILING COMPOSITE TOTALS: METHODS AND ASSUMPTIONS

Where the figures survive, each parish's Protestation returns and Compton Census results have been compared and a composite figure has been created, labelled 1660. As noted before, where one or other source does not survive, tax material has been taken into account. Where the two main sources offer very conflicting evidence, the 1641–2 figures have generally been given more weight, given the problems of interpreting the 1676 material. Thus if the 1676 figures suggested a population well above or below those of 1642, the possibility that they are actually a count of all inhabitants, including children, or, at the other extreme, only of adult males, has been allowed for. In some cases the 1676 figures are clearly guesses, often in round numbers. If a large increase is indicated before 1676, the likelihood of this has also been tested by looking at population figures from the eighteenth and nineteenth centuries. In a few cases, of course, a parish's population has undergone genuine, major change during these decades. For example, St Leonard's, just outside Exeter, apparently had a population of 1,053 in 1641–2, while the 1676 figures only suggest 150 inhabitants. The cause in this case is provided by the destruction of this suburb during the Civil War sieges of Exeter and thus the 1676 figures are a better guide to a notional 1660 total.

The creation of a composite figure is not without its dangers, but some of the risks of relying on a single source are removed and it enables a figure to be supplied for every parish. These can then be compared with the parish totals returned in the first two censuses, of 1801 and 1811. These also have their problems, in particular the distortions caused by the absence of men on military or maritime service on the census day in the

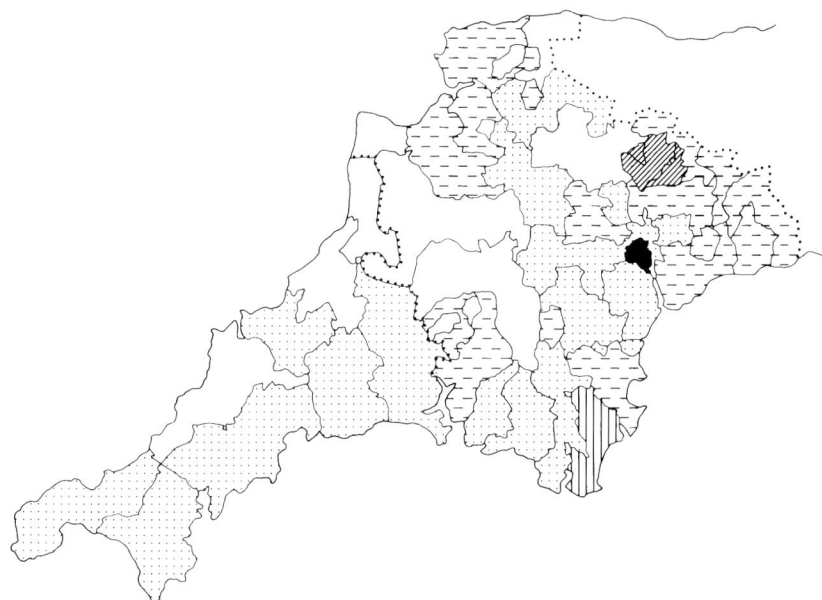

Map 16.1
Population density, 1660, all parishes, by hundreds.

Persons per 1000 acres

■	500
▨	300
▨	250
▥	200
∶	150
·	100

0 10 20 30km

Map 16.2
Population density, 1660, rural parishes, by hundreds.

Map 16.3
Population density
in Cornwall, 1750,
all parishes,
by hundreds.

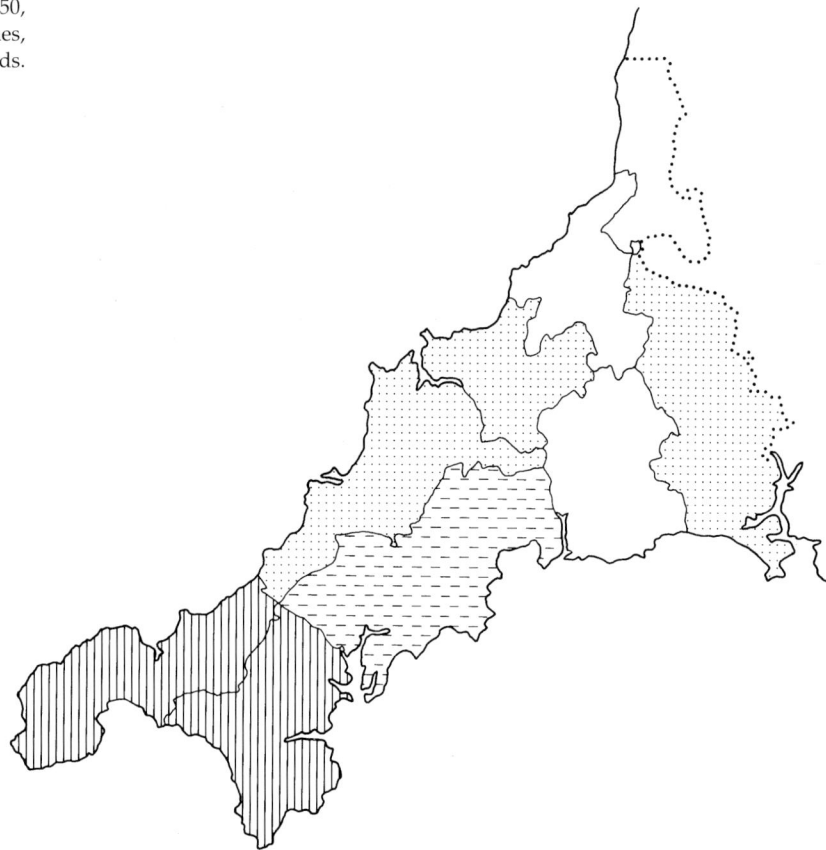

**Persons per
1000 acres**

500
300
250
200
150
100

0   10   20km

Map 16.4
Population density in
Cornwall, 1750,
rural parishes,
by hundreds.

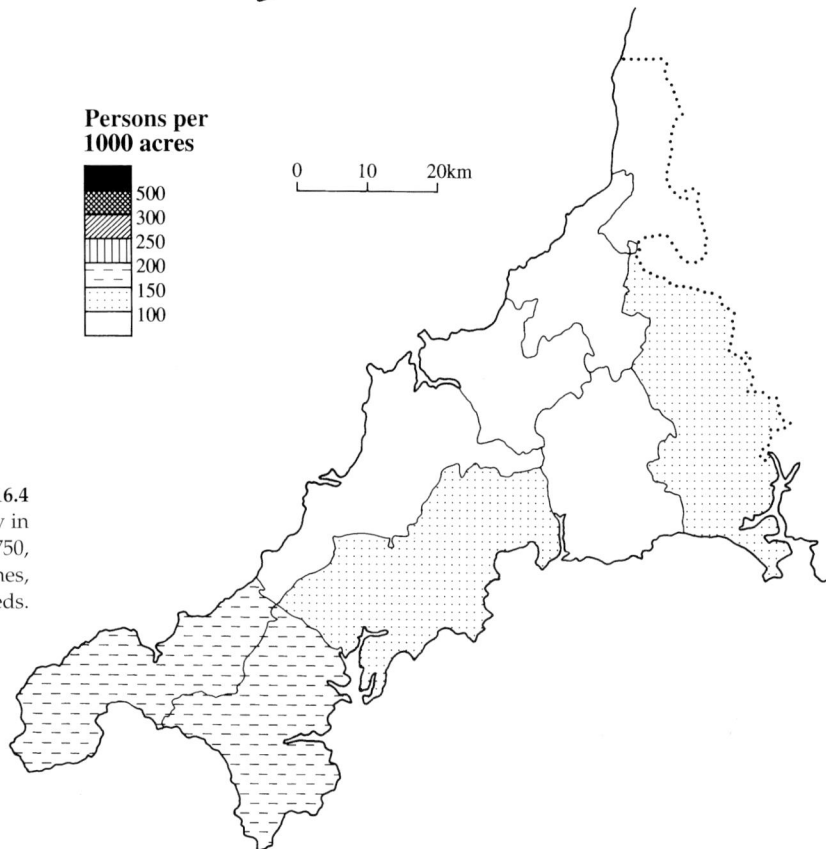

war years of 1801 and 1811.[13] The 1801 census is reckoned to be particularly imperfect and hence it was decided to iron out some of these problems by employing a notional intermediate, averaging 1801 and 1811, called in this section (and in chapter 53 on towns) 1805.

The gap from 1660 to 1805 is, of course, a major one. There are no obvious state-generated figures at a regional level to bridge this gap, but there is a series of further ecclesiastical surveys produced by the episcopal visitations of 1744–6, 1764–6 and 1779.[14] These asked incumbents to supply information about the numbers of families in their parish as well as about the level of dissent. A number of problems arise in relation to these surveys, including the absence of some parishes not part of the usual diocesan administration. The returns are often approximate and they also reflect very different understandings of the notion of 'family'. These issues have been well discussed in two studies of the Cornish returns. One of these, by Pounds, provides a map of the 1744–6 and 1779 returns by parish, but without subjecting the figures to any correction, while the other, by Thomas, provides a careful revision of the various figures, by comparing them with each other, with 1801, and with the marriage totals for problematic parishes.[15] It has not proved possible here to extend the same careful analysis to the much larger number of Devon parishes and no attempt has been made to map the Devon results, although the figures have informed some of the 1660 decisions. Moreover, the movements of population in Devon (excluding the urban developments analysed in chapter 53) are not so dramatic between 1660 and 1805 as to make it vital to establish what, if anything, had occurred by 1750. In Cornwall the rural changes are much greater and thus maps of Cornwall are presented (**Maps 16.3 and 16.4**), based on Thomas's figures for 1744–6 and 1764–6, averaged out to produce a composite figure for 1750, and so avoiding some of the pitfalls of relying on one figure.

Thus all three main data series used here are composite ones at one stage removed from the sources. The material presented in the maps has then been further aggregated to the level of hundreds, rather than parishes (only Ottery St Mary, being both hundred and parish, is directly represented). By using hundreds, many known data irregularities will be averaged out and it is also possible to reveal large-scale shifts in the geography of population within the region. On the other hand, there are dangers in using the hundred, an ancient administrative grouping of little immediate relevance to, or effect on, early modern society. One obvious issue concerns the distorting effects of urban populations, and especially of rapid urban growth, on hundred averages. Thus Roborough hundred's overall results become completely dominated, by 1805, by the Plymouth conurbation in its southern corner. In order to even out this urban effect, the results for each hundred have been given in two forms, one including all populations (**Maps 16.1, 16.3, 16.5 and 16.7**) and the other covering only the 'rural' population, derived by removing the urban populations (see chapter 53) from the hundred total (**Maps**

Persons per 1000 acres

500
300
250
200
150
100

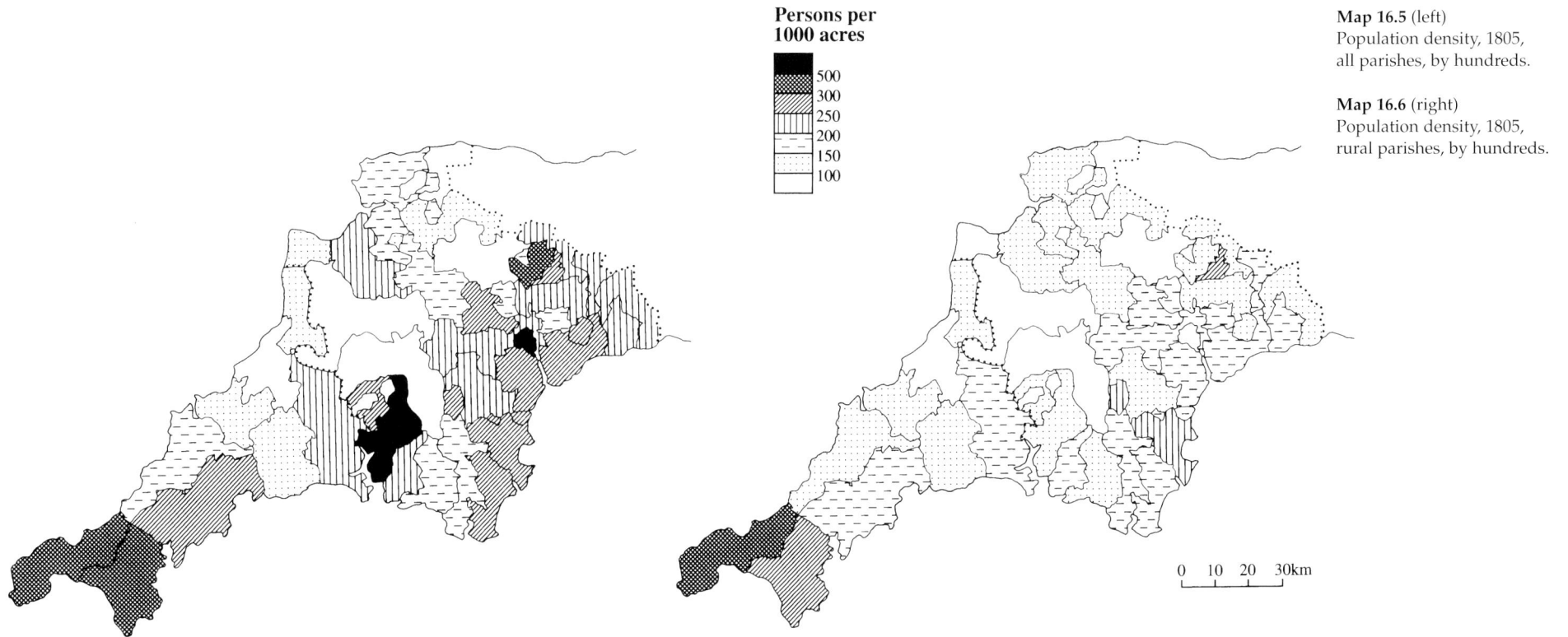

Map 16.5 (left)
Population density, 1805, all parishes, by hundreds.

Map 16.6 (right)
Population density, 1805, rural parishes, by hundreds.

0 10 20 30km

Percentage growth

200
100
70
55
40
25
10

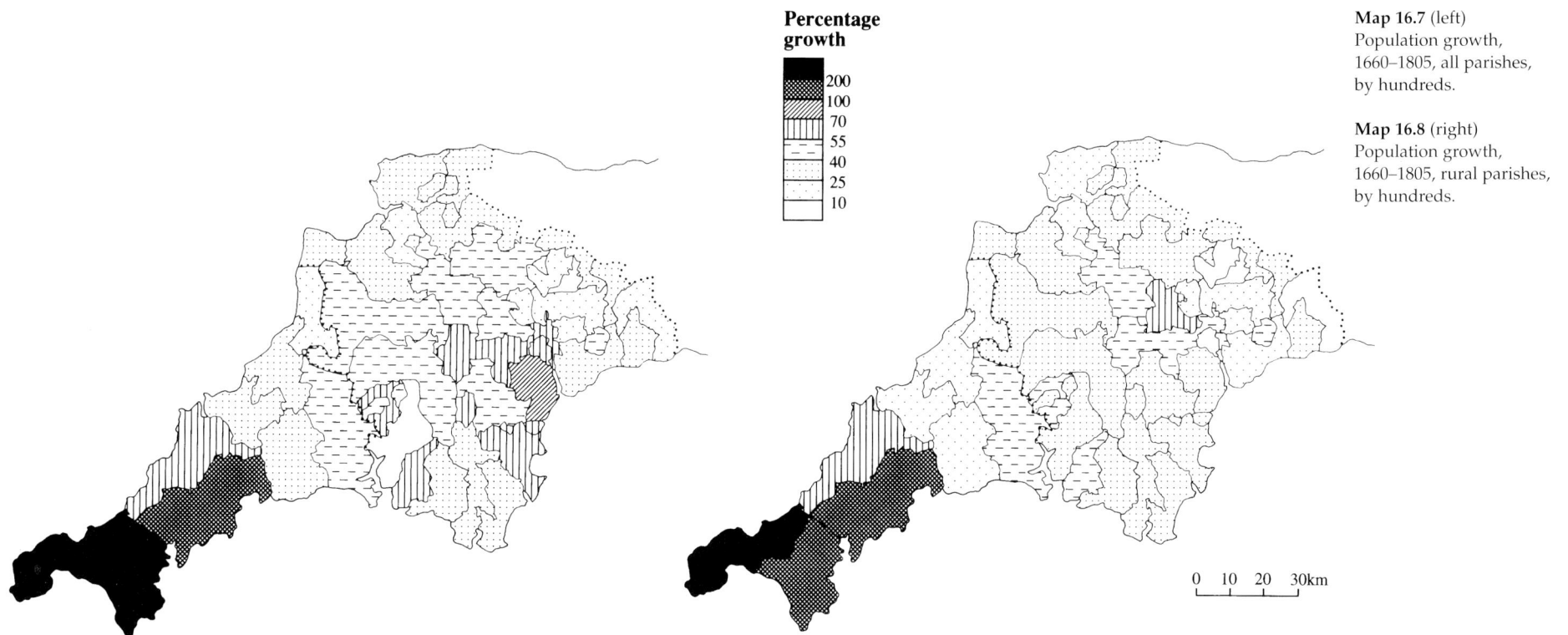

Map 16.7 (left)
Population growth, 1660–1805, all parishes, by hundreds.

Map 16.8 (right)
Population growth, 1660–1805, rural parishes, by hundreds.

0 10 20 30km

16.2, 16.4, 16.6 and 16.8). In the case of Roborough hundred, for example, this reduces the density per 1,000 acres in 1805 from 917 to just 100.

Hundred totals also undoubtedly conceal sub-hundred variations. However, examination of the individual parish figures does not suggest that the hundred averages are completely random. One indication of this is provided by **Map 16.9**, which shows the most densely populated rural parishes in 1660 and 1805 (those with over 200 people per 1,000 acres). These cluster in those parts of the region where the hundred averages are highest, such as Haytor in Devon and Penwith and Kerrier in Cornwall, and around a few of the large towns. Very few of them occur as isolated rural parishes within hundreds of much lower averages. A broadly similar conclusion can be drawn from the more detailed work on Cornish population undertaken by Professor Pounds, whose maps of parish density for 1642, '1672' (actually 1676), 1744, 1779 and 1801 confirm the broad movements sketched in this chapter by the hundred maps.[16]

POPULATION FIGURES FOR 1660: COMPARISONS WITH EARLIER ESTIMATES

Before discussing the trends shown on the maps, it is important to note that the overall population figures for 1660 presented here (summarized in **Table 16.1**) vary considerably from some of the best-known previous estimates of county populations. The figure for Cornwall of 98,104 is only slightly different from recent estimates and fits closely with Professor Pounds calculations based on the Protestation returns (99,000) and the hearth tax (*c*.95,000), although it is lower than his 1676 estimate of 105,000. He himself notes, however, that this may be an overestimate.[17] James Whetter's study of seventeenth-century Cornwall calculates a figure of 105,500 from the Protestation returns, but this is because he uses a more generous multiplier than is usual (3.333 rather than 3). The 1690 hearth tax total (multiplied by 4.333) yields a result of 107,517, but this is not incompatible with my 1660 estimate, given the growth in Cornwall's population evident between 1660 and 1750.[18] Thomas's figures of 115,710 for 1745, 123,200 for 1765 and 138,701 for 1779 are followed here in preference to Pounds's suggestions of 125,000 for 1745 and 149,000 for 1779, both of which are already downward revisions of earlier estimates based on Rickman's early-nineteenth-century attempts to use parish.register aggregations.[19]

The figures for Devon cannot easily be set against previous estimates, as these hardly exist. While W.G. Hoskins considered Devon's population in general terms and in comparison with other counties, he (perhaps wisely) eschewed attempts to fix a precise total. The nearest he comes is to suggest that Devon's 1674 hearth tax implies 39,353 households.[20] By my own estimates this is rather on the low side, with a figure of over 42,000 more likely, excluding Exeter. If Hoskins's figure is right, however, then he is

unlikely to have expected a total population of much above my estimate of 227,157 (**Table 16.1**), which would imply an average household of just over 5.75 persons. My own estimate of 42,301 households excluding Exeter would imply the more moderate, but still relatively high, average of 5.11 persons per household. For those hundreds where the hearth tax returns give us a precise and complete return of households, a comparison with my 1660 estimates yields the higher average of 5.58 (52,466 people in 9,398 households), which certainly does not suggest that the 1660 estimates are too generous.

The main problem with the estimate lies in its relationship to the Compton Census. The standard edition of this by Whiteman referred to above, derives a putative county total for Devon, based on an assumption of 33.33 per cent children, of 272,601 (and Cornwall 103,061). As this is a full 20 per cent above my calculation, it is important to explain this divergence. It should be noted at once that Dr Whiteman was not attempting to establish county totals, but merely establishing the consequences for her raw data if established multipliers were applied; she then tests these against various other assumptions. In fact, if we look at how the Devon figures fare in these other tests we can see that they begin to look too high. When she compares these totals with the numbers of houses recorded in two hearth tax totals, both give a much higher household size of people over sixteen in Devon (3.24/3.25) than in Cornwall (2.59/2.72), and the Devon figures are well above national averages, with only Northamptonshire and Rutland giving similar ratios in both cases. In the case of Northamptonshire/Rutland, Whiteman has documentary proof that many of the figures allowed for children and this, together with the hearth tax anomaly, leads her to conclude that the standard multiplier cannot be used in those counties, but she does not draw the same explicit conclusion for Devon, despite the same anomaly. Furthermore, there is documentary evidence, which she cites, to suggest that incumbents in Cornwall and Devon may well have interpreted their remit, in some cases, as including children. We have the original circular and returns for part of Penwith hundred and it could be read as implying that all inhabitants should be included. Four of the parishes whose original returns survive specified that they were returning inhabitants over sixteen, one returned communicants, but the majority merely referred to 'inhabitants' or 'persons'. In a considerable number of cases in both counties my comparison of the 1642 and 1676 figures has led me to conclude that the 1676 figure does indeed include children and this is the reason why the total arrived at here is well below the figure reached by applying the standard multiplier to all parishes.[21]

We can also test the 1660 figures by looking at their relationship to the 1569 musters and the fullest late-sixteenth-century tax record, the subsidy of 1581. Any such comparison is difficult, since variations might be caused either by errors in one or both sources or by genuine change over time, but certain broad conclusions emerge which encourage general trust in the 1660 figures. Except where the stannary factor makes the 1569 figures unreliable, most Devon hundreds fall within a fairly standard range of ten to fifteen people in 1660 per person listed in the 1569 muster, while the relationship to the 1581 assessment is also relatively stable at around twenty people in 1660 per 1581 assessment. Some of the most extreme variations (e.g. Exeter's apparent growth since 1569) almost certainly reflect real changes. Some other anomalous results cancel each other out. Thus the ratio 1660–1569 for Bampton hundred is a very low 6.95, suggesting a possible underestimate in 1660, yet the ratio of 1660 population to 1674 hearth tax households is a very high 6.72, making it unlikely that the 1660 figure is much exaggerated. The nearby hundred of Halberton also has a low ratio 1660–1569 of 7.63, yet its 1660 population already made it the densest of any Devon hundred!

Finally, we can test the 1660 figures against the total for the diocese of Exeter in 1603, which survives even though the detailed returns do not.[22] This gave 188,774 communicants and ninety-nine recusants, from an English total of 2,060,638 communicants and 7,782 recusants, making Devon and Cornwall 9.13 per cent of the total. However, a varying number of parishes are missing from the diocesan returns. It seems likely that the Exeter diocese is one of the more complete, but thirty-two fewer 'parishes' were included in the 1603 total than in the 1676 count, quite possibly most or all of either the twenty-eight dean and chapter peculiars or the thirty-six episcopal peculiars. Without knowing which, it is not possible to allow for their effect on the overall total. If we assume, however, that there is a 5 per cent shortfall in numbers, then 198,316 communicants are involved, which might imply a total population of about 297,500 people. This is just under 10 per cent below the population total for the two counties in 1660. This is reasonably close to Dr Whiteman's calculation of a 14.73 per cent growth between 1603 and 1676 for thirteen selected counties, given that these include a massive 122 per cent increase in Surrey. It is certainly a more plausible growth rate than one of 26 per cent, which would be the result of multiplying 1676 figures by 1.5, as no county studied except Surrey recorded above an 18 per cent increase. If Devon and Cornwall's respective populations grew at a similar rate (an untested assumption), then Cornwall in 1603 may have had about 90,000 people and Devon 207,500.

POPULATION TOTALS IN DEVON AND CORNWALL

There are no doubt many errors in either the sources or the assumptions made here about population totals, but, until rigorous research is done on all the sources and on register-based population movements, the figures provided, especially when aggregated to hundred and county level, are probably of the right order of magnitude. What, if we accept them, do they tell us? If we remain for a moment at the question of county population totals, we can compare the 1660 and 1805 figures with the best current estimates of English population, as given

Map 16.9
Rural parishes with a high population density in 1660 and 1805.

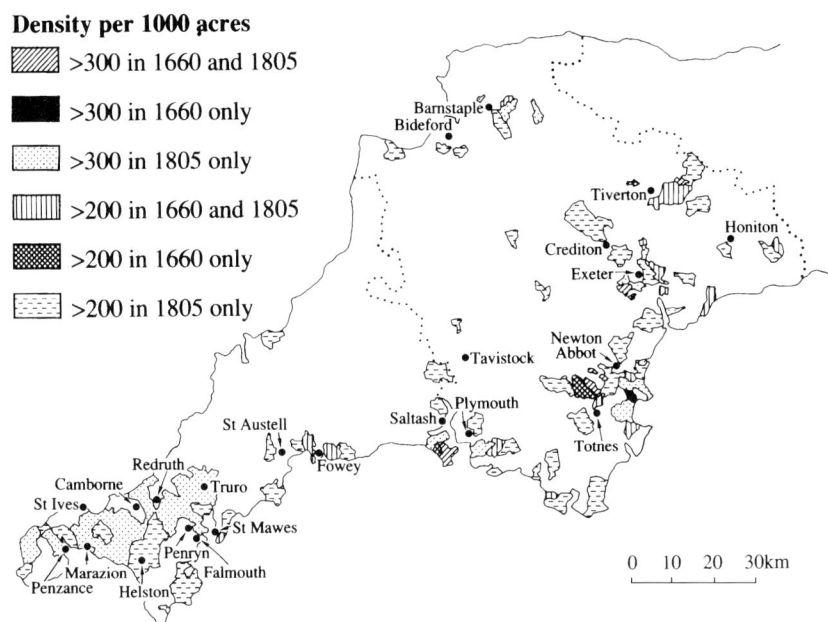

Density per 1000 acres
- >300 in 1660 and 1805
- >300 in 1660 only
- >300 in 1805 only
- >200 in 1660 and 1805
- >200 in 1660 only
- >200 in 1805 only

by Wrigley and Schofield. They estimate that in 1660 England had a population of 5,129,697, implying that 6.34 per cent of the nation lived in Devon and Cornwall (4.42 per cent in Devon and 1.91 per cent in Cornwall). It is harder to pick an 1805 comparison, since their 1805 total of 9,111,681 is lower than an average of their 1801 and 1811 estimates, namely 9,275,090. However, the latter seems more appropriate, given that our 1805 total is the average of those two years and on this basis Devon and Cornwall together had declined to only 6.07 per cent of the nation. However, this conceals a divergence between Devon, which had fallen substantially to 3.87 per cent, and Cornwall, which had risen to 2.2 per cent.[23] Thomas notes that by 1745–79 Cornwall had about 2 per cent of England's population.[24] Of course the national totals are themselves questionable, especially for 1660, as they are based on back projection from nineteenth-century figures using a sample of 404 parishes, whose behaviour is taken (with adjustments) to reflect national trends. It has often been noted that these parishes contain very few examples from the highland and pastoral zone of England and this is very clear in the case of Cornwall—no Cornish parish is included among the 404! Devon is more generously represented, with fourteen parishes in the sample, but only Hartland lies outside south and east Devon.[25] It is interesting to note that these sample parishes have a 37 per cent increase 1660–1805, compared to the Devon total of 58 per cent; they comprise 5.5 per cent of Devon in 1660 and only 4.8 per cent in 1805 (though if we exclude Plymouth, their experience is similar to that of the rest of Devon, which grew by 39 per cent).

Area	1660 all	1660 rural	1805 all	1805 rural	1660–1805 % growth	1660–1805 % rural growth
East Devon	85,182	55,067	118,427	69,032	39	25
East Devon excl. Exeter	73,682	55,067	97,927	69,032	33	25
North Devon	43,190	31,420	56,764	39,979	31	27
South Devon	69,381	50,306	141,475	66,525	104	32
South Devon excl. Plymouth	63,981	50,306	91,675	66,525	43	32
West and central Devon	29,404	24,494	42,321	34,006	44	39
Devon	227,157	161,367	358,987	209,622	58	30
Devon excl. Plymouth	221,757	161,367	309,187	209,622		39
West Cornwall	41,412	32,317	122,471	86,316	196	167
East Cornwall	56,692	47,652	81,242	64,847	43	36
Cornwall	98,104	79,969	203,713	151,163	108	89
Devon and Cornwall	325,261	241,336	562,700	360,785	73	49
England	5,129,697		9,275,090			

Table 16.1
Population growth 1660–1805 (as a percentage of 1660 figure).

Table 16.2
Population densities as number of people per thousand acres in 1660 and 1805.

NOTES:
East Devon contains the hundreds of Axminster, Bampton, East and West Budleigh, Cliston, Colyton, Crediton, Halberton, Hayridge, Hemyock, Ottery, Tiverton and Wonford, and the city of Exeter.
North Devon contains the hundreds of Braunton, Fremington, South Molton, Shebbear, Shirwell and Witheridge.
South Devon contains the hundreds of Coleridge, Ermington, Exminster, Haytor, Plympton, Roborough, Stanborough and Teignbridge.
West and central Devon contains the hundreds of Hartland, Lifton, Tavistock, North Tawton and Black Torrington.
West Cornwall contains the hundreds of Kerrier, Penwith and Powder.
East Cornwall contains the hundreds of East, Lesnewth, Pydar, Stratton, Trigg and West.

Area	1660	Rural	% urban	1805	Rural	% urban
East Devon	193	128	36	268	160	42
East Devon excl. Exeter	168	125	25	223	157	30
North Devon	114	84	27	150	107	29
South Devon	162	118	28	331	157	53
South Devon excl. Plymouth	151	119	22	216	157	28
West and central Devon	74	62	17	107	86	20
West and central Devon excl. Dartmoor	86	72	17	123	100	20
Devon	138	99	29	219	129	42
Devon excl. Plymouth	135	99	27	189	129	32
West Cornwall	120	94	22	354	251	30
East Cornwall	109	92	16	157	125	20
Cornwall	114	93	19	236	176	26
Devon and Cornwall	130	97	26	225	145	36

REGIONAL VARIATIONS IN POPULATION DENSITIES AND GROWTH

What then do the maps of 1660 and 1805 reveal? In 1660 the most densely populated parts of the region were all in Devon and in particular in east Devon, along the Exe and Culm valleys and along the southern coast round to Dartmouth. Both the Tamar and Taw/Torridge estuaries were relatively populous, but their densities fall sharply when urban populations are discounted, as do those in some parts of east Devon, but not all. Conversely, the least populous areas are in the central and moorland districts of both counties and along the north coast. These tendencies can be confirmed at an even larger scale by the figures in **Tables 16.1 and 16.2**, which divide Devon into four sections and Cornwall in half. Even without Dartmoor, west-central Devon emerges as the least populated region, followed by north Devon and then the two halves of Cornwall, which at that point were of similar density in rural population, although western Cornwall already had a higher urban proportion which raised its overall density.

How do these densities compare with the sixteenth-century situation? Pounds in his analysis of the 1569 material suggested that the Cornish population had already shifted to the coastal regions, but found quite high densities in north-east and south-east Cornwall and little sign of any westward movement of population.[26] If we compare the distribution of able men by hundreds in 1569 with the 1660 figures, we find that between 1569 and 1660 the six eastern hundreds increased their share of Cornwall's population by between 36 per cent (West) and 4 per cent (Stratton), increasing their overall share from 53.5 per cent to 58 per cent. Of the western hundreds, Kerrier declined most. While these figures are not entirely reliable, since the ratio of 'able' to 'unable' men may have varied, they should caution us against predating the 'westward shift' of Cornish population. In only three hundreds (Trigg, Stratton and Lesnewth, all in the eastern half) are both able and unable men probably given in full. In these cases the ratio of all men in 1569 to the Protestation return (itself extremely close, in aggregate, to the 1660 figure) is 1:1.6, suggesting that, here at least, population may have grown by 60 per cent during the ninety years from 1569 to 1660, almost exactly the same as the national average (59 per cent). If this is correct, Cornwall's population in 1569 may have been about 61,000. In the next ninety years (1660–1750) Cornish growth was to be much slower at 20 per cent. This, however, was well above the estimated national average of just 12 per cent, although the rural areas of Cornwall only grew by 11 per cent.

Turning to Devon, the broad picture is again one of continuity between 1569 and 1660, as far as the figures allow us to judge. The south Devon figures are largely unusable owing to the omission of tinners. It looks as though west and central Devon grew most slowly during this period, with a low ratio of able men to the 1660 figures, even though many of Lifton hundred's tinners were excluded. The figures for north Devon are quite normal for the county, as are those for the Torbay coast of south Devon, although here Haytor hundred may have been growing slowly. If so, given its high population density in 1660, it must have been especially densely populated in 1569. The greatest apparent growth between 1569 and 1660 occurred in parts of east Devon, namely Exeter, Crediton, Wonford, East Budleigh, Colyton, Ottery, Axminster and Tiverton hundreds. As we have seen, this was one of the most densely populated parts of Devon in 1660, but it would be wrong to suggest that this was necessarily a new phenomenon: the changes since 1569 were not that dramatic. With no contemporary count of 'unable men' it is not possible to relate these figures to those of 1660 at any absolute level and so no estimate of Devon's population growth over this period can be attempted. If we assume the same rate of growth as Cornwall (and the national average) then Devon in 1569 may have had about 142,000 people.

Turning to developments between 1660 and 1805, the most dramatic concern Cornwall. As **Table 16.2** shows, Cornwall moved from being less densely populated than Devon in 1660, to a considerable predominance by 1805; in rural areas Cornwall's population of 176 per 1,000 acres was not only higher than Devon's 129 but was higher than those of southern and eastern Devon. However, this outcome was almost entirely the result of the massive growth of population in western Cornwall, whose population density (both overall and in rural areas) had almost tripled. The whole of western Cornwall was now more densely populated than any part of Devon in 1660 except Exeter itself, and the rural population was much denser than any Devon equivalent in 1660 except the tiny hundred of Halberton in the midst of the east Devon cloth-making district. The concentration of this development in the tin-mining areas of Penwith and parts of Kerrier is clear both from **Maps 16.1 and 16.2** and from the individual parishes of high density identified on **Map 16.9**. As chapter 53 shows, this was also one of the areas of greatest urban growth, but here urban growth was merely an intensification of a process occurring in the countryside as well.

By contrast, the effects of major urban growth in Devon were much less evident in the countryside around. Some effects of Plymouth's growth may be discernible in the rising population of parts of southern and western Devon and the very south-east of Cornwall (in Plympton hundred and up the Tamar, rather than in Roborough itself), and the swelling populations of Haytor and Teignbridge hundreds may reflect the growth of such places as Brixham, Teignmouth and Newton Abbot, but otherwise changes in rural Devon are much less intense.

If we explore the Cornish changes in more detail, by dividing the period at 1750 we see an even more varied picture. Before 1750 the contrast between western and eastern Cornwall is even starker. Between 1660 and 1750 Penwith hundred grew by 89 per cent (its rural parts by 82 per cent) and Kerrier by 79 per cent (rural 64 per cent), while Powder grew by 29 per cent (rural 20 per cent) and Pydar by 6 per cent (its rural population remained static). The other five hundreds all experienced

absolute losses of population ranging from 7 per cent in Trigg to 26 per cent in Stratton, with rural populations falling between 13 per cent (Trigg) and 26 per cent (Stratton). The sharpness of this divergence of experience is somewhat concealed over the 1660–1805 period because in the next fifty-five years all the hundreds of Cornwall experienced strong growth, although it remained most intense in the west. During this period there was very little difference between overall growth and that in rural areas (see chapter 53), indeed in East hundred rural growth outstripped urban, thanks largely to the explosive growth of the parishes on the Rame peninsula opposite Dock.[27] Penwith led the growth table with 96 per cent, followed by Powder (82 per cent), East (74 per cent) and Kerrier (68 per cent), but no hundred fell below the 40 per cent of Trigg, and the other four all experienced between 56 per cent and 63 per cent growth.

CONCLUSIONS

The research needed to explain the patterns described here, notably the Cornish experience, is formidable. It is clear that attention for Cornwall needs to be centred on the 1660–1750 period rather than on the classic 'industrial age'. In explaining the 'westward shift' in population it is tempting to assume, as Pounds argued, that there was an actual movement of population from eastern to western parishes. Preliminary work on West Penwith casts some doubt on how far in-migration was responsible for growth there, drawing attention instead to the natural growth of the existing population, assisted by the types of family formation that tinning and fishing employment allowed.[28] Such an explanation is less likely to apply to urban growth, above all that of greater Plymouth, as urban and maritime mortality is unlikely to have left much scope for natural growth and it may have been to Plymouth, rather than to the west of the county, that east Cornwall's missing population moved before 1750. If so, then after 1750 Plymouth's recruitment area must have shifted elsewhere, perhaps to the more depressed parts of Devon, or possibly out of the region altogether. As David Souden has noted, during the eighteenth century Devon burials suggest an intensifying sexual imbalance, as women begin to outnumber men, so that by 1781–90 only ninety-four males were buried in Devon per hundred females. At the same period in Cornwall the ratio was 101 males per hundred females, although it had been lower at earlier dates.[29] This may indicate a shortfall in male employment in Devon, especially in rural areas. Towns traditionally employed more women, while Cornwall's growing mining and fishing opportunities for men contrasted with the decline in the cloth-making industry in Devon. As the work by Sharpe on Colyton and Cullum in West Penwith has shown, it is only when we can combine understanding of work, gender and demographic patterns that we can hope to unravel the full meaning of population levels.[30]

FURTHER READING

The fundamental study of English population history for this period is E.A. Wrigley and R.S. Schofield, *Population History of England, 1541–1871: A Reconstruction* (London, 1981). All the important studies of south-west England are listed in the notes to this chapter at the end of the book. There has been less work done on the population of Devon than on Cornwall.

POPULATION CHANGE
1811–1911

ANDREW ALEXANDER AND GARETH SHAW

Analysis of population trends for the nineteenth and early twentieth centuries provides important evidence of the considerable economic, social and demographic changes which were transforming Britain. The period was characterized by high rates of population growth and new patterns of distribution.[1] In Britain, as elsewhere, the degree of impact of change varied both geographically and temporally, revealing a patchwork of areas with very different experiences. In the case of south-west England, economic change was of comparatively minor importance during the surge of industrial activity which transformed England as a whole after the mid-eighteenth century. This contrasts with the region's more central role during earlier phases of development. However, increasing connectivity between areas, including political as well as economic linkages, meant the region could not remain immune from what was happening elsewhere. In response it displayed distinctive characteristics of change which are the main focus of this chapter. We begin with a brief discussion of national population trends before moving on to explore in more detail the patterns of population change within the region. Data are derived from the population censuses for 1811, 1851, 1881 and 1911, using the registration district as the basic level of analysis.[2]

NATIONAL TRENDS

National trends in population growth between 1811 and 1911 can be divided into two discrete chronological parts. The period up to 1871 witnessed a continuation of the unprecedented high population growth rates which emerged from the mid-eighteenth century onwards. Recorded growth rates in fact peaked at 1.7 per cent per annum in the census decade 1811–21.[3] An early driving force behind these population increases is now recognized to be high fertility trends.[4] Although real wages declined at the beginning of the nineteenth century, early and universal marriage was encouraged by rapid industrial growth, together with new employment opportunities for women and children. During the period 1783–1828, the crude birth rate in England increased from 35–36 per 1,000 to 44 per 1,000 in 1815 before levelling-off in the 1830s.[5] In contrast, mortality rates saw

little change between 1821 and 1871, although in many urban areas they increased in the face of poor living conditions that brought about epidemic diseases during the 1840s.

The decades between 1871 and 1901 witnessed fluctuating but overall similar population growth rates, averaging 12.4 per cent per decade. This represents the initial phase of the third stage of what is known as the 'British demographic transition model'. This model attempts to describe the general evolutionary trends in fertility and mortality. Within the context of Britain the model suggests changes relating to stages of industrialization and urbanization using a series of highly stylized phases. Towards the latter part of the nineteenth century the economic influences on population growth changed. As Lawton and Pooley argue, 'the dominant influence of food prices, previously the major element in real wages, on marriage rates and levels of fertility was no longer the key determinant'.[6] Whilst mortality rates continued to fall, especially infant mortality, the fertility rate also showed a downturn. However, annual birth rates did remain over one million between 1890 and 1911; these births allied to falling death rates would have produced large-scale population growth had it not been for the check imposed by emigration.[7] Some 10 million migrants moved from Britain in the century after 1815. In the 1880s alone, 1.8 million left England and Wales mainly from depressed rural areas.[8]

The geographical distribution of the nation's population was influenced by both national and international migration. Internally, population migration shifted the balance from rural to urban areas as well as redistributing people throughout different regions. However, to see the trend purely in terms of rural–urban is to ignore the complexity of the situation. Much migration was aimed towards regions experiencing the full force of industrialization and urbanization processes, but part of this was also between industrial areas.[9] As the movement to towns quickened during the nineteenth century, areas of concentrated urban development emerged in industrial Lancashire, Yorkshire, South Wales, central Scotland, the West Midlands and Greater London. By contrast, rural areas, like many parts of south-west England, were characterized by strong out-movements of population and falling birth rates, leading to an ageing population.

POPULATION DISTRIBUTION IN SOUTH-WEST ENGLAND

The emergence of highly urbanized, industrial regions, such as parts of northern England and South Wales, contrasts sharply with the situation in south-west England, where the urbanization process occurred far more slowly. In addition, within the South West some long-established rural industries suffered considerable decline. One consequence of these trends was a comparatively slow rate of population growth in the region from the 1830s onwards. In the case of Devon, population grew from 383,000 in 1811 to 604,000 by 1881, but failed to keep pace with other counties and as a consequence Devon declined from fourth to the ninth most populous English county.[10] The situation in Cornwall was even more severe, witnessing an absolute fall in population of more than 30,000 between 1851 and 1881, followed by a lesser but still significant decline up to 1901. The slow growth of population in Devon and the population decline experienced by Cornwall contrasts markedly with the national picture, and particularly with the growing industrial counties.

It is, however, the changing nature of population distribution that reveals the most interesting patterns. Thus, in Devon between 1841 and 1851 most rural parishes lost population to the county's major towns, especially Plymouth, Exeter and the growing seaside towns. The maps of population density for 1851, 1881 and 1911 (**Maps 17.1, 17.2 and 17.3** respectively) reveal the contrasts between the low densities recorded for the upland rural districts, which form the spine of the region, and the higher density values associated with the more urbanized districts of south Devon and the mining areas of west Cornwall. Not surprisingly, population densities were lowest in the moorland districts of both counties, the topography of many of which closely corresponded to Hoskins's description of 'the deep country of nothing but hamlets, farmsteads, and lonely cottages with little or no community life, where village settlement had never developed'.[11] Of the twenty-nine parishes that comprised South Molton registration district in north Devon in 1851, eighteen had fewer than 100 households and eight contained fewer than fifty. As a consequence population densities

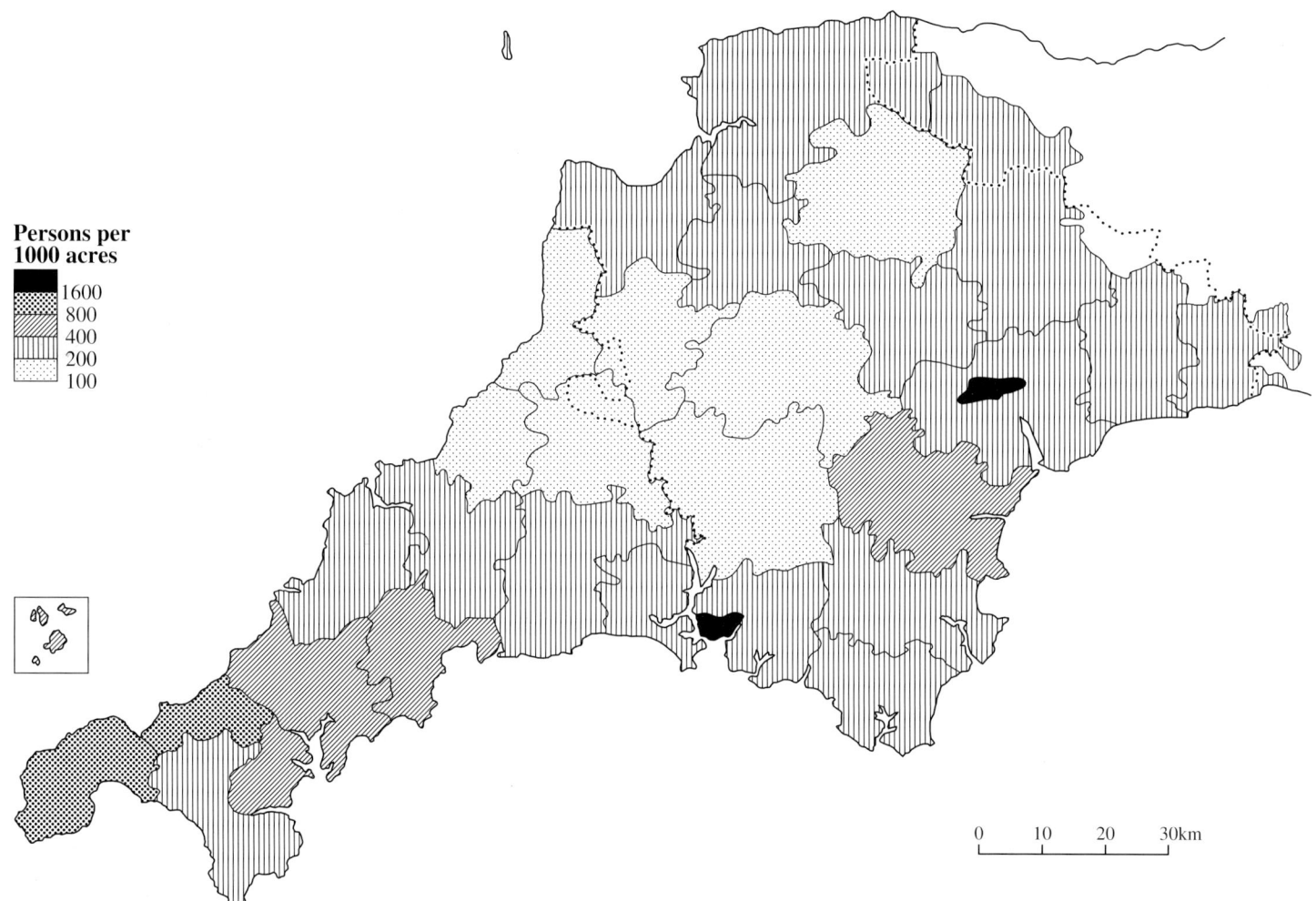

Persons per 1000 acres

1600
800
400
200
100

Map 17.1
Population density, 1851.

in such districts were frequently little more than one-third of the regional average.

Nonetheless, as **Map 17.2** indicates, these already sparsely populated districts endured prolonged depopulation during the nineteenth century as the drift of population to the towns gathered pace, stimulated by the decline of rural industries. The population densities of many rural districts throughout the region declined by more than 20 per cent between 1851 and 1911, and some in upland Devon suffered declines in excess of one-third. In the registration district of South Molton, for example, the reduction of copper-mining during the middle decades of the nineteenth century led to significant depopulation.[12] Similarly, the census explanatory notes for many of the upland districts, such as Holsworthy in Devon and St Columb in Cornwall, comment on the movement of agricultural labourers in search of better employment opportunities than those available in these depressed areas. Such population migration was generally rural to urban in nature and usually occurred over relatively short distances, typically less than 30 miles. In the case of Devon, of

those people recorded in the 1851 census as having migrated from the county, almost half settled in the neighbouring counties of Cornwall and Somerset. Many did go further, especially to London, although during the period 1870–90 there was increasing emigration to Britain's overseas colonies.[13] The effects of this persistent migration of the rural population were initially cushioned by the maintenance of comparatively high birth rates, but the continued loss of younger people who represented those most likely to migrate, together with the associated decline of rural birth rates from the 1860s onwards, reduced the fertility rates of rural populations.

In stark contrast to the situation in the moribund or decaying villages of the rural districts, many south Devon towns underwent a period of considerable growth during the century up to 1911. The most concentrated urban development in the region comprised the three adjacent towns of Plymouth, East Stonehouse and Dock (renamed Devonport in 1824). The population of the 'three towns' increased more than fourfold from 43,194 in 1801 to 193,184 by 1911.[14] Such growth was significant and reflected

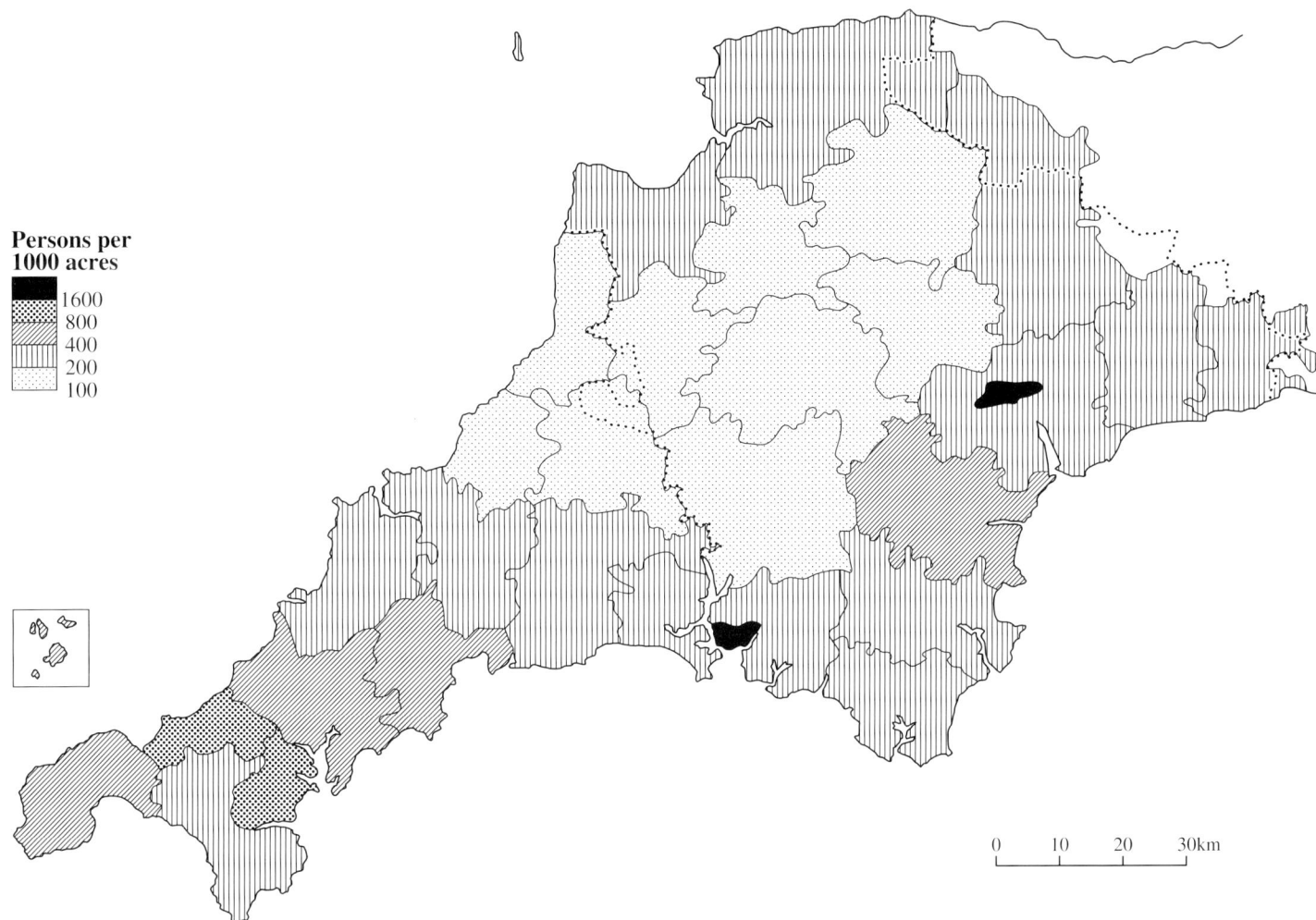

Persons per 1000 acres

1600
800
400
200
100

0   10   20   30km

Map 17.2
Population density, 1881.

Plymouth's increasing status as a naval port, as well as its increasing stature as a regional administrative, distribution and service centre. Migration was fundamental to this rapid growth and although most migrants were of local origin, the population of Plymouth and East Stonehouse included a significant and highly segregated Irish community.[15] Indeed, this community endured some of the worst conditions of overcrowding and this remained a severe problem throughout the nineteenth century as the rate of population growth outstripped that of house building. Similarly, Exeter's population grew between 1801 and 1911 from 17,398 to 95,621.

The other settlements of south-west England may be subdivided into three main types: older, established market towns; inland centres affected by limited industrialization; and the growing seaside towns. In Devon, the first group included settlements such as Cullompton where population declined from 3,138 in 1801 to 2,922 by 1901 and Ashburton which saw its population fall from 3,080 to 2,628 over the same period. Within the second group centres such as Tiverton in mid-Devon grew from a population of 6,505 in 1801 to 10,382 by 1901, largely in

response to increasing industrial growth associated with textiles. Other places in this group had more complex growth patterns, for example at Tavistock population grew very rapidly between 1801 and 1851 from 3,420 to 8,147 largely as a result of the mining activities encouraged by the Duke of Bedford. However, this growth was not sustained and by 1901 the population of the town had fallen to 5,841.

In general, mining operations provided an important stimulus to many settlements throughout the South West but its impact was greatest in the tin- and copper-mining districts of west Cornwall, which formed one of the most intensively mined areas in the world during the middle decades of the nineteenth century.[16] Mining and related industrial activity, such as pump engine production, was particularly concentrated around the towns of Redruth, Camborne and St Just. It was these areas that underwent rapid population growth during the first half of the nineteenth century as the region sought to meet the nation's enormous demand for tin and copper. As **Maps 17.1 and 17.2** indicate, these processes produced some of the highest population densities within the region outside of the major cities during the

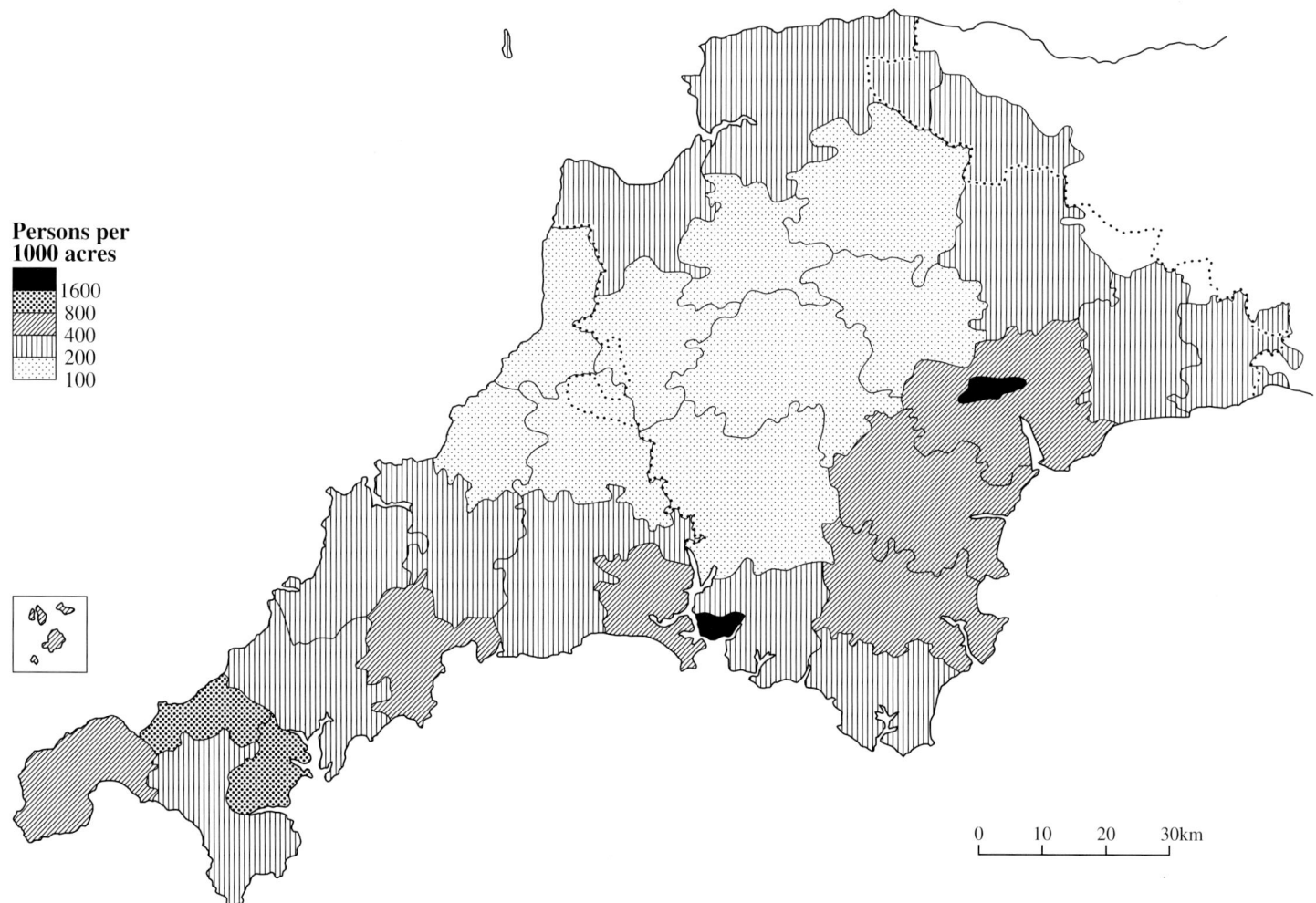

Map 17.3
Population density, 1911.

nineteenth century. The population of mining parishes such as Gwennap and Phillack, for example, doubled during the period 1811–51 as a result of the flourishing state of the industry and the opening of new mines.[17] The area's towns enjoyed growth as centres of settlement for miners and their families, and providing vital services to the industry. The recorded population of Redruth, for instance, increased from 5,903 in 1811 to 10,571 in 1851. Unfortunately, the mining boom was short lived and the collapse of copper-mining in the mid-1860s and of tin-mining in the 1870s dealt a sharp recessionary blow to the area.[18] As a consequence mining areas ceased to attract population from surrounding parishes and some began to decline in terms of population from as early as the mid-1860s. For example, the population of Redruth, the economy of which was heavily reliant on the trade in copper, declined by 7 per cent between 1861 and 1871. Redruth experienced faster population decline during subsequent decades along with other mining areas as is indicated in **Map 17.3**. As Brayshay notes, the demographic effects of the rapid decline of mining activity extended beyond the celebrated emigration of miners seeking new employment opportunities in rival mining areas in the southern hemisphere and included the collapse of remaining households that were stripped of their young adults and heads of household.[19]

The third group of settlements within the region, the seaside resorts, exhibited very different population trends. Most of these settlements witnessed significant population growth in the century up to 1911. This was especially so in Devon where resort development occurred earlier primarily because of the easier levels of accessibility (see also chapter 57). One such centre that witnessed the fastest rate of growth was Torquay in south Devon, where population grew from 1,639 in 1801 to 13,767 by 1851; growth continued throughout the century and then at a much slower rate after 1901. Similar, if less dramatic changes in population were experienced in all other coastal resorts. In general terms such patterns of growth are illustrated in **Map 17.4**, which shows increases in population density in many of the coastal registration districts. While the central rural registration districts of Devon were experiencing population decline, the majority of coastal districts showed increases as a result of tourism. In contrast the situation in Cornwall between 1851 and

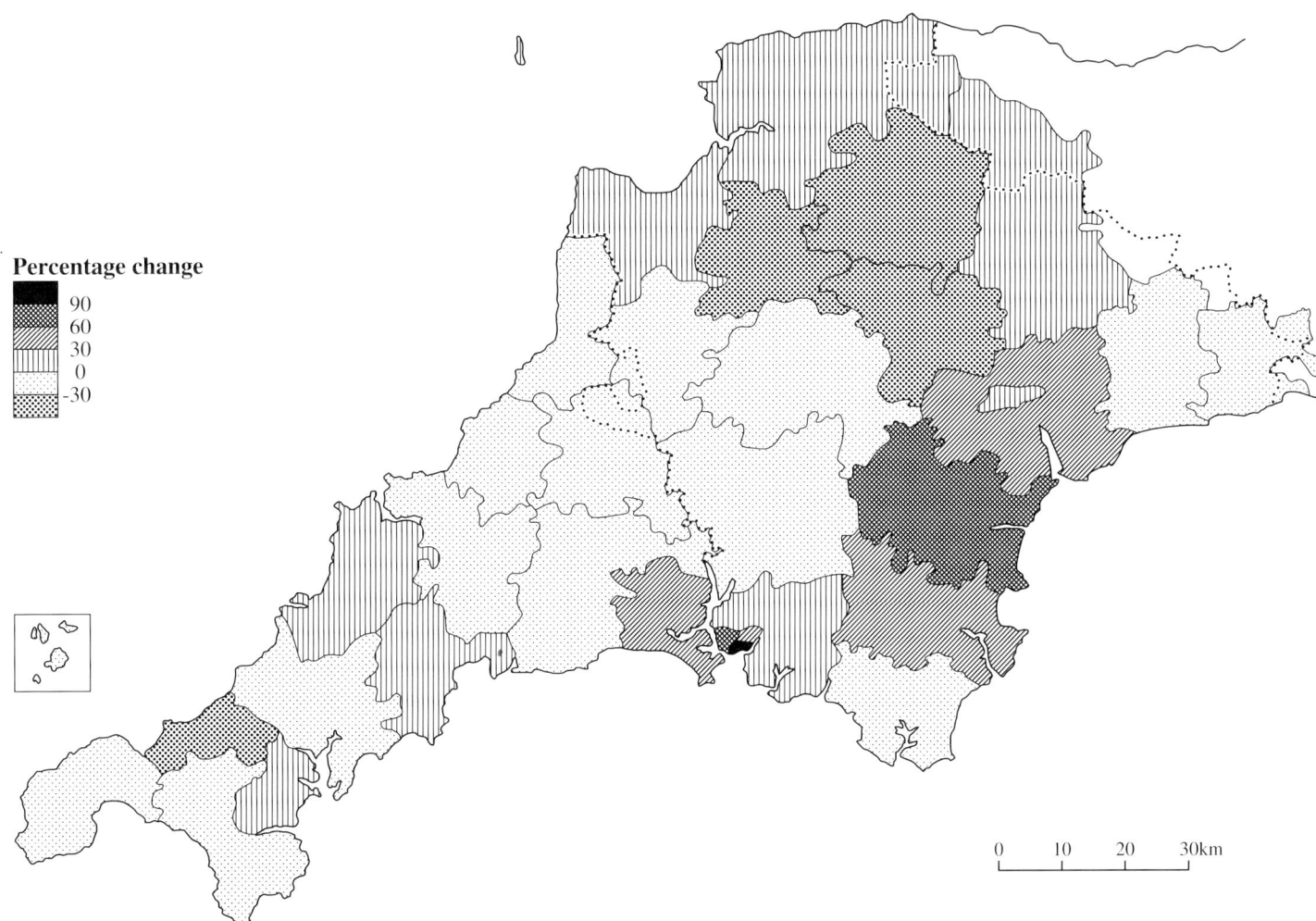

Percentage change

90
60
30
0
-30

Map 17.4
Percentage change in population density, 1851–1911.

1911 is much more complex and the influences of tourism growth are, in many registration districts, often counterbalanced by declines in mining activity.

In Cornwall, and to a lesser extent Devon, the activities of the Great Western Railway had an important influence in stimulating the tourism economy. Improved accessibility and heavy marketing, especially of the 'Cornish Riviera', enhanced the popularity of the region's resorts and enabled the development of larger hotel facilities.[20] Improved accessibility also served to accentuate the vulnerability of many of the region's agricultural parishes and traditional market towns by facilitating a further siphoning-off of younger inhabitants who moved to the larger towns and resort centres in search of better employment opportunities.

FURTHER READING

The national context is examined in R. Lawton and C.G. Pooley, *Britain 1740–1950: An Historical Geography* (London, 1992) and N. Tranter, *Population and Society, 1750–1940* (London, 1985). The principal books and articles on south-west England are listed in the notes to this chapter at the end of the book.

POPULATION

POPULATION CHANGES
IN THE
TWENTIETH CENTURY

ANDREW GILG

As **Table 18.1** shows, the populations of Devon and Cornwall in 1901 were 622,196 and 322,334 respectively. By 1991 their populations had grown to 1,015,969 and 470,912 respectively, or rises of 63 per cent and 46 per cent. However, these overall changes mask a number of significant trends:

(a) different rates of growth for the two counties over the ninety years, 1901–91;

(b) different patterns of growth within the counties over the period;

(c) the effects of some minor adjustments to the counties themselves (see chapter 24).

Because of boundary changes, it is impossible to portray population change exactly between any one census period and another. It is largely for this reason that this chapter does not analyse census data directly other than in **Table 18.1**. Instead, maps have been redrawn from a variety of published sources to depict the main population changes that have occurred. For convenience, the maps and analysis divide the ninety years up as follows: 1901–49, 1949–71 and 1971–91. Finally, the relative coverage of this chapter also reflects the greater importance of the Devon population (two-thirds of the total) and also the greater availability of published maps and literature for Devon.

POPULATION AND POPULATION CHANGE, 1901–1949

In terms of contemporary analysis very little work was carried out until the 1930s and most of the maps that we have date from the post-war period, when planners were forced to survey their areas under the duties given to them by the 1947 Town and Country Planning Act. For example, Cornwall County Council in their *Report of Survey* (1952) produced a map of population density for Cornwall in 1911 which shows densities of over 641 people per square mile in and around the main towns of Penzance, Redruth, Helston, Penryn, Truro, Newquay, Wadebridge, St Austell and Torpoint.[1] This is redrawn as **Map 18.1**.

Maps 18.2 and 18.3, obtained from a survey conducted by the University College of the South West in 1947, present 1931 data corrected to 1934 and 1935 boundaries for Cornwall and Devon respectively. **Maps 18.4 and 18.5** introduce the concepts of natural change and migration change and portray a very complex picture of change. There are, however, three main components to the pattern: first, in a broad strip along the north-western coast population increase by natural change was offset by a greater rate of out-migration to give a declining population; second, in a central band (with gaps) between St Austell and Tiverton natural increase counteracted the migration decrease; and third, in the southern coastal areas, notably of south-east Devon, population increased both by natural increase and by migration. This pattern of growth along the south coast and decline along the north coast of the south-west peninsula has been a recurring theme throughout this century.

Map 18.5 not only carries forward the trends of **Map 18.4** to 1938 but also attempts to explain the changes. In broad terms the 'decreasing areas' can be seen to be the main agricultural regions of the South West. In contrast, the areas of increase are due either to tourism, notably Newquay (which nearly doubled its population between 1901 and 1931), or to residential growth, mainly around Exeter, Torbay, Plymouth and along the south Cornish coast. It should be remembered at this point that the 1930s were the heyday of speculative building around towns before the introduction of a rigid town and country planning regime in 1947.

In the 1930s the so-called 'nadir' of agricultural fortunes was reached, and a number of maps have been produced that portray the demographic consequences of this process. For example, Vince in 1952 examined the so-called 'primary population' (that based on traditional rural employment, mainly agriculture) of the period, and the degree to which this was being replaced by an 'adventitious' population, notably commuters and retired people.[2] From these maps, redrawn here as **Map 18.6**, it can be seen that Devon and Cornwall had high percentages of primary population in 1931 (**18.6A**), but only moderate densities of primary population (**18.6B**) reflecting the pastoral nature of the area (see chapter 39 for further details). However, as **Map 18.6C** shows, north-west Devon and east Cornwall were experiencing not only depopulation of the primary population, but were also losing 'adventitious' population. In the

Map 18.1 (left)
Density of population
in Cornwall, 1911.

Source: Cornwall County
Council, *Development Plan:
Report of Survey* (Truro, 1952).

Map 18.2 (right)
Density of population
in Cornwall, 1931.

Source: University College of
the South West, *Survey*, 1947.

Persons per sq. mile

1000
640
352
96
48

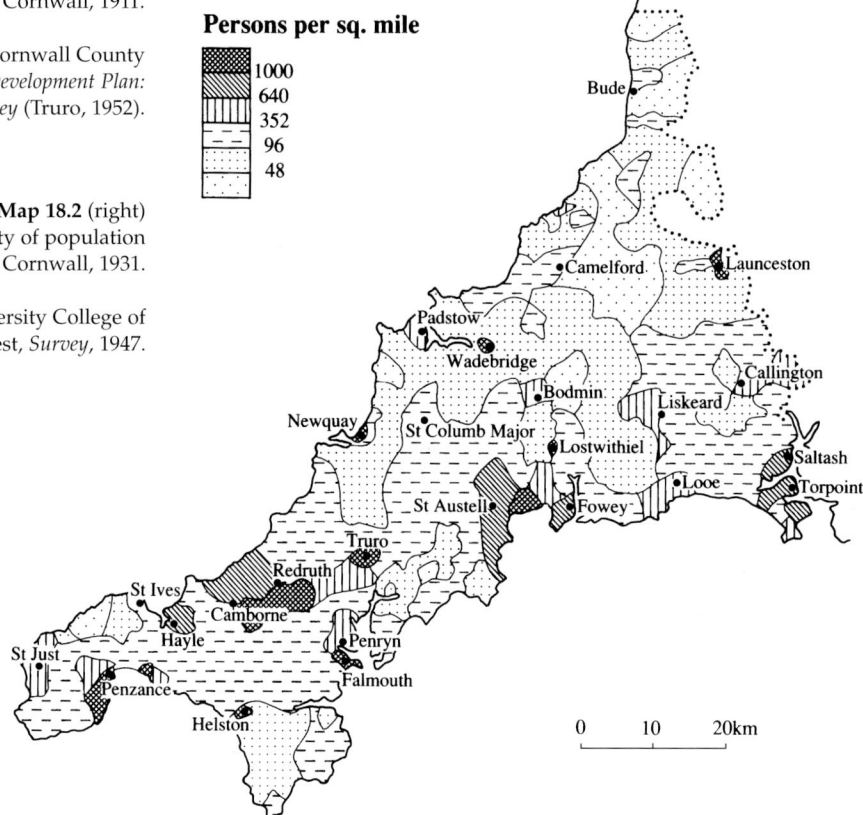

Persons per sq. mile

640
384
128
64

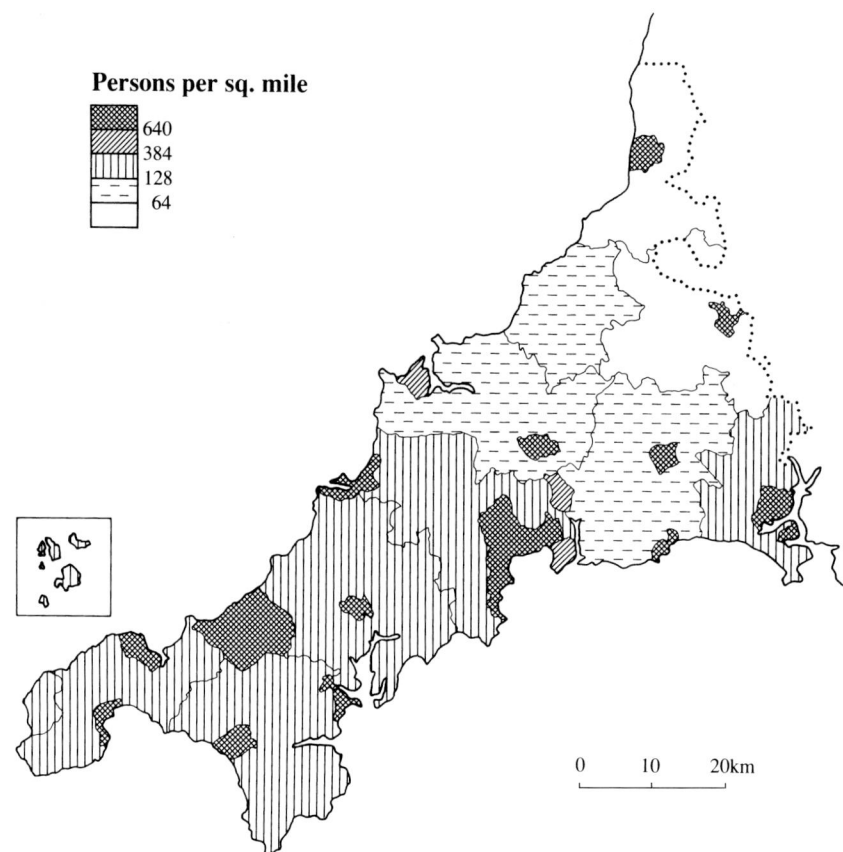

Map 18.3
Density of population
in Devon, 1931.

Source: University College of
the South West, *Survey*, 1947.

Persons per sq. mile

640
384
128
64

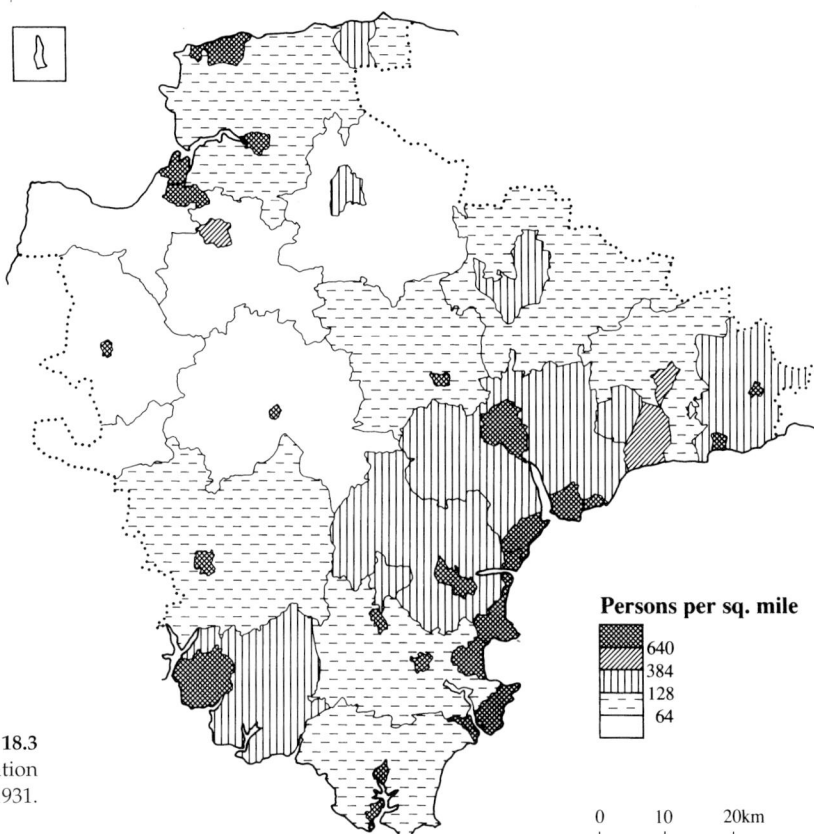

remaining areas, as **Map 18.6D** indicates, the primary component of the rural population was being moderately diluted, and in a few areas severely diluted.

Maps of population change in the period spanning the Second World War were produced by Willatts and Newsom in 1953 (**Map 18.7**).[3] **Map 18.7A** shows that the areas of rural depopulation already identified were gradually being eroded throughout the period and only a small area experienced depopulation throughout the entire period from 1921 to 1947. In contrast, south Devon, parts of north Devon and central Cornwall experienced continuous increase in the same period (**Map 18.7B**). By the immediate post-war period 1947–51 (**Map 18.7C**), nearly all the region was experiencing population increase, a trend which has been continued throughout the post-war years.

The main reason for population growth in Devon and Cornwall in the post-war period has been in-migration, and, as **Maps 18.7D and 18.7E** reveal, this was a trend already established before the war. The pattern of migration gain has, however, spread across the region in an uneven manner. For example, net migration gain was persistent only in south Devon, and only spread across most of the region after 1939. In the meantime, persistent migration loss was experienced in most of north Devon and parts of Cornwall throughout the period from 1921 to 1947.

The year 1947 marks a major watershed in population history because since that date planning controls and the supply of council housing have been major factors affecting the changing pattern of population. People have only been able to move to where planners have permitted new houses to be built, or to where councillors have decided to construct new council estates. This last is a major factor as more than half of the present housing stock of the South West has been been built since 1947.

POPULATION AND POPULATION CHANGE, 1949–1971

As **Table 18.1** shows, the 1949–71 period marks the transition from moderate to high rates of growth in Devon and from losses to rapid growth in Cornwall. It also marks the period during which Devon and Cornwall ceased to be truly rural counties and became increasingly urbanized under the influence of migration.

Examining Cornwall first, **Map 18.8** indicates that the 0.6 per cent decline in population experienced during the 1951–61 period was spread fairly evenly over the rural areas of the county. In contrast the main urban areas and their hinterlands experienced growth of up to 20 per cent. The resulting map of population density for 1961 (**Map 18.9**) shows a growing divide between

urban and rural densities, reflecting the introduction of tight planning controls (see chapter 55) and also the decline of agriculture as a major employer in the new technological age of farming ushered in by the 1947 Agriculture Act.

Year	Devon	Change since previous census (%)	Cornwall	Change since previous census (%)
1901	662,196	+4.6	322,334	−0.1
1911	699,703	+5.7	328,098	+1.8
1921	709,614	+1.4	320,705	−2.3
1931	732,869	+3.3	317,968	−0.9
1951	797,738	+8.8	345,442	+8.6
1961	822,699	+3.1	343,278	−0.6
1971	898,404	+9.2	381,672	+11.1
1981	958,745	+6.7	432,240	+13.2
1991	1,015,969	+6.0	470,912	+8.9

Table 18.1
Population of Devon and Cornwall, 1901–1991.

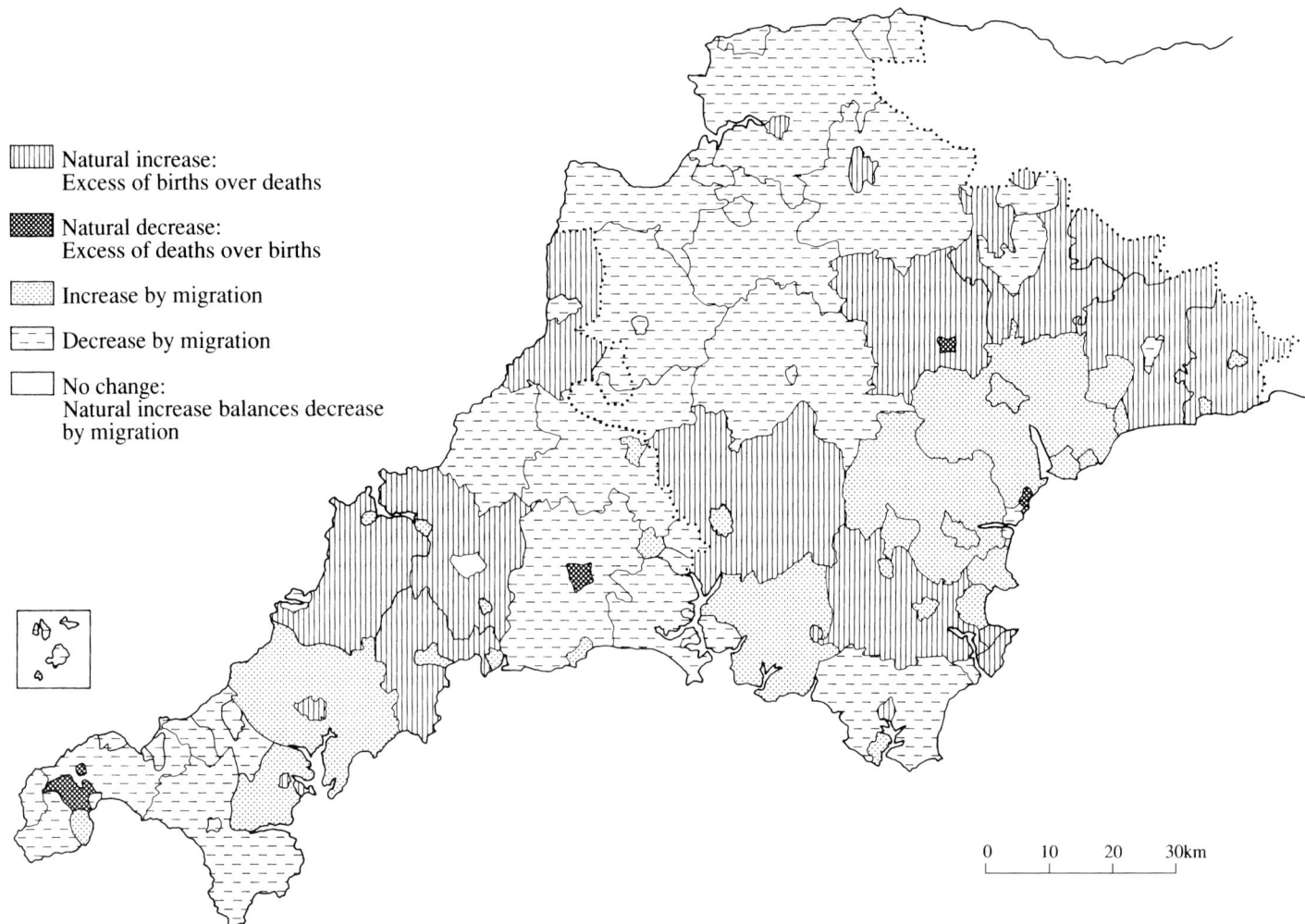

Natural increase:
Excess of births over deaths

Natural decrease:
Excess of deaths over births

Increase by migration

Decrease by migration

No change:
Natural increase balances decrease by migration

0   10   20   30km

Map 18.4
Population change, 1921–1931, by births, deaths and migration.

Map 18.5
Increasing, decreasing and
static areas, 1921–1938.

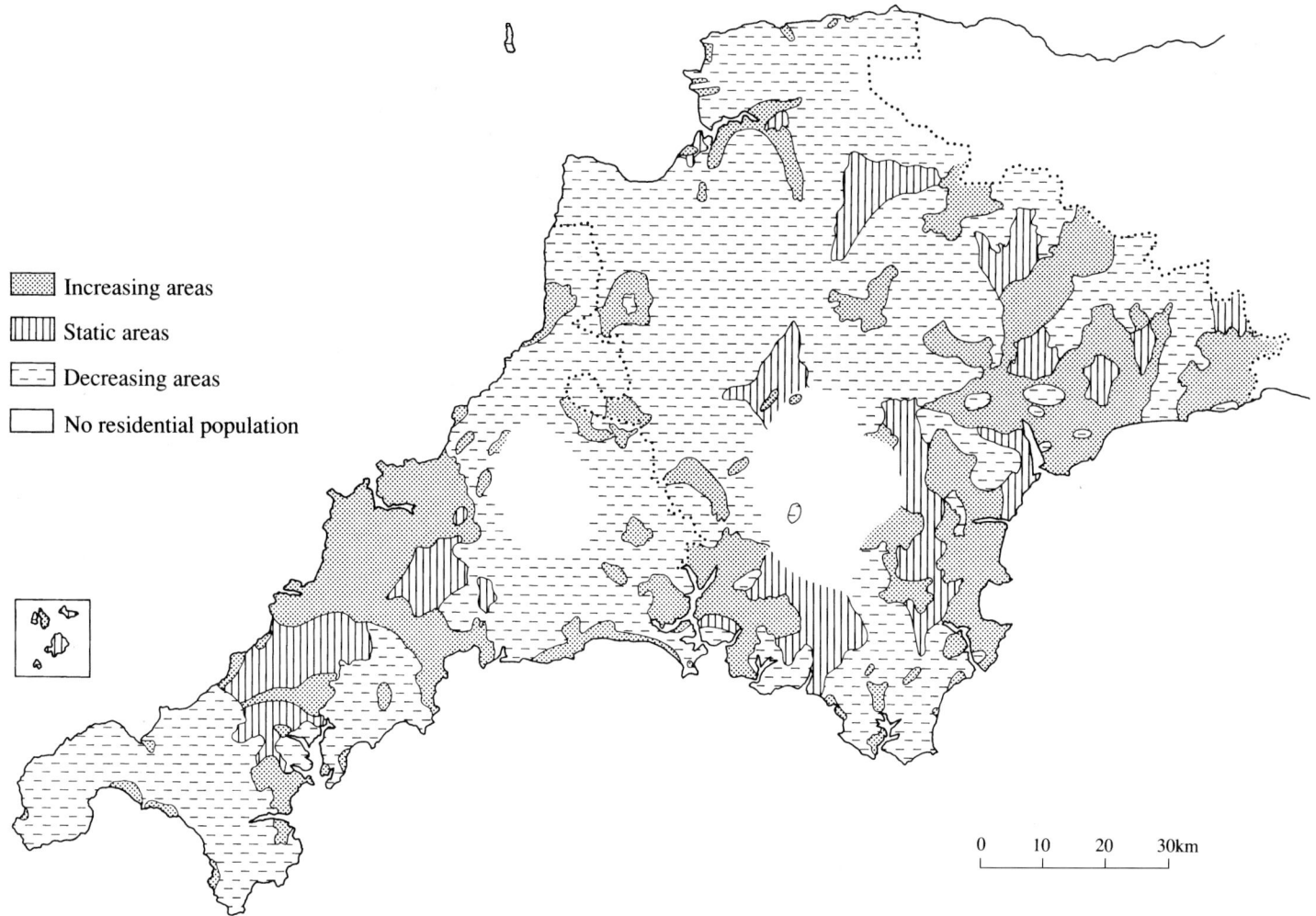

Increasing areas

Static areas

Decreasing areas

No residential population

0   10   20   30km

(A) Persons in agricultural
occupations 1931

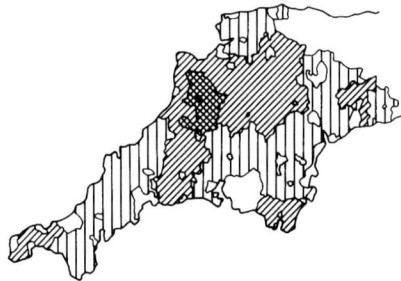

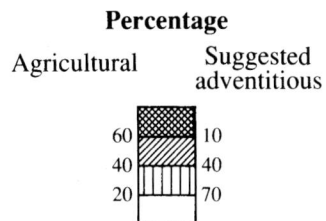

(B) Density of agricultural
population 1931

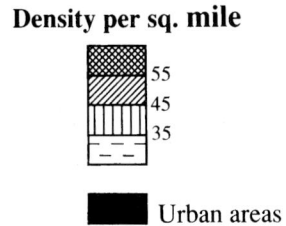

(C) Primary rural depopulation
1921-31

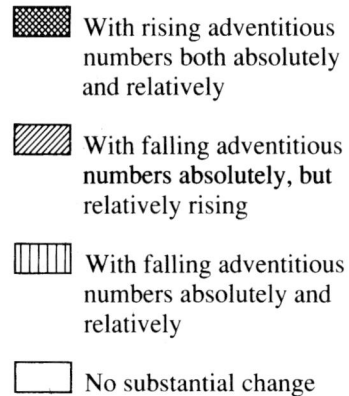

(D) Rural dilution 1921-31

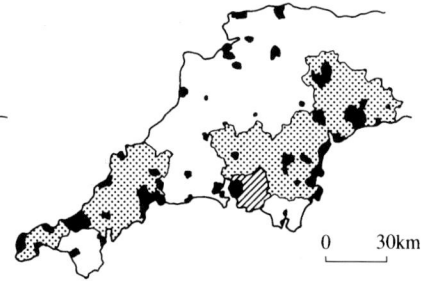

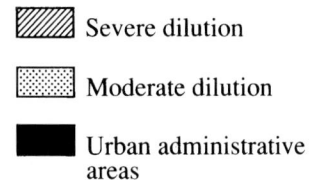

0   30km

Percentage

Agricultural   Suggested
adventitious

60	10
40	40
20	70

Density per sq. mile

55
45
35

Urban areas

With rising adventitious
numbers both absolutely
and relatively

With falling adventitious
numbers absolutely, but
relatively rising

With falling adventitious
numbers absolutely and
relatively

No substantial change

Severe dilution

Moderate dilution

Urban administrative
areas

Map 18.6
Rural population, 1921–1931.

SOURCE: S.W.E. Vince,
'Reflections on the structure
and distribution of rural
population in England and
Wales 1921–31', *Transactions
of the Institute of British
Geographers*, 18 (1952).

In Devon, **Map 18.8** shows that the overall 3.1 per cent increase in population between 1951 and 1961 was concentrated in and around the main towns, notably around Plymouth, Torbay, Exeter and Exmouth in the south, and Barnstaple–Bideford in the north. In common with Cornwall, widespread population decline was experienced in rural areas except in a zone parallel to the south coast. The net result of these changes as shown in **Map 18.9** is further affirmation of the growing importance of the main towns in the population structure of Devon, and in particular the growing importance of the Exeter–Newton Abbot–Torbay triangle. The dynamic nature of this area, and to some extent the Plymouth, Barnstaple–Bideford and Falmouth–Truro–St Austell areas, was, however, due entirely to in-migration as **Map 18.10** shows, except for the case of Plymouth which, as **Map 18.11** reveals, was one of the few areas to show any real natural increase in the 1951–61 period. Indeed some areas exhibited a natural decline in the decade.

As **Table 18.1** shows, the stagnation/decline of 1951–61 was followed by two decades of growth in the 1960s and 1970s, notably in Cornwall. Indeed, almost all the population growth of Cornwall in this century occurred after 1961. **Maps 18.12 and 18.13** indicate the differential experience within the region and in particular highlight growth in north Devon and the Helston, Bodmin and Saltash areas of Cornwall.

Overall, the 1960s marked the transition of the South West from a region with a fairly static population to one of the fastest growing regions of the country, mainly due to in-migration. This is a trend that has continued since as the next section demonstrates.

POPULATION AND POPULATION CHANGE, 1971–1991

A major problem of mapping population from 1974 is that in that year local government boundaries were radically redrawn, and the large number of local authorities that had been employed until that date were replaced by only sixteen areas, ten in Devon and six in Cornwall. This means that the spatial framework for population analysis becomes coarser, but on the positive side, the new areas are more even in size and population. These decades witnessed a reversal of previous trends in that Devon's rate of increase was less than that of Cornwall over the two decades. In order to aid comparison with earlier periods, population data for 1961–81 have been adjusted for post-1974 boundaries in **Map 18.14**. Everywhere in the region (as measured by the new boundaries) was experiencing population growth, but in a rather different way from before. For example in Cornwall, the most rapid rates of growth shifted from the west to the east, and notably to Caradon, west of Plymouth. As in the 1960s, the reasons for these changes are due to in-migration, since only seven of the twenty-seven districts of Cornwall showed any degree of natural growth between 1971 and 1975, as **Map 18.15** demonstrates.

(A) Decrease in Rural Districts 1921-47

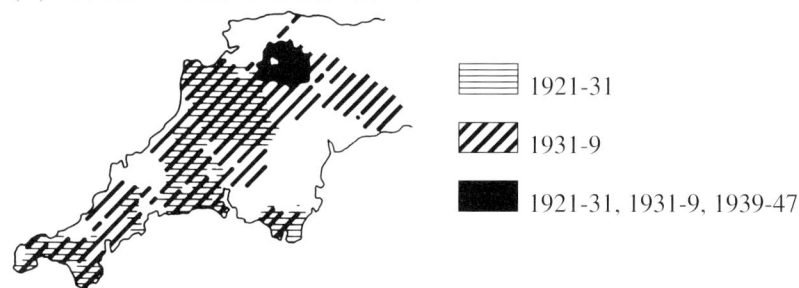

☰ 1921-31

▨ 1931-9

■ 1921-31, 1931-9, 1939-47

Map 18.7
Population change, 1921–1951.

SOURCE: G.C. Willatts and M.G. Newsom, 'The geographical pattern of population changes in England and Wales 1921–1951', *Geographical Journal*, 119 (1953).

(B) Persistent increase 1921-47

(C) Increase 1947-51

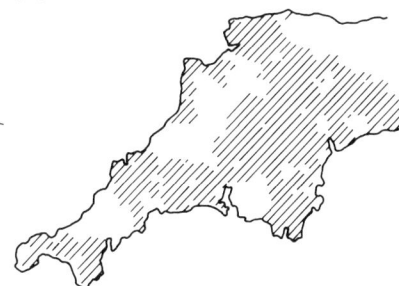

(Di) Persistent change 1921-47: Total population

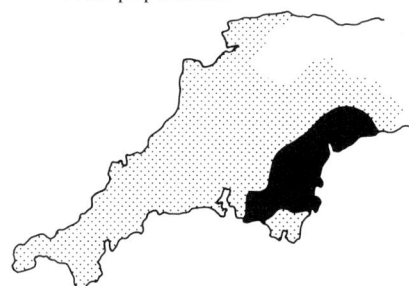

(Dii) Persistent change 1921-47: Migration only

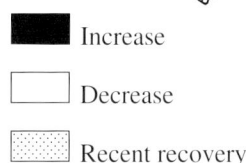

■ Increase

☐ Decrease

▨ Recent recovery

(Ei) Migration increase 1921-47

(Eii) Migration decrease 1921-47

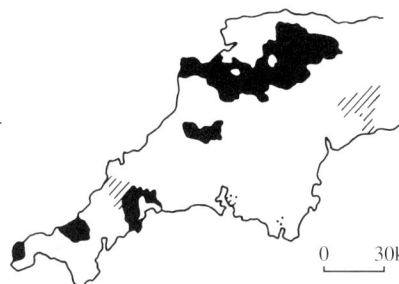

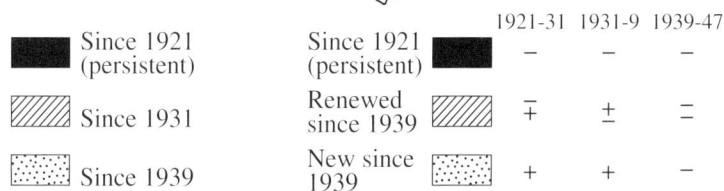

0   30km

		1921-31	1931-9	1939-47
■ Since 1921 (persistent)	Since 1921 (persistent) ■	–	–	–
▨ Since 1931	Renewed since 1939 ▨	∓	±	=
▦ Since 1939	New since 1939 ▦	+	+	–

Map 18.8
Population change, 1951–1961.

SOURCE: A. Shorter, W.L.D.
Ravenhill and K.J. Gregory,
South-West England
(London, 1969).

Percentage change

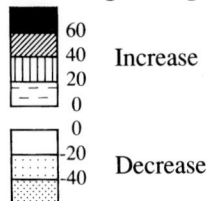

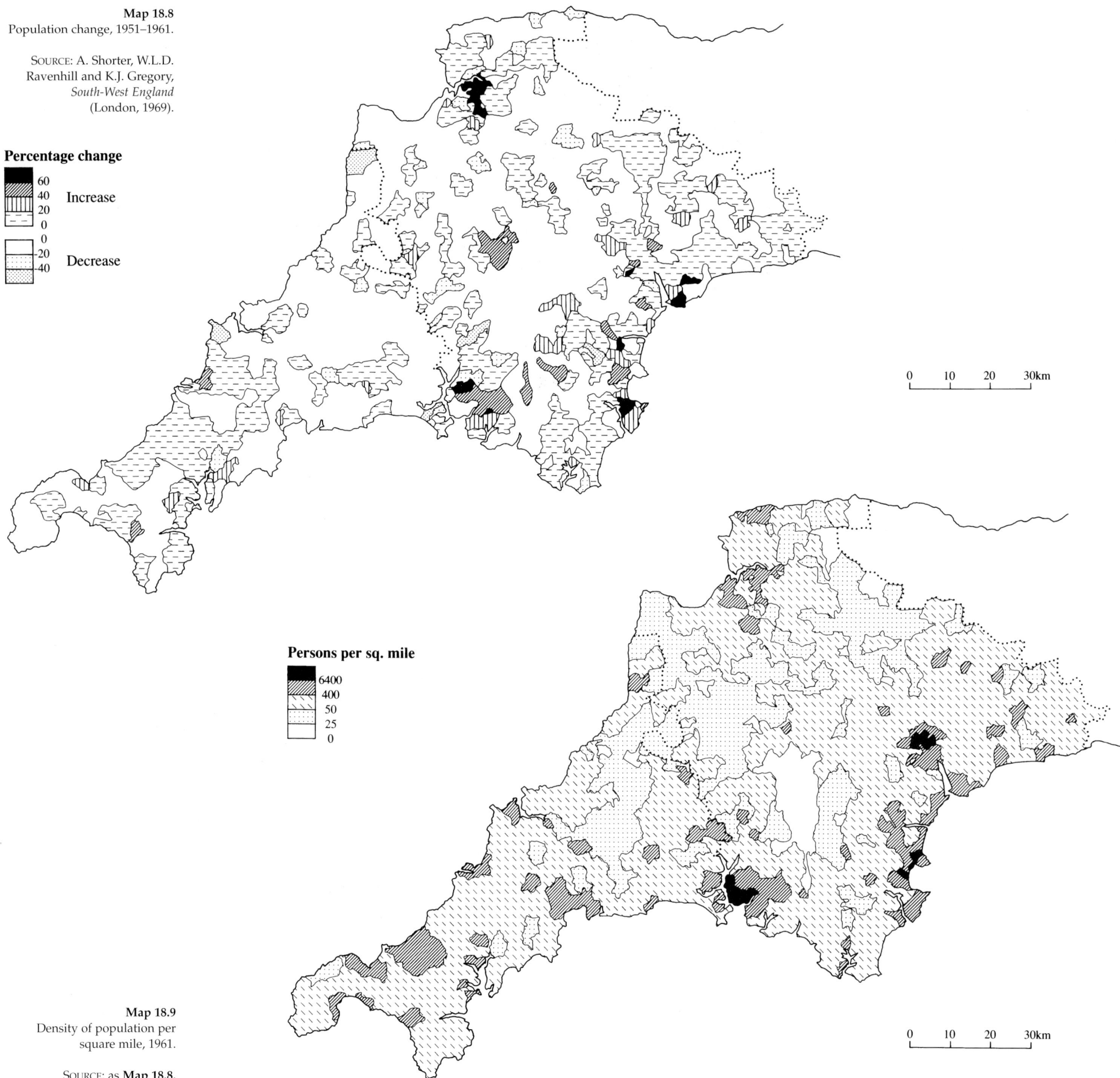

60
40   Increase
20
0
0
-20   Decrease
-40

Persons per sq. mile

6400
400
50
25
0

Map 18.9
Density of population per
square mile, 1961.

SOURCE: as **Map 18.8**.

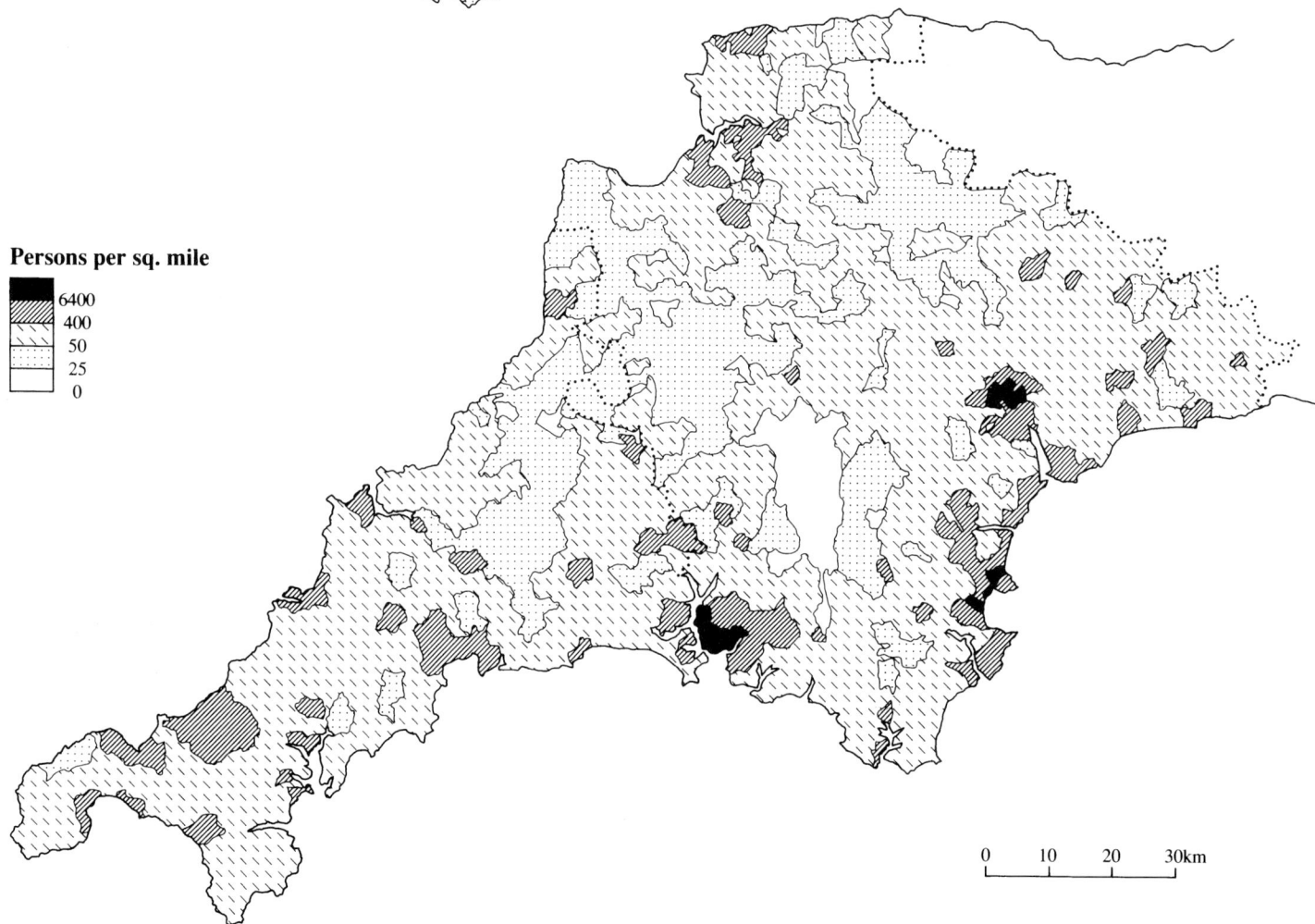

0   10   20   30km

0   10   20   30km

Percentage change

	8
	5
	2
	0
	-2
	-5
	-8

Map 18.10
Net migrational change,
1951–1961 (Registrar General
civilian mid-year estimates).

SOURCE: Department of the
Environment, *South West
Regional Atlas* (Bristol, 1971).

Percentage change

	4.5
	3.0
	1.5
	0
	-1.5
	-3.0
	-4.5

0   10   20   30km

Map 18.11
Natural change, 1951–1961
(Registrar General civilian
mid-year estimates).

SOURCE: as **Map 18.10.**

131

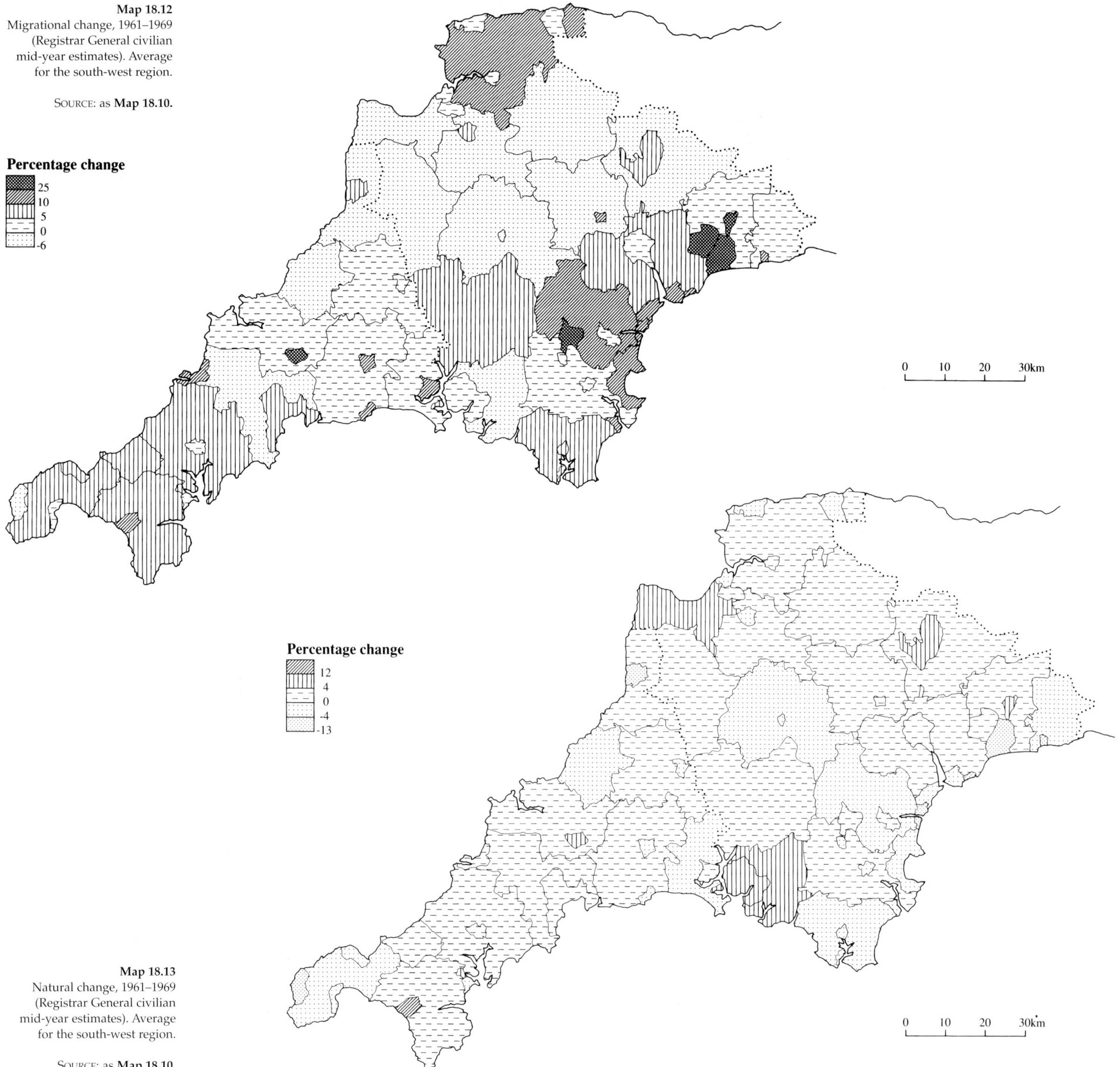

Map 18.12
Migrational change, 1961–1969
(Registrar General civilian
mid-year estimates). Average
for the south-west region.

SOURCE: as **Map 18.10.**

Percentage change

25
10
5
0
-6

Percentage change

12
4
0
-4
-13

Map 18.13
Natural change, 1961–1969
(Registrar General civilian
mid-year estimates). Average
for the south-west region.

SOURCE: as **Map 18.10.**

(A) 1961-71

(B) 1971-81

0   30km

Percentage change

20.0

10.0

0

Urban areas

(A) Population 1901-81

(C) Economic Planning Areas

Plymouth
Exeter/
E Devon
SE Devon
(Torbay)
N Devon
W Devon

(B) Development Control Areas

South

East

North

West

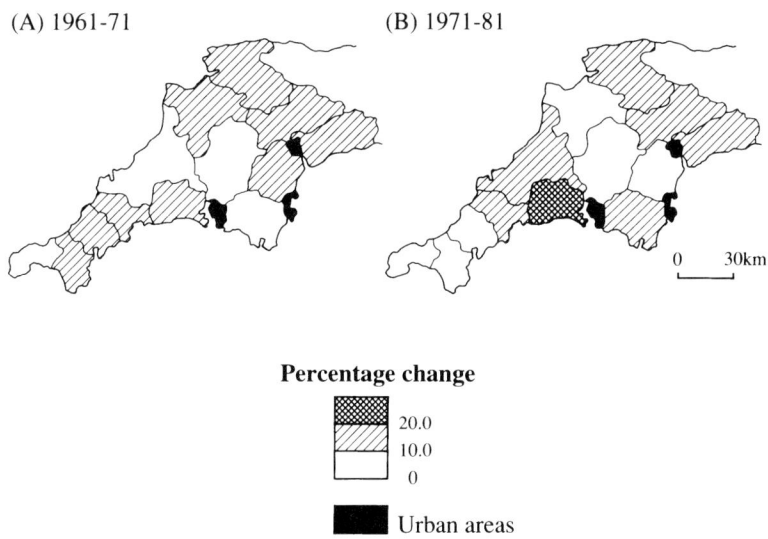

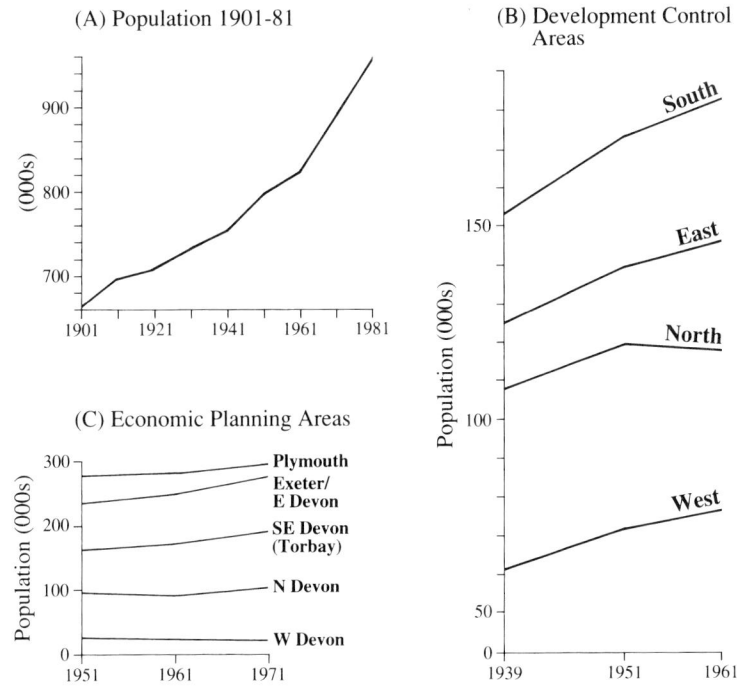

Map 18.14 (left)
Population change, 1961–1981, by 1974 local authority areas.

SOURCES: P.J. Cloke, 'An index of rurality for England and Wales', *Regional Studies*, 11 (1977), 31–46; P.J. Cloke and G. Edwards, *Rurality in England and Wales 1981: A Replication of the 1971 Index and a Contemporary Index of Rurality for England and Wales 1981* (Lampeter, 1985).

Fig. 18.1 (right)
Population change in Devon, 1901–1961.

SOURCES: (**A**) Devon County Council, *Devon in 2001: A Review of County Planning Policies* (Exeter, 1988); (**B**) Devon County Council, *Development Plan (First Review) Analysis of Survey* (Exeter, 1964); (**C**) Devon County Council, *Structure Plan* (Exeter, 1979).

THE TWENTIETH CENTURY SUMMARIZED

Overall, the population of Devon increased substantially during the century, as shown in **Fig. 18.1A**, but as **Figs 18.1B and 18.1C** indicate, the rates of growth have varied over the county since 1939, with the north and the west remaining static for most of the 1951–71 period. The most significant feature of the period since 1961 for Devon has been a demographic transition to one of natural losses being offset by in-migration gains, as shown in **Fig. 18.2**. The crucial date was 1969 since when births have remained well below deaths. In contrast, in-migration has shown a steady rise throughout the period, with an accelerated increase in the 1971–6 period, before a steady momentum was resumed. The period 1981–7 for Devon as a whole, however, witnessed an increase in the rate of growth compared with the 1970s from 5.9 per cent per annum to 7.4 per cent per annum. Only Exeter experienced a loss. Finally, as **Fig. 18.3** reveals, all these changes have produced an ageing population in Devon with fewer children but a lot more retired people. To some extent this is a national phenomenon, but in some parts of Devon around 50 per cent the population are retired, notably in the south coast resorts of Sidmouth and Seaton.

For the South West as a whole, a number of broad trends can be identified. First, the cities and large towns experienced growth; second, in the earlier half of the century the eastern regions experienced most growth, but since the 1970s the western regions have been growing fastest; third, cutting across both east and west, the spinal axis of both main line rail and the M5/A38 road has been a corridor of constant population attraction, notably to

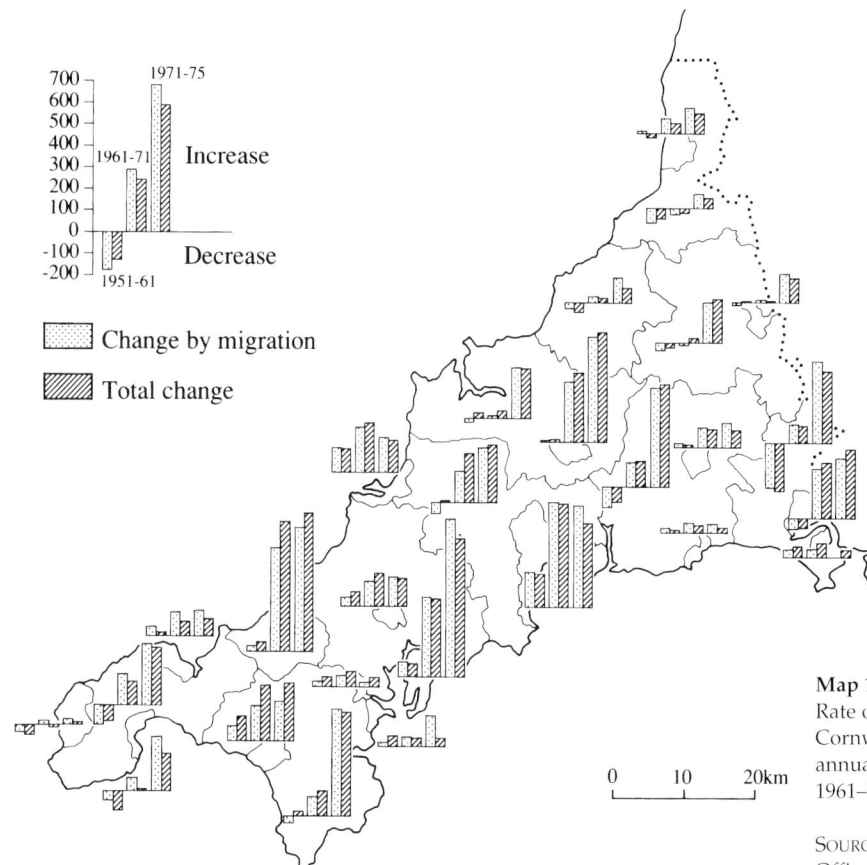

Increase

Decrease

Change by migration

Total change

Map 18.15
Rate of population change in Cornwall, 1951–1975: average annual change during 1951–61, 1961–71, 1971–5.

SOURCES: 1951–71 based on Office of Population Censuses and Surveys mid-year civilian population estimates; 1971–75 based on electoral population estimates and OPCS statistics.

Map 18.16
Rurality, 1961–1981.

SOURCES: as Map 18.14.

(A) 1961

(B) 1971

(C) 1971 using 1974 boundaries

(D) 1981 using 1971 data as starting point

(E) 1981 using 1981 data as starting point

Extreme rural

Intermediate rural

Intermediate non-rural

Extreme non-rural

Urban areas (excluded from the analysis)

0   30km

the south of the spine; fourth, and in contrast, some parishes in the remoter rural regions have experienced constant depopulation, though the overall trend of rural depopulation in south-west England was reversed in the 1970s and 1980s.

A CONTINUING RURAL REGION?

Is the South West still a rural region?[4] One of the most useful of the methods developed to determine the rurality of an area is that devised by Cloke in 1977 and subsequently refined by Cloke and Edwards.[5] **Map 18.16** reveals that in 1961 most of the region was classified as 'extreme rural', with the major exception of the Exeter–Plymouth axis and even most of this was 'intermediate or rural'. By 1971 little had changed in fact, but a number of areas in the 'extreme' category were becoming less rural, while in contrast the 'intermediate non-rural' area around Exeter was becoming more rural. By 1981 (**Map 18.16D**), the 'extreme rural' area had diminished to only about half of the region. When the classification is recalculated using 1981 data as the starting point as in **Map 18.16E**, then outside the three major cities only three districts fail to be classified as 'extreme rural'. These maps suggest, therefore, that the population of the South West remains essentially rural, and, therefore, that the twentieth century has

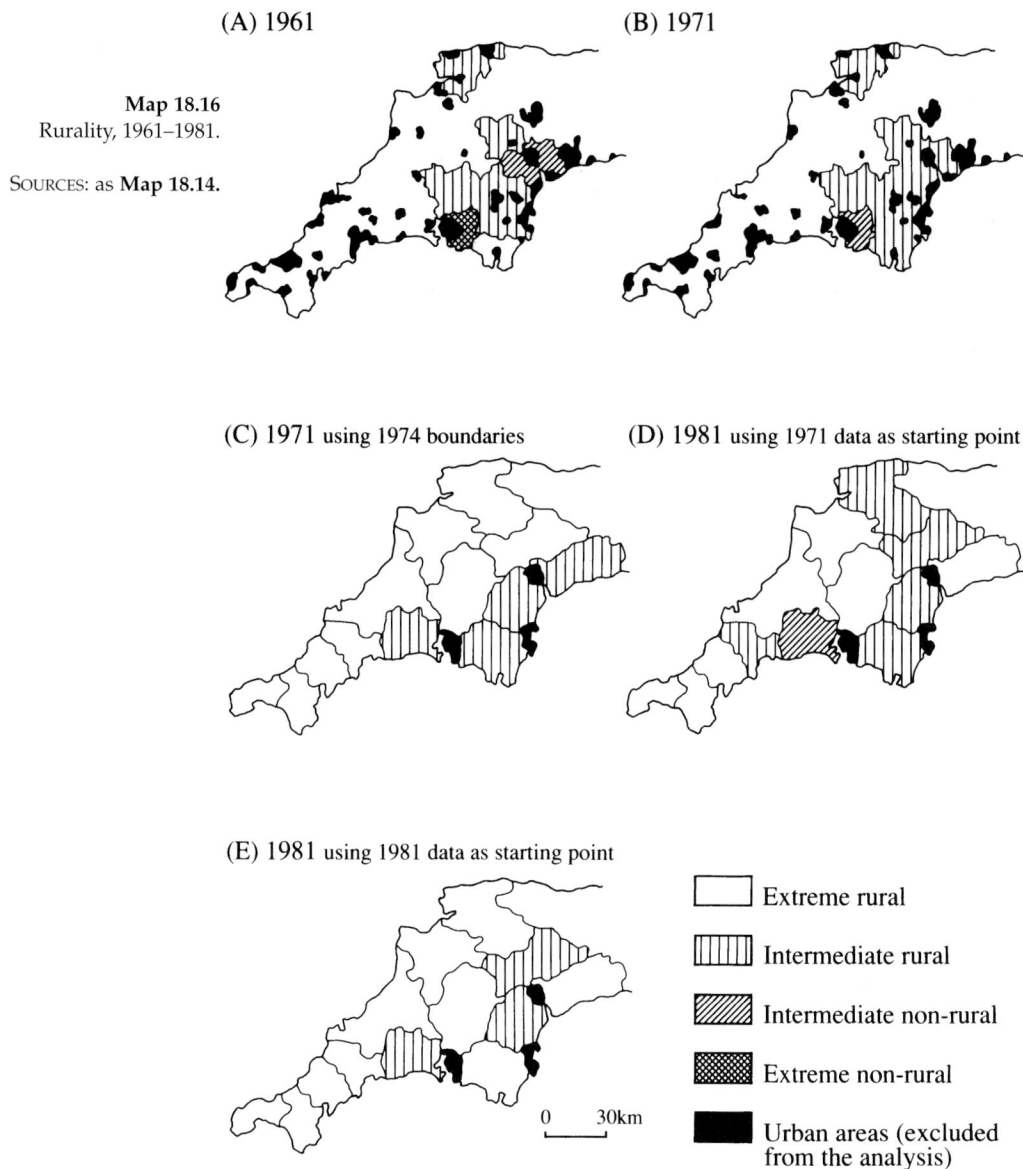

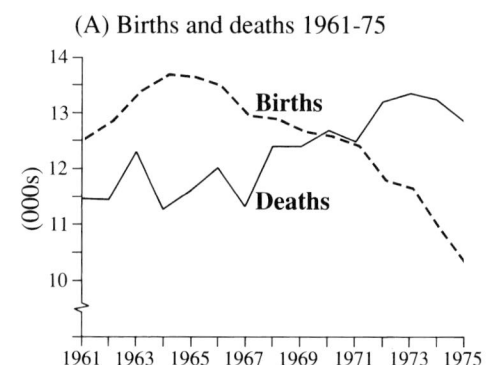

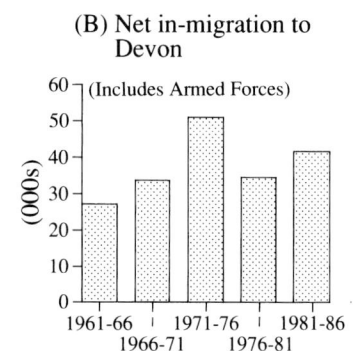

Fig. 18.2 (right)
Population change in Devon, 1961–1986.

SOURCES: the author and Devon County Council, *Structure Plan: Report of Survey* (Exeter, 1977).

Fig. 18.3 (left)
Age structure of Devon population, 1951 and 1981.

SOURCE: Devon County Council, *Devon in 2001: A Review of County Planning Policies* (Exeter, 1988).

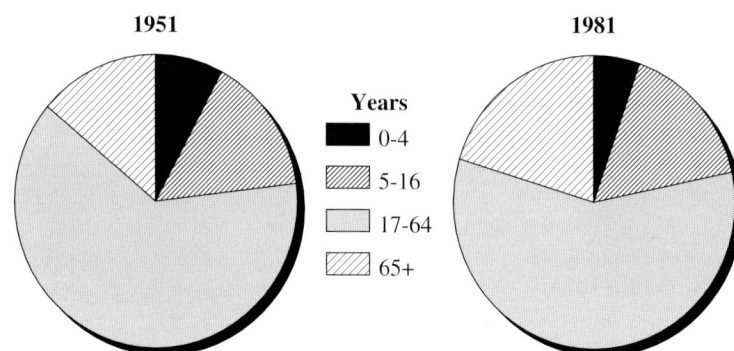

1951

1981

Years
0-4
5-16
17-64
65+

(A) Births and deaths 1961-75

(B) Net in-migration to Devon

only partially modified a centuries-long pattern of population in the region. The most recent data (**Map 18.17**) suggests that rural regions have experienced the fastest growth despite planning policies aimed at producing a contrary effect.

FURTHER READING

For further information on census data see D. Rhind, *A Census User's Handbook* (London, 1983) and B. Benjamin, *The Population Census* (London, 1969). An earlier atlas analysis of the socio-economic characteristics of south-west England and which contains demographic maps is Department of the Environment, *South West Regional Atlas* (Bristol, 1971).

Map 18.17
Population change, 1981–1991.

SOURCE: *1991 Census Newsletter.*

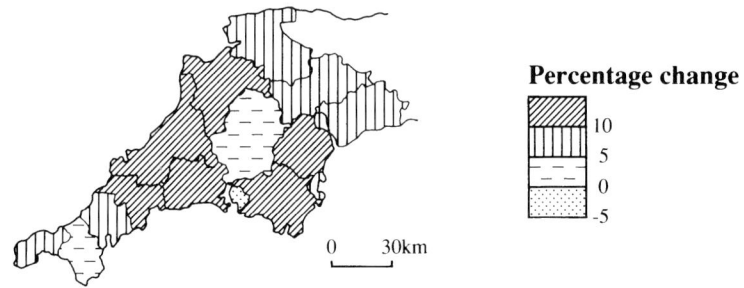

Percentage change

10
5
0
-5

0   30km

POLITICAL AND MILITARY HISTORY

CASTLES, FORTIFIED HOUSES AND FORTIFIED TOWNS IN THE MIDDLE AGES

ROBERT A. HIGHAM

Political thought, indeed much political action, leaves no physical trace amenable to mapping or any other sort of visual presentation. Showing on a map the location of places where important events occurred does not in itself illuminate political history except in a most rudimentary way. But in one important respect the political life of south-western society has left an unmistakable impression on the landscape, and one that can be mapped. In order to cope with threats both external and internal, and to display political power and social influence, those at the top of the medieval social ladder built fortifications.

Despite the controversy surrounding castle origins in Europe generally, there is no doubt that the Norman Conquest of England brought with it a rapid growth in castle-building. In the South West, castles appeared with the arrival of the Normans in the region early in 1068. Medieval castles were extremely adaptable. They could be built in a wide variety of forms, small or large, in timber or stone (or a mixture of both). They could be put to a similarly wide variety of purposes, in military campaign, in the consolidation of conquest, in the establishment of patterns of landownership, and in administration. Except where their purpose was purely military, and short-lived, they were also homes for their owners, or, in the case of kings, for their representatives. This residential function ensured that some castles stayed in use for centuries. This variety of purpose, as well as the long period over which castles continued to be built, resulted in a correspondingly wide array of castle designs. A brief indication of the form of each site shown on **Maps 19.1 and 19.2** is given in **Tables 19.1 and 19.2**, although the information presented there glosses over many difficulties.

Most English castles, unlike parish churches or vernacular architecture, do not reveal regional characteristics of design. They were built for an upper social class whose ideas and influences often transcended local limitations. Nevertheless, the South West is well known for its very large mottes, or castle mounds, such as those at Okehampton, Totnes, Launceston and Trematon. Circular buildings, as at Totnes, Barnstaple, Plympton, Launceston, Trematon and Restormel, were also popular, and this feature was shared with South Wales. Cornish castle-builders, for some reason, chose ringwork (enclosure) forms more frequently than did Devonian ones.[1]

Maps of castles share a weakness of all archaeological distribution maps. The quantity of contemporary activity can easily be exaggerated, since the precise date-range of occupation of individual sites is not always known. Even the dates of foundation for many of the sites shown on **Map 19.1** are obscure. Large numbers of the earlier castles, particularly those built of timber and which are now represented only by earthworks, are wholly undocumented.[2]

Map 19.1 shows castles established between the Norman Conquest and the thirteenth century. Some had gone out of use by c.1300, some indeed during the twelfth century. Others among these early foundations, such as Exeter, and Launceston, continued in use long after this. In contrast, although the licences to crenellate, which figure notably on **Map 19.2**, were predominantly a late medieval phenomenon, their origins stretch back, at Hartland and Axminster, into the early thirteenth century. There is, therefore, no meaningful division between the contents of **Maps 19.1 and 19.2**. The material has been divided in this way simply to provide a convenient visual distinction between origins in the earlier and later Middle Ages.

Listing of south-western castles was put on a firm footing with the Victoria County History's catalogue of earthworks, and there have since been other, more extensive surveys.[3] It must be stressed that no list of castles can be comprehensive. They are features that can disappear more or less without trace. A mutilated earthwork survives at Bratton Fleming in Devon; at Axminster the site has disappeared completely.[4] In Cornwall, the site of Tregony has all but disappeared, and that at Truro, completely wiped away in the nineteenth century, has only been glimpsed in excavation.[5] Later medieval fortified houses have been particularly vulnerable, and many places shown on **Map 19.2** no longer exist. On the other hand, new discoveries can (and do) still occur. The Royal Commission on the Historical Monuments of England recorded a newly observed motte at Poundstock in Cornwall as recently as 1993.

Smaller earthwork sites are not always easily distinguishable from sites of other periods, especially where the characteristically medieval motte is absent. Modestly-sized enclosures have been built from prehistoric times onwards. One such site, near Dunkeswell (Devon), excavated on the assumption that it was of Iron Age date, was actually occupied in the thirteenth century, though its

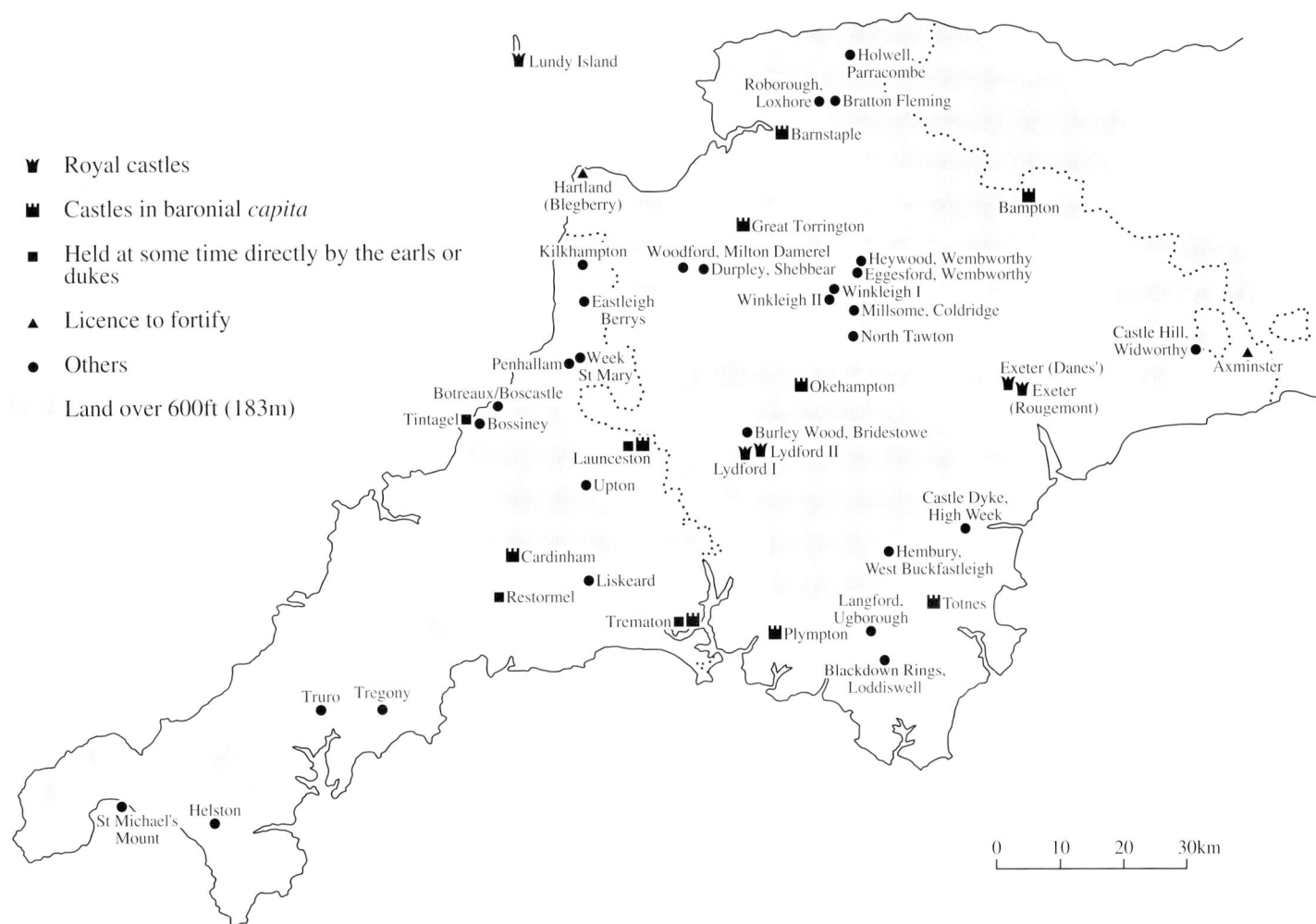

Map 19.1
Castles to c.1300.

SOURCES: this map is compiled from the most recent published listings, which are themselves based upon earlier listings; original fieldwork; documentary and map evidence; and the archives of the Sites and Monuments Registers for Devon and Cornwall. For Cornwall, A. Preston-Jones and P. Rose, 'Medieval Cornwall', *Cornish Archaeology*, 25 (1986); for Devon, R.A. Higham, 'Devon castles: an annotated list and bibliography', *Proceedings of the Devon Archaeological Society*, 46 (1988); for baronial estates of the thirteenth century, see I.J. Sanders, *English Baronies* (Oxford, 1960).

defences were too slight for it to be regarded as a castle.[6] In Cornwall, the 'rounds' of the late prehistoric and early historic periods may include examples which are in fact medieval ringwork castles.[7] Indeed, the problematic definition of a 'castle' is itself a barrier to the compilation of comprehensive lists. All are agreed that a reasonable level of defence is a necessary quality of a castle. But what is 'reasonable'? The medieval attitude to this issue varied, so it is not surprising that modern opinions waver in that difficult borderland between castle and house. **Map 19.2** is entitled 'Fortified houses', a useful phrase in that many of the sites were less heavily defensible than their predecessors. But it also obscures the issue, because all but the purely military sites of earlier times were themselves 'fortified houses'. Many late medieval houses with minor features of defence could no doubt have been listed here. Neither have the 'moated sites' of the region, upon which research has only recently begun, been included.[8]

Castles, unlike some more modern types of fortification, were not established in a strategic, pre-determined pattern. Although the maps reveal that obvious locations, such as major towns or estuaries, tended to have a castle, the eventual distribution of sites was not the result of a 'master plan'. The medieval attitude to the

siting of fortifications was rarely governed by general considerations, but rather by local ones, which might in any particular instance be military, or tenurial, or domestic or economic. The works at Dartmouth in the 1480s (see chapter 22) were eventually incorporated into a system of coastal defence which had a more unified 'national' character. But they had begun their development in the medieval context of local needs.

Castles were not generally isolated phenomena sited in strong positions by experienced generals intent on winning wars. Even when this motive lay behind the foundation of castles, and in the majority of cases it did not, their establishment rarely took place in an empty environment. A more rounded view of the subject sees castles, whether occupied for long or short periods, as elements in complex rural and urban landscapes. Sometimes they were destructive elements, as when houses were demolished to make room for them. But often they had a constructive influence, creating a new centre of activity in an existing town, giving rise to a wholly new town, enhancing the social and economic life of a rural manor, or stimulating the creation of new features such as hunting parks.

The castles of Devon and Cornwall are to be found in all sorts of topographical situations.[9] They were mostly accessible, and

Map 19.2
Fortified houses from *c*.1300.

SOURCES: as **Map 19.1**.

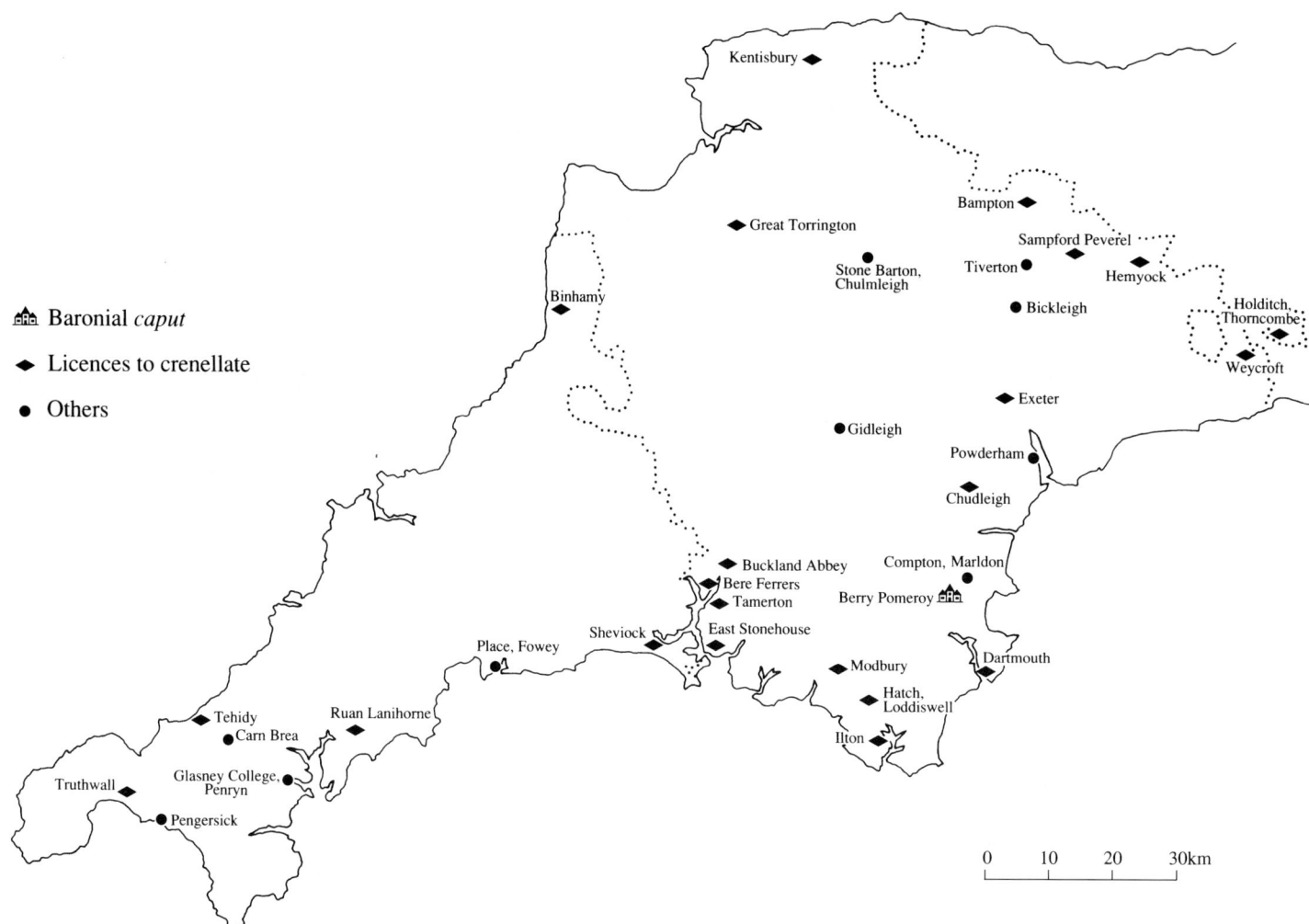

Kentisbury

Bampton

Great Torrington

Sampford Peverel

Stone Barton, Chulmleigh
Tiverton
Hemyock

Binhamy

Bickleigh

Holditch, Thorncombe

Weycroft

Exeter

Gidleigh

Powderham

Chudleigh

⌂⌂ Baronial *caput*

◆ Licences to crenellate

● Others

Buckland Abbey
Compton, Marldon
Bere Ferrers
Tamerton
Berry Pomeroy ⌂⌂

Sheviock
East Stonehouse

Place, Fowey

Modbury
Dartmouth

Hatch, Loddiswell

Tehidy
Carn Brea
Ruan Lanihorne

Ilton

Truthwall
Glasney College, Penryn

Pengersick

0   10   20   30km

well-drained spur-sites were much favoured for building. Major valleys were important for communications and good agricultural land. Estuarine locations were important for security and for coastal communications. Major road routes within the peninsula are also reflected in the pattern of castles, notably westwards from Exeter, for example, where roads divide north and south of Dartmoor to reach Cornwall.

Devon's four late Saxon towns received castles which affected the topography of these places permanently (for towns generally, see chapters 51 and 52). Elsewhere, the establishment of castles attracted urban growth, for example at Okehampton, Plympton and Great Torrington in Devon, or Launceston and Tregony in Cornwall. By the end of the fourteenth century both counties had large numbers of small market centres, many of which had borough status, and some of these, such as Bampton in Devon and Week St Mary in Cornwall, had also grown at places with castles. Up to *c*.1300, there was a high correlation between castles and towns of various sorts, both being features of an expanding society. Some castle/town associations, such Plympton and Launceston, were physically integrated settlements, but elsewhere borough and castle were separate, each taking advantage of a location suitable to their needs. Restormel castle in

Cornwall, for example, was separate from its associated borough of Lostwithiel, and at Okehampton in Devon, castle and borough stood apart.

In the countryside a similar variety of castle locations is found. There are a few, fairly remote sites. The motte and bailey at Burley Wood (Bridestowe, Devon) overlooks the route westwards to Cornwall, and the structure built in the corner of an Iron Age hillfort at Blackdown Rings (Loddiswell, Devon) may have been a campaign castle of the Norman Conquest. But many sites occupy fairly isolated locations without actually being inaccessible. They occupy more solitary locations simply because true rural nucleations were relatively rare in the two counties, whose settlement pattern was a dispersed one of hamlets and farms. Some apparent 'villages' with castles, for example Winkleigh in Devon, are in fact among the large numbers of very small south-western boroughs of the later Middle Ages. Where castles are truly rural and isolated, it may often be that they are early and did not survive into the period of borough and market expansion.

Except where castles had a short-lived military function, or were the centres of only small estates, they were normally the focal point of a much wider tenurial network. An example is the barony (to use the thirteenth-century terminology) of

A: Devon

** Royal castles
* Castles in thirteenth-century baronial *capita*
# Licence to fortify
+ Others

1. ** Exeter (Rougemont). Enclosure in angle of city walls
2. ** Exeter (Danes'). Siegework (ringwork) outside city.
3. ** Lydford I. Enclosure in angle of *burh* defences.
4. ** Lydford II. Keep (with added motte) and bailey.
5. * Okehampton. Motte and bailey.
6. * Totnes. Motte and bailey.
7. * Barnstaple. Motte and (destroyed) bailey.
8. * Plympton. Motte and bailey.
9. * Bampton. Motte and (destroyed) bailey.
10. * Great Torrington. Destroyed.
11. + Holwell, Parracombe. Motte and bailey.
12. + Roborough, Loxhore. Motte.
13. + Bratton Fleming. Fragmentary motte.
14. + Durpley, Shebbear. Motte and bailey.
15. + Woodford, Milton Damarel. (?) and bailey.
16. + Heywood, Wembworthy. Motte and bailey.
17. + Eggesford, Wembworthy. Ringwork and bailey
18. + Winkleigh I. Motte and bailey.
19. + Winkleigh II. Ringwork (partly destroyed).
20. + Millsome, Coldridge. Ringwork.
21. + North Tawton. Motte (?).
22. + Castle Hill, Widworthy. Motte (?).
23. + Burley Wood, Bridestowe. Motte and baileys.
24. + Castle Dyke, High Week. Motte and bailey.
25. + Hembury, West Buckfastleigh. Motte (?) and bailey.
26. + Blackdown Rings, Loddiswell. Motte (?) and bailey.
27. + Langford, Ugborough. Partly destroyed/unfinished motte?
28. # Axminster. Destroyed.
29. # Hartland (Blegberry). Destroyed.
30. ** Lundy Island. Keep and bailey.

B: Cornwall

† Held at some time directly by the earls or dukes
* Castles in thirteenth-century baronial *capita*
+ Others

31. †* Launceston. Motte and bailey.
32. †* Trematon. Motte and bailey.
33. * Cardinham. Motte and bailey.
34. + Week St Mary. Ringwork and bailey.
35. + Penhallam. Ringwork.
36. † Restormel. Ringwork and bailey.
37. + Tregony. Largely destroyed motte and bailey.
38. + Truro. Destroyed/ringwork (?).
39. + Botreaux/Boscastle. Largely destroyed/ringwork (?) and bailey.
40. + Kilkhampton. Motte and baileys.
41. + Eastleigh Berrys. Unfinished (?) motte and baileys.
42. + Bossiney. Ringwork/and bailey.
43. + Upton. Motte/ringwork (?).
44. + St. Michael's Mount. Destroyed.
45. † Tintagel. Three walled wards.
46. + Helston. Destroyed.
47. + Liskeard. Destroyed.

It is unfortunately impossible to portray in detail the political and tenurial context of all these castles by use of a series of symbols within a single map. Too little is known of the history of some sites, and the circumstances surrounding others changed through time. The *capita* of baronies can only be illustrated according to thirteenth-century and later definitions, though other castles may have fulfilled similar functions at an earlier date. Exeter was a royal foundation, but it became part of the Duchy of Cornwall in 1348 and also acted as the *caput* of the barony of Bradninch. Lydford, also a royal foundation, was part of the earldom of Cornwall from 1239 and part of the Duchy from its creation in 1337. In Cornwall, various castles were at different times held directly by an earl or duke, or by their tenants. This shifting pattern has not been depicted, and the symbols used on **Maps 19.1 and 19.2** cannot adequately illustrate the historical realities. Simple descriptions/classifications of many sites are also very difficult to convey in few words.

Table 19.1
Castles to *c.*1300.

A: Devon

1. + Tiverton. Walled enclosure and domestic buildings.
2. + Gidleigh. Tower house.
3. + Stone Barton, Chulmleigh. Destroyed.
4. + Bickleigh. Walled enclosure and domestic buildings.
5. + Powderham. House with towers and other works.
6. * Berry Pomeroy. Walled enclosure and domestic buildings.
7. + Compton, Marldon. Walled enclosure and domestic buildings.
8. # Exeter, Bishop's palace and Cathedral close (1290, 1322).
9. # Modbury (1334). Destroyed.
10. # Tamerton (1335). Destroyed.
11. # Ilton (1335). Destroyed.
12. # Bampton (1336). Destroyed.
13. # Buckland Abbey (1337). Precinct wall with slits may be related.
14. # Bere Ferrers (1337, 1340). Part of house/enclosure.
15. # Sampford Peverel (1337, 1339). Destroyed.
16. # Great Torrington (1328, 1340, 1347). Destroyed.
17. # Chudleigh, bishop's house (1379). Part of house/enclosure.
18. # Hemyock (1380). Walled enclosure with house.
19. # Holditch, Thorncombe (1397, now Dorset). House with tower.
20. # Dartmouth (? Gomerock, 1402). Partly destroyed.
21. # Weycroft (1427). Hall and tower porch in later building.
22. # Kentisbury (1457). Destroyed.
23. # Hatch, Loddiswell (1462). Destroyed.
24. # East Stonehouse (1515). Destroyed.

B: Cornwall

25. + Glasney College, Penryn. Defended enclosure. Destroyed.
26. + Place, Fowey. House with towers, within a nineteenth-century rebuilding.
27. + Pengersick. Tower house with (destroyed) hall.
28. + Carn Brea. Stone tower with later additions.
29. # Tehidy (1330). Destroyed
30. # Binhamy (1335). Destroyed. Moated enclosure survives.
31. # Ruan Lanihorne (1335). Destroyed.
32. # Truthwall (1335). Destroyed.
33. # Sheviock (1336). Destroyed.

* Baronial *caput*
# Licences to crenellate (dates in brackets)
+ Others

Table 19.2
Fortified houses from *c.*1300.

as Baldwin of Okehampton and Judhel of Totnes, there were other tenants-in-chief on whose lands castles are found: Parracombe (William of Falaise), Bampton (Walter of Douai), and Great Torrington (Odo fitzGamelin). But since such places are referred to for the first time only in the next century (if at all), their origin cannot be dated closely. In Devon, some sites are found in lands held by sub-tenants of the tenants-in-chief, such as Wembworthy and Loxhore, but these are never referred to and may be of later origin.

In Cornwall, by contrast, power was concentrated in the extensive landholding of Robert of Mortain, who was earl of this territory in all but name and had his own castle at Launceston. In addition to the Domesday castle at Trematon, held by his tenant Reginald of Vautort, other castles on his tenants' lands may have had an early origin: Cardinham, Week St Mary, Penhallam (Richard fitzTurold), and Restormel (Turstin the sheriff). Again, in the absence of documentation, distinguishing eleventh- from twelfth-century origins is very difficult.

THE TWELFTH CENTURY

The changing pattern of land tenure continued to encourage the building of castles. Plympton was established in the early 1100s by Richard de Redvers, a newcomer whom Henry I rewarded with many estates, and whose son, Baldwin, became first earl of Devon. In this century a problem arises from the 'adulterine' (i.e. lacking the king's permission) castle-building of the civil war in Stephen's reign (1135–54). This certainly took place in the South West, but distinguishing such sites from legitimately defended homes is not always easy. Bampton, Exeter, Plympton, Barnstaple and Great Torrington figure by name in narratives of the war, but they were presumably not the only sites occupied. The history of Devon landownership suggests some castle origins in this reign, for example at Winkleigh, but there are other sites, for example Shebbear and High Week, whose origins can be placed only somewhere in the late eleventh or twelfth century. In Cornwall, Kilkhampton and Truro were perhaps adulterine sites, but other

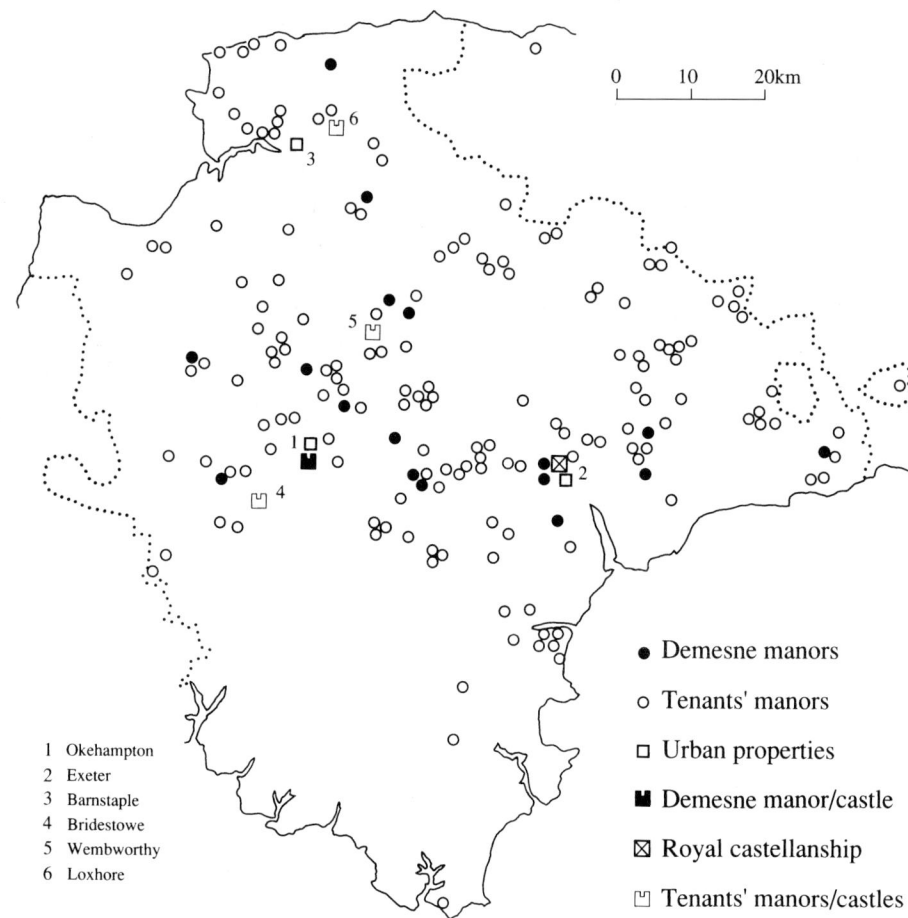

- Demesne manors
- Tenants' manors
- Urban properties
- Demesne manor/castle
- Royal castellanship
- Tenants' manors/castles

1 Okehampton
2 Exeter
3 Barnstaple
4 Bridestowe
5 Wembworthy
6 Loxhore

Map 19.3
The barony of Okehampton in the eleventh century.

SOURCE: R.A. Higham, 'Excavations at Okehampton Castle, Devon. Part I: the motte and keep', *Proceedings of the Devon Archaeological Society*, 35 (1977).

Okehampton, which at the time of the Domesday Survey comprised the new castle and developing borough at Okehampton itself, nearly 200 rural manors, mainly in the hands of tenants and including some lesser castles, and urban properties in Barnstaple and Exeter, where the lords of Okehampton were also castellans of the royal castle (**Maps 19.3 and 19.4**).[10]

THE NORMAN CONQUEST AND ITS AFTERMATH

Following the surrender of Exeter and the foundation of its castle early in 1068, William the Conqueror campaigned in Devon and Cornwall.[11] It is likely that other castles were also established at this time. Domesday Book (1086) mentions Okehampton, Launceston and Trematon by name, and there is evidence that Lydford, Totnes and perhaps Barnstaple belong to this period. How far castle-building had spread beyond the king and his immediate circle is not known, and circumstances were complicated by the different tenurial conditions which developed in the two counties. In Devon, in addition to powerful men such

Fig. 19.1
Drawing by Peter Orlando Hutchinson of the gateway of Rougemont Castle, 1893.

SOURCE: reproduced by permission of Devon Record Office (Z19/2/8A/2B).

castles were built by twelfth-century tenants of the earldom (which had been formally created in the civil war) with more residential interests: at Tregony (the Pomeroys), at Botreaux (the Botreaux), at Upton (the Uptons), and at Bossiney (the Mandevilles). In the civil war itself, the South West was largely against King Stephen. His only significant supporter was Henry Tracy, whose lands centred on Barnstaple castle.

THE THIRTEENTH CENTURY

In Devon, a castle may have been founded in 1200 at Axminster as a result of a royal licence granted to William Brewer, a royal servant of rising fortune. In 1201, Alan, lord of Hartland, received permission to fortify his north Devon property against raids of pirates from Lundy Island, where a royal castle was later established in 1242. Also now regarded as a thirteenth-century foundation is the castle at Tintagel. While this may also have had a role in protecting the north Cornish coast from similar problems, it has recently been suggested that, despite its defensible site, it was a non-military establishment, developed by Richard, earl of Cornwall (who deliberately acquired the site in 1233), as a prestigious symbol of the traditions of Cornish kingship which he inherited. The major works at Launceston castle were also part of this powerful man's building enterprises, as was the rebuilding of Lydford as a stannary court and prison. With the increasing definition of status and tenure which followed Magna Carta, several baronial estates emerged in Devon with existing castles at their centres (*capita*). There were fewer such estates in Cornwall, whose earldom was more all-embracing within the county, restricting the growth of a powerful gentry.

THE FOURTEENTH AND FIFTEENTH CENTURIES

It is too easily assumed that castles were less important in this period, the more lightly defended houses being significant in purely social terms. In fact, the situation was more complex. First, some earlier sites underwent major rebuildings, their political importance being enhanced rather than reduced. Okehampton, redesigned *c.*1300 by Hugh, the first Courtenay earl of Devon, is a good example, as is Restormel, rebuilt at the very end of the thirteenth century by Edmund, earl of Cornwall. Restormel stood outside Lostwithiel, which Edmund made the focus of the earldom: the so-called 'Duchy Palace' was also his creation. This was later adopted by the dukes as the administrative centre of the Duchy's lands in the West Country, but despite its name it was not their residence. Although the earls, and later the dukes, were rarely resident within Cornwall, they continued sporadic maintenance of their major castles: Restormel, Launceston, Trematon and Tintagel. Second, although the threats which castles might face were no longer the massive sieges of earlier centuries, the violence they could experience was real enough. In the Wars of the Roses Devon

and Cornwall became the most unstable areas outside the north of England. In 1455 Powderham castle withstood a siege of almost two months. Those who owned castles in coastal areas were also no doubt mindful of the possibility of French attack, which materialized at Plymouth and Dartmouth in the early 1400s. Third, and ensuring that the late medieval map of castles remained well filled, was the undoubted importance of the 'castle tradition'. The defended home had become so entrenched in society that landowners, great and small, sought to maintain this physical emblem of noble lifestyle. A reflection of this is the quantity of licences to crenellate (written grants of royal permission to fortify a house), which occurred especially in the fourteenth century. Devon experienced the general English popularity of these licences. Cornwall shared with Devon a fashion for licences in the 1330s, but thereafter the creation of the Duchy (1337) established conditions in which they were no longer appropriate. Such grants are not, in any case, a general guide to events, and important residences of established families, such as Berry Pomeroy and Powderham in Devon, could be built without them: in the majority of cases, those families who received (indeed sought) licences to crenellate were building castles for the first time and the licences enhanced their status. Tudor houses were built increasingly without any defence, though the castle tradition disappeared slowly. At Compton in Devon, the castellated facade was added only around 1500 and at Pengersick (near Helston) in Cornwall the early-sixteenth-century tower with gunloops was attached to a domestic hall. Nor was the tradition purely secular: the bishops of Exeter had strong residences including their Exeter palace, and the communities at Buckland (Devon) and Glasney (Penryn, Cornwall) also defended themselves.[12] Some Devonian and Cornish examples of ecclesiastical defence have been included to show that the theme of security extends beyond the 'castle' as traditionally construed. It is clear from the remains of Tavistock and Torre abbeys, for example, that strong gatehouses provided security for rich communities. In Plymouth, the strong enclosure wall of the Carmelite Friary, illustrated on a sixteenth-century map, has recently been identified in excavation. Remains of another enclosed episcopal palace survive at Paignton.

FORTIFIED TOWNS (**Map 19.5**)

The history of defended towns began in the Roman period, at Exeter, and continued in late Saxon times with the development of *burhs*. These were strongly maintained: in 997 Lydford resisted the Vikings and in 1001 Exeter also withstood them. Although Exeter fell to the Danes in 1003 (perhaps through treachery) it kept the

Fig. 19.2
Drawing by Peter Orlando Hutchinson of Lydford Castle, 2 August 1880.

SOURCE: reproduced by permission of Devon Record Office (Z19/2/8F/21).

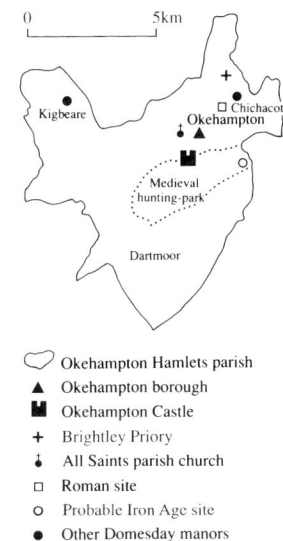

Okehampton Hamlets parish
▲ Okehampton borough
■ Okehampton Castle
+ Brightley Priory
⌁ All Saints parish church
□ Roman site
○ Probable Iron Age site
● Other Domesday manors

Map 19.4
Okehampton: castle and landscape.

SOURCE: R.A. Higham, *Okehampton Castle* (London, 1984; revised 1988).

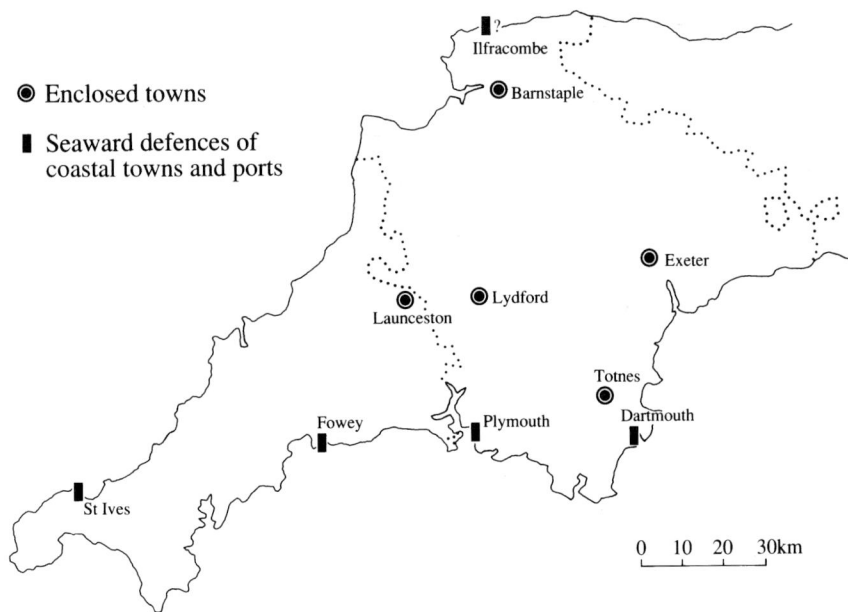

● Enclosed towns

▮ Seaward defences of
coastal towns and ports

Map 19.5
Fortified towns.

Enclosed towns: Exeter (Roman onwards), considerable remains of walls; Lydford (Saxon onwards), north-eastern earthworks visible; Barnstaple (Saxon onwards), destroyed, reflected in street plan; Totnes (Saxon onwards), destroyed, reflected in street plan, later gate; Launceston (twelfth century onwards; perhaps with late Saxon predecessor at St Stephen's by Launceston), south gate survives, course of wall known.

Seaward defences of coastal towns and ports: Dartmouth (late fourteenth century), parts of wall and tower survive; Plymouth (late fourteenth–early fifteenth century), destroyed, one fragment of tower; Ilfracombe (?) (fifteenth century), there is no evidence that anything substantial resulted from the grant of the murage for one year in 1418 'to the good men of Ilfracombe, Devon, who have begun to build a new stone tower on either side of the port there and to enclose the town with stone walls and high towers'; Fowey (mid-fifteenth century), remains of blockhouse, and another at Polruan, across the river; St Ives (late fifteenth century), blockhouse (destroyed).

SOURCES: based on H.L. Turner, *Town Defences in England and Wales* (London, 1971); R.A. Higham, 'Public and private defence in the medieval South West: town, castle and fort', in R.A. Higham (ed.), *Security and Defence in South-West England before 1800* (Exeter, 1987); P. Sheppard, *The Historic Towns of Cornwall* (Truro, 1980).

Fig. 19.3
Drawing of the East Gate, Exeter, mid-eighteenth century.

SOURCE: reproduced by permission of Devon Record Office (ECA, Exeter City Map Book).

century, stimulated new attention to urban defence in southern England generally. Piecemeal royal financial intervention soon led to the system of 'murage grants', whereby money levied as tolls on goods could be used by the burgesses for building and maintaining defences.[13]

Murage grants are known for Exeter (from the mid-thirteenth), Totnes (later thirteenth), Plymouth (later fourteenth), and Ilfracombe (early fifteenth century). But these grants are not a simple guide to works carried out. At Exeter expenditure also occurs in the civic records from the fourteenth to seventeenth centuries. There is no evidence that anything much was built at Ilfracombe. At Plymouth, little was achieved in enclosing the town itself. Instead, a strong fortification to protect its seaward side was constructed. Something similar was done at Dartmouth in the same period, and although there is documentary evidence for this work, it is not in the form of murage grants. In fact, some places had considerable urban defences built by means other than murage grants. South-western examples are Barnstaple, whose walls presumably followed the burghal line, and Launceston, whose walls enclosed the town which grew beside the Norman castle. At Fowey, whose history and rise to prosperity paralleled that of Dartmouth and Plymouth, blockhouses were built on either side of the estuary to protect the town after a French raid in about 1450. At St Ives, a fishing port in the far west, a blockhouse and ramparts across the neck of the headland were built by 1500.

Fear of attack was not, however, the only stimulus to town walls. In this period urban communities developed their own particular identities, and the social and economic distinction between town and country became more marked. Walls, and the

FRONT of *EAST-GATE* without.

Normans at bay for eighteen days in 1068. There is no reason to suppose that Barnstaple and Totnes were not also well defended. But the spread of castles, and the domination of warfare by castle-sieges, meant that town defences declined in importance from the later eleventh century. We do not know how long burghal defences continued to be maintained, and the history of town walls in the twelfth century is obscure. But the threat of French attack, more or less continuously present from the early thirteenth impressive gateways which punctuated them, became an important symbol of urban wealth and pride. One of the greatest tests of urban defences, however, came not from outside England at all, but from the five-week siege which Exeter suffered in the Prayer Book Rebellion of 1549 (see chapter 20).

FURTHER READING

Castles in the context of medieval building generally are best seen in the *Buildings of England* series, published by Harmondsworth: *Devon*, 2nd edn (London, 1989) is edited by N. Pevsner and B. Cherry, and *Cornwall*, 2nd edn (London, 1970) is edited by N. Pevsner and E. Radcliffe.

Castles in the context of defensive works from the immediately post-Roman period to the sixteenth century are discussed in R.A. Higham, 'Public and private defence in the medieval South West: town, castle and fort', in R.A. Higham (ed.), *Security and Defence in South-West England before 1800*, Exeter Studies in History 19 (Exeter, 1987), 27–49. This essay is mainly concerned with the social history of fortification, rather than its physical form, and discusses Devon and Cornwall in a wider south-western region.

Physical and settlement aspects of Devonian sites are dealt with in R.A. Higham, 'Castles in Devon', in S.C. Timms (ed.), *Archaeology of the Devon Landscape* (Exeter, 1980), 70–80. There is no comparable work for Cornwall, but the castles are discussed in the context of the county's medieval archaeology generally in A. Preston-Jones and P. Rose, 'Medieval Cornwall', *Cornish Archaeology*, 25 (1986), 135–85.

ACKNOWLEDGEMENTS

Generous assistance from Peter Rose (Cornwall Archaeological Unit) is acknowledged in the compilation of the Cornish information included in this chapter. Lesley Bryant helped in the compilation of the maps.

POLITICAL AND MILITARY HISTORY

REPRESENTATION AND REBELLION
IN THE
LATER MIDDLE AGES

NICHOLAS ORME

PARLIAMENTARY REPRESENTATION, 1254–1529

In 1254 the knights of each English county were ordered for the first time to elect two representatives to attend parliament, and in 1265 the practice was extended to some of the cities and boroughs. Knights and burgesses were not summoned to all the early parliaments, however, and their presence did not become regular until the early fourteenth century. Between 1295 and 1311, the crown sent summonses to large numbers of small boroughs, twenty-two in the South West being involved at one or more parliaments in this period, but the unimportance of some of these communities made their representation unnecessary or burdensome, and several ceased to be represented after one or two occasions (**Map 20.1**). The last to disappear, Great Torrington, formally petitioned the crown to be exonerated and dropped out in about 1369. By the 1370s only six boroughs in Devon and six in Cornwall were regularly represented, to which Plymouth (which had dropped out after 1313) was added in 1437. In the late Middle Ages, therefore, the South West normally sent four knights and 24–26 burgesses to parliament (each constituency sending two members) until 1529 when the number of boroughs began to be increased (see chapter 51).

REBELLION, 1450–1497

After the mid-twelfth century, the South West was not involved in national political conflicts for 300 years. Even royal visits were rare to a region which was off the king's accustomed routes and seldom important enough to merit a special journey. Edward I came three times, notably to Exeter in 1285 to see to the trial of the murderers of the cathedral precentor, the Black Prince (the first duke of Cornwall) visited south-west England on several occasions from 1345 to 1372, and Henry IV journeyed there to meet his second wife, the duchess of Brittany, in 1403 (**Map 20.2**). But these were exceptions. Not until the 1450s did the region again become part of the theatre of national events, in consequence of the disorders nowadays known as the Wars of the Roses. No major battle of the wars took place in Cornwall or Devon, but several of the campaigns traversed their territories, right down to the final one of 1497. The intervening years constituted a distinct political era, different from the more peaceful centuries which came before and afterwards.

In 1450 the success of Jack Cade's rising in Kent forced the king, Henry VI, to make a series of provincial tours to restore confidence in his government. This brought him to Exeter two years later. The South West had already become the scene of quarrels between the bellicose earl of Devon, Thomas Courtenay, whose centre of power was Tiverton, and his enemies James Butler, earl of Wiltshire, and Sir William (later Lord) Bonville of Shute near Axminster. In 1451 the earl of Devon gathered forces which he led against the earl of Wiltshire in that county and then against Bonville, who was holding Taunton Castle. The king's uncle, Richard duke of York, had to be sent to restore order, and the king's visit to Exeter in 1452 probably had a similar motive. If so, it was not effective for very long. In the autumn of 1455 the earl and his son Sir Thomas Courtenay waged a new campaign against Bonville and his supporters. On 24 October they murdered Nicholas Radford, an elderly lawyer and ally of Bonville, at his home at Upcott Barton near Tiverton. On 3 November a Courtenay army of over a thousand men entered Exeter, took charge of its gates and attacked the property of Bonville, Radford and their friends. Meanwhile the earl besieged the castle at Powderham, whose lord Sir Philip Courtenay was his kinsman but a Bonville supporter. Bonville and his men crossed the Exe to raise the siege, but they were driven back by the earl's forces and pursued into east Devon. On 15 December the earl's men won a pitched battle on Clyst Heath, and two days later they captured Shute and pillaged it (**Map 20.2**). Disturbances continued into the spring of 1456 and then subsided; two years later the earl died in his bed. His son succeeded him, but he and Bonville were both executed after battles elsewhere in England in 1461, and neither lived to see Edward IV become king in that year.

The 1460s were years of relative peace in the South West, but in March 1470 the earl of Warwick and the duke of Clarence, who had rebelled against Edward IV in the north, fled to France via Exeter and Dartmouth, closely followed by the king who reached Exeter on 14 April, staying two nights. A year later, on the very same day, Margaret of Anjou and her son Prince Edward, having landed at Weymouth, passed through the city on their way to Tewkesbury where Edward

IV defeated them. This made him undisputed king again, though one of his strongest opponents, John de Vere, earl of Oxford, seized and briefly held St Michael's Mount in 1473. The peace was broken for the third time in 1483 when Edward IV died and his son was driven from the throne by Richard III—an outrage which led to risings in several parts of southern England, including Devon and Cornwall. Opponents of the usurper Richard included three leading clergy (Peter Courtenay the bishop of Exeter, the archdeacon of Exeter and the abbot of Buckland) and several local gentry (notably Sir Thomas Arundell of Lanherne, Sir Edward Courtenay of Boconnoc, and Richard Edgecombe of Cotehele). Sir Henry Bodrugan of Bodrugan, on the other hand, supported Richard as he had previously done Edward IV. The rising in the South West did not prosper; Richard III moved hastily to Exeter with his forces, arriving on about 8 November, and his enemies had to take refuge in France. But within two years there was another revolution. Richard was overthrown at Bosworth, Bodrugan's power declined, and the rebels of 1483 were restored. The bishop was promoted to Winchester, and Peter Edgecombe became a man of importance in Cornish affairs.

These were mainly aristocratic conflicts, but in May 1497 Cornwall experienced a popular rising caused by resentment at new heavy royal taxation levied to finance war with Scotland. Local discontent was organized by a blacksmith from St Keverne, Michael Joseph, and a Bodmin lawyer, Thomas Flamank, into a march to London to present grievances to the government (**Map 20.2**). The Cornish entered Devon, gaining some recruits on the way, passed by Exeter and proceeded in a fairly orderly fashion into Surrey and Kent. The king refused to negotiate and confronted the protesters at Blackheath on 17 June 1497 where the Cornish were defeated. Later in the year, there was a final political rebellion. Perkin Warbeck landed at Whitesand Bay on 7 September, gathered supporters from some still disaffected Cornish, and had himself proclaimed at Bodmin as Richard IV, claiming to be the younger son of Edward IV. His forces attacked Exeter, but the city held out successfully, and they passed on to Somerset only to melt away (**Map 20.2**). The sequel was the arrival of Henry VII in Exeter on 7 October, where he stayed for nearly a month, pardoning rebels and presenting the city with a cap of maintenance (which it still possesses) in acknowledgement of its loyalty. This was the last military emergency in the region for half a century, and the last visit of a monarch until the Civil War in the 1640s.

FURTHER READING

Early parliaments are listed in E.B. Fryde, D.E. Greenway, S. Porter and I. Roy (eds), *Handbook of British Chronology*, 3rd edn (London, 1986), 525–81. The earliest lists of returns of members of Parliament are listed in *Parliamentary Writs*, ed. F. Palgrave, I (London, 1827) and II part i (1830), and in *Members of Parliament: Part I, Parliaments of England, 1213–1702* (London, 1878). There are

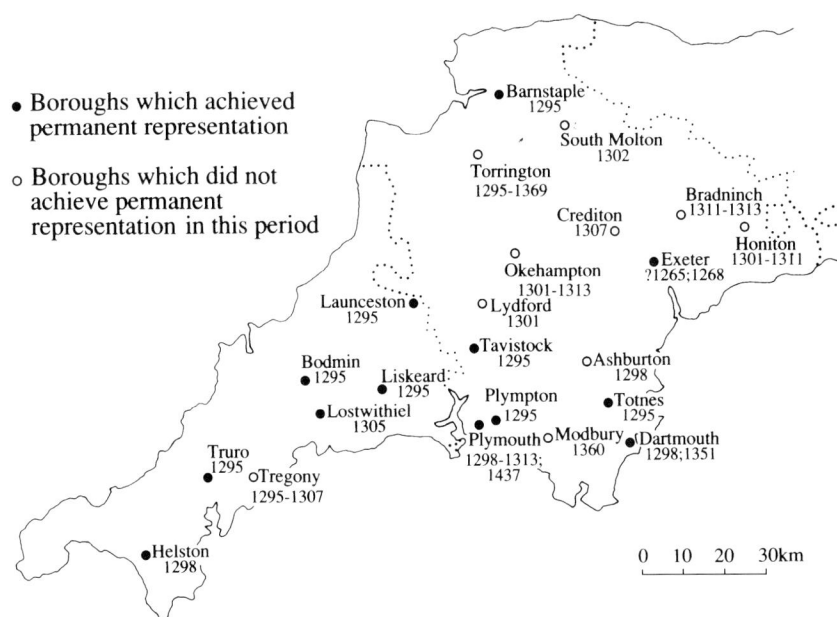

- ● Boroughs which achieved permanent representation
- ○ Boroughs which did not achieve permanent representation in this period

Map 20.1
Parliamentary representation, 1254–1529. Dates are those of the first known returns of members; and the last known in the case of boroughs which dropped out. Dartmouth and Plymouth were readmitted after long absences. County boundaries are those of 1254–1529.

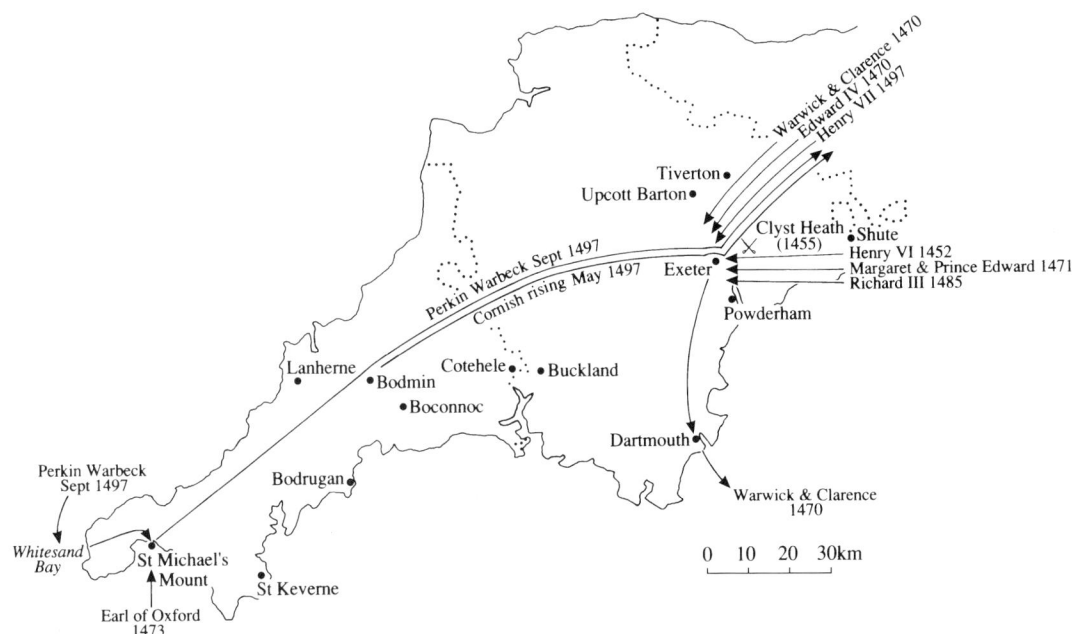

Map 20.2
Political history, 1450–1497. Routes are schematic not realistic.

studies of representation by A.F. Pollard, *The Evolution of Parliament*, 2nd edn (London, 1926), and May McKisack, *The Parliamentary Representation of the English Boroughs during the Middle Ages* (London, 1932).

N. Orme, *Exeter Cathedral As It Was: 1050–1550* (Exeter, 1986), 47–53, lists the royal visits. R.L. Storey, *The End of the House of Lancaster* (London, 1966), 84–92, 159–75, chronicles the Bonville and Courtenay conflict; and J.A.F. Thomson, 'The Courtenay family in the Yorkist period', *Bulletin of the Institute of Historical Research*, 14 (1972), 233–46, studies that family. There is also useful information in C.D. Ross, *Edward IV* (London, 1974) and *Richard III* (London, 1981). On the Cornish risings of 1497 see A.L. Rowse, *Tudor Cornwall* (London, 1941), 121–34 and Nicholas Orme, *The Cap and the Sword: Exeter and the Rebellions of 1497* (Exeter, 1997).

21

POLITICAL AND MILITARY HISTORY

CIVIL WARS
OF THE
SEVENTEENTH CENTURY

PETER GAUNT

A NATION AND REGION DIVIDE?

Following rebellions in Scotland and Ireland in the late 1630s and autumn 1641 respectively, England and Wales too slipped into civil war during 1642. By the end of that year, Devon and Cornwall were caught up in the war and the South West played a significant role in the unfolding events. But in this region, as elsewhere, contemporaries were unable to agree on what had caused the Civil War or on the factors which led them into support for one side or the other. If those who lived through the 1640s found these matters complex and confusing, it is hardly surprising that generations of historians have also disagreed over them.

Some historians of the Civil War have stressed long-term political and constitutional, or social and economic issues, seeing the war as an almost inevitable consequence of a power struggle between the crown and an ambitious parliament or as a class war of crown and nobility against the growing gentry middle classes. Other historians have given greater weight to short-term factors, such as the personal failings of Charles I which lost him the trust of a significant part of the political nation, the non-parliamentary financial levies which he imposed during the 1630s in an attempt to overcome the financial weakness of the crown and to rule without recourse to parliaments, his enforcement of a more elaborate, ceremonial, high-church form of worship within the Church of England, and his efforts to impose political and religious uniformity throughout the disparate Stuart inheritance of England, Scotland and Ireland. Indeed, some historians portray the English Civil War as just one part of a war of the three kingdoms, a British-wide reaction to the personality and policies of Charles I, which had ignited potential tensions within and between his kingdoms.

Similarly, many theories have been advanced to explain personal, local and regional allegiances once war broke out. Some would characterize the war as one of royalist aristocracy and their rural dependents against largely urban middle-class supporters of parliament. Others see more complex patterns of topographical and occupational divisions, principally between the small parishes of arable regions, portrayed as generally conservative, hierarchical and supportive of royal policies, and the larger parishes of woodland and pasture, which tended to be fluid, open to new ideas, less deferential in outlook and more likely to support the parliamentary cause.

In the South West, as in other regions, there were factors which may have led the population to support either the king or parliament in 1642. Devon and Cornwall were geographically distant from London and may, therefore, have seemed somewhat detached from the political controversies of Whitehall and Westminster. On the other hand, the crown was a powerful landowner in the area, for parts of Devon as well as much of Cornwall belonged to the Duchy of Cornwall. In this region, as elsewhere, the gentry had done well out of royal government in the century or so before the Civil War, acquiring property at the dissolution of the monasteries and power, status and money by office-holding. Thus a mixture of detachment, self-interest and innate loyalty ensured that many people in Devon and Cornwall felt strong ties to a divinely appointed monarch. The populace was also acutely aware of their area's vulnerability to sea-borne invasion, so reinforcing a natural inclination to rally to the crown in times of danger or unrest.

However, other factors might pull the region towards parliament. Devon and Cornwall had suffered from the heavy financial exactions imposed by the king during the 1630s, especially the repeated collection of Ship Money—even if, as maritime counties, both stood to gain from an improved navy and coastal guard, the ostensible ends of Ship Money. Most of this money was paid and there is little evidence of open ideological or constitutional opposition to the king's right to impose Ship Money, though rating disputes and pleas of poverty, of which there were many, may have overlain deeper discontent. The king's religious policies of the 1630s had clearly alienated many people, who viewed them as at best an attack upon the true Protestant Church of Elizabeth I and James I, and at worst as an attempt to reintroduce Roman Catholicism by stealth. The moderate and conciliatory approach of Joseph Hall, bishop of Exeter 1627–41, may have tempered religious tensions in the region, though there is evidence of discontent among active cliques of puritans in south-west England. Devon and especially Cornwall were very strongly represented in the House of Commons, and the many (mainly borough) MPs could keep their more politically aware constituents well-informed of deteriorating relations with the king.

Indeed, there is every sign that the gentry elite in the South West took a keen and informed interest in national politics and the policies of central government. In Devon and Cornwall there were also many small market and port towns, whose commercial and manufacturing sectors might be more sympathetic to parliament than to the king. Finally, because they felt particularly vulnerable to papist invasion, the inhabitants of the South West were deeply shaken by the 1641 rebellion of Irish Catholics and might have been inclined to support whichever side seemed to offer the best defence against Catholics and the Irish. Naturally, king and parliament each vigorously proclaimed their determination to defend Church and state during 1642.

Both king and parliament looked to the South West for support when they set about recruiting in the region during July and August 1642. However, both sides met with limited success, for each initially gathered only small groups of active supporters. For example, a royalist muster on Bodmin racecourse on 17 August produced less than 200 men. Some patterns of allegiance can be discerned at this opening stage of the conflict. From such evidence as survives, it seems probable that active parliamentarians outnumbered royalists in Devon—attempts by the royalist earl of Bath to recruit in Devon over the summer certainly proved very unpopular and aroused considerable hostility—while in contrast west of the Tamar royalism was stronger than support for parliament. Historians have also detected shades of allegiance within each county. In Cornwall, royalism seems to have been far stronger in the west than in the east, where there was some support for parliament. Although pro-parliamentary feeling was apparent in many parts of Devon, in the centre of the county—the area encompassing much of Dartmoor and the moorland fringes—there was some support for the king. However, in both counties open support was muted and activists comprised only a small if committed minority. At this stage, much of the population appeared unwilling to commit themselves and their counties to the imminent conflict. There was just a handful of resident peers within the two counties and no grand territorial magnates who might swing the region behind king or parliament. Instead, political leadership lay with the interconnecting web of several hundred gentry families, most of whom held back from actively supporting either side at this stage. The opening phase of the Civil War in the South West, as in most regions in England and Wales, was marked by a degree of hesitancy and uncertainty and a disinclination to become too involved.

DEVON AND CORNWALL SECURED FOR THE KING, 1642–1643

This mood of uneasy calm and neutrality was shattered in late September 1642 by the arrival of a 'foreigner'. Sir Ralph Hopton MP, an experienced soldier in his mid-forties, had declared for the king during the summer and had raised a cavalry regiment in his native Somerset. Because of his military experience and his wealth and social standing, he was given command of the horse in the embryonic army of the marquis of Hertford, the king's supreme commander in South Wales and southern England. Hertford tried with conspicuously little success to raise Somerset and north Dorset for the king during August and early September. By mid-September he had abandoned the attempt and, with parliamentary forces advancing from the east, crossed by boat with his infantry from Minehead to royalist South Wales. Hopton, meanwhile, led a body of 160 horse and dragoons westwards through north Devon and on into Cornwall, reaching Stowe, the house of the known royalist Sir Bevil Grenvile, without incident on 25 September. His presence in the area altered the course of events, galvanized local royalists and ended the period of uneasy peace. Hopton ensured first that Cornwall was swiftly secured for the king and then that Devon was overrun from the west. Despite the misgivings of many of the region's inhabitants, the South West could no longer avoid being caught up in the national civil war.

Hopton and his men entered Bodmin on 27 September, frightening Sir Richard Buller and other local parliamentarians into withdrawing to Launceston. Sir Ralph then voluntarily stood trial at Truro Assizes on charges of disturbing the peace but was triumphantly acquitted, perhaps thanks to a packed jury. The Cornish trained bands, the local, part-time self-defence force, declared their support for the king and on 4 October Hopton was able to review a body of over 3,000 foot. He marched them to Launceston and, for the second time within a fortnight, Buller and his fellow Cornish parliamentarians had to fall back before the superior royalist forces, this time retreating over the Tamar into Devon. Hopton now effectively held all Cornwall for the king, but he did not as yet attempt to push on into Devon with his motley self-defence force. Instead he sensibly spent several weeks in Cornwall raising new recruits and the money to pay them, actively assisted by Grenvile, Sir Richard Vyvyan, Lord Mohun and others. The county of Devon, particularly its major towns—Barnstaple, Plymouth and Exeter—was showing itself more sympathetic to the parliamentary cause. With the royalists now in firm control of Cornwall and raising an army there, the Devon parliamentarians, bolstered by their colleagues who had fled from Cornwall, also set about raising local regiments to cement their still precarious hold on the county and to meet the growing threat from across the Tamar. They also appealed to parliament to send reinforcements.

There followed six months of intermittent campaigning, from December 1642 to May 1643, comprising a series of raids and counter raids, several half-hearted sieges and at least three larger engagements which probably merit the label 'battle', as Cornish royalists clashed with Devon parliamentarians (**Map 21.1**). At the beginning and end of December, Hopton advanced as far as Exeter, but the city was well defended and on both occasions he soon fell back westwards. For much of the month Hopton concentrated instead on the parliamentary port of Plymouth. But Plymouth, too, was ably defended by Lord Robartes and Colonel

Map 21.1
The first Civil War in Devon
and Cornwall, 1642–1646.

Ruthin. They broke up a royalist rendezvous and muster at Modbury on 7 December and by early January were threatening Saltash, the royalists' main base for operations against Plymouth. Moreover, the first wave of parliamentary reinforcements now arrived from Somerset and Dorset and crossed the Tamar at Newbridge, near Callington. A second wave of reinforcements, under the earl of Stamford, was expected shortly. In the face of these setbacks, Hopton retreated into Cornwall in early January 1643 and by the 17th was quartered in Lord Mohun's park at Boconnoc, near Lostwithiel, preparing for the expected parliamentary attack. Without waiting for Stamford and the further reinforcements, Ruthin meanwhile gathered existing parliamentary forces at Liskeard. The two armies clashed on 19 January on open ground somewhere between Boconnoc and Liskeard.

Like many Civil War engagements, the resulting battle, generally known as the Battle of Braddock Down, occasionally as

the First Battle of Lostwithiel, was poorly recorded by contemporaries, whose accounts are often vague, even contradictory. In consequence, the numbers of troops involved on each side, the numbers killed, wounded or captured, and even the precise location of the battlefield are not now certain. We know that the armies drew up facing each other on two hillsides, separated by a marshy valley, perhaps the upper reaches of the western branch of the West Looe River, south of East and Middle Taphouse. The parliamentarians were superior in cavalry, but the royalists had more infantry. After a period of minor skirmishing and exchange of shot, the royalists charged. The parliamentary army quickly broke and fled back to Liskeard and on to Saltash. Three days later Hopton stormed Saltash, forcing Ruthin and his remaining troops to cross the river to Plymouth. Although casualties were probably quite light, Hopton took many prisoners and abundant equipment and supplies, and was confirmed in control of Cornwall. But he still lacked the strength to break into

Devon or seriously threaten Plymouth—a renewed blockade of the town during February was beaten off with heavy losses, and there was another mauling for the royalists in Modbury on the 21st. The sense of stalemate, with royalists supreme in Cornwall but parliamentarians holding Devon, was underlined when they signed a short cessation of hostilities at the end of February.

Both sides used the truce to reorganize and strengthen their armies. The parliamentarians, in particular, now under the command of the earl of Stamford, needed to recruit more men. When the truce expired on 22 April, Stamford sent young Major General James Chudleigh with a force of 2,000 horse and foot to test Hopton's base at Launceston. On the 23rd Hopton gathered his men in a defensive position on top of Beacon Hill. Chudleigh spent most of the day trying to fight his way up the hillside; both sides were reinforced during the day. Hopton held firm and in the evening Chudleigh conducted a well-organized retreat back over the Tamar. Hopton's men were too exhausted to pursue them immediately, but on the 25th he led over 2,500 royalists into Devon, marching on Chudleigh's headquarters at Okehampton. Warned in time of their advance, Chudleigh prepared an ambush on Sourton Down, into which the royalists carelessly stumbled in the late evening. The resulting action, somewhat flatteringly known as the Battle of Sourton Down, was a rather confused and scrappy affair, not least because it was fought in pitch darkness. The royalist force fell back in confusion when the ambush was sprung, but Hopton brought up a rearguard which restored some order in the royalist ranks and also compelled Chudleigh to pull back, while awaiting reinforcements from Okehampton. When these arrived Chudleigh renewed the attack, but by now the royalists were well prepared, lining ancient earthworks on Sourton Down. The forces disengaged, Hopton trudged back to Launceston through a heavy storm, and Chudleigh returned to Okehampton with captured arms and ammunition.

Some of Hopton's papers had also fallen to Chudleigh at Sourton. These included an order from the king that Hopton should march his army east to rendezvous with a larger royalist force under Prince Maurice in Somerset. Stamford was determined to prevent this and by mid-May had gathered a force of over 5,500 men at Torrington, effectively blocking the route from Cornwall to Somerset. Hopton, meanwhile, had gathered together almost 3,000 men at Launceston. He moved north on 13 May, well aware that Stamford was awaiting him. Indeed, the earl moved his army west to intercept the king's men, blocking their route at Stratton. Stamford deployed his troops behind helpful ancient earthworks on top of the hill, just north of the village, which now bears his name. Hopton, short of supplies but under orders to march to Somerset, felt that he had no alternative but to give battle, even though this meant attacking a numerically superior and well-deployed army.

The resulting Battle of Stratton was perhaps the largest and certainly the most decisive engagement of the opening phase of the Civil War in the South West. Hopton commenced battle at first light on 16 May, but time and again his men were repulsed as they struggled to fight their way up hill. Exhausted and running short of powder, Hopton ordered one final attempt in mid-afternoon. Against the odds, and showing the sort of determination that would make the king's Cornish infantry justly famous, the royalists carried the defences and took the hill. The royalists were exultant. Francis Bassett, writing to his wife, began: 'Oh Dear Soule, prayse God everlastingly. Reede ye inclosed. Ring out yor Bells. Rayse Bonefyers, publish these Joyfull Tydings, Beleeve these Truthes.' The parliamentarians had lost 300 dead and 1,700 captured, plus cannon, arms and cash. Stamford and the remnant of his infantry straggled back to Exeter. Although significant parliamentary pockets remained in Devon, especially in its main towns—Plymouth, Exeter, Bideford and Barnstaple all had parliamentary garrisons—in the wake of Stratton the local parliamentarians were a spent force. Hopton felt strong enough to march east with most of his Cornish army, joining the main royalist army in Somerset and contributing to a dazzlingly successful campaign which, by the end of the summer, left the king in control of virtually the whole of Somerset, Dorset and Wiltshire. Isolated now amongst a sea of royalism, most of the remaining Devon parliamentarians lost heart—Bideford, Barnstaple, Appledore, Exeter and Dartmouth surrendered without major fighting during the late summer and early autumn of 1643.

RELATIVE PEACE, 1643–1645

For the next two years, Devon and Cornwall largely escaped fighting, secure for the king and far removed from the theatres of military action. Minor parliamentary unrest in north, south and east Devon apart, there were just two significant exceptions to this period of undisturbed royalist rule, one long-standing, the other short-lived.

The former was the parliamentary town of Plymouth, the thriving port and commercial centre at the mouth of the Tamar. It was strongly defended by water on three sides and on the northern, landward side by a line of forts, linked by walls or embankments, from the Plym to the Tamar. Too strong for Hopton's half-hearted attempts during the winter of 1642–3, it held out even when the rest of Devon had fallen to the king (**Map 21.2** and see also **Map 65.4**). Although the royalists were able to encircle Plymouth with their own bases—including Mount Stamford (**Map 21.3**), Plymstock, Plympton, Egg Buckland, Tamerton and Mount Edgcumbe—and throw up a line of forts to the north of the town's landward defences, they could never take it. Parliamentary control of the navy ensured not only that supplies and reinforcements could always be shipped into Plymouth but also that the royalists could never mount effective amphibious operations from their bases on the other banks of the Plym and the Tamar Estuary—indeed these bases were themselves very vulnerable to counter-attack. The parliamentary forces always proved strong enough to repulse any direct, land-based attacks on the northern defences. At the end of September

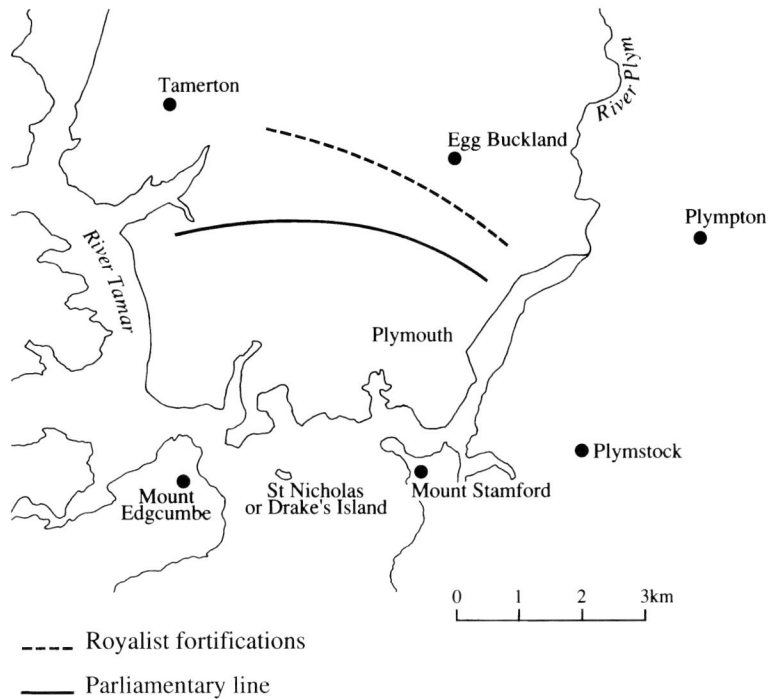

Royalist fortifications

Parliamentary line

Map 21.2
Operations around Plymouth.

Prince Maurice began a formal siege, but he abandoned the hopeless operation just before Christmas and for the next two years the royalists were content to blockade Plymouth, stationing troops in the area to prevent the parliamentarians straying far beyond their defences. Plymouth remained as parliament's only base in Devon and Cornwall, but its possession did not significantly disrupt royalist control of the region (**Map 21.4**).

The second and short-lived disturbance to royalist control was the arrival in the South West in summer 1644 of the main parliamentary army under the earl of Essex, the parliamentary Lord General. His original plan, to move west to tackle Prince Maurice's forces and so to relieve royalist pressure on parliamentary outposts, particularly Lyme Regis, was sound enough. But he advanced further west than was either wise or necessary and found himself deep in enemy territory. To make matters worse, the king's own army had brushed aside the parliamentary commander Sir William Waller in Oxfordshire and was pursuing Essex into the South West. Essex's army marched through Devon without meeting serious resistance and swept into Tavistock on 23 July, causing most of the royalist forces around Plymouth to fall back beyond the Tamar. Deciding to march on

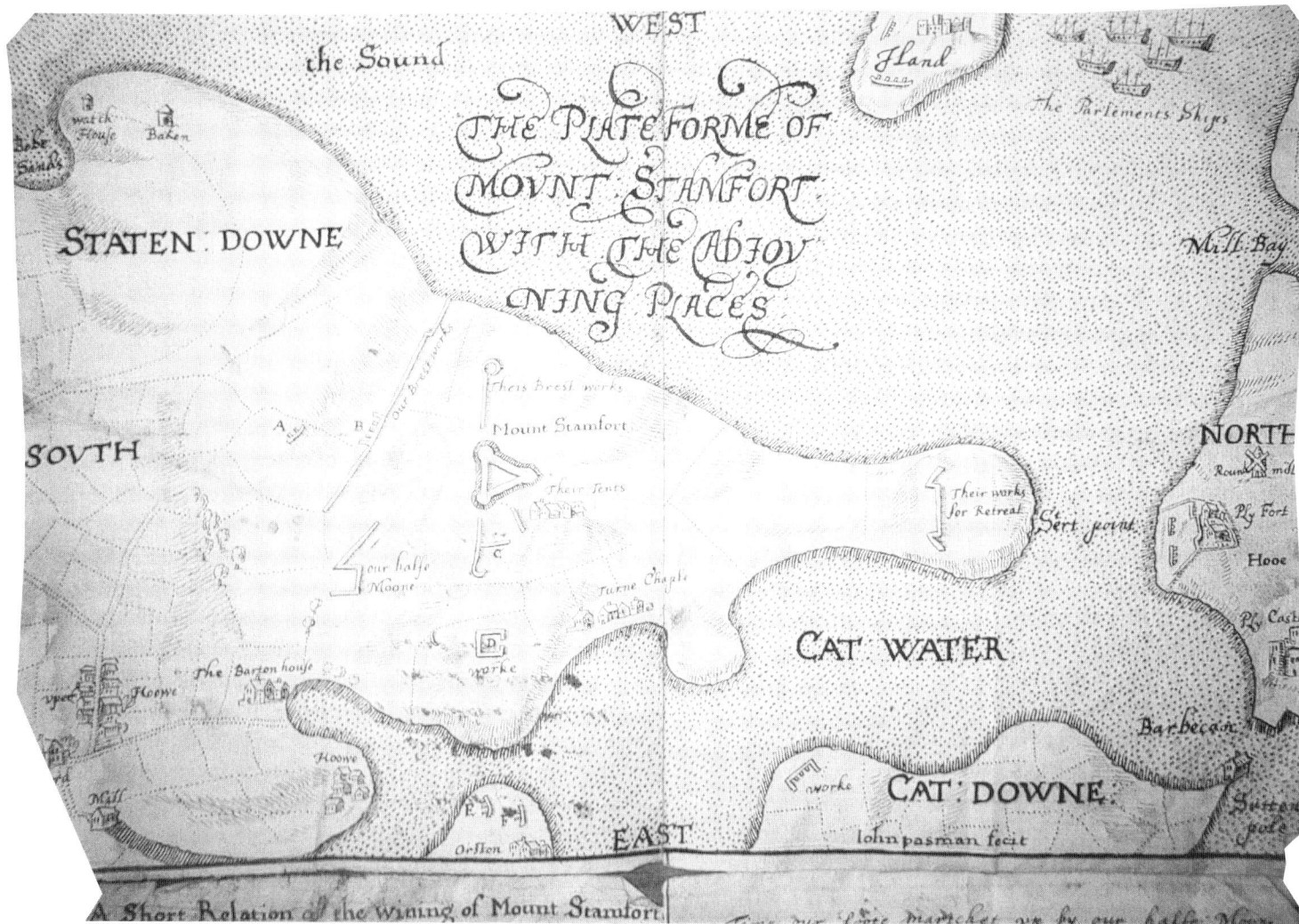

Map 21.3
Manuscript map by John Pasman of the royalist attack on Mount Stamford, Plymouth, 1643.

SOURCE: photograph reproduced by permission of Kent Archives Office (U269 0173).

behind. Fairfax occupied Launceston, Lostwithiel and Bodmin, encountering no serious resistance. From Truro, Hopton had nowhere else to run. He opened negotiations on 9 March at Tresillian. The treaty, signed on the 12th, led to the disbanding of his remaining troops. On 25 March 1646—old New Year's Day—Fairfax and Cromwell led a triumphant procession through the loyal town of Plymouth. Within a month both had left the South West, never to return.

IMPACT OF THE CIVIL WAR ON SOUTH-WEST ENGLAND

Despite the hesitancy of 1642, the South West had played a major part in the Civil War. A significant proportion of the adult male population of Devon and Cornwall, both the gentry elite and the common people, had entered military service and had fought for king or parliament, either within the South West itself or in campaigns further afield. In the South West, as in other regions, it is implausible to suggest that the common people blindly followed the lead of the elite, that those gentry who committed themselves were able, by exercising a mixture of deference, obligation and duress, to compel thousands of ordinary men to join up and fight. Nor does the suggestion that allegiance was determined by topographical and land-use patterns, with wood-pasture areas supporting parliament and arable areas the king, work well in the South West, though it has more credence in neighbouring Somerset and Dorset. Although some patterns emerge, with tin workers and the more remote rural areas inclined to support the king, and cloth-workers, seamen and large towns inclined to support parliament, there are significant exceptions to these patterns—the tinners around Plymouth supported parliament, some cloth-working communities of east Devon showed royalist sympathies, the town of Tavistock was pro-royalist and there was a significant royalist minority within Exeter.

Historians now tend to give more weight to religion as the key factor which not only motivated people to take up arms but also determined their allegiance. Those who favoured further reform of the state church, removing remaining ceremonial elements and creating a lower, more participatory, preaching church—those often labelled puritans or the godly—tended to support parliament. In the immediate pre-war decades, this more radical, reform-minded Protestantism was strongest in the ports, towns and cloth districts of Devon—particularly in the north Devon ports, in Plymouth and amongst a very active clique in Exeter—and had gained some support in south, particularly south-east, Cornwall. Those who feared that further religious reform would unleash heresy and social instability and who supported the Elizabethan and Jacobean Church of England tended to rally to the king in defence of the established Church. This more conservative, less reformist religious outlook was strongest in the more remote areas—the moorland interiors of Devon and Cornwall and western Cornwall—as well as amongst the Cornish tinning communities and the dean and chapter of Exeter Cathedral. Pre-war Exeter was thus torn between reformist and anti-reformist groups, which may explain why there were strong royalist and parliamentarian factions within the city during the Civil War.

No one living in the South West during the 1640s could have been unaware of, or unaffected by, the Civil War. It brought not only death and mutilation but also physical destruction to Devon and Cornwall. The long siege and blockade of Plymouth and the operations which saw Exeter fall to the king in 1643 and to parliament in 1646 had profoundly affected those two cities, with the construction of extensive offensive and defensive works, the clearance of the land and suburbs outside the defensive line and substantial damage within the city centre. But the physical impact spread well beyond these major centres. Even those who neither joined an army nor were directly caught up in the fighting would have been deeply affected by the war. Both sides, king and parliament, imposed novel, regular and very heavy taxes upon the people to finance their war effort.

Despite these social and economic costs, it would be wrong to overstress the direct military impact of the Civil War in the South West. Devon and Cornwall did not have to endure the sort of repeated campaigning and prolonged military action which many counties suffered during the war. They were secured for the king by spring 1643 and thereafter held by the royalists with little disturbance until the autumn of 1645, when, quite quickly and with limited bloodshed, the parliamentarians mopped up the remaining strongholds of an already defeated royalist cause. For much of the civil war, then, Devon and Cornwall were held secure for the king and without direct challenge, and saw only very minor military action within their borders—protracted operations around Plymouth and the short-lived presence of the earl of Essex's army. They were geographically separated from the main theatres of conflict by the sea and by the 'buffer zone' of Dorset, Somerset and Wiltshire, counties which suffered much heavier and more protracted fighting. For all but a few months of the war, they neither served as a frontier between the two forces nor formed part of a corridor linking royalist or parliamentary centres, and so escaped the intense garrisoning and incessant raiding and counter-raiding which afflicted many English counties. Whatever the price in terms of men, money and materials, in terms of military action within their borders Devon and Cornwall escaped relatively lightly.

THE UNEASY PEACE, 1646–1660

The collapse of royalism in Devon and Cornwall in 1645–6 was mirrored by the parliamentary conquest of the other royalist heartlands. By summer 1646 Charles I had to accept that militarily at least his cause was doomed. He surrendered to parliament's allies, the Scots, and on his orders most of the remaining royalist bases also surrendered. But following eighteen months of fruitless constitutional negotiations, the fragile peace collapsed. Charles fled to the Isle of Wight in the closing weeks of 1647 and from there he not only encouraged royalists in England and Wales to rebel but also concluded an alliance with the Scots. During 1648 there was a series of poorly coordinated royalist rebellions in many parts of England and Wales, the most serious in Kent, Essex and south-west Wales. During the summer a Scottish royalist army crossed the border and began marching south through England. Together these English and Welsh rebellions and the Scottish invasion are labelled the Second Civil War. However, parliamentary troops were able to put down the various rebellions and in August Cromwell destroyed the Scottish army of invasion in Lancashire. In the wake of these events, conservative MPs were purged from the Commons (December 1648) and, under army influence, Charles was tried and executed in January 1649 and both the monarchy and the House of Lords were abolished.

Devon and Cornwall played only a very small part in these disturbances. Only in the far west of the region was there a short-lived and rather feeble pro-royalist rebellion in 1648. During the third week of May a group of former junior royalist officers led a rebellion in the western tip of Cornwall. But there was no general rising, the Cornish gentry failed to lend active support and the insurrection remained very small-scale and geographically limited. One group of rebels, up to 300 strong, seized Penzance, but they were unable to take St Ives, Helston or St Michael's Mount. Parliamentary troops quickly moved west through Cornwall, and on 22 May Colonel Robert Bennett recaptured Penzance, killing or capturing many of the rebels and scattering the rest. A second group of rebels on the Lizard made a half-hearted march on Helston. On the evening of 23 May the parliamentary troops attacked them in Mawgan and after a fierce engagement drove them from the Gear, an ancient earthwork above Mawgan Creek. The rebels scattered, many seeking refuge in old mine workings or hiding in caves and cliffs around the Lizard coast; others in desperation threw themselves into the sea. The very minor Cornish rebellion was over.

Mainland Devon and Cornwall saw very little further military action after 1648. There was no sign of royalist insurrection during the summer of 1651, when another Scottish royalist army invaded England only to be destroyed by Cromwell at Worcester. During the 1650s there was a string of very minor risings in England and Wales, all efficiently and quickly crushed by the parliamentary army. But Devon and Cornwall remained quiet, and parliamentary fears of royalist insurrection in the South West proved unfounded. The largest home-grown rising of the 1650s, in March 1655, began in Wiltshire, where Sir Joseph Wagstaffe and John Penruddock raised a few hundred men and marched west. But Dorset and Somerset failed to rally to the king's banner and the rebels pushed on westwards, pursued by parliamentary forces. They entered east Devon, marching through Cullompton and Tiverton, but the south-western counties, too, showed no sign of rising in support. Instead of gaining further men, the rebel band was depleted by desertions. During the night of 13–14 March sixty parliamentary troopers out of Exeter under

Colonel Unton Crooke defeated and scattered the remaining 200 or so rebels at South Molton.

Although mainland Devon and Cornwall were secure for parliament by 1646 and remained so until the Restoration in 1660, the Isles of Scilly, like several other off-shore islands, followed a rather different course. The islanders rose up for the king in September 1648, not least because of the sufferings they had endured under the parliamentary garrison established there in summer 1646. For nearly three years the Scillies served as a sanctuary for disgruntled royalists and a base for pro-royalist piracy. The islands were under the command of Sir John Grenvile, while Hopton led the fleet of privateers on missions against the south-western coastline. In spring 1651 parliament moved to pluck this thorn. During the third week of May Admiral Blake led a fleet to the Scillies, landed 2,500 infantry on the islands and in a combined land and sea operation quickly took Tresco. Grenvile, with less than 500 men, was besieged on St Mary's and surrendered on 23 May.

THE 1650S

In the course of the 1650s a degree of normality returned to daily life and local administration in the South West. Garrisons remained in many of the larger towns (Exeter and Plymouth, for example) and in key fortresses (such as Pendennis and St Mawes). Often enough there was tension between the military and local civilians, and senior officers tended to oversee and to some extent control local and county administration. Colonel John Disbrowe, Cromwell's brother-in-law, took a particular interest in the West Country and in autumn 1655, as part of a scheme which covered the whole of England and Wales, he was appointed Major General of the region—Cornwall, Devon, Dorset, Somerset, Wiltshire and Gloucestershire—with special powers to levy money on former royalists and to impose order in the area, and with bodies of commissioners to assist him in each county. The system of the Major Generals, which lasted barely eighteen months and collapsed early in 1657, was an additional tier, over and above the processes of civilian local government and administration which were gradually returning in the wake of civil war.

In Devon and Cornwall, as elsewhere, many prominent royalists found themselves out of favour, deprived of local office and of some of their estates and forced to pay heavy fines. Even so, few royalist families were completely ruined by the experience and most came back into their own again at the Restoration. More broadly, from 1646 to 1660 many of the traditional greater gentry elite were partly or wholly excluded from local administration. Because (Plymouth excepted) Devon and Cornwall had been under royalist control for much of the Civil War, there was a need to reconstruct non-royalist county government anew when peace returned. Traditional forms re-emerged, particularly the wide-ranging roles of the justices of the peace and the control exercised by quarter sessions. However, they acted alongside post-war

Map 21.6
Parliamentary representation,
1646–1660.

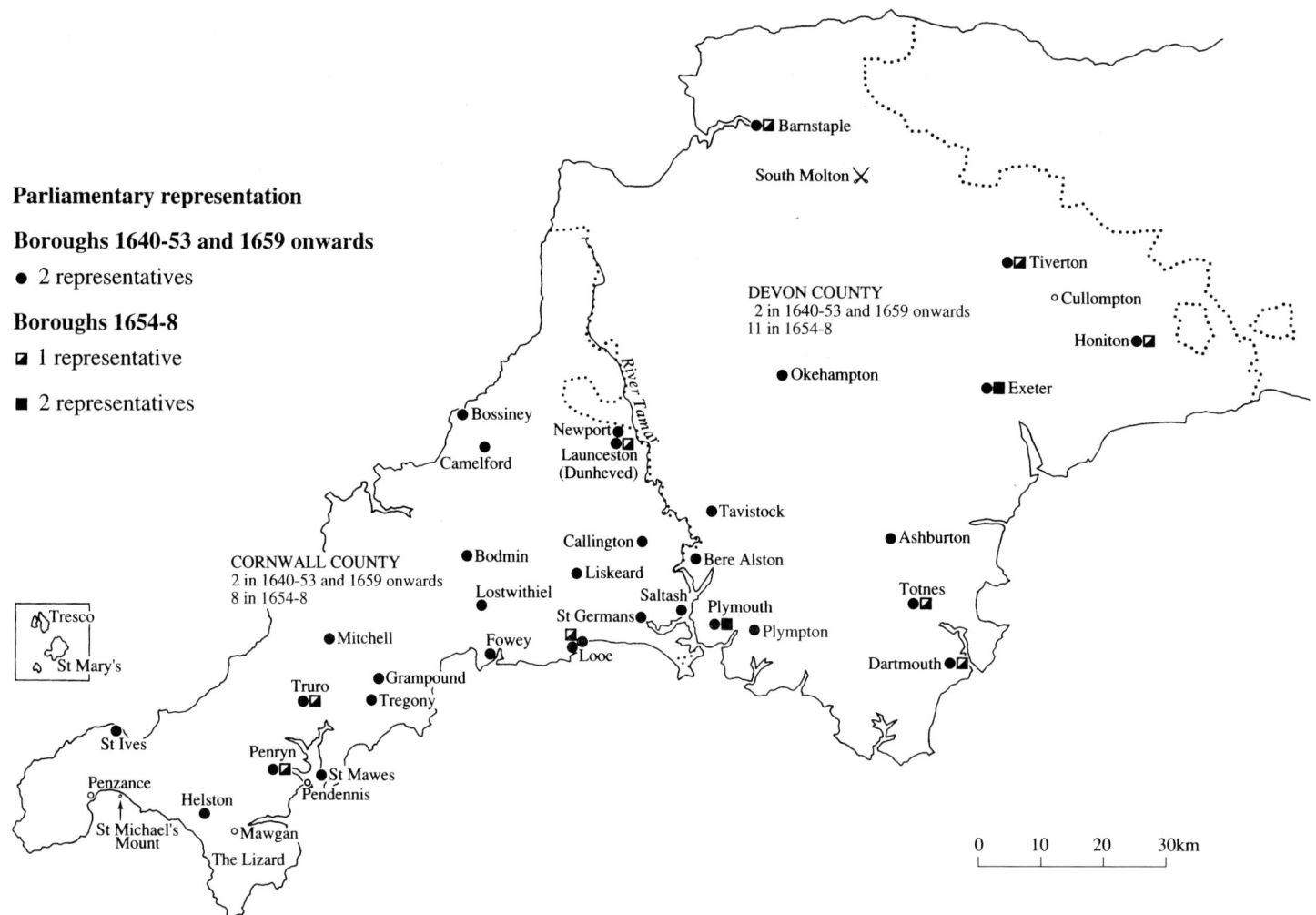

Map 21.6
Parliamentary representation, 1646–1660.

Parliamentary representation

Boroughs 1640-53 and 1659 onwards

● 2 representatives

Boroughs 1654-8

◩ 1 representative

■ 2 representatives

innovations, the various commissioners—a mixture of local civilians and military figures—appointed by parliament to oversee local defence and policing, to collect the regular and heavy taxes which continued to be levied, and to attempt to restore some order to the administration of religion. Between them, these traditional and new elements of local administration were charged with the task not only of rebuilding and maintaining good order and administrative efficiency in a post-war military, fiscal and religious context very different from that of the pre-war period, but also of imposing the legislative and executive initiatives of central government, including during the 1650s the fitful attempts to clamp down on intrigue and immoral behaviour.

During the 1650s central government, too, blended traditional and novel features. Until April 1653 power rested with the purged remnant—'The Rump'—of the House of Commons originally elected in 1640. Following its expulsion by Cromwell in April 1653 and a short-lived experiment entrusting power to a nominated assembly, a new form of government under a written constitution was established at the end of the year. During the Protectorate, which lasted from December 1653 to spring 1659, power was divided between a single head of state (Lord Protector Oliver Cromwell, succeeded in September 1658 by his eldest son Richard), a powerful and permanent executive council and an assured succession of triennial parliaments. One direct result of this scheme upon the South West was drastically to reduce its parliamentary representation (**Map 21.6**). By the early seventeenth century, Devon and Cornwall were very strongly represented in the House of Commons, with a proliferation of borough seats. In the Long Parliament elected in 1640 there were forty-four Cornish MPs (two for the county and two apiece for twenty-one boroughs) and twenty-six Devon MPs (again, two for the county and two apiece for twelve boroughs). Thus in a House of just over 500 MPs the two south-western counties accounted for seventy Members, almost 14 per cent of the total, a proportion well in excess of their quota had seats been distributed in proportion to the size, wealth or population of counties. Small places such as Bere Alston in Devon, and Bossiney, Grampound, Mitchell and Tregony in Cornwall, each returned two MPs (**Map 21.6**). Under the new Protectoral constitution of December 1653 the seats were radically redistributed, bringing parliamentary representation more in line with the distribution of wealth and population. In the parliaments elected in 1654 and 1656 England and Wales had 400 MPs (plus thirty apiece for Scotland and Ireland), the great majority of whom represented the counties; most boroughs were disfranchised. Cornwall was reduced to twelve MPs (eight for the county, one apiece for four boroughs) and Devon to twenty MPs (eleven for the county, plus two apiece for Exeter and Plymouth and one each for five other boroughs). With thirty-two out of 400 MPs, Devon and Cornwall now returned just 8 per cent of the total of English and Welsh MPs. The new arrangement was abandoned in 1659 and the pre-civil war distribution of seats returned, to survive, almost unchanged, until well into the nineteenth century.

Richard Cromwell's Protectorate lasted less than nine months, for in spring 1659 he resigned in the face of army pressure. There followed a year of constitutional uncertainty, as the restored Rump of the Long Parliament struggled to maintain control over both the country and the parliamentary army. In the opening weeks of 1660 the former royalist General George Monck (of Torrington) intervened with his Scottish-based army to restore order. It eventually became clear that this would be best done by restoring the monarchy in the person of Charles I's eldest son. Monck was actively supported in the South West by many gentry families, both royalist and disillusioned parliamentarians. The Moncks had property and kinsmen—including the Grenviles and Sir William Morice—in the South West, and George's own brother Nicholas, who held the living of Kilkhampton, acted as a go-between. In April 1660 the newly elected Convention Parliament recognized the late king's son as rightful king and Charles II returned in triumph and unopposed in May. Some of his leading supporters in the South West were rewarded with offices and titles, while prominent opponents of the royalist cause lost favour, office and, in some cases, life or liberty. The wheel had come almost, but not quite, full circle.

FURTHER READING

The best military histories of the Civil War nationally are Peter Young and Richard Holmes, *The English Civil War* (London, 1974), Charles Carlton, *Going to the Wars* (London, 1992) and Martyn Bennett, *The Civil Wars in Britain and Ireland* (Oxford, 1997).

On Cornwall, Mary Coate's *Cornwall in the Great Civil War and Interregnum* (Truro, 1933) is a masterly account of military, political, social and economic affairs; it should now be read with Anne Duffin's *Faction and Faith: Politics and Religion of the Cornish Gentry Before the Civil War* (Exeter, 1996), which explores the attitudes and actions of the Cornish elite in the decades leading up to the outbreak of Civil War.

On Devon, E.A. Andriette's *Devon and Exeter in the Civil Wars* (Newton Abbot, 1972) is a very sound but quite narrow military account, while M. Wolffe, *Gentry Leaders in Peace and War: The Gentry Governers of Devon in the Early Seventeenth Century* (Exeter, 1997) is a deeper examination of the Devon elite. But the most sophisticated recent work on Devon has been by Mark Stoyle. His *Loyalty and Locality: Popular Allegiance in Devon During the English Civil War* (Exeter, 1994) is a detailed and thoughtful examination of how and why ordinary people chose to support one side or the other during the Civil War—differing somewhat from I.R. Palfrey's interpretation in 'Devon and the outbreak of the English Civil War, 1642–43', *Southern History* 10 (1988)—while Stoyle's *From Deliverance to Destruction: Rebellion and Civil War in an English City* (Exeter, 1996) explores Exeter's Civil War and shows how the wartime divisions within the city stemmed from pre-war factionalism. Stoyle has written a much briefer account of 'Plymouth in the Civil War', *Devon Archaeology*, 7 (1998).

The best overall accounts of the period 1646–60 are contained within I.A. Roots, *The Great Rebellion* (London, 1966 and later editions) and G.E. Aylmer, *Rebellion or Revolution?* (London, 1986). Good studies of the republican period include A.H. Woolrych, *England Without a King* (London, 1983), T.C. Barnard, *The English Republic*, 2nd edn (Harlow, 1997) and R Hutton, *The British Republic* (Basingstoke, 1990), which covers local government and relations between the centre and the localities.

Turning to the South West, we need far more on Devon and (especially) Cornwall between the Civil War and the Restoration, to match the detailed studies now published on many other English counties and regions. The excellent accounts of the south-western gentry in the pre-war era—Duffin, *Faction and Faith*, and Wolffe, *Gentry Leaders in Peace and War*—provide no more than background to the post-war position, and Andriette's *Devon and Exeter in the Civil War* does not extend much beyond the end of the fighting. On Cornwall, the closing chapters of Coate's *Cornwall in the Great Civil War and Interregnum* briefly survey the period 1646–60. On Devon, S.K. Roberts has done a lot of work on county administration from the end of the war to the Restoration and beyond, most notably *Recovery and Restoration in an English County: Devon Local Administration 1646–70* (Exeter, 1985). However, we badly need more work on both Devon and Cornwall in the post-war era. Penruddock's Rising is examined in a 1955 Historical Association pamphlet of that name by A.H. Woolrych; although no major new sources or interpretations are presented, A.E. Button's article 'Penruddock's Rising, 1655', *Southern History* 19 (1997) is a sound, brief account.

ACKNOWLEDGEMENT

The author is indebted to Professor Ivan Roots for reading and commenting upon an earlier version of this text and for his many invaluable suggestions.

POLITICAL AND MILITARY HISTORY

COASTAL DEFENCES AND GARRISONS 1480–1914

MICHAEL DUFFY

BEFORE THE EIGHTEENTH CENTURY

Throughout this whole period the South West was either at the forefront of new developments in fortification or not very far behind them. Its strategic position opposite some of the country's most belligerent rivals and guarding the main trade routes to and from the west at a time of vast expansion of that trade, yet remote from the main centres of national power, exposed it to attack either by enemy raids or full-scale invasion. There was a clear need to keep its defences constantly modernized and sufficiently respectable either to deter attack or to be able to resist long enough for additional forces to be assembled to drive off attackers (**Map 22.1**).

Looe, Plymouth, Dartmouth and Fowey had all been raided in the fifteenth century before the citizens of Dartmouth with the aid of a crown subsidy constructed a new castle at the entrance to the Dart between 1481 and 1495, which is generally recognized as the first offensive artillery fortification in England. Whereas artillery had hitherto been used to defend castles or blockhouses (such as those built at Fowey after the 1457 Breton raid) against attack, the great rectangular blockhouse with its large gunports fronting the river at Dartmouth Castle was there to enable larger and more powerful cannon to attack shipping trying to enter the harbour (**Map 22.3**).[1] Between 1491 and 1502, Kingswear Castle was built on the opposite bank of the Dart to supplement it, and other south-western ports began to follow suit on a scale not imitated elsewhere. Henry Tudor's arrival at Cawsand Bay near Plymouth in 1483 and Perkin Warbeck's at Whitsand Bay near Lands End in 1497 marked the area as an attractive landing place for claimants to the throne, while the union of Brittany and France in 1491 combined Breton knowledge of the coast with the resources of a more powerful and persistent foe, and led Henry VIII to order extra coastal defences in the South West when he went to war with France in 1512–14—not unwisely, for the French raid which burned Brighton in 1513 had been driven by storms up-Channel from its intended targets of the Fal or Plymouth. Further war with France followed in 1522–3.

In consequence Plymouth constructed five artillery blockhouses (1491–1512), St Ives built two, perhaps in the 1490s, Dartmouth added another bulwark at Bayard's Cove on the town's southern edge in about 1509–10, Fowey built St Catherine's Castle in about 1520, and by 1539 there were others at Brixham and Salcombe in Devon, and at Looe, with batteries also at Penryn and St Michael's Mount, in Cornwall.[2]

When the next invasion scare developed in 1538–9 the South West was perhaps better prepared than elsewhere. Henry VIII's great coastal defence programme of 1538, therefore, gave priority to the south-east coast of England and ultimately only applied itself to filling in some gaps in the South West, where two circular artillery forts were built at Pendennis and St Mawes, on each side of the entrance to Carrick Roads at the mouth of the Fal.[3] Further scares in the 1540s led to further strengthening of the defences. In 1545 another blockhouse was added near Plymouth at Mount Edgcumbe, a bulwark was built at Exmouth to reinforce another at Powderham Castle to protect the Exe estuary, and a further battery was constructed at Lamberd's Bulwark near Dartmouth Castle. In 1548–9 St Nicholas's Island in Plymouth Sound was provided with artillery fortifications, and between 1548 and 1554 steps were also taken to prevent the Scilly Isles becoming a base for pirates or foreign warships. A blockhouse, later known as King Charles's Castle, similar to St Catherine's Castle, Fowey, or Salcombe Castle (a tower with one face projecting forward to give a wider field of fire), and two smaller blockhouses were built on Tresco. Based on English experience of the Flanders campaigns of 1543–4, a fortification of the very latest Italian bastion-trace style was designed for Harry's Walls on St Mary's—'the most advanced piece of military engineering for its date to be seen in this country'.[4] However, bad siting and a government economy drive meant that it was only partly completed (**Map 22.2**).[5]

These were the defences, reinforced by temporary batteries, which saw the South West through the year of the Spanish Armada, 1588. It is fortunate that the Spanish never attempted to land because, when they did so in the early 1590s, defences were found to be inadequate. In 1595 a raid burned Mousehole, Newlyn and Penzance, another in 1596 landed in Cawsand Bay, while a planned invasion through Falmouth in 1597 was wrecked by a storm only two days from Lands End. In consequence there was a new burst of refortification. In 1593 Star Castle was begun on the headland of the Hugh on St Mary's, Scilly,

Map 22.1
Defences, 1480–1800.

FRENCH 1797

■ Early Tudor artillery blockhouse or bulwark

▼ Artillery fort of Henry VIII's 1538-1539 programme

✳ Late Elizabethan bastioned fortification

⊙ Defences of the Dutch Wars

◗ Defences of the French Wars

➡ Foreign threats

Scilly
King Charles' Castle
Cromwell's Castle
Star Castle Harry's Walls
The Garrison

Powderham Castle ■ Exmouth

Teignmouth

FRENCH 1690

Brixham
Berry Head
Kingswear Castle
Dartmouth Castle
Maiden Fort

DUTCH 1667
DUTCH 1688
FRENCH 1690
PLANNED FRENCH INVASION 1692

Salcombe

DUTCH 1667
FRENCH 1690

St Catharine's Castle
Fowey
Looe
Plymouth
The Citadel
Mount Batten

Mevagissey

St Ives

Penryn St Mawes
Pendennis St Anthony's Head

Penzance St Michael's Mount

Helford

DUTCH 1667
SPANISH 1590
FRENCH & SPANISH ATTACK 1513
PLANNED FRENCH ATTACK 1779
DUTCH 1667
FRENCH 1690

0   10   20   30km

SPANISH 1595
SALEE CORSAIRS 1625

PLANNED FRENCH ATTACK 1513
PLANNED SPANISH ATTACK 1597

SPANISH ARMADA 1588 ➡

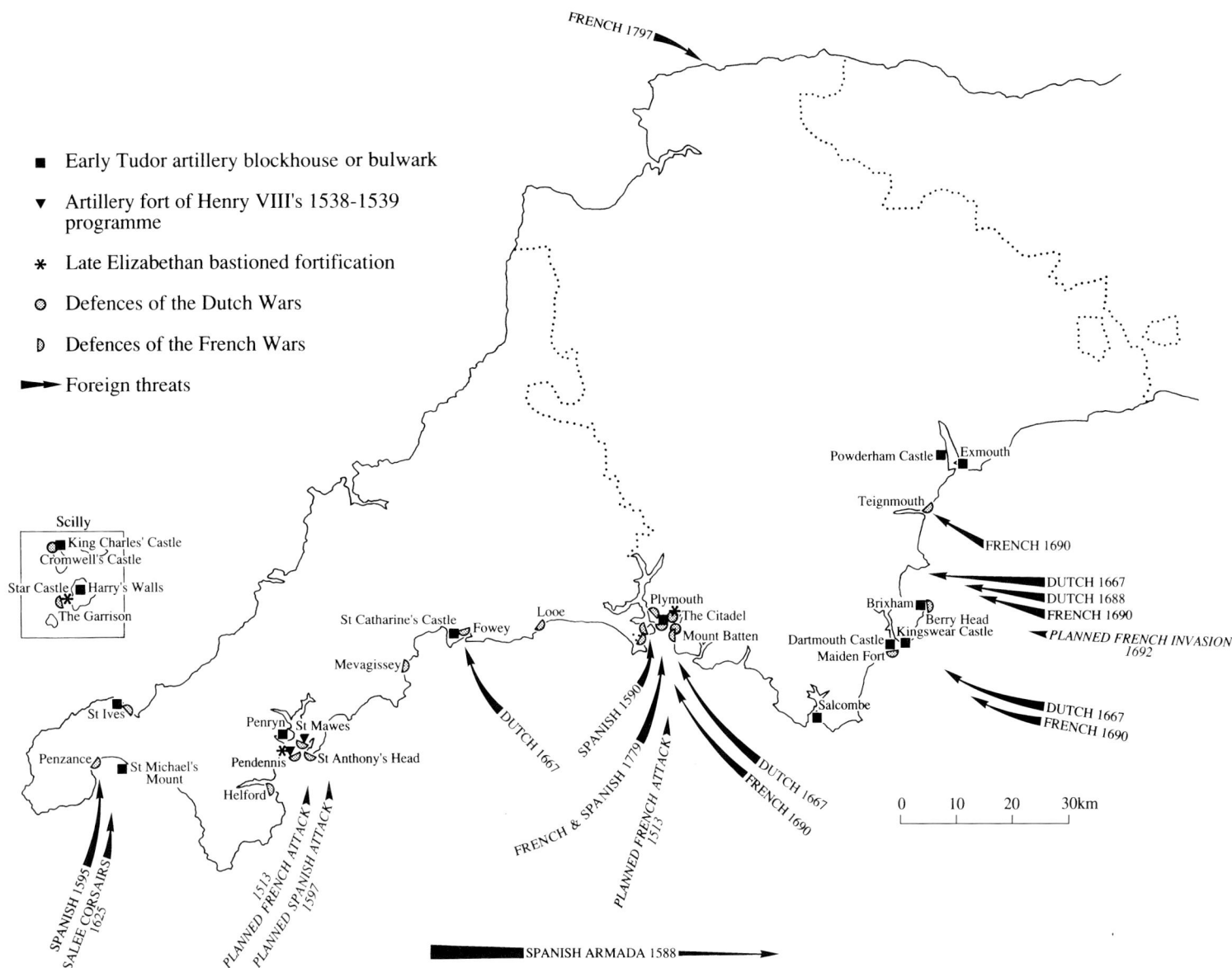

followed by a bastioned curtain wall across the neck of the Hugh itself. At Plymouth, St Nicholas's Island was refortified and between 1592 and 1596 a modern bastioned fort was built on the Hoe, which, with the increasing range of artillery, commanded both the entrance to Sutton Harbour and the town itself. Lastly, in recognition of the fact that Falmouth could not have resisted a Spanish landing, an extensive new bastioned *enceinte* was built between 1598 and 1611, subsequently extended by hornworks and advanced works, to cover the entire highground of Pendennis Head, converting Pendennis into a continental-standard fort of a size which was matched in England only by

the contemporary refortification of Carisbrooke Castle on the Isle of Wight (1597–1610), and whose strength was shown by its prolonged five-month resistance to the victorious New Model Army in 1646.[6]

The Civil War revealed the strengths and weaknesses of the South West's coastal defences. Against sea assault, blockhouses and temporary batteries proved sufficient to prevent the rather weak squadron of the parliamentary admiral, the earl of Warwick, from penetrating the Exe far enough to relieve Exeter in 1643 or to rescue Essex's army stranded up the Fowey in 1644, but the Scillies succumbed to attack by Blake's much stronger force in

Map 22.2 (right)
Harry's Walls, St Mary's, Isles
of Scilly: the original plan of
what remains the earliest
surviving example of a
bastion-trace defensive work
in Britain.

SOURCE: H.M. Colvin (ed.),
The History of the King's Works,
Vol. IV, Part 2
(London, 1982), 590,
based on a plan of 1551
at Hatfield House
(CPM II.34).

Map 22.3 (below)
The basement and main
gundeck of Dartmouth Castle:
the first offensive artillery
fortification in Britain.

SOURCE: A.D. Saunders,
Dartmouth Castle, 3rd edn
(London, 1991).

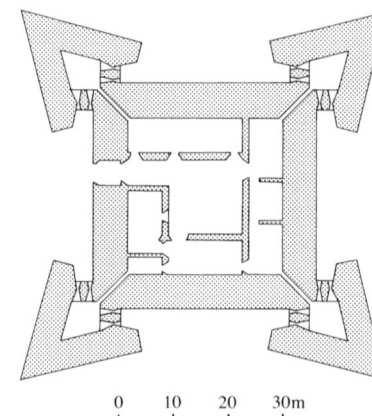

0 1 2 3 4 5m

Fig. 22.1
Engraving by Lowry of
Pendennis Castle, Falmouth,
*c.*1750.

SOURCE: reproduced by
permission of Westcountry
Studies Library, Exeter
(P&D 10282).

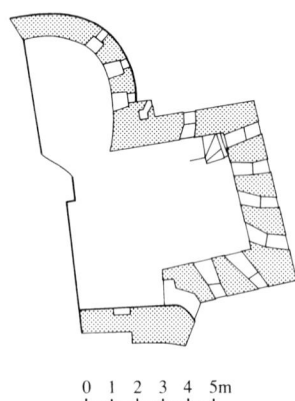

1651. However, only Plymouth and Pendennis showed themselves able to resist landward assault for any considerable length of time. There was still cause for anxiety, therefore, when from the 1650s England was faced by a powerful new maritime enemy, the Dutch. This threat produced two similar tall, round artillery towers in the early 1650s at Mount Batten, Plymouth, guarding the southern side of the entrance to the Cattewater, and Cromwell's Castle, Tresco. It then led to the construction, in place of the old fort on Plymouth Hoe, of the new Citadel (1665–77), to the designs of the engineer Sir Bernard du Gomme and of a standard approached elsewhere in England only by his *enceinte* to Portsmouth and his fort at Tilbury, and considered 'the best of the very few fortifications of its period in this country, to be set beside the masterpiece of Vauban in France'.[7]

The Citadel was only partially built, however, when the South West's defences were tested in 1667 by the fleet of Admiral de Ruyter, after the Dutch had won control of the Channel by their destructive raid on the English fleet in the Medway. De Ruyter anchored in Torbay, raided Torquay, tried to get into Dartmouth where the Mediterranean trade had taken shelter, went on to anchor in the Sound, and finally tried Fowey where the Virginia trade had sought refuge. In each case old defences were rearmed and temporary batteries established which kept the Dutch out. The South West was very much the focus of attacks in the late seventeenth century with invasions by the duke of Monmouth at Lyme, across the Dorset border, in 1685 and by William of Orange at Torbay in 1688, followed by a French raid in 1690 and a planned French invasion via Torbay in 1692. Of these the most traumatic was the appearance of the French Admiral Tourville in Torbay in 1690, having won control of the Channel by his victory at Beachy Head. He burned Teignmouth and tried to get into Dartmouth and Plymouth before withdrawing to Brest. Again temporary batteries were armed and Lambert's Bulwark at Dartmouth was rebuilt in stone as the main defence of the harbour entrance.[8]

0 10 20 30m

THE EIGHTEENTH CENTURY

Apart from the bastioned *enceinte* of Pendennis, which was repaired and put in a state of defence during a further invasion scare in the 1740s, and especially Plymouth Citadel whose defences were again upgraded in the 1750s and which remained one of the principal fortresses in the country, the tendency in the seventeenth and eighteenth centuries was to move out of the old stone castles and blockhouses, now inadequate against the weight of modern artillery fire, and to rely on batteries placed behind protective earthworks, sometimes faced with stone for more permanence. Many of the south-coast ports of Devon and Cornwall armed wartime batteries at their harbour entrances during the French wars of the mid-eighteenth century. More permanent fortified batteries existed at Falmouth, where the principal defence of the harbour became batteries established below Pendennis and St Mawes (**Fig. 22.1**), and at Dartmouth, where batteries were built on each side of the Castle, and Lambert's Bulwark became Maiden Fort, rebuilt as a two-tiered stone battery in 1747. Only on St Mary's, Scilly, however, was larger work undertaken when a bastioned curtain wall was built around most of the Hugh between 1715 and1746, leading to its renaming as the Garrison.[9]

The emphasis was on seaward defences rather than protection against attack from the landward side. Unlike the naval dockyard at Portsmouth, Plymouth Dock, begun in 1691, was given no landward defences until a continuous line of earthworks was thrown up around both dockyard and town during another invasion scare in 1756–7. Increasingly, defensive attention focused on the bases used by the Royal Navy in the region. A small 'Western Squadron' had been based at Plymouth since the 1650s and the building and dramatic growth in size of the dockyard, which rivalled Portsmouth by the 1780s, brought naval warships there in ever growing numbers and size as the Western Squadron became the main operational weapon of the navy in home waters. The fleet sought to maintain station in the western approaches watching the main French fleet base at Brest. It used Plymouth for repairs, but, because the Sound was unsafe except in the summer months until the Breakwater was built (1812–48), it used Torbay as

Map 22.4
Defensive installations,
1800–1914.

* Naval anchorages

► Napoleonic Wars coastal battery

● Napoleonic Wars signal station

○ Mid-Victorian coastal defences

St Mary's Late Victorian coastal defences

▣ Garrison towns

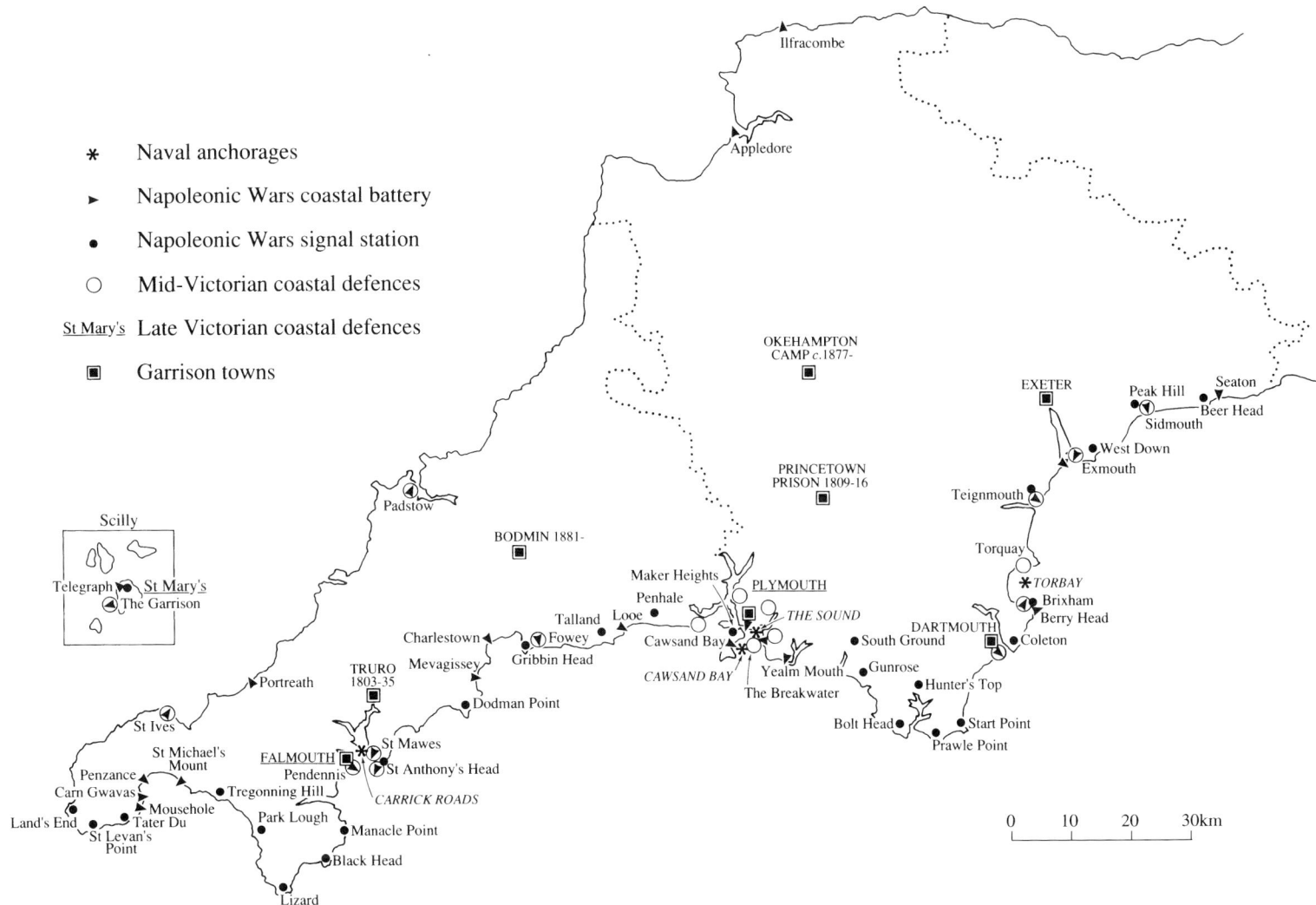

its principal refuge in bad weather, and a watering and stores base was created at Brixham where it could resupply while sheltering. Torbay was the largest and best sheltered anchorage along the whole coast, used not just by the Royal Navy but by the Dutch in 1667 and 1688 and the French in 1690, yet curiously, even though it had been identified as a gap in coastal defences as early as the great survey of 1538–9, little had been done to plug it except for the local defence of Brixham.[10]

The danger to both bases was highlighted in late-eighteenth century wars, when, in 1779, a Franco-Spanish fleet appeared in overwhelming force off Plymouth, and in 1781 and 1799 the proximity of further Franco-Spanish armadas sent smaller British fleets scurrying into Torbay as a defensive anchorage. At Plymouth it resulted in more batteries placed around the Sound, the Lines around the town were strengthened and detached redoubts were built as landward defences outside them, while batteries were erected against a French landing in Cawsand Bay, with defensive

redoubts on the Maker Heights (made permanent in the 1780s) to prevent a western landward assault getting into a position to bombard the dockyard from across the Hamoaze. At Torbay four new batteries were established between 1779 and 1783, three of them powerfully armed on Berry Head, to dominate the bay and its seaward approaches, and protected by an entrenchment across the neck of the headland against land assault. In 1794 Berry Head was purchased by the crown and work began to make these defences permanent. Elsewhere the 1779 scare saw new batteries established at Fowey, Looe and Mevagissey. A further scare in 1794 saw new batteries at Sidmouth, Exmouth and Teignmouth (**Fig. 22.2**), and a start was made on a system of watch and signalling stations that eventually covered the entire southern coast. The threat of invasion or raid was still a live one, and in 1797 a French raiding force even appeared on the less protected north coast, off Ilfracombe, before sailing on to Fishguard. In 1800 the South West was still as braced against attack as it had been in 1480![11]

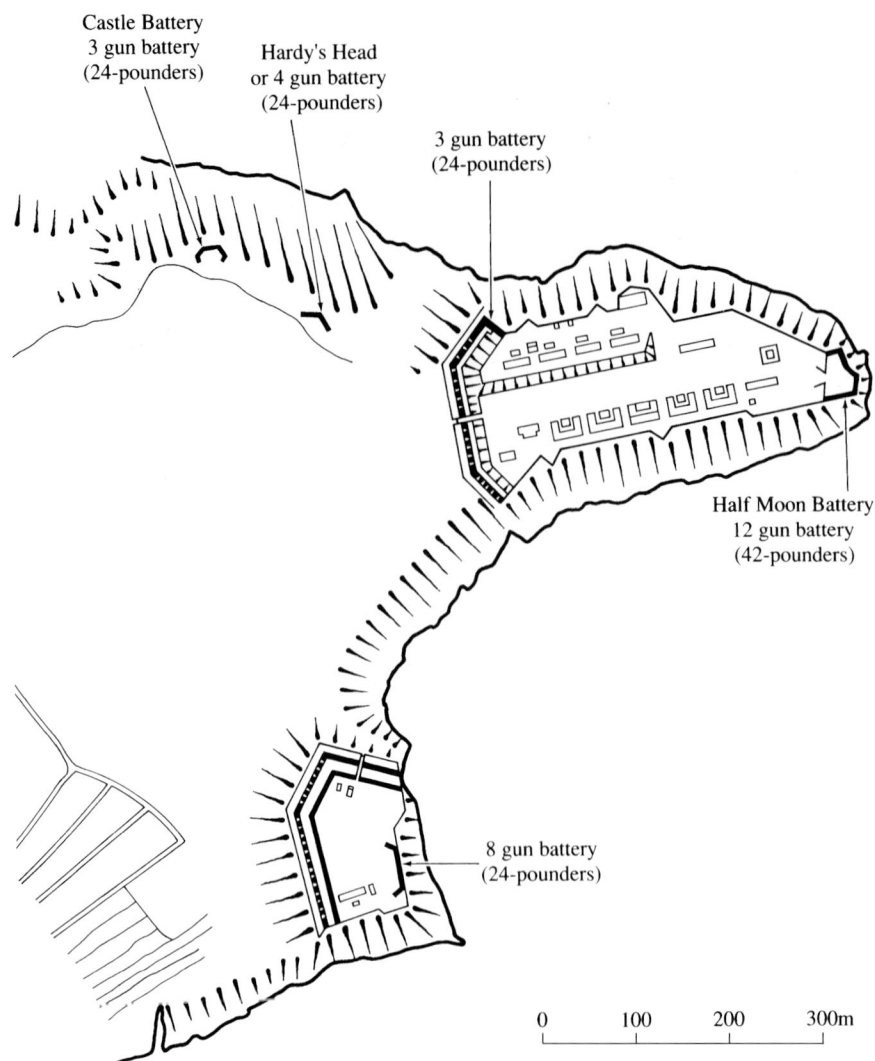

Castle Battery
3 gun battery
(24-pounders)

Hardy's Head
or 4 gun battery
(24-pounders)

3 gun battery
(24-pounders)

Half Moon Battery
12 gun battery
(42-pounders)

8 gun battery
(24-pounders)

0   100   200   300m

DEFENSIVE INSTALLATIONS AND GARRISONS, 1800–1914 (Map 22.4)

Map 22.5
The fortified batteries at Berry Head during the Napoleonic Wars.

SOURCES: based on a plan in the Public Record Office (MPH 233/11), reproduced in A.D. Saunders, *Fortress Britain* (Liphook, 1989) and the *Berry Head Official Guide Book* (Torquay, n.d.).

It is perhaps ironical that the high-tide of British power saw the nation at its most jittery about the possibility of invasion. The threat was real enough up to 1815 while Napoleon dominated the Continent and in 1801 and again in 1803–5 massed his army on the opposite coast. The entire country was mobilized against the danger, with local volunteer units formed for home defence and coastal batteries established, manned by volunteer artillery companies. The line of coastal signal stations, developed in the 1790s and intended to pass news of French invasion forces to the nearest naval warships or port, was reopened in 1803 and extended to the Scilly Isles, first on St Martin's in 1804, followed eventually by a still-surviving stone-built semaphore tower at Newford Down on St Mary's.[12]

The more major defensive works, as before, focused on the naval bases. At Torbay a small naval hospital was built at

Paignton in 1800, the watering wharf at Brixham was enlarged in 1801, and the victualling depot extended. The fortification of Berry Head (**Map 22.5**) was completed after 1803 as two stone-faced forts, fronted by ditches on their landward side, with massive heavy batteries facing out to sea, while two smaller adjacent batteries and another on the other side of Brixham near Finchcombe completed the defence of the base and the bay. At Plymouth the Lines around Plymouth Dock were reconstructed between 1801 and 1816 as a stone-faced rampart, fronted by a deep ditch, while the Maker defences protecting the dockyard against land assault from the west were strengthened by the building of barracks for the garrisons of the redoubts and the construction of a further redoubt at Empacombe. One particular defensive headache at Plymouth was the large number of prisoners of war held in hulks in the Hamoaze, a problem solved by their removal to a large prison, built at Princetown on Dartmoor between 1806 and 1809 and further extended in 1812 to house captives from an additional war with the United States. Defences were also strengthened at Falmouth where a naval stores depot was established, used by a frigate squadron based there in the 1790s and by the battleships of the Western Squadron from 1806.[13]

The ending of the Napoleonic Wars in the total defeat of France left the Royal Navy without a rival, and coastal defences were allowed to run down as part of the peace dividend. The navy concentrated its south-western operations into the Plymouth base. The Breakwater was built to provide a sheltered anchorage in the Sound between 1812 and 1848, obviating the need to use Torbay. All victualling operations were concentrated into the new Royal William Victualling Yard built at Stonehouse between 1825 and 1833. The naval depot at Falmouth was abandoned in 1815 and Berry Head was sold back into private hands in 1820. Such attention to coastal defence as continued thereafter was consequently focused on Plymouth, where naval investment further expanded with the building of the new Keyham Steam Yard between 1848 and 1853. The French navy was slowly rebuilt and extended its operational effectiveness by building a new base at Cherbourg. As the French moved enthusiastically into building steam warships and then ironclad warships, and a new Napoleon—Napoleon III—seized power in France in 1851 and made himself emperor in 1852, concern increased that the French might use the new mobility of steam power to slip past the British fleet and make a surprise attack, either to raid and destroy a British naval base or to effect a full-scale invasion. Successive invasion scares occurred in 1846–8, 1851–2 and 1858–9.[14]

The initial defensive response was to strengthen the batteries protecting the sea-approaches to Plymouth. In 1847–9 new batteries were built opposite the ends of the Breakwater. Concern for the land defences led between 1853 and 1863 to the further remodelling and completion of the Lines around Devonport (which Plymouth Dock had been renamed in 1824).[15] But these efforts were quickly overtaken by the rapid developments of

technology which now made all former defences obsolete and precipitated the most thorough examination of Britain's coastal defences and the most extensive refortification programme since Henry VIII's great survey of 1538–9.

THE ROYAL COMMISSION ON THE DEFENCES OF THE UNITED KINGDOM, 1859

The development of rifled and breech-loading cannon doubled the range of artillery and, with the replacement of solid shot by explosive shells, provided an attacker with the capacity to stand off at a distance, away from the existing fortifications, and inflict extensive damage. These same developments, combined with screw-propelled steam power and armour-cladding, further threatened to enable attacking warships to overcome the disadvantages that slow-moving, poorly protected and easily damageable wooden sailing ships had always experienced when facing artillery sea-defences ever since the building of Dartmouth Castle in the 1480s. The result was the enforced development of new and far more extensive defensive systems.[16] Modernization of the western defences of Plymouth began in 1858 with the building of two new distant detached forts at Tregantle and Scraesdon at the neck of the Rame peninsula, but the panic of 1858–9 accelerated and vastly expanded the work of refortification as a result of the 1859 Royal Commission on the

Defences of the United Kingdom. On the basis of its report in 1860, a ring of detached forts was built for landward defence of Plymouth, 4,000–7,000 yards away from the dockyard, in order to keep enemy artillery out of range. These forts abandoned the now discredited bastion-trace system in favour of a simpler polygonal form and were sunk into the ground in bomb-proof barrel-vaulted casemates, covered with concrete and topped with earth (**Map 22.6**). For seaward defence, batteries of massed cannon in armoured casemates (for protection against shrapnell and shellfire) were built in large numbers at the water's edge (**Map 22.7**), together with an armoured iron fort behind the middle of the Breakwater. With the breeches of breech-loaders proving initially unreliable, the heaviest, ship-stopping rounds had still to be fired by rifled muzzle-loaders, whose slow rate of fire meant that these batteries had to be armed with massed ranks of cannon (forty-two in two tiers at Picklecombe) in the hope that the sheer weight of their combined shellfire might stop attacking warships in the short period during which they could be fired on as they steamed past. It is some measure of Plymouth's

Fig. 22.2
Public notice regarding the erection of a Battery at Teignmouth, 27 July 1794.

SOURCE: reproduced by permission of Devon Record Office (1919Z/Z1/p. 60).

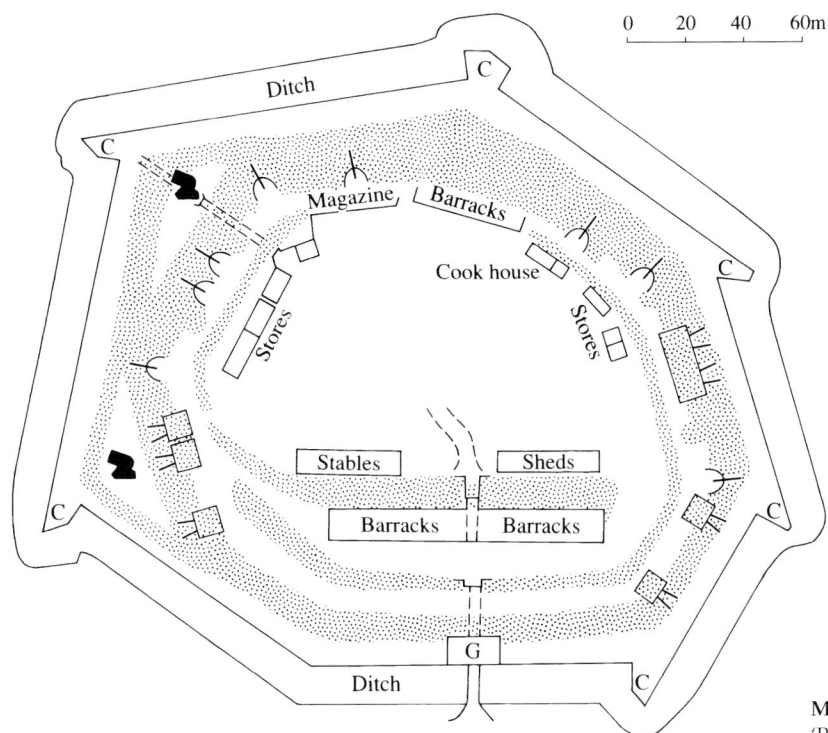

G   Guard house & cells

Mortar battery

Gun in covered 'Haxo' casemates

Open gun emplacement on rampart

C   Caponiers

Earth banked rampart

Map 22.6
'Palmerston's Follies': land defences—Crownhill Fort, the central pivot of the landward defences of Plymouth, showing the basic design of the new, detached, polygonal forts of the 1860s (armament arrangement of 1893).

SOURCES: special 1:2500 survey for the War Office, 1896; F.W. Woodward, *Plymouth's Defences* (Ivybridge, 1990).

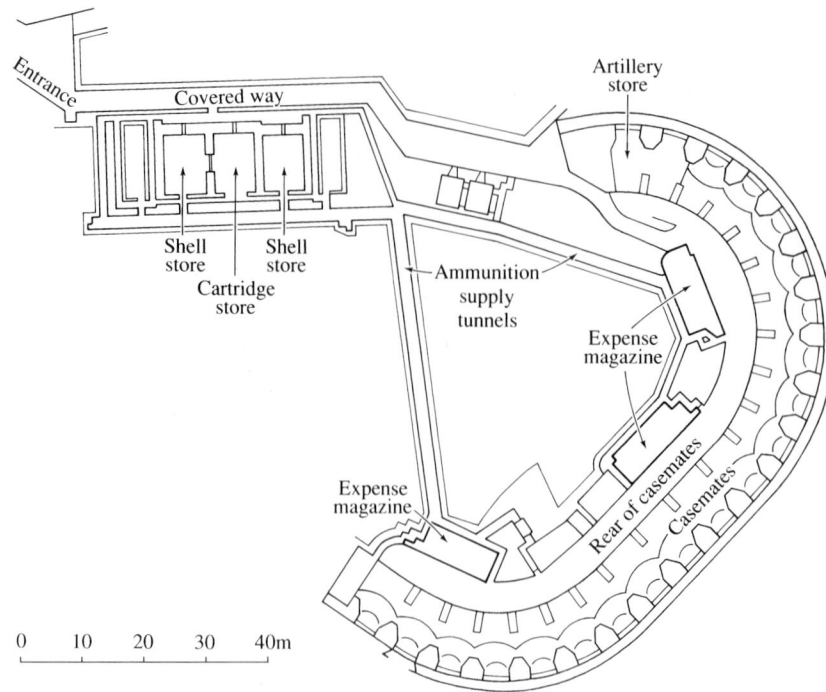

(two), Brixham, Fowey, St Ives, Padstow and Scilly (two) are among those recorded as arming batteries at this time.[18]

The 1859 Royal Commission concentrated on the defences of the naval dockyards, but it was followed by further enquiries by the Committee on Defence in the 1880s which looked at the commercial ports as it became clear that French strategic planning was turning more towards commerce-raiding and that British harbours were vulnerable to hit-and-run attacks by fast cruisers and the new small and highly manoeuvrable torpedo boats. Moreover the ever-increasing range of artillery and ever-increasing speed of warships was already exposing the limitations of the recent 1860s defences. Attention particularly concentrated on the sea-defences where the limited elevation and field of fire afforded by the narrow apertures of the casemated batteries no longer sufficed. The growing ability to hit at greater distances and eventually the perfection of breech-loading mechanisms removed the need for so many guns and saw the creation of much smaller, more open batteries both of long-range guns to keep more powerful warships at a distance, and of new quick-firing guns to stop torpedo boat attacks. Naturally the Plymouth defences were given priority, but the need to protect commerce passing along the south-western coast, and the need for advanced coaling and despatch and signals stations for the blockade of Brest now that steam had speeded-up the pace of naval warfare, led to Falmouth and St Mary's Road in the Isles of Scilly being raised to the status of defended anchorages in the 1890s.[19]

At Falmouth the bastions of the old *enceinte* of Pendennis were rebuilt and strengthened to protect batteries of quick-firer guns, and a long-range battery was established on the headland below the castle. On the opposite side of the harbour entrance, quick-firers were placed above St Mawes Castle and a long-range battery positioned to seaward at St Anthony's Head on the tip of

importance that the Commission recommended spending £3,020,000 on strengthening its defence, £220,000 more than on those of Portsmouth, and that these two accounted for more than half of the proposed £11,850,000 expenditure.[17]

In the South West only Plymouth was provided with detached forts in the 1860s, but at Dartmouth the former Maiden Fort was rebuilt as a two-tiered casemated battery. Elsewhere, volunteer artillery companies were again formed to man new open earthwork batteries which sprang up at towns and harbours along the coast: Sidmouth, Exmouth (two), Torquay

the St Just peninsula.[20] For the Isles of Scilly, two long-range batteries and one quick-firer battery were established on the Garrison at St Mary's (1898–1901) and a further quick-firer battery built at Bants Carn at the northern end of the island to cover Crow Sound. The latter, however, was not armed before the plan was aborted in 1906 by the replacement of France by Germany as Britain's main naval rival: a fleet anchorage was developed at Scapa Flow in the Orkneys instead.[21]

NINETEENTH-CENTURY GARRISONS

The defences of the South West were further strengthened by a growing military presence in the area. There had always been artillery garrisons at Dartmouth, Plymouth and Falmouth, and army regiments were regularly quartered at Exeter and Plymouth. Plymouth Dock had barracks built behind the new lines of the 1750s, and a Marine barracks was built at Stonehouse in 1779–85. Such early permanent accommodation was unusual, but from the revolutionary period of the 1790s there was a concerted move towards providing troops with permanent barrack accommodation in order to separate them from political agitators in the civilian population amongst whom they had formerly been quartered. At Exeter a major cavalry barracks for 400 men was built off New North Road in 1792 and the Wyvern Barracks constructed in 1804. Other small cavalry barracks for sixty men were built at Barnstaple, Honiton, Modbury and Totnes, and Truro had a wooden cavalry barracks from 1803 to 1835. Barracks also existed at the Berry Head defences at Torbay.[22] The third quarter of the nineteenth century saw a renewed expansion of the military presence. In 1853 the Raglan Barracks were built at Plymouth. After a trial firing in 1869, the army began to hold manoeuvres on southern Dartmoor in 1873. The artillery established a firing range and a camp near Okehampton after 1875, erecting permanent buildings in 1892–4, and in 1900 it purchased a further range at Willsworthy. In 1880–1 the Victoria Barracks were built at Bodmin as depot for the newly formed Duke of Cornwall's Light Infantry and in 1901 a new artillery barracks was built at Pendennis Castle. The Royal Navy too expanded the presence of its personnel with the move to Dartmouth in 1863 of the officer-cadet training hulk, HMS *Britannia*; the imposing buildings of the Royal Naval College were constructed in 1899–1905. There was also a build-up of shore-based ratings with the change to permanent naval service rather than recruitment for the commission of a particular ship, and this led to the building of the naval barracks, HMS *Drake*, at Keyham, Plymouth, between 1879 and 1886.[23] The people of the South West themselves also contributed to this heightened

military activity through the militia and the revived volunteer movement of 1859–60, which provided a further pool of manpower that would have been concentrated in the new Plymouth fortifications in the event of attack (**Fig. 22.3**). In 1882 the inspected strength of the Devon militia was 1,320 infantry (two battalions) and 418 artillery, and the Cornish militia numbered 688 Cornish Rangers and 348 Devon and Cornwall Miners, while the volunteers totalled 4,077 in Devon and 2,085 in Cornwall.[24] They were based on drill halls which were established in many towns across the two counties. With improved communications by rail and road, there was unprecedented confidence in being able to concentrate trained armed forces where they would be needed in any crisis in the area.

FURTHER READING

H.M. Colvin (ed.), *The History of the King's Works, Vol. IV, 1485–1660*, Part 2 (London, 1982) provides thorough background articles by J.R. Hale on defences from 1485–1558, J. Summerson on Elizabethan defences, as well as others on each of the particular defences built by the crown. A.D. Saunders, *Fortress Britain: Artillery Fortifications in the British Isles and Ireland* (Liphook, 1989) is the best general work on the subject.

For brief surveys of each county see A.D. Saunders, 'The coastal defences of Cornwall', *Archaeological Journal*, 130 (1973), 232–6, and R.A. Erskine, 'The military coast defences of Devon, 1500–1956', in M. Duffy *et al.* (eds), *The New Maritime History of Devon, Vol. I* (London, 1992), 119–29.

The best general surveys of threatened or actual invasions and raids are N. Longmate's two volumes, *Defending the Island: From Caesar to the Armada* (London, 1990) and *Island Fortress: The Defence of Great Britain 1603–1945* (London, 1991).

K.W. Maurice-Jones, *The History of Coast Artillery in the British Army* (London, 1959) provides details of local armaments. F.W. Woodward, in *Plymouth's Defences* (Ivybridge, 1990) and *Forts or Follies? A History of Plymouth's Palmerston forts* (Tiverton, 1998), gives a succinct summary of artillery developments in relation to Plymouth, while M.S. Partridge, *Military Planning for the Defense of the United Kingdom, 1814–1870* (New York, 1989) and I.V. Hogg, *Coast Defences of England and Wales 1856–1956* (Newton Abbot, 1974) set Plymouth in the national perspective.

Two useful articles on the late Victorian coastal defences of the South West can be found in Ian Hunt, 'Plymouth sound? The defence of a naval station', *Fort*, 11 (1983) 78–108, and I.V. Stevenson, 'Some West Country defences', *Fort*, 17 (1989) 11–26.

POLITICAL AND MILITARY HISTORY

DEFENCE AND DISRUPTION

THE FIRST AND SECOND WORLD WARS

MARK BLACKSELL

Whether or not a war is in progress, the military and their activities are ever-present realities in most parts of the country. The South West is no exception and has traditionally played a full part in the defence of the realm, with particular attention being accorded to the coastline of its long peninsula facing the Atlantic to the west and north, and the English Channel and continental Europe to the east and south. During the two World Wars in the first half of the twentieth century, this role was greatly intensified and once again put people to the test, deeply and at times cruelly disrupting the life of the region.

As in the rest of Great Britain, the Second World War from 1939 to 1945 marked an important turning-point for the region in military terms. For the first time since the Civil War nearly 300 years previously, Cornwall and Devon and the neighbouring counties were subject to sustained attack, albeit mainly in the novel form of bombing raids, and, as a result, were directly involved in the conflict in a way that was new and shocking to the people who lived there.[1]

How the south-west region moved from its role in the nineteenth and earlier centuries of providing coastal defence and essential support to national military operations, to one of more direct involvement, is the main theme of this chapter. Significant parts of the landscape were permanently maimed as a consequence and the relative luxury of languishing on the periphery of wartime conflicts was gone, probably for ever.

THE FIRST WORLD WAR

As far as the South West was concerned, the early years of the twentieth century, including the First World War itself, were a period of transition. The traditional role of ports such as Plymouth and Falmouth, key locations for maintenance and supply of the nation's navy, remained undiminished, but inland the moorland of Dartmoor and open heaths such as Woodbury Common in south-east Devon became increasingly important for training land-based forces as well (**Map 23.1**).

On Dartmoor the first formal move towards establishing permanent military training was taken in 1873, when the town clerk of Okehampton wrote to the War Office suggesting the northern part of the moor as an appropriate location.

Although there was no great initial enthusiasm in Whitehall, the idea gradually took official root and by the end of the century the firing ranges were well established, together with a permanent military camp (see also chapter 22). It is worth recording that this development was very much welcomed at the time in Okehampton itself. The arrival of the railway in 1871 had seen much of the economic activity on which the town depended siphoned away eastwards to Exeter, so that the new military investment represented a considerable fillip to the town. Subsequently, the extent of military training was greatly increased on Dartmoor to include sites across the whole of the open moor and encompassing almost 30,000ha by the end of the Second World War. Since then the area used for live firing has been progressively reduced, but the Okehampton ranges remain an important training facility for live firing.[2]

The improved accessibility of south and east Devon to other training grounds elsewhere in England occasioned by the building of the railways was also an important factor in causing other areas of open land to be reserved for military training manoeuvres. Woodbury, Colaton Raleigh and Aylesbeare Commons, a succession of sandy heaths inland from the Exe estuary and to the south-east of Exeter, were all requisitioned by the War Office for this purpose in 1893 and to some extent are still used for training.[3] The net result of all this interest by the military planners was to broaden and increase significantly the strategic importance of the south-west peninsula and the trend has continued, and subtly intensified, throughout the course of the twentieth century.

As far as the First World War itself is concerned, the years from 1914 to 1919 temporarily saw an actual reduction in training, because the troops were otherwise engaged in continental Europe, but although the region was for the most part removed from hostilities, the impact of the conflict was still very immediate. The evidence for this remains clearly visible in almost every town and village in the shape of memorials to those thousands of men who went away to fight and were killed (**Fig. 23.1**). There were many more, of course, who joined up and returned with the outbreak of peace, and the trauma and disruption to the social order of all the communities in the region that the loss of a large proportion of its male population caused, particularly during the war itself, are almost unimaginable.

One of the most obvious changes during the war was the way in which women took over jobs that had previously been the exclusive preserve of men. A graphic illustration of how disturbing this proved to be is provided by reactions to the Women's Land Army in Cornwall.[4] It was formally constituted in January 1917 and a year later there were forty-seven gangs of about 500 women in the country supervised by thirteen district committees. Most of the workers were part-time, but a core of about 1,000 were full-time and about 300 were trained to do specialized jobs, such as maintaining machinery. Initially the farmers for whom the members of the Woman's Land Army worked were very suspicious and sceptical, but they desperately needed the labour and seem to have been most pleasantly surprised at the ease with which the new recruits adapted to what was asked of them. Once the First World War was over, women's labour was for the most part quickly reconfined to domestic work in the house and home. The combination of demobilized troops returning and looking to reclaim their pre-war positions, the general depression in agriculture, and the substitution of manpower by machines on farms, meant that few of the jobs undertaken by the Women's Land Army were open permanently to the women concerned.

THE SECOND WORLD WAR

In the twenty years between the two World Wars, there was a gradual increase in the range and extent of military activity and facilities in the South West. Military training resumed on both

Fig. 23.1
The war memorial at Lympstone, Devon.

SOURCE: photograph by A. Teed.

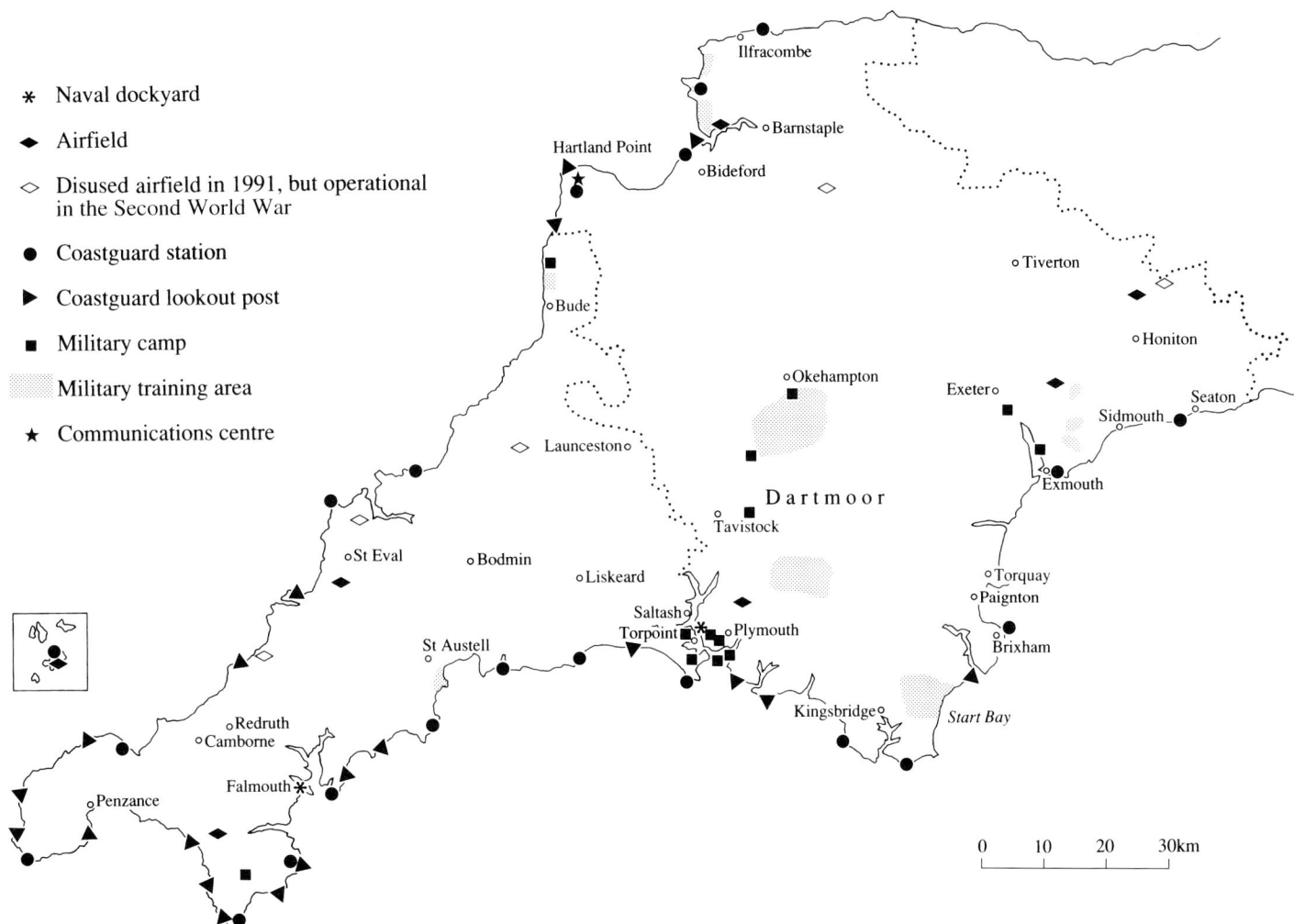

* Naval dockyard

◆ Airfield

◇ Disused airfield in 1991, but operational in the Second World War

● Coastguard station

▶ Coastguard lookout post

■ Military camp

Military training area

★ Communications centre

Map 23.1
Major military installations to the end of the Second World War.

SOURCES: Ordnance Survey maps and documents in Devon and Cornwall Record Offices.

later it had returns from all its district and borough councils showing that there was space for 242,091, so long as 30,000 extra mattresses and 80,000 extra blankets could be found.[8]

If evacuees were billeted at a particular house the occupant was obliged to accept them, but in the event the numbers involved were much smaller than had been anticipated, mainly because evacuation itself was voluntary. The borough of Okehampton in mid-Devon only received about 10 per cent of the 1,932 evacuees it had been deemed capable of housing. Nevertheless, there are many harrowing stories of children separated from their parents being hawked around houses, disorientated, bewildered and frightened. In some areas a virtual slave market developed, with farmers selecting the biggest and strongest to work on their farms. Equally, there are many stories of happiness and kindness with lasting bonds being formed between the families so randomly thrown together. In his seminal book about life in wartime Britain, Norman Longmate includes the following story from the village of Combe Raleigh in Devon:

> one billeting officer placed, with some misgivings, a three-year-old toddler with three elderly maiden ladies. On arrival he stood solemnly gazing round at his new home with tears pouring down his cheeks and announced 'My name is Robert: I am a big boy and I don't cry—well not often'. Within a few months he was idolised by the whole household and when his mother eventually came to see him she was horrified when at bedtime he knelt down and prayed 'O God don't let this woman take me away; she says she is my mother, but I want to stay here with my aunties'.[9]

In addition to the official, organized evacuation of children, there was also an exodus from the major towns and cities of people who had the means to take rooms in hotels in areas such as the South West, which were thought to be well away from the dangers of bombing. In the first months of the war Longmate recalls advertisements in *The Times* listing: 'Sanctuary Hotels recommended by Ashley Courtney … Torquay. You can sleep at the Grand Hotel for the drone of an aero engine is rare and sirens even more infrequent … Queen's Hotel, Penzance … for a sense of security that cannot be beaten.'[10] The business was of course welcomed in the South West where the holiday industry had been decimated by the outbreak of war, even if the visitors were frequently stigmatized for their unpatriotic behaviour.

Bombing was not the only spectre which plagued people in the early months of the Second World War. The threat of invasion was also seen as real and imminent, so that coastal lookouts and defences were accorded great importance. In the South West with its long coastline, coastguard stations and lookouts were key installations and very much in evidence, forming a continuous chain all around the peninsula, with each clearly visible by its neighbours on either side (**Map 23.1**).

Fig. 23.2
Exeter airport in the 1930s.

SOURCE: photograph reproduced by permission of Devon Record Office.

land and sea with the added ingredient of the air force, the scale of which grew steadily throughout the 1920s and 1930s. Airfields sprouted throughout Cornwall and Devon, though at that time most were little more than large, flat areas of mown grass where aeroplanes happened to land. A windsock and a few rather ramshackle buildings were all that distinguished them from the surrounding countryside.[5] The photograph (**Fig. 23.2**), taken in the late 1930s, of what is now Exeter Airport gives a flavour of the haphazard informality of these new airfields, which were used for military and civilian purposes.[6] By the outbreak of hostilities in 1939 there were five such airfields in Cornwall and six in Devon (**Map 23.1**).

Even before it actually began, the Second World War was perceived very differently from its predecessor in that the civilian population was assumed to be directly threatened by aerial bombardment. As a result elaborate plans were drawn up to evacuate women and children from the major towns and cities in the south of England, which were deemed to be especially at risk.[7] The South West was to be a major reception area and detailed preparations were made to establish how many people could be accommodated. The organization was hierarchical in that government worked through the county councils, who in turn required all the urban and rural district councils and the borough councils in their areas to provide inventories of the number of habitable rooms available, how many persons could be housed in them, and what amounts of extra bedding would be required.

The speed of the response was impressive. In Devon the county council was informed in May 1939 that it would be expected to receive and house 116,200 people and three months

Once the war in Europe was fully engaged, the pattern of hostilities turned out to be substantially different in a number of respects from what had been anticipated. As far as the South West was concerned, this meant that between 1940 and 1944 the region suffered heavy intermittent bombing, and, although the feared sea-borne invasion never materialized, with one disastrous exception in 1944 in Start Bay described below, there was still severe disruption as a result of the Allied preparation for the D-Day invasion of France in June 1944.

The logic behind the bombing by the Luftwaffe of the South West was complex, varied, and is still open to speculation. In the early part of the war, when planes were smaller and less sophisticated and also based at airfields further away from the northern coast of France, access was an important factor. Even flying to the south coast of England and back the bombers were operating at the extremes of their range. As far as the specific targets are concerned, it has frequently been claimed that historic towns like Exeter were deliberately chosen as revenge targets for the Allied bombing of Lübeck. These became known as the Baedecker raids, because the towns were supposedly chosen from those mentioned in the famous German tourist guides as being particularly attractive and historically interesting. Ultimately, however, the bulk of the bombing was almost certainly carried out for serious strategic reasons. Plymouth and, to a lesser extent, Falmouth were key naval dockyards; the airfields, particularly those in Cornwall, were important bases for the airborne coastal defence system; finally these same airfields almost certainly acted as covert communications centres, which became increasingly important as radar and other electronic surveillance systems were deployed with great effect in the later stages of the conflict.

Whatever the motives, the bombing itself was sustained and widespread in south Devon and Cornwall from July 1940 until May 1944, with more sporadic attacks occurring elsewhere in the region. The cities of Plymouth and Exeter were bombed repeatedly and large parts of the centres of both were almost completely destroyed. In Exeter there were eleven raids between 7 August 1940 and 30 December 1942, the heaviest occurring on the night of 4 August 1942. In that raid it is estimated that 157 bombs were dropped in and around the city centre, devastating many famous landmarks such as the Cathedral and the Georgian buildings of the Northernhay and Bedford Circus (**Map 23.2**).[11] It was one of two really heavy raids, the other occurring on 24 April 1942, and it was notable both for the scale of the attack and the way the bombs were concentrated in and around the centre of the city. Some indication of the sense of shock is shown by the photograph of the students sorting through the equipment salvaged from the ravaged shell of St Lukes Teacher Training College on the morning after the raid (**Fig. 23.3**).[12]

In Plymouth the bombing was even more severe, but it needs to be emphasized that the raids were by no means confined to these two cities. As early as 1941, fifty-six churches in Devon had been damaged by bombs and fourteen had been completely

destroyed, while in Cornwall most towns suffered to some degree and several were attacked repeatedly. Aside from Saltash and Torpoint, both of which were attacked several times during the major raids on Plymouth, Falmouth and Liskeard were each bombed on five separate occasions, Penzance on three, Redruth and Truro twice, and Camborne and St Austell once.[13]

It might be argued that these were all just sporadic attacks associated with the major raids on Plymouth, with off-target enemy planes just unloading their bombs where they could, but the evidence points to much more systematic planning. It is estimated that 4,001 bombs were dropped on Falmouth and Truro on the night of 24 September 1942, while in the case of Liskeard 810 bombs are thought to have been dropped on 14 March 1941 and 427 a week later on 21 March 1941. Even more convincing is the record of sustained attacks on the military airfield at St Eval near Newquay on the west coast of Cornwall. It was attacked four times: on 3 October 1940, 1 April 1941, 10

May 1941 and finally on 17 May 1941 when 1,096 bombs were dropped. The South West may not have been at the front line during the Second World War, but the widespread and repeated air raids ensured that it suffered a campaign of terror second only to London and a few other major cities. The last raid was on Falmouth on 30 May 1944 and by that time the sound of sirens and exploding bombs had become almost a way of life for much of the population in the region.

Fig. 23.3
St Luke's Teacher Training College after the air raid of May 1942.

SOURCE: photograph reproduced by permission of Devon Record Office.

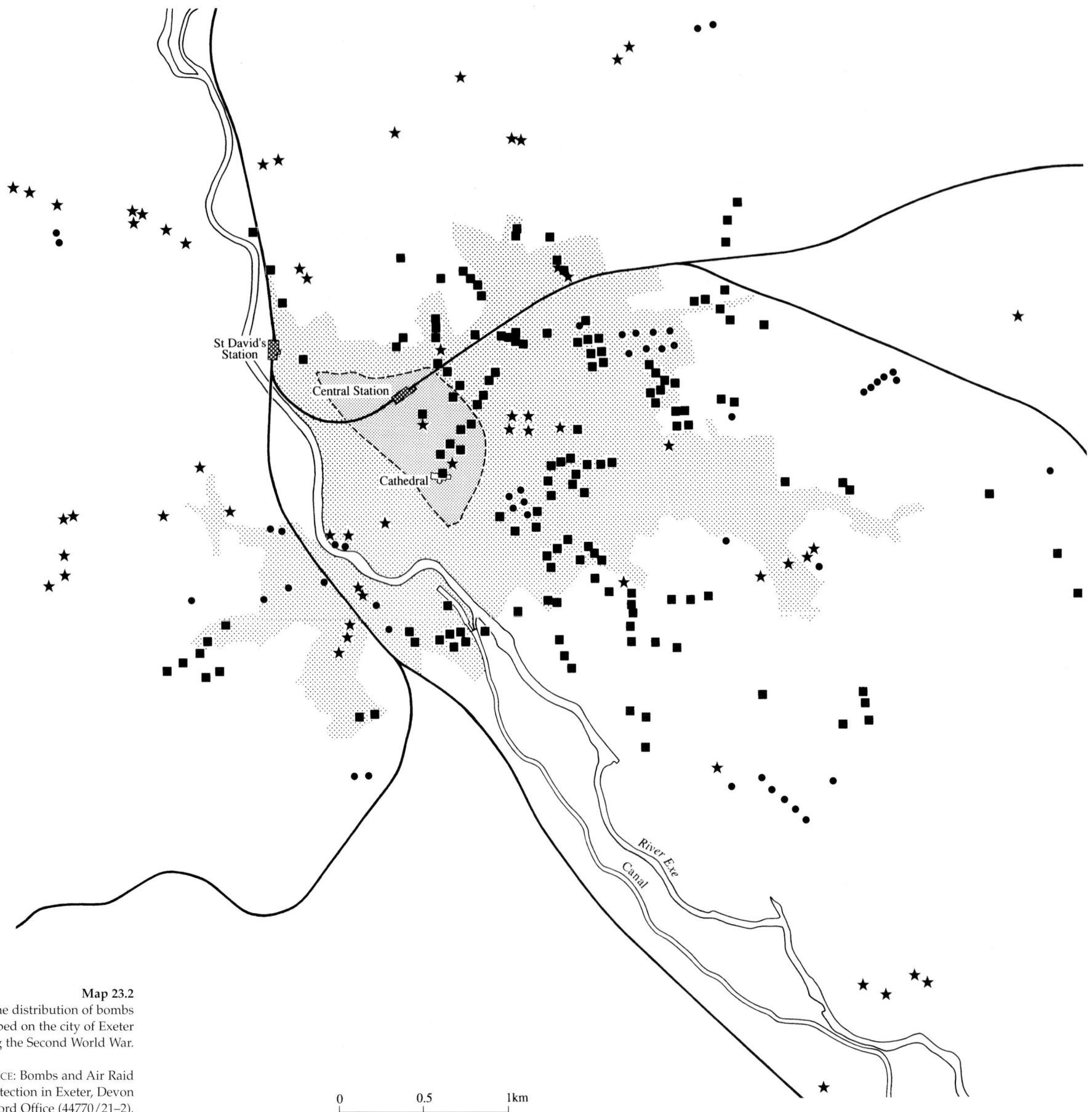

St David's
Station

Central Station

Cathedral

River Exe

Canal

Map 23.2
The distribution of bombs
dropped on the city of Exeter
during the Second World War.

Source: Bombs and Air Raid
Protection in Exeter, Devon
Record Office (44770/21–2).

0   0.5   1km

■	Major raid	4 May 1942	157 bombs
★	Major raid	24 April 1942	68 bombs
●	Minor raid	7 August 1940	5 bombs
		16 August 1940	3 bombs
		6 September 1940	1 bomb
		16 September 1940	2 bombs
		17 September 1940	9 bombs
		28 October 1940	1 bomb
		16 January 1941	5 bombs
		5 May 1941	3 bombs
		6 May 1941	9 bombs
		17 June 1941	6 bombs
		23 April 1942	6 bombs
		25 April 1942	1 bomb
		30 December 1942	5 bombs

Inner city centre

Built-up area 1942

Railway

PREPARATIONS FOR THE D-DAY LANDINGS

The other major disruptions to the life of the two counties, especially Devon, were the military activities by British and American troops in preparation for the D-Day landings in France on 6 June 1944. The dunes and beaches of Start Bay in south-east Devon, and Saunton, Croyde, Putsborough and Woolacombe in north Devon, were deemed to be ideal practice sites for simulating the planned invasion of Normandy and thousands of troops converged on the two locations to carry out the requisite training.

In the South Hams around Start Bay it was decided by the War Office that 12,000ha, involving parts of six parishes, 3,000 people, 750 families and 180 farms and other businesses, should be totally evacuated (**Map 23.1**). The request was received from Whitehall on 4 November 1943 by Devon County Council and the evacuation was completed on time by 20 December. The effect on the inhabitants of Slapton and the other villages affected was naturally traumatic; a whole network of communities were abandoned and left to the mercy of an unknown occupying force of American servicemen. In a vivid description of the whole episode, Grace Bradbeer relates how the following notice was left fastened to the main gate of each church in the evacuation zone:

> To our Allies of the U.S.A.
> This church has stood here for several hundred years. Around it has grown a community, which has lived in these houses and tilled these fields ever since there was a church. This church, this churchyard in which their loved ones lie at rest, these homes, these fields are as dear to those who have left them as are the homes and graves and fields which you, our Allies, have left behind you. They hope to return one day, as you hope to return to yours, to find them waiting to welcome you home. They entrust them to your care meanwhile, and pray that God's blessing may rest upon us all.
> Charles
> Bishop of Exeter[14]

It is now clear that the occupation of the whole area evacuated was complete and that the preparations for the sea-borne invasion, including the use of live ammunition, were comprehensive. Throughout the early months of 1944 a series of exercises was held to prepare the American and British troops for the invasion of Normandy the following June. The Allied command was under great pressure and this led to a number of serious mistakes being made in the mock landings at Slapton, which resulted in at least 100 casualties due to friendly fire. The culmination of these preparations was OperationTiger, centred on Start Bay but involving ships right along the south coast between Plymouth and Lyme Regis. Disastrously, on the night of 27–28 April a serious lapse in communication by Coastal Command allowed a flotilla of German E-boats to penetrate the naval defences and in the subsequent attack 946 American servicemen on board landing craft were killed.[15] This, and the other tragedies, brought the horrific reality of war very close to home in the South West, even though every effort was made to avoid the incident becoming public knowledge at the time. It is only now, more than half a century after the event, that the full extent of the carnage has been uncovered.

In north Devon the evacuation was less complicated, but the story was essentially the same, with a whole stretch of coastline requisitioned and occupied by Americans (**Map 23.1**). In addition, the shifting sands of the dunes between Saunton Sands and Braunton Burrows meant that much of the live ammunition and many of the mines laid as part of the operation were afterwards impossible to find and remained as a hazardous legacy after the war had ended.

THE POST-WORLD WAR LEGACY

Since 1945, the South West has reverted to its peacetime role of contributing to the national defence effort. The naval dockyard in Plymouth survives, though as the century comes to a close its long-term future is open to question. The continuing reduction in the UK's overall defence capability means that the yard has now been privatized and at the end of the millennium employs directly only 2,000 workers as compared with more than 20,000 at the end of the Second World War. Land-based training continues, and has been intensified on Dartmoor and in south-east Devon, assuming added importance since the Royal Marines established their permanent headquarters at Lympstone on the lower Exe estuary between Exeter and Exmouth, and overseas training grounds particularly in Germany have been relinquished. Some of the airfields continued to play a key role in the UK national defence strategy throughout the period of the Cold War. St Eval, renamed RAF St Mawgan, had a pivotal role in the national coastal defence system and Chivenor in north Devon was a major pilot training base. Here too, however, new technology and reduced defence expenditure have taken their toll and the levels of military activity have been much reduced. Nevertheless, Cornwall and Devon still house crucial activities. The communications centres at various locations, such as Goonhilly Down near the Lizard and Hartland at the entrance to the Bristol Channel, testify to the continuing importance of the south-west peninsula for early warning of impending invasion—a role famously embedded in the national psyche when the approach of the Spanish Armada was first seen from Plymouth Hoe by Sir Francis Drake more than 400 years ago.

FURTHER READING

A comprehensive and lively account of life in the UK during the two World Wars, including a number of detailed case studies from the South West, is provided by Norman Longmate in *How We Lived Then* (London, 1971). For a more detailed and well-illustrated account of Devon during the Second World War Gerald Wasley's *Devon at War 1939–45* (Tiverton, 1994) is very readable, while the compilation of contemporary documents assembled by Devon County Library Services, *Second World War: Home Front in Devon*, mimeo (Exeter, 1989), is an indispensable source of material on life in the county during the war. The most mysterious of the wartime events in the South West is the German E-boat attack on the American troops preparing for the Normandy landings in Start Bay in 1944. The events surrounding this and the whole history of the evacuation of the South Hams are included in Grace Bradbeer's *The Land Changed its Face: The Evacuation of Devon's South Hams 1943–44* (Newton Abbot, 1984), Edwin Hoyt's *The Invasion before Normandy: The Secret Battle of Slapton Sands* (New York, 1985), Ken Small and Mark Rogerson's *The Forgotten Dead* (London, 1989), and Nigel Lewis's *Channel Firing: The Tragedy of Exercise Tiger* (London, 1989).

POLITICAL AND MILITARY HISTORY

LOCAL GOVERNMENT AREAS
AND LOCAL AUTHORITIES
1801–1999

JEFFREY STANYER

[There are] certain occasions on which the little round of Devonshire business crossed or coincided with the larger orbit of the government of these islands. (A.H.A. Hamilton, 1873)[1]

Even educated men … often know more about the geography of remote countries than they know about the territorial subdivisions of their own country. (Census of Population, 1871)[2]

LOCAL GOVERNMENT HISTORY

The history of areas and authorities in Cornwall and Devon illustrates very clearly the ideas implicit in the two quotations at the head of this chapter. The quotation from Hamilton expresses exactly one theme of this chapter. The two counties have local government histories which are clear reflections of the operation of processes at national level. The distinctiveness of each county's history lies in the impact of these processes at county, district and parish levels. To the observer, therefore, both are a mixture of familiar and unfamiliar.

The ignorance of 'educated men' arose and still arises from the complexity of the historical processes that have created the successive 'landscapes' of local authorities. The complexity extends to minute details that are impossible to map on the scales used in this *Atlas*. These details are so numerous that even listing them in tabular form would take up a considerable amount of space and would not, of course, show their spatial characteristics.

Much of the work for this chapter was undertaken before it became common to talk about local governance in the widest sense rather than local government narrowly defined. Were it being planned today much greater emphasis would have been given to local administration by means of other forms of organization as well as the classical types of local authority. This would not, however, have changed the general verdict that local governance in the two counties has changed largely in an incremental fashion so that the present patterns are recognizably the descendents of a system already in place by 1801.

1801–1999

The processes by which the administrative 'map' of 1801 was transformed into its 1999 successor occurred on several geographical scales and these scales provide the framework for this chapter. First, the two counties were influenced by forces and events at national level. The structure of local government in Cornwall and Devon has been moulded by the same forces as the rest of England and Wales but at a relatively low level of controversy and conflict. Second, the two have distinctive features which differentiate them from the rest of the country and also from each other. Third, the highest level of local geographical process—the evolution of county boundaries—was represented by only very small changes compared with the next level down, the growth and consolidation of the major urban areas, all of which were in the 'ancient' county of Devon. The creation of modern Exeter, Plymouth and Torbay forms the fourth section. Fifth, the remainder of the counties were redivided into parts with different names at different times. At this level, however, the process ended in the second half of the twentieth century with mergers of the district areas that the nineteenth century had divorced. Finally, the lowest geographical level—the parish and its 'parts'—experienced a great variety and number of changes, many of which were generated within the localities themselves but some of which were a consequence of the higher level processes.

In addition to the processes above, there are also a number of general considerations which apply to the whole subject of areas and authorities and these are outlined first. The maps of present-day parish structure are printed at the front of the *Atlas* (front end papers), and keys to parish names at the end (rear end papers), rather than within the pages of this chapter. This is for the convenience of the user of the whole *Atlas* because so many social, economic and political activities are historically located by reference to the parish system.

GENERAL CONSIDERATIONS

In terms of the coherence of local government, Cornwall and Devon have benefited from two factors. First, their boundaries have been relatively trouble-free, partly because their 'frontier' runs through relatively unpopulated areas, and partly because this frontier is marked by a river of significant size.[3] Thus the present counties bear a very close relation to the 'ancient' counties of the centuries before 1830. Second, they do not contain large conurbations whose governing at a local level could pose considerable problems, as has been the case in London, the Midlands and the North of England. In fact urban and rural settlement patterns were tailor-made for application of the principles of local government structure in general use before the 1960s.

Period	Events and processes
Pre-1834	Continuation of the traditional system
1834–5	Creation of poor law unions and municipal corporations
1848–87	Creation of special purpose areas and authorities
1882–5	Implementation of the Divided Parishes Acts
1888–94	Creation of classical county, district and parish government
1894–1921	Expansion of urban government (annexation, 'promotion')
1930s	County review process: districts and parishes
1960s	Expansion of county boroughs
1971–4	Creation of modern county, district and parish government
1974–99	Reorganization of parishes
1991–8	The 'Heseltine' Review and its implementation

Types of 'county'	Types of 'district'	Types of 'parish'
'ancient'	'hundred'	'ancient'[a]
'county of a city'		ecclesiastical (1844) agricultural (1871)
parliamentary (1832)	'borough' 'improvement'[b] petty sessional (1828)	
poor law (1834)	poor law union (1834)	poor law (1834)
registration (1837)	registration (1837) ad hoc (1848–94)[b] sanitary (1872)	registration (1837)
geographical (1889)		
administrative (1889)	county (1894)	civil (1894)
county borough (1889)		rural borough (1958)
non-metropolitan (1974)	non-metropolitan (1974)[c]	non-metropolitan (1974)[d]
'unitary' (1998)		

NOTES:
a These include the established 'parts'.
b Improvement and ad hoc districts constitute categories, not specific types.
c This type also includes 'cities' and 'boroughs'.
d This type also includes 'towns'.

Though both counties were gradually urbanizing, they were doing so at a lower rate than many other parts of the country and their relative positions in the system of government, therefore, tended to decline in importance. The work of the Local Government Commission (1991–6) has left Cornwall unchanged but has partially restored the division between the main urban areas and the rest of Devon by granting 'unitary' status to Plymouth and Torbay from April 1998, though it rejected Exeter's case.[4] The county of 'new' Devon, therefore, remains recognizably the same.

Because the territorial patterns described here are a reflection of national events and processes, it is important for the reader to see them in their wider context. **Table 24.1** lists the main relevant episodes in English local government history. The detailed changes which were associated with them in the sub-region, however, were so numerous, and in many cases so small, that they cannot always be mapped in an intelligible manner. The maps and tables included in this chapter reflect these episodes during which the main features of the administrative landscape of the sub-region were laid down, but only general and summary references are made to most of the minute details.

Table 24.2 lists the different types of area that existed at 'top' (county), 'middle' (district) and 'lowest' (parish) levels. Because historical records are not always clear or accurate in this respect, researchers need to take great care to distinguish the different types of area. As their names suggest, the three levels of area were often closely connected.

OTHER AREAS UTILIZED FOR PUBLIC ADMINISTRATION PURPOSES

In addition to local government areas there are also other areas of administration too numerous to map in this volume. Many of these would now be called 'local quangos' and in some locations public services were provided by private agencies. The two most salient local government areas not dealt with in this chapter are electoral areas and areas for internal administrative purposes. Since 1894 all primary multi-purpose local authorities have been directly elected and this involves a system of electoral units called 'wards' in urban areas and 'divisions' for county councils.[5] The latter were delimited for the first time in 1888, and evolved slowly until the 1970s when they had to be re-thought as a result of the Local Government Act 1972. Some small towns had at-large elections, that is, they were not divided into wards, and rural area wards were and are closely related to the pattern of civil parishes. Maps of electoral divisions have not been included because they do have not special geographical significance and are changed more frequently than the authority's area itself.

Both county and district authorities divide up their territories for the purposes of service provision. As each service may adopt a different pattern and there are many separate local government activities, it is impossible to reproduce all of them in this volume.

The evidence to the Local Government Commission for England (1958–65) in its examination of the South Western General Review Area revealed that area administration was very important in both county councils. Local patriots might like to note the the Bains Working Group (1972) mentioned Devon's Social Services area committees as an example of decentralization within county council administration.[6]

OFFSHORE ISLANDS

Cornwall and Devon are also fortunate as counties because they do not have substantial islands or archipelagoes close to their coastlines. Note, however, must be made of the Isles of Scilly, Lundy Island and several smaller islands because their histories depart in various ways from the pattern of the mainland. In particular, the Isles of Scilly, which are 40 miles to the west of Land's End, cannot be administered in a straightforward way as part of Cornwall.[7] Since 1892 the archipelago has been a 'unitary' authority, combining the powers of county, district and parish councils, but purchasing some services from Cornwall.[8] It was also extra-parochial in respect of income tax until 1953.[9] It is governed by an elected council with members from each of the five inhabited islands.[10]

Lundy Island is situated off the north-west coast of Devon and was traditionally part of Braunton hundred. Because of its location it has not been *de facto* a full part of the mainland administrative system and there have been several unsuccesssful attempts on the part of its 'owners' in the nineteenth and twentieth centuries to claim that it is not part of the United Kingdom. It remains extra-parochial and was 'included for convenience' in the census county reports for Devon in 1961 and in 1971 in Bideford rural district.[11] The remaining islands, with the exception of St Michael's Mount, are so small and so close to the mainland that they do not contain the basis for claims of separation. St Michael's Mount, which had a population of 125 in 1811, was extra-parochial until 1858.

INTER-COUNTY BOUNDARIES

During the period covered by this chapter there were six main types of county in existence at various times. The 'ancient' counties of 1801 contained exclaves and unclear boundaries; the Parliamentary Boundaries Act 1832 (2&3 Will IV, c.64) created parliamentary counties by placing the enclaves in their surrounding county for national electoral purposes.[12] 'Registration' counties were created in the late 1830s on the basis of the areas of the newly introduced registration service.[13] The Local Government Act 1888 (51&52 Vict, c.41) introduced the 'administrative county' and 'county borough' (which were the local government areas) and these together were conventionally referred to as the 'geographical' county. Finally, the non-metropolitan county was created by the Local Government Act 1972.

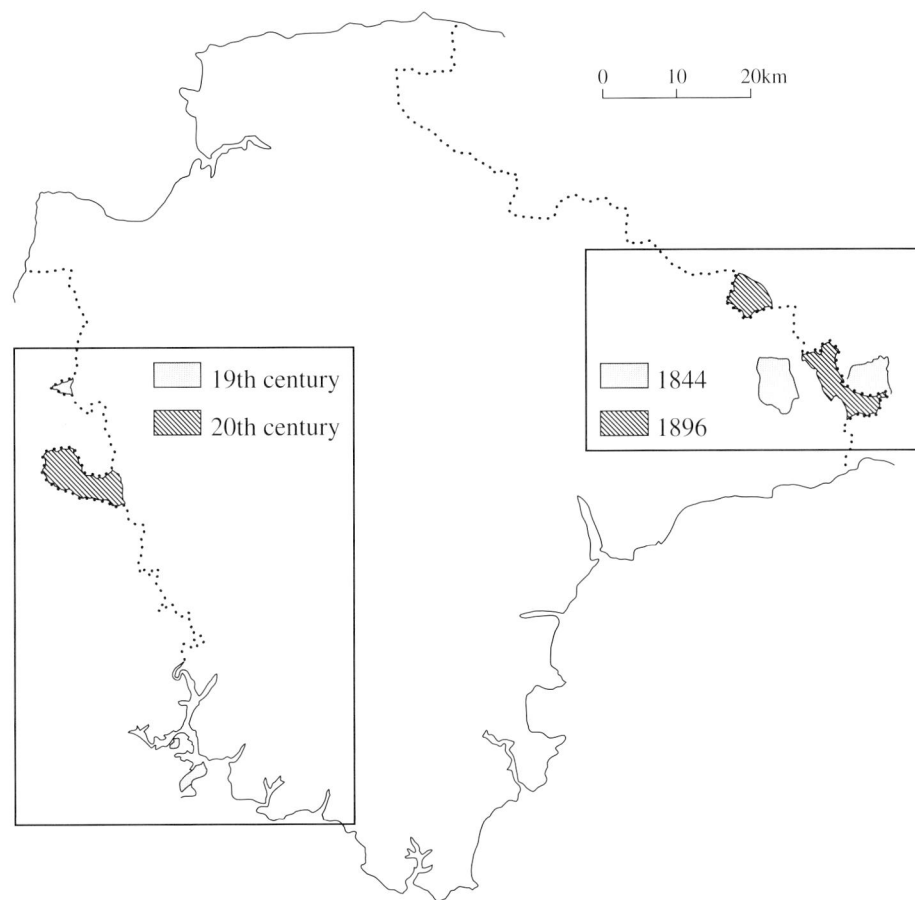

Map 24.1 County boundaries: elimination of exclaves and anomalies.

Map 24.1 depicts the main exclaves, enclaves and 'anomalies' in the county boundaries of 1801 and the main 'rectifications' made in 1844, 1896 and 1966.[14] The exclaves and enclaves were exchanged in 1844, parishes transferred in 1896 and a considerable number of the irregularities in the traditional boundaries were removed in the 1960s. Most of the changes of that decade involved such small areas that they cannot be mapped accurately on the scale used in this volume. The exception was the division of Broadwoodwidger rural district along the Tamar in 1966, and the transfer of its western portion to Cornwall. The eastern part was merged with Holsworthy rural district.

THE LARGE BOROUGHS

By national standards Cornwall had no large urban areas that could have been given or even reasonably sought exclusion from county administration. Devon, however, has three areas of substantial urban development, two of which—Exeter and Plymouth—have borough status dating back to the Middle Ages or earlier, the third being Torbay which is a creation of the nineteenth and twentieth centuries. In each case, as a result of urban growth, municipal boundaries have been extended frequently to take in suburban parishes, or parts thereof. The three

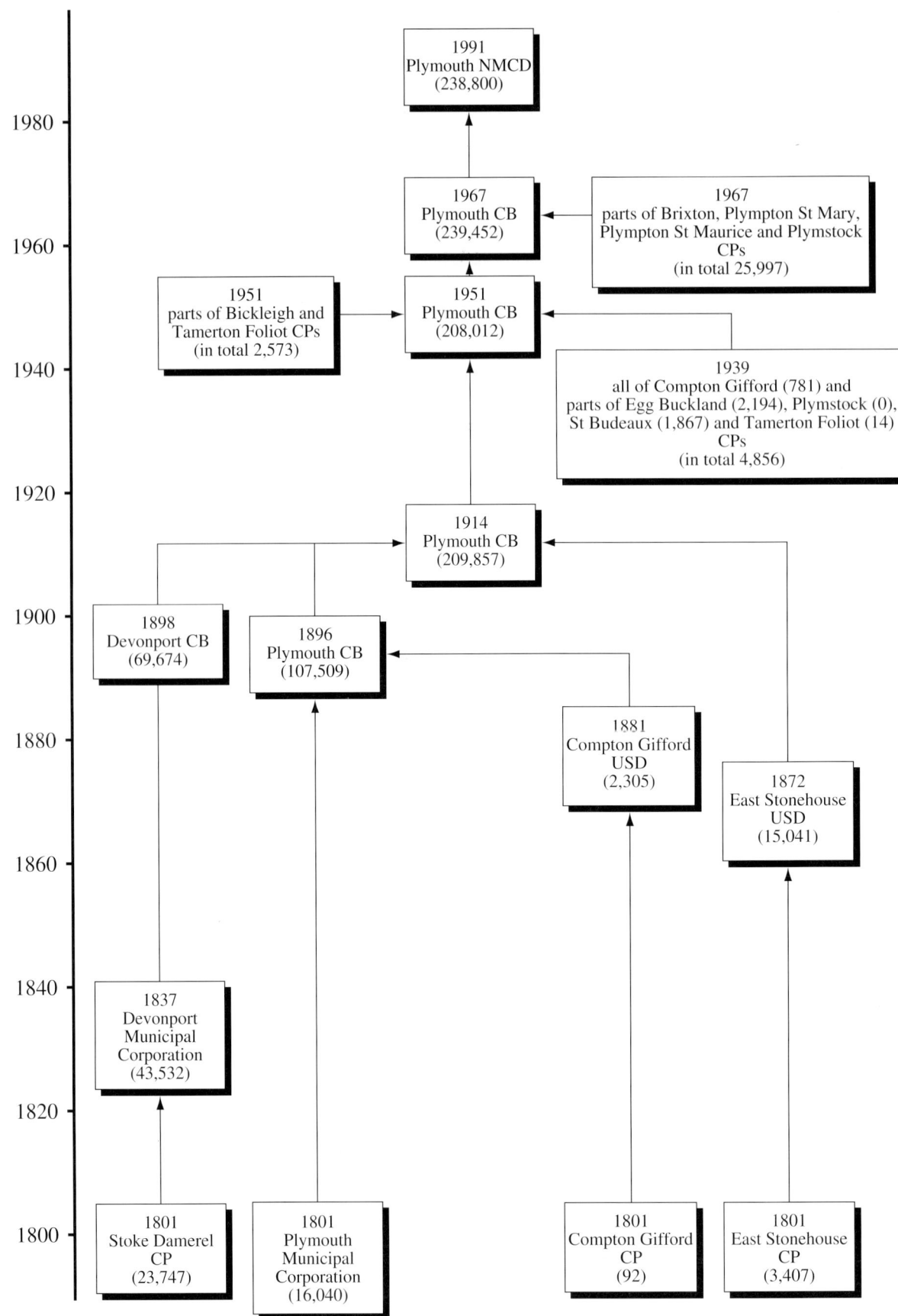

Fig. 24.2
The consolidation of the
Plymouth urban area.

1980
1960
1940
1920
1900
1880
1860
1840
1820
1800

1991
Plymouth NMCD
(238,800)

1967
Plymouth CB
(239,452)

1967
parts of Brixton, Plympton St Mary,
Plympton St Maurice and Plymstock
CPs
(in total 25,997)

1951
parts of Bickleigh and
Tamerton Foliot CPs
(in total 2,573)

1951
Plymouth CB
(208,012)

1939
all of Compton Gifford (781) and
parts of Egg Buckland (2,194), Plymstock (0),
St Budeaux (1,867) and Tamerton Foliot (14)
CPs
(in total 4,856)

1914
Plymouth CB
(209,857)

1898
Devonport CB
(69,674)

1896
Plymouth CB
(107,509)

1881
Compton Gifford
USD
(2,305)

1872
East Stonehouse
USD
(15,041)

1837
Devonport
Municipal
Corporation
(43,532)

1801
Stoke Damerel
CP
(23,747)

1801
Plymouth
Municipal
Corporation
(16,040)

1801
Compton Gifford
CP
(92)

1801
East Stonehouse
CP
(3,407)

1991
Torbay NMCD
(122,500)

1968
parts of Churston Ferrers (1,562),
Marldon (38), Coffinswell (6)
and Kerswells (35) CPs
(in total 1,641)

1968
Torbay CB
(109,257)

1951
Torquay NCB
(53,281)

1935
part of
Stokeinteignhead
CP
(187)

1928
Cockington CP
(279)

1900
Torquay NCB
(33,625)

1894
Cockington UD
(1,718)

1894
Brixham UD
(8,090)

1866
St Marychurch
LB
(4,472)

1863
Paignton LB
(3,590)

1862
Lower Brixham
LB
(4,941)

1850
Torquay LB
(7,903)

1801
Cockington CP
(294)

1801
St Marychurch
CP
(801)

1801
Tormoham CP
(838)

1801
Paignton CP
(1,575)

1801
Brixham CP
(3,671)

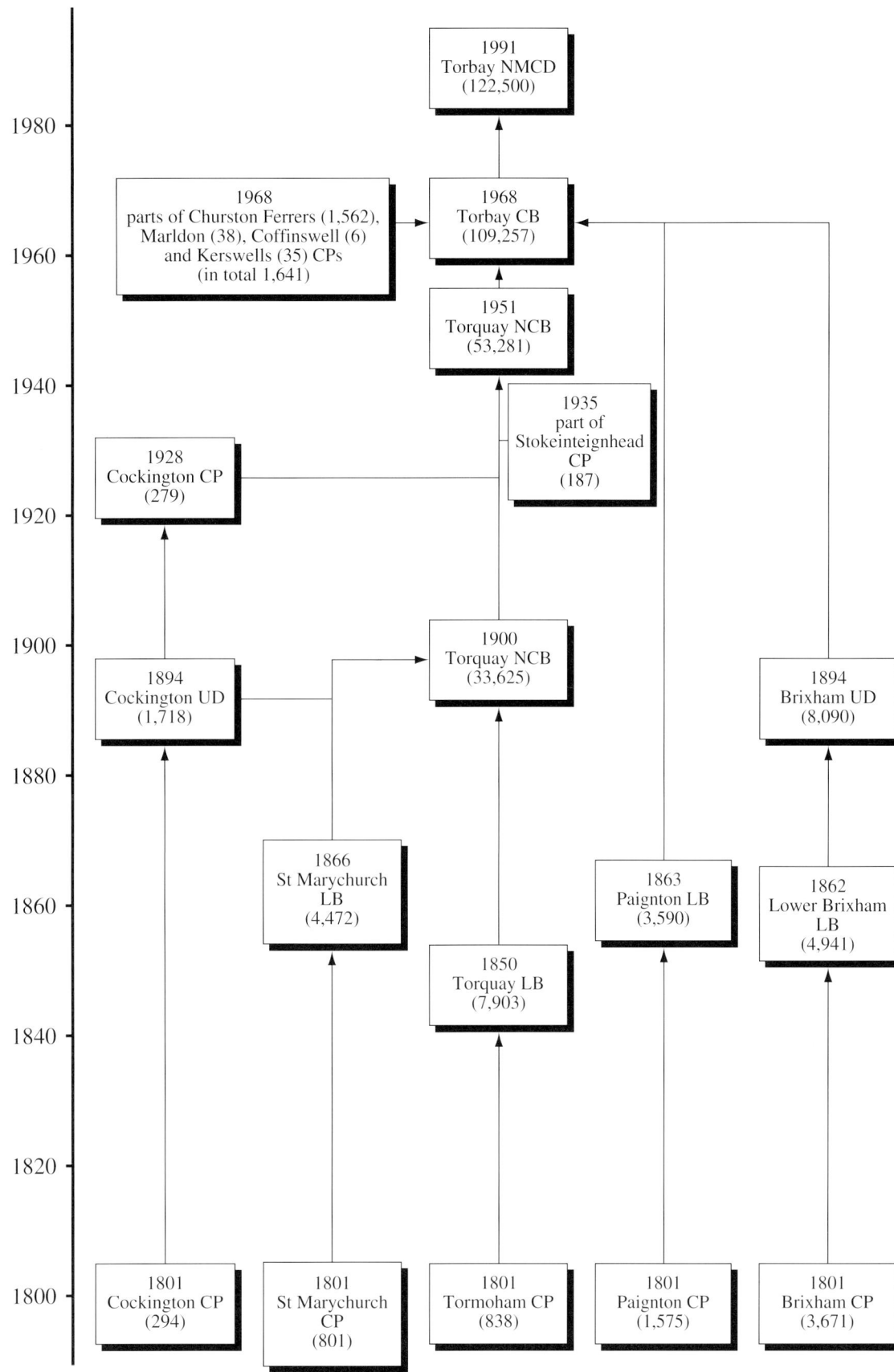

Fig. 24.3
The making of Torbay.

179

listed together as 'Plymouth Lighthouses' and sometimes separately.[19] Drake's Island was listed as extra-parochial in 1871 (population, 73) but included in St Andrew in 1891.

THE MAKING OF TORBAY (**Fig. 24.3**)

Torbay is from the point of view of the two counties unique; its early history is of small town progress towards an effective urban status. Torquay itself was a part of the parish of Tormoham in 1801 (population 838), thus smaller than Paignton (1,575) and Brixham (3,671), and not much bigger than Churston Ferrers (663) and St Marychurch (801). Torquay grew rapidly and this was marked by the stages of urban 'promotion'.[20] In 1835 a local public improvement act was passed; in 1850 wider powers were granted to a Local Board of Health, which took responsibility for the water supply in 1853; 1874 saw the end of the Turnpike Trust; in 1880 there was a public meeting of the Local Board to consider an

application for borough status or to divide town into wards; in 1890 there was a town poll and petition in favour of this and in 1892 there was an enquiry followed by the town's incorporation as a municipal borough. The borough then provided a focus for consolidation: in 1900 the boundaries were extended to include most of St Marychurch and Cockington, more was added in 1928, and in 1968 the borough merged with the other areas to form Torbay County Borough.

The two parishes adjacent to Tormohun—St Marychurch and Cockington—were also developing as urban areas. Each separately achieved urban status, St Marychurch in 1866 as a local

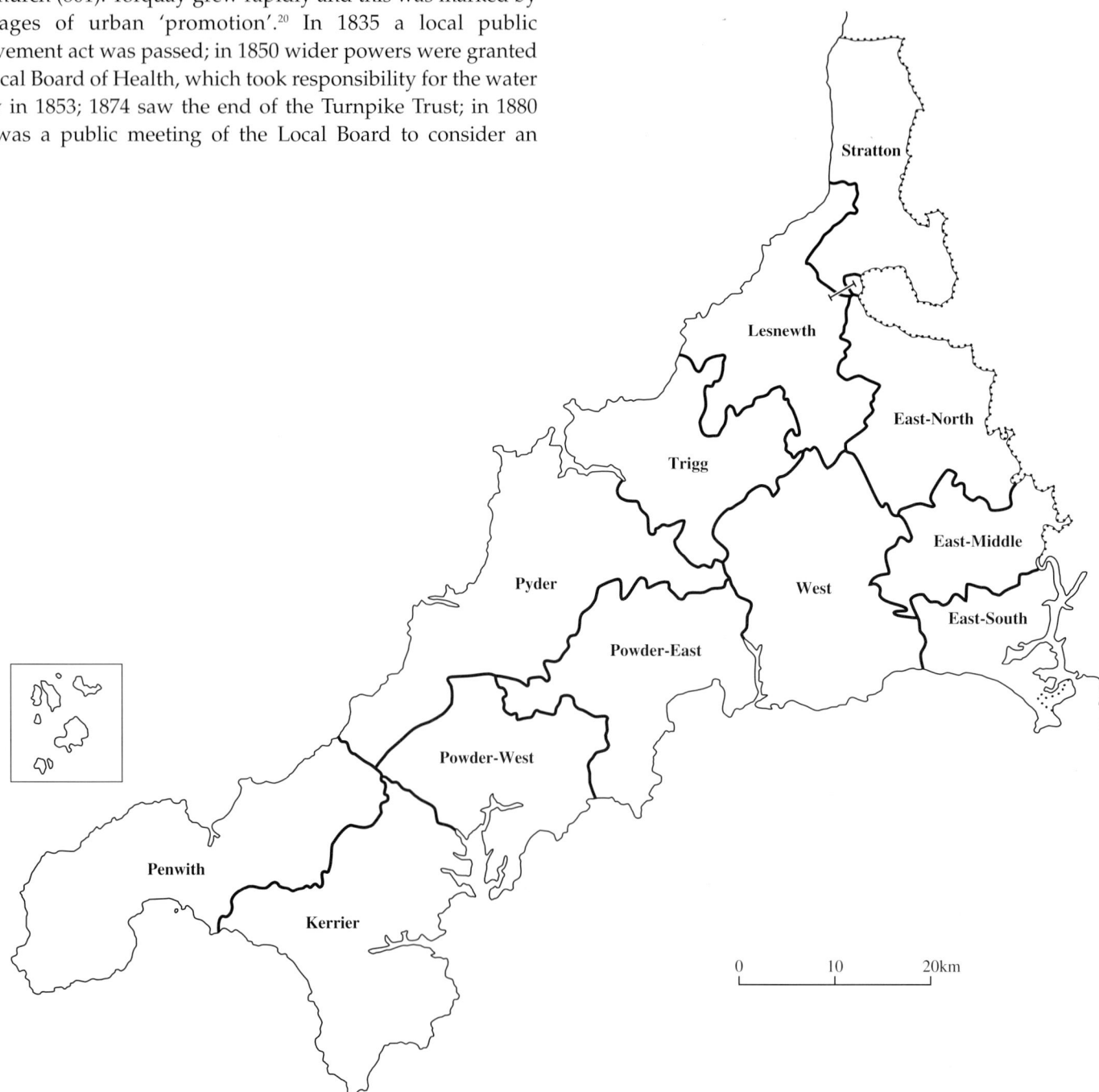

Map 24.2
The hundreds of Cornwall.

Map 24.3
The hundreds of Devon.

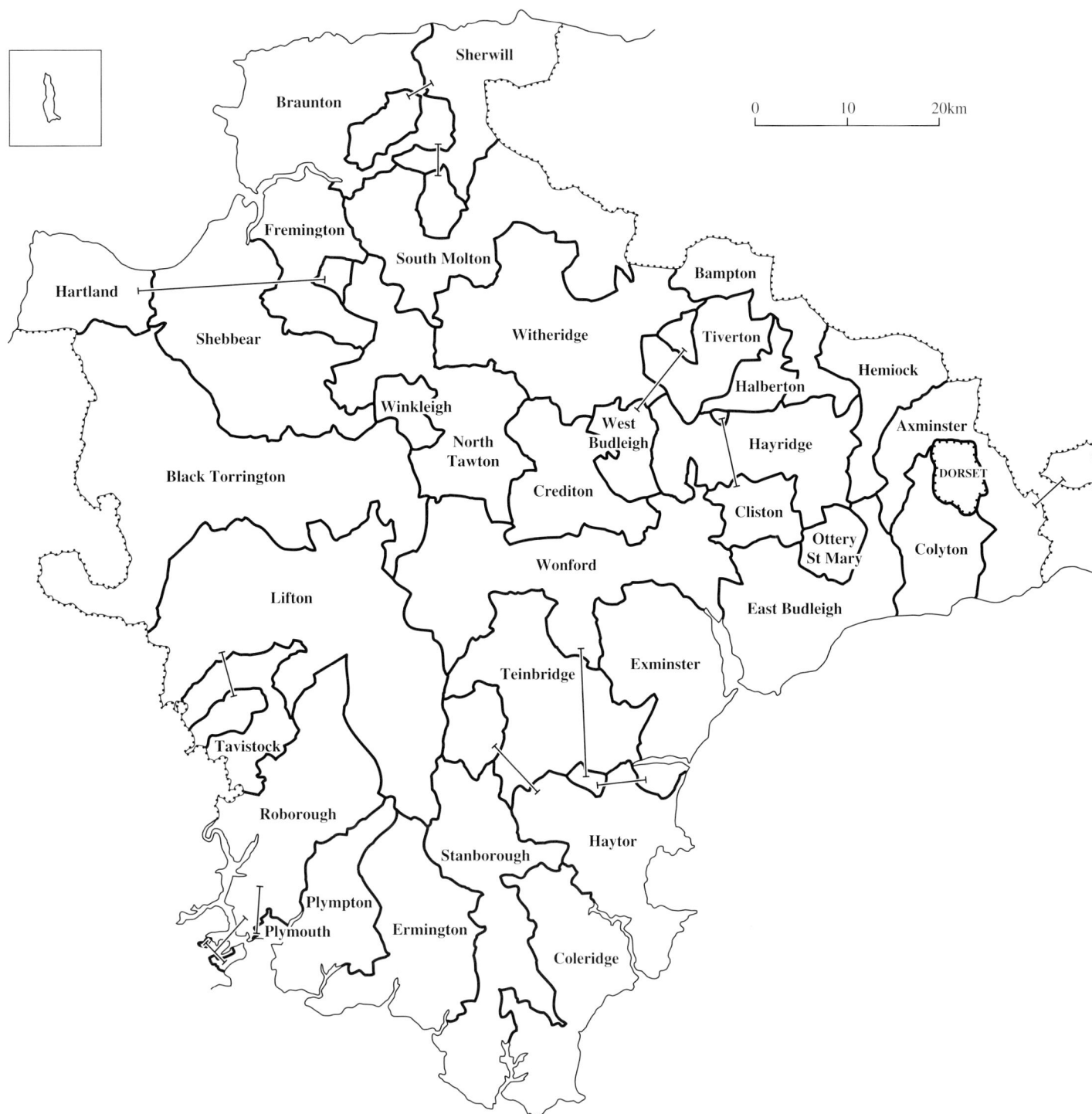

board and Cockington as an urban sanitary district in early 1894. Both disappeared in the first urban consolidation which took place in 1900. Small-scale rationalizations took place during the next four decades but further substantial change had to wait until the work of the Local Government Commission for England (1958–65) enabled Torquay to lead its smaller and somewhat distant neighbours into merger as a county borough in 1968. The borough regained its independence from Devon as a unitary authority in April 1998.

THE INTERMEDIATE LEVEL OF AREAS AND ADMINISTRATION

Prior to the 1830s the main division of the counties in England was most frequently called the hundred.[21] Hundreds were supposed to consist of whole parishes but this was not always the case. Because of their defects as areas the hundreds and their equivalents declined in importance, losing functions in the nineteenth century to registration districts, petty sessional

Map 24.4
The poor law unions of
Cornwall.

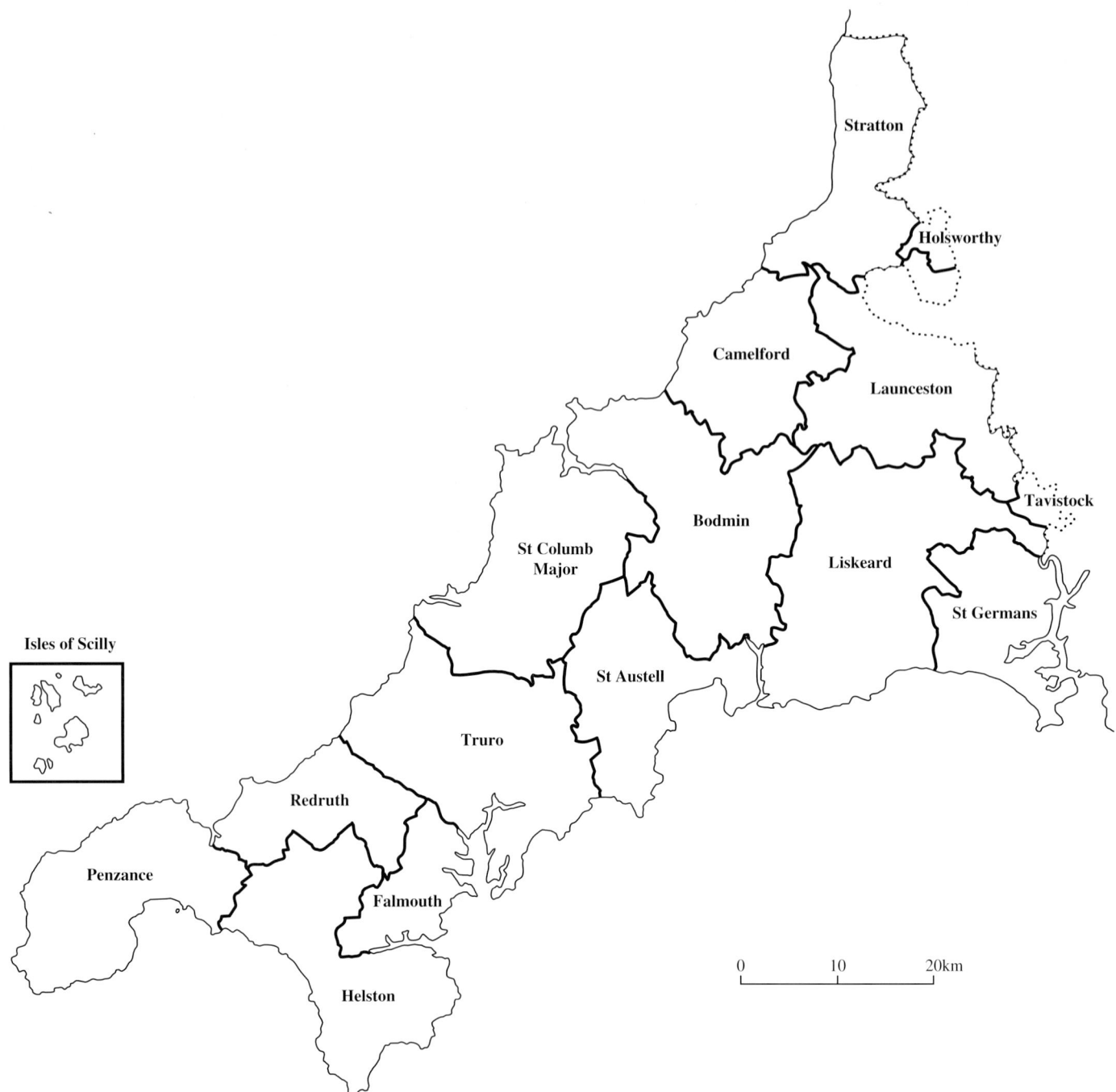

Stratton

Holsworthy

Camelford

Launceston

Bodmin

Tavistock

St Columb
Major

Liskeard

Isles of Scilly

St Germans

St Austell

Truro

Redruth

Penzance

Falmouth

0   10   20km

Helston

divisions and poor law unions. The latter category was the true inheritor of the intermediate role and evolved successively into registration districts (1837), sanitary districts (1872), county districts (1894) and non-metropolitan districts (1974). However, even during the times when the hundred was the main intermediate area there were complications caused by the granting or achieving of urban status by large villages and small towns. The history of the intermediate level is therefore a narrative of two interacting processes: the decline of the traditional hundred and the 'promotion' of small towns.

THE HUNDREDS

The hundred was a major statistical unit but had, by the nineteenth century, atrophied as a local authority. In 1801 it retained military and judicial functions but was not used for other public purposes. The inadequacies of the hundred were noted in 1801 and from the third decade of the nineteenth century it began to lose its judicial role to the petty sessional division, a new subdivision of the county which quarter sessions had the power to establish.[22] The hundreds remained statistical units until 1851 when they were in effect superseded by the national system of districts and subdistricts for purposes of registration.[23] Because of their antiquarian interest they continued to be mentioned in county directories until the Second World War.

The traditional hundreds of Cornwall (**Map 24.2**) were nine in total but two of these had well-established divisions which may have had some of the characteristics of hundreds themselves.[24] The towns of Launceston and Falmouth were 'extra-hundredal'; thus with the divisions there were fourteen

hundred-level areas in the county.[25] Devon had thirty-one ordinary hundreds (**Map 24.3**), and the parishes of Ottery St Mary and Winkleigh were hundreds in their own right, as were the boroughs of Exeter and Plymouth, making a total of thirty-five hundred-level areas in the county.[26]

The reasons for the rejection of the hundred as the modern intermediate level of local government were not so strong in Cornwall and Devon as they were in some other counties but their general defects were apparent. These defects included the fact that

Map 24.5
The poor law unions of Devon.

many of them, if they had any coherent geographical basis, reflected the urban hierarchy of centuries before. The parish that gave its name to the hundred had often been overtaken by one of its neighbours or had sunk into insignificance as a central place. But this fact was insignificant compared with the problems of irregular boundaries and exclaves or 'detached parts' as they were called in the nineteenth century.[27]

As **Maps 24.2 and 24.3** show, the hundreds of Cornwall were much more regular in shape than their counterparts in Devon. Cornwall appears to have had no detached parts to hundreds with the exception of Fowey, whose listing in 1801 and 1811 may be a mistake on the part of the census administration, and a detached part of Warbstow parish. The situation in Devon was very different. First, some Devon hundreds had detached parts.[28] Second, many of the hundreds of Devon were very oddly shaped, and third, the hundreds of Devon varied greatly in number of parishes, area and population. The contrast with Cornwall is marked (**Fig. 24.4**).[29] From 1830 onwards, the main drive at the intermediate level was to accommodate the urban areas and to eliminate inconvenient boundaries of the types mentioned above.

POOR LAW UNIONS

The 1830s saw the general acceptance of the demise of the hundred and the recognition that the parish as constituted was unable to carry out major public functions. The reforms of the decade, however, took the parish as their starting points. It was widely believed that a new intermediate level of administration was necessary and that urban areas needed a standard form through which they could develop further than the parish system allowed. These two aims were not fully achieved until the 1890s.

The Poor Law Amendment Act 1834 provided for parishes to be grouped for the purpose of the relief of destitution and for a workhouse to be built for each union. The intention was to base each grouping on a convenient socio-geographical centre. Though

the principle could not be easily applied in some parts of the country, Cornwall and Devon, with their pattern of market towns and rural catchment areas, were ideal (**Maps 24.4 and 24.5**). It is significant that most of the unions were given the name of the town which provided their centre and many of the lines drawn around groups of parishes to create a union have survived to the 1990s as administrative boundaries. The 1830s maps of the unions have a much more regular appearance than the hundreds they in effect replaced.

The boundaries were drawn in the first instance by peripatetic officers of the Poor Commission in the second half of the 1830s. In Cornwall and Devon their work was relatively straightforward, with two types of exception. First, the geographical principle implied that in some areas parishes should be grouped across county boundaries. The result was that the poor law county, which quickly became the registration county, did not have exactly the same boundaries as the geographical county (see **Table 24.3**). Second, the 1834 Act had been anticipated in Exeter, Plymouth and Devonport in local acts which established a corporation of the poor for the whole area, superseding parish responsibilities, and permitted the building of a workhouse.[30] These were allowed to continue as the equivalent of a Poor Law Commission-established board of guardians.[31]

THE SIGNIFICANCE OF THE UNION AREAS

The areas established by the Poor Law commissioners exerted a major influence on the development of the administrative landscape in provincial England and this is evident also in Cornwall and Devon. First, they were adopted by the registration service created in the late 1830s under the Births and Deaths Registration Act 1836 (6&7 Will IV, c.86), for its field office system. Second, they were the basis for the reform of sanitary administration in 1872, when the union areas became rural sanitary districts except for those urbanized parts which were declared to be urban sanitary districts. Third, the urban

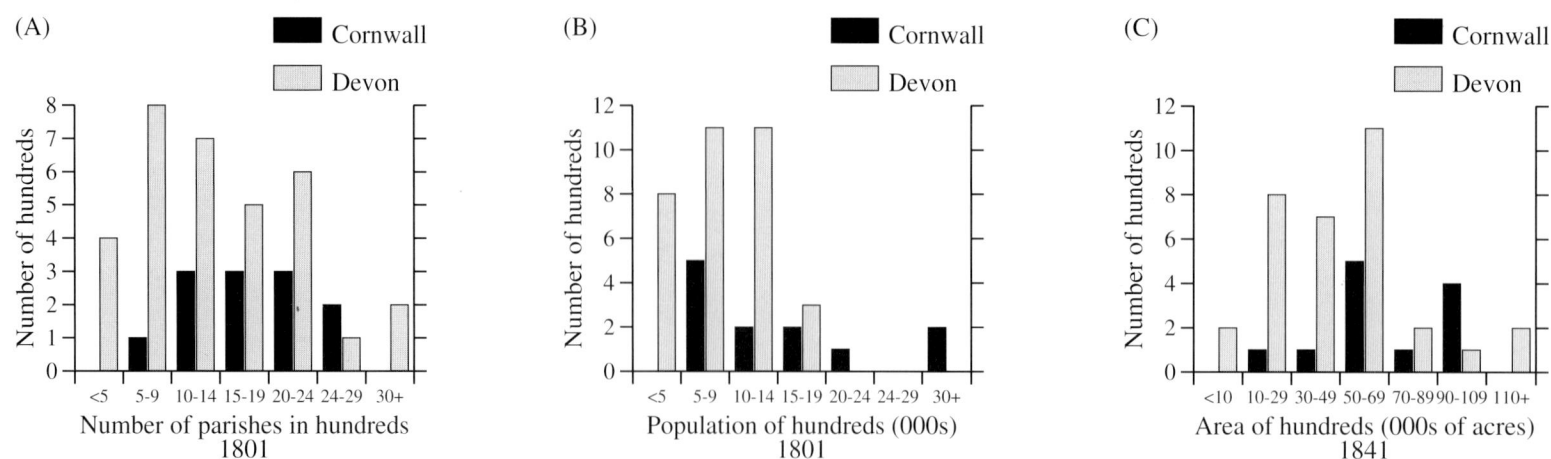

Fig. 24.4 Comparison of Devon hundreds with Cornwall hundreds.

(A) Number of parishes in hundreds 1801 — (B) Population of hundreds (000s) 1801 — (C) Area of hundreds (000s of acres) 1841

County	Equivalent
geographical Cornwall 1880	ancient county plus exclaves of Devon minus parts to Devon
geographical Devon 1880	ancient county minus exclaves in Cornwall and Dorset plus exclaves of Dorset plus parts of Cornwall
registration Cornwall	geographical county plus six parishes minus two parishes
registration Devon	geographical county plus eight parishes minus fourteen parishes
administrative Cornwall 1888	geographical Cornwall
administrative Devon 1888	geographical Devon minus three county boroughs
non-metropolitan Cornwall 1974	administrative Cornwall 1973
non-metropolitan Devon 1974	administrative Devon plus three county boroughs 1973
non-metropolitan Cornwall 1998	administrative Cornwall 1974
non-metropolitan Devon 1998	non-metropolitan Devon 1974 minus two districts

and rural sanitary districts were the basis for the creation of county districts in 1894, and as these were simply amalgamated in 1971–4, some of the boundary lines were carried through all successive changes of local authorities up to the present day. In Cornwall and Devon outside the main urban areas, therefore, many boundaries of the present-day district authorities are lines first drawn in the late 1830s.

In addition some parishes were assigned to unions that were primarily in other counties.[32] This situation continued throughout the nineteenth century. From time to time changes were made in the boundaries of the unions by the transfer of parishes from one to another, but the basic pattern remained the same until the Local Government Act 1929 abolished the unions and boards of guardians. In 1921 there were fourteen unions in Cornwall and twenty in Devon.

URBANIZATION OUTSIDE THE MAJOR TOWNS

Urbanization was the single most important influence on the development of the pattern of areas and authorities in the rest of

Table 24.3
Six types of county in south-west England, 1880–1998.

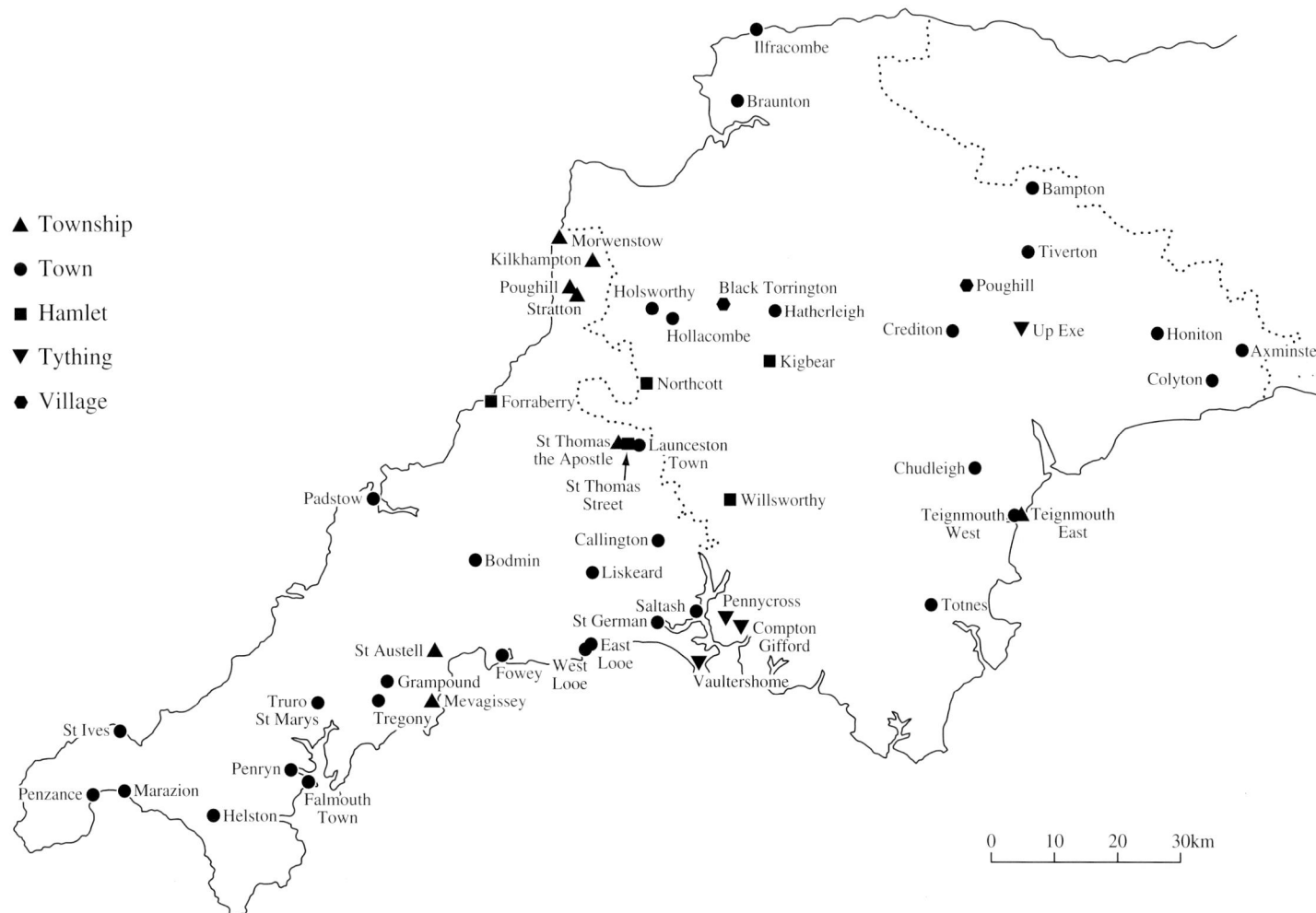

▲ Township
● Town
■ Hamlet
▼ Tything
● Village

Map 24.6
Places of non-parish status, 1801.

the two counties. At district level non-urban areas tend to be 'leftovers' or 'afterthoughts', solutions to problems created by the prior 'hiving-off' of towns from their hinterlands. The identification of urban areas and the allocation of appropriate powers, however, were complex but related processes, involving both national legislation which created types of local authority and local processes through which local areas interacted with these national developments.

Though it is necessary to describe all changes in terms of parish areas, or parts thereof, many developments were the product of pressures by growing urban areas for recognition as distinct territories and for the granting of a favourable local government status. The most striking examples of this have been described in relation to the largest urban areas—Plymouth, Exeter and Torbay—where the processes of annexation and unification eventually produced consolidated territorial units corresponding to a considerable extent to the urban area in socio-geographical terms (**Figs 24.1, 24.2 and 24.3**). The same pressures were felt on a smaller scale in the old inland market towns of both counties, in

the growing seaside resorts and in the small areas of industrialization.

Hindsight has been used to construct the list of small urban areas for the pre-1835 decades: areas are treated as towns if they had one or more of the urban statuses in 1851 or afterwards. Approximately fifty named territories in Cornwall and sixty in Devon were given one of the urban titles, or were described as a geographical 'town' at least once during the whole period of this chapter (**Maps 24.6, 24.7, 24.8 and 24.9**).[33] A handful of these places were not mentioned as urban during the first three decades and another handful were composite and/or renamed areas. Two-fifths of the Devon areas did not acquire an urban title until after 1831, but this was true of only a seventh of their Cornish counterparts.

The period 1832–35 provided the decisive dividing lines. Before this period urban status was partly a matter of legal status, sometimes self-description, and often unclear or controversial. Urban status could be claimed prescriptively, derived from a charter, or acquired by private act. The reform of the parliamentary boroughs (1832) and municipal boroughs (1835)

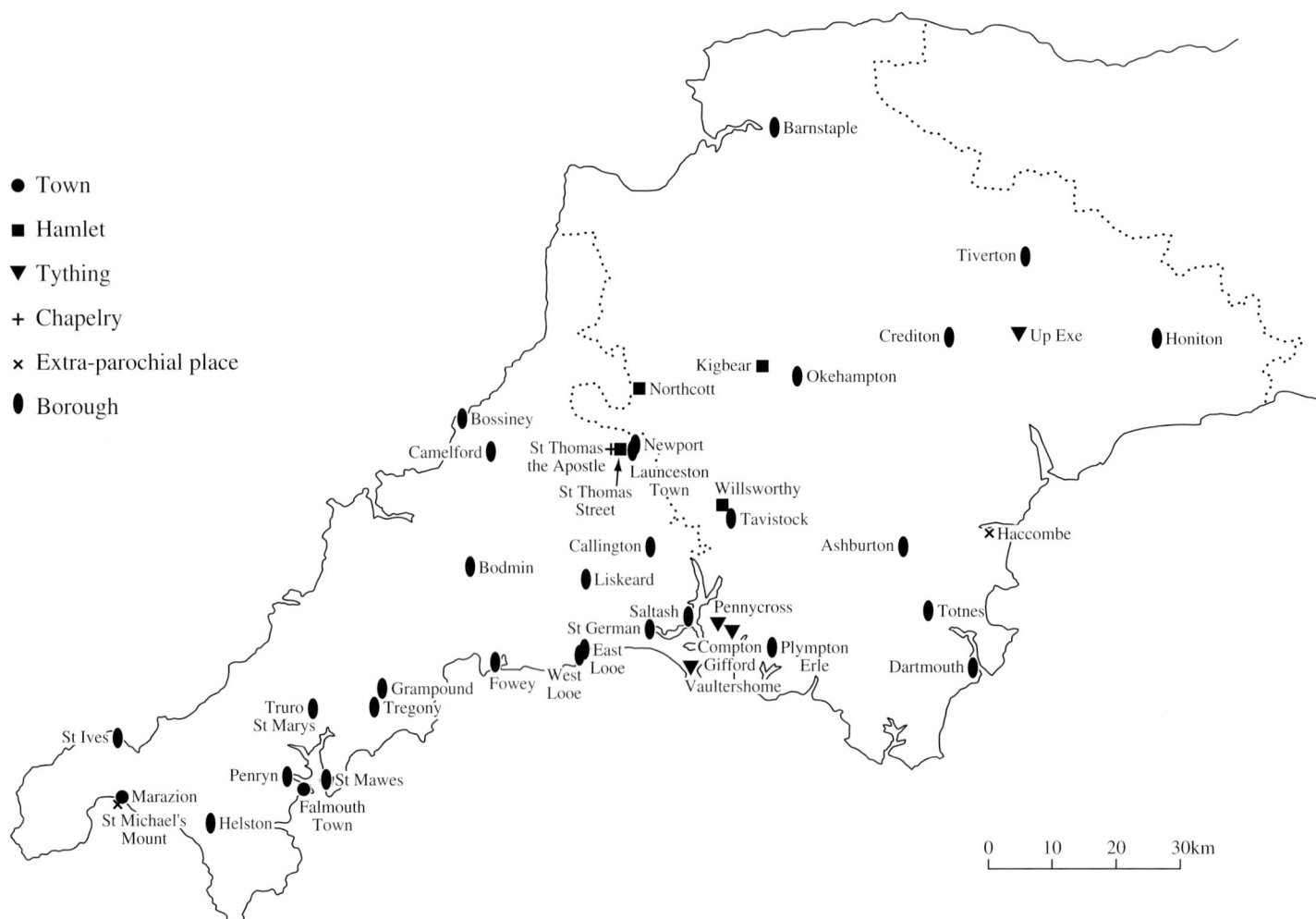

Map 24.7
Places of non-parish status, 1811.

created a new type of administrative landscape—one that was regulated by a set of general legal categories.

PARLIAMENTARY BOROUGHS (**Map 24.10**)

The first step was to reform the allocation of borough seats in the House of Commons. This was important because for the 'rotten boroughs' the benefit of parliamentary representation was the only consequence of this status. Once the connection between the municipal authority and parliament was broken it was possible for there to be a municipal corporation body focusing on local rather than national affairs. From 1832 onwards, therefore, the census drew attention to the differences in boundaries and population between parliamentary boroughs and municipal corporations where these had the same name. Some, of course, were identical (for instance, Barnstaple, Dartmouth, Honiton, Tiverton and Totnes in Devon and Truro in Cornwall); others had no more than a shadowy local government existence.

MUNICIPAL CORPORATIONS (**Map 24.11**)

The title 'borough' did not confer on an area a clear uniform local government status until all the possible confusions were dispelled by legislation in the last decade of the nineteenth century. In the period 1801–31, areas claimed the status 'by prescription' or 'by charter', but the claims were not always accepted by others and in many cases the governing body for the claimed municipal area was defunct or corrupt. The 1830s saw the beginning of the process of rationalizing and specifying the status of municipal corporation.

The Royal Commission of 1835 investigated 284 areas of which 178 were granted municipal status by the 1835 Act and a further five promoted soon after. Of the original 284, eighteen areas were in Cornwall and thirteen in Devon.[34] Of those in Cornwall, Bodmin, Falmouth, Helston, Launceston, Liskeard, Penryn, Penzance, St Ives and Truro were in the list of 178 towns that were given the new municipal corporation status, whilst Bossiney, Camelford, Fowey, Grampound, West Looe, Lostwithiel, Marazion, Saltash and Tregony were not. In Devon Barnstaple,

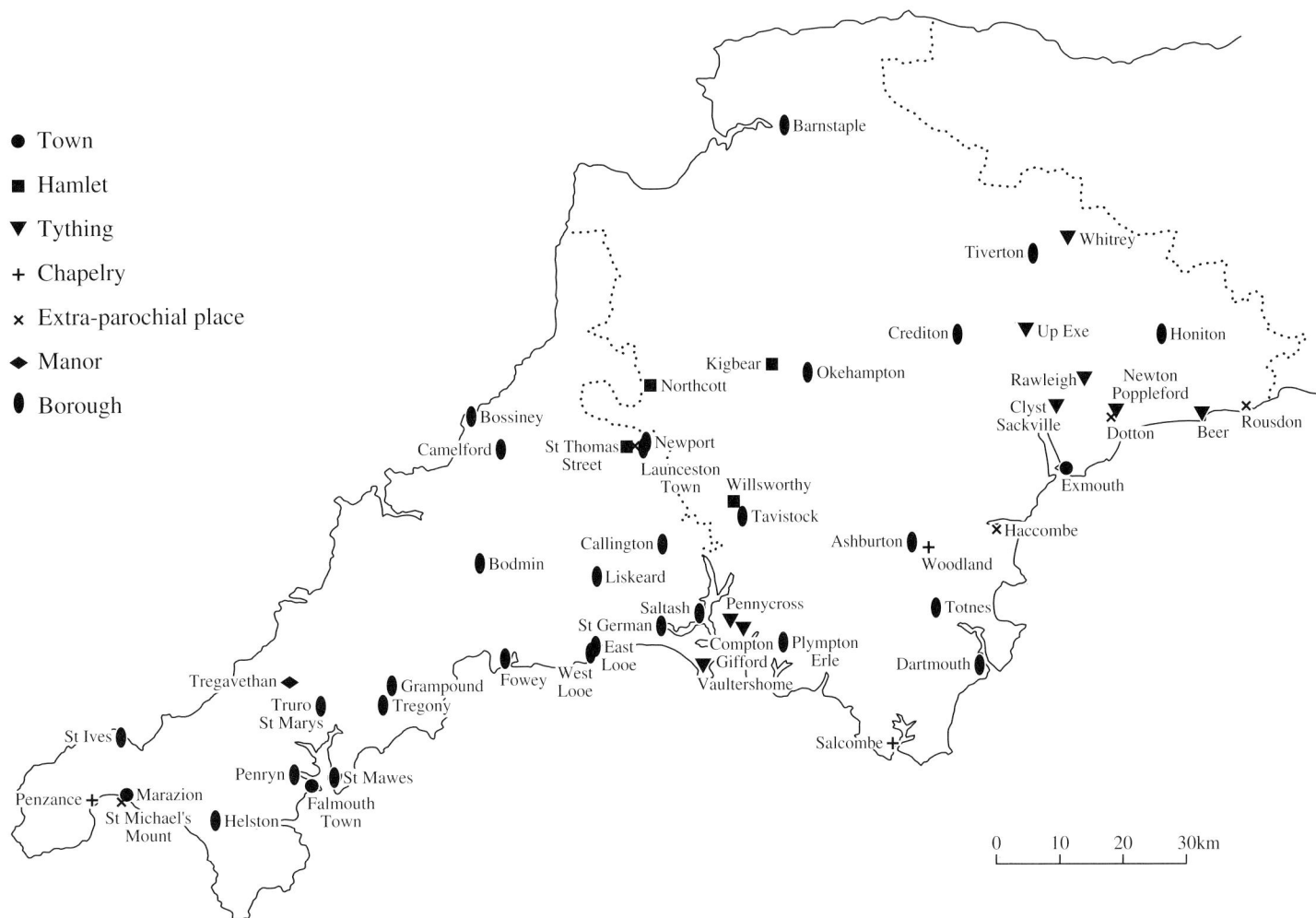

Map 24.8
Places of non-parish status, 1821.

Map 24.14
Cornwall: urban status in
the twentieth century.

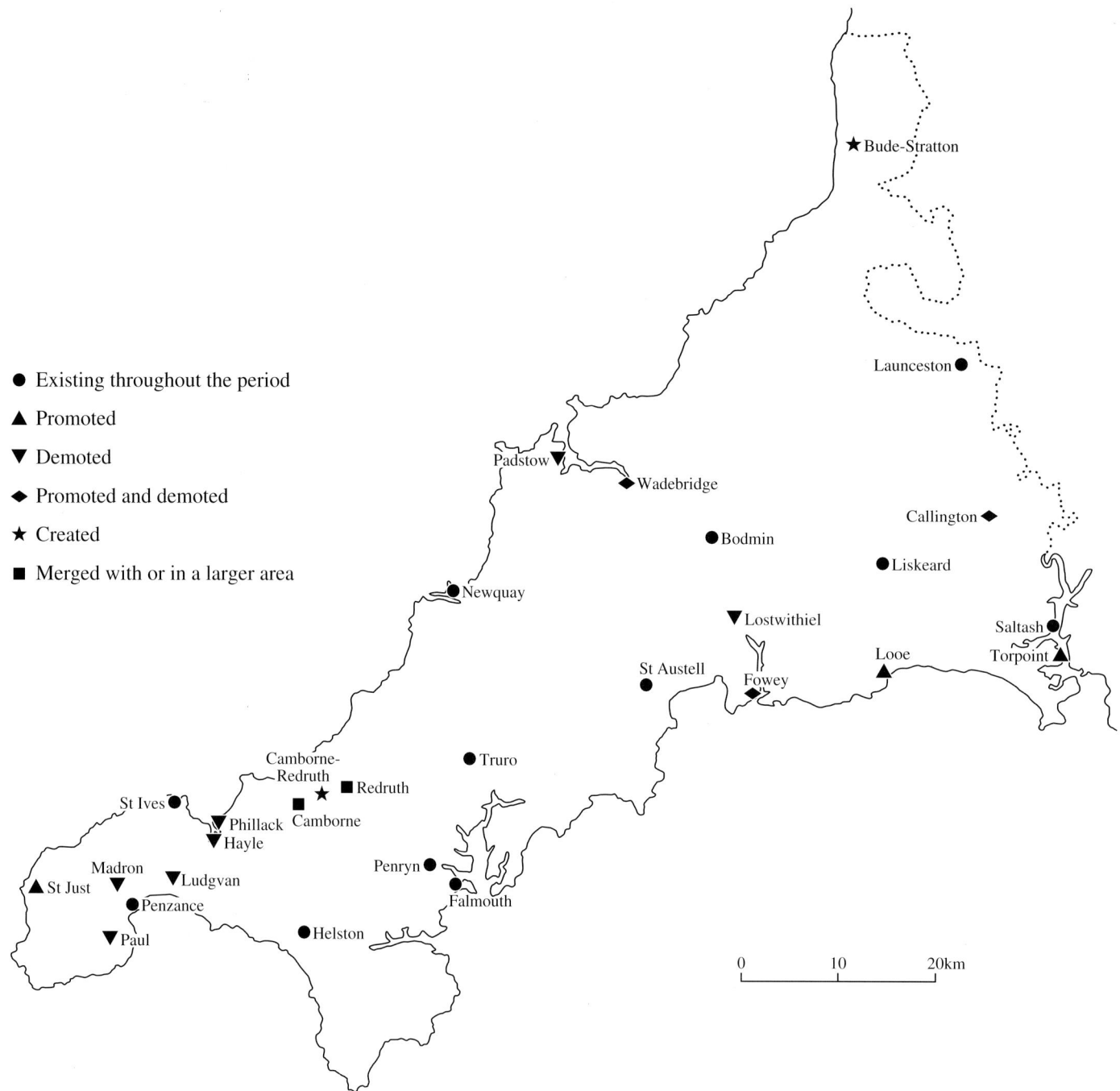

Existing throughout the period

Promoted

Demoted

Promoted and demoted

Created

Merged with or in a larger area

Bude-Stratton

Launceston

Padstow

Wadebridge

Callington

Bodmin

Liskeard

Newquay

Lostwithiel

Saltash

St Austell

Looe

Torpoint

Fowey

Camborne-Redruth

Truro

Redruth

St Ives

Camborne

Phillack

Hayle

Penryn

Madron

St Just

Ludgvan

Penzance

Falmouth

Paul

Helston

0   10   20km

24.6, 24.7, 24.8 and 24.9. As the period passed, however, there was a tendency for some simplification to take place. The experience of the two counties, however, was very different. The town and township statuses virtually disappeared, but in Cornwall towns became boroughs and in Devon they became parishes. Hamlets, tythings and extra-parochial places tended to maintain their status, whilst the village disappeared and manor appeared.

It is difficult to know how to interpret the differences between Cornwall and Devon during this period. They may reflect real variations of the administrative landscape or they might have resulted from different practices adopted by the local census administrators. The experience of the 1841–71 censuses confirms that there were substantial apparent differences between the two counties but provides no authoritative way of deciding

Map 24.17
Devon: parish status in 1998.

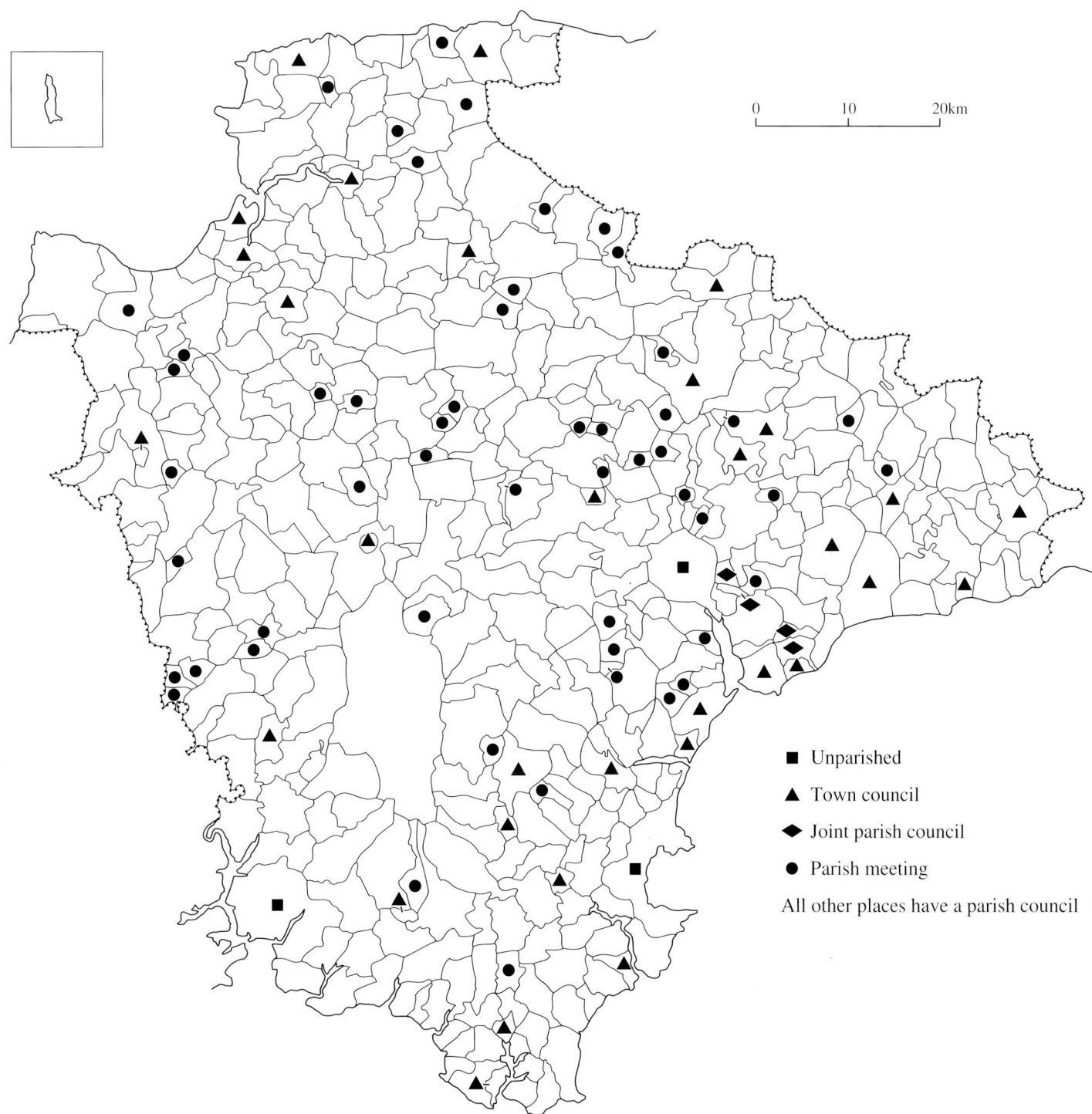

■ Unparished

▲ Town council

◆ Joint parish council

● Parish meeting

All other places have a parish council

impossible to avoid reasoning backwards from the later period to the former and concluding that many separate areas had been missed. Few of the problem areas were created after 1831 (or indeed after 1801); thus they must have existed in the earlier decades.

The 1841 census was the first organized by the newly created Registrar General's Office and exhibits a determination to identify

the territorial problems embodied in the local government structure. Some of the small units identified proved to be of no long-term significance but others played considerable parts in county administrative history in both Cornwall and Devon.

First, some parishes were divided between hundreds and hundredal level authorities, that is, they had parts in at least two hundreds or boroughs. This problem, however, immediately

Fig. 24.7
Devon election poster, 1812.

SOURCE: reproduced by
permission of Westcountry
Studies Library, Exeter.

Fig. 24.7
Devon election poster, 1812.

SOURCE: reproduced by
permission of Westcountry
Studies Library, Exeter.

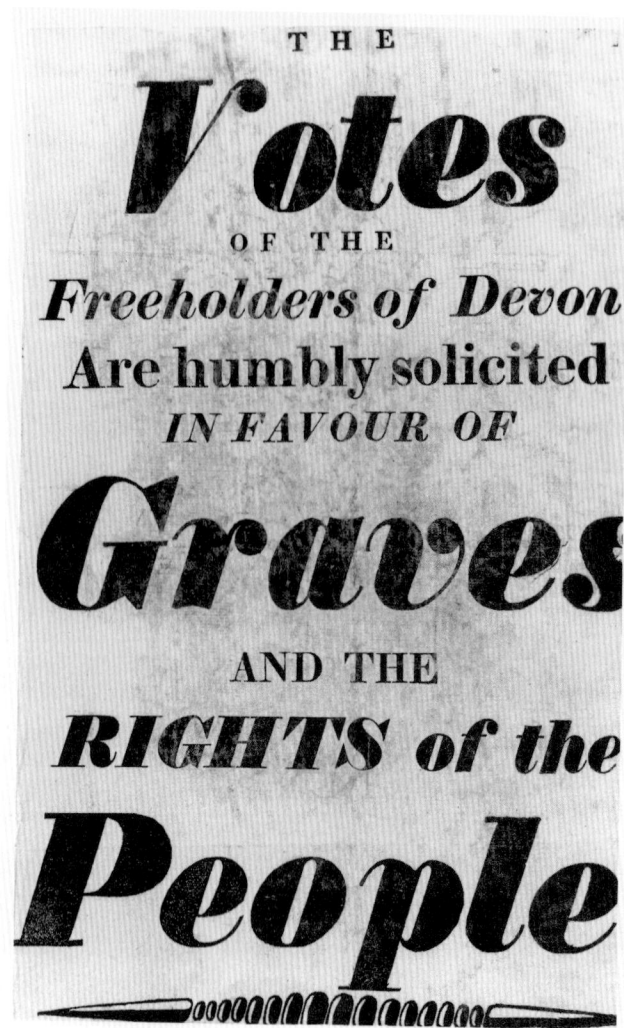

THE

Votes

OF THE

Freeholders of Devon
Are humbly solicited
IN FAVOUR OF

Graves

AND THE

RIGHTS of the

People

mainly because the existence of some extra-parochial places had not been perceived. It does not seem to have been applied to the small areas within the large municipalities and the situation in Exeter was not clarified until 1963.

ELIMINATING EXCLAVES AND INCONVENIENT BOUNDARIES

The problem of divided parishes was described in the Report of the Select Committee on the Areas of Parishes, Unions and Counties, 1873.[50] Cornwall had nine parishes in two parts and one in four parts whilst Devon had a much more fragmented pattern: forty-six in two parts, fourteen in three parts, five in four parts and two in five parts.[51] There were also detached parts of district level areas. The census regularly noted these and eventually most were eliminated as a result of other changes.

The application of the Divided Parishes Acts eliminated most of the exclaves at parish level. Cornwall Registration County was only slightly affected, with eight transfers of territory, including two in the 'ancient' county of Devon. In Devon between 1882 and 1886 five parishes disappeared completely, three gained whole parishes without changing their name, two entirely new parishes appeared and one rump was created. There were also 101 transfers of territory involving two parishes, many of which appeared in two or more such transfers.

IMPLEMENTING THE 1894 ACT

The processes described in respect of the Divided Parishes Acts were continued in 1894, largely because the whole structure was intended to be coherent and systematically rational. It was also necessary to integrate the parish pattern with that of urban authorities. Because parishes and parish equivalents and the various forms of urban status were different types of public entity before 1894, it was not necessary to have a standard form of relationship between the two. Changes had taken place but on an *ad hoc* basis. The 1894 Act, by differentiating between parishes in urban areas, that is in county boroughs, non-county boroughs and urban districts and in rural districts, made it necessary to rationalize parish boundaries within and overlapping the limits of the former.

In 1891 in thirty areas urban and parish boundaries were identical, fourteen urban areas consisted of only a part of a parish, and sixteen consisted of two or more parishes. No action was necessary for the first category and most instances of the second and third were dealt with by division or merger. There were, however, *sui generis* consolidations in Dartmouth, Exeter, Kingsbridge and Dodbrooke, the Launceston area, the Newton Abbot area, Plymouth, the Teign estuary, Ivybridge, the Torquay area, and Sidmouth.

By the late 1930s all urban areas consisted of only one civil parish, with the exceptions of St Austell UD (two), Barnstaple NCB (two), Dawlish UD (two), Exeter CB (three), Exmouth UD

disappeared because the use of hundreds as statistical units was discontinued, and the more rational registration districts and subdistricts were put in their place. Second, the most troublesome anomalies on the county boundaries were removed by legislation in 1844.[48] Third, the established subdivisions of parishes were listed, though for the most part only in footnotes. **Fig. 24.6** shows clearly that Devon had more types of subdivision and many more actual subdivisions. Sixty-seven of Devon's parishes had named parts compared with Cornwall's nineteen. Fourth, in the early years of the nineteenth century there were a number of small areas claiming, often controversially, to be outside the parish system or functionally divided between parishes. The areas depicted on **Maps 24.6, 24.7, 24.8 and 24.9** are only some of those identified after 1835 and additions were made erratically to the list of extra-parochial areas.[49] A determined effort was made to get rid of the whole category by the Extra-Parochial Places Act 1857 (20 Vict, c.19), which provided rather crudely that each area reported as extra-parochial by the clerks to the boards of guardians should become a poor law parish, except that very small areas might, with the consent of the owners and occupiers of land, be annexed to a neighbouring parish. The 1857 Act was not a complete success, partly because it was not properly implemented in all cases by local administrators, but (two), Newton Abbot UD (three), Plymouth CB (three), Sidmouth UD (three), and Teignmouth UD (two). The processes in the rural and mixed urban-rural areas were relatively straightforward. First, eight new parishes were created by merger, though only three had new names, leading to the dissolution of eleven old parishes. In the new urban districts fifteen old parishes in Cornwall and nine in Devon were divided into urban and rural parts, each a separate parish. Four Devon parishes experienced the opposite—the reunion of their urban and rural parts in an enlarged urban district. There was one transfer of area in Cornwall and three in Devon.

The processes under the 1894 Act operated until 1929. During this period there were four transfers of territory in Cornwall and eighteen in Devon. Four new Cornish parishes came into being by division, four by merger and one by division and merger, with the result that ten old parishes disappeared. In Devon nine new parishes were created by division and two by merger, with the disappearance of five old parishes.[52]

The county review process of the 1930s in Cornwall was much more substantial than in Devon. In the latter there were twenty-one transfers of territory, one simple and two complex mergers of parishes and parts of parishes, and a simple division into two parishes. Six parishes disappeared and five new ones appeared. The total number of changes in Cornwall was 103. Only twelve of these were straight transfers of territory from one parish to another. Forty were part of the restructuring of the town districts and are too varied to describe here.

A small amount of tidying up was perceived to be necessary between 1935 and 1939, and a very small number of changes in parish structure took place between 1945 and 1971. As a consequence of the consolidation of the major urban areas (**Figs. 24.1, 24.2 and 24.3**), a considerable number of parish areas disappeared completely, but on their fringes new rump parishes appeared.

The front and rear end papers are maps of parish and unparished areas in 1998. Prior to the Local Government Act 1972, civil parishes were to be found only in rural districts, but the vestiges of former parishes—or at least their names—were to be found in county boroughs, non-county boroughs and urban districts. After the 1971–4 reorganization of areas and authorities, the former urban areas were either given successor parish status or left without a parish level of government. In the case of the former county boroughs this did not matter because the new district authority had parish functions, but Camborne–Redruth UD, Penzance NCB, St Austell with Fowey NCB, Newquay UD, and Exmouth UD were temporarily left in 1974 in anomalous positions.

Two areas were divided before parish government was introduced. In the former Camborne–Redruth UD the separate parishes of Camborne, Carharrack, Carn Brea, Illogan, Lanner, Portreath, Redruth and St Day were created. In the former St Austell with Fowey NCB, a similar process created Fowey, Mevagissey, St Blaise, Treverbyn and Tywardreath parishes, leaving the St Austell area unparished. The areas of the former Penzance NCB and Newquay UD were granted successor parish status in 1980 and 1983 respectively, with part of Newquay becoming a new parish of Crantock. The area of the former Exmouth UD had to wait until 1996 before being promoted. In each case the status of 'town' was also adopted. Parish councils have the right to adopt the title of 'town', a dignity which does not confer additional executive powers but makes the presiding officer a mayor. **Maps 24.16 and 24.17** identify the parishes that have taken advantage of this legal provision.[53] In addition Exeter and Plymouth retained the title of city and Torbay that of borough, both honorific and not substantive statuses. The district councils of Restormel (Cornwall) and West Devon (Devon) also have the title borough and Truro town council retains the historic title of city. **Maps 24.16 and 24.17** also indicate those parishes that have joint councils or the parish meeting form of government. Cornwall has gained two new parishes by division and one by merger, thus losing three old parishes. There were also twenty-nine adjustments of boundaries. In contrast the changes in Devon have been much more substantial. Ten new parishes have been created by division, merger or more complex processes, resulting from

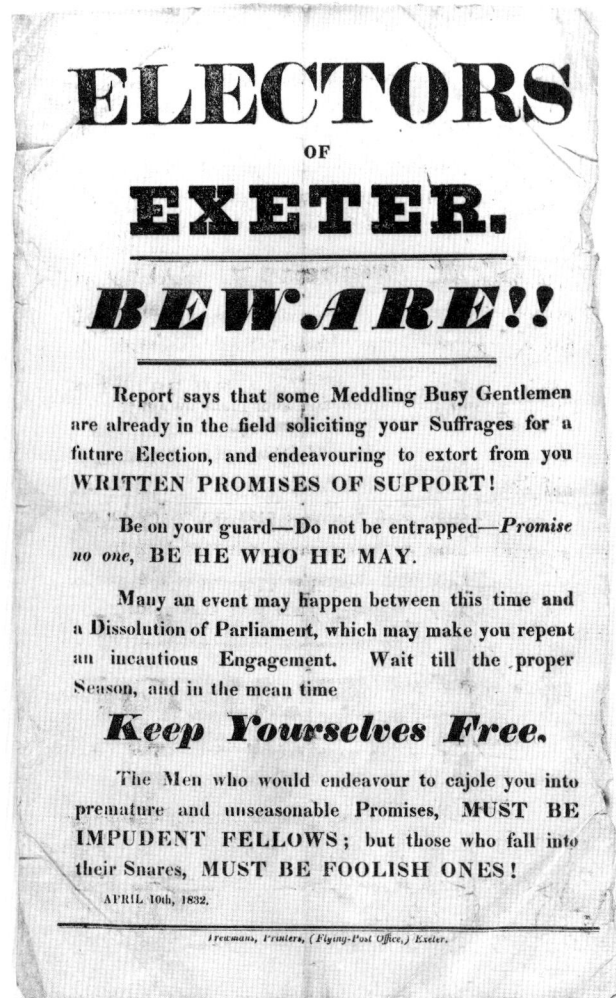

Fig. 24.8
Exeter election poster, April 1832.

SOURCE: reproduced by permission of Westcountry Studies Library, Exeter (LE 1832/04/10).

seventeen parishes disappearing, though in some cases the original name has survived. There were also 134 adjustments of boundaries but because of lack of detailed information the amount of double counting this involves is not known.

Though districts are considerably more important executive authorities than parishes, they are aggregates of the latter in geographical terms. This is the result of the simplicity of the reorganization process in 1971–4; every new area in Cornwall was equivalent to a combination of pre-existing areas, while areas of the new authorities in Devon were either identical to a single pre-existing one or a combination of several, with only the former St Thomas rural district being divided, though this was done on the basis of whole parishes.

also cause some surprise to the inhabitants of the two counties, but most of the inland towns and the rural parishes would be easily recognized. What has struck the author most strongly is the concern for the rule of law and the principles of natural justice embodied in the processes of change in areas and boundaries in England and Wales, of which this chapter is a case study. The reason why it has been impossible to record all the changes that have taken place is because those responsible for them tried to be accurate down to the last square yard and last individual. That this is what happened in local government reform has not been widely recognized and certainly not documented.

CONCLUSION

Present-day Cornwall and Devon are recognizably the successors of the counties of 1801. They have retained their shapes and in many of their parts the balance of urban and rural areas is familiar, though on a larger scale. Only the Torbay area would be completely unfamiliar to the Devonian of 1801. The development of seaside villages into small towns might

FURTHER READING

H.P.R. Finberg, 'The making of a boundary', in W.G. Hoskins and H.P.R. Finberg (eds), *Devonshire Studies* (London, 1952), 19–39; J. Stanyer, 'Devon—steadfast and changing: an essay in analytical local administrative history', *Devon Historian*, 15 (October 1977), 2–16; J. Stanyer, *A History of Devon County Council, 1889–1989* (Exeter, 1989).

POLITICAL AND MILITARY HISTORY

PARLIAMENTARY BOUNDARIES AND POLITICAL AFFILIATIONS 1918–1997

MICHAEL RUSH

In 1918 the combined electorates of Devon and Cornwall numbered over 487,000 (361,000 in Devon, 126,000 in Cornwall); in 1997 they numbered well over a million. The 1918 electorate itself was larger then ever before because of the extension of the franchise, notably to women aged 30 or over, by the Representation of the People Act 1918, and there was a further extension in 1928, when men and women had the vote on the same terms. In both counties the number of voters increased steadily after 1928, but after a major increase in 1945 Cornwall suffered a small decline and the electorate reached the 1945 level again only in 1964. Since then there has been significant growth (due partly to extending the vote to those of 18 years and over in 1969) to 375,788 in 1997, an increase of 198 per cent on 1918. The Devon electorate fell slightly after 1951, but later began to increase again, rising to 809,101 in 1997, an increase of 124 per cent on 1918.

Before the establishment of permanent parliamentary boundary commissions in 1948, boundary changes took place only when it was thought necessary, or in response to more limited changes in local government boundaries (**Map 25.1**). The latter practice continues and the inclusion of Pinhoe and Topsham in the Exeter constituency before the 1970 election is an example. Prior to 1918 major changes invariably accompanied an extension of the franchise. Thus in 1918 a major redistribution of constituencies took place, but there was no further major redistribution until 1948, when it was accepted that population movements required periodic reviews of constituency boundaries. The 1948 changes came into effect for the 1950 general election. Further, more limited changes took place in 1955, but there were no more redistributions until shortly before the general election of February 1974 and then again until 1997, from when the present boundaries date (**Map 25.2**).

Parliamentary representation of Devon and Cornwall since 1918 has been consistent in spite of the growth of the electorate, with Cornwall having five members of parliament throughout the period and Devon eleven members between 1918 and 1950 and since 1983, falling to ten between 1950 and 1983. This is markedly lower than the peak of nineteenth-century representation, when Cornwall had thirteen MPs (1832–68) and Devon twenty-two (1832–68). A number of the nineteenth-century constituencies were two-member seats, but since 1918 all seats in Devon and Cornwall have been single-member constituencies.

The constituency boundaries have been characterized by considerable historical continuity, although names have been changed from time to time, especially in Devon. Unlike Devon, which has three significant urban areas, Exeter, Plymouth and Torquay (Torbay), Cornwall has no entirely urban constituencies. Exeter and Torquay (now Torbay) have always been single-member constituencies since 1918, but Plymouth, as the largest and most populous urban area, has varied between two and three MPs, and before 1918 Devonport was a separate two-member constituency. Since 1918 Plymouth has had two or three single-member constituencies (Devonport, Drake and Sutton from 1918 to 1950 and from 1983 to 1997, and Devonport and Sutton between 1950 and 1983 and from 1997). The constituency boundaries of all three urban areas have been the appropriate local government boundaries (with a few minor exceptions) so that, as these have changed, so have the constituency boundaries. The boundaries with Plymouth have varied more than the constituency names suggest, although the core area in each has always remained the same.

The rural seats in Devon (six between 1918 and 1974 and between 1983 and 1997, five between 1974 and 1983, and seven from 1997) have undergone several major boundary changes since 1918, although Honiton survived largely unchanged and Tiverton relatively so, until they merged to form the Honiton and Tiverton constituency in 1997, with a new constituency of East Devon being created. The two or three remaining constituencies, sometimes with quite different names, have been based on variations and combinations of south Devon (Totnes, Teignbridge, South Hams, south-west Devon), west Devon (Tavistock, Torridge and west Devon), mid-Devon (South Molton, Torrington), and north Devon (Barnstaple, north Devon), as the maps illustrate.

The five Cornish constituencies all experienced boundary changes between 1918 and 1997, and in several cases different names have been used, but the main changes have affected the area in the centre of the county. Three constituencies—St Ives in the west (including the Isles of Scilly), Bodmin (now South-East Cornwall), and North (formerly Northern) Cornwall—have changed relatively little, whereas greater change has occurred in the two constituencies

Map 25.1
Parliamentary boundaries,
1918–1950.

SOURCE: Boundary
Commission for England,
*Report of Boundary Commission,
1917–1918,*
Cd. 8756.

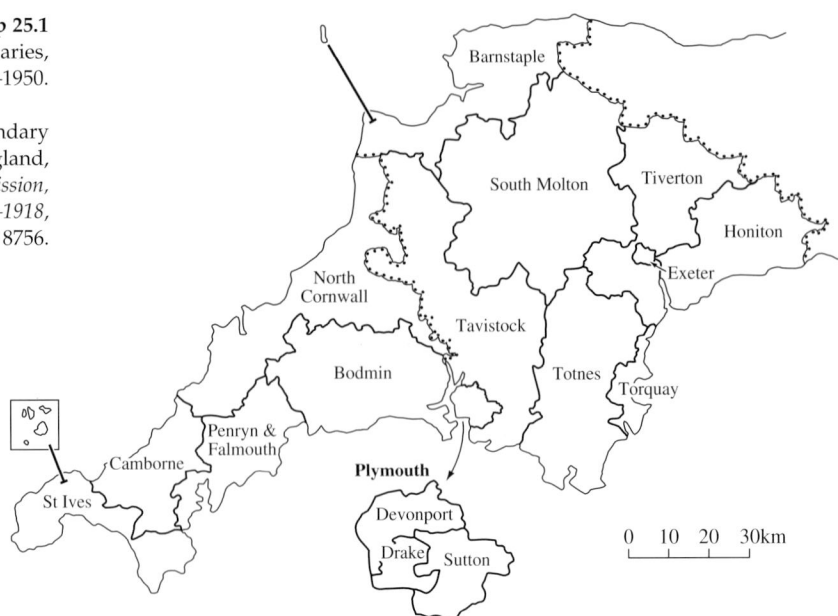

Map 25.2
Parliamentary boundaries,
1997.

SOURCE: Boundary
Commission for England,
Fourth Periodical Report, House
of Commons 433.iv, 1994–95.

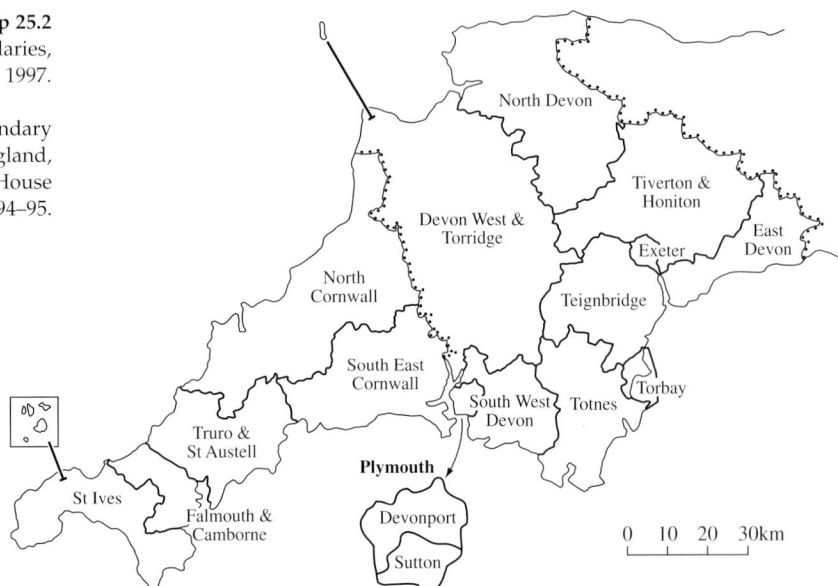

encompassing the towns of Falmouth, Penryn, Camborne and St Austell. Essentially, from 1918 to 1950 the dividing line ran down the peninsula and since 1950 it has run across it, as **Maps 25.1 and 25.2** indicate.

Urban areas apart, the result in both counties is a number of large constituencies in terms of area, with scattered electorates, the current North Cornwall and Devon West and Torridge being obvious examples.

POLITICAL AFFILIATIONS, 1918–1997
(Maps 25.3, 25.4 and 25.5)

An observer of general elections in Devon and Cornwall since 1918 could be forgiven for concluding that the Labour Party was a third party and that the historic clash between Conservatives and Liberals had continued more or less unabated since the middle of the nineteenth century. In terms of seats, in fact, Labour has only ever held one seat in Cornwall—from 1945 to 1970 (Penryn and Falmouth 1945–50 and Falmouth and Camborne 1950–70) and since 1997 (Falmouth and Camborne) (**Fig. 25.1**).[1] Indeed, Labour's electoral support exceeded that of the Liberals/Liberal Democrats in only three of the twenty-two general elections held between 1918 and 1997 (1950, 1951 and 1955) and has consistently been substantially below Labour's national level of support. Labour's weakness in Cornwall is further illustrated by the fact that, apart from its single seat, Labour has never been simultaneously second in more than two of the four remaining seats. Even in 1997, when it regained Falmouth and Camborne, Labour came third in Cornwall's other four seats, with only 17.1 per cent of the county's vote (**Fig. 25.2**). Labour has fared rather better in Devon, mainly though not entirely because of the presence of more electorally fertile urban seats. Labour held all three Plymouth seats from 1945 to 1950, retaining one Plymouth seat until the defection of David Owen to the Social Democratic Party in 1981, but regaining this in 1992 and winning both Plymouth seats in 1997. Exeter has also been in Labour hands, but only on two occasions—1966–70 and from 1997 (**Fig. 25.3**). In addition, Labour had more support than the Liberals in Devon from 1935 to February 1974, and between 1931 and 1959 was second in at least half the county's constituencies. Nonetheless, throughout the period Labour support has been consistently below its national level and its position in 1997 shows no change: in spite of winning Exeter and the two Plymouth seats, Labour's share of the county's vote was only 25.9 per cent (**Fig. 25.4**).

The broader picture that emerges is one in which the Conservative–Liberal clash of the nineteenth century has largely continued, seriously challenged by Labour only for a while during the 1950s—a period of Liberal weakness nationally. However, nineteenth-century Liberal strength was eroded, ushering in a long period of Conservative dominance, followed more recently by a Liberal and then Liberal Democrat resurgence.

Conservative

Labour

Liberal

Independent
Conservative

0 10 20 30km

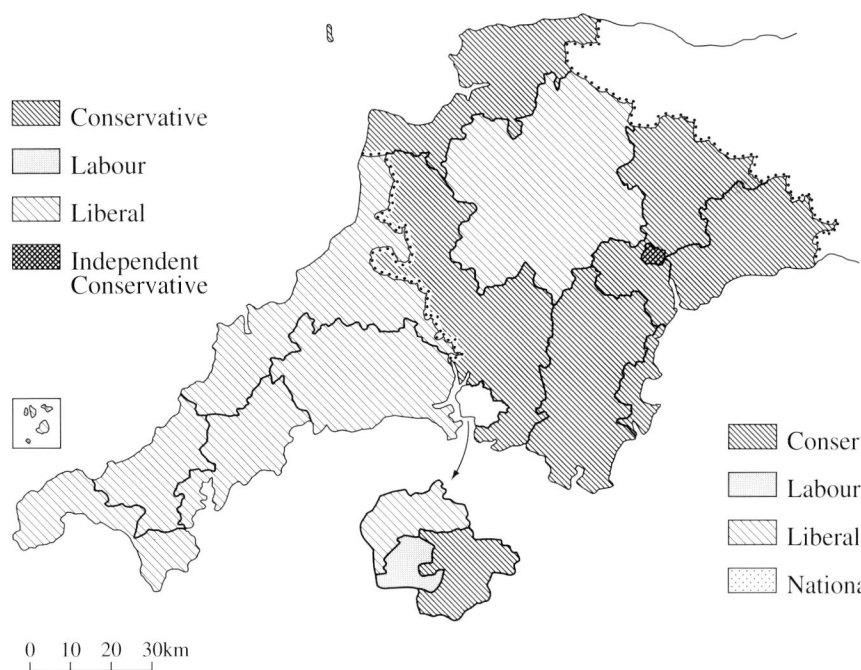

Map 25.3
Political affiliations, 1929.

SOURCE: F.W.S. Craig,
*British Parliamentary
Election Results, 1918–1949*
(Glasgow, 1969).

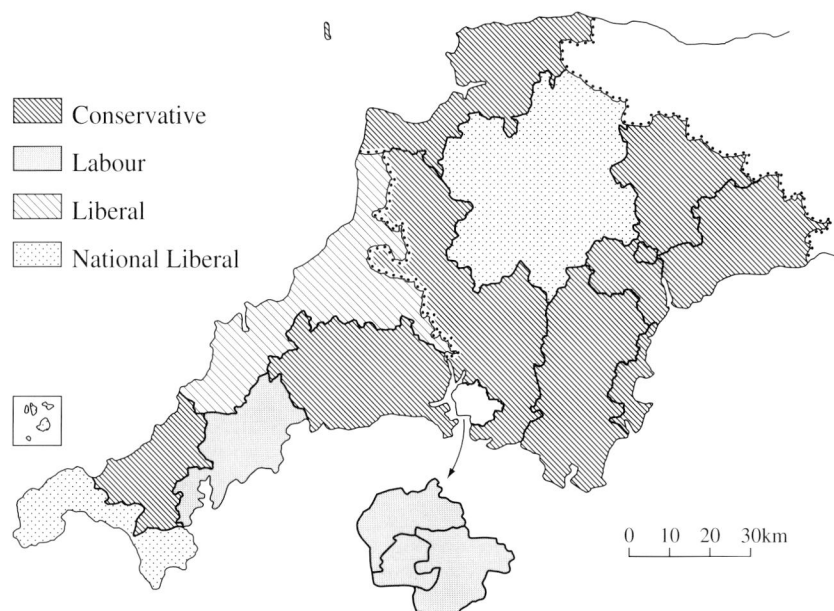

Conservative

Labour

Liberal

National Liberal

0 10 20 30km

Map 25.4
Political affiliations, 1945.

SOURCE: as **Map 25.3.**

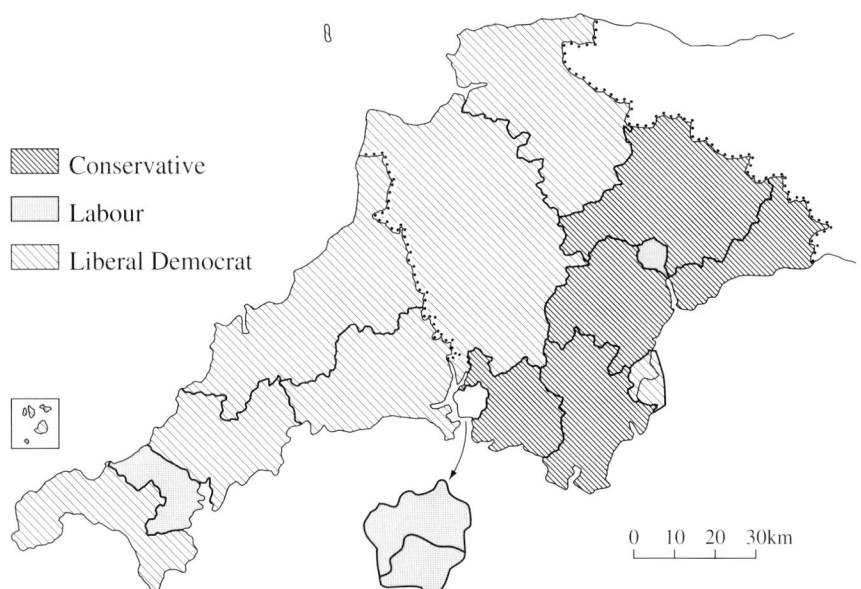

Conservative

Labour

Liberal Democrat

0 10 20 30km

Map 25.5
Political affiliations, 1997.

SOURCE: *The Times House
of Commons, May 1997*
(London, 1997)

Fig. 25.1
Seats won in Cornwall,
1918–1997.

SOURCES: F.W.S. Craig,
*British Parliamentary
Election Results, 1950–1970*
(Chichester, 1971) and
The Times House of Commons
(London, 1974–97).

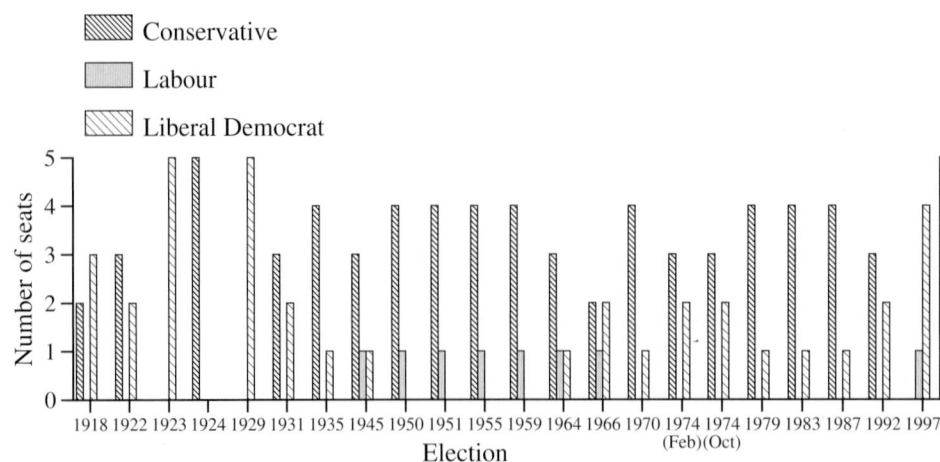

Fig. 25.2
Electoral support in Cornwall,
1918–1997.

SOURCES: as **Fig. 25.1**.

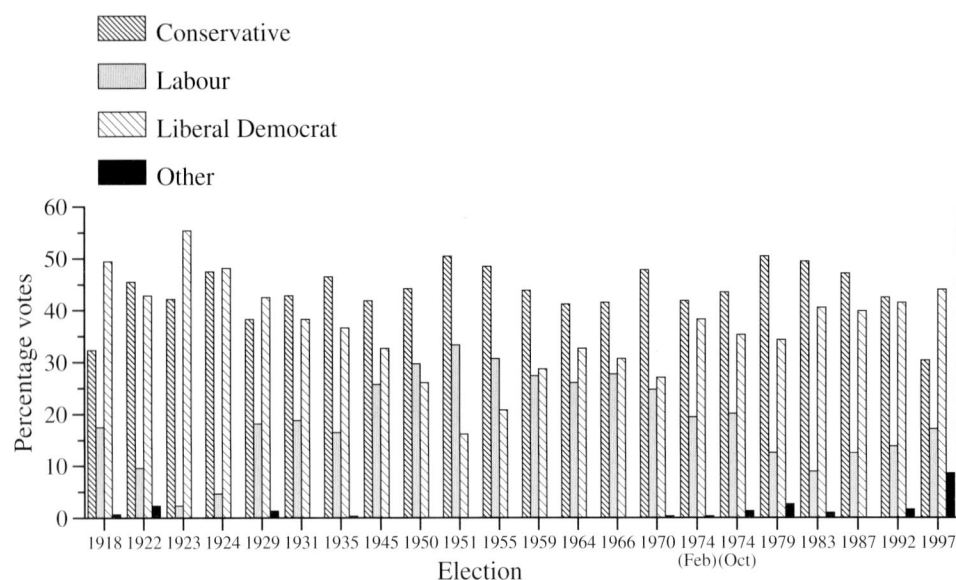

The Liberal erosion was assisted by the breakaway of the Liberal Unionists in 1886, the split between Coalition or Lloyd George Liberals and Asquith Liberals in 1918, and the defection of the National Liberals in 1931—the 'National Liberal and Conservative' label survived in Torrington until 1958 and St Ives until 1966. However, the Liberal revival, beginning in the late 1950s, has largely pushed Labour aside and increasingly made the Liberals and their Liberal-Democrat successors the main challengers to the Conservatives. Thus between 1832 and 1880, 62.7 per cent of the seats were Liberal and 37.3 per cent Conservative in Devon, and 66.0 per cent Liberal and 34.0 per cent Conservative in Cornwall; between 1918 and 1997 the figures for Devon were 80.6 per cent Conservative, 11.6 per cent Liberal/Liberal Democrat, 7.3 per cent Labour, and 0.5 per cent Social Democrat, and for Cornwall 60.0 per cent Conservative, 32.7 per cent Liberal, and 7.3 per cent Labour. The Liberals have therefore been more successful in retaining support in Cornwall than in Devon, where Liberal support is more concentrated in the north and west of the county. Indeed, in 1997 the Liberal Democrats won four of the five Cornish seats, which, with Labour's recapture of Falmouth and Camborne, deprived the Conservatives of representation in Cornwall and secured more votes than the Conservatives for the first time since 1929. In Devon, however, the Liberal Democrat vote increased by only 0.9 per cent, but the party benefited from the anti-Conservative swing and gained two more seats, which, with Labour's three seats, reduced the Conservatives to five, their lowest tally since 1923. In short, Devon and Cornwall do not fit the national pattern of Conservative versus Labour characteristic of much of the period since 1918, and, while not unique in this respect, they constitute the major regional exception to that pattern in England. That exception has not been significantly disturbed by Labour's victory in 1997.

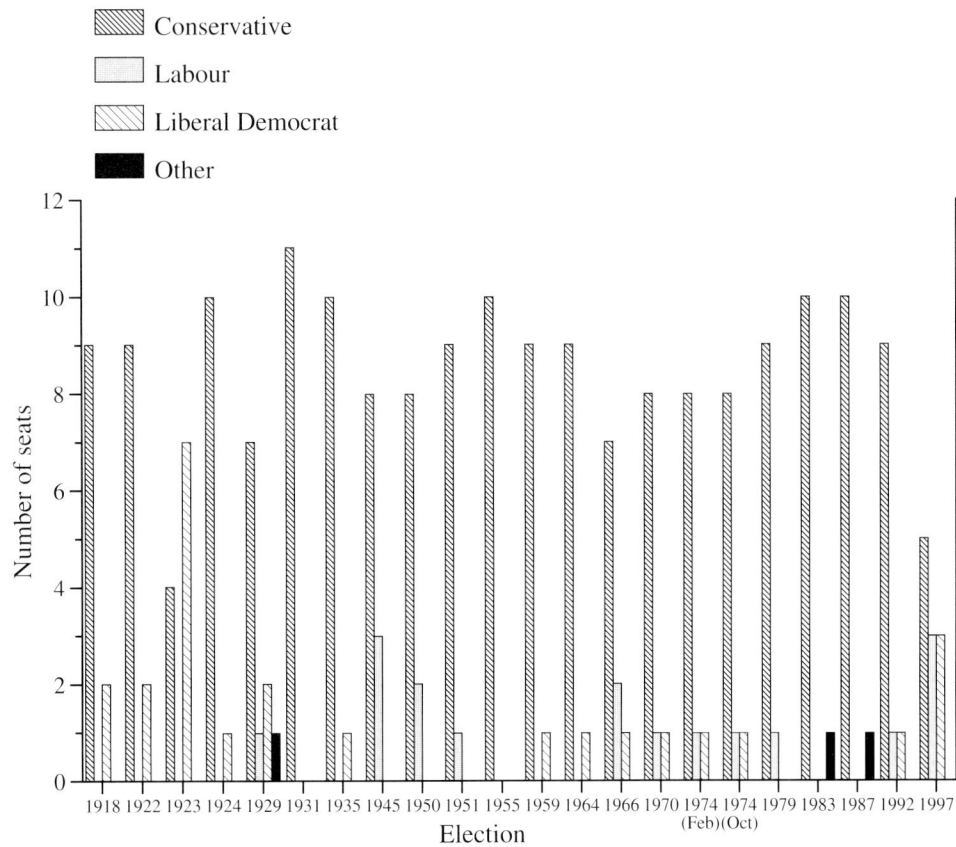

Fig. 25.3
Seats won in Devon, 1918–1997.

SOURCES: as **Fig. 25.1.**

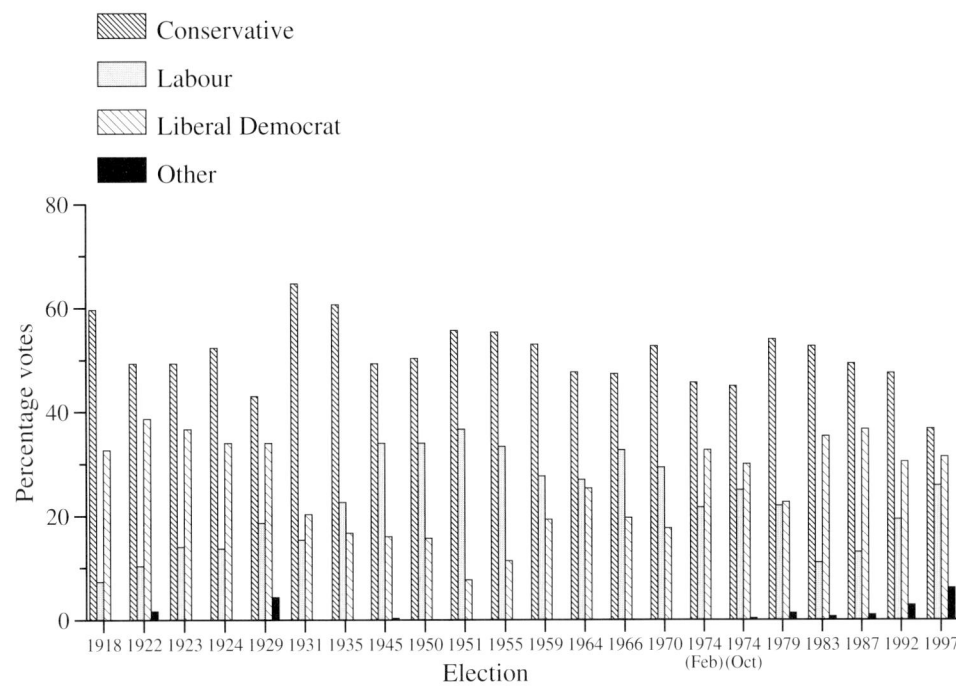

Fig. 25.4
Electoral support in Devon, 1918–1997.

SOURCES: as **Fig. 25.1.**

FURTHER READING

F.W.S. Craig, *British Parliamentary Election Results, 1918–1949* (Glasgow, 1969) and *British Parliamentary Election Results, 1950–1970* (Chichester, 1971). See also the serial publication: *The Times House of Commons* (London).

RELIGION AND RELIGIOUS INSTITUTIONS

ECCLESIASTICAL INSTITUTIONS
IN 1086 AND MONASTIC HOUSES
c.1300

CHRISTOPHER HOLDSWORTH

ECCLESIASTICAL INSTITUTIONS (**Map 26.1**)

By 1086 when William the Conqueror ordered that a survey be made of the kingdom which he had conquered twenty years before, the two counties of Devon and Cornwall had broadly speaking today's boundaries, as did the ecclesiastical organization for the region, the diocese of Exeter.[1] Two significant exceptions to this were both in the south-east. North-west of Axminster the parish of Stockland (and what was later called Dalwood) fell under the jurisdiction of the sheriff of Dorset in things temporal, and of the bishop of Sherborne in matters spiritual. To the east, across the shire boundary, the parish of Thorncombe was considered to belong to Devon, and to the see of Exeter. The explanation for these anomalies, not straightened out until 1842, lay well in the past in 1086.[2] Stockland had for some time belonged to a religious house within Dorset, Milton Abbey, whilst Thorncombe owed its position to the fact that it was part of a large royal estate centred upon Axminster.[3] A secular link seems to have determined Thorncombe's allegiances, whilst for Stockland a religious link was decisive.

By 1086 the authority of the bishop of Exeter was unchallenged everywhere else in the two counties, and **Map 26.1** shows how extensive were his estates.[4] Upon them he could draw for his own support, and for the maintenance of his own family of clerks, his cathedral, and those communities of clerks who still served the churches at Crediton and St Germans, which had served as seats of separate bishops for Devon and Cornwall until 1050 and *c*.1027 respectively.[5] These large holdings meant, too, that wherever the bishop travelled within his diocese he was never far from one of his manors. Some of the lands credited to the bishop in 1086 had been in the hands of his predecessors for a very long time: for example, Crediton had been given to the Church in 739, whilst Æthelstan was credited with the gift of Culmstock, Sidbury and Stoke Canon sometime between 925 and 939.[6] Other estates had been given much more recently by Leofric and Osbern, bishops from 1046 to 1072 and from 1072 to 1086 respectively.[7] Indeed two of the richest plums in the bishop's control, situated well beyond the compass of the map, were just such gifts: Brampton in Oxfordshire and Bosham in Sussex. Within Devon and Cornwall the estates held by the bishop meant that he controlled perhaps 100,000 acres, bringing in an income of about £377 a year,

making him the third wealthiest man in the region, after the king and Baldwin the sheriff of Devonshire.[8] Among bishops in England, although his see was the fourth largest, he was the sixth in wealth, reflecting probably the relative poverty of the two counties compared with other parts of England.

Map 26.1 also shows another dispersed pattern—that of the churches, which seem to have originally had rights and duties over wide lands around them, called in some legal texts 'superior churches' and now often known as minsters.[9] It will readily be seen that these were fairly evenly spread across the lowland parts of the diocese, but it would be hard to prove that this result came from planning. Some of these churches had links with holy men from the earliest days of Christianity in the South West, some of them seem to have been placed near centres of royal authority, but by the time that Domesday Book was made, they were being augmented by a network of local, or parish, churches.[10] Very few such churches were recorded for the area in Domesday, and it is thought that this reflects not so much their actual rarity, but the way that the panel of Domesday commissioners who worked there set about their task. Further research is needed both into surviving buildings and into the documentary sources before it would be meaningful to map this lower rung of churches.

One of the old minsters, Bodmin, which Domesday records as being served by canons, although there is evidence that at earlier times it had been a monastery, had a peculiarly rich endowment, reflecting the popularity of St Petroc to whom it was dedicated.[11] Lands given to him were mainly scattered across the northern edge of Cornwall, but with an interesting pair of outliers in Devon at Hollacombe and Newton St Petrock, themselves places where minsters may have existed in 1086.

Domesday only lists two monastic communities, Tavistock and Buckfast, both of them creations of the so-called Tenth-Century Reformation which had owed most to King Edgar (957–75).[12] Each Devonshire house was established by a local noble with close links to the royal family. Ealdorman Orbulf, the founder of Tavistock, was related by marriage to Edgar, whilst Buckfast was set up by Earl Ailward, a man close to Cnut. The two houses were, however, endowed on very different scales, Tavistock with lands worth around £75, Buckfast, £17.[13] Neither house was wealthy in comparison with places like Glastonbury or Christ Church,

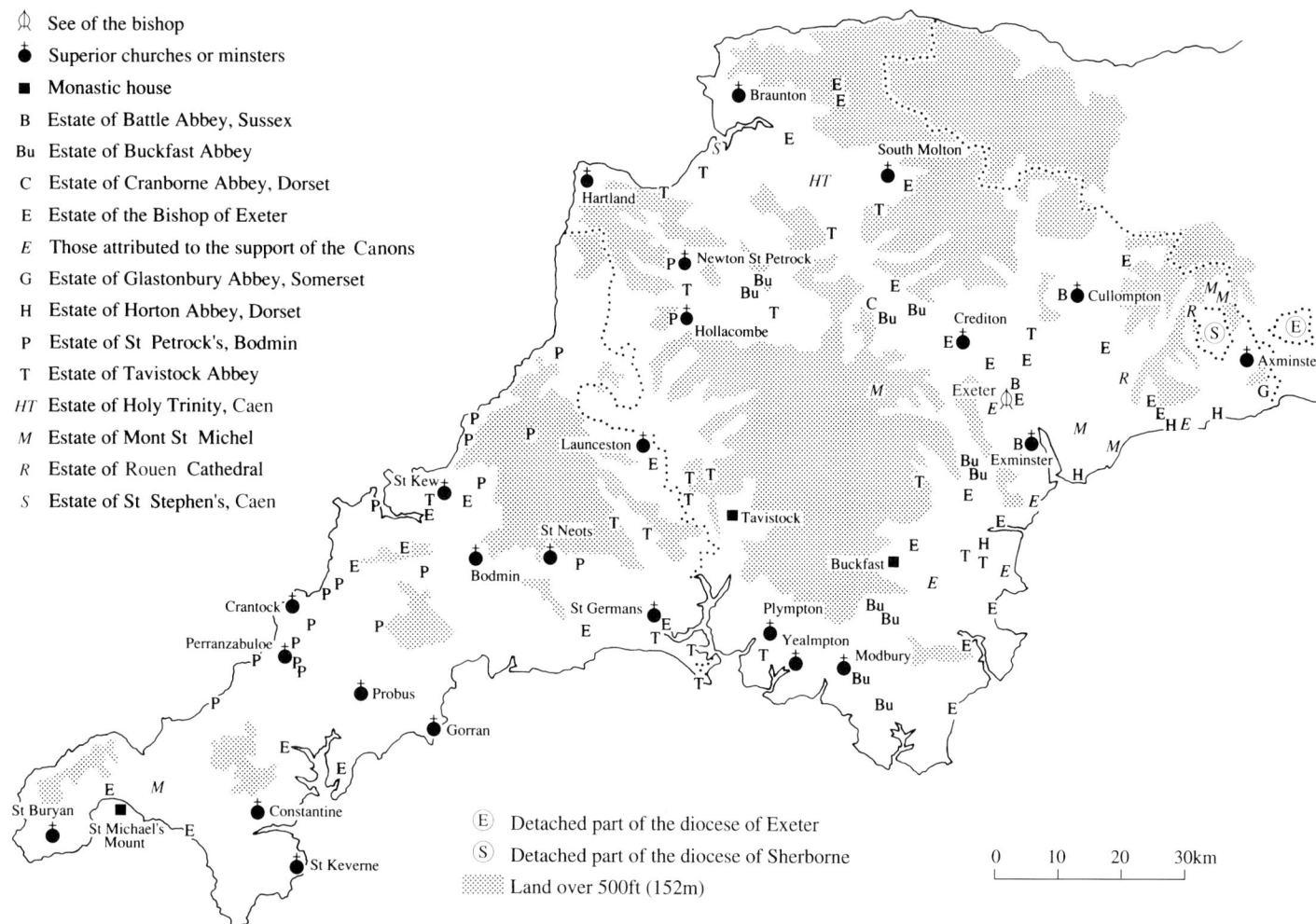

Map 26.1
Ecclesiastical institutions at the time of Domesday Book.

SOURCES: C. and F. Thorn, *Domesday Book: Devon*, 2 vols (Chichester, 1985) and *Domesday Book: Cornwall* (Chichester, 1979); H.C. Darby and G.R. Versey, *Domesday Gazetteer* (Cambridge, 1975).

Canterbury, enjoying Domesday incomes of about £828 and £688.[14] Again, we may see here another reflection of the relative poverty of the region. Already in 1086, however, each house had widely dispersed estates which must have required a good deal of time to exploit effectively. The impact of the Conquest on these two houses is rather hidden in the Survey, but it is clear that Tavistock had lost some of the land which it had held in 1066, and that it had had to settle knights upon about a third of its holdings, a fairly high proportion compared with most other monasteries, to fulfil the obligations which the new rulers had placed upon it, whereas Buckfast was not required to send any knights to the king's army.[15] Cornwall, in stark contrast, appears to have had no monastic community at all in 1086, and many of its old minster churches had lost part of their endowments to Robert count of Mortain, the king's half-brother.[16]

Some resources within the two counties had passed into the hands of two other groups of religious institutions: English monasteries based outside the South West, and four Norman communities, three monasteries and one cathedral. The first group began to profit from lands in the diocese well before the

Conquest, whereas benefaction to Norman institutions only began during the reign of Edward the Confessor, who gave the manor of Ottery to Rouen cathedral in 1061.[17] Three things are noteworthy about this income going to Normandy. In the first place, it surpassed in total the sum going to all the English houses, whether based in the diocese of Exeter or outside it. Second, it was derived from a few, fairly wealthy estates, and lastly it was diverted with relatively little cost to the Conqueror who was able to give away lands which had recently come into his hands from members of the Godwine family, or other nobles, whose lands had been forfeited to the crown.[18] He could get credit with religious groups, and in particular be assured of their prayers on his behalf, at relatively low cost to his own resources.

MAJOR MONASTIC HOUSES c.1300 (Map 26.2)

Map 26.2 shows that by around 1300 there had been a very dramatic increase from two to thirty-six in the number of religious communities within the region (that is to say of monks, canons,

Map 26.2
Major monastic houses, *c*.1300.

SOURCES: Ordnance Survey
Monastic Britain, South Sheet;
David Knowles and R. Neville
Hadcock, *Medieval Religious
Houses in England and Wales*,
2nd edn (London, 1971).

Monastic houses outside the diocese upon which local houses depended

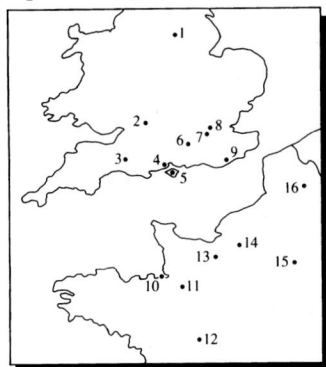

1 Welbeck, Notts.
2 Malmesbury, Wilts.
3 Montacute, Somerset
4 Beaulieu, Hants.
5 Quarr, Isle of Wight
6 Waverley, Surrey
7 Merton, Surrey
8 Holy Trinity, Aldgate, London
9 Battle, Sussex
10 Mont St Michel
11 Savigny, Normandy
12 St Serge, Angers
13 St Pierre sur Dives, Normandy
14 Le Bec, Normandy
15 St Martin des Champs, Paris
16 Arrouaise, near Arras

■ Benedictine
◪ Cistercian
◪ Cluniac
◆ Augustinian canons
◇ Premonstratensian canons
● Benedictine nuns
▲ Augustinian canonesses
Ⓒ Carmelite friars
Ⓓ Dominican friars
Ⓕ Franciscan friars
Ⓣ Trinitarian friars
◪ Preceptory of Knights Hospitallers
■ Founded pre-1100
■ Founded 1100-1149
■ Founded 1150-1199
■ Founded after 1200
▒ Land over 500ft (152m)

(K) Founded by the King
(b) Founded by a bishop
(B) Founded by William Brewer or descendant
(J) Founded by Juhel of Totnes
(fB) Founded by member of fitz Baldwin family
(R) Founded by member of de Redvers family
Ⓔ Detached part of the diocese of Exeter
Ⓢ Detached part of the diocese of Salisbury

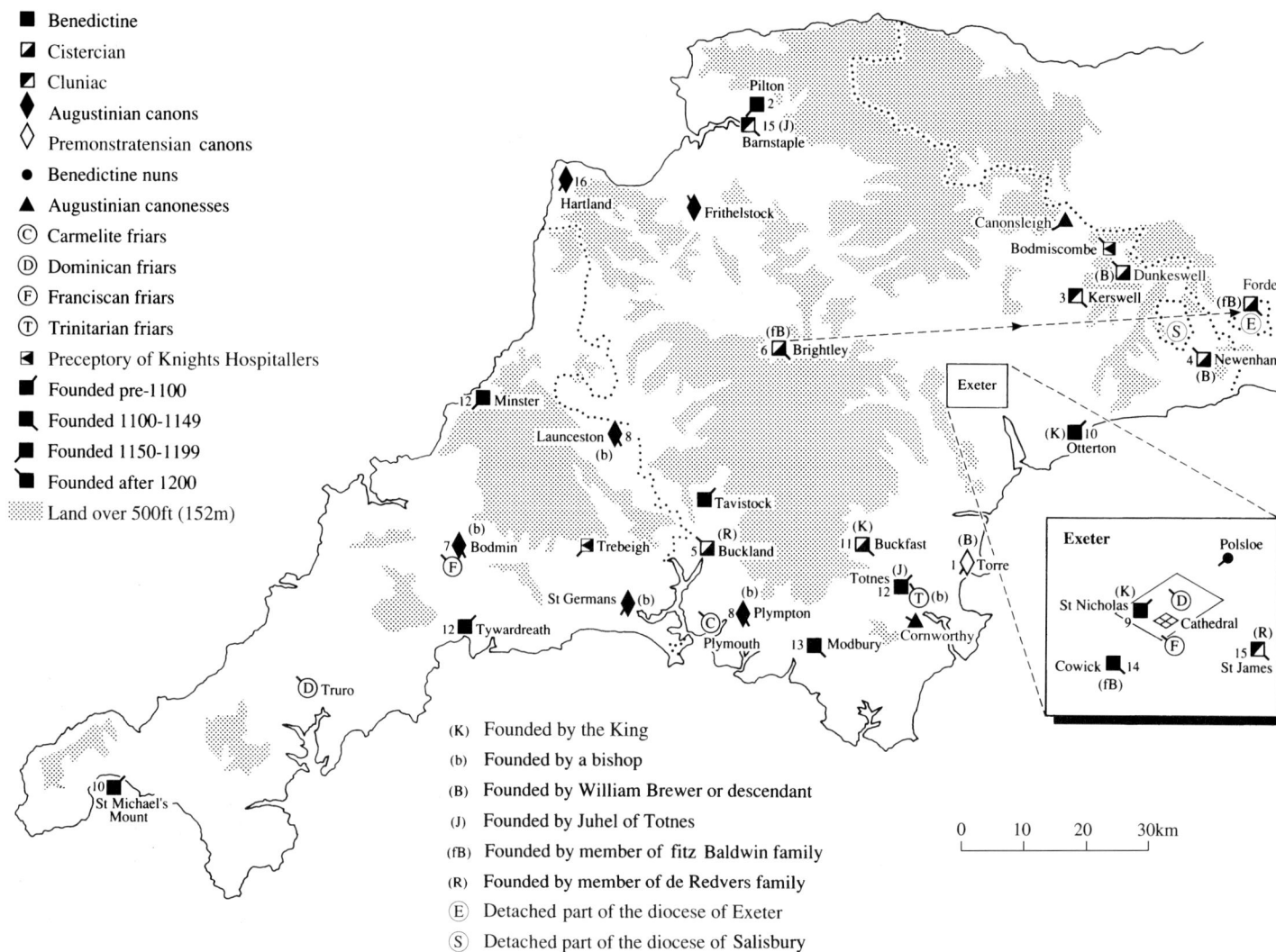

friars and nuns) compared with the Domesday situation shown in the preceding map.[19] Some of this growth was achieved through the transformation of old minster churches into stricter communities in accordance with new ideals, as at Bodmin or Hartland, but most of it came from entirely new foundations. These were made predominantly in the areas to the south of the major uplands of Bodmin Moor, Dartmoor and Exmoor, although five houses were planted not far from the northern seaboard. Comparison with the earlier map shows that with the one exception of Trebeigh, no monastery was created where the bishop already held land. The other striking thing from a spatial point of view is that only a quarter of the monasteries were situated west of the Tamar, and most of these were small houses living off relatively meagre resources. The probable explanation for this is that the major landholders in Cornwall preferred to support monasteries elsewhere on their estates. Successive earls of Cornwall, in particular, who dominated the county, gave Cornish lands and churches to monastic houses well outside the county. Henry III's brother Richard, for example, gave Cornish land and churches to his own Cistercian foundation at Hailes in Gloucestershire, as well as to Beaulieu in Hampshire, his father's creation.[20] One can only suppose that some Cornish men who wished to become monks travelled eastwards (the most famous is Peter of Cornwall who came from Launceston and became an Augustinian canon in London at Holy Trinity, Aldgate).

The growth in monasticism took place largely within the twelfth century when nineteen houses were created. The two houses of Domesday became seven by the end of the century, seventeen by about 1150, twenty-six by 1200, and thirty-six a century later. These houses attracted very different levels of endowment, and so could support very different numbers of religious. Until the refounding of Plympton, Bodmin and Launceston through the interest of Bishop William Warelwast, and the arrival of the Cistercians in 1136, few of these new houses held as many as a dozen religious.[21] Communities like that at St Nicholas, Exeter, or Totnes, were in this respect not a great deal more than centres at which rents from lands given to support a monastery in France could be collected.[22] In this sense they represented a logical development from the Domesday situation when, as we have already seen, a small number of such houses were already receiving income from English estates. Small houses continued to be founded right through the period, but even so the total number of religious within the diocese grew from perhaps about 50 to about 500, a tenfold growth. Why did this occur?

Some of the increase may reflect the steady growth in population, but this is only reckoned to have doubled, or conceivably trebled, between around 1050 and 1300.[23] Something more must have been involved. Throughout the whole western Christian world this was a time of unparalleled enthusiasm for monasticism, and a time in which new types of community life developed.[24] Previously the Rule of St Benedict (c.480–c.550) had dominated religious life; by c.1300 a number of changes had occurred based on that Rule, of which the Cistercian reform was the most successful, the so-called Rule of St Augustine (354–430) had been rediscovered as a guide for communities of canons, and a substantially new mendicant (or begging) way of life had been created by Francis Bernadone (1181–1226) and Dominic Guzman (c.1170–1221), each quickly canonized after his death. All of these forms took root here, gaining the interest of those who wanted to practice them themselves, and, just as crucially, the interest of people who were willing to devote some of their own resources to support them. Occasionally founders and later benefactors were moved to support one of their own relatives who wanted to become a monk or nun (the founder had a relative who was a monk at St James, Exeter), occasionally they wished to do this themselves (William Brewer entered Dunkeswell in 1224, two years before he died), but more frequently they gave so that others could flee the world.[25] They behaved like this because they had inherited a belief that the world was a dangerous place in which to try to save one's soul, and yet salvation was crucial, if one were not to suffer unending agonies in the world to come. The standards of behaviour which the Church commended in its teaching, and indeed tried to enforce through the coercive force of its own courts, were incompatible with the way most people lived, particularly with regard to recourse to violence and indulgence in sexual intercourse at times and in manners believed to be sinful.[26] Such fears lie behind the words of numberless charters, as, for example, when the founder of St James priory on the south side of Exeter stated that he gave Tiverton church to it 'for my soul and that of my wife Alice, and of my father Richard, my mother Alice, as well as that of the most noble King Henry'.[27]

The West Country evidence shows clearly that while only the wealthy could afford actually to found a monastery, a far wider circle could make some kind of gift. **Map 26.2** indicates the significance, even in a part of England where kings travelled infrequently, of royal intervention, which lay behind the establishment of Otterton and St Nicholas Priories, and Buckfast Abbey. Neither of the first two was particularly well endowed, and the last was achieved by King Stephen reforming an existing community by the imposition upon it of new stricter standards of observance pioneered by Savigny in Normandy, which he knew since it was situated within his own estates.[28] These things remind us of another feature of **Map 26.1**: the way that William I was able to endow Norman houses with English lands at relatively little cost to himself.

After the king, the bishop was the most significant figure in the growth of monasticism here, encouraging houses with his protection, sometimes with gifts, and, more rarely, with greater initiatives.[29] Besides William Warelwast's refoundations mentioned above, Bartholomew was involved in the reform of Hartland and St Germans between about 1165 and 1184, whilst much later Bronescombe helped a group of laity place one of the smaller new mendicant orders at Totnes in 1271.[30] On the whole, however, bishops, like kings, did not give much away.

Three great families dominate the scene around the middle of the twelfth century.[31] Two of the sons of Baldwin sheriff of Devon,

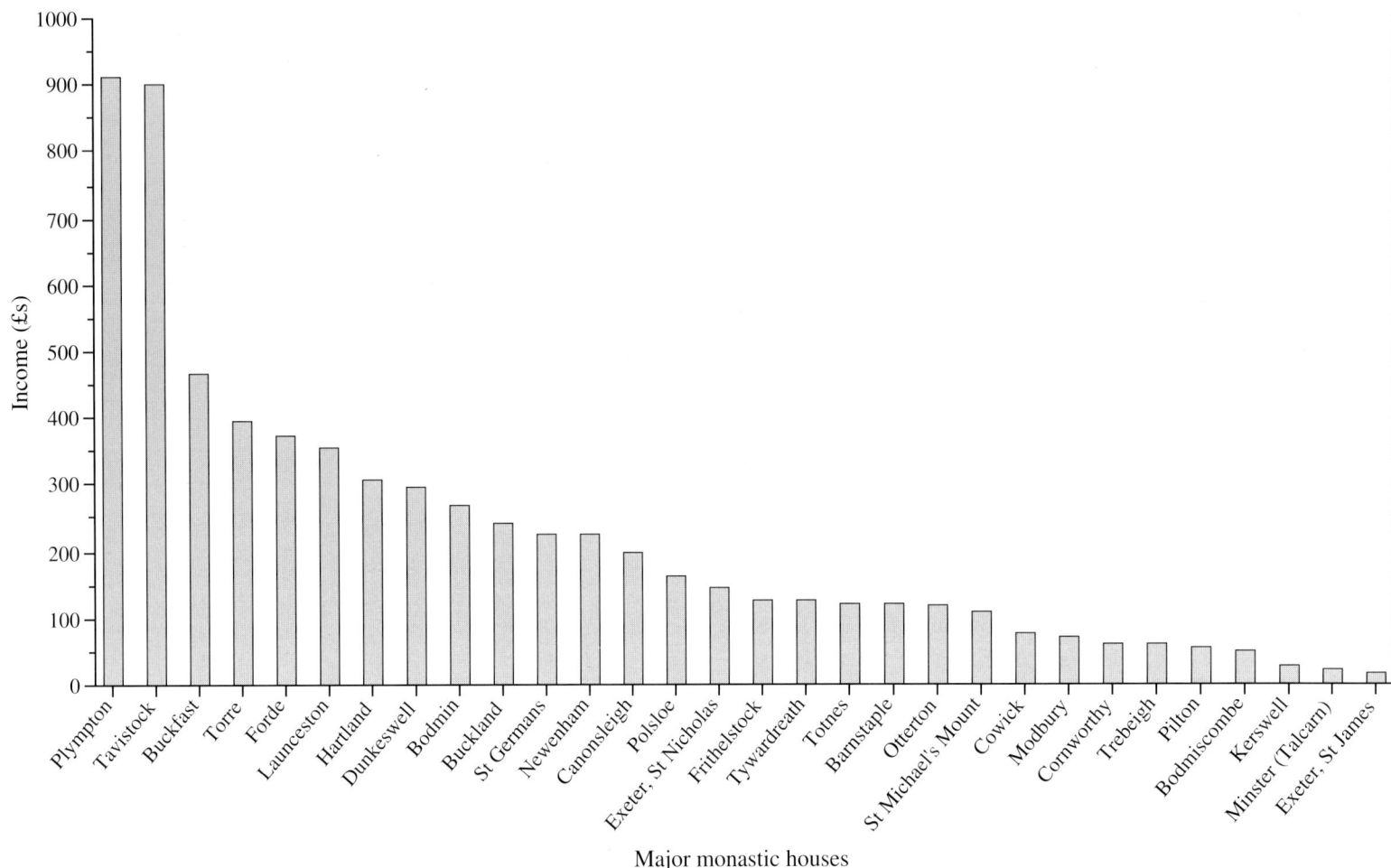

Fig. 26.1 Income of each house at the time of the Dissolution of the monasteries in the 1530s.

SOURCES: as **Map 26.1.**

son of Count Gilbert of Brionne, founded Brightley (better known as Forde, where it moved five years after its foundation) and Cowick, whilst Baldwin de Redvers established St James, Exeter, and Juhel of Totnes established Barnstaple and Totnes. Later on, at the turn of the twelfth and thirteenth centuries, one man, William Brewer, who had risen to wealth through service to the Angevins, created a monastery at Torre in 1196, near to his birthplace, and five years later in 1201 another at Dunkeswell, amongst the Blackdown Hills. In the first case the community housed canons following the customs of the north-eastern French house of Prémontré, whereas at the second there were Cistercian monks recruited from Forde. His and others' choice of groups to support may have been influenced by earlier experience, family connections and 'fashion'. William had come across Premonstratensians at Welbeck near Nottingham when he had been sheriff there, whilst Richard fitzBaldwin may have decided upon Cistercians for his foundation because cousins of his had recently planted them at Tintern. Continuity in patronage can be found at Cistercian Newenham, whose founder, Reginald de Mohun, was a son-in-law of William Brewer. If one looks beyond the county, however, one can find a surprising spread of interest: Baldwin de Redvers not only established Cluniac monks at St James, but Cistercians, who were in some senses then their critics,

at Quarr on the Isle of Wight, where he also had wide estates. Over 200 years later, the widow of one of his descendants, Amicia, lady of Devon, created a new house for Cistercians at Buckland, on the south-western edge of Dartmoor, and found monks to start it at Quarr, founded by her husband's ancestor.

Founders and benefactors had an enormous influence upon the development of individual houses. Often they retained some influence over the choice of superiors, or the running of the estates during vacancies, besides expecting to receive hospitality when they were in the neighbourhood, burial within the precinct, and, of course, regular remembrance in the prayers of their community.[32] **Fig. 26.1** shows one result of founders' activity, the widely varying economic resources which communities came to enjoy.[33] Firm figures are only available for most places from records made when the monasteries were closed down during the 1530s, but the general situation then was probably not very different from 200 years earlier, since neither donations nor losses were extensive after 1300. Not surprisingly there seems to have been a fairly close relationship between the size of endowment and the number of monks which could be supported. Only one of nine houses with a net income of under £100 ever had more than ten monks, and only four of nine houses with income between £100 and £199 surpassed that size, but all twelve richer houses were bigger.

Standards of life varied too. Kerswell, the poorest house, with £29 a year to share between two monks in the 1530s, must have been very different from the richest, Plympton, where the twenty-one canons had over £40 each. Such resources could be drawn, too, from very different kinds of estate. Cornworthy, for example, drew its income from lands and rights in five neighbouring parishes, whilst a middling-rank house, Canonsleigh, depended upon dispersed lands, in Dorset and Somerset, and even Suffolk. Such an estate would need a good deal of time and care if it were to be profitable.[34]

Behind the simple figures of total net incomes lies another difference, the degree to which a community depended upon the income drawn from the 'ownership' of a parish church (which brought with it substantial sums raised from tithes) or on payments from parishes. Among twenty-one houses for which the relevant information survives, three equal-sized groups can be distinguished. First, there was a group for whom this spiritual income was equal to, or greater than, income from temporal sources, i.e. that from land, rents etc.: Cornworthy, Hartland, Plympton, St Germans (Augustinian), and Pilton, Totnes and Tywardreath (Benedictine). Second, there was a group for whom spiritualities brought in between 25 and 49 per cent of their temporalities: Bodmin, Frithelstock, Launceston, Torre (Augustinian); St Nicholas, Exeter, and Tavistock (Benedictine); and Polsloe (Benedictine nuns). A third group consists of those for whom spiritualities were relatively insignificant: Barnstaple (Benedictine); Canonsleigh (Augustinian); and Buckfast, Buckland, Dunkeswell, Forde and Newenham (Cistercian). The presence of all the Cistercian houses within the diocese in this third group is an interesting reflection of the fact that one of the aims of the monks who founded Cîteaux in Burgundy, from which the group took their name, was to renounce income from churches.[35] Here the wishes of a group of monks seem to have been on the whole respected by those who patronized them. In general, however, it should be noted that a very substantial part of the economic resources given to the Church was in monastic hands by 1300, when there were far more secular clergy serving the wider Christian community than there were monks.[36] Prayer communities, apart from those of the mendicants, were not cheap.

Map 26.2 also illustrates another side of these establishments: their links with other monasteries beyond the region. As we have noted, a number owed regular payments to houses across the sea, but there were others linked to houses within England, and a sizeable group (all the Cistercian houses, and the canons of Hartland and Torre) had constitutional ties, involving them in attending general chapters of their order and being 'inspected' annually by the superior of the mother house from which their members had been originally drawn. The friars, too, were involved in such networks to maintain discipline. In such ways parts, at least, of the diocese were made aware of that wider cultural and political community which covered the whole of western Europe.

FURTHER READING

Frank Barlow, *The English Church 1000–1066*, 2nd edn (London, 1979) and *The English Church 1066–1154* (London, 1979) provide material to set the local situation against the national background. John Blair, 'Secular minster churches in Domesday Book', in Peter Sawyer (ed.), *Domesday Book: A Reassessment* (London, 1985), 104–42, provides a lucid entry to a complex subject. Christopher Holdsworth (ed.), *Domesday Essays*, Exeter Studies in History 14 (Exeter, 1986) gives a brief introduction using local, mostly Devon, examples. David Knowles, *The Monastic Order in England*, 2nd edn (Cambridge, 1963) is a masterly, readable account for the period before 1216. Robert J.E. Boggis, *A History of the Diocese of Exeter* (Exeter, 1922) is a clear narrative based on the materials then available. H.P.R. Finberg, *Tavistock Abbey: A Study in the Social and Economic History of Devon*, 2nd edn (Newton Abbot, 1969) is the finest study yet written about a west-country monastery. See Colin Morris, *The Papal Monarchy: The Western Church from 1050 to 1250* (Oxford, 1989), especially 237–62, 452–77, for a balanced account of monasticism across Western Europe, with full bibliography. Nicholas Orme (ed.), *Unity and Variety: A History of the Church in Devon and Cornwall* (Exeter, 1991), 1–80, brings together together recent work on the region in the Middle Ages.

RELIGION AND RELIGIOUS INSTITUTIONS

THE CHURCH FROM *c*.1300
TO THE
REFORMATION

NICHOLAS ORME

By 1300, the diocesan structure of archdeaconries, rural deaneries, peculiars and parishes (**Maps 27.1 and 27.2**) was complete, and would endure with little change (save for subdivision of parishes) for over 500 years. The foundation of religious houses was also virtually complete, the last monastery (Buckland) having been established in 1278. A fall in the population after 1300, much accelerated after the Black Death of 1348–9, meant that there were fewer vocations to become monks, friars or nuns, and the existing houses amply sufficed to cater for them.

THE DIOCESE OF EXETER IN 1291

In 1291 English clergy, acting on the pope's behalf, made a valuation of Church property throughout England, including that of bishops, cathedrals, religious houses and parish benefices (the incomes of the parish clergy). Though not the earliest such valuation, it is the most complete and remained in use as the basis of both papal and royal taxation of the clergy until it was superseded by Henry VIII's new *Valor Ecclesiasticus* in 1535. True, the 1291 valuation has its faults; it underestimated clerical incomes, perhaps by as much as half their value, so that its figures are better used for comparative purposes than as absolute guides. It also omits some parish benefices which certainly existed at the time. Still, it is a remarkably complete and helpful survey for its date, and an excellent framework with which to reconstruct the medieval diocese of Exeter, especially with regard to its system of parishes.

The parochial system which had grown up in Cornwall and Devon since Saxon times was well established by 1291 in the form that it would take down to the middle of the nineteenth century. Indeed, of the parishes listed in 1291 only three (Affeton, St James (Exeter) and Oldridge) were fated to disappear, and the main development in later times was the subdivision of some of the larger parishes into smaller units. **Maps 27.1 and 27.2** are based on the tithe surveys which followed the Tithe Commutation Act of 1836, and which provide the earliest comprehensive data on parish boundaries. The maps and gazetteers printed on the front and rear end papers enable almost all of these 1291 parishes

to be identified by name. To show the Church as it was in 1291, the parishes which came into existence after that date (as a result of subdivision) have been removed from **Maps 27.1 and 27.2** and put back into their original mother parishes. About a dozen parishes which were not listed in 1291 but which appear to have existed then, or did so by the middle of the fourteenth century, have been treated as if they were in the 1291 survey. The result is necessarily conjectural, because we do not know what other boundary changes were made between 1291 and 1836, and it is likely that there were some—perhaps many. Moreover, some of the parishes of 1291 probably had incipient divisions through the existence of parochial chapels which would one day achieve their independence, and some of the chapels may have had territories within the larger units, related to church attendance or tithe paying. The history of the English parochial system is a very complicated one, which needs to be studied parish by parish; no general map can do it adequate justice.

By the twelfth century the diocese of Exeter was divided into four archdeaconries: Exeter, Cornwall, Totnes and Barnstaple, in order of the rank of their occupants. Cornwall covered all the parishes west of the Tamar, Exeter east Devon, Barnstaple north Devon, and Totnes south and west Devon. The archdeaconries were further divided into rural deaneries: Barnstaple into six, Cornwall into eight, and the others into nine. Later, perhaps in the thirteenth century, a number of parishes were taken out of the system and became 'peculiars'. Most of these were places where the land and the patronage of the church belonged to the bishop or to the dean and chapter of Exeter Cathedral, each of whom made separate arrangements to govern their peculiars independently of the archdeacons and rural deans but subject, in the last resort, to the bishop. In Cornwall the bishop's peculiar parishes were organized into three rural deaneries of their own: St Germans, Pawton and Penryn. Some other parishes with different lords also claimed peculiar status, a claim usually contested by the bishop and archdeacons, but three or four such places in Cornwall managed to gain recognition in this way (St Buryan, Lanhydrock, St Michael's Mount and Temple), and in Devon two parishes did so after the Reformation (Templeton and Uffculme). These six peculiars claimed exemption even from the bishop's jurisdiction, as did certain religious houses until the

1530s, notably the Premonstratensian abbey of Torre, the six Cistercian abbeys, and the six friaries.

The parochial structure of Devon and Cornwall was varied, especially in Devon. Traces could still be seen in 1291 of some of the large Anglo-Saxon minster parishes, notably at Crediton, Hartland, Plympton and Tiverton. At the other extreme, there were some very small parishes like Dotton (214 acres) and Rousdon (255 acres). In Cornwall there was less variation and more uniformity, but some churches of early importance like Kea and Perranzabuloe continued to possess relatively large territories. Many parishes, especially in Devon, included outlying portions (at least, this was the case at the time of the tithe surveys), especially in the valley of the Little Dart. The maps show the larger of these outliers, but some of the lesser ones were too small and diverse to be reproduced. Finally, it should be noted that there was not an exact correspondence between the ecclesiastical system of the parishes and the secular units of the hundred, borough and county; some parishes lay in more than one such unit. Two in Cornwall (Boyton and Maker) extended their territories into Devon, and one in Devon (Bridgerule) did so in the opposite direction (see chapter 24). Such a system was ragged because it grew up rather than being planned, but it was to endure with little major change (except for parish subdivision) until the Victorians began to rationalize the units of local government in the 1840s (see chapter 24).

RELIGIOUS HOUSES

The only new religious houses to appear after 1300 (**Map 27.3**) were the Franciscan friary at Plymouth in 1383 (three attempts by the Augustinian friars to gain a foothold in Devon were unsuccessful) and five collegiate churches (Bere Ferrers, Haccombe, St Michael Penkevil, Ottery St Mary and Slapton). A sixth at Whitchurch was abortive. In addition, the vicars choral of Exeter Cathedral, who had existed as a group since the twelfth century, acquired collegiate buildings in 1387 and corporate status in 1400. There was a vogue for founding collegiate churches amongst wealthy noblemen and clergy in England during the fourteenth and fifteenth centuries, because they were cheap. They used an existing parish church for their worship, the rector usually becoming head of the college, and the rest of the staff consisting of poorly paid chaplains who cost less than monks but could as well perform the duties of prayer for the founders' souls. However, three of the new collegiate churches in Devon and Cornwall—Bere, Haccombe and St Michael Penkevil—were so poorly endowed that after the Black Death, when clergy numbers declined and clergy salaries rose, they were unable to support their staffing and returned to normal parochial status. They were not the only houses to disappear. The mid-fifteenth century saw the suppression by the crown of six small 'alien' priories at Ipplepen, St James (Exeter), St Michael's Mount, Minster, Modbury and Otterton. These houses had once been dependencies of monasteries in France, but the Hundred Years War first strained then broke this relationship and in the end their lands were transferred to endow new institutions in England, such as Syon Abbey (Middlesex) and Eton College (Buckinghamshire). The monks disappeared and the houses were closed, though St Michael's Mount was given a staff of chantry priests instead. Finally, in 1509 Bishop Oldham of Exeter suppressed the Trinitarian hospital at Warland (Totnes) and the moribund hospital for priests at Bishop's Clyst, and gave their endowments to the vicars choral of Exeter. Reorganizations of this kind look forward to the Reformation.

In its spirituality, the South West tended to be rather conservative in the later Middle Ages and remained so on the eve of the Reformation. Pilgrimages continued to flourish, not only to traditional shrines like Exeter, St Day (Redruth) and St Michael's Mount, but to new ones like Frithelstock (1351) and Whitstone (1359). Though there was a good deal of crime against clergy and churches, heresy was rare. Laurence Steven, one of John Wycliffe's Oxford disciples, made a preaching tour in Cornwall in 1382 but was soon caught and recanted. Drew Steyner was burnt at Exeter in 1431, probably in connection with a repression of Wycliffe's followers (also known as Lollards) that took place in southern England that year. Other instances of heresy occurred in the early sixteenth century at Axminster and Walkhampton, and at Exeter where a Protestant refugee from Cambridge (Thomas Benet) was convicted and burnt in 1532, but these were exceptions. Religious and cultural changes went on imperceptibly, rather than violently. The earliest publishing of printed books in the region was done at Exeter in about 1505–10 by Martin Coeffin, a Frenchman who had school textbooks printed for him at Rouen; the first printing was done at Tavistock Abbey by one of the monks, Thomas Richard, in 1525. Elsewhere, seven local guilds that had maintained a chantry priest to say mass at Barnstaple, Bodmin and other places, added school-teaching to his duties in the early sixteenth century—a sign of the interest in education characteristic of the Reformation period. Until the 1530s, however, there was no serious clash between old and new in the Church; both managed to coexist side by side. It was the coming of the Reformation that polarized attitudes and introduced an age of greater uniformity and intolerance.

THE REFORMATION

The Reformation affected Cornwall and Devon most dramatically, as it did England generally, with the dissolution of nearly all the religious communities. Under Henry VIII four smaller monasteries (Barnstaple, Frithelstock, St Nicholas (Exeter) and Tywardreath) were suppressed in 1536, the six friaries in 1538, and the remaining monasteries and nunneries in 1538–9. The collegiate churches began to be dissolved in 1545 (Crediton and Ottery), and the rest went under Edward VI in 1548, along with the chantries of one or more priests attached to parish churches and chapels. All that escaped were the cathedral

Map 27.1
Parishes in Cornwall,
1291, based on
the tithe surveys
of *c*.1840.

Peculiar Jurisdictions

Bishop of Exeter

Dean and Chapter of Exeter Cathedral

King (St Buryan)
Bodmin Priory (Lanhydrock)
Knights Templars (Temple)

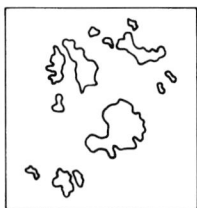

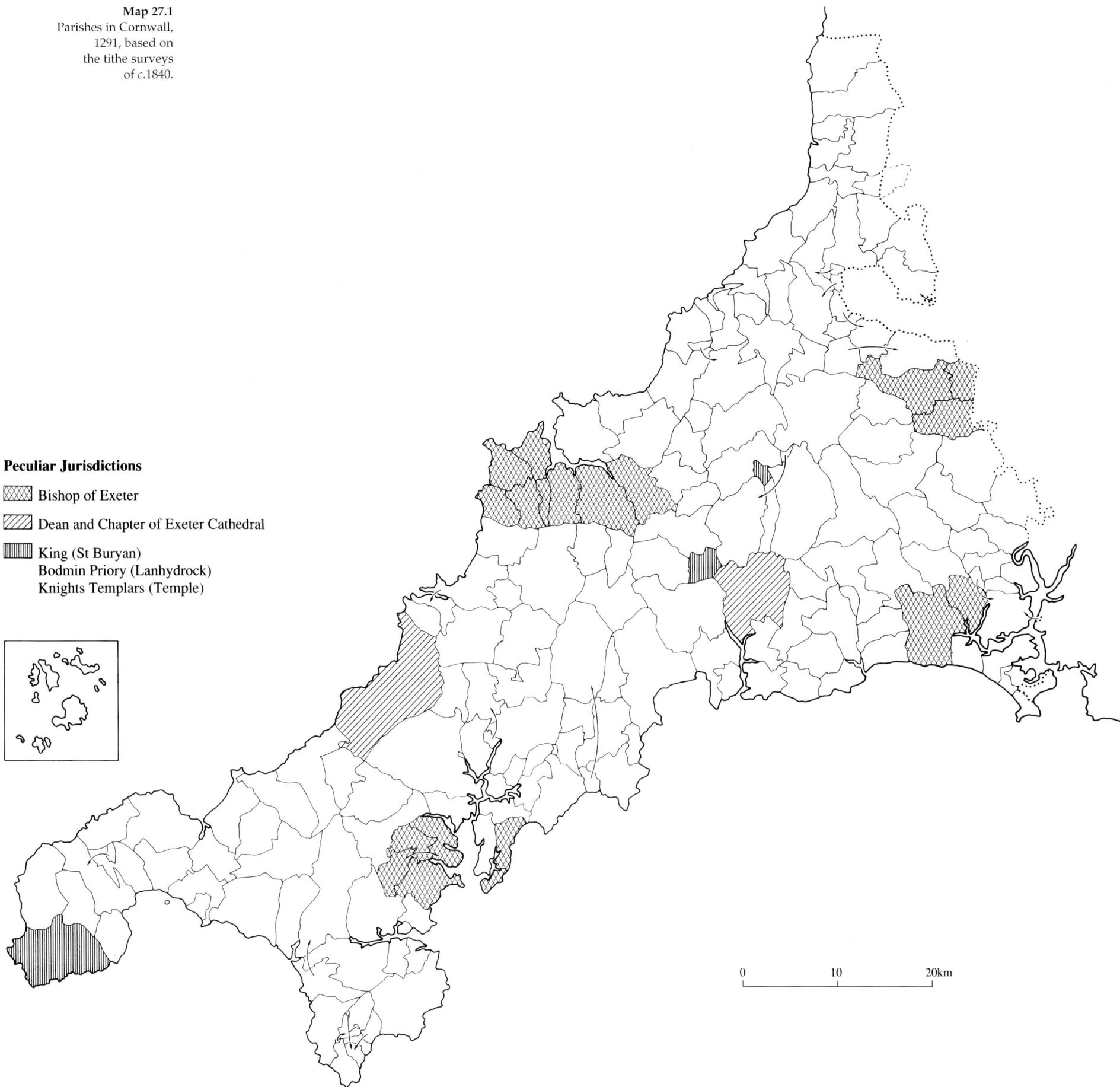

0   10   20km

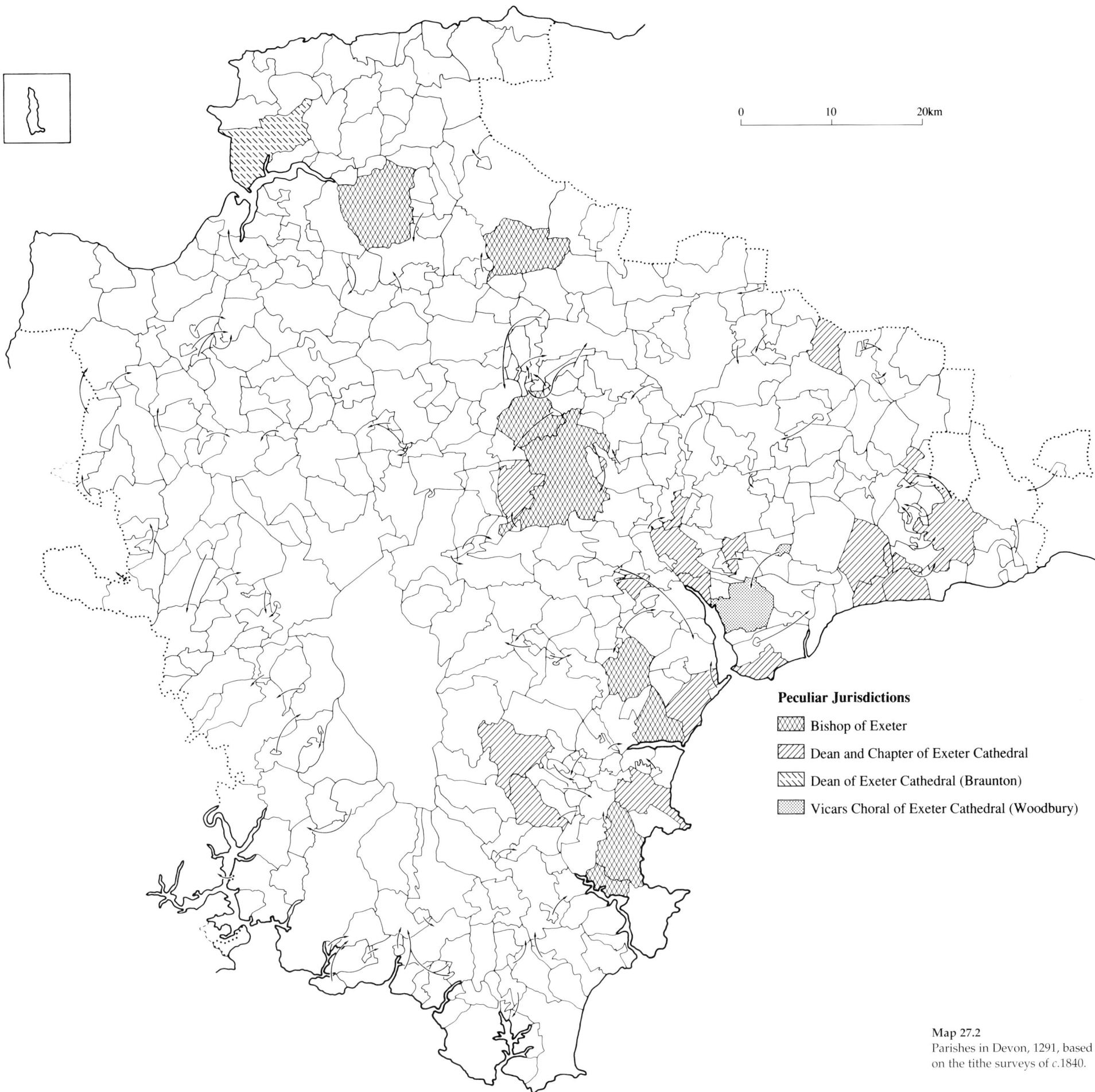

Peculiar Jurisdictions

- Bishop of Exeter
- Dean and Chapter of Exeter Cathedral
- Dean of Exeter Cathedral (Braunton)
- Vicars Choral of Exeter Cathedral (Woodbury)

0   10   20km

Map 27.2
Parishes in Devon, 1291, based
on the tithe surveys of c.1840.

(though its clergy were reduced in number), the associated college of vicars choral (also abridged in size), and three 'prebendal' churches: the chapel of Exeter Castle with three clergy, Tiverton with four, and Chulmleigh with six. A fourth such foundation, St Endellion, with four clergy, was suppressed in 1548 but revived in the late 1550s. These last four churches seem to have survived because they were not regarded as religious houses since their clergy did not live together in communities. In addition, the old collegiate church at Crediton remained for a time a shadow of its former self, with a vicar, two curates, an organist, four choristers and four almsmen.

The Reformation had an impact on the parishes too. In 1549 the medieval Latin services were superseded by the English ones of the first Book of Common Prayer, and in due course the mass, which had been the centre of Sunday morning worship, was replaced by mattins, the litany and the first part of the communion service, the communion itself being celebrated only three or four times a year. Church interiors were stripped of images and altars, while relics, shrines and votive lights were also abolished. Instead, the laity were directed towards Bible reading

in English, which was sanctioned in 1538—the same year that pilgrimages were forbidden. Some other institutions of the medieval Church—hermits, anchorites and indulgences—were not formally abolished (the bishop of Exeter issued an indulgence as late as December 1536), but simply faded away. Much changed, but perhaps even more continued unchanged. The laity remained obliged to be baptized at birth, to attend church regularly, to pay tithes and offerings, to fast during Lent, and to observe the Church's moral and social codes. Those who refused could still be summoned before the church courts of the bishop and archdeacon, which also kept the power to administer wills. The clergy, now permitted to marry, continued to possess courtesy titles ('sir', later 'reverend'), went on saying nearly the whole of the services in church, and even gained new powers and responsibilities in relation to the poor law and local public order.

The South West, as we have seen, lacked a tradition of heresy, and most people were probably still inclined to traditional Catholicism when the Reformation came, even if they were not always good Catholics. Several contemporaries referred to the region's conservatism. In 1536 there was a riot by Exeter women

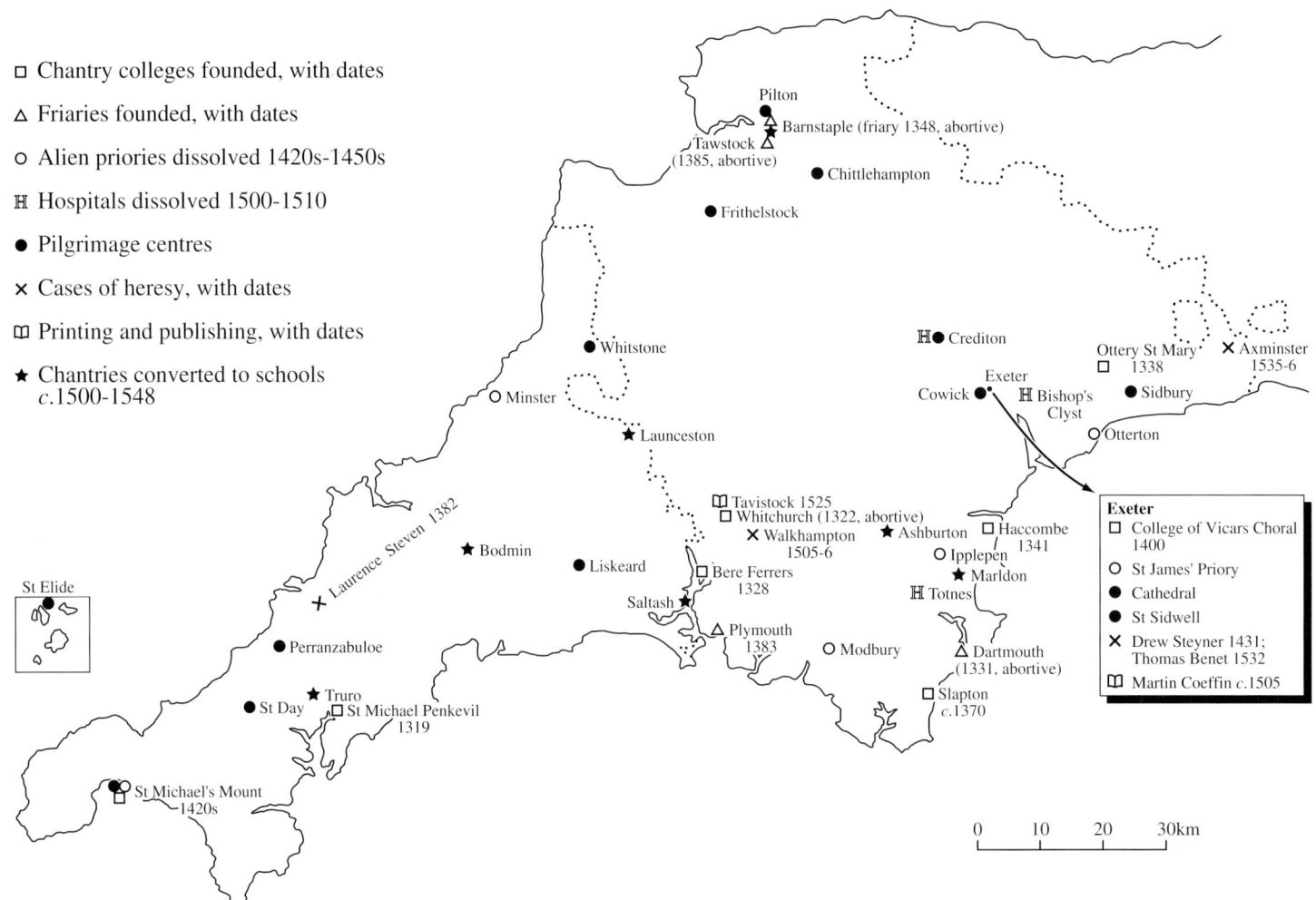

Chantry colleges founded, with dates

△ Friaries founded, with dates

○ Alien priories dissolved 1420s-1450s

H Hospitals dissolved 1500-1510

● Pilgrimage centres

✕ Cases of heresy, with dates

▥ Printing and publishing, with dates

★ Chantries converted to schools c.1500-1548

Map 27.3
Religious houses, 1300–1540.

when the priory church of St Nicholas had its images removed, and in 1547–8 other demonstrations took place in west Cornwall against Richard Body, deputy of the archdeacon of Cornwall, who was thought to be engaged in confiscating church plate. The region also produced one of the biggest popular protests of the sixteenth century against religious change: the 'Prayer Book' or 'Western' Rebellion of June to August 1549, sparked off by the introduction of the first English Prayer Book (**Map 27.4**). As with most early popular rebellions, our knowledge of its organization and events is relatively meagre and gathered from a few hostile contemporary writers. It was certainly a religious movement, and though all rebellions tended to gather up the various grievances of the day, recent attempts to cast doubt on the primary religious motive of the participants are probably misconceived. The protesters' demands, presented to the crown, were overwhelmingly concerned with religion, and imply hostility not only to the Prayer Book but to the recent abolitions of papal authority and monasticism. The rising was a serious one, involving both counties; the men of Cornwall and Devon joined up near Exeter, organized themselves in four camps, and besieged the city for five weeks. It took the king's forces two hard-fought battles to raise the siege, and a third at Sampford Courtenay to rout the rebels on the spot where the rising in Devon had begun.

HOSPITALS

Religion emphasized the duty of giving charity to the sick and the poor, and hospitals for this purpose were founded in Devon and Cornwall from about 1100 onwards (**Map 27.5**). Unlike modern hospitals which provide short-term medical treatment for the whole population, medieval ones generally centred their work on poor people with long-term disabilities: the blind, infirm, elderly, insane and especially lepers. Most of the early hospitals in the South West were for the latter, with only a few like St Alexius and St John (Exeter) for other infirm people. By the end of the fourteenth century, leprosy seems to have become less common, and some of the old leper hospitals were adapted for infirm people generally. By the sixteenth century, most houses were for the infirm and only a minority still for lepers.

Hospitals for the infirm were usually built in towns, and those for lepers in town suburbs or the countryside. Most hospitals of either kind were lowly institutions which have left few records, and **Map 27.5** plots only those which are known, not necessarily all that existed. A major source is the accounts of Bishop Bitton's executors in 1307; they tried to make a payment to each hospital in the diocese, but may have failed to reach all those in north Cornwall and north Devon. They also made payments to lepers in various other places, but these people may have lived dispersedly rather than in communities or institutions. A variety of people founded or helped to found hospitals, including bishops, monasteries, lesser clergy and citizens, and some institutions (not all) had a chapel, a chaplain and regular services. The hospital of Clyst Gabriel at

Map 27.4
The Western Rebellion and the Reformation.

The Western Rebellion 1549
☐ Initial outbreaks
✕ Chief engagements
--→ Rebel movements
—→ Royal forces

Religious houses surviving the Reformation
■ Cathedral
◉ College
⊕ Prebendal churches

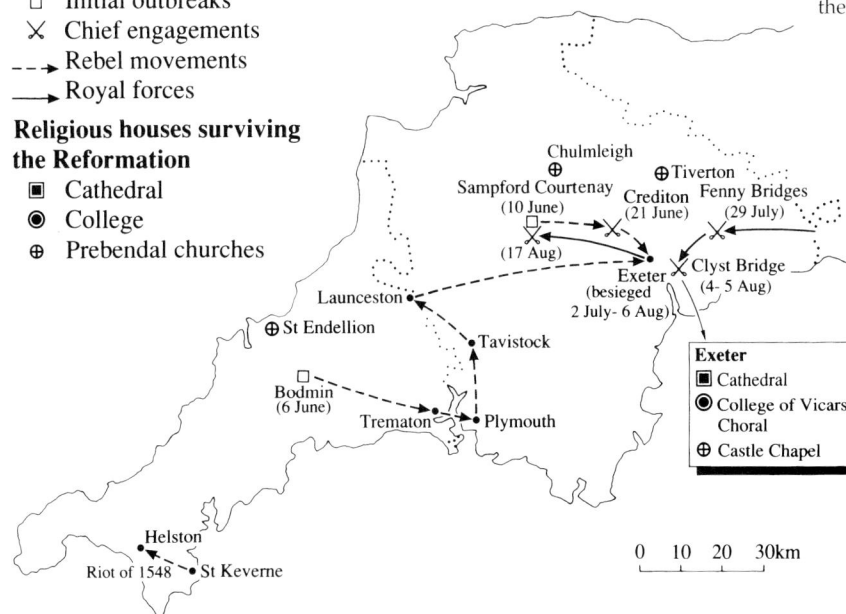

Bishop's Clyst (1309) was unique in the South West in providing a home for poor infirm priests of the diocese.

Some hospitals were endowed with property, producing an income from rents, but nearly all relied at least in part on voluntary donations: royal alms, or gifts and legacies by lesser people. Bishops often granted indulgences to encourage public giving. Each hospital was an individual entity, and though the bishops of Exeter sometimes intervened in their affairs, there was no regular episcopal supervision, and hospitals were not controlled by the Church in an effective way. Usually the founder of the hospital and his descendants also had rights of patronage, including the nomination of the hospital master or chaplain and the selection of inmates. City and borough councils sometimes issued regulations for their local hospitals, or acted to help them. Numbers of hospital inmates are rarely known, but many houses seem to have contained twelve people or less—sometimes only men, sometimes men and women. A few hospitals like Crediton, St John (Exeter) and Warland (Totnes) were staffed by clergy, but a more common arrangement was to have a resident lay prior or master who looked after the finances and discipline. Regulations were often strict in theory, with inmates being required to attend chapel services, and violence or immorality was punished by fines, the stocks or expulsion.

After 1400 a new wave of almshouse foundations began, catering for the elderly and infirm rather than for lepers. The typical almshouse accommodated its inmates in separate chambers or houses, rather than communally, and often provided an endowment to give the members regular weekly or monthly payments to live on. Some hospitals ceased to exist before the Reformation, due to the disappearance of their endowments or the decrease of leprosy; at the Reformation, the governments of Henry VIII and Edward VI dissolved religious houses and chantries, but not hospitals as such. This meant that the large hospital of St John (Exeter) came to an end, because it had been staffed by clergy and was deemed to be a religious house, and some others lost their chaplains, but the majority continued in existence with their premises, endowments and lay officers. They continued to fulfil their medieval functions into the seventeenth and eighteenth centuries.

CHAPELS

Parishes and parish churches were important institutions. Each parish boundary defined an area in which the clergyman had to minister and the laity accept his ministry, as well as paying him tithes and offerings (**see Maps 27.1 and 27.2**). The church was the main centre of parish worship, and parishioners were supposed to attend it, though it is not clear how far this happened and it is unlikely that church attendance was ever fully enforced. As early as the tenth century, however, some parishes possessed additional places of worship, subordinate to the parish church and eventually

described as chapels. Such places grew in number between 1100 and 1500, creating a first great age of chapels anticipating the later, nonconformist one; indeed, by the latter date there were far more chapels than parish churches. A sample area of nineteen parishes north-west of Exeter (**Map 27.6**) contained about twenty-six chapels and possibly more, since not all were documented.

Medieval chapels can be roughly categorized into three groups: domestic chapels, chapels of ease and cult chapels. Domestic chapels were established by the Normans in their castles at Exeter, Launceston and Okehampton, and later spread to the manor houses of the lay aristocracy as well as to some of the dwellings of wealthy clergy and citizens. The chapels at Ley, Mill and Pynes on **Map 27.6** belong to this category which became very common, hundreds of licences for such places being issued by the bishops of Exeter during the fourteenth and fifteenth centuries. Chapels of ease were built in larger parishes for communities living a long way away from the parish church. Kennerleigh and Sandford in Crediton parish fell under this heading, and Nether Exe originated as a chapel of ease to Brampford Speke. Cult chapels were built to honour a particular saint or shrine, and St

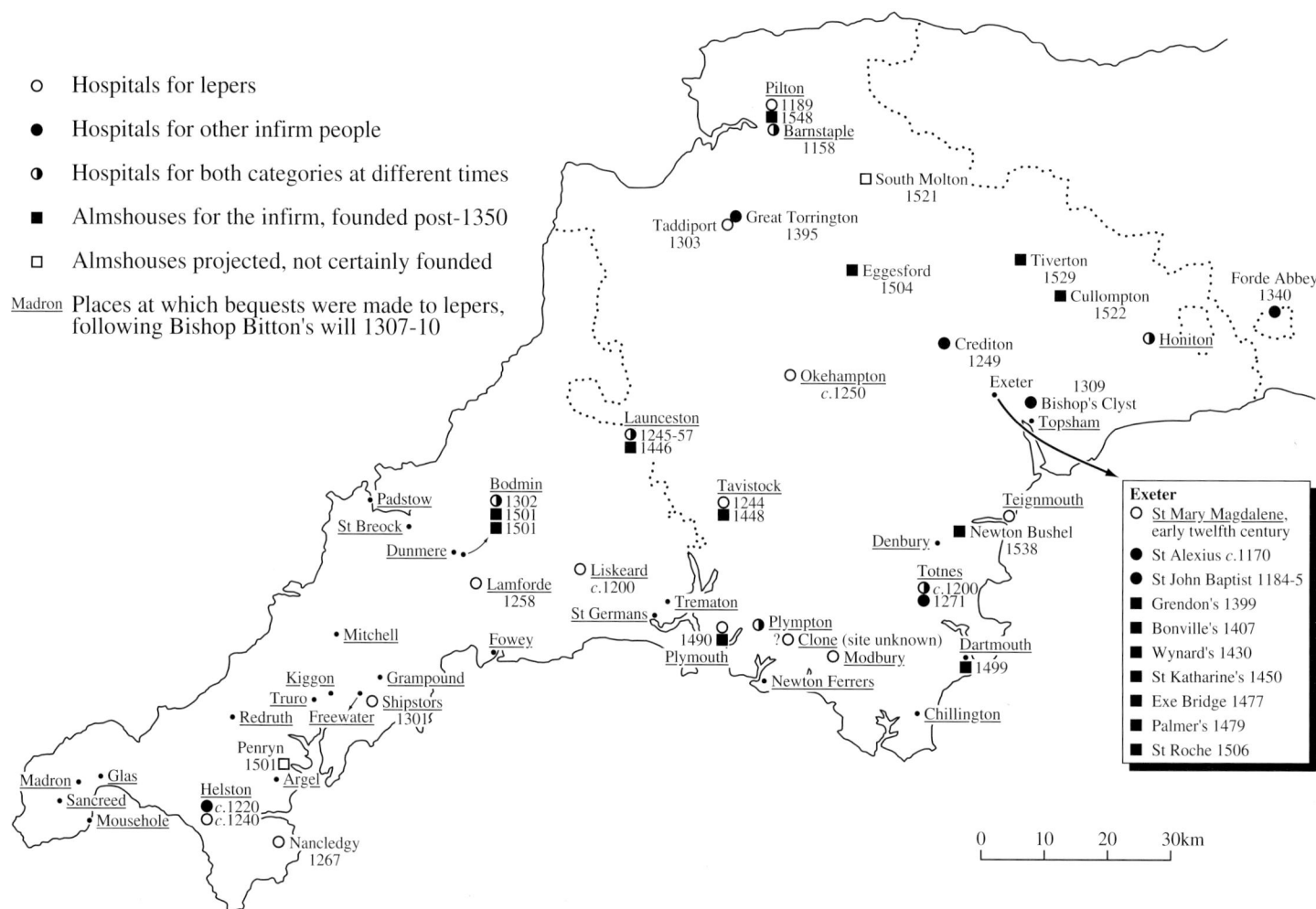

Map 27.5
Hospitals, 1100–1550.
Dates are of first mention,
not foundation.

SOURCE: Nicholas Orme
and Margaret Webster,
The English Hospital 1070–1570
(New Haven and
London, 1995).

George (Crediton) and St John (Poughill) may have been of this type. Elsewhere in the South West, Exeter had several cult chapels (St Anne, St Clair, St Roche and more than one of Our Lady), and the chapel of St Day (Redruth), with its famous image of the Trinity, was a major centre of Cornish pilgrimage.

Chapels posed a threat to parish churches by competing for congregations and offerings, but were tolerated as long as they supplemented, rather than replaced, people's obligations to the parish church. Sometimes their services were restricted to weekdays; money offered inside them was usually appropriated by the parish clergyman, and their devotees were expected to go on bearing all their usual parochial responsibilities. Still, tensions often existed, and the supporters of chapels of ease in particular developed ambitions to secede from the mother church and make their chapels parochial. This was resisted by the patrons and clergy of the parish churches, who stood to lose, and the bishop was often brought in to maintain the *status quo*, notably to prevent a secession by Sandford in 1437. In the long run, however, many chapels gained their independence, especially if backed by a local landlord; they gradually acquired rights of baptism and burial, and eventually their parochial standing was recognized. Most of the parishes which came into existence in Cornwall and Devon between 1291 and the early nineteenth century represent such developments by former chapels of ease.

The Reformation dealt a blow to chapels. They were not officially dissolved, but some of them lost the endowments which had maintained their chaplains, and their distinctive religious cults were discouraged. A shortage of clergy after the 1540s, coupled with a preference by the authorities to concentrate worship in the parish churches where it was more easily supervised, caused most of the chapels to fall into disuse and eventually ruin. Only the larger chapels of ease survived, and some of the other medieval chapels which exist today were rescued and restored to worship by the Victorians. They have not had a continuous history of use.

FURTHER READING

The best text of the 1291 valuation is to be found in *The Registers of Walter Bronescombe and Peter Quivil, Bishops of Exeter*, ed. F.C. Hingeston-Randolph (1889), 450–81. There is no complete list or study of medieval parishes, but N. Orme, 'The medieval parishes of Devon', *Devon Historian*, 33 (1986), 3–9, is an introduction to the subject. Hugh Peskett, *Guide to the Parish and Non-Parochial Registers of Devon and Cornwall, 1538–1837*, Devon and Cornwall Record Society, extra series, 2 (1979) is an excellent guide to parishes after the Reformation, and C.R. Humphrey-Smith (ed.), *The Phillimore Atlas and Index of Parish Registers* (Chichester, 1984) has useful maps (numbers 6, 9) of the post-Reformation archdeaconries, deaneries and peculiars. F.A. Youngs, *Guide to the Local Administrative Units of England, Vol. I, Southern England* (London, 1979) contains several errors and should be used cautiously.

Map 27.6
Parishes, rectories/vicarages, and chapels north-west of Exeter before 1530, based on the tithe surveys of *c*.1840.

The most recent general survey of late medieval religion in the South West is in N.I. Orme (ed.), *Unity and Variety: A History of the Church in Devon and Cornwall* (Exeter, 1991), chapter 3. In addition, on the Reformation period, see R.J. Whiting, *The Blind Devotion of the People: Popular Religion and the English Reformation* (a study of religion in the South West) (Cambridge, 1989). For Cornwall, see also A.L. Rowse, *Tudor Cornwall* (London, 1941). Two other useful studies are Joyce Youings, *The Dissolution of the Monasteries* (London, 1971) and Frances Rose-Troup, *The Western Rebellion of 1549* (London, 1913).

Hospitals are covered in detail in Nicholas Orme and Margaret Webster, *The English Hospital: 1070 to 1570* (New Haven and London, 1995) and chapels are reviewed in general by Orme, 'Church and chapel in medieval England', *Transactions of the Royal Historical Society*, 5th series, 6 (1996), 75–102. Chapels in west Cornwall are listed by C. Henderson, 'Ecclesiastical history of the 109 western parishes of Cornwall', *Journal of the Royal Institution of Cornwall*, new series, 2–3 (1955–60) and those in Devon by Jeanne James, 'Medieval chapels in Devon' (unpublished M.Phil. thesis, University of Exeter, 1997).

ACKNOWLEDGEMENT

The help of Margaret Webster with the section on medieval hospitals is gratefully acknowledged.

RELIGION AND THE SPREAD OF NONCONFORMITY BEFORE 1800

JONATHAN BARRY

The maps in this section portray the distribution of various forms of nonconformity in the later seventeenth and eighteenth centuries. In most cases they are based on information organized by parish, although no attempt has been made to portray the 650 or more parishes of the diocese of Exeter in their full complexity. The use of the parish reminds us that, while this map concentrates on nonconformity, the religious geography of the South West was fundamentally Anglican. It was the Church of England that not only had an official presence everywhere but also generated much of our information about nonconformity, through its various enquiries about the strength of dissent across the diocese. While omnipresent, however, the Church was neither uniform nor itself equally strong in its presence in every part of the region. The apparently straightforward equation between our two counties and the diocese of Exeter, with its hierarchy of bishop, archdeacons, deans and parochial clergy, each with a defined jurisdiction, was complicated in reality by the substantial number of parishes exempt from one or other jurisdiction by virtue of being a 'peculiar' subject to the dean and chapter of Exeter, Windsor or another cathedral, to the crown, or to the bishop directly. Since many parochial clergy held several livings and/or employed curates to perform their tasks and other parishes contained chapelries with their own ministers, ministration on the ground did not always follow the parochial pattern. Another linked variation concerned the patronage and remuneration of the clergy: who held the right to present and whether the clergyman was a rector (receiving the full tithes), a vicar (receiving lesser tithes) or merely a stipendiary clergyman on a set income. In every case much also depended on the wealth of the parishioners and how far they could or would supplement the minister's income through using his paid services or rewarding him by collections or gifts for his work. This created contrasts between town and countryside, large and small parishes, arable and pastoral parishes, which are not necessarily congruent with the official valuations of each living. Such complexities defy mapping: they changed constantly and could only be shown by detailed local studies, but they should be borne in mind in what follows.[1]

Anglican variation is also important because it is often used to explain the presence or absence of nonconformity. The simplest model of this correlates nonconformist strength with Anglican weakness, defined in terms either of pastoral provision or of jurisdictional control. In both cases it is assumed that nonconformists would flourish in towns and in pastoral parishes where population growth or scattered settlement made the parochial system less appropriate than in the classic nucleated village where the minister, supported perhaps by the 'squire', could maintain Anglican uniformity. One of the interesting questions to be asked of the maps that follow is whether this model, which has been applied most frequently to nonconformity of the late eighteenth and nineteenth centuries, also applies to so-called 'old dissent' in this earlier period.[2]

For nonconformity is itself a very misleading term, if it implies a monolithic movement. Nonconformists shared only the attribute of not conforming to the established Church, and varied widely in their reasons for this failure to conform, as well as in many other features. The maps distinguish the geographical spread of a variety of nonconformist groups. The most obvious contrast is that between the Roman Catholic nonconformists, usually labelled 'papists' by Protestants at this period, and the others. The spread of Protestantism in Devon and Cornwall, after an initial period of both conservatism and confusion, proceeded rapidly in the late sixteenth century, and by the early seventeenth century the number of papist 'recusants'—those refusing to attend Anglican services—was very small, although many suspected that 'church papists' (outward conformists who retained Catholic sympathies) were much more numerous. In the face of considerable persecution, Catholic priests found it hard to operate except where sheltered by important gentry families, of whom the most notable throughout our period were the Arundels of Lanherne. Once their numbers had stabilized, the Catholic laity's size seems to have remained fairly stable until the early eighteenth century, boosted occasionally by a notable convert such as Lord Clifford of Ugbrooke, whose family joined the Chichesters of Arlington Court and the Carys of Cockington and Torre Abbey as leading Devon Catholics. In the eighteenth century, and especially after the final defeat of Jacobitism in 1745, the threat to those practising Catholicism began to ease (although anti-Catholic prejudices remained rife) and it gradually became possible for priests to operate and Catholics to worship openly, with a *de facto* toleration only slowly matched by the repeal of the various statutes. Catholics remained barred from public life until 1829. Both Catholic sources and other evidence suggest that Catholic numbers in the region began to expand in the later eighteenth century. In addition to a regular influx of Catholics from Ireland and the Continent, the native population also seems to have been expanding, though the region remained one with a very small Catholic presence compared to the North West or to major cities like London or Bristol.[3]

THE COMPTON CENSUS AND SUBSEQUENT ENUMERATIONS OF CATHOLICS

Given the illegal status of Catholicism, sources that name or number Catholics are inevitably restricted and often suspect. During the seventeenth century, most government efforts were directed against the better-off Catholics, who were subject to heavy fines for recusancy and often to heavy taxation on their property. Such efforts were increased at moments of tension about the 'papist' threat nationally, such as during the late Elizabethan wars with Spain, before and during the Civil War and later during the anti-Jacobite wars and in face of Jacobite invasion. Many lists therefore survive of the leading Catholic figures in our region, but these will only correlate with the overall spread of Catholic laity if we assume that these clustered exclusively around their prominent leaders.[4] This assumption is largely but not entirely borne out by the occasional surveys we do have of overall numbers and their distribution. The so-called Compton Census of 1676, intended by Anglican bishops to show Charles II that neither 'papists' nor Protestant dissenters were any kind of numerical match for the established Church, asked incumbents to number the 'papists' in their parishes, and the results are indicated on **Maps 28.1 and 28.2**.[5]

In 1603 a total of 99 recusants, 44 men and 55 women, had been reported for the diocese.[6] This number was comfortably exceeded by the 297 recorded in 1676 (67 in Cornwall and 270 in Devon), but there is still reason to suppose that the figures are an underestimate, since two surveys in the decades around 1700 suggested that the diocese contained about 450 Catholics.[7] It may be that the period of toleration for Catholics in the reign of James II had revealed to official eyes many previously undetected Catholics, although it is also possible that some conversions had occurred during those years. However, the 1676 figures are more likely to be indicative of a Catholic presence, than to be accurate about its scale. The next attempt at a comprehensive survey by parish came in 1767, and these data are presented in **Maps 28.3 and 28.4**. This produced a reduced estimate, 291 in all, with 56 in Cornwall and 235 in Devon.[8] The balance between the two counties of about four Devon Catholics for one in Cornwall remains the same. Within Cornwall the pattern of distribution also remains similar, with the north-coast parishes around Lanherne still the centre of Cornish Catholicism, but with small numbers now registering near the Fal estuary and in Fowey. Apart from a growing presence in Exeter, Tiverton and the Plymouth conurbation, Devon's Catholicism also retained its traditional distribution based on such gentry centres as Arlington, Parkham, Ugbrooke and Torre Abbey.

All of these surveys were based on hostile sources—normally the Anglican clergyman—and it is not until 1773 that we have an estimate by the Catholics themselves of their strength. The vicar apostolic of the Western District then estimated that there were 485 Catholics in the two counties, 440 in Devon (considerably more than shown in 1767), but only 45 in Cornwall (less than in 1767). Devon had six priests operating, three in gentry households, two in rural missions and one in Exeter, while Cornwall just had two rural priests.[9] The discrepancies between the 1767 and 1773 estimates for Devon suggest that the 1767 figures on **Map 28.4** should be regarded with caution, at least in terms of absolute numbers, and it may also be that the number of Catholics in urban areas was under-recorded in 1767. From that period on we have an increasing number of Catholic sources for their missions and churches, which were now keeping registers and other records, and this enables us to plot with relative confidence the spread of Catholic chapels. How far this growth in open Catholicism can be equated with a growth in the number of practising Catholics is a question too complex to answer here.

HUGUENOTS, MORAVIANS AND JEWS

Before turning to the mainstream Protestant dissenting groups, it is worth noticing three other denominations with continental origins, each of whom owed their presence in the region in part to the effects of Roman Catholic activity in Europe. These are the Huguenots, Moravians and Jews. The South West was largely unaffected by the first wave of continental Protestant escapees from the Wars of Religion, in the sixteenth century, who settled in London and the east, but the campaigns of Louis XIV, notably those against Huguenots within France, did lead to the establishment of French churches in several ports. Central European conflicts, displacing Bohemian Protestants into Germany, where some of them formed the United Brethren, or Moravian church, led eventually to a movement, partly of exiles and partly of missionaries, which began to attract some converts by the mid-eighteenth century. Finally the Jewish community in the South West, which began to re-form after the centuries-long banishment of Jews was relaxed in the 1650s, depended heavily for its growth in the eighteenth century on immigrants. While most of these probably came for economic reasons, some may have been seeking religious freedom. The churches formed by these three communities are shown on **Maps 28.3 and 28.4**. As one might expect, they were basically urban and, in particular, port-based groups. During the eighteenth century the Huguenot communities became assimilated into English society and their churches faded, some moving into the Anglican fold and others into nonconformity. The Moravian movement, episcopalian yet reformed evangelical, can best be understood as part of the wider

Map 28.1
Cornwall: Roman Catholicism and nonconformity in the Compton Census of 1676.

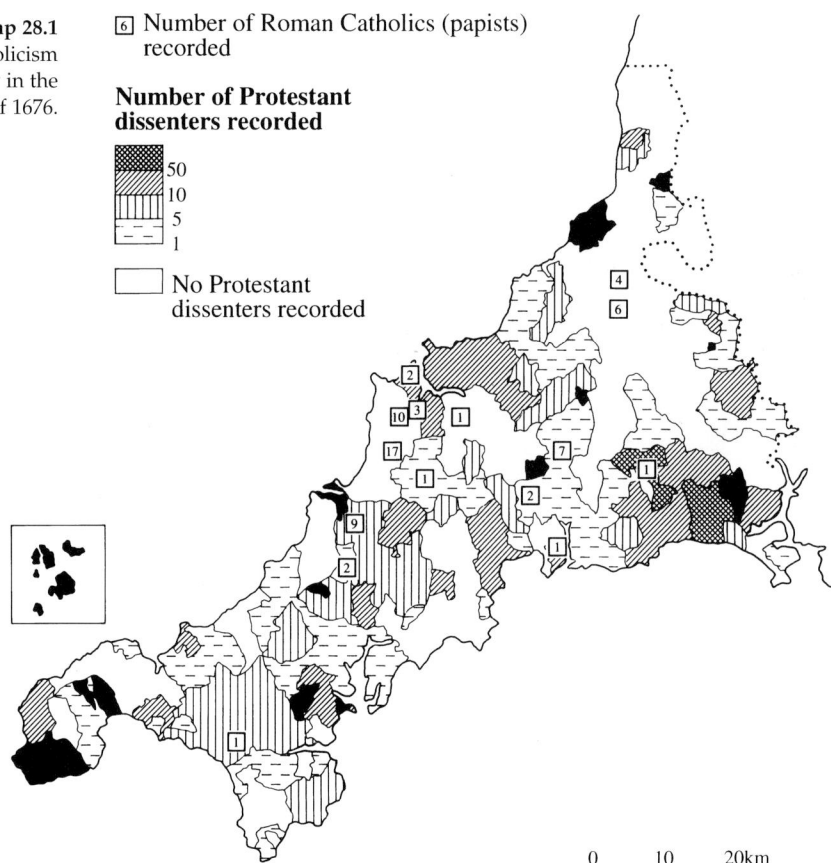

⬚6 Number of Roman Catholics (papists) recorded

Number of Protestant dissenters recorded

50
10
5
1

⬚ No Protestant dissenters recorded

0   10   20km

Map 28.2
Devon: Roman Catholicism and nonconformity in the Compton Census of 1676.

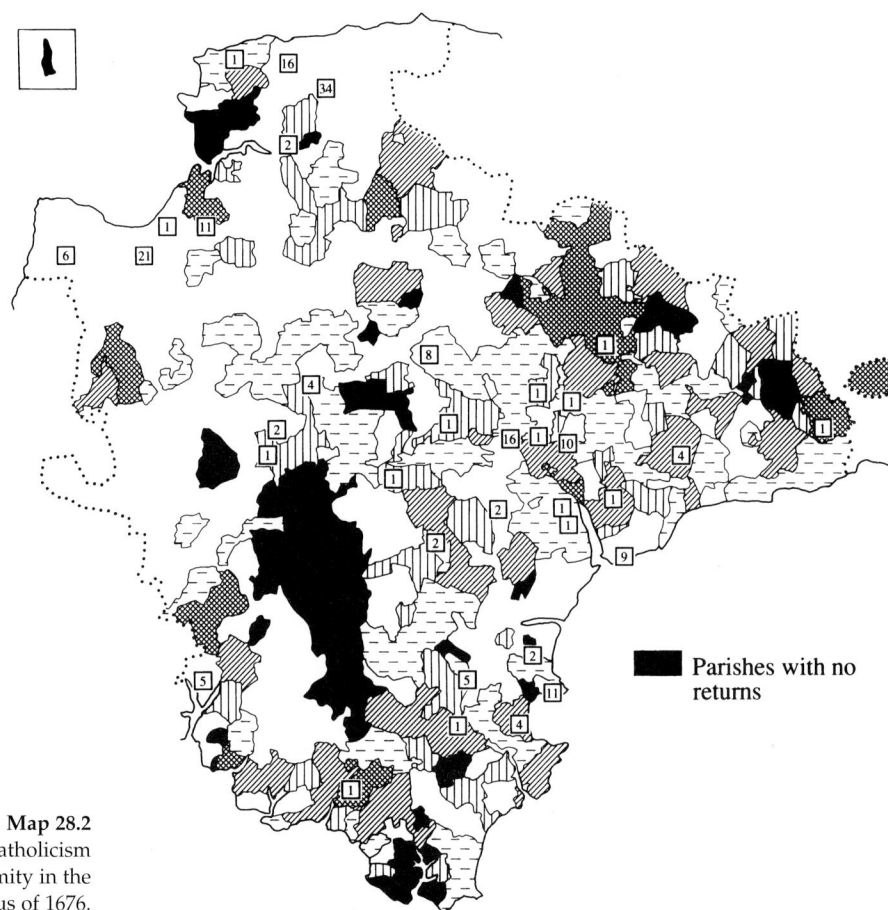

■ Parishes with no returns

evangelical revival, of which it formed an exotic component. The Jewish communities of Exeter and Plymouth added a new dimension to religious life in these already pluralistic communities.[10]

PROTESTANT NONCONFORMISM: 'OLD DISSENT'

All of these groups were, however, numerically tiny on a regional level, even if their concentration and distinctiveness sometimes gave them a significance beyond their numbers as examples of religious difference. The Catholic challenge to Anglican monopoly was always regarded with a seriousness beyond its local manifestation, partly because of its gentry leaders but mostly because local Catholics, and especially their priests, were regarded, understandably, as the agents of an immensely powerful international movement. It was the South West's geographic position in relation to Ireland, France and Spain, not its internal geography, that really shaped its anti-Catholic tradition.

This tradition played an important part in shaping Protestant nonconformity in the region as well. Within the Church of England, both before and after the cataclysm of 1640–60, there was indecision about how the national Church could best fulfil its historic role as the enemy of Rome. Should it extend the Reformation process by ridding itself of remaining elements of its medieval structure, or was its ability to fulfil its mission being threatened by another enemy, namely Protestant schismatics? This question informed the attitude taken to a range of church issues—the nature of church government, the role of the minister (or was he a priest?), the character of the liturgy and the balance between preaching, sacraments and other aspects of worship—which also had profound implications for how the church related to the local community. Tensions over these matters were considerable within the Elizabethan and early Stuart church. To some extent these can be simplified into a contrast between 'puritan' and 'conformist' wings and some efforts have been made to give these geographical precision and to relate them to the regional variations in allegiance which then appeared in the Civil War. Maps of 'puritanism' before 1640 would be largely maps of gentry and clergymen, but those for the 1640s can capture aspects of popular attitudes as well, though complicated by the other issues involved in wartime allegiances. No attempt has been made to present such maps here. One reason for this is that the sources available for building up a picture are highly fragmented and hence largely impressionistic, as might be expected given that we are discussing movements within a national church, fighting over its identity, not separate churches. Nevertheless, the existing evidence suggests very strong continuities with the pattern of late seventeenth-century dissent, with strongholds in north, east and south Devon and south-east Cornwall.[11]

It was only with the collapse of the Church in the 1640s and the splitting of 'puritanism' into a number of movements that we can begin to talk of 'nonconformists' in a later sense and to identify separate churches. Out of the confusion of these years emerged four dissenting traditions whose fortunes are traced in the maps: Presbyterians, Independents (otherwise known as Congregationalists), Baptists and Quakers. Together these constitute 'old dissent' and together they formed a Protestant challenge to Anglican monopoly of a much greater scale and intensity than the Catholic and other churches considered so far, at least in terms of numbers and activity at parochial level. As a consequence, successive Anglican leaderships sought to quantify the threat they posed, notably in the Compton Census of 1676 (shown in **Maps 28.1 and 28.2**) and then in questions posed prior to episcopal visitations. The responses to these survive for 1744–6, 1764–6 and 1779, and the first of these has been used in **Maps 28.3, 28.4, 28.5 and 28.6**. In 1676 incumbents were merely asked to number the Protestant 'nonconformists' as a whole, while in 1744 they were encouraged to distinguish the different types of Protestant dissenters and if possible to name their ministers and meeting-houses, stating if they were licensed or not.[12]

The differences between the questions posed in 1676 and 1744 reflect the changing status of Protestant dissent. In 1676 Protestant nonconformity was no more legal than Catholic: attendance at Anglican worship was compulsory for all under the recusancy laws and Protestant 'conventicles' and ministers were subject to prosecution. The extent to which this was enforced varied enormously, depending not just on local circumstances, as we shall see, but also on crown policy—both Charles II and James II proved unwilling to enforce Anglican uniformity consistently and at times offered toleration to dissent. Thus between 1660 and 1689 we can gauge the presence of dissent both from records of persecution (in church and secular courts) and by the appearance of congregations in times of toleration, notably in 1672. The lists of ministers and places of worship licensed in 1672 provides one of our best guides to the distribution of nonconformity, with the exception of the Quakers (who scorned the opportunity).[13] Following the revolution of 1688–9, an Act of Toleration was passed which permitted Trinitarian Protestant dissenters to attend an alternative worship to that of the parish church, provided that they did so in licensed meeting-houses, and from then on we have records of such licences, fairly complete for Devon but not for Cornwall, together with snapshots at such dates as 1744.[14] By 1744, therefore, Anglican ministers had become used (however reluctantly) to the fact that some of their parishioners could absent themselves legally from their services to attend a specified meeting of a given denomination. How well the ministers could judge the scale and nature of such dissent no doubt varied depending on their detailed grasp of parochial affairs.

However, the change between 1676 and 1744 involved more than the legal status of dissent. To understand this we need to consider the changing relationships of the four dissenting 'traditions' to the established Church. As emphasized above, the 'puritan' tradition was about reforming the national church, not disbanding it. During the Civil Wars and Interregnum what had

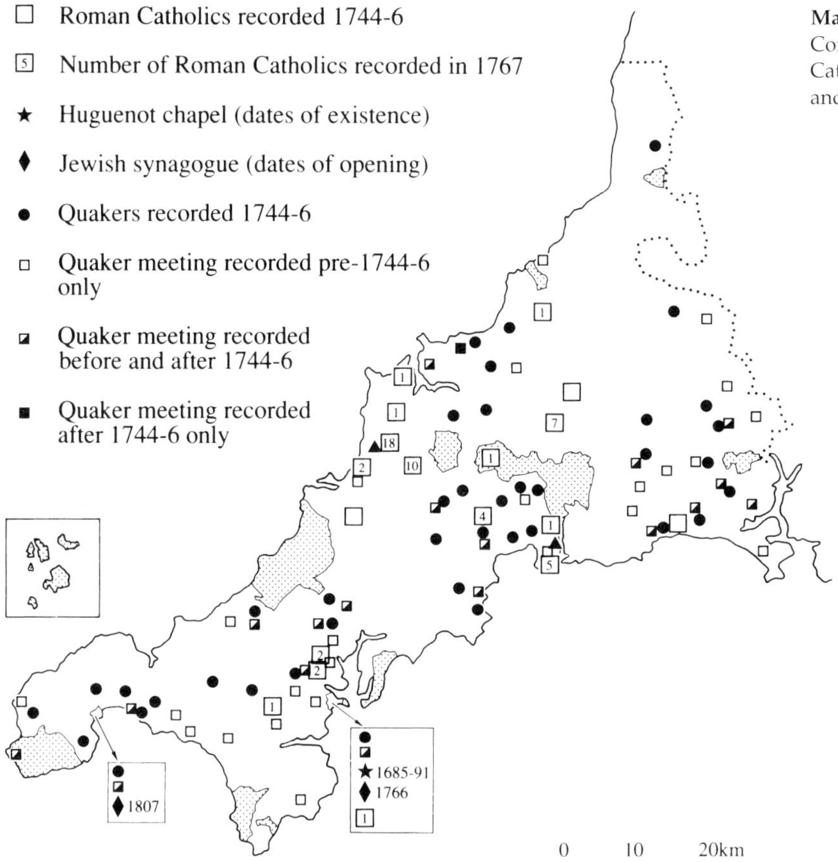

□ Roman Catholics recorded 1744-6

⑤ Number of Roman Catholics recorded in 1767

★ Huguenot chapel (dates of existence)

♦ Jewish synagogue (dates of opening)

● Quakers recorded 1744-6

□ Quaker meeting recorded pre-1744-6 only

◩ Quaker meeting recorded before and after 1744-6

■ Quaker meeting recorded after 1744-6 only

Map 28.3
Cornwall: Quakers, Roman Catholics, Huguenots, Jews and Moravians, 1672–1800.

0   10   20km

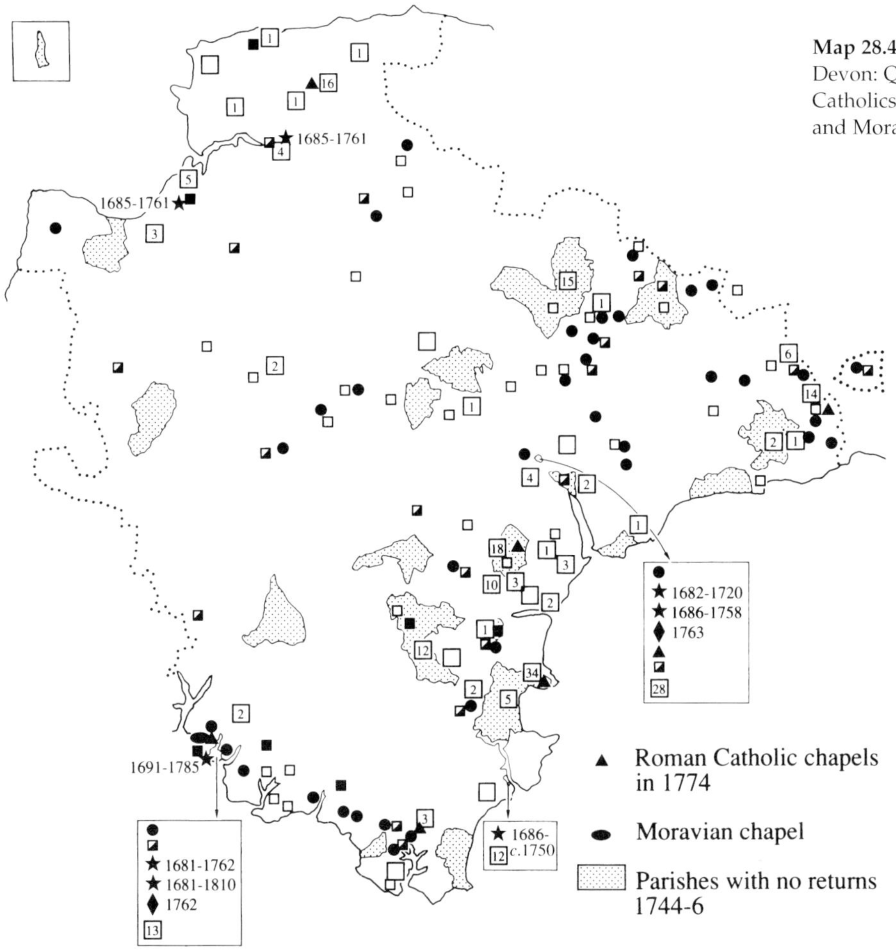

Map 28.4
Devon: Quakers, Roman Catholics, Huguenots, Jews and Moravians, 1672–1800.

▲ Roman Catholic chapels in 1774

● Moravian chapel

▨ Parishes with no returns 1744-6

previously been a tiny 'separatist' tradition began to grow in strength and led to the emergence of new churches which sought to establish a 'gathered church' of the godly and abandoned the aim of incorporating all people living in one place, as parishioners, in a single church. One way of signalling this 'gathering' was by adult baptism, but another was established by the Quakers with their close-knit churches who stood out from the ways of 'the world' and insisted that their members be married, for example, only within their movement. Similar aspirations characterized the 'independent' churches, since each congregation established itself, often by adhering to an agreed covenant, but since in other respects these churches followed practices identical to Presbyterians, their identity was less sharply defined. During the 1640s and 1650s Independents were generally content to see a national church with a parochial structure survive, provided that individual congregations were allowed to fashion their own form of worship. However, once the Church of England was restored in 1660–2 on pre-war lines, such groups moved firmly outside its orbit. The same cannot be said of the Presbyterians. As their name implies, those in this tradition wished to remodel the national church on a Presbyterian basis as both a mark and a means of further reform, but they had no desire to abandon a national church nor to permit the emergence of various Protestant denominations, much less to become such a denomination themselves. Their horror at the growth, first of Independent and Baptist churches and then in particular at the rise of the Quakers and other extreme sects, led them to acquiesce in, and often support, the restoration of Church and monarchy in 1660, hoping that a reformed Church of England would be developed in which they could participate. When this did not happen and the ministers of their persuasion were mostly forced to leave the

Church in 1662, these groups found themselves adrift between separatism and allegiance to their parish churches. It was only gradually, as they experienced intermittent persecution from the Church and had to evolve a separate congregational life, that the Presbyterians began to see themselves as a nonconformist denomination and only after 1689, when toleration was agreed and a more comprehensive national church again proved elusive, that they really settled down to a new status. Even then, many Presbyterians remained half-Anglican, participating in Anglican worship as well as Presbyterian and proving slow to develop the full range of alternative church practices, such as supplying rites of passage or charity for their members, independent of those provided by the parish.[15]

VARIETIES OF NONCONFORMITY

These variations in the type of nonconformity thus make it vital to offer, where possible, maps that distinguish the different traditions. This is the more essential because our Anglican sources for 1676 and 1744 are likely to have regarded the four dissenting traditions rather differently when asked to measure the strength of dissent. At one end of the spectrum will lie the Quakers and Baptists, whose separatism set them clearly in opposition to the parish church and whom all Anglican ministers are likely to have regarded as nonconformists. At the other end would be the Presbyterians, many of whom, at least in the seventeenth century, were partial conformists and on good terms with the substantial minority of Anglican clergy who, despite conforming in 1662, shared some of the attitudes of their erstwhile colleagues and parishioners, including their hostility to both Quakers and papists.[16] In 1676 many of these ministers may have failed to number such people among the nonconformists in their parish. This clearly raises a major problem about how far to trust the distribution pattern of dissent indicated on **Maps 28.1 and 28.2**. Not only may many dissenters have gone unrecorded, but this may be particularly true in areas where Presbyterians predominated. Peter Jackson has argued for the reliability of the figures, for east Devon at least, by showing their close correlation with what we know of dissenting strength from prosecutions in the courts 1660–88.[17] While encouraging, such correlations are not entirely conclusive, because the pattern of prosecution may itself be shaped by the sympathies felt by the clergy and churchwardens towards dissenters of various kinds: prosecution may have been rare in some communities because of general sympathy towards a very substantial Presbyterian grouping, but powerful in other parishes where the parish elite were hostile to, say, a pocket of Quakers. Hence the importance of confirming these guides by other evidence drawn from the nonconformist groups themselves.

The main sources here, unfortunately, concern the whereabouts of the ministers and of dissenting meeting-places, not the distribution of their congregations. Although the two

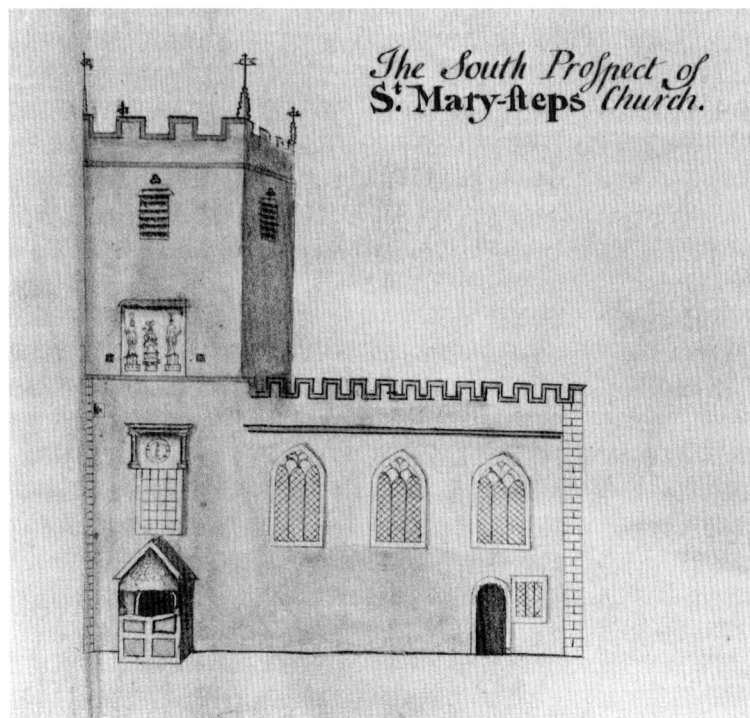

Fig. 28.1
Drawing of St Mary Steps Church, Exeter, mid-eighteenth century.

SOURCE: reproduced by permission of Devon Record Office (ECA, Exeter City Map Book).

The South Prospect of St Mary-steps Church.

Map 28.5
Devon: three
denominations,
1672–1800.

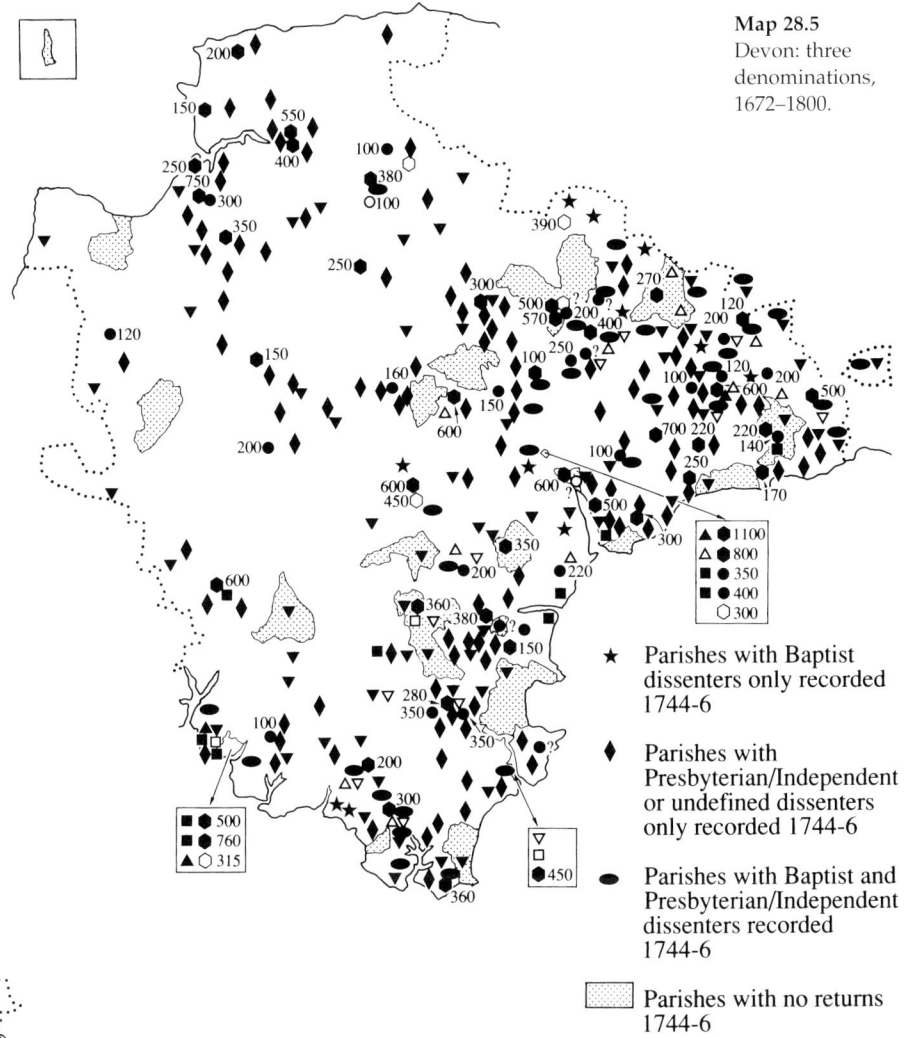

o Baptist meetings

Presbyterian/Independent

▼ Meetings recorded in 1672 or 1690 but not in 1715

⬤ Meetings recorded in 1715 and still extant in 1772
with number of 'hearers' in 1715

● Meetings recorded in 1715 but closed by 1772
with number of 'hearers' in 1715

▲ Meetings opened between 1715 and 1772

■ Meetings opened after 1772 and before
1800

★ Parishes with Baptist
dissenters only recorded
1744-6

◆ Parishes with
Presbyterian/Independent
or undefined dissenters
only recorded 1744-6

⬥ Parishes with Baptist and
Presbyterian/Independent
dissenters recorded
1744-6

▨ Parishes with no returns
1744-6

0   10   20km

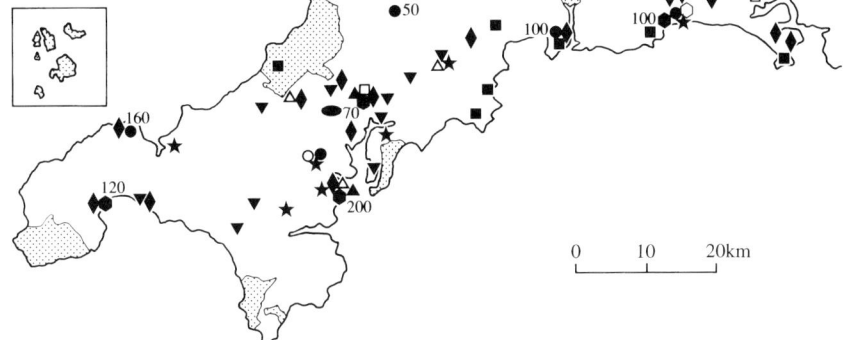

Map 28.6
Cornwall: three
denominations,
1672–1800.

might normally be expected to correlate, this was not necessarily so when, for example, ministers were particularly forbidden to preach in towns, and rural backwaters might attract less persecution, as was the case from 1665 to 1685. Many congregations formed around an ejected minister where he was living, which in turn might reflect the protection of a sympathetic gentlemen or incumbent: congregations also formed in places where Church jurisdictions were weakened by such factors as 'peculiar' status or on boundaries such as that between the three dioceses in east Devon and west Somerset and Dorset. Congregations might also vary enormously in size and wealth, as became clear in 1690 when the Presbyterians and Independents created funds to support their more marginal congregations. In this respect dissenters, like Anglicans, were affected by social geography. The conditions of the ministry were different in towns from countryside and in different areas of settlement. Furthermore, the different nonconformist traditions involved their membership in varied ways. The Quakers and Baptists both gave their laity a large role in worship, while developing a cadre of itinerant preaching leaders who helped to link congregations and bring the word to more scattered groups. The Independents and Presbyterians, by contrast, took their lead from their learned clergy. This placed on them the extra burden of both educating and then remunerating at a suitable rate such a clergy, which often proved a great burden on a small congregation. Such ministers often found it necessary to take on other work, such as schoolteaching, medicine or the like. The existence of such work, or of a wealthy core of adherents, thus became critical to the survival of their congregations.[18]

PATTERNS OF DISSENT

All of these social factors began to shape the distribution of nonconformist congregations, once the special conditions created by legal persecution had fallen away. The consequences are clearly visible on **Maps 28.5 and 28.6 (Maps 28.3 and 28.4** for the Quakers), which show the temporal pattern of church formation and survival across the region. What emerges for the period c.1689–1773 is that an early expansion of nonconformist meetings was followed from about 1715 onwards by consolidation and then contraction as many of the marginal congregations began to fail. The effects of this are already visible in 1744 and become clearer in subsequent visitations as the number of parishes that report a dissenting presence declines.[19] By and large dissent was shrinking back into the areas of its greatest initial strength and becoming, in setting at least, a largely urban movement, even though some of the attenders may have been from the countryside around.

The contrast is clear in the sources created by the two attempts of the dissenting congregations to measure their own strength nationally, undertaken in 1715 and 1773, resulting in what have become known as the Evans and Thompson lists.[20] The first of these was intended to indicate the strength of the dissenting presence in political life, by identifying the number of dissenting voters in county, borough and municipal electorates, but it also contains the names of all the ministers and congregations of the Presbyterians, Independents and Baptists, together with a crucial extra piece of information, namely the number of 'hearers' at each church. The category of 'hearers' may seem strange, but it reflects two facts. One is that the heart of the dissenting service (as of the Anglican, at this period) was the sermon and the second is that many people attended dissenting services without making the extra commitment of membership or even of using the church's services in such areas as marriages and burials, if these were offered. Thus the figures for hearers suggest a considerably larger public audience for dissent than that which emerges from surviving membership lists or registers.[21] There is clearly a danger that the figures may be exaggerated, given the aim of the survey, and they are usually in very round numbers, but they offer an unrivalled opportunity to get a sense of the relative strength of three of the four dissenting traditions at their high-point in the region. The 1773 material (again prepared as part of a lobbying movement for repeal of the remaining discrimination against Protestant dissenters) does not, unfortunately, contain the same detail, omitting any mention of hearers, and although a subsequent attempt was made in 1794 to update some of the Devon material, the column allowed for hearers was rarely filled in. As far as Presbyterian churches are concerned the 1794 material confirms the trend of 1773, namely the contraction of the range and size of congregations, except in a few growth areas such as Exmouth.[22]

The picture emerging from these surveys must be qualified. As noted, Quakers were not included, since they were not considered a genuine part of old dissent by the 'three denominations' of Presbyterians, Independents and Baptists. As it happens, the Quakers' own sources suggest that their meetings were suffering a very similar process of decline and urbanization as the other groups.[23] Secondly, during the eighteenth century a number of new issues brought new divisions within the ranks of old dissent, erasing some of the characteristic differences but creating others. One of these was the emergence of a strong Unitarian strand within Presbyterianism, following a controversy centred on Exeter in 1717. During the rest of the century, many Presbyterian congregations either split or renamed themselves as Independent to avoid association with what they saw as a doctrinal heresy. This tendency was strengthened by the other development, which was the evangelical revival of mid-century onwards. After a period of introspection and numerical decline, a substantial number of dissenting churches, mostly Independent and Baptist, began to reach out to new audiences, in part by the rediscovery of itinerant and lay preaching and other techniques of their seventeenth-century founders. This led, in Cornwall and the Plymouth area in particular, to the emergence of a substantial number of new congregations from the 1770s onwards. It also further eroded any distinction between Baptist and Independent churches.[24]

The evangelical revival was also having another profound effect on the religious geography of the region, which by the nineteenth century was almost to efface the traces of the conflict between old dissent and the Church. The rise of Methodism was a national movement, but one with particular implications for Devon and, especially, Cornwall. No attempt has been made here to map the early stages of this process, even though a mounting tide of references to Methodist activity can be found in the successive episcopal visitation returns. The main reason for this is the intellectual and geographical complexity of mapping Methodism before 1800. During this period Methodism was in a similar position to earlier 'puritanism' or to the early Presbyterians, in that it had no aspirations to be a separate denomination. Instead it was an evangelical tendency within both old dissent (symbolized by George Whitefield's followers) and the Church (notably the followers of John and Charles Wesley). The effect on old dissent, as we have seen, was to stimulate the growth of new 'independent' churches, only some of which explicitly identified themselves with Methodism as such. The effect of Wesleyan and other Anglican-based evangelicalism was more complex. In some cases it revitalized and modified parochial practices within the Church, but in others it led to the growth of Methodist societies which eventually grew into separate congregations. Where this latter process occurred it was often the result of the hostility or lack of pastoral care shown by the Anglican clergy. The Methodist recourse to itinerant and lay preaching and to small cells of lay members once again recalled the techniques of seventeenth-century sectarians. They proved particularly suitable to meeting the needs of the rapidly expanding mining populations of west Cornwall, but also the scattered rural settlements of Cornwall and north and west Devon. However, while it appears that their preaching reached a mass of 'hearers' neglected by the Church, they had not, before 1800, converted this audience into a mass membership or network of chapels such as characterized the region in the nineteenth century. Instead their characteristic institutions were the preaching circuit and the close-knit societies and bands of members.[25]

To return, finally, to the opening question about the geography of religion, it is far from clear that the geography of old dissent mirrored the weaknesses of the established Church. Instead dissenting groups owed their original geography to the power within that Church of a puritan tradition, strongest in parts of east, north and south Devon and leading to a largely Presbyterian or Independent dissenting movement. In 1676 and 1715 Devon dissent was much stronger than Cornish, with the partial exception of the Quakers. Once established, however, old dissent found it hard to survive, in the absence of the compulsory funding provided by the parish system, outside the towns and richer agricultural areas, where the Church was also a strong presence. It was not until the new methods of evangelical revival enabled a cheaper and more flexible system of religious provision for poorer and more scattered rural settlements to be provided outside the established churches, either of the Church or of old dissent, that a new pattern emerged in which nonconformity did begin to flourish where the Church was weak or unable to respond flexibly to population growth in mining or port areas. Ironically, the new dissent's massive success, largely in Methodist form, in Cornwall and west Devon may owe something to the complacency of an establishment which had defeated the challenge of old dissent.

FURTHER READING

H. Miles Brown, *The Church in Cornwall* (Truro, 1964); A. Warne, *Church and Society in Eighteenth-Century Devon* (Newton Abbot, 1969); J. Barry, 'The seventeenth and eighteenth centuries', in N. Orme (ed.), *Unity and Variety: A History of the Church in Devon and Cornwall* (Exeter, 1991).

29

RELIGION AND RELIGIOUS INSTITUTIONS

RELIGIOUS WORSHIP
IN 1851

BRUCE COLEMAN

Patterns of religious allegiance and practice in the South West have not been fixed over time. They have altered under many influences, some of them concentrated within the region itself while others had a wider significance and more national character. Yet there have also been continuities and from the mid-eighteenth century onwards the region's patterns of religion had recognizable characteristics that can still be seen in modified form in the late twentieth century.

From the Reformation to the eighteenth century the region had consisted of two reasonably distinct zones of religious culture. West Devon and Cornwall were characterized by a traditionalistic folk-religion which, though formally protestantized by the Reformation, had not been aggressively puritan in the early seventeenth century. Outside the largest town, Plymouth, it was shaped only modestly by those forms of Protestant dissent that would achieve recognition through the operation of the 1689 Toleration Act. The more advanced elements of English Protestantism tended to see the far west as backward, peripheral and superstitious rather than conspicuously pious.

East Devon, including Exeter, the cathedral city of the diocese covering both Devon and Cornwall, was ground better tilled and harvested. This was the part of the South West most responsive to the official faith of Court and Canterbury. It was more thickly studded with towns of size and significance than the western half of the peninsula and it had a flourishing woollens industry which was the westernmost extension of a belt of textile industrialism stretching from East Anglia through Wiltshire and Somerset. Protestant dissent was firmly established in this territory by the late seventeenth century and the towns of east Devon provided it with centres of influence of a kind matched only by Plymouth further west.

This picture was changing rapidly in the late eighteenth and early nineteenth centuries. The great expansion of deep mining and the associated growth of population in the west of the peninsula interacted with the spread of a new kind of nonconformity in the shape of Methodism (or rather Methodisms, as there were soon break-aways from the original Wesleyan connection). Methodism had originated within the established Church and here as elsewhere had drawn much upon the pieties and attitudes of traditional folk-religion. Its conservatism in moral teaching, its unpretentiousness and its organizational structures

(particularly local circuits and itinerant preachers) fitted it to grapple with the social conditions of the more western parts of the region. Perhaps too the eighteenth-century gentrification and better education of the Anglican clergy had made them less fitted to sway the hearts and minds of relatively unsophisticated people in the provincial backwaters. Whatever the reasons, by the mid-nineteenth century the western parts of the region—and Cornwall in particular—had come to rank among the most nonconformist and the most Methodist parts of England. Though it might be tempting to see Cornish distinctiveness, even separatism, as an explanation, west Devon had much the same experience as east Cornwall and the rapid spread of Methodist nonconformity in Cornwall represented the county's incorporation into one of the main contemporary national developments.

East Devon was much less affected by this tide of Methodism. Reasons included the predominant settlement patterns (rural settlement was less scattered here than further west), stronger landlord influence, a more numerous and strongly entrenched Anglican clergy, firmer episcopal supervision from Exeter and the near absence of significant mining developments. Here the long-established dissenting denominations served as more effective rivals and barriers to newer forms of nonconformity than they could ever have done further west. Though the old cloth industry had largely collapsed by this time, the old dissent remained influential in the towns alongside what seems to have been a majority Anglicanism. This eastern end of the region was its most metropolitan and least provincial part, well ordered from above and only marginally susceptible to the inroads of a Methodist *volkskirche*. Indeed Exeter and east Devon were Anglican enough to figure among the several models Anthony Trollope would use in the 1850s for his fictional Barchester and Barsetshire.

THE CENSUS OF RELIGIOUS WORSHIP, 1851

The one-and-only official Census of Religious Worship was held in 1851 and it provided a revealing snapshot of attendances at church and chapel across England and Wales. Attendances were recorded at morning, afternoon and evening services in all places of worship on census Sunday, 30 March. The original returns survive in the Public Record Office, Devon now among them, but historians have made most use of the summary statistics printed in a volume of the Census Report. The standard methodology has been to aggregate the attendances at the various services and to present the total as a percentage of the population of the area in question. This figure is called the index of attendance (IA). Though not a good guide to the numbers of individuals who attended worship that Sunday (some people attended more than once), it does provide a rough-and-ready basis for comparability between different denominations and different localities.

The census revealed that attendances at worship in the two counties ran rather above the national average: Devon's IA was 70.5 and Cornwall's 67.9 against a figure of 58.1 for England and Wales. The national average was, however, pulled down by the large cities and industrial districts and the figures for the South West were not remarkable among the predominantly rural counties of southern England (Wiltshire, for example, had an IA of 85.7). By contemporary standards there was nothing exceptional about the level of religiosity in south-west England. For the Church of England, however, a more interesting picture emerged. The IA for Anglican attendances in England and Wales was 29.5; in Devon it was 40.1 and in Cornwall only 19.2. For all nonconformity the national IA was 31.3; for Devon it was 30.4 and for Cornwall 48.7.

The suggestion that Cornwall was much the less 'normal' of the region's two counties was confirmed by the percentage shares of attendances attracted by the Church of England and by nonconformity respectively. For the whole of England and Wales the established Church returned 48.6 per cent of attendances and nonconformity 52.4 per cent. In Devon the equivalent figures were 56.9 per cent and 43.1 per cent and in Cornwall 28.2 per cent and 71.8 per cent. In fact Cornwall was the most nonconformist county in southern England in terms of the IA and showed the highest nonconformist share of attendances for all English (though not Welsh) counties.

It is, however, misleading to rely on whole-county figures and to stress simply the contrast between Devon's Anglican predominance and Cornwall's even clearer nonconformity majority. For religious purposes (and doubtless others too) the Tamar was not that significant a frontier. Neither county was homogeneous and the county figures quoted above are themselves averages which disguise significant variation amongst the component registration districts. Devon, the larger county in area and population and containing twenty registration districts to Cornwall's fourteen, was the more heterogeneous of the two, but Cornwall also displayed significant internal diversification. Analysis of the statistics for the separate districts suggests that the South West should be seen not as two sharply contrasted counties but as three discernible sub-regions. One would be Devon east of Dartmoor, where Anglicanism was strongly established, old dissent still moderately well entrenched and the Methodism

connections only modest in their following. The second, comprising west Devon and east Cornwall (though Plymouth might require separate treatment), provided a much weaker support for both the Church of England and old dissent but had Methodism strongly established in a variety of forms. The third area, the west of Cornwall, had seen the Church of England virtually overwhelmed by a rampant Methodism while older dissent had only a minimal presence. Clearly the main religious energies in the century before 1851 had gone into the establishment and rapid expansion of Methodism. The Church of England and the older dissenting denominations had shown much less capacity for growth and adaptation, had conceded ground to Methodism and had probably themselves been failing to keep up with the growth of population. But this pattern of change had been most clearly marked in the westernmost parts of the region and least so in the east.

PATTERNS OF CHURCH ATTENDANCE IN 1851

Maps 29.1, 29.2, 29.3 and 29.4 are based on the region's thirty-four registration districts. The Isles of Scilly constituted one district, and four others, three for the Plymouth/Devonport conurbation and one for Exeter, were almost entirely urban in character. All the others, though normally centred on a town of at least modest size, were predominantly rural in character, though some included significant industrial development, particularly mining and quarrying. The statistics represented through the maps are all derived from the 1851 Census of Religious Worship.

Map 29.1 shows the aggregate Index of Attendance. The highest levels of religious observance were found along Devon's eastern boundary, in the South Hams and in north Devon and in some, but not all, of the districts most influenced by Methodism.

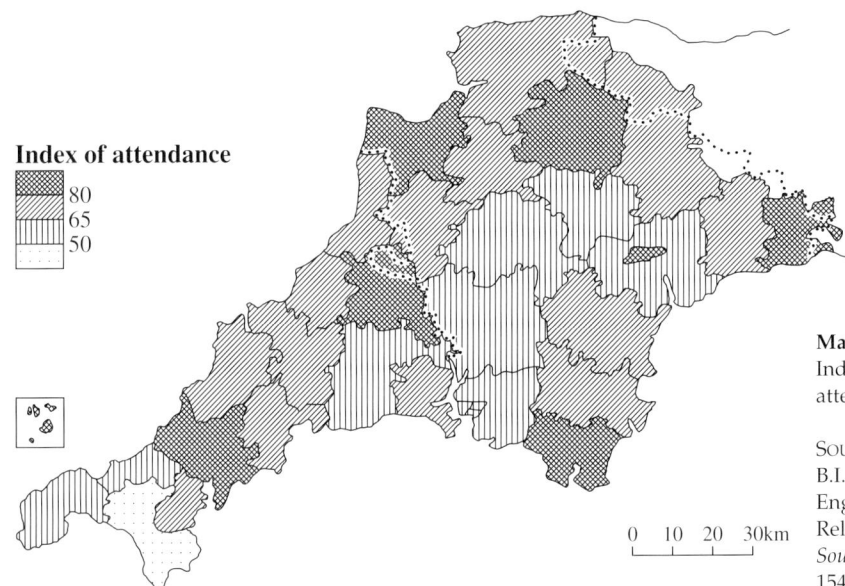

Index of attendance

- 80
- 65
- 50

Map 29.1
Index of attendance: total attendance.

SOURCE: redrawn from B.I. Coleman, 'Southern England in the Census of Religious Worship, 1851', *Southern History*, 5 (1981), 154–88.

0   10   20   30km

Map 29.2
Index of attendance: Anglican.

SOURCE: as **Map 29.1.**

Index of attendance

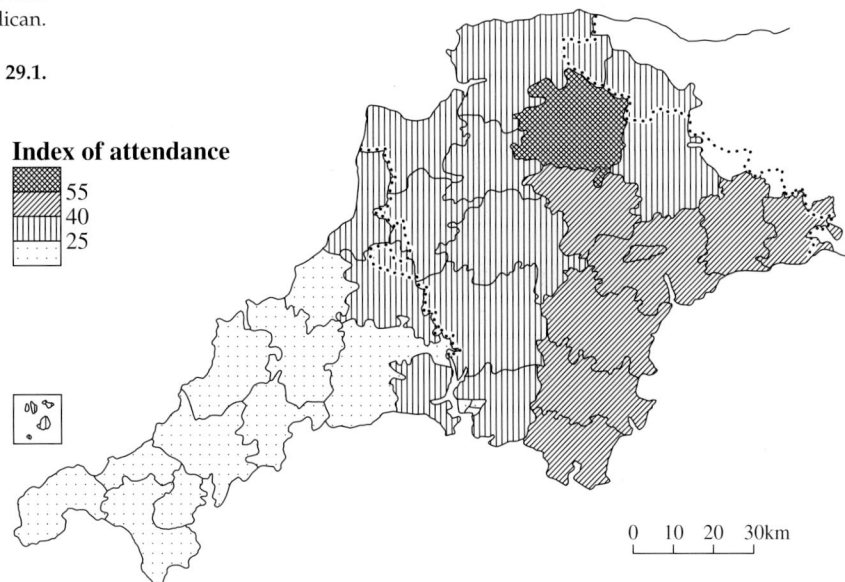

55
40
25

0   10   20   30km

Attendances were generally higher in east Cornwall than in the west. The lowest levels were to be found in the moorland areas and in parts of west Cornwall, but particularly in the large urban centres, Plymouth above all. The pattern is not a particularly neat one, but it is worth noting that the spread between the highest and the lowest levels of observance was much less than could be found elsewhere in England and Wales.

Map 29.2 shows the IA for the Church of England alone. It shows very clearly the high attendance attracted by the Church in the east of the region and the falling away of the level as one moved westwards along the peninsula. Though Exeter and its locality did not produce the very highest figures, the map conveys how much the Church's hold faded as distance from the cathedral

Map 29.3
Index of attendance:
non-Anglican.

SOURCE: as **Map 29.1.**

Index of attendance

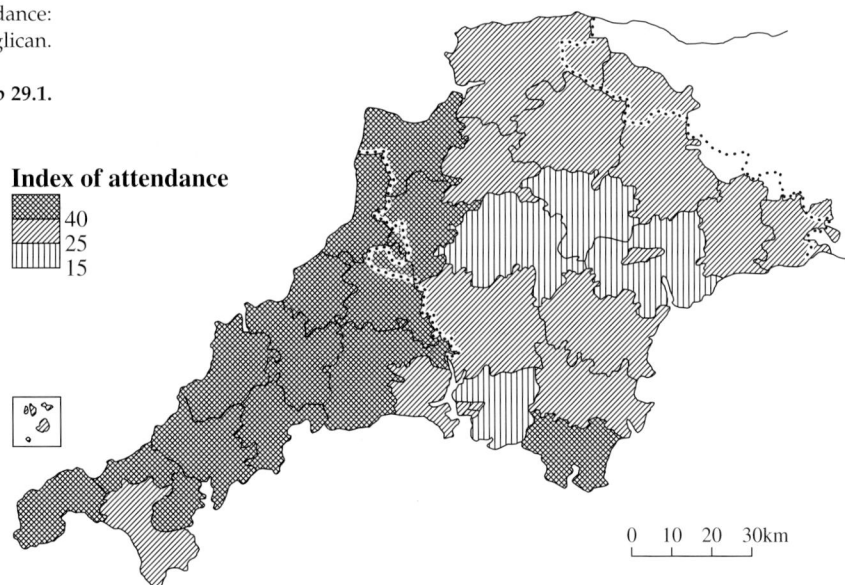

40
25
15

0   10   20   30km

city increased, though the Scillies (hardly a representative Cornish district) provided an exception to this tendency.

Map 29.3 shows the IA for all non-Anglican denominations in aggregate. The lowest figures came from the Exeter and Plymouth districts and from Dartmoor and its surrounds. The highest were, with the exception of the Kingsbridge and the north-western districts of Devon, to be found in Cornwall. The latter county lacked a very clear pattern, though the districts with the highest concentrations of mining tended not to return the highest figures. The highest IAs in the region were, however, all to be found in districts strongly influenced by Methodism. Had Methodism not inserted itself alongside the older denominations, religious observance would almost certainly have been markedly weaker in much of the region by 1851. At the same time the region's nonconformity, though uneven in its incidence, varied within narrower limits than in many other parts of southern England. Neither the lowest nonconformist IA (15.0 in the St Thomas district around Exeter) nor the highest (Truro's 61.7) was near the extreme of the national range.

Map 29.4 shows the percentage shares of attendance attracted by the Church of England and by nonconformity respectively. The establishment's highest figures occurred in east Devon and then fell away westwards along the peninsula, the Scillies again excepted. Some of the worst figures came from the Plymouth conurbation and the picture in west Cornwall was worse than in the eastern districts of the county. The range of variation within Cornwall was significant. In some eastern districts Anglicanism claimed around a third of attendances and in St Germans an actual majority; in the St Austell and Redruth districts nonconformity accounted for over four-fifths of attendances and they were the only districts in southern England where that was the case.

Several aspects which the maps fail to convey adequately are worthy of mention. First, there was little Roman Catholicism in Devon and Cornwall. Small congregations could be found on or near the estates of Catholic landowners and in ports trading with popish countries, but the only concentration of Catholic observance, one based on an Irish immigrant population, was in the Plymouth/Devonport conurbation. Consequently the South West was freer of the local tensions between Catholic immigrant and native communities that characterized some other parts of the country. Nevertheless the response of the region, particularly its Methodism, to national issues involving Ireland and Roman Catholicism down to the First World War showed a strong Protestant consciousness.

Second, **Map 29.3** shows only the strength of all nonconformity aggregated and does not reveal the contrasting patterns formed by the various forms of nonconformity within the region. The most striking contrast was that between Methodism, in its variety of forms, and other, largely older, types of Protestant dissent, amongst which the Independents and the Baptists were the most numerous. These two denominations, together with the Unitarians and the Quakers, had recorded 364 places of worship

attendances (the Church of England had 68,000), while the Bible Christians had 30,000, the Wesleyan Reformers 14,500 and the Primitive Methodists, who appeared not at all in Devon, over 8,000. In the Redruth district, one much affected by mining, Methodism returned about four-fifths of all attendances.

Third, the two large towns were not entirely at one with the rest of their region. The cathedral city and county town Exeter was the region's Barchester. Reasonably of-a-piece with its hinterland, it showed, nevertheless, higher levels of old dissent, Methodism and Roman Catholicism than were found generally in east Devon. Like any other town of its size (about 33,000 inhabitants) and history, it offered more religious liberty and consumer choice than the rural parishes and small towns could possibly do. Plymouth and its dockyard adjunct Devonport offered an even more remarkable spectacle. The nearest thing to an industrial city in the region and with a population around 100,000, the conurbation had some of the lowest levels of religious observance in the South West (only some of the mining districts of west Cornwall showed lower) and had a unique denominational character. The concentration of Roman Catholicism in one of its districts (over 15 per cent of attendances in East Stonehouse) has been mentioned already. Despite the local presence of significant numbers of the professional and officer classes, the Anglican attendances were among the region's lowest. One feature was the remarkably high level of attendances within what the Census Report labelled 'Isolated Congregations', that is those which could not be associated with a recognized denomination. This category, which accounted for nearly a quarter of all attendances within the main Plymouth registration district, almost certainly included places of worship associated with the Brethren (sometimes called the Plymouth Brethren), a movement which, though not originating in the town, had made it a major centre of its strength. How the impact of the Brethren related to the fortunes of other

Fig. 29.1
Charles Church, Plymouth, c.1855; lithograph by R. Groom after J. Shaw.

SOURCE: reproduced by permission of Westcountry Studies Library, Exeter (Somers Cocks 1909).

and nearly 102,000 attendances in the two counties on census Sunday 1851. Methodism in all its forms had returned 1,114 places of worship and some 230,000 attendances. But three-quarters of the non-Methodist dissent was in Devon and only one-quarter in Cornwall. With Methodism the position was nearly reversed: two-thirds in Cornwall, one-third in Devon. Again the county figures are too crude. The greatest concentration of old dissent was in east Devon, where in all the districts east of an imaginary line from Newton Abbot to Barnstaple it had a clear majority over Methodism and in some of these districts accounted for nearly a quarter of all attendances. But west of that line, except in Plymouth which had its own distinctive features, Methodism was in the ascendant for from a third to a half of all attendances at worship. In Cornwall the pattern was simpler. Though best established in east Cornwall, old dissent was modest in scale everywhere in the county and made no returns at all in four districts. Everywhere Methodism was the dominant form of nonconformity. The Wesleyan Methodism Original Connexion was the largest of all denominations, with about 100,000

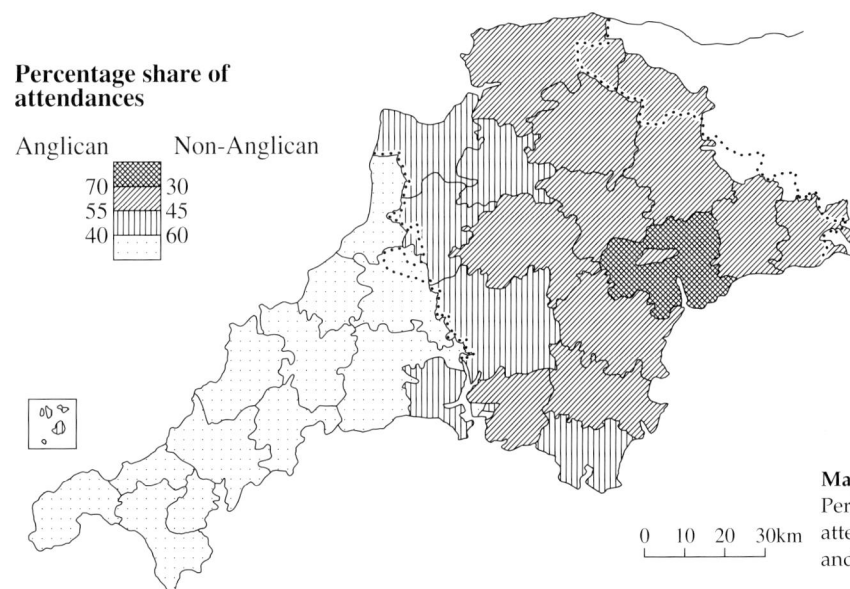

Percentage share of attendances

Anglican   Non-Anglican

70 / 30
55 / 45
40 / 60

Map 29.4
Percentage share of attendances between Anglican and non-Anglican.

0 10 20 30km

SOURCE: as **Map 29.1**.

denominations in Plymouth is a question as yet unexplored by historians.

The third feature of the region which merits some remark is the position of the Bible Christians, an offshoot of the Methodists originating in the South West in the early nineteenth century and entrenched by 1851 in a heartland either side of the Tamar. The connection, which achieved its greatest following in isolated agricultural areas, not in the mining districts where other Methodist denominations flourished, had relied heavily on cottage meetings in its formative years, and the chapels it was building by mid-century were mainly small and unpretentious in style. As a faith it was self-consciously simple and modest, suspicious of the pretentions of even mainstream Methodism, and with a character of inward-looking localism and moral conservatism. The Bible Christians, who would merge into a reunited Methodism in the early twentieth century, never spread on a great scale outside their locality of origin in Devon and Cornwall.

The local roots sunk by the Brethren and the Bible Christians suggest some capacity for religious dissidence and creativity in the region, particularly its more western (though not just Cornish) parts. The phenomenon of the prophetess Joanna Southcott, a Devon farmer's daughter, in the early years of the nineteenth century, had illustrated the scope for popular emotion and credulity in a period of social tensions. Most of the new faiths taking root in the South West were, however, less bizarre than

Southcott's and the region was anyway not the only one to experience surges of revivalist religion in the period. The region's experience, taken as a whole, cannot be regarded as unique. But one feature of the far South West was its somewhat peripheral relationship to much of national life. In many of the English regions certain parts were more obviously 'provincial' than others, with a distinctive character of their own. In contrast to the relatively metropolitan culture of east Devon, not very different from the rest of southern England to its east, west Devon and Cornwall had shown themselves to be dissident in matters of religious allegiance and practice, nonconformist in more than the technical sense, despite the achievement of Wesleyanism in binding so many of their inhabitants into the largest of the nation's nonconformist movements.

In contrast to some other regions, most of what had happened in the South West in that century of rapid change before 1851 had been small in social scale. The region now had little major industry outside mineral extraction and Plymouth dockyard. It also had little major urbanization outside Plymouth and Exeter, neither of which was large by the standards of the great contemporary conurbations. Even the ports of the region—Plymouth/Devonport again excepted—were now modest in scale. The religious life of the region was largely rural or semi-rural, characterized by hamlets and villages (and even more dispersed forms of settlement), by predominantly small towns, by modest-sized and rather isolated semi-industrial communities. Even the important maritime dimension of the South West's economy had left only a modest impression on its religious life, perhaps because the maritime economy outside the great naval dockyard had undergone no great surge of expansion during the recent period of religious evangelism.

The pattern of religion in the South West in 1851 was still, in part, a long-established one. The parochial structure of the established Church covered the whole peninsula, though better imposed and resourced and more effective in the east of the region than further west, and there remained significant residues of the old dissenting denominations, mainly in the once-prosperous textile districts at the eastern end of the region. The relative weakness of both established Church and old dissent in the west of the region was itself a reflection of their failings there in earlier centuries. But the rapid growth of new, particularly Methodist, forms of nonconformity in the decades before 1851 had ensured that the Census of Religious Worship would reveal a South West radically different from the region of even a century before.

Since 1851—and more so since the early twentieth century—nonconformity has declined. The most dramatic withering has been of some of the older dissenting denominations like the Congregationalists and the Baptists, though Methodism had particular problems in the South West with the late-nineteenth-century collapses of metalliferous mining and the losses of population from Cornwall and rural Devon. By that time the railway and, later still, road transport were reducing the isolation

Fig. 29.2
Providence Bible Christian Chapel, Exeter, photographed by Brown of Bedford Circus, 13 July 1898.

SOURCE: reproduced by permission of Westcountry Studies Library, Exeter (P&D 9240).

of much of the South West, a change evidenced by the rapid growth of the resort and holiday trades. The impact of these developments was to make the South West less isolated and less provincial in its cultural separateness in the twentieth century and more responsive to some of the main trends and influences of national life. One example of this tendency was the so-called Anglican revival in Cornwall, symbolized and spearheaded by the carving of a Cornish diocese out of the see of Exeter in 1876 and the building of Truro Cathedral. This revival was modest enough in scale but at least showed that the separateness of the far South West was now lessening rather than increasing further. But the luxuriant nonconformity of the region's westernmost parts would continue to shape its cultural and political life well into the twentieth century—witness the vigorous survival of the Liberal party in some of the very constituencies where a nonconformist-based Liberalism had achieved some of its greatest electoral triumphs after the 1884 extension of the county franchise. Witness too the failure of the Labour party to make much impact in the region outside Plymouth and Exeter. But Liberal survival as a major party is much less evident in the eastern half of Devon, a fact that underlines both the divided character of the region in religious terms and the continuing tendency of that part of the region to identify itself with the dominant national culture further east rather than with the dissidence of the region's western parts.

FURTHER READING

General discussions of patterns of religion in England are provided by R.B. Currie, A.Gilbert and L. Horsley, *Churches and Church-goers: Patterns of Church Growth in the British Isles since 1700* (Oxford, 1977), and by J.D. Gay, *The Geography of Religion in England* (London, 1971). B.I. Coleman, *The Church of England in the Mid-Nineteenth Century: A Social Geography* (London, 1980) focuses on the 1851 Census of Religious Worship and has statistical appendices which cover more than Anglicanism. That is also true of the same author's 'Southern England in the Census of Religious Worship, 1851', *Southern History*, 5 (1983), 154–88. Two other pieces by Coleman, 'Religious worship in Devon in 1851', *Devon Historian*, 23 (1981), 2–6, and 'The nineteenth century: nonconformity', chapter 6 of N.I. Orme (ed.), *Unity and Variety: A History of the Church in Devon and Cornwall* (Exeter, 1991), are relevant to this section. The original returns to the 1851 census for Devon have been helpfully presented by M.J.L. Wickes (ed.), *Devon in the Religious Census of 1851* (Appledore, 1990).

RELIGION AND ECCLESIASTICAL PRACTICES IN THE TWENTIETH CENTURY

GRACE DAVIE AND DEREK HEARL

The maps in this section have been determined at least in part by the availability of data.[1] They have, nonetheless, been selected to illustrate a persistent ambiguity concerning the religious life of the South West in the twentieth century—an ambiguity common to most of Britain, but with particular connotations in this region. On the one hand they demonstrate the undeniable decline in organized religion throughout the period in question, the principal conclusion to be drawn from **Maps 30.1 and 30.2**.[2] But other dimensions of religious life have shown themselves to be more persistent. One illustration of such persistence can be found in the continuing salience of the religious factor as a significant variable in the political life of the South West at least until the mid-1980s (see also chapter 25). A further point follows from this: the boundaries that emerge with respect to such politico-religious cultures are, substantially, those that have existed since the establishment of Methodism as the major religious alternative to the Church of England in this area. Such patterns need, moreover, careful delineation, for the cultural boundaries of the region, whether religious or political, do not coincide totally with county limits, a point already noted in the discussion of the nineteenth century (see chapter 29) which is confirmed in this chapter as well. The marked differences between Devon and Cornwall with respect to their religious life and its institutional manifestations are, however, important; the contrasts between the two counties provide a starting point for this chapter.

Before embarking on the substantive discussion, one methodological preliminary requires firm underlining. Reliable mapping depends to a considerable extent on an appropriate framework of statistics. But religious statistics are becoming increasingly difficult to handle for they reflect the ambiguity already noted above. To be more precise, the British of the late twentieth century, like many of their European counterparts, tend to manifest their religious allegiances by staying away from particular denominations rather than going to them, and there is little reason to suppose that the South West deviates from the norm in this respect. Two corollaries follow from this. The first is a practical one in that counting people who stay away from churches is bound to be a hazardous process since it involves, inevitably, a certain amount of guesswork. The second is a necessary reminder that nominal membership cannot

simply be ignored; nor should it be confused with no membership at all. For not only do these categories have quite different implications for the sociologist of religion, they also have widely different consequences for everyday living.

THE 1989 CHURCH CENSUS

Bearing these problems in mind, it is, nonetheless, possible to mark up the principal features of religious activity in the late twentieth century in the counties of Cornwall and Devon. A useful starting point for this exercise is the Marc Europe English Church Census of 15 October 1989 in that it provides moderately detailed information on church-going and on church-goers on a county-by-county basis, thus enabling some comparisons with the national position as well as between the two counties in question.[3] The Marc Europe information also includes trend data on recent growth and decline (a similar enquiry took place ten years earlier) and the figures are helpfully broken down by denomination as well as by a variety of socio-demographic factors. As a bench mark it provides a useful comparison with the 1851 figures already presented (see chapter 29). It would, however, be unwise to jump too quickly from one census to the other, for the overall figures mask a rather more complex picture, including for example a certain amount of adjustment between denominations as well as elements of growth in certain areas. The first part of this chapter will start by sketching the contemporary position. It will then work sideways in order to colour in some of the detail— including the variety within as well as between the counties of Devon and Cornwall and the trajectories, some more predictable than others, of particular denominations.

Nuances notwithstanding, the undeniable drop in regular church attendance over the last 150 years remains the dominant feature, both in the South West and in the nation as a whole. Only 10 per cent of the adult population of Cornwall attended either church or chapel on census Sunday 1989 and 12 per cent of adult Devonians, proportions very different from those discovered in 1851. In this respect the church-going activity of Cornwall in the late twentieth century is identical to the national average and that in Devon a little higher. A second point

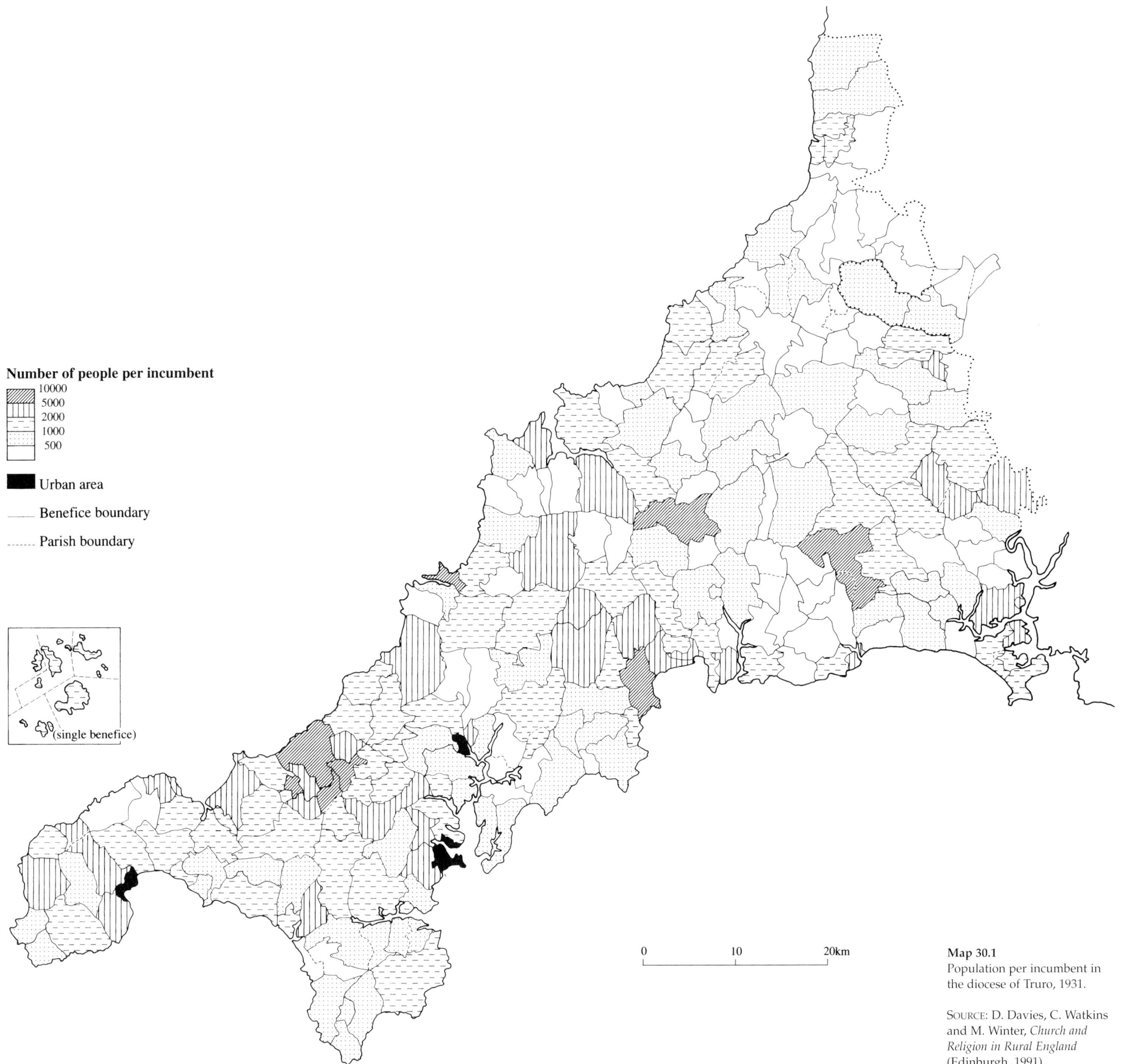

Number of people per incumbent

10000
5000
2000
1000
500

Urban area

Benefice boundary

Parish boundary

(single benefice)

0   10   20km

Map 30.1
Population per incumbent in
the diocese of Truro, 1931.

SOURCE: D. Davies, C. Watkins
and M. Winter, *Church and
Religion in Rural England*
(Edinburgh, 1991).

Map 30.2
Population per incumbent in
the diocese of Truro, 1988.

SOURCE: as **Map 30.1.**

Number of people per incumbent

10000
5000
2000
1000
500

Urban area

Benefice boundary

Parish boundary

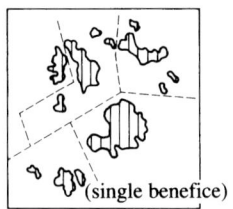

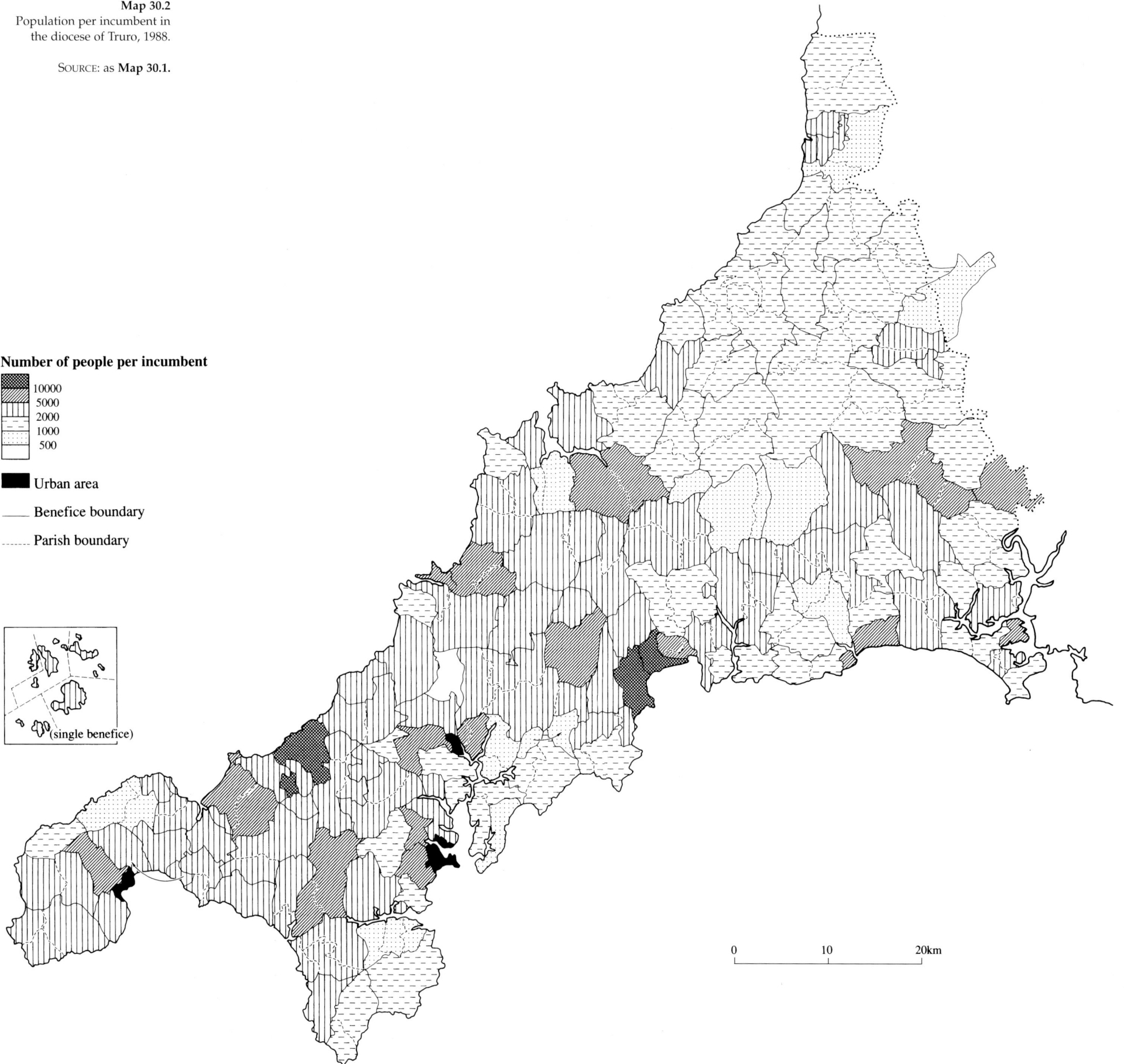

(single benefice)

0   10   20km

County	Methodists	Other free	Anglican churches	Roman Catholics
Cornwall	44	10	28	18
Devon	16	33	32	19
Avon	12	28	38	22
Dorset	7	26	45	22
Gloucestershire	8	27	55	10
Somerset	11	24	55	10
Wiltshire	12	23	45	20
Region average	15	26	41	18
National average	34		31	35

Christian or of other faiths. Once again this feature is explained by the predominantly rural nature of the counties in question and by the particularities of their economic and social history. The kind of labour that such incomers provided led them to the larger cities of post-war Britain, notably parts of London, the West Midlands and the industrial cities of northern Britain. Such explanations are straightforward enough; they have, however, important repercussions in the South West for the way in which terms such as 'pluralism' and 'multifaith' are understood, terms which are current in contemporary discourse. Children growing up in Devon or Cornwall, especially those from the country rather than the towns, are unlikely to have had first-hand experience of such phenomena. They must, however, be prepared for adult life in situations that may be very different—a formidable challenge for those with the responsibilities for teaching in the South West.[5]

Table 30.1
Adult church attendance in the South West in 1989 (each denomination as a percentage of total attendance).

SOURCE: Figures adapted from Peter Brierley, *Prospects for the Nineties: South West* (London, 1991), 6, and *Christian England* (London, 1991), 35.

must, however, be equally stressed. It concerns denominational difference rather than overall percentages, for the Marc Europe figures demonstrate unambiguously the disproportionate Free Church presence, both active and less active, throughout the South West alongside the relative absence of Roman Catholics in this region (see **Table 30.1**). In both respects, the South West continues to display a distinct rather than typical regional profile.

Most striking of all, however, is the distinctiveness of Cornwall, both within the region and within the country as a whole. The predominance of Methodism in Cornwall remains in the last decades of the twentieth century the defining feature of the county just as it was in 1851. Following Brierley's figures, for example, four in nine church attenders in Cornwall are Methodists, compared with one in nine in the population as a whole. It is hardly surprising, therefore, that the fortunes of church-going in Cornwall are largely bound up with the marked decline of Methodism in the country as a whole. The Free Church presence in Devon is, in contrast, far more varied and includes appreciable numbers from the Baptist, United Reformed, Brethren and Independent churches, once again a feature noted in the last century by Bruce Coleman in chapter 29.[4] Conversely, the South West contains correspondingly few Roman Catholics. This tendency is particularly marked in the counties of Somerset and Gloucestershire (see **Table 30.1**), but the feature pervades the extreme South West as well. It can, very largely, be explained by the relatively rural nature of the counties in question—indeed of the whole region—for Catholicism remains an urban phenomenon in contemporary Britain. The situation *is* beginning to change in the far South West due to in-migration into the area, but it will be some time before the figures rise to anything like the national average. Anglicanism, in contrast, flourishes in the South West, relatively speaking, though less so in the two most western counties than in their immediate neighbours. Cornwall, with its relatively weak Church of England, remains, as ever, an exception to the rule.

The far South West differs markedly from the national picture in one further respect. There is almost no representation south and west of Bristol of churches or places of worship which belong to the immigrant populations of contemporary Britain, whether

DIVERSITY OF WORSHIP IN SOUTH-WEST ENGLAND

The analysis so far has indicated county-by-county differences (particularly the contrasts between Devon and Cornwall); it has also drawn attention to the distinct nature of the far South West if it is compared to the rest of the country. A second set of differences now needs attention: those which relate to the internal diversity of the counties in question, and especially so in the relatively large county of Devon. Once again the patterns follow those established by Coleman for the mid-nineteenth century. North-west Devon, for example, (**Maps 30.3 and 30.4**) has more in common with north-east Cornwall than it does with the rest of Devon—a fact that can be explained geographically in that the natural dividing line of the Tamar loses its significance in the north of the two counties. From the point of view of religious history, the Bible Christian movement within Methodism developed on *both* sides of the county boundary, attracting adherents from the rural populations of this area and forming a distinct religious subculture within the region. The gradual reunification of the Methodist Church in the early part of the twentieth century has masked the internal divisions within Methodism which pervaded both counties in the previous century. Such patterns continue to show, however, in maps which indicate the *origins* of each Methodist chapel in the county in question. **Map 30.5** illustrates this for Devon. The spatial distributions are striking and—in the north at least—have little to do with county boundaries. Indeed the spatial distribution of the Free Church presence in Devon is sociologically unpredictable; or, to put this another way, it is most adequately explained using detailed and specific historical knowledge rather than trend data, for the trends are not easy to discern.

The varied nature of Devon's religiosity and the intricacy of its internal boundaries are undisputed. Cornwall's Methodist predominance, reinforced by Brierley's statistics, might seem rather more straightforward. A rather different picture for Cornwall in the twentieth century can, however, be found in the

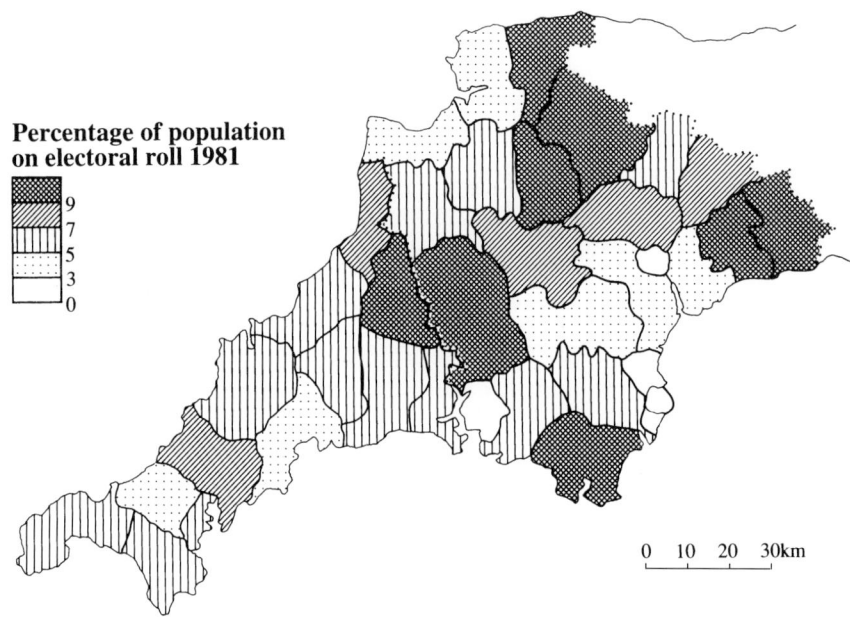

Percentage of population on electoral roll 1981

9
7
5
3
0

Map 30.3
Registered electoral roll members of the Church of England as a percentage of the total population of deaneries, 1981.

Map 30.4
Full members of the Methodist Church as a percentage of the estimated total population of Methodist Church circuits, 1981.

0 10 20 30km

tendency should not be overdrawn.[7] Nor should the contrast between high Anglicanism and the nonconformist churches, for they too have their internal differences, with the Wesleyan Methodists offering worship that is more sacramental in nature than the Bible Christians. Be that as it may, Winter's close analysis of the statistical sources available for both Methodism and the Church of England in Cornwall in the twentieth century points to the complexity of the picture which emerges, a picture in which Methodism seems to have suffered rather more than Anglicanism through the period in question. Such figures reflect, however, the changing nature of religiosity in Cornwall (like everywhere else in Britain). More particularly, they demonstrate relatively high levels of belief in a number of tenets of traditional Christian teaching—belief, however, which has become separated from regular religious commitment through activity and participation. Moreover, for many sociological and cultural reasons, not to mention doctrinal reasons, the Methodist Church—as a gathered church—is likely to suffer more than the Church of England from this changing situation.[8] Hence the particularly sharp decline in Cornwall, historically dominated by Methodism.[9] The term 'gathered church', however, needs to be considered with caution in the South West, for a significant proportion of the population in Cornwall and parts of north Devon still turn to the Free Churches rather than the established Church both for major festivals and for the occasional offices; indeed in many respects such churches have become the 'established' religion of the South West and continue—despite the fall in practice—to maintain a significant penumbra around themselves.

Map 30.3
Registered electoral roll members of the Church of England as a percentage of the total population of deaneries, 1981.

Map 30.4
Full members of the Methodist Church as a percentage of the estimated total population of Methodist Church circuits, 1981.

research of Michael Winter, who has emphasized the relative growth of Anglicanism in this area during the twentieth century, rather than the persistence of nonconformity.[6] Anglican renewal in Cornwall both reflected and encouraged institutional change in the area, notably the establishment of an Anglican diocese of Truro in 1877. The date itself is significant, for the diocese was founded at a time when the Anglo-Catholic movement was growing significantly within the Church of England, one reason perhaps why high churchmanship has been a prominent characteristic of Cornish Anglicanism since that time. Up to a point the same is true of the diocese of Exeter, though the

THE TRURO DIOCESE: PATTERNS OF DECLINE IN THE PROFESSIONAL MINISTRY, 1931–1988

A second and parallel measure of institutional change can be discovered in the falling number of religious professionals serving a particular area. For the Church of England in Cornwall, this reduction within the professional ministry, and the corresponding reorganization of parochial life of which it is part, was represented in map form as part of the material prepared for the Archbishops' Commission on Rural Areas, maps which are reproduced here (**Maps 30.1 and 30.2**).[10] They illustrate the numbers of people per incumbent in Cornwall for two dates, 1931 and 1988, revealing the extent of the changes that have occurred in this period. In 1931, for example, 52 per cent of the incumbencies within the Truro diocese contained fewer than 1,000 souls and more than a quarter of the benefices had less than 500. Parochial organization was dominated by the single-parish benefice, which comprised 88 per cent of the benefices. By 1988 the situation had been transformed. The number of incumbents declined from 222 to 146, only two of whom in 1988 held benefices with populations of less than 500. Not surprisingly, there had been a threefold rise in the number of benefices containing between 2,000 and 5,000 people. In terms of

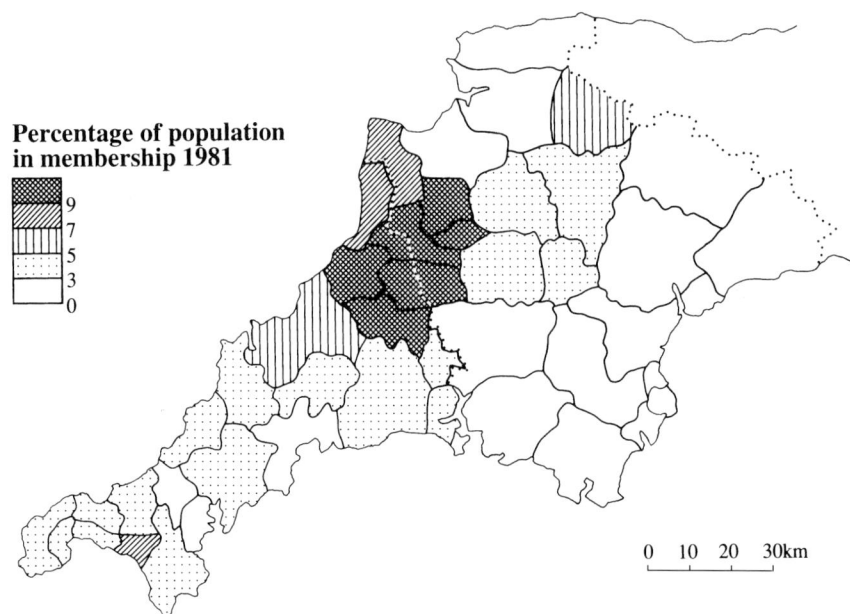

Percentage of population in membership 1981

9
7
5
3
0

0 10 20 30km

organization, only 49 per cent of the clergy now hold single-parish benefices and many of those have grown in size owing to population growth or parish amalgamation. Taken together, such changes have had a dramatic effect on the social visibility of clergy in rural areas and on the clergy as a group in terms of their own ability to network and encourage one another. The maps are a visible reminder of institutional shrinkage, though not of institutional collapse. Churches in the late twentieth century have proved themselves adaptable as well as resilient; diverse forms of lay ministry, together with appropriate theologies to underpin them, have emerged to maintain the system.

THE CHURCH OF ENGLAND AND METHODISM

Maps 30.3 and 30.4 indicate variations in Church of England and Methodist church membership over south-west England. This is measured by ascertaining the percentages of the total population of each of the thirty-seven Church of England deaneries which were registered electoral roll members of that Church, and by calculating the percentages of the estimated populations of each of the fifty-one Methodist circuits in full membership of the Methodist Church. Such figures have the great advantage of being readily available, though they are not the statistics most frequently used by political scientists who would prefer an indicator of church attenders. Sociologists of religion, on the other hand, (or some of them at least) would opt for a more generous indicator of cultural penetration, such as self-ascription, despite the difficulties involved in gathering such information. Such variations should be borne in mind in interpreting the results, for it may well be that the correlations between denomination and political allegiance would be stronger still had a broader indicator of church membership been utilized.

The results are, nonetheless, impressive. It is clear, first, that the Church of England is relatively much stronger in Devon than it is in Cornwall (**Map 30.3**), and second, that it is much weaker in urban areas than it is in the countryside. Since both findings were to be expected, such an encouraging sign helps instil confidence in the validity of the data. Even more striking, however, is the distribution of membership of the Methodist Church (**Map 30.4**). This time the pattern is, frankly, quite extraordinary, showing as it does the great sweep of Methodism across north Devon and down through the whole of Cornwall corresponding almost exactly with

● Wesleyan Methodist Connexion

▲ Bible Christian Connexion

□ Other Methodist denominations

the traditional Liberal 'heartland' of south-west England (see chapter 25). We can only speculate about how this apparent relationship might have been reinforced (or maybe attenuated) had other measures of religious attachment been used.[11]

FURTHER READING

John Gay, *The Geography of Religion in England* (London, 1971); Michael Havinden, Jean Quéniart and Jeff Stanyer (eds), *Centre and Periphery: Brittany and Cornwall and Devon compared* (Exeter, 1991), especially chapter 17; Michael Wickes, *The West Country Preachers: A History of the Bible Christians 1815–1907* (Hartland, 1907), copies available from the author, c/o Jamaica Press, Hartland, Bideford, Devon, EX39 6BE.

Map 30.5
Methodist churches in rural Devon: denominational origins.

SOURCE: compiled by Peter Davie from data in Roger Thorne, *Methodism in Devon: A Handlist of Chapels*, Devon Record Office Handlist No. 2, 2nd edn (1989).

EDUCATION, DISSEMINATION OF KNOWLEDGE, AND LANGUAGE

PRINTING, THE BOOK TRADE AND NEWSPAPERS
c.1500–1860

IAN MAXTED

THE BOOK TRADE, 1500–1800 (**Map 31.1**)

The arrival of a specialized book trade in the South West was a direct result of the growth of printing in the late fifteenth century. Exeter Cathedral Library included a few printed works in its 1506 inventory and at about the same time the region's first recorded publisher, Martin Coeffin, arrived from Normandy and issued at least two Latin schoolbooks which were printed for him in Rouen.[1] His pioneering activity was not regularly emulated until the following century, during which at least forty-seven works were printed in London for fourteen different booksellers in Exeter, as well as two each in Plymouth and Launceston and one each in Barnstaple, Totnes and Tiverton.[2] Booksellers could have considerable stocks in this early period. The inventory of Michael Harte of Exeter, made in 1615, suggests that he had more than 3,500 books in stock.[3]

Despite this, book collectors would frequently purchase books from London rather than use more local booksellers. The correspondence of Richard Coffin at Portledge near Bideford from 1687 to 1697 shows that he used Richard Lapthorne, a London agent, to attend auctions and search for items in the stock of booksellers.[4] The reader in the South West could be kept closely aware of London publications, as can be seen from surviving bound volumes which collect together series of pamphlets on religious and political controversies assembled by local individuals.[5] This reluctance of many to employ the local trade could perhaps be explained by the closed structure of the Exeter book trade which may have lead to a lack of dynamism. Analysis of freemans' records and other sources shows little injection of new blood in the sixteenth and seventeenth centuries.[6]

Probate inventories suggest that only one Devon household in three possessed books in the seventeenth century and these would normally be Bibles and prayer books, but in the following century the evidence of book subscription lists indicates a deep interest in local publications.[7] For example, in 1790 Martin Dunsford obtained 161 subscribers in Tiverton for his *Historical Memoirs of the Town and Parish of Tiverton*, perhaps 15 per cent of all households in the town.[8] Newspapers in the eighteenth century contain regular publicity for book auction sales and advertise the publication of catalogues of the stock of individual booksellers, although the catalogues themselves survive rarely.

The market was not, however, sufficient to permit the bookseller to depend on books alone and most businesses dealt in a range of other commodities. Perhaps the most popular was the sale of patent medicines, not such a surprising combination when it is considered that, like books, they were among the few branded goods produced during that period. Some booksellers in coastal towns sold navigational instruments and charts, and many booksellers acted as insurance or lottery agents.

From about 1780 an increasing number of booksellers ran circulating libraries. Subscription libraries were also established in some centres, for example the Cornwall County Library at Truro whose catalogue running to thirty-two pages was printed in 1794.[9] For towns where there was no circulating or subscription library, individuals grouped themselves to form book clubs. As these were informal associations evidence is difficult to come by, but Kaufman lists eighteenth-century book clubs in Penzance, Liskeard, St Ives, Exeter, Powderham and Tiverton.[10] Dunsford provides further evidence of the Tiverton Society. Formed in 1775, it consisted of twelve members 'from almost every sect, party, business and profession' who met monthly in each other's houses to choose books.[11]

By the 1790s there were specialist book dealers or stationers in the majority of centres with a population of 2,000 or more, although this is not so true of Cornwall where the population of several parishes was swollen by the presence of miners, who would not typically form a part of the book-buying public.

THE SPREAD OF PRINTING TO 1800 (**Map 31.2**)

Printing, invented by Gutenberg in Germany in the 1450s and introduced into England by William Caxton in 1476, did not reach the South West until about 1525 when a press attached to Tavistock Abbey is recorded.[12] No printing from that press is recorded after 1534 and the charter of incorporation of the Stationers' Company of London in 1557 effectively limited printing to London, with Oxford and Cambridge's right to presses being confirmed by the Star Chamber decree of 1586. As a result, the only presses recorded in the region before 1695 are the king's printers Robert Barker and John Bill, in Exeter 1645–6, and a printer with the initials J.B. (probably John Bringhurst) who landed with William of Orange in 1688.[13]

In 1695, with the lapse of the Licensing Act, the field was open for the spread of provincial printing. Plymouth seems to have led the way with the shadowy figure of D. Jourdaine in 1696, but a firmer foundation was laid in Exeter in 1698 when Samuel Darker arrived from London. Presses were busy for the next generation in Exeter, with some 170 surviving books and pamphlets printed there to 1720. Many of these early publications were highly ambitious, such as John Prince's *Damnonii Orientales Illustres*, printed in Exeter for two of the wealthiest London bookseller-publishers over a period of several years and finally appearing in a somewhat curtailed form in 1701. The middle of the century saw a slackening off, with less than half of this early level of production, a drop which can be paralleled in other parts of the country.[14] For some years in the 1730s and 1750s not a single Devon or Cornwall imprint is recorded, although much ephemeral material has disappeared. Only in the 1780s did production expand again and now it was swollen by presses in other towns. Andrew Brice of Exeter had briefly established a press in Truro in 1742 and the Jourdaine family appears to have operated in Millbrook in 1744, but Cornwall's first permanent press was established in Falmouth by Matthew Allison in 1753.[15] Printing then appears to have spread more rapidly in Cornwall than in Devon, as **Map 31.2** indicates, perhaps because of the isolation of much of the county from major centres of book production.

The scale of production in the region was never large, so too much cannot be read into yearly statistics. Individual events can grossly distort the figures. Examples are the Civil War in 1645, the landing of William of Orange in 1688, the Arian controversy in 1719, and elections in 1761 and 1790. For most of the eighteenth century output amounted to around 1 per cent of London titles, and London printers were preferred by most serious writers. Prince, for example, said that he was not willing to entrust further books to Exeter printers and it is perhaps significant that, apart from Prince, no county historians of Devon saw their works in print with local printers before the 1780s.[16]

Number of book trade firms c.1800

o Firms known from other sources

Map 31.1
Towns in which members of the book trade are recorded to 1800.

SOURCES: names mentioned in imprints and, for the end of the period, directories such as *The Universal British Directory*.

Printing firms established by

☐ 1720
◪ 1780
■ 1781-1800

Imprints recorded to 1800

■ Over 100
■ 10 - 99
■ 1 - 9
• None recorded

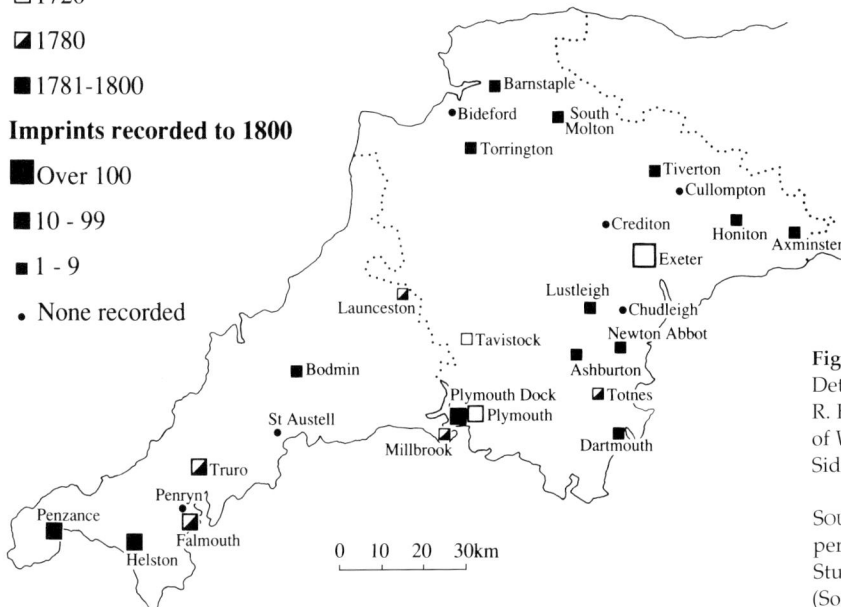

Map 31.2 (below right)
Towns where a printing press is known to have been located before 1801.

SOURCES: E.A. Clough, *A Short-Title catalogue, arranged Geographically, of Books printed and distributed … in the English Provincial Towns and in Scotland and Ireland … to … 1700* (London, 1969); Ian Maxted, *Books with Devon Imprints: A Chronological Checklist* (Exeter, 1989).

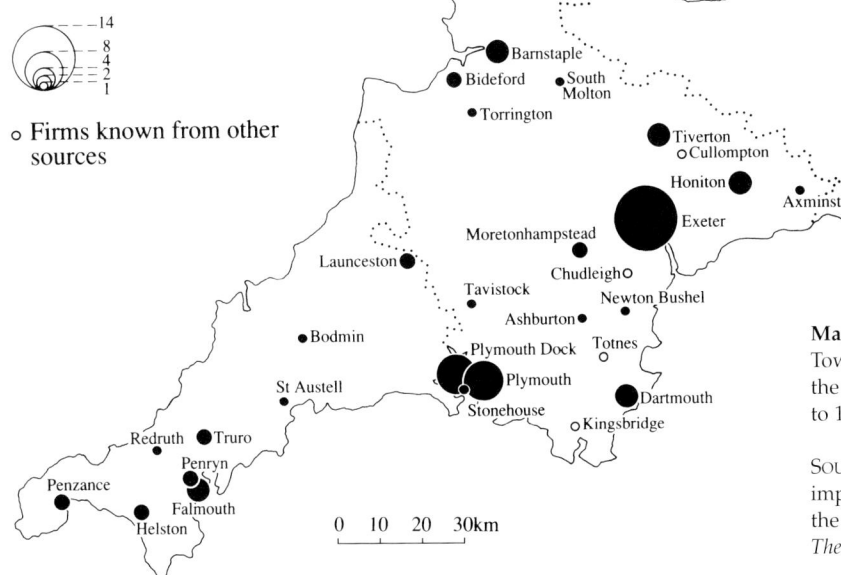

Fig. 31.1 (left)
Detail from panorama by R. Havell after H. Cornish of Wallis's Circulating Library, Sidmouth, 1815.

SOURCE: reproduced by permission of Westcountry Studies Library, Exeter (Somers Cocks 2473).

Fig. 31.2
Portrait of Tommy Osborne,
Exeter secondhand bookseller,
by unknown artist, *c.*1812.

SOURCE: reproduced by
permission of Westcountry
Studies Library, Exeter
(P&D 32608).

The staple fare of provincial presses was ephemeral material, such as printed forms, election squibs or gallows literature.[17] About a third of surviving Devon and Cornwall imprints are single-sheet items. The largest single subject grouping is religious literature, accounting for 34 per cent of titles, with sermons making up almost half of this. However, the proportion declined over the period, religious works making up 64 per cent of titles before 1720 and only 36 per cent after 1790. The popularity of verse as a medium is striking, with 10 per cent in this form, including many election and satirical items. Local guidebooks, a typical product of many provincial printers in the nineteenth-century, only began to appear in the 1790s. The most important category of all, however, was newspapers.

NEWSPAPERS TO 1860

While news books had been frequently produced in London from the 1590s and began to be numbered in series from 1622, the South West had to await the lapse of the Licensing Act in 1695 before its first newspaper could appear. The region vies with Bristol and Norwich for the honour of having the earliest English provincial newspaper. Counting backward from number 556, the only surviving issue of Sam Farley's *Exeter Post-Man*, we reach December 1700 as the date of the first issue, preceding the *Norwich Post* by almost one year and the *Bristol Postboy* by almost two, although this does assume that the Exeter paper appeared only once a week.

For the first century of its life the newspaper press in the South West was virtually limited to Exeter; newspapers founded in Plymouth in 1718, 1758 and 1780 all failed after a few years. In fact from the 1740s Sherborne newspapers were widely distributed in Devon and Cornwall through an extensive network of riders who also provided a means of distribution for other articles handled by the printer and were used as unofficial couriers by individuals.[18] The routes followed by about eight such riders who distributed *Trewman's Exeter Flying Post* in 1772 can be tentatively established from the information given in the newspaper (**Map 31.3**).

In Exeter there were probably at least two newspapers at any one time for most of the eighteenth century (**Map 31.4**), although

Map 31.3
The probable routes
of newsmen delivering
Trewman's *Exeter
Flying Post* in 1772.

SOURCE: Trewman's
Exeter Flying Post,
5–12 June 1772.

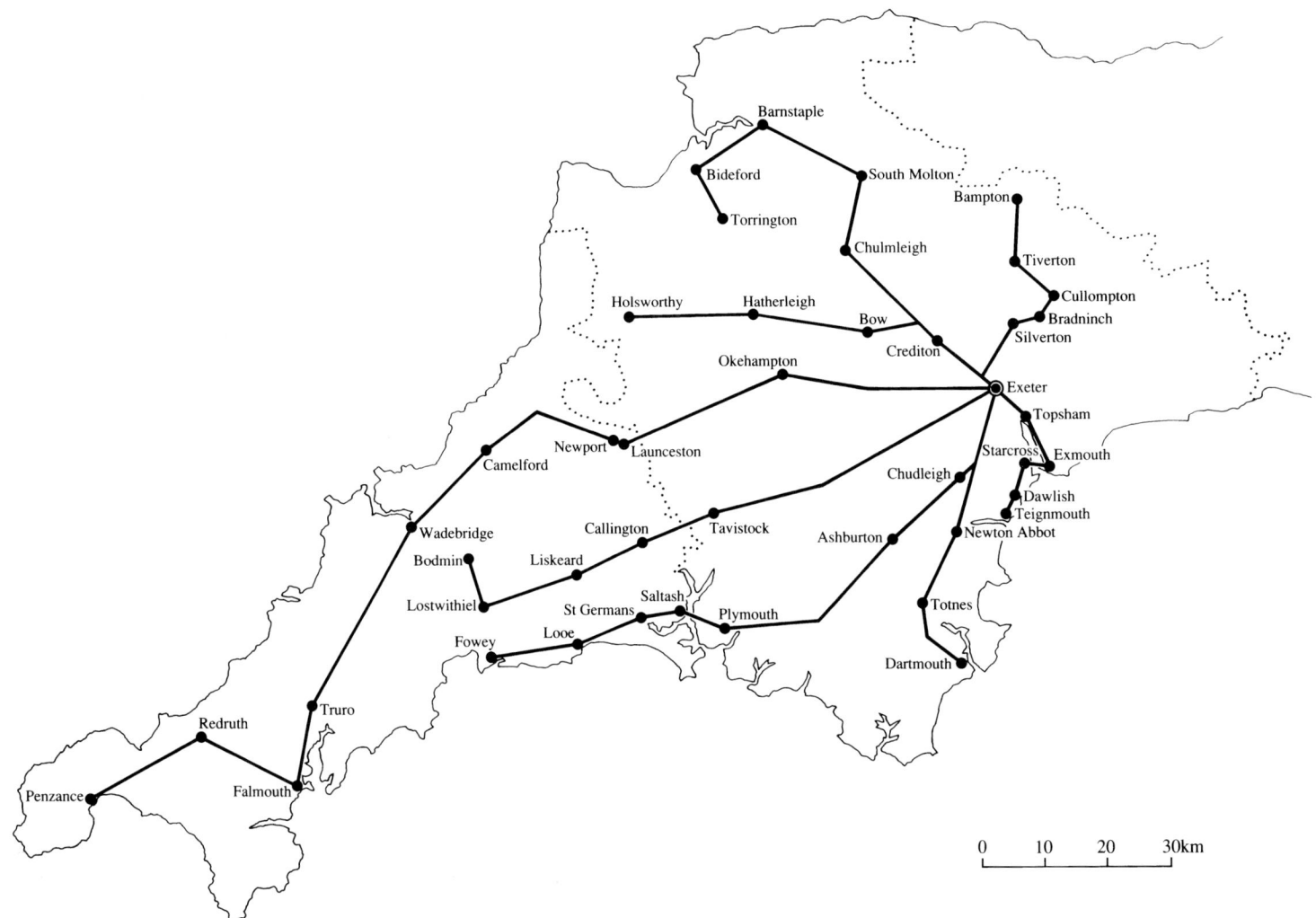

Fig. 31.3
Mastheads of
(**A**) *The Protestent Mercury* or *The Exeter Post Boy*,
(**B**) *Old Exeter Journal* or the *Weekly Advertiser*,
(**C**) *The Sherborne Mercury*
(**D**) *The Exeter Flying Post* or *Plymouth and Cornish Advertiser*,
(**E**) *The Western Times*,
(**F**) *The Exeter Independent*;
all 1717–1837.

SOURCE: reproduced by permission of Westcountry Studies Library, Exeter.

original copies survive infrequently. Some, such as a newspaper apparently published by Joseph Drew in the years around 1750, have disappeared entirely.[19] The earliest newspapers, often highly partisan, carried little local news, although as early as 1727 the acerbic Andrew Brice conducted a vigorous campaign on behalf of debtors in Exeter prisons.[20] Generally, the local newspapers copied news from the London press and there was an eagerness in the larger centres such as Exeter to read the London papers, the local publications providing a means of duplicating the news and relaying it to remoter areas.

As the eighteenth century neared its end, more local news gradually joined the local advertisements, and Cornwall's first newspaper, *The Cornwall Gazette and Falmouth Packet*, was established in Falmouth by Thomas Flindell in 1801. This failed but was re-established in 1803 in Truro under the title *Royal Cornwall Gazette*. In 1808 two rival newspapers were established in Plymouth, which finally laid a secure and permanent foundation for the newspaper press in the region's largest conurbation. Benefiting from the booming wartime economy and located in an important naval centre, these newspapers at times carried news of the Napoleonic campaigns before their London rivals.

The spread of the newspaper in the South West continued to be slow, as can be judged from **Map 31.4**. In 1824 Barnstaple obtained the first newspaper in the north of Devon and it was not

Newspapers and dates of first publication

◉ To 1800

● 1801-1840

○ 1841-1860

■ Cornish temperance journals

▼ Cornish mining journals

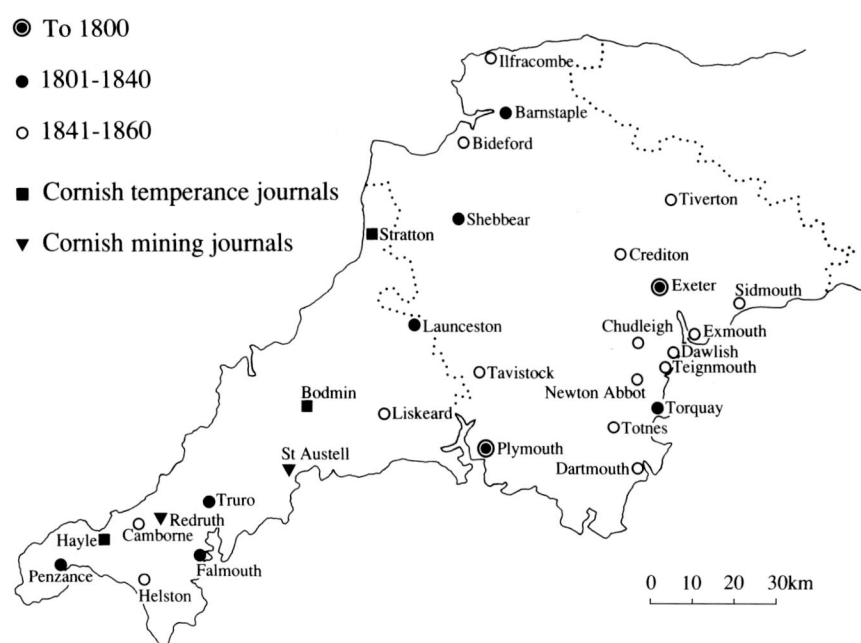

Map 31.4
Towns where newspapers were published to 1860.

SOURCE: *Bibliography of British Newspapers: Cornwall and Devon* (1991).

Map 31.5
Market penetration of Flindell's *Western Luminary*, 1815.

until 1839 that the first south Devon newspaper appeared in the newly developing coastal resort of Torquay. The first newspaper printed in Penzance also first appeared in that same year. From 1847 the spread accelerated, with newspapers appearing in Teignmouth and Tavistock in that year, in Sidmouth and Dawlish in 1850 and in no less than six centres in 1854: Kingsbridge, Dartmouth, Crediton, Bideford, Newton Abbot and Helston. Chudleigh and Camborne followed in 1855, in which year the stamp duty on newspapers was abolished. Ilfracombe and Liskeard had their first newspapers in 1856, Tiverton in 1858 and Totnes in 1860.

Copies sold as a percentage of households

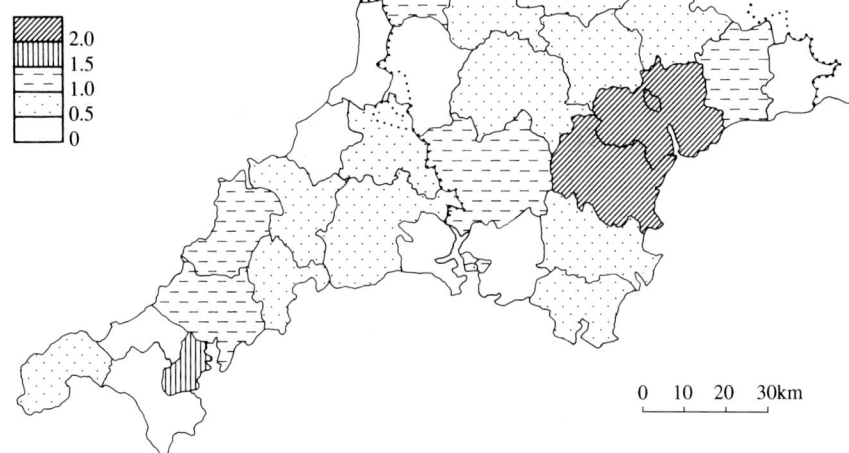

The tourist industry played a major part in the spread of newspapers in Devon, as can be seen from the large representation of coastal resorts in the list. Some titles appeared only during the summer season or were published less frequently in the winter. They would often include a local directory for the use of visitors and a list of the more notable holiday-makers. This catering for the tourist is reflected in such newspaper titles as *Wreford's Visitor's Guide for the Town and Neighbourhood of Torquay and Tor* or *Ilfracombe Gazette, Arrival List and General Advertiser*. In Cornwall there were journals for temperance and mining interests in several centres during the early nineteenth century.

In the larger towns rival newspapers had strong party affiliations and some of the liveliest personalities were to be found among the Liberal or more radical press. One example is Thomas Latimer, the campaigning editor of the *Western Times*, who conducted lengthy and vigorous disputes against what he saw as clerical abuses in the 1840s.[21] In 1835 Latimer introduced a steam-powered printing press and this made it possible to increase circulation and enabled the production of daily newspapers. It was in Plymouth that the first of these, the *Western Morning News*, appeared in 1860. It was soon joined by other daily titles, most of them developed from or associated with a weekly title.

Figures for the circulation of provincial newspapers are difficult to come by before the 1830s. In 1793 Woolmer claimed that 250–300 copies of the *Exeter Gazette* were distributed within Exeter, with 160 being despatched to Plymouth, 140 to Truro, 40 to Launceston, 60 to Penzance, nearly 30 to Dartmouth, 40 to Barnstaple, 'and Tiverton, Wellington, Tavistock, Falmouth, Camelford, Lostwithiel and every other opulent town … in like proportion', a statement which implies a total circulation of around 1,000.[22] In 1815 at least 1,720 copies of *Flindell's Western Luminary* were subscribed for, according to a list produced by the publisher.[23] **Map 31.5** indicates the market penetration of this title and **Map 31.6** its likely area of circulation in the year 1815. Even in the 1840s the influential *Western Times* could boast a circulation of little more than 3,000, although this was more than any other newspaper in Devon at that time.[24] In sharp contrast, by the 1980s and 1990s free distribution newspapers reached 45,000 households each week in Exeter alone, virtually saturation distribution in a city of 100,000 inhabitants.

FURTHER READING

On the book trade see Ian Maxted, *The Devon Book Trades: A Biographical Dictionary* (Exeter, 1991), which is arranged by town and gives brief details on personnel in all branches of the book-selling, printing and paper trades to about 1850. Ian Maxted, 'Four rotten cornbags and some old books', in Robin Myers and Michael Harris (eds), *Sale and Distribution of Books from 1700* (Oxford, 1982), 37–76, is a survey of the world of the book in Devon to about 1800. John Ingle Dredge, *Devon Booksellers and Printers in the Seventeenth*

Map 31.6
Circulation of Flindell's
Western Luminary,
3 January 1815.

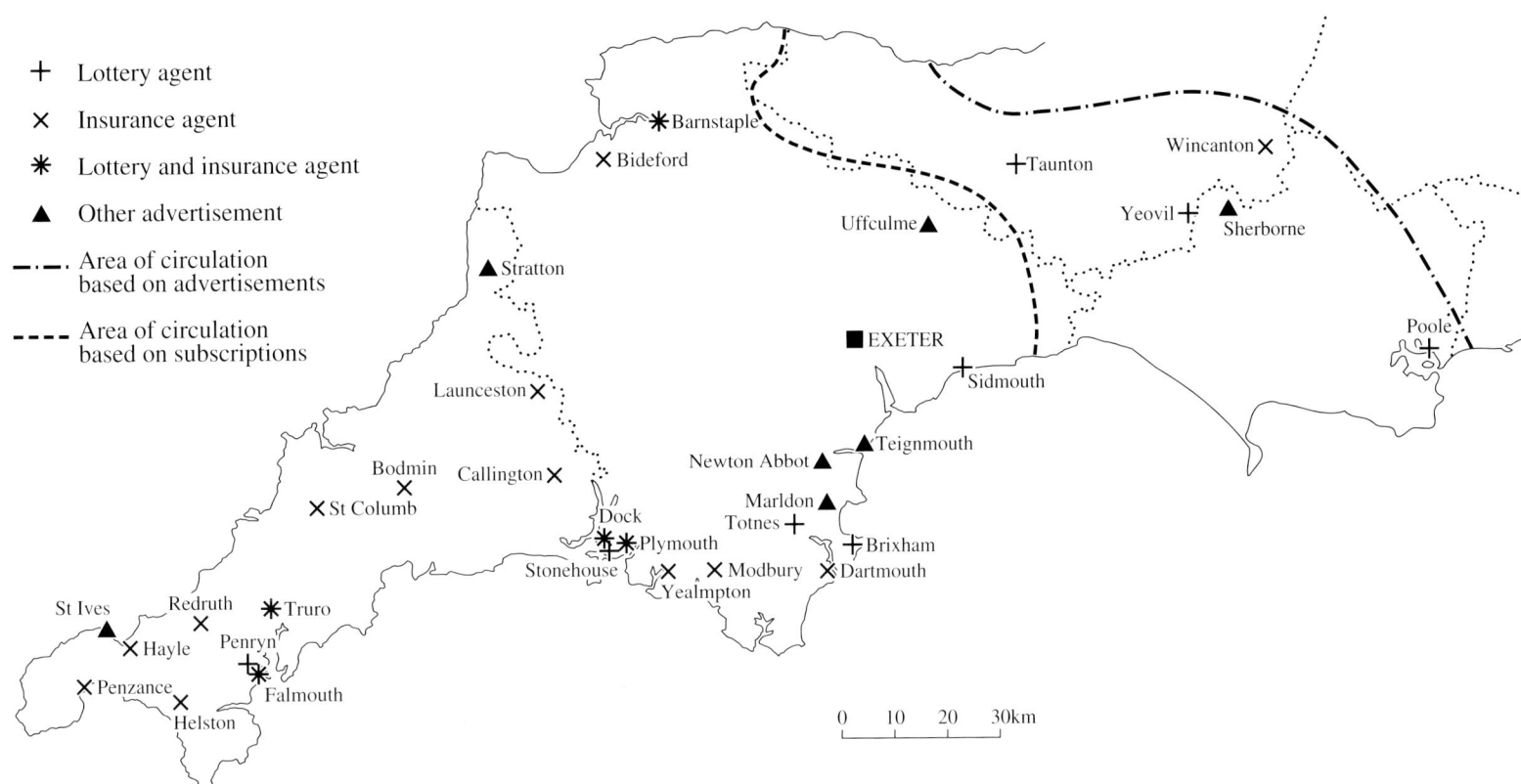

+ Lottery agent

✕ Insurance agent

✳ Lottery and insurance agent

▲ Other advertisement

—·—· Area of circulation
based on advertisements

— — — Area of circulation
based on subscriptions

and Eighteenth Centuries (Plymouth, 1885–91) contains articles which originally appeared in the *Western Antiquary*, listing books, advertisements and other details of booksellers and printers, and also covers much Cornish material. See also Ian Maxted, *A Good Face of Learning: Two Thousand Years of the Written Word in Exeter and Devon* (Exeter, 1999).

On the spread of printing see T.N. Brushfield, 'Andrew Brice and the early Exeter newspaper press', *Transactions of the Devonshire Association*, 20 (1888), 163–214, though this is now outdated for the earliest period. Ian Maxted, *Books with Devon Imprints: A Chronological Checklist* (Exeter, 1989) lists more than 1,000 titles to 1800. This can be supplemented by access to the Eighteenth Century Short Title Catalogue on-line or on CD-ROM. R.A.J. Potts, 'Early Cornish printers, 1740–1850', *Journal of the Royal Institution of Cornwall*, new series, 4 (1963), 264–325, is a detailed historical introduction with a biographical dictionary of printers arranged by town.

On newspapers the *Bibliography of British Newspapers*, general editor C. Toase, *Cornwall* edited by Jean Rowles and *Devon* edited by Ian Maxted (London, 1991) includes full details of surviving files and references to histories. H.R.P. Boorman, *Newspaper Society: 125 Years of Progress* (Maidstone, 1961) includes a chapter on the South West with short histories of individual newspapers.

Fig. 31.4
Engraving of Andrew Brice,
Exeter printer, published for
Barnabas Thorn of Exeter,
21 April 1774.

SOURCE: reproduced by
permission of Westcountry
Studies Library, Exeter
(P&D 32416).

EDUCATION, DISSEMINATION OF KNOWLEDGE, AND LANGUAGE

EDUCATION IN CORNWALL
IN THE
NINETEENTH AND TWENTIETH CENTURIES

L. BURGE AND F.L. HARRIS

The nineteenth century inherited from the eighteenth a provision of education essentially on the principles of the market: teachers of all kinds offered their wares and parents bought as they wished, according to their ability to pay the necessary pence, shillings or guineas. There were schools with endowments, usually very little, sometimes administered by parish or borough, but they could still allow master or mistress to make charges for pupils not entering free by the endowments. The result as a whole was a multitude of schools, from the dame's with a small group of infants in her kitchen to the best of the grammar schools and the (very few) first academies. But while there was some supply in every community, even often in the hamlet, size and quality cannot be measured by titles. Despite the number of schools there was not much effective schooling for working-class children, apart from the children of skilled workers; upbringing and education were on the whole traditionally left to the parents and the church. The apparent lavish provision of schooling shown on **Map 32.1** should be seen in this light.

With new dangers perceived from revolutions abroad and unrest at home, more extensive and strengthened elementary education was seen as an antidote. Philanthropy had also sprung forth: the education of 'the poor' became a public issue. In the second half of the century competition from abroad was seen as a threat and danger to trade and industry, demanding more elementary education in the 'three Rs' and training in special skills. New reforms were introduced to that end.

Reforms came in three stages of thirty to forty years. The first stage, in the earlier decades of the nineteenth century, was marked by action from the churches through their national societies—the National Society for the Education of the Poor in the Principles of the Established Church, and the British and Foreign Schools Society whose provision was on non-denominational, in effect nonconformist, lines. Compared with previous elementary schools, the schools introduced by the Societies were more uniform in their methods and standards, both in religious teaching and in the teaching of reading and writing. Sunday schools were also teaching reading (though not writing), and this contribution, especially as made by the Methodists, provided an alternative access to a degree of literacy.

By the mid-1830s the provision indicated on **Map 32.1** was seen as insufficient, and grant aid from the state was introduced, first on a limited scale for building, but later extended to the cost of training pupil teachers in schools and full-time students in training colleges. The twin aims were better accommodation and more efficient teaching. Progress was made, but increasingly seen to be inadequate to meet national requirements (**Map 32.2**). This map shows that over large areas of the county there was accommodation for 50 per cent or less of Cornish children. The Education Act of 1870 was calculated to provide the remedy—universal literacy through education to the age of 11 or 13. Every locality (normally the parish) was to provide sufficient places in efficient schools for every child whose parents would use the public provision. Where local voluntary bodies could not meet the need, elected boards would be charged with building and controlling schools, their funds supported by local rates. The result in Cornwall was that elementary education came to be supplied by school boards to an extent as great as in any county (**Map 32.3**). The new system was in full working order in the 1880s and the century closed on the double structure. For good measure the new county council began a county plan of technical education in the 1890s which was to be a useful foundation for development in the next century.

With public elementary schools available to all children, the private schools offering some degree of elementary education were greatly reduced. Secondary education, in the tradition of the grammar school, the academy and corresponding schools for girls, was still provided by private enterprise and by the few boarding schools founded mostly with firm religious purpose to meet the demands of middle-class parents. Examples were at Probus, the Cathedral School for boys and the High School for girls at Truro, all on Anglican principles, and the Wesleyan Middle-Class School at Truro.

THE TWENTIETH CENTURY

The twentieth century was to see a full system of secondary education to sixteen, universal, free and compulsory. The Education Act of 1902 made county councils the local education authorities for elementary education, taking over the functions of the school boards, of which there were 101 in Cornwall, and building upon what had been done. The Act also gave the councils the power to promote

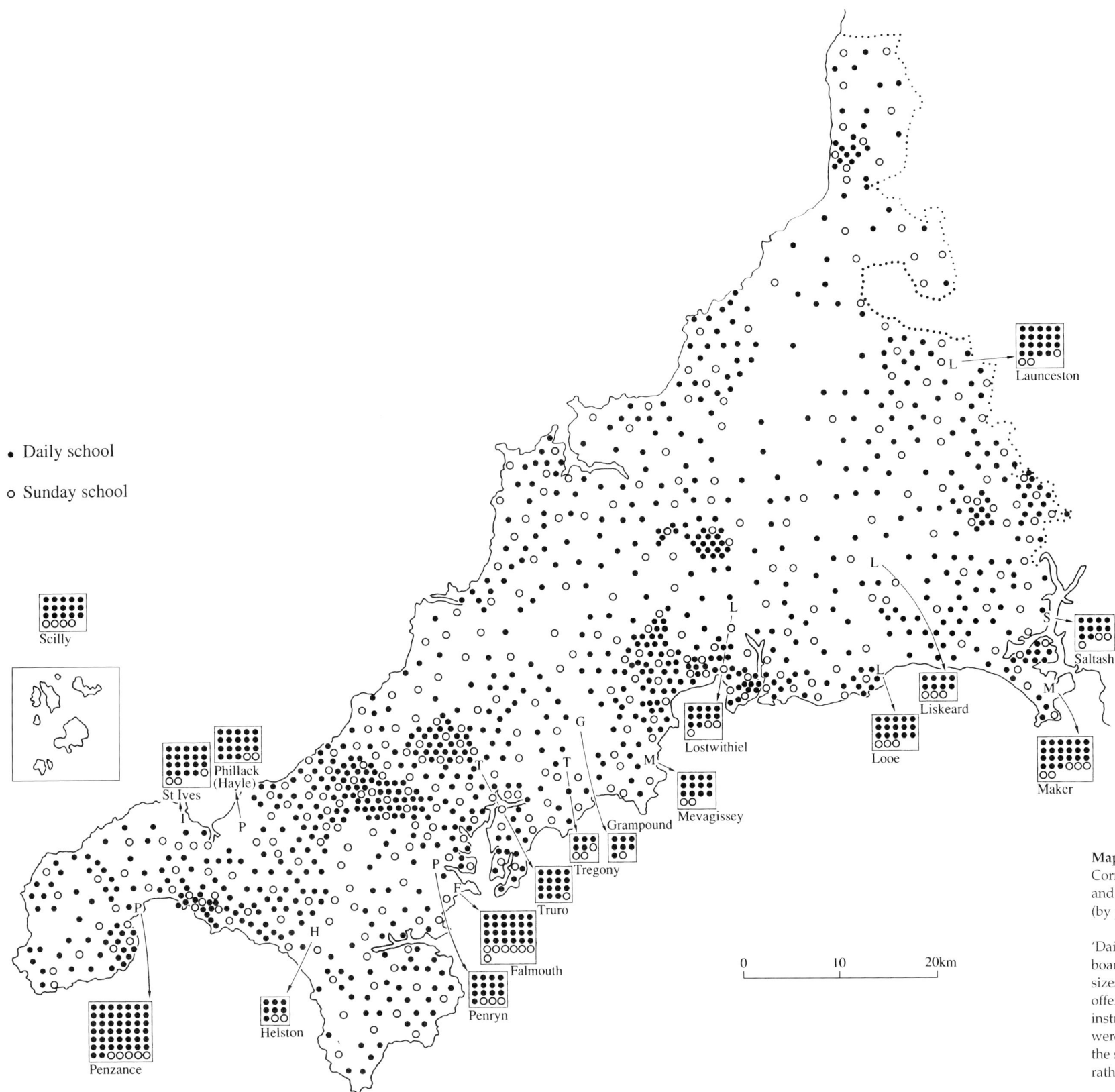

• Daily school

o Sunday school

Map 32.1
Cornwall: daily schools
and Sunday schools, 1833
(by parishes).

'Daily schools' includes a few
boarding schools, all types and
sizes, with the vast majority
offering only rudimentary
instruction. Sunday schools
were on average three times
the size of day schools and had
rather more scholars in total.

SOURCE: *British Parliamentary
Papers*, 1835, XLI: Abstract of
Education Returns, 1833,
96–119.

Map 32.2
Cornwall: grants and the supply of schools, 1870. Also shown is the estimated deficiency in school provision in 1870.

SOURCES: (grant data) *Report of Committee of Council on Education, 1870–1871: Schools Aided by Parliamentary Grants*, 439–41; (efficiency data) Education Department PRO 713 ED2 47–69; supplemented by *West Briton*, 25 September 1873, 6, Abstract of notices for the county of Cornwall; Isles of Scilly Museum Association.

Grants
Public elementary schools which had received grant (at some time) before 1870

B   British

C   Church of England

N   National

P   Parochial

W   Wesleyan

(N)   Former grant now withdrawn

Availability of schooling in the parishes at 1870
Fraction of children for whom accommodation existed in 'efficient' schools

Percentage

	76
	51
	26
	1

No data

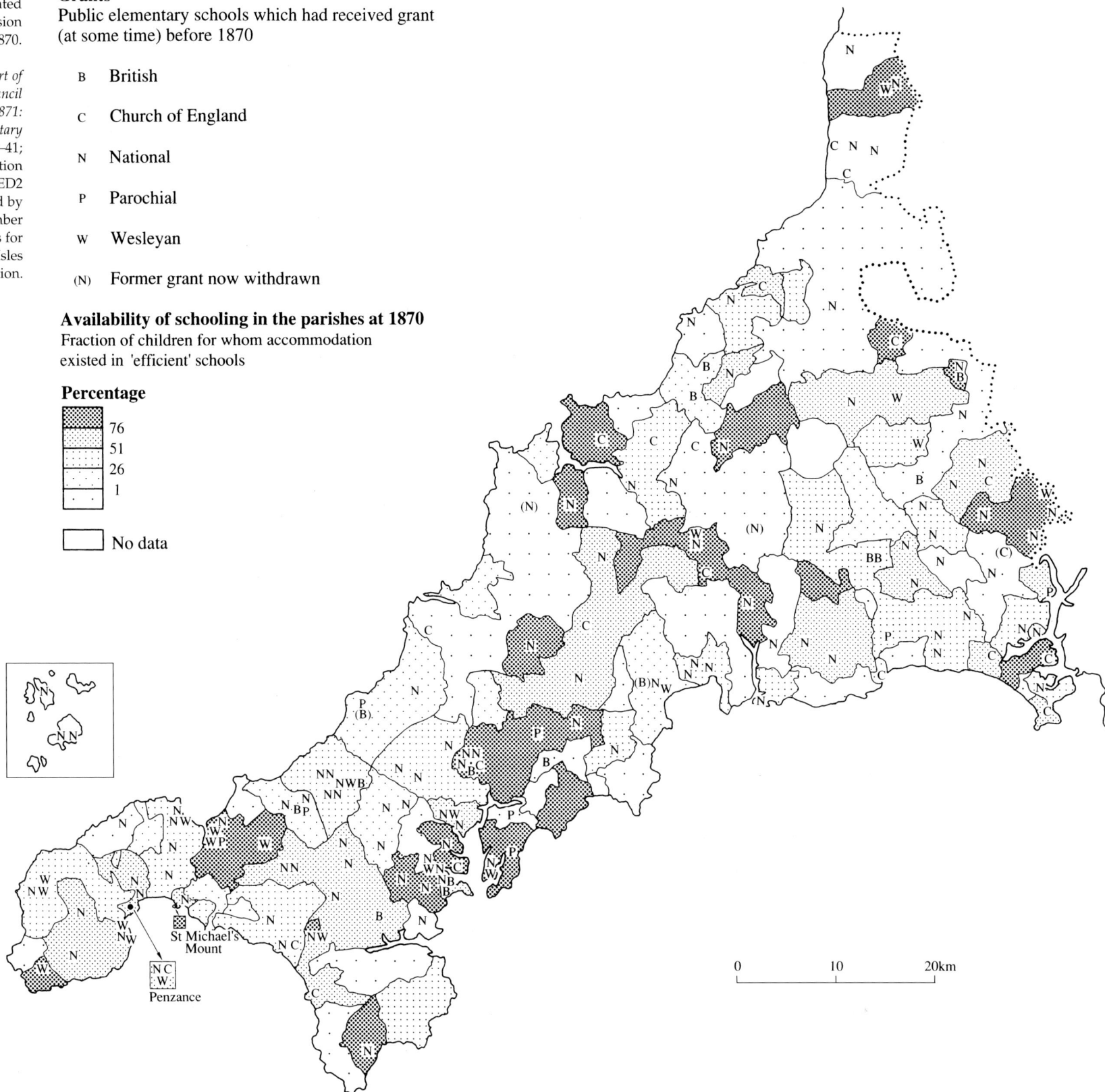

St Michael's Mount

Penzance

0   10   20km

School board district

Public elementary schools

* Board school

N National

B British

C Church of England

P Parochial

W Wesleyan

R Roman Catholic

O Other

Penzance

Helston

Falmouth

0 10 20km

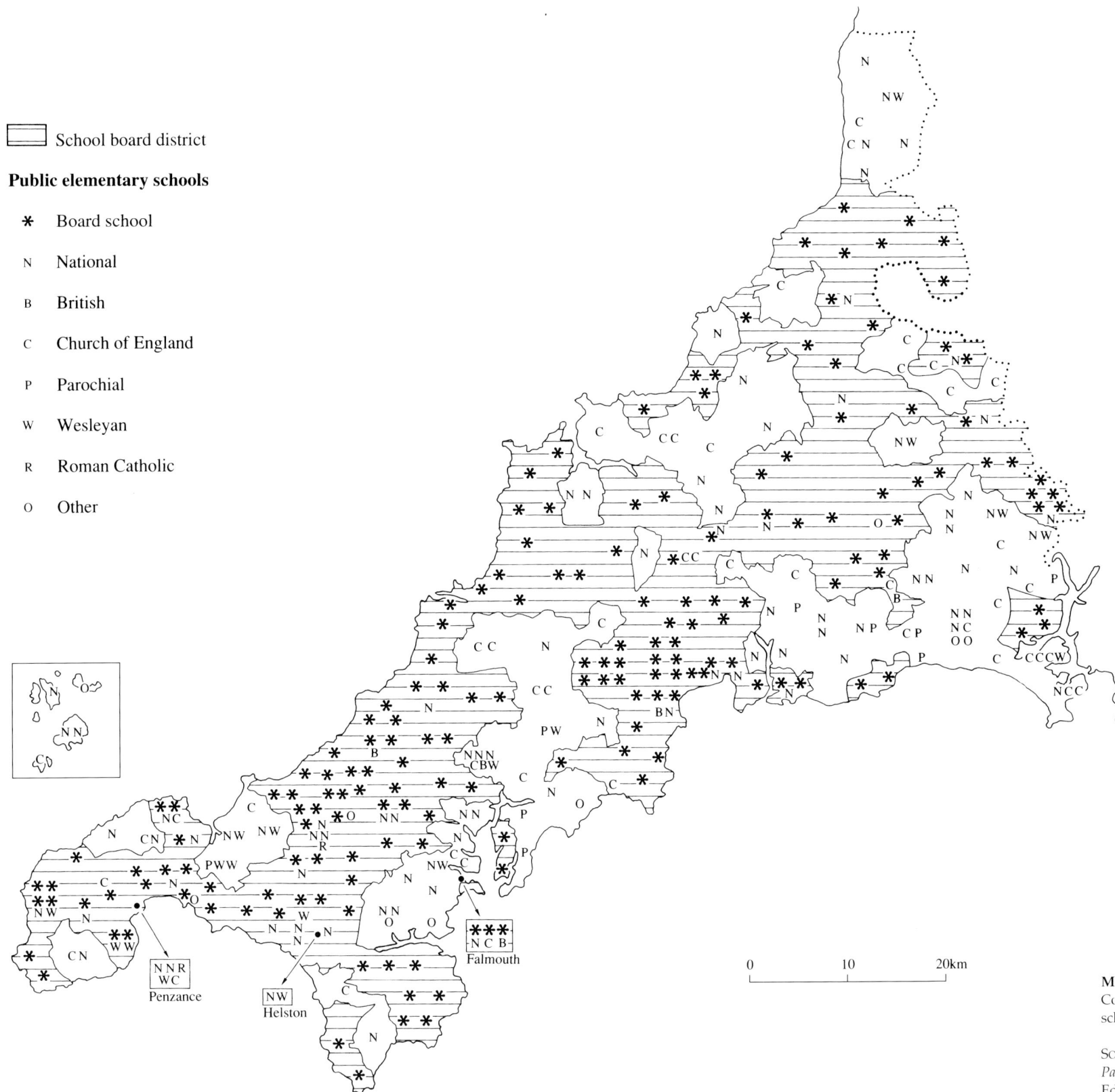

Map 32.3
Cornwall: boards and board schools, 1902.

SOURCES: *British Parliamentary Papers*, 1902, LI: Board of Education Lists of Public Elementary Schools and Training Colleges, 1902–1903, 26–31; Isles of Scilly Museum Association.

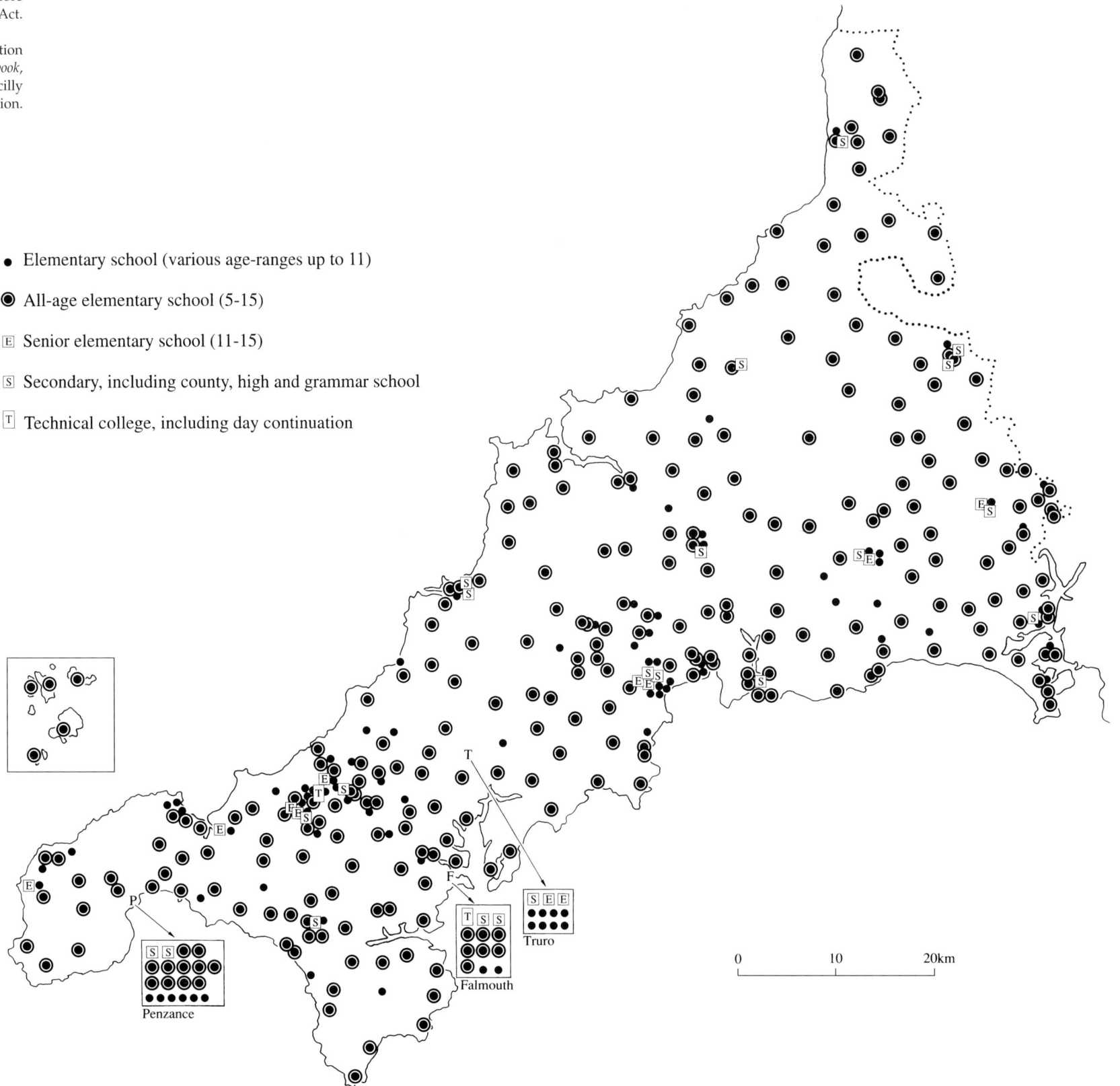

Map 32.4
Cornwall: schools maintained
by Cornwall Local Education
Authority in 1937, i.e. before
the Butler Act.

SOURCES: Cornwall Education
Committee, *Handbook*,
1937–1940; Isles of Scilly
Museum Association.

● Elementary school (various age-ranges up to 11)

◉ All-age elementary school (5-15)

E Senior elementary school (11-15)

S Secondary, including county, high and grammar school

T Technical college, including day continuation

Penzance

Falmouth

Truro

0   10   20km

M Modern school (11–15)

G Grammar school (11–18)

T Technical college (15+)

t Technical college outpost

A Art college (15+)

☐ Comprehensive school (11–18)

0   10   20km

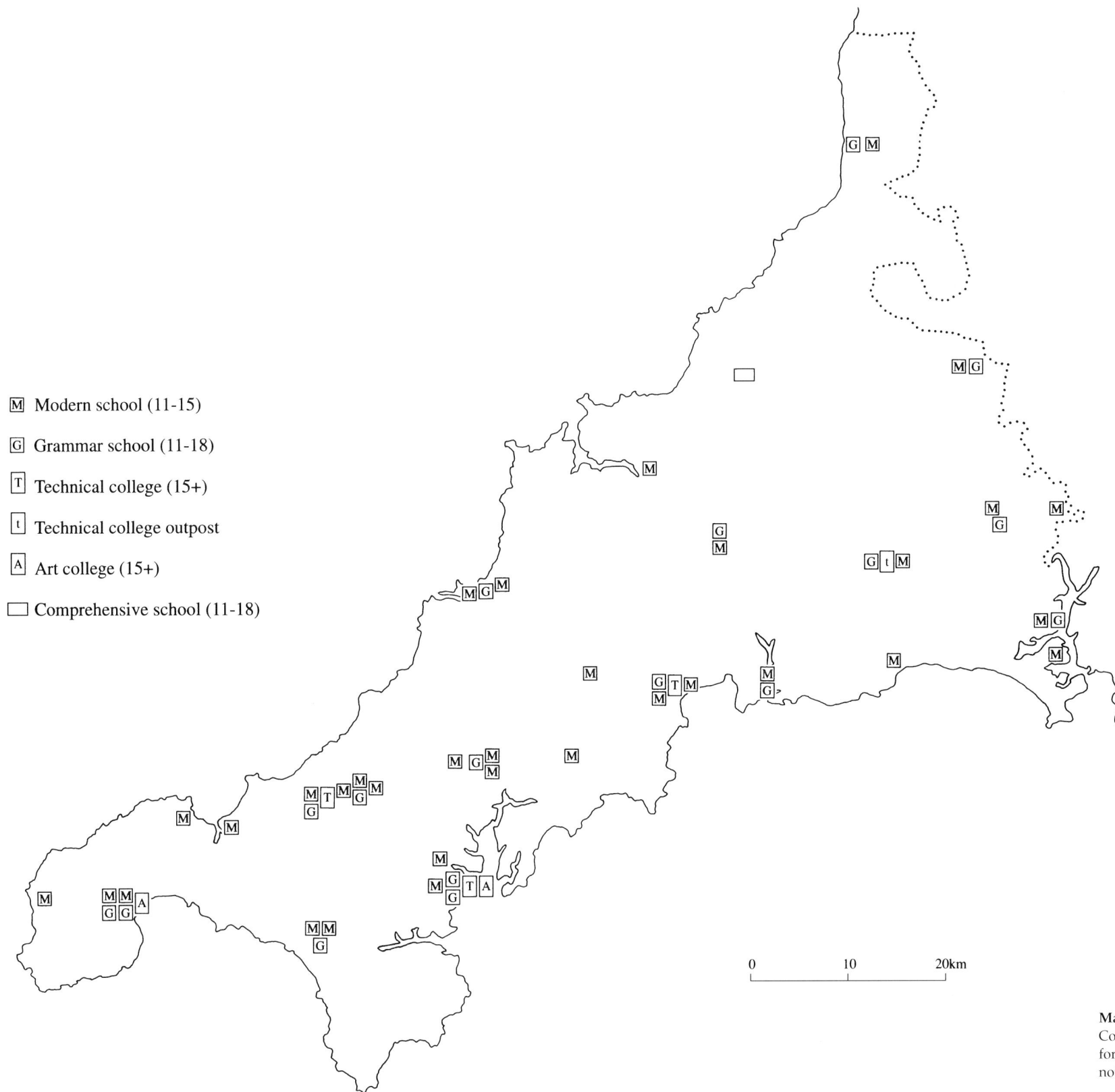

Map 32.5
Cornwall: secondary schooling
for all, 1964 (primary schools
not shown).

SOURCES: Cornwall Education
Committee, *List of Schools*,
September 1964; Isles of Scilly
Museum Association.

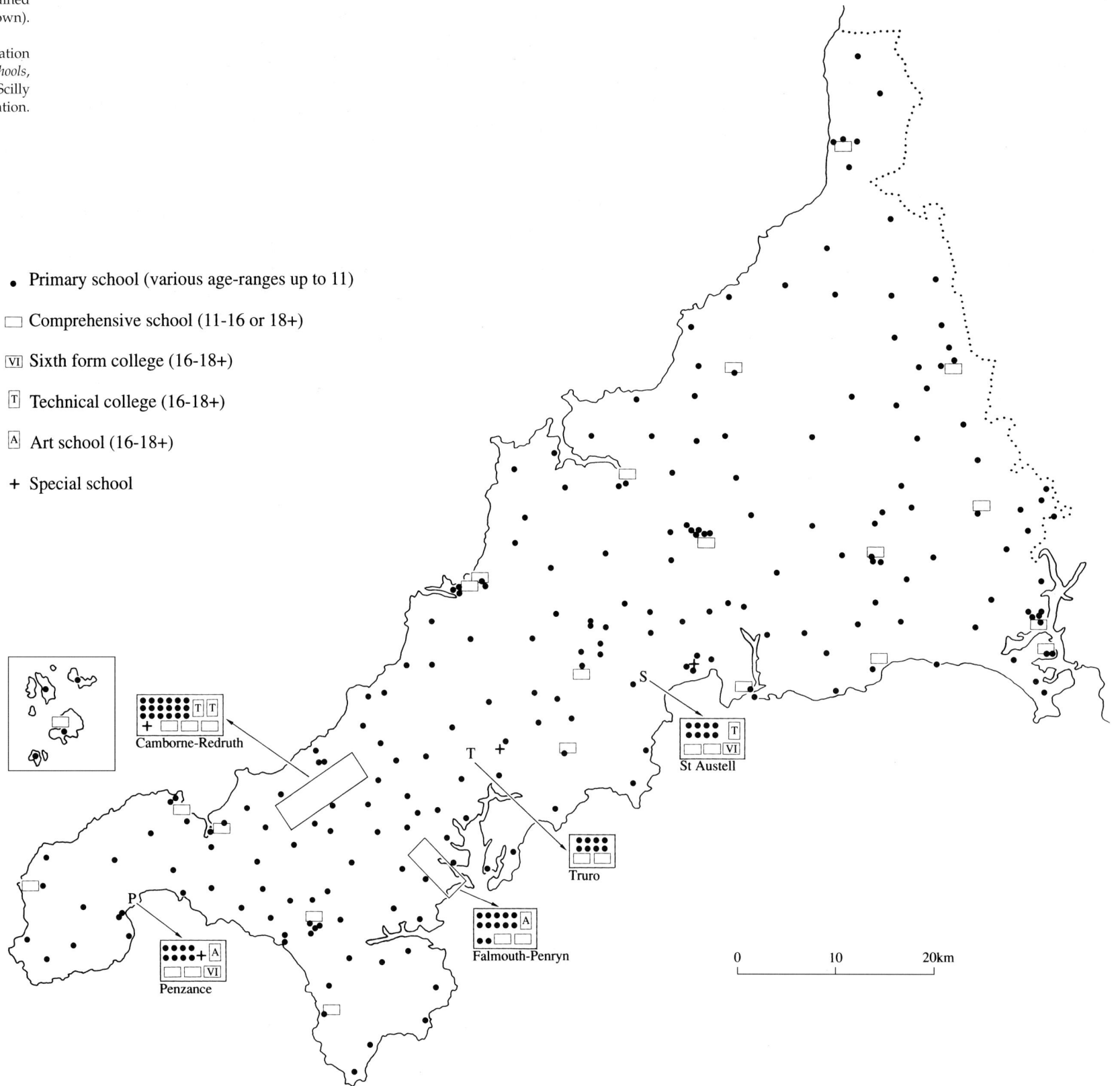

Map 32.6
Cornwall: comprehensives
complete, 1980
(all maintained
schools are shown).

SOURCES: Cornwall Education
Committee, *List of Schools*,
October 1980; Isles of Scilly
Museum Association.

• Primary school (various age-ranges up to 11)

☐ Comprehensive school (11-16 or 18+)

VI Sixth form college (16-18+)

T Technical college (16-18+)

A Art school (16-18+)

+ Special school

Camborne-Redruth

St Austell

Truro

Falmouth-Penryn

Penzance

0   10   20km

secondary education and in two decades the secondary 'county' schools were created to cover every area, catering for fee-paying pupils but with a limited number of free places to be won in competitive examinations. Curriculum and method in the county schools showed some continuity with the long tradition of the old grammar schools.

In the late 1920s, with more progressive ideas in play, the number of free places in the county schools was increased and a few senior elementary schools were formed for pupils aged 11–14. These were steps towards the new conception of Secondary education for all children; they moved away from the former limitation of Secondary to those who could pay, and also from the conception of Secondary as education in a superior field. **Map 32.4** shows the position achieved on the eve of the Second World War. By the end of the war the Butler Act had laid down plans for Secondary education for all children over 11, mainly in the two categories of grammar and secondary modern schools conceived as providing for different classes of ability and need. Selection of children from the primary schools at 11 was to be by examination.

The plan made to this order was implemented in Cornwall by the early 1960s (**Map 32.5**), but before it was completed the issue of comprehensive secondary schools had been introduced. Already, in 1960, one had been set up at Camelford and subsequently the policy of comprehensive schooling was adopted by the county. Reorganization on that basis was undertaken and completed by 1980. Meanwhile a further system of technical education had been constructed with centres at the principal college at Pool in Camborne–Redruth, at Falmouth and at St Austell, and their work and influence were spread over the county by extra-mural extension. **Map 32.6** shows that comprehensives are distributed much as were the post-war secondary modern schools. Primary schools are distributed not very differently from the elementary schools of a century before, often in identical premises. Nineteenth-century intentions for more neighbourhood infant schools have not been widely fulfilled.

NOTES ON THE MAPS

On the maps for 1833, 1870 and 1902, schools are shown by symbols placed schematically, usually within parish units, rather than being precisely located. On the maps for 1937 and 1980 each school is correctly placed, as far as the size of the symbol permits. Circular symbols are used for the 1833 schools and in general for elementary schools. The dot in a circle (1937) for the all-age elementary school is to remind us of the two stages under one roof that were eventually to be separated. Progress to secondary in the modern sense of post-primary is marked, when it appears, by the use of rectangular forms: squares first, for the various 11+ schools, stretching into horizontal oblongs for comprehensives and upright ones for the colleges of technology and art and their outposts.

EDUCATION, DISSEMINATION OF KNOWLEDGE, AND LANGUAGE

EDUCATION in DEVON
IN THE
NINETEENTH and TWENTIETH CENTURIES[1]

ROGER SELLMAN

ELEMENTARY SCHOOLS IN 1833 (**Map 33.1**)

The first date at which it is possible to obtain a list of all schools is 1833, when a return was made to parliament. However, this return, made by (often absentee) incumbents, cannot be fully relied on for accurate detail; and, as the printed report states, 'the maintenance of schools presents details incapable of exact classification'. All schools shown on **Map 33.1** had some element of outside support, but the extent to which this involved free places varied greatly. A minute endowment providing for two or three free places might be the only distinction between a school shown and a neighbouring dame school.

Endowments, if in the usual form of a fixed rent-charge, had declined in value; but those in land might well have kept up with inflation. Many schools were partly supported by 'subscriptions in aid of fees', some had subventions from the local squire and/or incumbent (which could vary or cease), some received money from parish funds, or a combination of two or more of these sources. Some schools were already connected with the National Society for Promoting Anglican Education, and a smaller number were 'Lancasterian' or British, officially undenominational but usually with nonconformist connections. Both Societies encouraged the 'monitorial' system whereby older pupils were instructed so that they could pass on their lessons to small groups and ease the burden on the teacher. It was difficult for this to be effective in rural schools.

As yet there was no system of teacher-training, apart from apprenticeship to an existing teacher which applied only in a few larger schools. Length of school attendance was still limited in rural parishes by parish apprenticeship at the age of 9 or 10, and in any case seldom exceeded three years. Better-supported schools generally offered all 'three Rs', but many attempted little more than reading, with the express object of reading the Bible.

Some schools, especially National schools, were 'day and Sunday', and many parishes without a day school had a Sunday school, mostly taught by unpaid volunteers and requiring very little expenditure. Nonconformists, seldom able in rural areas to support a day school, were well represented on Sundays. Sunday schools were open to children required to work on weekdays, and so had one notable if limited advantage. Rare indeed in the countryside were schools such as Sandford and Milton Abbot, handsomely built and supported by wealthy and interested patrons and offering a much better education than the norm. A few country schools such as Walkhampton and Silverton had land endowments sufficient to provide a similar standard, but these were apt to suffer from the extreme difficulty of displacing a master who proved negligent or inefficient—or, since the endowment made no provision for pension, too old to cope. Charity schools of the 'Bluecoat' type, providing clothing and often board as well as teaching, had been founded one or two centuries previously in Barnstaple, Crediton, Exeter, Plymouth, South Molton, Tiverton and Torrington, and most survived as such in 1833. Several other endowed schools also provided clothing for scholars.

While **Map 33.1** may give the impression of widespread educational facilities in 1833, this must be qualified by the fact that in most of the schools shown, the teaching was in every sense 'elementary', attendance was irregular, and teachers were often themselves of minimal education and capacity to teach. Classification of schools by type and size is hardly feasible at this date, particularly in towns where there were separate infants schools and separate schools for boys and girls; but in general it may be assumed that town schools were much larger, and length of attendance greater, and therefore that the map understates school provision in towns (and tends to overstate it in villages).

SCHOOLS ON THE GRANT LIST IN 1870 (**Map 33.2**)

Grants towards the cost of building schools were made by the Treasury from 1834 to 1840, but without assurance of subsequent upkeep. Only from 1840 were they made by the new Committee of Council for Education and came with assured maintenance in return for an acceptance of inspection. Her Majesty's Inspectorate (HMI) then began to visit schools and report deficiencies, and as a result positive steps were taken to improve the quality of education. State salaries were offered for pupil-teachers where HMI declared schools suitable for the apprenticeship teachers' certificate, examinations were introduced, and salary supplements were paid according to grade of certificate. Queen's scholarships to training colleges

Fig. 33.1
Lithograph by Maclure,
Macdonald and Macgregor of
Plymouth New Grammar
School, c.1830.

SOURCE: reproduced by
permission of Westcountry
Studies Library, Exeter
(Somers Cocks 2166).

inspectors reached any schools outside HMI's orbit, or which HMI was prevented from visiting through increasing obligations elsewhere. One result in Devon was the establishment of the Diocesan Training College (St Luke's), which sent out its trainees to schools in the county. Building grants to supplement those of the Committee of Council were also made and small but useful ones for books and equipment and towards the first-year salary of a certificated teacher, so as to qualify more schools for government grant.

These developments greatly increased the efficiency of those schools which had relatively good local support, but inevitably bypassed the majority of small rural schools where funds were so meagre (and often premises so poor) as to debar them from state assistance. There was, nevertheless, a steady increase in the number on the grant-list, until the famous 'Revised Code' of 1862 for a time checked the process. This abolished state salaries for pupil-teachers and salary augmentation for the certificated, and some Devon schools left the grant-list as a result, while the number and quality of male pupil-teachers and college trainees fell. From this time, the annual grant depended on individual passes in reading, writing and arithmetic, and was the only state payment apart from building grants. At first, too, the amount paid could not exceed the sum raised locally, from which endowment was excluded, with the result that some endowed schools were deterred from joining the list.

became available to pupil-teachers with a good final examination result, and in 1853 a capitation grant was offered for rural schools (and in 1856 for towns) where the headteacher was certificated, support was adequate, and HMI reported the school efficient.

Meanwhile, on the initiative of the National Society, diocesan boards of education were established in 1839 with the purpose of providing teacher-training, inspecting and advising schools, and promoting union with the Society. Diocesan

Fig. 33.2
Detail from engraving by
W. Spreat of St Sidwell's
Parochial School, 1854.

SOURCE: reproduced by
permission of Westcountry
Studies Library, Exeter
(Somers Cocks 1025).

The effect of the Revised Code was particularly hard on smaller schools, which required more income per head to reach comparable efficiency with the larger, and the six standards for examination (corresponding roughly with ages 7–13) seldom saw any above the fourth in country schools. Another result was concentration on the 'three Rs' at the expense of all else, though later Code changes went some way to correct this by introducing grant-earning 'specific' and 'class' subjects.

After the first shock, the number of schools on the grant-list again began to rise, and those shown on **Map 33.2** for 1870 may in general be taken to indicate the extent of 'efficiency' as then understood. Evident is the generous provision at Tiverton, largely due to the industrialists Heathcoat and Brewin, who preferred connection with the British Society. Plymouth and Devonport included naval and military schools.

ELEMENTARY SCHOOLS IN 1902 (**Maps 33.3 and 33.4**)

The Act of 1870 revolutionized state attitudes to the provision of schooling for the mass of the population. Previously, since 1833, it had grant-aided successful local effort but the great mass of rural schools had too little local support to come within reach of state aid. Compulsion was introduced in 1870, requiring existing voluntary schools to provide sufficient places for their district, to have a certificated headteacher, to work by the 'Code', and to submit to annual inspection. Otherwise a school board had to be elected, with power to raise a rate to meet the costs of building and maintenance, and to operate a school or schools to the same level of 'efficiency'. In towns, the board solution in general worked well but in the countryside the parish basis often meant a board of farmers with little interest in the education of their future employees and with much concern to keep down the rates. Some village boards had a respectable record, but others a deplorable one—as reports by HMI made very clear.

The great majority of rural Devon school boards were formed by compulsion, some four or five years after the Act, when the Department lost patience with local procrastination. The provision of schools took a further two or three years, and in rural areas the implementation of the 1870 Act took much longer than originally expected. But eventually 119 existing village schools were taken over by boards, and seventy-one of them given new buildings, while forty-seven new village or hamlet schools were founded (**Map 33.3**). The threat of the imposition of a board compelled managers of voluntary schools, if they were to survive as such, to meet the Department's requirements; and seventy-four new voluntary school buildings appeared in Devon villages with thirty entirely new foundations. (**Map 33.4**) The proportion of rural voluntary schools working by the Code and under inspection rose from 28 per cent in 1870 to 95 per cent by 1879.

Boards, and voluntary managers, varied greatly in the salaries they offered for schools of the same size, and so in their

ability to attract and keep headteachers of quality. Some such became respected and long-remembered pillars of the local community as well as the school, while other places had only a succession of young ex-pupil-teachers with provisional certificates who promptly left, or of older drifting failures who had to be dismissed. City and town school boards, particularly those of Plymouth and Exeter, mostly had a good record and effectively used their power to compel attendance, but in rural areas compulsion, whether by boards or the union, was entirely ineffective until the county took over in 1903. Even those authorities ready to prosecute soon abandoned the attempt when rural magistrates proved unwilling to do more than issue 'attendance orders', and when farmers were more concerned with cheap child-labour and parents with the small sums it raised to supplement their exiguous income.

In cities and towns, voluntary schools continued to flourish and both Barnstaple and Torquay were so liberally provided with them that the former, though having a school board, never had a board school, and the latter never had a board. In towns, nonconformity had financial backing and could continue to support schools, but in rural areas, though supporters might be numerous, they were mostly too poor to meet the 1870 requirements, and consequently nearly all surviving voluntary schools were 'Church'.

By 1902 there had undoubtedly been great educational advance, but far more in the towns than in the villages. Agricultural depression affected both the ability and the willingness of country parishes to pay for schools, and rural children were, with some exceptions, much less well provided for than their urban counterparts. A place in an at least nominally 'efficient' school had been provided for every child in the county, but an essential first step towards evening the urban–rural disparity was the wider county rating basis and the professional administration provided by the Act of 1902.

By this time educational practice had moved away from the passive and regimented mass-instruction of the nineteenth century, and required more space and more diverse facilities for a 'finding-out' approach. Corporate activities also needed reasonable numbers to be successful; the cost of running schools grew with the provision of meals facilities and staff, and with an entirely new dimension in equipment and a good reference library—all more easily provided and more fully used in a larger school. Reorganization to give the country child as good a primary school as his or her urban counterpart inevitably involved closures and grouping into larger units, either by enlarging and improving the better premises in one village or, preferably, by building new area schools with first-class facilities. An outstanding example is Stokenham, with a hall, and each classroom having access to separate 'quiet' and 'practical' rooms, which replaced five small existing schools (and another already closed).

Fig. 33.4
Barnstaple Grammar School, an anonymous etching, *c.*1820.

SOURCE: reproduced by permission of Westcountry Studies Library, Exeter (Somers Cocks 84).

RURAL SCHOOLS CLOSED, 1902–1983 (**Map 33.5**)

Nineteenth-century building effort, culminating in the post-1870 surge when all but the most minute parishes were required to provide their own school, endowed Devon with hundreds of small village and hamlet schools intended to cater for the whole 3–13 age group. Agricultural depression and associated rural depopulation after about 1879 meant that many schools had already declined in attendance, and a few even closed, before county administration in 1903. The process continued after this with some parishes eventually falling to below half their 1871 population, and many more by over a third. Moreover, those leaving rural parishes were mostly the young who had, or would have, children. Later recovery in agriculture was accompanied by mechanization, and the trend was not reversed except where villages became dormitories serving nearby towns or had some other source of local employment.

Meanwhile, educational and social developments entirely changed the circumstances for which school provision had been designed in the 1870s. Road transport made it no longer necessary to have a school within walking distance of every child, and rising living standards required water-supply, lighting and ventilation, clothes-drying facilities and cloak rooms, and some improvement on bucket lavatories. In 1925 the Board of Education 'blacklisted' 184 rural schools in Devon—over a third of the total. Though this number was reduced to fifty by 1939, by closure or improvement, contemporary concentration on new schools for seniors diverted resources from villages. It had also, where carried out, cut already diminished rolls by about a third; the raising of the entry age to 5 had further effect.

In 1945 a survey showed 103 schools with under twenty-five on roll, and therefore 'one-teacher', and forty-five of these were still 'all-age' (5–14)—including rolls as low as five or seven. Over thirty still had only one room; some had to be shut when the well ran dry, and about a quarter still relied on buckets.

GRAMMAR SCHOOLS (**Maps 33.6 and 33.7**)

Grammar schools had been founded in most Devon towns in the sixteenth and seventeenth centuries, and the bishop's licensing records show others at an early date in places as small as Thorverton. But by the early nineteenth century nearly all, in the form of a single-handed parson offering only Latin, had declined from the smallness of their endowment and the lack of local interest in classical education. An outstanding exception was Blundell's at Tiverton, which had developed into a boarding school for the sons of gentry and professional men from a distance, until a lawsuit in mid-century attempted to confine it to local boys—who soon proved not to want classics. New schemes by the Charity Commission allowed it again to take boarders, and by the time of the Endowed Schools Report of 1868 it was again

Map 33.1
Devon: elementary schools in 1833 with some external support besides fees (excluding those solely 'at parent's expense').

SOURCE: *British Parliamentary Papers*, 1835, XLIII: Summaries of Returns to an Address of the House of Commons on Education, 1833.

0   10   20km

Barnstaple

South Molton

Bideford

Tiverton

Holsworthy

Honiton

Crediton

Okehampton

Exeter

Seaton

Exmouth

Tavistock

Newton Abbot

Torquay

Brixham

Plymouth

◇ National

△ British/Nonconformist

○ Roman Catholic

● Others (mostly church-affiliated)

Map 33.2
Devon: schools on the grant list, 1870.

SOURCE: Committee of Council for Education, minutes and reports.

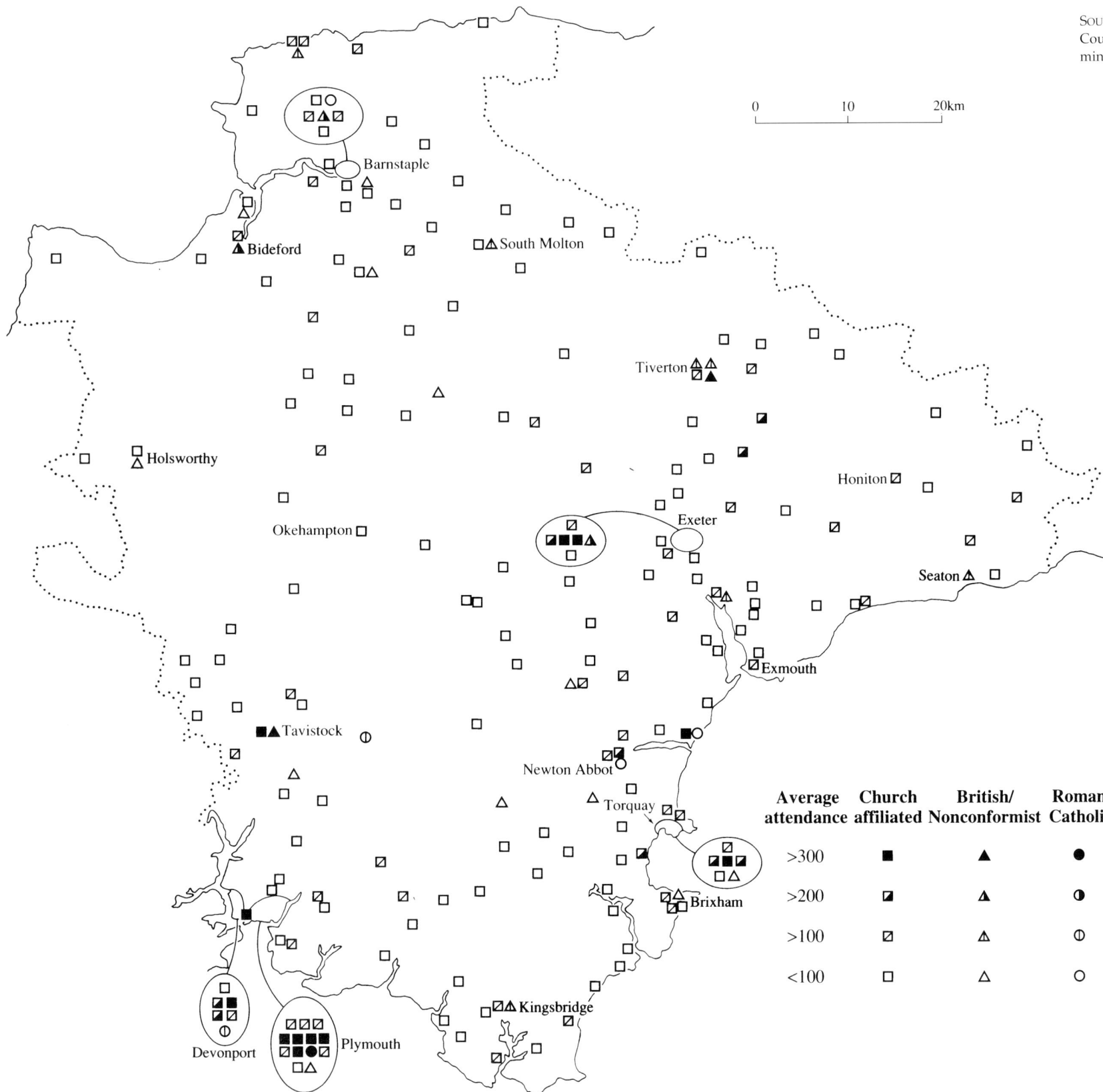

0   10   20km

Barnstaple

Bideford

South Molton

Holsworthy

Tiverton

Honiton

Okehampton

Exeter

Seaton

Exmouth

Tavistock

Newton Abbot

Torquay

Brixham

Kingsbridge

Devonport   Plymouth

Average attendance	Church affiliated	British/ Nonconformist	Roman Catholic
>300	■	▲	●
>200	◪	▲	◐
>100	◩	△	◑
<100	□	△	○

Map 33.3
Devon: school boards
and board schools, 1902.

SOURCE: Committee of
Council for Education,
minutes and reports,
directories, school board
minutes.

0   10   20km

Mortehoe
West Down
East Down
Marwood
Bratton Fleming
Braunton
Heanton
Punchardon
Stoke Rivers
Barnstaple
Instow
Bishops Tawton
North Molton
Westleigh
East Anstey
Hartland
BIDEFORD
Horwood & Newton Tracey
Clovelly
Littleham
Yarnscombe
Bishop's Nympton
Bampton
Clayhanger
Woolfardisworthy
Parkham
Buckland
Brewer
Mariansleigh &
Romansleigh
Rose Ash
Welcombe
Torrington
Clayhidon
Putford
Beaford
Chulmleigh
Worlington
Uffculme
Culmstock
Bradworthy
Chawleigh
TIVERTON
Hemyock
Sutcombe
Milton
Damerel
Shebbear
Washford
Pyne
Halberton
Willand
Winkleigh
Lapford
Cadeleigh
Coleridge
Cheriton Fitzpaine
Upottery
Hatherleigh
Zeal Monachorum
Shobrooke
Membury
Clawton
Halwill
Sampford Courtenay
Bow
Dalwood
North Tawton
Crediton
Newton St Cyres
Shute
Axminster
Inwardleigh
Colebrooke
Ashwater
OKEHAMPTON
Spreyton
EXETER
Ottery St Mary
Germansweek
Hittisleigh
Aylesbeare
Musbury
Broadwoodwidger
Cheriton Bishop
Tedburn
Southleigh
Colyton
North Petherwin
Bratton Clovelly
South Tawton
Drewsteignton
Ide
Alphington
St Giles on
the Heath
Throwleigh
Dunsford
Exminster
TOPSHAM
Lydford
Bridford
Moretonhampstead
Christow
Ashton
Kenton
EXMOUTH
Marytavy
Lustleigh
Hennock
Milton Abbot
Widecombe
DAWLISH
Tavistock
Whitchurch
Sampford Spiney
Highweek
TEIGNMOUTH
Shaldon
Ashburton
Stokeinteignhead
Denbury
Ogwell
Coffinswell
Shaugh Prior
BUCKFASTLEIGH
Abbotskerswell
Ipplepen
Broadhempston
DEVONPORT
South Brent
Totnes
PAIGNTON
STONEHOUSE
Ivybridge
Ugborough
Stoke Gabriel
Ermington
BRIXHAM
Plymstock
Halwell
Blackawton
Holbeton
Modbury
DARTMOUTH
Kingswear
Wembury
'Erme & Avon'
Stoke Fleming
Sherford
Slapton
PLYMOUTH
Kingsbridge
Stokenham
Chivelstone

Schools with average attendance

>300   ■

>200   ▲

>100   ●

<100   ·

Boards with total average attendance

>5000   PLYMOUTH

>2500   EXETER

>400   EXMOUTH

>50   Bow

<50   Halwell

Boards formed after 1875 shown with
names underlined

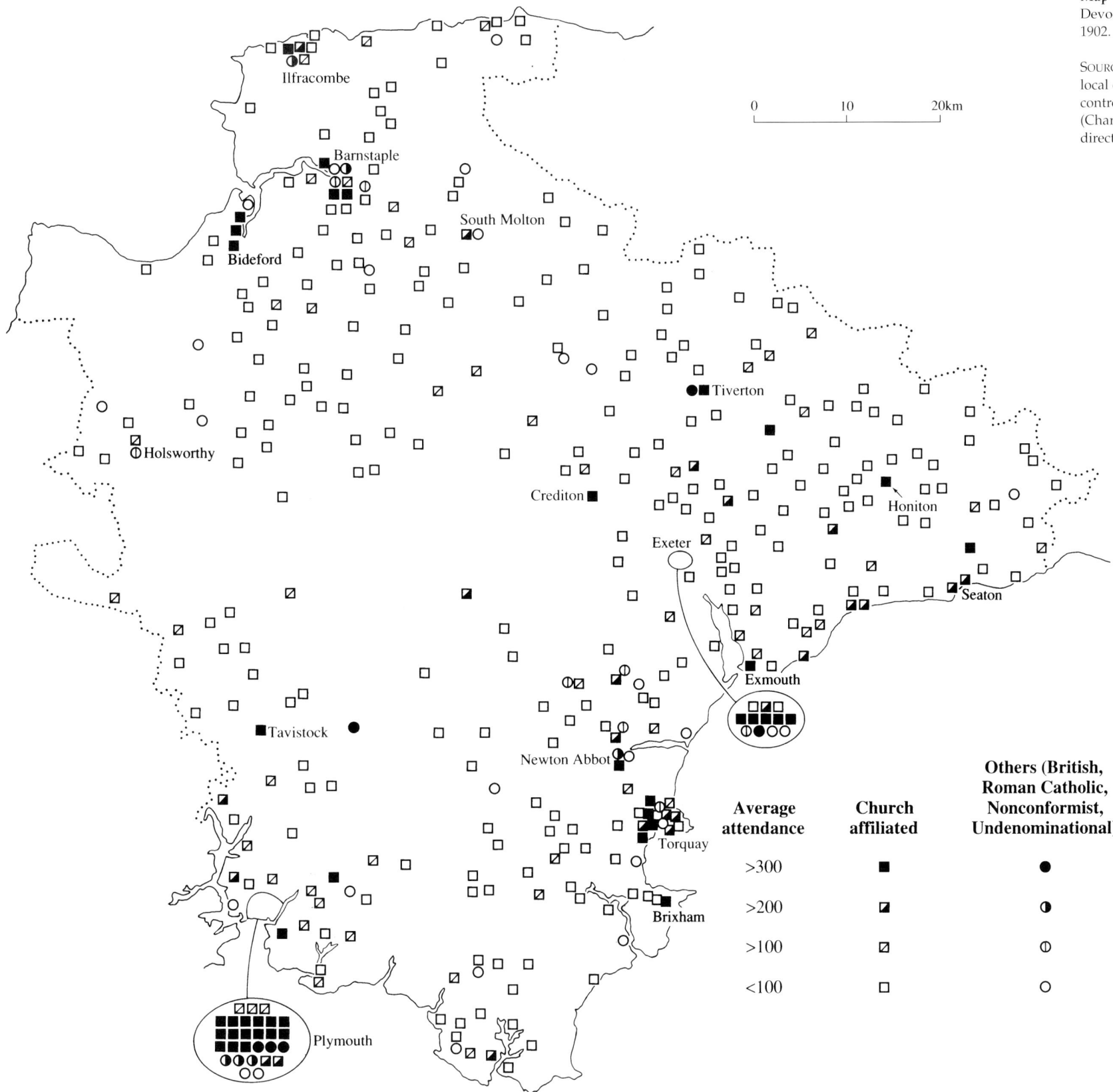

Map 33.4
Devon: voluntary schools, 1902.

SOURCE: list of schools under local education authority control by the Act of 1902 (Charity Commission) directories.

Map 33.5
Devon: rural schools
closed, 1902–1983
(omitting amalgamations
in one village and schools
for infants only).

SOURCE: Committee of
Council for Education,
minutes and reports.

0   10   20km

Countisbury

Lee Hele   Martinhoe   Barbrook
Mortehoe   Mill   Brendon

East Down

Challacombe

Loxhore

Stoke Rivers

Heasley Mill

Bickington   Twitchen   West Anstey
Westleigh   Molland

Arlington   Travellers Rest   Morebath
                                                        Clayhanger
Elmscott   Buck's Mill   Littleham   Atherington   Satterleigh   Newton   Knowstone
        Holwell   Rose Ash   Oakford   Huntsham
Welcombe   Frithelstock   Chittlehampton   Meshaw   Stoodleigh   Cove   Hockworthy
        St Giles   Roborough   Loxbeare   Clayhidon
Meddon   Little Torrington   Chevithorne
Bulkworthy   Peters Marland   Ashreigney   Templeton   Withleigh   Ash Thomas
Milton Damerel   Huish   Hollacombe   Cruwys Morchard   Puddington   Sheldon   Smeatharpe
Pancrasweek   Wembworthy   Cadeleigh   Butterleigh   Yarcombe
        Thornbury   Iddesleigh   Coldridge   East Village   Dunkeswell   Luppitt
Chilsworthy   Sheepwash   Bondleigh   Cadbury   Monkton   Membury
                                        Stockleigh Pomeroy   Combe Raleigh   Cotleigh
        Jacobstowe   Sampford   Colebrooke   Shobrooke   Buckerell
        Beaworthy   Courtenay   Rewe   Talaton   Gittisham
Tetcott   Hookway   Northleigh
        Germansweek   Hittisleigh   Upton Pyne   Poltimore
        Bratton   Whiddon Down   Whitestone   Marsh Green   Southleigh   Axmouth
        Clovelly   Throwleigh   Drewsteignton   Sowton   Salcombe   Rousdon
                                        Farringdown   Regis   Colaton
Stowford   Dunchideock   Exton   Raleigh
        Coryton   Bridford   Ashton   Powderham
Kelly   North Bovey   Trusham
                Manaton   Lustleigh   Ashcombe   Mamhead
                        Postbridge   Ideford   Luton
        Dartmeet   Bickington   Teigngrace

                By 1945   **Post-1945**

        Sampford   Buckland
        Spiney   Holne   East Ogwell
        Sheepstor   Coffinswell   ●   ○   Village
Bere Ferrers   Staverton
        Wotter   Lee Moor   Dean Prior   Littlehempston   ▼   ▽   Hamlet
        Harford   Avonwick   Rattery
                North Huish   Harberton
        Lee Mill   Cornworthy   Ashprington
        Brownston   Halwell   Dittisham
        Noss Mayo   Kingston   Woodleigh   Strete
        Ringmore   Bigbury   Goveton   Slapton
                        Sherford
        Galmpton   Chivelstone

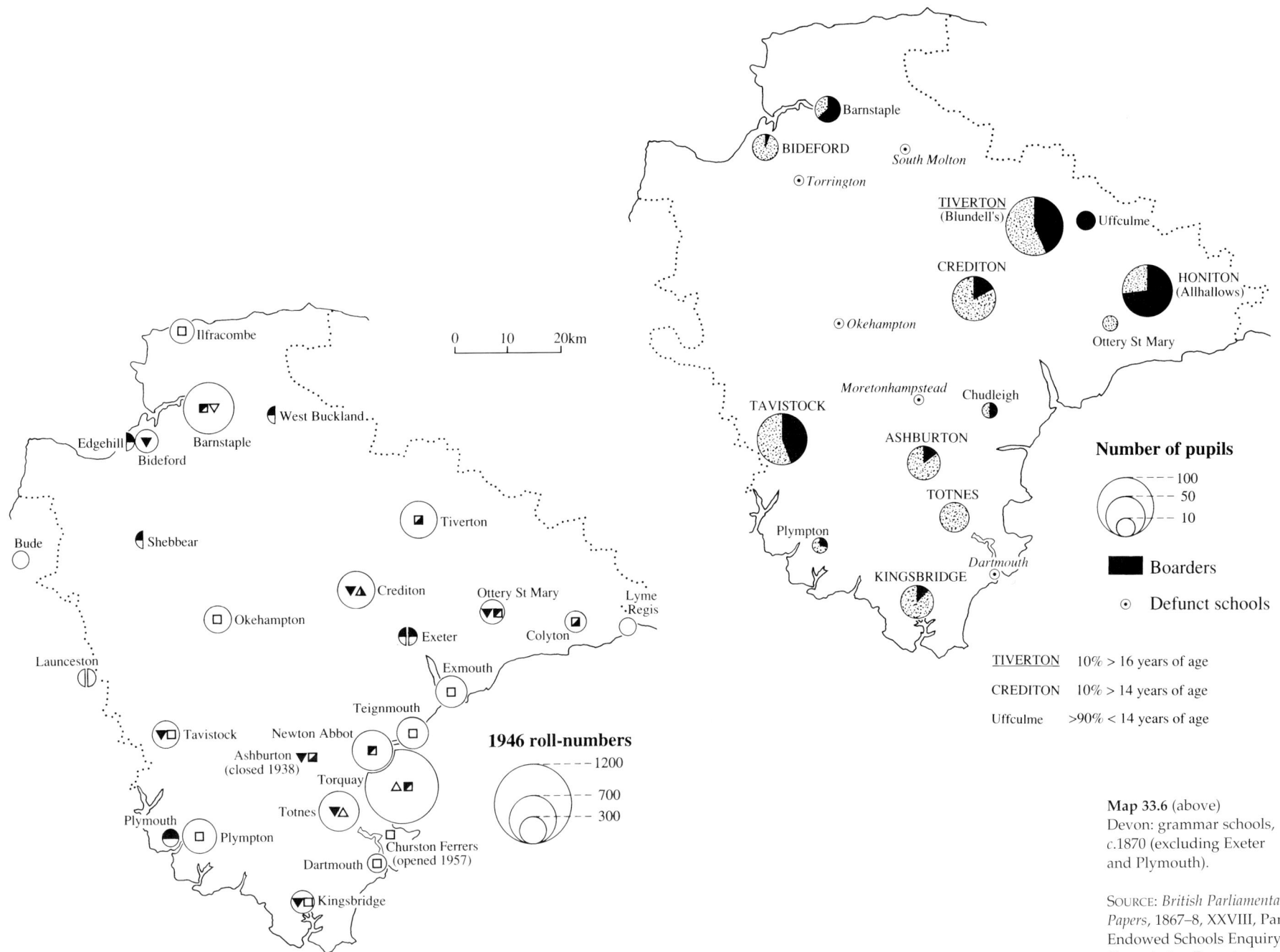

Barnstaple

BIDEFORD

South Molton

Torrington

TIVERTON
(Blundell's)   Uffculme

CREDITON

HONITON
(Allhallows)

Okehampton

Ottery St Mary

Moretonhampstead   Chudleigh

TAVISTOCK

ASHBURTON

TOTNES

Plympton

Dartmouth

KINGSBRIDGE

Number of pupils

— 100
— 50
— 10

◼ Boarders

⊙ Defunct schools

TIVERTON   10% > 16 years of age

CREDITON   10% > 14 years of age

Uffculme   >90% < 14 years of age

Map 33.6 (above)
Devon: grammar schools,
c.1870 (excluding Exeter
and Plymouth).

SOURCE: *British Parliamentary
Papers*, 1867–8, XXVIII, Part 1:
Endowed Schools Enquiry
Commission Report.

Ilfracombe

West Buckland

Edgehill
Barnstaple
Bideford

Bude

Shebbear

Tiverton

Crediton

Ottery St Mary

Lyme
Regis

Okehampton

Exeter

Colyton

Launceston

Exmouth

Teignmouth

Tavistock

Newton Abbot

Ashburton
(closed 1938)

Torquay

Plymouth

Totnes

Plympton

Churston Ferrers
(opened 1957)

Dartmouth

Kingsbridge

0   10   20km

1946 roll-numbers

— 1200
— 700
— 300

Map 33.7 (left)
Devon: grammar-school
provision, post-1903
(excluding Exeter
and Plymouth).

SOURCE: County Education
Committee records.

Boys	**Girls**	**Dual/mixed**	
			Aided (later maintained) schools
▼			Extant in 1904
	△	◪	Established by 1915
			Maintained schools provided
		◪	By 1915
▽	△	□	Post-1915
◐	◗	⬤	Independent schools providing places
◖	◗	○	Out-county schools providing places

Map 33.8
Devon: senior/modern and
comprehensive reorganization
(excluding Exeter
and Plymouth).

SOURCE: County Education
Committee records.

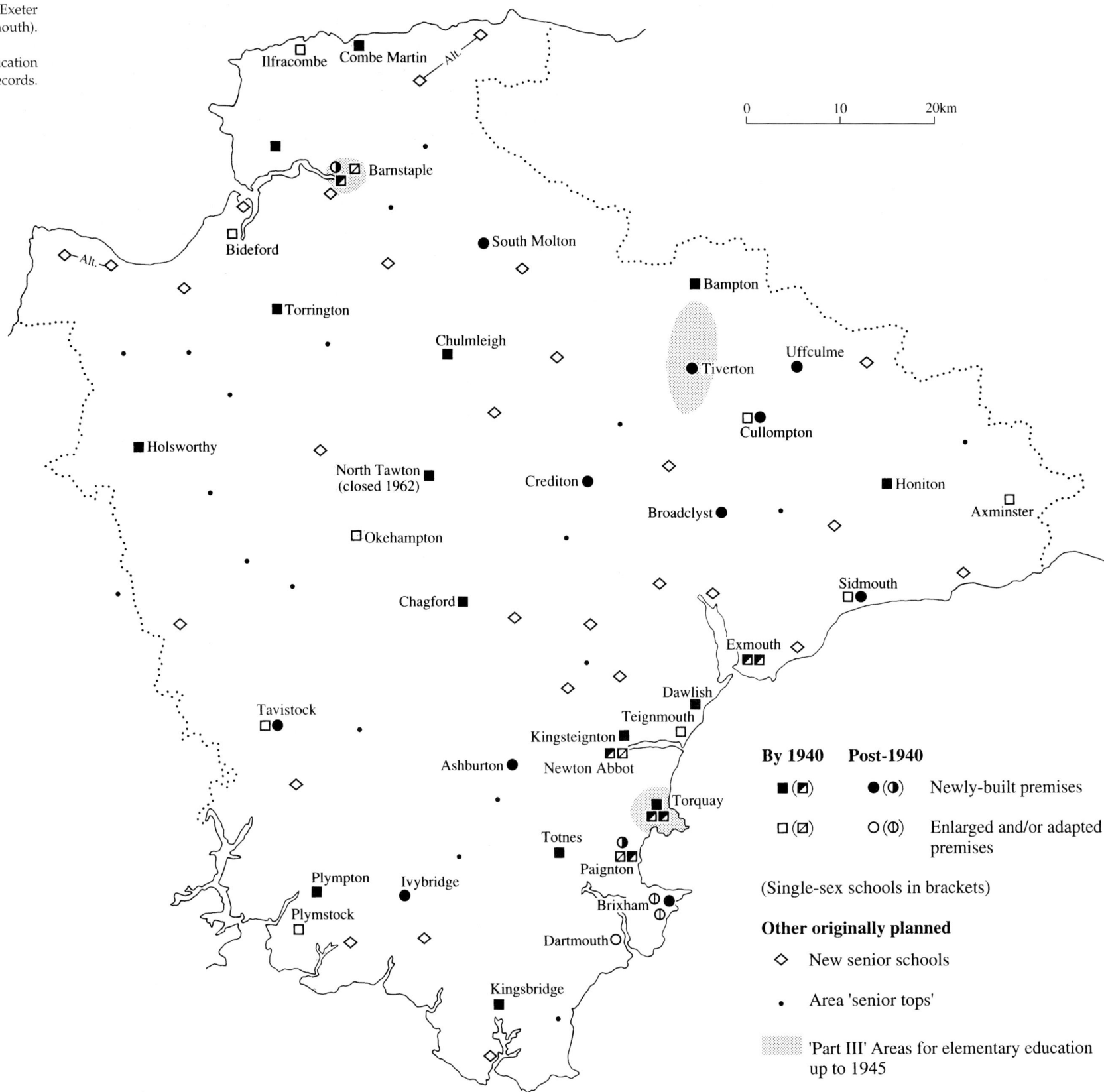

Ilfracombe   Combe Martin   Alt.

0   10   20km

Barnstaple

Bideford

South Molton

Alt.

Bampton

Torrington

Chulmleigh

Uffculme

Tiverton

Holsworthy

Cullompton

North Tawton
(closed 1962)

Crediton

Honiton

Broadclyst

Axminster

Okehampton

Sidmouth

Chagford

Exmouth

Dawlish

Tavistock

Teignmouth

Kingsteignton

Ashburton   Newton Abbot

Torquay

Totnes

Plympton   Ivybridge   Paignton

Plymstock   Brixham

Dartmouth

Kingsbridge

By 1940   Post-1940

■ (◪)   ● (◐)   Newly-built premises

□ (◨)   ○ (◑)   Enlarged and/or adapted
premises

(Single-sex schools in brackets)

Other originally planned

◇   New senior schools

•   Area 'senior tops'

'Part III' Areas for elementary education
up to 1945

emerging as a 'public school' with direct access to the universities—the only one in Devon outside the cities. Allhallows at Honiton had likewise become a mainly boarding school, though at that stage of only 'preparatory' status.

Four feebly endowed schools at Dartmouth, Okehampton, South Molton and Torrington ceased to exist in the earlier nineteenth century, as did the unendowed one at Moretonhampstead; but others continued, sometimes intermittently, despite minimal local demand for what they offered. By 1868 only two in country towns, Crediton and Tavistock, showed signs of developing into what would later be recognized as a grammar school, in each case as the result of enlightened patronage by respectively the town's trustees and the duke of Bedford. Blundell's and Allhallows flourished by cutting most of their local ties. Between them, these four mustered in 1868 130 boarders and 152 day boys; but the remaining nine in country towns had in all only 39 and 114 respectively—and no girls at all (**Map 33.6**). At this stage the more prosperous parents mostly preferred the 'middle-class' proprietary schools widely founded c.1860, such as West Buckland and Sampford Peverell, or the private 'academies' flourishing in larger towns, which offered a more relevant and practical form of education (and did not involve association with 'lower-class' children).

The newly founded county council used its powers under the Technical Instruction Act to make modest grants encouraging diversification and local relevance of curriculum, but large-scale reorganization had to await the 1902 Act, which made the county and the cities responsible for aiding or providing secondary education. Previously there had been virtually no link between elementary and grammar schools, and even where there were local free places in the latter, they had seldom been taken up. Now it was open to establish an 'educational ladder', and to provide equally for girls.

In 1901 the Board of Education offered grants to secondary schools, but on conditions which in effect doomed the smaller and feebly endowed: a course for children aged 12–16 including a foreign language, maths, science, and manual and physical training, adequately staffed, and implying a minimum roll of seventy. In 1907 the additional requirement of 15 per cent free places was made, and it was in this light that the county developed its policy, at first of aiding existing schools and shortly of providing new ones. By 1915, when the First World War temporarily halted progress, new maintained schools had been established in Barnstaple, Ottery St Mary, Newton Abbot and Torquay, and deficit-financing aid was given to Ashburton, Bideford, Crediton, Kingsbridge, Tiverton (Middle School), Tavistock, Totnes, and a revived school at Colyton which had long ago ceased to be 'grammar' (**Map 33.7**).

Wartime restriction built up pressure for further advance, and was quickly followed by the upgrading of the higher elementary schools at Ilfracombe and Okehampton, the opening of new schools at Teignmouth, Dartmouth and Exmouth, and the revival of Plympton in new premises. Provision was also made, where not already supplied, for girls. All aided schools in the county became 'maintained' by 1936; but two, Ashburton and Dartmouth, proved too small and too expensive per pupil to survive. The former closed in 1938, alternatives being available at Totnes and Newton Abbot; and the latter in 1957 when the last grammar school to be built in Devon was opened at Churston Ferrers on a compromise site to serve also Paignton and Brixham. The county also paid for free places at direct grant schools such as Edgehill (Bideford) and Shebbear, as well as at others under different authorities. In 1945 fee-paying ceased in grammar schools and all places became free to those selected in the '11+' examination. Meanwhile, curriculum development had made the grammar school a very different and more relevant institution than the 'parson and Latin' of a century before.

SECONDARY MODERN AND COMPREHENSIVE REORGANIZATION (**Map 33.8**)

The Hadow Report of 1926 recommended senior schools for older elementary pupils, and in 1929 Devon adopted this as a firm policy, opening its first at Holsworthy in that year and making a plan for no fewer than eighty-two separate centres, either purpose-built new senior schools or 'senior tops' taking older pupils from neighbouring schools into specially equipped central all-age schools. Hindsight shows the defects of this programme but in fact the delay imposed by the Depression of the 1930s mercifully prevented the county being saddled with far too many centres too small to be either effective or economical. Only one such was opened, in 1932 at North Tawton, for 120 pupils, and had later to be closed as inadequate. Many of the originally planned 'senior tops' could never have offered an education recognizable after 1945 as 'secondary', though a few served usefully as stop-gaps until a new school could be provided. Local rivalries, as between Chagford and Moretonhampstead, each demanding their own too-small school, further complicated the issue; there was at first much objection to moving seniors elsewhere.

With the worst of the Depression over, revised plans were made, cutting twenty of the original proposals (though leaving twenty more to be eliminated later). The offer of 50 per cent building grants in 1936 produced a spate of new senior buildings—and helped, together with rural closures, to raise the school transport bill to 240 times what it had been in 1904. More than half the reorganization was completed before 1939 brought a halt.

The raising of the leaving age to 15 in 1947 added urgency for completion. Reorganization, apart from replacement of premises, was completed with Crediton in 1962, by which time all remaining 'senior tops' had been eliminated. But by then the first step had been taken towards a further 'comprehensive' reorganization, with the building at Tavistock of a much-needed new school alongside the existing grammar school, which became the first comprehensive in the county when it opened in 1959. Comprehensive reorganization has continued, either with purpose-built new schools as at Ilfracombe, or by combining existing grammar and modern schools as at Okehampton, or the extension and development of a previous modern school with full sixth-form facilities, as at Plymstock. By 1988 separate grammar schools survived, for the time being, only at Colyton, Torbay (three), and Plymouth (four), the last alongside six 11–18 comprehensives. Exeter, while still an independent authority, reorganized into five 12–16 comprehensives with a large purpose-built sixth-form college.

EDUCATION, DISSEMINATION OF KNOWLEDGE, AND LANGUAGE

THE RETREAT
OF THE
CORNISH LANGUAGE

PHILIP PAYTON

Cornish is a member of the Brythonic branch of the Celtic group of languages, related closely to Breton and also akin to Welsh. It is related less closely to the Goidelic branch of Irish, Manx and Scots Gaelic. Historically, it is divided into three periods or phases. Old Cornish, for which there are few documentary remains, was the form of Cornish spoken from the ninth century (when the language had become distinguishable from other Brythonic tongues) to the thirteenth century. Middle Cornish, the medieval form of the language represented in the surprisingly large literary corpus of so-called miracle plays such *The Creacion of the World* and *Beunans Meriasek* (the Life of [St] Meriasek), dates from the fourteenth to the late sixteenth/early seventeenth centuries. Late Cornish covers the period from the seventeenth century to the language's demise in the late eighteenth or possibly nineteenth century.

Although, despite the efforts of a small group of dedicated revivalists, Cornish is no longer a spoken vernacular, it remains a powerful icon of separate identity and is evidenced across modern Cornwall in surnames and especially place-names (see chapter 13). The latter are an important clue to the long retreat of Cornish in the face of English intrusion, especially in eastern Cornwall where there is a paucity of other evidence. In western Cornwall, place-name evidence is complemented by contemporary accounts. **Map 34.1** is a composite, hypothetical delineation of the retreat of Cornish, based upon both place-name evidence and contemporary statements, and is drawn from the work of Holmes and George.[1]

The widespread presence of English place-names in north-eastern Cornwall suggests the early impact there of Anglo-Saxon settlement or influence, as does the more limited incidence of English place-names in the eastern fringe of the county. Cornish may have disappeared entirely from these areas as early as 1200 and 1300 respectively. Holmes's estimation of the retreat of Cornish in east Cornwall as a whole is based on his examination of the postulated sound change in Cornish represented in written form by '*t > s*', a change thought by him to have occurred *c*.1100. Holmes's assumption, therefore, is that places with names containing *s* rather than *t* may be supposed to have been Cornish-speaking at that date, a view accepted by George. Holmes has identified the number of places on modern maps where '*t > s*' appear, but which have known '*s*' forms for dates

considerably later than 1100, and argues that such forms indicate that Cornish was still spoken at those places at the time. On the basis of such evidence, Cornish is thought to have retreated by 1500 to a line running from just east of the Camel estuary on the north coast to the Fowey estuary on the south.

Although Holmes's analysis has been accepted by many commentators, notably George, a recent critic is Williams who asserts that 'there is good evidence that until the Reformation, i.e. the middle of the sixteenth century, Cornish was probably spoken as far as the Tamar'.[2] Noting that 'It is a common error of Anglophones to believe that Celtic languages have always been in retreat', Williams argues that Cornish enjoyed a territorial resurgence after the Norman Conquest, regaining lost ground as its socio-cultural status was enhanced during the medieval period.[3] However, Williams also argues that the increasing intervention of the Tudor state in Cornwall's affairs was an important vehicle of Anglicization, especially the Reformation (see chapter 27) with its English-language prayer book and Bible, and the repression of Cornwall after the 1549 rebellion (see chapter 20), resulting in the retreat of Cornish to the Fowey–Camel line by the late sixteenth century.

George has estimated that by 1600 there were probably some 22,000 Cornish speakers in a total Cornish population of about 84,000 (the greatest number of Cornish speakers at any one time would have been *c*.1300, when there may have been 38,000 speakers out of a population of about 52,000). Thereafter, the retreat quickened, with an estimated 14,000 speakers in 1650, 5,000 in 1700, and 'very few' by 1750. Although the general picture may indeed have been one of relentless westerly retreat, in detail the reality would have been more patchy (in 1587, for example, an interpreter had to be provided in a court case because two fishermen from Gorran Haven, on the south coast of mid-Cornwall, had no English).[4] Norden, however, writing in the reign of Elizabeth I, was insistent that 'of late the Cornish men have much conformed themselves to the use of the English tongue … from Truro eastward it is in manner wholly English. In the west part of the country … the Cornish language is most in use among the inhabitants'.[5] Carew, in his *Survey of Cornwall* of 1602, thought that 'the English speech doth still encroach upon it [Cornish] and have driven the same into the uttermost skirts of the shire', although he also noted that the inquisitive visitor

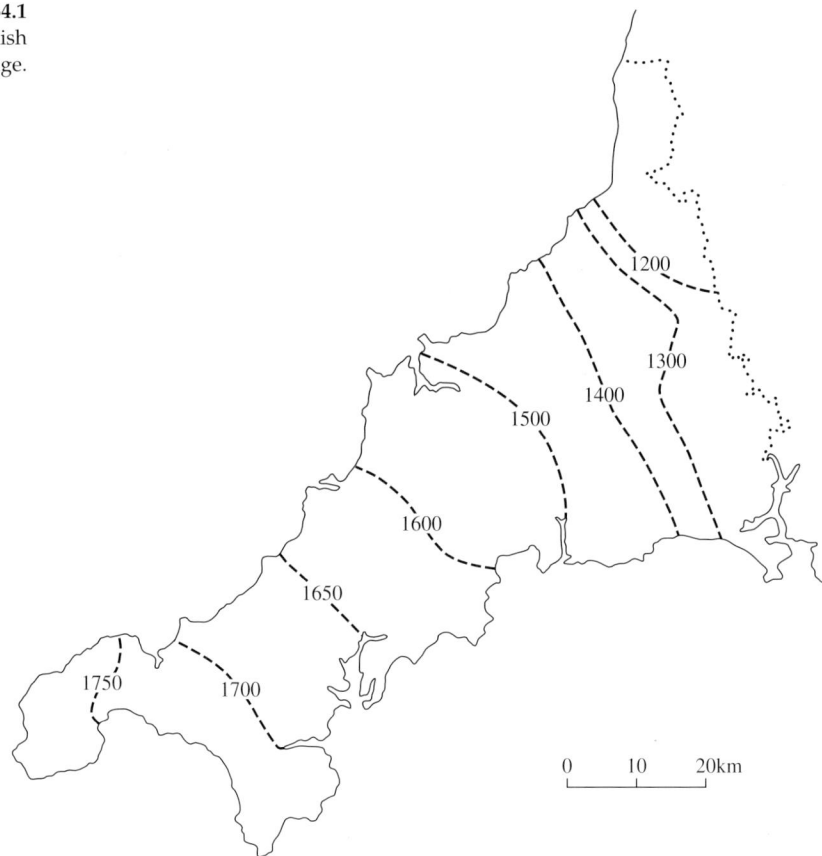

1200

1300

1400

1500

1600

1650

1750   1700

0   10   20km

only as occasional dialect words in English or possibly amongst Newlyn fishermen in obscure counting rituals which may have lingered into the early twentieth century.[8]

FURTHER READING

Peter Berresford Ellis, *The Cornish Language and its Literature* (London, 1974); Crysten Fudge, *The Life of Cornish* (Redruth, 1982); Brian Murdoch, *Cornish Literature* (Cambridge, 1993); Philip Payton, *Cornwall* (Fowey, 1996); P.A.S. Pool, *The Death of Cornish 1600–1800* (Penzance, 1975).

DOROTHY PENTREATH of MOUSEHOLE in CORNWALL
the last Person who could converse in the Cornish Language?

Fig. 34.1
Engraving by R. Soiddan
of Dolly Pentreath,
reputedly the last person
to speak Cornish in
daily life, n.d.

SOURCE: reproduced by
permission of Westcountry
Studies Library, Exeter
(P&D 32868).

from across the Tamar might be rebuffed with the retort *meea navidna cowzasawzneck*—I will not speak English.[6]

Edward Lhuyd, the Oxford scholar who had embarked upon an exhaustive comparative study of the Celtic languages, visited Cornwall *c*.1700 and found Cornish restricted to a string of largely coastal parishes in the Penwith and Lizard peninsulas of far-western Cornwall.[7] Popular belief insists that the very last speaker of Cornish was one Dolly Pentreath, a fishwife from Mousehole who died in 1777, but she was certainly survived by a small number of speakers who may have taken the language into the nineteenth century. But in the sweep of industrialization, the spoken remnants of the language fast disappeared, surviving

AGRICULTURE AND RURAL SETTLEMENT IN 1086

EVIDENCE OF THE DOMESDAY BOOK

F.R. THORN

Domesday Book, widely used as a source by archaeologists, place-name students, genealogists and several species of historian and geographer, is essentially a detailed survey of all the lands that constituted England in 1086 with especial emphasis on tenure, resources and value (**Maps 35.1 and 35.2**). The material in its finalized state is contained in the Exchequer Domesday Book arranged by county and within each county by individual landholder, generally beginning with the king, followed by a sequence of ecclesiastical, then lay tenants in descending order of importance.[1] Each tenant-in-chief has a chapter, or fief, containing a list of estates, usually grouped by hundred (the Saxon administrative subdivision of the shire) or by sub-tenants.

Planned after a meeting of the king's councillors and a synod at Gloucester at Christmas 1085 (= January 1086), the Inquest may well have been substantially completed by 1 August of that same year, although the writing up probably continued until William's death in October 1087 or even into 1088.[2] The major purposes of the Survey, apart from curiosity about a turbulent country which William had ruled for twenty years, frequently *in absentia*, were probably to ensure that taxation was levied efficiently and liabilities derived from agrarian capacity reassessed, to revalue estates as a whole in order to simplify transfer to new ownership, to check the legality of each individual's tenure, and to identify the current generation of major landholders on whose loyalty the king had to rely. The causes may well have been immediate: the difficulties with collecting the 1084 geld, the problems of paying for and billetting a large army brought to the country in 1085 to resist an expected Danish invasion, and the endless disputes involving laymen (including many of the king's relatives and close supporters) and churchmen concerning encroachment on and seizure of land.[3]

The *Inquisitio Eliensis* indicates that the questions to be asked in the course of the Inquest concerned the name of the 'manor', the identity of the tenants in 1066 and 1086, the hidage and number of ploughs in demesne and held by the men, the quantity and categories of population, the amount of woodland, meadow and pasture, the number of mills and fisheries, the values, the size of the subholding of any free man or sokeman, what additions or subtractions had been made to the 'manor', and the potential for greater revenue.[4] There may have been a separate survey of the king's land, and possibly of the boroughs, where included.

Information seems to have been collected from tenants-in-chief and sub-tenants, then compared with existing administrative records, before disputed matters at least were checked in the hundred and shire courts. Material may have been rearranged and examined both hundredally and feudally before being written up into 'circuit volumes' covering a group of counties, of which Exeter or Exon. Domesday is the south-western representative.[5] Though arranged by fief, Exon. is organized differently in detail and the hundred is much more prominent in its substructure. Now bound up with this circuit volume are tax returns (or geld rolls) for the hundreds of all five south-western counties (probably dating from the heavy 1084 geld-collection), and a list of *Terrae Occupatae* (lands in dispute or whose status had changed), now only surviving for Devon, Cornwall and Somerset. For the five south-western counties, Great Domesday Book is an abbreviation and reorganization of Exon., in which lengthy formulae were simplified, and all animals were omitted as were the details of villans' land, numerous bynames and often more precise dating of the values of 'manors' and how they were held. A few place-names were also passed over, sometimes by mistake.

The detail of Domesday Book, at first sight so vastly informative, is difficult to interpret. Information has to be teased out and surrounded by caveats. There are obvious discrepancies and omissions; some holdings are anonymous. It seems that certain questions (for example, concerning boroughs) were not asked at all, others concerning population and resources were not sufficiently inclusive, and were interpreted differently by those responsible for the Survey. Thus people of the same status and economic function are classified differently in diverse parts of the kingdom and even by jurors in adjacent hundreds. Only the working rural population is systematically included. The mensuration of woodland, meadow and pasture takes various forms and is somewhat difficult to interpret: each of these is normally treated as a single rectangle and no indication is given of their location and whether they were scattered. The questions mentioned only mills and fisheries, yet even these are not fully surveyed, and other manorial appurtenances such as salterns, quarries, markets and even castles and churches are only sporadically included. While the 1086 tenants-in-chief are normally clearly distinguished, the absence of most bynames makes it difficult to individualize sub-tenants and 1066 holders. The status of many of the last of

Map 35.1

Cornwall: estates and
hundreds, 1086.

SOURCE: reconstructed from
Great Domesday Book,
Exon. Domesday and the
Tax Returns (Geld Rolls).

Vills

Ecclesiastical	Royal	Lay	Tenure
▲	★	•	Estate < 1 hide
▲	★	●	Estate < 5 hides
▲	★	●	Estate < 10 hides
▲	★	●	Estate < 20 hides
▲	★	●	Estate > 20 hides
△	☆	○	Estate of unknown hidage
◇	◆	◆	Estate only named in Exon. Domesday Book. The extent is not indicated on the map
□	◨	■	Borough
⛉	⛉	⛉	Castle

Each estate is represented by a separate symbol. A lower
link ●● indicates that two or more form part of a vill.
A double link indicates that a single estate is subject
to a dispute or has a duplicate entry.

Domesday hundreds

Cornwall
1 Stratton
2 Rillaton
3 Pawton
4 Rialton/St Petroc's
5 Fawton
6 Tybesta
7 Connerton (Gwythian)
8 Winnianton

Devon
1 Braunton
2 Shirwell
3 Fremington
4 South Molton
5 Hartland
6 Merton
7 Witheridge
8 Bampton
9 Tiverton
10 Halberton
11 Uffculme
12 Hemyock
13 (Black) Torrington
14 (North) Tawton
15 Crediton
16 Silverton
17 Axminster
18 Lifton
19 Wonford
20 Cliston
21 Ottery (St Mary)
22 Colyton
23 Axmouth
24 Teignbridge (Teignton)
25 Exminster
26 Budleigh
27 Roborough (Walkhampton)
28 Plympton
29 Alleriga (Ermington)
30 Diptford
31 Kerswell
32 Chillington

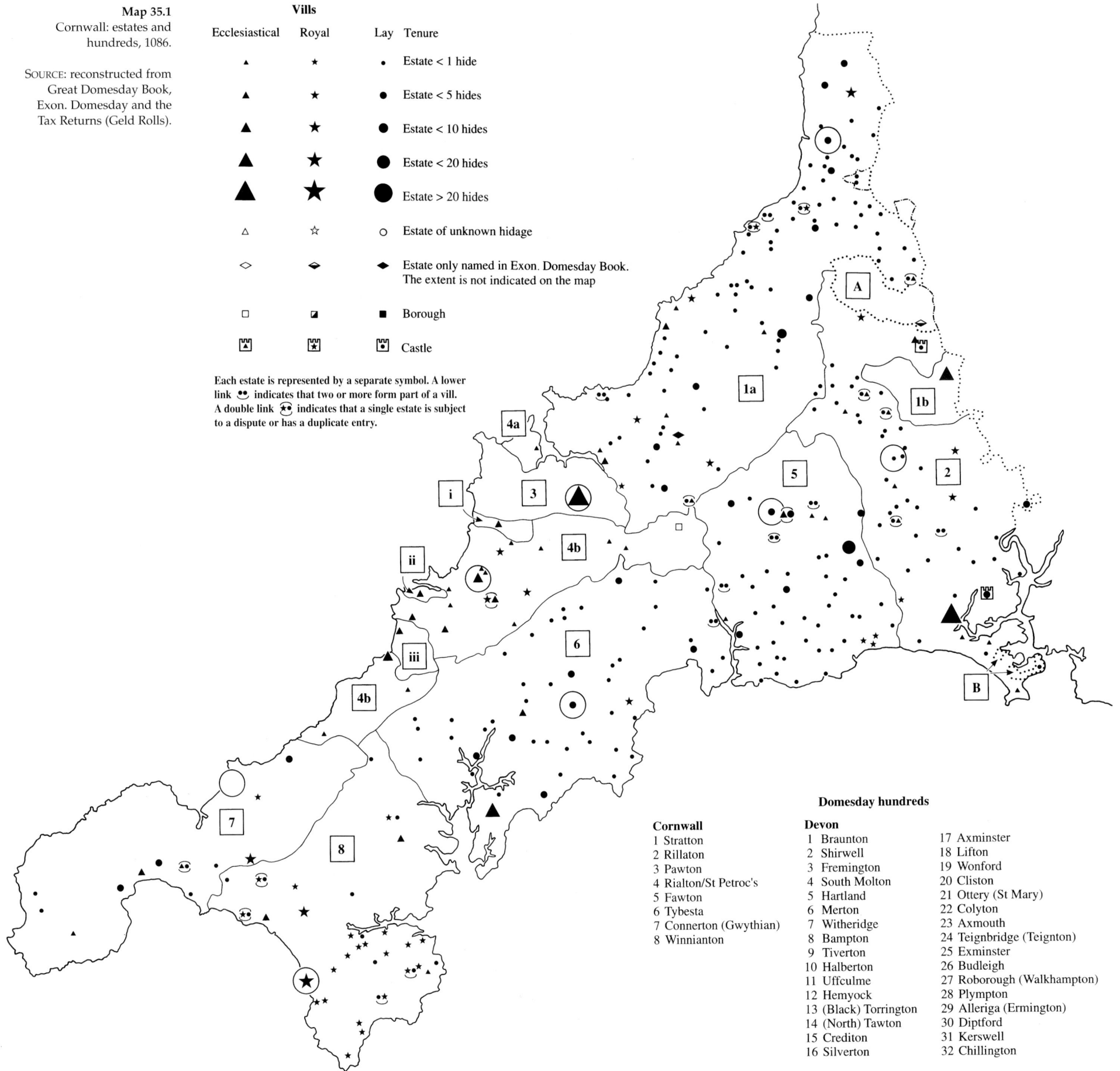

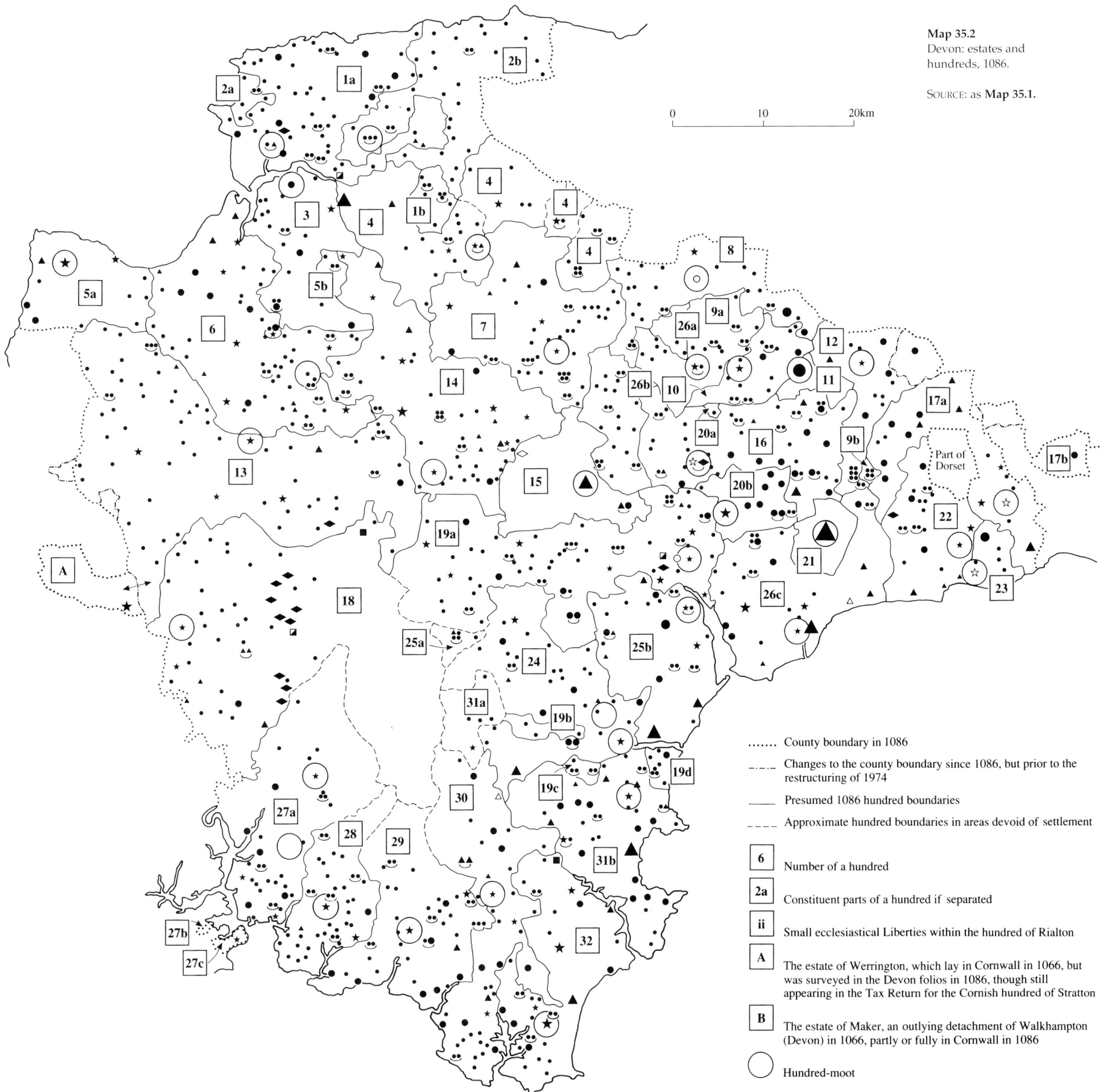

Map 35.2
Devon: estates and
hundreds, 1086.

SOURCE: as **Map 35.1**.

0   10   20km

6	Number of a hundred
2a	Constituent parts of a hundred if separated
ii	Small ecclesiastical Liberties within the hundred of Rialton
A	The estate of Werrington, which lay in Cornwall in 1066, but was surveyed in the Devon folios in 1086, though still appearing in the Tax Return for the Cornish hundred of Stratton
B	The estate of Maker, an outlying detachment of Walkhampton (Devon) in 1066, partly or fully in Cornwall in 1086

.... County boundary in 1086

—·—·— Changes to the county boundary since 1086, but prior to the restructuring of 1974

———— Presumed 1086 hundred boundaries

- - - - Approximate hundred boundaries in areas devoid of settlement

◯ Hundred-moot

Part of Dorset

these is equally hard to define: some may in fact have been sitting-tenants, holding from churches.

Both hundreds and vills are difficult to map satisfactorily. Hundreds are not named in the text of Domesday Book for either Devon or Cornwall, although hundredal groups of estates can be isolated in the text of Exon. and some tax returns allow a full reconstruction of the constituents of a hundred. The hundreds of Domesday Book are identical neither to the Saxon hundreds nor to those of the later Middle Ages, and some 'hundreds' seem to have had a short life as units existing only for the purpose of the Domesday Survey.[6]

Much of the interpretation of Domesday Book relies on place-name identifications. These in turn depend greatly on the quantity and quality of research devoted to them within individual counties. By no means all place-name identifications satisfy all the criteria necessary for certainty: palaeography, a satisfactory connection between the 1086 and modern name-form, documentary evidence of tenure both before and after 1086, location in the anticipated hundred, and harmonization of the proposed site with the manorial resources listed in Domesday Book. The Survey never provides the bounds of an estate and very rarely the distribution and location of resources within it. It does not indicate whether an estate was compact or was interlaced with others, nor always whether it had a number of detached members. The nature of the settlement (for example, dispersed, polyfocal, nucleated) is not stated, the exact location of the estate centre, if there was one, is not given, and there is no absolute measure of size such as is provided by a modern parochial acreage. Hidages are not always complete and by 1086 the hide had become a predominantly fiscal measure with the result that some estates, benefiting from reduced geld burdens, had their hidage reduced. The hidage assessment, ploughlands, ploughs, population and value, even when taken together, can give only a comparative size for a vill. The estates are themselves really taxation units: the places through which payments were made. They are not necessarily independent and self-supporting agricultural units, nor parishes, townships, villages, hamlets or farms. Small estates

may have contained several settlements, not named nor their existence indicated by Domesday Book. Even the name itself may not be that in local use in 1086, but the taxation or charter name. Where two or three adjacent estates share the same name in Domesday Book they had probably once been part of a larger unit or grant, and they may well actually have evolved quite different names on the ground or acquired differentiating suffixes. However, the conservative nature of Domesday Book's substructure of administrative records fails to record the fact. Thus, while there is abundant information for **Maps 35.1 and 35.2** to be drawn, they need to be interpreted with considerable caution.

FURTHER READING

No fewer than 1,757 items of Domesday bibliography (up to 1985) are listed and annotated in D. Bates, *A Bibliography of Domesday Book* (London, 1986). Modern approaches to Domesday are found in P. Sawyer (ed.), *Domesday Book: A Reassessment* (London, 1985); in J.C. Holt (ed.), *Domesday Studies* (Woodbridge, 1987); and in A. Williams and R.W.H. Erskine (eds), *Domesday Book Studies* (London, 1987). In the absence of a good general introduction to the complexity of Domesday Book, R. Welldon Finn, *Domesday Book: A Guide* (Chichester, 1973) will serve.

For Devon and Cornwall, readers can usefully consult the Introductions to Domesday Book in the *Victoria County Histories: Cornwall*, II, part 8 (1924), 45–59, and *Devon*, I (1906), 375–402; H.C. Darby and R. Welldon Finn (eds), *The Domesday Geography of South-West England* (London, 1967); C. and F. Thorn (eds), *Domesday Book: Cornwall* (Chichester, 1979) and *Domesday Book: Devon*, 2 vols (Chichester, 1985), which contain Farley's 1783 printed text, translation, notes, map and indices; and A. Williams and R.W.H. Erskine (eds), *The Cornwall Domesday*, and A. Williams and G.H. Martin (eds), *The Devonshire Domesday* Alecto County Editions of Domesday Book (London, 1988, 1991), which contain a facsimile, translation, extensive introduction, survey of the hundreds, indices and a county map at 1:100,000 scale.

MEDIEVAL FARMING
AND
RURAL SETTLEMENT

HAROLD FOX

FARMING

The quality of farming in the South West in the twelfth century and earlier sometimes received rather a bad press. William of Malmesbury, the twelfth-century chronicler, thought that soils in Devon were so poor that they could produce only oats, and Richard of Devizes noted contemptuously that Exeter people consumed the same food as their horses. In the seventh century, Aldhelm's visit inspired him to write of 'grim Devon and bare Cornwall'. These perceptions are understandable in travellers who probably struck westwards from Exeter along the old route which followed the northern flanks of Dartmoor, then across the Culm Measures (which, in the 1130s, the Cistercians of Brightly described as poor and sterile) and onwards towards Bodmin Moor. Medieval travellers who diverged from this route, sombre still today, received more favourable impressions. Such was John Leland, who sang in verse of 'fertile Cornwall' and noticed 'good corne' on the northern coastlands almost as soon as he entered that county, though he noted that the centre of the shire was 'rochel ground, very baren'. Leland's Devon in the early sixteenth century, likewise, had agricultural landscapes full of contrasts from 'morishe ground but very good for broode of catelle' on the Culm Measures to 'the frutefulest part', the South Hams. In the year 1001, the Anglo-Saxon chronicler refers to 'many goodly manors' destroyed by the Vikings in the neighbourhood of Kingsteignton.[1]

From any one of these perceptions it is not possible to generalize, as some historians have done, to the whole of the South West; each is specific to a particular region. Some idea of regional variation in crop cultivation may be gained by plotting information from medieval demesne accounts and related documents (**Map 36.1**).[2] The two most demanding crops, producing grain of highest value, were wheat and barley, and it is therefore not surprising to find that demesnes growing them were confined to regions of warm, well-drained soils, the South Hams and the coastlands of Cornwall.[3] A complete contrast is provided by cropping on the Culm Measures of mid-Devon and north-east Cornwall where, on poor often ill-drained soils and in a damp climate, the principal crops were the more tolerant (but low priced) oats and rye.[4] Local diet varied accordingly so that at St Columb on the Cornish coastlands mill accounts show that wheaten bread predominated in the diet of the people and that barley was replacing oats in brewing by the mid-fifteenth century; at Plymstock in the South Hams wheat and barley were used in baking. Quite different, less sophisticated and coarser was the fare of inhabitants of the Culm Measures. Mills at Bishop's Nympton and Sampford Courtenay ground rye predominantly and rye bread ('black loaves') featured in the rough diet of the demesne servants. The oats sown in this region went towards an ale so vile that, according to Hooker in the sixteenth century, it repulsed the stranger and made him vomit, although, presumably through acquired habit, 'the people of that country … do endure the same very well'.[5] Crop combinations influenced valuations of arable land in the Middle Ages (**Map 36.2**) so that where much wheat and barley were grown values per acre reached their highest, for example in the South Hams, the east Devon vales and on some manors on the coastlands of Cornwall where they could be 6d per acre (high by national standards) in places. The latter county was not a pastoral backwater as sometimes imagined, for where good soils, fair climate, closeness to markets (especially ports) and to supplies of sand for spreading on the land all combined together, arable farming was very profitable, 'as good and fair as any' according to a petition of 1361.[6]

Medieval livestock husbandry also varied regionally in its objectives. By the fifteenth century coastal demesnes in Cornwall and south Devon specialized in dairying, while farmers in mid and north Devon concentrated on extensive cattle-rearing, some stock being sent for a final fattening to the lush vales of east Devon (**Map 36.3**).[7] Other movements of cattle involved transhumance, that is seasonal movements to fresh pastures ('summering' on **Map 36.3**). On Bodmin Moor remnants of this system survived into the fourteenth century and later, while moorland farmers living on the flanks of Exmoor took in 'the beasts of strangers'; on Dartmoor, by the fifteenth century, as many as 10,000 head of cattle were brought yearly from lowland farms for the summer grazing.[8]

Four aspects of medieval farming practice in the South West stand out as distinctive. These were, first, convertible husbandry under which long periods of grass ley followed a few years of cropping; second, the associated practice of beat burning by which the matted grass sods of the ley were pared, dried and then burnt, a practice for which the South West was already famous by the

Map 36.1
Medieval crop combinations,
drawn largely from
fourteenth-century evidence.

SOURCE: medieval grange
accounts but with additional
information from
inventories, accounts of
multure and accounts
of tithe.

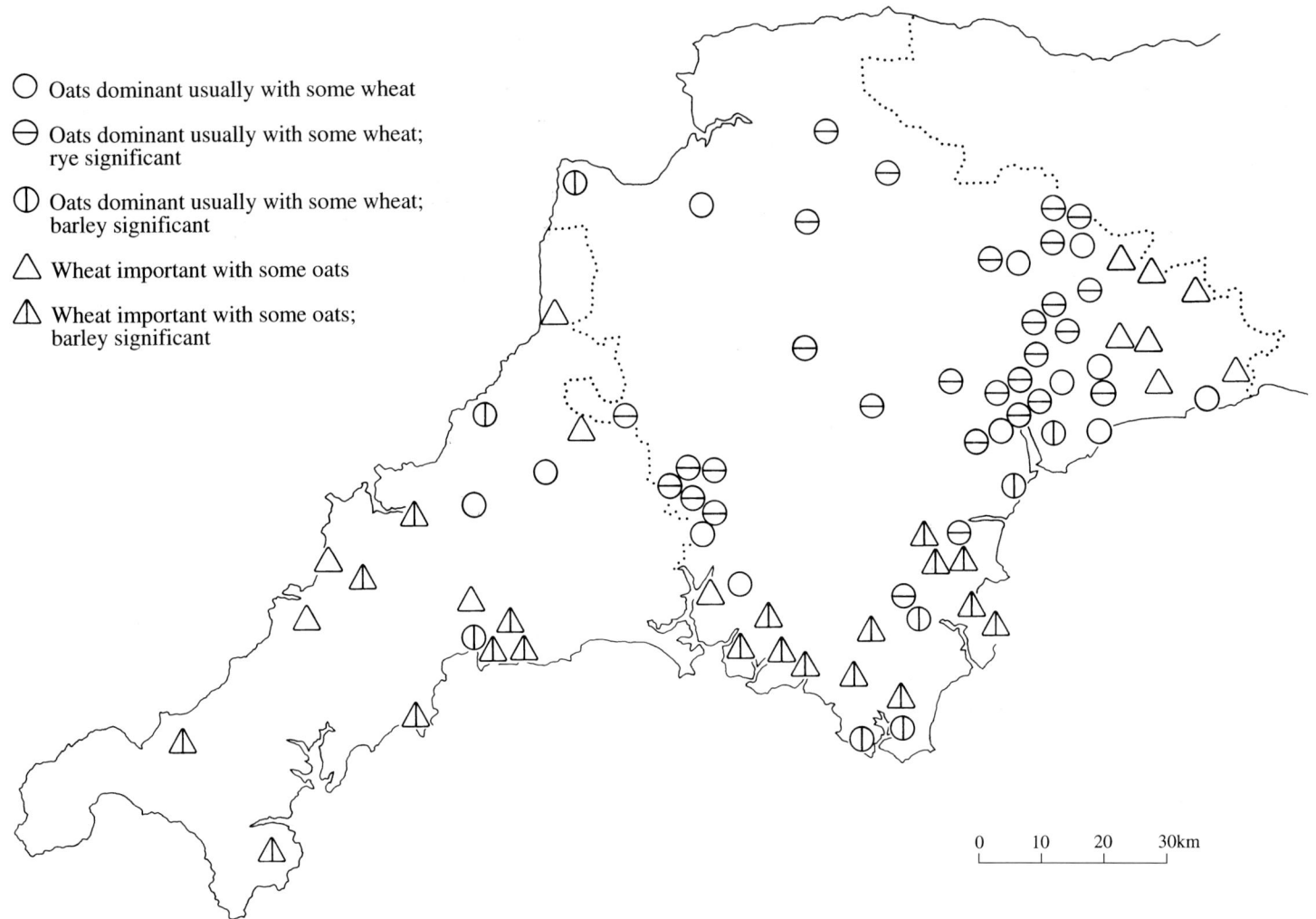

Map 36.1
Medieval crop combinations, drawn largely from fourteenth-century evidence.

SOURCE: medieval grange accounts but with additional information from inventories, accounts of multure and accounts of tithe.

○ Oats dominant usually with some wheat

⊖ Oats dominant usually with some wheat; rye significant

⊕ Oats dominant usually with some wheat; barley significant

△ Wheat important with some oats

◮ Wheat important with some oats; barley significant

fourteenth century when there are references to 'Devonshiring' in a document from Hampshire; third, the spreading of sand in order to lighten and lime the soil, which generated a lively traffic by barge and pack horse, as on the sandways leading inland from many Cornish coves; and finally periodic tillage of wastelands (outfield cultivation), part of a system of multiple exploitation of the rough pastures which are to be seen on the sky-line of a majority of parishes in the two counties.[9] These practices were ancient, clever adaptations to the physical make-up of the South West and are referred to at early dates; they are not indicators of agricultural improvement during the period covered by this chapter (late eleventh to late fifteenth centuries).[10]

For improvement we must look first and foremost to enclosure of the open-field strips which around 1300 would have covered far more of the landscape than it is easy to realize now, because so many south-western hamlets were once surrounded by strip-field systems in miniature (**Map 36.7**).[11] Piecemeal

enclosure speeded up during the late fourteenth century, rendered easy and uncontentious by the ever-present waste, which meant that common grazing rights on the arable were not strongly entrenched, and abetted by reduced pressure of population so that adjacent strips might easily fall into the hands of a single occupier.[12] There was also an active side to the movement: documents occasionally allow us to hear the voices of tenant farmers and landlords advocating exchange and amalgamation of strips, describing intermixture of parcels as 'a grave loss' (1394), and exchange as 'for profit and convenience' (1357), or 'to improve tenures' (1415).[13] The aim of enclosure was often to provide stoutly hedged closes for the better management of livestock (as stated in a sixteenth-century survey of Tinten) and it was often closely associated with another type of structural improvement, the amalgamation of smallholdings to create substantial farms.[14] Tenant farmers of the fifteenth century sought to supply markets from their newly enlarged acres, despite

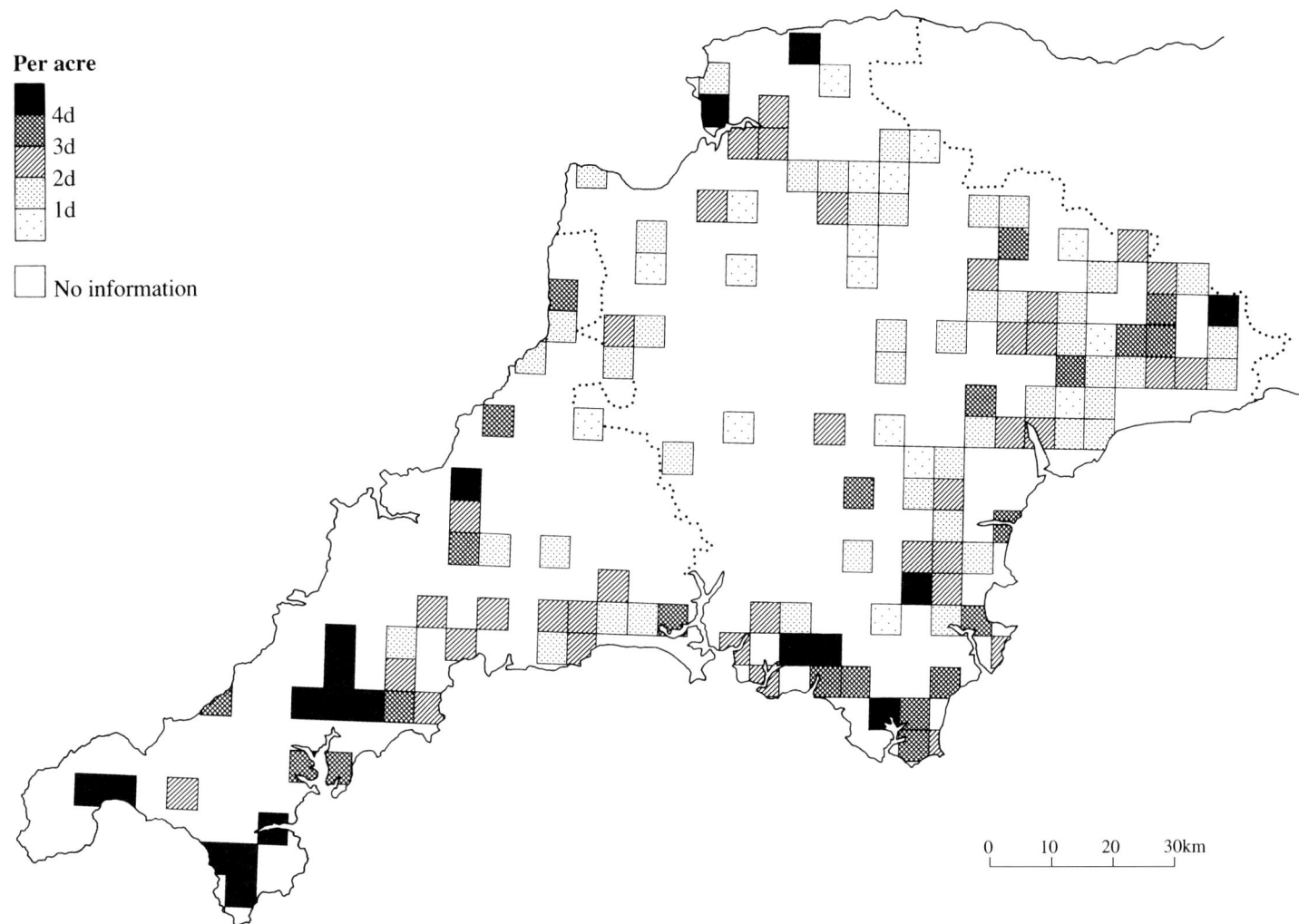

Map 36.2
The value of arable land, late thirteenth and early fourteenth centuries.

One shaded square represents the value from one manor or, in a few cases, the mean of values from two manors. Blanks indicate lack of data.

SOURCE: medieval extents, largely those in inquisitions *post mortem*.

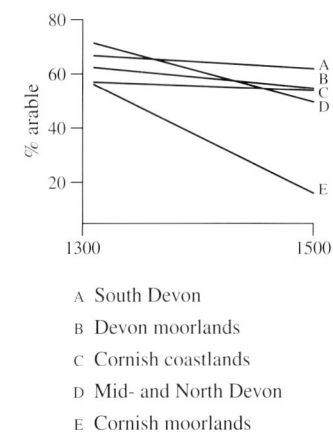

A South Devon
B Devon moorlands
C Cornish coastlands
D Mid- and North Devon
E Cornish moorlands

Fig. 36.1
Percentage of land under arable, *c.*1300–*c.*1500.

SOURCE: as **Map 36.2**.

labour and marketing difficulties. Types of farming which had once yielded specialisms for subsistence were now, by the fifteenth century, modified to produce specialities for sale and from an earlier germ true farming regions began to emerge. A crude indication of these developments is provided by **Fig. 36.1**, which shows regionally divergent trends in the amount of land kept under the plough. In Cornwall, for example, arable came to be concentrated on the coastlands where grain cultivation was most profitable, and these coastal tracts now came to supply the inhabitants, including tinners, of the moorland borders where evidence of decayed corn mills indicates that much former arable reverted to pasture or waste.[15]

Maps showing crop combinations and land use hint strongly at a mosaic of medieval farming regions—latent already by the thirteenth century, developing more strongly later—but they are blunt instruments for portraying the reality of regional differences.[16] The concept of a farming region certainly begins

with crops and livestock, so that, for example, a traveller through mid and north Devon in about 1500 would have seen much pasture, many cattle bred for fattening, some of them certainly of red colour, and small patches of oats and rye grown for subsistence. This type of farming engendered its own special type of cultural landscape, with few villages, enclosed fields, many patches of moorland, large farms, and farmyards with specialist steer houses, cow houses and linhays for shelter and hay storage. The concept also embraces quality of life, key features in mid-Devon being isolation, coarse diet, living-in servants (partly a product of isolation, partly of extensive pastoral farming) and, for some, the comforts of an improved farmhouse built with profits amassed from rearing.[17] By contrast, a traveller through the South Hams, from Dartmouth to Plymouth, at the end of the Middle Ages, would have been keenly aware that the basic type of farming, the farming landscape and all of the *minutiae* associated with each were very

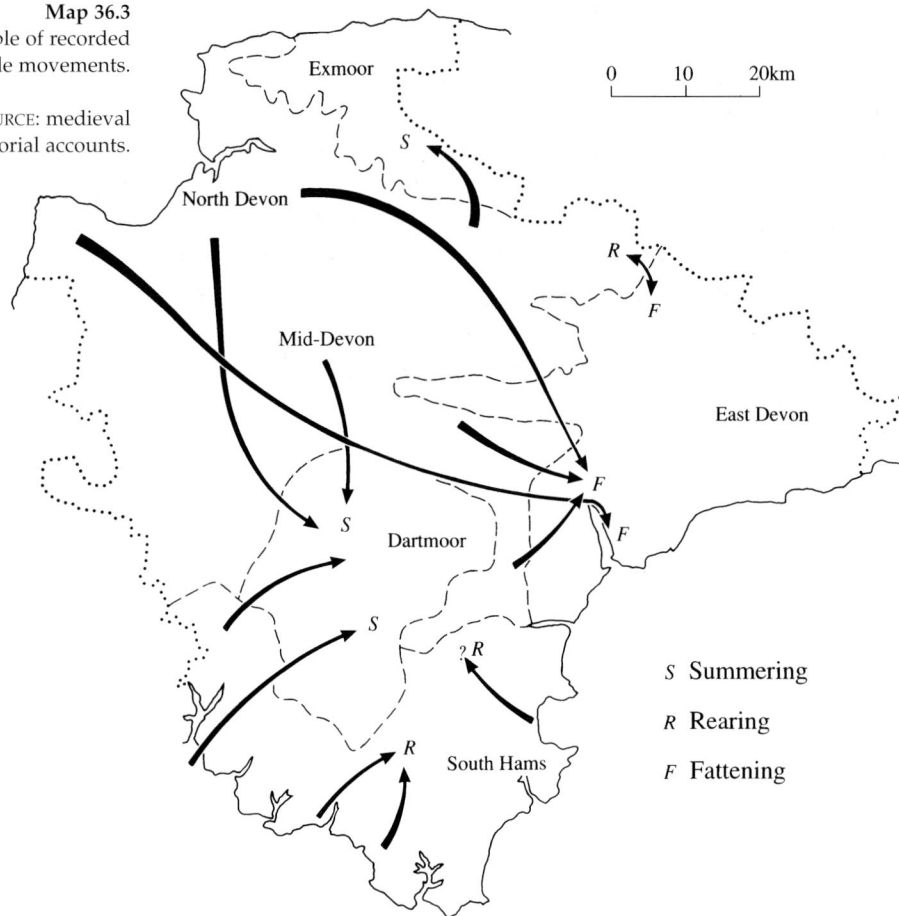

Map 36.3
A sample of recorded medieval cattle movements.

SOURCE: medieval manorial accounts.

s Summering
R Rearing
F Fattening

families, clearly a small hamlet.[18] Very occasionally twelfth-century documents shed a little light on the disposition of settlement sites. Thus one of the earliest fines for Devonshire, drawn up in 1198, mentions by name several isolated settlements in the high heathy country east of Combe Martin: Girt (named after its gravelly site), Truckham (once Trocumbe) at the head of its narrow valley, and Furswell. Already, and even in this relatively unpromising environment, we can see that characteristically south-western settlement pattern of dispersed, isolated sites.[19]

It is not until the late thirteenth or early fourteenth century that we have anything like a full picture of the rural settlement of any manor in Devon or Cornwall—derived from manorial surveys and rentals which list each tenant under the name of the site at which he or she lived. Officials sometimes enumerated tenants by trudging the lanes, for in some surveys the sites are listed in an order which reflects their position on the ground.[20] Such documents are not common but are susceptible, with much patience, to cartographic presentation; in the limited space available here we have mapped information from one Duchy manor, Helstone-in-Trigg, and from Sampford Courtenay in Devon.

On the manor of Helstone, surveyed in 1337 (**Map 36.4**),[21] there were no rural craftsmen; the churches of the two main parishes were isolated, Advent church having only a field track for access. The survey does not mention a manor house, although some tenants had to put up stabling when a duke came

different there. He would have seen more fields of grain, more wheat and barley, greater application of sand to the soil, dairy cattle not steers, persistence in places of unenclosed arable strips, less moorland. He would also have been aware of a diet including wheat, barley and cider (rarely drunk in mid-Devon at this time), cottages, sometimes clustered in villages, to supply labour, and relatively few large improved farmhouses, perhaps because of the variable profits from grain farming. Such details, and many more smaller ones, cannot be mapped for the Middle Ages because the information is so patchy, but they were of utmost importance for quality of life in the past and are important to us if we wish to understand the deep and complex nature of medieval regionalism.

RURAL SETTLEMENT

Map 36.4
Rural settlements in the manor of Helstone-in-Trigg, 1337.

SOURCE: Peter L. Hull, *The Caption of Seisin of the Duchy of Cornwall* (1337), Devon and Cornwall Record Society, new series, 17 (1971).

As Ravenhill points out in chapter 15, the fine details of rural settlement (the pattern of farmsteads from which farming, described above, was practised) cannot usually be reconstructed from Domesday Book, which provides a true glimpse of reality only in those (relatively rare) cases where a separately recorded manor or holding coincided with a single site. Such was Trehawke in eastern Cornwall where lived two bordars and their

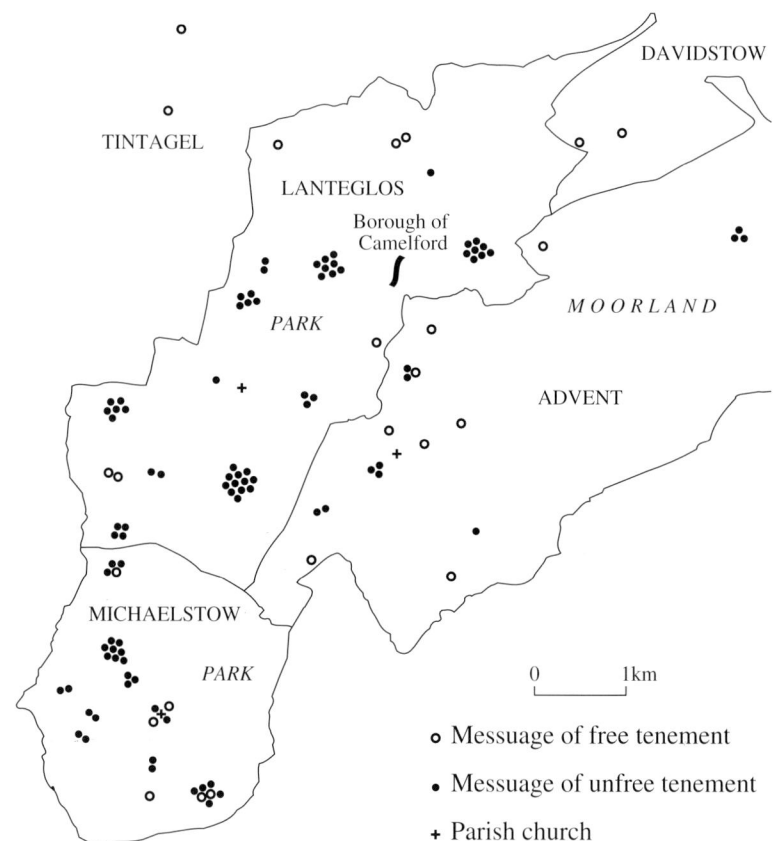

o Messuage of free tenement

• Messuage of unfree tenement

+ Parish church

to hunt. Farmsteads were dispersed among approximately forty settlement sites, some of them isolated farms but mostly hamlets. There was nothing on Helstone manor which matches our common conception of a medieval village, with church, manor house, a few craftspeople, some cottage labourers and, above all, a large number of nucleated farmsteads. Nevertheless the landscape of Helstone was very full of people in the early fourteenth century and farm size was on the small side, for example a mean of 16 acres at the hamlet of Trevia, and of 13 acres at Tregoodwell. Geographers and historians have never agreed a precise dividing line between village and hamlet, although none would doubt that the former is larger, and with more occupational and social heterogeneity, than the latter. Places like Trevia and Tregoodwell, small clusters of smallholding families, fall firmly on the hamlet side of the dividing line.[22]

Map 36.5 is a reconstruction of settlement on the Devonshire manor of Sampford Courtenay as it may have been in about 1300.[23] The pattern is similar to the Cornish one as exemplified by Helstone: hamlets and isolated farms with gaps in the pattern only where the demesne fields lay and in a moory middle tract. Towards the south, the hamlet clusters were smaller than on the manor of Helstone, and isolated, single farms commoner probably because lowland Helstone lay on good fine-textured soils whereas southern Sampford is inferior Culm Measures country inclined to moor. Towards the north, on a strip of good red land, hamlet clusters were larger, and in one of them, Sampford itself, was the manor's stock of landless cottages (not shown on the map), making this a different type of settlement from any on Helstone manor. But apart from that difference the basic pattern of settlement on the Devon manor was similar to that on the Cornish one.

The origins of a few hamlets are just late enough for us to be able to glimpse them, however imperfectly. Dunnabridge, high on Dartmoor, was not there in 1304 and was described as 'new' in 1306. Its five tenants held 96 acres, each claiming $19\frac{1}{5}$ acres precisely, a sure tell-tale sign of a communal reclamation divided equally into strips among pioneers. The hamlet of Brown Willy on Bodmin Moor began in a broadly similar way, but a little earlier. Both examples are from moorland environments where there was space for new settlement sites in the twelfth and thirteenth centuries, a period of population pressure in England generally.[24] At the same time, away from the moorlands, though under the same pressures, hamlets were still evolving in another way, through a process of internal fission. The Duchy survey of 1337 allows us to see one such subdivision in the making on Liskeard manor, where Luke Serle's holding was divided precisely into two, he himself (perhaps in old age) retaining 14.5 acres, the other 14.5 being let to Robert Faber.[25] Repeated fission of holdings like this, and associated subdivision of fields, could lead to the fragmentation of the land into strips. These few, very late examples help us to envisage hamlet origins but they do not help much in answering questions about the age of south-

western systems of dispersed settlement more generally. For Devon, W.G. Hoskins argued persuasively that, away from the moorlands, 'the pattern of settlement in 1086 was virtually what it is today'.[26] On some south-western manors this is almost certainly the case as far as the pattern of *sites* is concerned, but the picture is complicated by the 'pulsating' nature of the farmsteads at each site: a single farm could persist for centuries, then be split into several, then revert back to a single farm when holdings were amalgamated again, as often happened in the fourteenth and fifteenth centuries. We are not yet in a position to generalize about how many hamlets began as single farms, although there is a strong presumption of such an origin where a hamlet name incorporates the name of a single individual.[27]

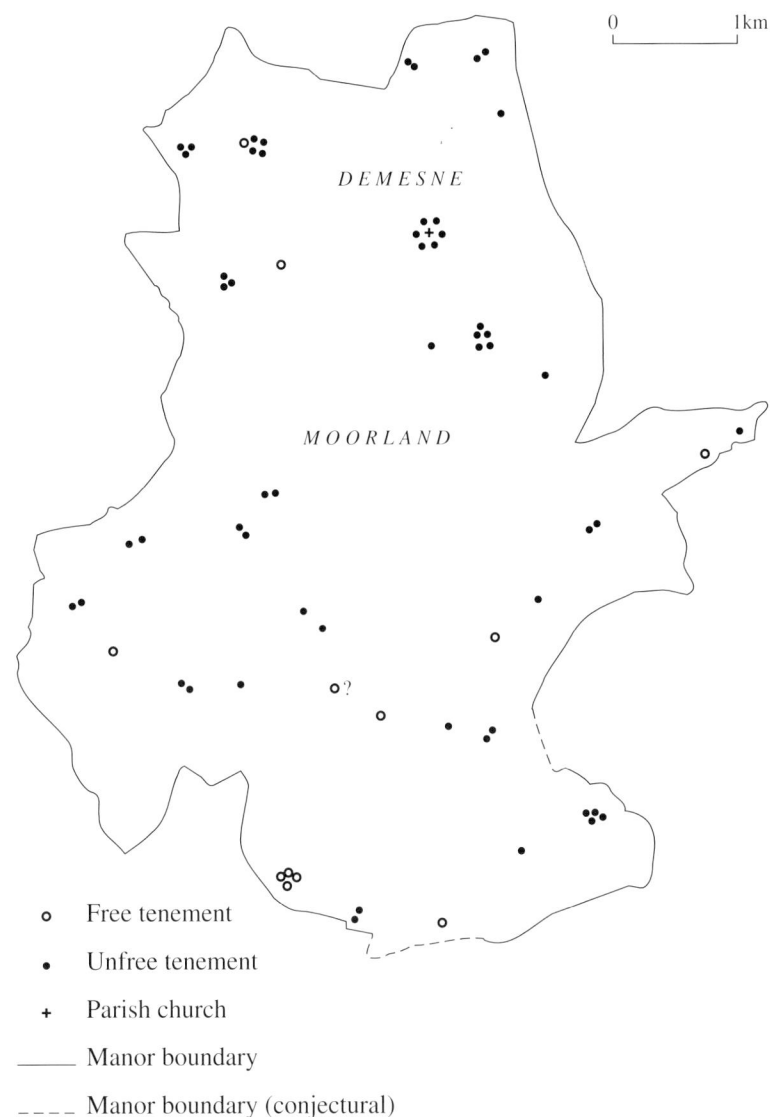

Map 36.5
Rural settlements in the manor of Sampford Courtenay, c.1300.

SOURCE: Public Record Office S.C. 12/22/19 collated with King's College Cambridge MSS SAC 67.

On neither of the manors mapped here (**Maps 36.4 and 36.5**) was there a true agricultural village, as defined above. Present-day appearances can be misleading, for a sizeable cluster of dwellings today sometimes turns out to be no more than a large group of

Map 36.6
Rural settlements in the manor
of Helstone-in-Trigg, 1498,
Lanteglos parish only.

SOURCE: P.R.O. E.306/2/16.

Fig. 36.2
Bench End,
East Budleigh Church,
showing shearing scissors.

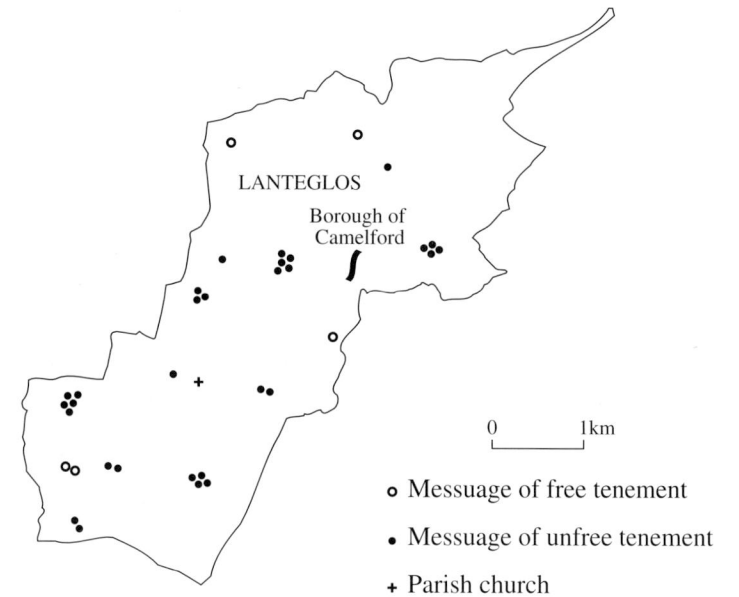

LANTEGLOS

Borough of
Camelford

0   1km

o Messuage of free tenement

• Messuage of unfree tenement

+ Parish church

landless cottages in medieval times. Sampford Courtenay 'village' is a good example, for landless cottages (perhaps twenty-four of them: not shown on Map 36.5) far outnumbered farms (six) here in the early fourteenth century; Stoke Fleming in the South Hams is another.[28] Many Cornish 'churchtowns' (the local term for the settlement next to the church) would have been far smaller in the Middle Ages than today and, indeed, we have Leland's word for it that St Keverne and St Buryan contained respectively about eleven and eight houses, and many of these were probably cottages without land.[29] The origins of these 'labourers' villages', certainly in existence in the Middle Ages, are obscure and the same applies to those relatively few true farming villages to be found in the South West, such as Sidbury, with about fifteen farmsteads and thirty cottages in 1350, or Axminster (a farming village with a minute urban borough attached), or Braunton with its surviving open field.[30] Large nucleated settlements of this kind were absent from most of Cornwall: those who know Cornish medieval records will agree wholeheartedly with Charles Henderson's view on this point.[31] A map of the distribution of such villages in Devon and eastern Cornwall, and detailed cartographic work on their morphology, are urgently needed if we are to explain their presence here and there in what Maitland typified as 'a land of hamlets'.[32]

More is known of south-western rural settlement between 1350 and 1500 than about earlier centuries simply because the documentation is fuller. The period was marked by a contraction, mirroring a falling population, from the very fully settled landscape which in a few places may be reconstructed for the period before the Black Death of 1348. Some settlements disappeared altogether, for example Hendon in Hartland parish, which had been a typical hamlet occupied by eight smallholders in the thirteenth century and was abandoned by its tenants between 1360 and 1400, its fields given over to pasture and infested with gorse. Folk memory of the settlement remained in the ditty: 'Hendon was a market town / When London was a vuzzy down'.[33] Far more common than the total abandonment of settlements was the severe shrinkage which occurred when surviving farmers acquired vacant holdings, as can be seen in that part of Helstone manor depicted in **Map 36.6**. Contraction of settlement was most severe in remote regions distant from markets, on Bodmin Moor for example,

where two moorland hamlets in the eastern part of Helstone manor were deserted by 1400 and many other settlements became decayed.[34] Also severely affected was the Culm Measures country of mid and north Devon (exemplified by Hartland), where contraction went hand in hand with the development of extensive cattle-rearing described above; in Cornwall's Stratton hundred, containing a good deal of Culm Measures land, a recent survey has revealed twenty-one deserted medieval sites and eighty-seven shrunken hamlets.[35] Here, and to varying degrees throughout the South West, tenant farmers of the fourteenth and fifteenth centuries were engaged in a quiet structural transformation which progressed with relatively little social dislocation—without riot or eviction—adding farm to farm, enclosing fields and helping to produce those lonely landscapes, with a few hamlets and more isolated farms, which we can still see today.

Settlement patterns and society are closely associated, each influencing the other. In a large parish, the church could be very distant from scattered settlement and reached over treacherous lanes (as witnessed by petitions to Exeter requesting local burial rights, with their stories of corpses falling from the bier on to the mud), and distance no doubt diminished a sense of parochial community.[36] Towards the end of the Middle Ages church houses were built and rural parish guilds were established and these would have strengthened parochial loyalty: Devon, it is claimed, has more church houses than any other county and the Cornish church-ales described by Carew had their counterparts in the Middle Ages; even a parish with rural settlement as scattered as that of Okehampton had its guild 'of the young countrymen on the land' (thus distinguished from the young men of the borough).[37] In areas of dispersed settlement the existence of such voluntary institutions set up in part for the strengthening of parochial ties helps us to understand how, on occasion, whole parishes might rise up in unison, as happened at St Keverne at the beginning of the 1497 rising against excessive taxation, or at Sampford Courtenay in the Prayer Book Rebellion of 1549. Both are parishes of excessively scattered settlement, but a guild existed in the former and the people of the latter had built, by 1549, the great church house which still survives.[38]

In landscapes of dispersed settlement, the usual world in which day-to-day contacts between countrypeople took place must have been the hamlet or a set of isolated farms sharing common boundaries. Especially before 1350 many hamlets had their own miniature open-field systems like the one described as 'divided into strips among neighbours' in a deed relating to Trevellion in Luxulyan (and see **Map 36.7**). At such places the most frequent contacts would have been between neighbours, who in some cases were also kin. There was a good deal of co-operation at the level of the hamlet, as in the 1520s when the five tenants who farmed at the hamlet of Bugford in Stoke Fleming came together to 'make a by-law' (something which, in other parts of England, was done at village level) or when communal corn-drying barns were constructed, as at some Cornish sites. Where families lived in completely isolated farmsteads there were frequent contacts with the occupiers of neighbouring properties: disputes about shared boundaries and, especially, about rights of way, but also co-operation as in 1285 when twelve tenants living in lonely farms on the eastern flank of Bodmin Moor came to an agreement with their lord about their ancient rights to a tract still known as Twelve Men's Moor today.[39] Dispersed settlement bred small-scale, intensely local loyalties, and, as well as the centripetal forces at a parish level mentioned above, there were also centrifugal tendencies, seen in the establishment of chapels of ease at the level of the hamlet and, more rarely, hamlet guilds.[40]

Another social repercussion of dispersed settlement was a tendency for farmers to retain servants in their households because pools of village labour, described above, were uncommon in the countryside of the South West. Living-in service was especially prevalent in Culm Measures country with its many remote farms and was probably widespread in Cornwall. It is almost certain, too, that dispersed settlement encouraged evasion of royal taxation. Isolated settlements might be missed altogether by the tax collectors, as was claimed for four hamlets in Dorset in the early fourteenth century, and historians now admit that the amount of tax collected would have varied according to the 'visibility' of the items which were taxed, so that, for example, livestock in distant fields or on the moors would have escaped notice. This flouting of authority was easier where farms were dispersed. We find it again in 1233 when the people of Cornwall are said to have refused to come to a royal eyre. For many countrypeople, too, manorial authority was distant and weak: most lived at some distance from a manor house and servile works were (as a consequence perhaps) exceptionally light in the South West, involving work on the demesne on only a few days yearly; manorial courts met with monotonous infrequency. Under these circumstances it is not difficult to see what lay behind the (English) archdeacon of Cornwall's charge in 1342 that, when faced with authority, his flock was of troublesome and obdurate character.[41]

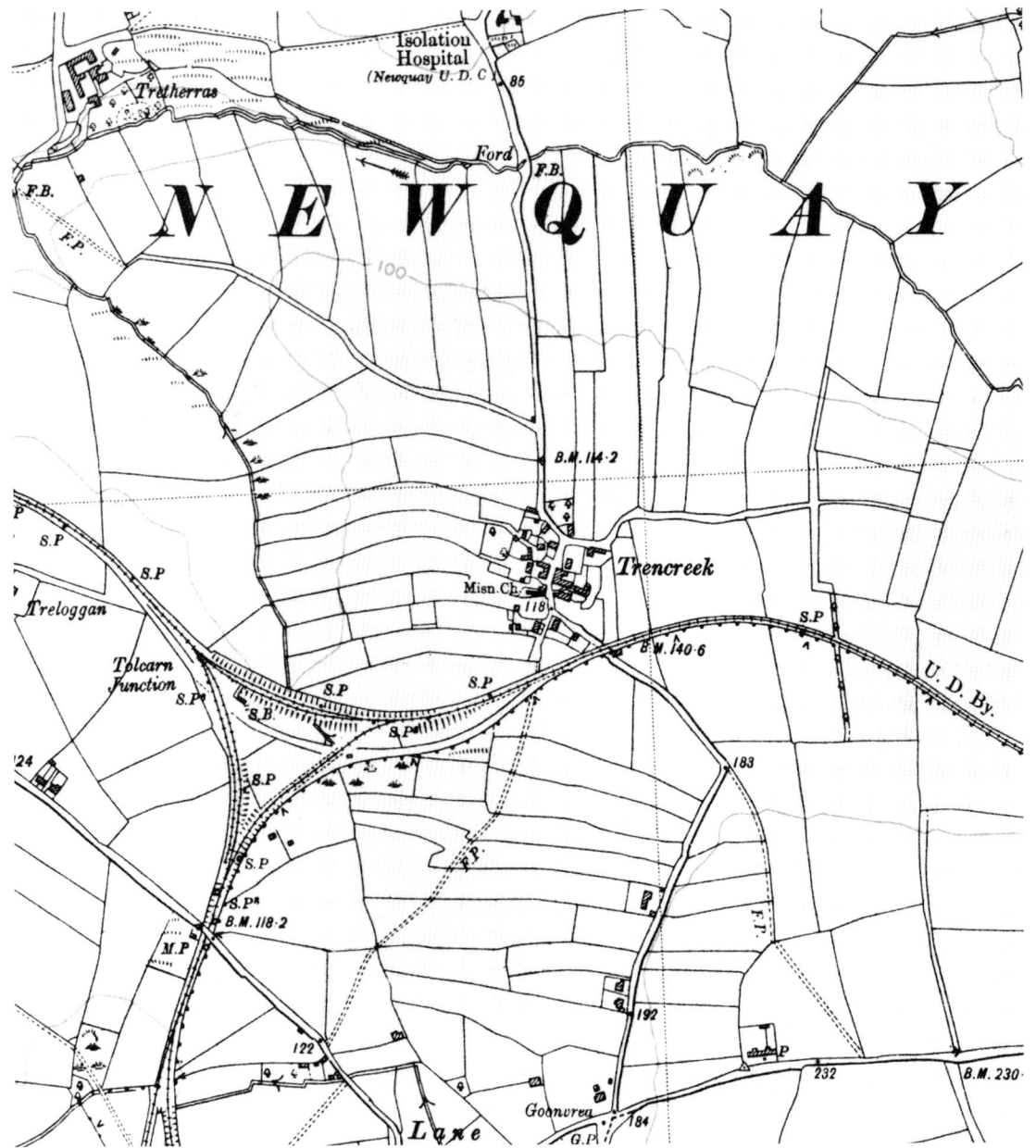

Map 36.7
Fields at Trencreek, near Newquay, Cornwall from Ordnance Survey six-inch map. The map shows the formerly unenclosed strips of farms belonging to the hamlet, now fossilized by hedgebanks. There is much documentary and cartographic evidence to show that many hamlets in this part of Cornwall were once associated with strip-field systems.

FURTHER READING

John Hatcher, 'New settlement', 'Farming techniques' and 'Social structure', covering Devon and Cornwall, in H.E. Hallam (ed.), *The Agrarian History of England and Wales, Vol. II, 1042–1350* (Cambridge, 1988), 234–45, 390–8, 675–85; Harold S.A. Fox, 'The occupation of the land', 'Farming practice and techniques' and 'Tenant farming and farmers', covering Devon and Cornwall, in Edward Miller (ed.), *The Agrarian History of England and Wales, Vol. III, 1348–1500* (Cambridge, 1991), 152–74, 303–23, 722–43.

Herbert P.R. Finberg, *Tavistock Abbey: A Study in the Social and Economic History of Devon* (Cambridge, 1951) is just what the subtitle claims, with much excellent detail on agriculture and rural settlement. John Hatcher, *Rural Economy and Society in the Duchy of Cornwall, 1300–1500* (Cambridge, 1970) is wider than its title suggests. William G. Hoskins, 'The highland zone in Domesday Book', in his *Provincial England* (London, 1963), 15–52, contains many insights. It was his last full statement on rural settlement in Devon and may be seen in part as a reassessment of his earlier work. See, though, the caution in note 26 of *this* chapter. Robin Stanes, *The Old Farm: A History of Farming Life in the West Country* (Exeter, 1990) is comprehensive and covers medieval and modern periods.

AGRICULTURE AND RURAL SETTLEMENT 1500–1800

MICHAEL HAVINDEN AND ROBIN STANES

In this period, although Devon and Cornwall could be generally assigned to the hilly, pastoral, 'highland zone' of Britain, there was a rough division into two regions. In the shaded region on **Map 37.1**, farming was basically concerned with livestock production, primarily cattle for beef and dairy products and sheep for wool and mutton, with corn grown only in small quantities for farm use. This region comprised the real upland moors like Dartmoor, Exmoor and Bodmin which provide only rough grazing, and the more fertile hill country of mid-Devon, and north and west Cornwall. In the unshaded region (south Cornwall and south and east Devon) some corn was produced for sale, especially in the fertile South Hams of Devon and along the Cornish coast, although livestock production was also important, and it was often combined with specialist production, such as the famous cabbages of the Paignton region and of course cider apples. Nearly every parish in Devon grew cider apples (**Map 37.2**) and they were also increasingly grown in Cornwall during the period, especially in the east. A Cornish writer, Kneebone, observed in 1684 that in the hundred of East: 'there is throughout the whole abundance of profitable Orchards whence syder as good as any mixture elsewhere in the land is produced'.[1]

In the south-west peninsula the mild, wet and constantly changing climate is well suited to the growth of grass and hence to livestock-rearing, with the massive hedges and ditches so characteristic of the region providing shelter from the buffeting winds and rain. This combination of climate and terrain discouraged the dense, village-based populations of arable regions, like East Anglia and the Midlands, and encouraged instead a sparser population settled in individual farms, hamlets, and small villages widely dispersed through the countryside.

The period saw an increasing diversification and commercialization of farming, especially after about 1650, when many new and unusual crops were introduced, such as onions and turnip seeds which were exported from Penzance as early as 1687.[2] This growing commercialization helped to link the region to other growth centres and to break down its isolation. The disadvantages of its peripheral location were compounded by a poor internal transport system and small local markets. Roads were only gradually improved

in the eighteenth century, and many remained the narrow, muddy tracks suitable only for pack-horses about which Celia Fiennes complained so bitterly on her journeys in 1698. From Exeter she wrote: 'went to Plymouth 24 long miles, and here the roades contracts and the lanes are exceeding narrow and so cover'd up you can see little about … the wayes now became so difficult that one could scarcely pass by each other, even the single horses, and so dirty in many places and just a track for one horses feete'.[3] The banks were so near together that she feared they would tumble down on her, especially as the roots of the trees which grew on top of them often loosened the stones and rocks which were meant to support them.

If the main roads were in this condition, it is not surprising that many of the farms, isolated off the beaten track, concentrated on self-sufficiency. This tendency was also strengthened by the small size of the local urban markets. Accurate figures for town sizes before 1801 are lacking, but at that date the largest town in Devon, Plymouth (with Devonport and other suburbs), had a population of only 43,532, compared with just over one million in greater London. Exeter, the capital of Devon, with its suburbs of St Thomas, Heavitree and St Leonards, had only 20,553 inhabitants. Tiverton was the largest town in mid-Devon with a population of 6,505, while the north Devon towns were much smaller. Barnstaple had 3,748 and Bideford 2,987 inhabitants. The Cornish towns were all small. Truro, the largest, with its suburbs in St Clement and Kenwyn had about 10,000 inhabitants, but Redruth, the next largest, had only 4,925.[4] St Austell with 3,788 people was the biggest town in mid-Cornwall (Bodmin had only 1,951 inhabitants) and the most populous town in east Cornwall was St Germans with 2,030 citizens. The important port of Falmouth boasted a mere 3,684 people.[5] Clearly, increased agricultural production was not going to receive much stimulus from such a limited hinterland, though exports to London and overseas provided some compensation.

Map 37.3 illustrates the farming systems of a sample of sixteen Devon farmers in the mid-seventeenth century based on probate inventories taken in the summer months, when crops were growing in the fields. Unfortunately only a few of the Devon inventories survived the air-raids on Exeter in 1942, and so this sample is smaller than desirable, but it covers quite a wide area in the

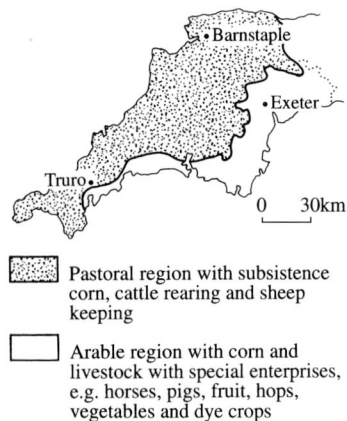

Pastoral region with subsistence corn, cattle rearing and sheep keeping

Arable region with corn and livestock with special enterprises, e.g. horses, pigs, fruit, hops, vegetables and dye crops

Map 37.1
Farming regions,
1640–1750.

SOURCE: based on map by Joan Thirsk in *The Agrarian History of England and Wales, Vol. V, Part 1, Regional Farming Systems* (Cambridge, 1984), xx and xxi.

county and a spread of farmers varying from the very large like Leonard Yeo, Esq., of Huish, whose movable possessions in 1641 were worth £1,140 13s 8d (a huge sum in the money of those days) to the very small like John Vallacott of Lynton in north Devon, whose possessions in 1645 were worth only £46 13s 2d.[6] Whereas Leonard Yeo had 59 acres of growing crops and 79 cattle and 118 sheep, John Vallacott had only 5.5 acres of crops, 12 cattle and 58 sheep. Leonard Yeo and John Lake of Brixham, whose movable possessions were worth £1,423 13s 4d in 1633, show that there were farmers who were operating on a large scale and producing for the market, but it is probable that they were very much in the minority.[7] The traditional Devon farm probably varied between 30 and 50 acres. In 1870 the mean average farm comprised 59 acres, and as late as 1907 55 per cent of Devon farms were smaller than 50 acres, and 17 per cent were smaller than five acres.[8]

The structure of farm sizes was a legacy from the past. Another was the size, shape and nature of the fields themselves, and equally important the farm buildings, yards, sheds (and indeed houses) from which they were worked, most of which were adapted to small farms. As so many south-western farmsteads are made from stone (often of massive granitic type), the cost of alterations could be heavy, and the character of past investments had a significant effect on the nature of farming practice.

EARLY ENCLOSURE OF FARMLAND IN THE SOUTH WEST

As far as the fields were concerned, their small average size and the often massive banks and hedges with which they were surrounded were all legacies from the very early date at which south-western enclosure occurred (there are exceptions, such as the famous Great field at Braunton in north Devon, where the remnants of common-field strip cultivation still survive). This early enclosure was a feature much commented on by past writers on agriculture. Much of it represented the original colonization from the waste, but some of it had resulted from enclosure of former common fields. Writing in 1796, William Marshall, an experienced observer of English agriculture, and then the manager of the Buckland Abbey home farm, near Plymouth, listed enclosure as one of the twenty-eight ways in which south-western agriculture differed from that of the rest of England.[9]

Two centuries before Marshall's time, the differences were probably even more marked, and if he had been writing in 1596 instead of 1796 Marshall would surely have remarked on the fact that the farmland of the two counties was almost entirely enclosed. By the time he wrote, enclosure in England was commonplace, but in 1596 it was exceptional and remarkable, although by that time much of Devon and Cornwall had been enclosed for at least a century and some parts for a great deal longer than that. Distinct 'pays', agrarian landscapes, are

recognizable in the two counties from at least the late Middle Ages, and the disappearance of open field derives largely from the nature of the farming in each of these. The change had taken place earliest in pastoral east Devon in the thirteenth and fourteenth centuries. Here, between the River Exe and the Dorset and Somerset borders, on the soils derived from greensands and marls, stock-rearing became the principal activity and the demands of this, in particular the need for enclosed pasture fields, led to the exchange and subsequent enclosure of open-field strips into small hedged plots banked and fenced in the traditional south-western fashion. Fields here were on average smaller than elsewhere in Devon, their enclosure dates from a time when the population pressure on land was considerable, and the accumulation and amalgamation of more than one or two strips at a time, for the purposes of enclosure, was probably difficult.

In the arable corn-growing South Hams, that part of Devon lying west of Dartmouth and between the Rivers Plym and Dart, the change was delayed until the fifteenth century and was facilitated by the reduction in population and the abandonment of land following the Black Death (*c.*1349). Here on the Devonian grits and shales, corn-growing was the dominant activity and for this open-field arable strips and furlongs were far less of a hindrance than they were for stock-raising. Hence there was less pressure to enclose, until a declining population made it easy. At Kenton in the fertile valley of the Exe the change was delayed even further as a consequence in part of a strong local market for corn from the rising population of nearby Exeter. In general, however, by 1500 Devon and Cornwall were enclosed counties, though here and there, particularly around the larger villages, Ottery St Mary, Woodbury, Otterton and of course at Braunton land was being worked in unenclosed strips.

The reasons for this early disappearance of what remained commonplace in much of England until the eighteenth century, varied with local conditions, but there were circumstances that applied to the South West as a whole that encouraged enclosure from an early date. The most obvious is that all over the South West in almost every parish there were considerable areas of common grazing; the rough moory hilltops were unenclosed and provided useful summer grazing for cattle and sheep. The abundance of common grazing of this kind in almost every parish or manor enabled farmers to summer their cattle and sheep without the need to graze the common arable after harvest or the common fallow field throughout the year. Hence in the South West it was easy to make private agreements, between tenant and tenant or tenant and landlord, to amalgamate and enclose neighbouring strips—agreements that would not be to the detriment of surviving open-field tenants. They had plenty of space on the commons for their livestock. It is clear from the field pattern as seen on a modern map that often all that was done as a result of these agreements was to put a hedge around existing strips or groups of strips which have thus become 'fossilized'.[10]

A second reason for early enclosure may be that open-field agriculture, as it is understood from studies of midland villages,

**Cider production in
hogsheads per parish**

1000
500
100

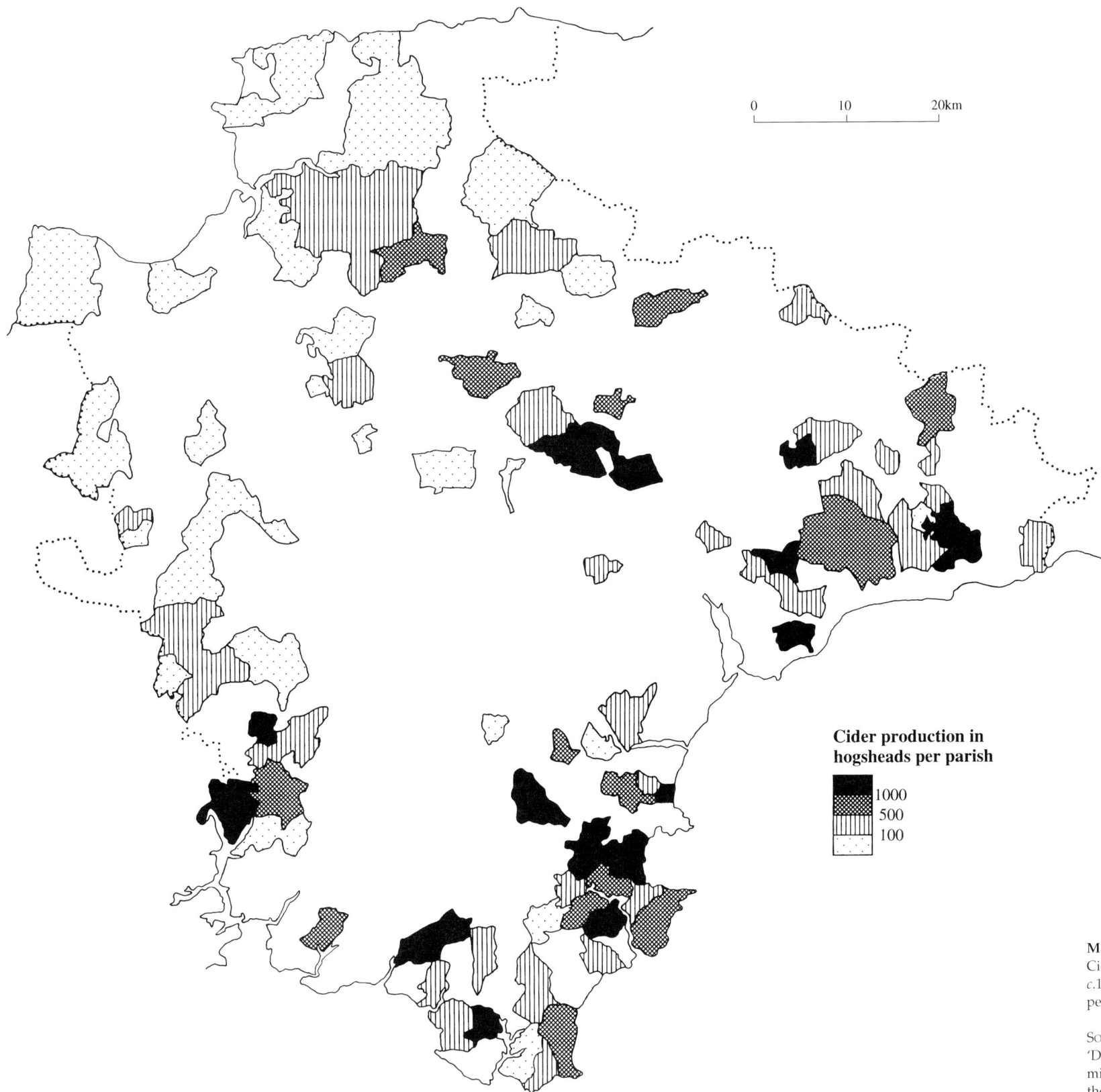

Map 37.2
Cider production in Devon,
*c.*1750, in hogsheads
per parish.

SOURCE: Robin Stanes,
'Devon agriculture in the
mid-eighteenth century:
the evidence of the Milles
enquiries', in M.A. Havinden
and Celia M. King (eds),
The South-West and the Land
(Exeter, 1969), 43–65.

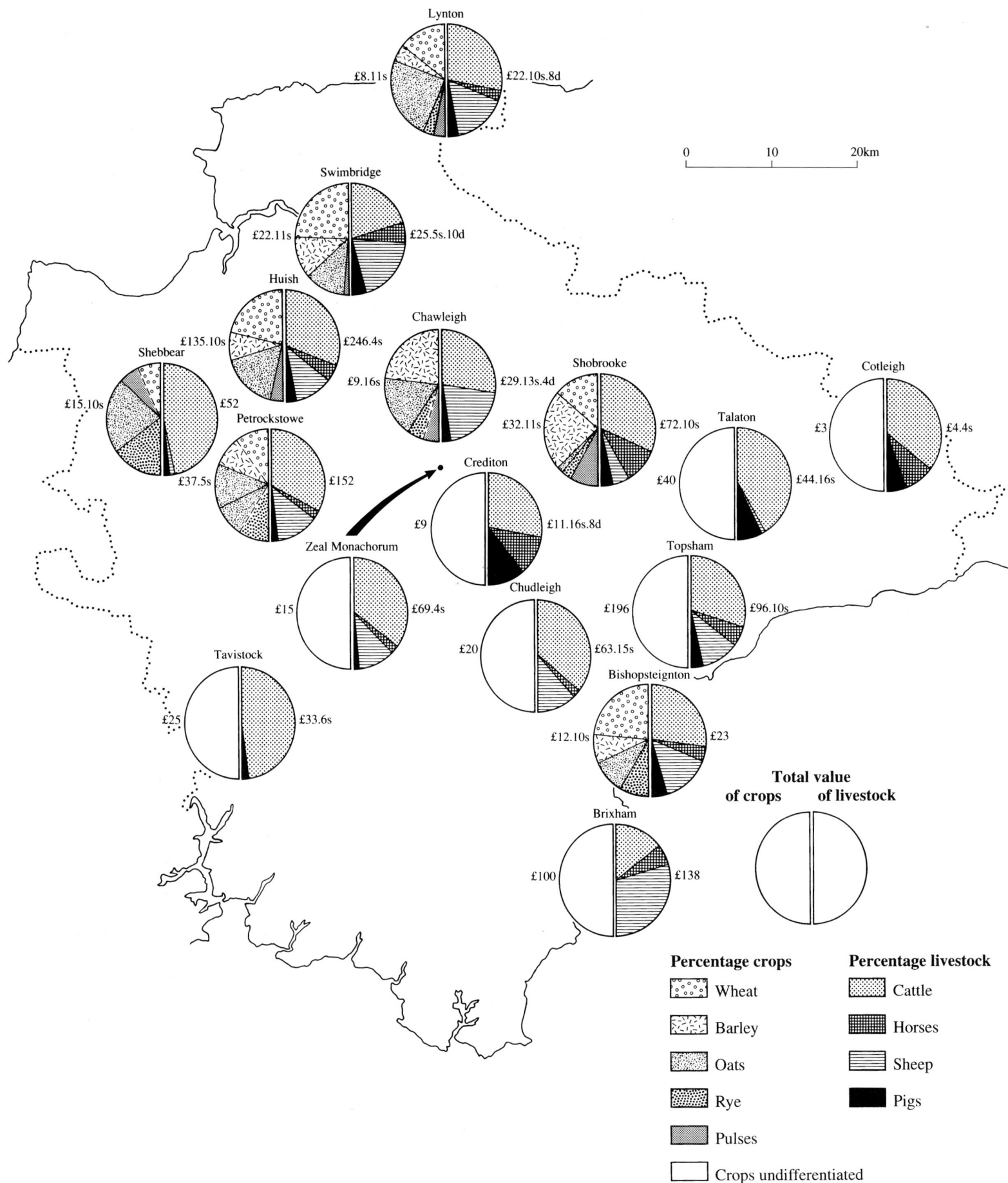

Lynton
£8.11s   £22.10s.8d

Swimbridge
£22.11s   £25.5s.10d

Huish
£135.10s   £246.4s
£9.16s

Chawleigh
£29.13s.4d
£32.11s

Shobrooke
£72.10s

Talaton
£40   £44.16s

Cotleigh
£3   £4.4s

Shebbear
£15.10s   £52
£37.5s

Petrockstowe
£152

Crediton
£9   £11.16s.8d

Zeal Monachorum
£15   £69.4s

Chudleigh
£20   £63.15s

Topsham
£196   £96.10s

Tavistock
£25   £33.6s

Bishopsteignton
£12.10s   £23

Brixham
£100   £138

0   10   20km

Total value
of crops   of livestock

Percentage crops

Wheat
Barley
Oats
Rye
Pulses
Crops undifferentiated

Percentage livestock

Cattle
Horses
Sheep
Pigs

Map 37.3
Proportions of crops and
livestock (by value) in
Devon probate inventories,
1631–1656.

SOURCE: Margaret Cash (ed.),
*Devon Inventories of the
Sixteenth and Seventeenth
Centuries*, Devon and Cornwall
Record Society, new series,
11 (1966).

with their two or three or more large arable fields, their fallow field and their prescribed rights to grazing after harvest and so on, never reached its full logical outcome in the South West. There is no evidence of a fixed, three-year, three-field rotation or of a fallow field in south-western records. Instead strips were grouped in an indeterminate number of 'culturas' or furlongs distributed around the village as the lie of the land dictated and cultivated according to no strict rotation. It may be that an initially scanty population and the existence of alternative occupations (fishing, mining, cloth-making), and the beginnings of a market economy as a consequence, all contributed to inhibit this full development. It may also be that manorial organization, historically strong in arable areas, never developed to the full in the more pastoral South West. Fully developed open-field agriculture seems to have resulted from pressure of population and a consequent need to regulate.[11] That pressure probably never existed in the South West, or, if it did, was absorbed from an early date by other economic opportunities, especially tin-mining, fishing and cloth-making.

Whatever the reasons, it is clear that by the early modern period the landscape of the two counties was much as it is today and that this was cultivated by individual farmers largely for their own gain unhindered by the rules of open-field communal husbandry. On their enclosed fields farmers were able to grow what crops or keep what livestock they pleased and were in a position to take advantage of any market opportunities that arose as the region began to develop towards mining and handicraft industries in the seventeenth century. It was this that probably prompted Oliver Cromwell's celebrated remark on a visit to the West Country (c.1650) that 'I have been in all the Counties of England and I think the husbandry of Devonshire to be the best'.[12]

CROPS AND LIVESTOCK

The farmers shown on **Map 37.3** were in every case mixed farmers and it is noticeable that their livestock were considerably more valuable than their crops. To some extent this was inevitable since livestock were a capital asset and crops only an annual increment, but the livestock usually exceeded the value of crops by a sufficient margin to indicate that the production of meat, dairy products and wool was the farmer's primary activity. An exception was John Weslake, yeoman, of Topsham in the fertile Exe valley, whose 66 acres of corn in 1646 were worth £196 compared with livestock worth £96 10s.

Generally amongst Devon livestock, cattle were the most important, including oxen for ploughing and dairy and beef cattle, with sheep usually coming second in value, although they were usually more numerous. For instance John Babb of Bishopsteignton in 1633 had seven cattle worth £12 10s and thirty-two sheep and lambs worth only £6 10s. Most farmers, though not all, kept a few horses and a few pigs, but poultry

were quite rare. Anne Parnacot, widow, of Petrockstow had poultry worth 15s in 1635, and Leonard Yeo had geese and other poultry worth £1 10s in 1641, but Humfry Harvie, yeoman, of Zeal Monochorum had none in 1643, although he was quite a wealthy man with livestock worth £69 4s.

Crop preferences varied quite widely. Of the seven farmers in the sample whose crops could be clearly differentiated, three had wheat as their most valuable crop, two had barley, two had oats and none had rye or pulses (beans and/or peas). Oats were the only crop that was grown by all seven farmers, and they were the second choice of four. Their prominence, especially in north Devon, is not surprising since they are more tolerant of cool, wet and acid soils than other crops, and were primarily fed to livestock—a further indication of the predominance of that sector of farming. Oats were also commonly malted to produce Devon's notorious beer. The relative popularity of wheat over rye on the other hand probably indicates that arable farming was improving because wheat requires better conditions and more care for its cultivation than rye. All in all, the inventories reveal a variegated and well-balanced farming system, partially adapted to a growing local and national market, and one which was no doubt profitable to its operators after the long rise in food prices during the sixteenth and early seventeenth centuries.

THE DEAN MILLES QUESTIONNAIRES, 1747–1756

Unfortunately very few inventories have survived for the eighteenth century, but a different source provides information about farming in Devon which is shown in **Maps 37.2, 37.4, 37.5 and 37.6**. These maps are based on the replies obtained between 1747 and 1756 by Jeremiah Milles, dean of Exeter, to a questionnaire sent to all the parishes in Devon. Dean Milles was intent on writing a history or description of Devon but never did so. His questionnaire was extensive, covering a great many subjects including detailed questions on farming practice. It was answered, sometimes rather perfunctorily, by the incumbents, or curates, or schoolmasters of 257 parishes in the county (57 per cent). It thus provides a very substantial sample and it is likely that conclusions drawn from it would have considerable validity for Devon as a whole. The maps illustrate the replies to the following questions. What is the usual value of arable, meadow, and pasture per acre? How manured—by lime, sand, dung, and in what proportions? What is the greatest produce of grain per acre? What quantity of cider is made yearly? What sorts of cattle are chiefly bred in the parish and are they remarkable for their size, shape, colour and breed?

Map 37.4 shows the variation in the annual value of land per acre (which probably in practice meant the annual rent) and the average yield of wheat per acre. The annual values have been divided into three categories and the yields of wheat into four. Poorish land, which was mainly in north and west Devon, was worth under 10s an acre; average land was worth between 10s

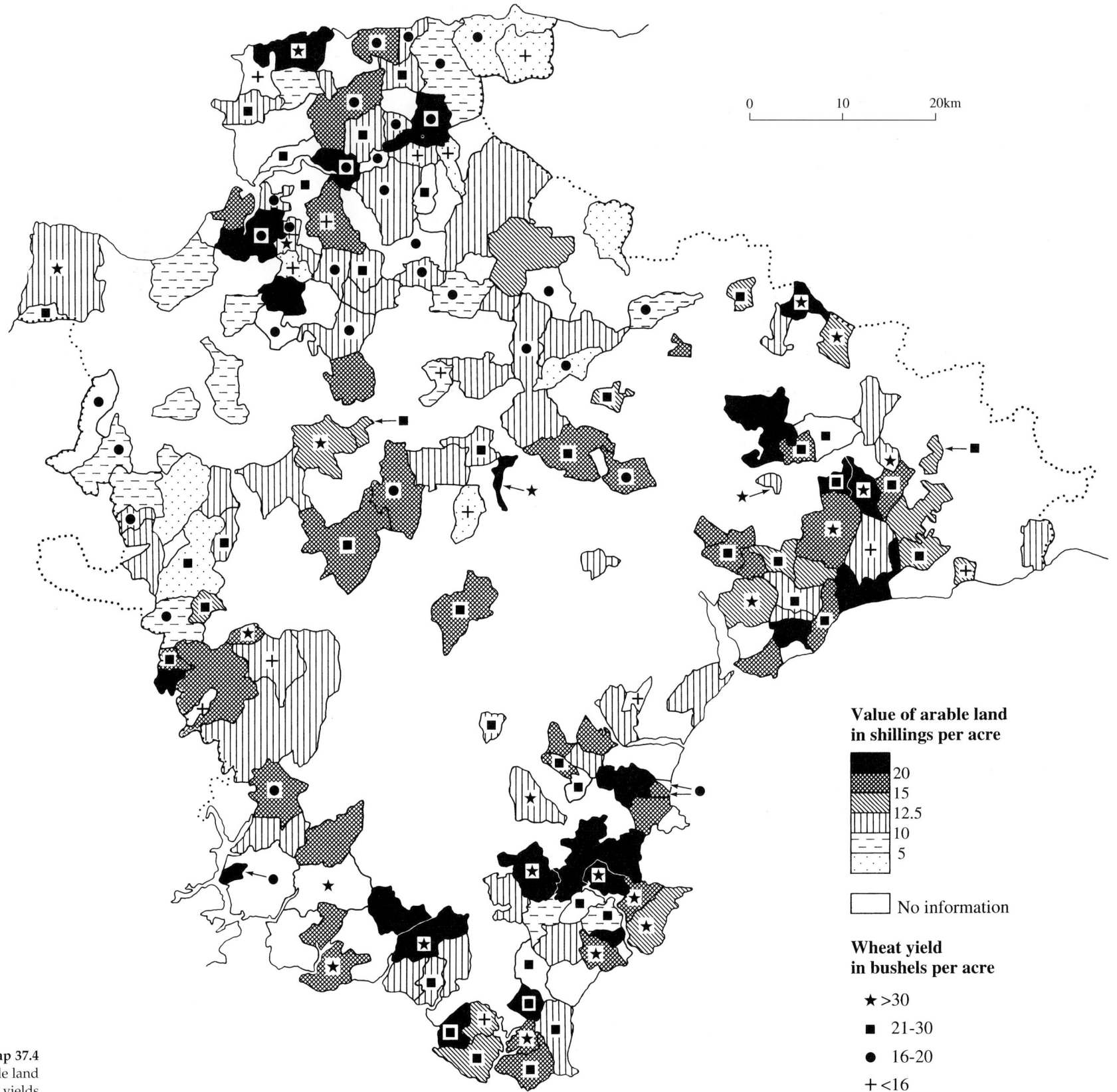

**Value of arable land
in shillings per acre**

20
15
12.5
10
5

No information

**Wheat yield
in bushels per acre**

★ >30

■ 21-30

● 16-20

+ <16

Map 37.4
The value of arable land
per acre and average yields
of wheat in bushels per acre
in Devon, c.1750.

SOURCE: as **Map 37.2.**

and £1, and good land over £1 an acre. The values indicated here are for arable land only, but in the South West 'arable' means land under temporary grass as well as land under crops. A range of figures is sometimes given in each parish and these have been averaged. The lowest figure for arable land was 1s 6d per acre at East and West Anstey on the fringes of Exmoor and the highest was £3 at Colyton in green and fertile east Devon. Meadow land—permanent grass—was generally twice the value of arable. Market garden land at Paignton for growing the huge flatpoll cabbages for which that town was famous fetched as much as £5 per acre. In contrast rough pasture at Buckland in the Moor was worth 6d per acre.

The wheat yields have been divided into poor, moderate, average and good and have been superimposed on the annual value data by means of symbols. This shows that in a general way wheat yields were related to annual values, although other factors, such as the quality of pasture and closeness to markets, prevent a perfect correlation. The most common answer to the question on yields was, 'twenty bushels of wheat, thirty bushels of barley and forty bushels of oats per acre'. Average yields, however, calculated from all answers, were 13cwt of wheat, 17cwt of barley and 13cwt of oats. Grain crops were grown normally for three years in succession without a break, and yields would lessen in the final year before the land was returned to grass. Using the seventeenth-century figure of six pecks to the acre as a usual sowing rate, yield ratios of wheat were about 15 bushels harvested for each one sown, compared to today's 32–40. The highest yield recorded was 60 bushels (33cwt) at Hatherleigh, the lowest was 8 bushels (4.5cwt) at Pancrasweek. Today's yields might be 50–60cwt. Generally the highest yields were obtained in the South Hams and east Devon, with mid and north Devon varying from average to poor.

The average yields obtained for cereals seem to represent evidence for a considerable improvement in cereal farming in Devon between the sixteenth and eighteenth centuries. For instance, Tavistock Abbey's Hurdwick Farm averaged 11.7 bushels of wheat per acre in thirteen assorted years between 1412 and 1537, while Samuel Colepresse's estimate of average wheat yields in 1667 was 15 bushels, and Milles's average was 22 bushels and William Marshall's in 1796 was the same. For barley there are no Hurdwick figures, but the 1667 estimate was 14 bushels, Milles's was 32 bushels and Marshall's in 1796 was 35 bushels per acre. For oats Hurdwick recorded 22 bushels, Colepresse only 13 (a suspiciously low figure) and Milles gave 42 bushels per acre. Marshall did not record a 1796 average.[13] No doubt the increased use of lime and other manures, and perhaps improved ploughs and other implements, were largely responsible for these quite significant improvements.

Map 37.5 shows the widespread use of lime throughout Devon and the areas where marl and sea-sand were used as substitutes. Almost all Devon soils are naturally acid and require a calcareous fertilizer like burnt lime, marl (a chalky sub-soil substance) or sea-sand derived from limestone or shells.

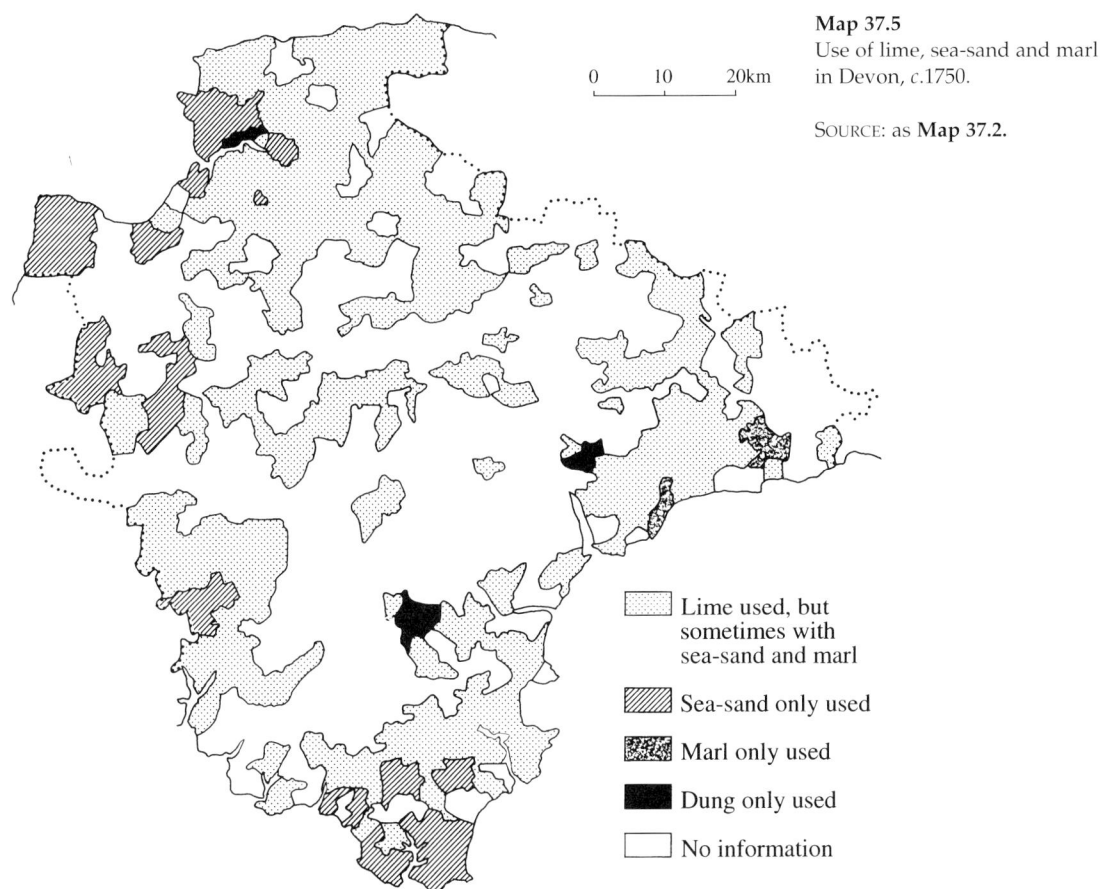

Map 37.5
Use of lime, sea-sand and marl in Devon, c.1750.

SOURCE: as **Map 37.2**.

Lime used, but sometimes with sea-sand and marl

Sea-sand only used

Marl only used

Dung only used

No information

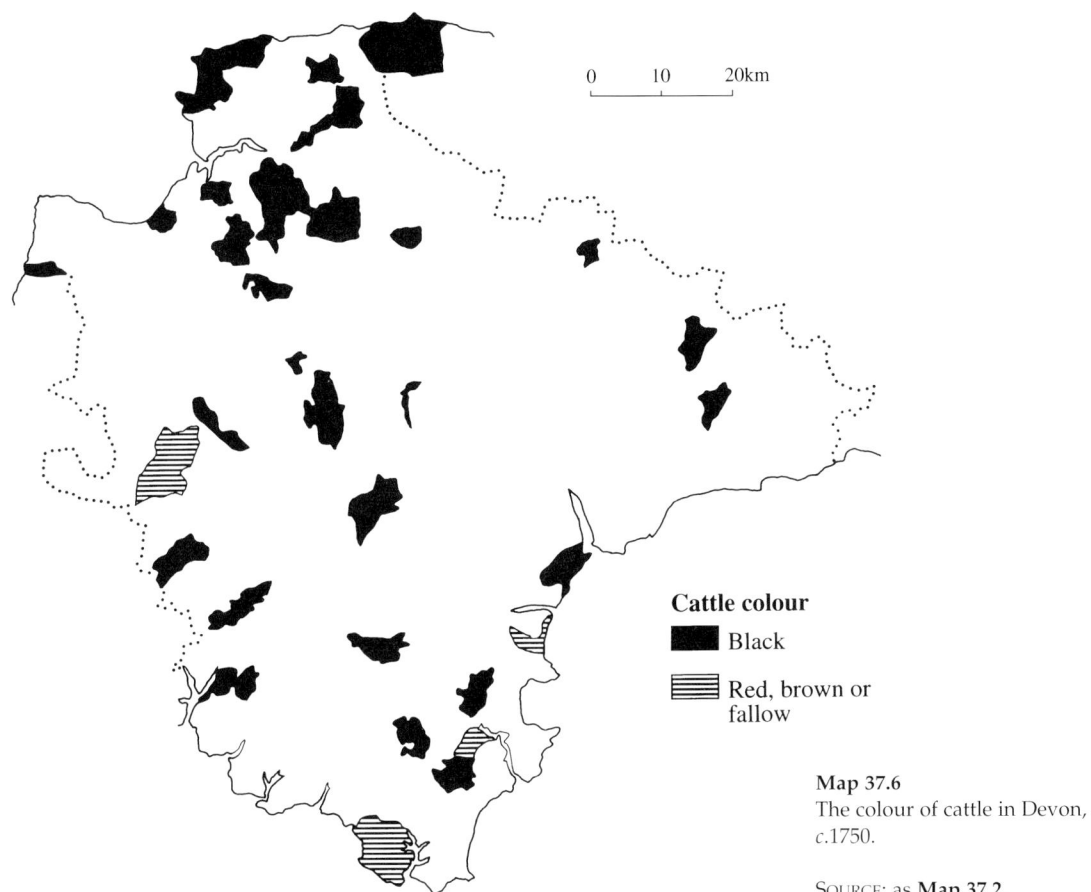

Cattle colour

Black

Red, brown or fallow

Map 37.6
The colour of cattle in Devon, c.1750.

SOURCE: as **Map 37.2**.

Eighteenth-century farmers did not know that these substances reduced soil acidity, allowing other manures like dung to work; they thought they were a plant food. Lime was used in 197 of the 234 parishes for which this question was answered (84 per cent). The remainder all applied calcareous sea-sand either dredged from estuaries or dug from beaches. Some forty-two parishes, mostly coastal, applied both lime and sand. By this date covenants in leases often expressly demanded the application of lime or dung or sand before land was ploughed for corn. Proximity to a river or to a beach clearly determined the use of sea-sand; sand was too far away and therefore too dear at Beaworthy and Bratton Clovelly, though it was used at Rose Ash, 24 miles from Barnstaple Bay. The use of marl was almost entirely confined to east Devon. This must reflect availability. Lime was also used in east Devon. Outside east Devon only St Marychurch used marl. This widespread use of lime, marl and sea-sand represented a considerable improvement in Devon agriculture since the sixteenth century. The first reference to the use of lime comes from John Hooker, a former chamberlain of the city of Exeter, who noted its use in 1599. It had probably been used earlier by the more advanced farmers, and its use was spreading rapidly. In 1630 Tristram Risdon of Winscott Barton near Torrington could write: 'of late, a new invention hath sprung up, and been practised amongst us, by burning of lime, and incorporating it for a season with earth, and then spread upon the arable land, hath produced a plentiful increase of all sorts of grain where formerly such never grew in any living man's memory'.[14] It seems probable that lime was first used by the better farmers between 1550 and 1650 and had spread widely by 1700, being virtually universal by 1750.[15]

Map 37.2 shows cider production in Devon c.1750. The parishes have been divided into three categories: those producing over 1,000 hogsheads (one hogshead = 52.5 Imperial gallons) a year, those producing between 500 and 1,000 hogsheads, and those making fewer than 500 hogsheads. Most Devon parishes could grow cider apples, but in much of north Devon, except in the lower Taw valley, production seems to have been for home consumption only. No cider could be made at East and West Anstey, while land at Frithelstock was too 'stiff and cold' for cider. In south Devon cider was made for sale. Paignton exported cider to London and across the Atlantic to Newfoundland. The greatest production was recorded at Berry Pomeroy where 2,000–3,000 hogsheads (c.105,000–160,000 gallons) were produced each year. East Budleigh, Bere Ferrers and Staverton produced 2,000 hogsheads. In contrast, Buckland in the Moor, Stoke Rivers, Newton Tracey and East Down, the last three in north Devon, could only manage between six and twelve hogsheads. This suggests a total production from all parishes of about 170,000 hogsheads or 10 million gallons, roughly 33 gallons per head of population, if no cider was exported.

Map 37.6 shows the colour of cattle in the rather small sample of forty-five parishes which responded to this question, but it produced the most unexpected responses. Devon cattle, by

the time the two breeds were recognized at the beginning of the last century, were red: the 'Red Rubies' of north Devon, bred by, among others, the Quartly family of Molland, and the large, yellower South Devons (also called South Hams or Malborough Reds) from around Kingsbridge. These dual-purpose cattle produced not only beef but the richest milk in England, which was the basis for the famous Devon and Cornwall clotted cream. The map reveals their presence clearly enough but gives no indication of the 'Red Rubies', though the Quartlys were at work in north Devon in the eighteenth century. In fact thirty-five of the forty-five respondents to this question recorded cattle that were black, totally unlike the present Devon breeds. There was a Black Cattle Fair at Torrington and black cattle were sold at Tavistock market in some numbers. There were still small black cattle in Cornwall in the early nineteenth century, perhaps rather like the Welsh Black, and it may be that it is to this sort of cattle to which the answers refer. The rector of Chivelstone in south Devon indeed indicated that there were two sorts of cattle, large cattle kept on the better land and smaller cattle 'kept on the commons'. These latter may have been black, perhaps descendants of ancient local stock. Black cattle were probably native to the South West and it is perhaps surprising that they have not survived in Devon and Cornwall. However, Devon cattle occasionally have black calves and this may be a clue to the fate of the old black breed—they were probably crossed with red cattle (perhaps South Devons) to produce the eventually new North Devon breed. Nevertheless the complete absence of red cattle in north Devon in the 1750s is an unexpected and puzzling aspect of the Milles enquiries, which suggests that the work of the late-eighteenth-century cattle breeders like the Quartlys of Molland and the Davys of Rose Ash may have been more innovatory than was previously believed. They probably did not merely improve a local breed but replaced it with a completely new one.

Generally the Milles enquiries show a Devon agriculture which was growing increasingly commercialized and in which significant improvements had occurred since the seventeenth century, notably in higher grain yields, more extensive use of lime, marl and sea-sand, and, less certainly, in improved livestock and extended cider production.

CORNWALL: SEVENTEENTH-CENTURY PROBATE INVENTORIES

In **Maps 37.7 and 37.8**, the focus shifts from Devon to Cornwall, where a much larger sample of probate inventories has survived and has been analysed by James Whetter in his *Cornwall in the Seventeenth Century*. **Map 37.7** shows the regional variation in the proportion (by average value) of crops grown in four regions of Cornwall in two periods, while **Map 37.8** shows changes in the proportions of livestock kept (again by value). The periods compared are 1600–20 and 1680–1700. The basic structure of Cornish agriculture was very similar to that of Devon, but owing to the greater remoteness of Cornwall, farming was probably less commercialized, and the average size of farms was even smaller than in Devon. For instance, in an analysis of several hundred Cornish probate inventories for the seventeenth century, James Whetter found that there was a clear relationship between the value of crops and livestock shown in an inventory, and the size of the testator's farm. This revealed that, both at the beginning and the end of the seventeenth century, well over half the farms were smaller than 15 acres and that 82 per cent of them were smaller than 40 acres.

Whetter notes that 'physical and cultural factors meant that in the main agriculture was not practised in open fields, as was still the case in much of England, but in enclosed land'. In the early part of the century, John Norden observed of the Duchy manor of Leigh Durrant, Pillaton, on the river Tamar 'some part of this Manor lieth in Comon Fields which is hardly found in any Manor of his Highness ells in Cornwall'. The inventories suggest that though there may have been some open fields in the early part of the century, little remained at the end.[16]

The fact that so much land was enclosed probably meant that Cornish farming, like Devon's was more flexible, and possibly more technically advanced than in other parts of England. John Taylor visiting Cornwall in 1647 shortly after the end of the Civil War observed: 'it is a wonder that such rugged Mountains do produce such fertility of Corn, and Cattle, for if the happy dayes and times of peace were once settled, Cornwall might compare with any County in England for quantity of all necessaries needfull'.[17]

Like their Devon neighbours, Cornish farmers concentrated in the main on livestock production and about 60 per cent on average of their total investment went into animals, although there was some regional variation, with corn being somewhat more important in the south. **Map 37.7** shows that in the period 1600–20 there was not much regional variation in the proportions of wheat and rye grown, but that with barley and oats the variation was greater. Wheat varied between 45 and 51 per cent of the value of all crops in all regions, while rye was insignificant at under 2 per cent in all regions. Barley was not a very popular crop, varying between 12 and 18 per cent in the centre and west and falling to a mere 4 per cent in the east. Oats were almost as popular as wheat in the east (at 45 per cent), but declined towards the west, being only 30 per cent in the far west.[18]

By the end of the century the picture had altered somewhat surprisingly, with a big advance in the amount of barley being grown at the expense of wheat and oats. Wheat fell as a proportion in all regions except the west. It declined by one-quarter to one-third in the east and south and by one-half in the north. Oats suffered similar sharp declines and rye disappeared altogether. It is hard to account for this decline in the popularity of wheat, as it was the essential bread crop, which was increasing nationally.[19] Whetter suggests three reasons. First, barley was better suited to Cornwall's soil and climate, and in an age of growing internal trade, perhaps wheat could have been imported more cheaply than it could have been grown, especially on the

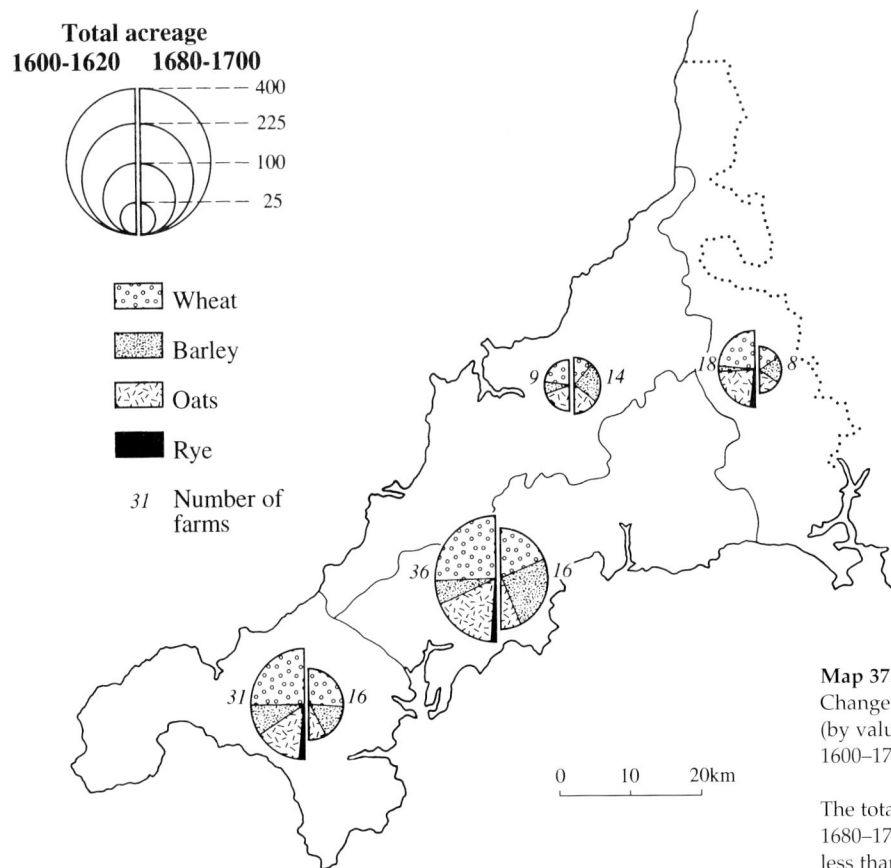

Map 37.7
Changes in crop proportions (by value) in Cornwall, 1600–1700, by region.

The total acreage sown in 1680–1700 was probably not less than in 1600–20, but the sample of farmers was smaller.

SOURCE: sample of probate inventories in James Whetter, *Cornwall in the Seventeenth Century: An Economic Survey of Kernow* (Padstow, 1974).

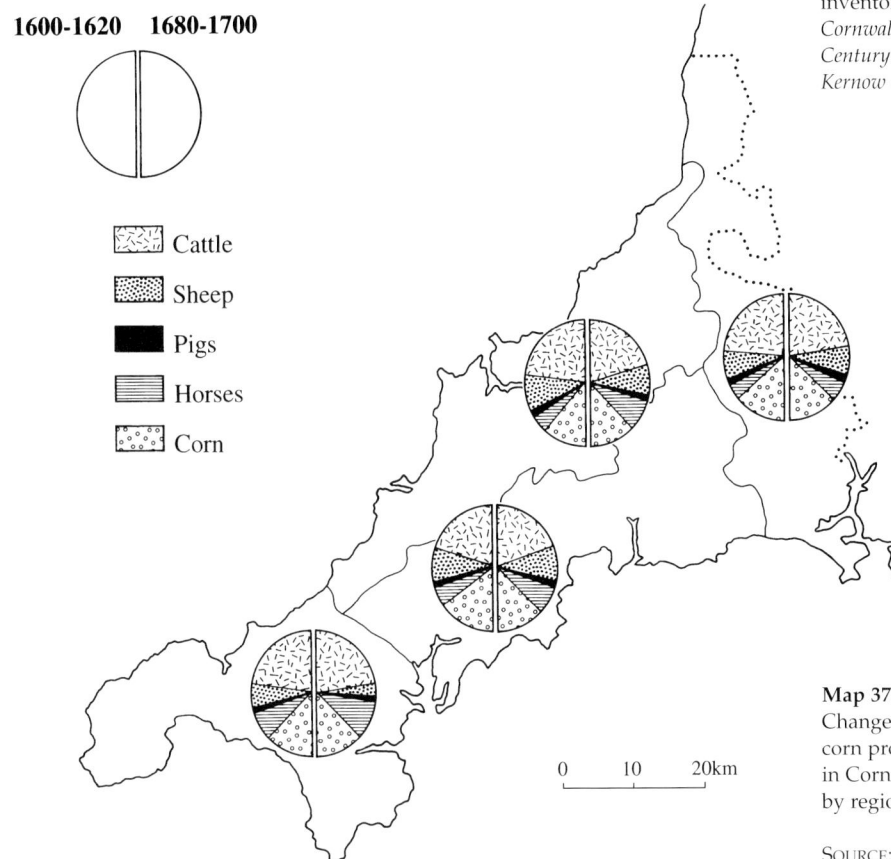

Map 37.8
Changes in livestock and corn production (by value) in Cornwall, 1600–1700, by region.

SOURCE: as **Map 37.7**.

poorer soils. Second, the demand for beer increased from the local population and the shipping trade. Third, barley was in demand from the growing mining population, not only for beer but also for food. At the end of the century it was noted that the diet of the tinners consisted of 'barley-bread (as coarse as Horse bread) with gruel thicken'd oftener with barley meal, than oatmeal'.[20] Whatever the reasons, Cornish grain farming was diverging quite markedly from the national and south-west regional trends.[21]

Map 37.8 shows changes in proportions of types of livestock kept (by value) and also shows the value of livestock compared with crops. It is divided into the same regions and time periods as **Map 37.7**. As in Devon, farmers in Cornwall kept relatively small numbers of all the basic livestock. Oxen and sometimes horses were kept for ploughing, as well as dairy and beef cattle. In all regions and in both periods cattle of various types comprised between 40 per cent and 50 per cent by value of all farm stock, including growing crops. This is what would be expected in a pastoral region, but their greater importance than sheep was quite marked. Sheep also showed a greater variation through the century. Averaging between 14 and 18 per cent in the earlier period, they maintained this position in eastern and central Cornwall in the later period, but declined to a mere 7.5 per cent in western Cornwall. Pigs always remained relatively unimportant, ranging between 2.5 per cent and 4 per cent in all regions at both periods. The growing popularity of horses was the greatest change during the century. Under 12 per cent by value in all regions except the west, where they were 15 per cent in 1600–20, they had risen to 10 per cent or over in all regions by 1680–1700 and had gone to just over 20 per cent in the west. Cornish horses and ponies were of course much in demand in the growing tin- and copper-mining industries, as well as for pack transport generally. Horses were, however, increasingly being used for ploughing and Whetter points out that it was a sign of advance in Cornish agriculture because horses could work more quickly and efficiently than oxen, although the latter were by no means yet displaced.[22]

Unfortunately, there is no Cornish equivalent of Milles's enquiries, nor many eighteenth-century inventories, so later developments cannot be mapped. There is some evidence that Cornish agriculture may have been somewhat stagnant, especially in the first half of the eighteenth century. With its mineral wealth and large fishing interests Cornwall was less absolutely dependent on its agriculture, and farming was frequently a part-time occupation and of second rank in the county. In discussing the profitability of his Cornish estates and the possibility of further purchases of land in the county, Thomas Pitt (of Boconnoc near Bodmin) ordered his agents in 1721 to 'remember to enquire into all ye other estates where there are mines working for I fear I am ill dealt with on that account for I hear of a great many Cornish people that have increased their estates very much by mines that have been discovered upon them, and never more need of looking out sharp than now'. This statement can be seen as a realistic verdict upon the status, progress, and relative profitability of Cornish agriculture at this time.[23]

John Rowe also suggests that Cornish agriculture remained fairly stagnant until about 1780, though he notes the earlier widespread cultivation of potatoes as a useful innovation. The early associations of the south-west with the New World might have led to the early cultivation of potatoes in Cornwall, but it seems that it was its utility in Ireland as a preparatory crop for barley on heath and bog soils that led to its wide cultivation in the south-west after 1700. By 1758 Borlase was writing that the potato was 'a more useful root' than the turnip, and that it was grown all over Cornwall, thriving best on shallow, poor lands.[24]

By the end of the century, though, there seems to have been considerable all-round progress, for William Marshall writing in 1796 could say: 'I am agreeably disappointed with respect to Cornwall. From what I had seen on the banks of the Tamar, I expected to have found as I went westward, a wretched county.... On the contrary the county whether in point of soil or cultivation—except the higher mountains and they are good in their kind—is above mediocrity.'[25]

LANDOWNERS

So far no mention has been made of tenures or landownership. Although a few farms were worked by the occupiers, most were let on various kinds of leases, frequently the long lease for three lives or ninety-nine years, whichever was the shorter. Such leases were of course purchased for a lump sum, and often continued for generations as new lives were bought in. Although the dominance of very large landowners was not so great in Devon and Cornwall as in some other counties, the influence of the great estates was still very strong, as can be seen from **Maps 37.9 and 37.10**, which indicate the location of properties of five of the largest landowners in each county in the eighteenth century. As there are no accurate contemporary figures it is not possible to show the acreages owned (as it is for 1873), but merely to indicate the parishes in which the leading owners possessed land. This shows the range of influence, and reveals that it was very extensive.

Map 37.9 shows five of the largest landowners in Devon based on catalogues and indexes in the Devon Record Office. These families (with their 1873 acreages) were the Rolles of Stevenstone (55,592 acres all in Devon), the dukes of Bedford of Woburn, Beds, and Tavistock Abbey (22,607 acres in Devon and 1,231 in Cornwall), the earls Fortescue of Castle Hill, Filleigh (20,171 acres in Devon and 571 in Cornwall), the earls of Devon of Powderham (20,049 acres in Devon) and the Aclands of Killerton, (15,018 acres in Devon and 4,578 in Cornwall). Lord Poltimore, Lord Clinton, the earl of Portsmouth and the prince of Wales as duke of Cornwall also had estates on the same scale. There were sixteen estates in Devon alone of over 10,000 acres. The process of building up these estates was already well under

Map 37.9
Five of the largest Devon estates in the eighteenth century.

SOURCES: estate records of Rolle, Russell, Fortescue, Courtenay and Acland families in Devon Record Office.

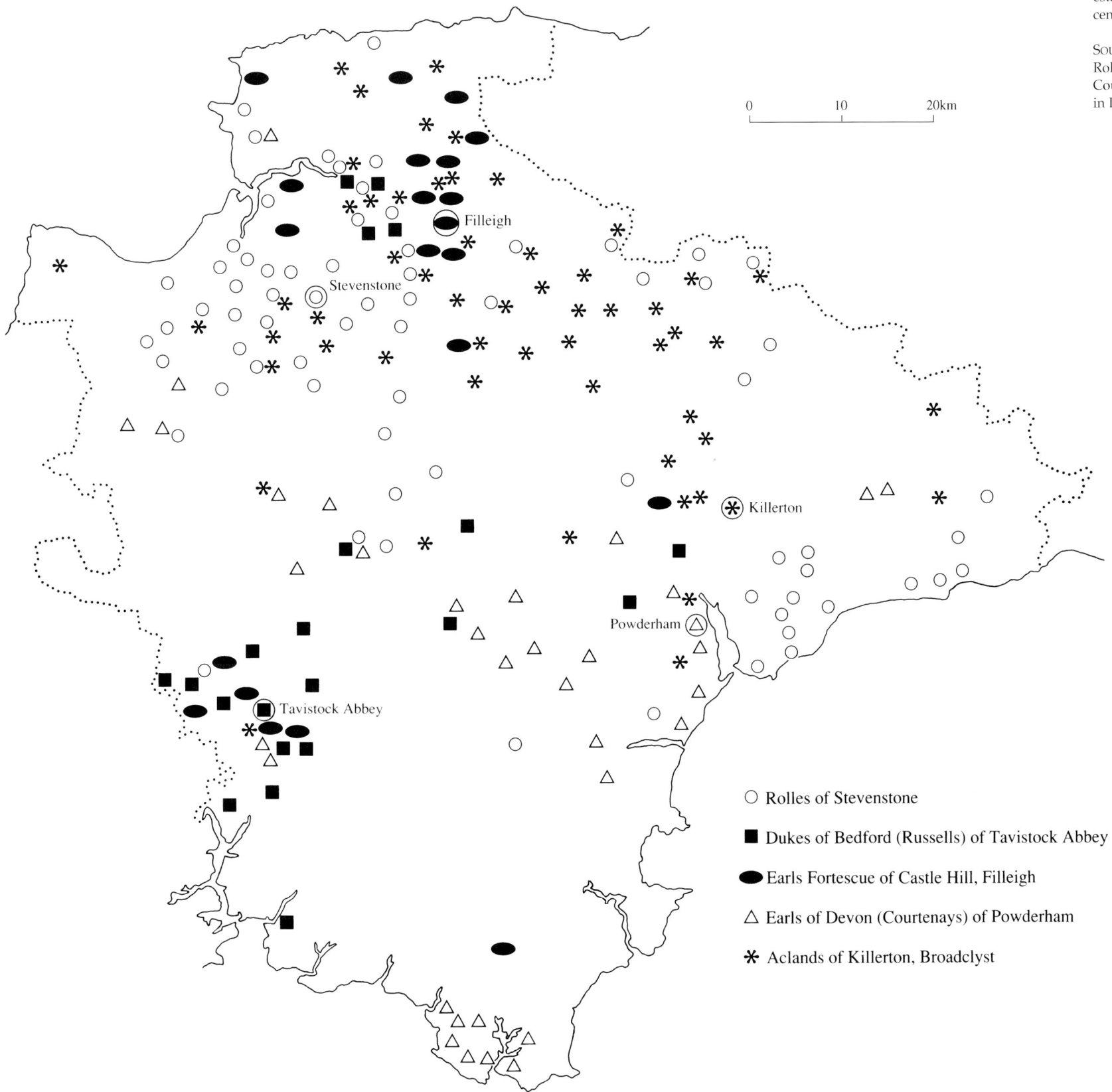

Filleigh

Stevenstone

Killerton

Powderham

Tavistock Abbey

○ Rolles of Stevenstone

■ Dukes of Bedford (Russells) of Tavistock Abbey

● Earls Fortescue of Castle Hill, Filleigh

△ Earls of Devon (Courtenays) of Powderham

✳ Aclands of Killerton, Broadclyst

Map 37.10
Five of the largest
Cornish estates in the
eighteenth century.

SOURCES: catalogues and
indexes in the Cornwall
Record Office (Rashleigh,
Robartes and Basset estates)
and David and Samuel Lysons,
Magna Britania, Vol. III,
Cornwall (London, 1814) for
Boscawen and Mount
Edgecumbe estates.

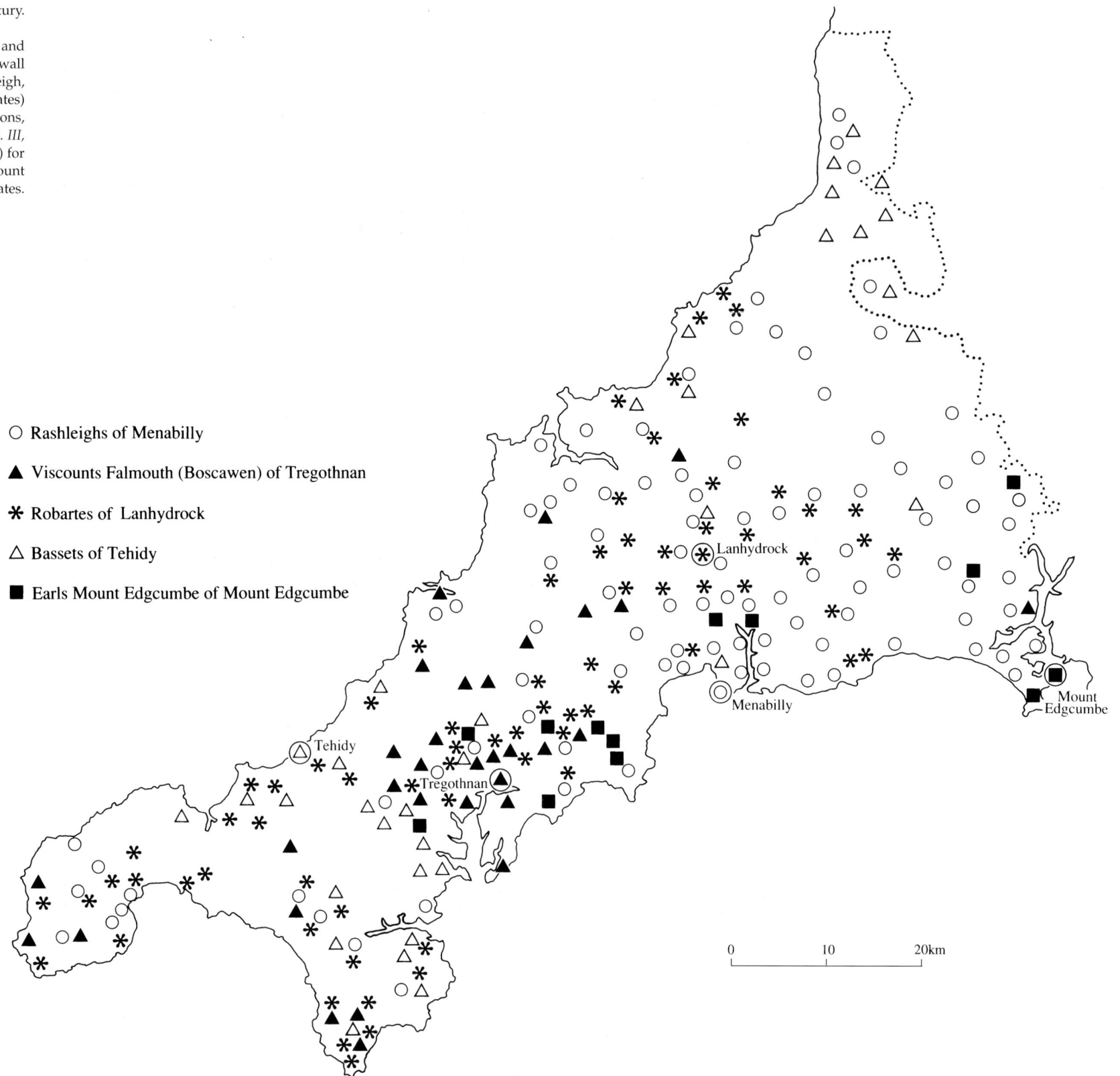

○ Rashleighs of Menabilly

▲ Viscounts Falmouth (Boscawen) of Tregothnan

✳ Robartes of Lanhydrock

△ Bassets of Tehidy

■ Earls Mount Edgcumbe of Mount Edgcumbe

way in the previous centuries. **Map 37.9** indicates the parishes in Devon in which five of these families had interests in land and property in the seventeenth and eighteenth centuries at some date, varying from very large to quite small holdings. So the Rolles owned almost all of Bicton and Otterton parishes but had only a farm or two or house property in some other parishes; and the same can be said of the other families. The map indicates only that these families had an interest in the parish in the form of freehold or leasehold property or as lords of the manor or as lay impropriators, entitled to collect the great tithes.

Rent from farmland was not the only concern of landowners; the dukes of Bedford were interested in the political control of the boroughs of Tavistock and Okehampton in Devon and Camelford and Launceston in Cornwall, and rented house property there gave them some control of their tenants' votes. The potentially vast profits of mining may also have influenced the acquisition of property, especially in west Devon along the Tamar valley with its rich copper mines. These families had differing origins. The Courtenays of Powderham, near Exeter, inherited the earldom of Devon ultimately from Baldwin de Brionne, the Conqueror's sheriff of Devon. Many of the numerous original estates of the Courtenays, Topsham and Tiverton for instance, were lost at the execution of Henry Courtenay marquess of Exeter in 1539, and were never recovered. The Aclands came from, and took their name from, Acland in Landkey parish near Barnstaple and were freeholders there in Henry II's day. The Fortescues of Castle Hill in Filleigh parish near South Molton were another medieval family, holding land near Modbury in the South Hams as early as 1140; one ancestor was chief justice and chancellor to King Henry VI. The first Rolle of note was a Tudor lawyer who took advantage of the sale of monastic lands at the Dissolution to add to an existing estate, centred on Stevenstone House in St Giles in the Wood parish, near Torrington. Monastic lands at Tavistock and at Woburn were also the basis of the wealth of the Russells, originally from Kingston Russell in Dorset, who served both Henry VII and Henry VIII well. Lord Russell was rewarded with all the Tavistock Abbey lands for his services to the king in the South West after the disgrace of the Courtenays. With all these families, it is probable that as much land was acquired by careful and fortunate marriages as by purchase.[26]

Map 37.10 (compiled on the same basis as **Map 37.9**) shows five of the largest estates in Cornwall in the eighteenth century, derived from catalogues and indexes in the Cornwall Record Office and Daniel and Samuel Lysons, *Magna Britannia, Vol. 3, Cornwall*, published in 1814 from materials of which many were gathered at, or referred to, earlier dates. These families (with their 1873 acreages) were the Rashleighs of Menabilly, Tywardreath, near Fowey (30,156 acres in Cornwall and 242 acres in Devon),

the viscounts Falmouth (Boscawens) of Tregothnan, St Michael Penkevil, near Truro (25,910 acres in Cornwall), the Robartes of Lanhydrock near Bodmin (22,234 acres in Cornwall), the Bassets of Tehidy, near Camborne (16,969 acres in Cornwall), and the earls Mount-Edgcumbe of Mount Edgcumbe, Maker, overlooking Plymouth Sound (13,288 acres in Cornwall and 4,935 acres in Devon). Other large estates were held by the Pitts of Boconnoc, the Hawkins of Trewithen, Probus, near Truro, and the Vyvyans of Trelowarren, near Helston.

The Rashleighs originated as merchants in Fowey and extended their property by astute purchases of monastic property in the Tudor period, with purchases like Bodmin Priory, bought for £360 in 1565.[27] They were voracious land accumulators through purchases and marriages. The Boscawens were a very ancient family. Hugh Boscawen of Tregothnan in 1620 was said to have been descended in the thirteenth generation from Henry de Boscawen.[28] Another Hugh Boscawen was created Viscount Falmouth in 1720. He drew on immense income from the copper mines of Chacewater and Gwennap where he was the principal landowner.[29] Sir Richard Robartes made a fortune from tin and banking in Truro. In 1620 he bought Lanhydrock, a former Bodmin Priory manor, and started building one of the grandest and most imposing houses in Cornwall (now owned by the National Trust). In 1624 he purchased a peerage for £10,000.[30] The Bassets were another ancient family who held the Tehidy estate for over 600 years. They were also prosperous mine owners in the eighteenth century. Their tin and copper mines at Pool, between Camborne and Redruth, yielded them an income of £10,000 a year and they also controlled Cook's kitchen and Dolcoath, two of the richest mines in Cornwall.[31] The Edgcumbes were also settled in Cornwall for a long time. William Edgcumbe, from the Devon side of the Tamar, founded the family's fortunes in the fourteenth century by marrying Hilaria de Cotehele, who brought him one of the most attractive old manors in Cornwall (now also owned by the National Trust), which overlooks the Tamar on the Cornish side.[32]

FURTHER READING

Robin Stanes, *The Old Farm: A History of Farming Life in the West Country* (Exeter, 1990); W.G. Hoskins, *Devon* (London, 1954); James Whetter, *Cornwall in the Seventeenth Century: An Economic Survey of Kernow* (Padstow, 1974); F.E. Halliday, *A History of Cornwall* (London, 1959); John Rowe, *Cornwall in the Age of the Industrial Revolution* (Liverpool, 1953); Joan Thirsk (ed.), *The Agrarian History of England and Wales, Vol. IV, 1500–1640* (Cambridge, 1967) and *Vol. V, 1640–1740* (Cambridge, Part 1, 1984 and Part II, 1985).

AGRICULTURE

FARMING
IN THE
NINETEENTH CENTURY

SARAH WILMOT

AGRICULTURE IN THE EARLY NINETEENTH CENTURY

The land-use and cropping systems which prevailed in the South West during the first half of the nineteenth century struck contemporary visitors as highly unusual. When William Marshall wrote his *Rural Economy of the West of England* in 1796 he listed no fewer than twenty-seven items of agricultural practice which he believed marked off the peninsula as distinct from the rest of England.[1] The peculiarities he listed are largely bound up with the practices and techniques associated with 'convertible husbandry' or 'ley farming' in Devon and Cornwall and its effects on landscape, yields and agricultural technology.

The essence of the convertible system was the breaking up of pasture for temporary cultivation to produce two or three crops of grain, before laying the land down to pasture again. This cycle would be repeated, with grass leys being broken up for arable at intervals of between two and fourteen years. The main effects on the landscape were two-fold. First, the landscape bore the marks of cultivation in even the remotest localities of Devon and Cornwall. 'Arable' and 'pasture' were in the main transitory states of land use, rather than fixed categories. Secondly, traditional rotations with long leys contributed to the multiplication of fields and hedges so deplored by agricultural improvers. A farm operated under a fourteen-year rotation, for example, would have required a minimum of fourteen enclosures.[2]

Traditionally, the convertible system practiced in the South West had involved using a rotation called the 'old Devon course', which entailed taking three corn crops in succession before sowing the land with grass seeds. Many agricultural writers were to assign to this rotation the cause of lower grain yields in the South West, and indeed the effect on fertility in terms of declining crop yields over a three-year period is readily apparent. The yield of oats or barley grown in the area depended on the place of the crop in the rotation. As a first crop, oats yielded 40 bushels per acre, as a second crop, barley and oats yielded 30–36 bushels per acre, and as a third crop neither oats nor barley exceeded 20 bushels per acre. In defence of the system it could be argued that the corn crops, in many parts of the region, were an 'accident', a step towards the production of grass leys, and not, as in the east of England, one of the main objects of the farmer. However, contemporary observers also lamented the condition of the pastures too.[3]

Charles Vancouver's survey of Devon and G.B. Worgan's survey of Cornwall in the early part of the century make it clear that convertible husbandry was practiced widely on a variety of soil types, as was the practice of taking two or three corn crops in succession. Where turnips had been introduced into rotations they were often grown in small quantities to provide winter feed. Turnips were also used in land reclamation as a pioneer crop. However, acreages were small and the crop was not usually managed intensively. In general, the western pastoral systems of the early nineteenth century did not depend upon root crops as the means of maintaining soil fertility, which was becoming the common pattern of agricultural systems in eastern England.[4]

The traditional tool/plough technology associated with convertible husbandry also remained an important feature of the region's agriculture in the early part of the century. Inaccessible fields, small enclosures and, in upland areas, steep gradients, ensured that modification of techniques proceeded slowly. The region's technology involved a heavy reliance on hand labour; particularly heavy demands were involved in initially spading or 'beating' the land to break up the old pasture where it was impossible to plough. Following clearing, rubbish would be collected for burning, after which the ashes would be mixed with lime and dung and spread over the field by hand. The central importance of lime in this system was reflected in the large number of lime kilns which dotted the landscape.[5] Finally, the crop was traditionally sown by a process of 'hacking in' which also involved a heavy input of hand labour, as the following description by Charles Vancouver in 1808 illustrates: 'the wheat is sown broadcast, and upon each ridge a single man proceeds with a broad bitted mattock, beginning on the lower side of the field, and advancing upwards, cutting down the edges of the furrows, and displacing so much of the soil as may afford a slight and imperfect covering to the grain'.[6] Although, wherever possible, horse technology had been substituted into these ancient operations, particularly in the lowland parts of Devon and Cornwall, the battery of hand-implements—the mattock, the shovel, the twibill or twobill and the breast plough—clearly remained essential to field cultivation in the early nineteenth century. In addition, in many areas, the heavy

Fig. 38.1
Drawing by Peter Orlando
Hutchinson: trimming
whetstones for scythes on
Punchey Hill, Kentisbeare,
Devon, c.1847.

SOURCE: reproduced by
permission of Devon Record
Office (Z19/2/8e/31).

The influence of mining was particularly important in shaping the character of agriculture in Cornwall, for it allowed the survival of smallholdings on a scale which was not reflected in the neighbouring county. Not only did the mining sector provide a ready market for the produce of smallholdings, it was also commonplace for miners to practice part-time farming themselves, particularly on marginal land. This process was encouraged by landlords who granted long leases to small tenants to encourage reclamation of moorland tracts. Additionally, retired miners, often returning from overseas, invested their savings in land.[10]

The result was the creation of a rural society which struck outsiders forcibly. Sir Thomas Dyke Acland, for example, commented of Penstrase Moor that: '[it] was a barren moor and now is a conglomeration of small leaseholds of 4–10 acres—the houses built by the people themselves, who have for the most part been miners; they regard the Estates like their own … They are interesting people, VERY independent—all Dissenters. Some very good farmers …'.[11]

The distinctiveness of Cornwall's farm structure is underlined by the census figures of 1851, which show that 28 per cent of Cornwall's holdings were below 20 acres, as compared with 11 per cent in Devon and 23 per cent in England and Wales as a whole (**Table 38.1**). With less than a quarter of its farms exceeding 100 acres in size, Cornwall may be seen as a 'refuge' for the small subsistence farmer. The pattern of farm size in Devon was much closer to the national average.[12]

Devonshire plough was retained in preference to newer designs. The overall result was that labour-productivity was considerably below the national agricultural average.[7]

Although convertible husbandry in south-west England met with much contemporary criticism from agricultural improvers, there was general agreement that cattle-breeding in Devon had attained a degree of excellence. Devon's breeds of beef cattle, and to a lesser extent dairy cattle, had built up a high reputation by the late eighteenth century, a reputation enhanced by the early patronage of such agricultural grandees as Thomas Coke of Holkham. By 1808 the North Devon breed was already one of Devon's most valuable exports, valued for both meat production and for draught. Charles Vancouver remarked on the 'very high prices they bring … either at home or among the Somersetshire graziers'. Devon's leading cattle-breeders, like Francis Quartly of Molland, achieved national notoriety, whilst at the market Devon cattle were said to receive the first attention of London's West End butchers.[8]

By the 1840s Devons represented the prevailing breed of cattle in Cornwall too, the original black cattle having rapidly diminished in numbers, although still surviving in some areas like the Scilly Isles. The best herds in Cornwall were said to have been built up from stock originally imported from North and South Molton. On Cornwall's richer pastures the Shorthorn breed of cattle was becoming established. Karkeek's mid-century survey of Cornwall indicates that there had also been considerable investment in the improvement of sheep and pig breeds during the early nineteenth century.[9]

Although the agriculture of Devon and Cornwall shared many features in common, there were some important differences.

Percentage of holdings in each size category				
Farm size (acres)	Devon	Cornwall	South West	England and Wales
Under 20	11	28	17	23
20–50	19	27	21	21
50–100	26	21	22	20
100–200	29	17	24	20
200–300	9	4	8	8
300–500	4	2	4	5
Over 500	1	<1	3	2
Acres not stated	1	1	1	1

Table 38.1
Farm size in Devon and
Cornwall, 1851.

SOURCE: Population Census of
Great Britain, 1851.

MAPPING AGRICULTURAL CHANGE

Although the regional surveys of Marshall, Worgan, Karkeek, Vancouver and others provide an invaluable insight into agricultural systems at the beginning of the century, they do not provide the region-wide coverage which an examination of extant agricultural statistics affords. Additionally, a series of agricultural maps, constructed at key intervals during the century, can provide a detailed picture of average change in land use over time. Contemporary observers, on the other hand, were not only restricted in terms of the geographical area which they were able to traverse, they often tended to focus on the unusual or the novel elements of agricultural practice, which may be misleading where an indication of *average* farming operations is required.

The agricultural statistics from the tithe files of the 1840s, and the government agricultural returns of 1801, 1875 and 1900, on which the maps in this chapter are based, also have their limitations.[13] Most important of all, livestock numbers were not collected with the same degree of accuracy, on the same basis, or at the same time of year in the different statistical surveys. The very mobility of livestock and seasonal patterns of grazing would make the mapping of livestock numbers highly misleading. It is for this reason that the mainstay of the region's pastoral economy is not represented in this chapter. Neither is it possible to map changes in the distribution of smallholdings or farm size. Statistics on agricultural holdings are only readily available for the counties as a whole, and not at a lower level of aggregation until the end of the century.

The information mapped here is restricted to land-use patterns, particularly the changing balance between arable and

pasture, and to crop combinations, which give an overview of rotations and agricultural specialization within the region.[14] Further limitations include the poor coverage of Devon in the 1801 crop returns, and the lack of sufficient data in the tithe files to enable the mapping of turnips, potatoes and other fallow crops for the period *c.*1836. Whilst these data undoubtedly fall short of revealing everything that we would like to know concerning agriculture in the South West during the last century, they do represent the most reliable information available for the period.

CROP COMBINATIONS IN 1801 (**Map 38.1**)

The scatter of data available for Devon suggests that most of the rotations practiced in the county were dominated by a combination of either two or three grain crops. Less than half of the parishes returning information mention root crops as a significant crop. This would seem to corroborate the early-

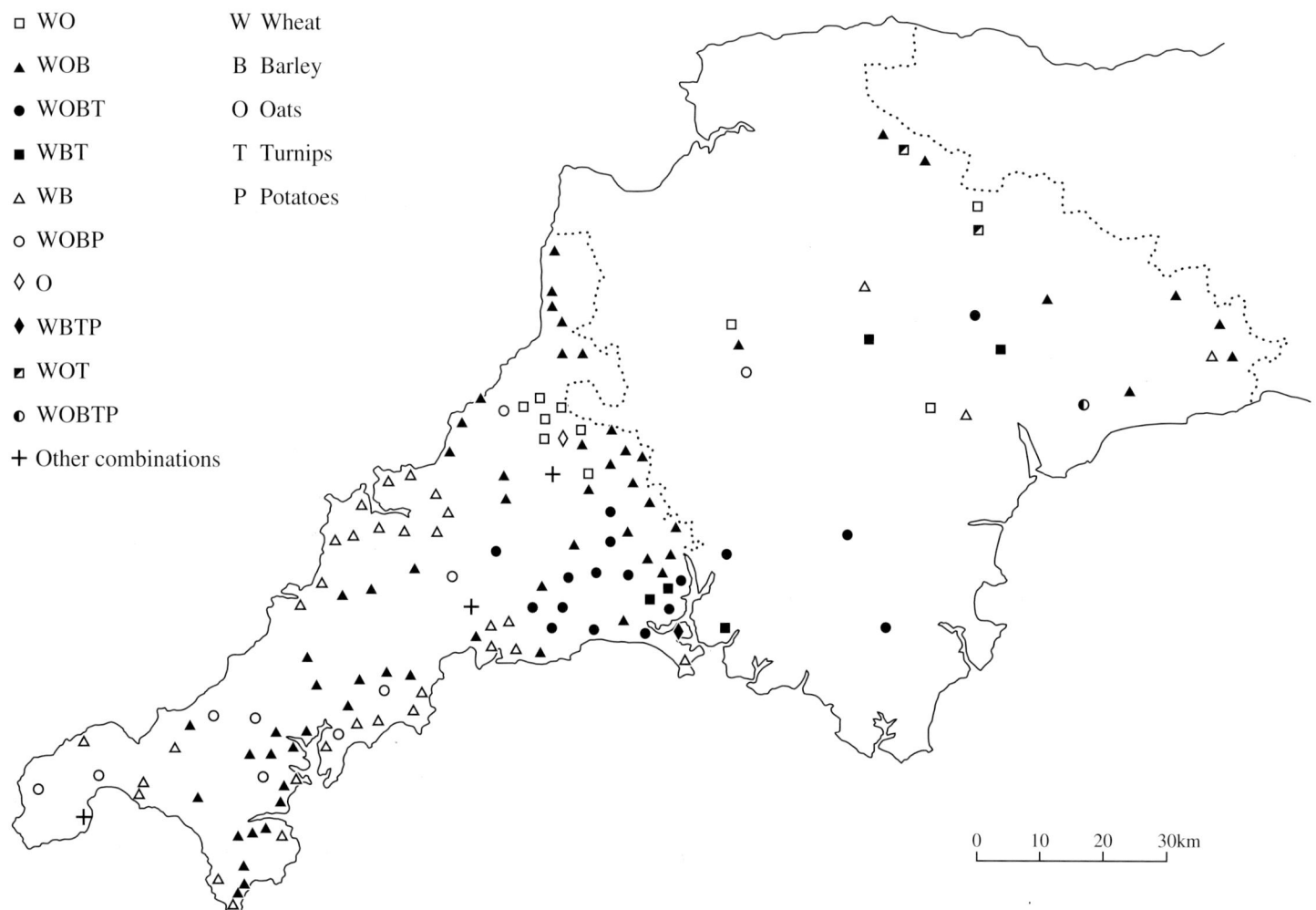

□	WO	W Wheat
▲	WOB	B Barley
●	WOBT	O Oats
■	WBT	T Turnips
△	WB	P Potatoes
○	WOBP	
◊	O	
◆	WBTP	
▨	WOT	
◑	WOBTP	
+	Other combinations	

Map 38.1
Crop combinations, 1801.

SOURCES: *Home Office Acreage Returns, 1801*; M. Turner (ed.), *List and Index Society* (1982).

nineteenth-century agriculturalists' comments on the prevalence of the 'old Devon course', whereby grain crops were sown in succession, without the intervention of a fallow crop. However, even in this very scant sample of parishes there is a considerable variety of crop combinations which makes further generalization unwise.

In Cornwall there is a similar predominance of crop combinations which exclude root crops at this date. Crop combinations which include turnips as a significant crop are concentrated in a limited area of south-east Cornwall. High prices for wheat during the Napoleonic Wars ensured that over 15 per cent of the cropped land of nearly every parish was sown with this grain. As Mark Overton pointed out in an earlier examination of the 1801 returns, the widespread distribution of wheat on the acidic soils of Cornwall is a tribute to the heavy investment by Cornish farmers in manures, including lime, sea-sand, sea-weed and the refuse of the pilchard industry.[15] Within this overall picture there were distinct regional variations. Wheat was an especially important crop in the Padstow area of the north coast and in the Falmouth area on the south coast of Cornwall. Barley also predominated in these areas, in the extreme west of the county, and in the area between Fowey and the Tamar valley. Oats became the most dominant grains where climate and soil factors limited the production of more valuable grains: on the fringes of Bodmin Moor and on the heavy clays. Potatoes were planted in large acreages in favourable soils in the more densely populated parts of west Cornwall.

CROP COMBINATIONS IN 1875 (Map 38.2)

A comparison of **Map 38.1** with **Map 38.2** indicates that cropping patterns in the South West had undergone considerable change by 1875, a date which marks the peak of arable production in the region. In Devon, the wheat, barley and turnips combination, which probably represents the 'Norfolk' four-course rotation,

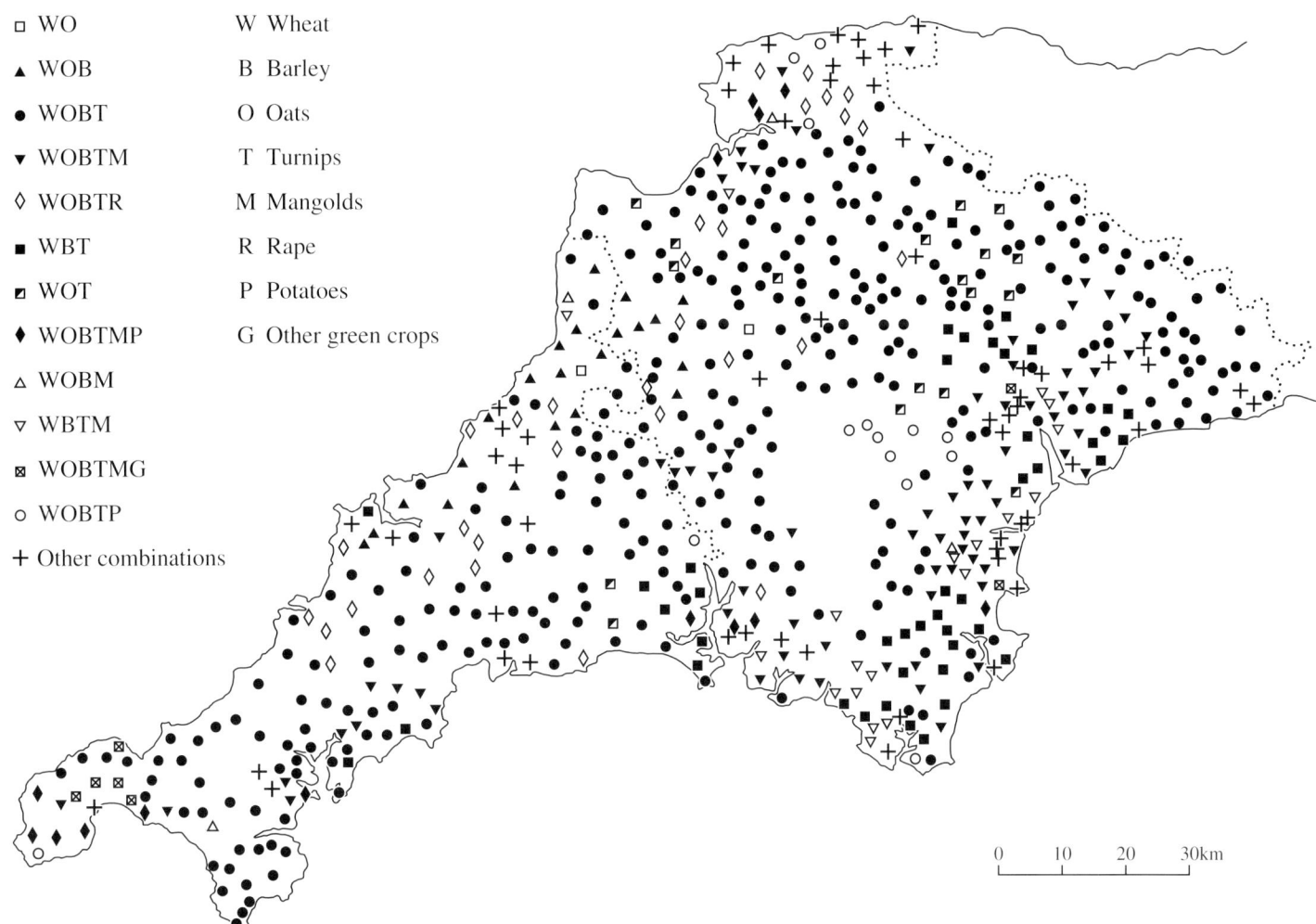

□ WO	W Wheat
▲ WOB	B Barley
● WOBT	O Oats
▼ WOBTM	T Turnips
◊ WOBTR	M Mangolds
■ WBT	R Rape
◪ WOT	P Potatoes
◆ WOBTMP	G Other green crops
△ WOBM	
▽ WBTM	
⊠ WOBTMG	
○ WOBTP	
+ Other combinations	

0   10   20   30km

Map 38.2
Crop combinations, 1875.

SOURCES: Agricultural Returns 1875: Parish Summaries; Public Record Office (MAF 68).

occurs in the barley-producing area of the South Hams and in part of the Exe Vale but is not common elsewhere. However, by this period the majority of crop combinations include root and green crops as dominant crops. The general importance of fodder crops in the county by this date suggests a widespread intensification of livestock farming. Crop combinations excluding significant acreages of root and green crops are confined to a small pocket in west Devon where two or three grain crop combinations occur. A further striking change is the evident importance of mangolds in crop combinations by this date, particularly in the South Hams and Exe Vale regions. This was a crop which was described as rare by Henry Tanner when he surveyed the progress of Devon's agriculture for the Royal Agricultural Society in 1849.[16] Rape is an important crop in combinations in north Devon and a scattering of west Devon parishes. Rape was valued for its high feeding qualities in sheep farming, and as a preparation for wheat. Potatoes formed a specialist crop on the lowland areas north of Barnstaple, in the Plymouth area and on the especially favoured granitic soils east of

Dartmoor. Despite the spread of root and green crops, there is still a strong continuity with traditional farming practices in that the most common crop combination in 1875 is that of wheat, oats, barley and turnips. This suggests that three corn crops were still being taken in succession after turnips, a practice which is particularly prevalent in mid-Devon, east Devon and on the fringes of Dartmoor. The distribution of this crop combination suggests that it was an adaptation to the heaviest and wettest soils of the county, although it is not confined to those soils alone.

In Cornwall, it is noticeable that crop combinations excluding root crops as significant crops are still fairly common in the north-coast area of Padstow, traditionally known as the 'granary' of Cornwall. This crop combination is also in evidence on the heavy soils of the northernmost tip of Cornwall whilst elsewhere in the county the practice has largely died out. As in Devon, the predominant combination of crops is that of wheat, oats, barley and turnips, and its distribution is widespread. In valley locations, particularly the Tamar valley and in isolated areas near

Map 38.3
Arable as a percentage of
tithe district area, *c.*1840.

SOURCE: R.J.P. Kain, *An Atlas
and Index of the Tithe Files
of Mid-Nineteenth-Century
England and Wales*
(Cambridge, 1985).

Padstow and Falmouth, the wheat, barley and turnip combination predominates, which indicates that the 'Norfolk' four-course rotation was being practiced. Mangolds are important crops in several parishes, particularly in the area between Mevagissey and Budock, a trend in cropping which was in progress when W.F. Karkeek was writing on Cornish agriculture in the mid-1840s.[17] In the Newlyn area and on upland margins, particularly around Bodmin, the appearance of rape as a dominant crop in combinations is a significant development.

The importance of intensive market-gardening in the Penzance to St Ives district is reflected on the map by the combinations of wheat, oats, barley, turnips and 'other green crops'. Combinations including potatoes as a dominant crop are particularly important in the Land's End district and, to a lesser extent, in the Tamar valley. It was principally from these areas that the export trade in fruit and vegetables was expanded in the late nineteenth century. An indication of the importance of the trade is given by the Great Western Railway's freight data: in 1889 the railway conveyed 300 tons of strawberries, 4,500 tons of new potatoes and 8,000 tons of broccoli from Devon and Cornwall to London, Manchester, Edinburgh and other large centres.[18]

The extent of arable and pasture, *c*.1840 (Maps 38.3 and 38.4)

Tithe file data indicate that in the south-west region as a whole, arable represented an average of only 23 per cent of the titheable acreage, pasture occupied 51 per cent, whilst the remaining 26 per cent of the titheable acreage was under woodland and common. Maps of the distribution of arable and pasture prepared by Roger Kain show considerable variations in land-use patterns within the region, reflecting the considerable diversity of soils and climatic conditions of the area.[19] Map 38.3 reveals that on the more fertile soils, particularly the light red marl soils of the Exe Vale, the percentage of arable rises to 40–60 per cent of the titheable

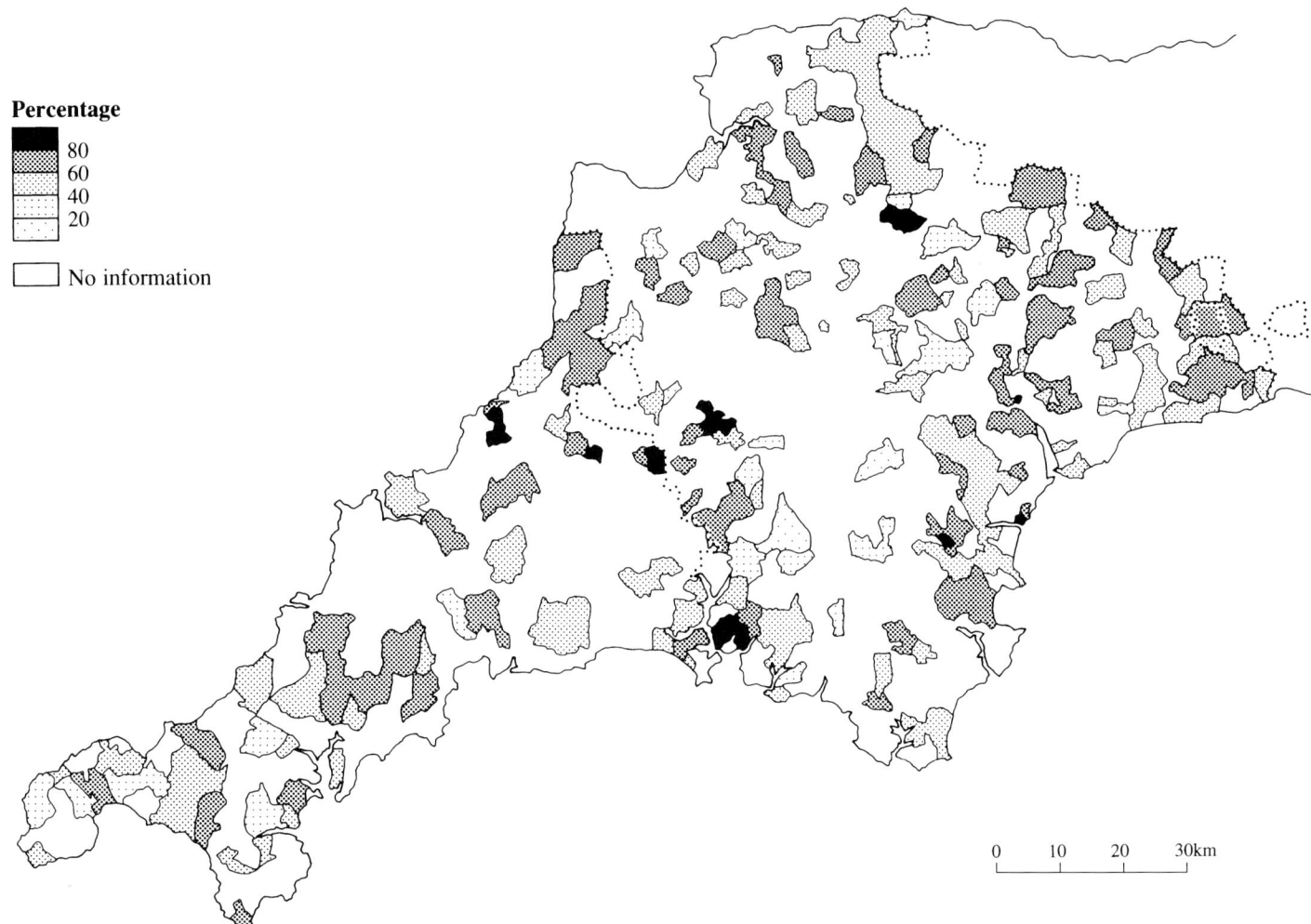

Percentage

- 80
- 60
- 40
- 20

☐ No information

0 10 20 30km

Map 38.4
Pasture as a percentage of
tithe district area, *c*.1840.

SOURCE: as **Map 38.3.**

acreage. In the area around Crediton the figure is as high as 60–80 per cent. In parishes bordering moorland areas, on heathy soils, and on the poorer soils of the Culm Measures, the proportion of arable falls below 20 per cent. In Cornwall, in all but the parish of St Minver on the north coast, arable accounted for less than 40 per cent of the titheable area, and for many parishes arable accounted for less than 20 per cent, especially in the Hartland Plateau and Bodmin Moor areas.

Map 38.4 shows the distribution of pasture *c*.1840. This category of land use includes land under temporary and permanent grass but excludes rough pasture classified as 'common'. This explains why the map shows lower than expected proportions of pasture in the upland areas of Dartmoor, the Exmoor borders, and in some upland parishes of Cornwall. Comparison of **Maps 38.3 and 38.4** makes clear the relatively pastoral character of east Devon, north Devon, and Cornwall, whilst the Exe Vale and South Hams emerge as mixed farming regions at this date.

THE EXTENT OF ARABLE IN 1875 (**Map 38.5**)

The mid-1870s saw arable acreages in Devon and Cornwall reach a peak of 702,812 acres. This represents a massive increase on the estimated extent of arable forty years earlier (**Table 38.2**). The increase in arable was not a result of the expansion of the acreage under wheat or barley, but of the extension of fodder crop cultivation: oats, roots and green crops.

Looking at the distribution of arable in Devon in 1875, it is apparent that the basic regional contrasts noted by agricultural observers at the beginning of the century are still very marked. The relative pastoral emphasis of north Devon, the keuper marls and sandstone area of east Devon, and of the heavier clay soils of west Devon remains evident. These areas specialized mainly in stock-rearing and dairying in the nineteenth century. The high concentration of arable in the western portion of the Exe Vale and in the South Hams, where a system of mixed farming was practiced, is also a pattern of long standing. The most striking

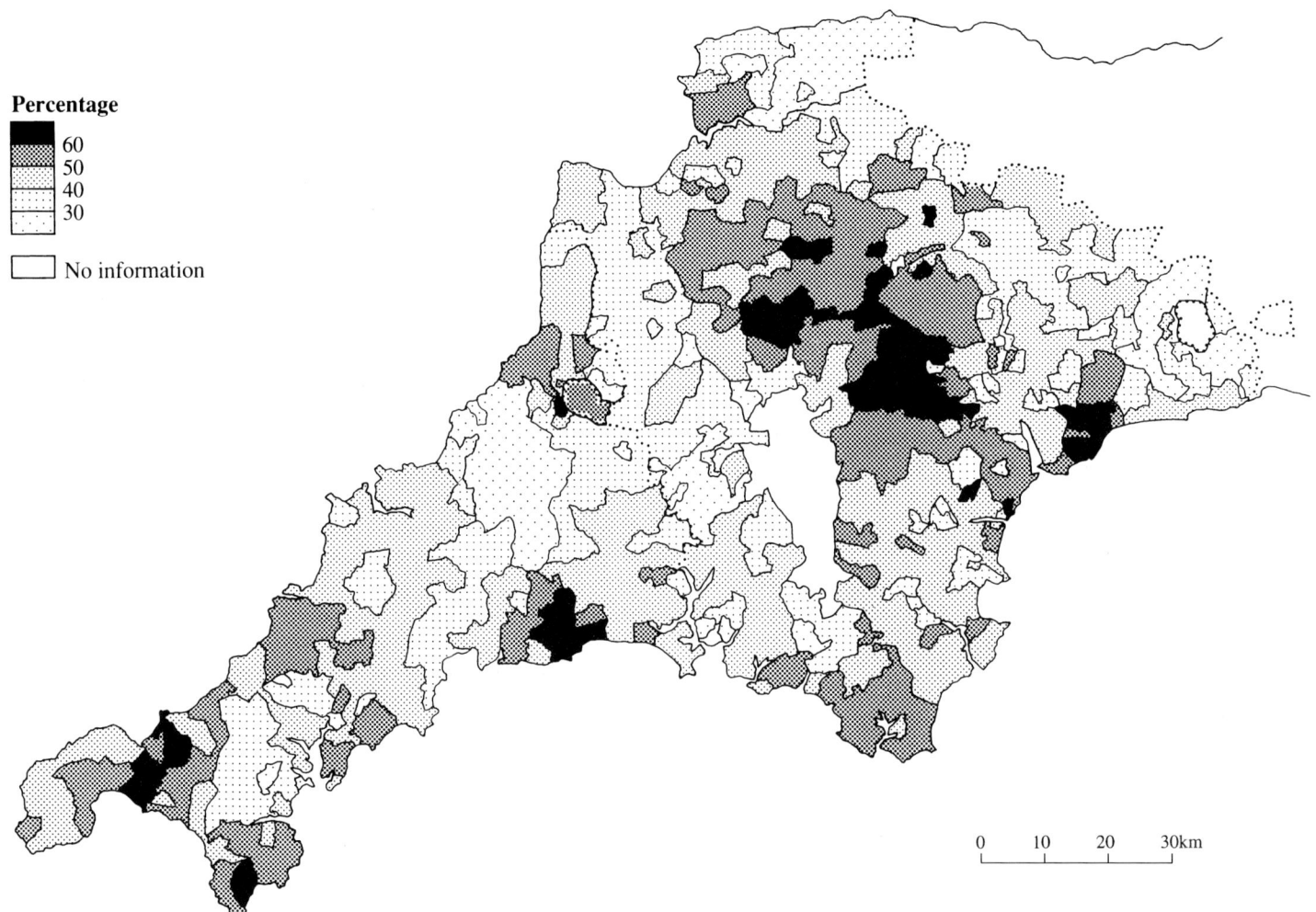

Percentage

- 60
- 50
- 40
- 30

☐ No information

Map 38.5
Arable as a percentage of crops and grass, 1875.

SOURCES: as **Map 38.2.**

0   10   20   30km

change since the early part of the century revealed by this map is the consolidation of the mid-Devon area as a major arable region. When Charles Vancouver described this area in 1808 he noted its greater inherent fertility than the Culm Measures to the west, but at that time remarked that the land gave the impression of a 'much worn and exhausted soil; fields bearing but a very thin and feeble plant of clover, light and indifferent crops of wheat and oats; and generally characterised by a tough, wiry and sour herbage'. On the inter-mixed soils of this district, Vancouver also noted considerable tracts of coarse moors which he maintained would repay improvement.[20]

The pattern of tillage in Cornwall in 1875 shows the heaviest concentrations of arable in the extreme west of Cornwall in a triangle of land, the three points of which are marked by Land's End in the west, Redruth in the north and the Lizard in the east; in the area north and south of Bude, including the parishes of Morwenstow, St Gennys, Poundstock, Jacobstowe and Whitstone; and in the central lowland parishes between Newquay in the

north and Falmouth in the south, and the area between Looe and Fowey on the south coast.

PERCENTAGE CHANGE IN ARABLE, 1875–1900 (Map 38.6)

The closing decades of the nineteenth century were difficult ones for the agricultural industry, although the pastoral South West appears to have suffered much less from the agricultural depression than many other farming regions. It nevertheless represented a period of great change in land use and landscape, as an examination of **Map 38.6** and **Table 38.2** reveals. Agriculture in the South West adjusted rapidly to the national decline in grain prices after 1875. Between 1875 and 1900 wheat and barley acreages contracted markedly, whilst the oats acreage was further extended. The acreage under root and green crops also fell in absolute terms, although continuing to represent approximately a third of the total arable acreage. By

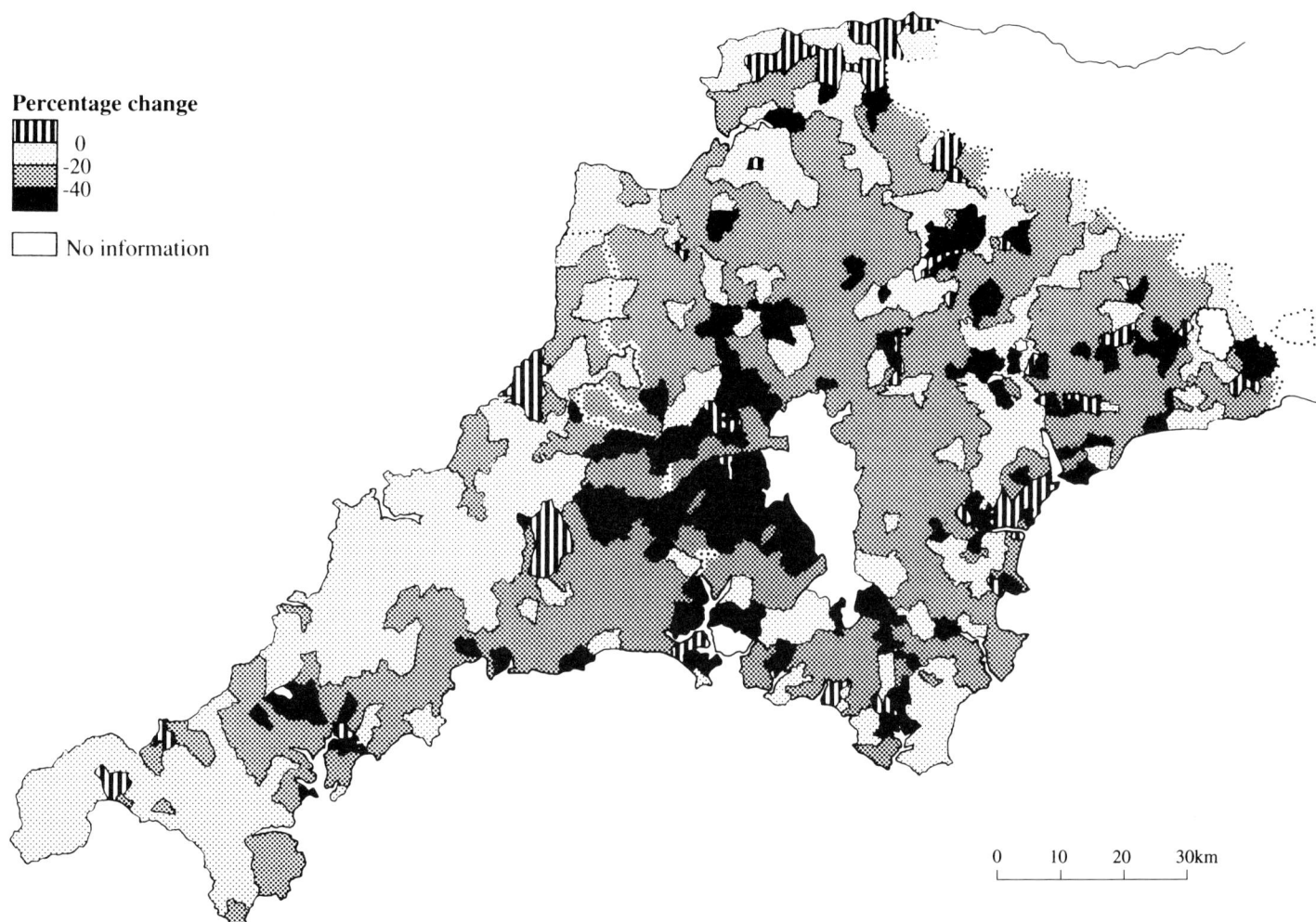

Map 38.6
Percentage change in arable, 1875–1900.

SOURCES: as **Map 38.2.**

301

Fig. 38.2
Livestock numbers in Devon
and Cornwall, 1870–1900.

SOURCE: *British Parliamentary
Papers*, 1870–1900:
Agricultural Returns of
Great Britain.

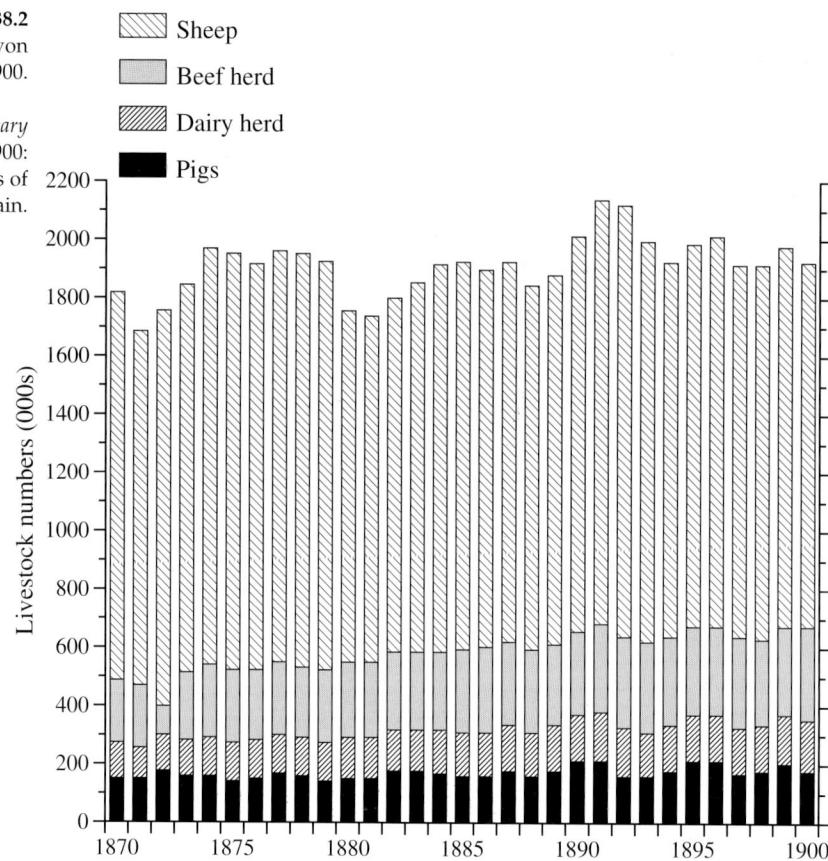

Table 38.2
Land use and crops in
Devon and Cornwall.

SOURCE: Figures for *c.*1836
from R.J.P. Kain, *An Atlas
and Index of the Tithe Files of
England and Wales* (Cambridge,
1986). Figures for 1875 and
1900 from Ministry of
Agriculture Statistics.

the county where agricultural systems were already strongly pastoral. East Devon, traditionally a dairying district, sustained similarly high rates of conversion to pasture. North Devon, in contrast, either shows little change in land-use patterns, or even some real increases in arable acreages in response to the growing markets provided by tourist towns like Ilfracombe. All along the south coast also, patterns of agrarian change are complicated by the effects of urban growth. The arable heartlands of the county—the mid-Devon parishes, the eastern Exe Vale and parts of the South Hams—also saw the contraction of tillage by over one-third, whilst most other parishes in the county saw a reduction of at least 11–20 per cent. In the 1900s the Devon landscape was beginning to take on its present appearance, with grassy and moory slopes replacing fields of wheat, barley, oats, rye grass, mangolds and turnips. An evocation of the change was given by J.G. Cornish in 1939 when he compared land use in his own day with land use in the 1870s: 'it seems strange to me now to look at a steep hillside covered with bracken or even gorse and brambles, and to think how I saw oats being sown by hand on it and the carter standing on the harrow at a perilous angle, to harrow in the grain'.[21]

In Cornwall the contraction of arable was concentrated in the south of the county. The western tip of Cornwall, with its increasing specialization in market-gardening and potato production, experienced relatively little change. Declines in arable acreage in north Cornwall were also relatively low (less than 20 per cent).

1900, wheat and barley output was down 39 per cent compared with 1875 levels, and the output of root crops had fallen by 18 per cent.

The overall margin of cultivation in Devon and Cornwall shrank at a more rapid rate than in England and Wales as a whole, and by 1900 the overall balance of arable to pasture was close to the balance which had existed in the late 1830s. The pattern of change within the region was complex. In Devon the contraction in the arable acreage was most marked in the west of

PASTURE IN 1900 (**Map 38.7**)

Overall by 1900 the great majority of parishes in the South West had seen an increase in the extent of their land laid to pasture. The general balance of rotational to permanent pasture had not changed significantly between 1875 and 1900. A broad contrast may be drawn between Devon and Cornwall in that Devon had a far lower ratio of rotational grass to permanent pasture. In Devon rotational grasses represented around a quarter of total pasture land, compared with over 40 per cent in Cornwall. Despite the region-wide increase in pasture, the overall contrasts between different farming districts had changed little since 1875. In 1900 the heavy pastoral emphasis of north west, and east Devon is still evident, whilst the South Hams, mid-Devon and the Exe Vale continued to practice mixed-farming regimes with between 30 to 49 per cent of the land under arable. The distribution of pasture in Cornwall closely reflects the pattern of uplands in the county, particularly the granite moorlands of Bodmin, Hensbarrow and Carnmenellis. The main tillage lands are situated in the lowland belt of central Cornwall, particularly in the hinterlands of Padstow and Newquay on the north coast, and at the western tip of the county. Other tillage areas include the parishes surrounding Bude in the north and the coastal area between Fowey and Torpoint in the south.

	c.1836	1875		1900	
A. Land use	Est. acreage	Acreage		Acreage	
Arable	580,016	702,812		531,557	
Grass	1,258,327	905,819		1,281,637	
Wood	100,458				
Common	537,203				
B. Crops	Acres	% Arable	Acres	% Arable	Acres	% Arable
Wheat	195,088	34	174,014	25	85,386	16
Barley	149,702	26	131,249	19	76,535	14
Oats	67,157	12	129,675	18	186,365	35
Turnips, swedes, mangolds			156,162	22	118,088	22
Other green crops			60,818	9	52,030	10
Other crops	168,069	29	50,894	7	13,153	3

LIVESTOCK NUMBERS IN DEVON AND CORNWALL, 1870–1900 (Fig. 38.1)

Fig. 38.2 and **Table 38.2** indicate that a dramatic increase in the acreage under fodder crops had taken place in the South West between the 1840s and the 1870s, suggesting that livestock production was becoming more intensive in the region during the 'High Farming' period. Unfortunately, annual statistics on livestock numbers are not available until the government Agricultural Returns began. **Fig. 38.2** charts trends in livestock numbers after 1870, a period in which the region was becoming increasingly pastoral (**Map 38.6**). Figures show a steady rise in the numbers of pigs, dairy cattle and beef cattle kept in the South West during the last three decades of the century. In general, dairying, pig, and beef production were growing in importance in relation to sheep farming.

Sheep numbers were erratic during this period: numbers peaked in the mid-1870s and the early 1890s, and slumped during the early 1880s. W.C. Little's reports on Devon and Cornwall in 1881 suggest possible causes for this slump. Little's report indicates that sheep rot was widespread in Cornwall, except on granitic soils, and was especially bad on the clays north of Launceston. A large number of sheep had also died from an unverified disease, possibly from anthrax. In north Devon, farmers had largely escaped the sheep rot, but in other parts of Devon, for example in the Exe Vale, sheep losses were heavy. Additionally, low prices for wool had made a serious dent in the profitability of sheep farming.[22]

Periodic fluctuations aside, the mainstay of the South West's agricultural economy during the late nineteenth century continued to be sheep and cattle breeding and rearing. New developments in livestock husbandry at the end of the century include an increase in the home-fattening of beef cattle for expanding markets in seaside resort towns, and an increasing focus on early lamb production. By the early twentieth century, pig production had become an acknowledged speciality of the region, particularly in

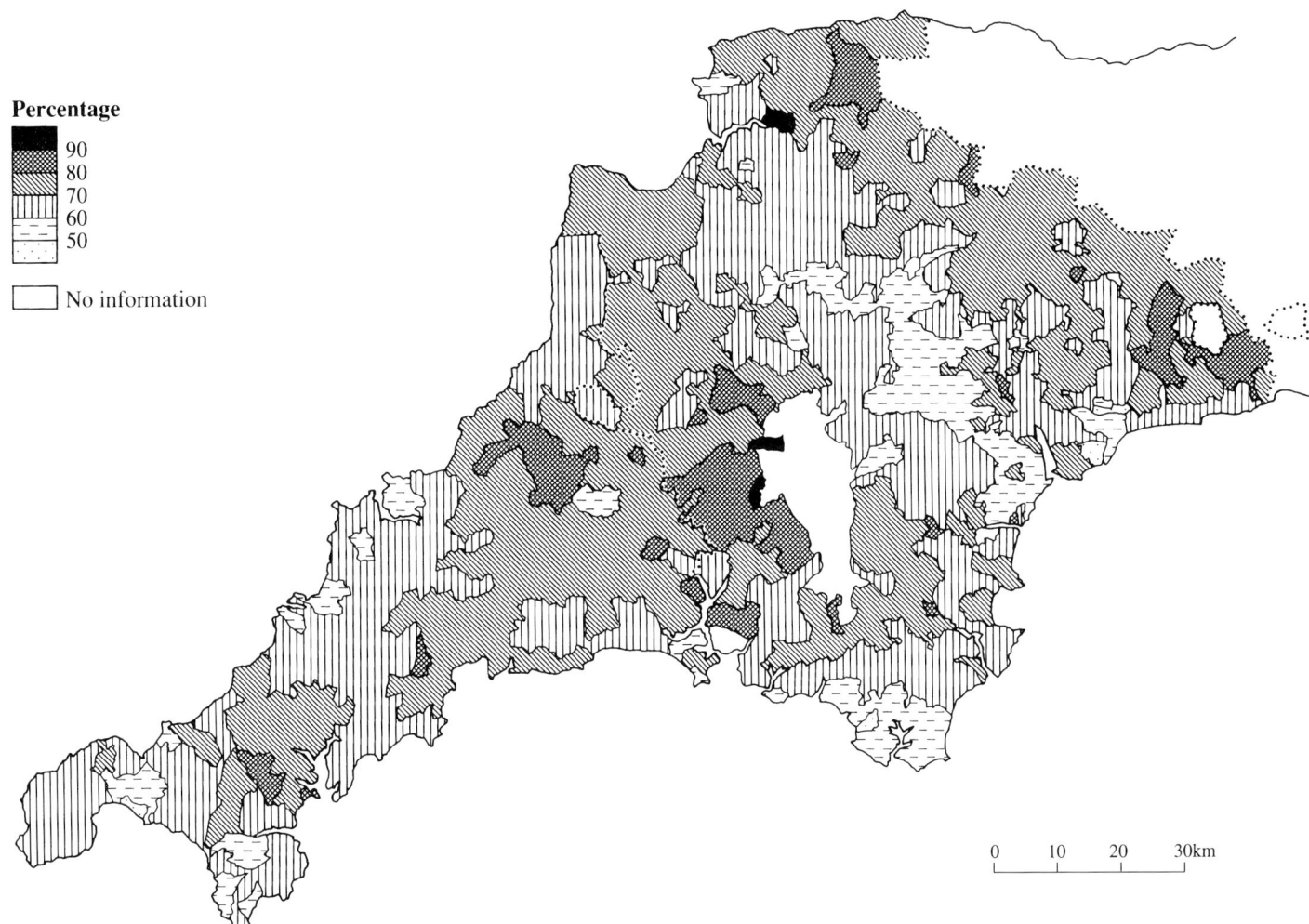

Map 38.7
Pasture as a percentage of crops and grass, 1900.

SOURCES: as **Map 38.2.**

303

Fig. 38.3
Bampton Pony Fair.

SOURCE: reproduced
by permission of
Devon Record Office
(Chapman 1578/11285).

the dairying districts of east Devon and west Cornwall. In the 1930s Devon and Cornwall accounted for nearly 10 per cent of the total pig population of England and Wales.[23]

In Devon, native breeds of cattle and sheep dominated for most of the century. The main cattle breeds were North Devons and South Hams. Attempts had been made to introduce Shorthorn, Jersey and Polled Angus cattle but these breeds had not been adopted widely by the county's farmers. The most important native breeds of sheep were the Devon Longwool, South Hams, Dartmoor and Exmoor. All of these breeds showed an aptitude to fatten and to carry heavy fleeces of wool of long staple. Dartmoor and Exmoor flocks were said to show better

wool, arrive at earlier maturity, and reach heavier weights than any other upland breed. Cornwall differed markedly from Devon in that by the late nineteenth century it boasted no native breeds and its farmers specialized in crossing the breeds of other districts. The major cattle breeds in Cornwall in the late nineteenth century were cross-bred Guernseys and Shorthorns.[24]

CROP YIELDS IN DEVON AND CORNWALL COMPARED WITH THE NATIONAL AVERAGE (**Fig. 38.4**)

Crop yields in the South West give some indication of the region's agricultural performance relative to the national average. Estimates for the Napoleonic War period suggest that yields of wheat in Devon, and of wheat, barley and oats in Cornwall, were substantially above the national average. Given that these figures are so much higher than estimates for the remainder of the century, and that intuitively we would not expect the region's climate to be particularly favourable to grain crops, it seems likely that these figures are an over-estimate. Yield estimates taken from the tithe file data *c.*1836 suggest, in contrast, that the region's performance, at least in grain crops, was lagging behind other counties. During this period figures indicate that wheat yields in Devon were only 73 per cent of the national average, barley yields 84 per cent, and oat yields 82 per cent.

Table 38.3
Crop yields per acre in Devon and Cornwall, 1800–1912.

SOURCES: as **Table 38.2**.

Crop	1800–16		c.1836		1861		1880s		1903–12	
	Devon	Cornwall	Devon	Cornwall	Devon	Cornwall	Devon	Cornwall	Devon	Cornwall
Wheat	25.0	24.0	16.0	18.0	22.0	24.0	22.0	24.0	27.0	30.0
Barley	24.0	36.0	26.0	28.0	32.0	32.0	29.0	32.0	33.0	33.0
Oats	32.0	40.0	28.0	31.0	37.0	38.0	32.0	37.0	38.0	40.0
Potatoes							5.5	5.4	5.4	6.3
Hay							1.2	1.5	1.2	1.4
Turnips							15.0	19.0	13.0	15.0
Mangolds							12.0	21.0	21.0	24.0

NOTE: Wheat, barley and oat yields in bushels. Potatoes, hay, turnips and mangolds in tons.

This picture of poor agricultural performance corroborates the impressionistic evidence of mid-nineteenth-century agricultural observers. For example, Charles Pym, an assistant tithe commissioner working in north Devon, concluded that the 'neighbourhood appears doomed not to participate in the improvement which has taken place in the Midland counties of England'. In similar vein, Henry Tanner assessed Devon's farming as 'inferior to that of most of the counties of England'. James Caird remarked that 'cumbersome and unskillful' practices were prevalent in every district of the county and had 'rendered Devonshire farming a by-word in the estimation of the great corn farmers of the eastern counties'.[25] Figures suggest that Cornwall's agriculture was delivering better grain yields, with yields of wheat at 82 per cent of the national average, barley 90 per cent, and oats 91 per cent.

Crop yields in both counties increased substantially between the late 1830s and the 1880s (**Table 38.3**). In Devon, wheat yields had risen from an average of 16 bushels per acre to 22 bushels per acre, barley from 26 to 29 bushels per acre, and oats from 28 to 32 bushels per acre. Cornish farmers consistently out-performed their neighbours with wheat yields rising from 18 to 24 bushels per acre, barley from 28 to 32 bushels per acre, and oats from 31 to 37 bushels per acre. Climatic disadvantages meant that grain yields in the South West were still generally below the national average, but the disparities had narrowed. Cornwall was achieving close to national average yields in wheat and oats, and was above average in barley yields. By the 1880s reliable statistics were being collected on root and hay yields. **Fig. 38.4** shows that Cornwall's yields were well above the national average in root and hay yields, whereas Devon lagged behind.

A number of factors contributed to the overall improvement in yields. **Maps 38.1 and 38.2** testify to the substantial changes in rotations practiced within the region during the first seventy-five years of the century. New manures, including guano and superphosphates, enabled improvements in root crops. Broadcast sowing and traditional hand-tool techniques progressively disappeared. In Devon, investment in underdrainage during the second half of the century also contributed to improve the productivity of grains, roots and grassland.[26]

Combining yield data with figures on crop acreages in **Table 38.2**, it is possible to estimate changes in the output of grains in the South West during the nineteenth century. Between *c.*1836 and *c.*1875 figures suggest that the output of wheat and oats increased by 21 per cent and 126 per cent respectively. Barley output remained static, the increase in yields being offset by the reduced acreage under barley. Between *c.*1875 and *c.*1900 wheat and barley output in the region declined by an estimated 38–40 per cent, whilst the output of oats rose by 60 per cent.

CONCLUSION

The maps, tables and graphs together tell a story of considerable change in the South West's agricultural systems during the

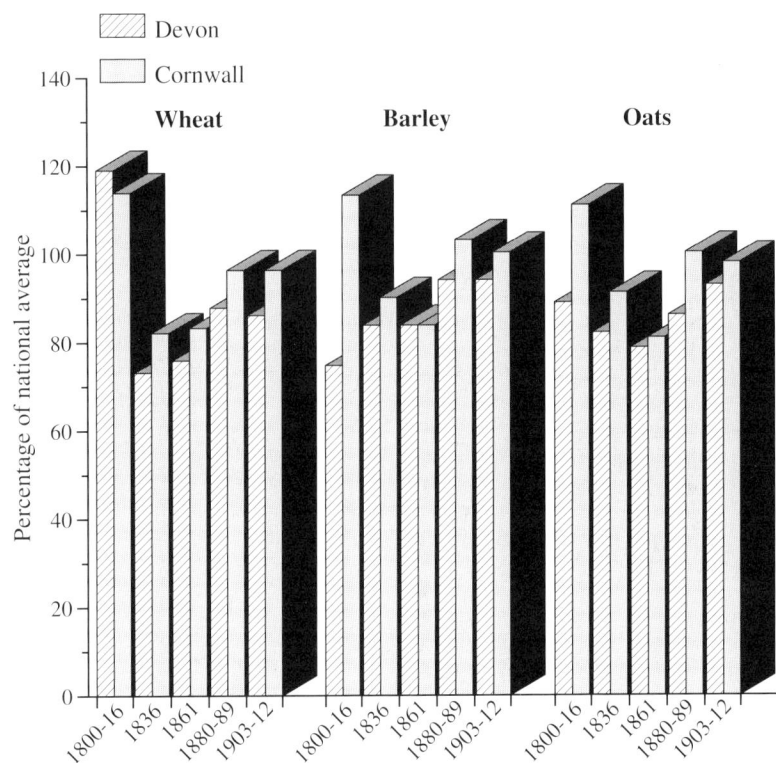

Fig. 38.4
Crop yields in Devon and Cornwall compared with the national average, 1800–1912.

SOURCES: 1836 figures as **Map 38.3**; 1800–16 and 1861 figures in P. Craigie, 'Statistics of agricultural production', *Journal of the Royal Statistical Society*, 46 (1833); 1880s figures in *British Parliamentary Papers*, 1884–1885 (C4537) LXXXIV: Agricultural Returns of Great Britain, 1885; 1903/12 figures in *British Parliamentary Papers*, 1914, Cd. 7325 XCVIII: Agricultural Returns of Great Britain, 1913.

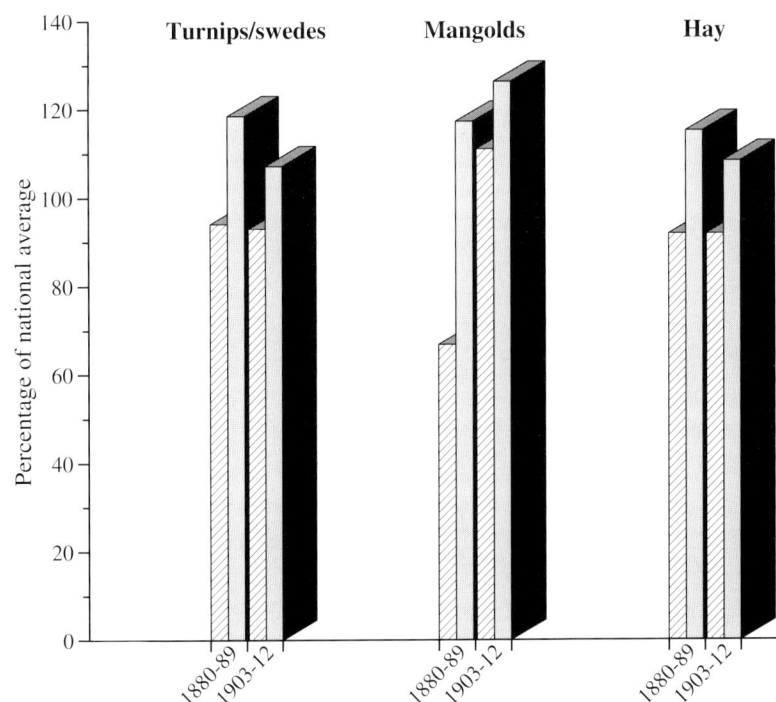

nineteenth century. The customary practices of traditional convertible husbandry and the 'old Devon course' were gradually modified or replaced. Most of the change was concentrated in the period after mid-century and was reflected in a widespread increase in the cultivation of root and green crops and in rising yields. This expansion of arable ended in the 1870s with the onset of agricultural depression, during which time much land was laid down to temporary and permanent pasture. Traditional strengths in livestock breeding and rearing were gradually supplemented by a growth in beef cattle and lamb fattening, whilst the burgeoning local tourist trade and improved railway links provided an expanding outlet for market-gardening, dairying, and pig production. By the end of the century the reputation of the South West's farmers had markedly improved, with F. Punchard commenting in 1890 that: 'Devonshire and Cornish farmers are to be congratulated upon their present position and upon many advances in local agriculture which have been made'.[27]

FURTHER READING

A.H. Shorter, W.L.D. Ravenhill and K.J. Gregory, *South-West England* (London, 1969) provides a useful summary of the economic and agricultural changes experienced in Devon and Cornwall in the nineteenth century. S.A.H. Wilmot, 'Landownership, farm structure and agrarian change in south-west England, 1800–1900: regional experience and national ideals' (unpublished Ph.D. thesis, University of Exeter, 1988) provides an overview of the economic and and agricultural history of Devon and west Somerset between 1800 and 1900, together with detailed studies of changes in farm size, tenancy and landownership in selected areas of these counties.

AGRICULTURE, FORESTRY AND LANDSCAPE CONSERVATION IN THE TWENTIETH CENTURY

ANDREW GILG

Although rural land use is one of the topics best supplied with published data, in this short chapter it is impossible to do any more than to highlight a number of key themes. The chosen themes reflect not only key or 'typical' periods in the land-use history of the century, but also the publication of major studies which the interested reader can follow up in more detail.

The fortunes of agriculture can be traced fairly easily since 1866, because since that date government has carried out an annual census of farm holdings which has counted virtually every field and animal in the country.[1] It is supremely ironic that Britain has a decennial population census but an annual farm census, and that we know more about the number of sheep than we do about the number of people. There are, though, some difficulties with these data. First, in the 1930s Dudley Stamp's Land Utilization Survey found severe under-recording by farmers in their census returns; second, there are problems of using the parish as the unit for publishing the data, and even in the 1980s, after over 110 years of data collection, researchers have questioned the accuracy of the annual agricultural census.[2] One alternative to the agricultural census is to use land-use survey data, notably the 1930s survey organized by Dudley Stamp which recorded land use in every field (but not livestock) and the resurvey in the 1960s organized by Alice Coleman. The merits of both these surveys are discussed by Peters and by Coppock and Gebbett.[3]

THE NADIR OF AGRICULTURE IN THE 1930S AND RECOVERY FROM THE 1950S TO THE 1980S

From the late nineteenth century to the 1930s, British agriculture suffered a long period of decline as cheap imports undermined the ability of British farmers to compete.[4] Three measures can be used to identify the decline into a depression which, by the 1930s, was as deep as that following the Black Death in the Middle Ages. These are the area of land under arable (the less land under arable the more depressed the state of agriculture), the area under permanent grass, and that under rough grazing (the more land under these last two uses the more depressed the state of agriculture). **Figs 39.1 and 39.2** show that both Devon and

Cornwall became increasingly depressed counties as arable land declined in total at the expense of increases in the areas of permanent grass and rough grazing. The end result was the agricultural pattern shown in **Map 39.1**, which indicates that most of the region was dominated either by 'dairying supplemented by other enterprises' or by 'mixed farming with a substantial dairying side'. The only other major activity was livestock-rearing along the west coast, in mid-Devon and on the uplands of Dartmoor and Exmoor. The whole region was effectively dominated by pastoral farming in the 1930s, based on permanent grass, a cheap form of farming in a depression. This is a situation similar to 1913 when permanent grass was identified as the main crop throughout the South West.[5] Dudley Stamp, in his essay on land use in Devon in the 1930s, illustrates graphically how the depression drastically reduced the area of arable in the upper Exe basin, while Robertson in his report on land use in Cornwall in the 1930s revealed by the Land Utilization Survey, provides a mass of detail concerning rural land use in the decade.[6]

The depression of the 1930s was rapidly lifted by the outbreak of the Second World War in 1939, when the need for Britain to feed itself became vitally important. **Fig. 39.3** shows how this was reflected in the amount of tillage, which virtually doubled between 1939 and 1945 as the 'ploughing-up campaign' of the war years had its impact. Although arable crops increased markedly, most of the rise was accounted for by temporary grass, as moribund permanent pasture was ploughed up to make way for sown grasses with greater vigour and feed value. This extra forage product was not, however, used to feed animals in the region, for as **Fig. 39.4** shows, sheep and pigs numbers fell during the war, but instead it was exported to other regions, presumably in the form of hay.

The war showed government how necessary it was for Britain to feed itself, and in 1947 the Agriculture Act provided the means and method of support which lifted the depression of the pre-war years. The result of expansion policies for a peacetime agriculture can clearly be seen in **Fig. 39.4** with marked increases in the numbers of sheep, pigs and cattle, and a very rapid increase in the acreage of feed barley. Wheat, not really suited to the South West or as a direct cattle feed, fell away as did oats as horses were replaced by tractors. Another change was the transformation of dairying, again dating from the 1930s when the depression in

Fig. 39.1 (left)
Land-use change in Cornwall,
1866–1938.

SOURCE: B.S. Robertson,
*Land of Britain, Part 91,
Cornwall* (London, 1951).

Fig. 39.2 (right)
Land-use change in Devon,
1866–1938.

SOURCE: L.D. Stamp,
*Land of Britain, Part 92,
Devonshire* (London, 1951).

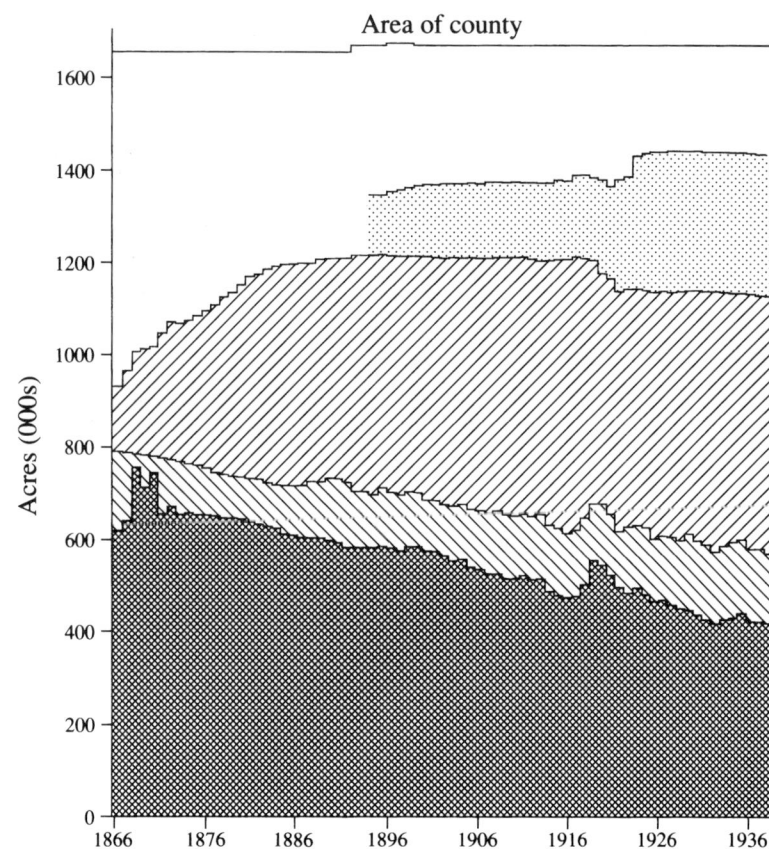

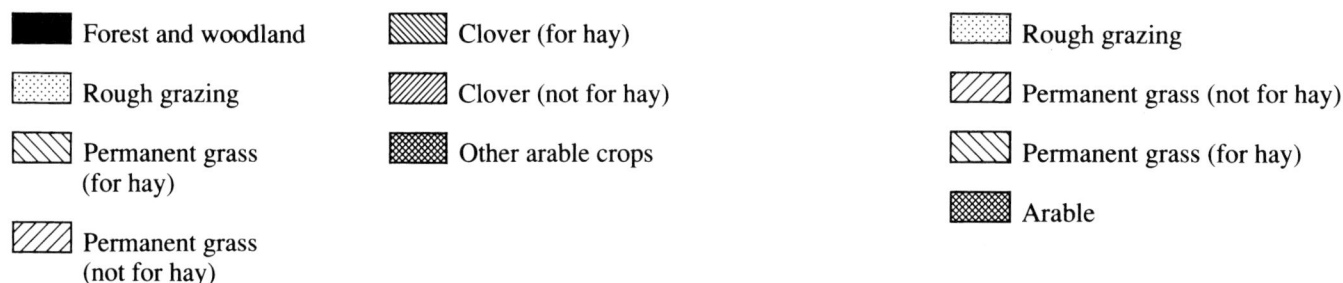

Fig. 39.1 — Land-use change in Cornwall, 1866–1938.

Legend:
- Forest and woodland
- Rough grazing
- Permanent grass (for hay)
- Permanent grass (not for hay)
- Clover (for hay)
- Clover (not for hay)
- Other arable crops

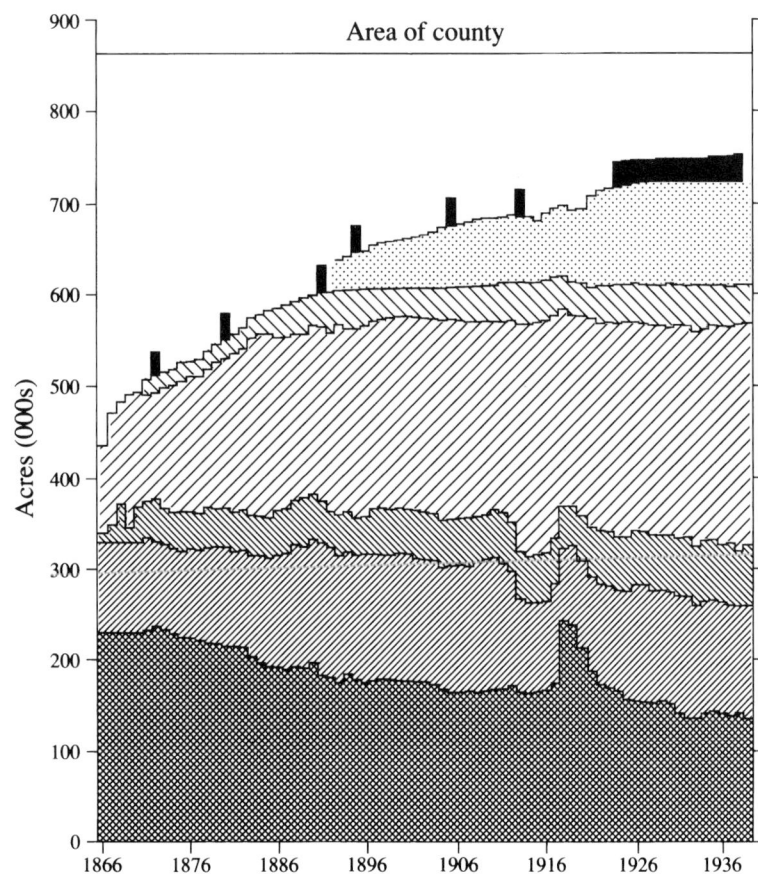

Fig. 39.2 — Land-use change in Devon, 1866–1938.

Legend:
- Rough grazing
- Permanent grass (not for hay)
- Permanent grass (for hay)
- Arable

farming was instrumental in bringing into being the Milk Marketing Board in 1933 in an attempt to provide dairy farms with an adequate income.

In many ways the thirty years from 1950 to 1980 will be looked upon as halcyon ones for agriculture in the South West, when not only were its products in demand but also prices were kept high and markets guaranteed by the UK government and, since the early 1970s, by the Common Agricultural Policy of the European Union. In 1964, the middle of this prosperous period, Shorter, Ravenhill and Gregory produced a series of detailed maps for a number of key agricultural indicators, three of which are reproduced here.[7]

Map 39.2 confirms the dominance of grass in the region, with virtually every area of the region recording 60 per cent or more, and substantial areas with over 80 per cent under grass. By definition, tillage is the mirror opposite of grass, and only in the Exe basin did tillage rise to more than 20 per cent of the total over

a wide area and with isolated pockets up to 55 per cent. Elsewhere there were small areas of tillage in excess of 30 per cent in the South Hams and the estuarine valleys of west Cornwall. Rough grazings as shown on **Map 39.3** dominated in the three moorland areas of Dartmoor, Exmoor and Bodmin Moor and in parts of west Cornwall on the bleak plateaux above the estuaries. The main end product of all this grassland and rough grazing land in 1964 was cattle production, with notably high numbers being recorded on Dartmoor and Bodmin Moor and in west Cornwall (**Map 39.4**). East Devon also had quite high numbers of cattle. Exmoor, however, had rather fewer.

Further evidence concerning agriculture in the three golden decades is provided by Coppock who has mapped agriculture using parishes and Agricultural Development and Advisory Service (ADAS) districts for both 1958 and 1970.[8] As **Map 39.5A** shows, livestock numbers, the main product of the region, varied widely in 1958, with concentrations in the west and the east, but fewer in the centre. The variations increase when dairy cows (**B**) and beef cows (**C**) are considered, with dairy cows being strongly represented in west Cornwall, and beef cows being dominant along the north-western coast, notably around Bude. Sheep are seen (**D**) to have been most important in 1958 in the Exmoor area (which accounts for the lack of cattle shown in **Map 39.4**), while pigs (**E**) are the inverse of the distribution of dairy cattle. In combination (**F**), cattle were the dominant animals in 1958 in all areas but north Devon, with sheep being the second most important animal elsewhere, except in west Cornwall.

Turning to the cereal crops, **Map 39.6A** shows that although less than a third of the crops and grass area was under tillage in 1958, most of the tillage area had more than 60 per cent of the area allocated to cereals (**B**), with barley being the main feeding cereal (**C**) except for the north-west coast, where mixed corn (a mixture of wheat, barley, oats and beans or peas, a peculiarity of the South West) and oats were the leading cereals. Taking these leading cereals in turn, barley (**D**) was concentrated in the South Hams and the Exe valley, mixed corn was concentrated in the western half of Cornwall (**E**), oats as in the 1930s in north-west Devon (**F**), and wheat (**G**) in south-east Devon. But all these crops were a small percentage of the total area. In combination (**H**), no one cereal was in a dominant second position in either the barley or mixed corn areas.

Map 39.1
Types of farming, *c*.1938.

SOURCE: as **Fig. 39.2**.

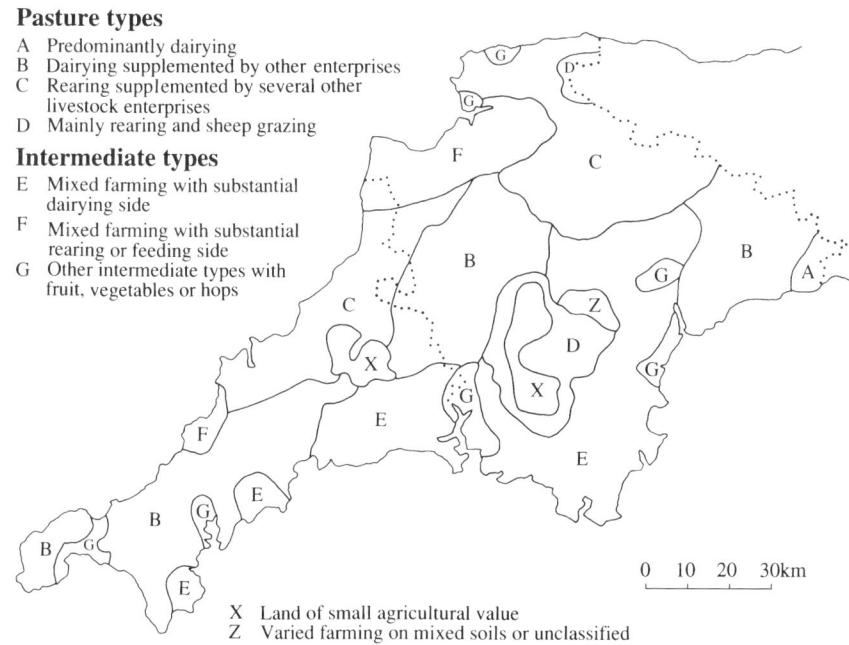

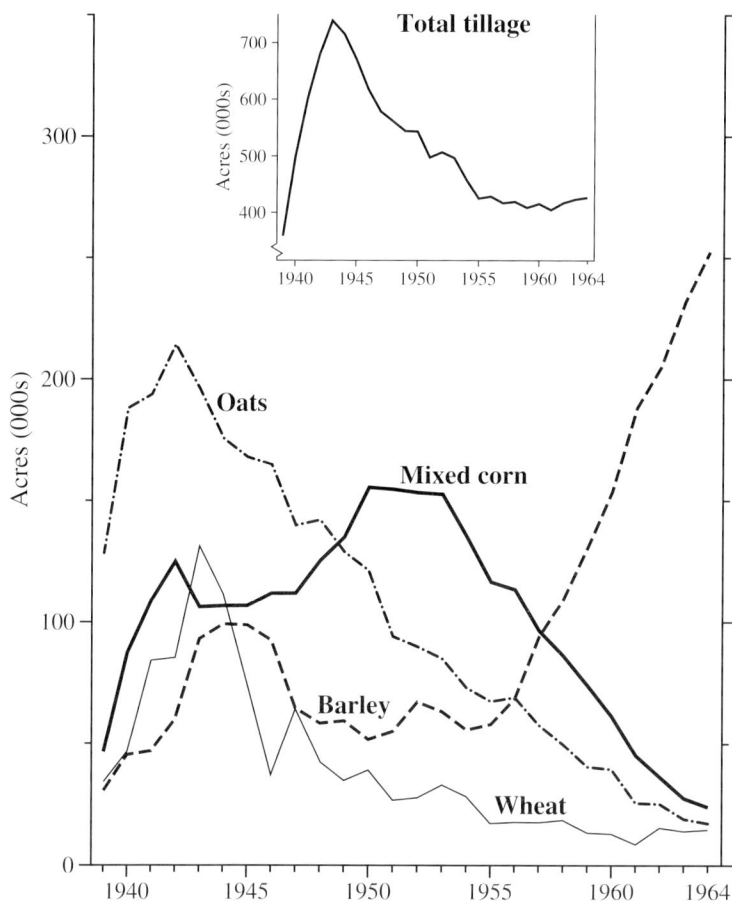

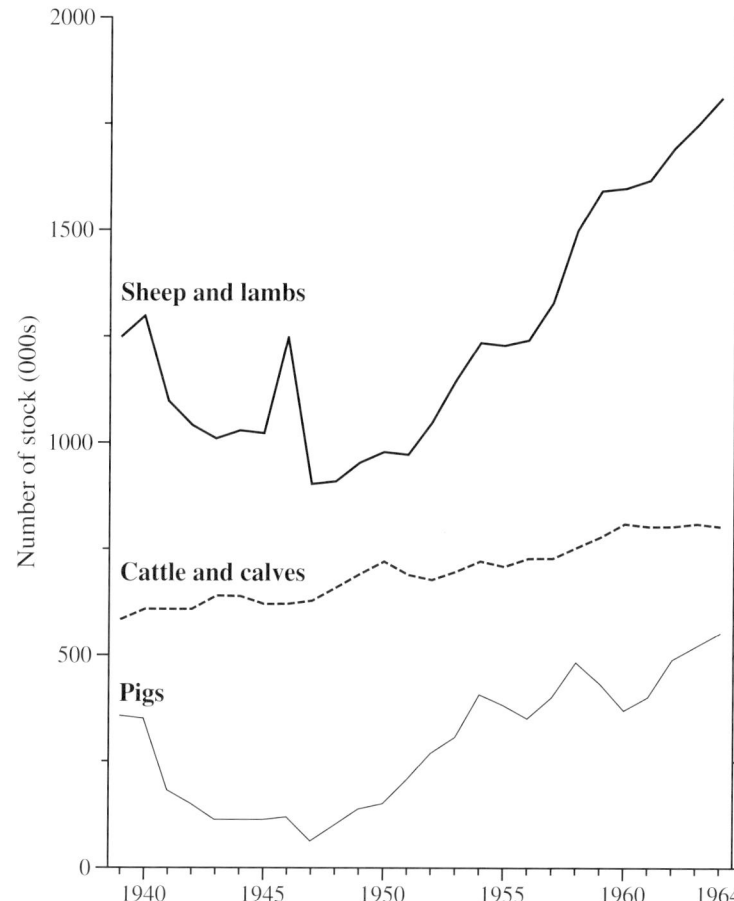

Pasture types
A   Predominantly dairying
B   Dairying supplemented by other enterprises
C   Rearing supplemented by several other livestock enterprises
D   Mainly rearing and sheep grazing

Intermediate types
E   Mixed farming with substantial dairying side
F   Mixed farming with substantial rearing or feeding side
G   Other intermediate types with fruit, vegetables or hops

X   Land of small agricultural value
Z   Varied farming on mixed soils or unclassified

Fig. 39.3 (left)
Tillage and cereal crops, 1939–1964.

SOURCE: A.H. Shorter, W.L.D. Ravenhill and K.J. Gregory, *South-West England* (London, 1969)

Fig. 39.4 (right)
Livestock numbers, 1939–1964.

SOURCE: as **Fig. 39.3**.

Map 39.2
Grass as a percentage of
total crops and grass, 1964.

S OURCE: as **Fig. 39.3**.

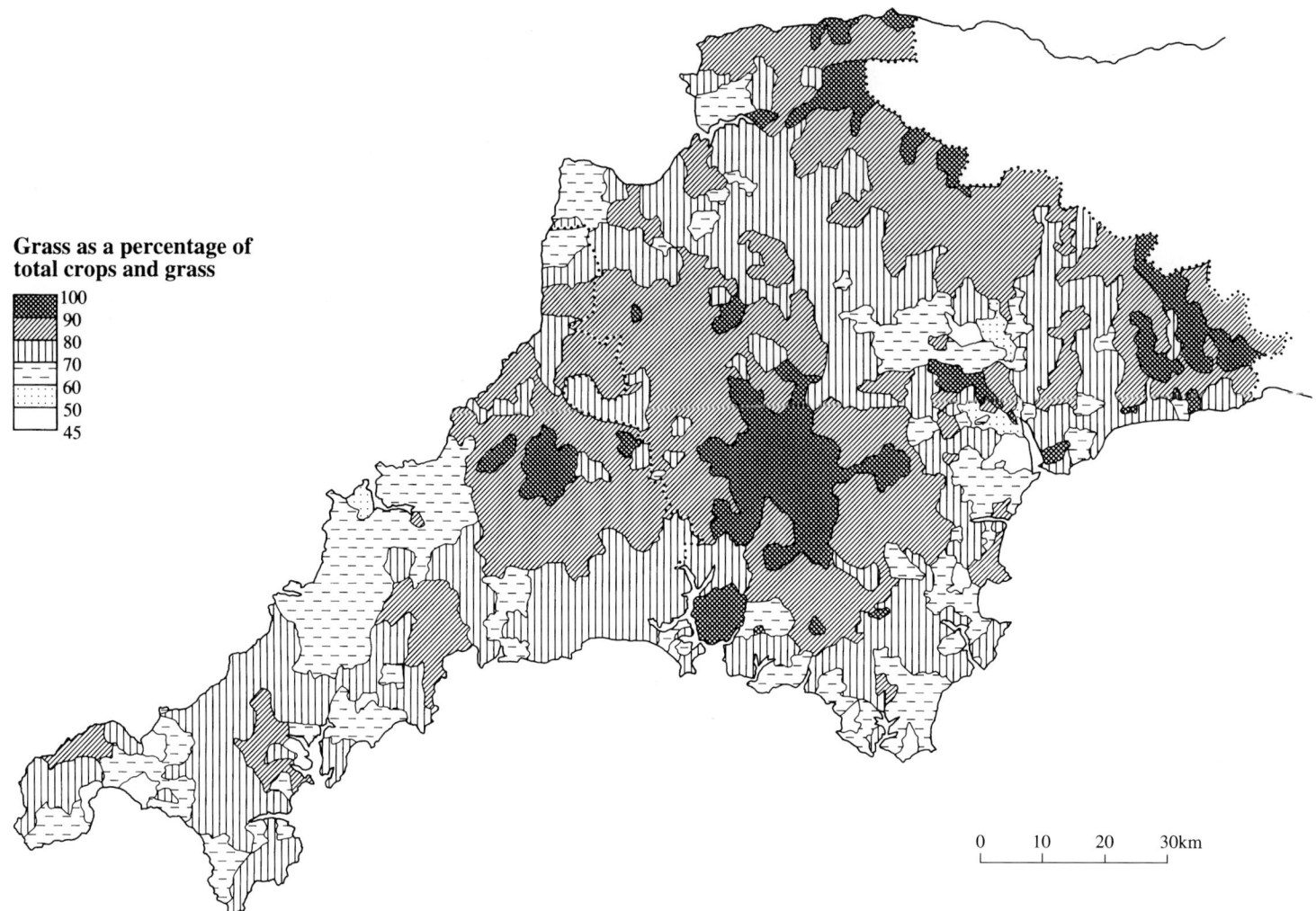

**Grass as a percentage of
total crops and grass**

100
90
80
70
60
50
45

0   10   20   30km

All other crops in Devon and Cornwall as a whole have traditionally been minority enterprises. Nevertheless, the detailed distribution reveals some interesting patterns (**Map 39.7**). For example, horticultural crops (**A**) were concentrated in west Cornwall, the Tamar valley and south-east Devon, as were vegetables (**B**) and potatoes (**C**). Orchards (**D**), however, were concentrated in south-east Devon with its sunnier climate, while green crops (**E**) and mangolds (**F**) for stock-feeding were concentrated in the milder southern coastal regions.

Combining crops with livestock, Coppock produced the map of leading enterprises based on ADAS districts (**Map 39.8**). This reveals that dairying was the leading enterprise except for north Devon and a small part of north Cornwall. Even in these areas dairying was the second most important enterprise. No other enterprise, except perhaps sheep and cattle, challenged dairying, and so for 1958 it can be concluded that the South West's agriculture was dominated by dairying and grassland in the lowlands, with sheep and beef cattle being locally important mainly in the uplands.

By 1970 little had changed, as **Map 39.9** reveals, with dairy farms accounting for over 50 per cent of all farms in the region, except for the Tamar valley and Exmoor.[9] Livestock farms were the dominant second type of farm and only horticultural farms in west Cornwall, the Tamar valley and east Devon accounted for over 20 per cent of farms. Nowhere in the region did cropping farms exceed 20 per cent, even in the Exe valley. Since the golden days of the three decades 1950–80, cutbacks in price support, the introduction of milk quotas in 1984, and general concerns about overproduction and farm pollution have led to a refinement in farming. In Devon and Cornwall, for example, the number of milk producers fell by 60 per cent between 1965 and 1985. Many of these changes have been influenced by British entry to the European Community in 1973 and so this section on agriculture concludes with an evaluation of the trends between 1973 and 1986 for Devon and Cornwall.

It has already been noted that dairying had increased its domination in the 1958–73 period, at the expense of sheep and

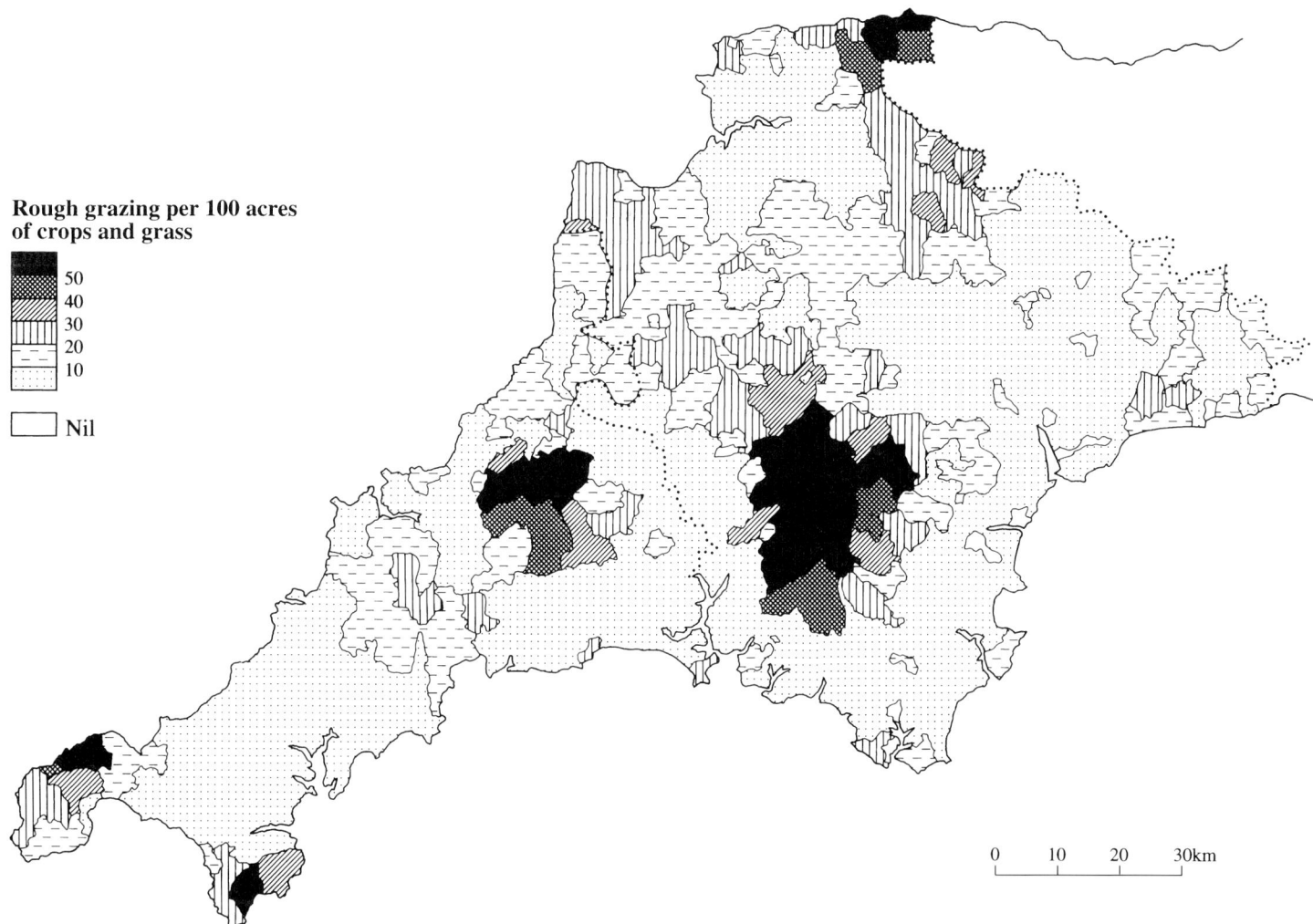

Map 39.3
Rough grazing (excluding commons) per 100 acres crops and grass, 1964.

SOURCE: as **Fig. 39.3**.

Rough grazing per 100 acres of crops and grass

50
40
30
20
10

Nil

0   10   20   30km

pigs and poultry. Nonetheless, between 1973 and 1986 there was a steady decline in the total number of cattle and calves. Within this decline, however, dairy cows have increased in total at the expense of beef cows, in spite of the imposition of milk quotas in 1984. This is because there is no realistic alternative to dairying, given the physical environment of the South West, and because South West dairying farmers have been able to buy quota. Indeed figures for farm incomes in 1988 showed that dairy farming was doing relatively well under the quota regime. However, since 1981 the numbers of other livestock have also increased: notably sheep and lambs in both Devon and Cornwall; breeding ewes in both Devon and Cornwall; pigs in Devon alone; and broilers in both Devon and Cornwall. In contrast, the numbers of other livestock have declined: notably pigs in Cornwall; and poultry kept for laying eggs in Devon and Cornwall.

It is difficult to see any obvious explanation for these changes either in underlying structural changes or peripheralization. For example, the British poultry industry has become heavily

concentrated into a few massive combines in recent years, mainly in the main cereal-growing areas near towns, but in Devon and Cornwall, although there has been a decline in the egg sector, the number of broilers has increased.

We have already seen that Devon and Cornwall are a marginal area for crops, other than horticulture. In both Devon and Cornwall the total area now devoted to grass as a percentage of the total cropped area is around 70 per cent, a figure that remained remarkably constant in the 1980s. In contrast, there have been significant relative (if not absolute) changes in the composition of crops other than grass. For example, crops for stock-feeding have increased by around 25 per cent since 1981, probably as a result of milk quotas releasing land which is then used to cut costs by growing grain rather than by purchasing it as an input. Less explicably in Devon, the barley area has fallen back since 1981 to the 1973 level at the expense of an increase in the wheat area. Finally, oil seed rape has made a dramatic entry owing to its relatively high intervention price prior to the cut in price in 1987.

Map 39.4
Total cattle and calves per 100 acres of total crops and grass, 1964.

SOURCE: as **Fig. 39.3**.

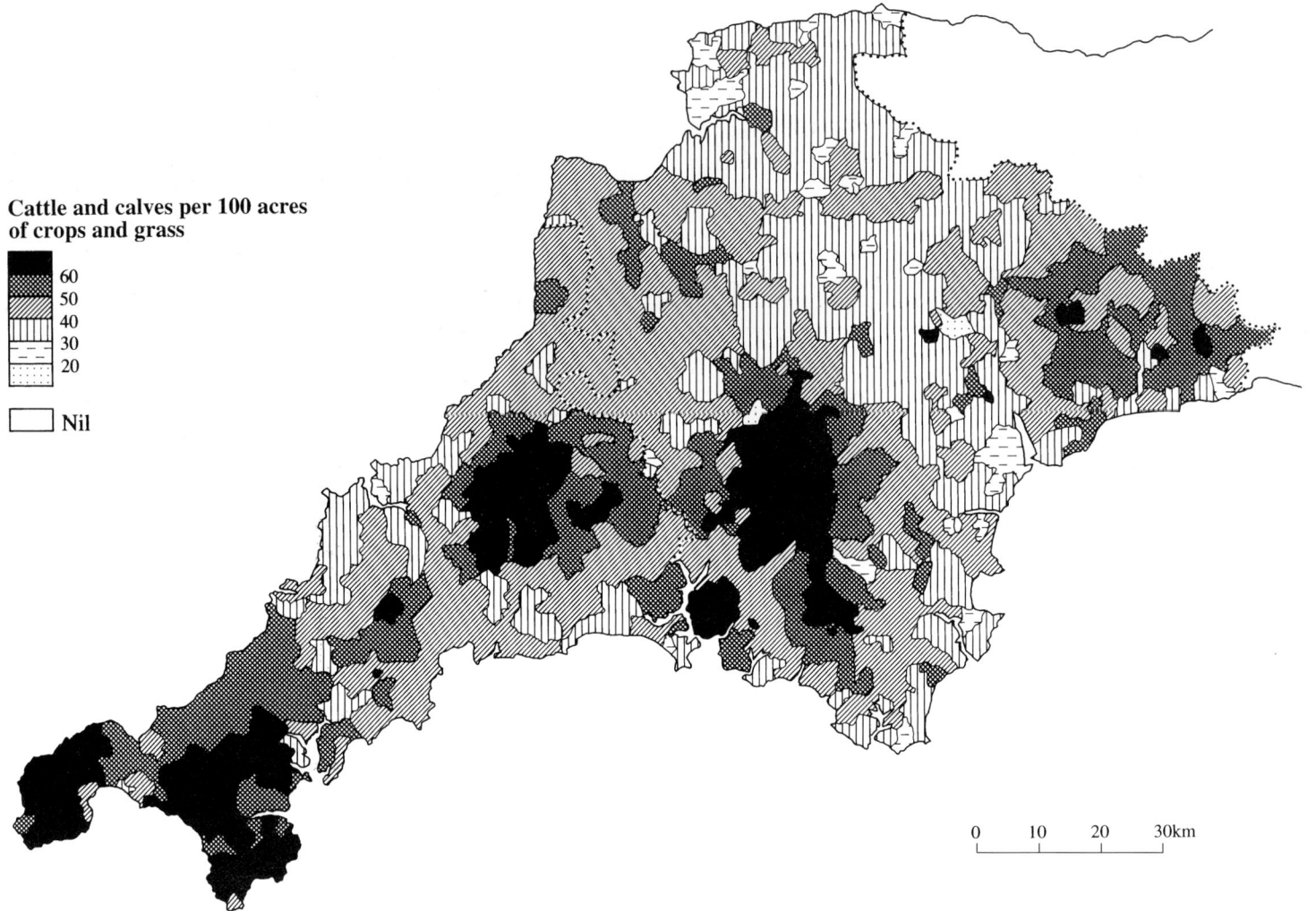

Cattle and calves per 100 acres of crops and grass

60
50
40
30
20

Nil

0   10   20   30km

Table 39.1
Type of farm in Devon and Cornwall, 1981 and 1986.

SOURCE: A. Gilg, 'The changing structure of agriculture in Cornwall and Devon, 1958–1986', in M. Havinden, J. Quéniart and J. Stanyer (eds), *Centre and Periphery: Brittany and Cornwall and Devon Compared* (Exeter, 1991), 151–63.

Turning to horticultural crops, the decline between 1973 and 1981 has been reversed in Cornwall and steadied in Devon. Within the sector, however, there have been significant changes with cabbage numbers growing at the expense of cauliflower and broccoli. This can largely be attributed to increased competition from Brittany with both accession to the European Community/Union and the introduction of the Roscoff–Plymouth ferry link. Most catastrophic, however, has been the decline of orchards in Devon, where the area has halved since 1973, with only a small offsetting increase in small fruit. This collapse is due to a concentration of cider production in a few centres, notably Hereford in the West Midlands, and to the high cost of replacing ageing and unproductive orchards with modern stock. Finally in the horticultural sector, flowers and bulbs have experienced a remarkable resurgence in Cornwall in contrast to a long decline in Devon. In this instance, competition from the European Union does not seem to have affected the mainly early flower trade centred on daffodils.

The end result of the changes outlined above is best summarized by analysis of changing farm type area as shown in **Table 39.1**. It is clear that horticulture as a type of farm in areal terms has suffered a severe decline in the 1980s, with pigs and

Type of farm	Hectares in 1981		Hectares in 1986	
	Cornwall	Devon	Cornwall	Devon
Dairying	107,017	194,115	103,311 (96)	183,517 (94)
Cattle and sheep:				
– Less favoured areas	13,979	58,454	12,614 (90)	55,840 (95)
– Lowland	90,866	161,246	97,492 (107)	183,622 (113)
Cropping	48,616	69,337	47,384 (97)	62,902 (90)
Pigs and poultry	5,762	15,191	4,409 (76)	10,716 (70)
Horticulture	5,352	4,790	3,709 (69)	2,979 (62)
Unclassified	7,775	15,894	10,007 (128)	19,355 (121)

NOTE: Extent is in hectares; figures in parentheses are 1986 figures as a percentage of the 1981 figures.

poultry suffering nearly as badly. In the area of less dramatic change it is clear that there have been some changes in enterprise patterns, notably Less Favoured Area cattle and sheep (as defined by the European Union) losing out in favour of lowland cattle and sheep, and cropping and dairying farms transferring to lowland cattle and sheep enterprises. The biggest change, however, has been in the unclassified category, namely those holdings returning only forage crops but no livestock at the June census. In more detail, dairying has increased its dominance as the leading entrprise, largely because Milk Marketing Board policies continued to encourage a pricing and manufacturing structure which to a large extent offsets peripherality from the main market, London. The introduction of milk quotas in 1984 caused a setback from which the industry gradually recovered by 1989. Second, fruit and horticulture lost their importance as locally important enterprises through a combination of increased competition from European Community countries, small scale of production rendering them uncompetitive, and because they do not have any major aid to offset their peripheral location, in contrast to the dairy sector. Third, mixed farming declined as dairy farming became more concentrated into single enterprise units, but has experienced a minor resurgence since the introduction of milk quotas in 1984, as farmers sought to diversify. Fourth, sheep farming has similarly declined, owing to a loss of market demand, but, for reasons which are not entirely clear, it staged a recovery in the 1980s.

In short, agriculture in the South West has been dominated by livestock farming in the twentieth century, notably dairy farming, and this was encouraged by the policies of the Milk Marketing Board between 1933 and the mid-1990s.

A CENTURY OF REAFFORESTATION

The South West is one of England's least wooded areas, and Cornwall is its least wooded county. This is partly a reflection of the windy climate but also of centuries of felling for boatbuilding and for agriculture. The lack of timber in the First World War, not only in the South West but elsewhere, was instrumental in the creation of the Forestry Commission in 1919 which has energetically set about its main task of reforesting Britain, though less spectacularly so in the South West than elsewhere. It is ironic then that some of the first Forestry Commission plantations in the 1920s were in Devon, on the Haldon Hills and in mid-Devon.

Tracing the history of woodland is not easy. First, it is not recorded consistently in the annual agricultural census, and Forestry Commission records only extend to their own woods and those private woodlands dedicated to them in return for grant aid. The main sources are published maps, land utilization surveys and surveys by the Forestry Commission in the late 1940s, mid 1960s and late 1970s.[10]

Map 39.10 shows the results of nearly fifty years of planting since 1919, notably the new forests of eastern Dartmoor and north-

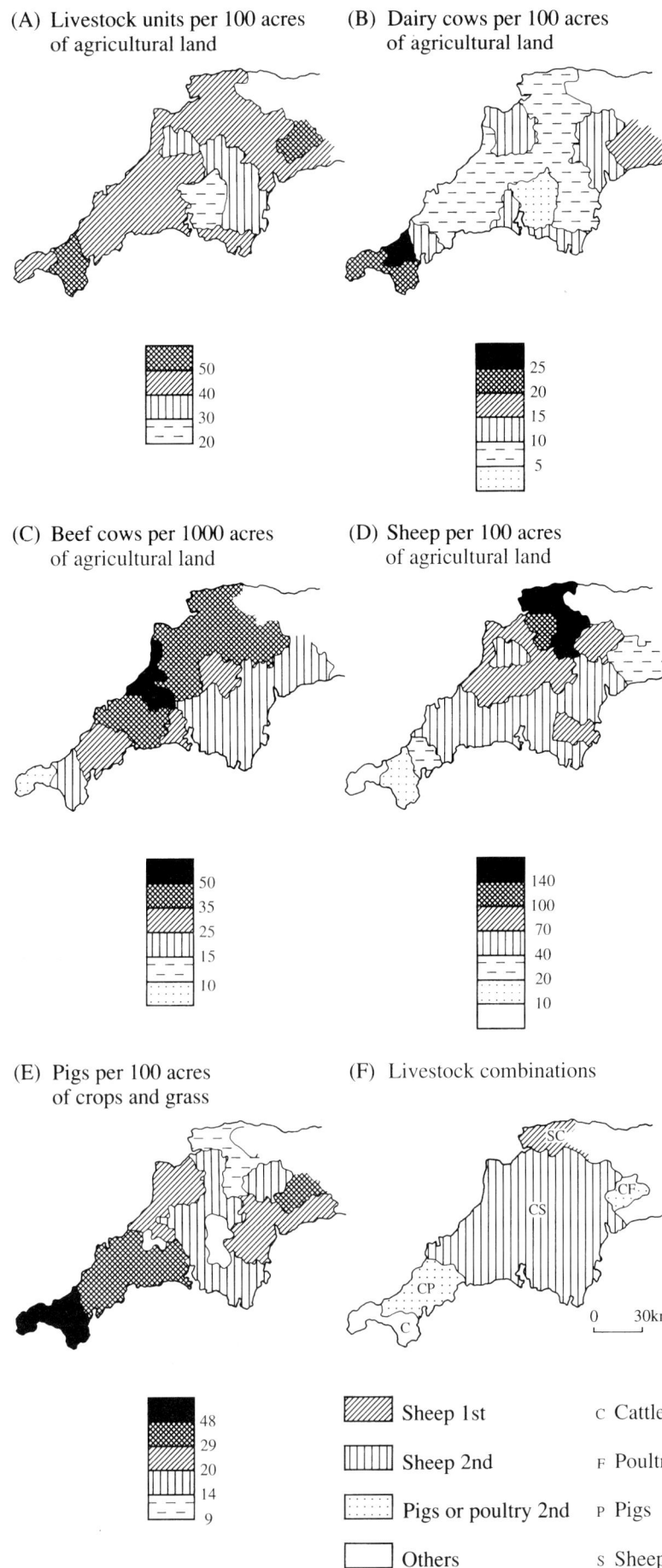

(A) Livestock units per 100 acres of agricultural land

(B) Dairy cows per 100 acres of agricultural land

(C) Beef cows per 1000 acres of agricultural land

(D) Sheep per 100 acres of agricultural land

(E) Pigs per 100 acres of crops and grass

(F) Livestock combinations

Map 39.5
Agricultural characteristics, 1958.

Source: J.T. Coppock, *An Agricultural Atlas of England and Wales* (London, 1964 and 1974).

Sheep 1st
Sheep 2nd
Pigs or poultry 2nd
Others

c Cattle
F Poultry
P Pigs
s Sheep

313

Map 39.6
Leading arable crops, 1958.

SOURCE: as **Map 39.5**.

(A) Percentage of crops and grass under tillage

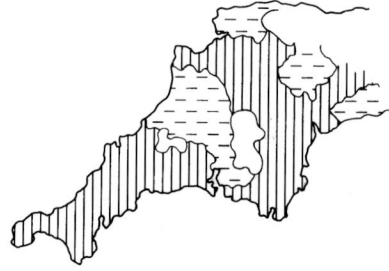

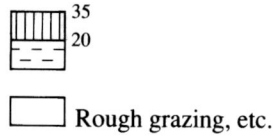

35
20

Rough grazing, etc.

(B) Percentage of tillage under cereals

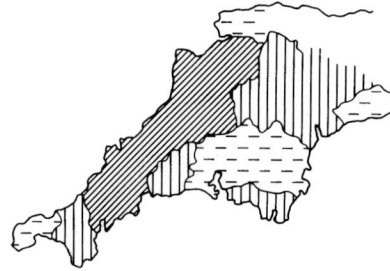

80
70
60

(C) Leading cereals

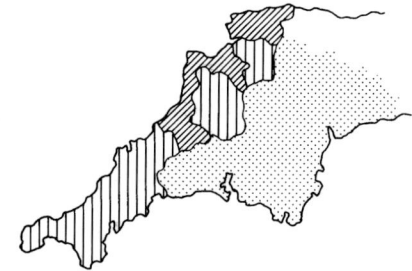

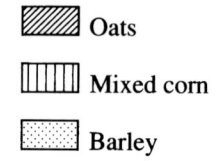

Oats

Mixed corn

Barley

(D) Percentage of crops and grass under barley

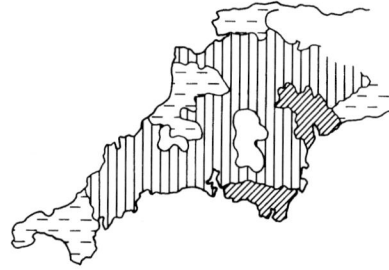

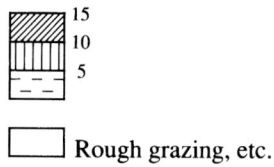

15
10
5

Rough grazing, etc.

(E) Acres of mixed corn per 1000 acres of crops and grass

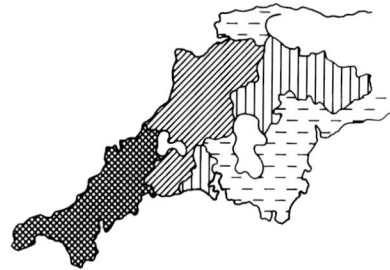

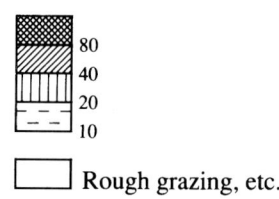

80
40
20
10

Rough grazing, etc.

(F) Percentage of crops and grass under oats

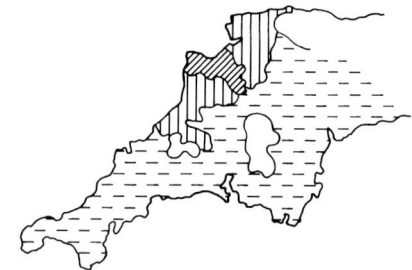

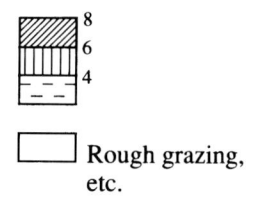

8
6
4

Rough grazing, etc.

(G) Percentage of crops and grass under wheat

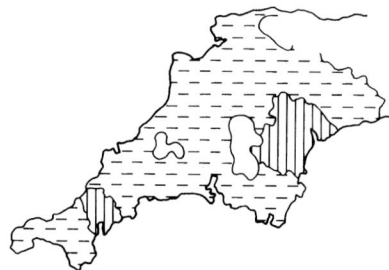

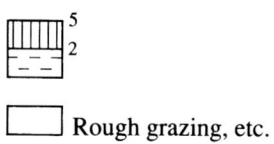

5
2

Rough grazing, etc.

(H) Leading crops

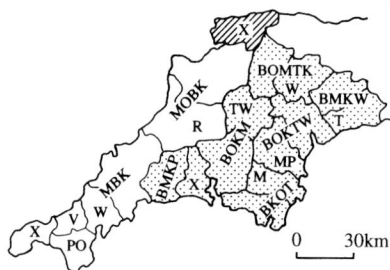

0   30km

Barley

Oats

Others

P   Potatoes
W   Wheat
B   Barley
O   Oats
M   Mixed corn
T   Turnips and swedes
R   Rape
K   Kale
V   Vegetables

west Devon. However, by the 1960s, opposition to further planting had increased on landscape and nature conservation grounds, notably in the most suitable areas of cheap rough grazing land on Dartmoor and Exmoor. As a result, voluntary agreements have reduced further planting since the 1960s. **Figs 39.5 and 39.6** indicate that the overall changes in Cornwall have followed a similar trend to that for Devon. The cutbacks in agriculture in the 1980s and the 1990s present an opportunity to reafforest the lowlands with traditional south-western broadleaved woodland.

LANDSCAPE CONSERVATION

The 1945 Town and Country Planning Act required local planning authorities to designate areas of landscape worthy of protection. Some of these are the so-called 'Areas of Great Landscape Value' as affirmed by Cornwall and Devon County Councils and depicted on **Map 39.11**. However, the 1949 National Parks and Access to the Countryside Act also provided powers for central government to designate National Parks and Areas of Outstanding Natural Beauty (AONBs) and the Nature Conservancy (Council was added to the title in 1973) to designate Nature Reserves and Sites of Special Scientific Interest (SSSIs). **Map 39.11** indicates that, by the mid-1960s, Devon had two National Parks and three AONBs and that in Cornwall Bodmin Moor was designated an AONB. However, none of these designations exerted any real power over change in the landscape in the same way that the 1947 and subsequent planning acts controlled the development of houses and factories. The speed of changes in the countryside outside planning control was considerable.

In an attempt to arrest these trends, the Nature Conservancy Council and Devon County Council have since 1949 designated various forms of Nature Reserve and SSSIs (**Map 39.12**). This shows a concentration of policies in the uplands, along the coast, and on the heathlands of east Devon and the Devon–Somerset

(A) Horticultural crops per 1000 acres of crops and grass

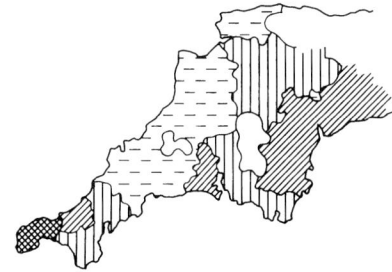

(B) Acres of vegetables per 1000 acres of crops and grass

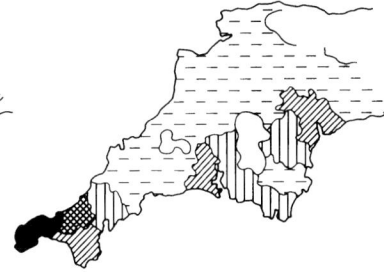

Map 39.7
Minor crops, 1958.

SOURCE: as **Map 39.5**.

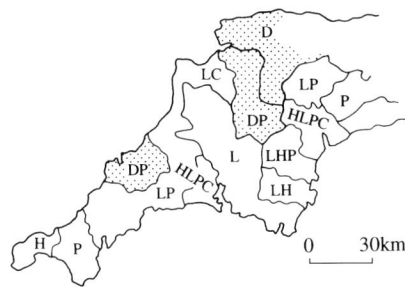

100 / 50 / 20 / 10

80 / 40 / 20 / 10 / 5

Rough grazing, etc.

Rough grazing, etc.

(C) Acres of potatoes per 1000 acres of crops and grass

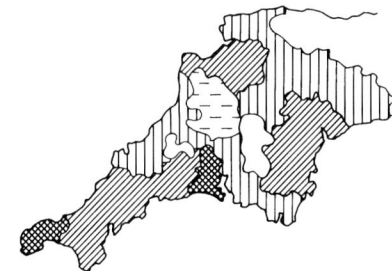

(D) Acres of orchards per 1000 acres of crops and grass

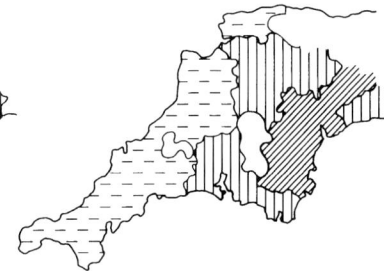

40 / 20 / 10 / 5

40 / 20 / 5 / 1

Rough grazing, etc.

Rough grazing, etc.

(E) Acres of cabbage, kale etc. per 1000 acres of crops and grass

(F) Acres of mangolds per 10000 acres of crops and grass

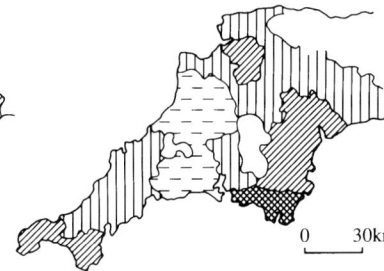

30 / 20 / 15

160 / 80 / 40 / 20

Rough grazing, etc.

Rough grazing, etc.

c Cash crops

p Pigs and poultry

H Horticultural crops

D Dairy cattle

L Sheep and beef cattle

0   30km

Leading enterprise

Sheep and beef cattle

Dairy cattle

Map 39.8 (above)
Enterprise combinations, 1958.

SOURCE: as **Map 39.5**.

Fig. 39.5 (left)
Devon: woodland planting,
pre-1861–1980.

SOURCE: Forestry Commission,
*Census of Woodlands and Trees,
1979–82: County of Devon*
(Edinburgh, 1983).

Fig. 39.6 (right)
Cornwall: woodland planting,
pre-1861–1980.

SOURCE: Forestry Commission,
*Census of Woodlands and Trees,
1979–82: County of Cornwall*
(Edinburgh, 1983).

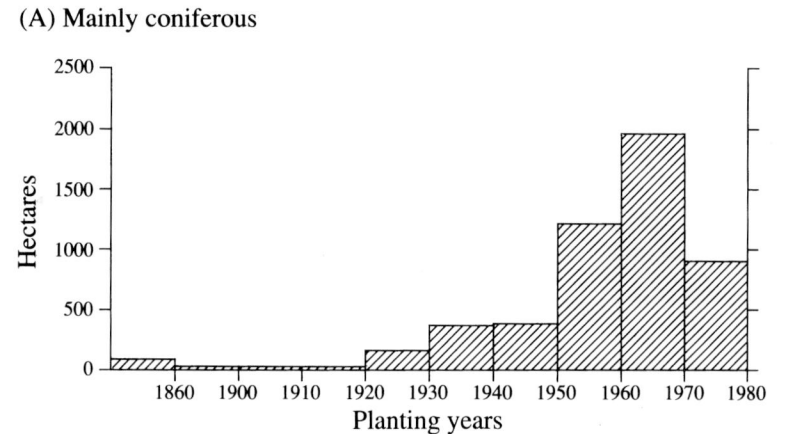

Fig. 39.5 (left)
Devon: woodland planting,
pre-1861–1980.

SOURCE: Forestry Commission,
*Census of Woodlands and Trees,
1979–82: County of Devon*
(Edinburgh, 1983).

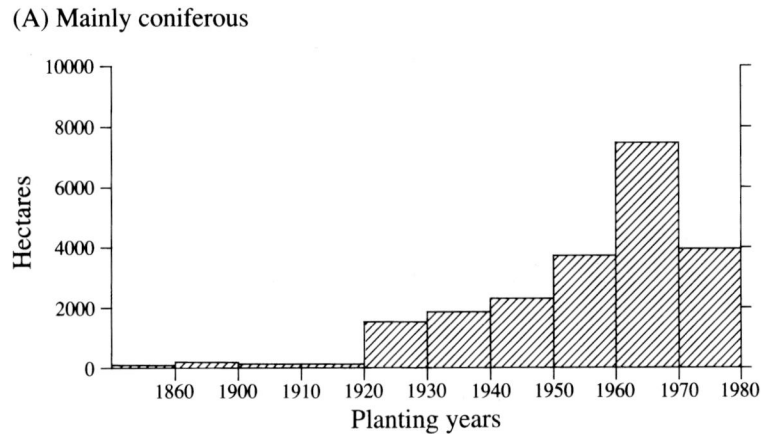

Fig. 39.6 (right)
Cornwall: woodland planting,
pre-1861–1980.

SOURCE: Forestry Commission,
*Census of Woodlands and Trees,
1979–82: County of Cornwall*
(Edinburgh, 1983).

(A) Mainly coniferous

(B) Mainly broadleaved

(A) Mainly coniferous

(B) Mainly broadleaved

Map 39.9
Types of farm, 1970.

SOURCE: as **Map 39.5**.

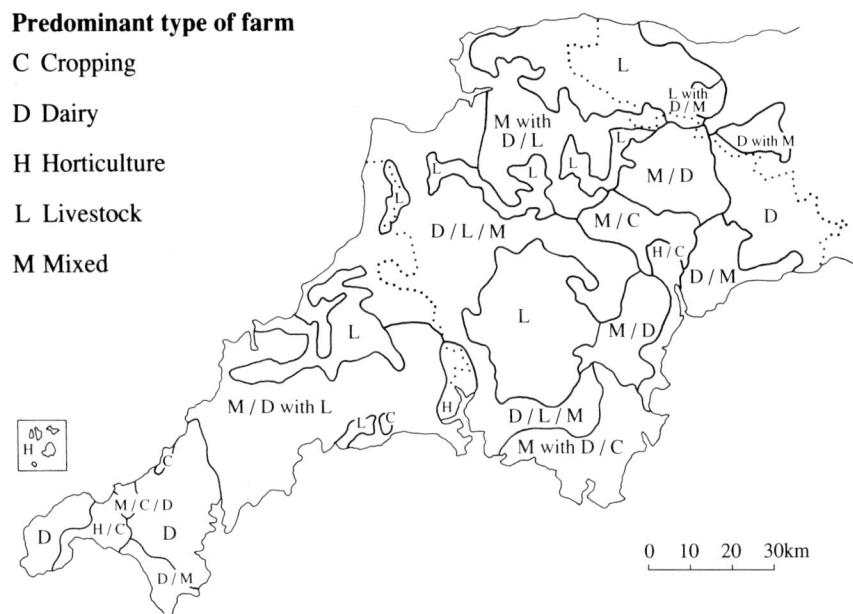

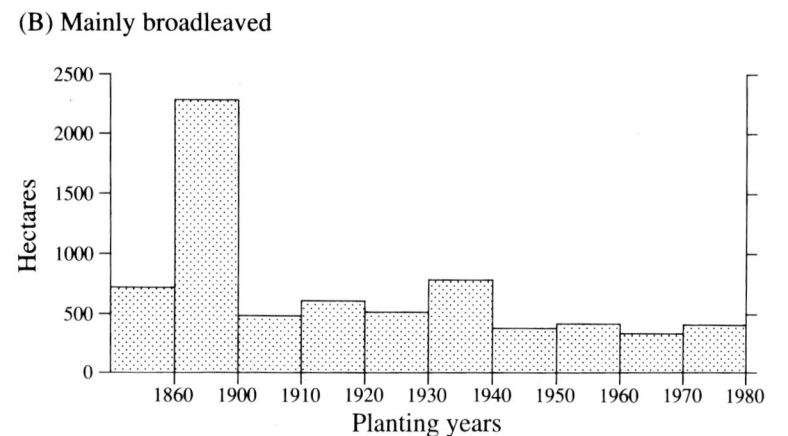

Predominant type of farm

C Cropping

D Dairy

H Horticulture

L Livestock

M Mixed

border. However, the policies are more apparent than real, as until the introduction of the 1981 Wildlife and Countryside Act SSSIs had no effective protection, and even since that date many SSSIs have been damaged.

Turning to the two prime conservation sites in the county, Dartmoor and Exmoor, both are National Parks, with similar landscape problems. In common with the lowlands, the main problem is conversion from rough land to farmland or forestry. Parry, who mapped land-use change from 1885 to 1958 in all the National Parks, makes the point that much change is temporary and reflects the centuries-long tradition of reclaiming marginal land in good times, but of then letting it revert in bad times.[11]

In the case of Dartmoor, land-use change has been analysed by Blacksell and Gilg, who concluded that most of the change in the 1950–70 period had occurred along the main river valleys of the Dart and Teign and at the fringe of the moorland.[12] The net effect was a loss of 680ha of heathland, offset by gains of 270ha of conifer woods, 130ha of deciduous woods, and 150ha of farmland.

Exmoor is a smaller and more fragile upland than Dartmoor. It is lower and with better soils it is more threatened by land-use conversion. Also it has far less common land and no Duchy of Cornwall to protect it. The threat to the very existence of the moor was recognized in the 1960s and a system of voluntary notification between farmers intending to reclaim land and the

Map 39.10
Distribution of woodland, 1967.

Source: as **Fig. 39.3**.

Map 39.11
Conserved areas, 1964.

Source: as **Fig. 39.3**.

National Parks

Areas of Outstanding Natural Beauty (AONB)

Areas of National Trust property

Areas of Great Landscape Value (AGLV)

0   10   20   30km

0   10   20   30km

Map 39.12
Devon: nature conservation policies and proposals, 1978.

SOURCE: Devon County Council, *Written Statement: 1978* (Exeter 1978).

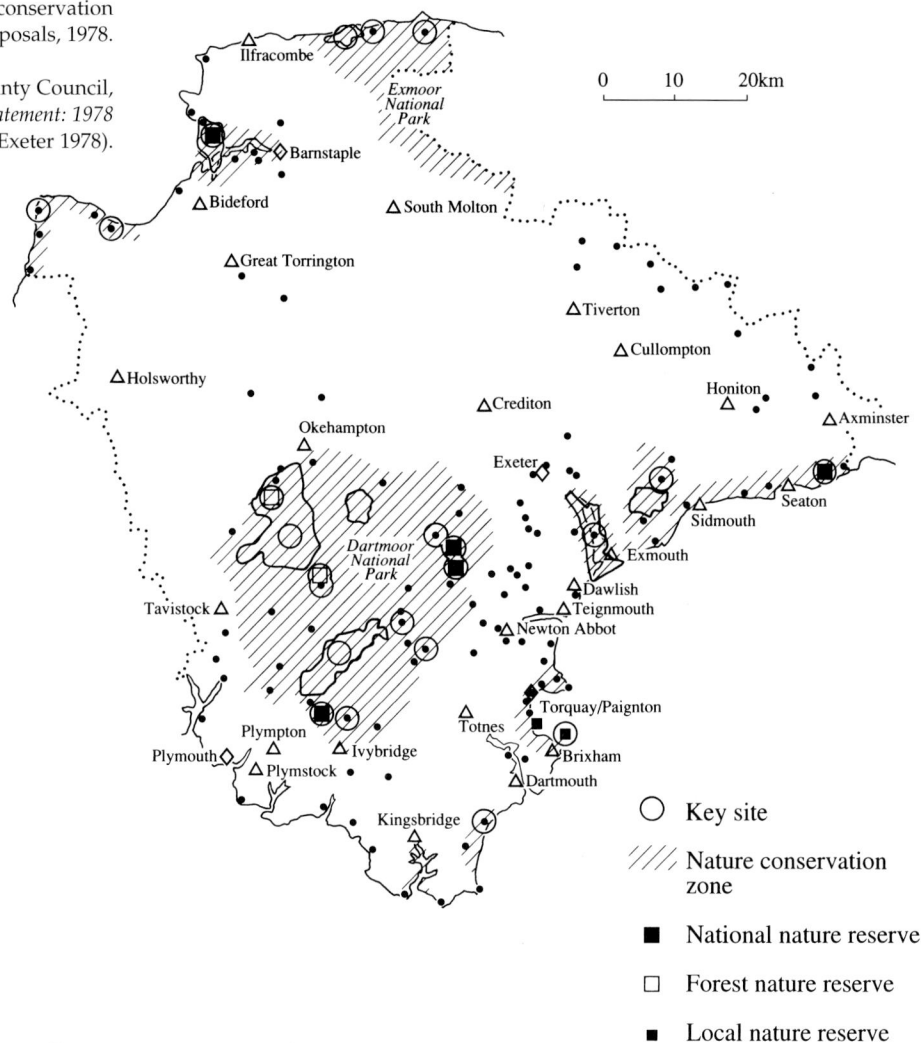

Ilfracombe

Exmoor National Park

0   10   20km

Barnstaple

Bideford   South Molton

Great Torrington

Tiverton

Cullompton

Holsworthy   Crediton   Honiton   Axminster

Okehampton   Exeter   Seaton

Dartmoor National Park   Sidmouth
Exmouth

Tavistock   Dawlish
Teignmouth
Newton Abbot

Plympton   Torquay/Paignton
Plymouth   Ivybridge   Totnes   Brixham
Plymstock   Dartmouth
Kingsbridge

○ Key site

/// Nature conservation zone

■ National nature reserve

☐ Forest nature reserve

■ Local nature reserve

Sites of Special Scientific Interest

⬭ >1000 acres

• <1000 acres

◇ Sub-regional centre

△ Area centre

National Park Committee was set in motion, based on a map of 'Critical Amenity Land'. A similar voluntary agreement was put into operation for Exmoor's woodlands not only to protect Exmoor's famous wooded combes from conversion to conifers, but also to protect the moor from afforestation. While the afforestation scheme had some success, the 'Critical Amenity Areas' (namely the moorland area) saw large tracts of heathland converted to farmland in the 1950s, 1960s and early 1970s.[13]

By the mid-1970s feelings had begun to run so high over the loss of moorland that the new National Park Authority failed to function for a while, so acrimonious did the meetings become. Accordingly government called in Lord Porchester, chairman of Hampshire County Council, to ascertain the facts and to find a solution. In his report, Porchester found that the rate of erosion had

indeed been rapid (**Map 39.13**) and that the concept of Exmoor as a moorland National Park was at risk. Porchester proposed a once-and-for-all payment to farmers who promised in perpetuity not to reclaim their land. Although accepted by the Labour government in 1978, this recommendation was rejected by the incoming Conservative administration in 1979 and replaced by yearly compensation payments in return for keeping land as moorland.

In contrast to Devon, landscape policies in Cornwall have had to deal with far less explosive issues, although the china clay industry continues to pose problems, as does derelict land from the tin-mining industry. Indeed, Cornwall has one of the highest proportions of derelict land of any county. The maps of conservation areas (**Maps 39.14 and 39.15**) also demonstrate that landscape conservation is basically a coastal issue in Cornwall, reflecting the nature of attractive fishing villages along the coast, especially along the more protected southern coast. In Cornwall there is a danger, though, that these conservation areas can become mere pastiches of their former selves as fishermen are driven out by dwindling fish stocks and second home owners. Buildings are preserved as a result, but not the communities.

CONCLUSION

This chapter has shown that Devon and Cornwall have been pastoral counties throughout the century, with a farm economy dominated by livestock and milk production. There have, however, been gradual changes in this pattern over the century and certain themes have emerged. These are:

(a) the agricultural area has expanded at the expense of rough grazing;
(b) the woodland area has expanded but mainly in coniferous plantations;
(c) rural land has come to be as valuable for conservation and recreation as it is for agriculture in many parts of the region, notably along the coast;
(d) the realization that the South West has to work out how productive farming can be reconciled with conservation and recreation so that a new type of rural landscape that combines the two main industries of the South West—farming and tourism—can be created and supported into the new millennium.

Land-use planners have formidable tasks ahead of them in the twenty-first century, but the lessons of the twentieth century as outlined in this chapter may help to guide them.

FURTHER READING

M. Blacksell and A. Gilg, *The Countryside: Planning and Change* (London, 1981).

Map 39.13
EXMOOR: extent of unimproved
hill in 1976 and hill converted
or improved, 1947–1976.

SOURCE: Lord Porchester,
A Study of Exmoor
(London, 1977).

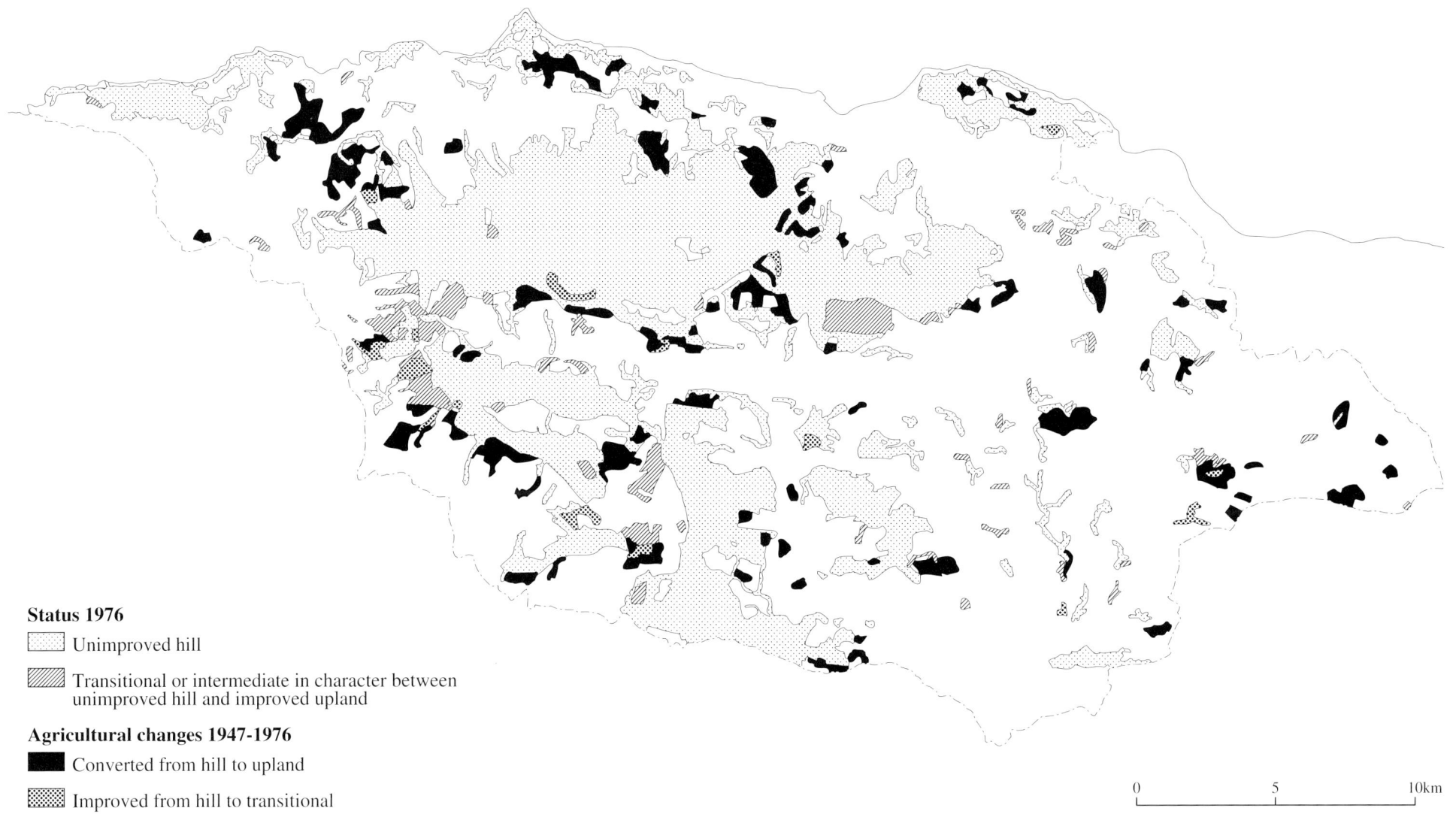

Status 1976

Unimproved hill

Transitional or intermediate in character between
unimproved hill and improved upland

Agricultural changes 1947-1976

Converted from hill to upland

Improved from hill to transitional

0   5   10km

Map 39.14
Cornwall: landscape
policies, 1979.

SOURCE: Cornwall County
Council, *Report of Survey: 1979*
(Truro 1979).

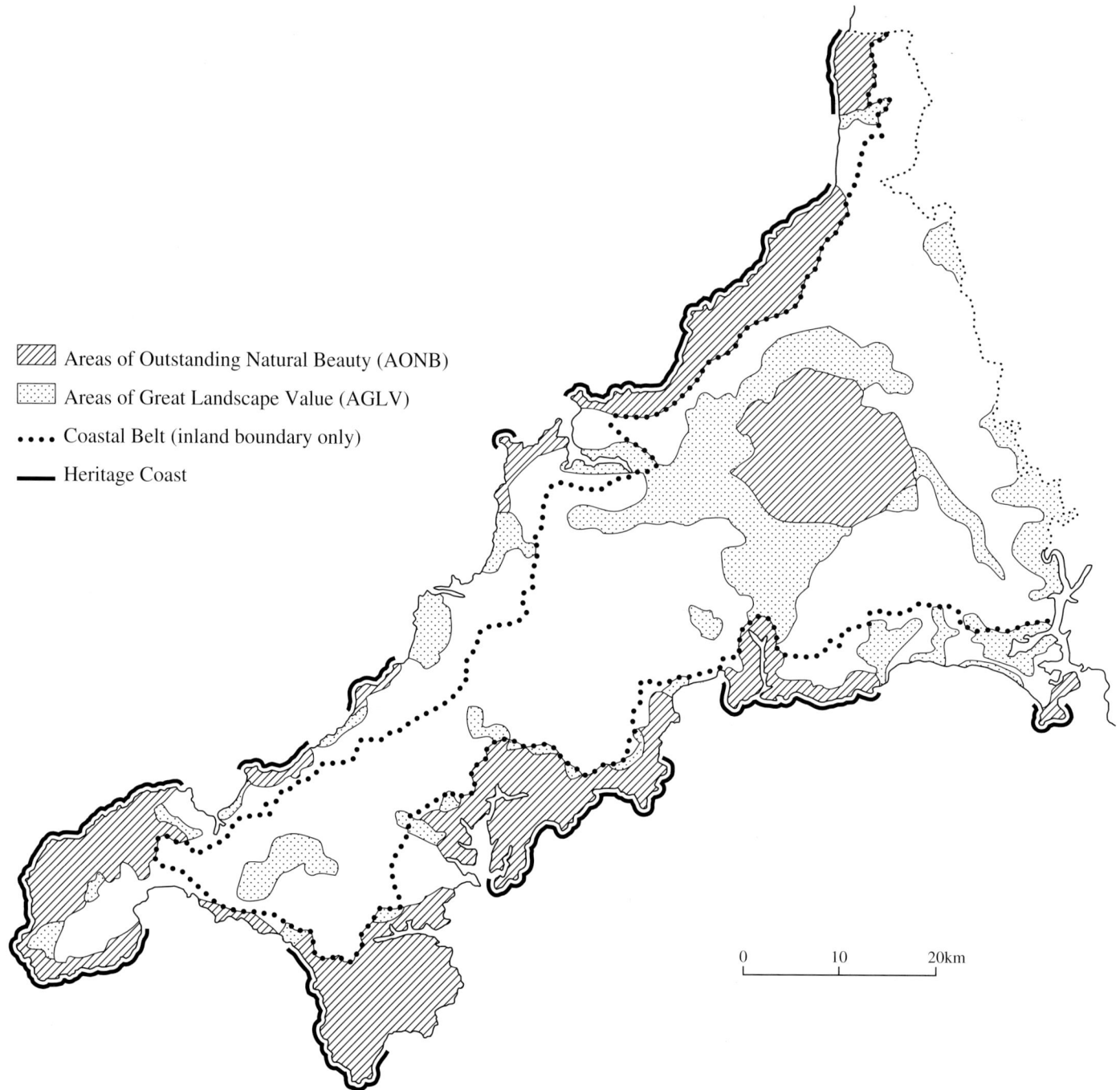

Areas of Outstanding Natural Beauty (AONB)

Areas of Great Landscape Value (AGLV)

•••• Coastal Belt (inland boundary only)

▬ Heritage Coast

Map 39.15
Cornwall: countryside
conservation.

SOURCE: W. Balchin,
The Cornish Landscape
(London, 1983).

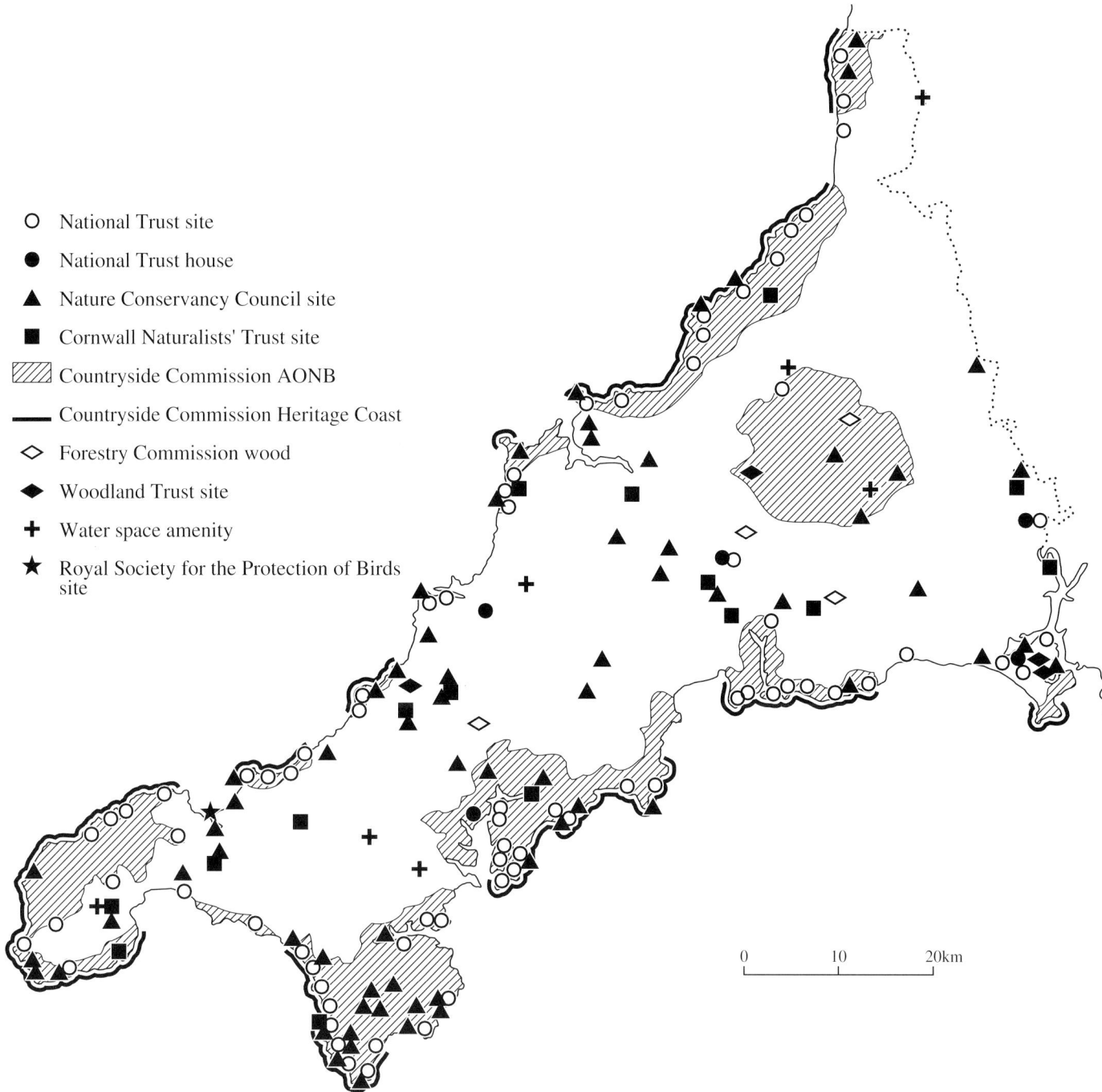

○   National Trust site

●   National Trust house

▲   Nature Conservancy Council site

■   Cornwall Naturalists' Trust site

▨   Countryside Commission AONB

▬   Countryside Commission Heritage Coast

◇   Forestry Commission wood

◆   Woodland Trust site

✛   Water space amenity

★   Royal Society for the Protection of Birds
    site

0   10   20km

INDUSTRY AND MINING

MEDIEVAL RURAL INDUSTRY

HAROLD FOX

When Defoe crossed the boundary between Somerset and Devon in the early eighteenth century he at once expressed delight, with a little exaggeration born of enthusiasm: he found the landscape 'so full of people … so universally employed in trade and manufactures' that it could not be equalled in all of Europe.[1] A traveller through Devon and Cornwall 400 years earlier would have been equally impressed by the rural industries of these two highly commercialized counties. Some medieval industries were rural because their bulky raw materials were located in the countryside; in other cases particular kinds of social structure and farming economy generated manufacturing. Rural industries complicated flows of commodities to and from towns. We often think simply of towns as supplying the countryside with manufactures and of country people supplying towns with food. But where industries were found in rural areas there were also flows in different directions: manufactured goods were brought from country to town and, under certain circumstances, country people engaged in industry came to town markets to purchase foodstuffs. Moreover, some rural industries had a considerable environmental impact. The smelting of silver and lead consumed great quantities of fuel, so that owners of woodland near the Tamar protested bitterly about demands made by the crown's controller of mines in the vicinity. Peat-digging to make peat charcoal—the standard fuel for smelting tin—denuded countless acres of moorland, the Cornish tinners complaining (with some exaggeration) in 1466 that their county was by then 'devoid of turf', while it was claimed in the early fourteenth century that digging for tin destroyed 300 acres yearly in Devon.[2]

THE TIN INDUSTRY

Among all the medieval rural industries of the South West, the getting of tin, largely by streaming, must be given high priority, for in the thirteenth century at least it undoubtedly created more wealth than any other. Maddicott has recently argued that the prosperity of Alfred's Wessex was in no small part derived from the tin trade and that pre-Norman urban development in the South West owes much to tin.[3] But it is not until the middle of the twelfth century that the royal pipe rolls give the historian a first taste of statistical sources relating to taxes on tin production which later expand into a vast body of material still in many ways untapped. Output grew rapidly during the twelfth and thirteenth centuries when tin from the South West commanded a large share of markets in Europe and the Middle East.[4] Population was increasing at this time, as was the wealth of those classes that purchased most items of pewter and bronze (the two alloys in which tin was largely used) for utilitarian, artistic, sartorial and cultural uses. Notable also is the way in which the production of tin suffered only about thirty-five years of acute depression during the centuries after the Black Death, a period of declining population. Clearly, objects made partly of tin (e.g. pewter plates and drinking vessels) were among the items increasingly used by a population which was experiencing a rise in standards of living at this time.[5]

The distribution of tin production is partially revealed by manorial sources, for lords made profits from their own tolls on tin.[6] More comprehensive are sources connected with royal taxation of the stannaries (tin-producing districts), which were, from west to east, Penwith with Kerrier (the far west), Tywarnhayle (environs of St Agnes), Blackmoor, Foweymoor (now Bodmin Moor) and the Dartmoor stannary districts around Tavistock, Plympton, Ashburton and Chagford.[7] Among several forms of royal taxation the most lucrative was coinage duty charged on the refined tin when the ingots were assayed at the chief town of each stannary district. Data in the earliest surviving coinage rolls, c.1300, and in rolls from around 1523 are presented in **Maps 40.1 and 40.2**.[8]

The maps show that by the beginning of the thirteenth century, Cornwall's tin production had outstripped that of Devon and that within Cornwall the tinners of Bodmin Moor were the most productive. At that time Bodmin and nearby Lostwithiel were, consequently, two of the wealthiest Cornish boroughs, the former a town of 2,000 people and more before the Black Death, the latter a port full of carousing sailors, many no doubt from ships exporting tin, who frequented taverns set up for them both on dry land and on deck.[9] There is also a connection between industrial activity on Bodmin Moor and the establishment of new moorland settlements during the thirteenth century as well as the more rigorous definition of manorial boundaries as lords sought to maximize their own private taxes on tinning. Thereafter there took place a quite marked westward shift in Cornish tin production, evident not only from the coinage rolls (**Map 40.2**) but also from a falling number of documentary references to tin works, stamping mills and blowing houses on Bodmin Moor.[10] The shift had an adverse effect on some east Cornish towns: in the sixteenth century John Norden noted many buildings 'decayed' in Bodmin and Lostwithiel, although in the latter case silting of the river, the environmental impact of centuries of tin-streaming inland, was an added cause of decay. By contrast, Helston in western Cornwall was described as a 'good market town' in the early sixteenth century.[11]

The westward shift of tinning in Cornwall has been explained by partial exhaustion of the most easily worked deposits on Bodmin Moor. On Dartmoor too, the fortunes of the different stannary districts ebbed and flowed as a result of what Lewis called the 'capricious' nature of the stream tin: it 'might be found … for a limited number of years and then disappear entirely'.[12] For example, in about 1300 Tavistock was the smallest Dartmoor stannary; by the early 1380s a boom had occurred; there then followed a slump and no recovery until the closing decades of the fifteenth century.[13]

The history of the workforce in the medieval tin fields will only be written when a complete study has been made of the records of the stannary courts. Far more easy to use, though surviving for Devon alone, are listings of people claiming exemption from regular royal taxation, for among the many liberties given by the crown to tinners was freedom from ordinary taxes. Any list of men trying to avoid tax is bound to contain some fraudulent individuals, and these lists almost certainly exclude a large number of the poorest tinners; but nevertheless a map based on this source is not without interest (**Map 40.3**). Tinners lived in most townships bordering Dartmoor but equally interesting are reputed tinners in coastal places and in two corridors, towards Exeter and Somerset and towards Barnstaple; these could possibly be men claiming exemption by virtue of involvement in the trade in tin.[14]

The nature of the working lives of medieval tinners has been debated. One view portrays the tinner as alienated from rural society, a 'poore hireling' in a tin works dependent on the vagaries of the labour market. The other view is of the 'free' farmer-tinner for whom streaming for ore was a part-time, seasonal occupation providing income to supplement husbandry.[15] Hatcher has rightly stressed that although medieval tin-streaming was not compelled for technological reasons towards a capitalistic organization of masters and men (for it needed a minimum of equipment), many tin works were nevertheless organized by entrepreneurs with dependent labour. The sometimes infrequent timing of the coinages tended to squeeze out the small man working for himself who was, therefore, obliged to seek advance payments from merchants; from that system it was a short step to reliance on wages alone.[16] Hatcher has provided many references to medieval mining enterprises run on capitalistic lines; to these may be added the tin works owned by the Arundell family who employed 'spaliards and diverse labourers' in 1488, and the seven tin mills at Newland on Bodmin Moor, leased to 'merchants' in 1451.[17]

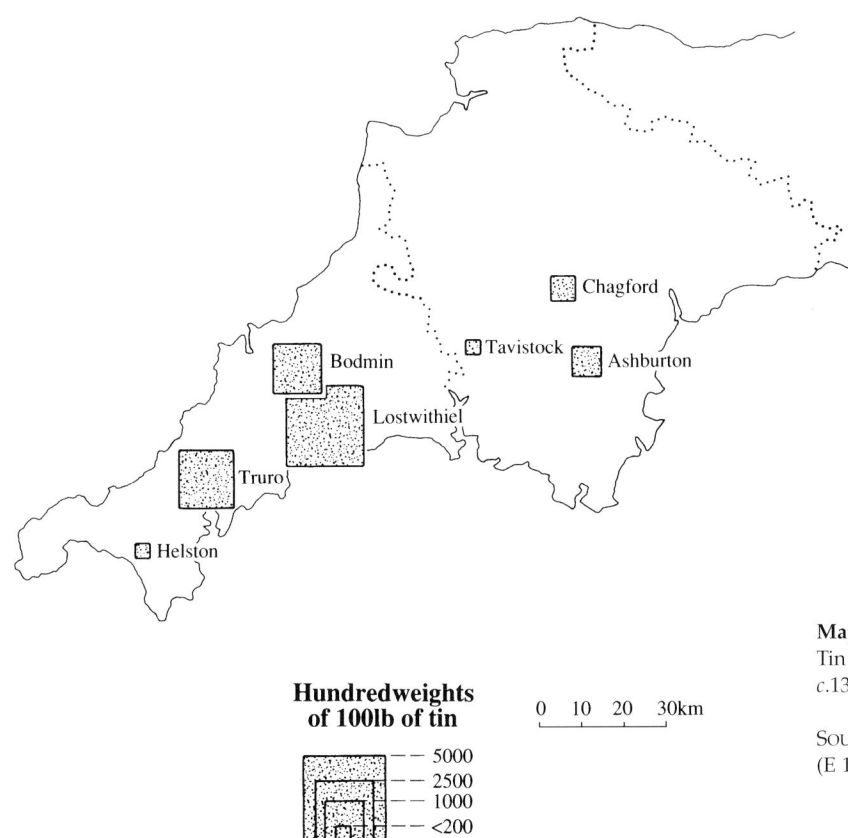

Map 40.1
Tin coined at stannary towns, c.1300.

SOURCE: Public Record Office (E 101/260/24 and 261/1).

Hundredweights of 100lb of tin

0 10 20 30km

5000
2500
1000
<200

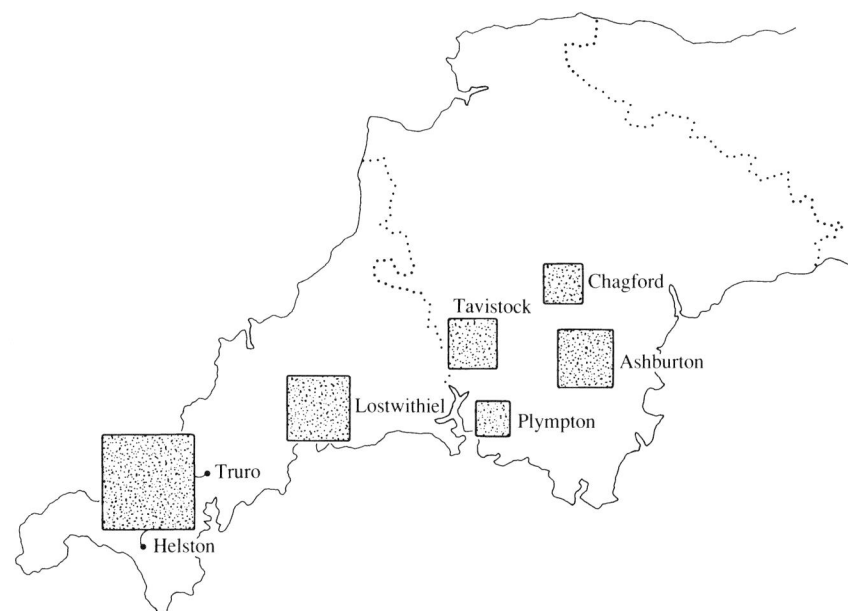

Map 40.2
Tin coined at stannary towns, c.1523.

SOURCE: Public Record Office (E 101/271/4–5 and 271/6–8).

Map 40.3
Reputed miners
in Devon, 1373.

SOURCE: Public Record Office
(E 179/95/28–32).

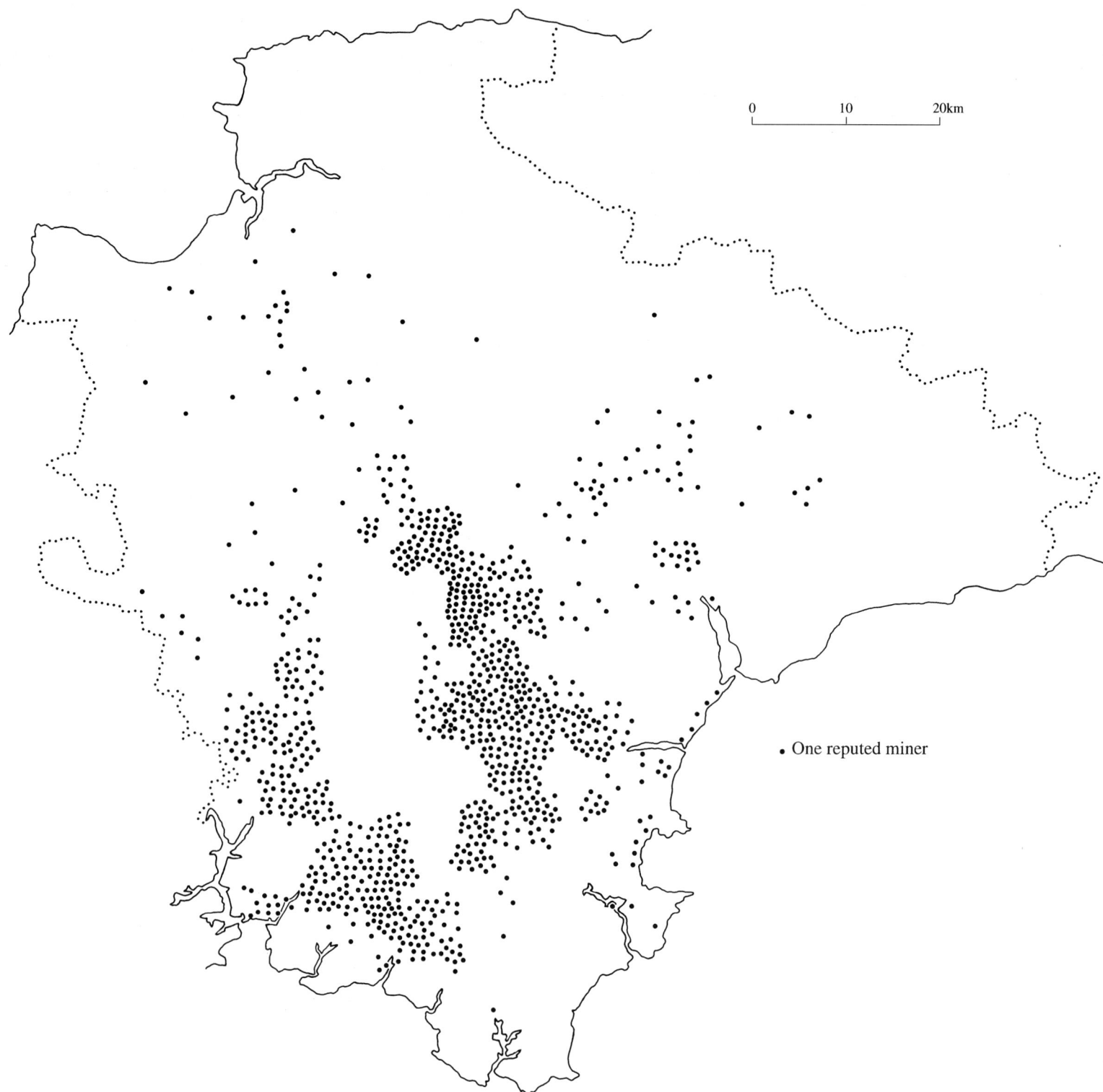

0   10   20km

• One reputed miner

One can also point to examples of men who certainly worked in the industry from an agricultural base: the 'sturdy farmer-miners' John Symon and John Kendal, with 25 and 7 acres respectively, who toiled in Blackmoor stannary in the 1380s; or Robert Richeman, who appears in the manor court rolls of Shaugh, having been killed in an accident in a tin works, and who also farmed a smallholding, for his chattels included several animals.[18] Both types of miner were always to be found in the stannaries but one can suggest, as a partial solution to the debate, that the relative sizes of the two groups changed during the course of the Middle Ages. Before the Black Death (1348), when farm holdings were on the small side, it may often have been necessary for needy smallholders to engage in tinning; after 1348, with pressure relaxed, there was less necessity for farmers, now with larger holdings, to engage in an uncongenial by-employment, and tinning, therefore, became an activity of the most needy of the labouring population and those without land, such as sons waiting to acquire a holding. Before 1348 relatively little tin was coined in summer and early autumn, but by the fifteenth century the majority was produced in those months. This trend is open to various interpretations, one of which is that the pattern before the Black Death reflects tinning by arable farmers who had little spare time during the hay and grain harvests; the later pattern could reflect tinning among a labouring population who by choice worked on the moors when they were less bleak in summer and who turned to other, perhaps more home-based, occupations during the winter.[19]

It is important to try to estimate the number of tinners because they formed a significant class of consumer in the countryside. In Devon we can very roughly estimate from the list of tinners exempt from taxation in 1373 and the poll tax total of 1377 that an absolute minimum of about one in twenty adult males was involved in the tin industry.[20] In Cornwall the figure may have been about one male in ten, with an additional contribution from women, which cannot be estimated but which cannot have been unimportant.[21] Ancillary industries, such as turf-cutting and the firing of turf charcoal, performed by *carbonarii*, gave employment to many more people.[22]

The cloth industry

Hooker, at the end of the sixteenth century, described tin as the 'third comoditie' of Devon, after farm produce and cloth; and indeed some time before he was born there were years (e.g. 1500) when cloth exports from Exeter exceeded those from any other English provincial port.[23] Customs statistics for the port of Exeter show a modest level of exports in the early fifteenth century, then a series of rapid surges from the 1470s onwards. There are, though, reasons for doubting that the development of the industry may be measured so exactly through the mirror of trends in exports. First, the figures exclude the coastal trades to, for example, Bristol and Southampton; it is significant that slumps in

exports from Exeter correspond with surges in those from Bristol and Southampton, suggesting changes in preferred ports among merchants rather than any sudden decline in manufacturing in the South West.[24] Second, export figures take no account of the overland trade in cloth, especially eastwards towards London, which was becoming the paramount cloth exporter during the second half of the fifteenth century, and also northwards into Somerset and beyond with Honiton acting as a distributive centre.[25]

There is, in fact, good evidence pointing to a clothing industry of respectable proportions in the South West well before the end of the fifteenth century. In the 1390s Devon and Cornwall could hold their own among some of England's leading cloth-producing regions such as Suffolk and rural Yorkshire.[26] Moreover, a considerable number of fulling mills for beating and cleaning cloth already existed by this time. If we guess that the total number of fulling mills in the 1390s was double the number (about forty) recorded in scattered sources and that a fuller needed to make no less than about £4 yearly (a minimal wage for a craftsman, but many fullers probably had other sources of income), then we can very speculatively calculate that the equivalent of about 3,600 cloths of assize were being processed. This is far more than the export figures for this decade would suggest, indicating that a good deal of the cloth produced went to the home market or was shipped in the coastal trade.[27] Earlier, in the 1350s, Devon's ports dispatched some 12 per cent of the nation's cloth exports and in 1315 the crown, concerned with the quality of certain cloths, listed Devon and Cornwall alongside other notable producers, Norfolk, Suffolk, the Mendips, Stamford and Beverley. Even earlier, at the end of the twelfth century, the royal almoner chose to buy cheap Cornish cloth for distribution to the poor.[28] Moreover, from 1225—when a financial account for the rural manor of Tiverton mentions rent from a fulling mill—there are numerous references to mechanical fulling in the South West. The number of rural fulling mills recorded during the thirteenth century suggest something more than a local clothing industry serving local needs; on balance one is tempted to agree with those historians who have concluded that rural cloth-making in England before 1300 had already developed in those counties which were to be the great exporters of cloth on international markets later on in the fifteenth century.[29]

Why did parts of Devon and Cornwall boast a medieval clothing industry of some importance whereas the countrysides of some English counties (e.g. Leicestershire) produced very little cloth? Technological reasons must be largely ruled out. If supplies of high-quality wool had been a locational factor, the two south-western counties should have had no cloth manufacturing at all, for the medieval fleece weights of local sheep were of puny proportions and the cloth produced was generally for the cheaper end of the market whether at home or abroad.[30] Contemporaries fully recognized this: inhabitants of the countryside around the south-western borders of Dartmoor made coarse 'Tavistocks', describing their wool as rough and

Map 40.4
Fulling mills recorded,
*c.*1350–1425.

SOURCE: miscellaneous sources
(mainly manorial).
For examples see text
and notes.

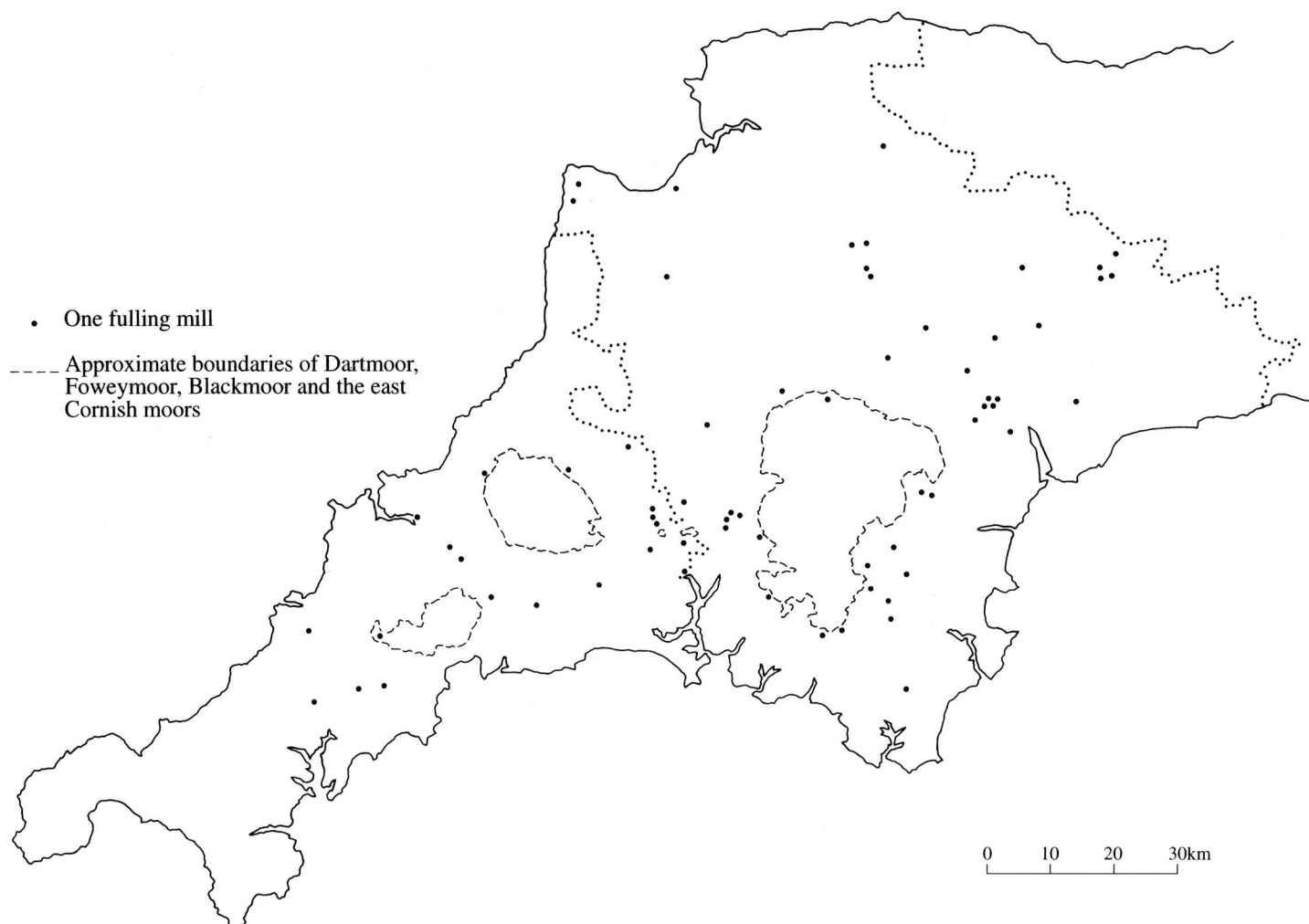

Map 40.4
Fulling mills recorded,
*c.*1350–1425.

SOURCE: miscellaneous sources
(mainly manorial).
For examples see text
and notes.

- • One fulling mill

- - - - Approximate boundaries of Dartmoor,
  Foweymoor, Blackmoor and the east
  Cornish moors

0   10   20   30km

stubborn in a late-fifteenth-century petition; an Elizabethan description refers contemptuously to Cornish wool as 'hair' which was coarse 'by the wildness of the country and evil chasing … [of] their sheep' (although today it is argued that genetic influences also affected fleece quality).[31] Nor can any force be given to the fulling mill as a locational factor. The principal investors in mills, lords and fullers, were usually seeking to exploit an existing activity: mills came to cloth-workers, not *vice versa*. That they were not a prime locational factor for the medieval cloth industry of the South West is shown by the fact that, although no parts of the peninsula were unsuitable for water-driven mills, their distribution is not random.

The distribution of fulling mills is, in fact, a good guide to the distribution of medieval cloth-making in the South West. References to fulling mills may be found in many types of medieval sources (especially manorial surveys and accounts) and these reveal the main concentrations (**Maps 40.4 and 40.5**).[32] There

were strings of mills around many of the granitic moorlands, so that we find them at, for example, Hamatethy and Cardinham on the flanks of Bodmin Moor and a great cluster driven by the gushing Tavy and its tributaries west of Dartmoor.[33] There was also a concentration in the rural vales of east Devon.[34]

The distributions reinforce the view that rural cloth-making was engendered not by technological factors but by certain features of the rural social structure and economy of particular regions. Cloth-making on the moorland borders developed upon relatively small farm holdings with unrewarding soils and some tendency towards pastoralism; here was a need for extra incomes as well as surplus time which could be occupied in a craft.[35] There were probably links between the getting of tin, unpredictable from year to year, and the making of cloth. The coincident general distribution of the two industries around the moors is suggestive and is sometimes thrown into sharp focus by local sources, as at South Tawton and South Zeal just north of

Map 40.5
Fulling mills recorded,
1426–c.1500.

SOURCE: as **Map 40.4**.

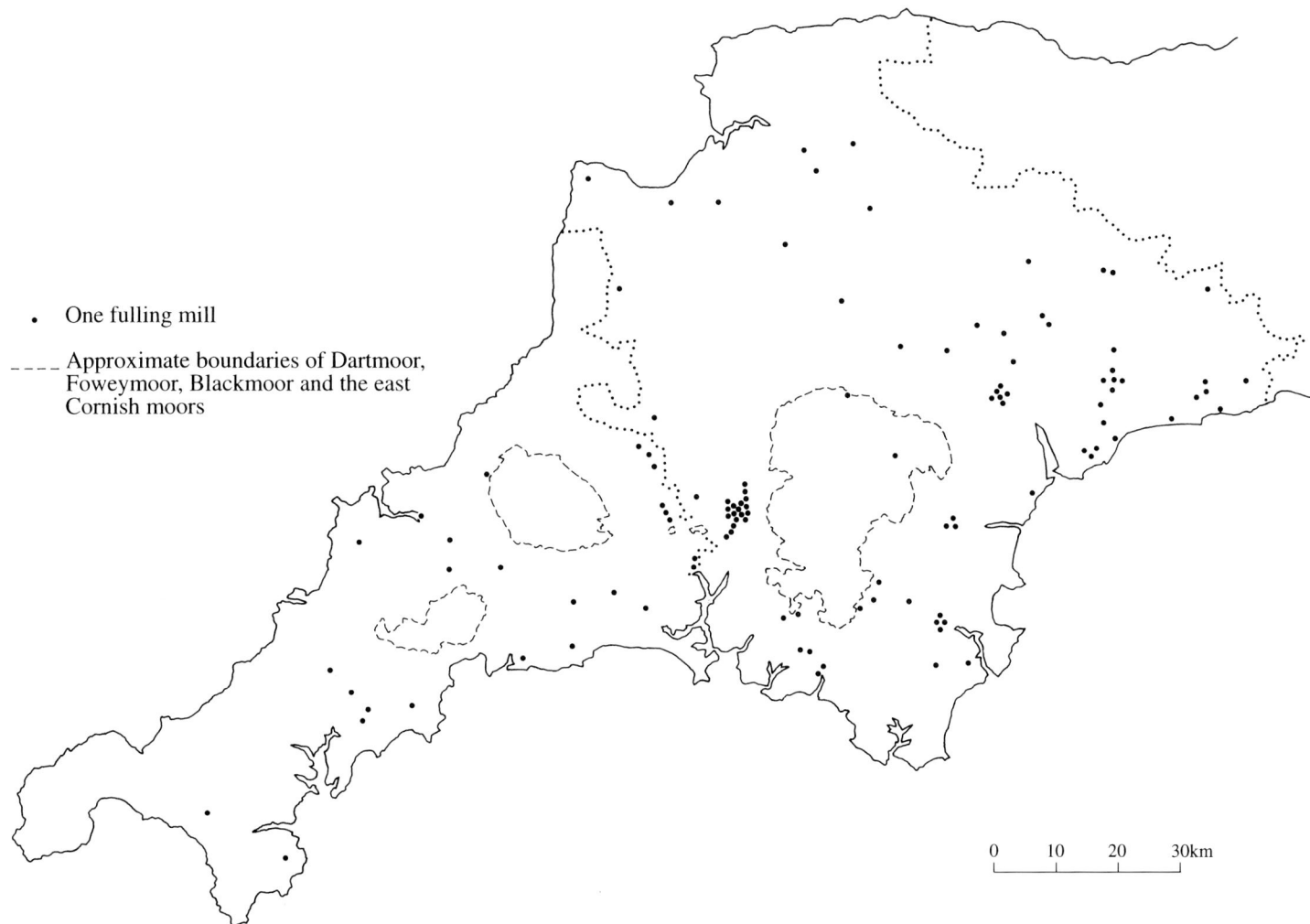

• One fulling mill

_ _ _ _ Approximate boundaries of Dartmoor,
Foweymoor, Blackmoor and the east
Cornish moors

```
0    10    20    30km
```

Dartmoor where, from the thirteenth century to the nineteenth, documents reveal cloth-making as well as tinning.[36] In such places some men were engaged in both industries, like the draper, the wool-beater and the fuller who were also Dartmoor tinners according to a list of 1339.[37]

The borders of Dartmoor in the fourteenth and fifteenth centuries provide to some degree an example of what economic historians now call a proto-industrial region, having many of the characteristics of later industrial zones but with dependent workers dispersed in the countryside rather than concentrated into factories.[38] Small-scale farmers and their families took to spinning, weaving and tin extraction out of need. Then their activities were much expanded by entrepreneurs who provided some capital, creating a degree of dependence, and who also opened up distant markets; such, for example, were the west-country merchants who bought up consignments of tin and cloth for shipping abroad. Local towns prospered, not because they

were the sole seats of industry but because of the trade created by the entrepreneurs and merchants who were responsible for articulating commerce in tin and cloth.[39] Within an industrial region of this kind an ebb in one industry might force a surge in another: when Tavistock stannary was in relative decline in the early fifteenth century, we find a boom in the construction of new fulling mills in the adjacent countryside as well as the building of a new cloth-workers' aisle in Tavistock church.[40] This example of what might be called 'industrial metabolism'—relative decline in one craft encouraging increased activity in another—can be paralleled in other places and periods: emergence of cloth-making in declining lead-mining settlements in medieval Derbyshire, seventeenth-century lace-making in decayed east Devon fishing villages, and, perhaps, the rise of thread-making among under-employed silver miners at Combe Martin.[41]

The other significant concentration of rural cloth-making was in east Devon. Here the industry grew within an agrarian society

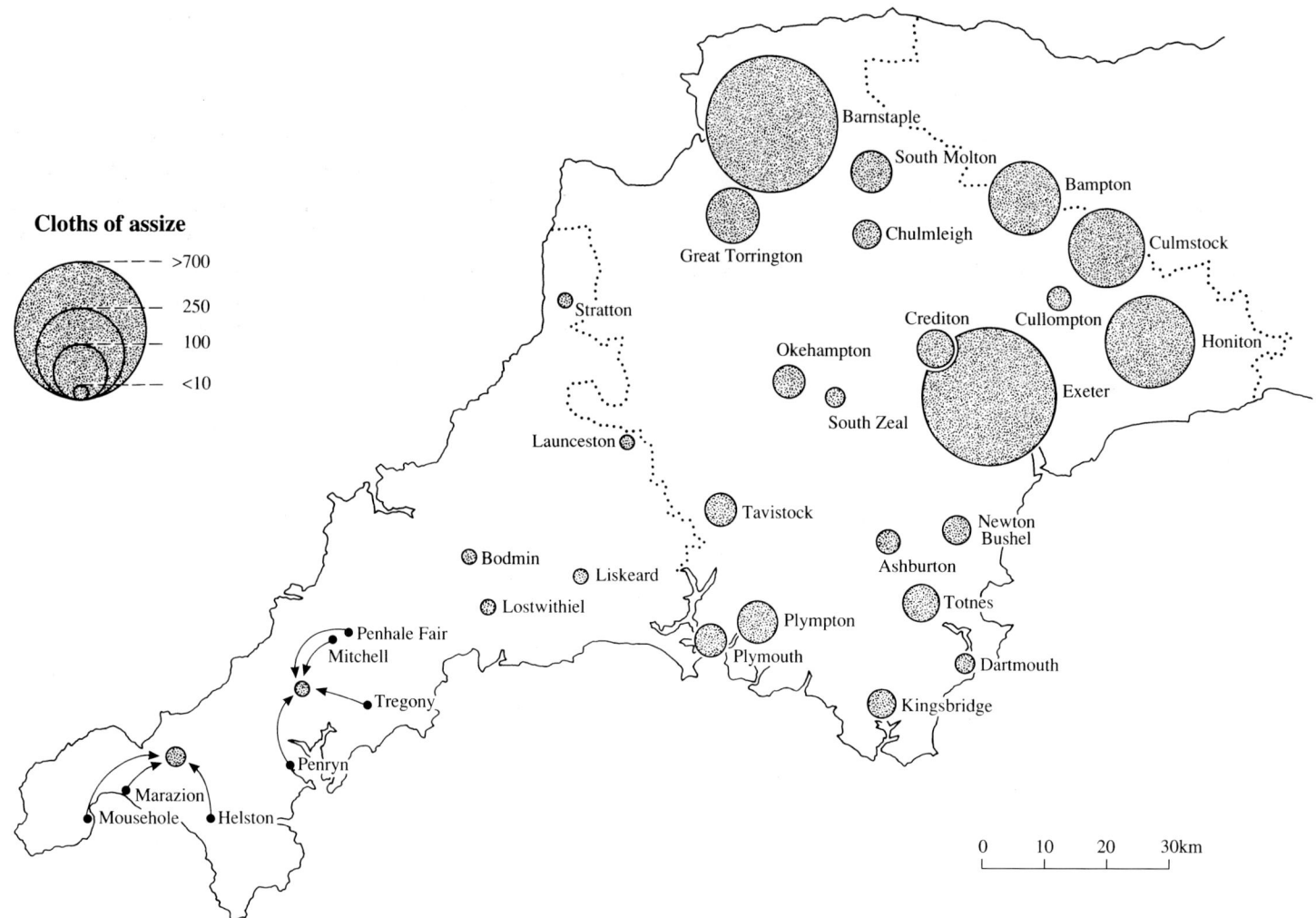

which had many smallholders and which practised intensive pastoral husbandry as early as the fourteenth century, if not earlier. Pastoral smallholders had surplus labour and a need for extra income: these factors, one often finds, were what lay behind the genesis of a rural industry.[42] Cloth-making in east Devon had roots which go back a long way, for rural fulling mills are recorded here well before 1300, while Exeter, at the hub of the region, had a textile industry of no small importance in the twelfth century.[43] Growth in demand for cloth in the later Middle Ages (after 1350) meant that smallholders here, and even landless cottagers, continued to survive during a period when both classes tended to decline in numbers in some other regions of the South West and in England at large.[44]

What reinforces the conclusion that cloth-making as a rural industry was associated (both in the stannaries and in east Devon) with small-scale pastoral farming is its absence in any large way from farming regions which did not have these two

characteristics, for example from the Culm Measures of mid-Devon and north-east Cornwall. During the fourteenth and fifteenth centuries, the populations of many Culm Measures parishes were dwindling; remaining husbandmen, now with large farms on their hands, were turning to extensive forms of stock-raising and were using living-in servants rather than cottage labourers as their source of labour.[45] This type of farm economy and society was not usually a seed-bed for rural industry and, consequently, fulling mills were few and far between in the region.

Before leaving the topic of distributions, one final map (**Map 40.6**) should be introduced. From the 1350s onwards those who exposed cloths for sale had to pay a fee to the royal quality-control officer, the aulnager. Historians now have most faith in the aulnage returns from the 1390s, though even then there was some fraud.[46] **Map 40.6**, based on the returns for 1395–6, confirms to a large degree what has already been said above about distributions.[47] Around Dartmoor a ring of moorside towns stands out and some of the towns of east Devon are also prominent: in both cases the cloths exposed for sale in the town markets were of rural as well as urban origin. There is, though, an apparent anomaly in the Culm Measures country and in north Devon between the distribution of rural fulling mills and the distribution of markets where cloths were produced for the aulnager. Rural fulling mills were uncommon in these parts, but there is a good deal of evidence for an *urban* cloth industry in towns such as Great Torrington, Pilton, South Molton, Barnstaple and Chulmleigh, which explains the aulnager's interest in those places.[48]

When one reads Hooker's charming description, probably from east Devon, of rural cloth-making in the sixteenth century— 'for wheresoever any man doth travell you shall fynde at the … foredore of the house … the wiffe, theire children and theire servantes at the turne spynninge or at theire cardes cardinge'—it is easy to conjure up a rural idyll of free and independent countrymen and women working up the clip of their own sheep.[49] Yet as Hooker goes on to stress, weavers in Elizabethan Devon were virtually in the pay of clothiers. The presence, already, in the countryside of clothiers who supplied capital and who organized marketing may be inferred from the reasonably reliable aulnage returns of the 1390s. To take just one example from east Devon: in 1396 seven vendors sold between 12 and 204 dozens (12 yards long) of cloth at the rural market of Culmstock, each disposing of far more than a single family could weave in a year. According to contemporary manor court rolls and accounts, six of these vendors were not Culmstock residents and were probably clothiers from other parishes in the Culm valley. The seventh, John Colyn, was a native of Culmstock. He saw to it that the cloth industry of this rural parish was serviced, acting as a pledge in the lease of a fulling mill there and importing dyestuffs; in several local court cases he was sued for debt by local men, perhaps weavers whom he had failed to pay; in his will he described himself as a 'merchant of Culmstock parish'.[50] We shall never know very much about the precise nature of the relationships between men such as Colyn and those who wove, or between spinners and those who supplied the wool (such as the syndicate of woolmongers who bought the clip of all the demesne flocks of Oliver de Dynham in 1347).[51] But there is little doubt that merchant capital had begun, first perhaps in a small way, to penetrate the industry long before the great days of Henry VIII's reign when men like Lane of Cullompton and Greenway of Tiverton articulated all stages of cloth processing and marketing in both town and surrounding countryside: Lane bequeathed money to 100 parish churches 'next about Cullompton', as if to emphasize his involvement with rural manufacture.[52]

The men and women who did the spinning and weaving are naturally less visible in the sources than the clothiers and merchants. From east Devon we find Sibil Pounde, 'spynster' and cottager of Sidbury, who in 1499 stole wool dyed a 'bloody colour' from a neighbour; Stephen Ham, another cottager of Sidbury, whose chattels included wool, a 'payre of cardez' and a 'turne'; and William Mochknap, a small farmer from Clayhidon, whose chattels in the autumn of 1381 included a little grain, twenty-one sheep and 8lb of wool perhaps stored away for winter work.[53] Here is a range of individuals employed in rural cloth-making: the cottagers Pounde and Ham, and Mochknap, a farmer. Where there was a fulling mill on a manor it was often found alongside a large number of cottages, which suggests that a proportion of cloth production was in the hands of the landless or near landless. To take but two examples, at Uffculme in the 1420s, alongside no less than three fulling mills, there were twenty-one cottagers (as many as there were farmers, an unusual situation by this date); at Menheniot in the cloth-producing region of east Cornwall, a fifteenth-century rental reveals a fulling mill and more cottage tenures than unfree farm tenures, again an unusual balance for a Cornish manor.[54] The presence of a clothing industry in a region thus ensured resilience, or even growth, in the settlement pattern at a time when decline was taking place elsewhere.

FURTHER READING

John Hatcher, *English Tin Production and Trade Before 1550* (Oxford, 1973) provides much new material, and revises some figures in George R. Lewis, *The Stannaries: A Study of the English Tin Miner* (Cambridge MA, 1924). Louis F. Salzman, 'Mines and stannaries', in James F. Willard, William A. Morris and William H. Dunham (eds), *The English Government at Work, 1327–1336, Vol. III* (Cambridge MA, 1950) has some material on tin extraction and the only printed account of lead and silver mining in west Devon. Eleanora M. Carus-Wilson, *The Expansion of Exeter at the Close of the Middle Ages* (Exeter, 1963) is an excellent account of the cloth industry in Exeter and its region. Maryanne Kowaleski, *Local Markets and Regional Trade in Medieval Exeter* (Cambridge, 1995) has much excellent detail on industries in both town and countryside, especially cloth and leather.

INDUSTRY AND MINING

THE TIN INDUSTRY
IN SIXTEENTH- AND SEVENTEENTH-
CENTURY CORNWALL

SANDY GERRARD

The history of the tin industry is complex, as the character of extraction techniques, capital and markets varied considerably through time.[1] The archaeological examination of the industry for all periods is in its infancy, particularly so for the period prior to 1700. By contrast historical research into the industry has been comprehensive, both for the prehistoric and for the historic periods.[2] The distribution maps in this chapter are the product of an extensive, but by no means complete, search of documentary sources, and published and unpublished literature. They represent the position of research at present and will inevitably be updated. The completeness of each map varies considerably. That for blowing houses (**Map 41.9**) is probably more or less complete since the bureaucracy associated with the valuable end product—the ingot—has meant that at least some records survive for each site. Stamping mills (**Map 41.7**) also generated large amounts of documentation, although some smaller examples may not have been recorded and the picture is thus probably less complete than that for blowing houses.

TIN WORKS

The tin work maps (**Maps 41.1 and 41.2**) are, however, the most incomplete and may only be used as a general indicator of the location of individual sites. Comparing the number of documented tin works per decade with the known output, it is clear that a large number of sites are unaccounted for. For the 1670s, if the total average annual output of 2.4 million pounds weight (lb) is divided by the known number of tin works (221), the average output per tin work would have been 10,873 lb. At the other extreme in 1530, 1.25 million lb were apparently produced by only four sites, meaning an unrealistic output of 311,474 lb per tin work. Establishing the true numbers of tin works is an exercise fraught with difficulties. The annual output for twenty-five Blackmoor tin works, where detailed accounts survive, is known to have been an average of 1,586 lb per tin work.[3] Tin works on Blackmoor were generally less productive than those in the west of the county and the true average production must be higher than this. There is thus a maxmimum figure of about 10,873 lb and a possible minimum of

1,586 lb, with the true figure probably somewhere between these two extremes. Using these figures as a guide, we can establish a minimum and maximum figure for the number of tin works per decade, although of course it is not possible to establish the total number since the average operational life of a site is not known. In the 1680s, the decade with most output, the number of tin works is likely to have been between 278 and 1,906, probably towards the upper end of the range. This compares with the 232 documented sites for this period and demonstrates the inadequacy of the record. These calculations cannot, of course, take into account the small-scale family operations producing just enough tin to augment family income. Throughout the period many small ventures left a physical trace but no documentary evidence. It will never be possible to quantify the scale of this activity, though it must have played a critical, although varying, role.

The two maps showing the distribution of tin works (**Maps 41.1 and 41.2**) illustrate only those sites for which at least a four figure grid reference has been identified. A large number of tin works from each stannary are represented, but because it was not possible to plot all documented sites, the total numbers are presented in **Fig. 41.1** to allow an examination of all available information. It is useful to compare the distribution maps and the 'pie graph' of **Fig. 41.1** for the two centuries. In general the seventeenth-century distribution is much more clustered. This may suggest that in the seventeenth century extraction was centralizing on the remaining rich lodes and that many of the smaller lodes and deposits which had been worked in earlier times were finally being exhausted and abandoned. This process continued into the twentieth century when fewer and fewer lodes were being exploited. This phenomenon is particularly marked on Foweymoor (as Bodmin Moor was known at this time) where a relatively random distribution in the sixteenth century is replaced by a clustered one in the seventeenth century centred on the Minions area and Trevenna. **Fig. 41.1** also shows fundamental differences in the number of tin works in each stannary. Most dramatically, Tywarnhayle rose from 2 per cent of the sixteenth-century total to 52 per cent in the seventeenth century. This large number of tin works (527) is a reflection of an extensive and complete set of documents and more particularly the ownership of a large number of sites by John Enys whose detailed records survive.[4] This situation must skew the distribution, but nevertheless reflects the level of activity in one of the busiest seventeenth-century stannaries. Surprisingly, the number of tin works on Blackmoor continued to rise despite falling output and increasing exhaustion of deposits. In Penwith with Kerrier, from which the bulk of sixteenth- and seventeenth-century Cornish tin came, there are only 39 per cent falling to 16 per cent of the total number of tin works. This may be explained either as a result of the western mines being larger and more productive or the vagaries of documentary survival. On Bodmin Moor, the marked reduction in the number of tin works accurately reflects the reduced role of this stannary (Foweymoor).

The surviving field evidence for the tin industry is still relatively abundant throughout the stannaries. Dating of the different elements is extremely difficult and it is rarely possible to equate documented activity with any particular surviving site. Many sites were exploited at different times during the prehistoric and historic periods and each phase of exploitation has left its own mark. The result is often a complex palimpsest which only detailed field survey can unravel. Work by the Cornwall

Archaeological Unit (as at Kit Hill and Minions, for example) and by the present writer is working towards identifying the many different features within the mining landscape.[5]

Water was essential to the industry, for powering the mills, blowing houses and other machinery and for streaming and dressing. Leats and reservoirs were constructed to harness available water supplies, as can be seen in **Map 41.3** which portrays the leat system at Penkestle Moor and bears witness to the sophistication of tinners' hydrological skills. On Bodmin Moor and

Map 41.1
Distribution of sixteenth-century tin works in Cornwall.

Penwith with Kerrier

Blackmoor

Foweymoor

Tywarnhayle

Sixteenth century

Seventeenth century

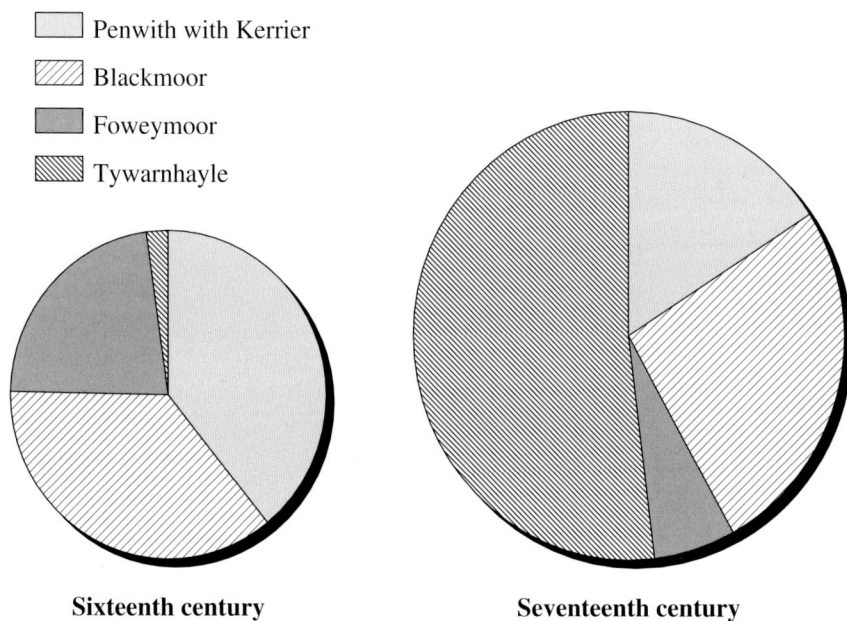

Fig. 41.1
Number of tin works
in each stannary
in sixteenth- and
seventeenth-century
Cornwall.

Blackmoor the most obvious physical legacy of the industry is the streamworks.[6] These are of two main types namely alluvial (those within the river valleys) and eluvial (those within dry valleys on the hillsides). There are many well-preserved examples of eluvial streamworks. At Harrowbridge on the northern slopes of Brown Gelly a group of three such works survive. The eastern one, in particular, demonstrates clearly the characteristic waste dumps associated with streamworking (**Map 41.4**). Their alignment reflects the water separation techniques employed by the tinners.

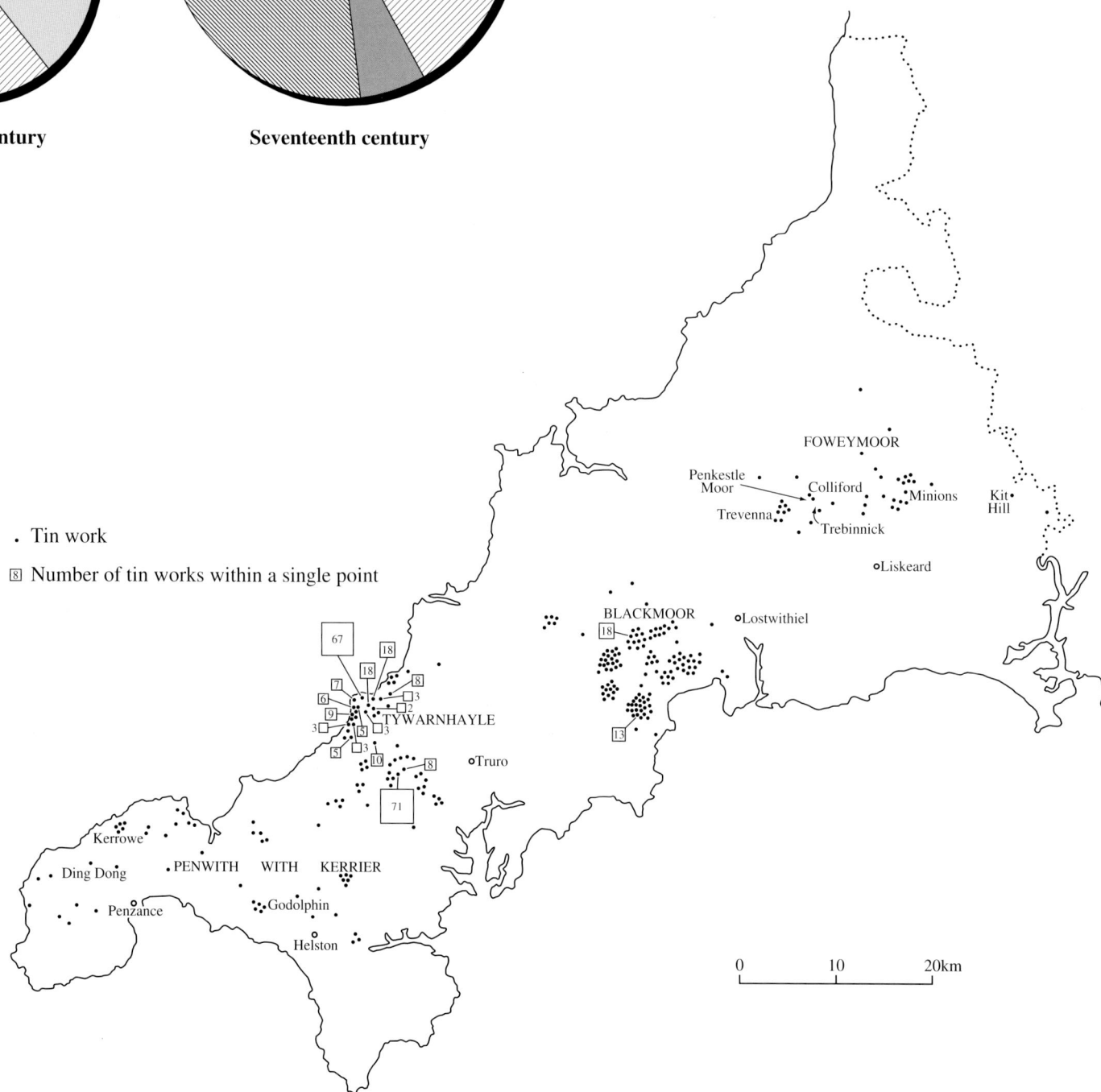

. Tin work

⊠ Number of tin works within a single point

Map 41.2
Distribution of
seventeenth-century
tin works in Cornwall.

In the three western stannaries the majority of the tin works were mines exploiting cassiterite directly from the lode. The archaeological record suggests that there were three different types. First are the shallow shafts exploiting the tin to a shallow depth (as at Kerrowe). These always occur close together following the line of the ore outcrop and are also known as lode-back pits. At Hobb's Hill (**Map 41.5**) earthworks of lode-back pits can be seen clearly following line of the lode and these may belong to the tin work known as 'Hobbys Worke' documented in 1516. Second are openworks (e.g. Godolphin), which are also known as beams or coffins (opencast quarries). Third are shafts exploiting the lodes to considerable depths (e.g. Pell Mine, Tywarnhayle). A small example of a shaft and adit mine survives at Trebinnick, Bodmin Moor (**Map 41.6**), where two shafts are clearly associated with a drainage adit situated immediately next to the river. Finally, associated with many tin works are small rectangular buildings which are generally interpreted as shelters.

STAMPING AND CRAZING MILLS

Tin was separated from its ore by crushing it into small particles and then using water to separate the lighter constituent parts from the metal oxide. Machinery was developed to crush the ore, and these devices were known as stamping mills, knocking/knacking mills, clash mills and tin mills. These consist of a set of water-powered trip hammers which crushed the ore placed below them. Documentation is of two types: descriptions of tin dressing processes and references to individual mills.[7]

Map 41.3
Plan of Penkestle Moor leat system (SX 200730).

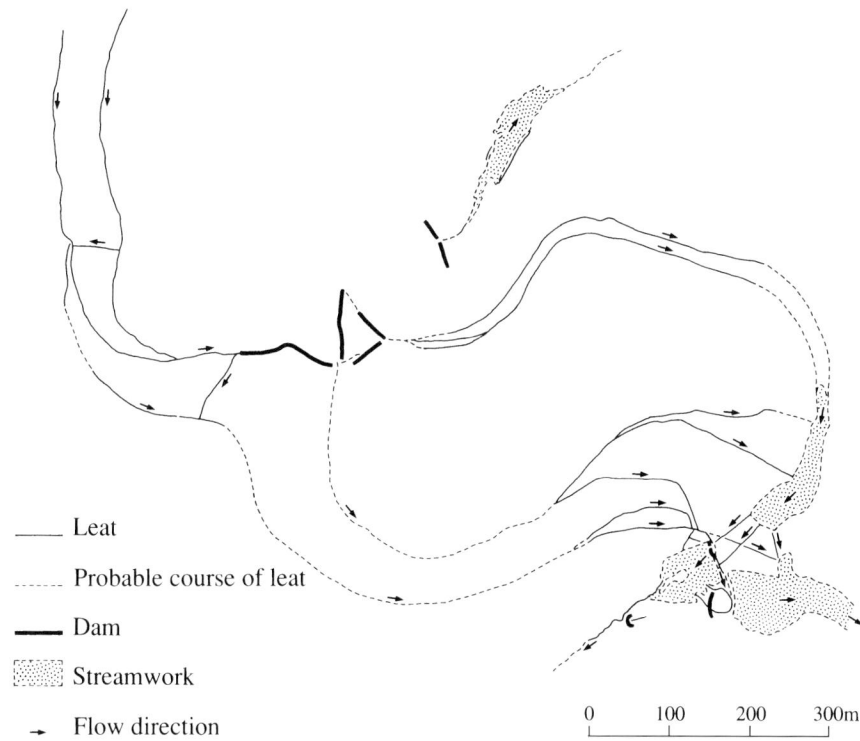

— Leat
····· Probable course of leat
▬ Dam
▒ Streamwork
→ Flow direction

0   100   200   300m

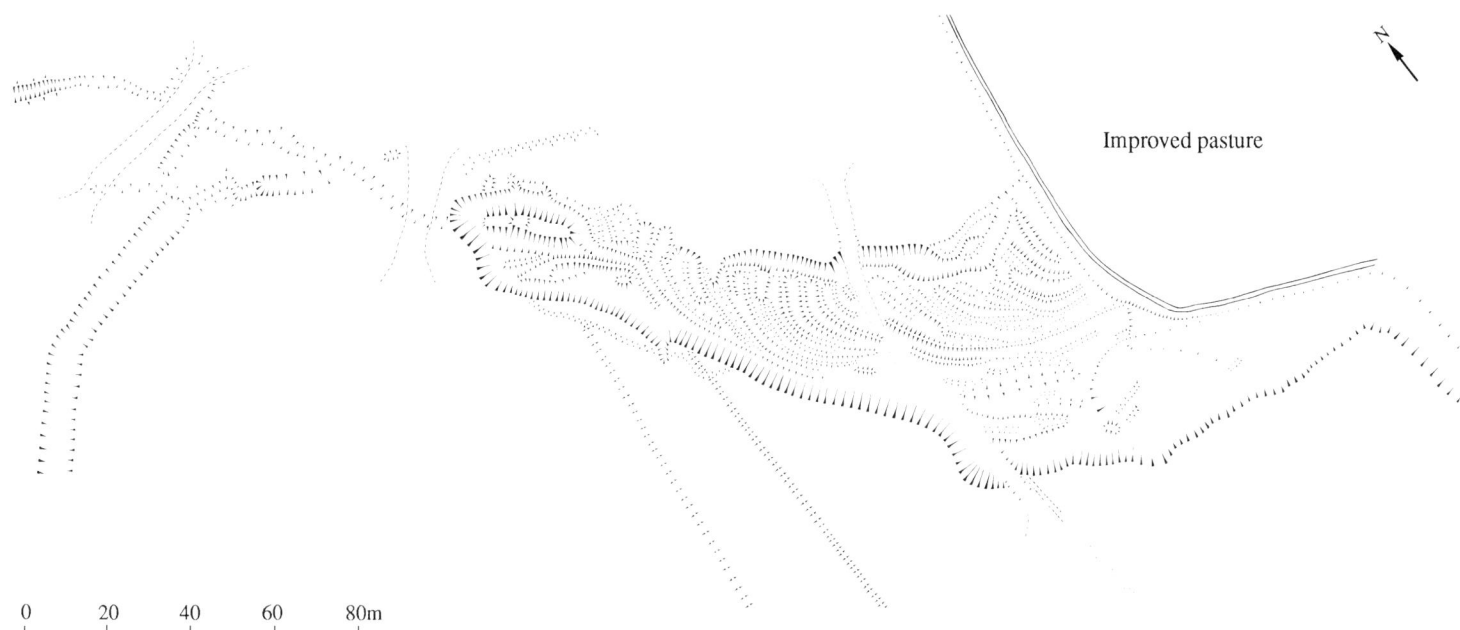

Improved pasture

0   20   40   60   80m

Map 41.4
Plan of Eastern Harrowbridge eluvial streamwork.

333

Map 41.5
Plan of lode-back works,
prospecting pits and
tinners' shelters at
Hobb's Hill
(SX 183694).

At least 141 stamping mills are known from the period. The distribution map (**Map 41.7**) shows that the majority of mills were in the three western stannaries. The stamping mills were primarily concerned with crushing lode-tin and thus they were necessarily situated in close proximity to the cassiterite lodes, and the scarcity of these in the east, compared with the west, accounts for the distribution indicated on the map. The tin output from the eastern stannaries was predominantly from alluvial and eluvial streamworks, and hence did not require stamping prior to dressing. It is possible to identify six separate clusters of mills, these being from east to west: Tregrehan, Polgooth (in Blackmoor), St Agnes (in Tywarnhayle), St Day–Creegbrawse, Tolgus and Porkellis (in Penwith with Kerrier). These probably represent the major tin ore production centres.

Very few surviving stamping mills are known from the archaeological record. In Devon thirty-two mills are known, but in Cornwall only three—two at Retallack and one at Newmill.[8]

Another group at Colliford was lost during the early 1980s. The shortage of known stamping mill sites in Cornwall is probably the result of a number of factors, not least that many of the earlier Cornish tin works have been reworked in modern times, frequently destroying the earlier evidence, and that there has been no systematic search for mills in Cornwall, whereas on Dartmoor much survey work has been carried out.[9]

A useful field indicator to surviving stamping mills is the distinctive mortar stones on which the ore was crushed. These are large slabs of stone with circular hollows on one or more faces. In Devon around 170 stones have been found whilst in Cornwall only fifty are known.[10] The distribution of these is shown on **Map 41.8**, and although generally there is no trace of the mill building, they do point to the one-time existence of a site. The majority of stones recovered are within close proximity to a documented site, though there are large areas of known intensive activity where no stones have been recovered (e.g. Porkellis).

In the earlier medieval period crazing mills with machinery similar to that in grist mills were employed to reduce the ore further. By the sixteenth century they were probably not so commonly used, although eleven documented examples testify to their continued use. Again their distribution is confined to known lode production areas (**Map 41.7**).

BLOWING HOUSES

Once the tin ore had been crushed, water was used to separate the tin from the waste utilizing the same basic procedures employed in streamworks. Within the immediate vicinity of every stamping mill there would have been dressing floors. The end product of this process was known as black tin. Black tin from the stamping mills was smelted in blowing houses (**Map 41.9**), which were substantial stone buildings containing a stone built furnace served by bellows operated by a water wheel. The most complete contemporary description of tin-blowing was given by Thomas Beare and a good review of much of the documentary evidence is provided by D.B. Barton.[11] Once the tin ingots were produced they were transported to the nearest coinage town where they were taxed prior to sale.

Compared to that of stamping mills (**Map 41.7**), the distribution of blowing houses on **Map 41.9** is more uniform and there are fewer clearly defined clusters. This is because both alluvial and lode tin were smelted, whereas the stamping mills were largely limited to the immediate vicinity of lodes. Thus on Bodmin Moor, where there are only five stamping mill sites, there are six blowing houses, whilst in Penwith with Kerrier there are sixty-four stamping mill sites and twenty-three blowing houses. Second, some blowing houses are situated in areas where there is no cassiterite. There are, for example, five houses in the immediate vicinity of Truro. The siting of these particular blowing houses is related to proximity to the coinage town. Black tin was probably less cumbersome to handle and transport than 'white' tin (ingots), and it therefore made sense to locate some blowing houses near to the place of coinage, as this would have considerably reduced the distance over which the tin ingots had to be carried before sale. Further, if blowing houses were close to the coinage towns, opportunities for smuggling and theft would have been reduced. This situation is confirmed by the fact that all blowing houses are within 8 miles of a coinage town and over 50 per cent are within 6 miles. Third, it must be emphasized that it is very unlikely that all the blowing houses represented on **Map 41.9** were contemporary, as

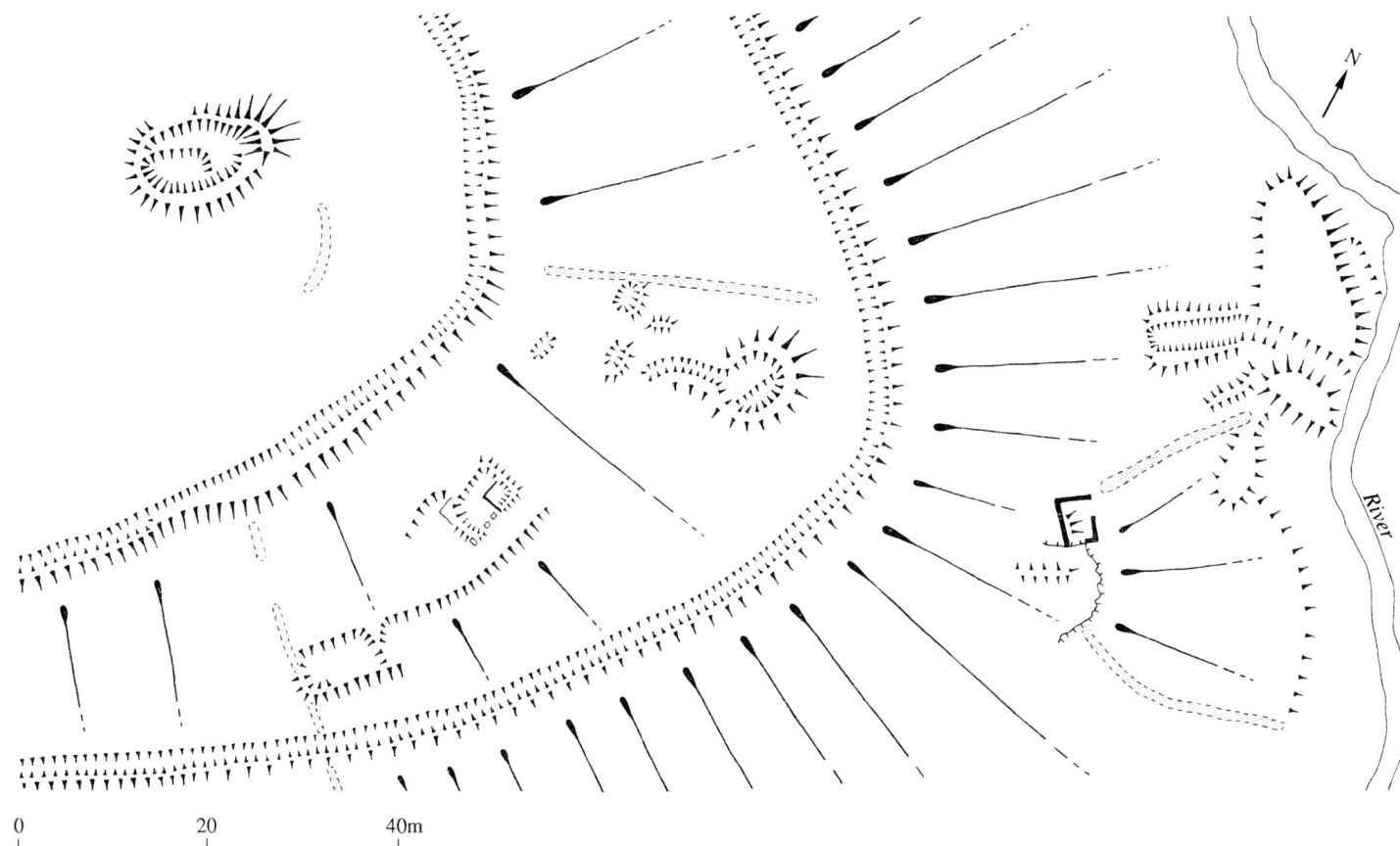

0   20   40m

Map 41.6
Plan of shaft and adit mine and tinners' shelters at Trebinnick (SX 183704).

Map 41.7
Sixteenth- and
seventeenth-century
stamping and crazing mills
in Cornwall.

. Stamping mill

● Several stamping mills

▽ Crazing mill

FOWEYMOOR

Colliford

BLACKMOOR

Tregrehan

St Agnes
TYWARNHAYLE

Polgooth

Tolgus
St Day

New Mill PENWITH WITH KERRIER
Porkellis

Retallack

0   10   20km

it is possible that some were built to replace others, particularly where examples exist in close proximity.

FURTHER READING

Georgius Agricola, *De Re Metallica*, translated from the first Latin edition of 1556 by Herbert Clark and Lou Henry Hoover (New York, 1950) describes in great detail the technology of sixteenth-century European mining. The descriptions of mining and processing technology are very relevant for students of tin-mining. The book is perhaps best known for the large number of its excellent quality illustrations.

D. Austin, G.A.M. Gerrard and T.A.P. Greeves, 'Tin and agriculture in the Middle Ages and beyond: landscape archaeology in St Neot Parish, Cornwall', *Cornish Archaeology*, 28 (1989), 5–251, reports on the most detailed and comprehensive fully published research project into the tin industry. J.A. Buckley,

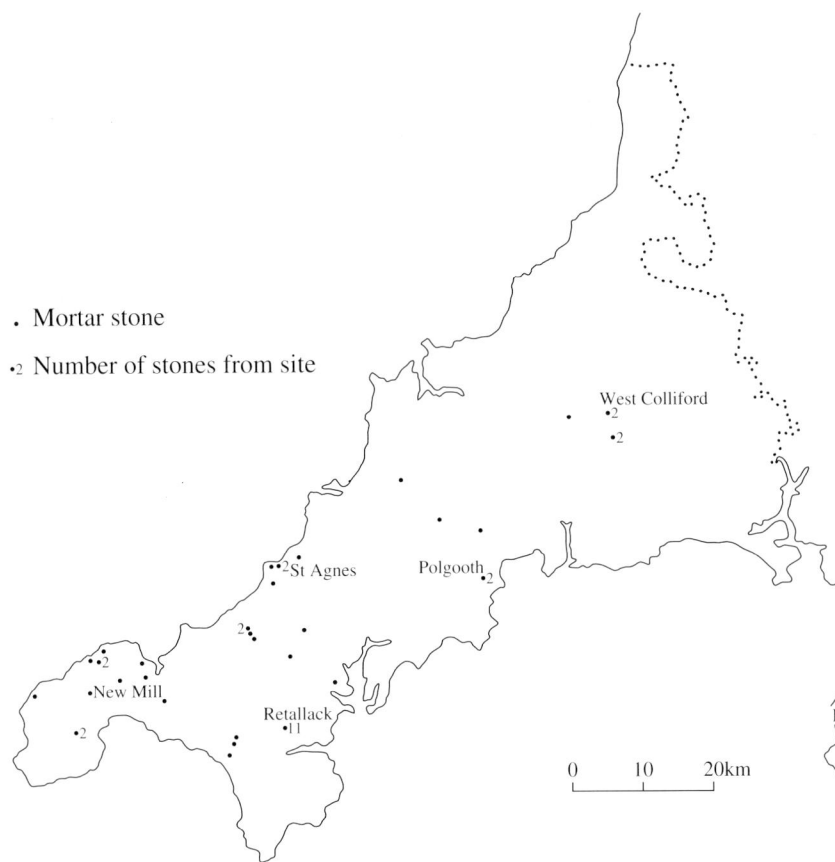

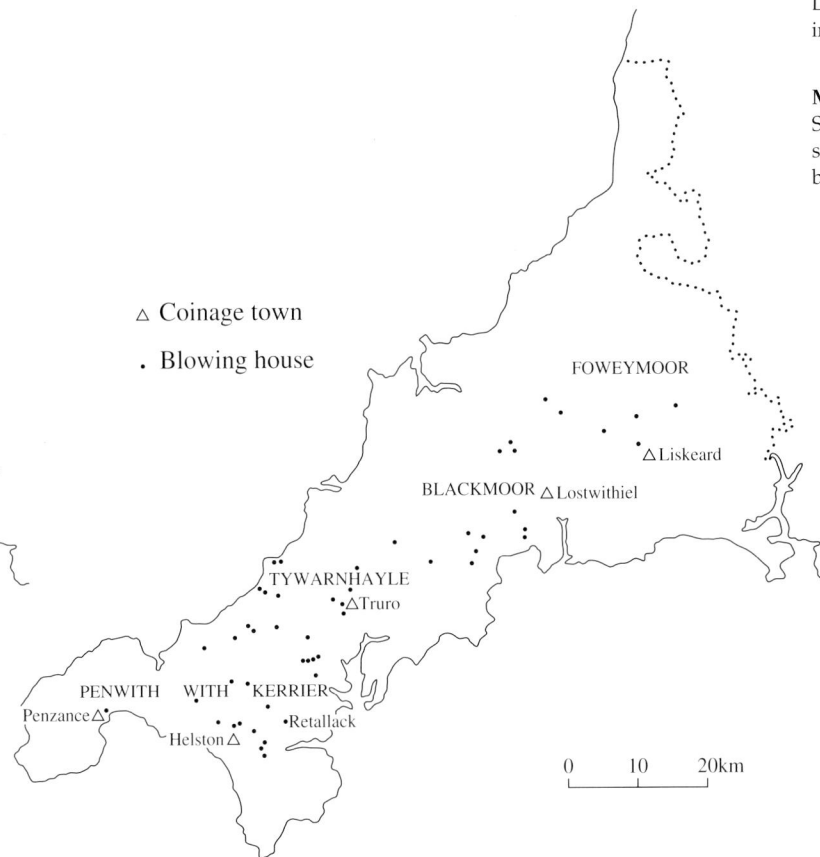

Map 41.8 (left)
Distribution of mortar stones in Cornwall.

Map 41.9 (right)
Sixteenth- and seventeenth-century blowing houses in Cornwall.

. Mortar stone

·2 Number of stones from site

West Colliford
·2
·2

·²St Agnes   Polgooth·2

2·
·²
·2
·New Mill
·2
Retallack
·11

△ Coinage town

. Blowing house

FOWEYMOOR

△Liskeard

BLACKMOOR △Lostwithiel

TYWARNHAYLE
·△Truro

PENWITH   WITH   ·KERRIER
Penzance△·
·Retallack
Helston △

0   10   20km

0   10   20km

Tudor Tin Bounds in West Penwith (Redruth, 1987) presents documentary evidence relating to a large number of sixteenth-century tin works. Sandy Gerrard, 'The early south-western tin industry: an archaeological view', in *The Archaeology of Mining and Metallurgy in South-West Britain*, Peak District Mines History Society, 13, 2 (1996), 67–83, is an examination of the archaeological evidence relating to the tin industry in the South West. J. Hatcher, *English Tin Production and Trade Before 1550* (Oxford, 1973) provides an economic historian's perspective on the tin industry. G.R. Lewis, *The Stannaries: A Study of the Medieval Tin Miners of Cornwall and Devon* (Truro, 1965), although somewhat dated, remains an easily digestible source of historical information.

THE WOOLLEN, LIME, TANNING AND LEATHER-WORKING, AND PAPER-MAKING INDUSTRIES c.1500–c.1800

MICHAEL HAVINDEN

THE WOOLLEN INDUSTRY

Between 1500 and 1800 south-western England rose to be one of the most important regions in Britain for the production of woollen cloth. The abundant pastures nourished many sheep and the spinning of wool into yarn and the weaving of cloth were carried out in a large number of villages and small towns, and even as a part-time occupation in farmhouses. Because of the scattered, domestic nature of these occupations the production areas cannot be mapped, but most south-western cloth subsequently underwent fulling (beating in mills with water-powered hammers, to thicken and shrink it) and this process is more conducive to mapping.[1] The mills were sited on rivers and streams, and most of their sites were identified and mapped by Alfred Shorter in 1969.[2] **Map 42.1** is a slightly modified version of Shorter's with the post-1800 woollen factory sites removed and three new mills in Cornwall added. Most of the fulling-mill sites shown on the map were active between 1500 and 1800, although a few of the pre-1600 mills may have been abandoned before or during the sixteenth century.

Exeter was the largest centre of the fulling industry, followed by other south Devon towns like Tiverton, Cullompton, Ashburton and Buckfastleigh, and in Cornwall the area around Truro and Falmouth. Generally speaking most of the cloth was woven in the south and centre of the region, with the northern villages concentrating on spinning wool into yarn. This tendency was reinforced after about 1600, when the expansion of the industry called for imported wool. Much of this came from Ireland into north-coast ports like Barnstaple, Bideford and Padstow.

Important changes in the cloth industry occurred in the years around 1600. Prior to that date south-western cloth had been of a type called kerseys, made from the local wools, which were short staple and of only medium quality, and they were thickened by fulling.[3] They were 12 yards long (shorter than many English cloths) and hence came to be called Devon and Cornish dozens.[4] The wool or yarn was usually dyed before being woven. The production of kerseys began in the 1430s, and after a lull in the 1450s, expanded rapidly, especially between 1480 and 1520. By about 1505 over 10,000 kersey cloths were being exported annually to Europe.[5] The region around Tiverton, Cullompton and

Uffculme was the centre of the industry, and initially the Cornish and west Devon cloths (often known as 'Tavistocks') were poorly valued, being made from low quality Cornish wool, known as 'Cornish hair'.[6] However, during the course of the sixteenth century, improved wool was grown in eastern Cornwall, between Bodmin and Tavistock, and by 1582 Cornish wool was said to be 'verye good and much bettere and plenty fullere than in tymes past'.[7] Even so, local supplies of wool were not adequate and the import of long wools from other parts of England and abroad enabled the next development to take place around 1600— the manufacture of serges (the best of which later came to be called 'perpetuanas' because they were so long-wearing).[8] These were cloths woven from two types of yarn, the short-wool carded variety and the long-wool combed variety (which was a worsted, lighter yarn). Consequently serges were lighter than kerseys, though not so light as the pure East Anglian worsteds. Like kerseys, serges were fulled, and were thus very strong as well as light. They were popular, not only at home and in the traditional Dutch and German markets, but also in the newer Spanish, Italian and Mediterranean markets being developed after about 1570.[9]

The introduction of serges led to a huge expansion in the cloth industry. This cannot be measured precisely because there are no figures for domestic consumption, but export statistics survive. It was generally thought at the time that about a quarter of the total production was exported.[10] In the seventeenth century Exeter became a leading fulling and finishing centre, and with its outport, Topsham, handled most of the west-country serge exports, including those from Somerset centres like Wellington and Taunton, though some went out from ports like Dartmouth, Plymouth, Fowey, Truro and Falmouth, and a considerable quantity were exported from London.[11] However, the Exeter exports well illustrate the general trend. Serge exports were well established in the early seventeenth century, when 9,500 pieces were exported, but later development was disrupted by the Civil War, and other wars between 1638 and 1680, so that by 1666 exports were still only 10,000 pieces. Then began a very rapid rise to 120,000 pieces in the 1680s and to a peak of 365,000 pieces in 1710, when Exeter was exporting over a quarter of all England's cloth exports.[12] In 1710 the official value of serge exports from Exeter and Topsham was £463,000 and another £302,000 worth of west-country serges went out from London. The total of

Map 42.1
Fulling mills, *c.*1500–1800.

SOURCE: A.H. Shorter,
W.L.D. Ravenhill and
K.J. Gregory, *South-West
England* (London, 1969).

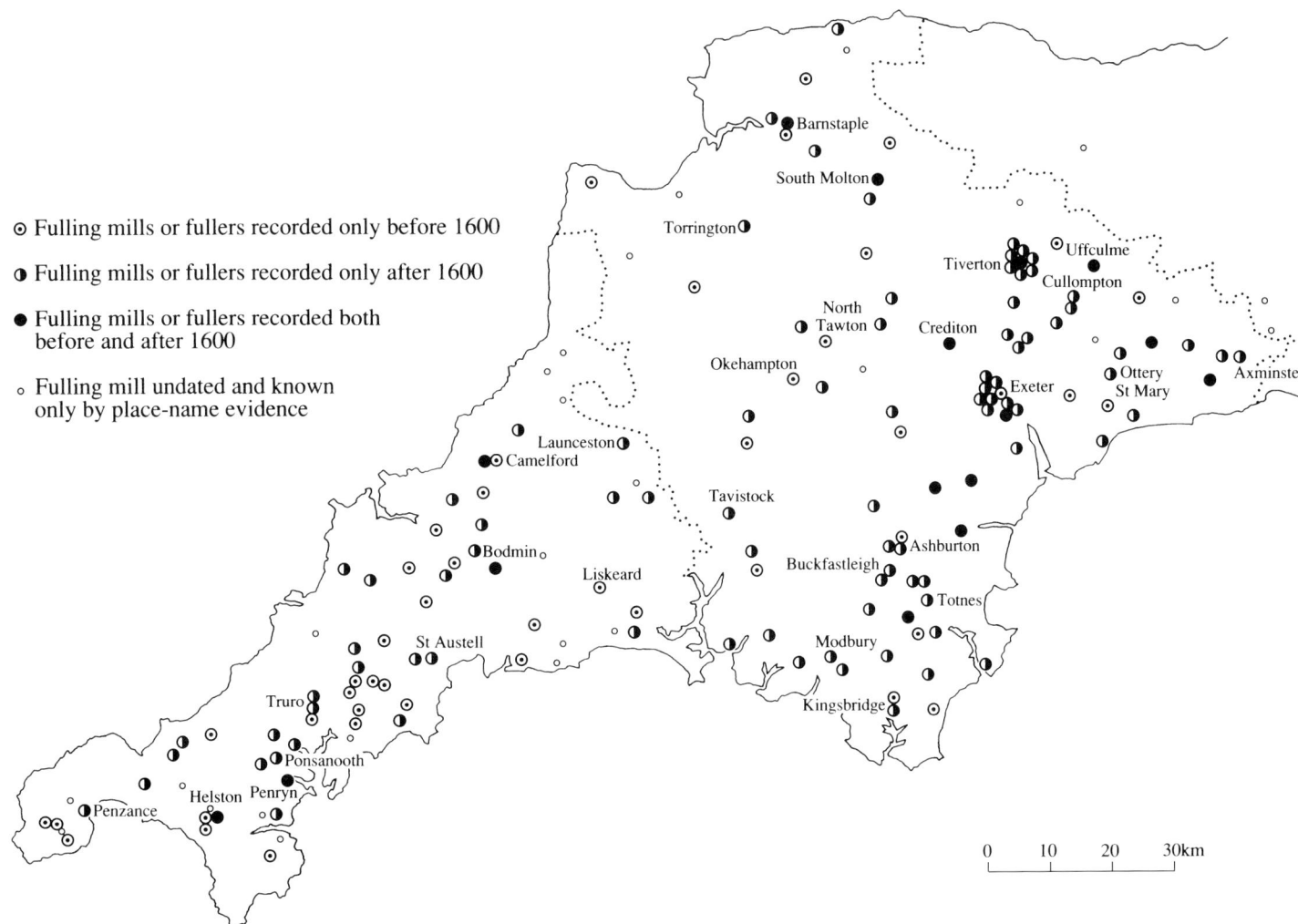

⊙ Fulling mills or fullers recorded only before 1600

◑ Fulling mills or fullers recorded only after 1600

● Fulling mills or fullers recorded both
  before and after 1600

○ Fulling mill undated and known
  only by place-name evidence

£765,000 represented 16 per cent of the value of *all* English and Welsh exports (not just cloth).[13] The reign of Queen Anne (1702–14) saw the epitome of the south-western cloth industry. A period of decline then set in with exports from Exeter down to 162,802 pieces in 1745. This was followed by a rather artificial revival based on exclusive contracts with the East India Company, with exports rising to a new peak of 390,000 pieces in 1777. However, the domestic and European markets had been largely lost, to East Anglian and Yorkshire competitors, and when the French Revolutionary and Napoleonic Wars disrupted English trade in the 1790s, there was a catastrophic decline. In 1800 serge exports from Exeter fell to a mere 8,126 pieces (nearly all sold to Portugal). The East India Company trade enabled a partial revival to take place in the early nineteenth century, but when the Company's monopoly was ended in 1833, the south-western serge industry virtually died out.[14]

Although woollens dominated south-western textiles, there were others, particularly the making of fine lace in east Devon, of which Honiton and Ottery St Mary were the centres. A largely female, domestic industry, lace-making began in the late sixteenth century and spread rapidly in the seventeenth century. A petition to the House of Commons in 1699 against the repeal of the customs duties on foreign lace showed that over 4,700 people were employed in lace-making in east Devon. There were more than 300 in Colyton, Beer with Seaton, Sidbury and Sidmouth, but Ottery St Mary with 814 and Honiton with 1,341 were the leading places. The trade flourished throughout the eighteenth century, but languished after John Heathcote established his machine-made net and lace factory at Tiverton in 1815.[15]

THE LIME INDUSTRY

Map 42.2 shows the distribution of the old kilns in Devon and Cornwall, where limestone was burnt and turned into quicklime (later to be slaked by water) before the introduction of large-scale 'factory' kilns in the later nineteenth century. All the lime kilns in Cornwall, and most of those in Devon, were originally mapped by Alfred Shorter, but a considerable number of Devon kilns were later added by Michael Havinden.[16] The great majority of these kilns were built between 1550 and 1850 (probably mainly in the eighteenth century) and the kilns shown on the map were not all working at the same time, but the map accurately emphasizes the main areas of lime production.

There were two main consumers of burnt lime: the building industry, for lime mortar, and agriculture, which was by far the largest consumer. Lime mortar had been in use from Roman times, and possibly even earlier, but lime was not used in farming, so far as is known, before about 1500.[17] The earliest lime kilns were probably near, or on, the sites of cathedrals and large churches, or in the larger towns like Exeter, Plymouth and Truro. They were probably fairly few in number and have not survived. The big expansion in kiln-building began with the discovery of the importance of lime for raising agricultural production.[18] What lime actually does is neutralize soil acidity (and Cornwall and Devon both have a great deal of naturally acid soil), but this function was not known before the mid-nineteenth century; what was observed at the time was that when lime (or ground chalk, or calcareous sea-sand) was applied, production of crops and grass increased, sometimes dramatically. This was because farmyard manure and other fertilizers are only effective when the soil acidity is neutralized. Not surprisingly, farmers before 1850 thought lime was a plant food, but this error was unimportant, since it acted like one.

Fig. 42.1
Drawing by Peter Orlando Hutchinson of the converted lime kiln, Sidmouth, 3 June 1851.

SOURCE: reproduced by permission of Devon Record Office (Z19/2/8D/245).

Map 42.2
Old lime kilns in Devon and Cornwall, c.1550–c.1850.

SOURCES: as **Map 42.1** and M.A. Havinden, 'Lime as a means of agricultural improvement: the Devon example', in C.W. Chalklin and M.A. Havinden (eds), *Rural Change and Urban Growth, 1500–1800: Essays in Regional History in Honour of W.G. Hoskins* (London, 1974).

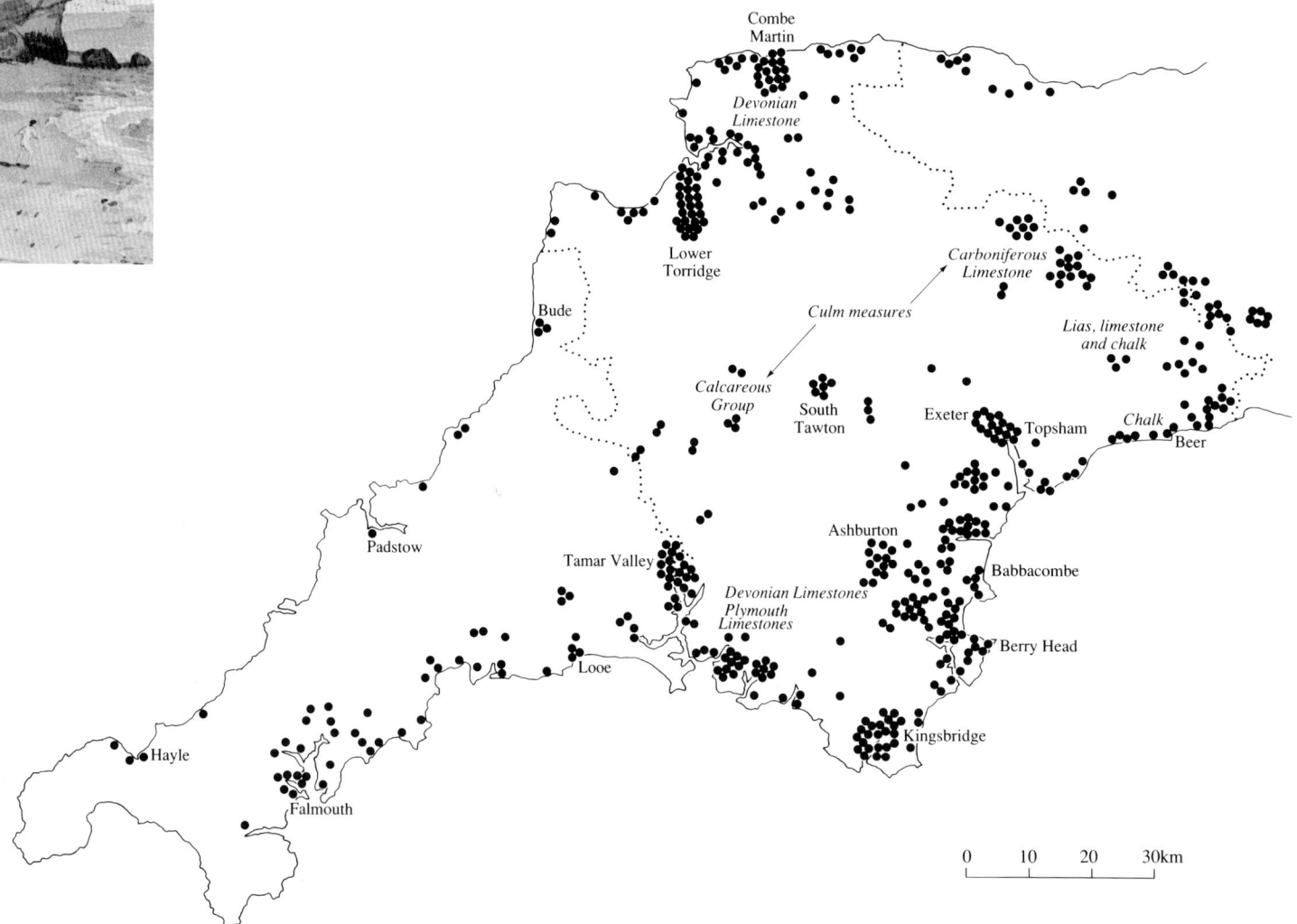

The earliest introduction of lime into south-western agriculture is not known. Its importance for 'mendyng of errable lands' was known to Fitzherbert in Derbyshire in 1523, and it seems to have been introduced to Devon and Cornwall from South Wales in the mid-sixteenth century.[19] In 1578 R. Merrick wrote of Glamorgan that: 'now of late years, since the knowledge or use of lymmige was found, there groweth more plenty of grayne'.[20] Emery has found that: 'despite a general belief that the convertible farmers of Devonshire and elsewhere began liming about 1600, it can be shown that in North Devon at least kilns were burning Glamorgan limestone in the mid-sixteenth century. It was exported from places like Porteynon in Gower with a story of cross-Channel trade in stone right through to the 1870s.'[21]

Once it was established, the use of lime in farming spread rapidly (see also chapter 37). Tristram Risdon, of Winscott Barton near Torrington in north Devon, described it as a 'new invention' in 1630, but by 1667 Samuel Colepresse described its familiar use in most of Devon, though he said it had only recently been introduced in west Devon and Cornwall.[22] By the time that Dean Milles of Exeter carried out a parochial survey of Devon in 1750, the use of lime seems to have been almost universal, and it was widely used in Cornwall as well, where it stimulated projects for canal-building after 1770.[23] Rising prices for cereals were an important cause, especially between 1550 and 1650, and again between 1685 and 1700, when Exeter wheat prices rose rapidly.[24]

The high costs of transporting limestone and coal dictated a coastal, or estuarine, location for the kilns. It also meant that a large number of small kilns, serving local markets, was more efficient than a smaller number of larger ones. In addition kilns could often be built cheaply using local labour, though sometimes they were grander affairs. In 1796 William Marshall wrote that: 'The LIME KILNS of Devonshire are large, and of an expensive construction; … But their duration is in proportion: one that has been built thirty years is still firm and sound on the outside. The walls are of extraordinary thickness; wide enough, on top, for horses to pass round the kiln, and deliver stones.… The stones are brought up from the water side on horseback or upon asses; and, being distributed round the top of the kiln, are there broken, and thrown into the kiln with shovels.'[25]

Coal was fed into the bottom of the kilns, and when the limestone reached a temperature of about 550°C it began to decompose and release its carbon dioxide and to turn into quicklime (calcium oxide). It took about two tons of limestone to produce one ton of quicklime—a hot unpleasant substance, which quickly absorbed water when spread on the land, to produce slaked lime (calcium hydroxide), a fine, soluble, alkaline powder.[26] This method of producing lime continued to thrive during the early nineteenth century, but gradually declined after 1850 in the face of competition from the new chemical fertilizers, which many farmers believed, erroneously, were a substitute for lime. Many of the old kilns have been destroyed, but others may still be found peeping out from undergrowth, often in picturesque coastal sites.

TANNING AND LEATHER-WORKING

Prior to 1800 the production of leather was widely dispersed throughout the South West in a multitude of small tanyards (**Map 42.3**), though there was some concentration in south-east Devon and south-east Cornwall near to the urban markets in Exeter and Plymouth.[27] Tanning was well established in Exeter by 1500. William Crugge, a wealthy tanner and merchant, was mayor four times between 1506 and 1518.[28] The South West, with its abundant pastures, nourishing many cattle and sheep, and with reasonable quantities of the oak bark and lime needed for tanning, was a good location for the industry; but although it was important for the local market, it was overshadowed nationally by larger centres of production, especially in London, Kent and the western region between Bristol and Chester.[29] The leather industry was divided into two branches: tanning, which produced heavy leather for shoes and saddles etc; and dressing, which produced lighter leather for gloves and bags etc. (including chamois leather). Tanning, in which the hides were treated by lime and manure, before being 'tanned' (e.g. beaten) in baths of water and oak-bark, was a laborious process, and could take from six months to two years to complete; whereas dressing, in which the skins were smoked, oiled and soaked in a mixture of water and alum (salts of aluminium and potassium, mostly imported from the Papal States), could be completed in about three weeks.[30] The equipment needed for tanning was not very expensive, and many tanners operated on a small scale, sometimes engaging in the trade as a supplement to farming; but by the seventeenth century some tanners were operating on quite a large scale and were relatively wealthy. For instance in Cornwall, Anthony Wills of St Stephens-by-Saltash left a personal estate (excluding land and houses) of £250 (quite large when a horse could be bought for £1) when he died in 1619, of which tubs and vats of leather were worth no less than £160—or 64 per cent of his personal estate. Farther west, Richard White of Truro left £239 in the same year, of which £60 was the value of his leather, hides and skins. In the extreme west of Cornwall at Penzance in 1688, the tanner Richard Cunnack left a personal estate of £247, of which £125 represented the value of leather in production and finished.[31]

Glove-making from dressed leather was a craft widely dispersed, especially in north Devon and throughout Cornwall. In the seventeenth century, probate inventories reveal glovers in the following towns in Cornwall: Launceston, Bodmin, Truro, Stratton, Camelford, Penzance, Falmouth and Looe, and also in many villages.[32]

As the population of the region grew in the eighteenth century, the leather industry grew with it. In the early eighteenth century Daniel Defoe recorded on his tours of the West Country (probably c.1705–6) that Liskeard was 'remarkable for a very great trade in all manufacturers of leather, such as boots, shoes, gloves, purses, breeches, etc'. Farther west in Redruth he noted that 'they have a great market of provisions once a week, and a great sale of shoes brought from all parts round for near thirty miles'.[33] In addition to the Exeter and Plymouth regions, Tavistock was also

Map 42.3
The tanning industry, sixteenth to nineteenth centuries.

SOURCE: as **Map 42.1**.

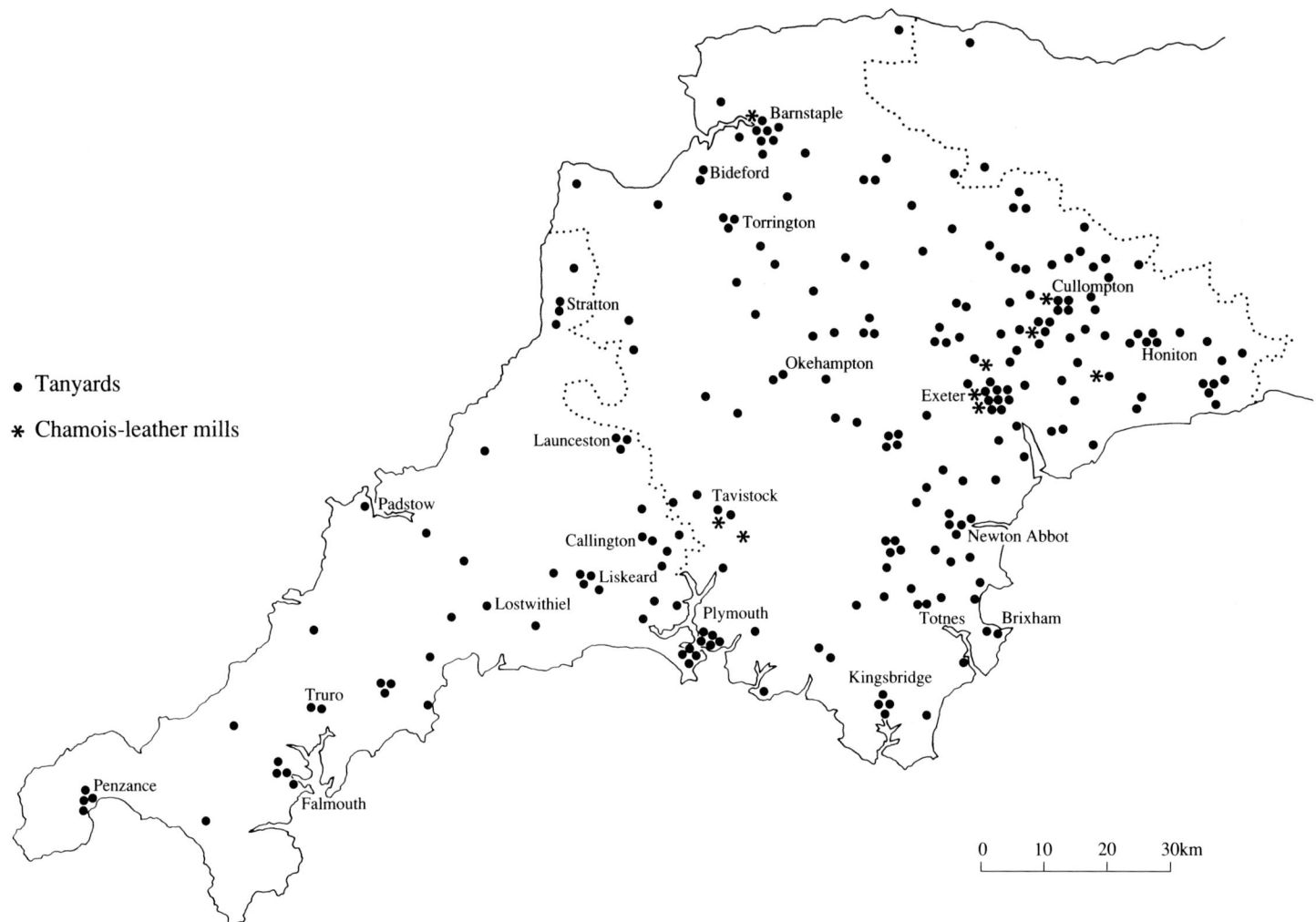

• Tanyards

✳ Chamois-leather mills

an important leather town in the eighteenth century.[34] Stock farmers benefited from the leather industry, as did the owners of oak woods, who sometimes operated on quite a large scale. For instance in 1748 an owner of oak coppice at Linkinhorne in eastern Cornwall sold the bark from five acres of coppice (a considerable quantity of bark) to Mr Avery, a Liskeard tanner, seven miles away.[35] Another large deal took place in 1691, when the Chamber of the city of Exeter sold 500 oaks to Mathew Frost, tanner, of Shobroke, near Crediton, for £900. The oaks came from the ancient manor of Duryard, an estate belonging to the city, on its northern edge.[36] The leather industry continued to be widespread well into the nineteenth century.[37]

PAPER-MAKING

Maps 42.4 and 42.5 (which are based on the work of Alfred Shorter) show the location, and date of foundation, of the small,

water-powered vat mills in which paper was laboriously made by hand before the invention of a paper-making machine in France in the late 1790s.[38] Only a handful of people, most of whom were women and children, worked in these mills, which were not capable of producing more than a few hundredweights of paper a week.[39] There are no regional production statistics, but by the 1720s, English and Welsh production was around 3,000 tons compared with about 1,000 tons of imported paper. By 1800 imports had ceased and domestic production was over 10,000 tons.[40]

The location of paper mills depended partly on a reliable supply of clean water (which Devon and Cornwall had in abundance) and on a good supply of linen rags; but more importantly upon the size of the local market (many of the early skilled workers being Huguenot refugees from France). The small size of west-country towns was a disincentive initially, but by about 1638 Exeter was large enough to support the first paper mills in the region. These were two mills erected at Countess

Wear on the Exe just south of Exeter. Little is known about them, but in a law case of 1670, one Peter Trenchard, a paper-maker of St Thomas, Exeter, stated that in about 1656 he had served an apprenticeship at 'Wear Paper Mills' under Abraham Langdon, who is thus the first recorded paper-maker in the South West.[41] However, the process spread slowly and by 1700 there were still only a few mills in Exeter and Plymouth, and none elsewhere in Devon or in Cornwall (**Map 42.4**). Progress was much more rapid, though, in the eighteenth century, and by 1800 there were thirty-one paper mills in Devon and six in Cornwall (**Map 42.5**), though progress prior to 1750 was still slow. By 1750 there were six mills in Exeter and east Devon, one at Plymouth (at Millbay founded in 1710) and one in north Devon, at Barnstaple, founded in 1746. The first paper mill in Cornwall was not established until 1784 (Coosebean Mill, near Truro); but five more followed in fairly quick succession, stretching across Cornwall from Danescombe in the east (on the upper Tamar estuary) to Stithians in the west (north-west of Falmouth). In Devon the Exe and Culm valleys remained the most favoured locations with a cluster of mills around Exeter, Crediton and Cullompton, and others at Colyton and Moretonhampstead. The second concentration of mills in the late eighteenth century was around Plymouth, and

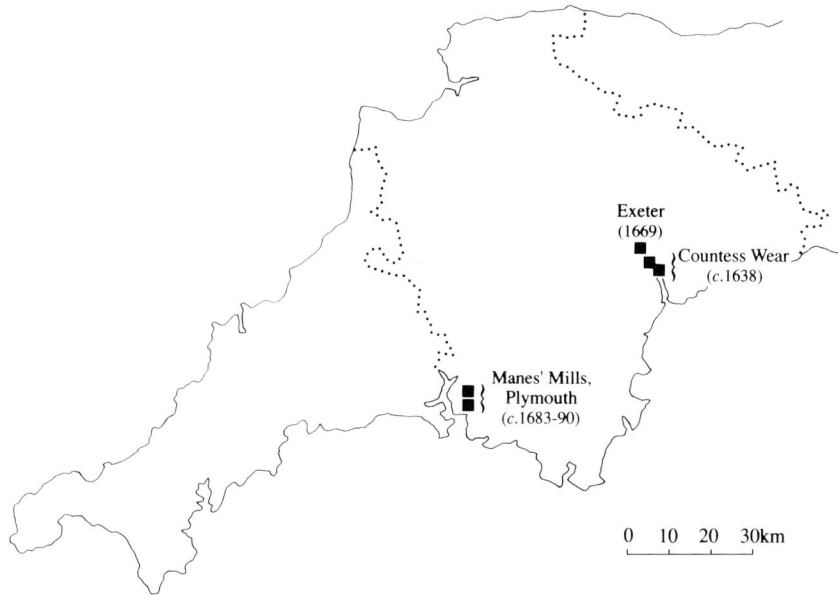

Map 42.4
Working paper mills, 1676–1700, with starting dates.

SOURCE: A.H. Shorter, *Paper Mills and Paper Makers in England 1495–1800* (Hilversum, 1957).

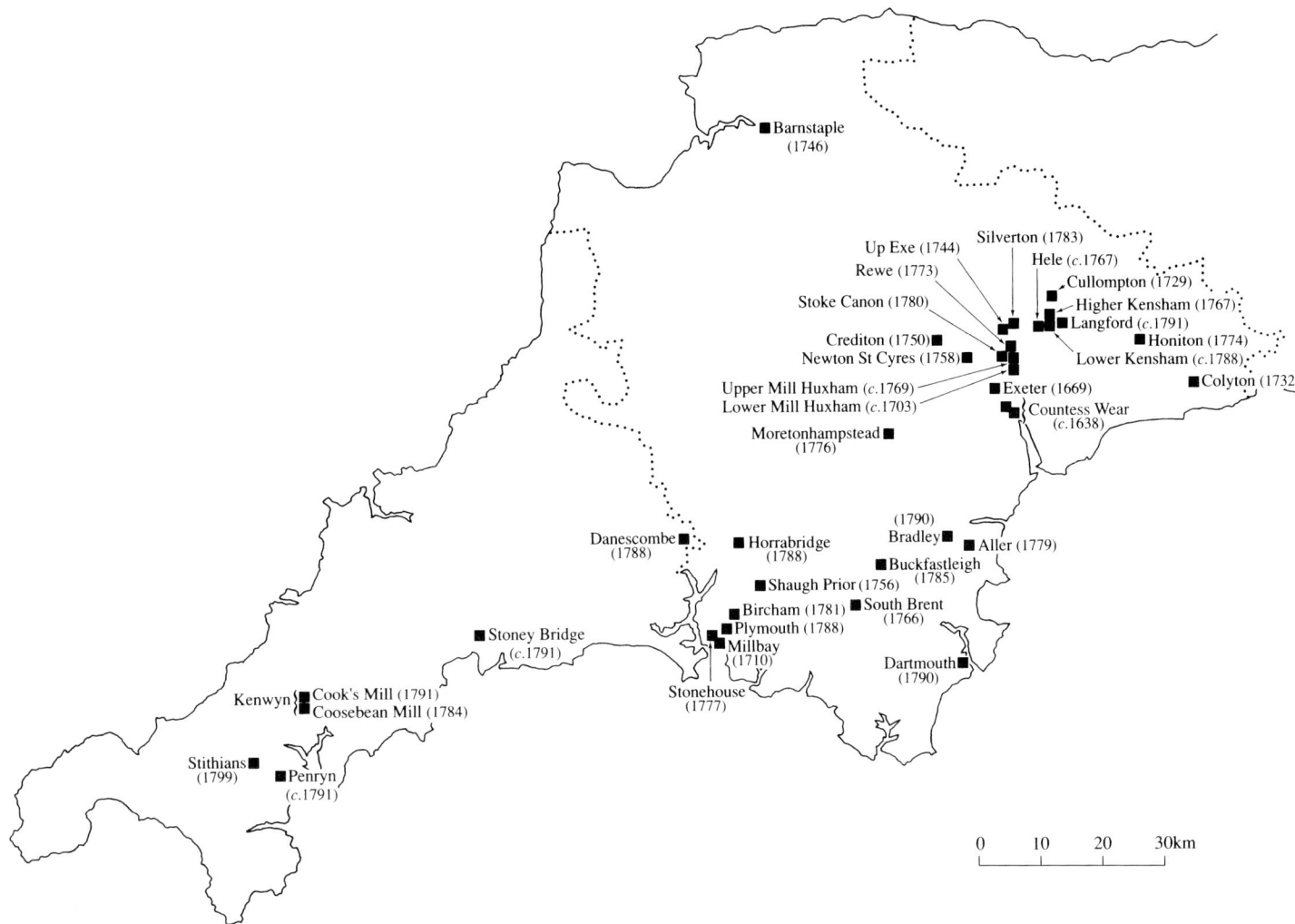

Map 42.5
Working paper mills, 1776–1800, with starting dates.

SOURCE: as **Map 42.4**.

Fig. 42.2
The interior of a
water-powered
paper mill, *c*.1662.

SOURCE: G.A. Böckler,
Theatrum Machinarum Novum
(Nuremberg, 1662).

there were five more mills in south Devon between Exeter and Plymouth. West and central Devon were without mills, and the Barnstaple mill remained a lonely outpost in north Devon. It was referred to by P. Luckcombe in *England's Gazeteer*, I (1790), but its later history is unknown, though paper-making was carried on near Barnstaple in the nineteenth century at Playford and Blatchford mills.[42]

Vat-made paper reached its climax in the South West in the early nineteenth century, and then declined steadily in the face of competition from machine-made paper (mainly made by large firms in the London area, and in the northern and midland industrial regions) and from imported paper.[43] Devon and Cornwall's share of production cannot be calculated, but in the decade 1801–10 there were 434 paper mills in England, of which thirty-seven (8.5 per cent) were in Devon and Cornwall.[44]

FURTHER READING

A.H. Shorter, W.L.D. Ravenhill and K.J. Gregory, *South-West England* (London, 1969); Joyce Youings, *Tuckers Hall: The History of a Provincial City Company through Five Centuries* (Exeter, 1968); W.G. Hoskins, *Industry, Trade and People in Exeter, 1688–1800* (Manchester, 1935); M.A. Havinden, J. Quéniart and J. Stanyer (eds), *Centre and Periphery: Brittany and Cornwall and Devon Compared* (Exeter, 1991); Daniel Defoe, *A Tour through Great Britain* (1724); Alfred H. Shorter, *Paper Making in the British Isles* (Newton Abbot, 1971).

INDUSTRY AND MINING

METAL-MINING
SINCE THE
EIGHTEENTH CENTURY

ROGER BURT

The mapping of mine sites is at best an imprecise exercise and at worst a thoroughly misleading one. Most maps are concerned with the indication of surface features, while mines, by their very nature, are invisible underground labyrinths. A map of mine sites becomes simply a map of mine surface features—the points at which the products of underground working are disgorged for processing and sale. For modern workings it is possible to show the margins of the area of the mine's lease and/or to delineate a plan of the extent of the workings. For historical times, however, this information is rarely available and never sufficiently comprehensive for any particular date to produce a complete and detailed map. No historical map of mines can be more than a composite of approximations—points of general reference taken from the workings' main shaft, principal adit, or dressing floors.

These problems also create obstacles for demonstrating the size and extent of the industry. In very general terms, underground workings started off in the sixteenth and seventeenth centuries as small-scale, independent workings. There was a multiplicity of separate leases, separate mines and separate companies. During the course of the eighteenth and nineteenth centuries, however, the extent of mineral exploitation and underground working increased, individual operations became larger, and leases and companies were progressively 'balled' or consolidated into larger units. At Dolcoath, for example, what might have been shown on the map as seven different mines in the first half of the eighteenth century had become just one mine by the end of the century. Underground operations had also expanded but a simple comparison of maps would suggest a significant contraction of activity. This process of amalgamation accelerated over time and creates particular problems when the indicated level of activity in the late twentieth century is compared with that of the nineteenth century. A direct comparison of the number of mines working in the 1980s with those working in the 1880s, for example, would suggest that the industry had shrunk from several hundred to just five active producers. However, the mines working in the 1980s included the leases/sets of around seventy of the most productive workings of the 1880s and their total level of production remained little changed.

With these reservations about the interpretation of mining maps firmly in mind, what can be said about changes in the spatial distribution of mining in Cornwall and Devon during the last 300 years? The rapid development of copper-mining after the 1690s and the greater exploitation of lead deposits started a gradual diversification of mining away from the established tin areas of south-west England (see also chapters 40 and 41). Also, the publication of contemporary reviews of the industry by Borlase, Pryce and others began to provide a clearer view of the more productive workings. From these and other sources, Hamilton Jenkin has recorded the existence of over forty separate copper mines between Redruth and Camborne alone, during the eighteenth century, and Morrison has calculated from ore sales returns that there were ninety active copper producers in the county as a whole in 1770.[1] Many of these mines were small and shallow workings whose location might never be known with accuracy, but a few larger enterprises clearly stood out. In the mid-1750s, Borlase noticed twenty-six major tin, copper and lead workings in the county, and twenty years later Pryce referred to thirty-eight separate mines. Fourteen of these mines were the same, which provides some further indication of their size and continuity. Many of the fifty separate sites noticed by these authors lay very close together and were later amalgamated. Their distribution is shown in **Map 43.1**. This may be taken as a general guide to some of the principal mine sites in Cornwall during the third quarter of the eighteenth century.

Unfortunately, there is no equivalent data available for Devon in this period. The first clear picture of the distribution of the industry in that county appears with the publication of Lysons's *Devonshire*. That volume contains details, provided by the very knowledgeable John Taylor, of mines at work in the period 1780–1810.[2] More than a hundred different sites were referred to, including thirty-one tin, twenty-one copper, thirty-nine lead and eleven manganese mines. Those that can be located are shown on **Map 43.1**, which may be regarded as providing a view of Devon mining *c*.1800. By comparison with the Cornish mines, most of these workings were relatively small, though Lysons commented that three of the ventures operating at the beginning of the 1820s—Vitifer, Ailsborough and Whiteworks—'are upon a large scale'.[3] The explanation of the predominance of small-scale working is probably to be found in the shallow development of the tin deposits, which had been mainly worked out by earlier streaming and opencast operations, and the slow development of copper-mining in Devon. Unlike

Map 43.1
Tin, copper, lead and
manganese mines working,
*c.*1750–1780.

SOURCES: W. Borlase,
*The Natural History
of Cornwall* (1758);
William Pryce, *Mineralogia
Cornubiensis* (1778);
D. and S. Lysons,
*Magna Britannia, Vol. VI,
Devonshire* (London, 1822).

- Mine

░░ Land over 600ft (183m)

0   10   20   30km

Fig. 43.1
Copper production,
1720–1900.

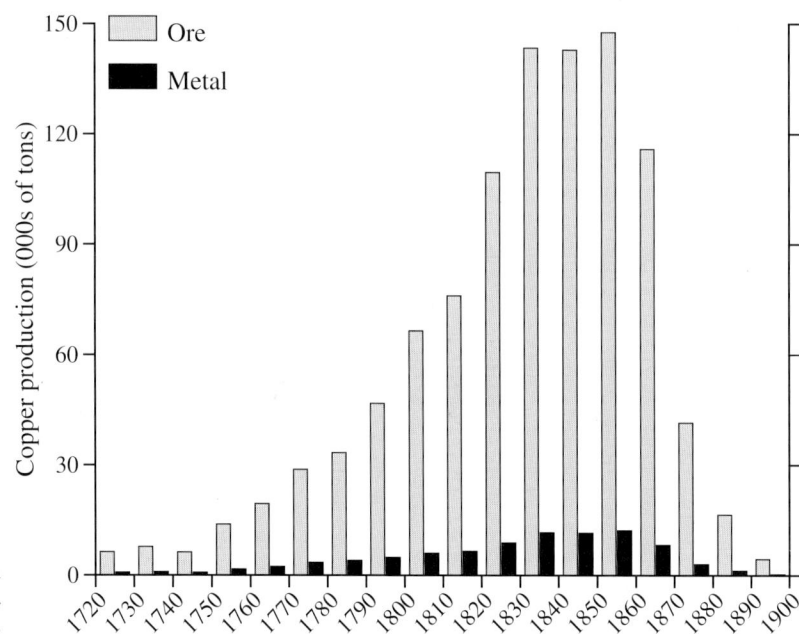

Cornwall, where most of the copper and tin came from the same mines, in Devon the output of these minerals was spatially separate. The tin mines were all on, or very close to, the Dartmoor granite, while copper was worked in the surrounding metamorphic auriol and particularly the Tamar valley. Lead-mining was widely distributed across the county, though most of the rich argentiferous ores—for which the county was nationally famed—were derived from a small group of lodes around Bere Ferrers and Bere Alston, in the southern Tamar valley, and Combe Martin on the north coast. Manganese production was also important around the turn of the century, with the county becoming one of Europe's principal sources of supply.[4] The most productive mines at this time were a few miles to the north-west of Exeter, around Upton Pyne and Newton St Cyres. Like Cornwall, Devon mines also produced small quantities of a wide range of other minerals, from iron and cobalt to clay, slate, marble, granite and other building stone. There was even some small production of coal, particularly near Tavistock, Bovey Tracy and Bideford.[5]

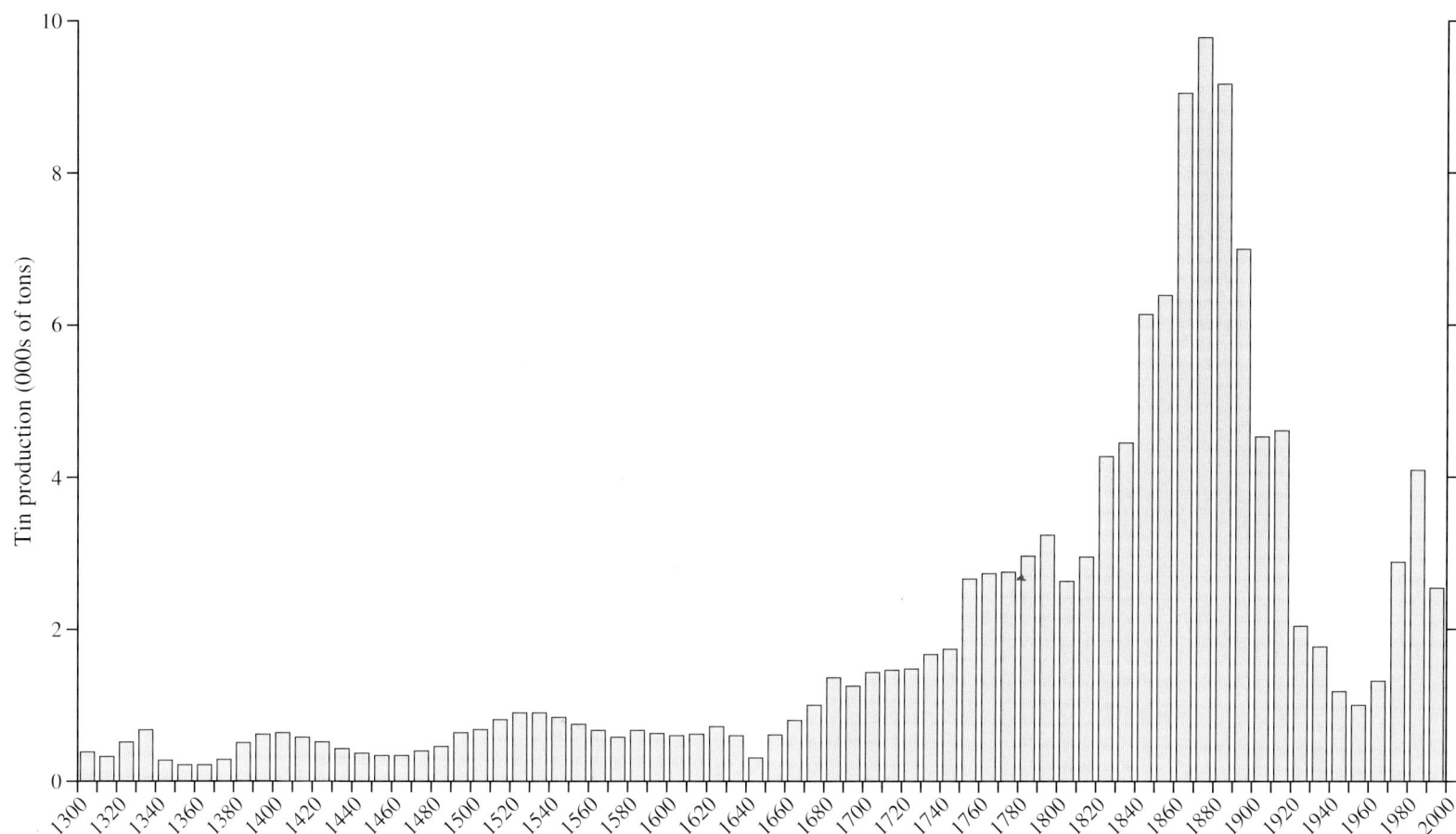

Fig. 43.2
Decennial average of white tin production, 1300–2000.

During the early nineteenth century, the story of south-western metal-mining was dominated by copper. Tin output, still derived largely in association with copper, recovered from a low point at the end of the Napoleonic Wars but in the 1820s and 1830s averaged only a little more than the level of the 1780s and early 1790s (**Fig. 43.1**). There was more rapid expansion thereafter but in the-mid 1850s the value of copper ore output in the South West still stood at twice that of tin ore.[6] This sustained prosperity and expansion of copper was the result of a combination of fortuitous circumstance and good business practice (**Fig. 43.2**). On the one hand, the Anglesey producers, who had flooded the market with cheap copper during the latter part of the eighteenth century, were now effectively spent and alternative sources of international supply were not forthcoming. Prices of metallic copper were nearly twice their early 1770s level in 1820 and remained generally buoyant thereafter.[7] On the other hand, the industry invested wisely and heavily in state-of-the-art steam technology to exploit deepening lodes, built new transport facilities to cheapen the cost of moving ore out and fuel in to the region, and above all developed major new copper deposits. In this latter respect, the focus of the industry gradually widened. Until the end of the eighteenth century, the general belief in Cornwall that there was little worth mining east of Truro bridge was hardly challenged. The early years of the nineteenth century, however, saw first the development of rich new mines around St Austell and then, from

the 1830s, a major new mining district around Caradon Hill and on eastwards, into the Tamar valley and west Devon. During the same period, a less important, but nevertheless significant, westward-moving frontier carried copper-mining down into the far St Just peninsula. The long-established dominance of the central Cornwall Redruth–Camborne–Gwennap mining district was gradually challenged and overthrown. Many of its mines, however, like Dolcoath, United and Poldice, remained among the largest industrial operations in Britain, as their deepening lodes gradually turned from copper to tin and gave a new lease of life to the production of that latter metal.[8]

One of the principal aspects of the eastward expansion of copper-mining was gradually to raise the profile of Devon mines within the region. Notwithstanding the early development of some of the Tamar valley deposits, Devon copper output was just a small fraction of that of Cornwall in the years following the Napoleonic Wars. It expanded rapidly, however, in the 1820s, 1830s and early 1840s from just a few hundred tons to over 4,000 tons annually. The really big breakthrough came in the later 1840s with the development of the Devon Great Consols deposits. This one mine was producing around 25,000 tons of copper ore a year by the mid-1850s, valued at almost £150,000. It had rocketed into first place as the region's, and the world's, leading producer. Together with Bedford United, Wheal Maria, Wheal Friendship and Wheal Crebor, and several other lesser producers, it enabled

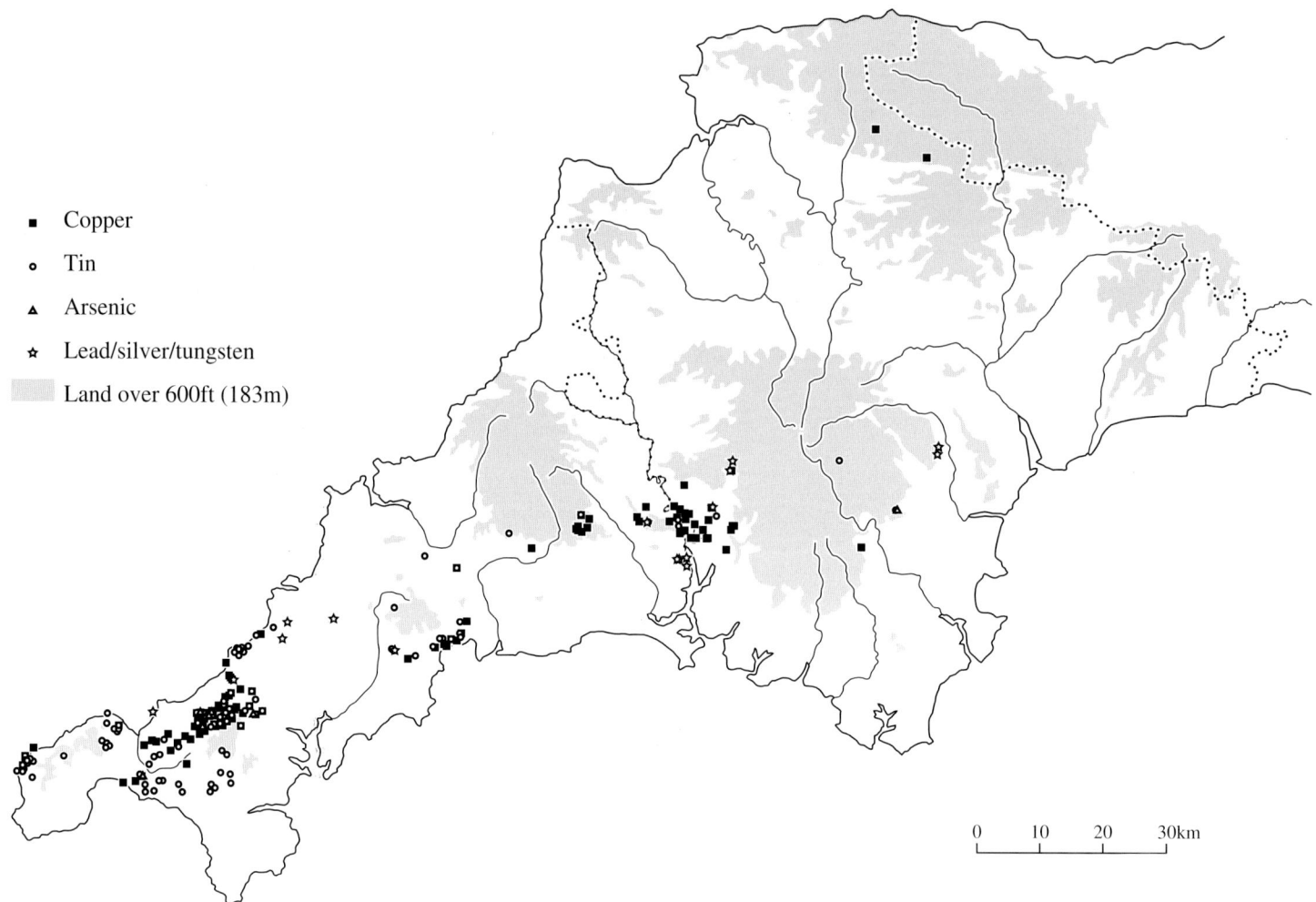

■ Copper

o Tin

▲ Arsenic

☆ Lead/silver/tungsten

▨ Land over 600ft (183m)

0   10   20   30km

the eastern side of the Tamar valley to produce up to a quarter of the country's total copper output during the third quarter of the century.[9]

The early and mid nineteenth century also saw the first large-scale development of some 'new' minerals and mining districts in the South West. Iron deposits, for example, which had been worked sporadically in a small way since the sixteenth century and earlier, began to attract increasing attention as the rapidly expanding iron and steel industry of South Wales depleted its own local resources and searched for new sources of supply.[10] In Cornwall, the Perran Iron Lode on the north coast was mined near St Agnes, and in the south-east of the county near St Austell a number of mines were worked on a significant scale during the third quarter of the century. In Devon, small iron mines were opened in Brixham/Torquay and around Ilsington and Hennock, and there was particularly intensive exploration on Exmoor, mainly in the parish of North Molton, and in the neighbouring Brendon Hills. On a national scale, however, none of these workings ever came to much. Their lodes were too small and difficult to work and the two counties taken together never produced much more that 1 per cent of annual UK output. There

was a very different picture for manganese. As the important mines near Exeter in the east of the region were exhausted, major new deposits were developed in west Devon around Milton Abbot and Marystow. Together with some small mines in adjacent east Cornwall, these became the principal domestic source of UK supply during the third quarter of the century and competed effectively with rising imports.[11] For the production of lead and silver, the declining output of established centres of production in Cornwall and the Tamar valley was compensated by the development of important new mines in the Teign valley of central Devon. In the 1850s and 1860s these mines often had an annual output exceeding 1,000 tons of lead and 30,000 ounces of silver. As by-products they sometimes produced small quantities of zinc and barytes and in the twentieth century became primarily dependent on the output of that latter mineral.[12]

The distribution of the south-western mining industry in the middle of the nineteenth century is shown in **Map 43.2**. For the first time, the map can be regarded as reasonably comprehensive and accurate. This is made possible by reference to the detailed annual production statistics for the mines which began to be published from the late 1840s. The industry at this time was at its all-time

Map 43.3
Mines in the early twentieth century.

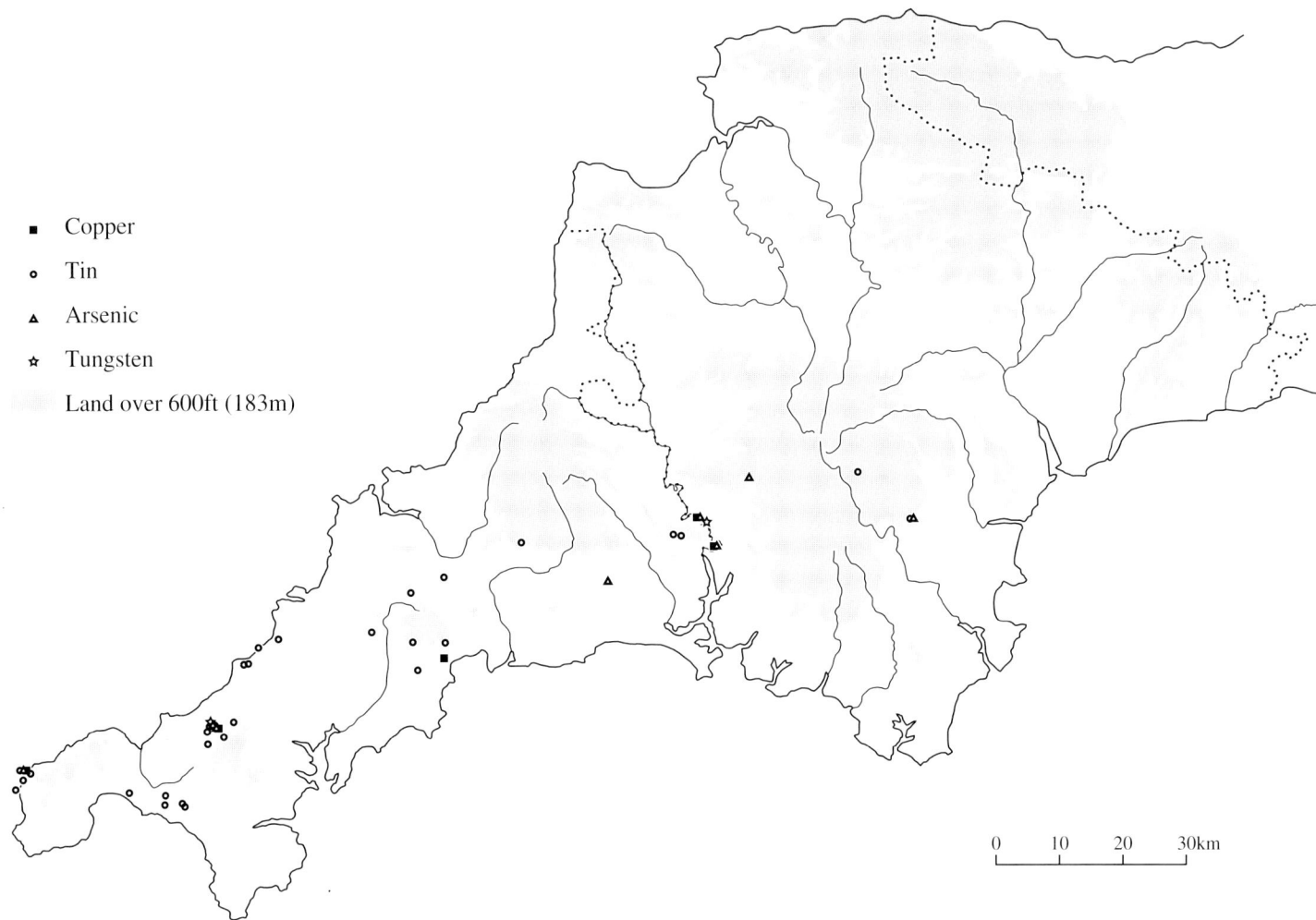

■ Copper

○ Tin

▲ Arsenic

☆ Tungsten

▨ Land over 600ft (183m)

0   10   20   30km

peak. It produced a wider range of minerals, from more mines, in greater quantities and for higher returns, than ever before or since. Few could have guessed that the prosperity was to be so fragile and so short-lived. A rising wave of cheap imports of all minerals was about to break on unprotected British shores and within a generation an industry that had developed over two millennia would be in ruins. First copper, then lead and manganese, and finally tin prices were to decline sharply, and production from deep, high cost 'old' mines went down with them. A boy of 12 joining his father and elder brothers underground in the booming conditions of the early 1860s would have seen the industry's near total collapse well before his fiftieth birthday. In just thirty-five years, between the early 1860s and the late 1890s, the region's copper ore output fell from over 180,000 tons annually to less than 6,000 tons; lead and silver production had entirely disappeared; manganese production had risen to a peak in the early 1870s and equally rapidly sunk into oblivion; tin ore production had been halved, from more than 14,000 tons to less than 7,000 tons; and only arsenic output stood at significantly more than its earlier position. The much diminished industry, as it stood at the beginning of the twentieth century, is shown in **Map 43.3**. Thereafter there was little

relief from seemingly inexorable decline. Depressions following the First World War, the 1929 crash, and the Second World War gradually whittled the industry away to almost nothing. When effective cartelization of the tin market finally restored high prices to that section of the industry in the late 1970s and early 1980s, it was restricted to a small area of central Cornwall and the St Just peninsula. By this date, underground metal-mining in Devon had entirely ceased and the prospect of developing a major new underground tungsten/tin working at Hermerdon Ball, near Plymouth, was dashed by the collapse of the Sixth International Tin Syndicate in 1985, just before it was due to come into production. With the cessation of production at South Crofty mine near Camborne in 1998, the mapping of south-western mining can now only be conducted as an historical exercise.

FURTHER READING

On Devon see R. Burt, P. Waite and R. Burnley, *Devon and Somerset Mines* (Exeter, 1984) and for Cornwall see R. Burt, P. Waite and R. Burnley, *Cornish Mines* (Exeter, 1987).

EMPLOYMENT
IN THE
TWENTIETH CENTURY

ANDREW GILG

In Great Britain the employment history of the twentieth century has been one of radical transition from extractive and manufacturing employment to one of service employment. It may well be thought to have been especially marked in the South West, which in the popular imagination has traditionally relied on agriculture, fishing and mining for employment. However, by as early as 1931, Devon and Cornwall already had above average percentages employed in most sectors of the service industries, notably in public administration and personal service. Within the extractive sector, agriculture was a dominant employer in 1931, but mining and quarrying accounted for less than 3 per cent and fishing less than 1 per cent of employment. Every sector of industrial employment in south-west England was below the national average at this time.

THE INTER-WAR PERIOD

The overall employment structure in the inter-war period for the region is shown in **Fig. 44.1** and the sub-regional breakdown in **Map 44.1**. **Map 44.1** shows that services were locally important in Exeter, Plymouth and Torquay, that extractive industries were important in all areas except Torquay and Plymouth, and that manufacturing was only important in Exeter and Plymouth. In terms of a balanced employment structure, Exeter and Truro stand out as having roughly equal shares in the three major sectors, but Plymouth, although unbalanced in its structure, was nearer the national norm for the time.

The 1930s were of course a time of the great economic depression in both manufacturing and agriculture, but to some extent Devon with its economy already firmly based in the service sector was able to survive the depression fairly well, while Cornwall, where in 1935 unemployment reached over 40 per cent in Redruth and St Columb and over 60 per cent in Gunnislake, fared far worse.[1] By the end of the 1930s unemployment had fallen back to the 1929 levels except in Cornwall, but as **Map 44.2** shows, every sub-region increased its number of insured persons in employment between 1929 and 1939, notably Exeter, Torquay and Plymouth, while in contrast north Devon and all of Cornwall registered lower rates of increase. In explanation, Plymouth benefited from a

rapid expansion in manufacturing and related building, Torquay from increased service activity, and Exeter from increased jobs in distribution. For the region as a whole, the increase was due entirely to the service sector with mining and quarrying in contrast showing a rapid decline and manufacturing a small decline. Only one sub-region, Exeter, enjoyed employment growth across its whole area, while in every other sub-region serious losses of employment offset to some extent the gains made. Notable gainers (above 40 per cent) were Penzance, Falmouth, Bude, St Columb, Saltash and Brixham. In contrast notable losses of employment (above 30 per cent) were recorded in St Ives, Gunnislake and Ivybridge.

Most of the material for the 1929–39 period used in this chapter derives from a survey carried out by the University College of the South West published in 1947 to coincide with the advent of statutory planning under the 1947 Town and Country Planning Act.[2] For the more recent period I rely mainly on the *Reports of Survey* produced since 1947 by Devon and Cornwall Councils as part of their Development and Structure Plans process. Unfortunately, the Devon *Report of Survey* did not produce any maps of the overall distribution of employment, but in contrast the Cornwall *Report of Survey* contains the map reproduced here as **Map 44.3**.

THE POST-WAR RISE OF SERVICE INDUSTRY EMPLOYMENT

The post-war period witnessed first a recovery from the distortions of the war, which in the South West had been strongly influenced by the build-up of defence industries in the Plymouth area, and the need to expand agriculture, so it was not until the 1960s that a new genuine, post-war employment picture emerged. **Map 44.4** reveals the situation by 1965. By this date, primary employment only accounted for a significant percentage in the St Austell area with its china clay workings, and in one or two smaller rural areas of north and west Devon and east Cornwall and the horticultural areas of west Cornwall. In contrast service employment accounted for well over half of all employment in virtually every area, and in some areas reached figures in excess of 75 per cent, notably in the

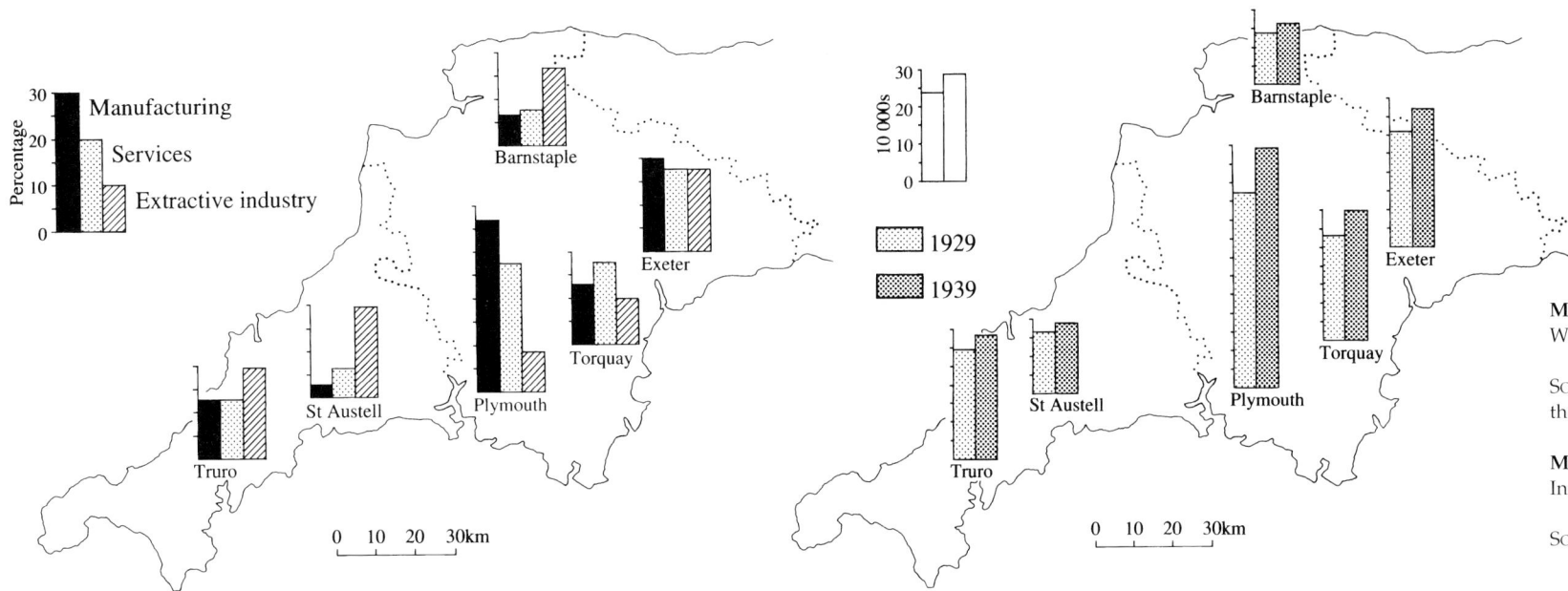

Map 44.1 (left)
Workers in sub-regions, 1931.

SOURCE: University College of the South West, *Survey*, 1947.

Map 44.2 (right)
Insured persons, 1929–1939.

SOURCE: as Map 44.1.

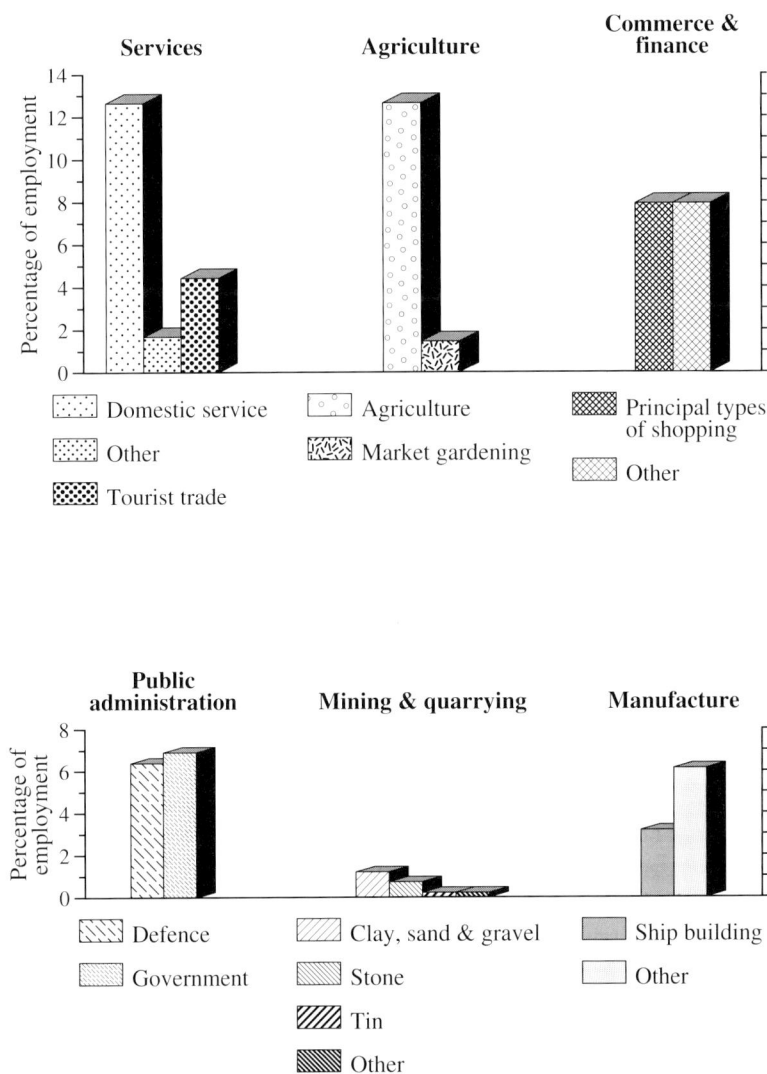

Fig. 44.1
Structure of employment, 1931.

SOURCE: as Map 44.1.

coastal areas of north Devon and Torbay, where tourism added to the totals. Manufacturing across the region was only significant in Camborne–Redruth, Plymouth, Tiverton and Newton Abbot, but in absolute terms the large numbers employed in Plymouth made manufacturing more important than the map suggests. South Devon experienced an economic boom in the 1950s, but north and east Devon experienced sluggish growth, albeit more palatable than the decline experienced in south-west Devon. Primary employment in agriculture fell rapidly, notably in east Devon, while service employment grew strongly everywhere except in south-west Devon. The most patchy performance was, however, in manufacturing, where the very fast growth of south Devon was mirrored by sharp decline in south-west Devon.

The second review of the Devon Development Plan revealed an expanding number of firms overall between 1961 and 1968, with notable growth in food, drink, engineering, electrical goods, textiles, paper and printing, and building materials, but a number of other areas with a static number of firms.[3] Overall, growth was almost uniform across the county. In north Devon, nearly all of the growth was due to an influx of new firms, while in east Devon, there was widespread growth of existing firms. The role of incoming employment is discussed in more detail at the end of this chapter.

Employment in Cornwall in the decade 1961–71 has been analysed by the County Council in their *Report of Survey of the Structure Plan*.[4] During the decade the county experienced overall growth of 8.3 per cent, but this concealed a major restructuring not only of the pattern shown in Map 44.4, but also within more detailed sub-sectors. For example, agriculture, forestry and fishing experienced a loss of over a quarter of all jobs in the decade, and utilities and transport suffered nearly as badly. The

Map 44.3
Manufacturing industry,
agriculture and fishing
in Cornwall.

SOURCE: Cornwall County
Council, *Report of Survey*
(Truro, 1952).

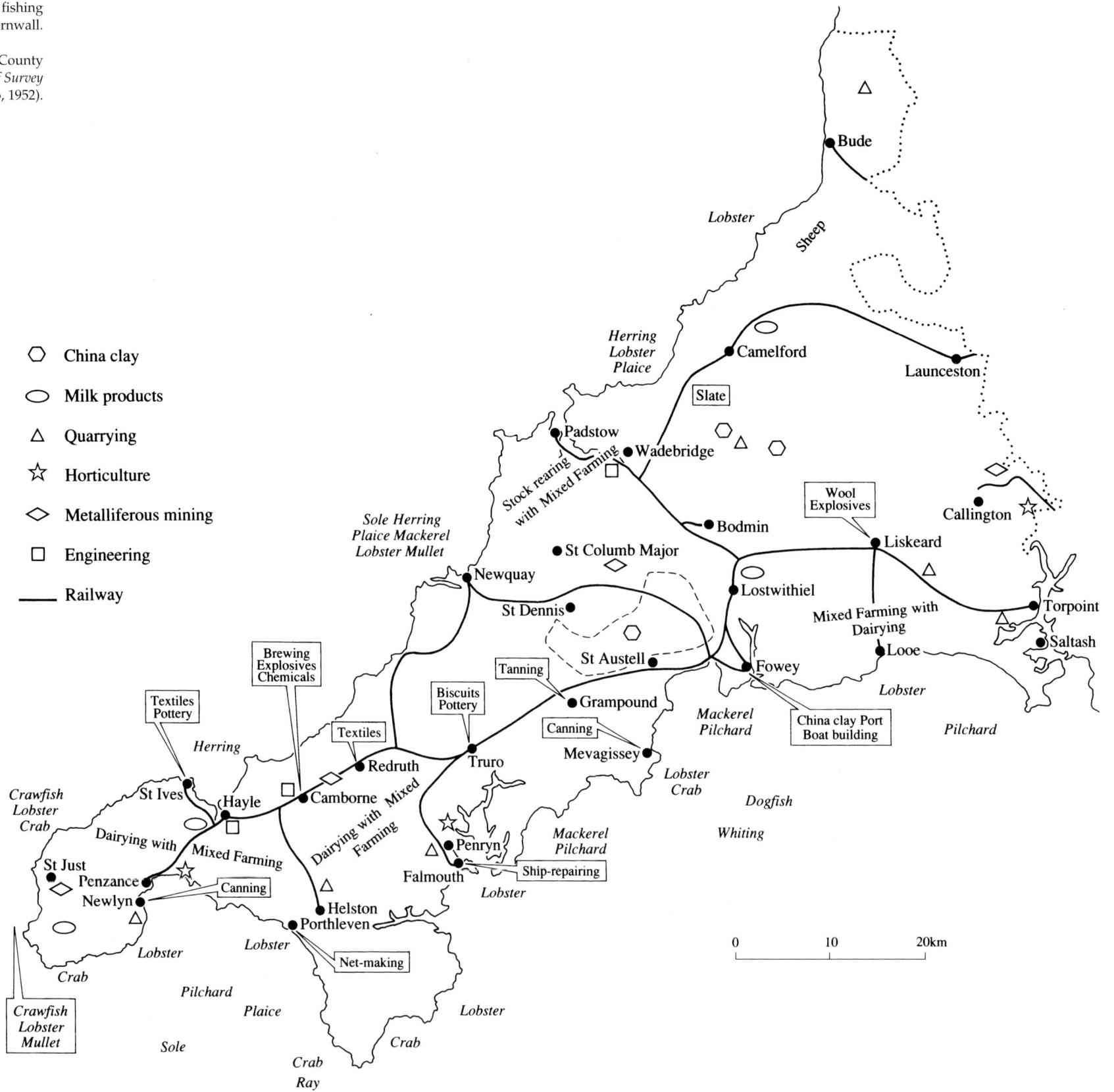

◯ China clay

⬭ Milk products

△ Quarrying

☆ Horticulture

◇ Metalliferous mining

▢ Engineering

—— Railway

Bude

Lobster

Sheep

*Herring
Lobster
Plaice*

Camelford

Launceston

Slate

Padstow

Wadebridge

Stock rearing
with Mixed Farming

Bodmin

Wool
Explosives

Callington

Liskeard

*Sole Herring
Plaice Mackerel
Lobster Mullet*

St Columb Major

Newquay

Lostwithiel

St Dennis

Mixed Farming with
Dairying

Torpoint

St Austell

Fowey

Looe

Saltash

Tanning

Lobster

Brewing
Explosives
Chemicals

Biscuits
Pottery

Canning

*Mackerel
Pilchard*

China clay Port
Boat building

Pilchard

Textiles
Pottery

Textiles

Grampound

Mevagissey

*Lobster
Crab*

Dogfish

Herring

Redruth

Truro

*Mackerel
Pilchard*

Whiting

*Crawfish
Lobster
Crab*

St Ives

Hayle

Camborne

Dairying with Mixed
Farming

Penryn

Ship-repairing

St Just

Dairying with Mixed Farming

Falmouth

Lobster

Penzance

Canning

Newlyn

Helston
Porthleven

Net-making

Lobster

Lobster

Lobster

Crab

*Crawfish
Lobster
Mullet*

Pilchard

Plaice

Crab

Lobster

Sole

*Crab
Ray*

0   10   20km

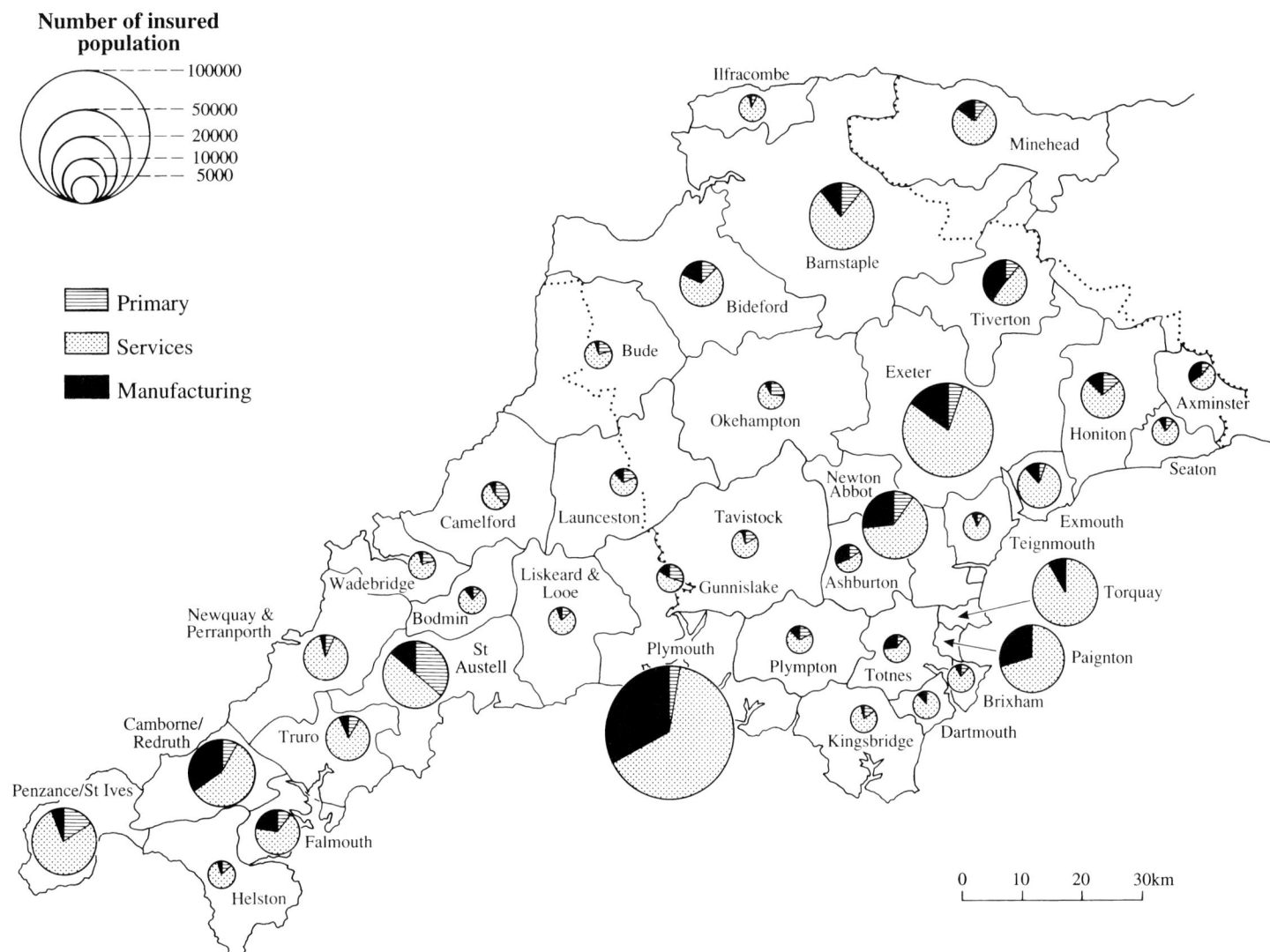

Number of insured population

- —————— 100000
- —————— 50000
- —————— 20000
- —————— 10000
- —————— 5000

Primary
Services
Manufacturing

Map 44.4
Employment structure, 1965: numbers of insured population.

SOURCE: A.L. Shorter, W.L.D. Ravenhill and K.J. Gregory, *South-West England* (London, 1969).

rundown of defence was also highlighted by a 7 per cent fall. In contrast, manufacturing showed a rapid rate of growth of nearly 50 per cent in the decade, which probably also accounted for the 37 per cent rise in the number of construction jobs. Other areas of strong growth were mining and professional services. These trends were not, however, continued into the 1970s; although the growth rate remained high, this was a result of the service sector doubling its rapid growth rate, distributive trades accelerating their growth trend, and public administration reversing a decade of decline to slow substantial growth. Elsewhere, however, decline, albeit slow, set in across the former boom areas of manufacturing and mining, and the primary sector continued to decline but at a slower rate than before.

Fig. 44.3 shows the pattern of employment change for the county of Devon (including Exeter, Plymouth and Torbay) for the 1961–74 period. It shows that services of all kinds, professional, financial, governmental and distributional, all increased their share considerably in the period. The only other area of major

growth was in engineering and food industries. Declines were registered across a wide spectrum of employment areas, but, except for leather goods, the percentage cut was not very large. The net result of all these changes for employment in the South West as a whole by the mid-1970s is shown in **Map 44.5**. This confirms the trend to service employment already established, and a comparison with **Map 44.3** clearly confirms that by 1974 Devon was an economy dominated by service employment, except in Plymouth, and to a lesser extent Barnstaple–Bideford, Newton Abbot, Exeter, Tiverton and parts of east Devon. In Cornwall, as in Devon, service employment accounted for well over half of all jobs, except in the St Austell area and north of Padstow. Manufacturing was locally important in Camborne–Redruth, Falmouth and St Austell, where mining and quarrying also remained important. The resurgence of tin-mining in the 1970s is not reflected in the data, as, although it attracted a lot of attention, it created very few jobs. Agriculture, forestry and fishing remained important in a number of rural areas across the

Map 44.5
Number of employees
and industrial structure
in Cornwall (1975)
and Devon (1974).

SOURCES: Cornwall County
Council, *Structure Plan:
Report of Survey* (Truro, 1979)
and Devon County Council,
*Structure Plan: Report and
Survey* (Exeter, 1977).

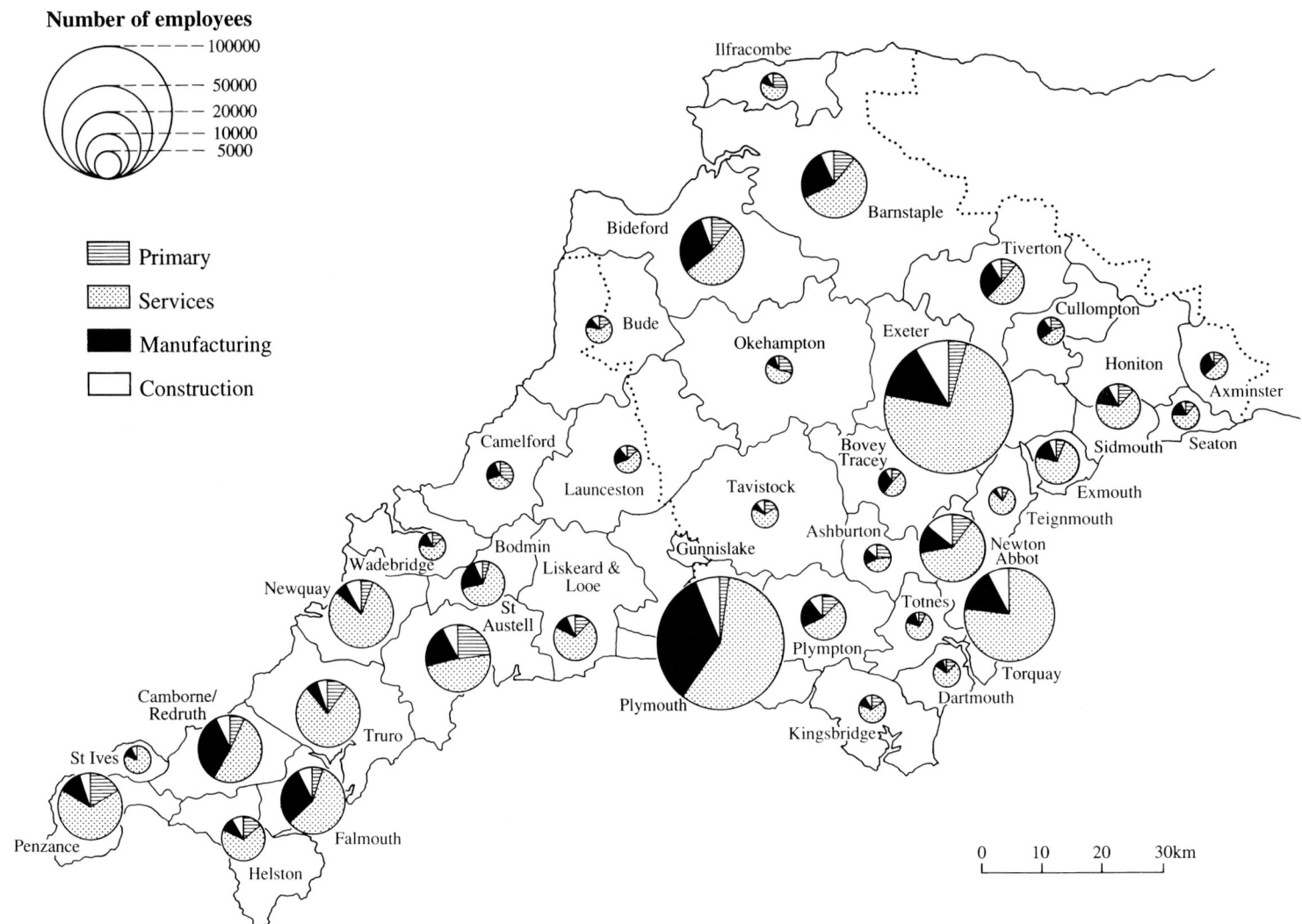

Number of employees

- Primary
- Services
- Manufacturing
- Construction

Fig. 44.2
Rate of movement of firms
to Cornwall, 1961–1974.

SOURCE: Cornwall County
Council, *Structure Plan:
Report of Survey*
(Truro, 1979).

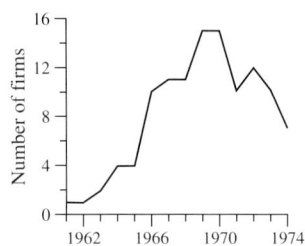

county but always well below a quarter of the total. For Devon, **Map 44.6** updates employment to 1986. Manufacturing accounts for only a small share of jobs except in Plymouth, the only area where it exceeds one-quarter. Even in Plymouth, though, service employment accounts for over half of all jobs.

GOVERNMENT EMPLOYMENT POLICIES AND THEIR EFFECTS IN SOUTH-WEST ENGLAND

Because Devon and Cornwall are fairly remote areas, they have at times suffered from greater rates of unemployment during the twentieth century, as peripheral enterprises have been shut or run down by firms in order to save core activities in London or the Midlands. In order to counteract this trend, the county and other councils have adopted plans to attract employment, and these have often been allied to central government policies to induce employers to move to the region. Cornwall is remote both

apparently and in real terms. Nonetheless in the late 1960s, a boom time in the economy and for regional planning, Cornwall managed to attract up to fifteen firms a year (**Fig. 44.2**). However, not all parts of the county benefited equally from this inward migration of firms; Falmouth and Camborne–Redruth in the west, St Austell and Bodmin in the centre and Saltash in the east were the main beneficiaries. In the Bodmin area the new jobs accounted for over 15 per cent of jobs in industries and services.

Between the 1930s and the early 1980s it was assumed by both central and local government that employers would only move from dynamic to depressed areas of the country in response to government intervention. This normally took the form of a carrot in terms of financial inducements of various kinds and the provision of services and supplies, for example advance factories and trading estates. In the post-war years various sticks were also tried by making it difficult to obtain permission for new premises in the more prosperous regions.

Regional assistance from central government did not arrive

in the South West until 1958 when the so-called Development Areas Training Advisory Committee grants or loans were made available to assist industrial firms.[5] Cornwall benefited most at first from the policy, and then a year later parts of Devon and finally Plymouth were included in the scheme. The scheme was wound up in 1960, however, and replaced by Development Districts with a wider range of powers, grants and subsidies under the Local Employment Act 1960. **Map 44.7** shows that the areas under which various grants, subsidies, advance factories etc. could be provided under the 1960 Act and subsequent acts have changed considerably between 1966 and 1984. In the 1980s there was the shift away from encouraging manufacturing firms to move, which had become increasingly costly in terms of jobs created per pound invested (as industry became more labour extensive), towards encouraging offices and service employment. This policy shift was obviously of some benefit to the South West with its dominant service economy, and also because many opponents of regional policy had always argued that increasing manufacturing would jeopardize a more important part of the area's economy, tourism, by spoiling the attractions of the region. A final change in this decade was the creation of Rural Development Areas by the Development Commission, which, in collaboration with programmes by the individual counties, has provided a new layer of assistance for small firms to set up factory and workshop projects in both the Rural Development Areas and

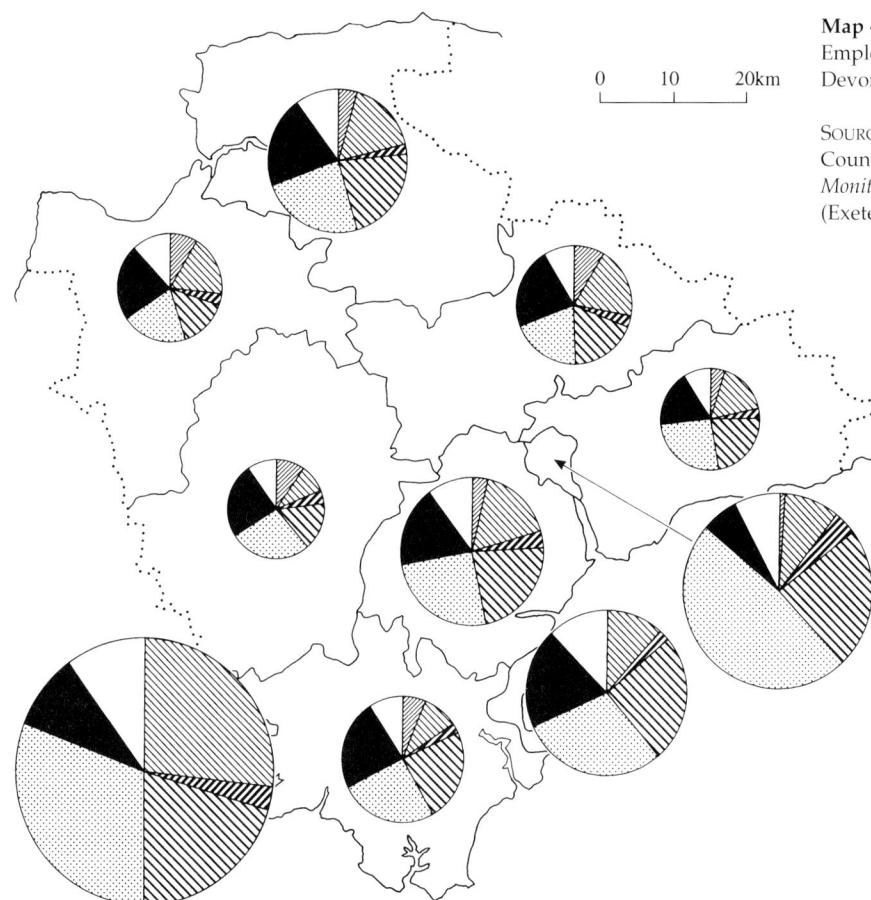

Map 44.6
Employment structure in Devon by district, 1986.

SOURCE: Devon County Council, *Devon Structure Plan Monitoring Report* (Exeter, 1988).

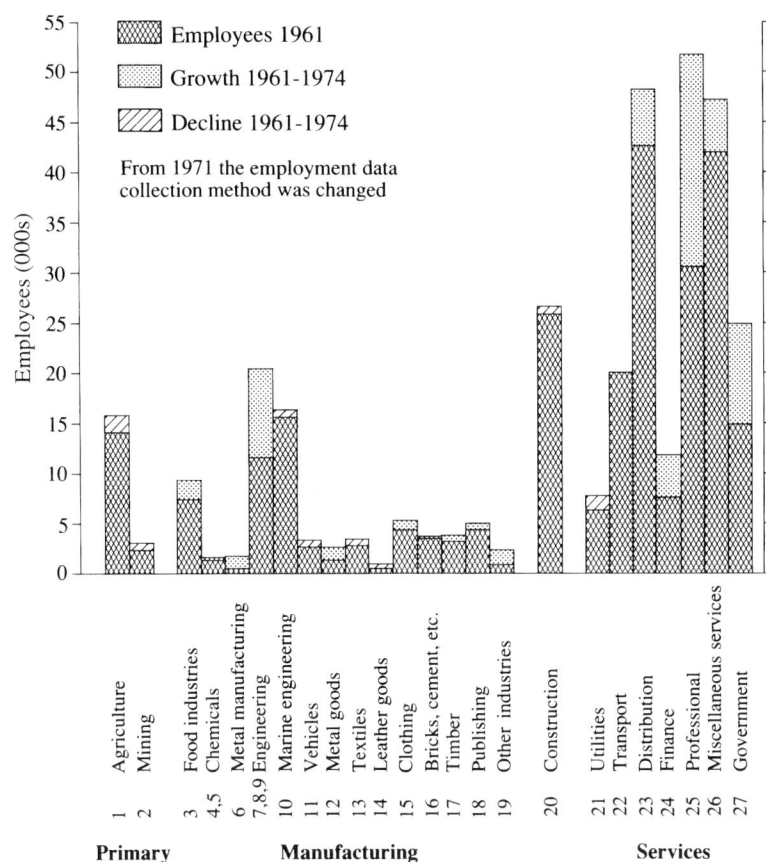

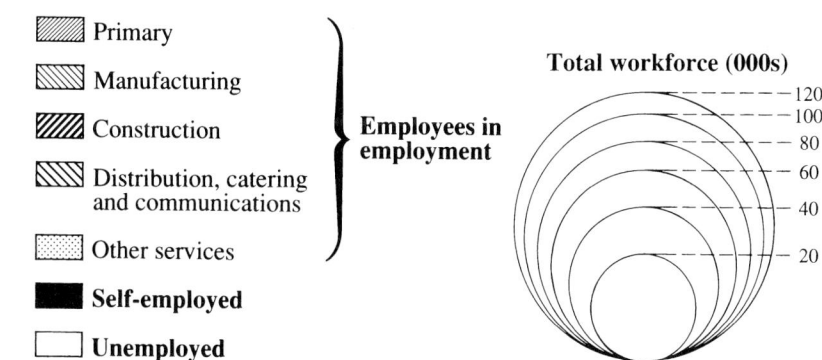

Fig. 44.3
Employment growth and decline by industry in Devon, 1961–1974.

SOURCE: Devon County Council, *Structure Plan: Report and Survey* (Exeter, 1977).

outside. In the 1990s the South West has fared quite well, with lower than average rates of unemployment. However, Devon and in particular the Exeter sub-region have fared best, while Plymouth, with the run-down of the naval dockyards after the end of the Cold War, and Cornwall continue to be unemployment problem areas with some pockets of severe depression.

CONCLUSION

The twentieth century has been a period of steady employment change in the South West, with service employment moving from a majority to a dominant position as both extractive and

(A) 1945-60   (B) 1966   (C) 1967   (D) 1972

(E) 1979   (F) 1982   (G) 1984

■ Special Development Areas

▨ Development Areas (Development Districts in 1966)

▦ Intermediate Areas

0   30km

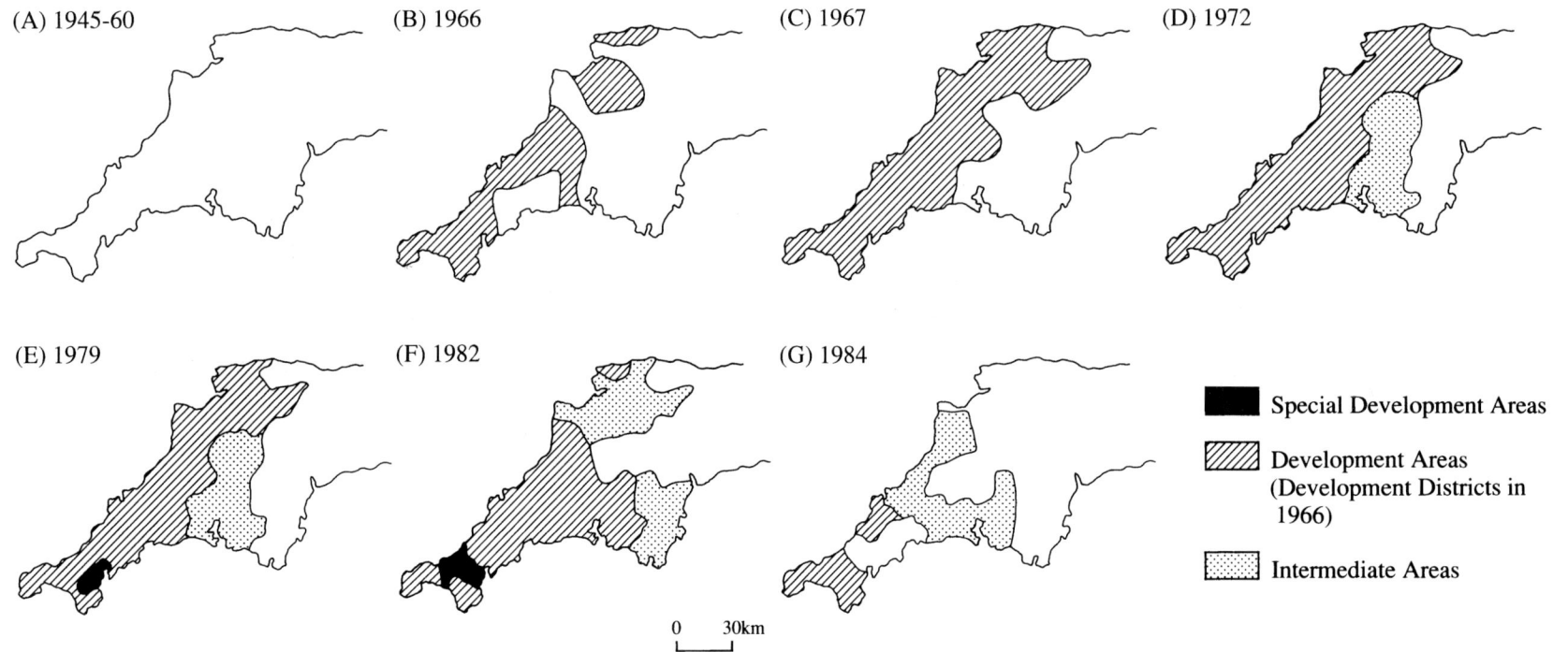

Map 44.7
Regional planning areas, 1945–1984.

SOURCE: R. Hudson and A. Williams, *The United Kingdom* (London, 1986).

manufacturing employment declined. Throughout most of the century, the South West has fared relatively well in employment terms, and, in the service economy of the 1980s, its attractive environment and improving communications have brought it to the fore as a major destination for service relocation, notably the Exe valley where several London firms have relocated. Structural problems do remain. One lesson from the history of this century is that the peninsularity of Cornwall, and its remoteness, combine to make it an unattractive area for employers to relocate from outside the region. However, in an age increasingly dominated by telecommunications, the attractive environment of Cornwall could lead to an employment surge in the next millennium.

FURTHER READING

On twentieth-century rural planning issues in general see G.E. Cherry (ed.), *Rural Planning Problems* (Glasgow, 1976) and specfically on rural employment the chapter in that book by A.W. Gilg, 110–25. See also B. Ilbery (ed.), *The Geography of Rural Change* (Harlow, 1998).

INLAND TRANSPORT

TURNPIKE ROADS

JOHN KANEFSKY

South-west England was not well served with roads before the coming of the turnpikes after 1750. In the middle of the eighteenth century, wheeled vehicles were not common outside towns, the horse being the main means of inland transport over long distances—packhorses for goods, and riding horses for personal journeys. Journeys of any distance were in fact rarely undertaken, except by the wealthy, and many people never left the locality in which they were born. Wherever possible, goods were moved by sea and transhipped to carts for local delivery or to packhorses for destinations farther inland. Few coach routes existed, although there were some reasonably regular local cart services. There were no regular and reliable coach schedules linking even Exeter and Plymouth, let alone destinations further afield such as Falmouth, Truro, Bristol or London. A traffic census in Tiverton over two weeks in 1757 logged only thirty-two wheeled vehicles but nearly 3,000 horses.[1]

What few roads there were serving the longer-distance traveller were in fact little more than bridleways, invariably in very bad condition, especially in winter. These avoided marshy ground, since this was soon churned up and became impassable. Given the terrain of Devon and Cornwall this inevitably meant frequent steep climbs, which further restricted the usefulness of wheeled vehicles and the rate of progress they could achieve.[2]

A.K. Hamilton Jenkin summed up the state of roads in Cornwall at this time as 'for the most part in their primitive state, still following the highland ridges, as in ancient times, and descending only to the valleys by breakneck hill where some river or creek had of necessity to be crossed'.[3] The first recorded coach in Falmouth, for example, was in the late 1760s, and it was not until about 1790 that there was a scheduled coach from there to Exeter via Bodmin. Similarly, the first stagecoach from the Torpoint ferry in Plymouth to Exeter was not introduced until 1762. The Falmouth–Exeter journey of about 100 miles was supposed to take twenty-one hours, but in fact usually took two days, with an overnight stop in Bodmin, until well into the nineteenth century. In 1750 the typical journey time from London to Exeter was three days.[4] The flight of Dona St Columb from London to Launceston—some 200 miles—in two days in the 1660s in Du Maurier's *Frenchman's Creek* seems improbable, despite all the guineas Sir Harry had paid for the horses! Even in 1808, when the first London–Penzance mail was introduced and most of the journey was on improved turnpike roads, it took three full days with numerous changes of horse to cover that journey.[5]

The principal roads in 1750 shown on **Map 45.1** should, therefore, be regarded with some caution: the designation of a 'road' at that time does not necessarily convey the same meaning as the modern concept, particularly in the winter months. Celia Fiennes oft-quoted description of Devon lanes as 'exceeding narrow … so difficult that one could scarcely pass by, even the single horses' referred to the road from Exeter to Plymouth, and there is no evidence that things had improved much by the mid-eighteenth century.[6] All that can be reliably said of the road network in 1750 as delineated on contemporary maps and described in literature is that it encompassed the least impassable roads over which coaches and carts could sometimes pass.[7]

TURNPIKE TRUSTS

From the middle of the eighteenth century, however, a gradual process of improving the roads in the South West began with the introduction of the first turnpike trusts. Such trusts had begun to be introduced in the more populous parts of England in the late seventeenth century, and by 1750, 146 such Acts had been passed, mostly in the south and west of England, but none had yet reached the South West.[8]

Turnpike trusts were very much local initiatives, whereby a group of people sponsored an Act of Parliament to maintain and improve one or more roads by levying tolls and using the money to pay for work on the roads. Although the trusts were (loosely) overseen by Parliament, they were never controlled or planned from London; the development of the turnpike road system into a national network was more by accident than design. Once established, the trusts began to improve the roads and sometimes to construct new ones, although it generally took many years for all the mileage under their control to be developed.

The earliest trusts in the South West were principally those turnpiking the pre-existing roads around the towns: Exeter, Axminster, Honiton, Ashburton,

Map 45.1
Principal roads, 1750.

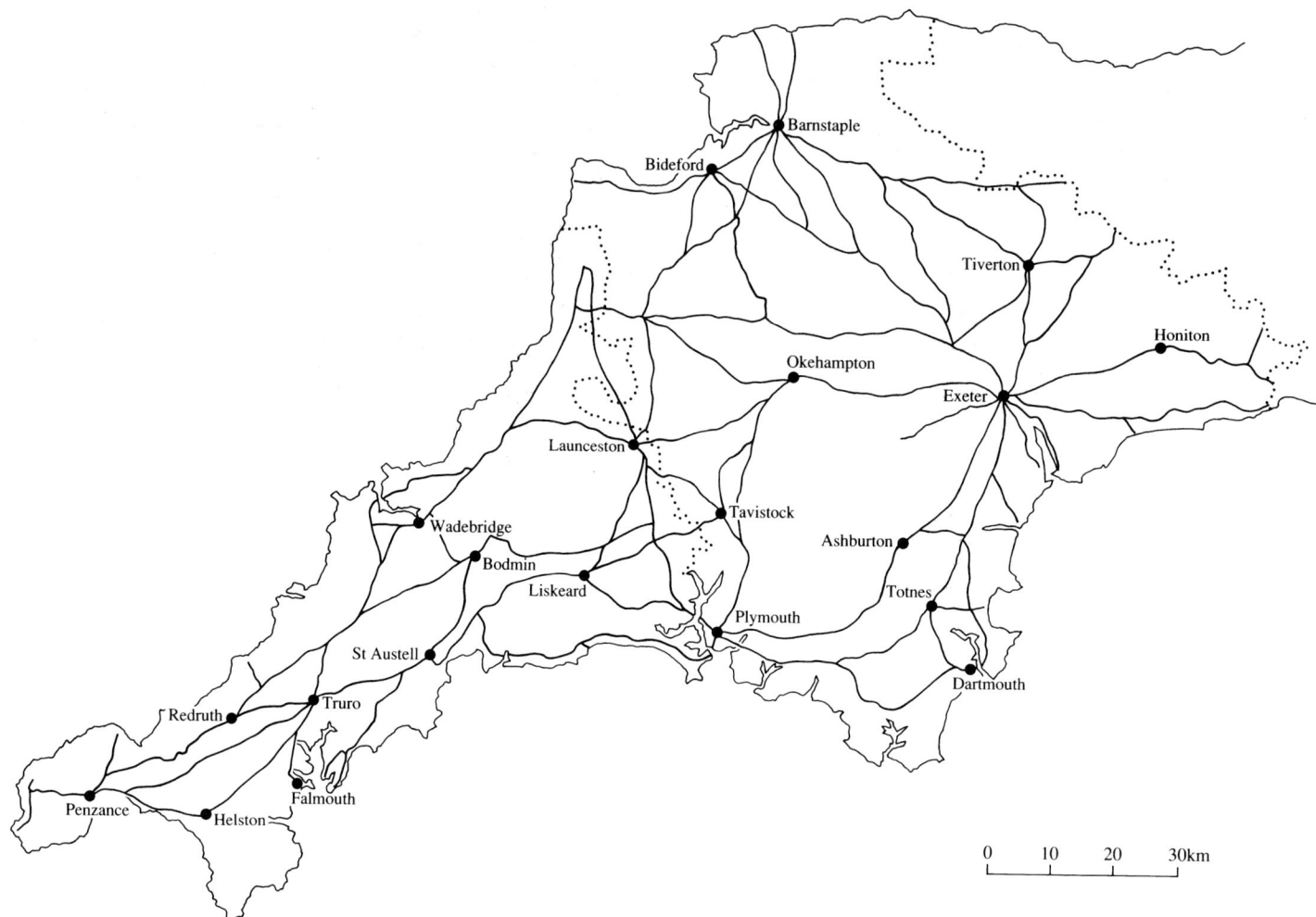

Totnes, Plymouth and Tiverton all had Acts between 1753 and 1758, and by 1772 there were twenty turnpike trusts in Devon, plus portions of roads from three Cornish and four Somerset and Dorset trusts. Many of the early Devon trusts, unusually, covered a web of roads radiating from a town and were designed to serve local markets; benefits to longer distance traffic were a secondary consideration.[9] The first trust in Cornwall, however, was unsurprisingly to turnpike the road between Truro and Falmouth, the main port for the packet ships bringing mail from Spain, America and Africa, in 1754. Thereafter there was a gap until the 1760s, when ten more trusts were started covering the main towns. By 1770 the principal long distance routes in Cornwall were covered by turnpike trusts, including a new route to the Torpoint ferry, although they stopped short of Penzance. Some earlier routes, such as that from Mitchell to Lands End via Redruth, became virtually disused.[10]

This burst of enthusiasm for setting up turnpike trusts, which was mainly driven by a wish to improve existing roads and make

them at least passable to wheeled traffic in all but the worst of the winter weather, culminated with the commencement of two new roads to replace what had previously been very doubtful tracks, those across Dartmoor in 1772 and Bodmin Moor in 1769, although it took many years for both to be completed because of the difficult terrain. Thereafter there was a period of consolidation, during which only one new trust—Stonehouse in Plymouth in 1785, turnpiking the roads through the town to the Cornish ferries—was established, although several new lines of road were added to existing trusts and some new roads were constructed, principally short lengths to cut off the worst gradients or to straighten bad bends.[11]

Tolls were high in real terms—up to a shilling (5p) for a coach and a penny (0.4p) for a horse were typical, in a period when the tollgate keeper's wages were normally less than 10 shillings a week. Foot travellers did not pay tolls—it would have been impossible to enforce them—but payment had to be paid for animals on their way to market. Tolls were usually payable only once a day, and covered all nearby gates. Given that any lengthy journey attracted several tolls, it is not surprising that coach travel was expensive, at about £2.00 for the journey from Exeter to London, plus the cost of meals etc. *en route*—perhaps a month's wages. Just to go from Exeter to Plymouth cost about a week's wages for a labourer.[12]

As **Map 45.2** shows, there were nonetheless some significant gaps in the turnpike road network in 1800. In Cornwall, improved roads had not yet linked Redruth, Camborne and Penzance, nor Bodmin and St Austell. In Devon, large parts of the south-east coast and the South Hams were untouched by the turnpikes, and there were many gaps in the cross routes between towns. The sparsely populated area to the west of Okehampton and north of Launceston was also devoid of turnpike roads, and remained so throughout the nineteenth century; there was simply not sufficient wheeled traffic to prompt local people to sponsor turnpike acts.

Fig. 45.1
Detail from an engraving by L.E. Reed of Teignmouth showing a carriage and horses, n.d.

SOURCE: reproduced by permission of Devon Record Office (1919Z/Z1/p. 55).

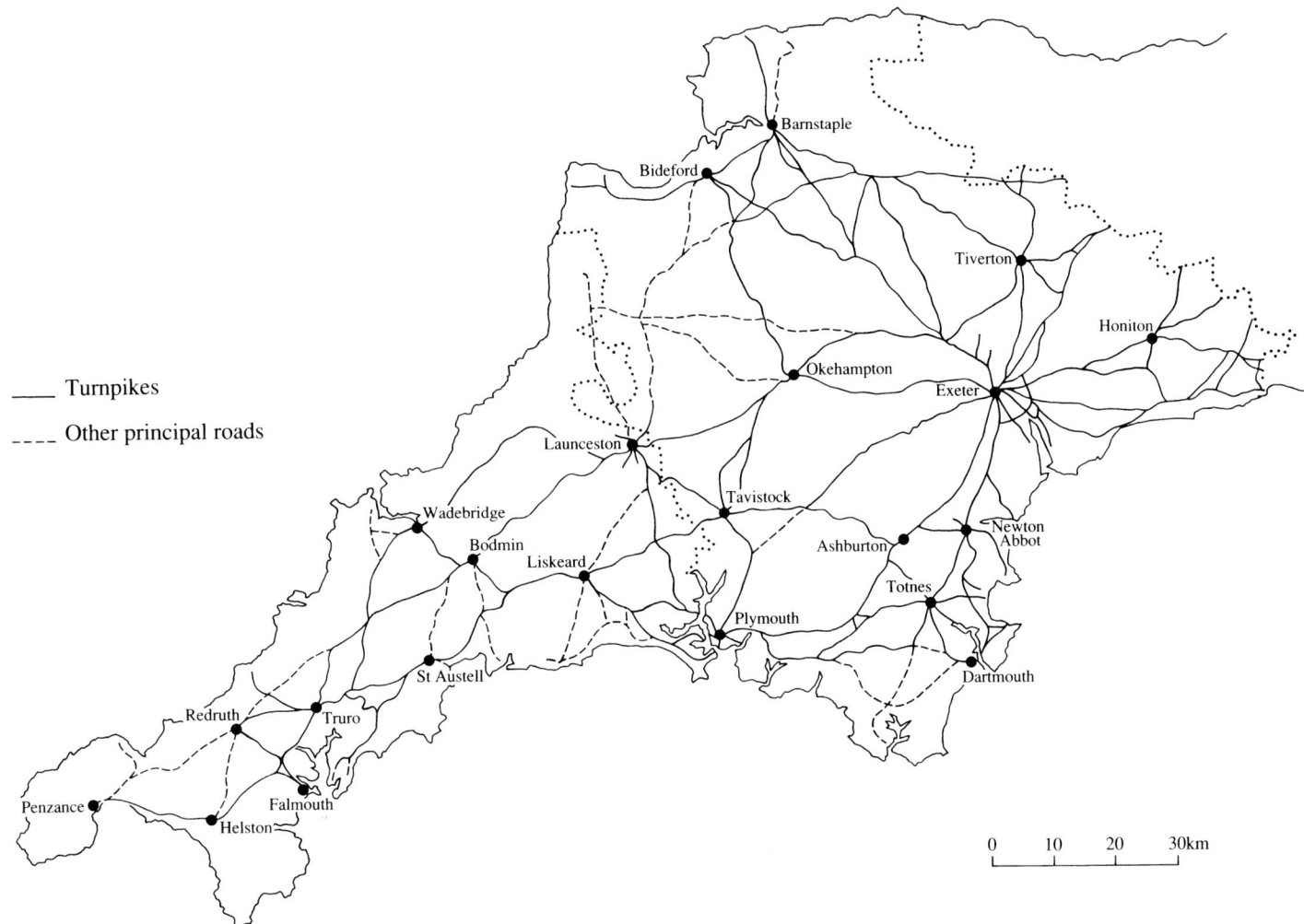

_____ Turnpikes

_ _ _ _ Other principal roads

Map 45.2
Turnpike roads, *c.*1800.

Map 45.3
Turnpike roads in the
nineteenth century.

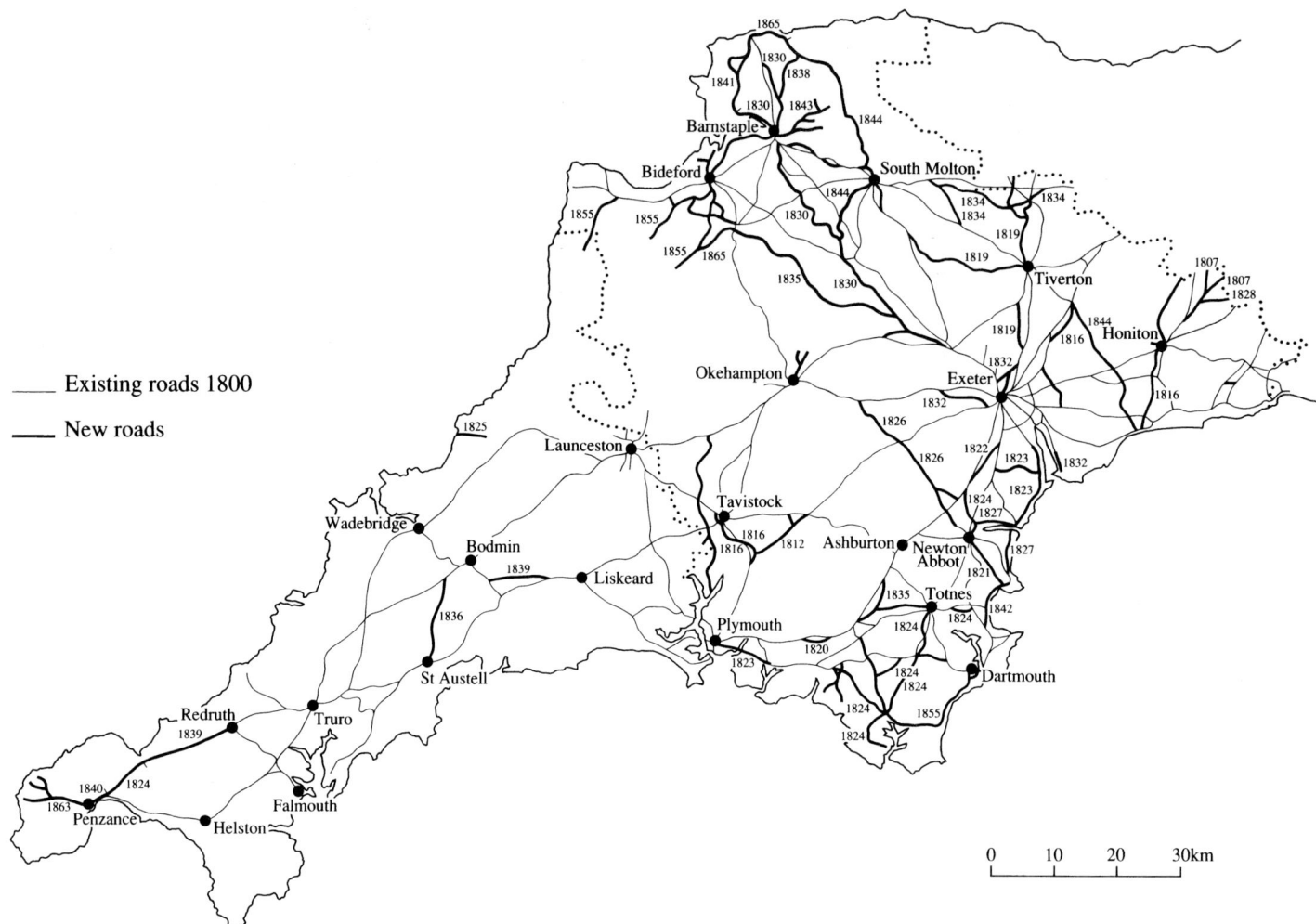

Existing roads 1800

New roads

NINETEENTH-CENTURY IMPROVEMENTS

After 1800 there was a new impetus to formation of turnpike trusts, and later the improvement of existing trusts by the creation of new routes. The Napoleonic Wars created new demands for road improvements, and in the distressed economic conditions that followed the end of the wartime boom in 1815 there was both the perceived need to provide new roads to stimulate the economy and an abundant supply of labour to work on them.[13] As important were advances in the technology and organization of road-building which provided an undeniable boost to the expansion of the turnpike network. As has already been noted, the early turnpikes mostly followed old routes, which avoided the river valleys wherever possible because of the difficult winter conditions. New road-building techniques and materials introduced in the early nineteenth century remedied this problem, however, particularly the system introduced by John McAdam and rapidly adopted

throughout Britain. His son William also became surveyor to several Devon trusts from 1820 onwards.[14]

The principle and practice of making roads from layers of well-graded materials with proper drainage ditches, together with improved methods of bridge-building, levelling and surveying, meant that the river valleys were no longer an impediment to road travel. Because of their more even gradients they became the preferred routes, allowing larger, better-sprung vehicles with fewer horses, substantial improvements in journey times, and making carts rather than packhorses the most effective way of transporting most goods. A horse could pull about five times the load in a cart as it could carry on its back.[15]

Beginning in 1807, therefore, eleven new turnpike trusts were created in Devon and four in Cornwall in the period up to the mid-1840s. There was also one very late-starting trust in Cornwall, from St Just to Penzance in 1863—railways never reached west of Penzance. Most of these new trusts were primarily designed to fill in gaps in the coverage of the earlier

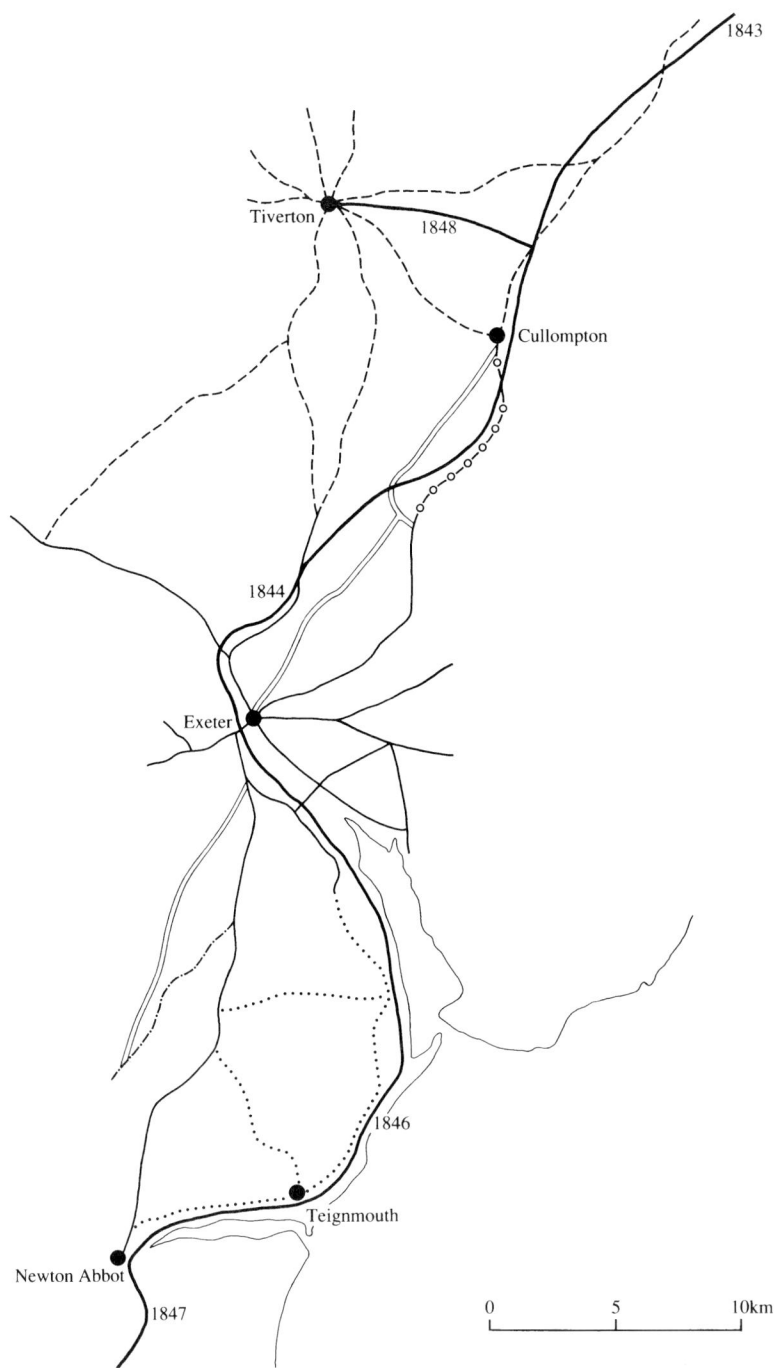

Fig. 45.2
Income of turnpike trusts in east Devon, 1830–1865.

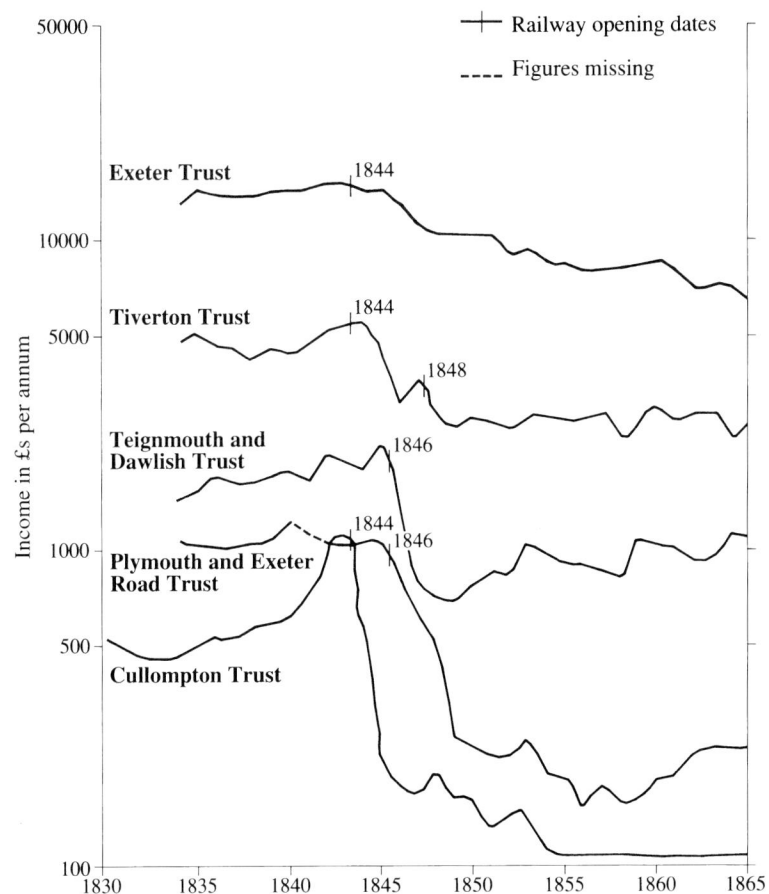

Map 45.4
Turnpike roads and railways in east Devon, 1843–1848.

trusts, particularly in east, north and south-east Devon (where tourism was building up in places such as Dawlish and Exmouth) and between Redruth and Penzance in Cornwall. One curiosity was the Trebarwith Sands Trust, commenced in 1825, which turnpiked a country lane from a small beach south of Tintagel to a point a couple of miles to the east but did not carry on to the road from Launceston to Wadebridge; its toll income was less than £150 a year.[16]

Some of the new trusts were to build new roads to replace bad routes, most notably the new roads from Exeter to Cullompton avoiding Stoke Hill in 1816 and that over Haldon Hill in 1822–3. Other new roads were built along the river valleys by existing trusts to replace hilly routes, some of which were later 'disturnpiked'. The most important of these were the roads to Barnstaple (1830), Bideford (1835), various improved stretches of the Exeter–Plymouth route between Ashburton and Lee Mill, the Laira Bridge road east of Plymouth in 1823, and the new Liskeard–Bodmin road through the Fowey valley in 1839, which was described as the fastest stretch in Cornwall. The South Molton Trust also built a completely new road to Combe Martin in 1844, and as late as 1865 a new road was built linking the latter with Ilfracombe, after the opening of the railway (**Map 45.3**).[17]

Map 45.5
Exeter turnpike gates, *c.*1850.

SOURCES: John Kanefsky,
Devon Tollhouses (Exeter, 1984);
W. Buckingham,
A Turnpike Key (Exeter, 1885).

Fig. 45.3 (right)
Shaldon Bridge Tollhouse,
Teignmouth, *c.*1943.

SOURCE: author's collection.

Fig. 45.4 (below)
Okehampton Gate,
Exeter, 1884.

SOURCE: Mark Searle,
Turnpikes and Toll-bars
(London 1930).

COMPETITION FROM RAILWAY COMPANIES

All this activity was, however, overshadowed by the coming of competition from the railways. The new roads and better vehicles with innovations such as improved springs were reducing journey times considerably—many were halved by the 1830s compared to eighty years earlier—but they could not hope to match the ability of the railways to provide speedy and cheap transport for both passengers and goods. The Bristol and Exeter Railway reached the Devon/Somerset border in 1843 and progressed rapidly: to Exeter in 1844, Plymouth in 1849 and, once the Royal Albert Bridge over the Tamar had been opened in 1859, linked up the existing Cornwall Railway and West Cornwall Railway into one main line connecting Penzance to London.[18]

Some of the Devon trusts were very hard hit by the railway. After a period while construction traffic swelled their coffers, the railway took traffic away from trusts in direct proportion to the degree to which it paralleled their routes (**Fig. 45.2**). Thus the income of the Cullompton Trust, whose line was followed exactly, went down from £500 a year to just over £150 (**Map 45.4**), and much the same happened to the Plymouth and Exeter Road Trust, whereas the Exeter Trust (**Map 45.5**), with a web of roads radiating from the city, only lost a quarter of its tolls, from £14,000 to £10,000 per annum.[19] The effect was much less pronounced in Cornwall, where journeys were more local and feeder traffic presumably made up to a greater extent for the loss of the longer-distance business. Only the Liskeard Trust suffered a severe drop in income, from approaching £3,000 to just over £1,500 a year. Other, more rural trusts in both counties were barely affected until branch lines reached them in the 1860s and 1870s, and it has already been noted that new roads were built in some places to feed the railheads.[20]

By the 1870s it was nonetheless clear that the turnpike trusts had outlived their usefulness, and legislation was enacted which provided for local administrations to take over their duties. Starting with the Okehampton Trust as early as 1863, trusts were wound up one by one and the new county councils took over their function. The Cornish trusts were mostly ended between the late 1870s and mid-1880s; the last in Devon lingered until 1889. Their property was sold off, but as most had substantial debts there was little to pass on to the highways authorities apart from land and tollhouses (**Figs 45.3 and 45.4**).[21] Most local people did not mourn the passing of the trusts, whose tolls were generally regarded as iniquitous, but they had performed a substantial service to the infrastructure of the South West for over a hundred years, at a time when there was effectively no national government body which could have performed the function they exercised.

INLAND TRANSPORT

CANALS AND RAILWAYS
IN THE
NINETEENTH CENTURY

RICHARD OLIVER

CANALS IN THE SOUTH WEST (Map 46.1)

Inland navigation was never of more than modest importance in the South West. Two categories can be distinguished: canals built for short-haul traffic down to tidal water and canals intended to form a part of a national network.

The Exeter Canal, built in 1564–6, was the first artificial waterway for transport purposes to be built since Roman times, and was intended purely to improve access to Exeter from the sea. In this it was less than wholly successful, as extensions to its length and widening and deepening were carried out in 1676 and 1825–7. The canal then assumed its present form as a ship canal bypassing Topsham and the upper tidal part of the Exe. In one sense it was too good for its purpose, as the lower Exe estuary remained unimproved, and so the ship canal never reached its full potential. For all that, it is the only canal in the South West to remain in full working order.

The next canal to be built in the South West was a short one near St Austell, serving clay pits and mostly in tunnel. It came to a premature end in 1731, after about eleven years' service, when the tunnel collapsed. More successful, and built for a similar purpose, were the Stover and Hackney canals near Newton Abbot, which remained in use into the twentieth century. The Tavistock Canal, completed in 1817, connected that town with Morwellham Quay on the Tamar, the considerable difference in level being overcome by an inclined plane down to the quay. The necessary transhipment increased operating costs, and the canal's days were numbered once Tavistock gained its first railway, in 1859. The coming of the railway also closed the Torrington Canal, which was built because of the enthusiasm of a local landowner, Lord Rolle (Map 46.2).

Possibly the most successful of all the south-western canals was the Liskeard and Looe Union Canal, opened in 1827–8; indeed, it was a victim of its own success. Since 1844 traffic had reached it at the Liskeard end by a railway from the tin mines beyond, and in 1860 the railway was continued down to Looe. The original idea was that the railway would relieve congestion on the canal, but in the event use of the canal quickly dwindled; there was no point in a needless transhipment of goods from railway to canal at Liskeard.

The most ambitious of the local canals in the South West was the Bude Canal, built in the 1820s partly to facilitate agricultural development by bringing sea-sand from the coast to the uplands inland for manure. In this it was fairly successful, though the system was never completed, being intended to reach Holsworthy and Launceston. The inclined planes on which the boats were raised and lowered also gave trouble, and no doubt contributed to most of the system closing down in 1891, seven years before Bude was reached by railway.

There were two features in common between the Bude Canal and the Grand Western Canal: neither was completed as originally envisaged, and both used mechanical rather than hydraulic means to raise and lower boats from one level to another. The Grand Western was a far less successful canal than was the Bude, but it was the only canal in the South West which was built with a view to forming part of a regional or national network rather than with the limited aim of serving a particular place or industry. There were numerous schemes canvassed from the 1790s onwards for canals to connect the Bristol Channel and the English Channel, some of which were for conventional 'narrow' canals and some—the last of which was promoted as late as 1869—for ship canals. The two main ship canal routes were from Bridgwater through Taunton, Chard and Axminster to Seaton and from Bridgwater through Taunton and Wellington, and thence along the Culm valley to Topsham. Taunton was served by both the navigable River Tone and a canal to Bridgwater, and there was a prospect around 1800 of its being connected by inland waterway to Bristol and thus to the national canal network. The Grand Western scheme, first authorized for a canal from Taunton to Topsham with a branch to Tiverton, was on paper sound and practicable, but it did not prosper. The Tiverton branch was completed first, and opened in 1814; the section onwards to Taunton was only completed in 1837, by which time the Bristol and Exeter Railway, which ran very close to the canal, had been begun. The railway was completed in 1844, and its Tiverton branch in 1848. The railway company's purchase of the Grand Western in 1867 was the logical outcome. The eastern part of the canal was closed immediately; the western part continued to carry some local traffic into the twentieth century.

THE DEVELOPMENT OF THE RAILWAY NETWORK (**Maps 46.3 and 46.4**)

The north-east of England is often treated as 'the cradle of railways', but South Wales and Cornwall also have claims to be considered. Richard Trevethick, who built the first working steam railway locomotive, was a Cornishman, though it was South Wales which first had the benefit of his invention. In the early nineteenth century Cornwall was both relatively and absolutely more of an 'industrial' area than it is today, and thus the sort of place where railways might be expected to develop, particularly as the terrain was not at all favourable for canal-building. Much the same applied in South Wales and the North East, and in all three areas the early railways were built for conveying minerals from inland mines to the nearest sea port. There was no thought of them forming part of a larger regional or national network. Thus Cornwall saw its first railway in 1812, but was only connected to the national network comparatively late, in 1859, whereas predominantly non-industrial Devon's first railways appeared in the early 1820s, on the fringes of Dartmoor for quarry traffic, and the county was penetrated by a 'main line' railway, the Bristol and Exeter, by 1844. Even in the last third of the nineteenth century, short hauls of minerals to nearby ports remained important in Cornwall, and the extensive Cornwall Mineral Railways system was built in 1873–4.

One reason which explains in part why Devon was connected to the national network before Cornwall is the nature of the terrain. Railway-building was much easier and, therefore, cheaper east of Exeter than to the west; the Bristol and Exeter had merely to follow the Tone and Culm valleys, whereas the South Devon Railway, which ran onwards to Plymouth, had first to cling to the seashore (scenically splendid, but an engineering headache), and then clamber up and down the hills of south Devon. Admittedly, the gradients were rather steeper than they might have been, as it was intended that the line would be operated by atmospheric propulsion, which offered better hill-climbing possibilities than did steam operation. This proved a practical failure and left the steep gradients as its most tangible memento. Worse was avoided by circumventing Dartmoor, at the cost of increasing the distance between Exeter and Plymouth by a quarter. Once the Tamar had been crossed by the Royal Albert Bridge, the Cornwall Railway presented a continuous succession of cuttings, embankments, viaducts and steep gradients all the way to Truro, where it joined a network of previously constructed lines. Cornwall thus became the last English county to be connected to the national railway network, and Penzance town clock was one of the last to be set to London rather than to local time.[1] Even today, the fastest trains take nearly half as long again to cover the 132 miles from Exeter to Penzance as they do to cover the 173 miles from London to Exeter. Exeter gained its second main line, from London *via* Salisbury and Yeovil, in 1860; the problem of finding a route through the hilly terrain of east Devon contributed to the long delays in building this line, first proposed in the mid-1840s. This

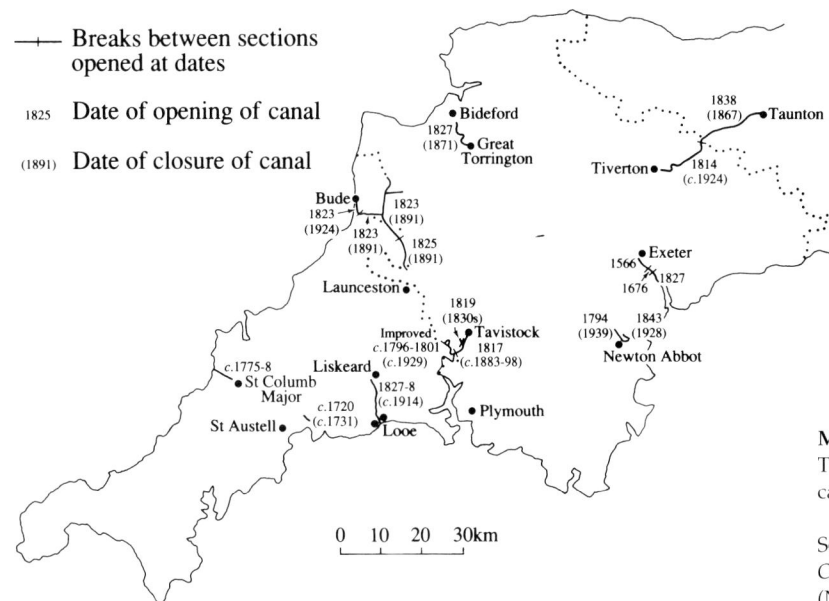

Map 46.1
The growth and contraction of canals.

SOURCE: Charles Hadfield, *Canals of South West England* (Newton Abbot, 1967).

rival route pushed westwards and eventually reached Plymouth in 1876. Thus whilst the South West had railways almost before the railway age had begun, they developed much more slowly than in most other parts of Britain. It was one thing for a railway to be desirable for local or regional economic development, which was the motive behind most of the railways which were actually built; it was another to engineer a route and raise capital, and the

Map 46.2
Rolle canal, Torrington.

SOURCE: reproduced by permission of Devon Record Office (DQS, 91/3).

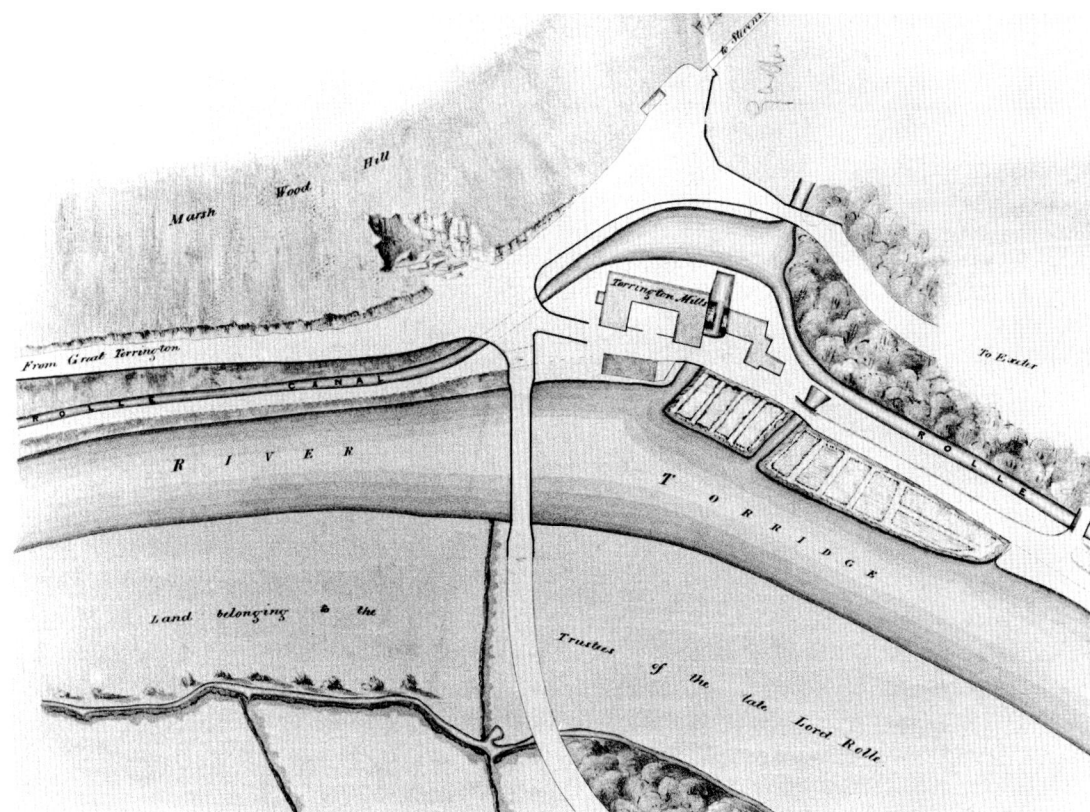

Map 46.3
The development of
the railway network,
1812–1925.

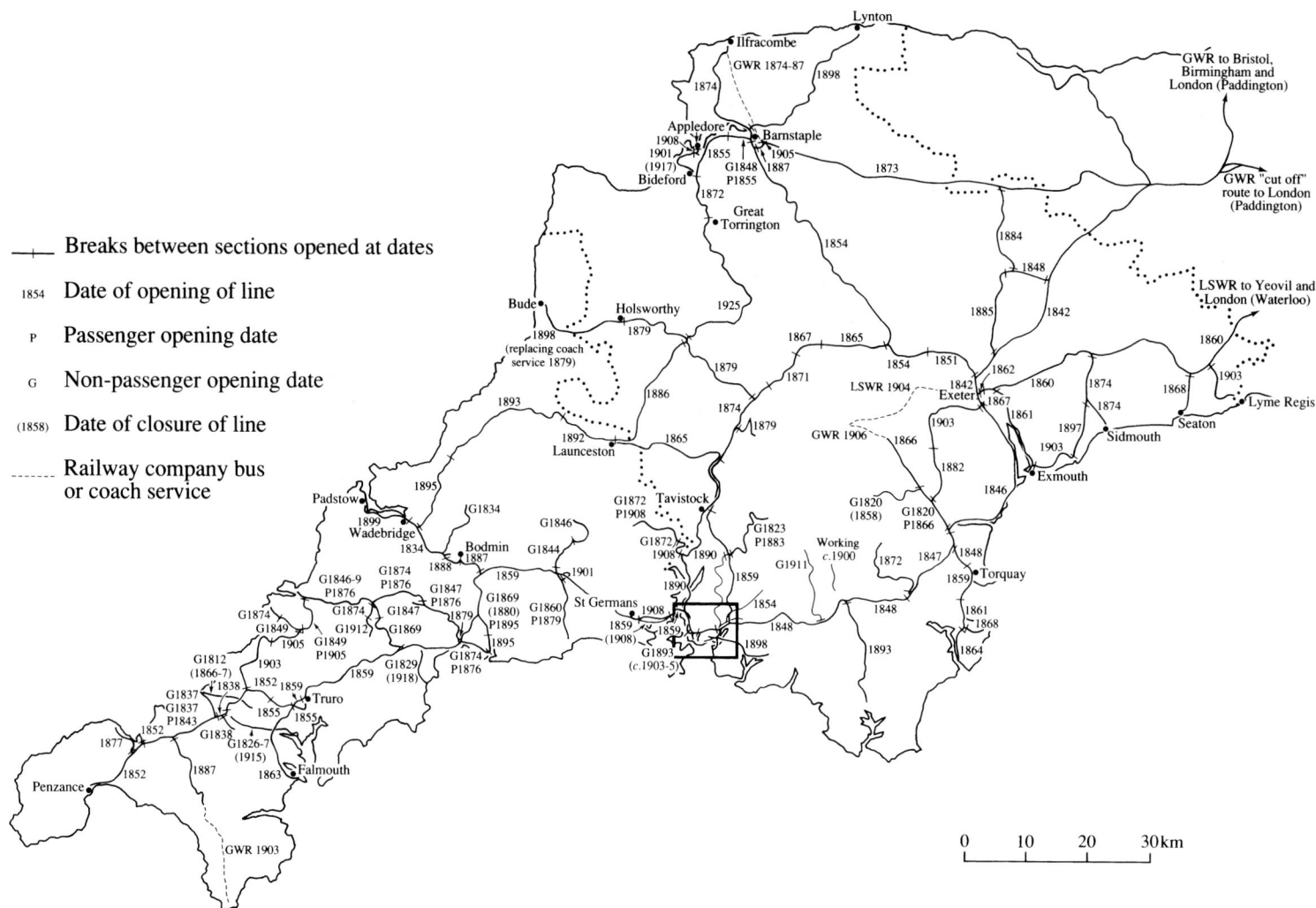

Breaks between sections opened at dates

1854 Date of opening of line

P Passenger opening date

G Non-passenger opening date

(1858) Date of closure of line

Railway company bus
or coach service

0   10   20   30km

raising of capital was particularly difficult in the late 1840s and early 1850s, when the Cornwall Railway and the Salisbury route to Exeter were endeavouring to get started.

Terrain also influenced the building of branch lines. The branches to Torquay (opened 1848), Bideford (1855), Exmouth (1861) and Seaton (1868) were mostly across level country and easy to construct; those to the smaller towns of Ilfracombe (1874), Sidmouth (1874), Lynton (1898) and Lyme Regis (1903) were all steeply graded, and all suffered the additional disdavantage that the terminus was several hundred feet above the town, which was only tolerable as long as there was no alternative means of mechanized transport.[2]

BROAD AND STANDARD GAUGES; SINGLE AND DOUBLE TRACKS

From the 1840s until 1892 the railways of the South West were built to two gauges: the standard gauge of 4ft 8.5in (1.435m) and

the broad gauge of 7ft 0.25in (2.140m).[3] The standard gauge was used for most of the early lines in west Cornwall, but it was the broad gauge which first connected Devon and Cornwall to the outside world. It was the brainchild of Isambard Kingdom Brunel, who claimed various advantages for it; it was adopted as standard by the Great Western Railway and its three allies, the Bristol and Exeter, the South Devon and the Cornwall companies, which it absorbed in 1876. The London and South Western Railway, which took over the Bodmin and Wadebridge Railway in 1847, and reached Exeter by way of the Salisbury and Exeter company in 1860, used the standard gauge, and could thus offer the advantage of being able to convey traffic to and from any part of the country without transhipment. The Great Western could not, and whilst this could be mitigated by laying a standard gauge inner rail on broad gauge lines, producing a 'mixed gauge' (or, in west Cornwall, adding a broad gauge outer rail to an existing standard gauge line, as between Truro and Penzance in 1866), this created operating problems (running broad and standard gauge stock in

Fig. 46.1
The railway line near Teignmouth breached by a storm in March 1855, printed in the *Illustrated London News* from a drawing by F.W.L. Stockdale.

SOURCE: reproduced by permission of Devon Record Office (1919Z/Z1/p. 84).

the same train was hazardous) and the broad gauge was finally abolished in 1892, over a single weekend in which 132 miles of railway in Cornwall were 'narrowed' in one of the most remarkable of British civil engineering operations.

Although to its devotees the abolition of the broad gauge in 1892 may have been a retrograde step, it did facilitate the upgrading of the main line of the former Cornwall Railway. Almost all the South West's railways—the Bristol and Exeter and the direct Tavistock–Plymouth line of 1890 are the main exceptions—were built as single track, in order to keep down construction costs. Single-track main lines were difficult to operate, particularly when steeply graded and carrying significant freight traffic, and in easier country the South Devon and Cornwall railways would almost certainly have been built as double-track from the start. As with the basic expansion of the system, terrain hampered the work. For example, the Exeter–Salisbury route was converted from single to double track by 1870, within ten years of its opening, but doubling of the former Cornwall Railway only began in 1893 and was only completed belatedly in 1930. Doubling involved more than just earth-moving as bridges and viaducts had to be reconstructed. Both the South Devon and the Cornwall railways were initially furnished with timber viaducts, which were cheap to construct, and therefore kept down initial capital costs, but expensive to maintain, and thereby depressed subsequent dividends. The last of them, on the Falmouth branch, was replaced in 1934. Timber viaducts were common in North America and were used

occasionally elsewhere in Britain, but by far the greatest concentration was in the South West.

The London and South Western may have reached Exeter in 1860, but it was 1895 before its outpost at Wadebridge was connected to the main system, and 1899 before it reached its westernmost outpost, at Padstow. Its system in north Cornwall

Map 46.4
The development of the railway network in the Plymouth area outlined on **Map 46.3**.

SOURCE: as **Map 46.3**.

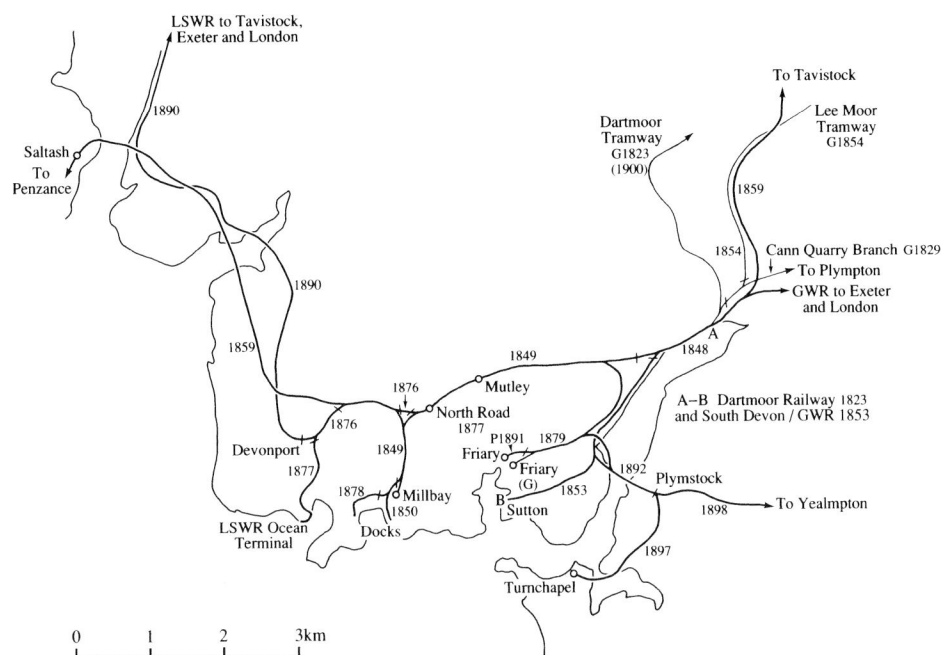

was built, and remained, as single track; holiday traffic was not sufficient to offset the much more meagre traffic offering for the other eight or nine months of the year. It had reached Plymouth, with the help of running powers over the South Devon/Great Western branch from Tavistock, in 1876, and thereafter Plymouth as well as Exeter had two routes to London. The London and South Western service was usually slightly slower than the Great Western's, but it had the advantage at Exeter of a more conveniently situated station (Queen Street, now Exeter Central Station) for the city centre.

EXPANSION ENDS AND CLOSURES BEGIN

The railway system continued to expand even after 1900, when it was first threatened by mechanized road transport, and the South West's railway network only reached its maximum extent in 1925, with the completion of the Torrington–Halwill Junction line. This had been proposed in 1907, was eventually built in 1922–5 as an unemployment relief measure, and was something of an anachronism even then, when, elsewhere, the retreat of the railway network had begun. A number of the early railways had closed some time before—the earliest of all, the Poldice Railway, had closed in the 1860s—and 1917 saw the closure of the Bideford, Westward Ho! and Appledore Railway, mostly opened in 1901 but only completed in 1908. This line was curious in that, though worked by steam, it was owned by a larger company whose main interests were in electric tramways, and it was perhaps the ultimate example in the South West of the influence of terrain. Connecting Bideford to Appledore by way of Westward Ho! ensured a fairly easy route, but at the expense of covering nearly three times the direct distance.

An alternative after 1900 to building a new railway was a motor bus service. As **Map 46.3** shows, some of the later branch lines, such as those to Bude and Lynton, had been preceded by horse bus services run in conjunction with one of the main line railway companies. In 1903 the Great Western Railway started a motor bus service from Helston to the Lizard as an alternative to

building a new length of railway. It was the first rural motor bus service in Britain and, by 1914, a number of similar services were operating.

FURTHER READING

For canals this chapter draws largely on Charles Hadfield, *The Canals of South West England* (Newton Abbot, 1967), and also E.A.G. Clark, *The Ports of the Exe Estuary 1660–1860* (Exeter, 1960). A recent study which explains the development of the national rail network is David Turnock, *An Historical Geography of Railways in Great Britain and Ireland* (Cambridge, 1998). The literature on railways in the South West is extensive, and there is only space here to mention some of the standard texts, to which the present chapter is indebted. The leading example remains David St John Thomas, *A Regional History of the Railways of Great Britain, Vol. I, The West Country* (Newton Abbot, 1966).

Those seeking greater depth of coverage are referred to several important company histories: E.T. MacDermot and C.R. Clinker, *History of the Great Western Railway*, 2 vols (London, 1964); C.F. Dendy Marshall, revised by R.W. Kidner, *A History of the Southern Railway* (London, 1968); R.A. Williams, *The London and South Western Railway*, 2 vols (Newton Abbot, 1968, 1973); J.W. Faulkner and R.A. Williams, *The L.S.W.R. in the Twentieth Century* (Newton Abbot, 1988); and C.R. Clinker, *The Railways of Cornwall 1809–1963* (Dawlish, 1963). Examples of the numerous more local histories are R H. Gregory, *The South Devon Railway* (Tisbury, 1982); M.J.T. Lewis, *The Pentewan Railway, 1829–1918* (Truro, 1960); Philip Payton, *Tregantle and Scraesdon: Their Forts and Railway* (Redruth, 1988); and Edwin Welch, 'Camm Quarry canal and railway', *Report and Transactions of the Devonshire Association*, 100 (1968), 111–23. Studies of the effects of the railways include G. Finch, 'Railways and the balance of trade in Victorian Devon', *Devon and Cornwall Notes and Queries*, 36 (1988), 85–8, and J.V. Somers-Cocks, 'The Great Western Railway and the development of Devon and Cornwall', *Devon and Cornwall Notes and Queries*, 36 (1987), 9–20.

INLAND TRANSPORT

RAILWAYS AND ROADS IN THE TWENTIETH CENTURY

MALYN NEWITT

RAILWAYS (Map 47.1)

The railway network of the South West has evolved over a period of nearly 200 years and with the opening of the Channel Tunnel in 1994 with a terminus at Waterloo it is clear that its evolution is far from over.

The coming of the motor car and motorbus at first stimulated a new phase of railway development with the Great Western Railway (GWR) building a large number of unmanned halts and small manned platforms to try to make the railway more responsive to local transport needs. Altogether the company built 420 halts and platforms between 1903 and 1947.[1] These were often served by rail cars and many of them had an ephemeral life. The rail companies were also among the pioneers of bus travel. In 1903 the GWR decided to run a motorbus service from Helston to the Lizard instead of building a branch line and from that time developed as a major bus operator, introducing double deckers on rural routes as early as 1908.[2]

During the First World War railways were taken under government administration and the advantages of operating a single national network were made apparent. Although it was eventually decided to return the companies to private ownership in 1921, it was on condition that they amalgamated into four large groupings. The railways in the South West were already effectively divided between two big companies, the GWR and the London and South-West Railway (L&SWR). After 1923 this division continued, with the GWR running the mainlines from Paddington and the railways in south Cornwall and south Devon and the new Southern Railway running the Waterloo to Plymouth mainline and the railways in east and north Devon and north Cornwall.[3]

In the twentieth century no major new lines were opened in the South West, though two light railways were built, the Bideford to Appledore line being opened in 1908 and the Torrington to Halwill Junction in 1925. The biggest change to the system came in 1906 when the direct route from Taunton to Westbury was completed and GWR trains from Paddington could bypass Bristol and save twenty minutes on the journey.[4] This line was built to give the GWR a competitive edge over the L&SWR line from Waterloo but in the second half of the century it was to enable the railway to compete with the motorway which

was opened in 1976 and was routed via Bristol. Connecting the South West to London was always the priority of the GWR. It signally failed to develop rail services with the Midlands and the North, a factor of some significance in the orientation of south-west England's economy.[5]

In the early part of the century the railway companies, and notably the GWR, diversified their investment policies. While investment in the railways was limited to minor improvements, a few branch extensions and some station-building, large investments were made in steamers, docks, hotels and motorbuses. The Great Western even started an air service between Cardiff and Plymouth in 1933.[6] Although this horizontal integration made good business sense in an era when, for example, there was strong competition for transatlantic liner traffic, there is little doubt that the railways suffered neglect. Although main line electrification had been recommended as early as 1931, the Great Western, with its heavy, and in the end debilitating, involvement in the South Wales coal industry, was slow to adopt new railway technology—unlike the Southern.[7] Although the GWR began to run diesel units after 1933, when it was nationalized in 1947 it had 3,856 steam locomotives and only a single diesel electric shunting engine.[8]

All the railway companies were slow to respond to the threat of the motor car. Speed and comfort had always been the railways' greatest asset, and remained so at the end of the twentieth century. Yet the railway companies refused to compete with each other in speed and it was not until after the line closures masterminded on behalf of government by Lord Beeching that a serious attempt was made to make the railways competitive again. As David Smith put it in his book, *The Railway and its Passengers*, 'it remains the case that the railways under private enterprise drifted through the period without any clear or coherent strategy for the future'.[9]

The organic life of the railway system dictated that while lines continued to be built into the twentieth century, some lines would close, their purpose completed. Mineral lines in Cornwall and Devon were closed when the mines ceased operation and many of them were of no use for passenger traffic. The Appledore line closed in 1917, the first passenger-carrying railway to go out of business. By the end of the 1930s, lack of investment was already suggesting that closure was a better policy than reinvestment. The Lynton railway closed in 1935 after only thirty-seven years of life, but this was an isolated example in the inter-war years, and the companies, rather unimaginatively, maintained their network without developing it in any significant way. It was after the war that line closures really began. The line to Moretonhampstead closed in 1952, the Princetown line in 1956, and the Teign valley line in 1958. However, the bulk of the railway network survived until the early 1960s when the Beeching axe fell heavily on the South West.

During the 1950s passenger miles travelled on British Railways reached record peacetime levels as did freight tonne miles—hardly the sign of a business in terminal decline.[10] However, in the South West the railways experienced to the full the problem that the roads were just beginning to experience—the high concentration of traffic on Saturdays, and in particular on summer Saturdays. 'Many stations in Devon and Cornwall despatched 90 per cent of their long-distance passengers, not only in a single day of the week, but within a few hours', and the Beeching report listed 2,000 coaches in British Railways' rolling stock that were only used for ten days of the year.[11]

The controversy surrounding the Beeching closures remains fresh. The neglect of rural lines, the lack of investment and the failure to adapt the speed and type of service to changing needs created an impression of a loss-making transport system whose time was over. The accounting techniques employed by Beeching, whereby the profitability of each line was tested independently of the network as a whole, were guaranteed to highlight the unprofitability of parts of what should have been seen as a single network. However, as branch lines closed, the core railway did not become more profitable but sank further into loss—a phenomenon which the builders of the original network would clearly have understood. Just as the road system of the 1960s urgently needed investment, so did the railways, and had they received even a fraction of the investment that was poured into

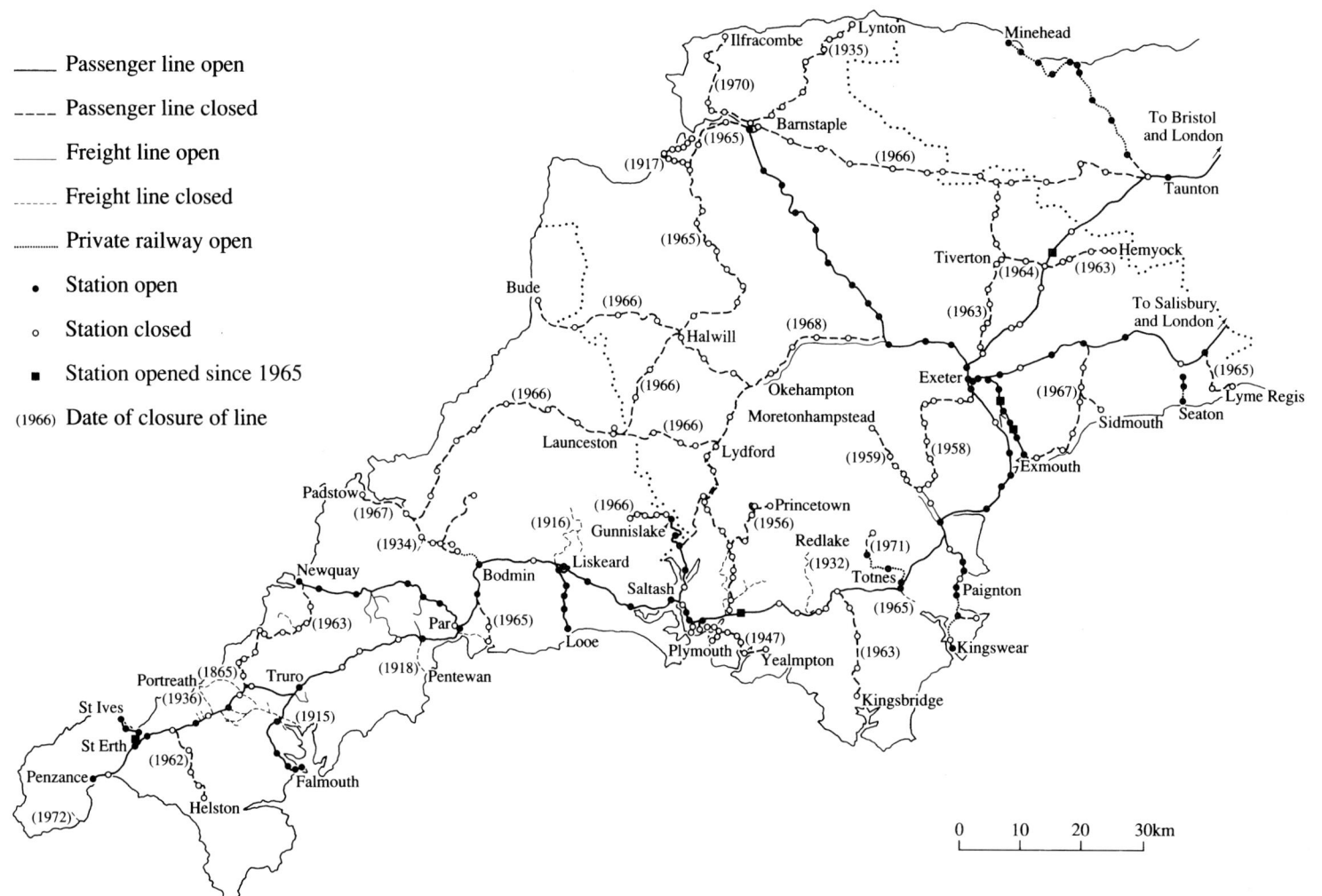

Map 47.1
Railways.

SOURCES: O.S. Nock, *London and South Western Railway* (London, 1965); E.T. Macdermott and O.S. Nock (revised by C.S. Clinker), *History of the Great Western Railway*, 3 vols (London, 1964); R.H. Clark, *An Historical Survey of Selected Great Western Stations*, 3 vols (Oxford, 1976–81).

the motorway system, the rail network of the South West might have survived to continue to serve the region effectively.

The Beeching cuts aimed to close all the lines in north and east Devon apart from the main lines to Paddington and Waterloo and the line to Barnstaple that was to become single track (**Map 47.2**). All the lines in Cornwall, except the mainline and the branches to Falmouth and Newquay, were to close. In the event the cuts were stopped before the rail network could be quite destroyed and the St Ives, Looe, Exmouth and Gunnislake branches were reprieved. Plymouth suffered worse than Exeter (**Maps 47.3 and 47.4**). The port was cut off from the railway and North Road station became little more than a stop on the main line to Cornwall. Exeter, however, remained a major railway centre with five lines feeding into the city providing an opportunity for a rail-based transport policy in the 1990s. Cornwall retained four of its branches and the old Southern line from Exeter to Waterloo (condemned to single line operation by the narrow vision of Dr Beeching) also survived to assume a wholly new importance with the building of the Channel Tunnel and the operation of continental trains out of Waterloo. It is known that Beeching planned a second round of closures that would have limited rail in the South West to a single mainline to Plymouth—a plan that briefly resurfaced in the Serpell Report of 1982.[12]

The post-Beeching era saw the rail network evolve in minor but important ways. The railways remained crippled by government policies which continued to favour road-building and which starved them of investment while demanding exaggerated standards of safety and maintenance and wholly inappropriate levels of return on investment. Nevertheless the introduction of diesel locomotives in the 1960s and the high speed trains in 1977 showed, a generation late, the potential that rail had to compete with the motor car. Journey times from Exeter to Paddington were cut to under two hours, over an hour faster than the journey could be done by road. The fastest times to Penzance were now under five hours. At the same time the Transport Users Consultative Committee began to monitor the deplorable condition of many of the stations in the region, persuading British Rail to initiate a policy of station refurbishment.[13] Supported after 1974 by county councils, British Rail also began to construct new stations to tap new markets. 'Parkway' stations were built at Tiverton Parkway and Bodmin, suburban stations opened at Feniton, Pinhoe, and Digby and Sowton outside Exeter, and at Ivybridge outside Plymouth, while tourists for St Ives were lured on to trains at Lelant Saltings and marines were given their own station at Lympstone Commando.[14] Plans were discussed for rail-based transport strategies in Exeter and studies were undertaken to reopen the line to Tavistock. Meanwhile rail's share of the transport market was retained and the rail network remained, truncated but still a network, to provide the basis for a modernized and relatively clean and safe system of transport for the twenty-first century. In 1996 British Rail track and stations were transferred to Railtrack, while train operation was placed in the hands of twenty-five private companies.

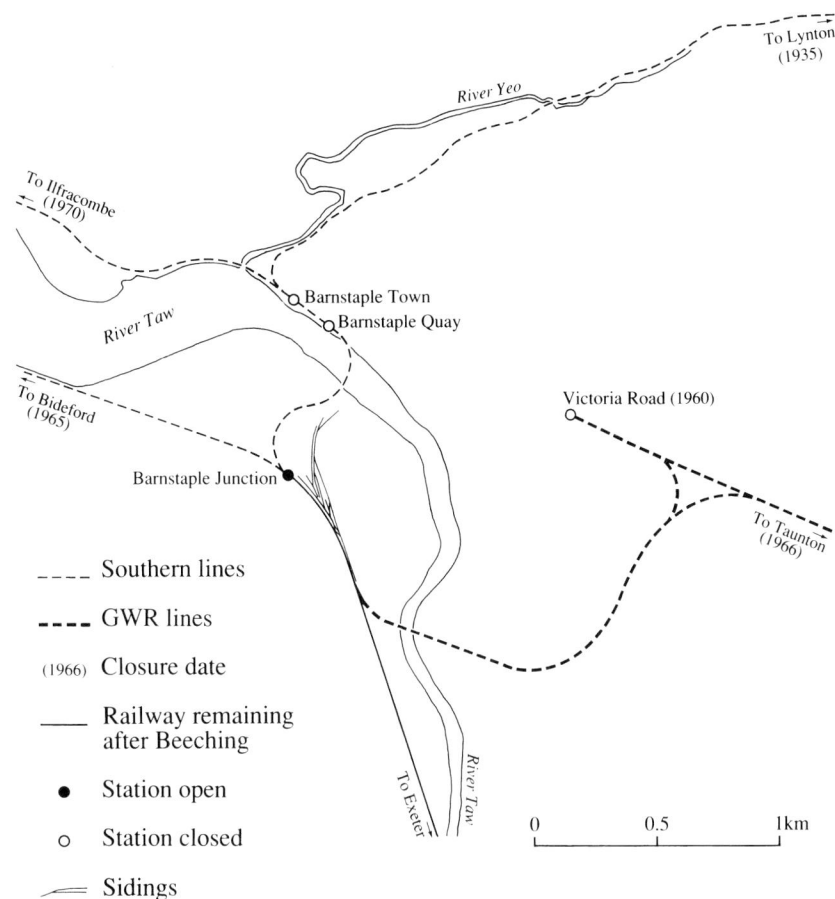

ROADS (**Maps 47.5, 47.6, 47.7, 47.8 and 47.9**)

The maps showing the growth of traffic on the roads of the South West are to some extent self-explanatory. However, the exact nature of this traffic growth, the reasons for it and the consequences of it for the south-west region are more complex.

Detailed information about traffic only exists for the second half of the twentieth century. Although some local authorities carried out regular traffic censuses before the Second World War (Cornwall carried out a census every three years), the practice was discontinued during the war and development plans in the 1950s were frequently based on out-of-date information. The *County of Cornwall Development Plan: Report of Survey* of 1952 based its traffic information on a census taken in 1938. In Devon it was only in the 1960s that the first Land Use Transportation Studies were undertaken.[15] However, the Act that created the new county councils in 1974 laid on transportation departments an overall responsibility to monitor and plan transportation within their counties and to publish an annual statement of Transport Policies and Programmes (TPP). This responsibility was enhanced by the 1974 regulations providing for the drawing up of structure and local plans in which transport policy was to be one of the major issues addressed. These documents allow a

Map 47.2
Railways in the Barnstaple area of north Devon.

SOURCE: G.A. Brown, J.D. Prideaux and H.G. Radcliffe, *The Lynton and Barnstaple Railway* (Newton Abbot, 1964).

Map 47.3
Railways in the Exeter area.

SOURCE: Geoffrey Body,
Railways of the Western Region
(Cambridge, 1983).

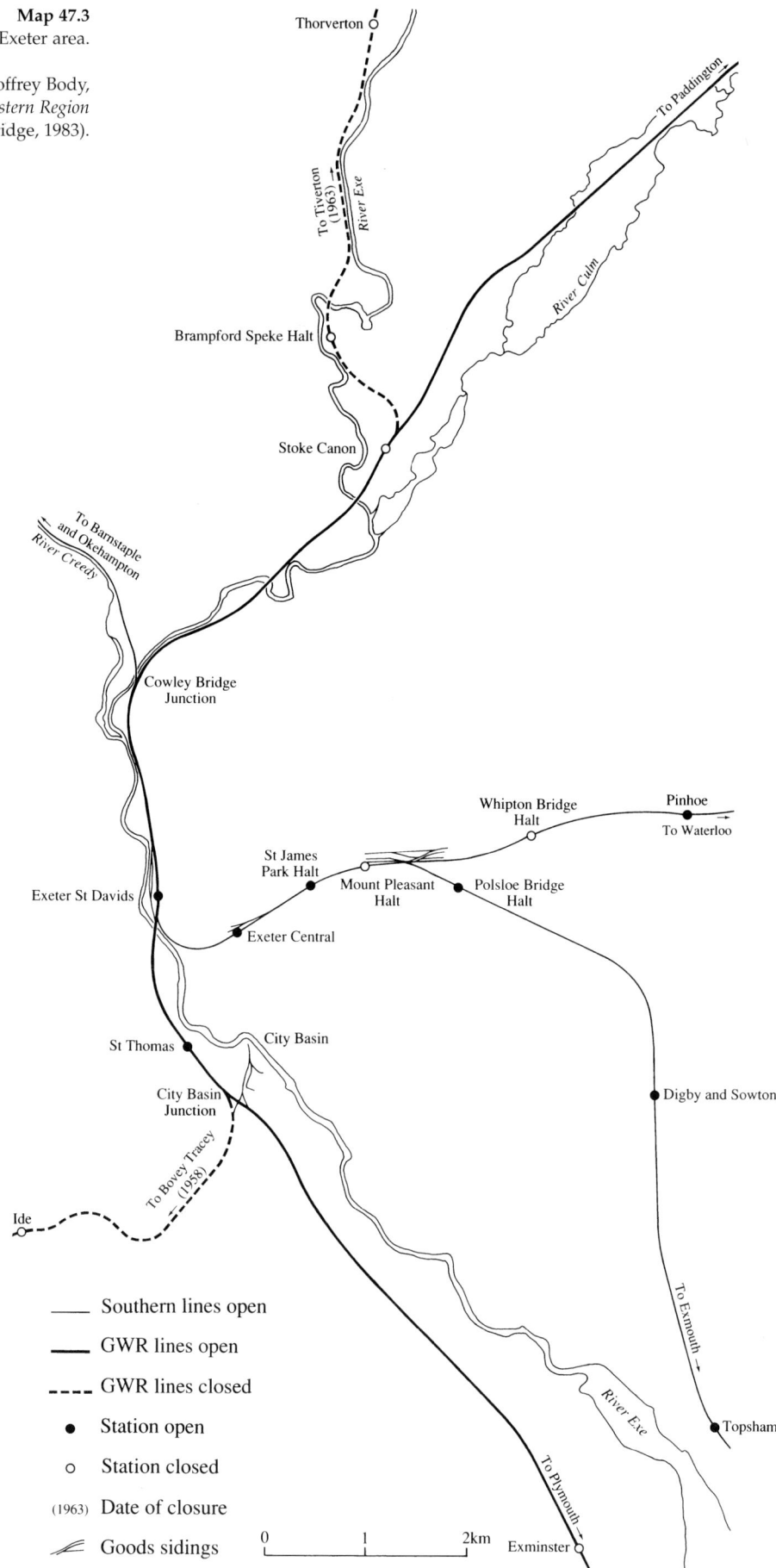

Thorverton

To Paddington

River Exe

River Culm

To Tiverton (1963)

Brampford Speke Halt

Stoke Canon

To Barnstaple
and Okehampton
River Creedy

Cowley Bridge
Junction

Whipton Bridge
Halt

Pinhoe

To Waterloo

St James
Park Halt

Mount Pleasant
Halt

Polsloe Bridge
Halt

Exeter St Davids

Exeter Central

City Basin

St Thomas

City Basin
Junction

Digby and Sowton

To Bovey Tracey (1958)

Ide

To Exmouth

River Exe

Topsham

To Plymouth

Exminster

—— Southern lines open

━━ GWR lines open

---- GWR lines closed

● Station open

○ Station closed

(1963) Date of closure

⫽ Goods sidings

0   1   2km

reasonably clear picture of transport development in the South West to be obtained.

The first motor vehicles appeared in Devon in 1897, but the lorry and the private car did not dominate the transport scene until after the Second World War. In the inter-war years road traffic was very varied. A relatively dense network of local buses grew up and bicycles were a common mode of transport. By 1990 traffic volumes in Britain were more than six times the level they had been in 1950.[16] In Cornwall numbers of licensed motor vehicles rose from 33,231 in 1938 to 52,943 in 1952, an increase of 59 per cent. In 1990 there were 237,102 vehicles, a figure which had grown by 31 per cent in the previous decade. In Devon there were 105,000 private cars licensed in 1960, a figure which rose to 270,000 in 1979.[17]

English roads were first classified in 1919 and in the following two decades all such roads were given a tarmac surface. By the Second World War, Cornwall and Devon had one of the largest networks of roads in Britain. In the inter-war years few new roads were built and the growing number of vehicles had to be accommodated within the historic pattern of existing roads. The only significant road-building was the construction of the Exeter bypass in the 1930s. Immediately after the war little road construction took place in either Devon or Cornwall except for some improvements to the main road to Plymouth.[18]

After the war the patterns of transport began to change rapidly. Whereas passenger miles travelled by rail remained static, there was a dramatic increase in car and lorry ownership and use. Between 1950 and 1979 the volume of motor traffic increased 600 per cent, while the number of licensed road vehicles rose by 450 per cent. Bus travel declined from 39 per cent of passenger miles in 1955 to only 11 per cent, with a corresponding decline in rail freight.[19] There is some dispute over which factors have contributed most to this revolution in transport. It appears that national population increase has been less significant than the rise in personal incomes, although in the South West the scattered settlement pattern has been more important than either. Despite lower than average personal incomes, Devon in the 1980s had higher car ownership than the national average.

The policies pursued by the government greatly promoted this traffic revolution. Real fuel prices for motor vehicles remained exactly the same (with some intermediate fluctuations) over the period from 1960 to 1990 in spite of the growth of personal incomes.[20] Taxation policies encouraged the use of company cars which by the 1980s accounted for two-thirds of all new registrations, while the taxation of motor vehicles declined in real terms and ceased to reflect the social and environmental costs of motoring. At the same time the government created a favourable framework for investment in roads.[21]

The road-based transport strategies pursued by various governments from the late 1950s, but particularly by the Thatcher government in the 1980s, were based on the assumption that roads contributed to strengthening the economy. This belief was

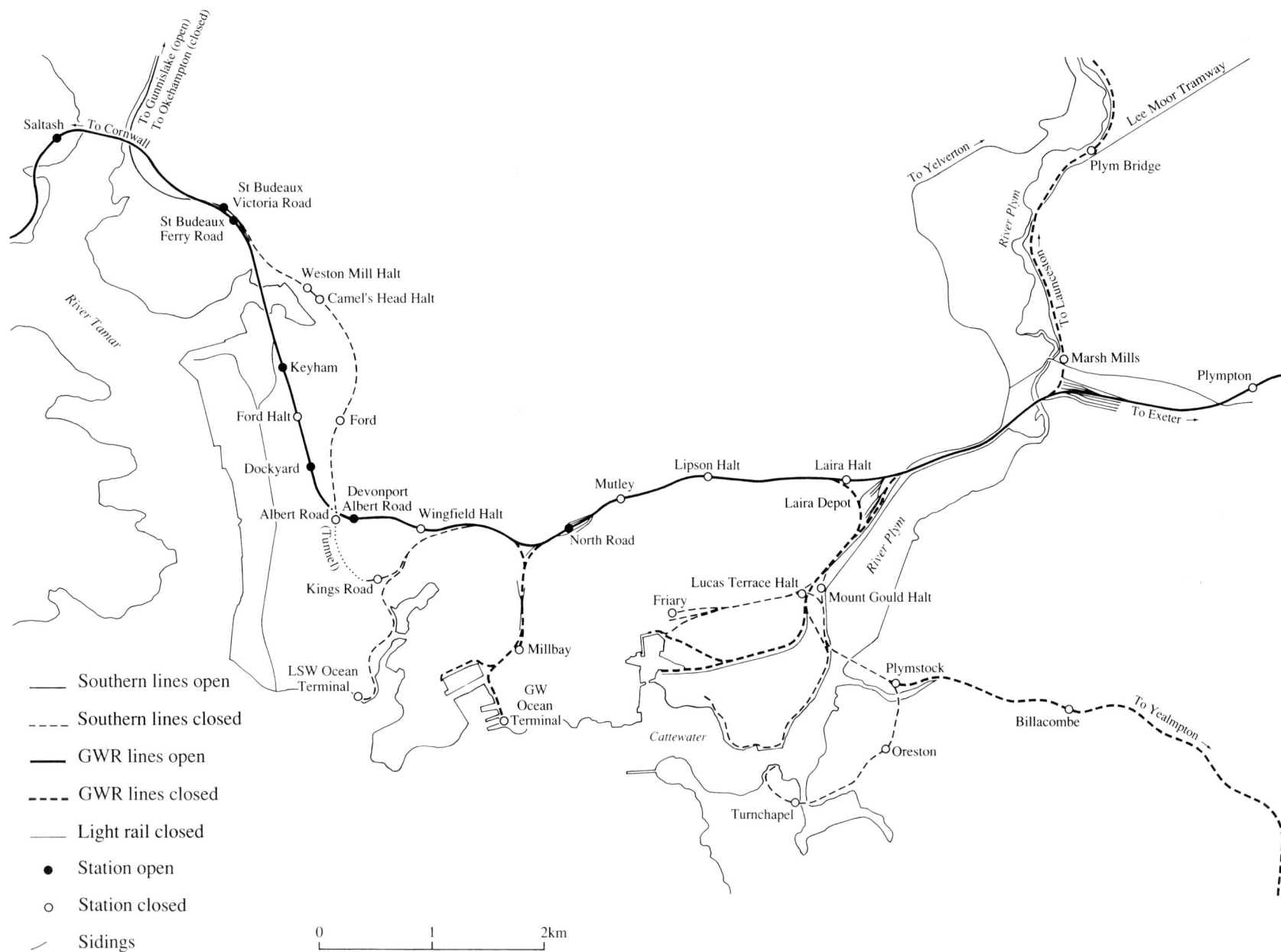

Southern lines open
- - - Southern lines closed
GWR lines open
- - - GWR lines closed
Light rail closed
● Station open
○ Station closed
╱ Sidings

0   1   2km

summed up in the title of a pamphlet published in 1983 advocating a dual carriageway through Cornwall—*The Road to Prosperity for Cornwall*.[22] This policy culminated in the White Paper *Roads to Prosperity*, published in May 1989, which forecast further traffic growth of between 83 per cent and 142 per cent by 2025 and described the main objectives of transport policy as 'measures to improve the country's economic geography, increasing opportunities for the less favoured areas [and] assisting urban regeneration'.[23] As late as 1992 the Devon County Council paper, *Towards a Sustainable Transport Policy for Devon*, in many ways a far-sighted and innovative document, could still make the crudely simplistic comment, 'the economy of the County very largely depends on efficient and cost effective transportation connections within and outside Devon'.[24] If this and other similar statements

were unexceptionable as general propositions, it did not follow that improving roads would encourage industry or other forms of economic activity to locate in peripheral parts of the country like the South West.

Already by this time major new trunk road schemes had been built in the South West. In Devon the A38 dual carriageway to Plymouth had been completed in 1974 and the M5 to Exeter by 1976. The North Devon Link Road was opened in 1987 and the highly controversial Okehampton bypass in 1988. Devon's system of 8,200 miles of classified road was by far the largest of any shire county in England, its nearest rivals being North Yorkshire and Norfolk.[25] This network was made up of 24 miles of motorway, 180 miles of trunk road, 571 miles of main roads and a massive 7,351 miles of minor roads and lanes.

Map 47.4
Railways in the Plymouth area.

SOURCE: David St John Thomas, *A Regional History of the Railways of Great Britain, Vol. 1, The West Country* (Newton Abbot, 1960).

Map 47.5 (right)
Traffic flows on
Cornwall roads,
1938–1939.

SOURCE: Cornwall
County Council, *Development
Plan: Report of Survey*
(Truro, 1952).

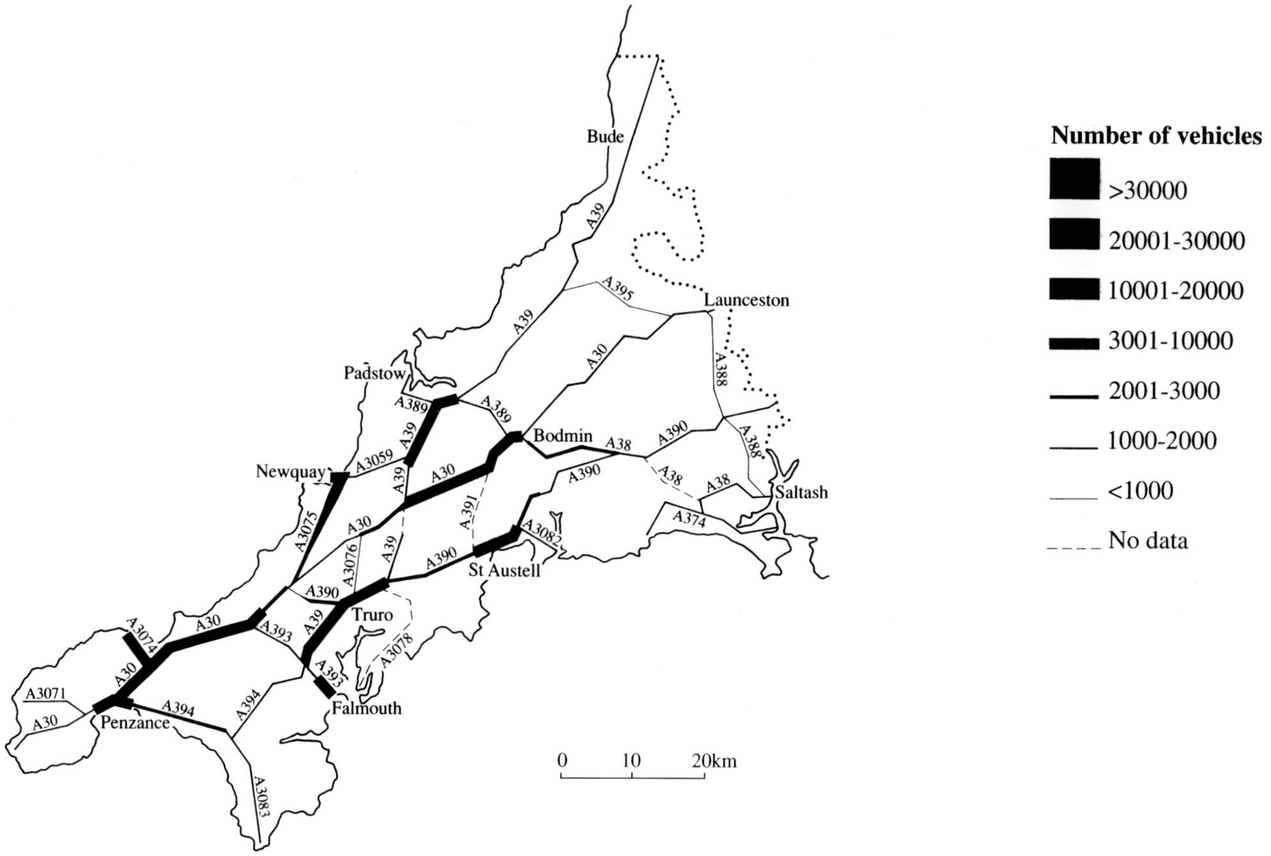

Map 47.6 (below left)
Traffic flows on
Devon roads, 1938–1939.

SOURCE: Devon Record Office
(86/47/7).

Map 47.7 (below right)
Traffic flows on
Devon roads, 1954.

SOURCE: as **Map 47.6.**

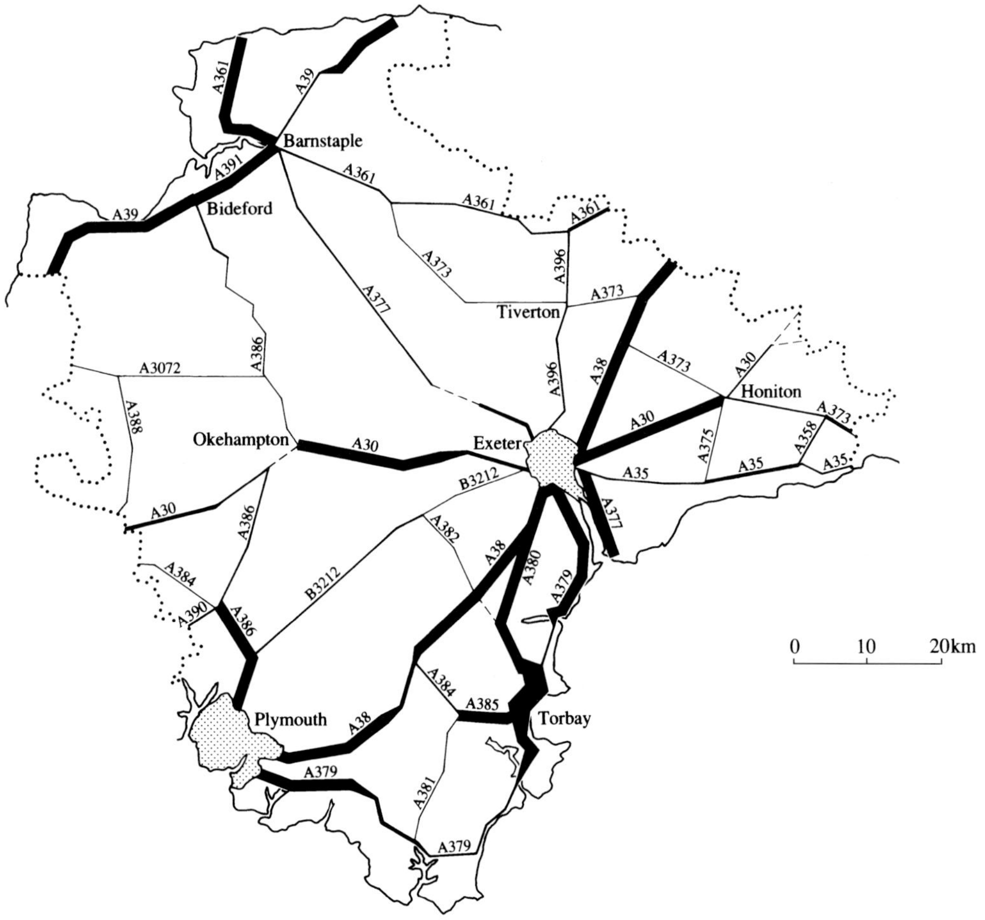

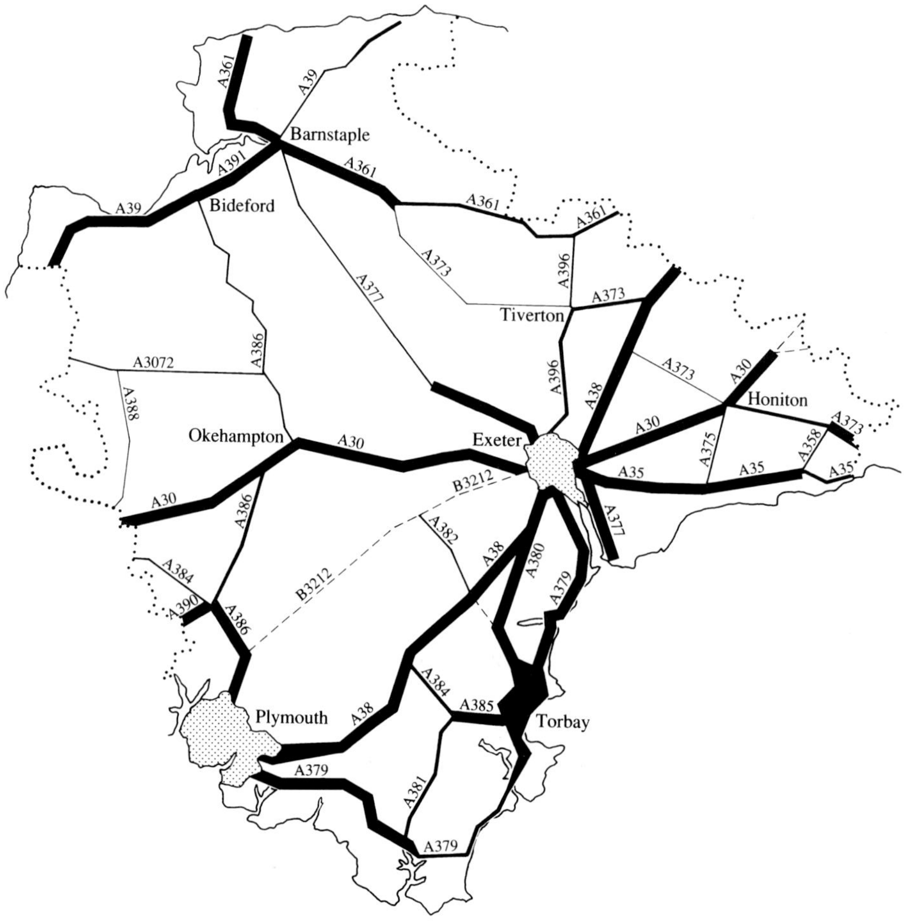

Number of vehicles

>30000
20001-30000
10001-20000
3001-10000
2001-3000
1000-2000
<1000
No data

At the end of the Second World War, Cornwall was still to an extraordinary extent cut off from the rest of Britain. Although good rail connections existed, traffic surveys showed that circulation within Cornwall was always much heavier than traffic entering the county from Devon. In 1938 only 2,494 vehicles daily entered via the Torpoint ferry, where in summer there could be a two-hour wait, and 1,900 via the A30 at Launceston.[26] The 1952 Development Plan outlined far-reaching proposals for new road works, including a Tamar road bridge, a widening of stretches of the A30, and bypasses or relief roads for Liskeard, Bodmin, Fraddon, Penzance, St Ives, Redruth, Hayle, Launceston and Wadebridge. It drew attention to the particular problem which transport planning in Cornwall faces in making sense of the inheritance from the past of a scattered settlement pattern and numerous small ports, difficult of access because of the indented geography of the county's coastline. The Tamar bridge was opened in 1962 and a tunnel under Saltash in 1987, while transformation of the A30 into a dual carriageway remained high on the government's list of trunk schemes throughout the 1980s and 1990s. In 1990 Cornwall had 159 miles of trunk road (47 miles of it dual carriageway), a further 306 miles of main roads and 1,959 miles of minor roads. In addition it had 2,366 miles of unclassified road.[27] The Cornwall Structure Plan Report of Survey in 1979 was able to record that average journey times to Penzance by road from selected starting points in Britain had for the first time become as fast by road as they were by rail.[28]

Few assessments of the consequences of this massive road-building programme have been made to see if it has fulfilled the economic expectations of the planners. The construction programme coincided with economic downturn not economic prosperity. The opening of fast roads to the West Country was followed by a decline in tourism and a steady contraction of local industries and distribution centres. Instead of reviving the local economy during a period of recession, the roads encouraged business to retrench by locating away from the south-west peninsula and delivering to local markets by road. The decline in Cornwall's economy meant that by the early 1990s, traffic growth in the county had almost ceased and a sharp decline was taking place on minor roads.[29]

The factors affecting traffic growth in Devon and Cornwall were always somewhat different from those in Britain as a whole. Since the war the South West has enjoyed a greater population growth than the rest of Britain.[30] The scattered nature of the population also led to a higher than average car ownership pattern—in 1980 Cornwall ranked ninth among shire counties in England and Wales in households with a car, while between 1979 and 1988 vehicle ownership in Devon increased by 35 per cent against a national increase of 27 per cent. In 1990 the South West had 410 cars per 1,000 population, greater than the figure for England as a whole which was 370 cars per 1000, but comparable to the national figures for France and Italy.[31]

Traffic growth in Cornwall, as opposed to car ownership, however, has been slightly below the national average. Between

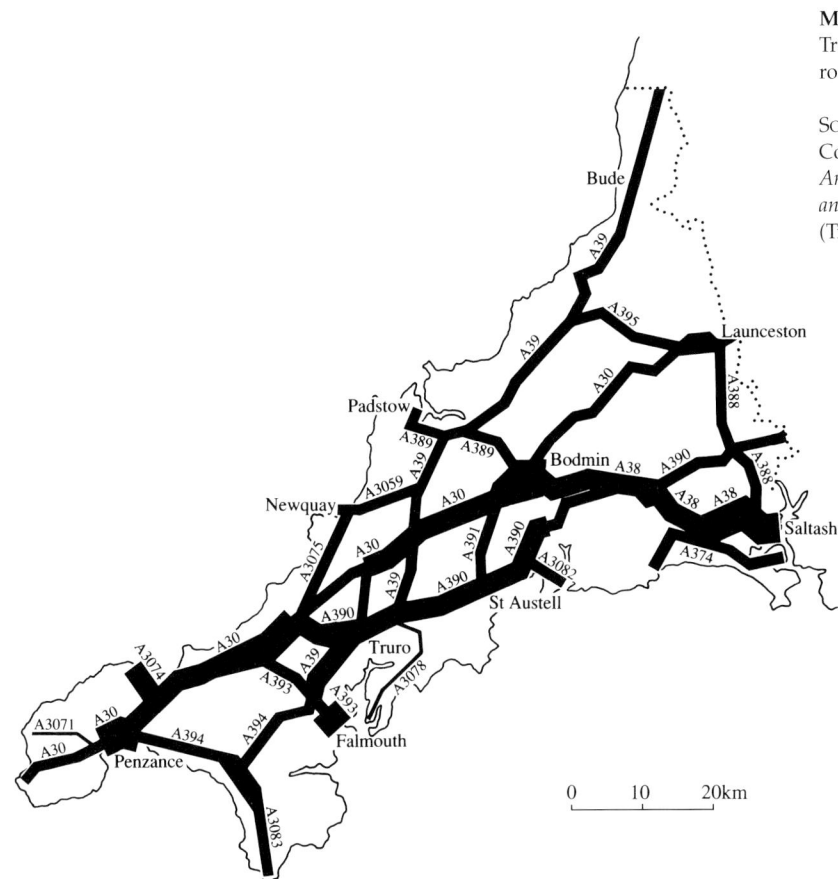

Map 47.8
Traffic flows on Cornwall roads, 1991.

SOURCE: Cornwall County Council, *Traffic in Cornwall: Annual Report of Traffic and Accident Statistics* (Truro, 1992).

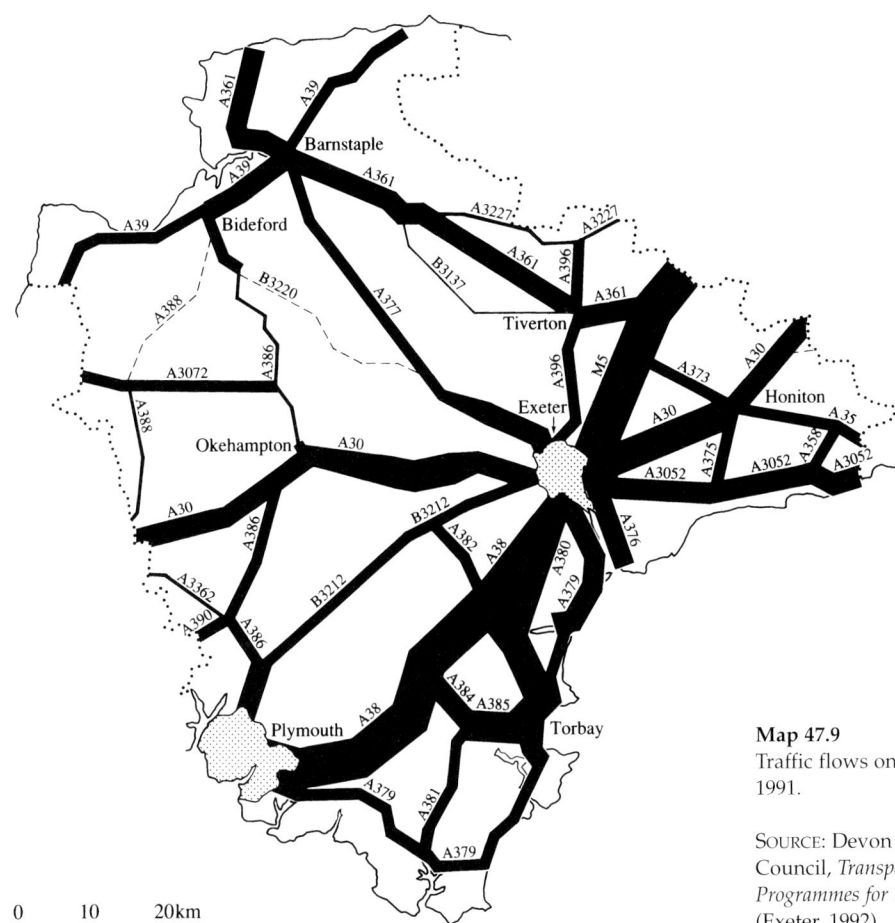

Map 47.9
Traffic flows on Devon roads, 1991.

SOURCE: Devon County Council, *Transport Policies and Programmes for 1993–4* (Exeter, 1992).

1970 and 1990 the national index of traffic growth rose from 100 to 203.3, while in Cornwall it only rose to 192.7. Although growth in demand in Cornwall was strong during the period 1985–9, there was decline in traffic on minor roads after 1990 and no increase on the major roads. This development was important as it showed that traffic growth was sensitive to a variety of government policies and was not some inexorable law of nature.[32]

Among other factors that affect traffic in the South West are the seasonal fluctuations. Throughout Britain as a whole traffic levels differ little between winter and summer, the variation being less than 10 per cent. In Devon, however, variations on major routes may be considerable and August traffic flows can be 50 per cent above the daily average and 80 per cent above the winter low points. On Saturdays the problem is accentuated with flows on the M5 six times higher than in winter.[33] Indeed the building of the M5 had the effect of concentrating into Saturdays traffic that previously was distributed throughout the week.[34] In Cornwall in 1994, variations in August could be 40 per cent above the daily average and 100 per cent above the winter low periods. Ten years earlier this seasonal variation had been much more marked with an August peak 60 per cent above the average—a difference mostly accounted for by a decline in summer traffic.[35]

By the 1990s the period of road-building, which had lasted from the late 1950s, was clearly coming to an end. As the *Plymouth and South East Cornwall Environs Transportation Study* put it in 1992, 'in many places where new highways have been constructed, traffic demands immediately expand to fill the space available and the increased traffic can place an even greater burden on the surrounding environment'.[36] Devon County Council recognised this issue in 1992 when it wrote that it was necessary to 'recognise that the longer term demand for road space cannot, and should not, be met and that this demand needs to be managed'.[37] The idea that road-building can actually make the problems of accommodating traffic worse not better informed the whole of the county's 1993 Structure Plan Review.

FURTHER READING

For a general history of the railways in the South West see David St John Thomas, *A Regional History of the Railways of Great Britain, Vol. 1, The West Country* (Newton Abbot, 1960). On roads see Geoffrey Hindley, *A History of Roads* (London, 1971) and Michael Hawkins, *Roads in Devon* (Exeter, 1987).

MARITIME ACTIVITIES

FISHERIES, EXPLORATION, SHIPPING AND MARINERS IN THE SIXTEENTH AND SEVENTEENTH CENTURIES

TODD GRAY

THE FISHERIES AND EXPLORATION, 1500–1650

Although the West Country is noted for its Elizabethan sea adventurers, such as Drake, Raleigh, Hawkins and Grenville, there was a less assuming and long-standing industry which was annually worth many tens of thousands of pounds and employed thousands of men and women. Moreover, while the Sea Dogs' world of piracy and privateering lay at the whim of political fortune, the fisheries remained a constant presence throughout. Drake's circumvention of the globe from 1577 to 1580 may have captured the country's imagination, but the trade in the unglamorous pilchard and cod persisted to prove of more lasting value.

The demand for fish arose from great practicality: throughout this period there was a harvest failure in the South West at least once every decade and consequently an inexpensive and easily preserved source of protein was of pressing necessity. The famous Cornish topographical writer Richard Carew noted that the poor utilized shellfish to save them 'from dread of starving, for every day they may gather sufficient to preserve their life, though not to please their appetites'.[1] Occasionally scarcities of grain and fish coincided, such as in 1594 when there were few pilchards available in the markets.[2] It must be assumed that fish comprised a fair portion of the general diet in the coastal parishes and probably in many inland ones. The early Stuart household accounts for the Reynell family of Forde near Newton Abbot show a consistent diet of local seafood such as pilchards, salmon, oysters, lobsters and crabs. The only resident noble family in Devon at the time, that of the earls of Bath at Tawstock in north Devon, also purchased considerable amounts of fish, including barrels of Irish herring.[3]

Fishing also provided the South West with one of its principal exports. Carew and his Devon colleague John Hooker both recognized the importance of the trade.[4] It is paradoxical that following the Reformation English Protestant fish merchants derived their greatest profits from trading with the Catholic countries. The peculiarity of the situation was recognized in a Cornish toast to the Pope:

Here's a health to the Pope,
and may he repent

And lengthen by six months
the term of his Lent.
It's always declared
betwixt the two poles,
there's nothing like pilchards
for saving of soles.[5]

There was also a great market for fish in the South West, not only amongst the local population, but also amongst foreign buyers: the papers of the Corsini brothers, Italian merchants of the late sixteenth century, show highly competitive buying at Plymouth with their Dutch and French counterparts.[6]

COASTAL FISHING

All of the topographical writers of the sixteenth and seventeenth centuries noted that the coasts of the South West were abundant with fish; there was a great number and variety of local fish to be found. However, the most well-known regional fishery is that for pilchards, the mature sardine which arrived off the coasts generally at the end of the summer. Pilchard fishing was especially attractive to fishermen because of the highly valued train oil, a by-product which resulted from squeezing the livers. The best general account of the fishery comes from Richard Carew in 1602. He described several methods for catching 'the least fish in bigness, greatest for gain and most in number' and for their preservation by smoking, pressing and pickling.[7] In about 1600 John Norden depicted the seine fishing for pilchards at Polkerris near Fowey and observed a single draught of pilchards filling sixteen boats as well as a great number which were carried away in baskets (which he also illustrated).[8] It is probably because of the importance of the Cornish pilchard fishing in the eighteenth and nineteenth centuries at Newlyn, St Ives, Mount's Bay and Mevagissey among other places that there is little general identification of this fish with Devon.[9] However, Thomas Westcote noted in his *View of Devonshire* in 1630 that one of Devon's two

Map 48.1
South Devon: seine, tuck and hake nets, 1566.

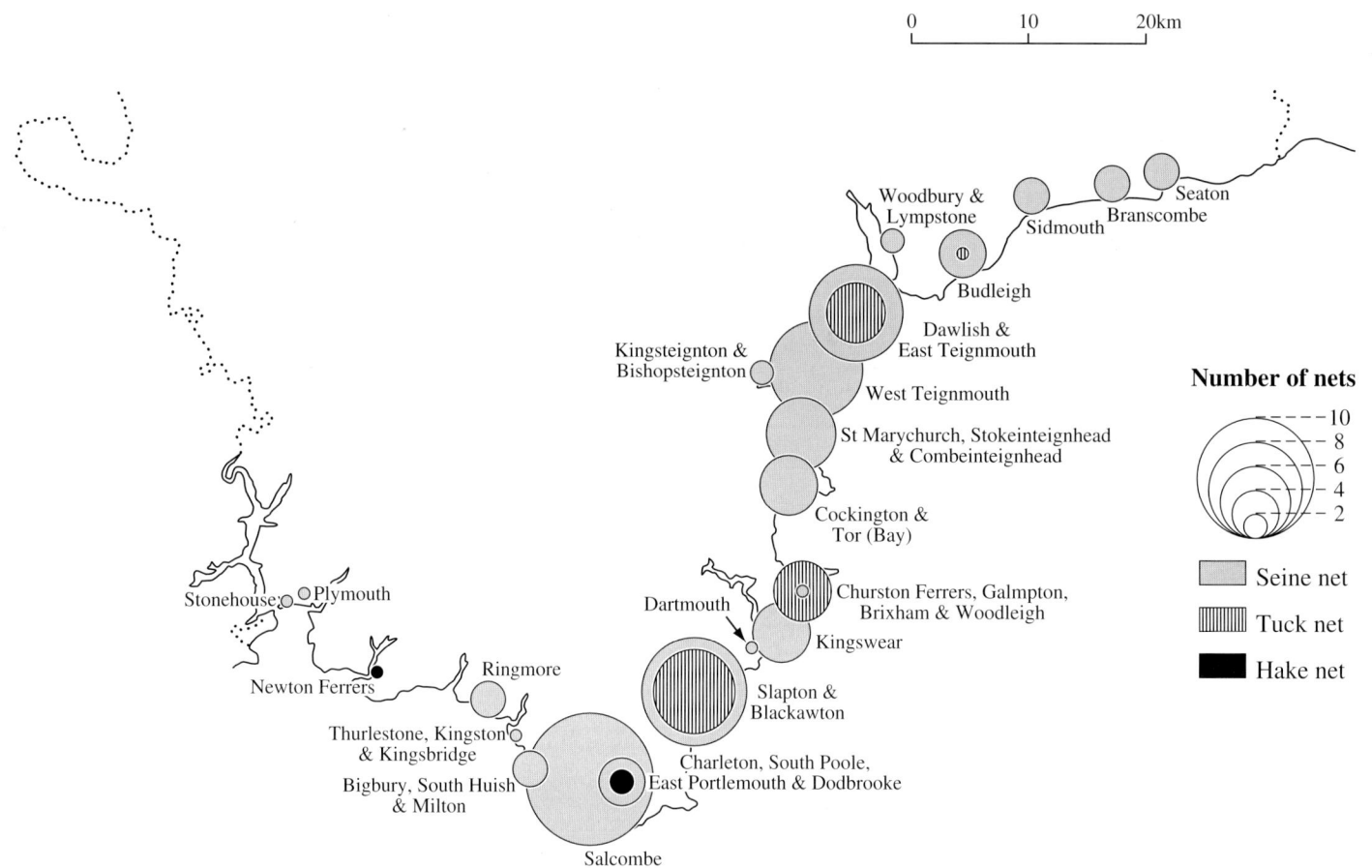

principal fisheries was for pilchards on the south coast.[10] In 1566 a survey was conducted by Sir Arthur Champernowne, vice-admiral of South Devon, of the coast's fishing shore nets.[11] It established that there were eighty-seven seine, eighteen tuck and three hake nets scattered along the length of the coast and demonstrates the extent of the local fisheries; while coves such as Hope Cove were noted by others as being an important place for pilchard fishing, the survey demonstrates that the fishery extended along the length of the coast (**Map 48.1**). The fishery continued until 1649 when, it was claimed, the fish withdrew to the western coast of the region in protest at the execution of Charles I.[12]

The second noteworthy fishery was that for herring. It was Tristram Risdon who noted in his *Chorographical Description or Survey of the County of Devon* that herring was 'a kind of fish which in our forefather's day kept, as it were, their station about Norway but in our time … take their course round this isle of Great Britain by shoals in great numbers'.[13] It was probably due to the herring fishery that there was a burst of building activity in the late sixteenth century at Porthilly across the River Camel from Padstow. At this time John Norden wrote of it that 'of late years there were few or no houses, now by their industrious fishing and the blessing of God, the inhabitants so increase in ability as their

prosperity allureth others to resort to the place and daily increase the buildings'.[14]

In 1583 herring suddenly appeared off Lynmouth in north Devon. The season took place during the last three months of the year. It is now realized that it was to Lynmouth, and probably to many other places not yet appreciated, that local and other fishermen migrated in great numbers for the season. The numbers were substantial enough to attract a cleric who administered to the fishermen every Sunday, alternating each week to the east or west sides of the River Lynn. In addition it was to Padstow that fishermen from Salcombe in the 1580s travelled annually for the herring fishery. While there they rented cottages and storehouses for their tackle. Likewise it is known that forty years earlier men from Paignton sailed west into Cornwall to fish at Mount's Bay and the Isles of Scilly.[15]

The Paignton fishermen also travelled to a number of other places for fish. This great migration was mostly due to the unreliable nature of fish stocks, demonstrated by the sudden arrival of herring on the north coasts and disappearance of pilchard from south Devon. **Map 48.2** shows the fisheries outlined by Robert Hitchcock in his *Pollitique Platt* of 1580. A later copy noted three additional fisheries, including the pilchard fishery in the west of England and the fishery for herring along the Welsh coast.[16] The erratic and uncertain character of fish stocks is further underlined by these changes and it was because of this characteristic that there was a continual search for new sources of fish.

The local shore-based fishing was only a small portion of west-country fishing activity. The fishermen of the South West established what became a network of fisheries through the sixteenth and seventeenth centuries (**Map 48.3**). In the first part of the fifteenth century boats from Fowey and Dartmouth are known to have travelled to Iceland for the fish there.[17] This was most likely an extension of what was a traditional fishery off the Irish coasts. The region's fishermen sailed to Ireland for fish throughout this period. The Cornish ports of Fowey, and especially St Ives and Padstow, had a natural advantage because of their closer proximity. Carew noted of the latter that it 'reapeth greatest thrift by trafficking with Ireland'. Plymouth was a centre for the selling of Irish hake in the 1530s and other Devon parishes such as Newton Ferrers and Stoke Gabriel continued this involvement.[18] By 1623 it was claimed that as many as 400 vessels from Devon were fishing in Ireland.[19] The arrival of pilchards off the Irish coast in the first few years of the seventeenth century resulted in further migration, and in some cases emigration, of fishermen from the region. They have even been credited with introducing the seine net to Ireland.[20]

This western migration coincided with a movement in the opposite direction. It is known that in the middle of the sixteenth century the region's fishermen, particularly those from south Devon, were travelling to the Sussex coast for fish. By 1540 men from Torbay, and later men from parishes along the Teign and Exe estuaries, were regularly fishing off Winchelsea. The migration extended into the North Sea: for example, south Devon fishermen were at Lowestoft in the 1560s and other mariners, including from Kenton, are now known to have fished not only at Great Yarmouth but further north to Scarborough.[21] This migration to the North Sea escaped the attention of the region's contemporary topographers and later of its historians. It is the movement across the Atlantic that has attracted the greatest attention.

THE NORTH ATLANTIC FISHERIES

In 1497 John Cabot conducted his famous voyage in the *Mathew* from Bristol to the 'new found land'. The discovery of fish stocks there proved to be of great interest to west-country merchants, although it is uncertain whether there was much traffic there for another fifty years. It was claimed in 1563 that Dartmouth men were only recently voyaging to Newfoundland. Certainly by 1583, when Sir Humphrey Gilbert claimed the island as England's first colony, English fishermen were there in considerable numbers. The fishery became increasingly popular through Elizabeth's reign; so much so that Sir Walter Raleigh famously claimed that it was the 'main stay' of the South West. By the early seventeenth century several hundred English ships, many of them from Devon

or Cornwall, were sailing to Newfoundland and it was claimed that 10,000 sailors were regularly employed there.[22] One of the most well-known of the fish merchants was John Rashleigh, a resident of Fowey whose family was originally from north Devon, and of whom Carew wrote 'his industrious judgement and adventuring in trade and merchandise first opened a light and way to the townsmen's new thriving'. The fish sales to continental markets, particularly to Spain, Portugal and Italy, established a pattern which persisted for several hundred years. Rashleigh, like many others, exported mixed cargoes of local fish, especially

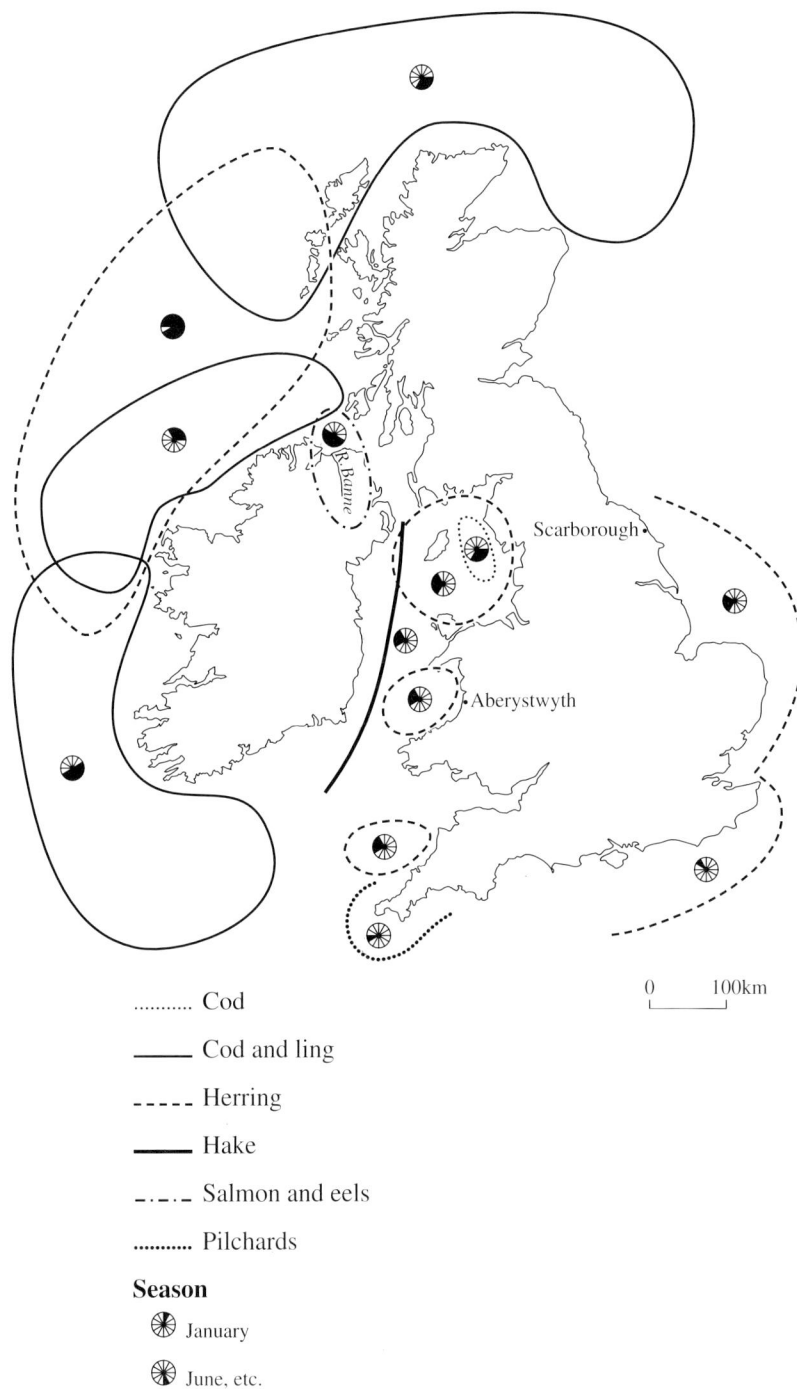

.... Cod
_____ Cod and ling
----- Herring
_____ Hake
_._._ Salmon and eels
.... Pilchards
Season
✳ January
✳ June, etc.

Map 48.2
British fisheries, 1580.

Map 48.3
Devon's overseas fisheries.

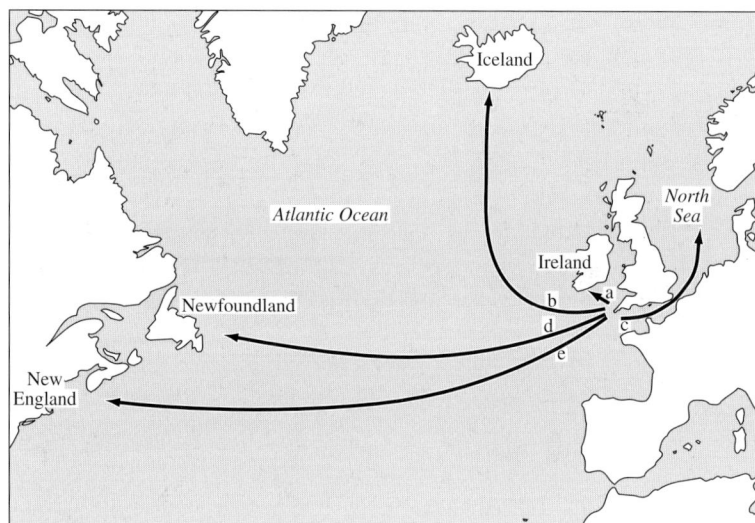

a Ireland

b Iceland, early fifteenth century

c Channel and North Sea, sixteenth century

d Newfoundland, from the early and mid-sixteenth century

e New England, from early seventeenth century

pilchards, with fish from overseas, particularly Newfoundland cod.[23] An additional factor for success was that the Newfoundland fishery was also substantially aided by the crown, which perceived it to be a 'nursery' for sailors.[24] The region's network of fisheries was extended at the beginning of the seventeenth century, and possibly as early as 1597, by the discovery of fish stocks off the coast of what is now known as New England. Early exploration was funded in part by fishing merchants and the first English settlement, at Sagadahoc in Maine, in 1608 was led in part by George Waymouth of Cockington in Torbay. The failure of that colony left the region to the migratory west-country fishermen who continued there until the arrival of the Pilgrims sailing from Plymouth in 1620. While southern New England became associated with religious nonconformists, the northern part of New England continued to be dominated by fishermen. The best documented of the fishing 'plantations' or 'stations' was that established by Robert Trelawny. He was a Plymouth merchant, although his family was a recent arrival from east Cornwall. Trelawny maintained a substantial enterprise at Richmond Island from the late 1620s until his death in 1644. He employed a Plymouth man as his manager and principally recruited his workforce from the Cornish and Devon parishes around Plymouth. Amongst these migratory workers was Priscilla Bickford, a young girl also from south Devon, who was employed to cook and look after the men. She found conditions too primitive, fled the island and eventually married a fisherman from another settlement.[25]

Fishermen from the south-west Devon parish of Newton Ferrers featured greatly at Richmond Island and the parish records reveal the diversity and range of the region's North

Map 48.4
Detail from map of Ruan Lanihorne, Cornwall, depicting small fishing boat, early seventeenth century.

SOURCE: reproduced by permission of Cornwall Record Office.

Atlantic fisheries. In the first years of the seventeenth century not only did they engage in local shore-based fishing but they also sailed to the fisheries off Ireland, Newfoundland and northern New England.[26] The parish was typical of many others throughout Devon and Cornwall during this period when the need for fish compelled fishermen to explore new shores and worlds.

SHIPPING IN THE SIXTEENTH AND SEVENTEENTH CENTURIES

It is most fortunate that several surveys of mariners and ships are extant for south-west England. In this period such occupational censuses are extremely unusual. These reports were made for the purposes of the national government, nearly always in order to assess those sectors of the maritime economy which could potentially be useful for naval purposes. Consequently, the surveys generally did not include the smallest of boats nor did they ever report the number of women or children who were involved in maritime activity. Those undertaking the surveys were commissioned exclusively to list those men and ships which could be drawn temporarily into the navy.

At least eleven maritime surveys were taken in Devon and Cornwall between 1560 and 1635. In 1572 there was a total of 51 vessels in Cornwall and 131 in Devon, in 1582 there were 66 and

124, and in 1626 the figures rose to 83 and 202 vessels. The precision suggested by the surveys masks important differences: they seldom reported the same categories of information, for instance the 1582 survey omitted Plymouth, and there are differences in the classes of vessels listed, which makes it difficult to make clear comparisons. Nevertheless it is clear that there was an overall increase in the total number of ships, although it must be assumed that these figures fluctuated widely. One uncertain factor is the incidence of the loss of ships. There are no central records of shipwrecks and while parish registers sometimes recorded burials of drowned men and women, they failed to note the loss of parishioners' lives in other places. This is problematic given that the majority of drownings occurred far from home, and it is therefore unknown how frequently ships and their crew were lost at sea.

In 1602 Richard Carew provided a useful explanation of the character of Cornish shipping in his *Survey of Cornwall* when he wrote that the county was 'stored with many sorts of shipping (for that is the *genus* to them all) namely, they have cock-boats for passengers, seine-boats for taking of pilchards, fisher-boats for the coast, barges for sand, lighters for burthen and barks and ships for traffic'.[27] The surveys show that Carew's last catergory, ships with the greatest tonnage, were located in 'greater' Plymouth, which included Saltash and Millbrook, as well as Dartmouth, the Exe estuary and to a lesser extent Barnstaple, Bideford and Fowey. By the earlier seventeenth century not only were the Cornish vessels smaller but the county had less than half as many ships as Devon, which also had four times as much in tonnage. The greatest number of Cornish vessels were in the eastern part of the county and could be considered within the economic sphere of Plymouth. East Devon was also a poorer cousin: all of the twelve vessels at Beer and Seaton were fishing boats of between 5 and 10 tons (**Maps 48.4 and 48.5**).

To a certain extent, the lower share of ships in west Cornwall and east Devon are comparable: lesser economic activity in west Cornwall, particularly with regard to the cloth industry, reduced the demand for ships of greater burthen, while in east Devon, an area of considerable cloth activity, the lack of any decent harbourage boosted the use of shipping elsewhere, particularly in the Exe estuary. Presumably the more generally vibrant economies of Devon and east Cornwall generated a surplus of cash among individuals who without any necessary maritime expertise could easily invest in short-term voyages, whether in trade, fishing or privateering. Richard Carew recognized this distinction between Devon and Cornwall and the importance of personal enterprise when he wrote of the afore-mentioned leading merchant and ship-owner of Fowey: 'I may not pass in silence the commendable deserts of Master Rashleigh the elder, descended from a younger brother of an ancient house in Devon, for his industrious judgement and adventuring in trade of merchandise, [who] first opened a light and way to the townsmen's new thriving'.[28] John Rashleigh established voyages to Newfoundland and, like the Devon merchants but largely unlike his fellow Cornishmen, traded with the Italian ports. He overcame a lacklustre economy by taking advantage of the natural topography.

MARINERS (**Maps 48.6 and 48.7**)

Three sixteenth-century surveys provide what appear to be exact figures for the number of seamen in the two counties. In 1560 it was recorded that there were 1,703 mariners in Cornwall and 1,268 in Devon; by 1582 these numbers had increased to 1,918 and 2,166 men, respectively. The last county survey was made in about 1626 and shows the difficulty in relying too strongly on the surveys: it listed 1,594 mariners for Cornwall but only 567 for Devon. The sharp fall for Devon resulted from the survey only reporting part of the county when it was noted that in 1626 'this survey was taken about January/February towards the ends of the wars when shipping was at its lowest and the seamen were most abroad at sea looking for prizes'.[29] Also, seamen fled from being listed as they recognized that the surveys were intended as reports for subsequent impressment for the navy. In the 1620s complaints were made of the mayors of both Saltash and Dartmouth that they warned local seamen to be absent when the official surveys were to be made.[30]

Map 48.5
Detail from map of Dartmouth, 1619, showing the village of Hardness.

SOURCE: reproduced by permission of Devon Record Office (R9/1/Z33).

Map 48.6
Cornwall mariners, *c.*1626.

SOURCE: Todd Gray (ed.),
*Early Stuart Mariners and
Shipping*, Devon and Cornwall
Record Society, 33 (1990).

Parishes with resident
mariners *c.*1626

0   10   20km

However, significantly there is also an earlier survey, made for south Devon in 1619 for the Duke of Buckingham, which provides considerable clues into the nature of the maritime sector. It reveals, first, that there were 3,653 seamen residing along Devon's south coast, a considerable increase on the 567 men recorded a few years later and a significant number in comparison with the 15,905 seamen estimated nationally in 1635.[31] But just as important are the details on the categories of men that were recorded. In certain regards this survey provides great insights into activity for which there are no other surviving documents. For example, twenty-three shipwrights were recorded at Plymouth and others were at Plymstock (ten), Ringmore (four), Malborough (five), Chivelstone (three), Stokenham (thirteen), Blackawton (ten), Stoke Fleming (fifteen), Dartmouth (forty), Kingswear (five), Paignton (two), West Teignmouth (three), Woodbury (seven), Colaton Raleigh (five), Withycombe Raleigh (two) and Otterton (twenty-one). No doubt many were employed merely repairing existing shipping or in the building of boats.

The majority of the men were listed as fishermen, sailors or mariners. The description 'sailor' referred to a member of a ship's company below the rank of officer, while the term 'mariner' described the most senior rank aboard a vessel as well as being a generic term used in the same sense as 'seaman' or 'seafarer'. It is

important to recognize the distinctions between the terms but another lesson is learned by comparing them with other details listed in the 1626 survey. This shows the importance of dual employment, with such descriptions given as 'bargeman fisherman', 'mariner fisherman', 'sailor fisherman' and 'fisherman sailor', indicating that annual income was derived from various forms of labour.

The area immediately around Dartmouth and Torbay comprised some 42 per cent of the mariners in south Devon in 1619, with the remainder at Plymouth (17 per cent), Salcombe (12), the Teign (13), the Exe (8) and east Devon (8). Dartmouth was clearly the centre of the maritime population. What is uncertain about the surveys is to what extent there was reporting of men who were engaged in maritime employment on a temporary or irregular basis. Many men who resided along the coast would have taken part in seasonal fishing, particularly for pilchards and to a lesser extent for herring, and many others travelled to Newfoundland for the occasional fishing season. It does not appear that these men were included in the surveys.

FURTHER READING

On fishing see Gillian Cell, *English Enterprise in Newfoundland* (Toronto, 1969); M.G. Dickinson, *A Living from the Sea: Devon's Fishing Industry and its Fishermen* (Exeter, 1987); Todd Gray, 'Devon's fisheries and early Stuart northern New England', in Michael Duffy *et al.* (eds), *The New Maritime History of Devon, Vol. I* (London, 1992), 139–44; Cyril Noall, *Cornish Seines and Seiners: A History of the Pilchard Fishing Industry* (Truro, 1972).

For discussion of the maritime surveys see M. Oppenheim, 'Maritime history', in William Page (ed.), *The Victoria History of the Counties of England* (1906), I, 492, and also his *The Maritime History of Devon* (Exeter, 1968), 38–40, 61–2. For more recent work see Todd Gray, 'The duke of Buckingham's survey of south Devon mariners and shipping, 1619', in Duffy *et al.* (eds), *The New Maritime History of Devon, Vol. I*, 117–18, and for the sixteenth century see Joyce Youings, 'Raleigh's country and the sea', *Proceedings of the British Academy*, LXXV (1989), 267–90, and her chapter, with Peter Cornford, 'Seafaring and maritime trade in sixteenth-century Devon', in Duffy *et al.* (eds), *The New Maritime History of Devon, Vol. I*, 98–107, as well as Alison Grant's 'Devon shipping, trade and ports, 1600–1689', 130–8, which has good discussion of the overall character of maritime Devon.

The early-seventeenth-century surveys have been printed in Todd Gray (ed.), *Early-Stuart Mariners and Shipping: The Maritime Surveys of Devon and Cornwall, 1619–35*, Devon and Cornwall Record Society, new series, 33 (1990). For recent work on the history of one port see Todd Gray, 'Fishing and the commercial world of early Stuart Dartmouth', in Todd Gray, Margery Rowe and Audrey Erskine (eds), *Tudor and Stuart Devon: The Common Estate and Government* (Exeter, 1992), 173–99.

0   10   20km

Parishes with resident mariners in 1619

Parishes with resident mariners c.1626

Parishes surveyed with no resident mariners c.1626

Map 48.7
Devon mariners,
1619 and c.1626.

SOURCE: as **Map 48.6.**

MARITIME ACTIVITIES

MARITIME DEVON
1660–1815

STEPHEN FISHER

The period from the Restoration of Charles II to the end of the Napoleonic Wars witnessed striking innovation and expansion in England's maritime trading and shipping activities, to the extent that some historians have regarded the century itself after 1660 as a time of 'commercial revolution'. The changes may be briefly summarized as a consolidation of the earlier widening of the geographical areas of trade; marked transaction growth and composition changes with the rise of colonial and Asian imports and re-exports of these products, as well as a far greater variety of English manufactured exports; and a distinct rise of the English mercantile marine. Devon's merchants, shipowners and ports also experienced a notable expansion of maritime activity in the period 1660–1815, particularly in the decades to 1730 or so, which may be regarded as the heyday of Devon's maritime activity on the national stage.[1]

Devon's noteworthiness as a maritime county sprang from a variety of factors: economic, social, geographic and political, or more strictly naval. Devon had an extensive woollen and worsted textile industry particularly located around Exeter, as well as copper and other mining interests in west Devon, while the county's agriculture was flourishing, and each of these activities generated important coastal and foreign trades, inwards and outwards. The county's towns, particularly Exeter, were becoming more populous and prosperous, with resident gentry and a rising middle class sustaining new patterns of consumption. Devon's strategic location on the English Channel, and its proximity to the Western Approaches and the wide Atlantic, also benefited its maritime business. So too did England's rise as a great power in this period and the relatively frequent wars fought against France and Spain. Finally, Devon's maritime importance was promoted by the proximity of good local fishing grounds, and the strong traditions of skill and enterprise found amongst the county's seafarers, merchants and shipowners (see chapter 48).

The maps presented here provide a view of Devon's maritime activities in the years from 1660 to 1815, specifically concerning foreign and coastal trade, the Newfoundland fishery, and naval and privateering activity. The chief ports involved in these activities were Exeter, Dartmouth and Plymouth on the south coast and Barnstaple, Bideford and Ilfracombe on the north coast. Each port presented scenes of varied maritime business, and especially those on the south coast, in both peacetime as well as wartime, with Plymouth assuming an extraordinarily hectic level of activity in the war years.

Map 49.1 displays customs figures of the total tonnages of shipping for the years 1759–63 in the trades, both foreign and coastal, of the principal Devon ports. In these latter years of the Seven Years War Plymouth clearly was the busiest of all the Devon ports, with some 515 vessels altogether engaged in the port's foreign trade, totalling nearly 60,000 tons with an average vessel size of 116 tons. Some 36 per cent of these vessels were foreign-owned, reflecting the use of neutral vessels in wartime in bringing naval stores and other vital supplies to this great naval port and base. Exeter, including Topsham, was the second Devon port, according to this measure, with a substantial coastal trade, reflecting the port's role as a major local entrepôt. Dartmouth came third, with the north Devon ports employing sizeable if less significant volumes of shipping in their trades.

These customs figures, of course, relate to the legal trades carried on and do not include the vessels engaged in contraband commerce. But the official national customs report, from which the figures were extracted, contains some interesting comments on the extent and character of smuggling.[2] At Plymouth this was chiefly centred on Cawsand and Kingsand and 'principally with Guernsey', employing eight 'sailing boats of about 8 tons each' as well as 'several rowing boats'. The articles involved fitted the conventional pattern of contemporary illicit imports, namely 'tea, brandy, rum, geneva [gin]', all high dutiable commodities, and interestingly enough 'china', i.e. porcelain. At Dartmouth it was thought 'there are several cutters and other small vessels employed' in contraband commerce, 'which trade has rather increased than decreased, notwithstanding the utmost diligence and alertness of the [customs] officers'. At Exeter such commerce was also thought to have 'rather increased' in these war years, it being centred on the Channel Islands with nine vessels being so employed as well as 'a great number of open boats which trade to and from Guernsey, [and] serve to take goods out of larger vessels on the coast'. At Barnstaple contraband trade was considered by the customs to be 'rather decreased' in these years, at Bideford tea and brandy smuggling was thought to be in 'greater quantities … than heretofore', while at Ilfracombe the illicit trade was reported to have decreased.

Maps 49.2 and 49.3 relate to Devon ports' involvement in the long-distance Newfoundland cod fishing trade. The Devon ports had a disproportionate national role in the Newfoundland fishing industry in this period, reflecting perhaps their historic links with the island and prior decades of contact and settlement.[3] **Map 49.2** shows that in certain years in the 1670s and 1680s Dartmouth led the other Devon ports in both the fishing vessels sent out and the dispatch of supply and carrying-to-market 'sack ships'. **Map 49.3**, which gives the average annual number of ships clearing from the Devon ports for Newfoundland for six years between 1770 and 1792, again shows Dartmouth to be the dominant centre of this trade, with Exeter a notable second. Of the other English ports involved in the Newfoundland cod fisheries and trade, only Poole can be said to rival the two leading Devon ports with seventy-four average annual clearances compared to Dartmouth's eighty-two and Exeter's fifty-three, the Devon ports' total vessels clearing for Newfoundland in these years constituting well over half of all the English vessels involved.[4]

The remaining maps relate to Devon's participation in English naval activity in the years 1660–1815 and also the rather curious practice of privateering, that is the legally sanctioned business of English merchant vessels in wartime taking out Admiralty letters of marque and being refitted as men-of-war to attack and seize enemy merchant vessels as prize.

After 1690, Plymouth became the great 'western dockyard' for the English Royal Navy in the frequent wars fought with France and Spain (the seven Anglo-French wars of the period 1689–1815 occupying sixty-seven years of the period), culminating in the long drawn-out Revolutionary and Napoleonic Wars with France.[5] In 1782 in the closing stages of the War of American Independence, in which France had earlier intervened, the dockyard labour force at Plymouth numbered 2,438 exceeding all the other royal naval

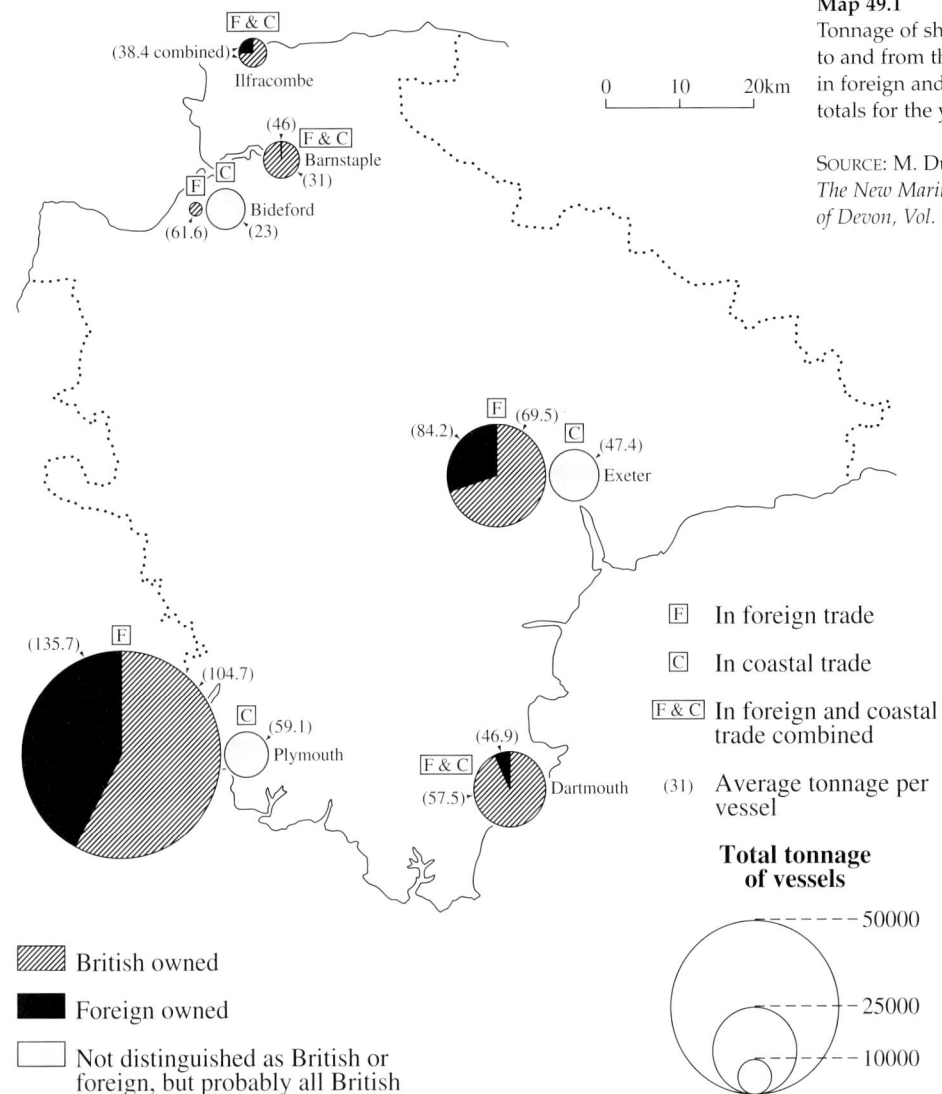

Map 49.1
Tonnage of shipping trading to and from the Devon ports, in foreign and coastal trades, totals for the years 1759–1763.

SOURCE: M. Duffy *et al.* (eds), *The New Maritime History of Devon, Vol. 1* (London, 1992).

F — In foreign trade

C — In coastal trade

F & C — In foreign and coastal trade combined

(31) — Average tonnage per vessel

Total tonnage of vessels

British owned

Foreign owned

Not distinguished as British or foreign, but probably all British

Fig. 49.1
View of Plymouth Dock, *c.*1760; detail from copper line engraving by C. Bowles after J. Clevely jr.

SOURCE: reproduced by permission of Westcountry Studies Library, Exeter (Somers Cocks 1922).

Type of ship
Fishing   Sack

Number of ships
— 25
— 15
— 10
— <5

Number of ships
— 80
— 50
— 20
— <5

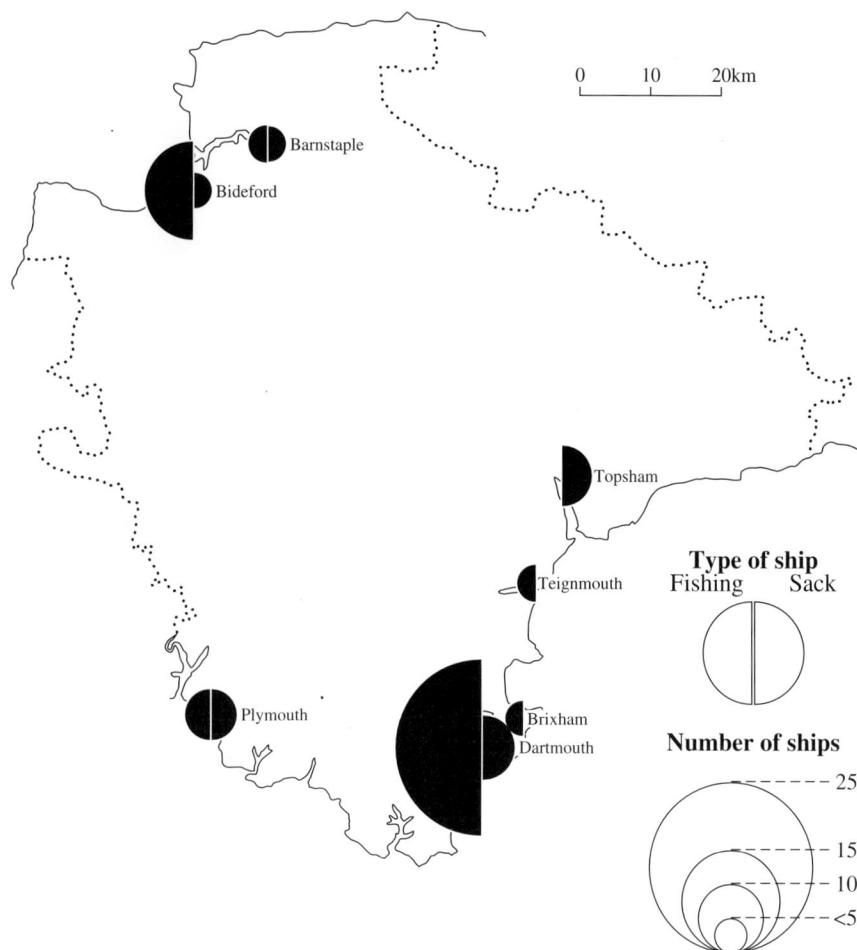

Map 49.2 (above left) Devon 'fishing' and 'sack' ships at Newfoundland by home port, annual average for 1675, 1681 and 1684.

SOURCE: as **Map 49.1**

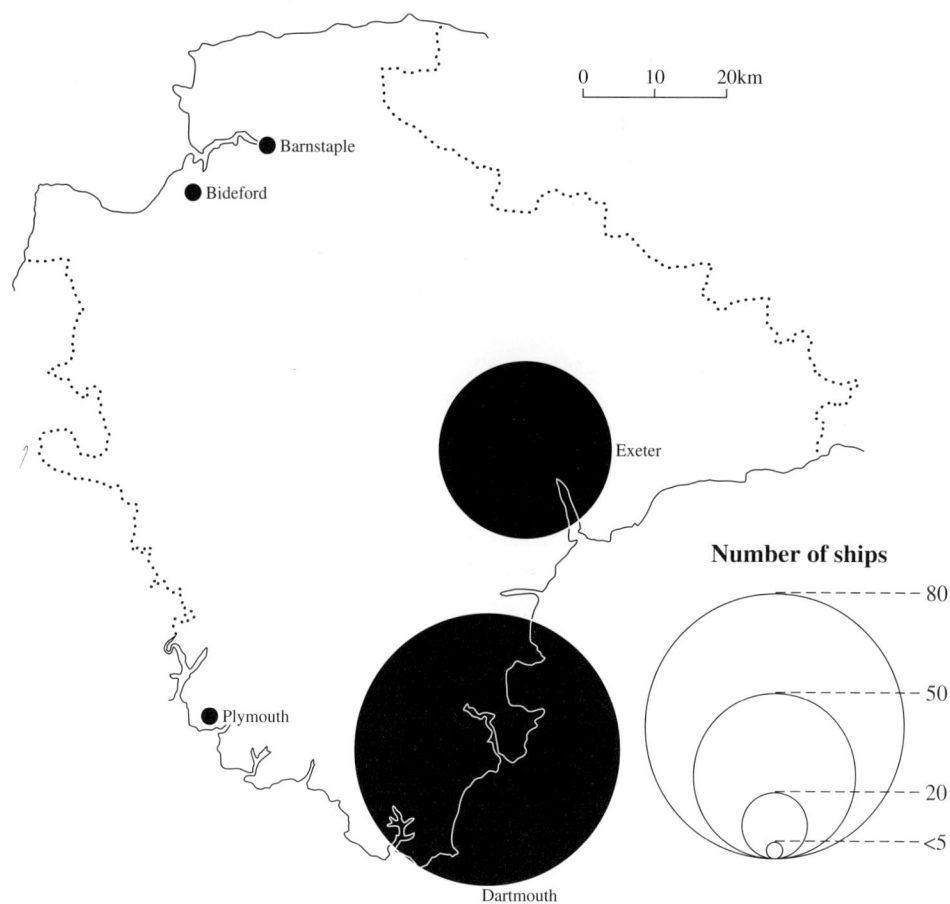

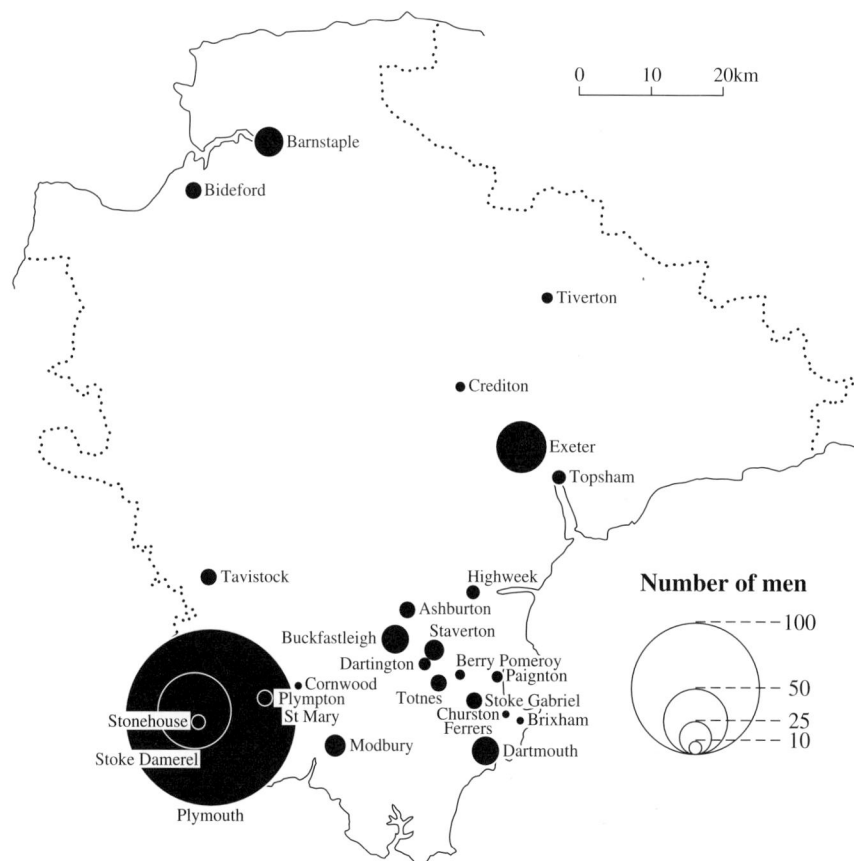

Map 49.3 (above right) Ships clearing for Newfoundland from the Devon ports, annual average for 1770, 1772, 1774, 1788, 1790 and 1792.

SOURCE: as **Map 49.1**.

Map 49.4 Devon parishes (place of birth) supplying five or more men to twelve Royal Navy vessels commissioned at Plymouth between 1770 and 1779.

SOURCE: as **Map 49.1**.

Number of men
— 100
— 50
— 25
— 10

dockyards, with Portsmouth coming second with 2,385 workers.[6] About 1807, in the midst of the great struggle with Napoleonic France, in terms of moorings for laying up warships in the national naval dockyards Plymouth had fifty-nine such moorings for battleships compared to Portsmouth's forty-eight, although when the additional moorings for frigates and sloops are taken into account Portsmouth just surpassed Plymouth with total moorings of eighty-four compared to seventy-nine. Over the wars of the eighteenth century the 'western squadron' based at Plymouth escalated, and Torbay became, as Dr Michael Duffy has written, 'the essential complement to Plymouth for naval operations to the westward in the French wars: a ready refuge for admirals careful of their fleets [and] a vital last safety line for admirals determined to hang in close to Brest [the principal French Atlantic naval base] as long as possible'.[7]

Using the surviving muster books of twelve of the warships commissioning at Plymouth between 1770 and 1779, it has been shown by Dr Nicholas Rodger that of the 2,923 ratings whose place of birth was recorded, 693 (rather less than a quarter) were born in Devon.[8] The Devon-born seamen came from 128 parishes, of which twenty-six yielded five men or more, as shown in **Map 49.4**. As will be noted, most of the parishes were in the south of the county and especially in and near to Plymouth (whose parishes themselves supplied 133 men) with the Exeter parishes accounting for thirty-nine and inland Tiverton and Crediton eight and seven native-born Devon naval seamen respectively.

The final two maps in this chapter relate to Devon's privateering activity, a maritime business which has been most ably explored both nationally and regionally by Dr David Starkey.[9] **Map 49.5** indicates the privateering commissions taken out by twelve Devon ports and havens in the wars between 1689 and 1815. Altogether 423 vessels were so licensed, with Plymouth and Dartmouth leading the fleet with 152 and 123 vessels respectively. The lesser centres of Teignmouth, Brixham and Exmouth commissioned seven, four and three vessels respectively, while Appledore and Totnes each commissioned a single privateer. The raison d'être of privateering was the profit distributed to shipowners, masters, officers and crew derived from the sale of captured enemy vessels and their cargoes condemned as legal prize. **Map 49.6** shows the successful seven Devon ports (some of whose prizes were taken in consort with non-Devon captors) in the war years between 1702 and 1809. Dartmouth interests were the most fortunate with fifty-eight prize vessels followed by Plymouth with thirty-four, while, of the lesser ports, Teignmouth gained three prizes and Ilfracombe one.

FURTHER READING

The discussion summarized in this chapter is elaborated in Stephen Fisher, 'Devon's maritime trade and shipping, 1680–1780', in Michael Duffy et al. (eds), *The New Maritime History of Devon, Vol. I* (London, 1992), 232–41.

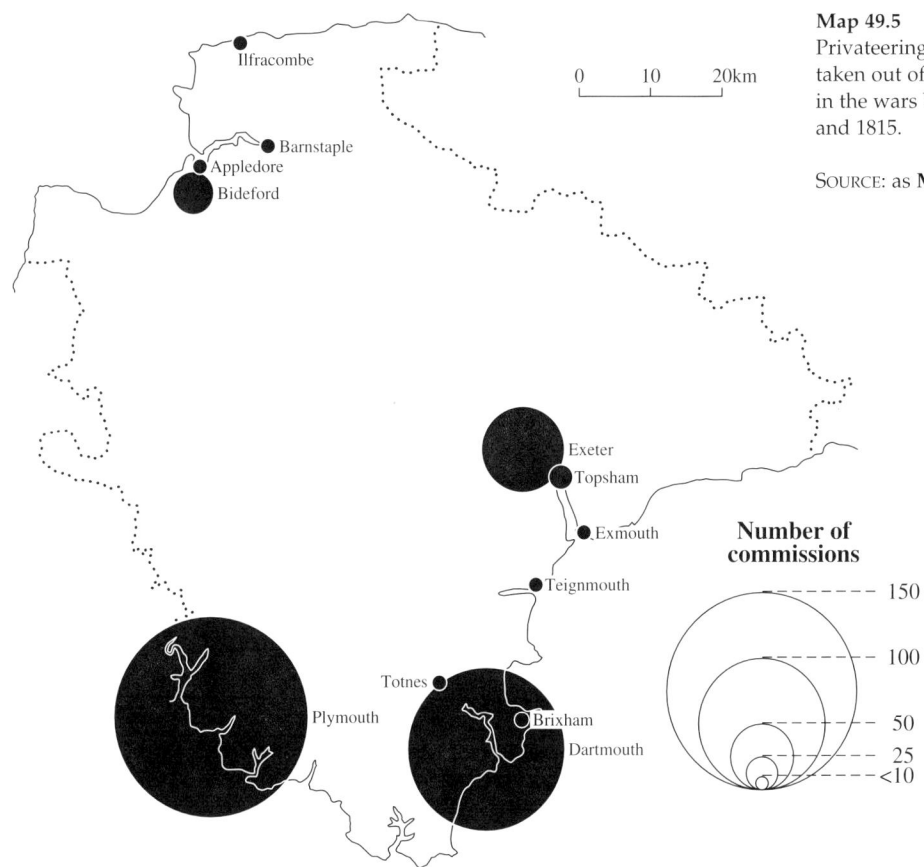

Map 49.5
Privateering commissions taken out of the Devon ports in the wars between 1689 and 1815.

SOURCE: as **Map 49.1**.

Number of commissions

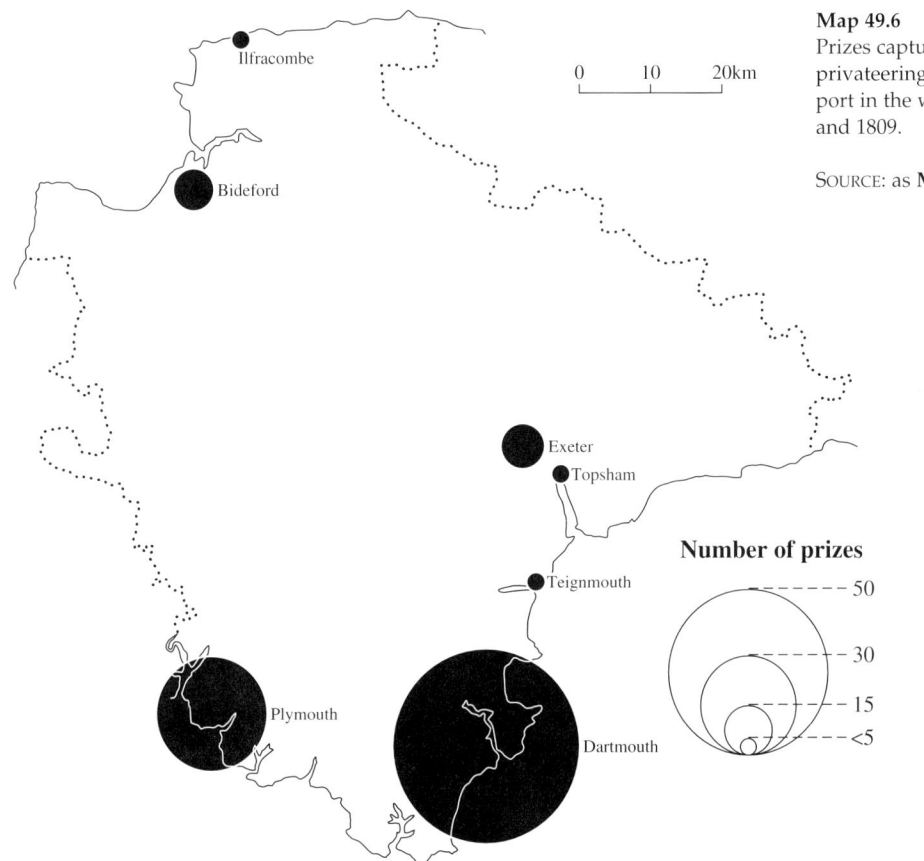

Map 49.6
Prizes captured by Devon's privateering vessels by home port in the wars between 1702 and 1809.

SOURCE: as **Map 49.1**.

Number of prizes

MARITIME ACTIVITIES

SEA-BORNE TRADE, FISHING AND MARINE RECREATION SINCE 1800

DAVID J. STARKEY

THE SEA TRANSPORT INDUSTRIES OF SOUTH-WEST ENGLAND, 1800–1914

The transportation of goods and people by sea gives rise to various maritime activities. Cargoes and passengers require safe and convenient waterside facilities at which to embark and discharge, demands which are met by the harbours, wharves, docks and other amenities operated by port authorities. The conveyance of goods and people over the oceans or along the coasts is the business of shipowners who, in turn, purchase and utilize the products of the shipbuilding industry. Discrete, yet closely related, these sea transport industries are supported and supplied by an ancillary sector which ranges from marine insurance and ship agency to warehousing, and from salvage to marine engineering.

During the nineteenth century, population growth and the spread of industrialization stimulated a vast increase in international trade, a large proportion of which passed through British ports. While the bulk of this traffic was conducted in home-owned vessels, Britain's shipping industry also dominated the world's carrying trades. Thus, by 1912 nearly one in three of the cargoes shipped between foreign ports, 55 per cent of those conveyed between the Empire and foreign countries, and some 92 per cent of inter-imperial shipments were carried in British holds.[1] Confirming this maritime pre-eminence, a substantial proportion of the world's tonnage—peaking at over four-fifths in the early 1890s—was launched from British shipyards.[2]

Trade, particularly overseas trade, was largely channelled through London, Liverpool, Glasgow, Hull and the coal ports of the North East and South Wales— the major ports which served the principal industrial hinterlands. It was in these places that the massive docks of the Victorian age were built, and it was here also that the main shipping firms and their satellites gravitated.[3] The shipbuilding industry, responding to different locational factors, such as the pull of coal, iron and engineering, centred increasingly in the North East, the Clyde, and other northern districts from the 1850s.[4]

While spatial concentration marked Britain's sea transport industries as the nineteenth century wore on, regions distant from the industrial areas continued to engage significantly in port, shipping and shipbuilding activity. In south-west England, for instance, the absolute level of traffic handled by the region's ports grew steadily down to the early twentieth century, with coastwise trade proving notably buoyant.

PORTS AND SEA-BORNE TRADE

The term 'port' relates to both administrative and business units. It refers to a specified stretch of coastline administered for revenue purposes by customs personnel based at a designated 'head' port. It is likewise applied to the places— which are generally the customs head ports—where shipowners are obliged to register their vessels. Ports in the sense of commercial undertakings established for loading and discharging vessels function within these customs and registry ports. Comprising docks, harbours, piers, wharves and other forms of infrastructure, such concerns are operated by various types of port authority.[5]

In 1881, as **Map 50.1** shows, the shores of the South West were divided into thirteen customs ports, seven of which were centred in Cornwall, and six in Devon. The limits to these administrative units had changed in certain respects since 1800. For instance, the customs port of Ilfracombe was abolished in 1839 and redesignated as part of Barnstaple, while Teignmouth was granted 'independence' from Exeter in 1852, and St Ives lost its customs port status in 1860, forming part of Penzance until 1864, and then part of the newly created customs port of Hayle. Moreover, they were to alter by 1914, with St Ives reappointed in the place of Hayle in 1882, and Bideford 'integrated' into Barnstaple in 1886.[6] Though these head ports invariably functioned as registry ports, the reverse was not always true. Thus, registries were established at Brixham in 1863 and Salcombe in 1864, though both ports remained part of Dartmouth for customs purposes, while St Ives maintained its registration role even when it was part of the customs ports of Penzance or Hayle between 1860 and 1882.

With regard to the volume of traffic recorded at these customs ports, certain clear trends were evident between 1800 and 1914. At the aggregate level, activity tended to increase as the period progressed. Though the figures are not entirely compatible, it would seem that the tonnage entering the South West's ports, which stood at 48,349 tons in 1789, had more than doubled by 1840.[7] Thereafter,

Map 50.1
Ports, 1881.

SOURCE: *British Parliamentary Papers*, 1882, LXII, A return of the names of the port and harbour authorities in the United Kingdom in August 1881, 409.

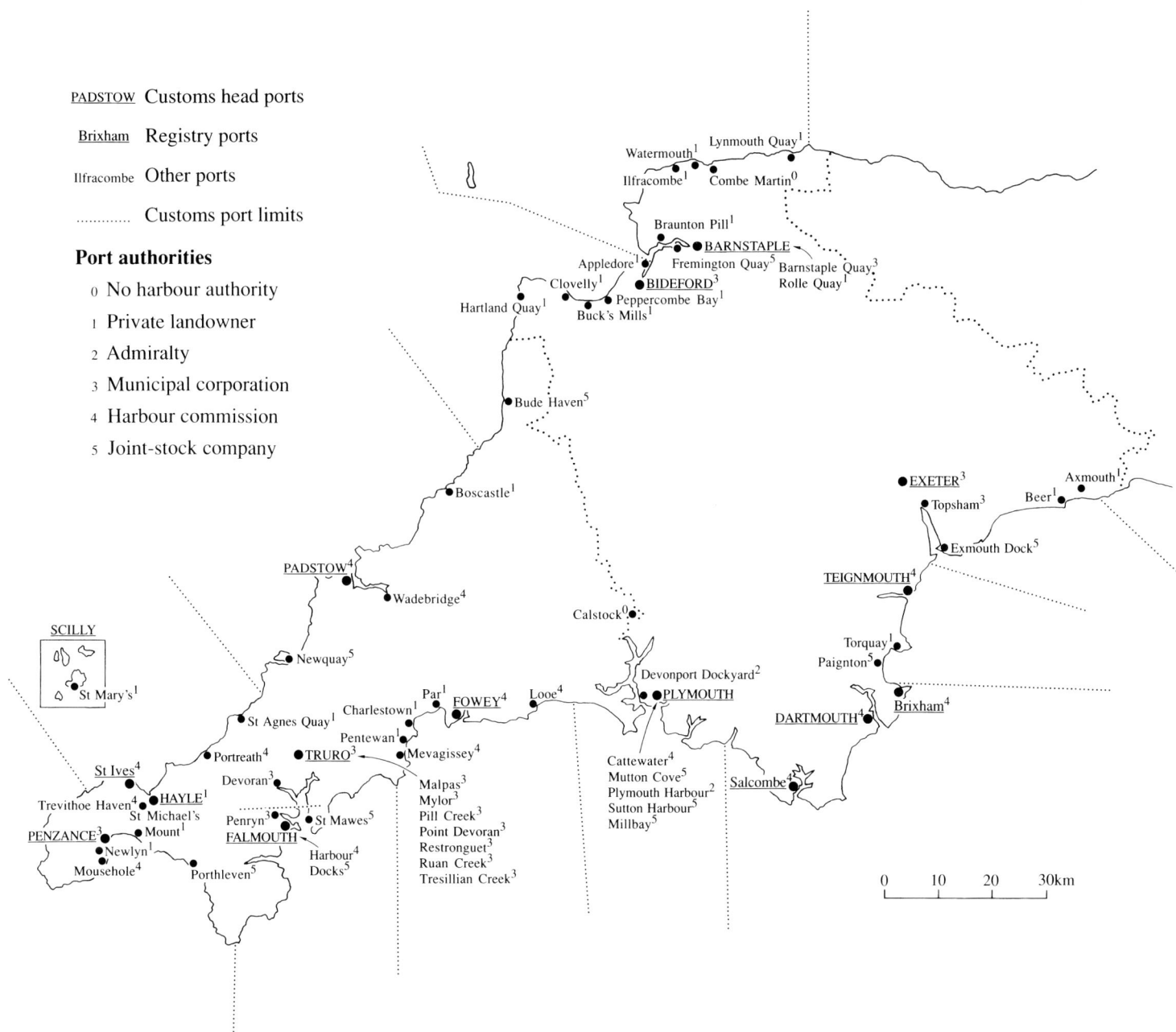

PADSTOW Customs head ports

Brixham Registry ports

Ilfracombe Other ports

.... Customs port limits

Port authorities

0 No harbour authority

1 Private landowner

2 Admiralty

3 Municipal corporation

4 Harbour commission

5 Joint-stock company

Watermouth[1]
Lynmouth Quay[1]
Ilfracombe[1]   Combe Martin[0]
Braunton Pill[1]
BARNSTAPLE
Appledore[1]   Fremington Quay[5]   Barnstaple Quay[3]
Clovelly[1]   BIDEFORD[3]   Rolle Quay[1]
Hartland Quay[1]   Peppercombe Bay[1]
Buck's Mills[1]

Bude Haven[5]

EXETER[3]   Axmouth[1]
Topsham[3]   Beer[1]
Boscastle[1]
Exmouth Dock[5]

PADSTOW[4]   TEIGNMOUTH[4]
Wadebridge[4]
Calstock[0]   Torquay[1]
SCILLY   Paignton[5]
Newquay[5]
St Mary's[1]   Par[1]   Devonport Dockyard[2]   Brixham[4]
Charlestown[1]   FOWEY[4]   Looe[4]   PLYMOUTH
St Agnes Quay[1]   Pentewan[1]   DARTMOUTH[4]
Portreath[4]   TRURO[3]   Mevagissey[4]   Cattewater[4]
St Ives[4]   Devoran[3]   Mutton Cove[5]   Salcombe[4]
Trevithoe Haven[4]   HAYLE[1]   Malpas[3]   Plymouth Harbour[2]
St Michael's   Mylor[3]   Sutton Harbour[5]
PENZANCE[3]   Mount[3]   Penryn[3]   St Mawes[5]   Pill Creek[3]   Millbay[5]
Newlyn[1]   Point Devoran[3]
Mousehole[4]   Porthleven[5]   Harbour[4]   Restronguet[3]
Docks[5]   Ruan Creek[3]
Tresillian Creek[3]

0   10   20   30km

the increasingly regular and reliable shipping statistics indicate that growth continued. Whereas in 1841 a total of 114,386 tons of shipping, with cargo and in ballast, entered the region's ports in the foreign trade, the equivalent figures for 1881 and 1910 were 277,553 and 1.62 million tons respectively. In relative terms, however, the South West's performance was less impressive, for these totals represented a small proportion—2.9 per cent in 1841, 1.1 per cent in 1881, and 2.6 per cent in 1910—of the tonnage entering England and Wales from abroad. By volume, the South West's coastal traffic was invariably and considerably greater than that engaged in overseas commerce; thus, coastwise entrances, with cargo only, measured 803,804 tons in 1841, 1.3 million tons in 1881, and 1.8 million tons in 1910, over five times its foreign trade equivalent in the latter year. Moreover, the South West tended to maintain its share of the nation's coastal trade, accounting for 9

per cent of entries in 1910 as against 8.3 per cent in 1881 and 8.9 per cent in 1841.[8]

As the tonnage entering the ports of Devon and Cornwall expanded during the nineteenth and early twentieth centuries, so the volume of outward traffic increased, with coastwise clearances again eclipsing those in the foreign trade. However, the volume of traffic clearing the region's ports was never as great as that entering, especially if the vessels sailing in ballast are eliminated from the figures. In 1895, for example, laden vessels measuring 286,466 tons entered the South West's ports from overseas, while only 190,344 tons—over 90 per cent passing through Fowey and Plymouth—cleared thence with cargoes for foreign destinations. Coastwise, a similar imbalance was apparent, with 1.6 million tons of shipping delivering goods to the region as opposed to 1.1 million tons clearing with cargoes.[9]

While this pattern was typical of most areas of southern England, it is significant that in the ports serving the industrial regions, particularly those sited near the coalfields of the North East and South Wales, clearances with cargo vastly outnumbered laden vessels entering. Underlying these contrasts, of course, was the character and productivity of the respective regional economies. Ports exhibiting a marked surplus of clearances—like Cardiff and Newport, Newcastle and Sunderland—generally served the more productive hinterlands. On the other hand, a preponderance of entrances, together with a declining share in the nation's foreign trade, as occurred in the South West, implied that lower than average rates of economic growth pertained in the region.

Fig. 50.1
'Picturesque Version' by Peter Orlando Hutchinson of a whale washed up at Beer, 1876.

SOURCE: reproduced by permission of Devon Record Office (Z19/2/8E/221).

The location of Britain's coal deposits explains many of these regional differentials. In general, industrial activity focused increasingly on the coalfields during the nineteenth century, a process reflected in the burgeoning export of manufactured goods through Liverpool, Hull, and Glasgow—a proportion of which 'entered' the ports of less productive areas like Devon and Cornwall. More specifically, in volume terms, coal itself was the single most important commodity handled in British ports. Though it was not until the 1870s that coal was carried overseas in significant quantities, coastwise deliveries from coalfield to consumer formed the backbone of the coastal trade throughout the nineteenth century.[10] The most important facet of this business, as it had been for centuries, was the shipment of coal from the North East to London. However, there were other significant

channels.[11] The South West, for instance, received large quantities of sea-borne coal from both the North East and South Wales. Thus, in 1811 vessels carrying culm and coal from Neath and Swansea formed a vital part of the import trade of Fowey, Plymouth and Dartmouth; in 1829 Devon was the fourth most important coal-importing county, after London, Norfolk and Kent, while Cornwall was seventh; in the 1850s north-eastern coal constituted the mainstay of Teignmouth's inward trade, while coal from South Wales likewise dominated Barnstaple's imports; moreover, from the 1870s, Plymouth, Penzance and Dartmouth were consistently amongst the nation's most important recipients of coastal coal, largely because they accommodated bunkering facilities for ocean-going steamships.[12]

If, at the regional level, there was a tendency for sea-borne traffic to expand between 1800 and 1914, with coal imports underpinning the dominance of the coastwise, inwards flow, not all of the South West's customs ports conformed to the overall pattern, as **Map 50.2** demonstrates. Shipping activity at ports such as Ilfracombe, Hayle and Bideford, for example, declined to such an extent that their appointment as customs ports was terminated, while elsewhere, notably at Plymouth, the volume of trade expanded at a much higher rate than that apparent regionally. Such contrasting fortunes were further reflected in the commercial viability and physical development of the various units of the region's port industry. As **Map 50.1** indicates, there were sixty-seven separate 'ports' in Cornwall and Devon according to a customs return of August 1881, rights over each being exercised by private landowners, the Admiralty, municipal corporations, harbour commissions, or joint-stock companies.[13]

While some of the places identified in the 1881 return—Peppercombe Bay and Watermouth, for instance—were ports in name only, others, like Beer and Mousehole, were principally fishing harbours which handled little, if any, sea-borne trade. Many were working ports, however, and as such their prosperity depended upon a complex of inter-acting factors. At base, there were the physical features of the various ports—depth of water, navigational hazards, space to landward—which might facilitate or hinder the handling of ships and cargoes, an especially important consideration in an age when vessel size was increasing dramatically. The expansion of Falmouth's harbour and dock provisions from the 1870s, for example, reflected the significance of a deep-water site, while, on the other hand, the construction of Exmouth Dock in the 1860s was designed to obviate the navigational difficulties posed by the Exe.[14] Inland, the resource endowments of the hinterland were critical, for a local abundance of bulky, low-value commodities generally stimulated export-oriented port growth. Harbour facilities at Charlestown and Pentewan, for instance, were specifically built as outlets for the china clay of the St Austell district, while the port of Teignmouth's prosperity hinged to a large degree on the export of ball clay from the Bovey basin.[15]

Communications with the interior influenced port development, with railway linkages proving especially significant in the collection and distribution of sea-borne cargoes. Thus, the marked growth in Fowey's export trade from the 1870s was largely inspired by the marriage of a deep-water site and a productive hinterland effected by the construction of the mineral railway from Lostwithiel. Entrepreneurship was a further influence. In this respect, individuals like Joseph Thomas Treffry, founder of harbour facilities at Par and Newquay, and members of the Harvey family, proprietors of Hayle Harbour, played a notable part in the expansion of the South West's port capacity, while the capital and initiative of railway companies underlay the establishment of cargo-handling provisions at Fremington Quay, Exmouth Dock, and elsewhere.[16]

With the opening of Millbay Docks in 1857, and the development of substantial cargo-handling facilities in the Cattewater during the 1870s and 1880s, the port of Plymouth acquired the modern infrastructure which set it apart from the region's other landing places. Accordingly, by 1910, as **Table 50.1**

	Coastwise	Foreign	Total
Barnstaple	414.5	8.8	423.3
Dartmouth	321.9	1,205.0	1,526.9
Exeter	129.2	13.5	142.7
Plymouth	1,851.6	6,738.1	8,589.7
Teignmouth	386.2	36.8	423.0
Falmouth	776.8	1,230.7	2,007.5
Fowey	485.9	341.9	827.8
Padstow	46.0	0.8	46.8
Penzance	489.3	55.3	544.6
Scilly	38.4	43.5	81.9
Devon	3,103.4	8,002.2	11,105.6
Cornwall	1,836.4	1,672.2	3,508.6
South West	4,939.8	9,674.4	14,614.2

Table 50.1
Total shipping movements* in south-western ports, 1910 ('000 tons).

SOURCE: *Annual Statement of Navigation and Shipping,* 1910.

NOTE: * Total shipping movements comprise arrivals and departures in the foreign trade, and entrances and clearances in the coastal trade, with cargo and in ballast in both cases. It provides a measure of shipping activity rather than the trade passing through each port, for it includes vessels which called at ports without entering them. Thus, Plymouth's figures include the liners which disembarked passengers in the Sound, while departures from Dartmouth include the steamships which called to pick up bunker coal

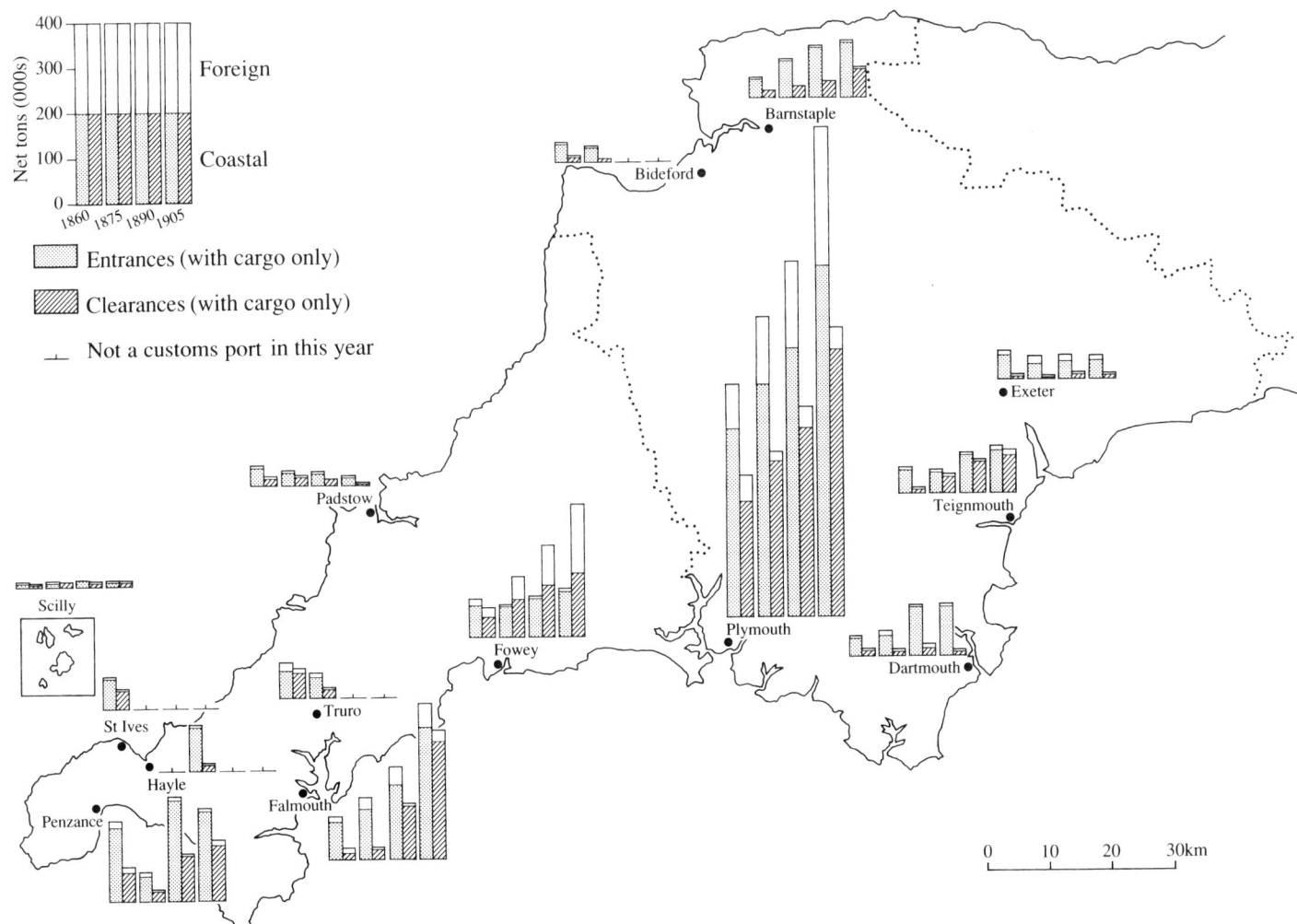

Map 50.2
Sea-borne trade, 1860–1905.

SOURCE: *Annual Statements of Navigation and Shipping.*

indicates, over 58 per cent of the South West's total shipping movements (arrivals and departures, with cargo and in ballast) occurred at Plymouth, including over 70 per cent of those in the foreign trade. Prominent amongst this traffic were the great transatlantic liners which regularly called in the Sound so that passengers could disembark and board the London train. The bedrocks of the port's business, however, were the mundane imports of coal, Baltic timber, Black Sea grain, and South American fertilizers—some of which were disseminated coastwise to the region's lesser harbours—and the export of locally won clay, stone and minerals.[17]

Nearly all of these shipments were carried in steamships. Requiring deep water and extensive waterside provisions, the growing utilization of the steamer from the 1860s had been a vital factor in Plymouth's expansion and, conversely, in the contraction of activity at less well-endowed ports like Exeter, Salcombe and Padstow.[18] It also had a significant bearing on the shipping and shipbuilding industries of the South West.

SHIPOWNING IN THE SOUTH WEST

While the growth in the volume of Britain's sea-borne trade was accompanied by an expansion in the nation's shipping industry during the 1800–1914 period, this was not altogether true in the South West. Here, the tonnage registered increased steadily until the third quarter of the nineteenth century when a sharp decline set in at most, but by no means all, of the region's registry ports. That the South West's maritime traffic continued to grow in the half century prior to the First World War clearly suggests that the greater part of this business was conducted in vessels registered elsewhere.[19] In essence, this divergence

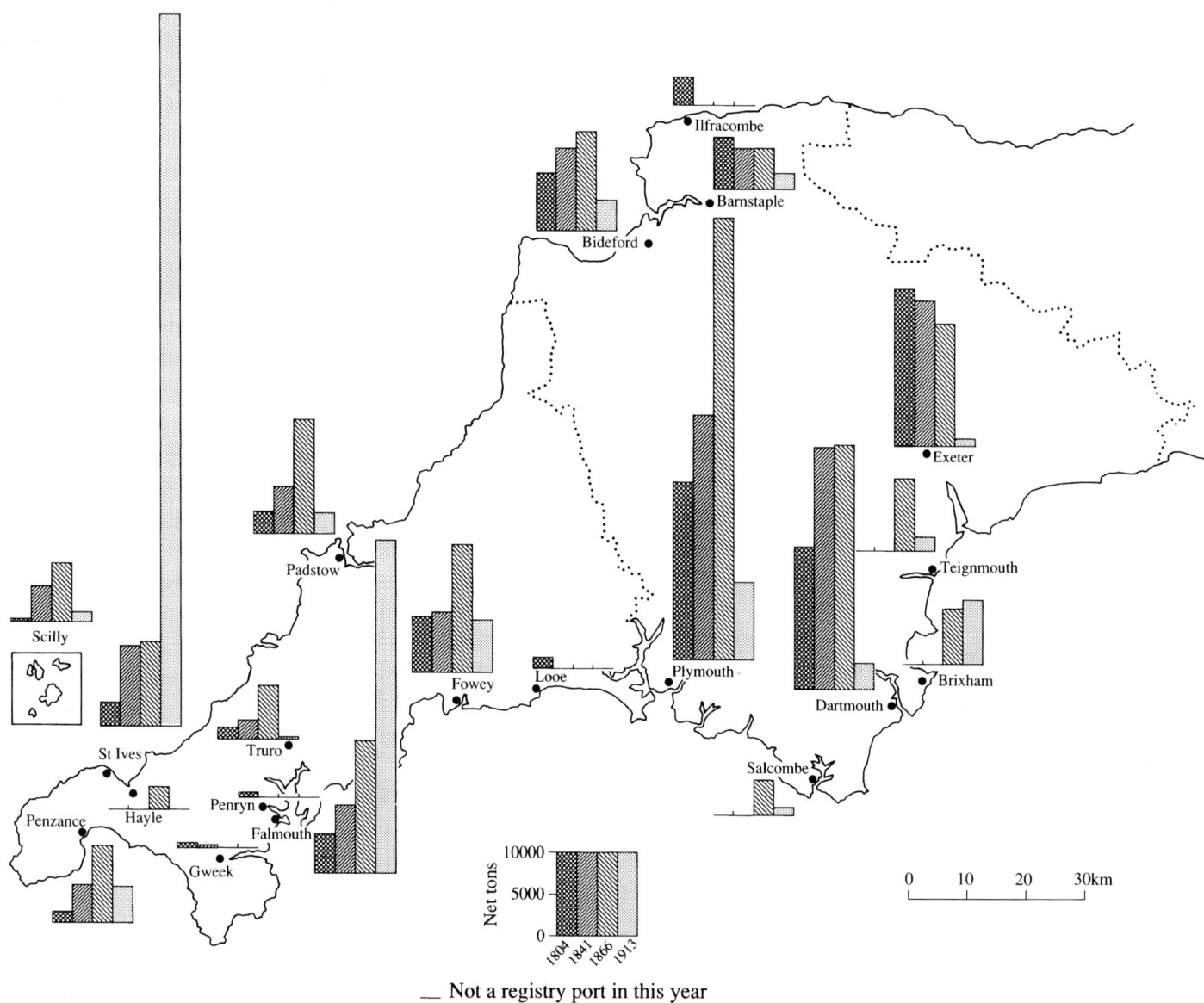

Map 50.3
Tonnage of ships registered in the South West, 1804–1913 (the tonnages owned by the New Zealand Shipping Company and Reardon-Smith Line, registered at Plymouth and Bideford respectively in 1913, are excluded).

SOURCE: 1804: PRO, CUST 17/26; 1841: *British Parliamentary Papers, 1843, LII*, Accounts Relating to Shipping; 1866: *Annual Statement of Trade and Navigation*; 1913: *Annual Statement of Navigation and Shipping*.

__ Not a registry port in this year

between local shipping and local trade was a reflection of the growing concentration of the British shipping industry in the major ports. Central to these developments was the shift from sail to steam which gathered an increasingly irresistible momentum from the 1860s onwards.[20]

The pattern of shipowning in the South West between 1804 and 1913 is indicated in **Map 50.3**. In 1804 some 95,733 tons of shipping were registered in the region, a fleet constituting 5.4 per cent of the tonnage owned in England and Wales. A slightly higher than average growth rate was apparent in the South West's stock of shipping in the first half of the nineteenth century, so that in 1841 nearly 132,000 tons, 5.9 per cent of the national aggregate, were registered in Devon and Cornwall. This expansion continued during the next two decades, the South West's shipping industry reaching its greatest extent in the mid-1860s, with 2,108 vessels, measuring a total of 212,610 tons, belonging to the region's ports in 1865. Already, however, there were signs that the local rate of increase was beginning to lag behind that of other areas, for this fleet represented just 4.6 per cent of the tonnage registered in England and Wales.

Thereafter, the South West's shipping industry declined both absolutely and relatively, though the official returns mask the true extent of the demise. Thus 267,311 tons were registered in south-western ports in 1913, just 3 per cent of the national aggregate. However, these figures include the vessels of the Cardiff-based Reardon Smith line, which measured 12,998 tons and were registered for sentimental reasons at Bideford, and the New Zealand Shipping Company's steamers, which totalled 86,379 tons and were registered at Plymouth, one of their ports of call.[21] Excluding these extra-regional tonnages from the calculation, the South West's shipping industry comprised 167,934 tons in 1913, just 1.9 per cent of the total for England and Wales. Over half of the regional aggregate (84,911 tons) belonged to a single firm, the Hain Steamship Co. Ltd of St Ives, and a further third (56,026 tons) was registered at Falmouth by R.B. Chellew.[22] Shipowning in Cornwall and Devon was clearly much less pervasive on the eve of the First World War than it had been during the Napoleonic Wars.

Vessels belonged to twenty registry ports in the South West during the course of the nineteenth century, as **Map 50.3** illustrates. In 1804 south Devon dominated the region's shipping industry, with over 56,000 tons of shipping belonging to Dartmouth, Exeter and Plymouth, as against 16,000 tons in north Devon and 22,460 tons dispersed amongst ten Cornish ports. Growth in the shipping stock occurred at most of the region's ports down to the mid-1860s, though the rate of increase was most marked in Cornwall, notably at Falmouth, Fowey, Padstow and St Ives, and at Plymouth, where 52,525 tons, roughly a quarter of the South West's tonnage, was registered in 1866. By 1913, Cornwall had clearly eclipsed its eastern neighbour, accounting for 82.8 per cent of the tonnage owned in the South West. While St Ives (which meant Hain's) and Falmouth (which meant Chellew) were now the centres of the south-western

shipping industry, shipowning had declined to negligible proportions throughout Devon.

Local investment in steamships was atypical. Large quantities of capital were needed to purchase, maintain and operate steam-driven vessels, while deep-water ports, with extensive docking and cargo-handling facilities, and abundant supplies of locally produced goods for shipment, were also desirable. Such prerequisites were largely lacking in the South West. This is not to say that south-western interest in the new technology was entirely lacking. Indeed, there were numerous ventures into steamship ownership, though these tended to be the exceptions that proved the general rule. Thus, west-countrymen like J. Anning, W.J. Tatem and William Reardon Smith left the region to establish significant shipping enterprises in London, Cardiff, Liverpool and other vibrant centres of maritime activity. There were others who stayed in the locality, though their investments in steamships were usually secondary to their other maritime interests. For instance, J.A. Bellamy of Plymouth, who managed twelve steamers between 1883 and 1905, the largest being the *Endsleigh* of 3,079 gross tons, and R.S. Triplett, who owned five Plymouth-registered steamships in the late 1880s, both derived the bulk of their income from their ship agency businesses. Likewise, Holman & Son of Topsham— owners of eleven Exeter-registered steamships between 1870 and 1884—were shipbuilders, ship agents and marine insurers as well as shipowners, before they, too, shifted their operations to London.[23]

Fig. 50.2
Drawing by Peter Orlando Hutchinson of the wreck of the *Sarah* at Sidmouth, 1877.

SOURCE: reproduced by permission of Devon Record Office (Z19/2/8E/247).

Overshadowing these Devonian steamship owners, however, were two Cornish firms, Chellew of Truro and Hain of St Ives. While R.B. Chellew was essentially a steamship manager, Edward Hain had the foresight to perceive in the 1860s that long-term survival as a shipowner entailed extensive investment in steam shipping. With the unstinting support of the Bolitho Bank, and an unusually long association with Readheads, shipbuilders of South Shields, Hain established a shipping line which comprised thirty-seven modern steamers of 84,911 tons by 1913. Based in St Ives throughout, though its vessels were deployed in the general tramp market, never entering their home port, the firm was eventually purchased by P&O in 1917.[24] Though extraordinary in a south-western context, Edward Hain and R.B. Chellew were living proof that entrepreneurial ability was not one of the shortcomings of the region's shipping industry.

THE SOUTH WEST'S SHIPBUILDING INDUSTRY

Shipbuilding in the South West was largely conditioned by the extent and character of the region's shipping industry during the nineteenth century. Accordingly, as the stock of locally registered shipping expanded down to the 1860s, so the output of south-western shipyards, both in terms of new building and repair work, remained buoyant. Subsequently, the contraction apparent in the region's fleet was reflected in a decline in the number of vessels launched and repaired, and in the closure of many shipbuilding firms. This pattern is evident in the data for new building presented in **Map 50.4**. Thus, in 1804, 7,224 tons of shipping were launched from the South West's yards, production rising to over 10,000 tons by 1866 before falling precipitously to less than 2,000 tons in 1913.

In 1804, the most considerable of the South West's shipbuilding enterprises, employing sixty-eight shipwrights and

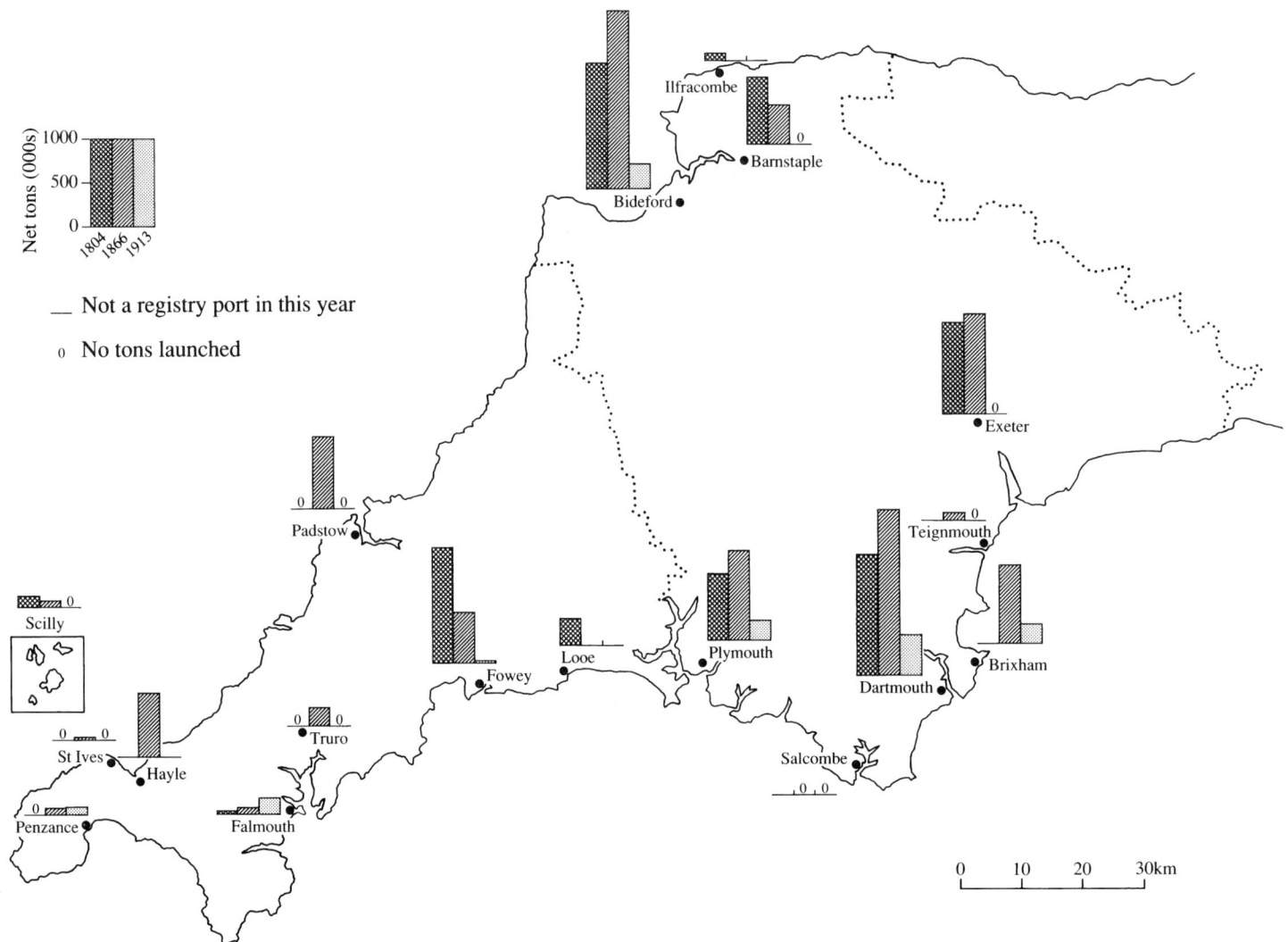

Map 50.4
Tonnage of ships built and registered in the South West, 1804–1913.

SOURCE: as **Map 50.3**.

thirteen apprentices, was that owned by Benjamin Tanner of Dartmouth. This was one of the largest workforces in the country according to one official return. But it was exceptional in the South West, where most shipbuilders employed less than twenty workers and apprentices.[25] The land and capital requirements of these producers were also modest. A flat, dry, waterside site, equipped with a shed for storing cut and treated timber, a saw pit, and a steam box for bending wood were the principal fixed assets of the typical nineteenth-century west-country shipyard, with timber the most important item of working capital.[26] The ready availability of such productive factors facilitated entry into the shipbuilding industry, while the fluctuating nature of demand, even at times of prolonged growth in the shipping stock, meant that many enterprises were short-lived. Accordingly, a host of firms contributed to the South West's shipbuilding output. At Brixham, for instance, twenty-four different builders launched vessels which were subsequently registered in the registry port of Dartmouth between 1824 and 1854, while Plymouth's registers indicate that no fewer than forty-three different builders operated in the port during the same period. Similarly, to the west, the products of at least eighteen local shipyards were registered at Fowey during the first half of the nineteenth century.[27]

Though the local shipping industry was a vital market for the South West's shipbuilders, it was not the sole source of demand. In the Napoleonic Wars, for instance, many merchant shipyards were kept busy fulfilling orders from the Admiralty. While the Falmouth firm of Symons built the *Dispatch* in 1804, few other warships were launched in Cornwall. In contrast, at least seventy naval vessels were constructed in Devon yards between 1803 and 1814.[28] Demand for vessels built in the South West also came from other parts of the country. In the 1850s, for example, shipowners in South Wales, Liverpool and London purchased a number of relatively large barques and fully rigged ships built at the yards of Westacotts of Barnstaple, Cox and Evans of Bideford, and Holmans of Topsham.[29]

With such firms deploying extensive facilities—dry docks, covered slips, power-driven machinery—for large-scale wooden shipbuilding, the South West's shipbuilding industry reached a peak.[30] Already by this time, however, the inexorable shift in the location of the industry had commenced. Exploiting the immense advantages accruing from the proximity of supplies of iron and steel, coal, and engineering expertise, shipbuilding firms on the Clyde, the Mersey, in the North East, Belfast, and Barrow became dominant in the production of metal-hulled, steam-driven vessels. This shift was associated with the changing structure of the shipping industry which saw the decline of shipowning in the South West. Losing its principal source of demand, the shipbuilding industry of Devon and Cornwall contracted rapidly from the late 1870s.

PORTS AND SEA-BORNE TRADE SINCE 1914 (Map 50.5)

On the eve of the First World War, the sea-borne trade of the South West was divided for government revenue purposes into twelve

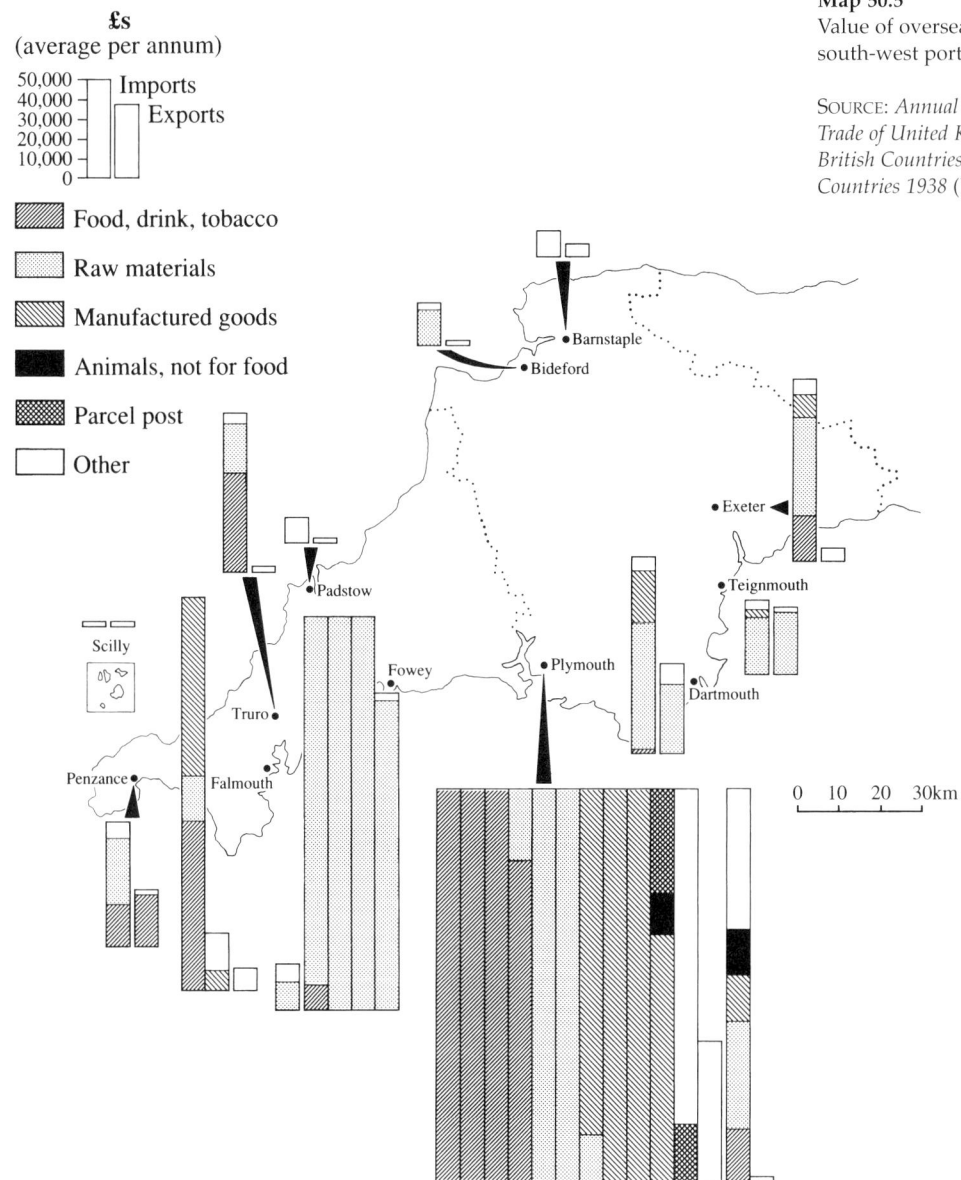

Map 50.5
Value of overseas trade at south-west ports, 1934–1938.

SOURCE: *Annual Statement of the Trade of United Kingdom with British Countries and Foreign Countries 1938* (London, 1939).

£s
(average per annum)

Imports
Exports

50,000
40,000
30,000
20,000
10,000
0

Food, drink, tobacco

Raw materials

Manufactured goods

Animals, not for food

Parcel post

Other

customs ports, seven in Cornwall and five in Devon. A similar number existed in the region in 1938, though by this time there were just six Cornish head ports, St Ives having been subsumed into Penzance, while Devon's complement also stood at six, Bideford having been detached once more from Barnstaple. Half a century later, in 1990, only seven of these administrative units served the region—Exeter, Teignmouth and Plymouth in Devon, and Fowey, Par, Penzance and Scilly in Cornwall—though now they were described as customs trade ports and defined as 'shipping places where trade is large enough to justify separate analysis of customs data'.[31] Within the customs port or customs trade port, of course, there were usually distinct port units where goods were actually loaded or discharged. In the twentieth century, the number of these shipping places in operation in the South West has tended to decline. Whereas in 1881 some sixty-seven west-country shipping places were identified by the

Customs Board, in 1990 the Department of Transport compiled returns from just twenty-one port authorities and undertakings operating in the South West.[32]

The volume and character of the trade handled by the South West's ports changed both in degree and kind during the twentieth century. Before the First World War, the pattern of the region's sea-borne commerce exhibited several clear characteristics: in absolute terms, the level of the South West's trade was increasing steadily, though relative to the UK total it tended to fall; imports greatly exceeded exports; coastal trade exceeded foreign; coal arriving coastwise was the single most important commodity flow; clay was the region's only significant bulk export; and Plymouth's business tended to expand as the trade of other south-western ports stagnated or declined. Though well established, this trading pattern was greatly disrupted by the First World War. In particular, the coastal coal trade, the bedrock of the traffic handled by most of Britain's smaller ports, especially in southern England, contracted massively during the war years as enemy action took a heavy toll and the government, anxious to maintain supplies, subsidized the carriage of coal by rail.[33]

Activity levels at many south-western ports never really recovered from these hammer blows, a fate common to many minor ports.[34] In Devon, for instance, the volume of coastal shipping entering the county's ports declined to just a third of the 1913 level in 1918, and clearances to less than a fifth. Moreover, pre-1914 traffic levels were not attained during the inter-war years, the tonnage entering Devon's ports coastwise reaching a peak for the 1919–39 period in 1929, but at a level that represented just 60 per cent of the 1913 figure. The Second World War further disrupted coastal trade in the South West, though some ports benefited from a government-directed increase in coal imports. Again, the post-war recovery was generally slow and modest, particularly in south-western havens such as Bideford which relied on the traffic in domestic coal. The long-term demise of the coal trade, instigated by the onset of war in 1914, largely explains the contraction and virtual closure of ports such as Exmouth, Dartmouth, Bideford, Barnstaple, Hayle, Padstow and Penzance during the last twenty years.[35]

As **Map 50.5** indicates, imports predominated at every south-western port with the exception of Fowey and its sub-ports (Charlestown, Mevagissey, Par, Pentuan and Polkerris), whence commodities with a mean value of £762,118 per annum cleared for overseas destinations. Fresh, cured and canned fish comprised a small part of this business, with clay accounting for 90 per cent of the port's export trade. Peaking at £913,743 in 1937, the shipment of clay through Fowey was the single most valuable commodity flow in the South West between the two World Wars.

Plymouth's overseas trade, however, was much more significant in aggregate terms. Here an average of £2.27 million worth of goods was imported from overseas each year, some ten times the value of commodity exports (**Map 50.5**). Compared to other south-western ports, the range of goods handled at Plymouth was immense, with clay the leading export, and large

quantities of food imported, most notably maize, as well as metal ores, timber of all varieties, chemical goods, paper, animals and, most significantly of all, refined petroleum. Another category of import, parcel post, valued at £108,000 in 1937, points to a further aspect of Plymouth's traffic, the ocean liner trade. Commencing at Plymouth in the mid-nineteenth century, the business was initially dependent for its viability on goverment mail contracts. Once the railway reached the port in 1850, Plymouth could exploit its westerly location, both to embark later mails than was possible at London or Southampton, and to land homeward-bound mail, which could be in London long before the ship even reached Southampton. These advantages were soon appreciated by passengers, and long before 1914 Plymouth was a port of call for the great ocean liners of the day. This high-profile business peaked in the inter-war years, with at least 500 vessels and 29,000 passengers annually arriving at or leaving Plymouth between 1925 and 1935, a total that comprised 'the great and the good, the famous and glamorous people of the day, as well as thousand upon thousand of less exotic but still worthy folk'.[36]

The ocean liner was one of the casualties of the transport developments of the second half of the twentieth century. While the advent of the jet aircraft fatally wounded Plymouth's prestigious passenger liner business, the unerring development of the road network accentuated the difficulties of the coastal traders, whose business had been increasingly undermined by competition from the railways since the First World War. With traffic, notably the coal trade, transferring to the rail network at an accelerating rate from 1945, the intensification of road competition from the 1960s effectively brought to an end commercial activity at minor coastal trade ports such as Exmouth, Dartmouth, Padstow and Bideford in the 1980s.

In the 1990s, sea-borne trade and port activity is relatively less significant to the economies of Cornwall and Devon than it had been in the Edwardian era. In the face of this contraction— perhaps even because of it—the maritime communities of the South West have continued to exploit the resources of the sea and the recreational potential of the coast and inshore waters.

THE SEA FISHERIES (**Map 50.6**)

The twentieth century has seen the South West retaining its fishing interests despite the prolonged depression that afflicted its fisheries between and beyond the two World Wars. Indeed, with the massive contraction of distant-water trawling activity at ports such as Hull, Grimsby, Aberdeen and Fleetwood during the 1970s, the South West has again emerged as one of the UK's leading fishing regions.

On the eve of the First World War, the South West's fishing industry formed a relatively important part of the nation's fisheries, with Cornwall and Devon the home of 5,218 fishermen, nearly 21 per cent of the 25,239 fishermen enumerated in the 1911 census of England and Wales. However, this sizeable regional workforce was not matched by the extent of the South West's stock of fishing vessels, which stood at 21,803 net tons, just 11.5 per cent of the national total. This was essentially due to the simple fact that the average south-western craft, at 13.4 net tons, was barely half the tonnage of the typical fishing vessel of the day. More significantly, perhaps, only 6.4 per cent of the regional fleet was steam-propelled in 1911 as against a national figure of 62.7 per cent.[37] Thus, while Cornwall and Devon possessed a relatively large fishing workforce, the region's vessels were comparatively small and predominantly sail-driven at a time when distant-water steam trawling was increasingly the leading sector of the UK fishing industry.

That the South West was responding slowly to technological change was recognized as early as 1913 when the Devon and Cornwall Local Fisheries Committee applied to the Board of Agriculture and Fisheries for a grant of £10,000 to fund a modernization scheme for fishing vessels in the South West. Numerous south-western vessels were motorized under the terms

of this scheme during the First World War, increasing the catching capacity of each craft by up to three times. But government assistance of this type only marginally mitigated the depression of the South West's fishing industry that commenced in earnest in 1918 and did not abate until the 1970s. This depression was reflected in the decline of the region's fishing fleet, both in terms of its relative quality and in terms of its absolute size. The seeds of the revival in the South West's fishing interests that was to take place in the 1970s were sown in the immediate post-war decades. While some of the region's fishermen moved out of the industry, others took up government grants and loans to purchase and modernize second-hand vessels such as the motor fishing vessels de-commissioned by the Admiralty after the Second World War, and the European vessels which came on to the market when territorial limits were extended to 12 miles in 1965. Diversification in catching methods also took place, with south-western fishermen investing increasingly in lining rather than trawling so that species at the expensive end of the market formed a growing

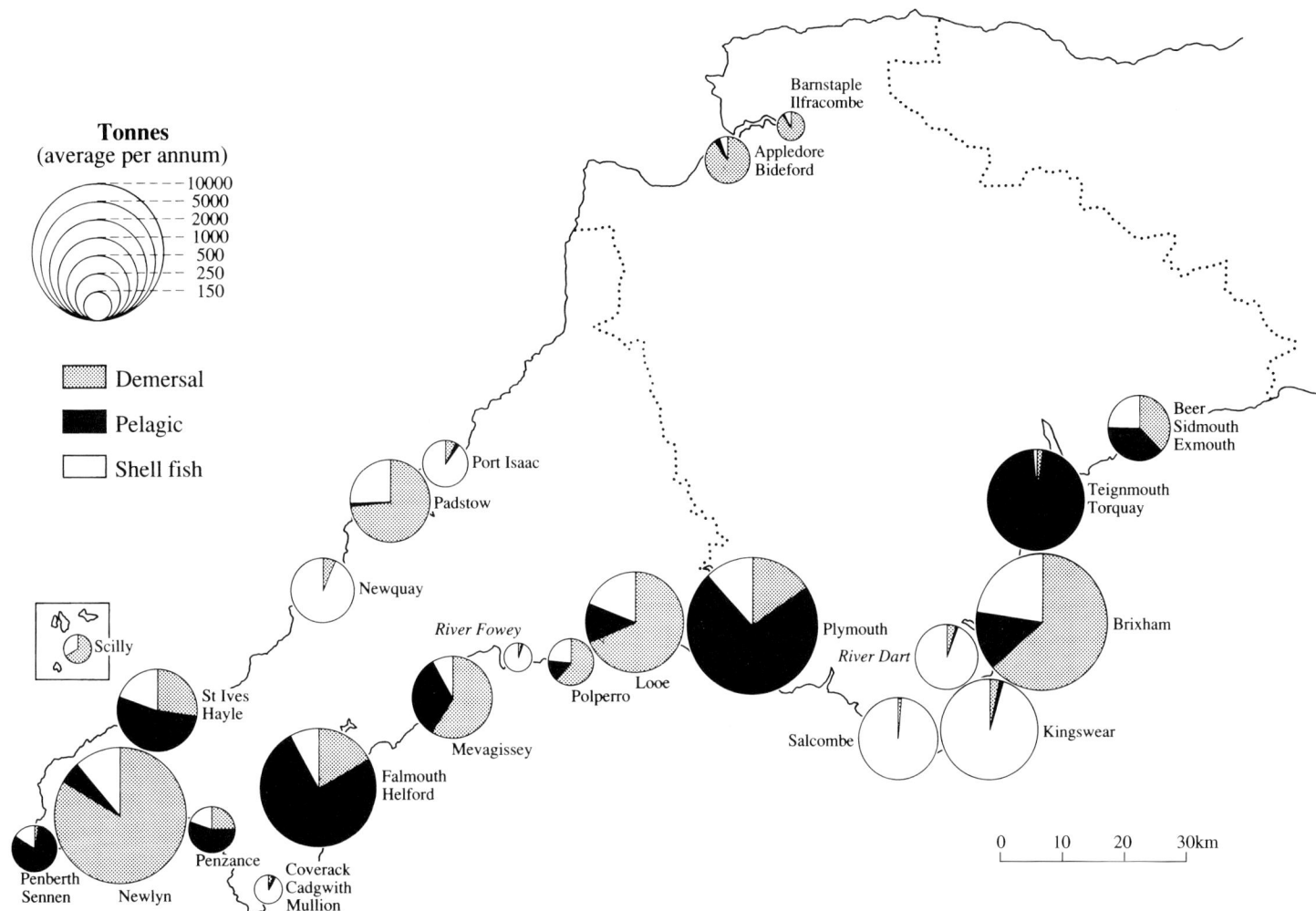

Map 50.6
Fish landed by British vessels at south-west ports, 1986–1990 (small quantities of fish were also landed at Paignton, Cawsand, Portloe, Flushing and Penryn).

SOURCE: Ministry of Agriculture, Fisheries and Food.

part of the region's catch. Marketing also benefited from the stabilization of prices and lower distribution costs deriving from the operations of fish co-operatives such as Brixham and Torbay Fish Limited founded in the late 1960s. As a consequence of these various developments, the South West retained its fishing interests during a period in which the prospects facing the industry were far from propitious. When such conditions evaporated in the early 1970s, with the exclusion of British trawlers from Icelandic grounds and the imposition of quotas under the terms of the European Union's Common Fisheries Policy, the renewed and intensified exploitation of home waters helped to revive the region's fisheries.[38]

Thus, by 1990, 52,000 tonnes of fish were landed in the ports of the South West, a total that exceeded all other regional aggregates in England and Wales.[39] As **Map 50.6** shows, Newlyn and Brixham were the leading centres of the South West's resurgent fishing industry in the 1986–90 period, particularly with regard to the taking of demersal (mainly flat) fish. Plymouth and Teignmouth/Torbay accounted for relatively large catches of pelagic fish, chiefly mackerel, while in Cornwall, Falmouth and Helford were significant in the pilchard fishery. The shell fisheries also constituted an important facet of the South West's fisheries, most notably the harvest of crabs and lobsters off the South Hams.

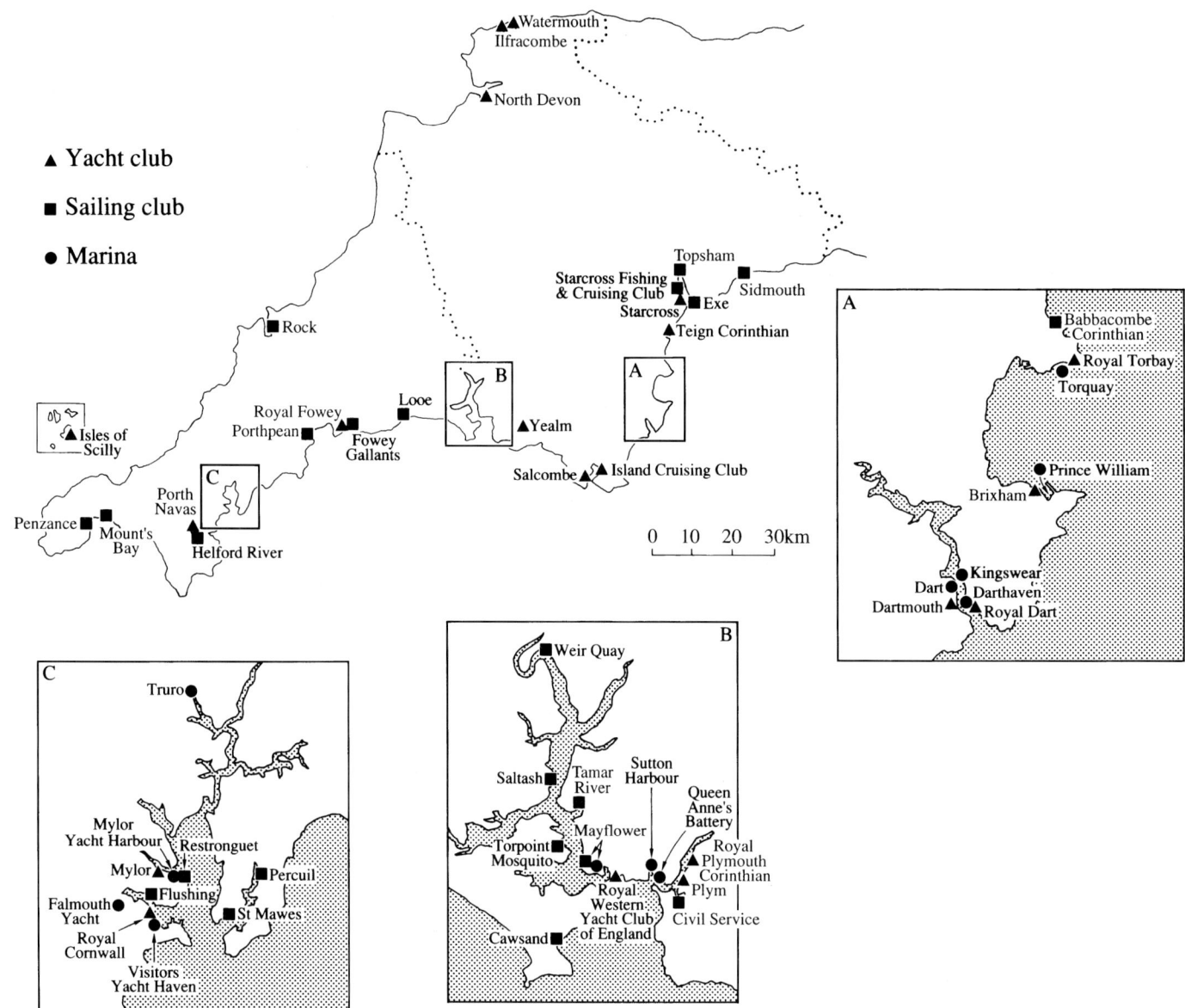

Map 50.7
Yacht clubs, sailing clubs and marinas.

SOURCES: Brian Goulder (ed.), *Pilot Pack, Vol. 3, Bridport to the Scilly Isles* (1989); *Reed's Nautical Almanac* (1988).

MARINE RECREATION (MAP 50.7)

The estuaries and inshore waters of the south-west peninsula have long been used for recreational purposes. Aquatic pastimes such as rowing, boating and swimming have for centuries featured in the lives of the region's coastal communities, such leisure activities becoming more formalized and therefore more visible in the mid-eighteenth century with the establishment of regattas. Teignmouth, in the 1740s, was probably the first west-country port to host such an event, and over time regattas came to occupy a prominent position in the social calendars of many local havens. Yachting, like aquatics, has a long pedigree, the earliest accounts of vessels constructed and deployed specifically for recreational use occurring in the early seventeenth century. Generally the preserve of the aristocratic or gentleman seafarer, sailing for pleasure was not confined to local waters: Lord Courtenay, for instance, embarked for London and the Channel Islands in his yacht *Dolphin* in 1790, while in 1809 Sir William Curtis sailed for the Iberian peninsula to 'see the war with the special permission' of the Admiralty.[40]

Such recreational activities developed rapidly in the final quarter of the nineteenth century as wealth accumulated and the real incomes of much of the population expanded. Aquatic sports and regattas served to attract visitors to Exmouth, Torquay, Teignmouth, Dartmouth and other seaside towns, thereby enhancing the growing resort function and diversifying the economic base of numerous erstwhile ports and fishing villages. Yachting became increasingly institutionalized at this time, with clubs such as the Royal Dart, the Royal Torbay, the Royal Western and the Royal Plymouth Corinthian established to cater for the social and recreational needs of the well-to-do. By 1900, in south Devon alone, there were six yacht clubs, with a membership of 282 boat owners and a 'fleet' totalling 17,512 tons (Thames measurement), and a further thirteen sailing clubs for those with more modest vessels, purses and social aspirations.[41] Mirroring this growth, the late nineteenth century witnessed an increase in yacht construction, as the more astute of the South West's shipbuilders shed their interests in merchant shipbuilding and began to concentrate on the production of wooden yachts and, in cases such as Cox of Falmouth and Simpson Strickland of Dartmouth, steel-hulled steam launches and cruisers.[42]

In the twentieth century, the South West's interests in marine recreation have expanded greatly, generally in line with the growth of the region's holiday industry. As purchasing power has increased, especially since the mid-1950s, the demand for yachts, motor cruisers and, more recently, dinghies and multi-hulls, has mushroomed, a trend accentuated by the declining costs of mass-produced small boats and the improvement in road links which has made the region's coasts accessible to the urban centres of the Home Counties, the Midlands and the North. As a result, the character of the South West's ports and estuaries has changed considerably, with recreational craft of all descriptions dominating the anchorages, moorings and docks that were once the preserve of the merchant ship and the fishing vessel.

This process of transition has proceeded furthest on the south coast of the peninsula; thus, as **Map 50.7** indicates, comparatively few facilities are available for the recreational sailor along the rugged coasts of north Cornwall and north Devon, while substantial yacht and sailing clubs are established in all the principal estuaries to the south, with many other small clubs, as well as individual vessels, operating also in these waters. In the port areas of the region, most harbours and docks cater for leisure vessels, some accommodating marina developments of some size. Torquay and Brixham harbours now have substantial marinas, while Sutton Harbour and Queen Anne's Battery in Plymouth, and three marinas on the Fal cater for the cruising business. Encapsulating the transition from sea-borne trade to marine recreation, which has so changed the economic and social profile, and the seascapes, of many south-western coastal communities, is the firm of Uphams. Famed for building merchant schooners, ketches and sail trawlers in Brixham between the 1850s and the 1930s, the company now owns the Darthaven marina on the Kingswear side of the Dart.

Such developments might not please the 'old salts' and the romantics, but marine recreation generates considerable business for the South West's maritime communities, both in terms of serving the needs of the recreational sailors and in constructing, repairing and maintaining their vessels.[43] Indeed, though shipping and shipbuilding, the region's 'traditional' maritime industries, have largely died and the naval presence at Devonport is much reduced, the South West, by dint of its sea-borne trade, fisheries and marine leisure interests, probably depends as much on the sea in the late 1990s as it did a century ago.

FURTHER READING

Michael Duffy *et al.* (eds), *The New Maritime History of Devon, Vol. II* (London, 1994) is a collection of essays concerned with Devon's maritime interests in the nineteenth and twentieth centuries. Ranging from traditional sea-related industries, such as shipping, shipbuilding and fishing, to the more recent developments in seaside tourism and naval facilities, this highly illustrated volume provides a comprehensive guide to Devon's rich maritime history. Gordon Jackson, *The History and Archaeology of Ports* (Tadworth, 1983) is an overview of British port development from medieval times to the 1980s. This well illustrated book is particularly informative on the physical and commercial growth of the nation's ports in the Victorian age. Sidney Pollard and Paul Robertson, *The British Shipbuilding Industry, 1870–1914* (Cambridge MA, 1979) is an analysis of the British shipbuilding industry in its most successful era, and includes a chapter on the regional distribution of activity. See also Stephen Fisher (ed.), *Recreation and the Sea* (Exeter, 1997).

TOWNS

MEDIEVAL
URBAN DEVELOPMENT

HAROLD FOX

For 200 years following the Norman Conquest the people of England witnessed what Finberg has described as a 'golden age' of town foundation when many new towns were added to the previously existing pattern of Anglo-Saxon or Danish urban places. A rapidly growing rural population needed supplies of those simple commodities which they could not produce themselves, like salt, for example, which was to be had in Bodmin, or horse-shoes traded at Marazion market. They needed cash to buy such commodities and to meet the financial demands of lords; in local market towns money could be raised by country dwellers, like the Pinhoe tenants who we glimpse in a document selling cheese and eggs in Exeter. Landlords needed towns in order to buy simple commodities and also luxuries, for example items made of gold which could be bought in places the size of Bodmin or Barnstaple. Moreover, these 200 years witnessed an expansion in foreign trade which was bound to have repercussions for the urban development of the South West.[1]

TOWN FOUNDATION

Town foundation involved a combination of physical and legal changes at a designated site, often a site where some commercial activity, albeit informal, small-scale and perhaps periodic, already existed. In physical terms it meant the laying out of a skeleton of urban house plots fronting on to a street or streets and forming a pattern often still observable today. Very occasionally there are documentary references to this physical planning, as at Pilton in the thirteenth century where the lord granted out a defined area to be divided up into plots of standard size or at the new town of Newton Bushel for which an order of 1246 authorized the lord 'to make plots and to let them to anyone he pleases'. Lords also provided market places, sometimes later furnishing them with market houses and booths at which tolls were collected. The building of houses and workshops was left to a town's new inhabitants: Henry de Tracey's foundation charter for Bovey Tracey ordered that plots should be built up within two years.[2]

Legally the foundation of a town by a lord gave certain privileges to its new inhabitants, as typified by the charter (about 1230) by which Reginald de Valletort confirmed the borough liberties of Saltash. He confirmed that property belonging to a deceased person could be retained by the heir; there would be no arbitrary seigneurial taxes or detention in the lord's gaol; townspeople were to be judged only in their own courts and would pay fines for misdemeanours at a fixed rate. Other freedoms commonly granted in town charters were permission for unregulated circulation of urban property and freedom from market toll (e.g. in Okehampton's charter, about 1200).[3] All of these privileges reduced some of the burdens normally placed upon unfree people in the Middle Ages, thereby allowing commercial activity with fewer fetters.

Some town charters also conferred limited degrees of self-government. For example, Helston was allowed to have a guild merchant, a self-governing association to regulate trade, by its charter of 1201, and Launceston was allowed a guildhall by a charter of the 1260s.[4] Further early developments towards self-government included urban courts presided over by an elected mayor (e.g. Exeter and Barnstaple) and arrangements whereby the people of a town compounded with their lord to pay him a fixed sum for the urban revenues (e.g. Helston and Plympton).[5] But for many of the smaller towns in the South West municipal development went no further than acquisition by townspeople of the right to elect their own town reeve or portreeve who had responsibility for collecting rents and for overseeing the market in urban property. The presence of a chief townsman responsible to his fellows and in charge of financial management gave the people of a town that degree of social cohesion which historians expect in their definition of 'urban'. The society of most towns was no doubt riven by fault lines, but some urban dwellers nevertheless thought of themselves as members of a commune. The self-styled 'portreeve and commonality' of Ashburton clubbed together in 1314 to pay for their urban chapel of St Lawrence, appending their own common seal to the document which confirmed their obligation; in 1356 the 'commonality of the town of Helston' petitioned the Black Prince in order to be allowed independence in the affairs of their own chapel of ease.[6] Among some people sentiments of urban identity persisted even if a place went downhill and lost most of the characteristics of a town: decayed towns in the South West were still jealously guarding the right to elect portreeves in the nineteenth century, as described in *The Portreeve* (1906), a novel by Eden Phillpotts.[7]

Before leaving the topic of town foundation it must be stressed that in many places some kind of commercial activity pre-dates the charter, the granting of privileges and the formal urban lay-out. For example, the large fleets which assembled in the deep anchorage of the Dart in 1147 and 1189 in preparation for the Second and Third Crusades would have encouraged the development of victualling, outfitting and ship-repair trades on the shoreline, and, sure enough, in the early thirteenth century there are references to the occupations of taverner, cook, smith and miller here and to men, almost certainly merchants, who had migrated to the nascent port from Hastings and Lostwithiel; these developments almost certainly took place before any charter was granted to Dartmouth. In the same way Helston was well enough developed by 1177 to have a separate identity, distinct from the manor of the same name, as perceived by the royal officials who collected levies known as aids and to have merchants engaged in the export trades—all before the first charter was granted in 1201. One might expect a degree of trade and manufacture at some old-established religious or administrative centres—such as Crediton, Tavistock and Liskeard—some time before a formal town was established.[8]

MEDIEVAL BOROUGHS

A very basic definition of a medieval town is a compact place where most of the inhabitants got their livings from the manufacture of a diverse range of items and from marketing rather than from farming or extracting natural resources. Unfortunately there is no comprehensive medieval source relating to occupations and so we have to fall back upon contemporary terminology and make do with a map of all places which were described as a borough or at which burgesses or burgages were recorded (a *burgus* or borough was a place whose people, burgesses, had some of the privileges described above and a burgage was a house plot within a borough). **Map 51.1**, which

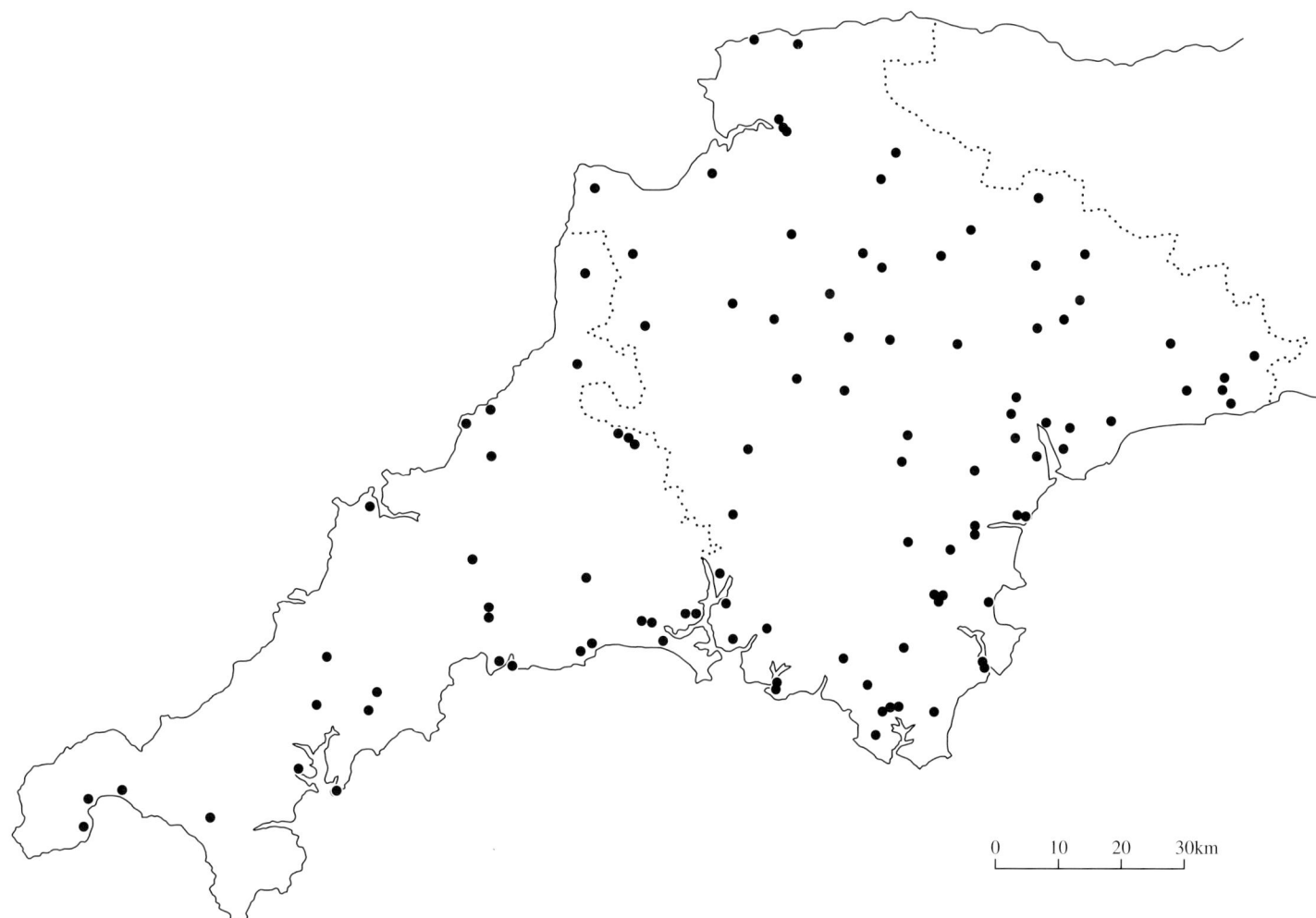

Map 51.1
Places styled boroughs or with evidence of burgages or burgesses, all probably in existence before 1300.

relates to around 1300, is based upon those criteria following the indispensable list compiled by Beresford and Finberg, with a few modifications.[9]

What is immediately striking about **Map 51.1** is the very large number of places shown. Among all English counties Devon ranked first in having the largest number of boroughs per 1,000 acres and Cornwall ranked third.[10] Why so many? Three basic reasons can be suggested, the first having to do with the pattern of lordship. Where the manors of a single lord lay adjacent in a great, compact territorial fief it was usual for only one borough to be established, a good example being the core estate of Glastonbury Abbey in Somerset which extended west to east for about 20 miles with no towns apart from Glastonbury itself. Another example comes from the Liberty of St Edmunds in Suffolk where the Abbey of St Edmund was able to prevent establishment of rival commercial centres within a great swathe of country (over 75 square miles) centred on its own market town of Bury.[11] In Devon and Cornwall there were no such compact estates or liberties. The home manors of Tavistock Abbey stretched for no more than 5 miles from the abbey gate and its other properties were widely dispersed, as were those of the see of Exeter, the earls of Devon and the earls (later dukes) of Cornwall.[12] In any district, therefore, many different lords were represented and the scope for establishment of boroughs was duly magnified.

A second reason for proliferation of boroughs was the physical make-up of the South West. As Beresford put it, 'difficult inland transport ... encouraged more frequent ... resting places'.[13]

Few journeys were on the flat; branching estuaries complicated movements and tended to divide the land up into a series of compartments, each with its local market town.

Third, and most important, it was the highly diversified economy of most parts of the medieval South West which gave rise to establishment of towns in such numbers.[14] Chagford, at the centre of its tin-producing territory, may be used as an example. A market is recorded there in 1220, described as a gathering (*congregatio*) of people buying bread and meat. This tells us a lot: that farmers attended the place in order to sell produce and that it was also frequented by people who were not producers of foodstuffs, in this case tinners, as we know from another source. Moreover, by the late thirteenth century Chagford was perceived as being urban enough to be selected as an official stannary town, that is a place to which tinners brought tin to be assayed, taxed and then sold. Buyers came to Chagford from as far afield as Coventry and Bristol to purchase tin for making girdles and from the south Devon ports through which the metal was shipped, ultimately for export.[15] We can thus see how a town is brought into being by one element of the diversified economy of the South West, the tin industry. It set up local flows of commodities of all kinds and long-distance flows in tin itself, all centred upon a town which also acted as an administrative centre in a much administered industry, attracting at times the warden of the stannary (or his deputies) with a sizeable entourage. What has been said here of Chagford would apply to all of the other stannary towns of the South West (named on **Map 40.2** in chapter 40): they

A

B

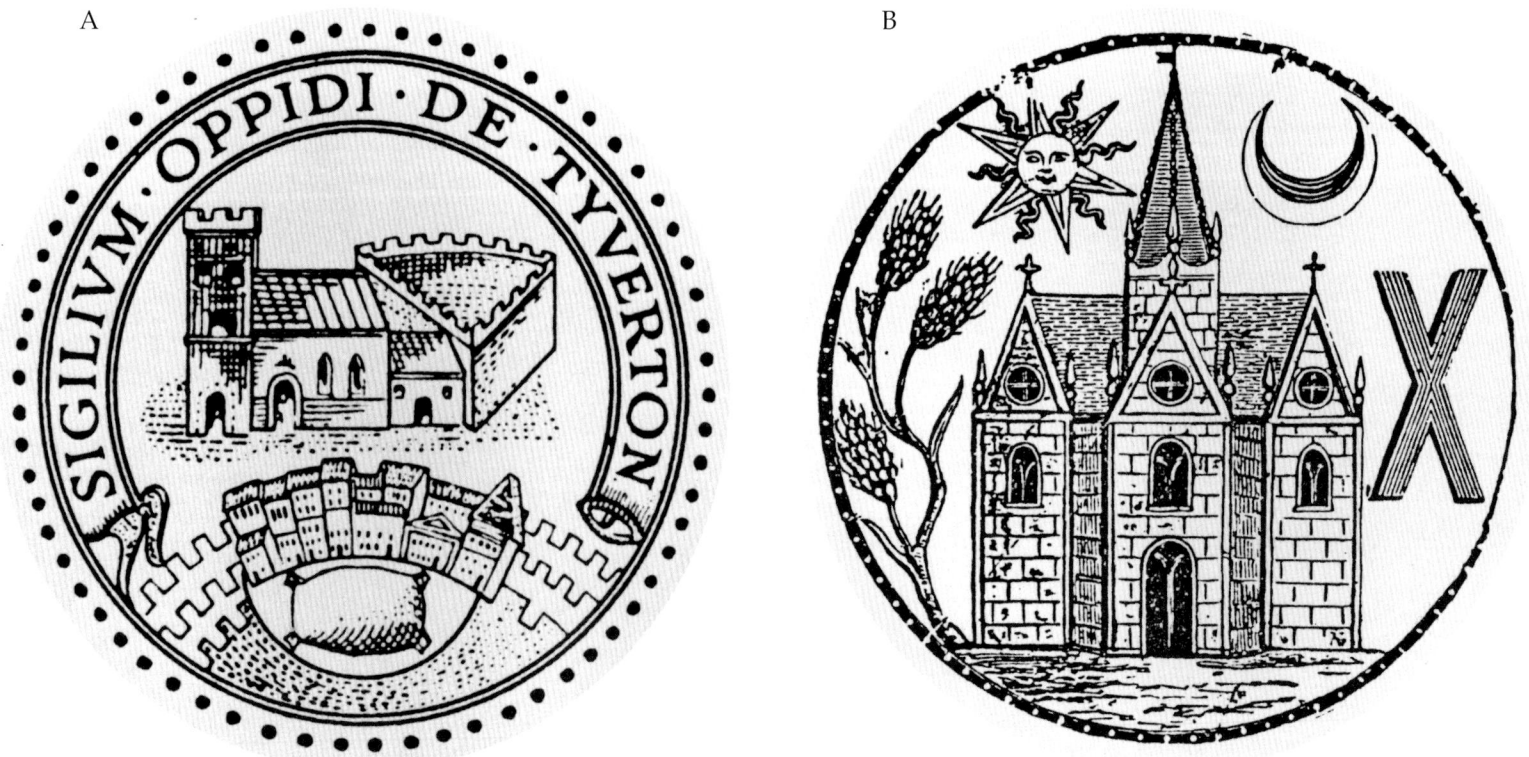

Fig. 51.1 Town seals of (**A**) Tiverton and (**B**) Ashburton.

Tiverton's seal shows the two bridges, over the Exe and the Lowman (some versions of the seal have a wavy line, for water, between the two); the town strung out in between; a wool sack; the parish church and the Courtenay's castle. The images seem to show that, in popular perception, the town owed most to cloth manufacture and to the two rivers, which drove so many fulling mills.

Ashburton's seal shows, in stylized fashion, what is probably the chapel of St Lawrence, given to the town in 1314, and, to the left, teazles used for raising the nap of cloth.

did not all owe their origins to tinning, but their economic development was dependent upon it as well as upon local marketing unrelated to the industry. One other connection between tinning and urban growth must be mentioned. Those Coventry and Bristol girdlers, and other merchants, making the long journey to Chagford and other stannary towns, would need accommodation and provender at intervals along the way, so that towns were established to serve them and other travellers. Trade in a commodity, which began in the scrapings of tinners on the bare moors, therefore helped to encourage urban development on routeways some distance away.

The diversified medieval economy of the South West produced other commodities which were traded over relatively long distances, giving rise to much marketing and commercial movement and thereby stimulating urban development. Fish, salted or dried, was one of these. Markets and towns on or just behind the coast had an active trade in fish. From these it was carried overland by packhorse and by the coasting trade to the great redistributive centre of Exeter and thence to points of sale at inland Devon towns and at markets in Somerset, Dorset and land-locked Wiltshire. Much of the evidence for this trade comes from the later fourteenth century. Sources are more patchy before then but a fish trade, both local and overseas, was certainly in existence in the twelfth century and it could be argued that the volume of traffic should theoretically have been even greater then, for people were more numerous and poorer so demand for this cheap and plentiful foodstuff should have been higher.[16]

Two long coastlines endowed with valuable fisheries and the tin fields of the interior gave Devon and Cornwall a combination of non-agrarian resources enjoyed by few other English counties. Other commodities added to the diversified economic base: there was some domestic trade, overland or coastal, in slate and coarse cloth by the end of the twelfth century and in silver by the thirteenth.[17] Production of all these commodities was bound to lead to development of trade and hence of towns. Their production was partly (not wholly because of the participation of farmers engaged in by-employments) in the hands of specialist workers who stimulated food markets and thereby fostered market towns.[18]

The two coastlines made an additional contribution to commercialization because they were national frontiers for imports and exports. During the twelfth and thirteenth centuries, when so many boroughs were established in the South West, evidence on foreign trade is patchy, but there is enough to show that tin, wine and fish and probably salt—all very important later on—already featured at that time. Tin was traded in, and almost certainly exported from, Exeter in the twelfth century. By the first decades of the thirteenth century, Lostwithiel was the point of export for Cornish tin and, as a consequence, had a foreign trade of high value by national standards.[19] There were abundant supplies of wine in Exeter during the anarchy of Stephen's reign (1130s), by 1178 there was a locally imposed custom on imported wine, and by 1199 a recognized body of vintners was established

there; imported French pottery appears to have been associated with this trade.[20] By then a good proportion of the wine came from Gascony after Eleanor of Acquitaine brought that province to the English crown in 1152, thereby at a stroke giving places along the south coast of the peninsula a new role as closest points of contact with a developing colony.[21] The earliest mention of export of cured fish, to Gascony, is in 1202.[22] This sample of early references to overseas trade comes from the period which witnessed the height of the programme of borough foundation in the South West. Foreign trade was one impetus behind this programme: coastal concentrations of people engaged in shipping, transhipping and ancillary crafts—merchants, smiths, carpenters, shipwrights, porters, bakers, taverners and other victuallers—were elevated to borough status; carriage of goods to and from ports stimulated borough foundation inland on the most used routeways.

Many of the boroughs marked on **Map 51.1** developed partly as a result of commerce in the diverse commodities mentioned in the preceding paragraphs and partly as local market towns. Ports, the most sucessful of which by the thirteenth century were near river mouths with deep anchorages, were obviously the creations of trade, overseas and coastal. In the South West there were also very numerous 'thoroughfare' boroughs founded on routeways, partly the creations of overland commerce.[23] The siting of these is, indeed, (along with evidence relating to medieval bridges) almost the only guide we have to the precise routes used by commercial travellers in the Middle Ages. For example, one important overland route between Exeter and north Devon was through Chulmleigh and Chawleigh, bypassed backwaters today but both boroughs in the Middle Ages. In north Cornwall one may reconstruct an east–west routeway crossing into the county on the Norman or pre-Norman bridge at Bridgerule, then passing through Canworthy, Camelford and Wadebridge all with medieval bridges, then curving southwards towards Truro: five thoroughfare towns were strung out along it.[24]

What has been said above helps our interpretation of the distribution shown in **Map 51.1**. There are four salient points. First, the southern coastlands supported a greater density of towns than the north as a result of closeness to the stannaries, greater volume of foreign trade and superior productivity of fisheries and agriculture here. Second, the lowest possible bridging points on deeply penetrating estuaries attracted urban development of special importance. These were prime sites for towns: they had a magnetic effect on routeways, they had access downstream for small craft and they had larger rural catchment areas than the river-mouth ports where size of hinterland is always halved by the presence of the coast. This last point was developed by Carew with reference to Truro: towns 'that stand highest up in the country afford therethrough … a fitter opportunity of access from all quarters' and, sure enough, we have a lively picture of that town's role in attracting country people with their farm produce in a case of 1238 which describes how a new deer park enclosed by the bishop of Exeter at Lanner obstructed local men and women coming there 'with carts, on

Size of borough (no. of people)	Number of boroughs	Total population	Percentage of urban population
More than 1,000	19	19,006	48.6
500–1,000	15	8,250	20.9
200–499	23	7,360	18.6
Less than 200	47	4,700	11.9

NOTE: Statistical base of **Table 51.1** and **Map 51.2**

Derivation of urban population on basis of 1377 poll tax figures.
I have only used figures where the source clearly indicates that the population enumerated was that of the borough, not borough plus surrounds. The poll tax is said by most historians to greatly underestimate the population over age 14 which it purports to record, although Maryanne Kowaleski, *Local Markets and Regional Trade in Medieval Exeter* (Cambridge, 1995), 371–5, has good evidence to show that the figure for Exeter is reasonably satisfactory. In order to take account of children under 14 and other unrecorded people, historians usually inflate the recorded figures by a multiple of between 1.65 and 2.0: Kowaleski, *Local Markets,* 371; Alan Dyer, *Decline and Growth in English Towns* (Basingstoke and London, 1991), 72; Richard Holt and Gervase Rosser, 'Introduction' in their *Medieval Towns, 1200–1540* (London, 1990), 22. I use a very unrealistically low multiplier of 1.5 in order not to be seen to be over-stating my case for a high level of urbanization. In order to extrapolate backwards to about 1300 (i.e. to cross the period of famines and pestilence) I also use the unrealistically low multiplier of 1.5. This, in effect, is to assume a drop of population between 1300 and 1377 of 33 per cent, again far too low, for demographic historians now consider that 50 per cent is more reasonable: Richard M. Smith, 'Human resources', in Grenville Astill and Annie Grant (eds), *The Countryside of Medieval England* (Oxford, 1988), 190–1. For a variety of reasons, therefore, my estimates for 1300 based upon poll tax figures are very low.

Derivation of urban population on the basis of numbers of burgage plots.
I have striven only to use surveys from the period 1270–1340. The multiplier used is 4.5, which begs many questions relating to spatial variation within towns in the size of households, the possibility of vacant plots, the fact that some burgages contained more than one household and the existence of sub-letting (variables which in some ways cancel one another out). A similar multiplier is acceptable to other historians: Rodney H. Hilton, 'Low-level urbanization: the seigneurial borough of Thornbury in the Middle Ages', in Zvi Razi and Richard Smith (eds), *Medieval Society and the Manor Court* (Oxford, 1996), 490.

For the purposes of **Table 51.1** and **Map 51.2**, populations of towns which physically adjoin have been amalgamated, because there was no doubt much economic interaction between two adjacent boroughs.

horseback or on foot'.[25] So propitious for urban growth were these lowest bridging points that in some cases adjacent boroughs were founded on adjoining lordships near the bridgehead, such as Lostwithiel and Penknight, Kingsbridge and Dodbrooke or the veritable constellation of boroughs at the head of the Dart estuary. Third, each of the two moorlands, Dartmoor and Bodmin Moor, were surrounded by a ring of moorside boroughs. These served in part as thoroughfare towns, for medieval routeways tended to skirt the moorlands. But their development was also partly due to the growing economies of the moors in the twelfth and thirteenth centuries when there was expansion in farmland, in tin output and in demand for other moorland products such as granite and gorse: here the frontiers of colonization, industrialization, commercialization and urban growth all marched together.[26] Fourth, the sites of some boroughs make little sense in terms of the commercial forces of the types discussed above: these were towns

next to prominent residences on great estates, like the borough of Tintagel. It was on a route to nowhere but it served a famous castle built by Richard earl of Cornwall in the late 1220s on a spot which already had Arthurian associations. The domestic establishment of one of the most extravagant men of his age would have needed servicing and this was the reason for the borough that Richard founded at more or less the same time as the castle. Trematon borough, probably adjacent to the Norman motte and bailey of the same name, and the minute borough at Wiscombe, at the seat of the Bonvilles, but in a little combe which goes nowhere, are two other examples.[27]

SIZE OF TOWNS

The highly diversified and commercialized economy of the South West was thus the principal reason for the very large number of towns. This view runs counter to often-quoted statements suggesting that many of these boroughs were tiny 'still-born' failures or can hardly be considered urban, so we must now address the question of their populations. Tentative estimates of the sizes of many towns *c.*1300 may be made by extrapolating backwards from the figures given in poll tax returns for 1377 or by working with the number of burgesses recorded in early medieval urban surveys. For a few towns, especially the smaller ones, sources are lacking for either method, but these may usually be placed in a size band on the basis of taxed wealth, or of value of rents or (in a few cases) of numbers of burgage plots shown on early maps. Full details of these difficult calculations, in which conservative estimates have always been preferred, are set out in the note to **Table 51.1** and the results of the calculations are presented in the table itself. The South West boasted no urban giants: Exeter with a population of about 3,000–4,000 in 1300 was dwarfed by places such as Norwich (25,000) or Winchester (10,000).[28] But what is remarkable about Devon and Cornwall is the number of towns with well over 1,000 people, relatively large places by the national standards of the thirteenth century: Plymouth, Barnstaple, Totnes, Dartmouth and Great Torrington in Devon, and Bodmin, Lostwithiel and Truro in Cornwall, all (except Torrington) having connections with tinning or the port trades. Also remarkable was the large number of towns in the next two size bands, between 500 and 1,000 people, a good size for a town in the thirteenth century, and between 200 and 500 people, a very respectable size for a small market town at this time. These comprised a heterogeneous collection of places: inland towns enlivened by the tin trade such as Helston (about 420 people in 1300) and Tavistock (about 540); ports not among the top-ranking coastal towns such as Saltash (about 530) and Penryn (about 680); and the largest of the thoroughfare boroughs such as Camelford (about 306) and Colyford (about 500).[29] Towns of around 500 people might contain about thirty different occupations, such as (from Bradninch and Tavistock sources *c.*1300 or a little later) bakers, weavers, smiths, tailors, tanners, painters, carpenters, cordwainers, cobblers, glovers and sellers of salt, coal and pottery. They provided a heterogeneous though fairly basic range of services for the surrounding countryside and for travellers, though not the more specialized trades such as those of goldsmith or furrier or more exotic commodities such as figs and skylarks, probably pickled (both to be had in Great Torrington), which were only to be found in the larger places.[30]

Altogether, as **Table 51.1** shows, the combined population of the boroughs with over 200 people in about 1300 accounted for almost 88 per cent of the total urban population at that time. It was not an urbanization of small fry or failures but of towns of considerable or respectable size, most of which are still recognizable as urban places today.

The other boroughs, with below 200 people and in the lowest size band in **Table 51.1**, must not be written off as urban failures because they were small. Axminster may be taken as an example. The borough was created by charter in 1209 and its liberties confirmed in the middle of the thirteenth century. A survey of 1334 shows that it was a single street tacked on to an existing village with eighteen burgage plots (nine on each side), some occupied by burgesses with trade-related surnames; there were market stalls and buildings of two stories, an urban trait at this time.[31] The borough, therefore, had thoroughly urban characteristics, legal, topographical and occupational. Yet if we calculate its size by using a multiplier of 4.5 people for each burgage (see note to **Table 51.1**) it comes out at about eighty. For many of the smallest towns there is similar evidence to indicate that in their heydays they were legally created, discrete places of trade, important within their small local hinterlands and, in some cases, thoroughfares as well. If a few did indeed fail before the middle of the fourteenth century, this will not affect our description of the South West as thoroughly urbanized: the population of this group of towns comprised no more than 11.9 per cent of the total urban population (**Table 51.1**).

Map 51.2, which depicts town sizes, is of great interest. It shows, as we should expect from what has been said earlier about

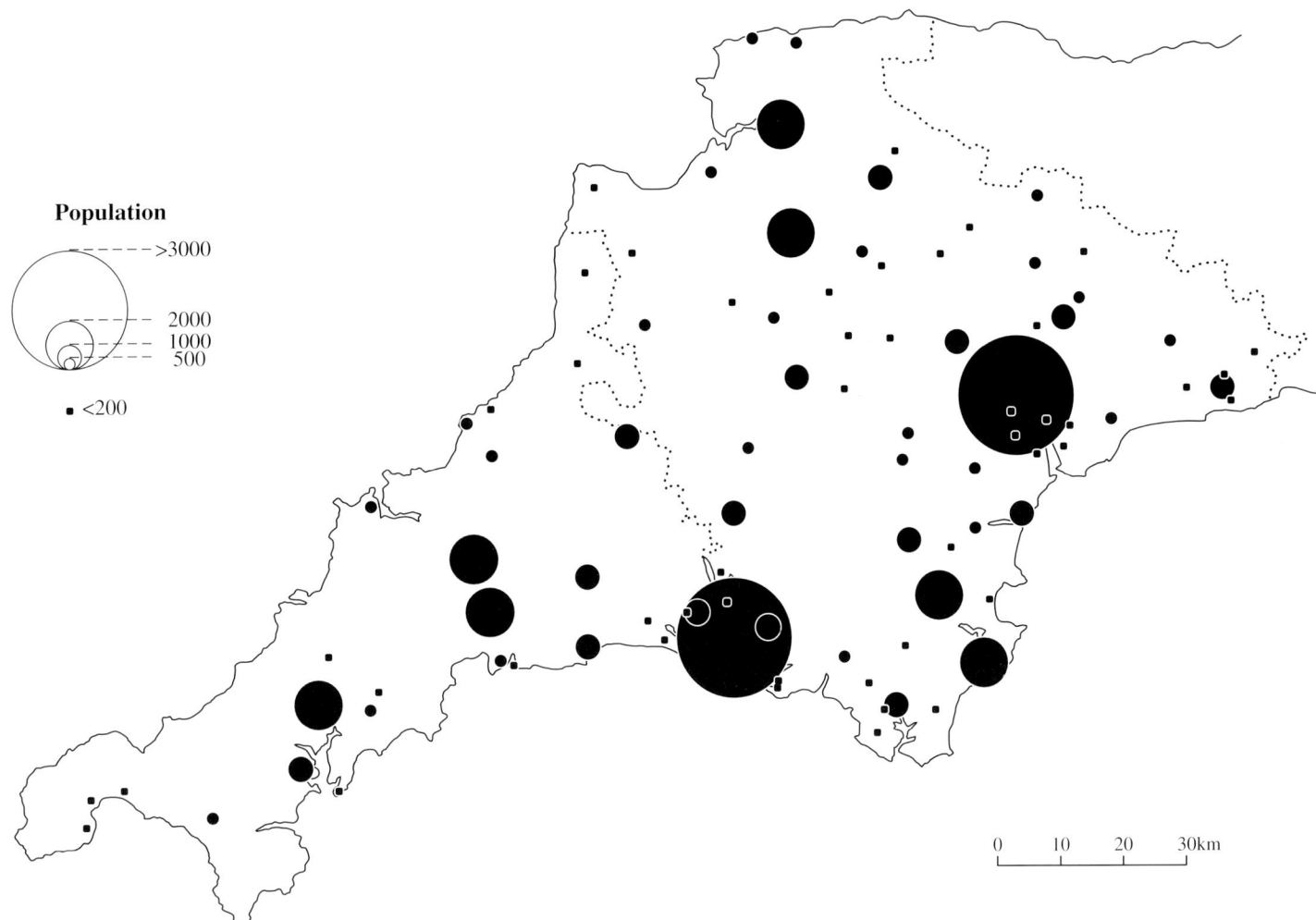

Population

>3000

2000
1000
500

■ <200

0   10   20   30km

Map 51.2
Size of towns c.1300.

There is a discussion of the data for this map in the note to **Table 51.1**.

the overall distribution, that the greatest concentrations of the largest towns were in the south and that few people would have been without easy access to a town of moderate size.

For Devon alone (because the Cornish figures are complicated by large numbers of tinners and by doubts about their taxable status) we can move analytically further forwards from **Table 51.1** and estimate the proportion of the total population living in boroughs around 1300.[32] The figure, an approximation only, is 26 per cent. Historians now consider that a national figure for this time might be about 20 per cent but that of course includes London.[33] The South West, therefore, was especially well urbanized, albeit after its own particular, dispersed fashion, but this should come as no surprise given that its location and resources, both inland and coastal, greatly stimulated commercial development. A high degree of commercialization is further indicated by the existence in Devon (the Cornish figure has never been properly researched) of thirty-six rural markets with crown charters, that is places which were not boroughs but where periodic trading was licensed (in some cases optimistically) to take place once a week.[34] Moreover, some places had a diversified non-agrarian base yet for some reason were never styled boroughs and had no charters, especially minor ports such as Exmouth, Kingswear, St Ives (about 400 people in 1300) and Landulph. Two significant Cornish thoroughfare towns also come into this category: Wadebridge and St Columb Major.[35] If only we could be sure of the total number of such places they would significantly add to the urban density around 1300 shown on **Map 51.2**.

URBAN DECLINE IN THE LATE MIDDLE AGES?

There has been much debate about whether or not the close of the Middle Ages (roughly between 1350 and 1550) witnessed urban decline along with the population decline which was undoubtedly a feature of much of this period.[36] For Devon and

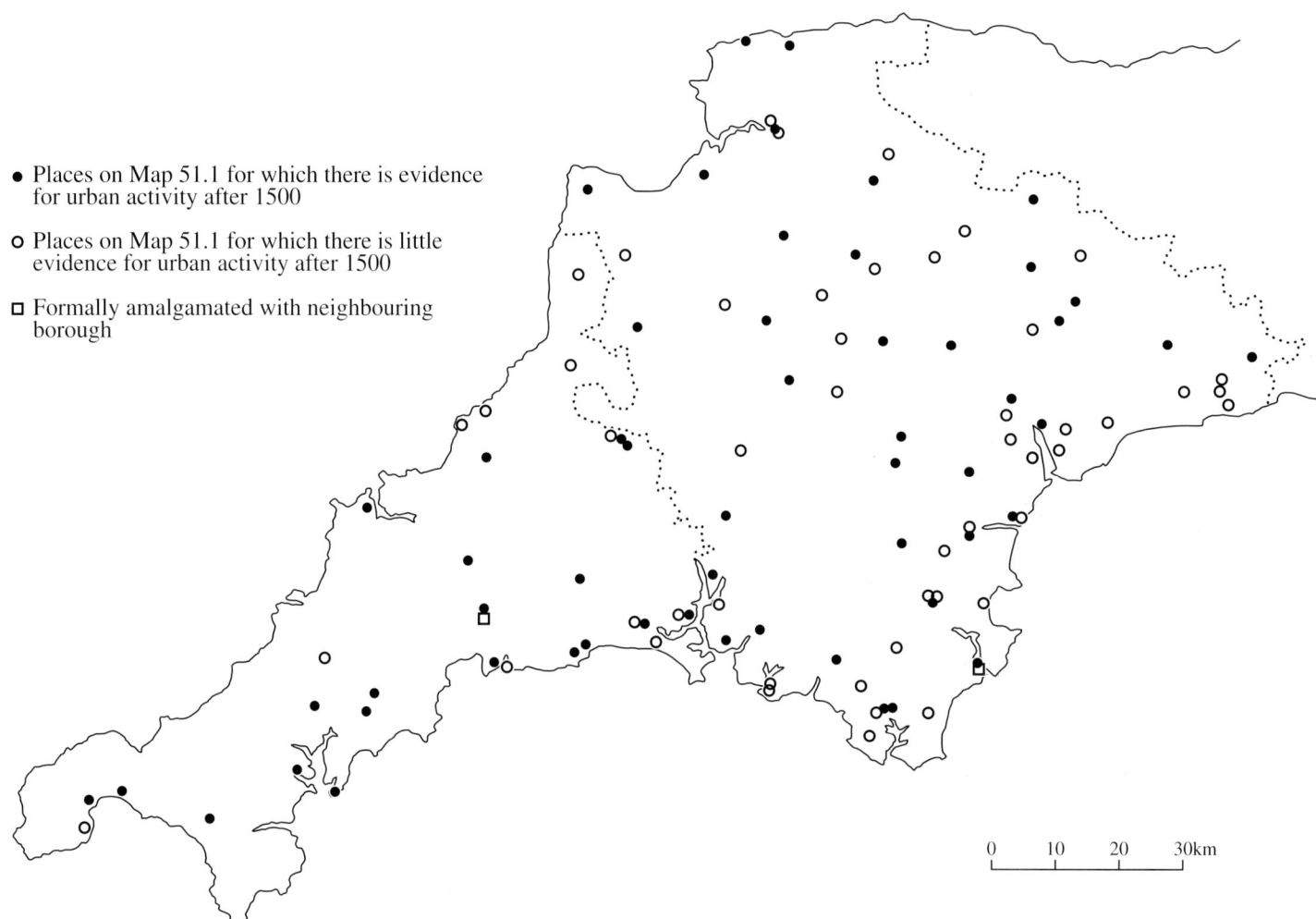

• Places on Map 51.1 for which there is evidence for urban activity after 1500

○ Places on Map 51.1 for which there is little evidence for urban activity after 1500

□ Formally amalgamated with neighbouring borough

Map 51.3
Urban resilience and decline during the later Middle Ages.

Cornwall it is difficult to find comprehensive sources with which to address this topic because the lay subsidy of 1524–5 frequently fails to distinguish borough from parish. All that we can do in a rapid survey is to examine the works of topographical writers of the sixteenth and seventeenth centuries to see whether the early medieval boroughs still passed muster as towns then.[37]

Evidence of this kind has been used for **Map 51.3** showing boroughs which ceased to have many urban characteristics during the later Middle Ages (perhaps a little later in a few cases where the evidence is rather uncertain). Boroughs which owed their siting to the existence of an estate centre failed if the importance of the seat declined: such were Wiscombe, with no traces on the ground today, or Tintagel described by Tonkin as a mean cluster of mere cottages.[38] Many boroughs which were both thoroughfares and local markets were very close to other urban places and it is probable that their selective decline reflected a drop in the population of their small hinterlands, so that they lost one of their sources of income; **Map 51.3** shows that such 'failures' were especially numerous in mid-Devon, a region of severe population loss in the fifteenth century. Another common (but not universal) trait among declining boroughs was their small size, for very few of the towns whose populations had reached over 400 by the end of the thirteenth century subsequently decayed. Decline is of course a relative term. Small decayed thoroughfare boroughs such as Colyford and South Zeal remained in the landscape as settlements, but without markets, without much individual wealth and without a diversity of occupations; in their later days they can best be described not as towns but as industrial villages specializing in cloth-making (later lace) in the first case and the home of cloth-makers and tinners in the second.[39]

Because towns in the lowest size band predominated among those that failed, it may well be that the urban population of the South West was not much smaller in the early sixteenth century than it had been in about 1300; indeed it may have been larger. For while some of the smaller boroughs failed, some other towns certainly contained more people in the early sixteenth century than they had in the thirteenth. The Black Death of 1348 hit them all: the population of Bodmin is said to have fallen by 1,500 people, while a petition of 1369, calamitous in tone, speaks of desolation in the bishop's boroughs of Crediton, St Germans and Penryn.[40] But from some point, probably around the middle of the fifteenth century, growth began to occur again, as can be seen from the example of Tavistock. Around 1300 the town had about 540 people, its population then fell back as a result of the pestilences of the middle of the fourteenth century, but by the 1520s it stood, at a very conservative estimate, at around 860, growth being associated with a flourishing clothing industry in the town and its surrounds and with a high level of activity in its

Fig. 51.2
Drawing of Exe Bridge, mid-eighteenth century.

SOURCE: reproduced by permission of Devon Record Office (ECA, Exeter City Map Book).

stannary district.[41] It is very unlikely that this degree of population growth could have been concentrated into the first decades of the sixteenth century: it must have had earlier roots and the same can be said of other towns for which this kind of calculation can be made.[42] The diversified economy of the South West which brought so many towns into being in the twelfth and thirteenth centuries continued to serve them well later on.

FURTHER READING

Maurice Beresford, *New Towns of the Middle Ages* (London, 1967) provides much background and gives many examples from Devon and Cornwall. A gazetteer at the end of this book discusses all 'planted towns'. Maryanne Kowaleski, *Local Markets and Regional Trade in Medieval Exeter* (Cambridge, 1995) concentrates on the fourteenth century but with many earlier and later references and contains much excellent detail on towns other than Exeter.

See also Charles Henderson, 'A note on the origins of Cornish towns', in his *Essays in Cornish History* (Oxford, 1935), 19–25. The same volume contains Henderson's essays on medieval Truro, Fowey, Lostwithiel, Mitchell and Helston. Herbert P.R. Finberg, 'The borough of Tavistock: its origins and early history', in W.G. Hoskins and H.P.R. Finberg, *Devonshire Studies* (London, 1952), 172–97, is a classic local study with many general points. Peter Sheppard, *The Historic Towns of Cornwall: An Archaeological Survey* (Truro, 1980) provides an annotated plan, brief history and discussion of medieval topography for each town. There is no equivalent for Devon.

Towns

MEDIEVAL
TOWN PLANS

T.R. SLATER

CLASSIFYING MEDIEVAL TOWN PLANS

The plan characteristics of medieval towns in the South West are determined by a wide range of factors reflecting their long history and many functions. First, the physical site often gives a strong frame to the plan and to the appearance of a town, right through to the present. It is one means by which medieval plans can be classified. There is, for example, a distinctive group of towns developed on beaches, particularly on sand and gravel spits at the mouths of estuaries along the southern coasts. They include boroughs such as Exmouth and East and West Teignmouth in Devon, and East Looe and Marazion in Cornwall. Another distinctive group of medieval towns is those which developed on either side of a spine road running down a ridge to a river crossing or harbour, such as Totnes and Kingsbridge in Devon, or Saltash and Penryn in Cornwall. A third group, often wrongly characterized as most typical of the South West, are the picturesque fishing harbour towns, with houses clinging to steep cliffsides around a sheltered harbour or inlet, such as Dartmouth and Brixham in Devon, West Looe and Fowey in Cornwall. The most common plan type, however, has nothing to do with physical site. It consists of the street borough, usually deliberately laid out along a main road to take advantage of trading opportunities from passing travellers.

A second set of factors which influence town plans are those related to the development of central-place functions. Three such functions can be identified in particular. First is the location of early minster churches which, since they served a wide area of the countryside, often acquired a marketing function as a matter of convenience at the gates of the churchyard. Towns which developed around minster churches have distinctive plans: the church is large, usually set in an extensive precinct, and the street pattern focuses on the church like the spokes of a wheel, with the market place close to the churchyard. Axminster and Ottery St Mary are good east Devon examples; Launceston and Bodmin are Cornish cases.[1]

Another important central-place function is derived from lordship. Places which were the centres of feudal estate administration have plans distinguished by lordly residences. After 1066 this usually meant a motte and bailey and, later, a stone castle. In towns such as Okehampton and Lostwithiel, the castles were a little distance from the borough and did not impinge on their plans, but at Totnes, Exeter and Barnstaple the castles were intended to overawe the townspeople and were superimposed on to the built fabric of these Anglo-Saxon *burhs*. By contrast, at Plympton and Winkleigh in Devon, and Boscastle and Tregony in Cornwall, town and castle enjoy a more symbiotic relationship.

The third set of factors which enables classification to be undertaken is to do with planning. Many medieval towns were carefully planned, and the way in which surveyors undertook their task distinguishes one plan from another. Some towns are fully planned new towns laid out in a single operation. Some of the street boroughs are the best examples; Honiton was the most spectacularly successful, whereas Bow, Colyford and Newton Poppleford in Devon and Grampound and Mitchell in Cornwall are tiny planned street boroughs that later lapsed back to rurality.[2] Other planned towns were grafted on to earlier central-place settlements. Crediton, in Devon, is a good example, with a carefully laid out street borough beside the earlier minster settlement. Thirdly, some towns which were particularly successful have several planned areas. At Totnes, for example, and its Bridgetown suburb (though this was an administratively separate medieval borough), there were at least five phases of planned development.

The administrative division of a settlement between two lordships, as at Totnes, is especially common in the South West and can almost always be recognized in the plan. East and West Looe are obviously separated by the estuary, but East and West Teignmouth occupy the same sand spit. Newton Abbot and Newton Bushell are divided only by the stream of the little River Lemon. Most complex of all was Dartmouth, in legal terms the three-fold borough of Clifton–Dartmouth–Hardness, to which could be added Southtown in Stoke Fleming parish, and Kingswear, across the estuary of the Dart, in Brixham parish!

PLAN ANALYSIS: ESTABLISHING PHASES OF MEDIEVAL DEVELOPMENT

These generalities are perhaps relatively familiar to those who know something of the medieval towns of the region. Less familiar is the technique of plan analysis which allows for more detailed categorization and understanding of the built form of towns. The principal plan elements in plan analysis are buildings, the plots within which buildings are located, and the combination of these plots into plot series and street blocks, divided from each other by the public spaces of streets and market place.[3] It is the analysis of these elements and the recognition of homogeneous plan units that enable the phases of medieval urban development to be recognized, since the vast majority of medieval town plans are composite in nature, consisting of more than one plan unit.

Taking each of these elements in turn, relatively few medieval buildings, with the exception of churches, chapels and castles, survive in towns today. Where there are survivals of domestic buildings they tend to be late-medieval in date and of high income households. Despite modern development pressures they tend to be in the largest towns since it was these towns which were prosperous at the end of the medieval period, as they still are today. There are fine courtyard houses in the cathedral close at Exeter, for example, and interesting burghers' houses in Totnes, Dartmouth, Barnstaple, and some very early houses in Silverton.[4] Other relict buildings include fragments of medieval hospitals, bridges (notably the 'Long Bridges' of Bideford and Barnstaple), and town walls, which, again, are a feature of the larger towns, namely Exeter, Totnes, Barnstaple and Launceston.[5] Another important relic is the site, though not the building, of water mills complete with their leats, ponds and yards. The mill leat is still a major feature of the plan of Totnes, whilst the site of the pond of the great medieval tide mill in the centre of Dartmouth is still easily discerned since it was filled in only in the mid-nineteenth century.[6] The majority of urban domestic buildings were built across the plot with their long axis parallel to the street.[7] In much of Devon they were normally of cob with thatch roofs, hence the large number of disastrous fires that marked both the eighteenth century and the medieval period.[8] In a few towns the site was sufficiently restricted, and the demand for properties sufficiently great, that houses were built gable end to the street, often with stone walls separating the timber frames.

Plot dimensions in towns in the South West are more varied than in many regions, partly because of the particularities of site. Towns crowded on to sandy beaches or clinging to steep cliffsides normally have relatively short, narrow plots, sometimes little more than room for the house and a small yard; East Looe is a particularly good example. In contrast, those laid out along the crest of a ridge often have very long narrow plots stretching down the sides of the hill to the valley streams on either side; Kingsbridge and Penryn are fine examples of this type of layout. The majority of new planned towns, particularly those laid out on either side of a main road, normally have substantial rectangular plots of up to two acres in size initially, but if the town prospered, then the majority of these plots would be subdivided lengthways in halves, thirds, or quarters, so that the burgage holder would be able to sub-let and make a good income from his property.

Market places in medieval towns in the South West were comparatively small. The broadened street or a long narrow triangle are probably the most common shapes. There are few of the spacious rectangles or triangles found elsewhere in southern England. Indeed market space was so restricted in many towns that a newly inserted market area is a common feature of nineteenth-century improvement. Totnes is notable for having a livestock market outside its gates as well as a provisions market within, but this was unusual in smaller towns.[9] Streets too were often extraordinarily narrow and steeply graded. As a consequence, many had to be widened, or new streets driven through in the early nineteenth century as stagecoaches were unable to pass. There are examples of new roads being laid down in the medieval period too, often as a means of encouraging traffic to pass through a new town or because a new bridge replaced an ancient fording place. Such a diversion is what makes the grid-like plan of Lostwithiel.

The detailed town plans in **Map 52.1** illustrate some of these features in selected towns in the region. They are derived from the first edition of the 1:2500 Ordnance Survey plans dating from the 1880s. The plans have been analysed, the principal plot boundaries extracted, and plan units superimposed. The paragraphs that follow provide brief interpretations of the particular places, drawing out wider comparisons.

LOOE (**Map 52.1**)

The separate boroughs of East and West Looe are both first documented in 1201.[10] West Looe, often known as Porbuan, was in the parish of Talland. It was always inferior in size and prosperity to East Looe and was probably founded to try to take advantage of some of the developing trade across the river. Its plan is simple. There is a single series of burgages on each side of the road to Talland which runs up a steep-sided valley. The houses are crowded at the foot of the slopes. A broad triangular market space adjoins the quayside. The community had sufficient land for crofts in the upper part of the valley (unit IIC), and for some pasture on the hill top (unit IIA, B), but little more, as is so often the case with medieval new towns. The borough expanded up the valley over its own garden ground (units IA, B) and may then have contracted in the late-medieval decline (unit IC).

East Looe is morphologically more complex. It was founded in St Martin's parish, the church being two miles to the north east.[11] It probably represents a case of a lord giving legal confirmation to a place that was already flourishing so that he could draw rents from it. The close-packed, narrow blocks of houses, warehouses and repair sheds, with almost no open ground (unit III), is typical of such informal trading settlements on beaches, and is probably the earliest development in East Looe.[12] The more regular plots along Fore Street, with the later Guildhall (unit IV), represent the first formal stage of development as the lord began to take control. East Looe, too, had garden crofts (units VII, VIB) and pastures (unit VB) stretching up the steep hillside. As the town expanded, a new plot series was planned facing the river (unit VA) with a narrow back lane behind

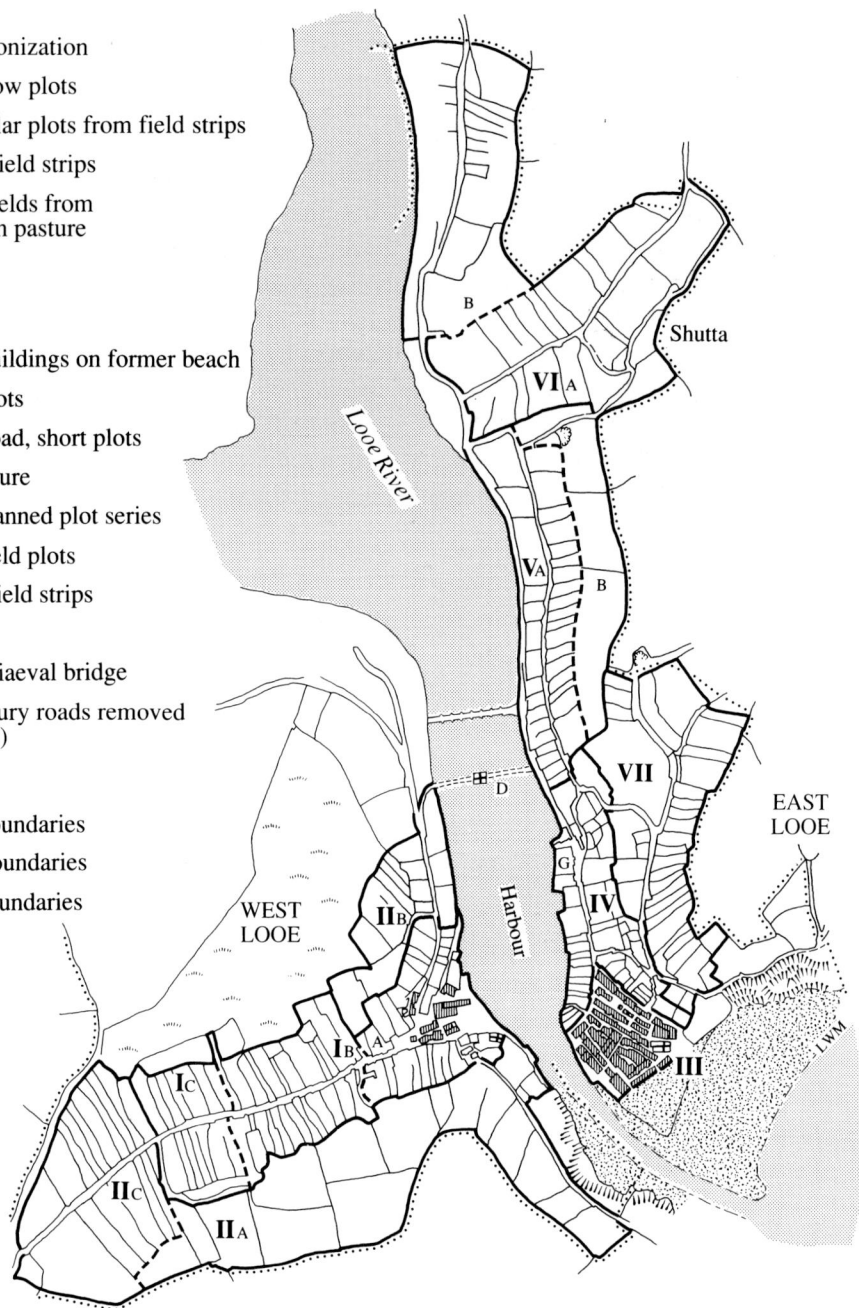

WEST LOOE

▨ Market colonization

IA Short, narrow plots

B Short, regular plots from field strips

C Surviving field strips

IIA Enclosed fields from
B former open pasture

C Croft land

EAST LOOE

III Irregular buildings on former beach

IV Irregular plots

VA Regular broad, short plots

B Hilltop pasture

VIA Regular, planned plot series

B Irregular field plots

VII Surviving field strips

G Guildhall

D Site of mediaeval bridge

(Nineteenth-century roads removed
except for bridge)

⊞ Chapel

.... Borough boundaries

——— Plan unit boundaries

- - - - Sub-unit boundaries

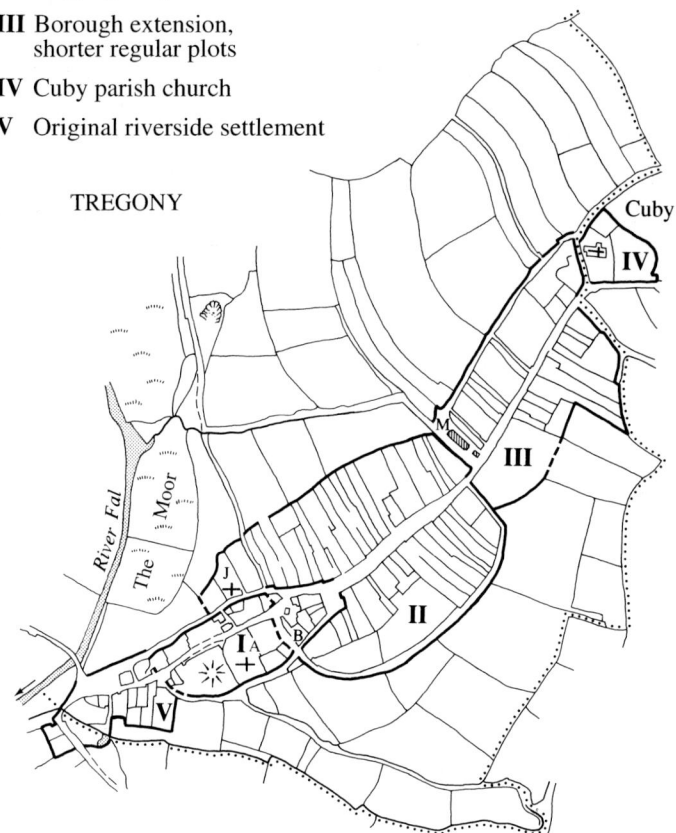

IA Site of castle and priory

B Original market at castle gate

II First planned borough,
long regular plots

III Borough extension,
shorter regular plots

IV Cuby parish church

V Original riverside settlement

J Ruins of St James Church

M Secondary market space

TREGONY

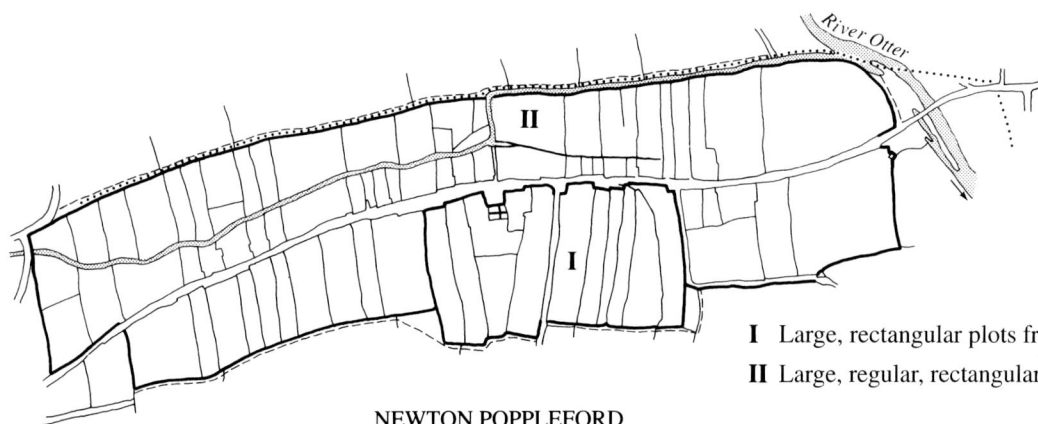

NEWTON POPPLEFORD

I Large, rectangular plots from fields

II Large, regular, rectangular plots

AXMINSTER

I Minster church enclosure

II Early market space and radial roads,
irregular plots

III Castle site, short, more regular plots

IV Later developed, more regular plots

V Mill complex

the tofts and then garden crofts stretching up the slope, most of which were divided transversely to terrace the steep slope. Houses reached little further than the present bridge in the 1880s, but in the medieval period it seems probable that they filled the whole plot series since there is a third phase of expansion represented by the relict plots on each side of the road up to St Martins (unit VIA). These plots are much more typical of late-medieval burgages in their dimensions and, again, there is a back lane running around the southern series. This area, too, was probably depopulated in the late medieval decline as inhabitants of the borough concentrated in the town centre. In the later medieval period a bridge, with a bridge chapel, linked the two boroughs together a little below the site of the present bridge.

TREGONY (Map 52.1)

This small Cornish borough is a classic 'castle town' of a kind that is particularly common in the Welsh borders. The castle stood on a prominent ridge overlooking the upper reaches of the Fal and was held by the de Pomeroy family. Beside it was a small Augustinian priory founded by Goscelin de Pomeroy sometime before 1112, so it must be presumed that the castle is older still (unit IA). Below the castle, a small group of irregular plots (unit V) probably represents the settlement that gives name to the place ('dwellings by the riverside common').[13] The borough is carved from St Cuby parish and St Cuby church stands beside the road at the top of the ridge (unit IV) attached to, but outside, the borough, which was served by St James's chapel. The early borough (unit II) had been laid out by 1197. An irregular rectangle, later much infilled, at the junction of castle and borough served as the market place (unit IB). The borough has steep slopes on all sides, but the curved back lane behind the southern burgage series suggests that the founder intended to surround his new borough with at least a ditch and fence.

Later, this small borough had flourished sufficiently to encourage its lord to lay out an extension. The new burgages were laid out in two series on either side of the main road, infilling the space between borough and St Cuby church (unit III). In this plan unit it is the northern plot series which is most regular, with a back lane giving access to the rear of the plots and a narrow footway dividing the street block in half. A second marked space seems to have been provided at the junction of the two parts of the borough. A historical context for the borough extension is the building of Tregony bridge in 1300. It seems probable that the Pomeroys were attempting to compete with the lord of Grampound, only 3 miles to the north, but better situated on the main road to Truro.

NEWTON POPPLEFORD (Map 52.1)

This borough is a classic thirteenth-century new-founded street town. It stands on the west bank of the Otter on the Exeter–Dorchester main road. A market had been granted to William Brewer, lord of Aylesbeare, within which parish the town is situated, in 1226, and a fair in 1239.[14] The first record of the borough is in 1274, when 55½ burgages are recorded, each holder paying the standard medieval burgage rent of 12d.

The plan is almost a unitary one with a plot series laid out on each side of the main road, stretching for more than half a mile from the bridge over the Otter. A chapel was built for the community halfway along the street. A footpath marks the back of both plot series and the northern plots slope steeply down to a stream which runs along the back of plots to the east and bisects the western group. The only subdivision of the plan that can be recognized is in the centre of the southern plot series (unit I) where pre-existing field boundaries have made a group of plots somewhat longer. They may represent a group of initial plots laid out at the time of the market grant in 1226, to which the more regularly planned remainder were added soon after. There is no record of the original dimensions of the plots, but many seemingly remained unaltered through to the late nineteenth century. Modern analysis suggests a burgage width of 66 feet (four statute perches) which, again, is common in other planned new towns.

AXMINSTER (Map 52.1)

The settlement history of Axminster dates back to Roman times. Two Roman roads cross just to the south of the medieval town; indeed, it may be that they were linked across their north-western quadrant by a road which was to form the axis of the borough.[15] The minster church, which gives name to the settlement, had been founded on what was an extensive royal estate by the eighth century.[16] As a royal estate centre, head of a hundred, and with a minster church, Axminster is an excellent example of a central place with proto-urban characteristics in the Anglo-Saxon period. Thus, when William Brewer was granted a market charter by King John in 1204, it was confirmation of an existing Sunday market held at the churchyard gate, not for a new market.[17]

The town plan is complex and difficult to interpret. The sub-circular pattern of boundaries to the south of the church (unit I) may represent the enclosure of the minster. At the north door of the church, now partly taken into the churchyard, is the irregular triangular market place with all the major roads running into it (unit II). On the hill top, probably where the later market hall stands (unit III), was a small castle looking out over the Axe.[18] The river powered a number of mills (unit V). Two are documented in 1086 and others were added later. The town continued to expand in the high medieval period along the main roads and the lanes linking them, such as South Street (unit IV), and there are references to Newenham, a nearby Cistercian abbey to which the manor had been granted in 1236, granting new burgages for the standard 12d rent on the fringe of the town.[19] The irregularity of the plot pattern contrasts markedly with the other towns considered.

Map 52.1 (facing page)
Plan analyses of West and East Looe, Newton Poppleford, Tregony and Axminster.

SOURCE: base plans are derived from the first edition 1:2500 county series plans of Devon and Cornwall.

FURTHER READING

Little work has been undertaken on the physical development of towns in either Devon or Cornwall. M.W. Beresford, *New Towns of the Middle Ages* (London, 1967) provides detailed documentation for the many medieval new towns in the South West, and some plans. J. Haslam (ed.), *Anglo-Saxon Towns in Southern England* (Chichester, 1984) gives plans of some Devon towns and somewhat controversial interpretations for the early medieval period. P. Sheppard, *The Historic Towns of Cornwall: An Archaeological Survey* (Truro, 1980) contains plans and some basic interpretations for Cornwall.

TOWNS AND PROCESSES OF URBANIZATION IN THE EARLY MODERN PERIOD

JONATHAN BARRY

Many problems, both conceptual and evidential, arise from the mapping of towns in the early modern South West. Consequently, this section will be devoted largely to explaining the decisions made in creating these maps and indicating their reliability, with only a brief evaluation of their implications, in particular for the scale and nature of urbanization in this period.

The first three maps (**Maps 53.1, 53.2 and 53.3**) present population estimates for ninety-two 'towns' at three dates (c.1660, c.1750 and c.1805), based on **Table 53.1**, which gives the population estimates in full and explains some of the 'urban areas' used in the maps. The inclusion of ninety-two places may appear generous, given the limited size of many of these settlements, but the intention is to reflect contemporary ideas of a town, not impose a modern definition. Thus any place which was recognized as a market town, incorporated or parliamentary borough during the period, or was regularly referred to as a town, or was on the threshold of such recognition in 1800, has been included, even if its population falls below normal thresholds for urban status.[1] One advantage of this approach is that it makes it possible to follow both the loss and gain of urban status, which was, as we shall see, a vital feature of this period in this region, especially in Cornwall and along the coast.

SOURCES AND ASSUMPTIONS

Two major issues are raised by the population estimates for towns. The first concerns the survival of evidence for population and the second concerns its attribution to specific urban areas. Before the 1801 census the first is the more serious, but from 1801 the latter still renders the population figures a matter of educated guesswork. The basic unit of nearly all sources used for population was the parish and this remained the case with the census. Many towns formed either a part of a parish or straddled several parishes. With increasing frequency from 1801, and sometimes retrospectively after 1831, the census reports include a separate figure for the 'town' population, but this is provided only for larger towns or boroughs with established jurisdictions. The urban limits of many small towns can only be recovered by painstaking work on individual census enumerators' returns. Considerable use has been made of preliminary work of

this type for Cornwall by Dr Tom Arkell, using the 1851 census.[2] However this research, like many of the published figures, leaves open the question of changes in the urban proportion of the parish since 1801, let alone since the seventeenth century. Here the historian has to turn to other sources that may, at some point, distinguish town from parish, or, if these do not survive, to the density of settlement in the census period, although other variables, such as parish size and the nature of rural settlement, make the latter an inexact guide. The figures presented here represent the result of such calculations for each individual town; where the process has been more than usually uncertain this is indicated.

The other problem concerns sources for population totals, even at parish level. From 1801 this is relatively unproblematic, although the wartime conditions in 1801 and 1811, plus the difficulty of capturing a mobile population on one night (10 March 1801 and 27 May 1811), especially in maritime communities, make even these sources slippery; this is one reason why an average of 1801 and 1811 (notionally called c.1805) is used here, rather than either date alone. Before 1801 no national effort was made to capture population totals, although the growing habit of counting population produced some individual town or parish 'censuses', used here where appropriate.[3] Instead we have to rely on two main types of record, which are discussed in more detail in chapter 16. The first and most important are ecclesiastical records. Of these, parish registers are potentially most useful but are little used here. They have problems of omission, especially as nonconformity and absenteeism grew in urban settings, but, conversely, the very scale of such records has prevented their systematic use to generate population estimates. Where aggregate figures are easily available they have been used to check other figures (e.g. for Exeter and Tiverton), employing standard multipliers to derive population totals from baptisms (annual average × 40) and burials (annual average × 31), although these are hard to specify for urban populations with fluctuating mortality and high immigration.[4]

Three other religious sources have been used systematically. The Protestation returns of 1641–2 survive for most areas except much of east Devon; it has been assumed that the adult males listed will represent one-third of the total population.[5] The 'Compton Census' of 1676 (see chapter 16) also omits certain parishes, especially chapelries or special jurisdictions, and it is often

Map 53.1
Urban population, *c*.1660.

SOURCE: data from **Table 53.1**.

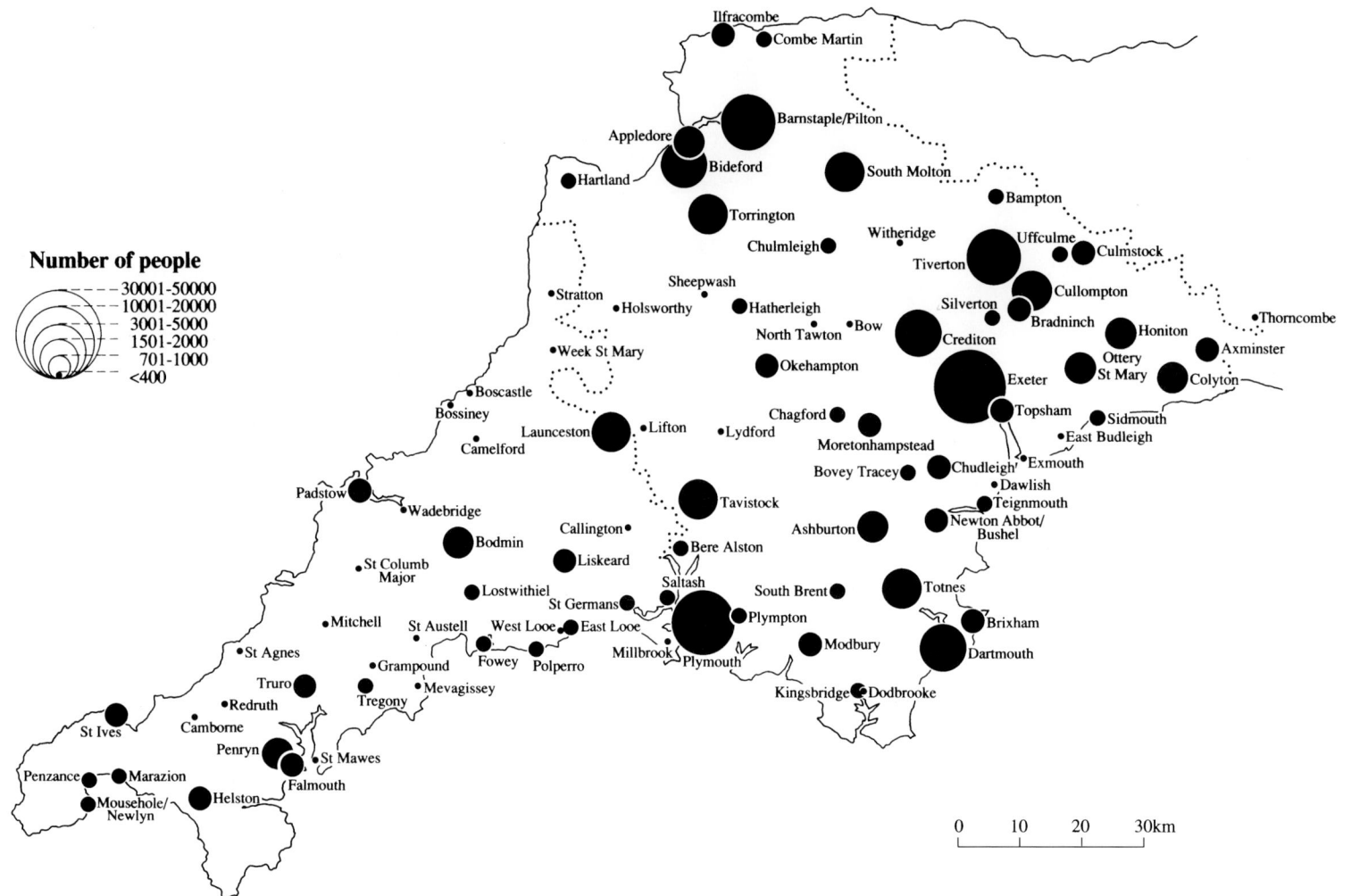

Number of people

30001-50000
10001-20000
3001-5000
1501-2000
701-1000
<400

Ilfracombe
Combe Martin
Barnstaple/Pilton
Appledore
Bideford
Hartland
South Molton
Bampton
Torrington
Witheridge
Uffculme
Chulmleigh
Tiverton
Culmstock
Sheepwash
Cullompton
Stratton
Holsworthy
Hatherleigh
Silverton
Bradninch
Honiton
Thorncombe
North Tawton
Bow
Crediton
Week St Mary
Axminster
Okehampton
Exeter
Ottery
St Mary
Colyton
Boscastle
Topsham
Sidmouth
Bossiney
Chagford
East Budleigh
Launceston
Lifton
Lydford
Moretonhampstead
Exmouth
Camelford
Bovey Tracey
Chudleigh
Dawlish
Padstow
Teignmouth
Wadebridge
Tavistock
Newton Abbot/
Bushel
Bodmin
Callington
Ashburton
Liskeard
Bere Alston
Lostwithiel
Saltash
South Brent
Totnes
St Germans
Mitchell
West Looe
East Looe
Brixham
St Austell
Millbrook
Plympton
Dartmouth
St Agnes
Fowey
Polperro
Plymouth
Modbury
Grampound
Truro
Mevagissey
Kingsbridge
Dodbrooke
Redruth
Tregony
St Ives
Camborne
Penryn
St Mawes
Penzance
Marazion
Falmouth
Mousehole/
Newlyn
Helston

0   10   20   30km

unclear whether the incumbent is returning all parishioners or only communicants or even (in a few cases) only male communicants. Generally this study has followed the advice given in the standard edition, except where other evidence suggested a better reading. A multiplier has to be applied to transform adult communicants to total population, in this case 1.5.[6] Finally, we have a series of visitation returns in 1744, 1764 and 1779, and in some cases also 1768 and 1771, recording the number of families within the parish.[7]

The answers given vary enormously in their precision and as far as possible (again, not all parishes made returns, especially the parishes outside episcopal jurisdiction) the successive returns have been compared to select the most plausible. Some incumbents appear to have excluded individuals or small, often poor households from their definition of 'family', while others counted each separate household unit. If the latter is assumed, then the standard multiplier adopted here is 4.25, but this may be an underestimate.

The other main sources are taxation returns. Of these, only the hearth tax returns of the 1660s and 1670s survive (often imperfectly) for most communities, but some use has been made of poll tax material for the 1660s and 1690s. The hearth taxes measure hearths and households and sometimes omit the non-exempt, but multipliers between 4.25 and 5.75 were employed to cross-check other data. Taxation material quite regularly distinguishes town or borough from parish, so enabling estimates of town size in relation to the parish at this period.[8]

A final source used to establish town size is estimates of numbers of houses, and sometimes of population, provided by visitors and guidebooks. Often these are very approximate, with 200 houses, for example, a frequent guess for a moderate-sized town, but they can be used to give an order of magnitude, employing the same guide multipliers as for the hearth tax.[9] One particularly important source of such estimates is Browne Willis's *Notitia Parliamentaria* of 1716, a two-volume gazetteer of parliamentary boroughs based partly on personal travel but

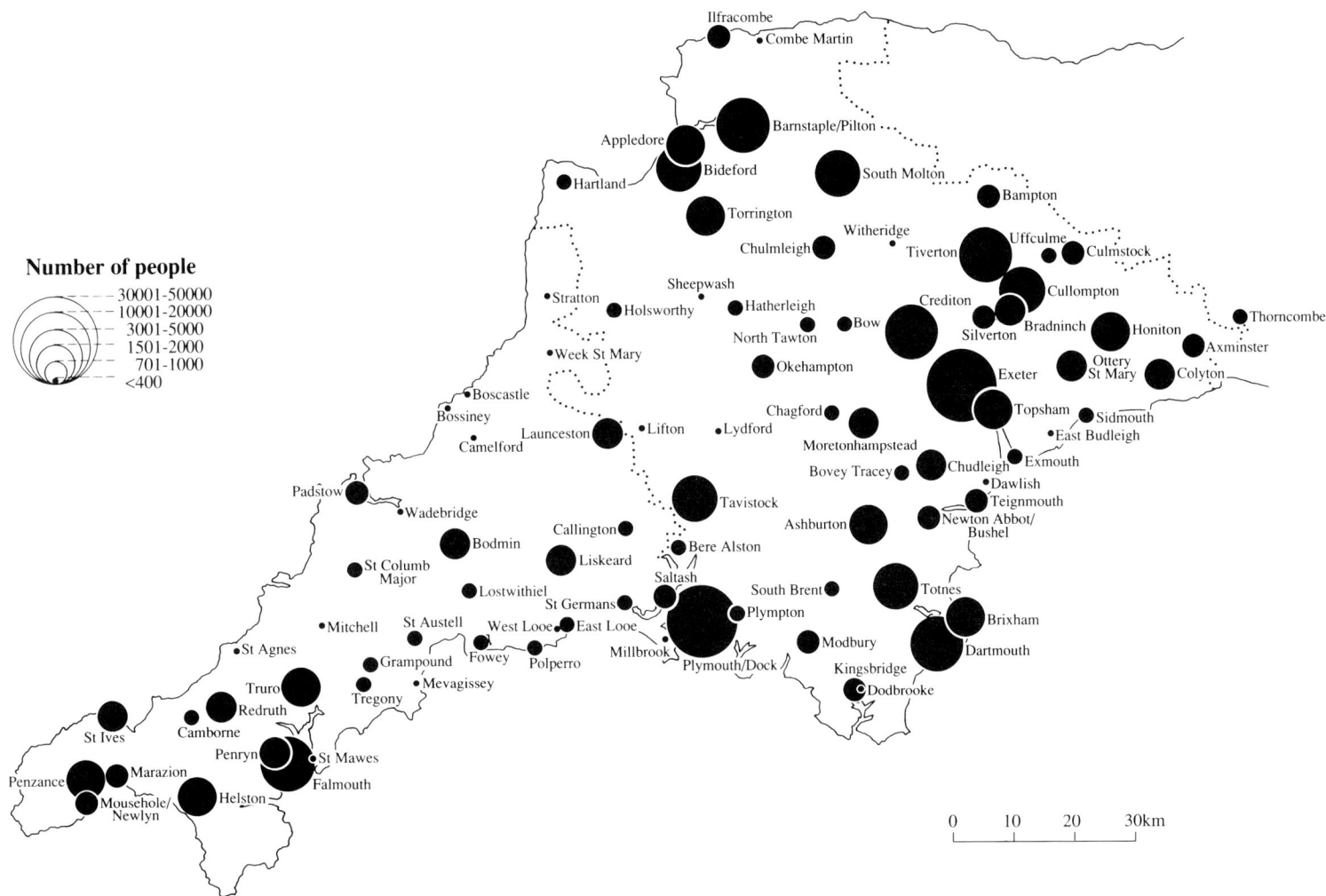

Map 53.2
Urban population, *c.*1750.

SOURCE: as **Map 53.1**.

mainly on antiquarian correspondence. Given the high proportion of boroughs, especially in Cornwall, this provides an important insight into town size, as well as other features of town life discussed later.

The imperfections and gaps in all these pre-census sources explains the decision to produce population estimates for a particular period, rather than deriving figures from a single source. The three main mid-seventeenth century sources (Protestation returns, hearth tax or poll tax and Compton Census) have been used to derive a composite figure for *c.*1660. Where these sources imply considerable population growth between 1642 and 1676 (as for example at Plymouth) a figure for 1660 has been estimated, allowing for differential growth within those decades. In Plymouth's case it has been assumed that Civil War destruction on the one hand, and the growth of naval facilities and building of the Citadel during the 1660s on the other hand, would have skewed Plymouth's growth so that a 1660 figure of 5,400 has been chosen between the 4,500 suggested in 1642 and a figure up

to 7,900 indicated by 1676. The figure *c.*1750 is based on the various visitation returns, giving most weight to the 1744–6 return unless it seems implausible and again allowing for change in the next few years where appropriate. Given the problem of multiplying up from families, this estimate has often been altered to take account of sources like Willis or later eighteenth-century estimates, while throughout the exercise the 1801 and 1811 figures are used to help place the earlier figures in context. This is, however, of only limited use given the often quite rapid changes in population after 1750, with many towns growing rapidly by 1800, but a significant minority stagnant or even declining.

URBAN GROWTH AND ITS REGIONAL IMPACT

Enough has been said to show the caution with which these figures should be used; each is open to detailed local research and subsequent revision. The bands shown on the map are thus more meaningful, as approximations, than the apparently precise figures given in **Table 53.1**. However, these have been offered to suggest a guide figure and to enable some quantification of changes in urban status and urbanization in this period. As **Table 53.2** indicates, the period in question was one of considerable urban growth, with the urban population as a whole increasing by 141 per cent to almost two and a half times its 1660 level in 1805. The bulk of this growth was concentrated in the post-1750 period, with a 73 per cent growth in fifty-five years, compared

to 39 per cent growth in the previous ninety years. Whereas the average town in 1660 had just over 900 inhabitants, in 1750 it was just over 1,250 and in 1805 nearly 2,200. Within this picture of growth, a considerable difference emerges between Devon and Cornwall. Overall Cornish towns grew 190 per cent, 59 per cent by 1750 and another 81 per cent thereafter, while Devon towns grew a little less than the regional average at 34 per cent before 1750 and 70 per cent thereafter, making 127 per cent in all. The 'average' Devon town (Ashburton is a close approximation) grew from 1,200

Map 53.3
Urban population, *c*.1805.

SOURCE: as **Map 53.1**.

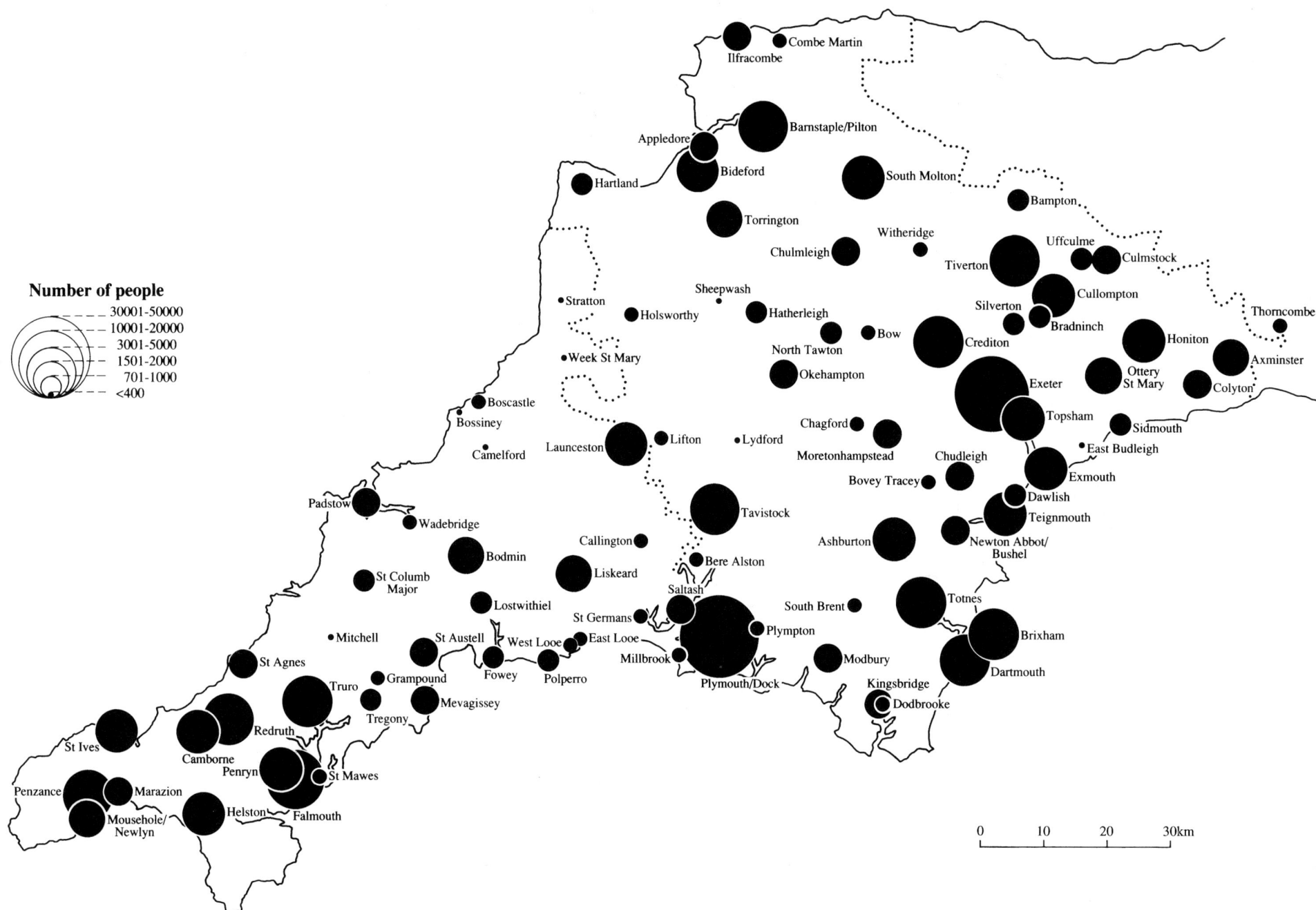

Town	c.1660	c.1750	c.1805
1. Plymouth/Dock*	? 5,400	11,500	49,800
2. Exeter*	11,500	15,500	20,500
3. Falmouth*	850	3,500	5,550
4. Tiverton	3,500	4,250	4,965
5. Barnstaple/Pilton*	3,100	3,600	4,755
6. Truro*	860	1,950	4,750
7. Penzance	695	1,610	3,925
8. Dartmouth*	2,900	3,350	3,590
9. Brixham	720	1,510	3,500
10. Crediton	2,500	3,350	3,275
11. Redruth	295	1,500	3,200
12. Tavistock	1,800	2,250	3,050
13. Totnes*	1,900	2,475	3,005
14. Bideford	2,100	2,050	2,850
15. St Ives	950	1,375	2,550
16. Penryn*	1,125	1,140	2,505
17. Camborne	? 330	? 640	? 2,400
18. Teignmouth*	480	950	2,350
19. S. Molton	1,625	2,400	2,300
20. Honiton	1,260	1,760	2,300
21. Ashburton	1,200	1,600	2,300
22. Helston	720	1,600	2,275
23. Cullompton	1,800	2,750	2,270
24. Topsham	900	? 1,700	2,250
25. Exmouth	? 250	? 500	2,250
26. Launceston*	1,550	1,485	2,220
27. Bodmin	1,050	1,275	2,000
28. Liskeard	900	1,200	1,920
29. Torrington	1,600	1,600	1,800
30. Mousehole/Newlyn*	? 500	? 950	? 1,575
31. Axminster	900	950	1,560
32. Ottery St Mary	1,100	1,200	1,510
33. Chudleigh	? 800	? 1,100	? 1,500
34. Ilfracombe	800	800	1,415
35. Saltash	480	750	1,310
36. Colyton	? 1,250	? 1,250	? 1,280
37. Appledore	1,100	1,530	? 1,275
38. Newton Abbot/Bushel*	710	800	1,275
39. St Austell	? 305	? 500	? 1,250
40. Padstow	705	740	1,200
41. Moretonhampstead	850	1,350	1,200
42. Kingsbridge	550	720	1,180
43. St Agnes	? 125	?? 375	? 1,150
44. Modbury	825	840	1,110
45. Okehampton	800	850	1,100
46. Mevagissey	105	375	1,070

Town	c.1660	c.1750	c.1805
47. Culmstock	? 750	?? 750	? 1,065
48. Marazion	500	725	1,015
49. Chulmleigh	? 600	? 800	? 1,005
50. Polperro	? 450	? 450	? 995
51. Fowey	510	580	990
52. Hatherleigh	? 450	? 690	? 975
53. Lostwithiel*	425	625	950
54. Sidmouth	? 510	? 510	? 950
55. Dawlish	? 400	? 400	? 950
56. Bradninch	? 825	? 1,015	? 940
57. Tregony	450	600	925
58. Silverton	? 525	? 900	? 925
59. Bampton	675	825	870
60. Uffculme	? 640	?? 750	? 850
61. Hartland	? 600	? 625	? 820
62. St Columb Major	325	450	715
63. North Tawton	? 325	? 475	? 710
64. Bovey Tracey	? 700	? 700	? 700
65. Millbrook	350	340	? 675
66. Grampound	300	410	675
67. Chagford	500	520	660
68. Plympton	500	450	660
69. Bere Alston	? 405	? 405	? 650
70. St Germans	450	550	605
71. South Brent	? 420	? 420	? 600
72. Callington	310	460	585
73. Combe Martin	? 450	? 350	? 580
74. Dodbrooke	? 315	? 230	? 580
75. Thorncombe	? 400	? 500	? 575
76. St Mawes	175	250	? 550
77. East Looe	640	550	540
78. Bow	? 260	? 450	? 525
79. Holsworthy	300	405	475
80. Lifton	? 275	? 300	? 460
81. Boscastle	300	300	450
82. Wadebridge	? 250	? 350	? 450
83. Witheridge	? 230	? 375	? 450
84. West Looe	330	375	405
85. Camelford	125	250	400
86. East Budleigh	250	250	? 400
87. Stratton	? 340	250	330
88. Week St Mary	? 200	? 150	? 295
89. Sheepwash	? 165	? 195	? 275
90. Lydford	? 100	70	? 200
91. Mitchell	115	115	105
92. Bossiney	45	45	50

Table 53.1
Town population (ranked by size in c.1805).

? implies serious uncertainty in calculating population of town (as opposed to parish, or because of conflicting data);

?? implies absence of data and consequent extrapolation from other periods.

In all cases only the 'urban area' of the town/towns is included. Suburban areas outside formal jurisdiction are included (see numbered notes below flagged by an '*' for most important cases).

NOTES:

1. Plymouth/Dock. These figures include both Plymouth parishes together with Stoke Damerel, and thus the three towns of Plymouth, Dock (later Devonport) and East Stonehouse. The 1750 figure includes 6,700, 4,200 and 600 for Plymouth, Dock and East Stonehouse respectively and the 1805 figure 18,750, 26,800 and 4,250 respectively.
2. Exeter. These figures include St Thomas and St Leonard (but not Heavitree).
3. Falmouth. These figures include the Falmouth parts of St Budock.
5. Barnstaple/Pilton. The urban area of Pilton parish has been included here, although not the suburbs of Barnstaple in Bishop's Tawton mentioned in the 1851 census.
6. Truro. These figures include all of St Mary and the suburbs in St Clement and Kenwyn.
8. Dartmouth. These figures include all of the parishes of St Petrox, St Saviour and Kingswear and the urban parts of Townstall.

13. Totnes. The Bridgend suburb of Totnes in Berry Pomeroy parish is included here.
16. Penryn. The figures include parts of Penryn in St Budock as well as St Gluvias.
18. Teignmouth. The two parishes (East and West) are treated as one here; West Teignmouth was about three times the size of East Teignmouth throughout.
26. Launceston. The figures here include the urban areas of St Stephen and St Thomas parishes as well as St Mary Magdalene, so including the borough of Newport as well as that of Launceston.
30. Mousehole/Newlyn. Both these quasi-urban areas were in Paul parish and although not contiguous they are hard to separate; Newlyn probably grew from being the smaller in 1650 to twice the size of Mousehole by 1805.
38. Newton Abbot/Bushel. The urban parts of Wolborough and Highweek respectively formed the two towns which were usually treated as one, referred to mostly as Newton Bushel in early sources and Newton Abbot later; they are combined here.
53. Lostwithiel. The figures include the suburb of Bridgend in St Winnow.

people in 1660 to 1,600 in 1750 and just over 2,700 in 1805, while its Cornish equivalent (Saltash is the nearest to the model) grew from a mere 490 to 780 by 1750 and 1,420 in 1805. Despite the differential rates of growth, Devon towns were much larger at the start and remained larger throughout. Only 21.6 per cent of the region's urban population lived in Cornwall in 1660 and even by 1805 the figure was only 26 per cent.

To some extent these figures are misleading. As we shall see, they ignore major shifts *within* each county, especially in Cornwall, but they also place the Plymouth conurbation exclusively within Devon when, given its position, as well as its national character as a naval and shipping centre, its growth was hardly a Devonian matter alone.[10] If we exclude the Plymouth

figures from the equation then Devon's urban growth was more sluggish, at 26 per cent to 1750 and 31 per cent thereafter, making a mere 65 per cent in all, one-third of Cornwall's growth rate. From holding 23 per cent of the non-Plymouth urban population in 1660, by 1750 Cornwall had 27 per cent and by 1805 35 per cent. To set this in context, in the two censuses of 1801 and 1811 Cornwall's population was only 56 per cent that of Devon as a whole, or 65 per cent that of Devon without the Plymouth conurbation. Only 26 per cent of Cornwall's population lived in towns c.1805, compared to 42 per cent of Devon's including Plymouth and 32 per cent without Plymouth; for the region as a whole the figures are 36 per cent with Plymouth and 29.5 per cent without.

As **Table 53.3** shows, overall the urban population of the two counties increased from 26 to 36 per cent between 1660 and 1805. Excluding Plymouth, the rate of growth was greater in Cornwall (from 19 to 26 per cent) than Devon (from 27 to 32 per cent), though Devon remained more urbanized. Cornwall's greater urban growth reflected the much steeper rise in its general population during this period, although urban growth outstripped rural, even in densely populated west Cornwall, which was, by 1805, as urbanized as any part of the region, if we exclude the two major conurbations, equalling east Devon with 30 per cent. But by that date no part of the region was less than 20 per cent urban, east Cornwall and west-central Devon having both risen to that proportion (from 16 and 17 per cent respectively in 1660).

Table 53.2
Comparative rates of urban growth, c.1660–1805.

Region	Total c.1660	% growth 1660–1750	Total c.1750	% growth 1750–1805	Total c.1805	% growth 1660–1805
A. All towns	83,925	39	116,635	73	201,920	141
B. Devon towns (B as % of A)	65,790 (78)	34	87,845 (75)	70	149,365 (74)	127
C. B – Plymouth (C as % of A)	60,390 (72)	26	76,345 (65)	31	99,865 (49)	65
D. Cornish towns (D as % of A)	18,135 (22)	59	28,790 (25)	81	52,555 (26)	190
E. Inland Devon* (E as % of B)	34,330 (52)	24	42,720 (49)	20	51,430 (34)	50
F. Maritime Devon* (F as % of B)	14,560 (22)	24	18,125 (21)	54	27,935 (19)	92
G. Exeter/Plymouth (G as % of B)	16,900 (26)	60	27,000 (30)	160	70,300 (47)	316
H. West Cornwall* (H as % of D)	7,125 (39)	119	15,615 (54)	101	31,445 (60)	341
I. East Cornwall* (I as % of D)	11,010 (61)	20	13,175 (46)	60	21,110 (40)	92
J. Inland Cornwall* (J as % of D)	8,440 (47)	47	12,410 (43)	72	21,300 (41)	152
K. Maritime C'wall* (K as % of D)	9,695 (53)	69	16,380 (57)	91	31,255 (59)	222

NOTES:
E. all Devon towns not in F or G (36 towns)
F. Barnstaple/Pilton, Dartmouth, Brixham, Bideford, Teignmouth, Topsham, Exmouth, Ilfracombe, Appledore, Sidmouth, Dawlish, Hartland, Combe Martin, East Budleigh (14 towns)
H. towns west of Truro and St Mawes (inclusive) (12 towns)
I. Cornish towns not in H (25 towns)
J. Redruth, Camborne, Helston, Launceston, Bodmin, Liskeard, St Austell, Lostwithiel, Tregony, St Columb Major, Grampound, St Germans, Callington, Wadebridge, Camelford, Stratton, Week St Mary, Mitchell (18 towns)
K. Cornish towns not in J (19 towns)

URBAN HIERARCHIES OF CORNWALL AND DEVON

However, regional and county trends are only part of the story and they are comprised of a series of individual town histories, only some of which can be related closely to the rest of the urban network in the region. A brief consideration is needed of the individual successes and, to a lesser extent, failures. The most obvious is the growth of the Plymouth conurbation, replacing Exeter as the region's largest town after 1750 and by 1805 dwarfing all the other towns—with a population larger than the next seven towns put together, or the bottom sixty-two towns combined. Most of this growth occurred after 1750 and it was concentrated not in historic Plymouth but in the new towns of East Stonehouse (the ninth largest town in 1805 by itself) and, in particular, Dock, which was the largest town in Devon in its own right by 1805, with Plymouth proper still below Exeter.

While the naval connections of the Plymouth conurbation gave it a unique stimulus, its spectacular growth can be paralleled by that of a number of other maritime communities.[11] These include the ports of Falmouth and Penzance, the fishing towns of Brixham, Mevagissey, Newlyn and Mousehole and the seaside resorts of Exmouth and Teignmouth. All of these towns rose spectacularly in urban status, especially Falmouth, which had only emerged as a town a few years before 1660 but was already the largest town in Cornwall and seventh town in the region by 1750, and by 1805 was the largest town outside Plymouth conurbation and Exeter. Like the Plymouth conurbation, it owed its growth largely to external factors, namely its role in the packet business and in sheltering and victualling for the Atlantic trades, rather than as an outport for regional products. Penzance, by contrast, grew in part from fishing but mainly after it became a coinage centre for tin in 1663. Its growth and rise in urban status paralleled that of western Cornwall in general, with tin and other extractive industries proving the key, along with fishery. The Cornish urban network was turned upside down, with the old towns of eastern Cornwall such as Bodmin, Launceston, Liskeard and Lostwithiel losing rank to Falmouth and Penzance, together with the old centres of Truro, Helston, Penryn and St Ives now growing rapidly, and the mushroom new towns of Redruth, Camborne and St Agnes, plus St Austell further east. Truro's growth was only slightly less spectacular than Falmouth's and from 1750 it was rapidly catching up its neighbour, while Penryn, with two rivals so close, did well to grow at a standard rate. The figures for the new mining towns are inevitably very uncertain, with no clear evidence of their exact share of population within what were becoming the most populous parishes in Cornwall, but their growth pattern is evident.

Compared to these Cornish changes, the Devon urban hierarchy was much more stable, save at the very top. The case nearest to the Cornish is that of Brixham, whose fisheries brought it from nowhere to within the region's top ten, challenging Barnstaple and Dartmouth (**Map 53.4**) and eclipsing Bideford. The other maritime winners were Teignmouth and Exmouth where the fishery was significant but, especially in the latter case, the tourist trade was more important. In the early nineteenth century Dawlish, Budleigh Salterton, Sidmouth and Torquay were to follow their lead, along with Ilfracombe and Appledore on the north coast. Nevertheless, Devon's top fifteen towns in 1805 included only two newcomers from 1660 (Brixham and Teignmouth, replacing Torrington and Colyton) and considerable internal stability, notably Tiverton and Barnstaple as third and fourth, then Dartmouth, Crediton, Totnes, Tavistock and Bideford, followed by Cullompton, South Molton, Honiton and Ashburton. Taken together, however, these eleven towns only grew by 26 per cent to 1750 and a mere 16 per cent thereafter, and whereas they had formed more than a third of Devon's urban population in 1660 and in 1750, they provided less than a quarter in 1805. The main reason for this, without doubt, was the declining role of the cloth trade and the effect this had on the economy of north and east Devon, which hit the towns both directly, as manufacturing and finishing centres, and indirectly as centres of distribution, marketing and services for the countryside around. Exeter, of course, was affected by the same trends.

It is important to remember, however, that population as such is not necessarily an index of wealth or economic activity; its significance can vary according to the nature of a town and its activities. Before 1750, in particular, many thriving towns had

Area	1660	Urban 1660	%	1805	Urban 1805	%
Devon	227,157	65,790	29	358,987	149,365	42
Devon excl. Plymouth	221,757	60,390	27	309,187	99,565	32
East Devon	85,182	30,035	35	118,427	49,395	42
East Devon excl. Exeter	73,682	18,535	25	97,927	28,895	30
North Devon	43,190	11,770	27	56,764	16,705	29
South Devon	69,381	19,075	27	141,475	74,950	53
South Devon excl. Plymouth	63,981	13,675	21	91,675	25,150	27
West and central Devon	29,404	4,910	17	42,321	8,315	20
Cornwall	98,104	18,135	19	203,713	52,550	26
West Cornwall	41,412	9,095	22	122,471	36,155	30
East Cornwall	56,692	9,040	16	81,242	16,395	20
Devon and Cornwall	325,261	84,105	26	562,700	201,915	36

NOTES:

East Devon contains the hundreds of Axminster, Bampton, East and West Budleigh, Cliston, Colyton, Crediton, Halberton, Hayridge, Hemyock, Ottery, Tiverton and Wonford, and the city of Exeter.

North Devon contains the hundreds of Braunton, Fremington, South Molton, Shebbear, Shirwell and Witheridge.

South Devon contains the hundreds of Coleridge, Ermington, Exminster, Haytor, Plympton, Roborough, Stanborough and Teignbridge.

West and central Devon contains the hundreds of Hartland, Lifton, Tavistock, North Tawton and Black Torrington.

West Cornwall contains the hundreds of Kerrier, Penwith and Powder.

East Cornwall contains the hundreds of East, Lesnewth, Pyder, Stratton, Trigg and West.

Table 53.3
Levels of urbanization in 1660 and 1805.

their population levels curbed by repeated fires, destroying a great deal of urban property, and by very high levels of mortality, with smallpox, typhus and other diseases, especially of children, replacing plague as urban killers. Many of the older towns in cloth districts had also served highly populated rural areas, just as the new mining towns were to do, with many of those dependent on the town economy not actually living there. In this they contrast with shipping or fishing centres or resorts where, by the later period at least, most of those involved in the town's main activity were necessarily resident; the same could be said for the new factory towns. The sex ratio of towns is also an important variable. During the seventeenth and eighteenth centuries most towns were attracting more females than males, though this effect is magnified by male absenteeism on military service and on ship on the census days.[12]

MARKET FUNCTIONS

Population alone, in other words, is not a reliable guide to urban status. While contemporaries certainly noticed population sizes, often very vaguely, they also measured towns by other criteria. Perhaps the most traditional mark of a town was its provision of a market. During the early modern period this market function

improved transport links to other towns and the growth of alternative retail outlets finally pushed marginal markets out of business. In several cases these towns shifted to holding monthly or quarterly cattle markets or annual fairs, which were also held by many villages with no urban status.

Several problems arise in the identification of markets. Without detailed local research on each community, it is necessary to rely on lists of market towns provided by county historians, notably those of the early seventeenth century and the decades around 1800, plus various national and local directories or gazetteers of towns which list markets, together with market towns marked on county maps. References to markets in travel accounts and guides, notably Willis's 1716 account, are also helpful.[13] However, all of these sources raise two issues. One is the tendency for a town, once established on the list of markets, to remain there, either because of the copying of earlier sources or from a sense of tradition; some attempt has been made to identify the most obvious examples of towns where markets had become marginal though surviving.[14] The second issue concerns what constituted a market. A market town was generally conceived as

Map 53.4
Detail from map of Dartmouth, 1619.

SOURCE: reproduced by permission of Devon Record Office (R9/1/Z33).

Fig. 53.1
Drawing of 227 High Street, Exeter.

SOURCE: J. Crocker, *Sketches of Old Exeter* (London, 1886), plate 29.

faced challenges, from private dealing in inns, from the emergence of retail shops, at first for imported goods but later for a much wider range of products, and finally from pedlars and chapmen. By 1800 the market was only one manifestation of a town's retailing, distributive and service function, but it remained a vital focus of town life and, in particular, for the regular use of the town by its rural hinterland. As **Map 53.5** suggests, no parts of the South West, save for the furthest tips of Lands End and the Lizard and some small tracts of Dartmoor and Bodmin Moor, were more than 7 miles from a market town for most, usually all, of this period, and in some areas, such as the Culm valley, market towns were even more densely concentrated. No well-established town lacked a market and many of the growing towns obtained one, although some very small towns on the margins of urbanity lost their markets during the period, especially in the last fifty years when, it seems likely,

Map 53.5
Market towns, 1600–1800.

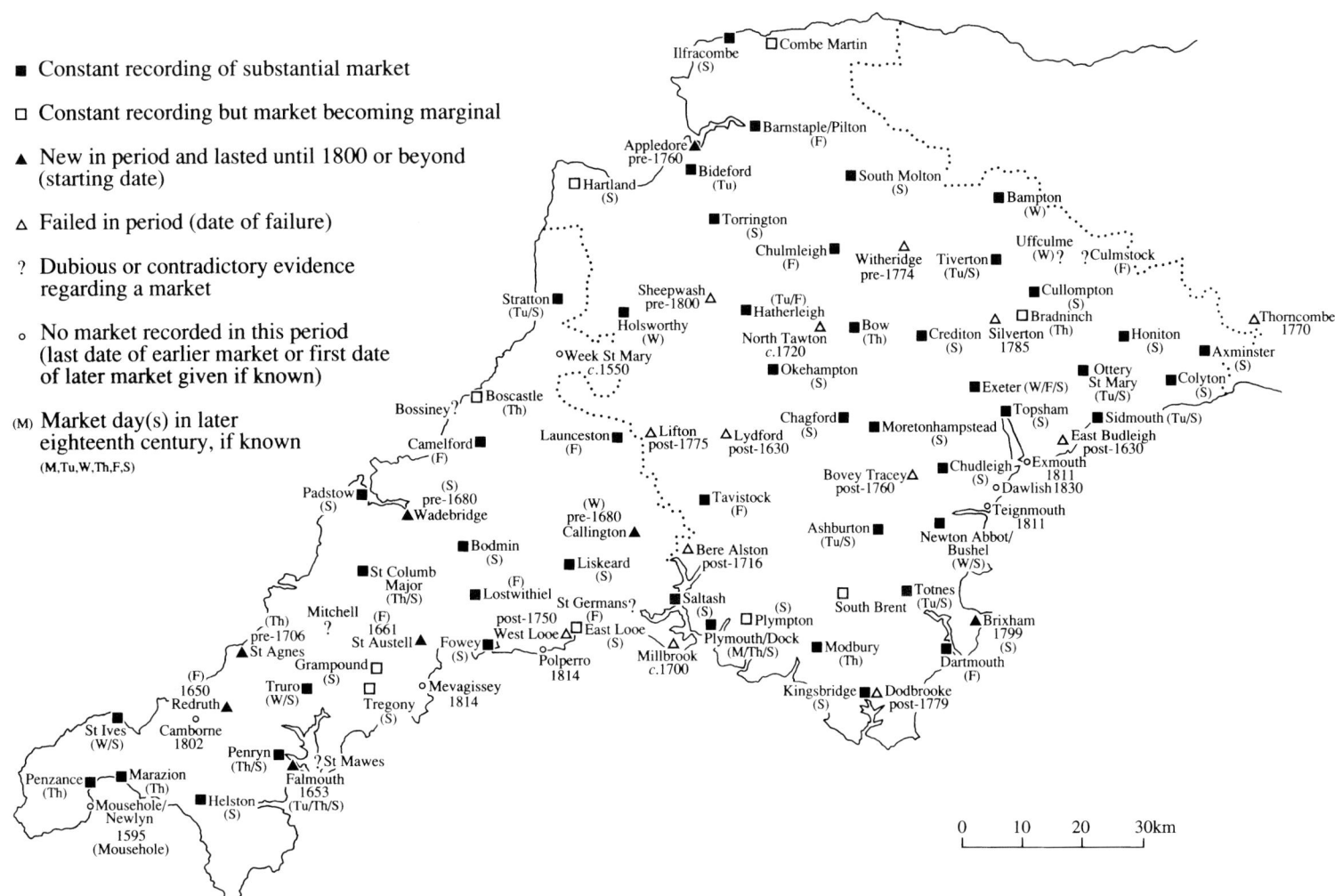

Legend:

■ Constant recording of substantial market

□ Constant recording but market becoming marginal

▲ New in period and lasted until 1800 or beyond (starting date)

△ Failed in period (date of failure)

? Dubious or contradictory evidence regarding a market

○ No market recorded in this period (last date of earlier market or first date of later market given if known)

(M) Market day(s) in later eighteenth century, if known
(M,Tu,W,Th,F,S)

0 10 20 30km

somewhere which provided a market not so much for its own inhabitants as for the exchange of goods within its hinterland, together with the sale of goods moving into or out of the district. Thus a proper market town would deal in cloth, wool, leather, cattle or corn on a considerable scale, as well as provisions and craft products from the district. While many of the larger towns continued to act in this way, many smaller ones increasingly had a market solely for local exchange and even, in some cases, only for the supply of the town's own needs. By the early nineteenth century some towns had a market only for butchers' meat and other provisions or even just for a few trifles: clearly such places were not proper market towns, but still had some kind of market.

Overall, it is the continuity of market towns (above the smallest) that strikes one, especially in Devon. There the towns that failed were all, save those on or around Dartmoor, very close to a stronger rival, and the only new markets to emerge were in two thriving ports—Appledore in the seventeenth century and Brixham in 1799. The only other trend is the loss of market towns

north of Exeter after 1750, as Silverton and Witheridge lost their markets and Culmstock and Uffculme may well have done the same, while Bradninch market declined—all evidence of the faltering prosperity of the cloth trade. In Cornwall, the market system was rather less complete in 1600 and the shifts in population changed market requirements. Gaps in coverage allowed Wadebridge, Callington and, in particular, St Austell, to establish new markets. John Norden had identified the Duchy manor of St Austell as a good market prospect in 1615 because its nearest rivals (Grampound, Tregony, Fowey, Lostwithiel and Bodmin) were all approximately 7 miles away—this being then seen as the maximum viable distance to attend a market. The first three rivals all blamed their subsequent decline as markets on St Austell, although Tregony also suffered from Truro's rapid expansion.[15] Meanwhile in west Cornwall there was room for several new markets, first and most significantly Redruth but then St Agnes and finally, in 1802, Camborne, to cater to the mining populations of what had been sparsely populated districts.

INCORPORATED TOWNS AND PARLIAMENTARY BOROUGHS

Another vital factor in urban status was institutional complexity, autonomy and political power, expressed above all in borough status. **Maps 53.6 and 53.7** display two forms of borough, incorporated towns and parliamentary boroughs, but it should be stressed that many other towns (plus a number of places not considered here) had some claim, often residual, to borough status. Medieval grants of borough status (often together with markets) were often long remembered, for example in the continued nomination of mayors or portreeves to conduct manorial business, distinguishing such places from mere villages. Real power to run local affairs (beyond that offered to any parish) was by now limited in the main to those towns that had obtained a charter of incorporation in the Tudor and Stuart period. However, success in this often expensive process depended more

on friendly courtiers than on urban standing. In Cornwall, with Duchy influence, all the major Cornish towns in 1660 had obtained charters (although Padstow had allowed its charter to lapse), as had several quite small ones, with Fowey and Marazion the largest unincorporated towns; the new town of Falmouth obtained a charter in 1661. In Devon, by contrast, the smallest incorporated towns (apart from the ancient borough of Plympton) were (Duchy-owned) Bradninch and Okehampton, but twelve of the towns larger than these two were unincorporated, including Crediton, Tavistock and Cullompton. This situation was remedied, temporarily, for Tavistock, Honiton and Ashburton, as well as Fowey, Grampound and Callington, during the extraordinary burst of charter granting and revocation during the 1680s, when Charles II and James II used charters to remodel town government to their (rapidly changing) political will. Save for Fowey, which obtained a new charter in 1690, none of these

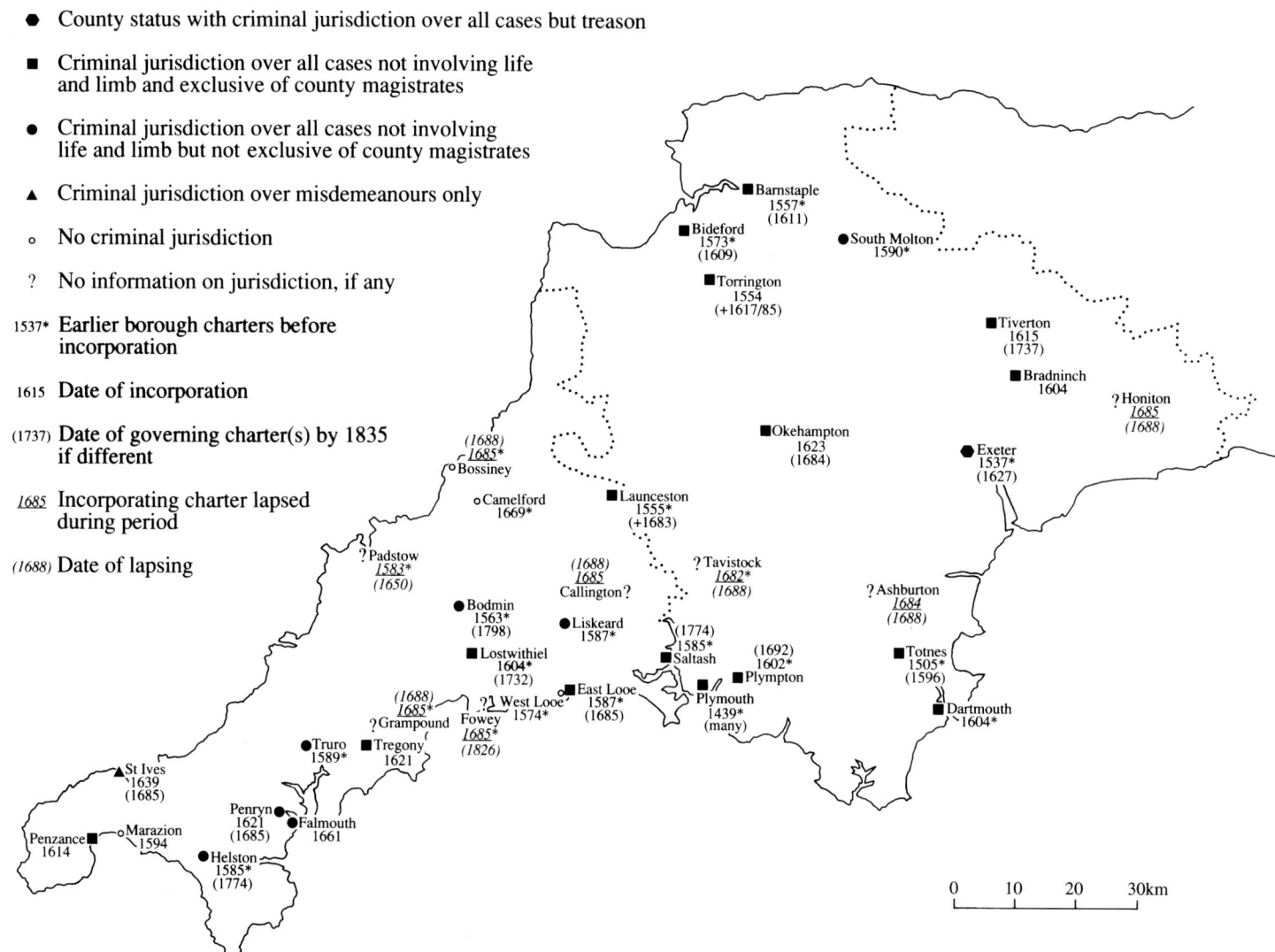

- ● County status with criminal jurisdiction over all cases but treason

- ■ Criminal jurisdiction over all cases not involving life and limb and exclusive of county magistrates

- ● Criminal jurisdiction over all cases not involving life and limb but not exclusive of county magistrates

- ▲ Criminal jurisdiction over misdemeanours only

- o No criminal jurisdiction

- ? No information on jurisdiction, if any

- 1537* Earlier borough charters before incorporation

- 1615 Date of incorporation

- (1737) Date of governing charter(s) by 1835 if different

- 1685 Incorporating charter lapsed during period

- (1688) Date of lapsing

Map 53.6
Incorporated towns,
1500–1800.

parliamentary boroughs retained its incorporated status when, after 1688, parliament declared most of these new charters invalid (despite this, several towns continued to use these charters, as the dates for the governing charters in 1835 indicate). After this fiasco, new charters were few and far between and were all granted to old towns to clear up confusions in their constitution, so that none of the expanding towns thereafter was incorporated before 1800.[16]

Exactly what difference incorporation made is hard to measure. The most tangible right conferred was judicial, with rights to hold both criminal and civic courts. **Map 53.6** shows the rights of criminal jurisdiction conveyed and whether this was exclusive of the county bench or not (whether these powers were exercised or not at different periods is another question, requiring detailed research); generally speaking the larger communities had the greatest rights. However, in daily life the right to hold weekly courts to settle local business or a court of record to settle debts and other business affairs may have been more crucial; towns valued the speed and cheapness of not having to deal with gentry magistrates or irregular and distant county sessions. The other main effects were political, with charters laying down the form of municipal government, usually one with a restricted and self-perpetuating inner council, though sometimes with a wider freemen body. In practice, however, this did not so much create an oligarchy as recognize the oligarchical tendencies inherent in local government when this was unpaid and time-consuming—the richer and more leisured always dominated whether legally supported or not. Nevertheless, the formal trappings and conditions of power did matter and were certainly valued by townsmen, and often viewed jealously by the country gentry. From Carew's strictures onwards, the pretensions to power, especially of the governors of the smaller boroughs, irritated and amused gentry and visitors.[17]

Such apparently petty matters took on a wider importance, however, when it came to parliamentary boroughs.[18] As **Map 53.7** shows, the region, and south-eastern Cornwall in particular, had more than its fair share of (two-seat) constituencies, notably the four seats allocated to Launceston (Dunheved and Newport) and East and West Looe. In Devon, only Plympton (a medieval constituency) and Bere Alston (enfranchised in 1584) were particularly small, although the late re-enfranchisement of Honiton, Ashburton and Okehampton in 1640 looks rather arbitrary when Crediton, Bideford, Cullompton, Torrington and South Molton were all somewhat larger. However, the criterion for gaining a seat in this period again owed more to aristocratic politics than to urban standing, and this is very clear in Cornwall.[19] Here a respectable group of medieval constituencies was supplemented, thanks to Duchy and gentry influence, by a series of new seats of which only St Ives, Penryn and Saltash retained any real urban credibility. At the same time some of the older towns in the east also lost ground, leaving two of the top three Cornish towns in 1750 and 1805 (Falmouth and Penzance) unrepresented. However, given the ability of urban interests to find representation indirectly before 1800, this mattered less than

what seemed, to many in the eighteenth century, to be an indefensible power given to the corporations or townspeople of various obscure Cornish settlements. Two of these (Bossiney and Mitchell) were not urban in any other respect, while St Mawes and St Germans were little better and none of the other places displayed any signs of real vitality during the period.

Although such seats became proverbial for corruption during the period, the contrast between them and the larger towns should not be pushed too far. Except in the largest towns, members of parliament were nearly always drawn from the gentry not from townspeople, and in towns of every sort they were then expected to cater to the constituents' needs. How assiduously they did this depended considerably on how secure their control of the seat was and this depended more on the nature of the franchise and local property ownership than on urban status. **Maps 53.7 and 53.8** attempt to indicate some of the factors influencing constituency politics by the eighteenth century through a five-fold franchise classification (**Map 53.7**), a note of the number of general election contests 1701–1800 and an indication of the maximum size of the electorate between 1700 and 1784 (**Map 53.8**).[20]

None of these classifications is entirely reliable, either in itself or as an indicator of how politics operated. The choice of a single franchise type or number of voters conceals the fact that both changed. Before 1700 franchises often altered so much and so drastically that it would be impossible to represent the situation on any single map; we often cannot find out what

Fig. 53.2
'A view of the Hamoaze and Plymouth Dock, &c. from Mount Edgecumbe'.

SOURCE: reproduced by permission of Westcountry Studies Library, Exeter.

Map 53.7
Parliamentary boroughs:
nature of franchise, 1500–1800.

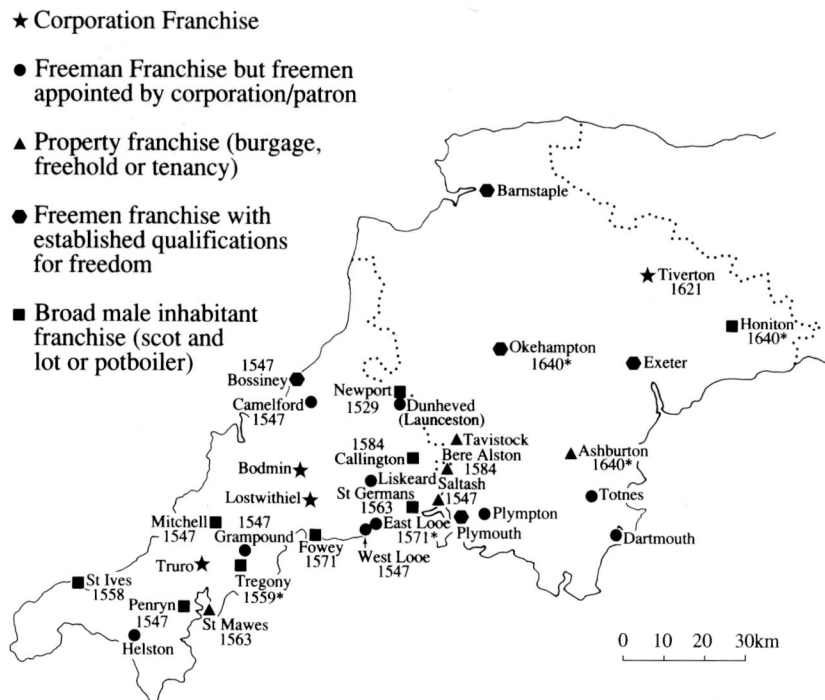

★ Corporation Franchise

● Freeman Franchise but freemen
appointed by corporation/patron

▲ Property franchise (burgage,
freehold or tenancy)

● Freemen franchise with
established qualifications
for freedom

■ Broad male inhabitant
franchise (scot and
lot or potboiler)

● Barnstaple

★ Tiverton
1621

■ Honiton
1640*

● Okehampton
1640*   ● Exeter

1547
Bossiney●
Camelford●   Newport
1547   1529 ■ Dunheved
(Launceston)
1584   ▲ Tavistock
Bodmin ★   Callington   Bere Alston
St Germans ■ Liskeard 1584   ▲ Ashburton
Lostwithiel ★   1563   ▲ Saltash   1640*
Mitchell ■ 1547   1547   ● Plympton   ● Totnes
1547 Grampound ● East Looe   ● Dartmouth
Truro ★   ● Fowey 1571* Plymouth
■ St Ives   1571 West Looe
1558 Penryn ○   1547
1547 ▲ St Mawes
● Helston   1563

0   10   20   30km

1621 Date of enfranchisement (none given if continuous from before 1500)

1640* Earlier right restored after 1500

**Size of electorate
1700-1784**

(maximum at any point
in period)

– 1281-3560
– 321-640
– 81-160
– 1-40

(8) **Number of contests
known in general
elections 1701-1800**

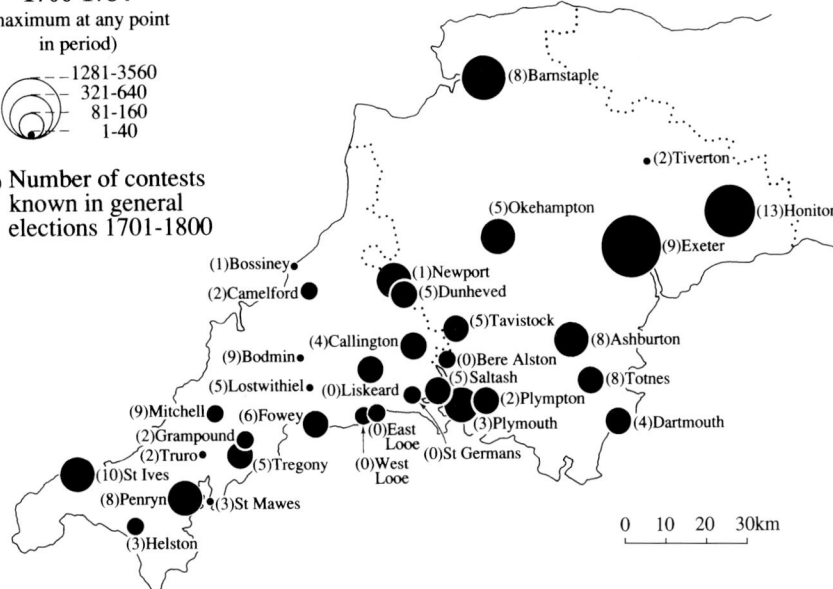

(8)Barnstaple

● (2)Tiverton

(5)Okehampton
(13)Honiton
(9)Exeter

(1)Bossiney●
(2)Camelford●
(1)Newport
(5)Dunheved
(5)Tavistock
(9)Bodmin●   (8)Ashburton
(4)Callington   (0)Bere Alston
(5)Lostwithiel (5)Saltash   (8)Totnes
(0)Liskeard   (2)Plympton
(9)Mitchell●   (6)Fowey   (3)Plymouth (4)Dartmouth
(2)Grampound   (0)East
(2)Truro●   Looe
(10)St Ives   (5)Tregony (0)West (0)St Germans
(8)Penryn   Looe
(3)St Mawes

(3)Helston

0   10   20   30km

Map 53.8
Parliamentary boroughs:
size of electorate, 1700–1784.

contemporaries used as a franchise and in many cases they did not know or care until more frequent and contested elections for a parliament of increasing importance made the right to vote a valued one. Between 1679 and 1715 rival claims about the franchise constituted a vital element of party struggle, resolved by Commons decisions that were often dictated by party advantage, and this remained the case even in the more stable conditions after 1715. When rival franchises operated, or some hybrid was selected, **Map 53.7** shows the more broadly based variant. Equally, the number of voters could vary widely, partly in accord with the franchise and partly depending on how that franchise was operated. For example, freemen's lists could be expanded or narrowed, especially when they were appointed by corporation or patron, and property qualifications could be manipulated, not just by creating and destroying legal rights such as burgages and tenancies, but by physically building or destroying houses within the borough, as happened at St Germans (gradually dismantled by the Eliots) and Tregony (filled with dismal cottages for the 'potboiler' electors).[21] The number of contests (**Map 53.8**) can also be misleading, since the convention then was only to vote if the outcome was in question, so that a lengthy pre-election canvas and debate might not result in any 'contest' strictly identified, while a clear majority over time for a particular party might also prevent the rival side from bothering to challenge. Nevertheless all these factors taken together can indicate a spectrum of political types from the open constituencies like Honiton, St Ives and Exeter (wide electoral franchise, large electorate, many contests) to the completely closed seats like Bere Alston and West Looe. Whatever the type of seat, however, the ever more complex and expensive preparations for elections played an important part in the life of such towns, often adding an extra dimension to their municipal politics given the crucial role of corporations and especially mayors as voters and returning officers. For the smaller seats, in particular, being a parliamentary borough fundamentally altered their whole character.

Leaving aside their political significance, the parliamentary status of such insignificant towns had a considerable contemporary impact on views of urbanization in the region. One cause of the tendency to contrast Devon's vigorous urbanization with Cornwall's weak urban system, especially before 1750, was a fascination with these small seats, although, as we have seen, there was some truth in the claim. It did not help Cornwall's image that the traveller west passed from Devon into declining east Cornwall, forming an image hard to dispel when he or she reached the expanding western towns.[22] By 1800, this contrast began to lessen, not just as Cornwall caught up with Devon but as many of Devon's old-established towns began to slip down the urban league. Here both changing evaluations of urban character and the changing basis for urban life interacted to alter the impact of the region's towns on contemporaries.

The 'urban renaissance' after 1660 changed expectations of urban form, lessening the weight of old criteria, such as the

partly a matter of changing taste but also of real shifts in the south-western economy, especially after 1750, with the rise of resorts and decline of some cloth towns. Meanwhile the urban population was increasingly concentrated in ports. These had always (partly due to dense settlement and partly from their non-agricultural nature) been regarded as a type of town, however small, as Leland's frequent references to 'fisher-towns' suggest, but in other respects, such as institutional complexity or architectural dignity, such towns often seemed non-urban.[24] The same uncertainty surrounded the new mining towns. During the early modern period, therefore, in Devon but even more in Cornwall, we see changes in the type and character of urbanization as a matter not just of hard data but also of perceptions and expectations. The maps in this chapter must be read in this light, not just because the evidence on which they are built reflects such changing ideas, but also because the classifications adopted are themselves an uneasy effort to capture the shifting meanings of urban life.

Fig. 53.3
Lithograph by M.H. of the Butterwalk, Dartmouth, 1845.

SOURCE: reproduced by permission of Westcountry Studies Library, Exeter (Somers Cocks 542).

FURTHER READING

For the general issues discussed here see also P. Clark and J. Hosking (eds), *Population Estimates of Small Towns 1550–1851*, revised edn (Leicester, 1993). Although this analysis extends and corrects the material presented in the latter volume, it (and the earlier edition) formed an invaluable basis for this work. For parish boundaries see H. Peskett, *Guide to the Parish and Non-Parochial Registers of Devon and Cornwall 1538–1837*, Devon and Cornwall Record Society, extra series 2 (1979) and for the census returns E. Higgs, *Making Sense of the Census* (London, 1989). Earlier estimates for Devon and Cornwall can be found in W.G. Hoskins, *Devon* (London, 1954), 104–22, and J. Whetter, *Cornwall in the Seventeenth Century: An Economic History of Kernow* (Padstow, 1974), 8–13.

church or the market, and giving new emphasis to 'urban' styles of building such as street, square and classical frontage (churches and market houses might of course be rebuilt or fronted to meet these new demands).[23] Towns where timber or thatched houses in irregular and narrow roads predominated were not just liable to disastrous fires but also to denigration as old-fashioned. It was the retailing and service sectors of urban life that best expressed this new urbanity, and thus towns were measured by shops, banks or leisure facilities more than by market or manufacturing, for example in the new guidebooks and directories. This change is

ACKNOWLEDGEMENT

The author thanks Dr Tom Arkell for his assistance with the figures, for sharing data and for his general advice.

MORPHOLOGICAL
DEVELOPMENT OF TOWNS
SINCE THE
MIDDLE AGES

MARK BRAYSHAY

Despite considerable growth both in the areal extent and in the total population of the towns of south-west England, especially since the beginning of the nineteenth century, urbanization in the region has remained below the national average. In 1981 the census showed that in Cornwall and the Isles of Scilly only 56.1 per cent of the total population resided in towns, while in Devon the figure was 70.6 per cent.[1] But in England and Wales as a whole, 76.9 per cent of the population were living in towns or cities. Yet while Devon and particularly Cornwall have remained less urbanized than elsewhere, for many centuries the towns of the two counties have nonetheless played a vital economic, social and political role in the life of the region. They have been the 'essential cogs in the machinery' of rural society, providing not only organization and focus, but also a range of central services and functions for the surrounding countryside.[2] There is, moreover, a great variety of towns in the South West. They range in size from Plymouth, which by 1991 had almost a quarter of a million inhabitants, to the numerous small market towns like Liskeard with 7,559, Okehampton with 4,572, or Holsworthy with only 1,890.[3] Each town possesses its own individual characteristics reflecting its particular historical development. Indeed, the diversity exhibited by the urban settlements of the region makes any attempt at generalization extremely difficult. This is especially true with regard to urban morphology.

Although no attempt is made here to provide a comprehensive discussion, three themes in the urban morphology of the South West will be considered. First, the historical extension of the built-up area of towns will be explored by examining urban growth in Tavistock. Second, an application of town-plan analysis will be undertaken for Holsworthy. Third, the influence of the pre-urban cadastre on the urban morphology of part of Plymouth will be investigated. Thereafter, some of the particular morphological features of coastal resorts will be discussed in the context of Sidmouth and Ilfracombe, while aspects of nineteenth-century industrial towns are considered in the case of Redruth.

THE ANALYSIS OF THE AREAL GROWTH OF TOWNS: TAVISTOCK (**Maps 54.1 and 54.2**)

A consideration of the manner in which town growth is expressed in physical terms, as the gradual extension of the built-up area, has traditionally formed the major, sometimes the only, component in morphological analysis.[4] Urban extension is, however, never a continuous process, but rather tends to be erratic or cyclical with periods of rapid growth alternating with periods of standstill. Inevitably, the character and timing of these phases will vary from one town to another. Moreover, town growth is usually a twin process of both outward extension and internal reorganization.[5] Maps which depict only the spread of a town from its original core by means of gradual accretions cannot, therefore, convey fully the complexities of the processes involved. Indeed, because such analyses depend critically on the availability of a series of accurate historical maps, which will not in any case provide for continuous monitoring of change, their value in presenting an authentic image of the dynamic of urban extension is somewhat questionable. The value of an urban growth map lies instead in revealing the spatial relationship between accretions to the built-up area at different dates, the areal spread of various phases of development, and the general configuration of the street system which forms the essential framework of the town plan. While no two towns will exhibit an identical pattern of spatial growth, the case study presented here, Tavistock, not only demonstrates something of the broad chronology of urban extension in south-west towns, but also indicates the usefulness of a cartographic approach in studying historical changes in a specific town (**Maps 54.1 and 54.2**).

The original core of Tavistock developed around the Tavy in 974.[6] In 1105 the abbot established a weekly market and by 1116 there was an annual three-day fair. Later in the twelfth century 325 acres (132ha) of the monastery's land was legally designated as a borough, but only approximately 47 acres (19ha) was occupied by the market and the burgage plots.[7] In 1305 Tavistock became a stannary town, though when Dartmoor tinning declined, local prosperity increasingly relied upon the cloth trade. The town was transferred to the Russell family (later the earls then dukes of Bedford) at the Dissolution in 1539, but significant urban growth occurred only when copper-mining began to flourish in the 1790s.[8] Rapid population increase was underway by the early nineteenth century and the first major redevelopment of the town centre since medieval times was then carried out. Moreover, a significant expansion of the built-up area took place to the south-west of the original core. Between 1845 and 1866 the dukes of Bedford also financed the construction of workers' cottages on the edges of the town.[9] The broad-gauge railway line (later taken over by the Great Western Railway and converted to standard gauge) was completed by 1859 and a station was erected near Abbey Bridge. But urban extension in the final decades of the nineteenth century was focused mainly on the other side of the river towards the western margin of the town where there was vacant land adjacent to Tavistock's second railway, the London & South Western line, which was completed in 1890.

The decline of copper-mining in the Tamar and Tavy region effectively halted Tavistock's population growth in the later nineteenth century, but the built-up area depicted on Ordnance Survey maps of the town published in 1906, 1955 and 1982 indicates that following a hiatus in development during the early part of the twentieth century, expansion proceeded rapidly during the inter-war and post-war periods. Residential estates were established on a significant scale to the east of the River Tavy in the period before 1955 and these have since been 'infilled' and extended. Indeed new suburbs have spread as far south as the village of Whitchurch which is now no longer physically separate from its urban neighbour. Today the town occupies more than 642 acres (260ha) representing a fourteen-fold increase in its urban area since its foundation as a borough five centuries ago.[10]

TOWN PLAN ANALYSIS: HOLSWORTHY, DEVON (Map 54.3)

While simple growth maps can thus be related in a general way to the factors which have shaped and influenced the pattern of urban expansion, their capacity to reveal the processes of change is rather limited. But since the 1950s geographers and historians have taken much more interest in unravelling and interpreting the detailed complexities of the evolution of the morphological structure of urban areas by means of town-plan analysis (see also chapter 52). Before exploring its application to the study of the urban morphology of the Devon town of Holsworthy, it may be noted that pioneering methodological work on town-plan analysis was published by A.E. Smailes in 1953 and 1955, and was substantially developed during the 1960s by M.R.G. Conzen, not only in his classic monograph on the Northumbrian town of Alnwick, but also in his subsequent studies of Newcastle upon Tyne, Ludlow, Whitby and Conway.[11] Conzen's work demonstrated that techniques of plan analysis can illuminate the historical development of a town even in cases where other documentary evidence is scarce. He argued that urban morphology comprises three basic elements, namely the town plan, the form and fabric of the buildings, and the uses made of land and buildings. In studying Alnwick, Conzen in fact focused primarily on the town plan and used detailed cartographic and documentary analysis to explore the development of the streets, the plots of land, and the block plans of buildings. He developed an analytical method which researchers working elsewhere in Britain have sought to replicate, and he established a flexible classification system to describe the key elements in a town plan.[12] However, apart from a brief consideration of the morphology of Dawlish and Chelston by Ewart Johns in 1971, there have been no detailed 'Conzenian' studies of town plans in south-west England.[13]

Holsworthy in Devon has been selected as an example where Conzen's approach to plan analysis can be explored (**Map 54.3**). Although something of the initial plot pattern survives to a greater or lesser degree in many south-western towns, few escaped substantial alteration during the nineteenth century and medieval cores have in any case often been dwarfed by expansion occurring during the past 150 years. But as one of the smallest towns in the region, isolated in the cold-clay country of west Devon, the pace of change in Holsworthy has been rather more modest.[14] When the Ordnance Survey published its 1:2500 map of the town in 1906, the built-up area had still not spread much beyond the limits of the medieval burgage plots. Holsworthy had thus escaped any wholesale reorganization of its central layout during Victorian times. Thereafter major change occurred only through expansion on the periphery, and the core of the town remained largely unaltered.

Although Holsworthy is thought to date back to the early eighth century, the legal right to hold a weekly market and collect tolls from traders who came to sell their produce on the triangular-shaped plot of land in the centre of the settlement was granted only in 1155.[15] An annual three-day fair was established in 1185. According to Conzen's typology, Holsworthy's central market place represents the town's core and is a separately identifiable 'plan unit'. Moreover, in view of their position and because they occupy well over 85 per cent of the plot area, the two blocks of buildings located in the middle of the square may be categorized as 'market concretions'. Temporary market stalls in medieval towns were frequently made more permanent (often for the sale of perishable goods) or converted into regular shops. Holsworthy's market concretions eventually included a covered market hall (rebuilt in 1857), a public house and a hotel.

Surrounding the market place Holsworthy's burgesses established their homes on burgage plots. T.R. Slater has argued that such plots represent the basic 'cells' of a medieval town (see chapter 52).[16] They were generally long and narrow strips of land aligned roughly parallel to one another at right angles to the street. Medieval plans rarely show a strictly geometrical arrangement of elements but instead were adapted to local site conditions. Although some plots have been amalgamated and

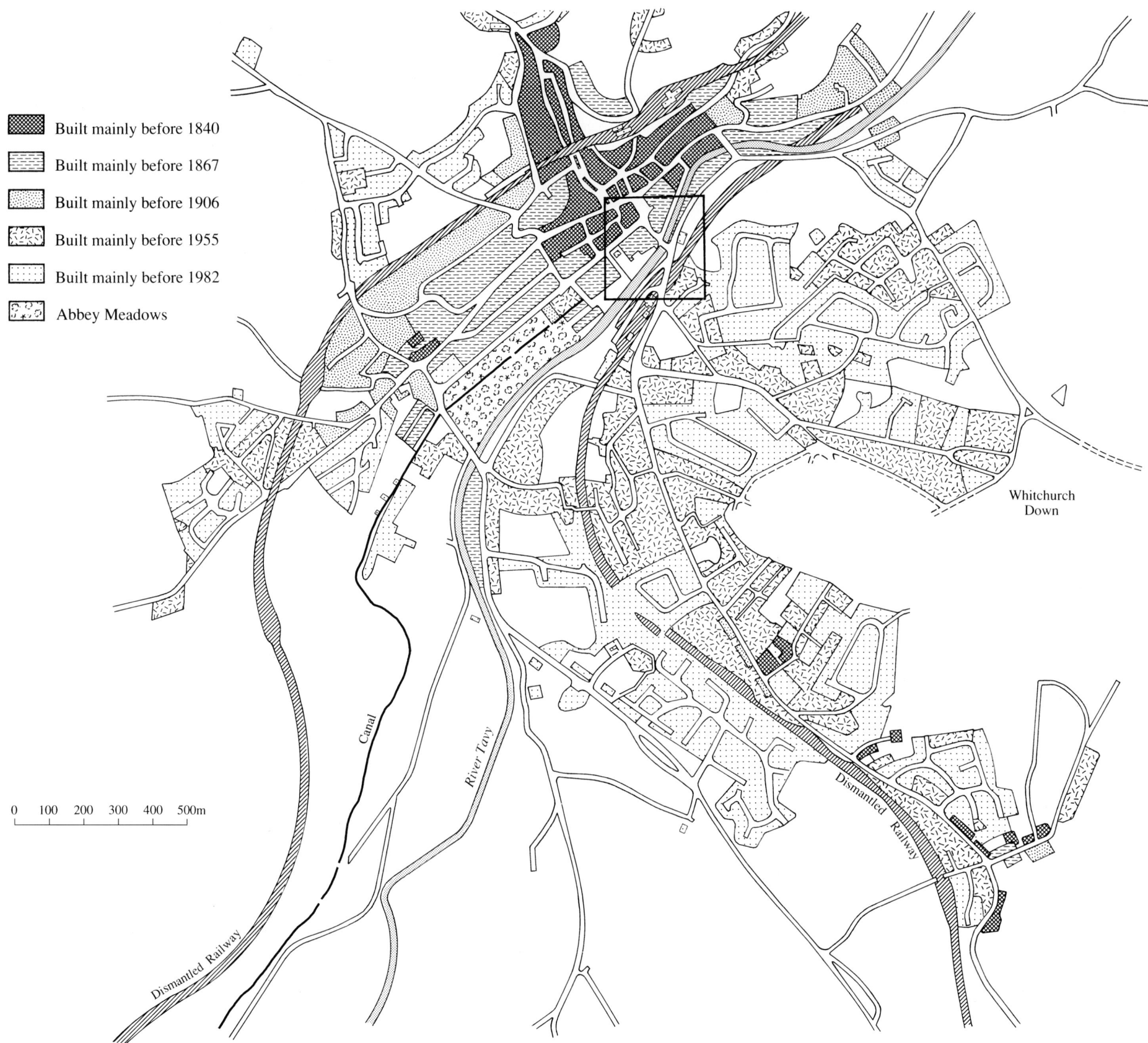

Built mainly before 1840
Built mainly before 1867
Built mainly before 1906
Built mainly before 1955
Built mainly before 1982
Abbey Meadows

0   100   200   300   400   500m

Whitchurch Down

Canal

River Tavy

Dismantled Railway

Dismantled Railway

others divided, the burgages of Holsworthy appear to have remained remarkably stable. This long-term survival of plots reflects in part the difficulties of realigning land boundaries in a crowded urban environment, but it is also related to the legal attributes of burgages in medieval towns. The essential privilege enjoyed by a medieval burgess was that his manorial labour services were commuted to an annual fixed money rent. But to qualify, it was necessary to hold a burgage plot. Important political and legal rights were thus linked to these purposely defined strips of urban land and the boundaries were thereafter carefully protected. Burgesses were, however, often allowed to sub-let their plots and even in medieval times it was common for burgages to be divided internally. Employing Conzen's nomenclature to describe the morphology of Holsworthy's medieval core, it may be noted that the majority of the plots were of the 'long burgage' type where the ratio of depth-to-width exceeds 6:1. Moreover, clusters of similar burgages can be identified which shed some light on the sequence of medieval accretion. Thus the Fore Street long burgages flanking the market place were probably established first. It is possible that the Bodmin Street long burgages (and possibly those at the southern end of Fore Street) were later additions, while the 'standard burgages' of High Street and Victoria Street appear to represent a further, separate accretion. In the latter, the depth-to-width ratio lies between 3:1 and 6:1.

The conjectural classification of Holsworthy's burgage plots presented in **Map 54.3** requires detailed verification by means of field work comparable with that already carried out by T.R. Slater in a small area of Totnes. His work has shown that even in apparently unplanned medieval towns, groups of new burgages added to the existing built-up area display remarkable regularity. In analysing the plot series below the East Gate in Fore Street, Totnes, he showed that each burgage was either 7 perches or 3.5 perches (116ft or 58ft 9in) in width (**Map 54.4**).[17] Despite subsequent alterations, Slater was able to pick out these essential characteristics of a town plan deliberately formed centuries ago as the physical expression of well-understood legal rights and tax liabilities.

In Holsworthy (**Map 54.3**) the tails of the plots aligned along Victoria Street and Bodmin Street were clearly subject to substantial subdivision and encroachment; in the latter case the rearrangement extended across five of the burgages with their heads on Fore Street or Stanhope Square. Elsewhere, however, even those plots accessible via Croft Road were not built on until recent years. Indeed, the entire complex of the market place and the medieval burgages was contained over many centuries within a notional boundary which, in Conzen's terminology, is a 'fixation line'. Fixation lines are easily identified in towns enclosed by walls or where there is a physical barrier to growth such as a river or a steep slope.[18] The medieval fixation line defined for Holsworthy, however, marks the separation of visibly contrasting patterns of landholding and building. It indicates a long period of quiescence in urban

development when economic dynamism was lacking and the town failed to expand. J.W.R. Whitehand has argued that in most towns, particularly during periods of stability, a 'fringe belt' often developed beyond the fixation line where a range of marginalized functions such as industries, cemeteries, isolation hospitals, or slaughter houses might locate.[19] When urban extension resumed, such fringe belts were partially replaced or infilled with residential or commercial development, producing a curious, often chaotic, mix of uses until fresh areas of unoccupied land beyond were reached. But no significant fringe belt appears to have developed in Holsworthy, although arguably the graveyard to the north of the church might have been so classified, along with the two smithies originally located in Sanders Lane, the timber yard near the railway, and the cattle market established at Greenway to the south of the town in 1905.

Traditional post-medieval ribbon development, exhibiting no evidence of linear burgages, characterizes both Chapel Street (north of the line of the railway) and North Road, though map evidence indicates that some of the buildings in the latter were demolished and replaced between 1843 and 1906 and several boundaries were altered as a result. Beyond the medieval fixation line, the exceptionally modest residential accretions belonging to the eighteenth and nineteenth centuries stand out noticeably in the town plan. A rail link to Holsworthy was

Map 54.1 (facing page) The growth of Tavistock, Devon.

SOURCES: E.R. Dymoke's plan of Tavistock, 1867 (Devon Record Office, Bedford Estate Maps & Plans, T1258M/E12); Ordnance Survey 1:2,500 map, 1906 (sheet CV.8); Ordnance Survey 1:10,560 map, 1955; Ordnance Survey 1:10,000 map, 1982.

Map 54.2 The town centre of Tavistock, Devon as outlined on **Map 54.1**. The darker lines show the position of the main boundary walls of the monastic precincts of Tavistock Abbey. Much of the site of the monastery is now obscured by the modern town.

SOURCES: E.R. Dymoke's plan of Tavistock, 1867 (Devon Record Office, Bedford Estate Maps & Plans, T1258M/E12) with the ground plan of the abbey adapted from H.P.R. Finberg, *Tavistock Abbey*, 2nd edn (Cambridge, 1969).

Map 54.3
Town plan analysis of
Holsworthy, Devon.

SOURCES: tithe map, 1843;
Ordnance Survey 1:2,500 map,
1904; Ordnance Survey 1:2,500
map, 1954; Ordnance Survey
1:10,000 map, 1964.

established only in 1898 and the town experienced no great population increase. The housing stock was actually reduced when the alignment of Under Lane was altered to accommodate the railway. Early in the present century some building occurred in Bodmin Street and Croft Road, but the modern estates of Glebelands and West Croft are typical low-density post-war developments.

THE IMPRINT OF THE PRE-URBAN CADASTRE ON URBAN MORPHOLOGY: PLYMOUTH (**Maps 54.5, 54.6, 54.7 and 54.8**)

Studying the relationship between morphological elements and the process of historical town development clearly offers considerable scope for a more thorough evaluation of urban settlements in the South West than has so far been attempted. Indeed townscape elements may perhaps be viewed as a kind of 'text' or a series of encoded messages which, when properly deciphered, shed light on the origin and evolution of urban forms. However, although Conzen's approach is a powerful descriptive tool, it does not facilitate a detailed understanding of the context in which town plans developed. Towns and cities are, after all, human creations and are the physical product of innumerable decisions made in the past by individuals or institutions.[20] Thus Harold Carter has argued that town plans ought to be interpreted in terms of the values and decision-making characteristics of the society that gave rise to them.[21] While such a holistic approach to urban morphology lies beyond the scope of this chapter, nevertheless it is possible to demonstrate in the case of part of Plymouth the influence of the 'pre-urban cadastre' on the form of subsequent urban development. We can seek to understand urban morphology by exploring the pre-urban pattern of fields, the ownership of land, and the manner of its release for building

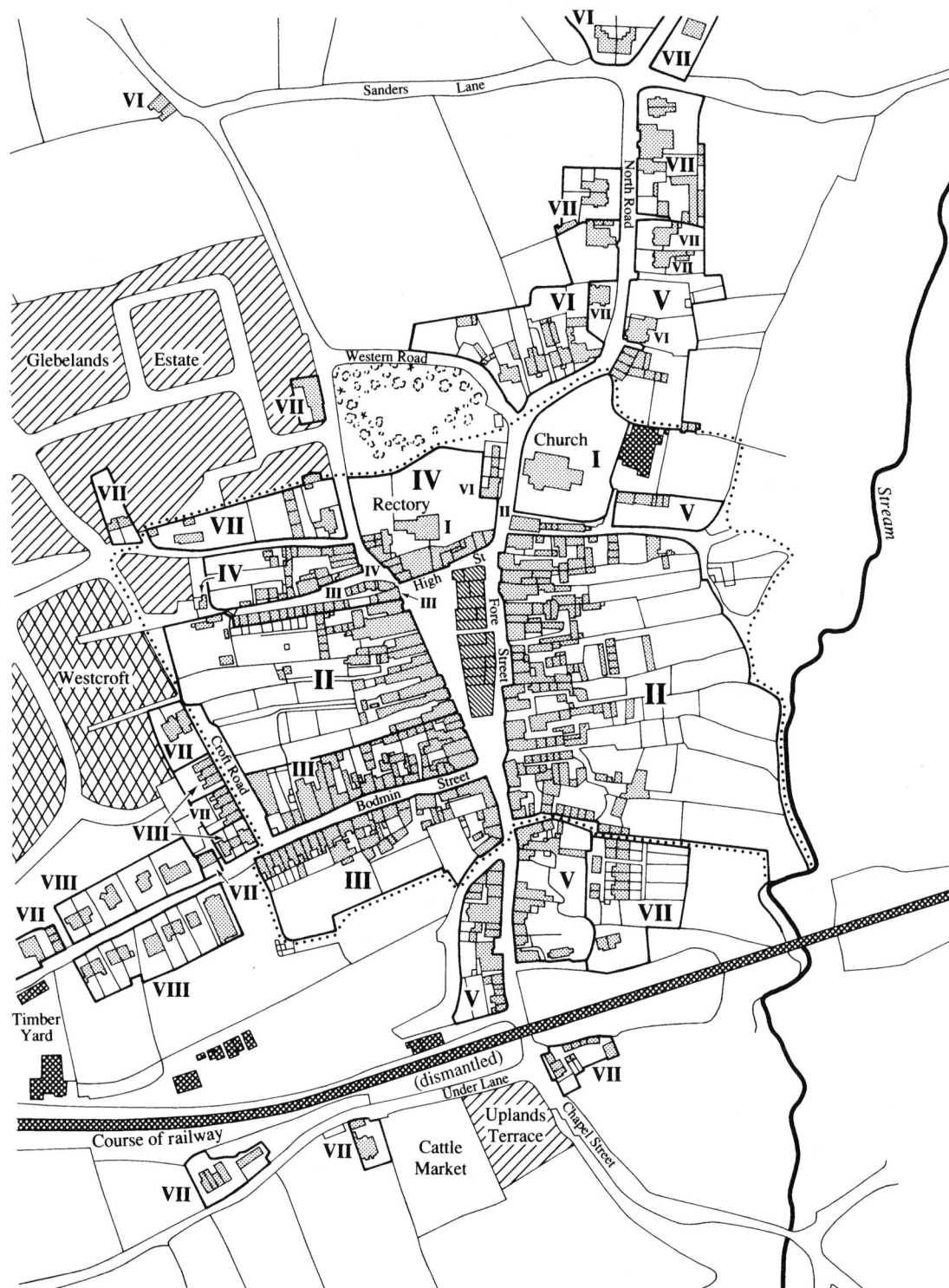

I	Church and rectory complex
II	Long burgages (Fore Street type)
III	Long burgages (Bodmin Street type)
IV	Standard burgages (High Street/Victoria Street type)
V	Traditional arterial ribbons
VI	Late Georgian and early Victorian 'accretions'
VII	Late Victorian 'accretions'
VIII	Early twentieth-century ribbon and infill
	Market 'concretions'
	Modern estate development (Pre-1955)
	Modern estate development (Post-1955)
	Industrial buildings
....	Conjectural 'fixation line' of medieval town

0   100   200m

purposes. Research has shown that development has commonly been fitted into the framework of existing roads, tracks, field and rural plot boundaries. Moreover, the decisions of landowners as to whether or not to sell land, under what precise terms, and for what purposes, have emerged as fundamental factors shaping the form of urban growth.

It may be noted that the first major attempt to relate urban expansion to the underlying pre-urban cadastre was made by J.D. Chambers in his 1951 study of Nottingham where he demonstrated that the configuration of new streets added after 1845 was determined largely by the alignment of the footpaths and furlongs of the former open fields, and that the character of development was strongly influenced by the pattern of ownership established by the enclosure commissioners.[22] In examining the role of the pre-urban cadastre of Leeds, David Ward found that although the strength of the relationship varied from one part of the city to another, the perimeter boundaries of rural properties had often been perfectly preserved in the pattern of subsequent building development.[23] In general, Ward found that the release for building of the larger holdings produced considerable coherence in the subsequent layout of streets and terraces. Although similar work has been carried out for example in Bradford, Bromley and Huddersfield, no studies on a comparable scale have so far been attempted for the towns of south-west England.[24] Indeed, in examining urban design in Dawlish and Chelston, E. Johns found that street patterns show only a 'partial adaptation to the former network of fields'.[25] But Ward's work in Leeds did not in fact point to a detailed correspondence between old field boundaries and later roads; instead he revealed how former landownership patterns lay behind the homogeneity displayed in the internal layout of streets and buildings within discrete areas of the city. There is evidence that private landowners in both Dawlish and Chelston also contrived to influence the layout of certain areas of the town, but Johns pointed to other factors, including local topography, shaping the precise designs which were eventually adopted.

In Plymouth the pre-urban cadastre appears to have exerted a particularly strong influence on the modern street pattern which often clearly preserves the position of former field boundaries (**Maps 54.5 and 54.6**). In some cases the alignment of earlier enclosures is followed by particular roads over considerable distances. Linkages between the pre-urban landscape and subsequent development can be established in Plymouth because the town was chosen in the mid-1780s by the newly formed Corps of Surveyors of the Board of Ordnance as the focus of a remarkably accurate survey at a scale of six inches to the mile.[26] Plymouth was still then largely confined to its medieval site and a network of roads and trackways threaded through the neighbouring fields towards isolated farms and estates in a predominantly rural landscape (**Map 54.5**). Unfortunately, there has been no thorough analysis of the release of this agricultural land for building development, yet

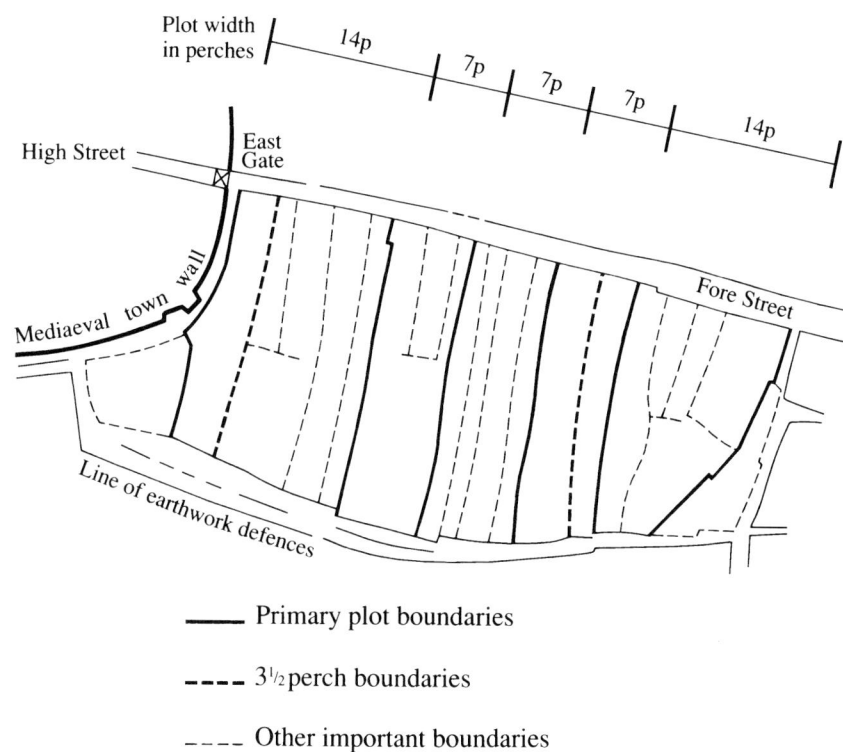

— Primary plot boundaries

---- 3½ perch boundaries

---- Other important boundaries

the clear preservation of key elements of the former agrarian landscape in the urban morphology of modern Plymouth (**Map 54.6**) suggests that in future such research could prove fruitful. Detailed work is required on the precise correspondence between coherent areas of townscape and the shapes of the original landholdings, including for example the Mannamead Estate of the Culme-Seymours, the Beaumont estate of the Bewes family and the Barley Estate of the Elliots.[27] **Maps 54.7 and 54.8** show matching portions of the Ordnance Survey 1:2500 map of the Mutley area of Plymouth (first surveyed in the mid-1850s), as revised in 1888 and in 1907 respectively. While the construction of terraced housing clearly proceeded in a piecemeal fashion as speculative building firms leased and developed plots of land of varying size, the overall plan of the roads was pre-determined years earlier.[28] In fact the alignment of Ford Park Road follows field boundaries appearing on the 1784 six-inch drawing (**Map 54.5**), and the 'fan-shaped' arrangement of terraces to the south echoes the pattern of elongated enclosures over which they were built. The tithe apportionment of 1840 indicates an intricate pre-urban pattern of landownership in this small area of Mutley. Land to the east of Mutley Plain was owned by the Culme-Seymours. The Buller family owned the fields to the north of Ford Park Road, while those lying to the south, as far as the line later taken by the railway, formed a well-defined but detached portion of the Beaumont estate owned by Thomas Bewes. Beyond lay the core of the Houndiscombe Estate of the Derrys.[29] Even today the shapes of these once discretely owned tracts of land are clearly visible in the layout of the modern townscape.

Map 54.4
An analysis of the medieval burgages on the south side of Fore Street, Totnes.

SOURCE: after T.R. Slater, 'The analysis of burgage patterns in medieval towns', *Area*, 13 (1981).

Mutley

Efford

Lipson

Tothill

Plymouth

Urban

Sutton
Pool

Area

The Hoe

Mount Batten

Central Park

Higher
Compton

Lower Compton

Mutley

Cemetery

Ford Park

Lipson

Reclaimed land

Central
Station

Freedom
Fields

Mount
Gould

Post-war
city centre

Beaumont
Park

St Jude's

Reclaimed

Friary
Station

land

Area built

Sutton
Pool

up in

1780s

Prince Rock

The Hoe

Cattedown

Mount Batten

——— Field boundaries

===== Roads and tracks in 1780s

0   20   40   60   80m

===== Streets in 1950s/1960s

▬▬▬ Modern roads following course of those on 1780s map

===== Modern roads exactly following alignment of 1780s field boundaries

Map 54.5
The pre-urban cadaster of Plymouth.

SOURCE: unpublished 'six-inch drawing' of Plymouth (1:10,560) made by the surveyors of the Board of Ordnance in 1784–86 (Public Record Office, MR 1385 and MR 1199).

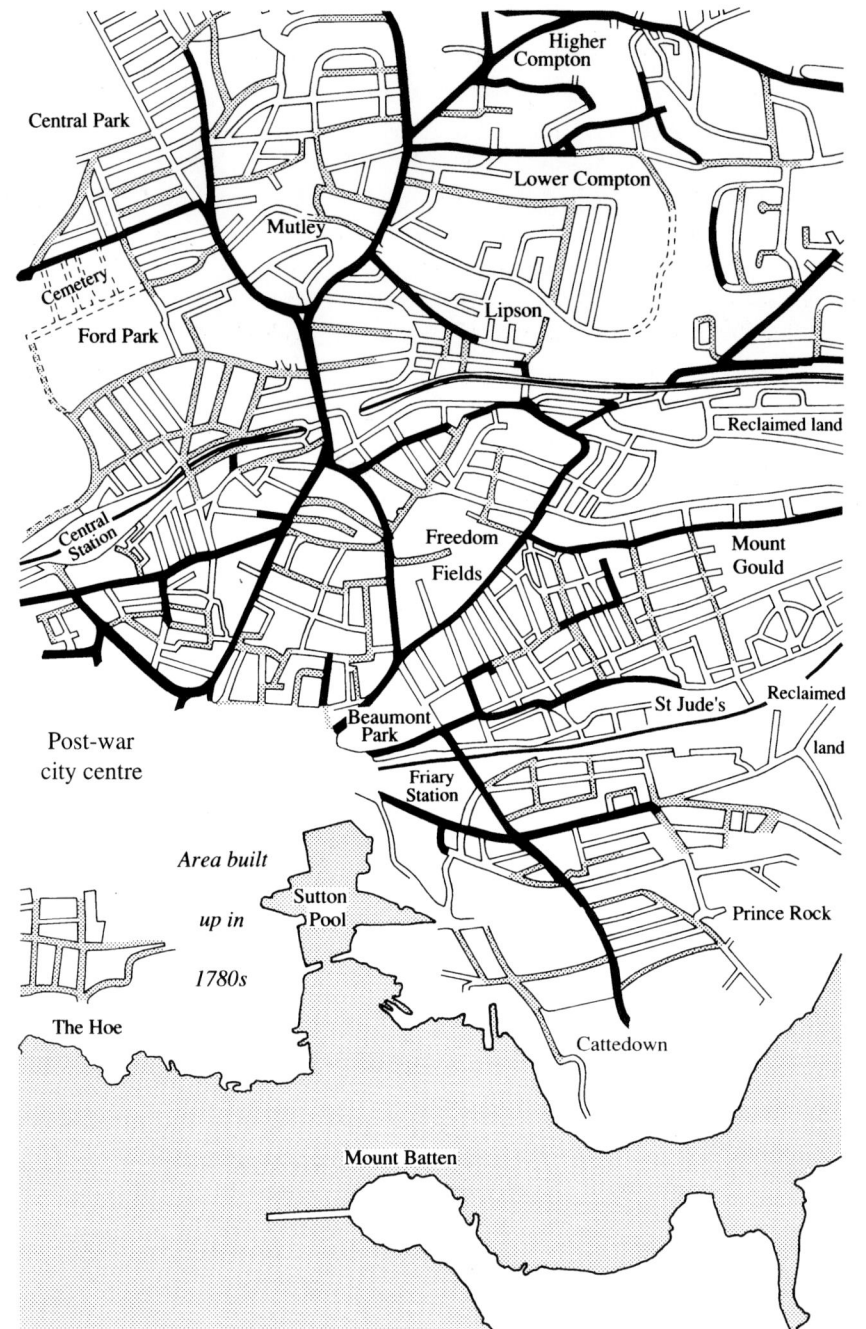

Map 54.6
The survival of the pre-urban cadastre in the modern townscape of Plymouth.

SOURCES: Ordnance Survey 1:10,560 maps, 1957 and 1966.

THE MORPHOLOGY OF COASTAL RESORTS: SIDMOUTH AND ILFRACOMBE (Maps 54.9, 54.10 and 54.11)

Having examined aspects of urban morphology from the point of view of urban growth, town-plan analysis, and the role of the pre-urban cadastre, consideration is now given to the morphology of two coastal resorts and an industrial town. To a certain degree, towns of a particular type tend to possess distinct morphological characteristics which are immediately recognisable in the town plan. The layout of a coastal resort, for example, is likely to exhibit an urban design aimed at pleasure and recreation where seaside promenades, ornamental gardens, parkland and generally low-building densities are the morphological hallmarks. By contrast, industrial towns tend to be more prosaic, utilitarian and close-built.

The development of coastal settlements as holiday resorts in south-west England is a relatively recent phenomenon in the urban history of the region. Yet their growth since the end of the eighteenth century has been prodigious. Many, like Sidmouth and Ilfracombe, originally depended on fishing for their prosperity, but were well-endowed with scenic attractions which appealed to visitors when seaside summer holidays became fashionable. Sidmouth began to expand in the 1790s partly because war with France had closed off the Continent as a destination for English tourists.[30] When a map of the manor of Sidmouth was made by William Day in 1789 (**Map 54.9**), the settlement was still a tiny village at the mouth of the River Sid with open fields behind. By the end of the Regency period a fashionable resort had been created. While lodging houses were constructed in the centre of the village, on the outskirts the wealthy built their summer 'cottages' surrounded by parkland.[31] Names like Myrtle Cottage, Violet Bank Cottage, Woodlands Cottage and Elysian Fields eloquently convey the atmosphere and social tone of early nineteenth-century Sidmouth. The majority of the 'cottages' are now hotels and, as **Map 54.10** shows, despite the enormous growth of Sidmouth's built-up area, many are still surrounded by their delightful landscaped gardens. The core of the resort was well-established when the tithe map was made in 1839. Indeed Sidmouth's growth rate peaked in the early 1840s and expansion thereafter was extremely modest. At the census of 1851 the population was 1376. By 1901 there had been a reduction to 1188. When a railway branch was finally brought from Sidmouth Junction in 1874, it terminated well short of the town centre and thus provided only a relatively weak stimulus to further growth. Rival resorts such as Torquay were more accessible and prospered at Sidmouth's expense. In morphological terms the stagnation of the town's economy in the later nineteenth and early twentieth centuries helped to preserve

Map 54.7 (above right)
Part of the Mutley area of Plymouth as shown on the Ordnance Survey 1:2,500 map published 1888 and amended by the local authority to show the alignment of planned new streets.

Map 54.8 (below right)
The same area of Mutley as in **Map 54.7** as shown on the Ordnance Survey 1:2,500 map published 1907 and bearing amendments made by the local authority.

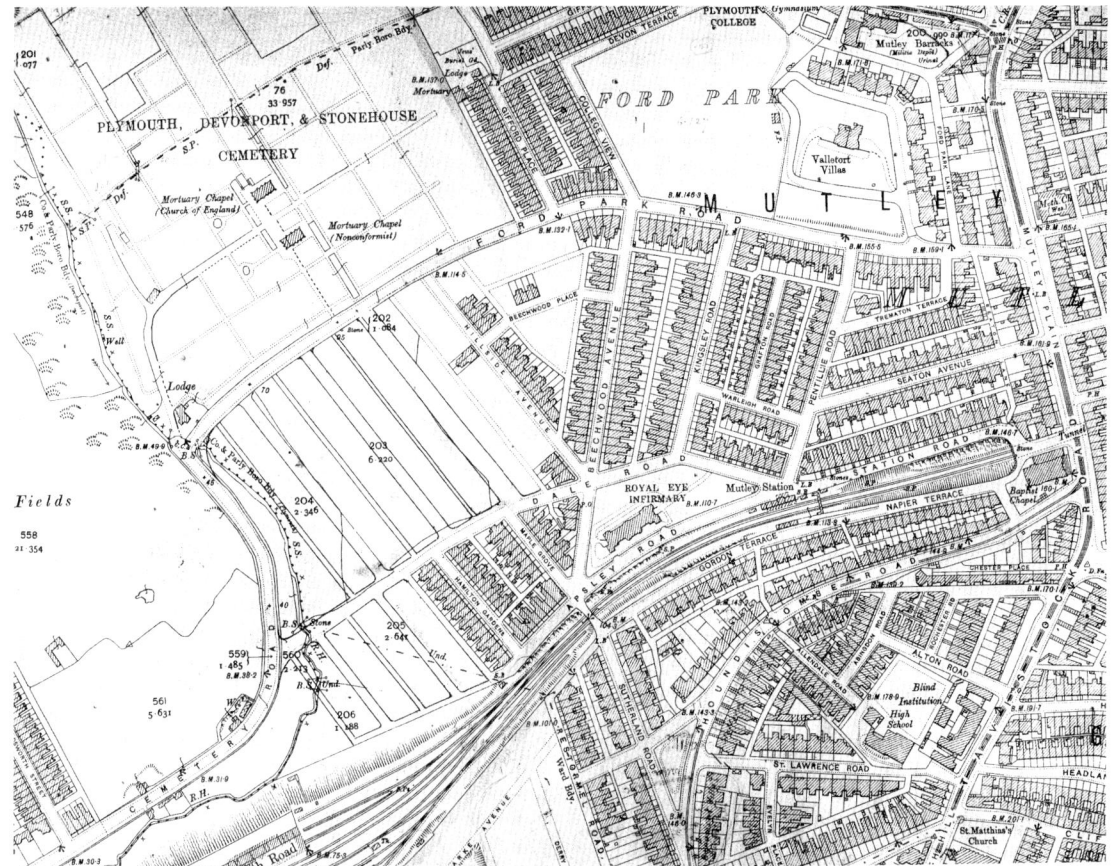

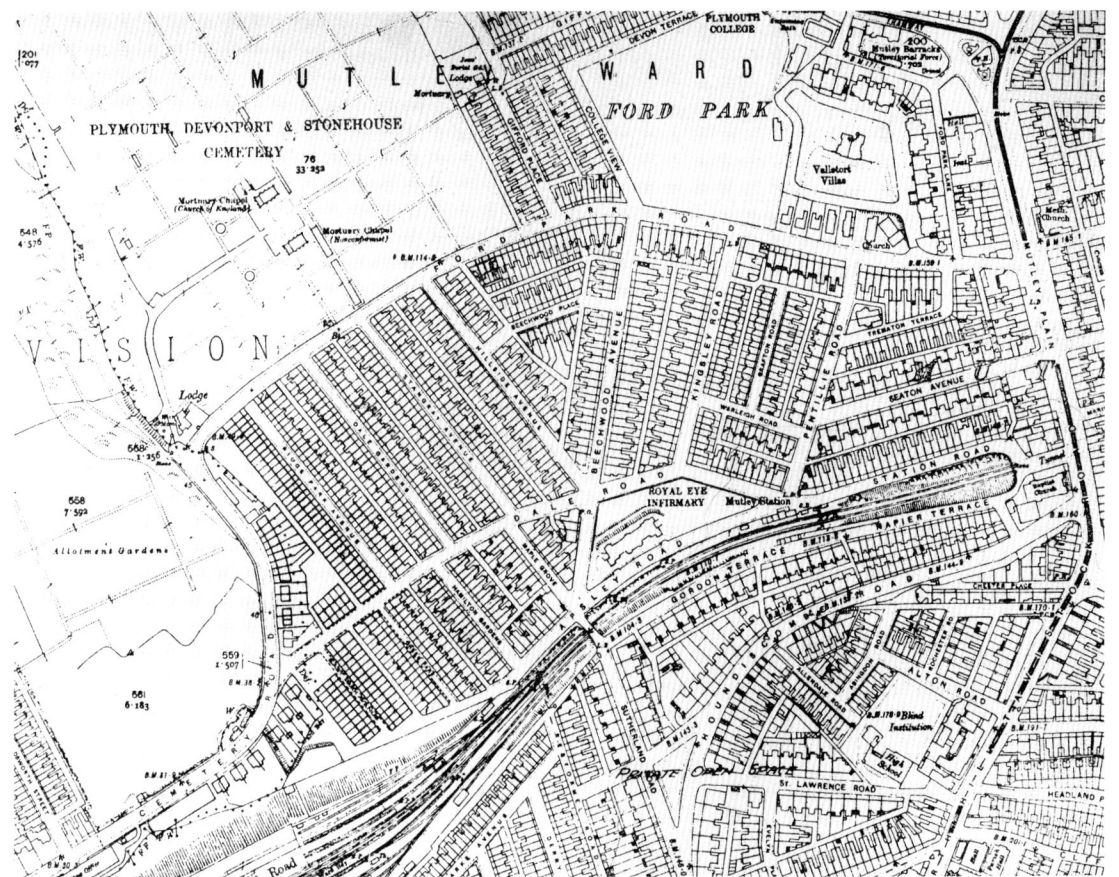

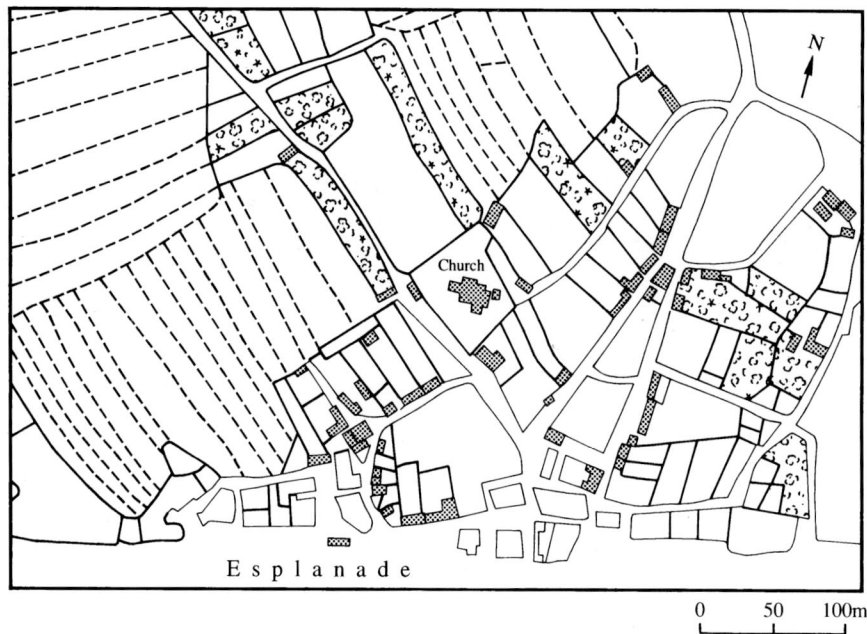

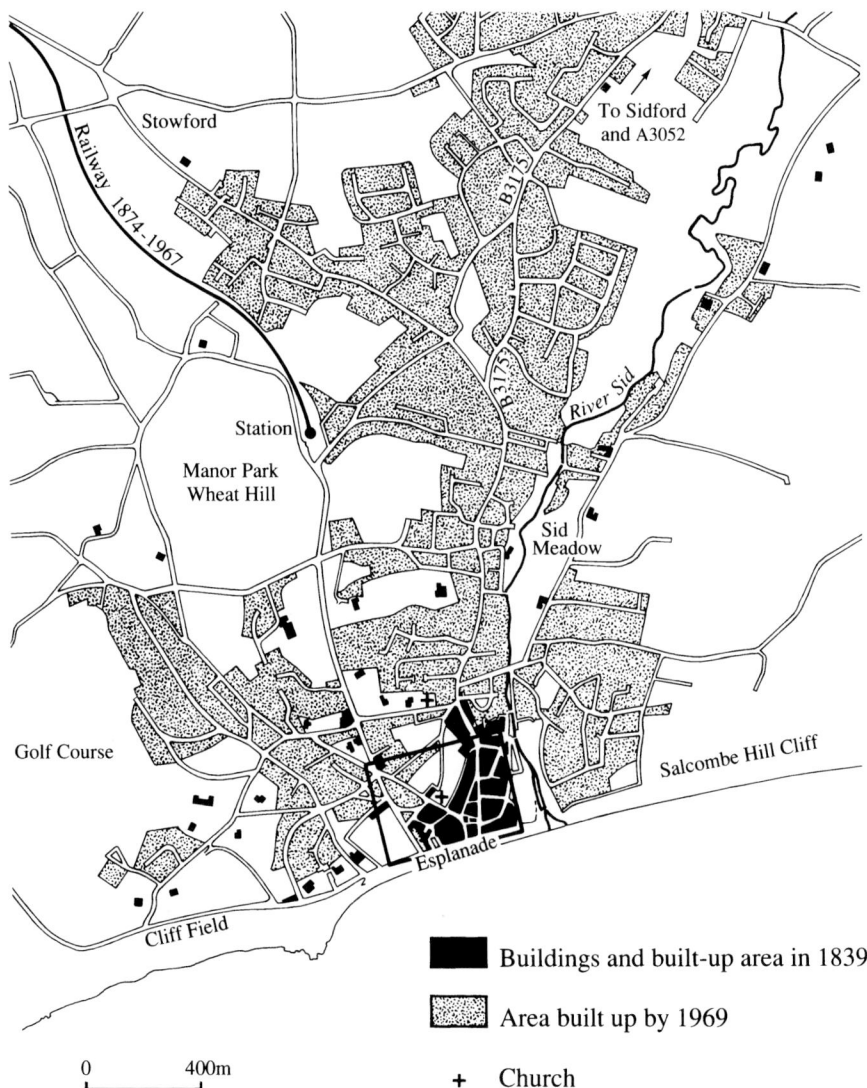

much of its late Georgian and Regency core. But on the slopes of the Sid valley beyond, large-scale expansion has occurred in the present century, especially during the post-war period. The population increased ten-fold between 1931 and 1981, and these extra residents live in the pleasant cul-de-sacs and crescents of low-density housing which now extend as an unbroken ribbon inland from the coast to the village of Sidford almost 3.5km away (**Map 54.10**). Sidmouth remains a holiday resort, but it is also now a 'retirement town' where more than 47 per cent of the permanent inhabitants are of pensionable age.[32]

The morphological character of central Ilfracombe indicates its rather different history compared with Sidmouth (**Map 54.11**). In the 1820s and 1830s the town plan still reflected Ilfracombe's role over almost seven centuries as a market town and fishing port. A weekly market and an annual fair were granted in 1278.[33] However, Ilfracombe's growth as the 'Brighton of north Devon' commenced at the end of the Napoleonic Wars. The resident population doubled between 1801 and 1851. Former medieval burgages were infilled by new terraces erected behind Fore Street and High Street. Coronation Terrace and Hillsborough Terrace were completed by 1820. Montpelier Terrace came a decade later, while Adelaide Terrace was built in the 1830s. Hotels and public houses were established in increasing numbers in the town centre. Unlike Sidmouth, expansion in Ilfracombe continued in the later nineteenth century. The railway arrived in 1870 and the population doubled again between 1851 and 1901. Narrow passages formerly leading to the rear of medieval burgage plots were widened into roads, and the lanes running out to the fields surrounding Ilfracombe, which had marked the urban 'fixation line' for centuries, were paved and flanked by new housing. While late-Victorian architecture is scarce in Sidmouth, it dominates Ilfracombe and the morphological development of the town as a whole appears much more even and balanced.

THE MORPHOLOGY OF INDUSTRIAL SETTLEMENTS: REDRUTH AND BOSCASTLE, CORNWALL (Maps 54.12 and 54.13)

Victorian population growth also occurred in towns where the local economy depended not on tourism, but on industry. However, the urban morphology of industrial towns exhibits

Map 54.9 (above left)
Sidmouth, *c.* 1789 (the area outlined on **Map 54.10**).

SOURCE: manorial survey of Sidmouth by William Day, 1789 (copy in Sidmouth Museum).

Map 54.10 (below left)
The growth and morphological character of Sidmouth, Devon.

SOURCES: tithe map of Sidmouth and Salcombe Regis, 1839; Ordnance Survey 1:2,500 map, 1889; Ordnance Survey 1:10,560 map, 1969.

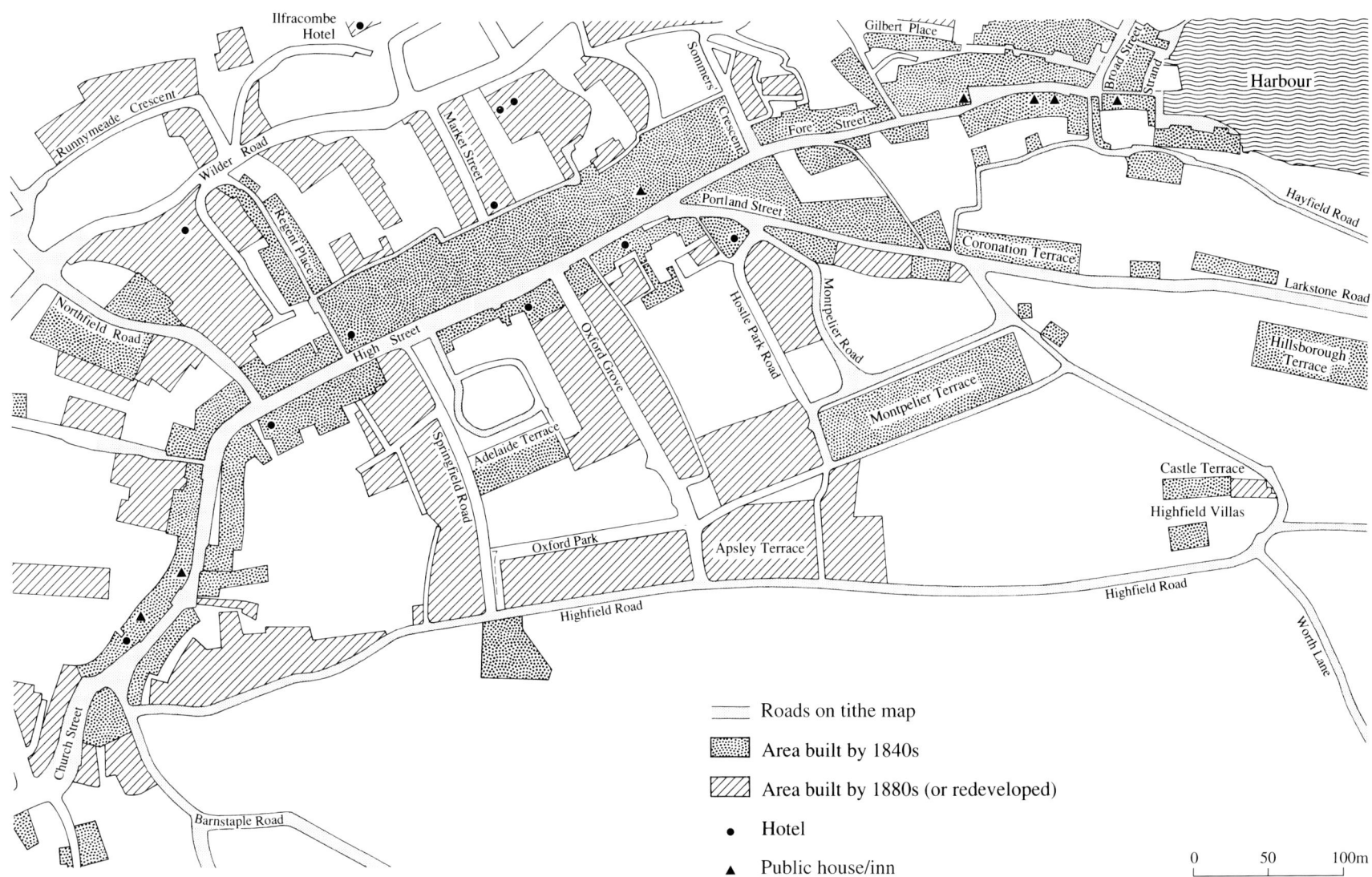

Roads on tithe map

Area built by 1840s

Area built by 1880s (or redeveloped)

● Hotel

▲ Public house/inn

0   50   100m

some rather different features. High-density terraced housing was a common feature of Britain's rapidly growing manufacturing towns.[34] It was often arranged in courts where houses might be built to share a common rear wall (back-to-back housing), thereby suffering restricted light and ventilation.[35] Such dwellings were unhealthy, often overcrowded and unsalubrious. They were the result of a deliberate attempt to pack as many houses as possible into a limited site. In south-western towns experiencing rapid growth in the early nineteenth century, court housing was built, often in considerable quantities, on the 'tails' of old burgage plots. Few dwellings were of the classic back-to-back type, but many were 'blind-backs' simply because they were erected against the perimeter wall of a neighbouring plot. Thus, while court dwellings were less numerous as a proportion of the South West's total urban housing stock than was the case in some other areas of Britain, they were no less prejudicial to health and well-being. Their presence is unmistakeable as a morphological element in the

town plans of many of the region's industrial towns. The court housing behind Fore Street in Redruth provides a good example (**Map 54.12**). The map shows the extensive early Victorian encroachment of rear court housing on the former medieval burgage plots of this Cornish town. Dwellings were often built along one side of the plot facing a high wall. Thus in Pearce's Court the yard was only 10ft wide. Occasionally, the rear space was entirely enclosed, as in Leggo's Court, which was just 15ft across at its widest point and in places less than 8ft.

As local copper- and tin-mining flourished, Redruth's population increased from 4,924 in 1801, to 11,504 in 1861.[36] Thereafter, the industry collapsed, large-scale emigration ensued and the population total fell by almost 20 per cent.[37] Thus while the morphological legacy of the years of expansion remained in the townscape, by the time the large-scale Ordnance Survey map was published in the late 1870s, there had been a four-fold increase in the number of vacant houses. But the social-status

Map 54.11
The growth and morphological character of central Ilfracombe, Devon.

SOURCES: tithe map, 1840; Ordnance Survey 1:2,500 map, 1888; and town plan in J. Banfield, *A Guide to Ilfracombe and the Neighbouring Towns* (Ilfracombe, n.d., but probably before 1834 when 2nd edn was published).

	Bank		Police station
	Hotel		Church
	Public house		Trees
	Post office		

0   50   100m

Map 54.12
Mid-nineteenth-century 'rear-court housing' in Redruth, Cornwall.

SOURCES: Ordnance Survey 1:500 map, 1870, with the names of the courts from the census of 1871 (Public Record Office, RG10/2315/4).

gradient between families occupying houses in Fore Street, and those in the tiny court dwellings crowded at the rear, in fact became more pronounced than ever. In earlier enumerations the occupants of the courts were predominantly miners—many employed in the Tolgus mines located near the outskirts of the town. By 1870 all these local mines had closed and a year later only seven of the ninety-nine residents of the courts labelled on the map were still miners. Three of the twenty-four households

were headed by paupers receiving parish relief, while fifteen of the household heads were women. Typical of many was Eliza Francis, who lived in Pearce's Court with her year-old son, Joseph. The census recorded that her husband was 'in America' and that her widowed mother supported the family by taking in laundry.[38] A clear symptom of the Cornish mining recession and the emigration which resulted was the movement of many married women with children to live with their parents when left behind

Garden/field boundaries

Trees

Post office

Engine house mine buildings

Public house

Church

Mine shaft

0   100   200m

by husbands who went abroad to seek work. And it was in the cheap, court housing of towns like Redruth that recession victims such as Eliza Francis were most frequently encountered.

Industrial communities in nineteenth-century Devon and Cornwall were not confined to the towns. Mining and quarrying villages dominated the landscape in some districts. But while the setting was rural, some of the morphological elements which comprised these settlements were essentially urban in character.

As mines were opened in remote locations in the early nineteenth century and local populations increased, high-density terraced cottages sprang up like neat rows of urban 'bedding-plants', self-conscious and out-of-place on the borders between farmland and the uncultivated moorland commons. In the midst of so much space these dwellings appear incongruous and alien. Numerous examples existed in the mining district of West Penwith. **Map 54.13** shows the nineteenth-century mining hamlet of Higher

Map 54.13
Urban 'plan units' in a rural setting: the hamlet of Higher Boscaswell, St Just in Penwith, Cornwall.

SOURCES: Ordnance Survey 1:2,500 map, 1878; Census of 1871 (Public Record Office, RG10/2343/6).

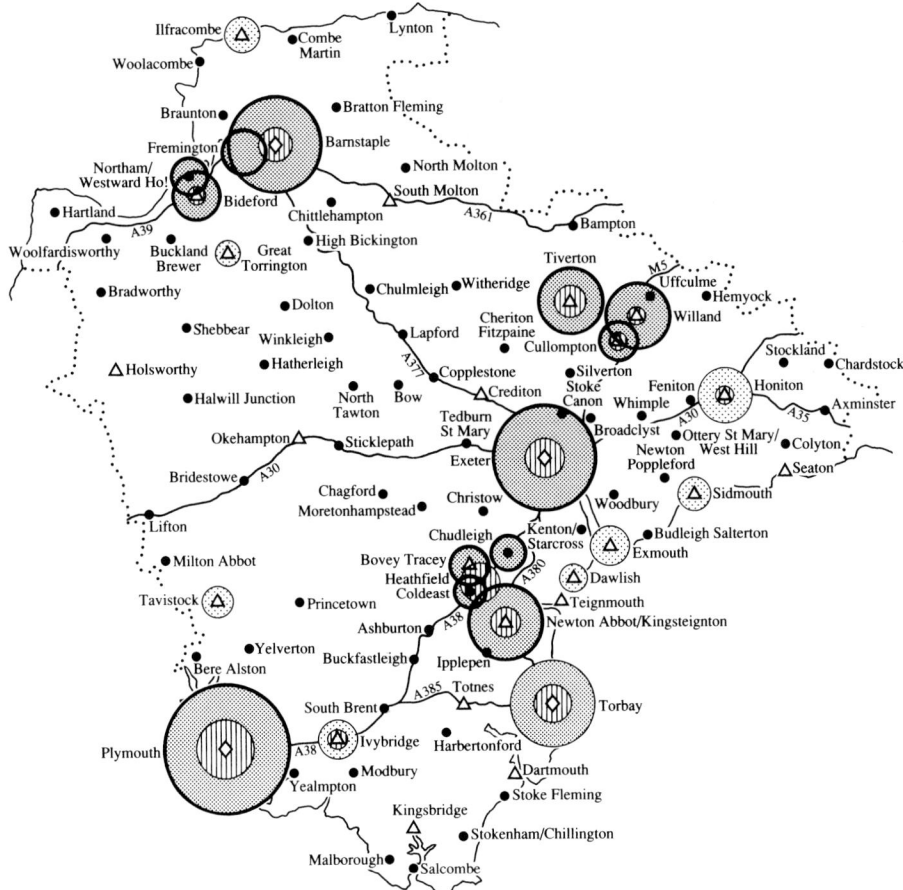

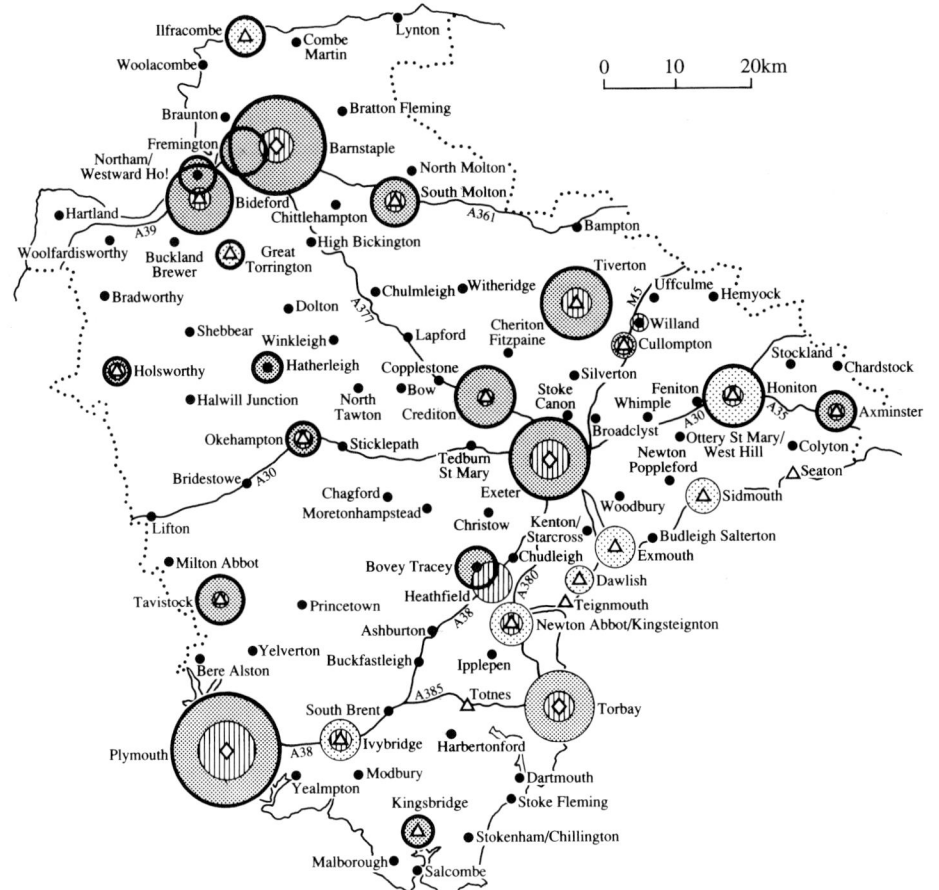

Map 55.3 (above left)
Devon Structure Plan 1977,
Option 1.

Map 55.4 (above right)
Devon Structure Plan 1977,
Option 4.

Map 55.5 (below right)
Devon Structure Plan 1977,
Option 5.

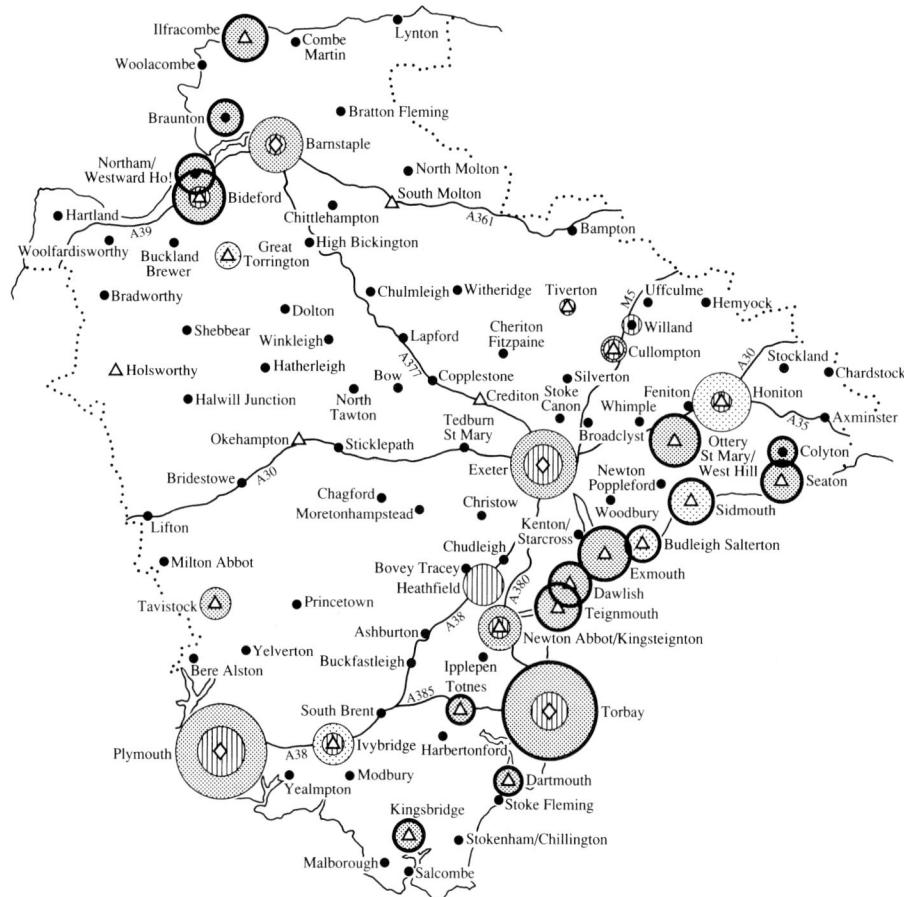

Residential land

▨ Settlements where new infrastructure would be required

▨ Settlements where no new major items of infrastructure are required

◯ Places of particular importance to the implementation of this option (outside acreage circle)

▥ Employment land

◇ Sub-regional centre

△ Area centre

● Local centre

Acres
— 1000
— 500
— 200
— 100
— 50
— 10

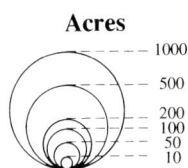

The area of each circle is proportional to the amount of land required for development

focused development in some thirty settlements, the majority of them in the south of the county. In addition the amount of land to be allocated for new housing was overwhelmingly concentrated in the south, notably between Exeter and Torbay, as shown in **Map 55.7**. Mid and west Devon in contrast were allocated virtually no housing land, in recognition of their century-long history of population stagnation (see chapter 18).

In 1985, the county council published the *First Alteration* to the structure plan, and in 1987 this was approved with amendments by the Secretary of State for the Environment.[5] While the strategy remained essentially unaltered, the policy for the provision of growth in villages was relaxed, to include not only more selected local centres, for example Woolfardisworthy and Halwill Junction, but also a less strict policy with regard to buildings in the open countryside. Nonetheless, the basic policy remained one of concentrating growth in a few centres, with only small-scale growth elsewhere on the grounds of economies of scale, to preserve the character of villages, and for the efficient provision of services.

PLANNING IN CORNWALL (Maps 55.8, 55.9 and 55.10)

In Cornwall, the 1930s also witnessed a survey of the county instigated by the Council for the Preservation of Rural England and carried out by W.H. Thompson.[6] He found that very few local authorities had passed a resolution to prepare plans under the 1925 Act, possibly because no town in the county then exceeded the 20,000 population threshold to make the production of a plan

mandatory. However, St Austell, Newquay, Truro and St Ives councils had passed resolutions to prepare plans. Thompson also discussed plans to create national parks in Cornwall in three areas: near Padstow, St Ives and Lands End. The need for planning must have been apparent for all to see, for as **Maps 55.8, 55.9 and 55.10** show, many towns in Cornwall were experiencing a rapid expansion of their area. Just as importantly, this expansion was often unrelated to the existing area of the town, as at Bude (**Map 55.8**) or Fowey (**Map 55.10**), or was sterilizing large areas of farmland, as at Newquay (**Map 55.9**) and also at Bude.

CORNWALL: DEVELOPMENT, STRUCTURE AND LOCAL PLANS (Map 55.11 and Fig. 55.2)

In common with Devon, the first Cornish development plan in 1952 was little more than a statement of existing patterns and trends, though the two large cloth-bound volumes of the 1952 *Report of Survey* are well worth reading for the detail they contain about the Cornwall of the 1930s and 1940s.[7] The plan itself assumed a population growth of 7,000 or 2 per cent by the year 1971, with increases of 5 per cent in the towns and decreases of

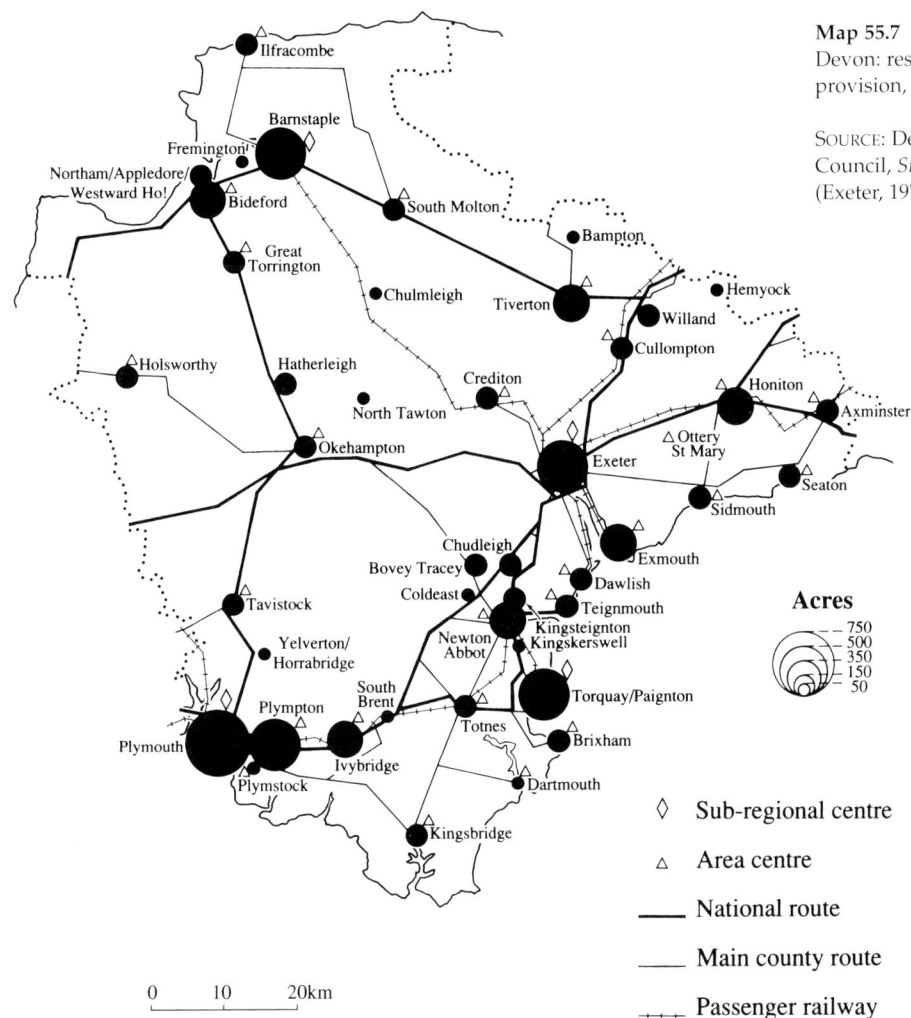

For **Maps 55.6 and 55.11** see pages 13–14 in the colour section.

Map 55.7
Devon: residential land provision, 1976–1991.

SOURCE: Devon County Council, *Structure Plan* (Exeter, 1979).

Map 55.8
Growth of Bude-Stratton
in the nineteenth and
twentieth centuries.

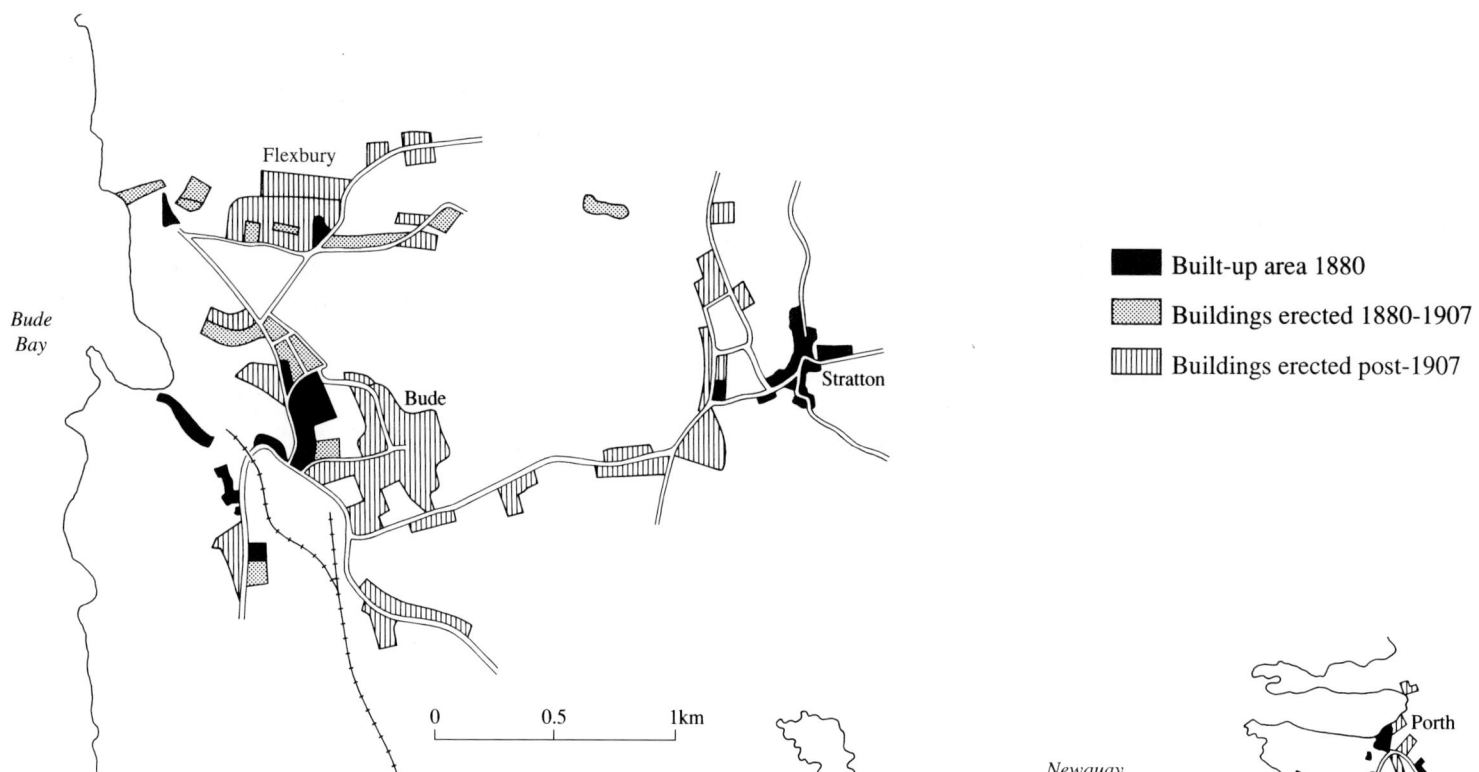

Flexbury

*Bude
Bay*

Bude

Stratton

Built-up area 1880

Buildings erected 1880-1907

Buildings erected post-1907

0   0.5   1km

Map 55.9
Growth of Newquay
in the nineteenth and
twentieth centuries.

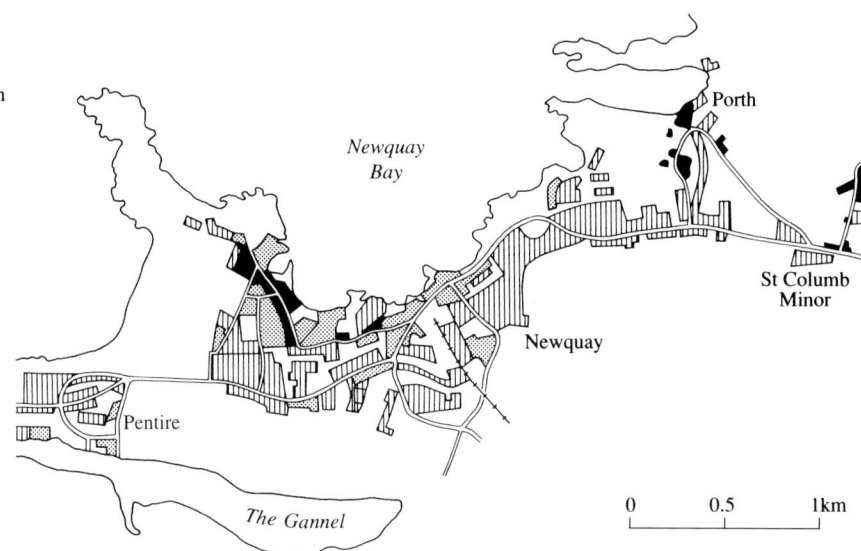

*Newquay
Bay*

Porth

St Columb
Minor

Newquay

Pentire

The Gannel

0   0.5   1km

Map 55.10
Growth of Fowey in
the nineteenth and
twentieth centuries.

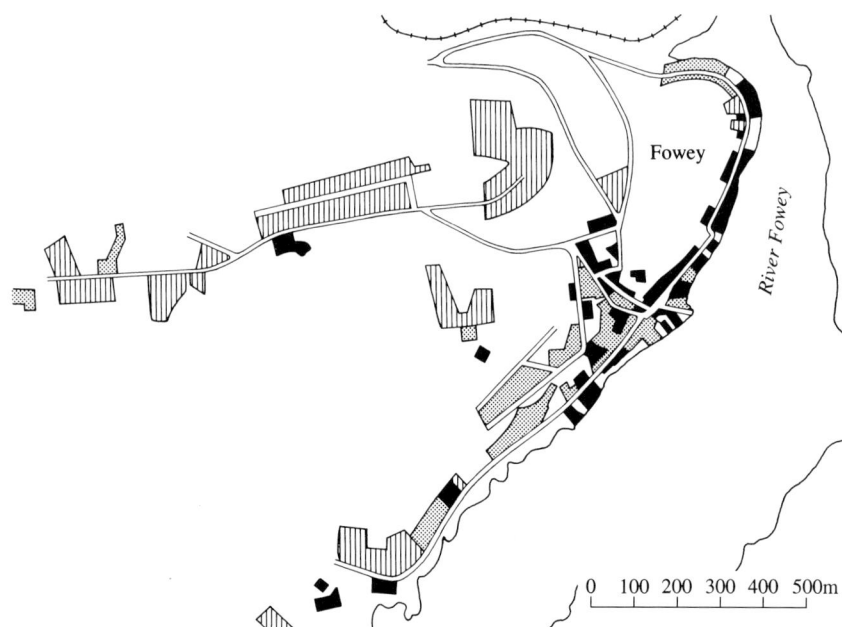

Fowey

River Fowey

0   100 200 300 400 500m

1 per cent in the rural areas, to give a population of 337,000 by 1971. In 1969 the *First Review* of the development plan did not seek to amend the plan significantly but simply to take into account recent trends, notably proposals to reserve land for certain purposes and the effect of planning permissions given contrary to the plan. In terms of population forecasts it increased the rate of growth to 4 per cent, divided into a 7 per cent growth rate in urban areas but a 0.5 per cent decrease in rural areas to give a population of 343,000 by 1971 compared to the 1952 forecast of 337,000.

As a result of the public participation exercise on the structure plan which succeeded the development plan, the county council decided to follow a compromise policy which placed some emphasis on economic growth combined with efforts to maintain the physical character of the county. The draft plan was published in 1979 and after public examination and amendment by the Secretary of State for the Environment it became operative on 15 August 1981.[8] The 'key diagram' is reproduced as Map 55.11. Unlike Devon there was no specific settlement plan, but there were housing allocations for each of the six districts and for the Falmouth and Penryn areas. Emphasis was placed on the provision of housing in urban areas, and outside these, development would normally be permitted only in villages with high levels of service provision. The rate of house-building forecast for 1976–91 in the 1981 plan was similar to the rate achieved in the early 1980s, but markedly lower than the rate of over 3,000 houses a year achieved in the mid-1970s (**Fig. 55.2**).

In common with Devon, planners in Cornwall have had to balance the demands for growth mainly from an incoming population (see chapter 18) with the demands for Cornwall to be kept attractive and to turn away from the grim industrialization of the past towards a more arcadian landscape. In both counties there is a fundamental dilemma between the need to provide factories and jobs for local residents and houses and scenery for migrants, frequently cash-rich and retired. Planning policies, if too restrictive, force up property prices and turn the area into a

For **Maps 55.6 and 55.11** see pages 13–14 in the colour section.

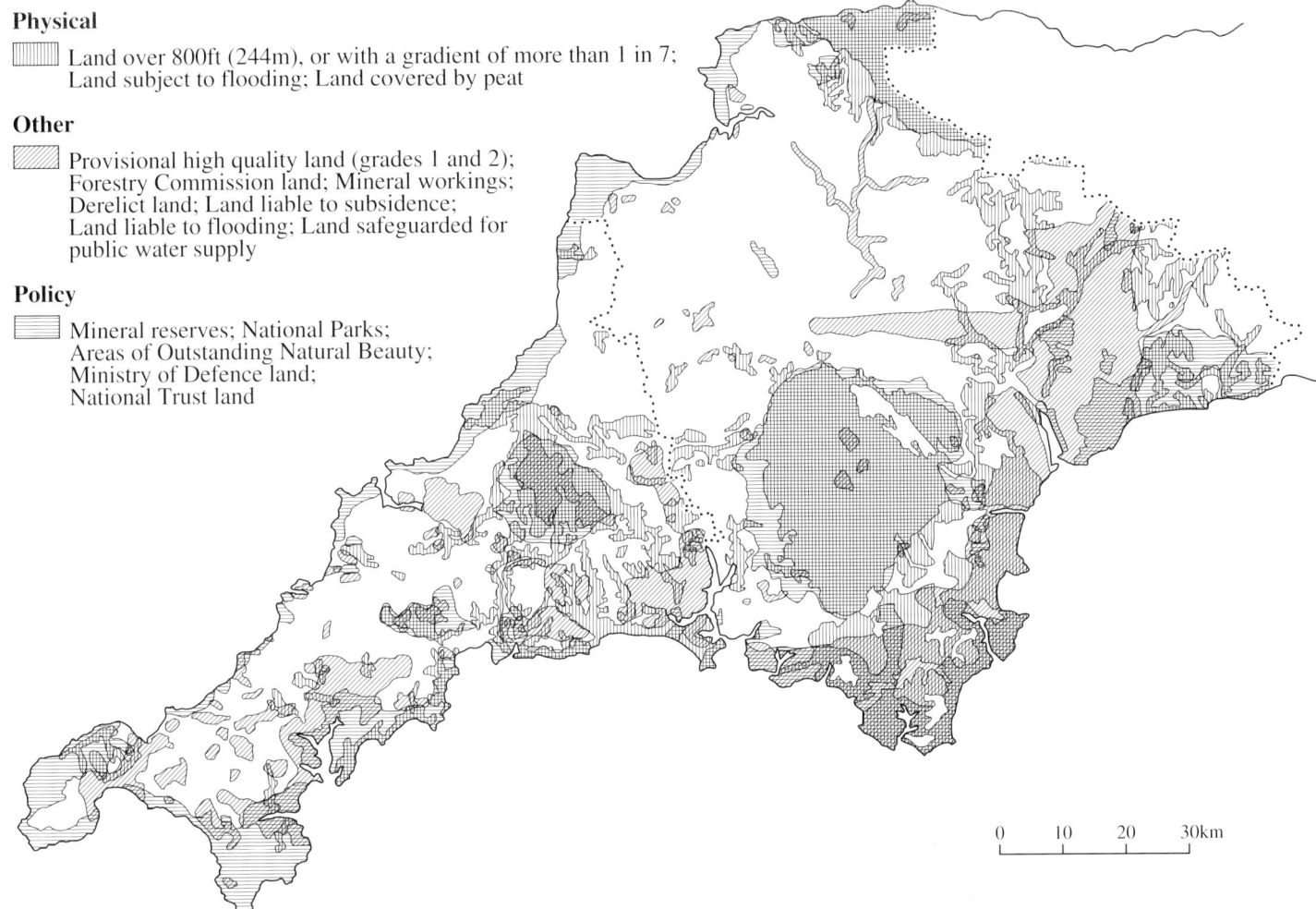

Physical

Land over 800ft (244m), or with a gradient of more than 1 in 7; Land subject to flooding; Land covered by peat

Other

Provisional high quality land (grades 1 and 2); Forestry Commission land; Mineral workings; Derelict land; Land liable to subsidence; Land liable to flooding; Land safeguarded for public water supply

Policy

Mineral reserves; National Parks; Areas of Outstanding Natural Beauty; Ministry of Defence land; National Trust land

0   10   20   30km

Map 55.12
Restraints to development.

Fig. 55.2
House-building in
Cornwall, 1961–1985.

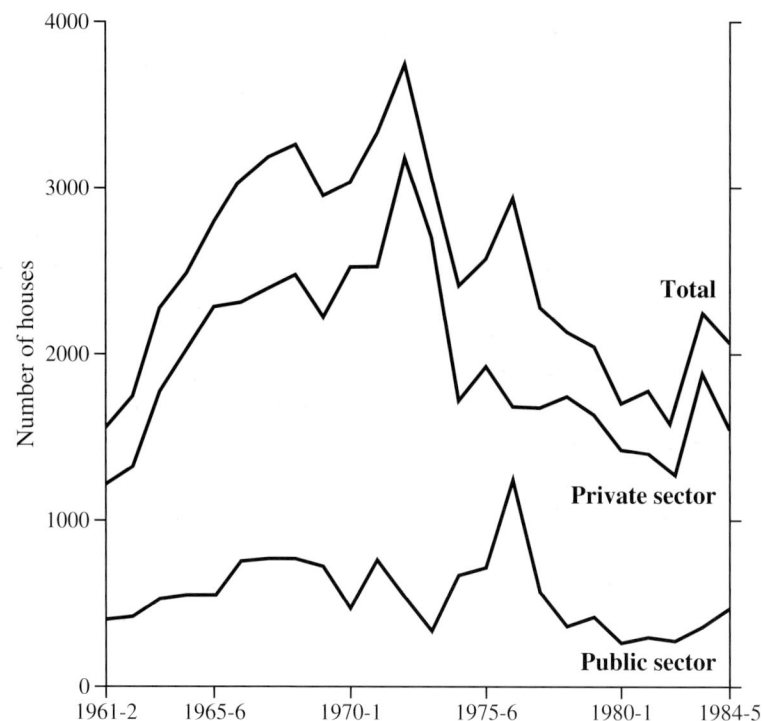

middle-class enclave. If, however, policies are too flexible, then the scenic beauty of the area may be spoilt by too much development. In order to balance the two demands and to produce policies that are neither too flexible nor too restrictive, planners in both Devon and Cornwall would ideally like to have turned to national and regional plans which would have set out the environmental role of the South West, as either a holiday playground, the scenario of the next industrial/service revolution, or a 'costa geriatrica'. Unfortunately, as the next section shows, such regional planning policies have either been weak or short-lived.

REGIONAL PLANNING IN THE SOUTH WEST (**Maps 55.12 and 55.13**)

In the 1960s the reform of plan-making that replaced development plans with structure plans was also accompanied

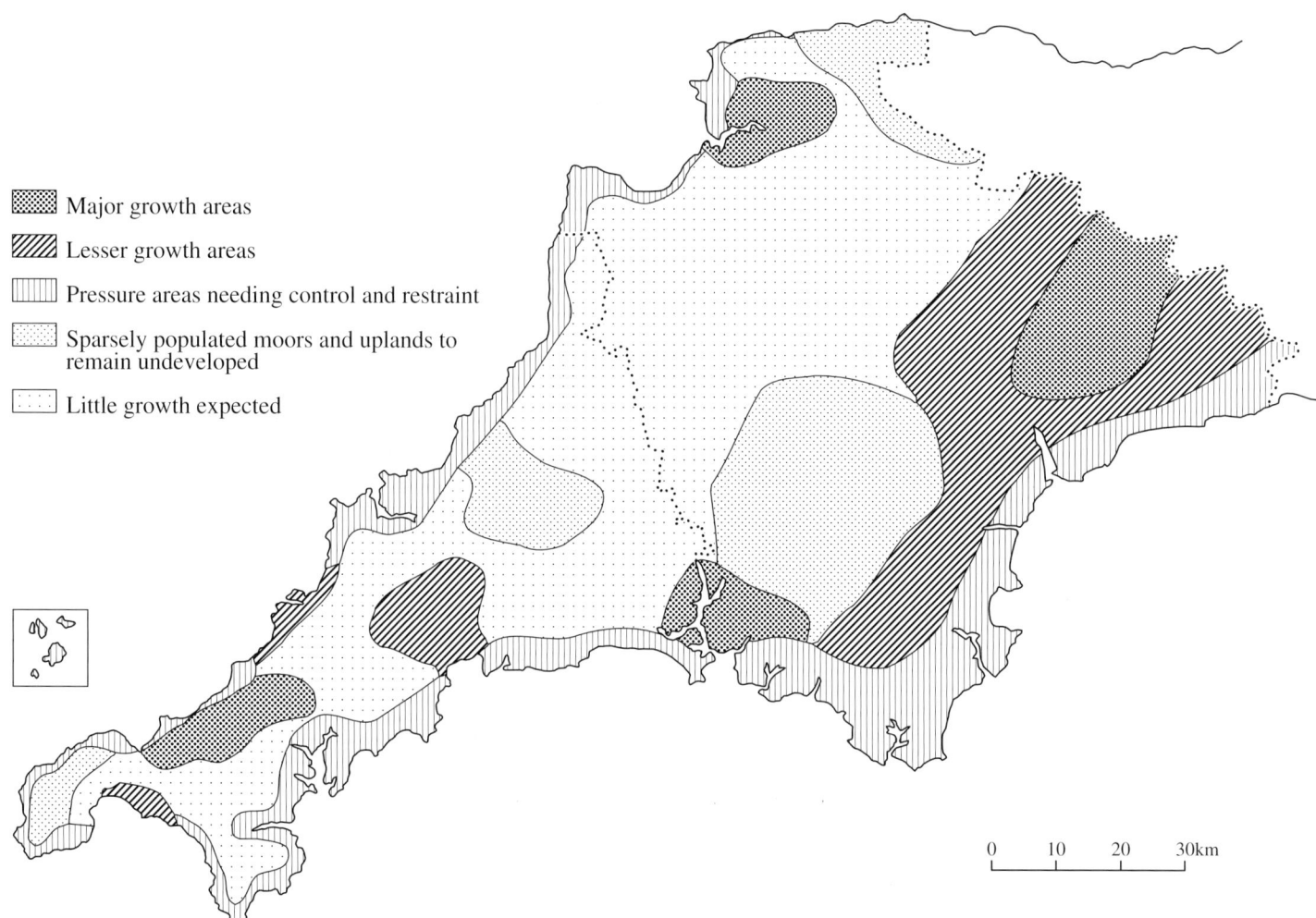

Major growth areas

Lesser growth areas

Pressure areas needing control and restraint

Sparsely populated moors and uplands to remain undeveloped

Little growth expected

Map 55.13
Strategic settlement pattern:
expected pattern of change.

SOURCE: South West Economic
Planning Council, *A Strategic
Settlements Pattern for the
South West* (London, 1974).

by the creation of a new type of planning, at the regional level. The Labour government of 1964–70 was committed to a national plan (short-lived) which would then filter down to regional plans and thus provide a regional framework for structure plans.

In the South West, two regional planning documents were produced.[9] The first of these, *A Region with a Future*, was merely an inventory of existing resources and problems and in particular of the need for a new 'spine road'. Its main impact was to persuade Whitehall to fund the M5 from its then end-point at Bridgwater to its current end-point Exeter, and also to agree to dual carriageway the A38 and A30 links to Cornwall as a matter of urgency.

As part of the inventory process, the regional planning office in Bristol which embraced the six pre-1974 counties of Gloucestershire, Wiltshire, Somerset, Dorset, Devon and Cornwall produced a series of regional planning maps in an *Atlas for the South West*. One of the most enduring of these is the map of restraints to development reproduced as **Map 55.12**.

Unfortunately, most of the land free from restraint is shown to be in the areas of least population, namely mid and west Devon. A second regional planning document largely ignored these restraints and predicted major population growth in the restraint areas. Published in 1974 under the title of *A Strategic Settlement Pattern for the South West*, it used projections of migration to the South West to predict very rapid rates of population growth between 1971 and 2001.[10] Even in the Exmoor area a growth of over 11 per cent was predicted, and in the Exeter region, population growth equivalent to the size of Exeter was forecast, namely 90,000 more people or a growth rate of 32 per cent.

Not surprisingly, the document had a very critical response in the two counties of Devon and Cornwall, most often from recent migrants. Even the formal responses were critical.[11] Although the overall forecasts in the plan were open to severe criticism, the plan translated these forecasts into more realistic sub-regional expected patterns of change as shown in **Map 55.13**. This accepted the fact that most growth had occurred

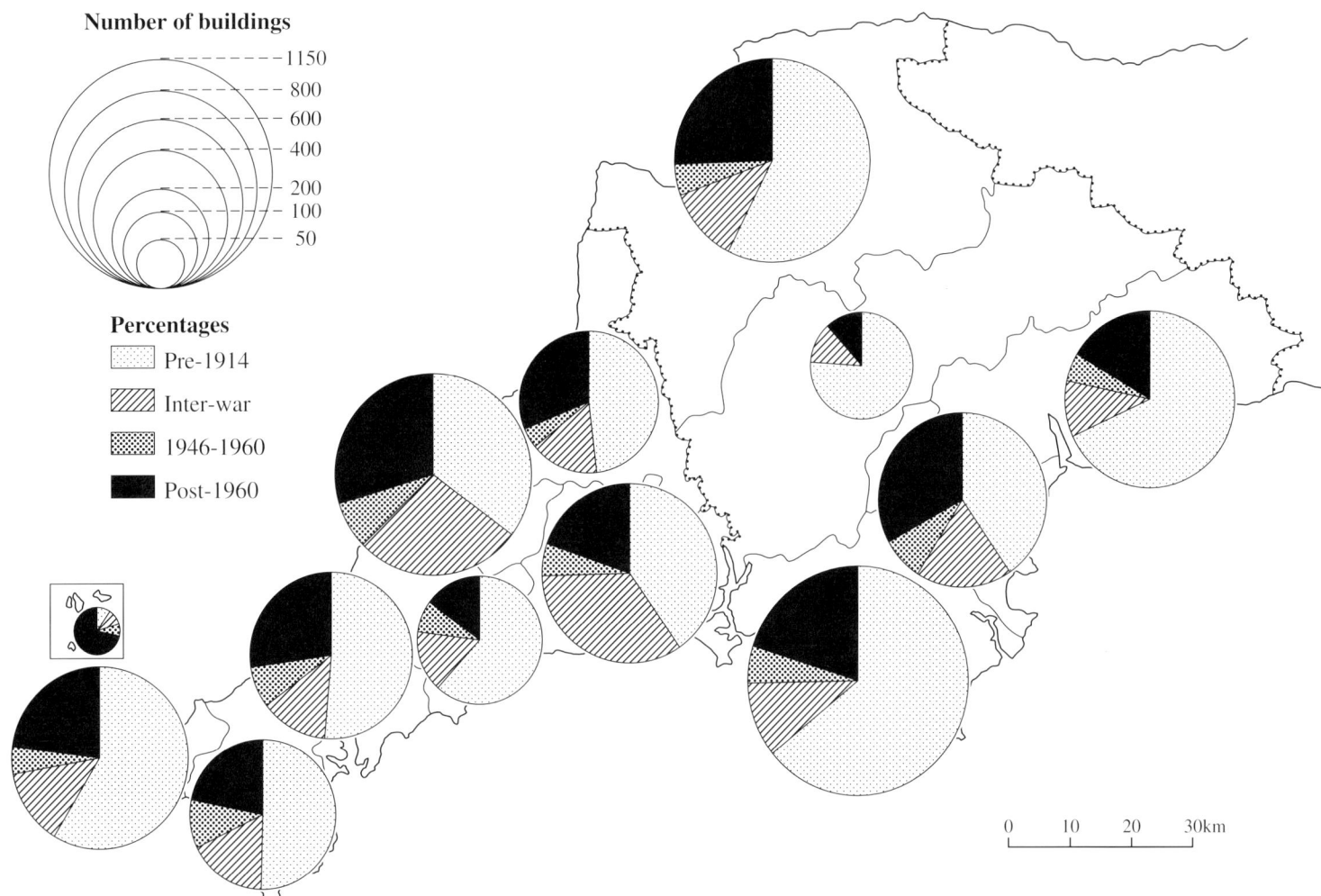

Number of buildings

1150
800
600
400
200
100
50

Percentages
Pre-1914
Inter-war
1946-1960
Post-1960

0 10 20 30km

Map 55.14
Dates of buildings used as second homes.

Map 55.15
Growth of non-permanently
occupied dwellings,
1931–1971.

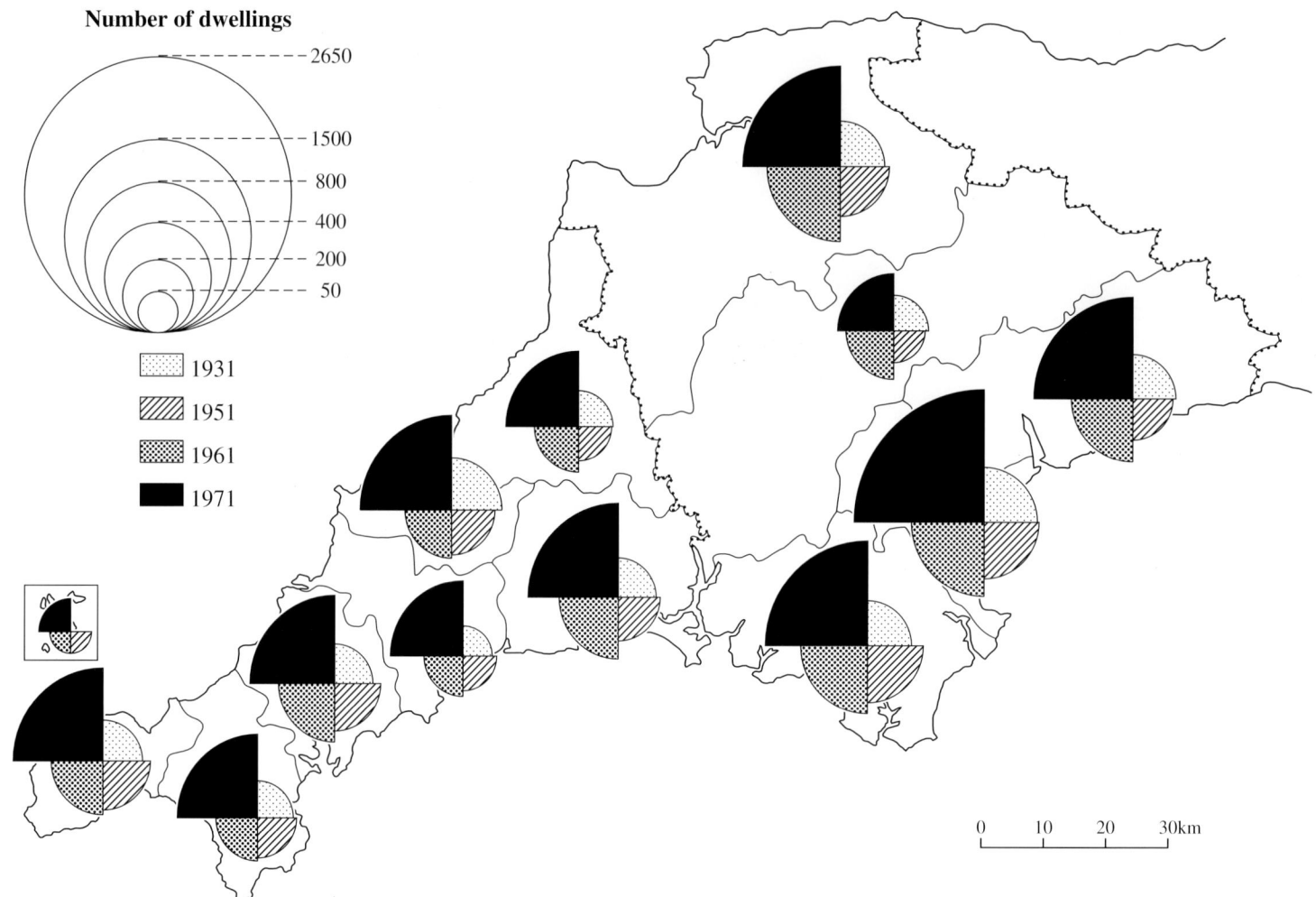

Map 55.15
Growth of non-permanently
occupied dwellings,
1931–1971.

Number of dwellings

along the Taunton–Exeter–Plymouth–Truro spine in the past, and that the planned development of the M5 and A38 in the 1970s would not only continue but also accelerate this trend. The plan, therefore, predicted in effect four main regions. First, the spinal corridor of growth along the M5 and A38 roads; second, a parallel corridor to the north where there would be growth but to a lesser degree; third, a ribbon of restraint around the entire coast; and fourth, low growth in the upland areas, not because people did not want to live there, but because of planning restraints.

In spite of severe criticisms, the government accepted the report in 1975 and the Minister for Planning and Local Government, John Silkin, recommended that: 'the region should in general plan to accommodate growth at the trend rate' and that 'the report offers useful regional guidelines to be tested in the structure planning process'. To the extent that the forecasts in both the Devon and Cornwall structure plans have proved to be underestimates, the *Strategic Settlement Pattern* document was

with hindsight more accurate than it was thought to be at the time. Nonetheless, it is still hard to see the population of Devon and Cornwall reaching the predicted 1.63 million by 2001.

SECOND HOMES (**Maps 55.14, 55.15 and 55.16**)

One final theme of planning interest in the South West is the growth of second homes. As second homes are mainly buildings constructed before the advent of planning (**Map 55.14**), the change from primary to secondary residence occurs outside the planning system, since a change of use from a main to a second home does not need planning permission. Most second homes date from before 1914, indeed that is part of their attraction. The growth in their use as second homes dates from the 1960s as shown in **Map 55.15**. However, in spite of this growth, only a few areas of Devon and Cornwall have percentages of second homes above 7.5 per cent of the housing stock (**Map 55.16**). It seems, therefore, that

Map 55.16
Second homes as a percentage
of dwelling stock.

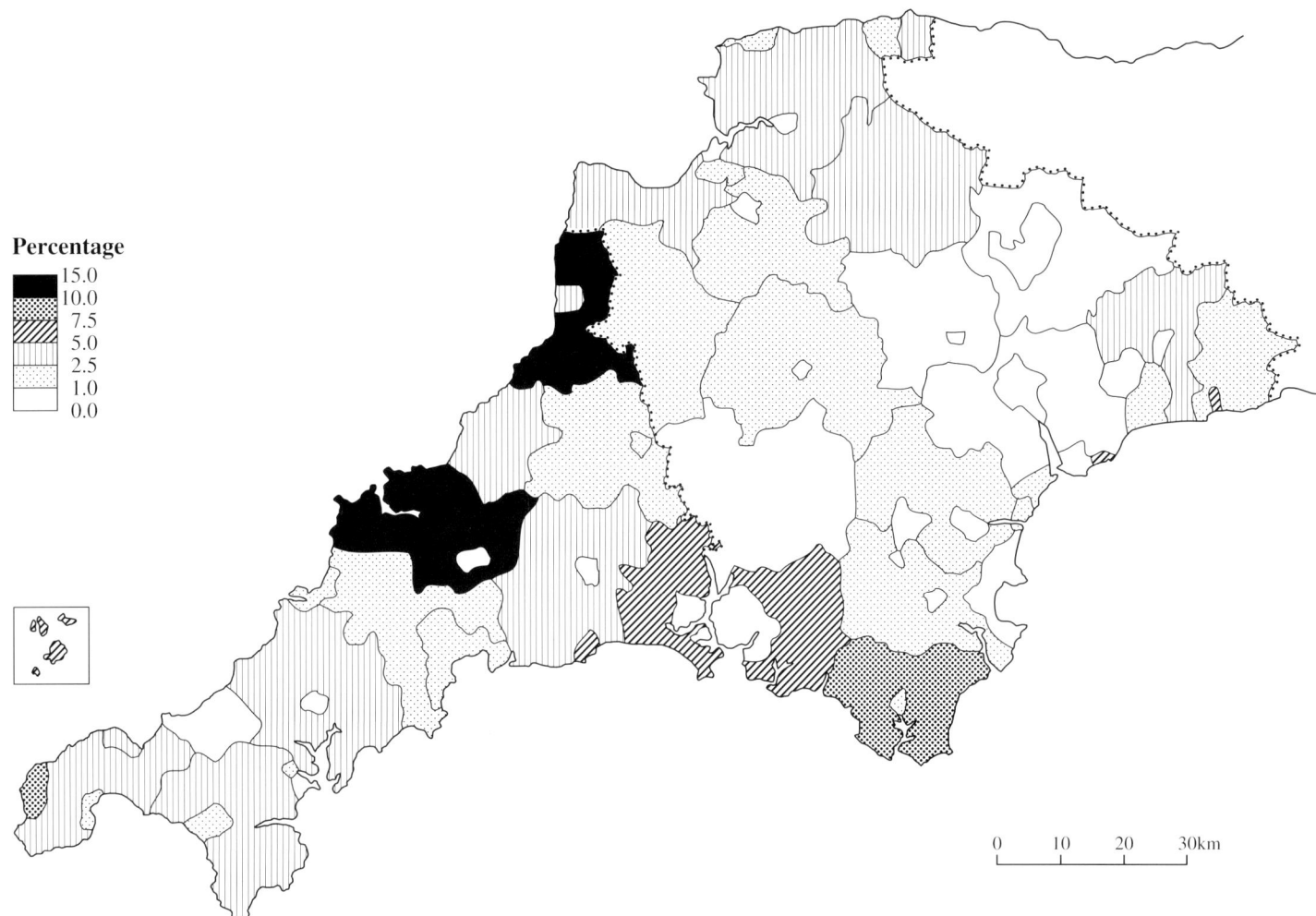

Percentage

15.0
10.0
7.5
5.0
2.5
1.0
0.0

0   10   20   30km

the second home issue is more apparent than real, except for a few coastal villages and other attractive locations near the main lines of communication.

CONCLUSION

In conclusion, no plan for Devon and Cornwall has yet had time to make a real impact on the region. The plans of the 1950s were ineffectual, the development plans of the 1960s were hardly approved before their days were numbered by the advent of structure planning, and the structure plans of the 1980s have been constantly modified. Plans, if they are to work and to have the confidence of the public and developers, must be long-term and

be seen to be long-term. All the plans for the South West have done little more than recognize the existing distribution of population and have merely sought to allow for its expansion. The main issue has been one of where precisely this expansion should be accommodated.

FURTHER READING

A.W. Gilg, 'Regional planning in the South West', *Social and Economic Administration*, 9 (1975), 220–5; G. Shaw and A. Williams, 'The regional structure of structure plans', *Planning Outlook*, 23 (1981), 2–7.

SERVICE INDUSTRIES: TOURISM AND RETAILING

EARLY TOURIST DESTINATIONS
THE INFLUENCE OF ARTISTS' CHANGING LANDSCAPE PREFERENCES

PETER HOWARD

Pictures are produced in such numbers that they can be used to measure quite accurately the changes and fashions in landscape preferences, and artistic fashions influence others significantly. **Maps 56.1 and 56.2** show the locations of a sample of landscape pictures exhibited at the Royal Academy Summer Exhibition before and after 1870. That date does seem to mark a significant change nationally in the kind of landscapes preferred by artists, and this change is clearly reflected in the very different distributions shown on the maps. **Fig. 56.1** shows the changing popularity of Devon and Cornwall to artists. Devon's early popularity was much concerned with its country houses, but the great fashion for Devon in the 1840s, when it was the most depicted county in England, largely reflects the popularity of its river scenes. By the time Dartmoor itself was in fashion, after 1870, Devon had been overtaken by Cornwall, as coastal views displaced rivers. The decline in Cornish popularity since 1940 has been dramatic, though there are some hints of an upturn in both counties in the last few years.

Map 56.1 includes many pictures of major buildings and country houses—Exeter Cathedral, St Michael's Mount, Eddystone Lighthouse, Okehampton Castle are named, and Mount Edgcumbe, Kitley, Powderham and Berry Pomeroy may be identified. Such places seem to attract for several reasons: some are of obvious antiquarian interest, others attract because they fit into aesthetic concerns of the time, for example Berry Pomeroy castle is a picturesque study, while the Eddystone, usually seen in storm, illustrates the taste of the sublime. However, many of the country houses were only recently constructed, and although many such pictures were commissioned, there is little sign of a distaste for the modern before 1850.

The most obvious feature, however, is the importance of river valleys—very popular in mid-century and especially in Devon, with the Teign, Dart, Lyn, Avon, Tavy and Plym all much visited, usually in those sections with steep, but densely wooded sides, and rocky streams. A bridge or a water mill was commonly included. Fingle Bridge and Holy Street Mill, both on the Teign, and Watersmeet on the Lyn, are reminders of this fashion. Chudleigh Glen in the grounds of Ugbrooke House is a special case of the romantic little glen, immensely popular as a day trip from the growing resort of Torquay.

The coastal depictions are largely of major resorts. Sidmouth is typical of the earlier preference for long sweeping views over the town, and this was followed later by more rocky and intimate views of Torquay and Ilfracombe. There are signs too of the growing interest in the fishing harbour, which was to become the greatest feature after 1870. This is seen as Clovelly began its rapid rise to fame in the 1850s and at several places in Cornwall, though most Cornish pictures are of uninhabited and rugged parts of the coast—notably Kynance and the Lizard.

Map 56.2 shows the dominance of the fishing village as Cornwall rose to popularity. While the opening of the railway clearly helped this move, which relates to a similar move to Brittany on the other side of the Channel, artists had long struggled to reach equally inaccessible places such as North Wales. In an urban England, rural labour became heroic, and the fisherman was a notable example. Painters flocked to such villages, with St Ives heading a long list including Brixham, Newlyn, Tintagel, Polperro, Falmouth and Appledore. However, rugged coastal scenes also increased as at the Lizard, Land's End and Morte Point, for example.

The lack of interest in the rivers or in major buildings was very marked in this period, in comparison with the years before 1870. The pictures of Dartmoor were by now usually of the high moor, perhaps enlivened by a tor, pool or pony, rather than the rivers. It reflects a fascination for bleak landscapes which is seen elsewhere, in the northern fells of England, and in the East Anglian fens. Bodmin Moor was smaller and lower than Dartmoor and further from London so was largely ignored, but Exmoor was fashionable after the publication of *Lorna Doone*.

Twentieth-century discoveries, such as the thatched village, the farmstead and the orchard, add to the pattern of views pictured in south-west England, though these tend to be very widely distributed, especially after the coming of the car.

Maps 56.1, 56.2 and Fig. 56.1 demonstrate the speed with which landscape fashions change. The changes shown here often presage movements in tourist destinations, and act as a reminder that our own preferences may be only temporary.

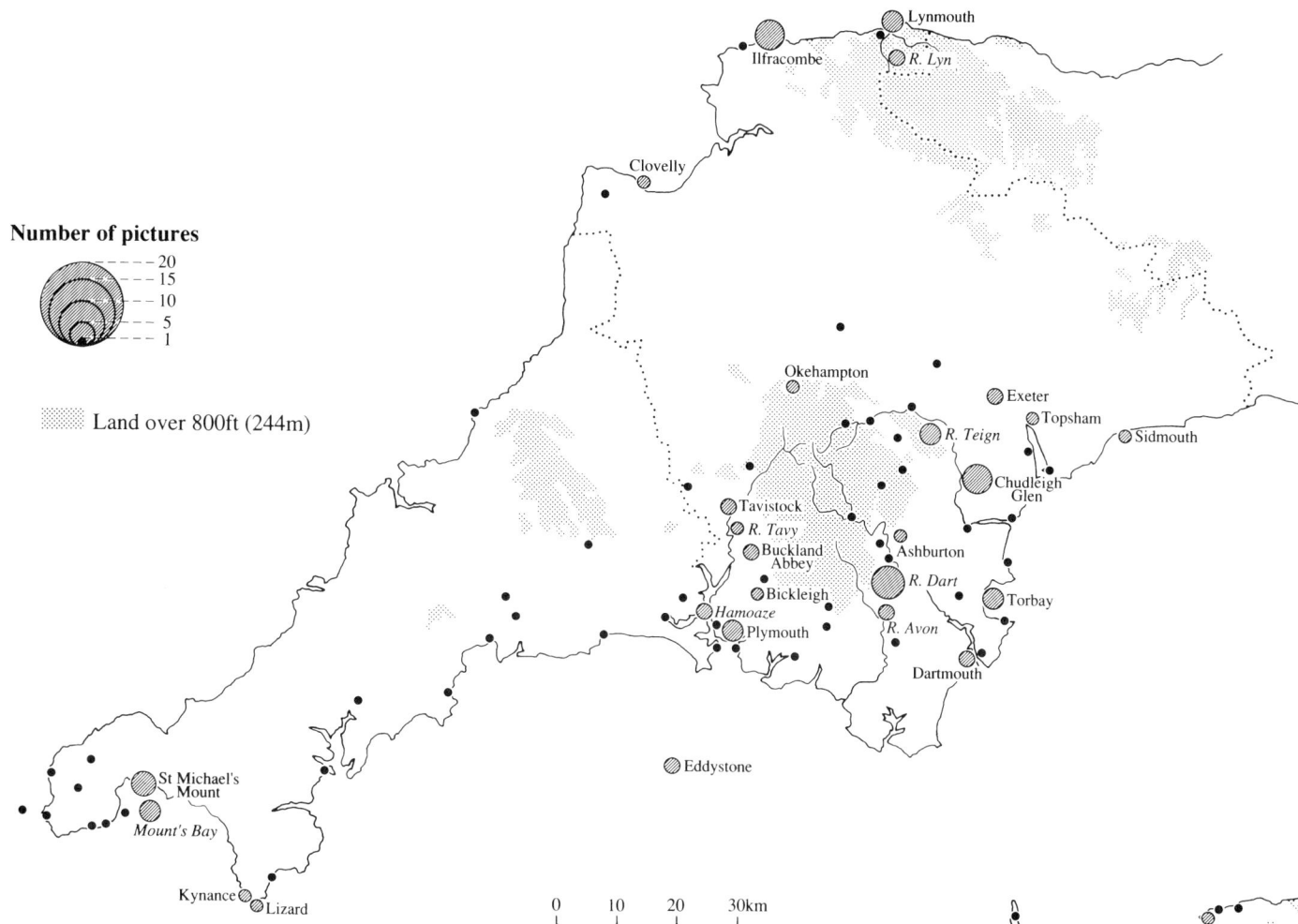

Map 56.1
Locations of south-western
landscape paintings,
1800–1865.

SOURCE: a random sample of
pictures exhibited at the
Summer Exhibition of the
Royal Academy of Arts, where
the title gives a south-west
location (places with two or
more pictures in the sample
are named).

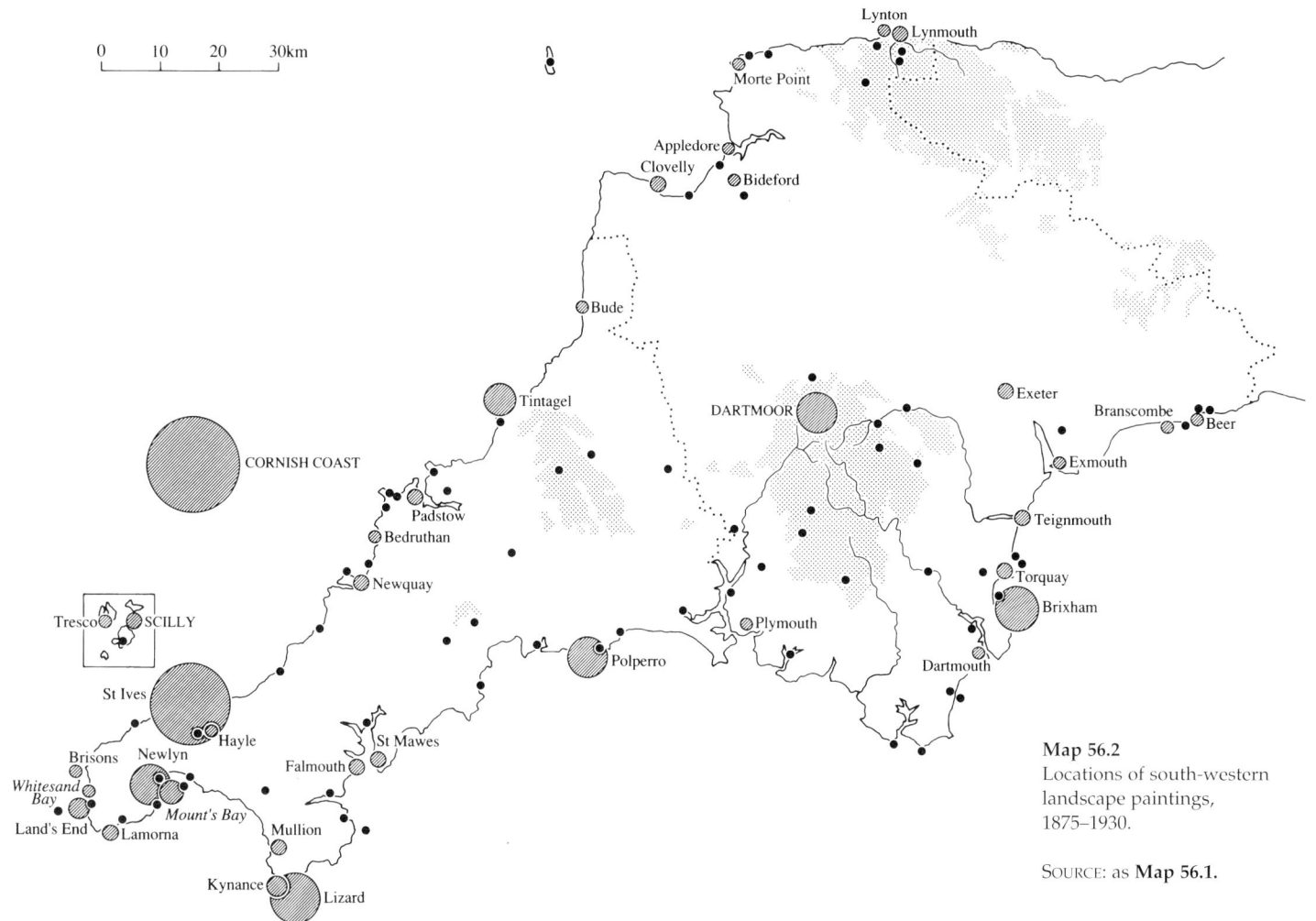

Lynmouth
Ilfracombe
R. Lyn
Clovelly
Okehampton
Exeter
Topsham
R. Teign
Sidmouth
Chudleigh
Glen
Tavistock
R. Tavy
Buckland
Abbey
Ashburton
Bickleigh
R. Dart
Hamoaze
Torbay
Plymouth
R. Avon
Dartmouth
St Michael's
Mount
Eddystone
Mount's Bay
Kynance
Lizard

0   10   20   30km

Map 56.2
Locations of south-western
landscape paintings,
1875–1930.

SOURCE: as **Map 56.1.**

Lynton
Lynmouth
Morte Point
Appledore
Clovelly
Bideford
Bude
Tintagel
DARTMOOR
Exeter
Branscombe
Beer
Exmouth
CORNISH COAST
Teignmouth
Padstow
Torquay
Bedruthan
Brixham
Newquay
Tresco
SCILLY
Plymouth
Dartmouth
St Ives
Polperro
Hayle
St Mawes
Brisons
Newlyn
Falmouth
Whitesand
Bay
Mullion
Land's End
Lamorna
Mount's Bay
Kynance
Lizard

451

Fig. 56.1
Artistic importance of
south-west England
landscapes, 1869–1980.

SOURCE: the numbers of
pictures of Devon and
Cornish landscapes
exhibited in the Summer
Exhibition of the
Royal Academy of Arts
are expressed as percentages
of the total number of
English landscapes (as
decades running averages).

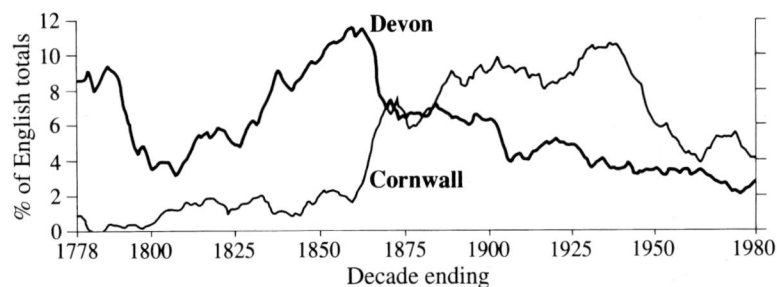

FURTHER READING

Peter Howard, *Landscapes: The Artists' Vision* (London, 1991)
presents the material for England and Wales from which the maps
in this chapter are derived. The same author's 'Painters' preferred
places', *Journal of Historical Geography*, 11 (1985), 138–54, discusses
the methodology of quantitative studies applied to pictures, and
'Changing taste in landscape art' (unpublished Ph.D. thesis,
University of Exeter, 1983) gives much more detail of changes in
artistic taste, including a detailed case study of Devon. Ian Ousby,
The Englishman's England: Taste, Travel and the Rise of Tourism
(Cambridge, 1990) allows the artist's situation to be set in the
general context of the history of tourism in England.

SERVICE INDUSTRIES: TOURISM AND RETAILING

THE GROWTH OF TOURISM
IN THE
NINETEENTH AND TWENTIETH
CENTURIES

GARETH SHAW, JUSTIN GREENWOOD AND ALLAN WILLIAMS

'TAKING THE WATERS': EARLY TOURISM IN DEVON AND CORNWALL

The growth of tourism in Devon and Cornwall relates to a number of factors, some operating at a national level and others at a more local or regional level.[1] Important influences on the region's coastal settlements were the fashion for sea-bathing stimulated by George III and the health-giving properties ascribed to the south coast by the medical profession. In the latter respect medical writers such as Granville played a dual role, first through publicizing the health-giving properties of coastal environs and secondly by recommending particular resorts to the wealthy.[2] Granville gave Bournemouth his wholehearted preference as a resort benefiting the invalid and consumptive; whilst in Torquay he found the Strand 'filled in general with respiratory-bearing people, who look like muzzled ghosts'.[3] Granville was writing in the 1840s, but from the second half of the eighteenth century travel books 'fairly deluged from the presses, topping the best-seller lists of Georgian Britain'.[4] It was just such publications that stimulated travel in parts of Devon and Cornwall, together with the fact that the French Revolution and wartime disruption made large areas of Europe difficult for recreational travel, thereby putting greater emphasis on British-based trips.[5]

Such changes opened up Devon and Cornwall to early tourists, leading to a fairly rapid growth in coastal resorts. As **Map 57.1** shows, there was a strong geographical shift in coastal resort development during the nineteenth century, with the earliest growth occurring in parts of south Devon. In fact resorts such as Exmouth and Teignmouth had developed as localized watering places for Exeter merchants as early as the mid-eighteenth century. By the late eighteenth century not only had the pace of change quickened, but more facilities were being provided for visitors as local communities became involved in developing a tourist industry. It was at this time that Sidmouth, Dawlish and Torquay emerged as small tourist resorts. Furthermore, the splendid scenery and mild climate had made this part of Devon an attractive retirement area for ex-officials of the East India Company.[6] In the case of Torquay naval officers also encouraged the development of a small resort through the practice of having their families join them when naval ships sheltered in Tor Bay. This

quickly led to the building of houses near the quay of the adjoining village of Torre.[7] Indeed, from 1797 onwards advertisements for houses within Torquay stressed both the warm climate and the convenience for bathing, and early town development plans show provision for a bathing beach and 'bathing machines'.[8]

Cornwall was opened up to visitors at a much later date than Devon, mainly because of its greater remoteness from London and the fashionable centres such as Bath. However, some early tourism developments can certainly be found in Cornwall, especially in and around Penzance. This was the earliest Cornish settlement to benefit from the growth of visitors towards the end of the eighteenth century, largely as a result of publicity surrounding its mild climate. Its early importance is reflected by the fact that the first guidebook on Cornwall was published for Penzance and district in 1815 by a distinguished physician, who gave great emphasis to the benefits of the local climate, comparing it to Mediterranean resorts.[9] But for many potential visitors Cornwall, and certainly the far west and north of the county, remained a remote place with difficult transport links even in the mid-nineteenth century.

Such difficulties, and the sense of adventure enjoyed by these early tourists, can occasionally be glimpsed in the written accounts of their journeys. One such account is the holiday diary kept by Mariana Tuckett, who journeyed from Bristol to Falmouth with her two sisters in 1858.[10] Her diary gives insights into their holiday activities and places visited, as is shown in **Map 57.2**. The three sisters were members of a Quaker family and they stayed with relatives in and around Falmouth. The holiday was a mixture of social visits, bathing and exploration of the surrounding countryside, with the longest excursion being to Penzance and Land's End. They made this trip, which necessitated an overnight stay, by omnibus. This, according to Mariana, 'was pretty full when we got in and it fell to my lot to sit very close to a tipsy man'.[11] Mariana Tuckett's diary reflects the more wealthy type of tourist being attracted to Cornwall at that time, but they were few in number and concentrated in one or two resorts. Over much of the rest of the county, fishing and mining still dominated, and the development of places like Newquay had to await the arrival of the railways.

Map 57.1
The growth of resorts:
population change, 1801–1911.

SOURCE: J.K. Walton,
*The English Seaside Resort:
A Social History, 1750–1914*
(Leicester 1983).

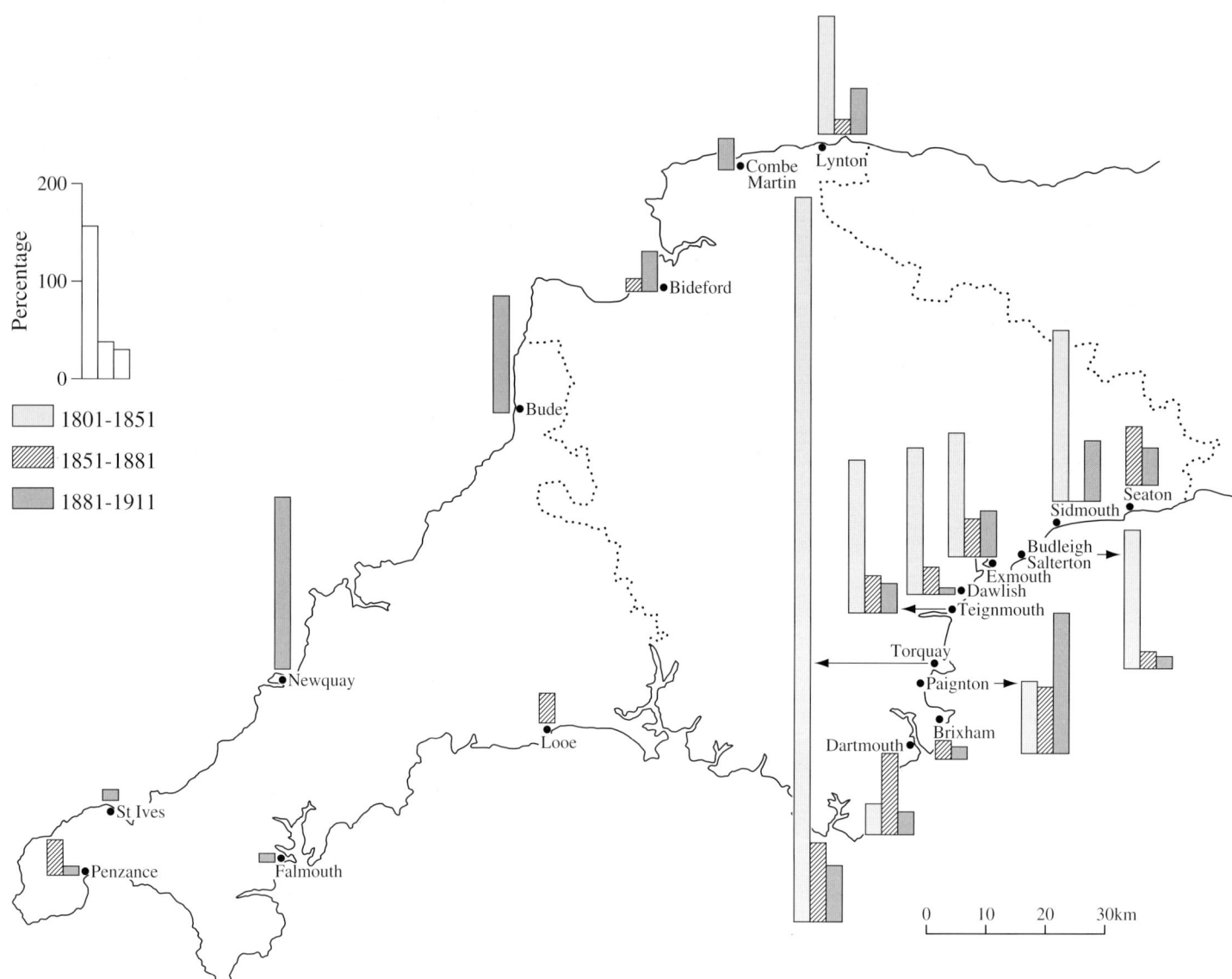

Fig. 57.1
Detail from lithograph by
W. Spreat of the Butterwalk,
Dartmouth, *c.*1850, showing
a crowd gathered for a
Punch & Judy performance.

SOURCE: reproduced by
permission of Westcountry
Studies Library, Exeter
(Somers Cocks 543).

SELLING THE SOUTH WEST IN THE RAILWAY AGE

According to Walton, 'a few resorts were called into being by railways, but almost invariably these were incidental by-products of schemes for commercial ports'.[12] Whilst this view holds good for parts of Devon and Cornwall, it tells only a partial story. To a considerable extent the Great Western Railway shaped an enduring image of Cornwall that lasts through to the present day, as well as building up resorts like Newquay. The story of Cornish railways is told by Simmons, who has highlighted the increasing significance of passenger traffic as Cornwall became the last English county to become connected to the main railway system in 1859 (and see chapter 46).[13] It was from this period that tourism started to expand. In 1860, for example, Thomas Cook organized the first 'package' tour to Cornwall, chartering a special train from Bristol; this proved an instant success paving the way for further tours later the same year.[14]

The impact of the railways took different forms, as may be illustrated in Penzance, Newquay and St Ives. In Penzance the railways added to the already important local tourism economy through new hotel developments during the 1860s, but essentially this resort was not a product of the railway age. In contrast, Newquay and St Ives owed their success as tourist resorts more closely to the coming of the Great Western Railway. St Ives had fallen on hard times in the 1860s, with a decline in mining and pilchard fishing, only to see its economy revived by the Great Western, which took a lease on Tregenna Castle and opened it as a hotel in 1878.[15] Similarly, new hotels were developed in Newquay from the late 1870s onwards in response to the railways, culminating in majestic facilities such as the Atlantic Hotel, opened in 1898.

The full blast of change associated with the railways came during the early twentieth century when the Great Western Railway introduced the concept of the 'Cornish Riviera',

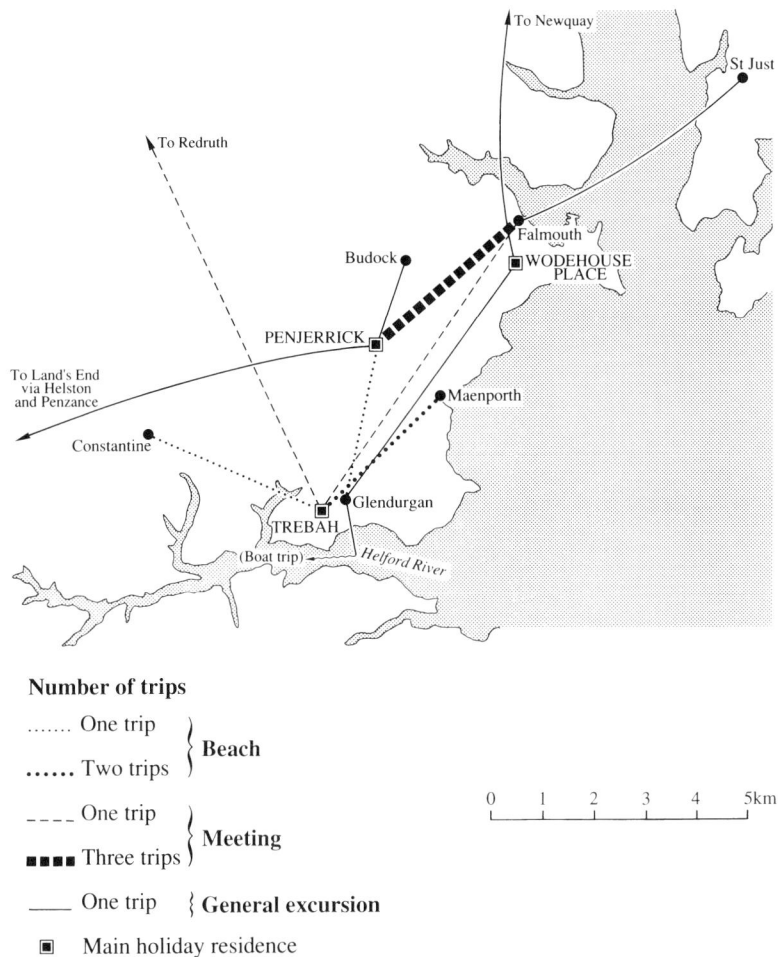

Number of trips

.... One trip ⎱
⎰ **Beach**
...... Two trips ⎰

- - - One trip ⎱
⎰ **Meeting**
▪▪▪▪ Three trips ⎰

———— One trip ⎰ **General excursion**

▣ Main holiday residence

0 1 2 3 4 5km

Map 57.2
Patterns of early tourist
behaviour in Cornwall.

SOURCE: H. Fox (ed.),
*Mariana's Diary: A Record
of a Holiday at Falmouth*
(Falmouth, n.d.).

brought changes, the most conspicuous being large-scale developments of villas. In Exmouth, as in many other ex-fishing villages, the original community remained clustered in the 'lower or old part of the town [where] the streets are narrow and irregularly built, but all the modern parts are composed of terraces surmounted by good houses, mansions of considerable size and villas, pleasantly detached'.[19]

By the second half of the nineteenth century the railways had increased the pace of change in Devon's resorts, but not to the same extent as in Cornwall. Leaving aside the early failures associated with the South Devon Railway Company, the first passenger service was opened in 1848, although, once again, it was absorption into the Great Western Railway in 1876, which proved to be a major turning point.[20] Improvements to the doubling of the track, with the ending of the broad gauge in 1892, stimulated frequent and faster trains bringing in more tourists. In Torquay these improvements led to changing attitudes towards tourism by local businessmen and the local authority. By 1900 the town had decided to advertise itself as a major tourist resort and to popularize its attractions. In 1902 a Corporation Bill was approved which included spending a penny rate on advertising, and the establishment of the first visitor information bureau.[21] This marked the start of Torquay's development as a tourist resort rather than the fashionable watering place of the nineteenth century.

Fig. 57.2
Great Western Railway
Company advertising poster
(Cornwall in the image
of Italy).

associated with its new Cornish Riviera Express which had a journey time of only seven hours from London to Penzance. The 'Riviera' was used as a successful advertising device to market holidays in the West Country and Cornwall from 1904 onwards. In effect it extended the emphasis previously given to the county's environmental qualities by making direct comparisons with the Italian Riviera (**Fig. 57.2**). Every Cornish resort was promoted in this context, as exemplified by St Ives, which became the 'Naples of the West', whilst Newquay was promoted through the message that 'the lamps are alight in London before the sun has set at Newquay'.[16]

In Devon the relationship between railways and tourism was more subtle than in Cornwall, primarily because most of the resorts were already well established by the mid-nineteenth century. Exmouth, for example, was already 'one of the handsomest and most fashionable sea bathing places on the southern coast of Devonshire', according to White's Directory of 1850.[17] The resort had acquired, during the 1840s, all the trappings associated with fashionable tourism, since 'the promenades are numerous … and the situation for bathing is excellent' with many bathing machines.[18] By the 1840s, therefore, most south Devon resorts had been transformed into settlements for fashionable tourists and wealthy retired classes. Both groups

CONSOLIDATING THE TOURISM MAP IN THE EARLY TWENTIETH CENTURY

The late nineteenth and early twentieth centuries were a period of growth and increasing maturity in the British tourist industry. The main source of this expansion was the growing number of lower-middle-class and working-class families able and wishing to enjoy holidays or day outings, and this demand was fostered by the promotional activities of a rapidly maturing industry. Resorts in Devon and Cornwall did benefit from this expansion, but places such as Torquay failed to capitalize on their early advantages. By 1911 Torquay was ranked by population only in twelfth position amongst English seaside resorts, and no other Devon or Cornwall resort figured amongst the top twenty-five.[22] The reason for this was quite simple: they were beyond easy access from the major centres of population in the Midlands and the North.

While the major tourist resorts grew no more than steadily, tourism growth became far more diffuse in these years in both England as a whole and the South West in particular. Walton writes that 'the new foundations of the late nineteenth and early twentieth centuries were more modest adaptions of existing village economies, with little new building beyond the occasional new hotel, villa or terrace of lodging-houses'.[23] Places such as Newquay and Bude in Cornwall, and Mortehoe in north Devon, doubled their resident populations during the years 1881–1911, although from small base numbers. They were especially attractive to middle-class tourists seeking alternatives to the increasingly commercialized attractions of larger resorts such as Blackpool or Weston-Super-Mare. This was the period when the fundamental map of English and south-western seaside tourist resorts was firmly established.

During the 1920s and 1930s the tourism industry continued to expand despite the difficult economic conditions of those years.

The polarization of incomes meant that there were growing numbers of lower-middle-class tourists in the Midlands and the South East, even during the Depression. Paid-holiday entitlement was also being extended to more of the population: from 1.5 million in 1925 to 11 million in 1939.[24] While the established resorts derived some benefits from this growth, 'some of the newcomers and, indeed, some of the established visitors were being pulled in new directions, away from the resort-centred holiday and away from the seaside landlady'.[25]

One of the main reasons for this geographical re-orientation was greater middle-class mobility. This was fuelled partly by improvements to the public transport system, but more significantly by the tenfold increase in motor-vehicle ownership between 1919 and 1939. This reduced the dominance of the railheads and allowed increasing numbers to seek out remoter coastal villages and the countryside, often as parts of touring holidays. Indeed, as early as 1914, the preface to Crossing's third edition of the *Guide to Dartmoor* had stated that 'the claims of Dartmoor as a holiday and health resort have become widely recognized'.[26] Hence, the framework of coastal resorts which emerged before the First World War was gradually filled out during the inter-war period. For example, the population in Cornwall's coastal parishes increased by 3.2 per cent between 1921 and 1931, while the county average growth rate was only 0.7 per cent.[27]

Another change in the industry was the emergence of rival types of resorts. The most innovative were the holiday camps offering full board and entertainment, although they were still far from common in the 1930s. Far more numerous were the camping grounds, chalet parks and caravan sites that began to sprout along the coastline, especially in south Devon. There were also growing numbers of retired in-migrants, drawn to the new villages of bungalows located outside the main resorts. Both these developments contributed to the continued

Fig. 57.3
Detail from lithograph by Newman & Co. of Exmouth showing bathing machines, *c.*1850.

SOURCE: reproduced by permission of Westcountry Studies Library, Exeter (Somers Cocks 1081C).

Map 57.3
Women employed as lodging
and boarding house keepers.

SOURCE: Census of England
and Wales, 1931.

Number of women

550
200
100
50
25
10

0   10   20   30km

dispersion of the holiday industry and were also potent indicators of post-1945 trends.

The geographical distribution of the tourism industry in 1931 is indicated in **Map 57.3**. This shows the number of women employed as lodging-house or boarding-house keepers; similar data are not available for men, at this spatial scale. While some of the newer forms of tourism are under-represented in this map, it does give a general picture of the state of the industry. The most striking feature is the concentration of tourism in the coastal areas (excluding Exeter and one or two other inland towns). However, there is also polarization within the coastal areas, despite the dispersion previously referred to. At the top of the hierarchy are Torquay and Paignton, followed by the other Devon resorts of Exmouth, Ilfracombe, Barnstaple and Teignmouth. Cornwall lagged considerably behind Devon, and its main resorts were Falmouth and rapidly expanding Newquay.

The flavour of tourism at this time is caught in *Holiday Haunts 1935*, published by the Great Western Railway.[28] Torquay was still seen as the queen of the South West resorts, 'With palm trees lining its roadways, and its all-pervading air of carefree joyousness, Torbay more readily resembles one of the famous Mediterranean resorts than any other English town'.[29] Its attractions included more than a thousand acres of parks, bathing, yachting, thirty hard and grass public tennis courts, two golf courses and other sports, concert parties, a covered seawater swimming bath and a vita-glass sun-lounge. For those who sought less commercialized attractions, there were resorts such as Looe: 'in the twin towns of East and West Looe the holiday-maker can enjoy all the pleasures of a country holiday without sacrificing any of the joys of the seaside'.[30] Alternatively, for those with a car and a desire for real solitude, there were myriads of unspoilt villages awaiting discovery. Deep in the South Hams, for example, was Hope Cove, 'a remote and wholly delightful fishing village'.[31] In a sense, this was the culmination of a golden age of leisurely tourism development, for great changes lay in store after the Second World War.

THE 1950S AND 1960S: MASS TOURISM COMES TO THE SOUTH WEST

Fig. 57.4
Use of non-traditional units of
holiday accommodation.

SOURCE: Devon County
Council, *County Structure Plan,
First Alteration* (Exeter, 1983).

Fig. 57.5
Detail from engraving
published by L.D. Westcott
of Teignmouth beach
showing bathing machines
and swimmers, n.d.

SOURCE: reproduced by
permission of
Devon Record Office
(1919Z/Z1/p. 92).

The outbreak of the Second World War did not mean the end of tourism in the South West. Indeed the closure of the southern coastline east of Dorset to holidaymakers gave a boost to tourism in Devon and Cornwall. This was to provide the platform for a major expansion of the industry in the 1950s and 1960s. During these decades there were growing real incomes and an extension of paid holidays to virtually the whole of the working population. These demand-side changes were matched on the supply side by the expansion of new and cheaper forms of self-catering accommodation, so that the net result was a virtuous circle of growth. Devon and Cornwall took a major share of this expansion, as the extension of mass car-ownership to the working class finally broke down the access constraints on the summer holiday industry. The 1950s and 1960s were to be an era of growth and prosperity for South West tourism. Meanwhile, the development of the modern foreign holiday 'package' industry meant the emergence of strong new competition. Between 1951 and 1972 the proportion of British people taking their main holiday abroad increased from 6 per cent to 25 per cent. This was to dampen growth in domestic tourism and by the late 1970s the South West was facing largely static demand.

The aggregate growth of tourism was remarkable, and numbers of visitors doubled within fifteen years. Cornwall, for

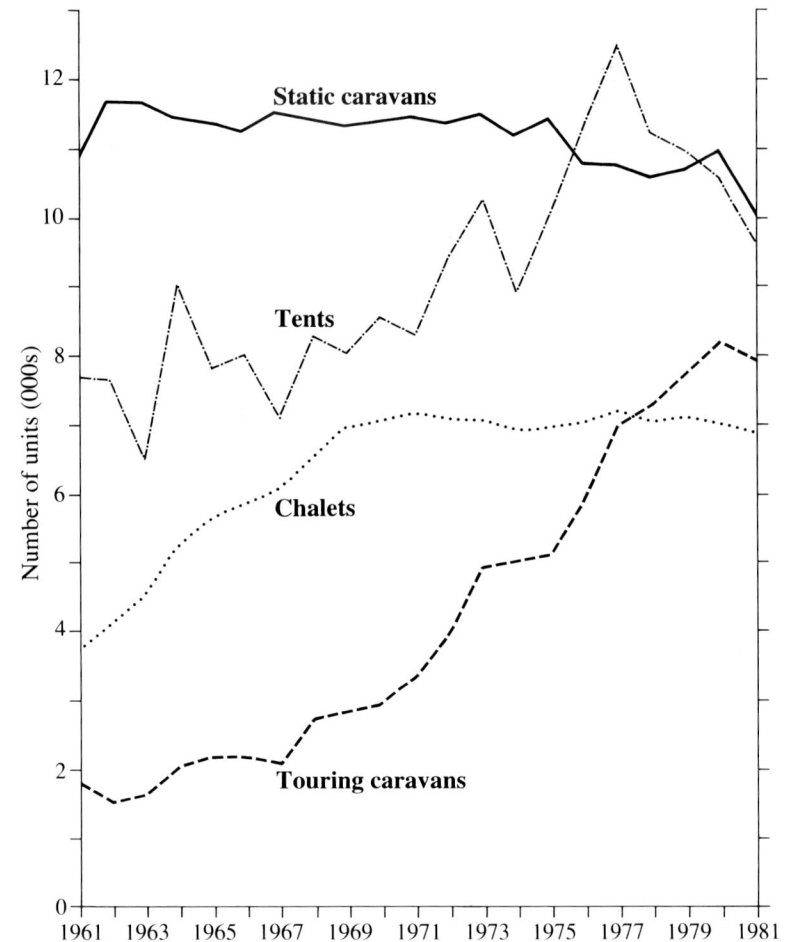

instance, had 1.25 million visitors in 1954, 2.09 million in 1964 and 3.25 million in 1974.[32] In Devon numbers increased from an estimated 2.5 million in 1960 to approximately 3.5 million in the mid-1970s.[33] In terms of numbers of tourists attracted, therefore, Cornwall almost caught up with Devon during the post-war boom years. The major resorts continued to expand, but much of the new growth was diverted to smaller towns and villages, as tourists became more locationally flexible.

Perhaps the most remarkable change in this period was the growth of self-catering. Between 1962 and 1980 the proportion of holidaymakers who self-catered increased from 41 per cent to 57 per cent, while there was a corresponding decrease in traditional serviced accommodation in hotels and guesthouses. The motives were partly those of cost—especially as real labour costs and prices in guesthouses and hotels increased—and partly the desire for greater freedom and flexibility.[34] **Fig. 57.4** shows the changes which occurred in some of the major types of self-catering. The greatest increases were in camping and touring caravans, which were the most flexible and cheapest types of accommodation. The growth of chalets was strong in the 1960s but had peaked in the 1970s, as had also the demand for static caravans. There were similar trends in Cornwall: between 1964 and 1974 the 10 per cent increase in the number of visitors in serviced accommodation was considerably surpassed by a 91 per cent increase in self-catering.[35] Again the largest increases were in touring caravans (37 per cent) and camping (96 per cent) rather than in static caravans and chalets (40 per cent). The growth of the self-catering sectors was also reinforced in both counties by the increasing popularity of second homes and of rented holiday cottages.

By the 1970s the tourist industry in the South West was in a relatively mature phase. Excepting London, it was the strongest single regional market in the UK, but it was facing a largely static level of demand. Most holidaymakers came for a one-week or two-week visit, and the average length of stay in Devon was 9.5 nights. Most came by car, and 40 per cent came from London and the South East.[36] Within the South West, the coastal areas were still the great attractions, and resorts such as Torbay, Ilfracombe and Newquay continued to provide a large proportion of the accommodation available (**Map 57.4**). Torbay, for example, attracted 39 per cent of all visitors to Devon in 1975.[37] There had, however, been an intensification of tourism in many smaller coastal towns and villages, especially in Cornwall and in north Devon.

The long evolution of tourism in Devon and Cornwall during the nineteenth and twentieth centuries had produced a very distinctive settlement morphology by the 1960s. In Cornwall there were three main types of tourism areas.[38] First, there were old coastal towns and villages such as Penzance and Padstow, which had to some extent retained their original character. Secondly, in places such as Bude and Newquay the original settlement was so small that large-scale tourism growth had transformed the character of the resort. Thirdly, there were the new centres of the holiday industry which had emerged since the 1920s. These were diverse and included chalet and caravan townships outside existing settlements at places such as Whitsand Bay and Hayle Towans, as well as estates of holiday homes as at Constantine Bay and Mawgan Porth. There have been similar developments in Devon.[39]

With the growth of tourism, especially outside the major resorts, there were increasing land-use conflicts between farmers, tourist businesses and conservationists. The balance of interest has shifted over time, but the tourist industry has been subject to increasing constraints. For example, Cornwall County Council

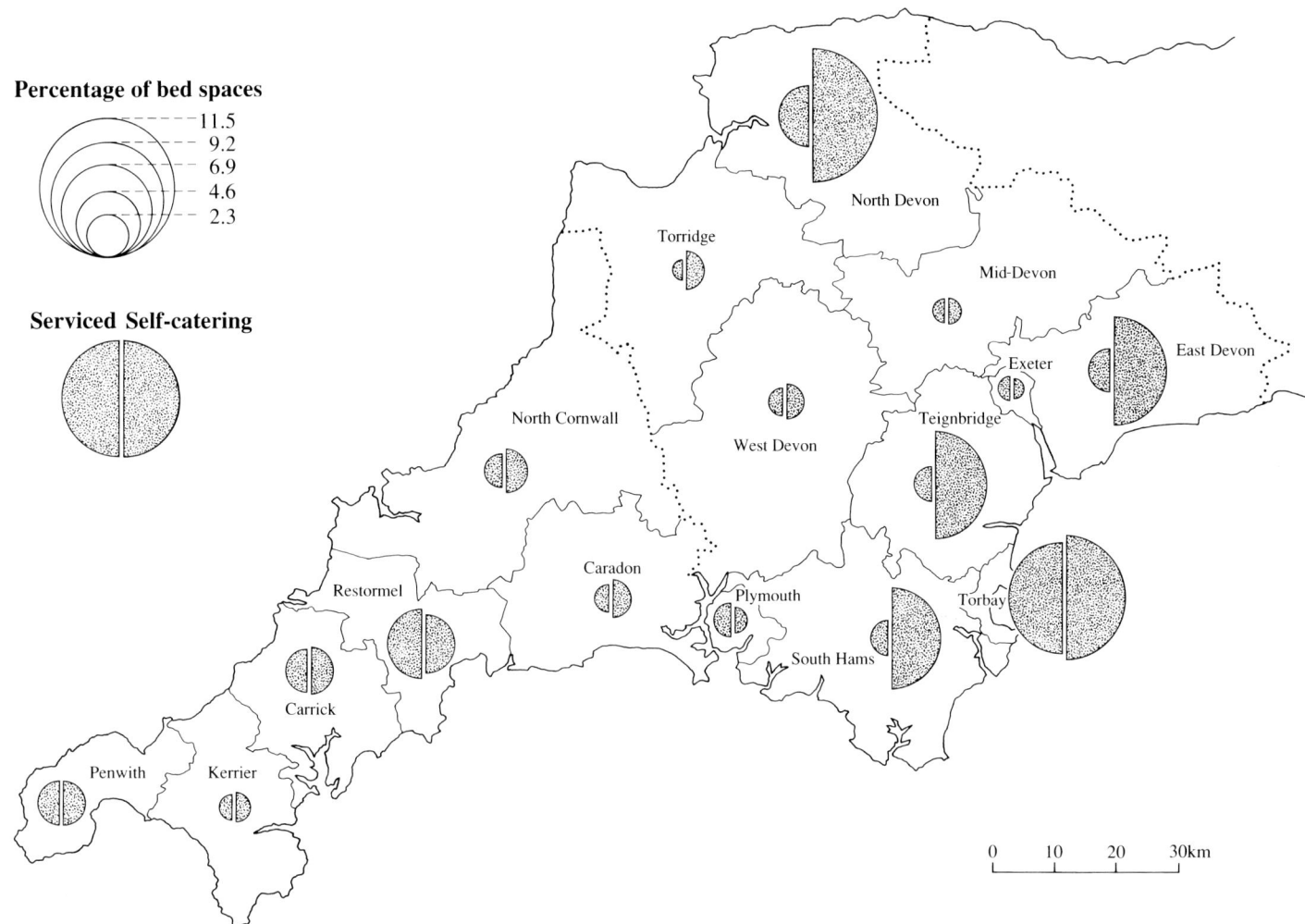

Map 57.4
Accommodation stock: number of bedspaces, 1990, by district.

SOURCE: West Country Tourist Board.

declared 'saturation areas' in the 1960s within which 'no new sites or extensions to existing sites for static holiday caravans or chalets are permitted'.[40] Similarly, Devon in 1961 identified areas which were 'approaching capacity for holiday development' and in which further camping and caravan sites were discouraged; seven of the eight designated areas were, not surprisingly, on the south coast.[41] Against this, however, there was growing realization of the economic importance of tourism to the South West.[42] This was to be one of the most important issues facing the industry and indeed the region in the 1980s.

TOURISM IN THE LATE TWENTIETH CENTURY: ADJUSTMENT AND CHANGE IN THE AGE OF LEISURE

By the 1980s, it was argued that Devon and Cornwall represented the 'jewel in the crown' of British domestic holiday tourism. Although tourism figures are always difficult to calculate, it is estimated that around one in eight of all trips made by British tourists in the 1980s and 1990s was to Devon and Cornwall, while annual tourism expenditure in the two counties is around £1,250 million.[43] Although the annual family holiday by the seaside is still a favourite, there have been considerable changes in visitor patterns and behaviour. Visitors to Devon and Cornwall in the 1990s were increasingly likely to be on a second or subsidiary holiday, and often as not on a short break. Tourists are drawn to the region for many reasons; the beaches still hold great attraction, but above all it is the beautiful landscape which is the magnet.[44]

So great is the attraction of the landscape and of the lifestyles of the South West, that these have also attracted tourism business owners to the region. Today's visitors to the area often become tomorrow's entrepreneurs in the region's tourist industry. It is no accident that the South East region, which supplies more visitors to Cornwall than any other in Britain, is also the birthplace of one-third of the owners of hotels and guesthouses in Cornwall. This has contributed to a relatively slow adjustment of the tourism industry to the demands of modern visitors for improved standards of catering and accommodation. Many business owners are simply not interested in investment in modern tourist facilities, such as en suite bathrooms, yet there is clear demand for these.[45] Occupancy rates are highest in the exclusive end of the serviced accommodation market.[46] This—and changing tastes—has contributed to a long-term decline in the serviced accommodation sector.

In contrast, self-catering accommodation has grown in recent years, although here again the growth is in quality provision with retrenchment occurring in cheaper accommodation types. Increasingly, sites which have not been updated have been unable to attract guests and, consequently, have been unable to generate investment capital. As a result some chalet units, static caravans and tent sites have been closed in recent years. This shift in accommodation preferences has important economic implications.

Self-catering accommodation is less labour-intensive than other forms, and the jobs it provides are more likely to be part-time and low-skilled. In comparison, serviced accommodation generates more tourist spending; for instance, in 1988 25 per cent of all tourist nights in the region were spent in hotels and guesthouses, but they accounted for 35 per cent of all the spending on accommodation.

Another important segment of tourist accommodation is second homes. These are notoriously difficult to estimate, although census data suggest that there are around 13,000 second homes and holiday lets in Cornwall, particularly in the north Cornwall area. Whilst they generate some jobs and incomes in the county, they also inflate house prices and can cause considerable housing difficulties for families in some towns and villages (see chapter 55).

The total breakdown of accommodation stock throughout Devon and Cornwall for both hotels and guesthouses and self-catering accommodation is shown in **Map 57.4**. The character of tourism varies throughout Devon and Cornwall, with serviced and self-catering accommodation predominating in different areas. For instance, of the six districts in Cornwall, Penwith is only ranked fourth in terms of bed spaces but has the largest number of self-catering establishments, while self-catering overwhelmingly predominates in both north Devon and east Devon. Overall, Torbay and Newquay are the most popular holiday destinations, with serviced accommodation still being relatively strong. Indeed, despite the shift to self-catering, the West Country accounts for a quarter of all hotels and guesthouses in the UK.[47] This confirms that the traditional seaside holiday is far from dead and, indeed, still provides the mainstay of tourism in the region. However, there has been a loss of bedspaces in traditional resort areas, with some hotels and guesthouses being converted into nursing or residential retirement homes. This is particularly true of Torquay, Paignton and Teignmouth. In Devon alone in 1989, some 4,500 bedspaces were lost in this way.[48]

Facilities for tourists tend to be available only during the peak holiday times of April to September, and Devon and Cornwall's tourist industry is characterized by a marked seasonality. Only in Plymouth and Exeter, where there is significant business tourism, does occupancy exceed 20 per cent in the winter months. Similarly, almost one-third of restaurants and cafés close during the winter months. This also serves to emphasize the benefits and drawbacks that tourism brings to local economies in Devon and Cornwall, where almost all sectors of economic activity benefit from tourist spending. In Newquay, for instance, the majority of manufacturing firms are to some extent dependent upon tourism for their sales.[49]

Although visitors are well catered for on the tourist attraction side, repeated surveys of visitors show that tourists seek attractions which blend well with the heritage and environment of Devon and Cornwall.[50] The most popular holiday activities involve enjoying the natural beauty of the area, such as beach, sightseeing and walking activities. The attraction of the area's

Fig. 57.6
Swimmers at Dawlish, n.d.

SOURCE: reproduced by permission of Devon Record Office (Chapman 8154).

natural heritage has also resulted in a significant growth of rural tourism in recent years, particularly in Devon. This is another indicator of the relative shift away from traditional seaside holidays.

The rich variety of visitor attractions in Devon and Cornwall clearly owes its continued existence to tourism. Attractions range from historic homes and gardens run by the National Trust, to privately owned theme and leisure parks, local-authority run museums and small visitor centres, and this range offers a diversity of activities and facilities. The most successful attractions are, by and large, those situated close to major visitor centres, and which offer a range of indoor and outdoor facilities. These include Buckfast Abbey (551,413 visitors in 1989), Dartington Cider Press (512,000), Cornwall Coliseum (490,000) and Babbacombe Model Village (472,032). A new generation of attractions is emerging to challenge these, such as Crealy Adventure Park in Clyst St Mary, east Devon, and the Plymouth Dome. In recent years, emphasis has also been placed upon attractions which integrate the themes of conservation, recreation for local residents, and tourism, through the provision of countryside facilities. Examples of these include the development of the Tarka Trail footpath route in north Devon, the Templar Way in south Devon, the promotion of rural railway lines with British Rail such as the Tamar Valley route, and the creation of country parks.

The emphasis upon rural and countryside tourism has not been to the detriment of developing facilities in the cities. Major developments have occurred in Exeter, Plymouth and Truro. Indeed, urban tourism has been one of the major growth points of UK tourism in the late twentieth century, and although an overwhelming majority of tourists still come to Devon and Cornwall for holidays, business tourism in the area is of increasing importance. In 1989, for example, new four-star hotels opened in Exeter and Plymouth. Small conferences are also being attracted to the area, particularly to Torbay, Plymouth and more recently Exeter.

Plymouth, Exeter and Truro have been major success stories of tourism in Devon and Cornwall in recent years. In 1989 Plymouth won both the British Tourist Authority 'UK Marketing Authority of the Year' award and the English Tourist Board's 'England for Excellence' award. Work is also continuing on the Waterfront Strategic Development Initiative and the Mast projects, which aim to transform the city's river and sea frontages. Exeter, too, has engaged on major waterfront development schemes, and both Exeter and Truro have developed into regional shopping centres of some importance. Part of the attraction of these two cities is their historical heritage, including their cathedrals. 'Selling the past' is, however, a more general theme. Cornwall's industrial tin-mining heritage, for example, is now a visitor attraction, as is the 'Celtic revival culture' of the county.[51]

FURTHER READING

G. Shaw and A. Williams (eds) *The Rise and Fall of British Coastal Resorts* (London, 1997); J.K. Walton, *The English Seaside Resort: A Social History, 1750–1914* (Leicester, 1983).

SERVICE INDUSTRIES: TOURISM AND RETAILING

RETAIL TRADING
1850–1939

ANDREW ALEXANDER AND GARETH SHAW

STRUCTURAL AND REGIONAL TRENDS IN BRITISH RETAILING

The middle decades of the nineteenth century witnessed the beginning of a fundamental transition to the structure of the British retail industry. Central to this change was the emergence of an increasing number of large-scale retail firms which sought to reduce distributive costs through mass-merchandising. There were three main elements to this retail revolution: the multiple, the co-operative and the department store. Co-operative retailing is considered in chapter 59; in this chapter, attention focuses on the emergence of the department store and more particularly the multiple retailer.[1]

Both these types of business operation sought to enhance economies of scale and scope beyond those typically achieved in retailing. In the case of the multiples, such economies were most usually secured through the opening of additional branches in new locations, a process hastened by retailers' perceptions of an emerging national market. In relation to the menswear trade, for instance, the founder of the Leeds-based multiple bespoke tailor Montague Burton Ltd noted: 'the whole of Britain is practically like one city today. What sells in Penzance will also sell in Putney and Pontypridd. The changes date from the Great War when the miner, bricklayer and bank clerk were all thrown into one melting pot.'[2]

DEPARTMENT STORES

Department store entrepreneurs adopted a different method of achieving such economies, concentrating instead on expanding individual store size and thus retail floor space. This was usually accomplished in the earliest years by the acquisition of properties adjoining existing stores. Subsequent developments, however, epitomized by the opening of Selfridge's in 1909, reveal a trend toward purpose-built stores which were better able to offer the latest in retail amenity. Other strategies adopted by multiples and department stores included centralized buying, and, more occasionally, backward integration into the production process.[3]

While evidence exists of incipient department store and multiple shop retailing during the first half of the nineteenth century, such developments were of very limited significance to overall retail provision. In contrast, the half century after 1850 saw a notable increase in the number of both types of organization, a widening of their geographical distribution, and a significant increase in the market share of each. Available data reveals London to have been the primary focus for early department store trading, with evidence for a provincial department store trade before 1850 restricted to the development of Kendal Milne's in Manchester (1836) and Bainbridge's in Newcastle (1838).[4] In the subsequent half century or so, however, there followed a wave of provincial openings such that Birmingham alone could boast eight full department stores by 1910 and Walsall five. It is estimated that there were as many as 200 department stores by early Edwardian times, situated mainly in the nation's cities but extending in some instances to lesser provincial towns and seaside resorts.[5]

MULTIPLE RETAILERS

The expansion of multiple retailing in the period after 1850 was, by way of comparison, even more rapid. While firms such as the Singer Manufacturing Company, W.H. Smith & Son and J. Menzies were among the first modern multiple retailers in Britain, the emergence of multiple retailing in the grocery and provisions and footwear trades during the 1870s and the clothing, chemists' goods, meat and variety store trades in the 1890s was more typical of the pace and style of development of the ensuing half century or so.[6] In each of these trades large, capital-rich firms operating extensive chains of geographically dispersed fixed-shop outlets began to develop. Leading businesses such as Lipton, Freeman Hardy Willis, Hepworth, Boots, Eastmans and Marks & Spencer flourished in this period, extending their burgeoning retail networks beyond original localized 'heartlands' toward the point at which they would achieve national market coverage. Lipton, for instance, who opened the first branch of his grocery and provisions chain in Glasgow in 1871, had shops as far south as Birmingham by 1885. The company's first London branch opened in 1888 and provincial towns such as Ipswich and Plymouth gained branches around the turn of the century.[7]

The effects of these changes were felt nation-wide, but the timing and extent of their impact varied across space. It is possible, therefore, to identify distinct historical-geographies of retail change at the regional and local level which had important social and political, as well as economic, ramifications. In a study of retail trading in Devon and Cornwall in the period 1800–70, for instance, Phillips detects a surprisingly high level of institutional stability in the region's food trades.[8] Despite the expansion of the region's railway network, he can find little evidence of the expected centralization of the region's retail system or of the rapid decline of market-based trading. Instead, he reveals a high level of stability which he explains by reference to a series of social relations in the marketplace. Further evidence of the particular nature of the development of the region's retail system can be seen in relation to the emergence of multiple retailing. Calculations of branch per population ratios reveal marked and enduring contrasts between the English regions. Analysis of the distribution of multiple grocery branches in 1911, for example, reveals above average per capita representation in the industrial north and London and the Home Counties. In contrast, the West of England, East Anglia and other less industrialized regions recorded below average representation.[9] Similar analyses for the year 1919 suggest a national average of 1.7 multiple grocery shops per 10,000 population. The industrial North and London and the Home Counties are again calculated as having above average representation, while the West of England, with an estimated 1.1 multiple grocery shops per 10,000 population, is calculated to have had one of the lowest levels of representation in Britain.[10]

Yet even regional comparisons such as these mask the real paucity of early large-scale retail development in Devon and Cornwall. Alexander's more detailed analysis reveals considerably lower levels of representation in these counties than is suggested by the regional averages.[11] Taking the case of the grocery and provisions trades in 1911, Alexander calculates that per capita multiple branch representation in Devon was 19 per cent below the regional average and that in Cornwall some 30 per cent below. These inter-regional disparities had lessened somewhat by 1931, but the South West as a whole continued to have one of the lowest

Fig. 58.1
Drawing from an original sketch by Henry Ellis of his shopfront, Exeter, showing a traditional shop of the type being replaced by the new multiples discussed in this chapter.

SOURCE: reproduced by permission of Devon Record Office (76/20/4).

per capita rates of multiple retail branch representation. Similar trends are identified in relation to the menswear trade, revealing the pattern of under-representation recorded for the grocery and provisions trade to be unexceptional. The belated emergence and subsequent slow growth of this new retail type can be attributed in part to the region's remoteness from the seedbeds of the emerging multiple retail system and to its comparatively limited urbanization.[12] Many of the earliest multiple retailers in the grocery and provisions trade originated in large commercial port locations, such as London and Glasgow, where the goods in which they dealt were landed and sold. Similarly, the seedbed of the multiple trade in menswear was concentrated around the northern textile towns, particularly Leeds. Second, multiple retail firms tended to develop first in large cities offering a concentrated source of working-class demand. Devon and Cornwall, peripheral to the main thrust of Victorian industrialization, generated very few indigenous multiples and were instead reliant on the diffusion into the region of the branch networks of exogenous firms.

INTER-URBAN PATTERNS (**Maps 58.1, 58.2 and 58.3**)

Some of the earliest examples of multiple retailing in south-west England are provided by the grocery and provisions trade. Evidence exists of such practices during the 1850s, but these were largely unconsolidated efforts with few implications for the longer-term evolution of the region's retail trades. Consequently, the 1883 edition of Kelly's county directory of Devon and Cornwall records the operation of only two multiple grocery businesses within the South West. Six branches of the London-based national multiple the Star Tea Company were recorded, their location concentrated in the larger towns of Devon, with one branch in each of the 'three towns', Plymouth, Stonehouse and Devonport, and further outlets in Exeter, Torquay and Newton Abbot. The directory also lists the business of John Lethbridge, a local grocery multiple based in east Devon with outlets in Sidmouth, Exeter, Cullompton and Barnstaple. Offering a mixture of wholesale and retail services, Lethbridge's business was one of the small number of local multiple firms which developed in the region during the study period.

The subsequent edition of Kelly's county directory for Devon and Cornwall, published in 1889, reveals little further evidence of the development of multiple retailing. Perhaps most significant was the opening of branches of the London-based multiple the International Tea Company, which had outlets in Tiverton, Barnstaple and Truro by this date. These few cases apart, there is little indication of change and it would seem that the majority of consumers within the region continued to rely upon existing co-operative ventures, traditional fixed-shop outlets or markets for their groceries and provisions. The relative unimportance of the multiples in grocery and provisions retailing is confirmed by their limited geographical distribution (**Map 58.1**).[13] Largely confined to towns at the upper end of the region's urban hierarchy, which

offered the largest concentrations of working-class demand to which the trade of the grocery and provision multiples was most suited, the region had yet to witness the full implications of the revolutionary changes in the structure and organization of retailing that had started almost twenty years earlier.

By the turn of the century the situation had begun to change quite rapidly. The county directory for 1902 records more than fifty multiple grocery branches within the region. These additional branches came in part from the expansion of representation of the national multiples already trading in the region. The Star Tea Company, for instance, was recorded as operating fifteen branches in Devon and Cornwall by this date. More significant, however, was the emergence of a number of rival national multiples. By 1902 the branch networks of the Maypole Dairy Company, Lipton, Pearks Dairies and the Home and Colonial Stores had diffused into the region and these firms opened new branches with characteristic rapidity. Nonetheless, network development programmes continued to concentrate on opportunities in the larger towns of the region and were restricted almost entirely to locations in Devon.

This increased pace of development was sustained throughout the following decades as the multiples consolidated their foothold within the region. Such consolidation was achieved through a two-phased widening of branch networks. First, there was more comprehensive development in the major towns of Cornwall. Second, and more significant, was the process of branch diffusion down the urban hierarchy, with the opening of branches in towns smaller than those represented in the initial phase of colonization (**Map 58.2**). The pace of expansion recorded for the grocery and provisions multiples reflects the distinctive nature of their trade. Encouraged in particular by rising demand from an increasingly urbanized working class, a number of national grocery multiples expanded with striking rapidity and grew to considerable size. Home and Colonial Stores, for instance, formed in the middle years of the 1880s, had more than 200 branches in Britain by 1895 and more than 400 by the turn of the century. Similarly, Thomas Lipton's business empire included some 500 shops by 1899.[14] Such growth rates were made possible by the small and often rudimentary nature of early branches, particularly in the provisions trade where limited branch size represented less of an impediment to growth simply because retailers commonly stocked fewer lines. Consequently, limited capital represented less of a brake on growth rates during the initial years of development, while subsequent share flotations enabled often ambitious branch opening programmes. This can be seen in the case of Home and Colonial's rapid expansion into the towns of Devon around the turn of the century, which was part of a larger nation-wide branch opening programme made possible by the capital raised from share and debenture issues during the 1890s.

The final map in this series (**Map 58.3**) illustrates the maximum extent of coverage achieved by grocery and provision multiples by the close of the inter-war period. It reveals a quite dense pattern of representation, with branches present in towns throughout the region and at all levels of the urban hierarchy. Virtually no urban borough with a population of more than 5,000 was neglected by the multiple retailer, and larger towns enjoyed considerably enhanced representation. In the case of Exeter, for instance, the number of national multiple grocery and provisions branches recorded increased three-fold in the period between 1911 and 1939.

One particularly interesting trend is the concentration of branch outlets in the coastal resort towns of the region. Initial evidence of this can be seen in the map of branch location for 1910 (**Map 58.2**) which reveals a cluster of openings in resort towns of south Devon such as Dartmouth, Teignmouth, Brixham and Exmouth. This pattern was consolidated by 1939 (**Map 58.3**) and extended to include the resort towns of Cornwall. Consequently, small towns such as Hayle enjoyed the representation of multiple grocery firms attracted by the increasing number of visitors taking paid leave in the region's resorts. A similar pattern can be identified in other trades, most particularly in the variety store trade in which retailers such as Marks & Spencer afforded special

status to seaside stores, and represents part of a wider process of transformation of the structure of these communities initiated by the expanding tourism industry.

T.J. LIPTON (**Map 58.4**)

Many of the trends highlighted in the preceding overview can also be identified in the pattern of branch development of the multiple provisions dealer T.J. Lipton. The earliest references to Lipton's business found in the Kelly's county directories of the region come from the directory of Devon and Cornwall for 1902, which records branches in Plymouth, Devonport, Exeter, Torquay and Barnstaple. The choice of these locations, together with those chosen for the initial development in Cornwall (branches were operating in Redruth and Falmouth by 1910), form part of a wider process of expansion into the West of England based on a hierarchical pattern of diffusion. Shops were opened first in the

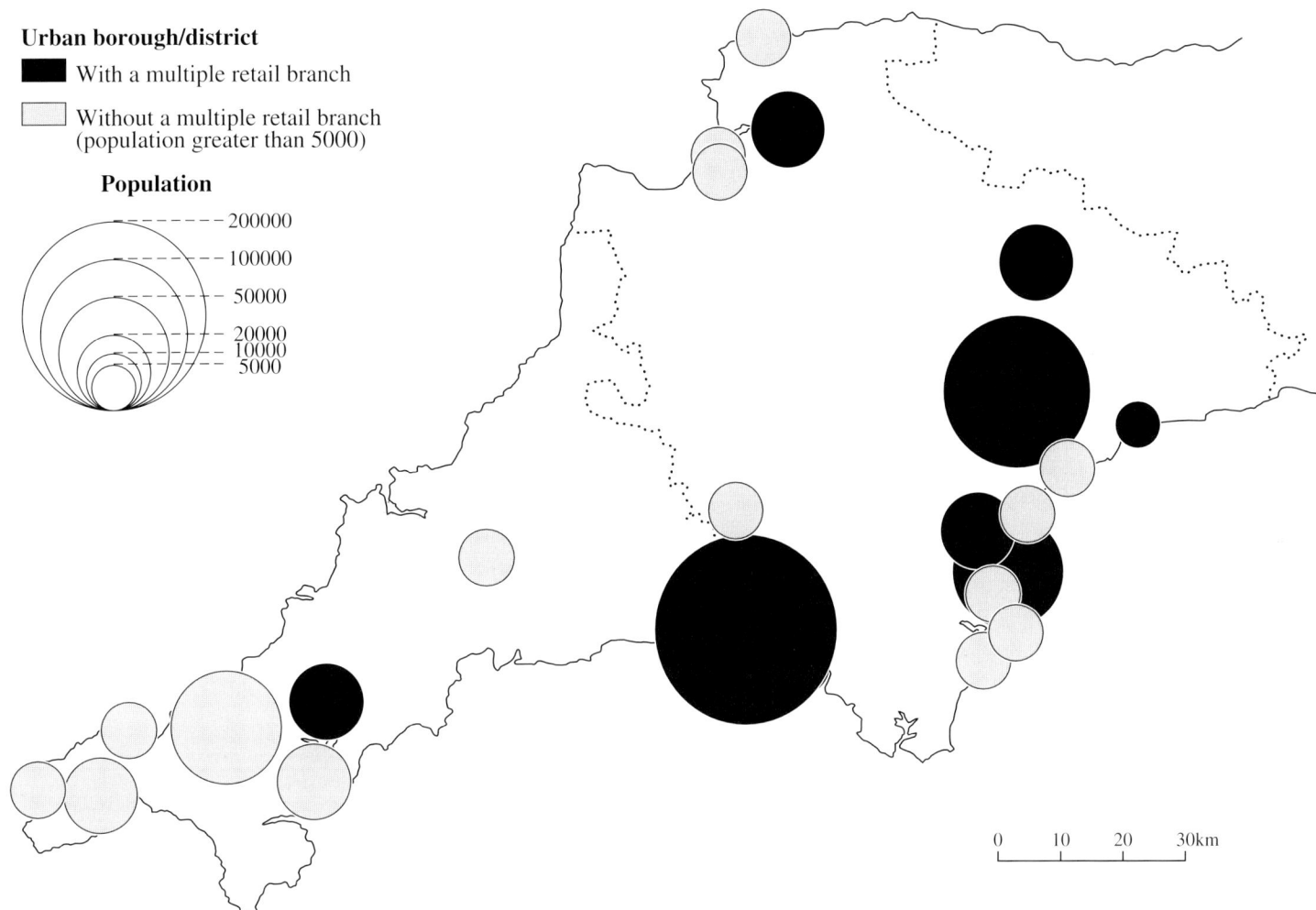

Urban borough/district

■ With a multiple retail branch

☐ Without a multiple retail branch (population greater than 5000)

Population

200000
100000
50000
20000
10000
5000

Map 58.1
The location of multiple retailers in the grocery and provisions trade, 1889.

SOURCES: population data from the Census of Population 1891; retail data from Kelly's *Directory of Devonshire and Cornwall*, 1889.

larger towns of the region, which offered greatest potential demand for the products and style of trade pursued by Lipton's shops and, perhaps most importantly, which were to act as bridgeheads for subsequent developments. The first stage of this hierarchical diffusion into the West of England, which had begun with the opening of a branch in Bristol in 1886, was essentially complete by 1902 (**Map 58.4**). The market and resort towns of south Devon formed the focus of the next phase of development in the South West, with attention turned initially to larger towns of this type bypassed in the original diffusion process. Such towns offered attractive opportunities for Lipton in the increasingly competitive grocery and provisions trade. Thus branches were opened in rapid succession at Tiverton, Ilfracombe and Newton Abbot in Devon, and subsequently at Penzance, St Austell and Truro in Cornwall.

The stock turnover rates of the firm's branches in relatively affluent towns such as these compared well with those of its older branches in the working-class districts of large cities.[15] Yet, despite

attempts to rationalize the existing branch network by closing some of the least successful branches, it is estimated that by 1926 as many as half of Lipton's English shops were making losses.[16] This failing profitability, combined as it was with gross inefficiencies in other parts of Lipton's business empire, resulted in the purchase of a controlling share of the company by Van den Berghs, which controlled the rival Meadow Dairy Company and Pearks Dairies, in 1927.

The newly appointed senior management team determined to close as many as 100 of the worst loss-making branches in the Lipton chain. As noted, these were heavily concentrated in lower-working-class districts of large urban areas, where the company was increasingly vulnerable to competition from both rival multiples and co-operative societies, and the branch network in the West of England seems to have been less severely affected. Indeed, in the case of Devon and Cornwall the county directories reveal no evidence of branch closure whatsoever. Instead available evidence suggests that efforts were concentrated upon securing important

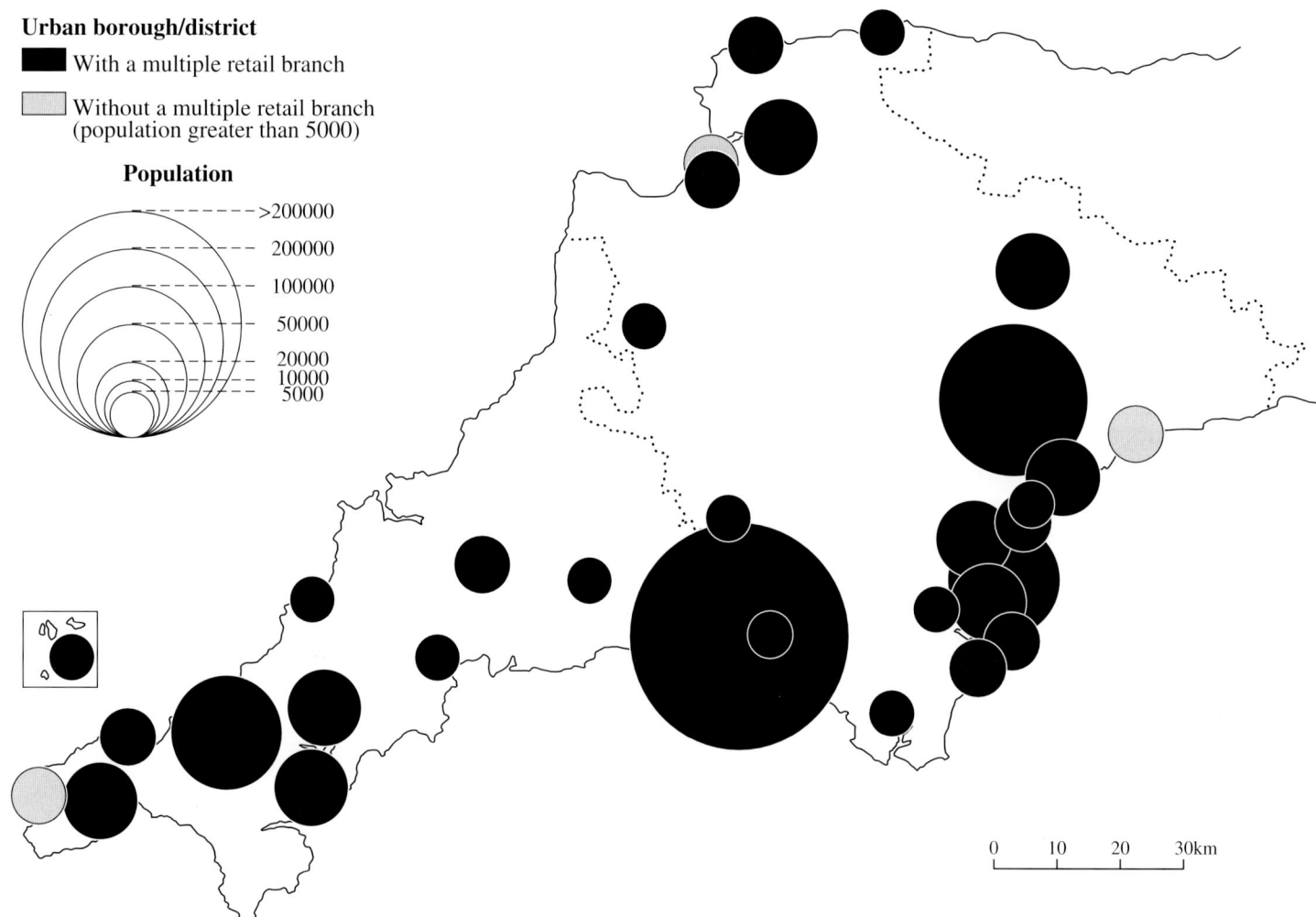

Urban borough/district

■ With a multiple retail branch

▨ Without a multiple retail branch (population greater than 5000)

Population

>200000
200000
100000
50000
20000
10000
5000

Map 58.2
The location of multiple retailers in the grocery and provisions trade, 1910.

SOURCES: population data from the Census of Population 1911; retail data from Kelly's *Directory of Devonshire and Cornwall*, 1910.

0   10   20   30km

economies of scope made possible by the acquisition of the Lipton chain. This was apparently achieved by augmenting and complementing the selling space of some existing Lipton branches by opening branches of Pearks Dairies in close proximity. In Barnstaple, Tiverton and Torquay, for instance, branches of Pearks were opened in the same street as existing Lipton outlets during the final years of the 1920s and, it would appear from directory data, continued to trade in tandem with them throughout the remainder of the inter-war years. Van den Berghs' acquisition of the Lipton chain also required the rationalization of existing junior management structures. This included a thorough reorganization of the shop inspectorate schemes of the previously competing Lipton and Pearks Dairies chains in the region. As a result, the majority of the Lipton network was subsumed into a reorganized south-western division of Pearks' shop inspectorate. The areal extent of this inspectorate had originally encompassed all branches between Penzance and Swindon, but now, although including about the same number of branches, extended only as far east as Teignmouth.[17]

INTRA-URBAN TRENDS

The study of retail change at the intra-urban level offers a different perspective on the development of multiple retailing in the South West. Such analysis reveals the precise locational trends of multiple retail firms and permits an alternative assessment of their impact on the fixed-shop independent sector. It also provides information on the development of local and regional multiples which provided a valuable contribution to their particular retail system but restricted their expansion plans to local and regional horizons.

Many of the largest British cities enjoyed the representation of local multiples. Notable amongst these was the Glasgow-based tea and provisions multiple Cochranes, the branch network of which totalled 110 branches by 1920, all of which remained within the Glasgow city region.[18] Local multiples of this scale were clearly the exception, however, Bristol based firms like Bendall's Stores (ten branches in 1939), Butt's Stores (nine branches in 1939) and

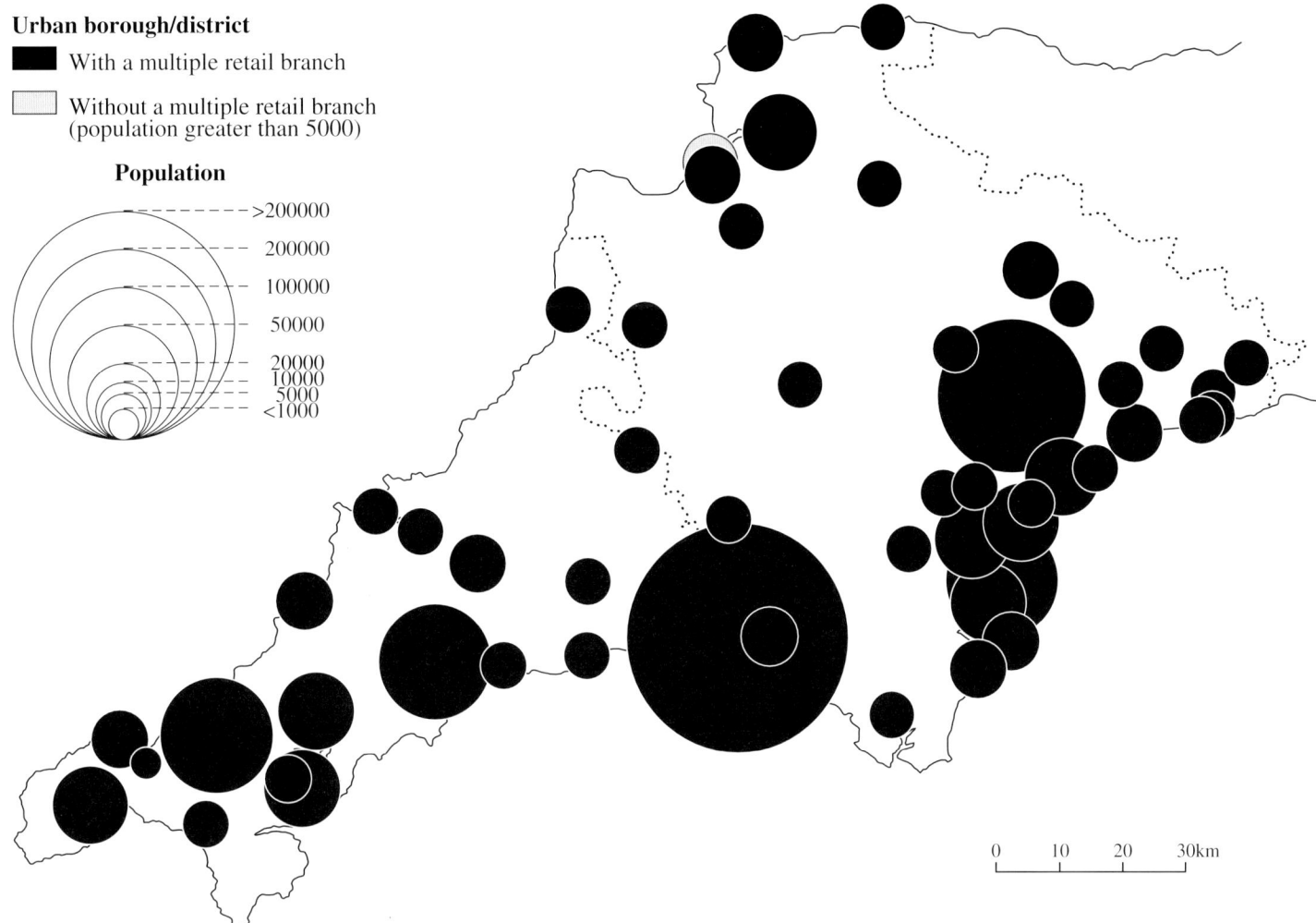

Urban borough/district

■ With a multiple retail branch

☐ Without a multiple retail branch
(population greater than 5000)

Population

>200000
200000
100000
50000
20000
10000
5000
<1000

0   10   20   30km

Map 58.3
The location of multiple retailers in the grocery and provisions trade, 1939.

SOURCES: population data from the Census of Population 1931; retail data from Kelly's *Directory of Devonshire and Cornwall*, 1939.

Map 58.4
The location and opening dates of branches of Lipton Ltd, pre-1939.

SOURCE: Kelly's *Directories of Devonshire and Cornwall*, 1902–35.

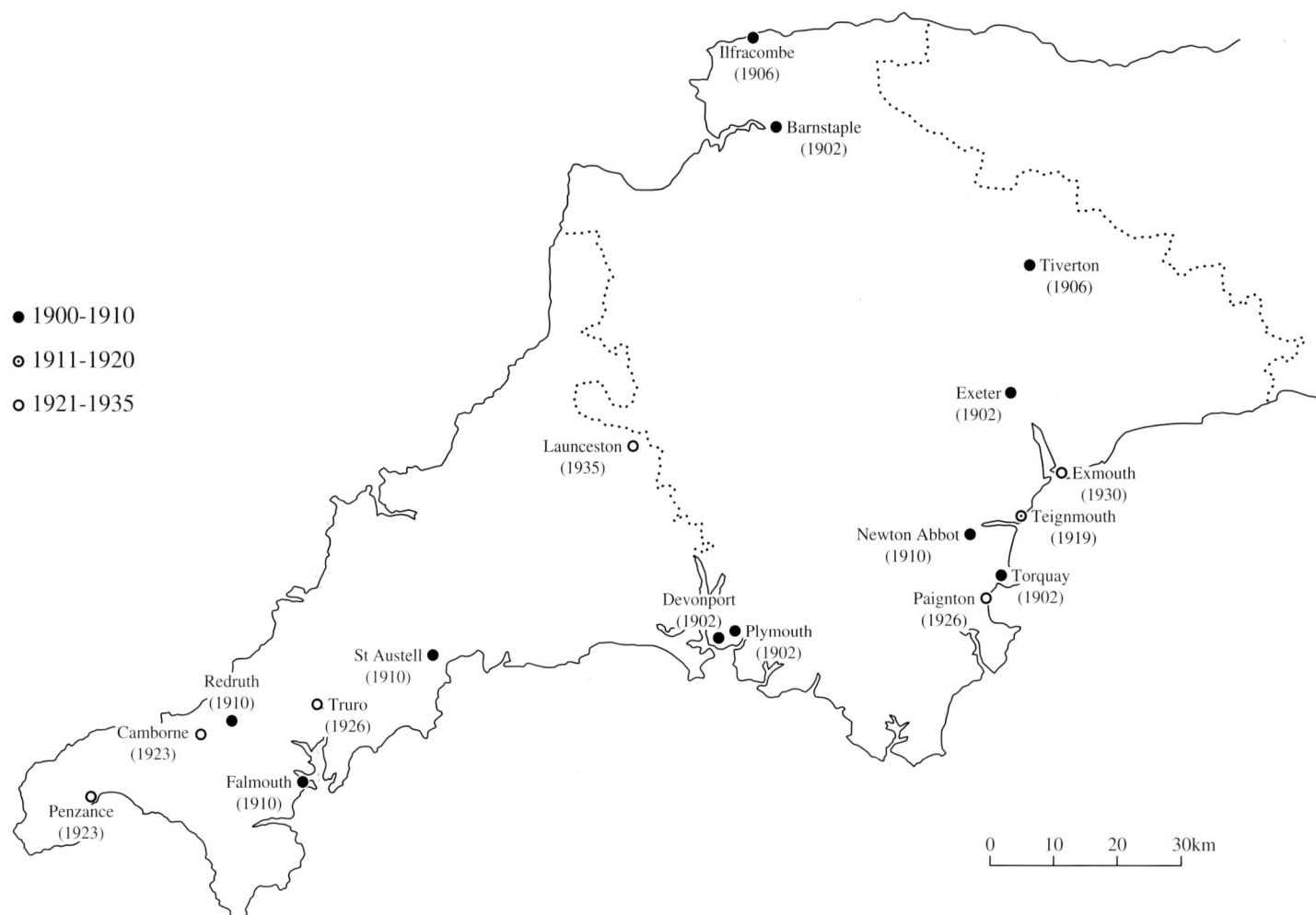

- ● 1900-1910
- ◉ 1911-1920
- ○ 1921-1935

J.H. Mills (seventeen branches in 1939) being more typical. By way of comparison, the largest town in Devon, Plymouth, enjoyed only minimal representation. In 1910, for instance, there were at least six local or regional grocery multiples operating a total of thirty-eight branches in Bristol. These firms augmented the representation provided by national firms and provided a valuable service to customers in the city's expanding suburbs. Available directory data suggest that Plymouth had only one such firm, Underwood & Co. (Plymouth) Ltd, with six branches in the three towns making up the city and two further branches in surrounding towns. This low level of representation was perhaps the result of the city's strong co-operative movement (see chapter 59) and its role as a western bridgehead for expanding national multiples. By the beginning of the 1930s another local grocery multiple had emerged in the city, George S. Dilleigh & Co. Ltd. Although interpretation of the pattern of the firm's branch location is complicated by the unusual history of the city, branches being located in each of the formerly independent 'three towns', outlets at Stoke and Mutley reflect the increasing momentum of branch decentralization in the grocery multiple trade. This

decentralization became a characteristic feature of the evolution of multiple retailing in larger towns and cities during the latter part of the inter-war period. The main emphasis, however, remained on the development and redevelopment of outlets in principal shopping streets, and the business records of multiple retailers reveal a continued concern for the acquisition of prime sites and for economies of retail agglomeration.[19]

GROCERY AND PROVISION DEALERS IN CENTRAL EXETER (Maps 58.5, 58.6 and 58.7)

In the case of smaller towns and cities development remained almost entirely concentrated on centralized sites. **Maps 58.5, 58.6 and 58.7** reveal the identity and location of grocery and provision dealers in five of Exeter's most important commercial streets for the years 1883, 1902 and 1939, respectively.[20] Comparison of the mix of retailers in each street reveals the intensification of multiple retailing and suggests a damaging effect on the city's independent grocers and provisions dealers. Taking the issue of intensification first, data from Kelly's Devon and Cornwall county directories record an increase in the number of multiple retail grocery branches in the city from only three in 1883 to fourteen by 1939. This increase in branch numbers resulted from the expansion into the city of national multiples. Lipton, Home and Colonial, Pearks Dairies and the Maypole Dairy Company all opened their first branches in the city by 1902, the International Tea Company following by 1926. Firms such as David Greig and John Bull Stores also targeted the

city for development in the later years of the period covered by this chapter. An Exeter branch of Greig's was first recorded in the Devon directory of 1923 and John Bull Stores in the 1939 edition.

A further aspect of this intensification process which the directory does not directly record was the simultaneous increase in the size and turnover of branches. In the case of grocery and provisions retailing the inter-war years saw a lessening of the distinction between the two parts of the trade and a consequent

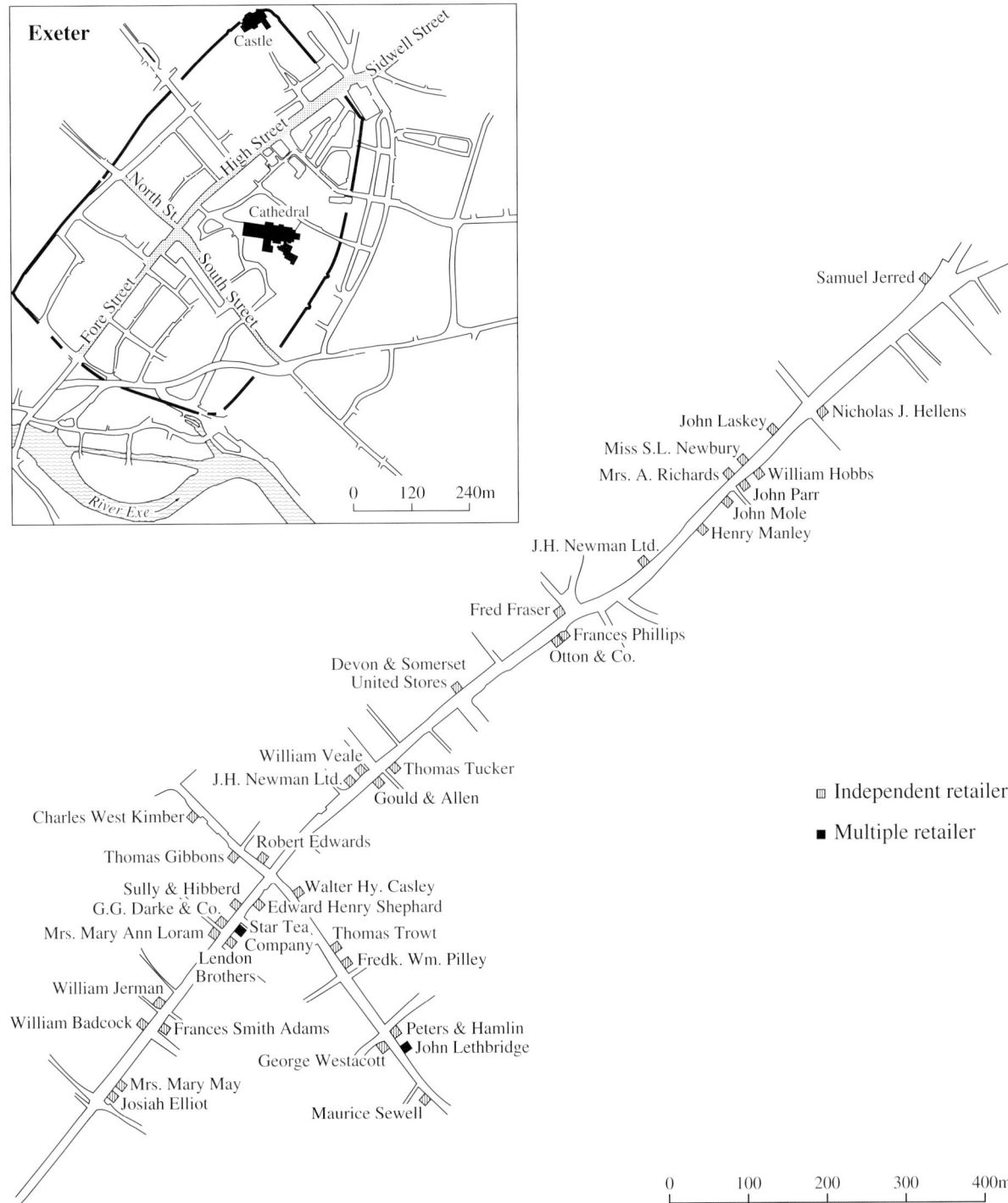

Map 58.5
Independent and multiple grocery and provisions retailers in the main commercial streets of Exeter, 1883.

SOURCE: Kelly's *Directory of Devonshire and Cornwall*, 1883.

Map 58.6
Independent and multiple grocery and provisions retailers in the main commercial streets of Exeter, 1902.

SOURCE: Kelly's *Directory of Devonshire and Cornwall*, 1902.

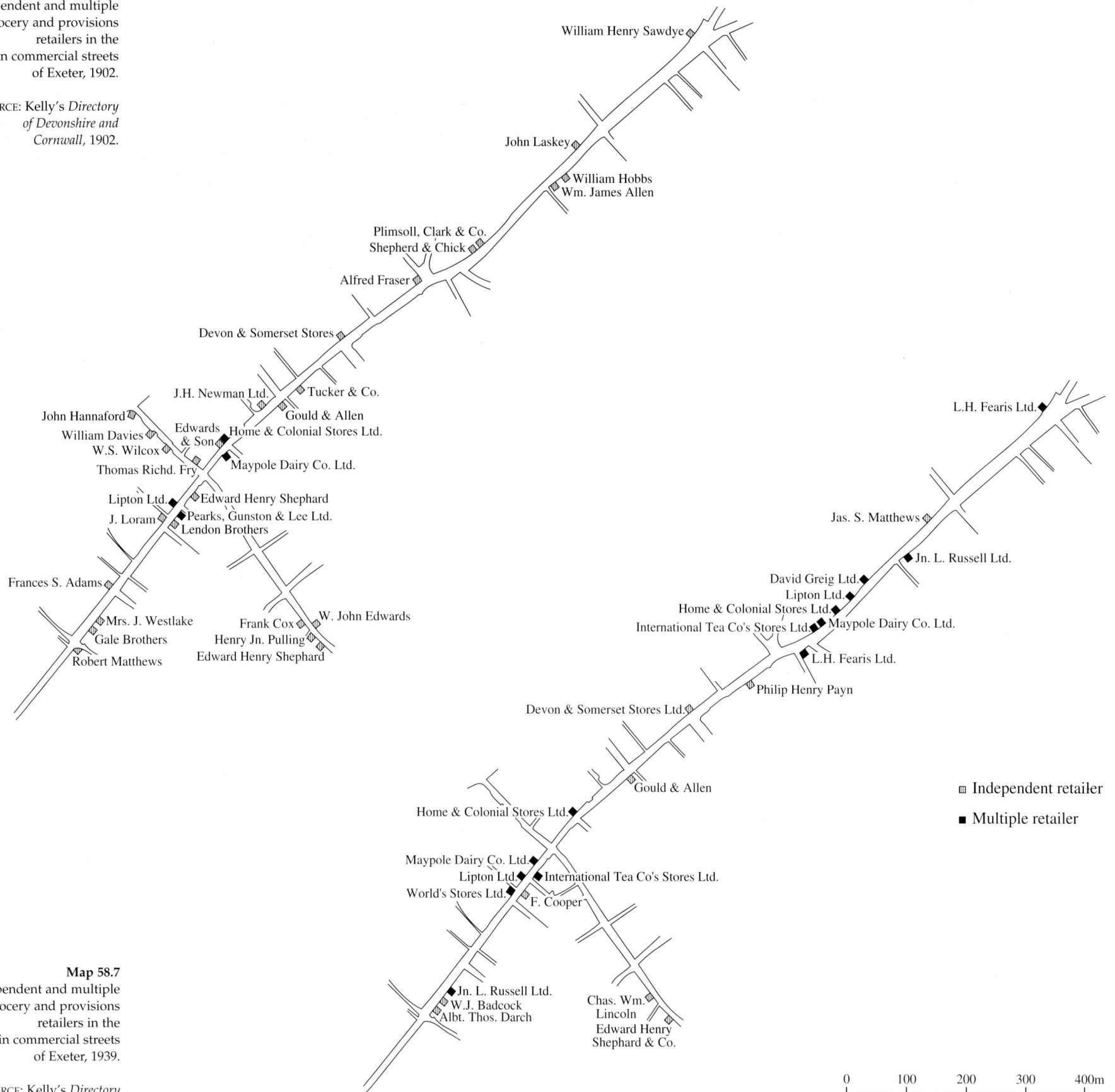

William Henry Sawdye

John Laskey

William Hobbs
Wm. James Allen

Plimsoll, Clark & Co.
Shepherd & Chick

Alfred Fraser

Devon & Somerset Stores

J.H. Newman Ltd.   Tucker & Co.
Gould & Allen
John Hannaford
William Davies   Edwards   Home & Colonial Stores Ltd.
W.S. Wilcox   & Son
Thomas Richd. Fry   Maypole Dairy Co. Ltd.

L.H. Fearis Ltd.

Lipton Ltd.   Edward Henry Shephard
J. Loram   Pearks, Gunston & Lee Ltd.
Lendon Brothers

Jas. S. Matthews

Frances S. Adams

Jn. L. Russell Ltd.
David Greig Ltd.
Mrs. J. Westlake   Lipton Ltd.
Frank Cox   W. John Edwards   Home & Colonial Stores Ltd.
Gale Brothers   Henry Jn. Pulling   International Tea Co's Stores Ltd.   Maypole Dairy Co. Ltd.
Robert Matthews   Edward Henry Shephard   L.H. Fearis Ltd.

Philip Henry Payn

Devon & Somerset Stores Ltd.

Gould & Allen

Home & Colonial Stores Ltd.

Maypole Dairy Co. Ltd.
Lipton Ltd.   International Tea Co's Stores Ltd.
World's Stores Ltd.   F. Cooper

◪ Independent retailer

◆ Multiple retailer

Jn. L. Russell Ltd.
W.J. Badcock   Chas. Wm.
Albt. Thos. Darch   Lincoln
Edward Henry
Shephard & Co.

Map 58.7
Independent and multiple grocery and provisions retailers in the main commercial streets of Exeter, 1939.

SOURCE: Kelly's *Directory of Devonshire and Cornwall*, 1939.

0   100   200   300   400m

decline in the number of specialist multiple retailers. Most firms sought to widen the number of lines they carried in a bid to ensure continued profitability in a time of declining proportional expenditure on foodstuffs. These additional lines required additional selling space and consequently many firms began to replace their existing single frontage outlets with new deeper, double frontage units, using one window and counter for the display and sale of groceries and the other for provisions. Combined with the continued increase in the number of multiple retail branches these measures enabled multiple retailers to extend their share of total national retail sales of grocery and provisions from about 13 per cent in 1915 to about 24 per cent in 1939. In contrast, the share of sales by all independents fell by around one-fifth to about 52 per cent.[21]

Maps 58.5, 58.6 and 58.7 also reveal the continued concentration of multiples in the main commercial streets of the city throughout most of the study period, Sidwell Street in particular being heavily colonized by multiple grocery branches during the course of the inter-war years. The opening of a branch of Pearks Dairies on Cowick Street, St Thomas, and the John Bull Stores branch on New Bridge Street in the late 1930s elongated the existing ribbon-like pattern of development but it remained one of essentially centralized development. This locational pattern, contrasting with the evidence of multiple grocery branch decentralization which has been gathered from the analysis of patterns of retail change in larger cities, reflects the comparative proximity of most suburbs, and hence potential customers, to the Exeter city centre area. One consequence of this continued concentration was an acute reduction in the number and variety of fixed-shop independent grocers and provision dealers trading in the city centre. The extent of the reduction is revealed by comparing **Maps 58.5 and 58.7**.

CONCLUSIONS

Using the multiple retail shop as an example, this chapter has explored the patterns and processes related to the establishment of large-scale retailing in Devon and Cornwall. As such it provides an insight into the nature of retail change in the periphery, a theme which has as yet to receive adequate attention within the literature on retail change. Chief among the particular characteristics identified for the region is the belated emergence and subsequent slow growth of multiple retailing. This resulted from a scarcity of indigenous multiple firms, the region's isolation from the seedbeds of national multiple retailer growth, and the comparatively low levels of urbanization. Consequently much of the region's retail system enjoyed a period of extended institutional stability.

Fig. 58.2
Detail of engraving by W. Deeble after A. Glennie of Molls Coffee House and Spark's Bookshop and Circulating Library, Cathedral Yard, Exeter.

SOURCE: reproduced by permission of Westcountry Studies Library, Exeter (Somers Cocks 1031).

Representation of multiples in the larger towns and cities of the region expanded quite rapidly during the initial wave of development, but towns lower down the urban hierarchy frequently gained their first branch representation as much as forty years later. Similarly, representation in Cornwall compared poorly with that in Devon. The absence of a large number of major urban markets precluded concentrated development during the initial phase of diffusion and, despite the boost given by the tourist trade, resulted in restricted growth throughout the study period. Such variations have important implications for the nature and pace of economic and social change in the periphery, not least because the multiples were agents in the forging of ideologies and patterns of mass consumption. Furthermore, large-scale retail developments were rarely simply passive elements in the process of urban development, more usually playing a dynamic role in the restructuring process.

FURTHER READING

M. Winstanley, *The Shopkeepers World 1830–1914* (Manchester, 1983); J. Benson and G. Shaw (eds), *The Evolution of Retail Systems, c.1800–1914* (Leicester, 1992).

COMMERCE AND MARKETING
1800–1914

CO-OPERATIVE RETAILING

MARTIN PURVIS

As a formal embodiment of self-help initiatives by groups of working people, the activities of nineteenth-century co-operative societies were often varied. Indeed they formed but one part of a wider range of popular initiatives, including friendly societies, trades unions, working men's clubs, religious and educational groups, within which 'connections were made between spheres of activity kept separate elsewhere, such as education and material consumption and production, or sociality and insurance, or entertainment and collective self-help'.[1] All of these were in some way the concern of co-operatives, but in practice many local societies focused principally on retailing. Certainly trading activities offered the best prospect of attracting popular support and generating finance for other initiatives providing education, welfare and entertainment.

Co-operative retailing, a combination by consumers to establish their own stores and distributive channels, is most often associated with northern industrial England. But in fact interest in such developments was widespread throughout nineteenth-century Britain.[2] Thus while Devon and Cornwall were not particularly fertile territory for co-operation, they saw a number of attempts to found societies for this purpose. Indeed, the counties fostered some long-lived associations and, at Plymouth, one of the largest of contemporary British co-operatives.

EARLY INITIATIVES IN SOUTH-WEST ENGLAND

The earliest recorded initiatives concentrated on the supply of bread and flour—a popular response to high prices and heightened fears of adulteration during and immediately after the Napoleonic Wars. These prompted a local friendly society, the Unity and Amity Society of Brixham Quay, to extend its operations. By 1798 it had built its own windmill to produce flour for its membership.[3] More substantial was the flour and milling society formed at Devonport in 1817. This initiative with a retail store in the town and a mill at Ivybridge was established by dockyard workers, probably emulating similar projects in other naval towns.[4] Devonport also had a co-operative coal-purchasing association first registered in 1837. Both survived as independent ventures until the 1890s.[5] Less enduring was

the Stonehouse bread and flour co-operative established in 1847, but it did achieve annual sales of around £12,000 in the early 1850s.[6]

Other early co-operatives were not numerous, with limited south-western participation in the first national wave of enthusiasm for co-operative storekeeping during the late 1820s and early 1830s. However, record survives of local initiatives to establish stores at St Columb and Barnstaple in 1829, and an Owenite co-operative community planned near Exeter during 1826–7.[7] All these projects proved abortive. Subsequent revivals of interest in co-operation also had relatively little impact in the South West. Chartists in the region do not seem to have followed some of their northern counterparts in espousing the principle. Further efforts to stimulate co-operative developments during the early 1850s did prompt interest in the establishment of stores at Ashburton, Barnstaple, Tavistock and Plymouth.[8] It was also reported that several stores were operating in Cornwall.[9] However, as was often the case elsewhere, the revival around mid-century was short-lived.

ACCELERATED GROWTH FROM c.1860 (Map 59.1, Fig. 59.1 and Map 59.2)

The real starting point of modern co-operation in the South West dates, therefore, from the national upturn in co-operative fortunes around 1860. New societies were established in many of the larger centres; there were several separate efforts to add general retail stores to the existing co-operative provision in Plymouth and there were also foundations in Falmouth, Truro, Barnstaple, Tiverton, Exeter and Torquay (Map 59.1). Although not all were successful, the establishment of the Plymouth Mutual Co-operative Society in 1859 initiated a continuous record of co-operative storekeeping in the region. While most of the initial foundations were in the larger towns, the later 1860s saw a cluster of co-operatives established in the smaller metal-mining communities of east Cornwall and more societies were added in this area during the 1870s. Relatively few new foundations were recorded in the late 1870s and 1880s, although co-operation was reintroduced to Exeter in 1884 after an absence of over ten years. New societies were more numerous during the 1890s, especially in south Devon.

But it was not until the 1900s that the proliferation of co-operatives in medium-sized and smaller communities spread into the previously relatively untouched territories of northern Devon and western Cornwall.

Despite the presence of early societies, the growth of co-operation in Devon and Cornwall lagged behind developments in more northerly counties. Membership and sales grew during the initial decades from the 1860s to the 1880s, allowing for fluctuations chiefly associated with cyclical change in the general health of the economy. Yet this early growth was from a low base and was dwarfed by subsequent increases (**Fig. 59.1**).

This growth was not, however, a reflection of significant co-operative progress everywhere. As the map of co-operative membership in 1913 shows, most societies were still of modest size (**Map 59.2**). In some cases this reflected a recent origin and/or the small size of the communities in which they were based. But there were other barriers to co-operative success. There were few counterparts in the South West to the occupationally homogeneous and economically dynamic manufacturing and mining communities amidst which northern co-operation flourished. Even the cluster of co-operatives in the metal-mining villages of east Cornwall remained individually small. This reflected stagnation and decline in the local economy and population, and practices such as the long interval, in some cases up to two months, between payments to miners.[10] This last tended to perpetuate debt, binding consumers to established private retailers, and worked against the 'prudence' in household expenditure on which co-operative success frequently rested. Elsewhere, from Buckfastleigh, there were reports of the survival of truck payments amongst textile workers inhibiting co-operative progress in the early 1870s.[11]

Overall, there were few urban centres or industrial villages focused around a single workplace breeding the collectivity amongst workers that sustained many of the strongest co-operatives in the northern counties. The relative demographic and economic stability of the South West also limited the growth

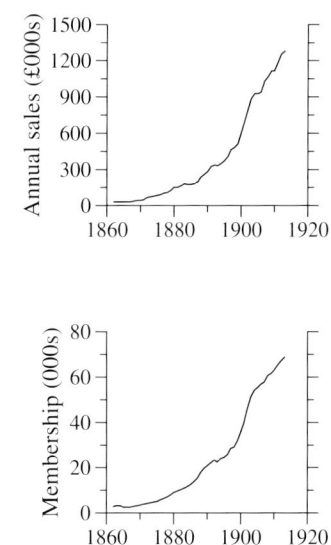

Fig. 59.1
Co-operative membership
and sales, 1862–1913.

SOURCE: as **Map 59.1.**

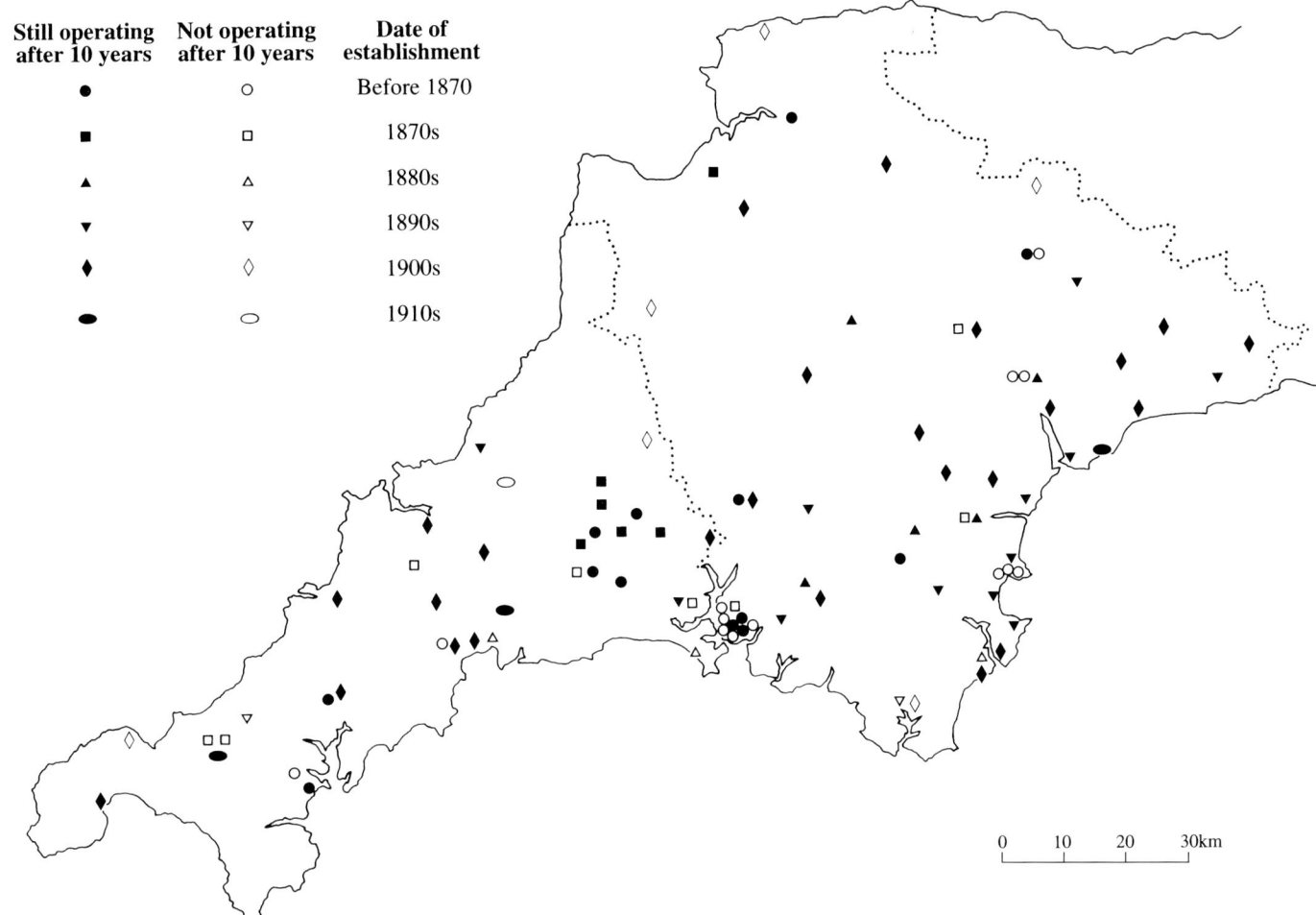

Map 59.1
Co-operative retail societies,
c.1860–1914.

SOURCE: *British Parliamentary
Papers*: Reports of the Chief
Registrar of Friendly Societies,
Industrial and Provident
Societies for the years 1862
to 1914.

of retail demand, so reducing openings for new entrants into the trade. In some northern communities co-operatives could become strongly established relatively quickly through meeting consumer demand that would otherwise have been only inadequately fulfilled; in many southern centres, however, there were not such easy pickings and any co-operative expansion was rendered difficult due to more direct competition with established private shopkeepers. In some towns established traders wielded not only commercial power, but also a degree of social control that could be used to discourage independent initiatives by workers. Thus the overall balance of power between workers and the middle-class establishment was often less favourable to the former in the South West, as compared with districts in the industrial north.

It is clear that experimentation with co-operation was not precluded, but most early societies found it difficult to generate any significant support, several dissolving within a few years of foundation. Thus, co-operation was held in check for much of the nineteenth century. Its later relative success reflects at least in part a new assertiveness amongst workers, perhaps linked to a strengthening of other forms of collectivity including trade unionism.[12] Co-operation in the South West also benefited from the increasing national strength of the movement, which generated a new willingness and capacity to foster the foundation and expansion of societies in areas which had previously resisted co-operative advance. The national federations of the Co-operative Union and Co-operative Wholesale Society were involved in supporting and advising some of the new local societies formed in Devon and Cornwall towards the end of the century.[13] Yet even at a late date co-operation was not invulnerable: for example the St Ives society formed in 1902 had been wound up by 1905 and the long-established co-operative at Barnstaple, never particularly large or successful, lost nearly half its membership in the decade to 1913.[14]

The largest Devon societies in 1913 included those based in significant urban centres: Exeter, Torquay, Newton Abbot and

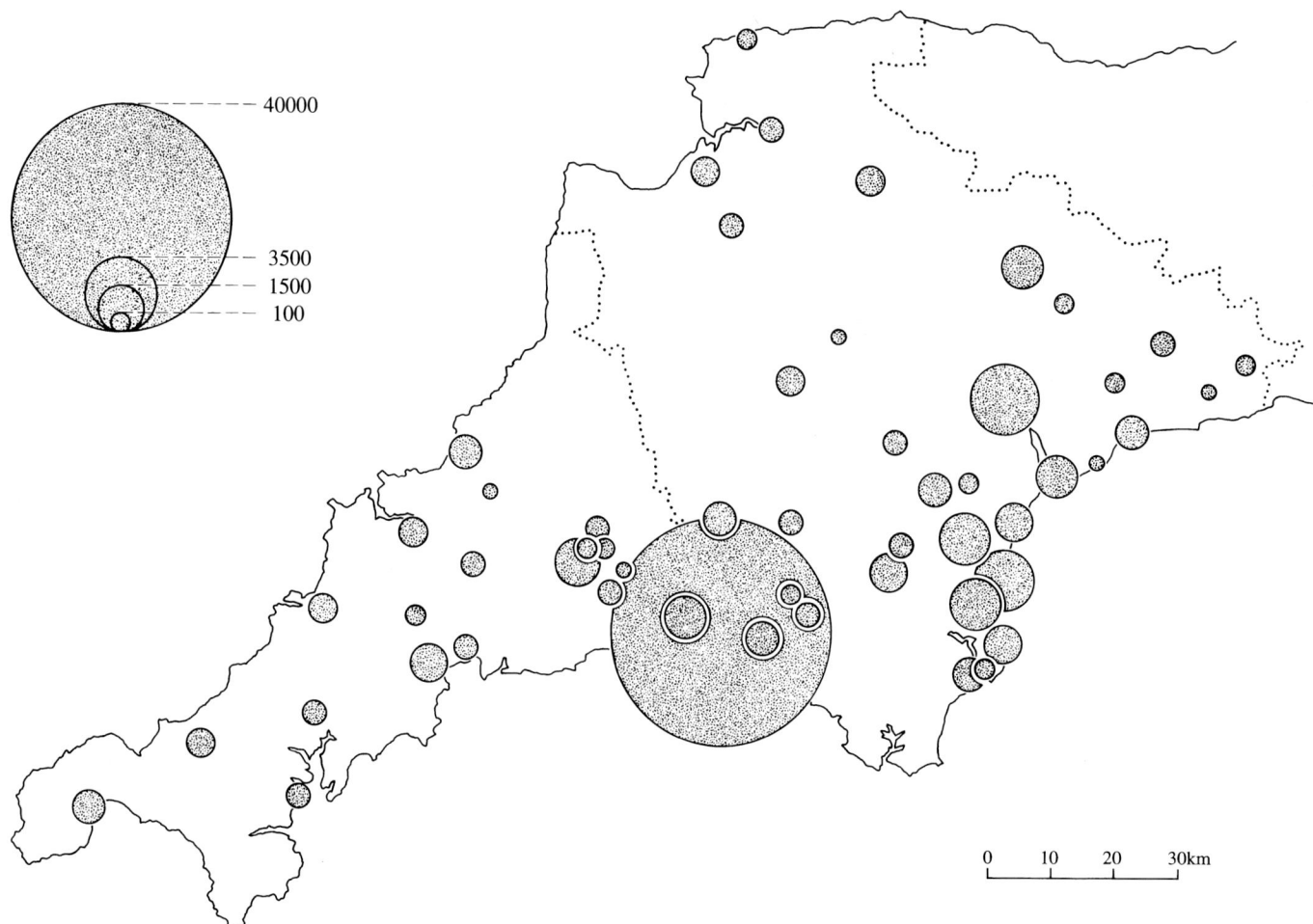

Map 59.2
Co-operative Retail Society membership, December 1913.

SOURCE: *British Parliamentary Papers*, 1914, LXXVI, 422–427: Reports of the Chief Registrar of Friendly Societies for the year ending 31 December 1913, Part B Industrial and Provident Societies.

Paignton (**Map 59.2**). Cornish societies were generally smaller, the most substantial co-operative based in the county being at St Cleer. The dominant presence in south-western co-operation was, however, the Plymouth Mutual Society, whose growth was an extension of the strong local co-operative tradition originally fostered by dockyard workers and other artisans.

THE PLYMOUTH MUTUAL SOCIETY (**Map 59.3**)

The Plymouth society alone accounted for 58 per cent of regional membership and 61 per cent of sales in 1913. Indeed it was of national significance with a place amongst the five largest British retail co-operatives at this date. Its size also rendered the Plymouth society a significant element within the overall retail structure of the town. By 1900 its central premises on Frankfort Street constituted a department store selling a wide range of goods, including tailoring, millinery and furnishings. This contrasted with most of the smaller south-western societies, which concentrated on a more basic range of goods—chiefly groceries and provisions, with footwear, hardware and drapery as the most frequent additions.[15]

Plymouth was the only co-operative in the region to develop a major network of branch stores, again rivalling the larger northern societies. In 1900 most co-operatives in Devon and Cornwall operated a single store only; Torquay and Exeter had respectively one and two branches in addition to their central premises. Plymouth, however, supplemented its substantial main store with twenty-one branches throughout the built-up area, including Devonport, Ford, St Budeaux, Stoke, Stone and Torpoint (**Map 59.3**). It was also the best regional example of co-operative achievement beyond the sphere of retailing; the associated activities included a library and reading room, educational classes, a building society, a penny savings bank, and subscriptions to local hospitals and dispensaries to provide medical care for members. Thus, while overall co-operative activity formed only a relatively small element within the total regional economic activity in the South West in the period up to the First World War, it was locally of some significance and for some individuals its impact touched many different areas of their lives.

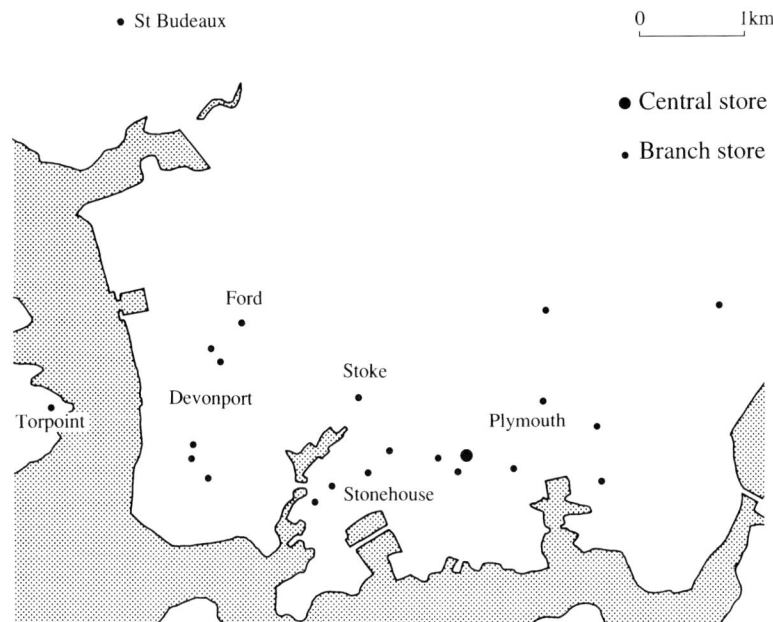

Map 59.3
Branch stores of the Plymouth Mutual Co-operative Society, 1900.

SOURCE: Co-operative Union, *Co-operative Directory for 1900* (Manchester, 1900).

FURTHER READING

Published material relating to co-operation in the South West consists chiefly of the histories of individual societies; these habitually give detailed narrative accounts of local progress but offer little explanatory analysis. Histories of leading regional societies include Exeter Co-operative and Industrial Society Ltd, *History of the Society 1885–1935* (Plymouth, 1935) and R. Briscoe, *Centenary History: A Hundred Years of the Plymouth Co-operative Society. A Hundred Years of Progress, 1860–1960* (Manchester, 1960). A wider range of brief histories of societies throughout the South West is given in Co-operative Union, *Souvenir of the Thirty-Fourth Co-operative Congress, held at Exeter 1902* (Manchester, 1902) and Co-operative Wholesale Society Handbook Committee, *Plymouth: A Handbook of the Forty-Second Annual Co-operative Congress, Whitsuntide 1910* (Manchester, 1910). For a more recent account of developments in Devon and especially Exeter see J.H. Porter, 'Co-operative experience in nineteenth-century Exeter and Devon' Part I, *Devon and Cornwall Notes and Queries*, 36 (1988), 113–18; Part II, 36 (1989), 159–66.

RETAIL DEVELOPMENT
AND THE
CHANGING SHOPPING HIERARCHY
IN THE LATE TWENTIETH CENTURY

GARETH SHAW AND ANDREW ALEXANDER

RETAIL CHANGE

Retail development has been characterized by rapid and dramatic change during the second part of the twentieth century. In terms of the geography of such changes two main locational themes can be identified. One is associated with the development and refurbishment of town centre shopping environments, while the other is most clearly represented by demands for new retail locations out of town. Significantly the pace of these locational trends has varied over time and space, affecting different regions in different ways. In this context the South West has often been relatively late in receiving the products of such changes, a fact that has important repercussions on both the nature of retail developments and their impacts on existing shopping centres.

Before discussing these developments and their spatial patterns, it is important to recognize that changes in retailing have been brought about through a complex interplay of different forces. It is not necessary to discuss all of these in great detail, but we should at least be aware of the three major ones, namely: changes in consumer demand, structural and organizational changes in retailing, and the nature of retail planning. Of the many changes in shopping habits perhaps one of the most significant has been the increased use of the car.[1] National surveys have revealed the importance of car-based shopping trips, with some showing that almost a third of shoppers used out-of-town retailers and cited cheaper and easier parking as important factors.[2] Of course such processes are selective and tend to discriminate against the elderly and unemployed, whose use of out-of-town retail facilities is far less than that of other shoppers.[3]

The most dynamic element of the retail scene has been within the structural and organizational elements of retailing itself. At a very broad scale there has been a shift from a production-led to a market-led economy which has changed and expanded consumer-related activities. As Dawson points out, this has produced new space for retailers through a strong process of property development which has geared particular developments to specific retailers.[4] Such processes have brought into play large financial institutions that have provided capital for new types of retail schemes.[5] Within retailing itself, structural changes have occurred as large organizations have taken a stronger hold on the British market, thus between 1981 and 1986 the number of small, independent retailers fell from 375,000 to 217,000. Over the same time the share of retail sales from large organizations (those with over fifty stores) grew from 24 per cent to 52 per cent. Such large organizations have introduced change at the store level, first in the 1950s with self-service methods and since the late 1960s with larger stores. The superstore started in grocery retailing, but now encompasses DIY, furniture and electrical goods. In all retail sectors the share of sales through such large stores has increased from 3 per cent in 1981 to 11 per cent by 1990. In grocery retailing, superstores account for 25 per cent of all sales.[6]

The third and least dynamic of the forces acting on retail change is that of the planning system, which has operated in both positive and negative ways. In general the most positive aspects of planning have been directed at creating new and attractive retail environments in town centres, while more negative mechanisms have attempted to control the growth of out-of-town stores. Planning perspectives have also been strongly influenced by the attitudes of central government, which has moved in the last twenty years from a position of strong control over out-of-town retailing, through to a recent period of relative laxity, back most recently to one of control. Such changing attitudes have had an influence on the way that retail developers have perceived the merits of different locations.

RETAIL DEVELOPMENT AND PRESSURES ON THE SHOPPING HIERARCHY
(Maps 60.1 and 60.2)

Throughout the post-war period, but more especially in the 1960s and 1970s, retail planning has largely focused on two main themes. The first is the desire to relate retail supply to changes in consumer demand, whilst the second is the preservation of a clear recognizable hierarchy of shopping centres.[7] It is the second of these aims that has proved most problematic in the face of new forms of retail development within the South West. The county structure plans provide

Map 60.1
Shopping hierarchy, *c.*1980.

SOURCE: structure plans for Cornwall and Devon.

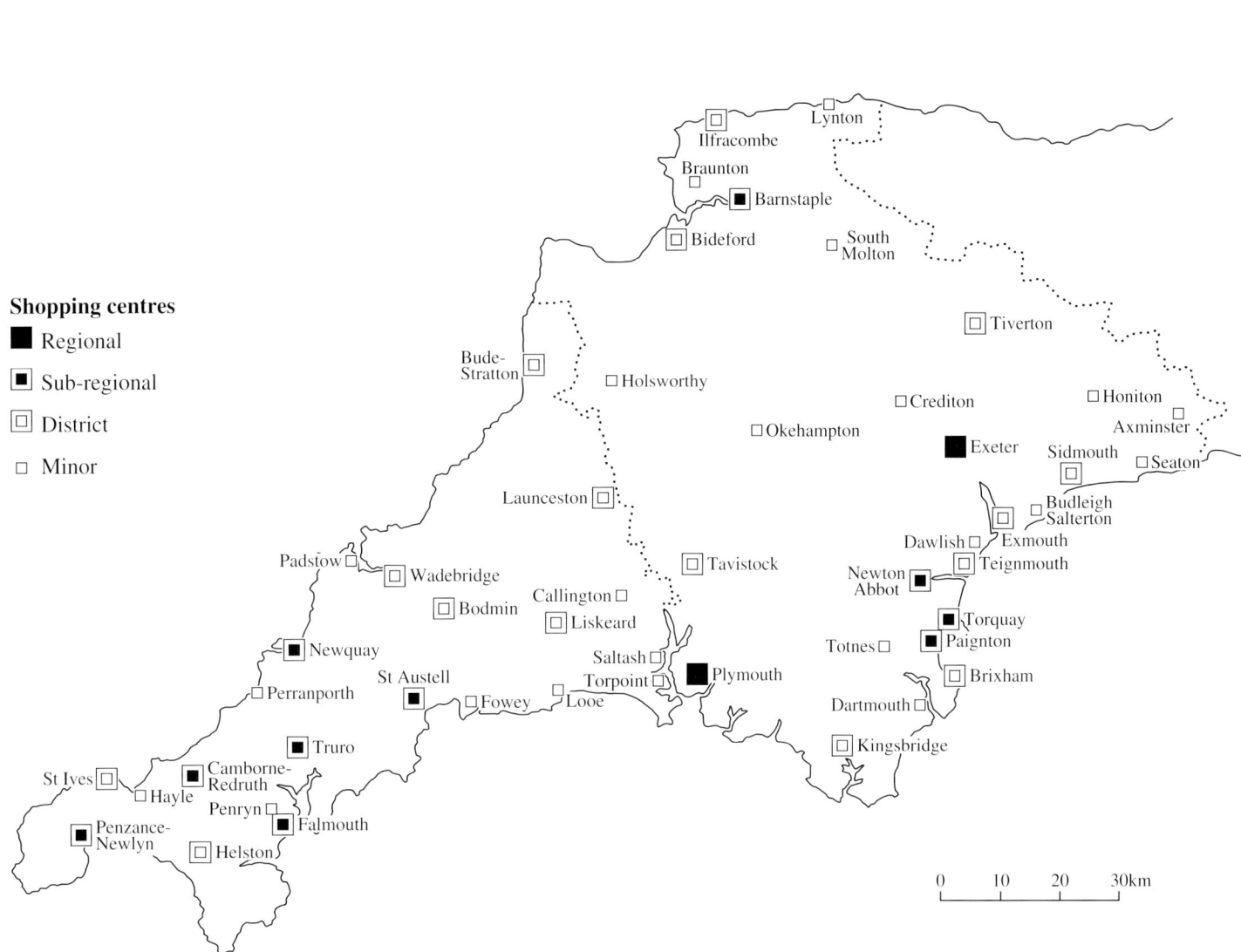

Shopping centres

■ Regional

■ Sub-regional

▣ District

□ Minor

a clear expression of the established shopping hierarchy as shown in **Map 60.1**. From this we can see the region encompasses a wide range of shopping environments, ranging from regional centres such as Exeter and Plymouth, through to a wide range of district and minor shopping centres. The region therefore presents both the planner and the developer with a variety of planning problems and development opportunities. It is also an area that has recently experienced large increases in consumer demand as a result of both social and demographic changes. Such factors coupled with the increasing scale of retailing have placed strong pressures on the traditional shopping centres.

The pressure on these centres has come through what Schiller has described as a series of 'waves' of retail decentralization.[8] The first of these started in the 1960s and concerned the demands from food retailers for out-of-town sites for new superstores. The second wave was more associated with stores selling household goods, including DIY and electrical products, and began during the 1970s. The third and latest wave started in the mid-1980s and

concerned the sale of comparison goods, such as clothing, footwear and toys.

Within the South West each of these decentralization waves has been experienced, though often at some time after the main national trends. For example, most of the early developments for large grocery superstores came during the 1970s, when the core metropolitan regions had already started to experience the onset of the second wave. Such a time lag may also be detected within the region, with superstores being initially attracted to locations around the major centres of Plymouth and Exeter, before taking up locations around smaller towns or moving into the far west of the region (**Map 60.2**). Thus, the earliest out-of-town superstores in Devon were a 30,000 sq. ft (net floorspace) development at Lee Mills near Plymouth and a 30,000 sq. ft (net floorspace) at Broadclyst east of Exeter. As **Map 60.1** suggests, many of the smaller shopping centres within the region were targeted by superstores at a time when the debate on 'out-of-town' retailing had moved to focus more on the implications of the second and third waves of change. This spatial trend was partly associated with successful planning policies by Devon and Cornwall that attempted to integrate the demand for grocery superstores into the needs of local areas by making them act as district shopping centres, and partly because of the expansion programmes employed by large-scale retailers. The use of superstores as so-called district centres has been most successfully employed within Plymouth, where such stores are located at Roborough, Estover, Plympton and Plymstock.

In the period after 1980, the region has attracted far more attention from large retail organizations as other core regions became somewhat overprovided with outlets and companies were looking for new development opportunities. Such expansion focused not only on the large centres such as Exeter and Plymouth, but also targeted smaller towns. During this period Cornwall, for example, experienced a rash of 'out-of-town' superstores at all points of the shopping hierarchy as shown in **Map 60.2**.

THE SECOND WAVE OF RETAIL DECENTRALIZATION

The late 1980s and early 1990s also saw the region experience the full impact of the second wave of retail decentralization associated with DIY, electrical goods and furniture. This again has largely focused on the larger urban areas, with large-scale retail warehouses being developed on the edges of Plymouth, Exeter, Torbay, Newton Abbot, Truro, Camborne–Redruth and St Austell. As **Map 60.2** shows, the timing and spatial extent of these developments has been far quicker than those of food retailers, in that they have spread to small centres in a much shorter period of time. The main reason for this is that such retail warehouse developments have largely faced much less opposition from local planning departments. This is because the site and locational requirements for these types of stores were very similar to those earmarked for employment development by local planners. Consequently, there was little resistance from planners when retailers wanted to move into those sites where employment

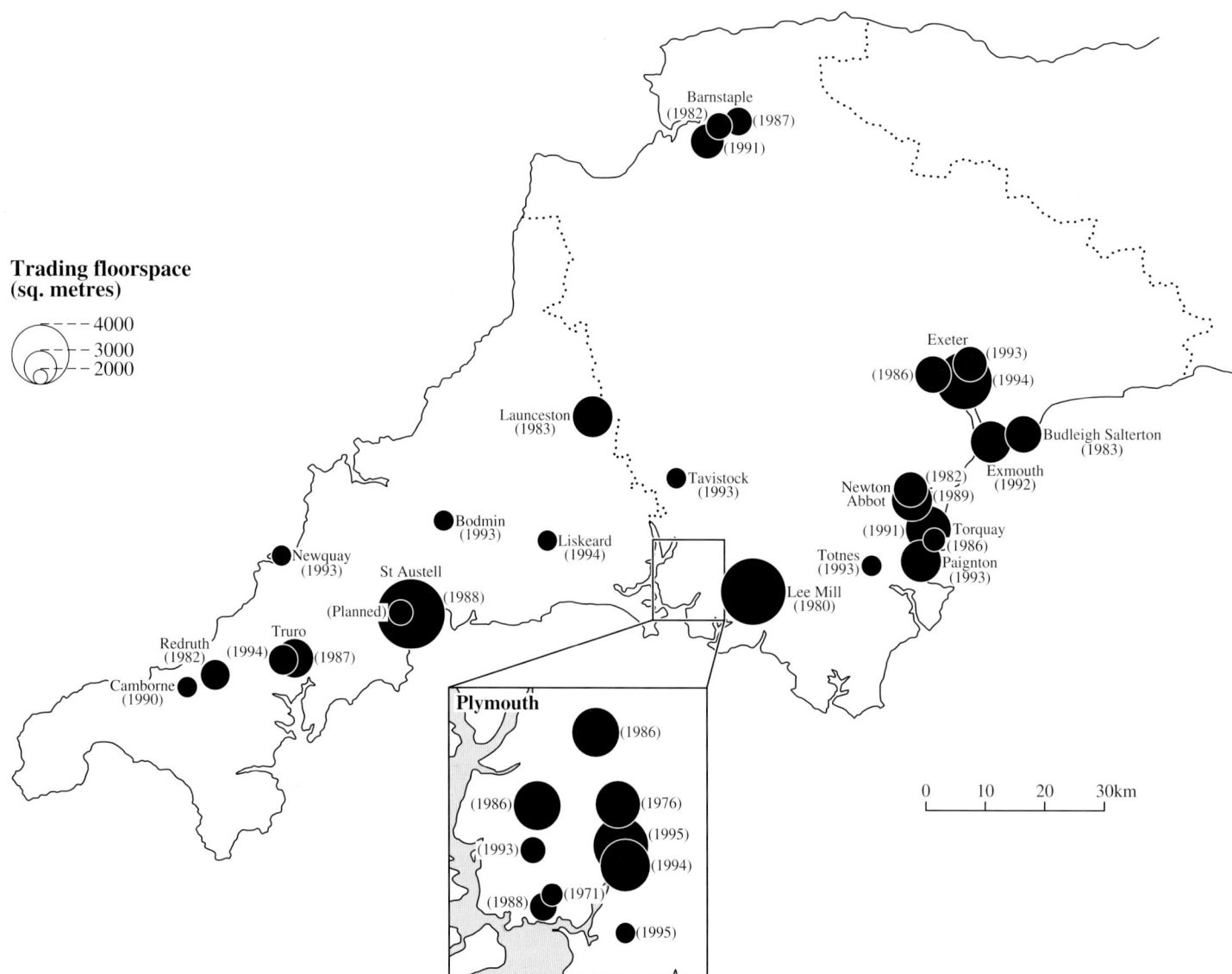

Map 60.2
Early superstore development.

SOURCE: *1994 Register of UK Hypermakets and Superstores.*

land allocations could not be achieved by manufacturing or office developments. In addition, planners had increasingly come to recognize that town centres could no longer act as feasible locations for these bulk goods type stores and that edge-of-town sites provided a better solution.

Such views have prevailed generally, although within the region there are important examples of local authority opposition to these retail warehouses. A key case is that at Sowton, east of Exeter, where a proposal for a large CRS Homeworld superstore was opposed by the district and county planning authorities in 1985. At the public inquiry the local authorities lost their case and the store was approved on the grounds that Sowton provided an acceptable site for such developments. This came at a critical time in the region's retail development in that the third wave of retail decentralization, that of comparison goods, was focusing on the South West. The developments associated with this wave of change can only work if new out-of-town regional shopping centres are constructed, which of course hold much larger threats for town centres than any previous retail developments. More significantly, such proposals were taking place against a weakening planning background, if not locally, then certainly nationally as central government appeared to give a green light to these regional centres. In addition, there was a changed economic climate which weakened the fiscal base of many local authorities, who in turn were searching around for new ways of raising money. In such circumstances some local authorities were being tempted into considering large out-of-town retail schemes to solve some of their financial problems and improve local employment.

RETAIL DEVELOPMENT PROPOSALS IN THE EXETER AREA
(Map 60.3 and Fig. 60.1)

Exeter provides a focal point of proposals for a new retail centre following the approval of the CRS development in the mid-1980s. The judgment had the effect of stimulating a wide interest in the area as a potential location for a regional centre and it was only a short time after this that such interest was translated into a number of formal applications. In all there were some eleven applications for out-of-town centres around Exeter, ranging from 650,000 sq. ft of floorspace through to smaller developments of around 70,000 sq. ft (**Table 60.1 and Map 60.3**). Before discussing the case of Exeter in a little more detail, it is worthwhile considering the speculation process that produces the type of spatial clustering so evident in **Map 60.2**.

This process may best be described as one of 'cumulative speculation'. Evidence from Exeter suggests that the process starts when first one and then several developers become interested in sites within the same area (**Fig. 60.1**). The process of speculation takes off extremely quickly as different developers follow each other in submitting planning proposals. The net result is to produce a strong focus of development interest around one urban area, as is shown for Exeter during the late 1980s (**Map 60.3**).

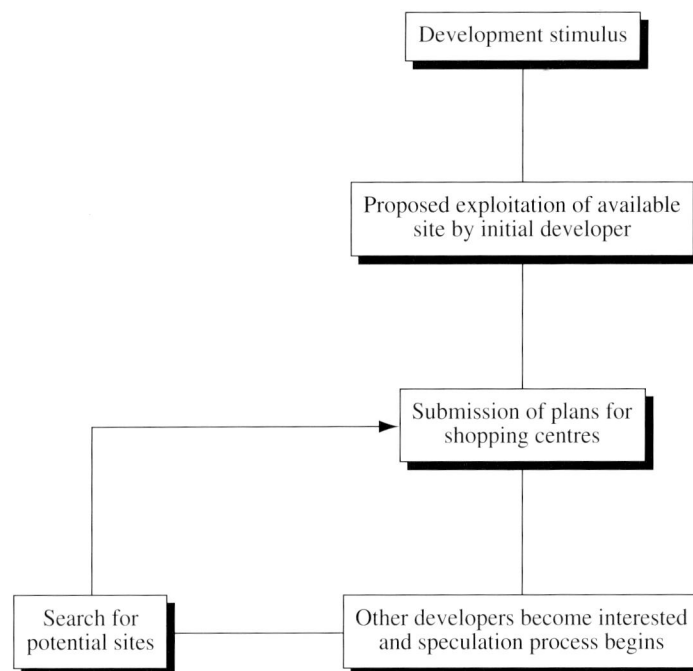

Fig. 60.1
Model of shopping centre development decisions.

Developer	Site	Size (000 sq. ft)
Cameron Hall (Metro West)	Marsh Barton	400
E.B.C.	Sowton	406
Deepdene	Exeter Airport	650
Shearwater/South West Water Authority	Digby	650
Carkeek Developments	Alphington	550
Baker Property Services	Pinhoe	350–400

Table 60.1
The main proposals for large shopping centres around Exeter, 1985–1986.

SOURCE: Planning applications.

Centre size	Estimated trade diversion (%)	Commercial viability
1 million sq. ft	24–28.5*	not feasible
550,000–600,000 sq. ft	15–19.5	sufficient demand
450,000–550,000 sq. ft	11.5–16	feasible
350,000 sq. ft	8.5–12.5	feasible
below 350,000 sq. ft	5–8.5	not viable

Table 60.2
Viability and impact of proposed regional shopping centres in Exeter.

SOURCE: * Estimated by authors; all other data from Exeter Planning Department, 1986.

Such a process also greatly undermined the existing planning policies in Exeter and caused the local authority to reassess its attitude towards out-of-town retailing. Thus, the original thrust of the planning policy, as expressed by the City's Development Committee, stated that 'shopping policies in Exeter be summarized as promoting city centre expansion and opposing out-of-town superstores'.[9] However, under the pressure of a wave of planning applications for out-of-town centres following the CRS appeal, the local authority decided to adopt a more pragmatic approach to retail planning. The new strategy involved the local authority deciding, on the basis of a consultant's report, to support one of the applications for a shopping centre. This, it

Map 60.3
Retail development
proposals in the Exeter area.
See also **Table 60.1**

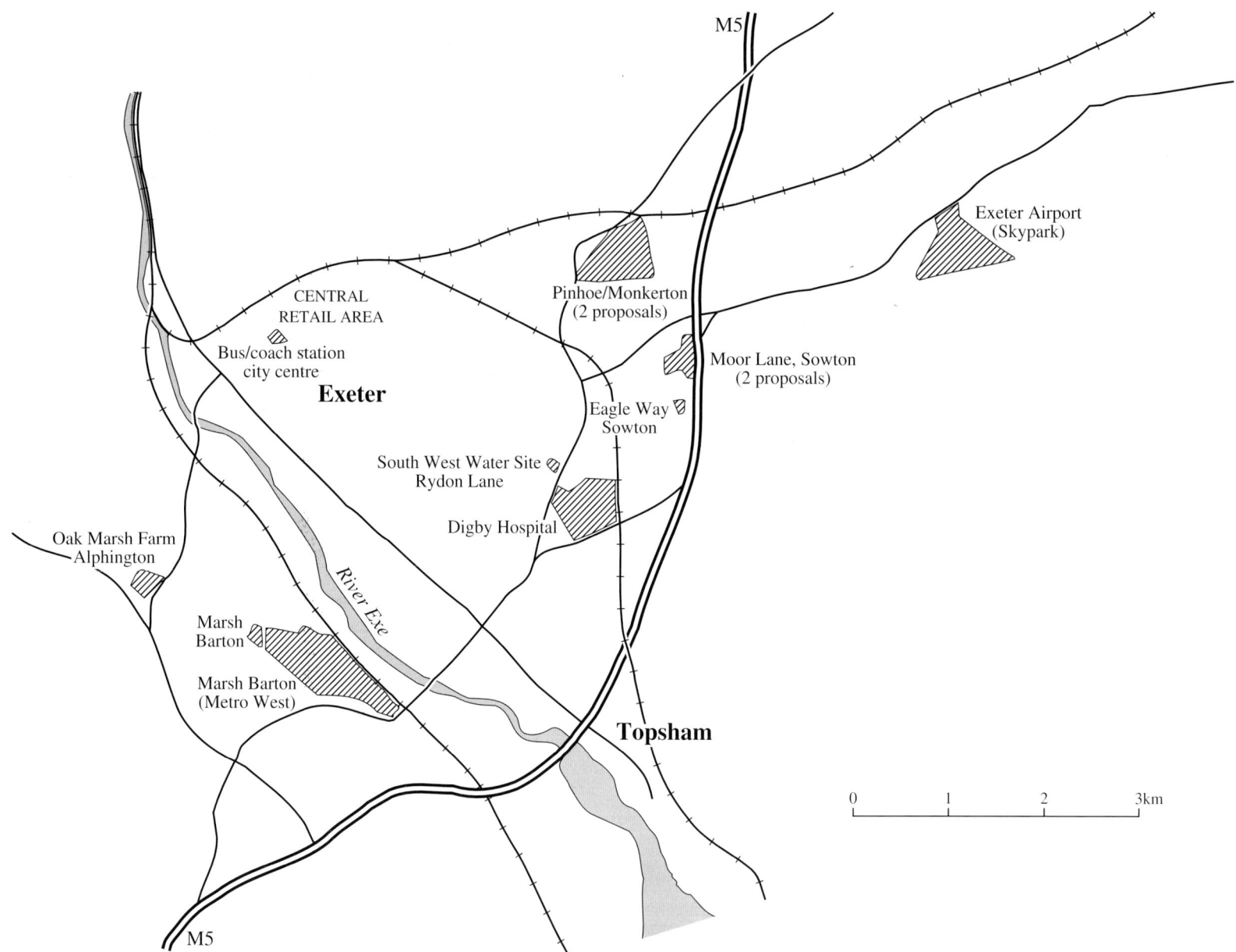

was hoped, would have two important effects. First, it should allow some local control over future shopping developments, rather than leaving any decision to a planning appeal process; and second, it should halt further applications.

As a prelude to this changed approach, Exeter City Council commissioned a survey to examine the commercial viability and likely impact of different types of out-of-town centres. The survey encompassed a range of options and proposed developments as shown in **Table 60.2**. The consultants were given the task of identifying an acceptable shopping centre, both in terms of minimal impact on Exeter's central area and the centre's commercial viability. This type of approach suggests that there is an ideal type and location for a large, out-of-town shopping centre in an environment such as Exeter, a suggestion that seems to fly in the face of existing evidence.[10] In the view of the consultants, a development proposal of between 350,000 sq. ft and 400,000 sq. ft of retail floorspace would have the least impact on the city whilst at the same time be commercially viable. This partly explains the local authority's support for the Metro-West scheme, which fitted some of the criteria (**Table 60.1 and Map 60.3**).

By following this approach local authorities like Exeter are suggesting that large, out-of-town centres can complement existing shopping areas in at least three broad ways. The first concerns transport links, and the view that such new developments should be sited to complement rather than compete with the town centre and that out-of-town centres should provide adequate public transport links. The arguments over complementarity are strongly related to the second area of concern, that of tenant mix. This is perhaps the most problematic for local authorities to control, and past evidence on tenant mix policies in major shopping developments show strong tendencies toward major multiples. Finally, complementarity would also have involved shifting the nature of Exeter's central area more towards so-called leisure shopping, as the out-of-town centre would have taken on more of the city centre's traditional retail functions.

Fortunately for the long-term health of Exeter's city centre a national recession, a decline in retail property speculation, and a change in central government's attitude to out-of-town retailing all served to halt such proposals. Perhaps of greatest significance for Exeter and indeed other town centres within the region is the government's recognition of the need to support city centre retailing and to start to redress the balance between out-of-town developments and the traditional shopping hierarchy.[11] The full implications of this changed attitude will be discussed in the concluding part of this chapter.

TOWN CENTRE RETAILING

Between 1981 and 1986 surveys within Devon have shown that retail floorspace in central areas increased by 120,000 sq. ft, from 5.1 million sq. ft to over 5.2 million sq. ft. However, a closer inspection of such data reveals strong structural differences, with DIY, furniture and carpet retailers experiencing a decline in floorspace in most town centres. In Exeter, for example, there was a decline of 18,000 sq. ft of DIY goods floorspace between 1981 and 1986, while floorspace for food retailing only remained static in the face of competition from out-of-town stores. Perhaps most worrying for the health of the region's town centres is the fact that of their lost retail floorspace, some 47 per cent did not stay in retail use but was either vacant or converted to office activities.[12]

However, set against these shifts in retail floorspace and the overall competition from edge-of-town and out-of-town retailing, many town centres within the region have experienced investment in new shopping facilities. In general these schemes have tended to filter down the urban hierarchy, with the earliest ones being in Plymouth and Exeter. Once again such purpose-built shopping centres came relatively late to the region when compared with national trends, which according to Davies and Bennison started in the early 1960s.[13] Thus, in Plymouth since 1971 there have been three new major shopping schemes developed within the central area.[14]

Exeter, in considerable contrast to Plymouth, utilized pedestrianization at an early date in its central area planning with the development of Princesshay. The city also constructed a much larger purpose-built shopping centre, the Guildhall Centre (some 180,000 sq. ft of retail floorspace), in 1976. This is located adjacent to the city's major central conservation area and as such it acted as a powerful magnet attracting other retailers to locate in that part of the city.

Perhaps some of the most significant improvements to town centre shopping have taken place in the region's medium-sized towns. Thus, in Cornwall, Truro has greatly enhanced its retail potential through the construction of a new shopping complex based around a large Marks & Spencer store and new in-town

grocery superstores accounting for some 20,000 sq. ft of retail floorspace.[15] Elsewhere within Cornwall superstores have found ready locations as part of town centre development schemes, the largest ones being in St Ives (20,000 sq. ft), Newquay (20,000 sq. ft) and Bodmin (22,000 sq. ft). In many cases such stores have formed the focus of shopping improvement schemes that have included other smaller units and new parking facilities.

Within Devon towns there has been a similar development of grocery superstores, although during the last decade there has been an increasing number of shopping centres built. Within the sub-regional level shopping centres identified in **Map 60.1**, there has been a rash of such schemes. Thus, Torquay with its major tourist function has seen the construction of the Haldon Centre and the more tourist orientated 50,000 sq. ft Fleet Walk development. Similarly, nearby Paignton has opened the Victoria Centre adding mainly to the tourist goods floorspace. At the district and minor centre level of shopping provision, new developments have also occurred, though these tend to be characterized by small, open-air shopping complexes containing mainly small units, as in the Brutus Centre, Totnes.

There are exceptions to these in-town developments in that significant sub-regional level centres such as Barnstaple and Newton Abbot have not had any major new facilities added. In the latter town the main retail developments that have taken place have all been outside the central shopping area and the centre is now under considerable competitive pressures from the newer centres in nearby Torbay.

CONCLUSIONS: TOWN CENTRE OR OUT OF TOWN?

Quite clearly the retail geography of the region has changed markedly during the last forty years. In very broad terms the forces of change have swung away from established retail cores and the most dynamic trends have been for retail decentralization as large retailers have sought to find more 'economic' locations. The pressures for such developments have swept across the region from east to west and down the urban hierarchy usually some time after national trends. Even so, the impacts on shopping patterns and established retail centres have been significant. Planning policies within the region have in the main supported the established shopping hierarchy, although during the mid to late 1980s a combination of circumstances sought to undermine such policies. Had such trends continued then the region would have witnessed even greater changes in the way many of its town centres operated.

FURTHER READING

G. Shaw, 'Retail development strategies and the British shopping hierarchy', in G. Johnson (ed.), *Business Strategy and Retailing* (Chichester, 1987); C. Guy, *The Retail Development Process* (London, 1964).

61

EXETER AND PLYMOUTH

THE CITY OF EXETER
FROM AD 50 TO THE
EARLY NINETEENTH CENTURY

C.G. HENDERSON

ISCA ROMAN FORTRESS AD c.50–75 (Maps 61.1 and 61.2 and Figs 61.1 and 61.2)

Little evidence of pre-Roman activity has come to light in the area of the Roman and later city, although High Street may approximate to the line of a prehistoric trackway leading down to a ford across the River Exe. The first settlement of any size at Exeter dates from around AD 50–55, when the Roman army founded a fortress (probably named *Isca*, from the Celtic name for the River Exe) on a rectangular site of about 16ha (40 acres) occupying a broad, sloping spur overlooking the early river crossing. This served as the main winter base of the Second Augustan Legion for a period of about twenty-five years. The fortress contained accommodation for perhaps 6,000 legionaries and cavalrymen, making it one of the largest settlements in Britain at this period.

With the single exception of the stone bath-house, discovered in 1971 to the west of the cathedral, the buildings in the fortress were all constructed in timber, wattle and daub. The base contained at least seventy barrack blocks, each housing about eighty men, as well as stables, store buildings, smiths' and carpenters' workshops, a hospital, headquarters building, commanding officer's residence, and houses for other senior officers. Stone for the bath-house was obtained from basalt quarries in the flanks of Rougemont hill, immediately to the north of the base. An aqueduct brought water to the baths from a spring to the north-east of the fortress in the St Sidwells area.

The defences comprised a timber-revetted rampart surrounded by a large ditch; watch-towers stood at 30m intervals along the rampart, and imposing timber gatehouses marked the entry points on each of the four sides. A wide zone outside the fortress was occupied by ancillary compounds, civilian settlement areas, cemeteries, and extractive and industrial sites. The main streets in the fortress were metalled with gravel, as were the approach roads outside. To either side of the road leading to the north-east gate there is thought to have been a settlement for civilians who provided goods and services to the soldiers; this area also contained the legionary tilery and pottery workshops. The main road leading to *Isca* from the east ran up to the eastern corner of the defences before turning to run alongside the ditch to the north-east gate and beyond. The north-west defences overlooked the steep-sided Longbrook valley, beyond which lay a relatively level hilltop (the modern St David's Hill area) which was possibly the site of a cemetery. An annexe, enclosed by a bank and ditch, was appended to the south-west side of the fortress, overlooking the river, whilst on its south-east side an area between a minor combe and the Shutebrook valley was occupied by an embanked and palisaded compound containing timber buildings.

The main port facilities serving the fortress and its hinterland lay at the head of the estuary on the north-west side of the later port of Topsham. It is likely that sea-going ships unloaded their cargoes here for onward carriage by road; Holloway Street and Topsham Road probably date from this period. Other roads must have approached the river crossing at Exeter from the west and south, through modern St Thomas and Alphington.

THE EARLY ROMAN TOWN ISCA DUMNONIORUM c.75–150 (Map 61.3)

Following the redeployment of the Second Legion to South Wales, around AD 75, their base at Exeter was dismantled and its site appropriated for the foundation of a new town, *Isca Dumnoniorum*, which served until the early fifth century as the commercial and administrative capital of the south-west peninsula. This was the only major Roman town in the Dumnonian tribal territory—represented by Cornwall, most of Devon, and part of western Somerset. The timber buildings in the fortress were probably all taken down when the legion vacated the base, but the stone bath-house, initially reduced in size, continued in use for a few years. In the early to mid 80s the bath-house was demolished in order to provide a site for the forum (civic centre) of the new town, although some of its external walls were retained for incorporation into the basilica (town hall). This latter building contained a council chamber, a large market hall, offices for civic officials, the town shrine and the magistrates' court. In front of the basilica was the forum-piazza, a wide courtyard enclosed by long ranges of shops and offices, and beyond this, to the south-west, an open market place, probably for livestock. Construction of the forum was probably completed in the early 90s, when work would have started on the public baths, which occupied a site to the south of the forum. An open contour leat, built in 100–101, conveyed spring water to the baths.

The rampart enclosing the site of the former legionary fortress was retained for up to a century after the town's foundation, albeit only in the form of a grassy bank marking the formal limit of the urban area. There seems to have been no need for the town to be defended at this period, and the timber revetment of the fortress rampart must rapidly have fallen into decay, while the ditch became partially silted up. By the middle of the second century, scattered extra-mural settlement is attested on all sides of the early Roman town. Within the town proper, most of the main streets of the legionary base were retained, while some new ones were laid out in the late first century to complete a simple grid pattern. Private houses and workshops were constructed in timber at this time, which meant that there was a serious risk from fire, with the buildings on some town centre sites being destroyed and rebuilt on the same plot as many as three times over a 100-year period.

The only townsman whose name has come down to us from this period is an individual named *Vitanus*, a late first-century earthenware potter whose kiln lay in the south-western extra-mural zone overlooking the river. Around the middle of the second century the old fortress ditch began to be used as a rubbish disposal site, and not long afterwards the disused rampart was levelled and spread over the rubbish in the ditch. It appears that for a few years the town lay unenclosed, as were most of the other Romano-British tribal capitals in the earlier second century.

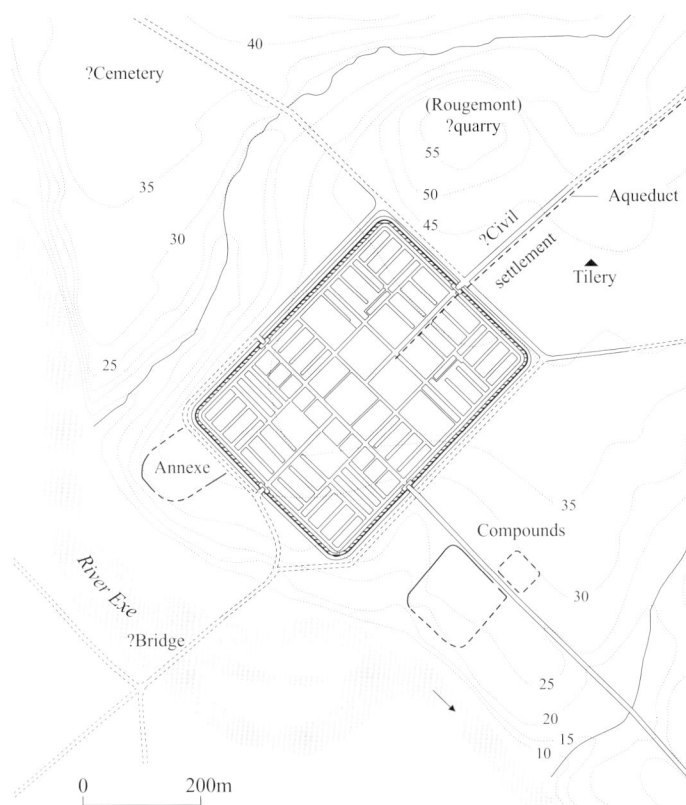

Fig. 61.1
The *caldarium* hypocaust (underfloor heating system) of the Exeter fortress bath-house underlying the front wall and steps of the late first-century basilica (town hall) at the head of the Roman forum.

SOURCE: reproduced by permission of Exeter Archaeology.

THE LATER ROMAN TOWN, *c*.150 TO THE EARLY FIFTH CENTURY (**Map 61.4 and Fig. 61.3**)

The town did not remain unenclosed for long. During the third quarter of the second century, or a little later, many urban centres in Roman Britain were provided with earthwork defences. At Exeter, these comprised a broad, low rampart, probably with a wattlework front revetment, protected externally by two deep ditches. These defences, which enclosed an area of about 37ha, were apparently erected very rapidly, over just a few weeks or months, with large numbers of the local inhabitants assisting in their construction. Relatively impermanent defences of this sort were generally thrown up as an emergency measure (as for

Map 61.1 (left)
The topographical setting of *Isca* Roman fortress, *c*.50–75.

Fig. 61.2 (below)
Model of the *caldarium* (hot room) of the mid-first-century bath-house of the Roman legionary fortress at Exeter.

SOURCE: reproduced by permission of Exeter Museums.

Map 61.2
The internal layout of *Isca* Roman fortress, *c.*50–75.

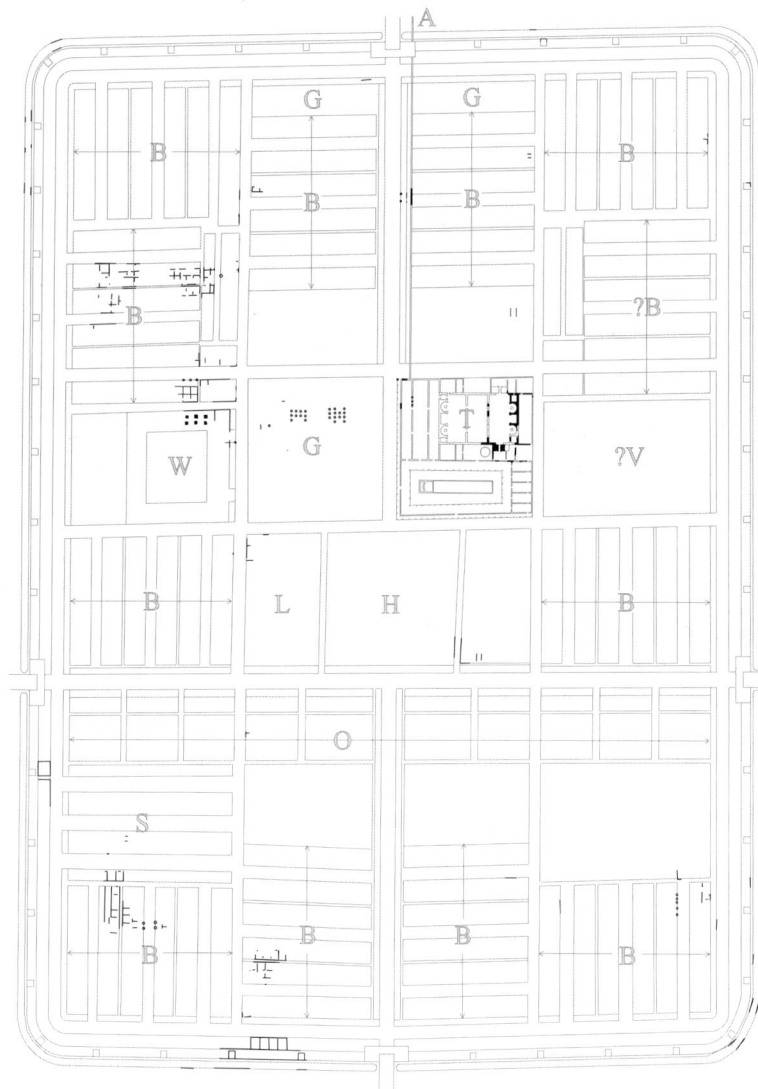

```
0                    200m
```

A Aqueduct
B Barracks
W Workshops
G Granaries
T Baths

V Hospital
L Legate's residence
H Headquarters
O Officers' houses
S Stables

The second-century earthwork defences are likely to have fallen into decay within a year or two of being erected, and were probably never refurbished after the initial threat that occasioned their construction had passed. They were superseded in the earlier third century, when work commenced on the construction of stone gatehouses and a substantial town wall at the front of the second-century rampart. This enterprise must have been very expensive and probably took many years to complete. Few new streets were built in the outer areas of the newly walled town, and within the original street grid a number of old streets had already fallen out of use by the end of the second century, in several cases subsequently being built over. From around the middle of the third century, the character of the town began to change quite markedly, with extensive masonry houses now taking the place of the much smaller timber buildings that had characterized the second- and earlier third-century settlement. Such 'urban villas', with their subsidiary yards and ranges of outbuildings, were associated with barns and stock enclosures indicative of a primarily agricultural economy. The basilica was remodelled extensively in the mid-fourth century, and some of its floors were relaid on at least one occasion subsequently; the make-up of the latest surviving floor incorporated a worn coin of Valens (364–78), suggesting that the building remained in use until at least the end of the century. By the last decade of the fourth century fresh supplies of coinage had ceased to reach Exeter: the latest Roman coin from the town is an issue of the house of Theodosius, dated 388–92. In the late fourth and early fifth centuries

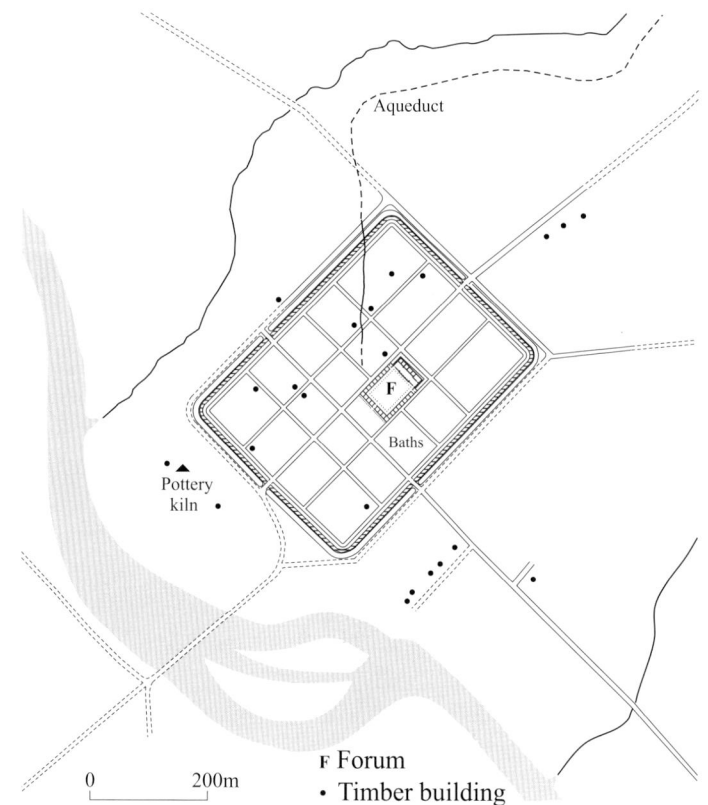

```
0        200m
```

F Forum
• Timber building

example 1,500 years later, in the English Civil War). These defensive precautions must have been prompted by a serious military threat, such as civil insurrection or barbarian incursions, but no historical record has come down to us of such an episode at this time. The earthwork defences enclosed a considerably larger area than was strictly necessary for the defence of the core settlement. This is probably because the defended enclosure was intended as a temporary refuge for the surrounding rural population, their animals and agricultural produce.

Map 61.3
The early Roman town *Isca Dumnoniorum*, *c.*75–150.

there occurred a permanent breakdown in the relatively sophisticated economic and administrative system that had sustained the towns of the Roman province. For the next five centuries urban settlement was absent in the south-west peninsula.

THE SUB-ROMAN AND EARLY MEDIEVAL PERIOD, EARLY FIFTH CENTURY TO *c.*880 (**Map 61.5**)

The Roman town was probably largely deserted by the middle of the fifth century, if not before. The streets and houses would quickly have become overgrown with vegetation, as indicated by the blanket of dark humus-rich soil that covers their remains. Except possibly in the forum area, there is nothing to suggest that any Roman buildings survived to influence the topography of medieval Exeter. The Roman town wall, over 3m thick at its base, was apparently the only major Roman structure left standing throughout the sub-Roman and subsequent periods. Of the intra-mural streets, only a 170m section of the street immediately inside the north-east gate is thought to have continued in use as a thoroughfare throughout the early Middle Ages (and down to the present day, as the upper part of High Street). The line followed by South Street and North Street probably came into existence at this period. These streets cut across the Roman street grid, taking the shortest route between the south-east and north-west gates. The course taken by the route leading up from the south-west gate

Fig. 61.3
The interior of the later Roman city wall (*c.*200) at Post Office Street, Exeter.

SOURCE: reproduced by permission of Exeter Archaeology.

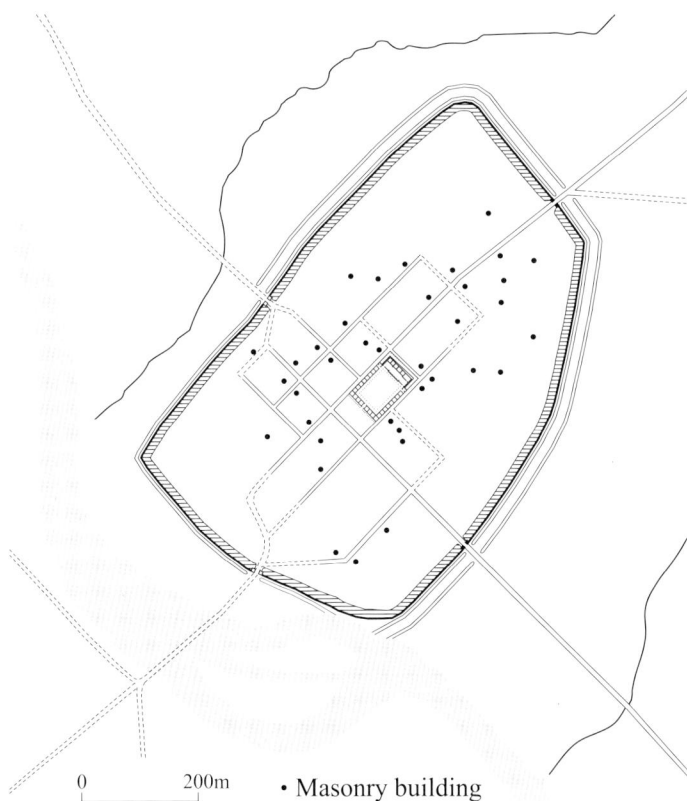

0   200m   • Masonry building

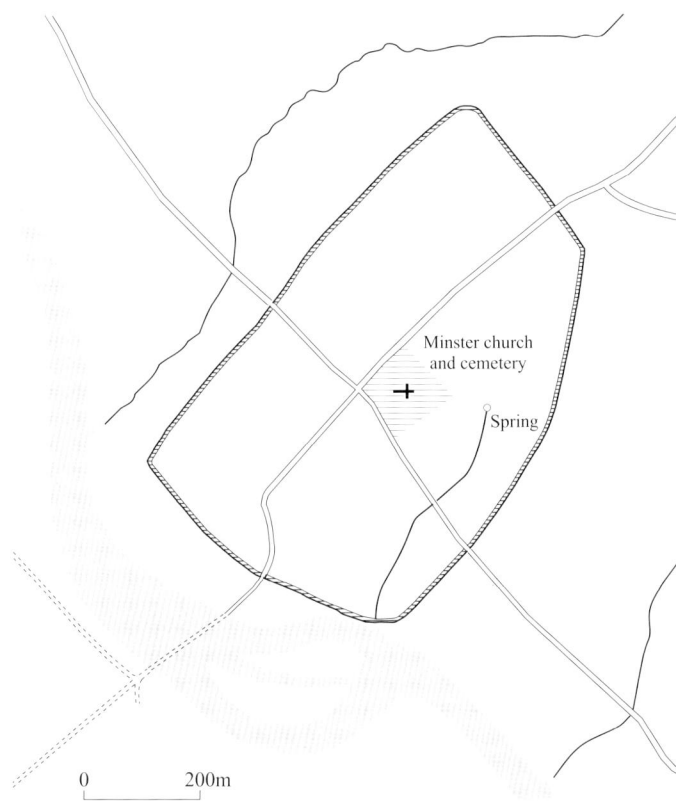

0   200m

Map 61.4 (left)
The later Roman town, *c.*150 to the early fifth century.

Map 61.5 (right)
The sub-Roman and early medieval period, *c.*420–880.

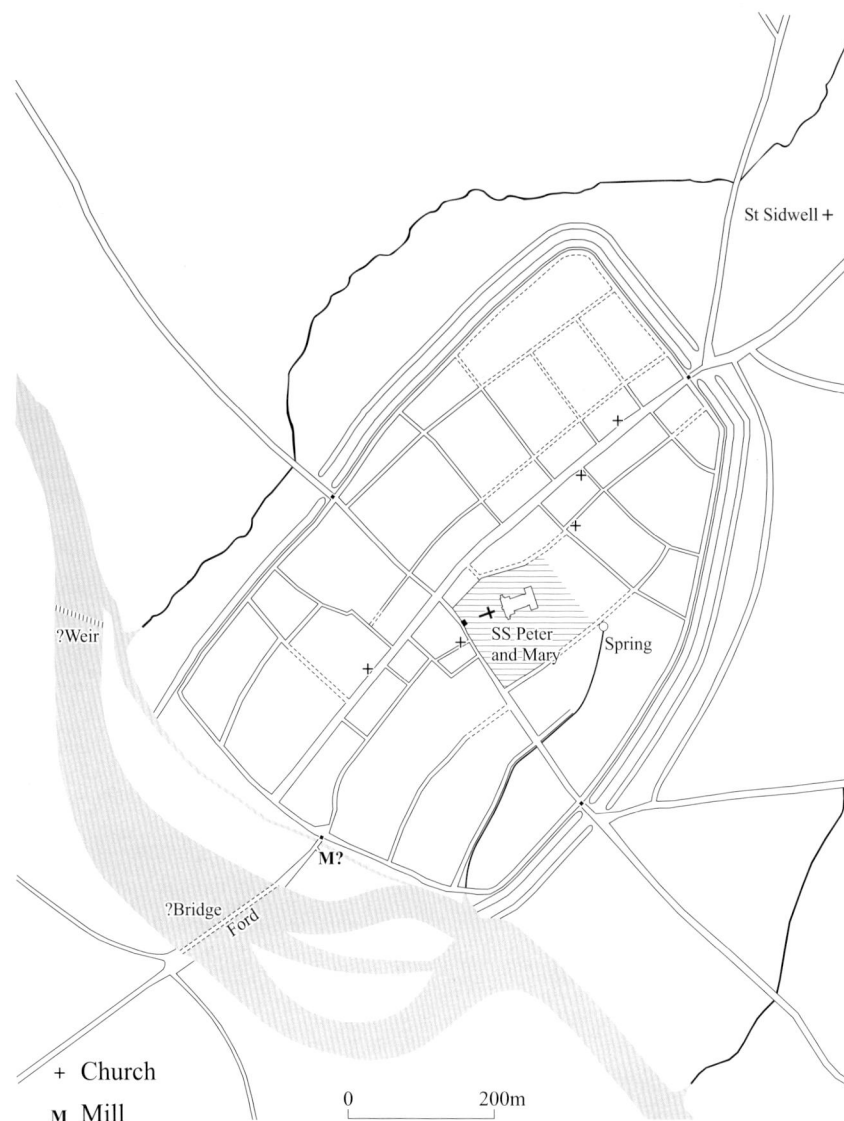

Map 61.6
The late Saxon town,
c.880–1068.

established within the area of the early Christian cemetery; it was here that St Boniface, Apostle to the Germans, received his early education around 690. The alignment of the graves in the middle Saxon cemetery which succeeded the early Christian one suggests that a new minster church was erected in the later seventh or eighth century which was oriented close to east–west. Once again, no remains of a church were found.

It is likely that on occasion during the eighth and ninth centuries the walled *enceinte* was used for defensive purposes; a Danish army overwintered here in 877, when the Anglo-Saxon Chronicle described the place as a stronghold (*faesten*).

THE LATE SAXON TOWN, *c*.880–1068 (**Map 61.6**)

Exeter was refounded as an urban settlement on the orders of King Alfred, probably in the 880s. At this time the Roman wall would have been refurbished and new streets laid out to create a planned defended town or *burh*. It is likely that the main elements of the medieval street system, much of which survives today, date from around this time. The streets provided a framework for the subdivision of the land within the walls to create large tenement plots (*hagae*) for allocation amongst the holders of rural estates over a wide area in Devon. The backbone of the new street system was a broad market street, the later Fore Street/High Street, which extended uninterrupted from the north-east gate to a point near the south-west gate. On each side of this axial street, parallel lanes were laid out which ran along the backs of the tenements fronting on the main street; it is not known whether these back lanes (which included Catherine Street) were originally continuous to either side of the main street. A series of long lanes ran back from the main street to meet a track running along the top of the Roman bank at the rear of the town wall. Of these lanes at right angles to the main street, only St Martins Lane/The Close now survives to its full original length. The wall was protected externally by a large ditch, supplemented by a second ditch on the more vulnerable north-east and south-east sides.

In the late ninth and tenth centuries many of the tenements within the walls were comparatively large—in some cases an acre or more in extent. The population increased rapidly from the late tenth century onwards, and the large primary tenements laid out in the time of King Alfred became progressively subdivided. The process of successive subdivision would have been especially marked along the frontages of the main street, but much less evident in the areas nearer the walls, behind the back lanes. In the late tenth or early eleventh century an immigrant earthenware potter established a workshop in the eastern corner of the town. The products from this kiln, some of them glazed (a rarity at this period), have been found on contemporary sites in Exeter and in other late Saxon towns in Devon. Other pottery current in Exeter in the late Saxon period was imported from Normandy.

Coins were minted at Exeter from Alfred's reign, starting around 895. In a ranking of Anglo-Saxon mints operating between

in sub-Roman times is less certain, though Stepcote Hill and Smythen Street could date from this period.

A possible thread of continuity between Roman and medieval Exeter is to be found in the area of the Roman civic centre. By the earlier sixth century a Christian burial ground had been established on the site of the Roman basilica (to the west of the later medieval cathedral). This apparently fairly extensive cemetery contained burials without grave goods lying with their heads to the north-west, on the Roman alignment, suggesting the survival nearby of some Roman building (a church?) or boundary feature. This early Christian burial ground could perhaps have been associated with a Dumnonian religious community but no church or other buildings of this period were located in the excavations carried out on the site in the 1970s.

By the late seventh century the eastern part of Dumnonia had come under West Saxon control and a Saxon monastery had been

Map 61.7
Norman Exeter, 1068–c.1220.

Churches and chapels		18 St David
1	St Lawrence	19 St Sidwell
2	St Stephen	20 St Leonard
3	All Hallows Goldsmith St	21 St Thomas
4	St Paul	22 St Mary Minor
5	St Pancras	23 St Bartholomew
6	St Kerian	24 Christ Church
7	St Petrock	25 Holy Trinity
8	St Martin	26 St Cuthbert
9	St Mary Major	27 SS Simon and Jude
10	St George	28 St Peter Minor
11	St Mary Arches	29 St Radegund
12	St Olave	30 St James
13	St John	31 St Clement
14	All Hallows on the wall	**Hospitals**
15	St Mary Steps	H1 St Mary Magdalen (leper)
16	Holy Trinity	H2 St Alexius
17	St Edmund	H3 St John

973 and 1066 Exeter is placed seventh, and at the beginning of the eleventh century it ranked in fifth position. The names of over twenty moneyers are known from the legends of coins minted in Exeter. These men were prominent Anglo-Saxon merchants who belonged to the wealthiest class of the town's inhabitants. Many of the local population must have been of British extraction. According to William of Malmesbury, writing in the early twelfth century, Alfred's grandson Æthelstan compelled the British population to leave Exeter 'which they had until that time occupied with equal rights with the English'.

Of Exeter's parish churches, St Sidwell's, situated 280m outside the north-east gate, is the earliest attested. Its appearance in an early eleventh-century list of saint's resting places implies that the church was a major place of pilgrimage at this time, presumably possessing a shrine housing the holy relics of the martyr herself.

NORMAN EXETER, 1068–c.1220 (**Map 61.7**)

By the Norman Conquest, Exeter was one of the ten largest towns in England. The total of 399 houses recorded in Domesday Book implies a population in 1066 of around 2,000 inhabitants. William the Conqueror besieged the city for eighteen days, and upon receiving the surrender of the citizens early in 1068 built a large earthwork castle on Rougemont hill within the northern corner of the walls. Certain regions of the town witnessed major changes in the aftermath of the Conquest; the imposition of the castle obliterated earlier streets and houses over a wide area, and the building of a much larger cathedral on a new site from 1114 necessitated a considerable expansion in the size of the Close. St Nicholas Priory was founded about 1090 as a daughter house of Battle Abbey. Its precinct eventually came to occupy a considerable area in the western quarter of the city, requiring the

Map 61.8
Late medieval Exeter,
*c.*1220–1540.

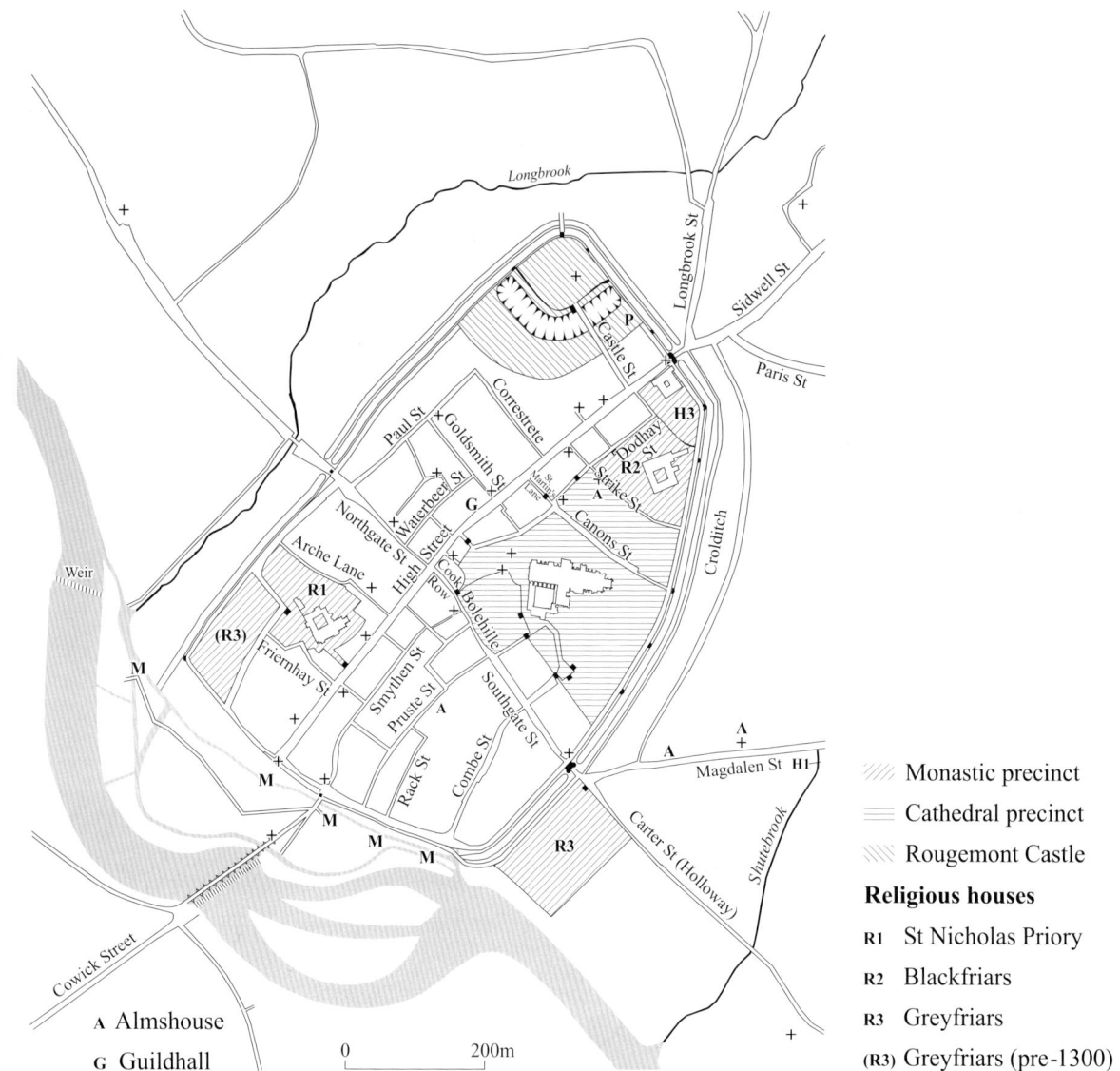

Monastic precinct

Cathedral precinct

Rougemont Castle

Religious houses

R1   St Nicholas Priory

R2   Blackfriars

R3   Greyfriars

(R3)   Greyfriars (pre-1300)

A   Almshouse

G   Guildhall

0   200m

closure of one or more public lanes running between blocks of tenements acquired by the monastery. St John's Hospital for the relief of the sick and the poor was established on a site just inside the East Gate around 1180; while about twenty years earlier, St Mary Magdalen's leper hospital had been built on a site 370m outside the South Gate. Over the next century St John's acquired most of the land in the eastern corner of the walls between High Street and Dodhay Street. A third hospital founded in the twelfth century, that of St Alexius, which failed early in the thirteenth, was located in the western corner of the city on a site later taken over by the Greyfriars.

Burial in Saxon and Norman Exeter was confined to the cathedral cemetery, since the lesser churches were technically only parochial chapels without the right of sepulture, which was reserved to the city's mother church, the Anglo-Saxon minster and its successor the medieval cathedral. By the early thirteenth

century there were at least thirty-two chapels in the city and its suburbs. For most of these no foundation date is recorded, although the great majority are likely to have originated in the tenth to twelfth centuries as proprietary chapels serving as places of worship for the inhabitants of a single wealthy household or a limited local area. St Martin's was consecrated in 1065 and the dedication of St Olave's suggests a foundation date in the 1030s. After the consecration of the east end of the Norman cathedral in 1133, the chancel, eastern transept and lateral elements of the Anglo-Saxon minster were demolished, leaving the nave to form the body of what became St Mary Major Church, whose congregation was drawn from an extensive parish lying mainly in the thinly populated southern quarter of the city where there were few chapels. Parish boundaries were delineated in 1222, when certain chapels were designated as parish churches (although still without burial rights).

By the late twelfth century, a narrow timber bridge had been built alongside the ancient ford at Exeter. Around 1200, this was replaced by a stone bridge of seventeen arches joined to a raised causeway leading to the West Gate. A chantry chapel stood on the bridge near its north-east end; while on the other side of the carriageway St Edmund's Church was built to serve the inhabitants of Exe Island, the city's low-lying western suburb which was developing rapidly at this period. Close to the south-west end of the bridge was St Thomas's Church, destroyed by flooding in the early fifteenth century and subsequently rebuilt on its present site in Cowick Street, 400m from the river. A weir under the bridge diverted water into a leat which led to a mill for fulling cloth (in existence by 1291). Immediately beneath the city, the Higher Leat followed a course hard against the eastern edge of the floodplain, leaving the river at a weir 400m above the Exe Bridge and rejoining it at Cricklepit about 130m below the bridge. This leat powered corn- and fulling-mills (the later City Mills) located just outside the West Gate, as well as Cricklepit corn mill, built about 1220.

Late medieval Exeter c.1220–1540 (Map 61.8)

In the thirteenth and fourteenth centuries Exeter ranked as a modest provincial town, exceeded in wealth and population by more than twenty other places in England. The city was badly affected by the Black Death in 1348–9, probably losing at least a third of its population; by 1377 it contained perhaps 3,000 inhabitants, just half as many again as in 1066.

Progressive enlargement of the cathedral precinct, commencing in the early twelfth century (see chapter 62), and the establishment in the thirteenth century of monastic houses belonging to two of the mendicant religious orders, the Franciscan and Dominican friars, together wrought major changes in the city's topography. A large area to the south-east of Catherine Street, between Dodhay Street and Strike Street (the later Chapel Street), was by stages acquired as the site of the Dominican friary (or Blackfriars), founded by 1232 to accommodate thirty-six friars. By the beginning of the fourteenth century, the section of Dodhay Street running between St John's Hospital and the Blackfriars had ceased to be a public highway, and it was eventually absorbed into the Blackfriars precinct. Other public streets and lanes also went out of use at this period, as for example the upper section of Rack Street in the southern quarter.

The Franciscans (or Greyfriars) probably reached Exeter during the 1230s, establishing their house within the western corner of the walls. Their site proving damp and unhealthy, however, the friars obtained royal consent to acquire land overlooking the river outside the South Gate, to which site they removed around 1300. Other medieval monastic houses established within a mile or two of the city were St Andrew's Priory, Cowick; St Katherine's Priory, Polsloe; St Mary's Priory, Marsh Barton; and St James's Priory, which lay next to the river a mile below Exe Bridge.

By the early fourteenth century, the cathedral dean and chapter had acquired almost all of the tenements on the north-east frontage of St Martin's Lane, as well as other properties, for example along the south-east side of their precinct. In 1286 they obtained royal licence for the erection of gates at each of the seven entry points to the Close. A gate already existed near the top of South Street (the *porta lata*, or broad gate—the later Little Stile) which had probably served as the main ceremonial entrance to the Anglo-Saxon and Norman precinct; its name (Broadgate) and ceremonial function now transferred to a new gate facing onto the High Street near the merchant's Guildhall (itself first documented in the twelfth century). Most of the houses in the cathedral precinct were occupied by members of the cathedral clergy, ranging from the bishop in his palace, through the senior clergy, including the dean, treasurer and the archdeacons, in their substantial hall-houses, down to priests, vicars and other lesser clerics living in lodgings or collegiate accommodation. The vicars choral occupied their own college in Kalendarhay from 1387, and the annuellars (chantry priests) lived in purpose-built premises from about 1529 until their suppression in 1548.

Exeter's medieval and early modern water supply (Map 61.9 and Fig. 61.4)

Until the later twelfth century the inhabitants of the city obtained most of their water from wells and springs. Exeter's first piped water supply was engineered by the cathedral authorities in the third quarter of the twelfth century. A lead pipe conveyed water from St Sidwell's Well in the north-eastern suburb to St Peter's Conduit in the Cathedral Close, and thence to St Nicholas Priory, with a further extension being laid in the early thirteenth century to Town Well in Fore Street. In the thirteenth century, the water supply to this lengthy system was augmented by the laying of a subsidiary pipe from St Sidwell's Well up to Head Well.

Each of the monastic houses possessed its own piped water supply. The Blackfriars aqueduct, built in the 1240s, was the longest at about 1,400m, with an offshoot being laid to the Bishop's Palace around 1260; the second Greyfriars house obtained a supply in 1347 from a spring in the town ditch in Southernhay; and from 1380 the monks of St Nicholas piped water to the priory from their own well in Paul Street. In 1346–9 the cathedral decommissioned its first aqueduct and built an entirely new system. The original aqueduct had flowed on a relatively gentle gradient, passing through suburban fields and gardens before running along Southernhay to enter the city at the head of St Martin's Lane. The new aqueduct followed a more undulating course from a new well house at Head Well: it twice crossed the Longbrook valley before running up Longbrook Street to pass through the city wall into the grounds of St John's Hospital. A section of the pipe in the hospital grounds lay on the

Map 61.9
Exeter's medieval and early
modern water supply,
*c.*1270, *c.*1500 and *c.*1590.

1270

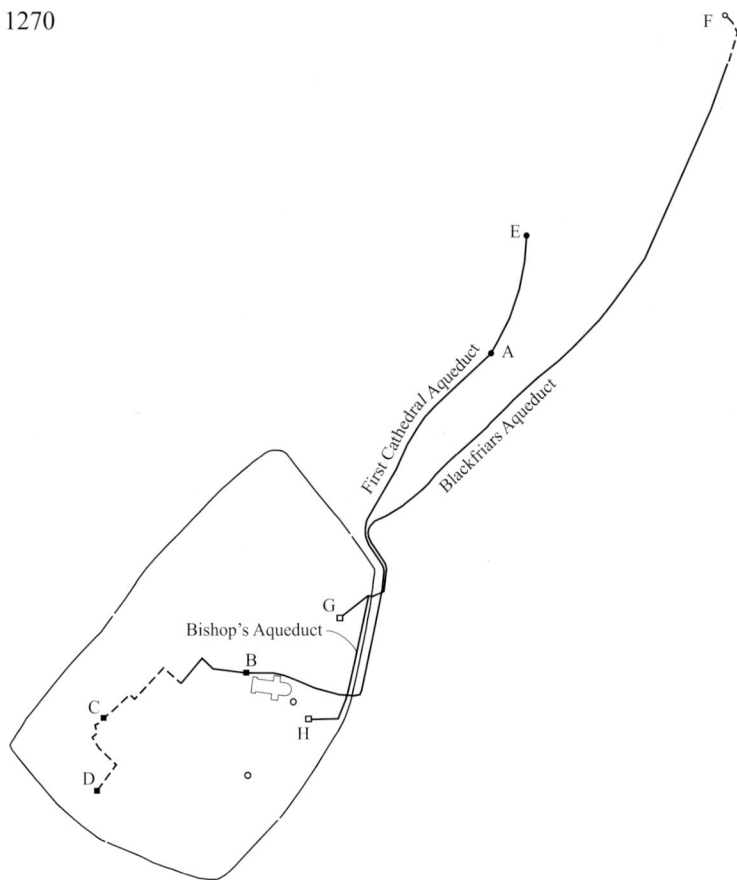

1590

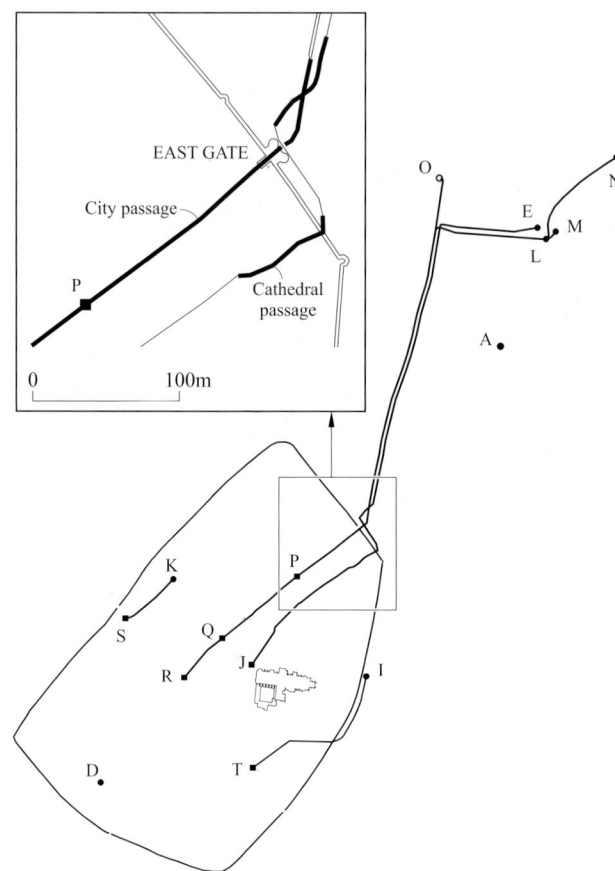

1500

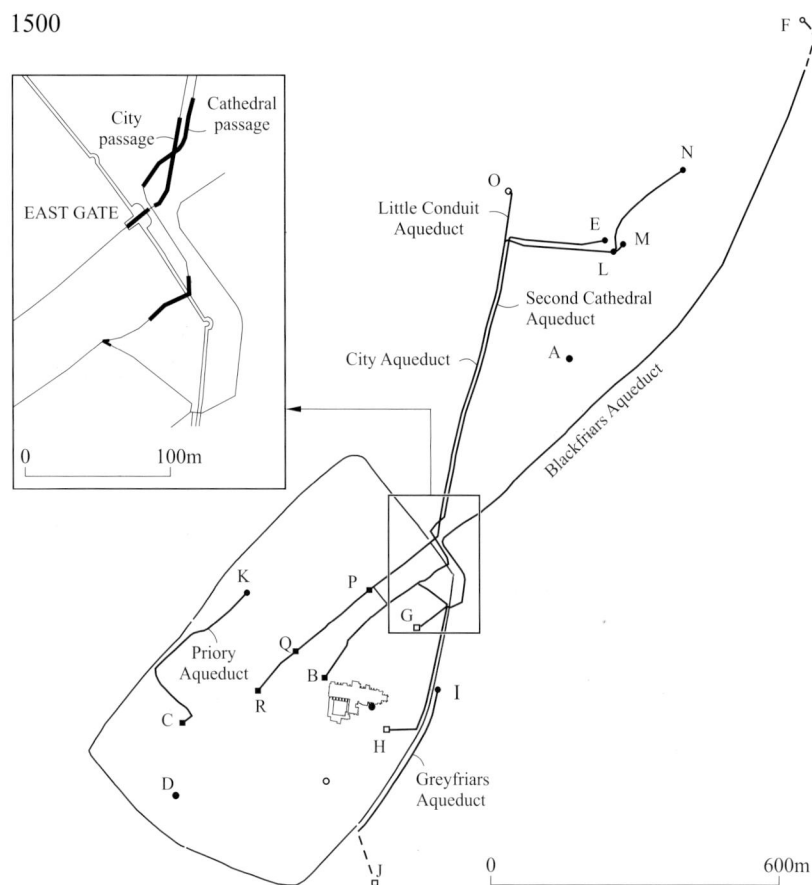

A St Sidwell's Well

B St Peter's Conduit

C St Nicholas Priory

D Town Well

E Cathedral Headwell

F Podwell

G Blackfriars

H Bishop's Palace

I Southernhay Well

J Greyfriars

K St Paul's Well

L City Headwell

M Salter's Well

N Higher Well

O Marpool Well

P Little Conduit

Q Guildhall Stoup

R Great Conduit

S North Gate Conduit

T South Gate Conduit

nineteenth century. From 1694, however, the city's supply was augmented by water pumped from the Exe by a water-powered 'engine' on the Higher Leat. The old city aqueduct was finally superseded when the Danes Castle reservoir opened in 1832.

EXETER QUAY AND CANAL (**Map 61.10 and Fig. 61.5**)

The Exe was probably navigable for small craft as far up as Exeter in Roman and early medieval times, but by the twelfth century, if not before, vessels importing wine were unloading their cargoes at Topsham. Three weirs were later constructed below the city (including, in the thirteenth century, the notorious Countess Weir), rendering the river quite impassable. In the 1540s attempts were made by the city council to remove all obstructions from the river bed, but it proved impossible to make the waterway navigable even for small boats. In 1564, therefore, the engineer John Trew was engaged by the council to cut a canal through the marshes on the west side of the floodplain, so as to bypass the worst hazards. Completed in 1566, the Exeter Canal was 1.75 miles long, 3ft (1m) deep and 16ft (5m) wide. It was the first English navigation to employ 'mitre gates' and pound locks, of which there were three. There was a difference in level of only 2m between the upper and lower ends of the canal, but the three locks were necessary in order to maintain the water level and provide basins for boats to pass one another or lie up in when the river was in flood. The upper entrance lay 400m from the southern corner of the walls, where a new gate and quay were constructed; a weir, built of stakes and stones, diverted water into the mouth of the canal, which was furnished with a single pair of gates to guard against flooding when the river was high. The lower entrance opened into a tidal pool navigable only at high

Fig. 61.4
View of the Exeter Cathedral aqueduct passage built in 1346–1349.

SOURCE: photograph reproduced by permission of Exeter Museums.

Fig. 61.5
The Custom House at Exeter Quay erected 1680–1681.

SOURCE: photograph reproduced by permission of Exeter Archaeology.

floor of a masonry tunnel, which allowed inspection and mending without the need to dig up the ground. This is the earliest part of the system of underground service tunnels known today as the Underground Passages. Beyond the hospital, the pipe ran beneath streets and paths to St Peter's Conduit.

In the 1420s the city council obtained their own well, close to the cathedral's, and over the next decade built an aqueduct that entered the city beneath the East Gate and ran down the High Street to a conduit at St Stephen's Church and thence to the Great Conduit at Carfax. Around 1500 the council built a second aqueduct to supply the Little Conduit, which stood in the High Street within East Gate. Between 1490 and 1540 further sections of service tunnels were added to the cathedral and city aqueducts, creating most of the passage system that is visible today. After the Dissolution, water from the Greyfriars' springs was piped to a conduit in South Street, while St Nicholas Priory's well in Paul Street supplied a conduit in North Street. The city and cathedral aqueducts continued in use down to the

Map 61.10
Exeter Quay and canal from the mid-sixteenth century to c.1830.

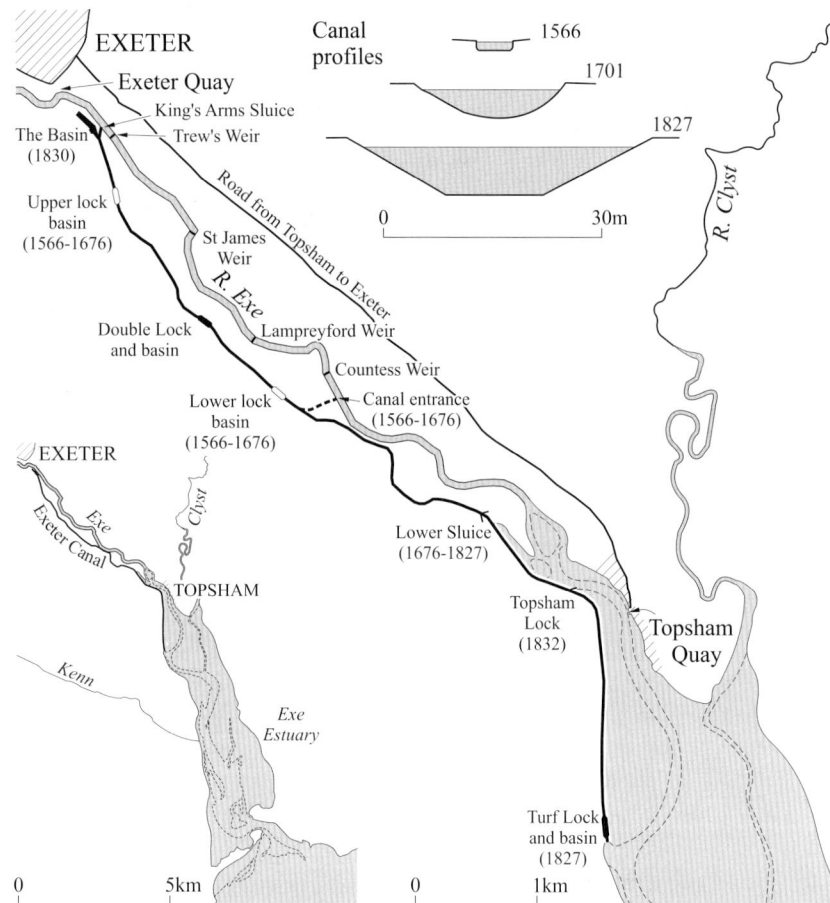

spring tides. The canal was suitable for lighters of 8–10 tons, which mainly carried lower-value, bulk commodities such as coal, roofing slates and limestone from sea-going vessels anchored within the estuary mouth at Starcross. The lighters could be sailed in the estuary but had to be hauled up the canal. Excavations at Exeter Quay have revealed the site of the first warehouse, known as the Crane Cellar, which was erected about 1574 and contained two small lock-up units for the storage of higher-value goods in transit. Around 1600 the warehouse was rebuilt on a larger scale; the new building, called the Quay House, incorporated a projecting roof canopy to keep cargoes dry during loading. The canal was extended to Lower Sluice, a little above Topsham, in 1676. A major transformation was initiated at Exeter Quay in 1680, with the erection of a much larger Quay House and the Custom House (the oldest surviving in Britain). Around 1700, the canal was widened and deepened to permit sea-going vessels of up to 200 tons to reach the Quay, via a single pound-lock basin at Double Locks. Other additions were made to the quays in the eighteenth century. Finally, in the 1820s, the canal was further enlarged and lengthened to accommodate ships of 500 tons, which entered at Turf Lock, 5.25 miles below the Kings Arms Sluice at the upper entrance, whilst in 1830 a 300m long dock basin was opened at the head of the canal. The Exeter Canal was at that time the second deepest inland waterway in Britain, after the Caledonian Canal.

EXETER IN 1600 (**Maps 61.11 and 61.12 and Fig. 61.6**)

The population of Exeter increased rapidly in the later fifteenth and early sixteenth century. By 1525, the city contained around 8,000 inhabitants and it now ranked fifth or sixth amongst English provincial towns, a position it was to maintain for almost two centuries. The city's prosperity at this period was closely linked to growth in the Devon cloth industry. Woollen cloths manufactured in the surrounding area were finished in Exeter and exported to foreign markets, particularly northern France, through Topsham. By 1580 there were three fulling mills in Exe Island. These can be seen on the Hooker map of Exeter, published in 1587 (**Map 61.12**), which also depicts a pair of cloth-drying racks.

Starting in the late thirteenth century, amalgamations had taken place between several of the more impoverished parishes, resulting in losses amongst the churches, so that by the close of the Middle Ages there remained only fifteen parish churches in the city. In the late fifteenth and sixteenth centuries many of the churches were enlarged, generally by the addition of one or more aisles. Chapels which were forcibly closed at the Reformation, during the 1540s, include the guild chapel built in 1471 by the Weavers, Tuckers and Shearmen (now Tucker's Hall in Fore Street) and the chapel of St Catherine's Almshouses, built about 1450. Almshouses had been a feature of the city since the early fifteenth century, when a number were founded by wealthy

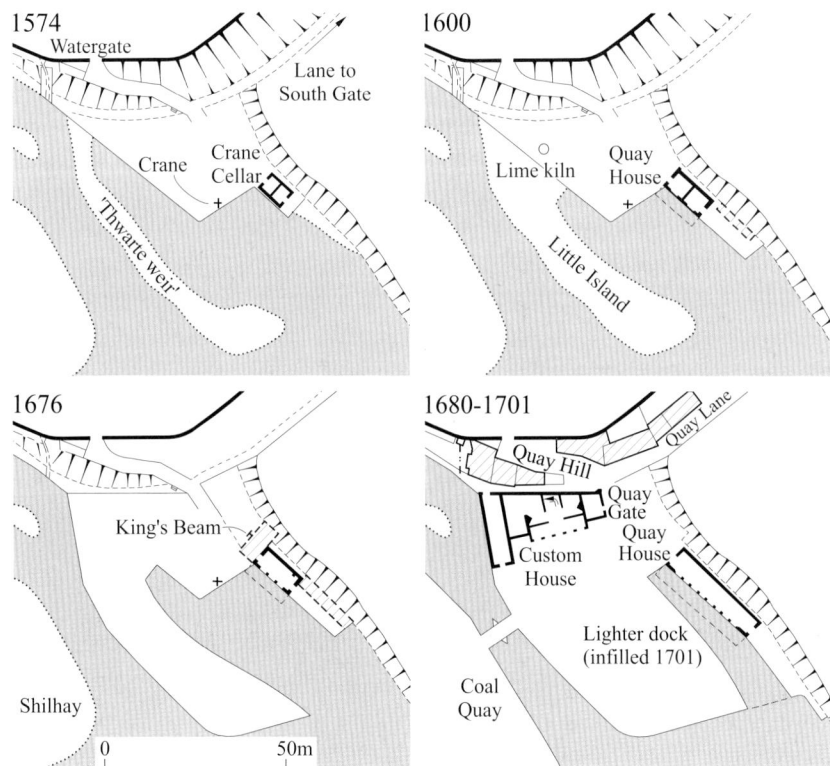

Map 61.11
Exeter, c.1600.

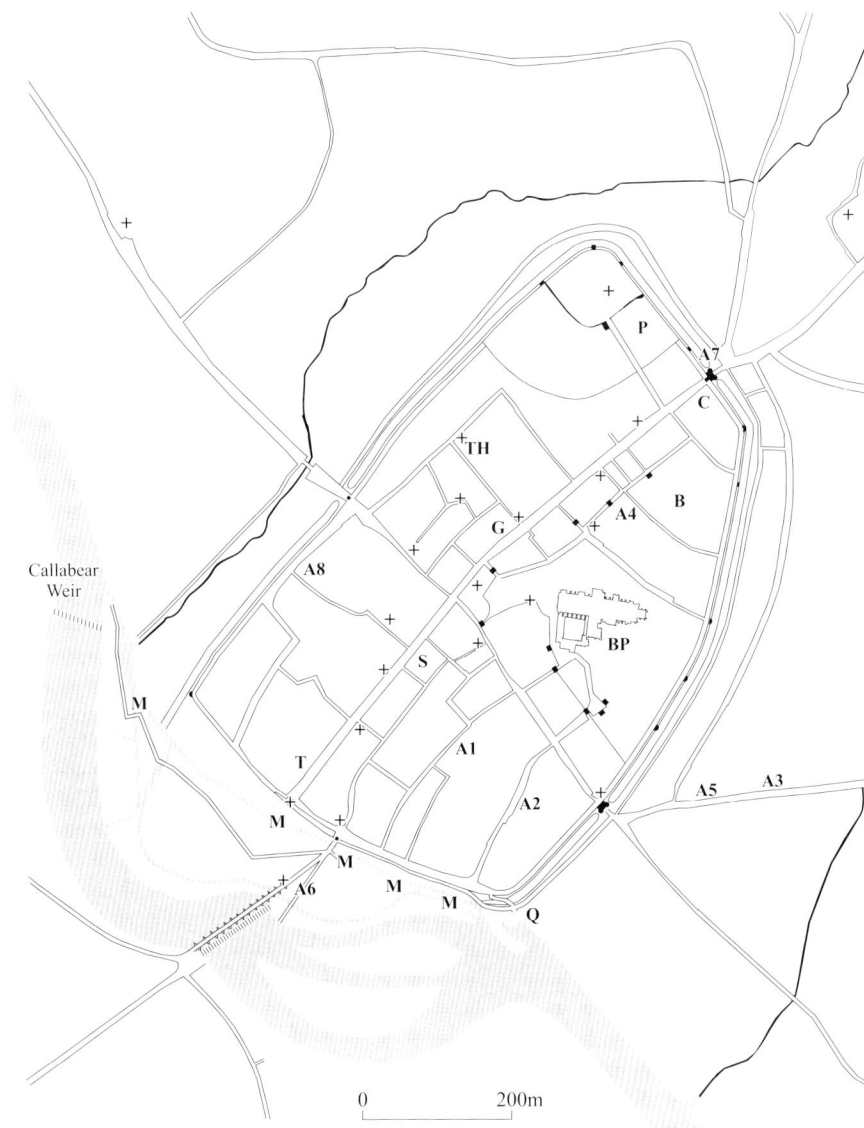

TH Taylor's Hall

G Guildhall

T Tucker's Hall

B Bedford House

S Shambles

C Cloth Hall

Q Quay

M Mill

P County Gaol

BP Bishop's Palace

Almshouses

A1 Grendon's

A2 Bonville's

A3 Wynard's

A4 St Catherine's

A5 Palmer's

A6 More's and Fortescue's

A7 Hurst's

A8 Davy's

citizens; new foundations continued to be made well into the seventeenth century.

An important influence on the topography of the city was the dissolution of the four urban monastic houses in 1536–9. Their precincts can still be recognized as open spaces on Hooker's map of 1587 (**Map 61.12**). At St Nicholas Priory and the Blackfriars the church and the eastern claustral range were demolished, leaving the other main ranges to be converted into spacious domestic accommodation. The priory became the house of successive merchants, while the friary was acquired in 1539 by Lord John Russell, later duke of Bedford, who built himself a substantial mansion, Bedford House, containing over seventy rooms; it was here that Princess Henrietta, daughter of Charles I, was born in 1644. St John's Hospital had a rather different history. Some of its buildings were pulled down at the Dissolution while others were converted into dwelling houses. The church was divided into two

storeys, the ground floor of the nave becoming the city's Cloth Hall. The Free Grammar School (predecessor to Exeter School) and the Bluecoat School later occupied the premises.

From the later twelfth century, a single defensive ditch had extended around the walls from the East Gate to the South Gate. The extensive strip of ground outside the walls between these gates, known as Crolditch or Southernhay, was for centuries the site of the Lammas fair (which also occupied St David's Down). From the early thirteenth century onwards, considerable sums of money were expended in successive building campaigns aimed at strengthening the walls. Projecting towers were constructed at intervals around the more vulnerable parts of the circuit, and the four gates were progressively enlarged, the most impressive being the South Gate and the East Gate. A fifth gate was constructed in 1565 at the southern corner of the walls which gave access to the new Quay.

For **Map 61.12** see page 15 in the colour section.

EXETER IN THE CIVIL WAR (**Map 61.13**)

Fig. 61.6
The projecting arcade of the
Elizabethan front block
of Exeter Guildhall built
1593–1594.

SOURCE: photograph
reproduced by permission of
Exeter Archaeology.

Map 61.13
Exeter defences in
the Civil War.

Exeter grew steadily over the century up to 1640, the population reaching perhaps 12,000 by the eve of the Civil War. In the early sixteenth century about a quarter of the inhabitants lived outside the walls; a hundred years later the proportion of suburban households had risen to around a third. The most populous of the extra-mural parishes was St Sidwell's, outside the East Gate. Exeter declared for parliament in 1642 and over the next four years the citizens were to endure a number of sieges. In 1642–3 the ancient walls were strengthened by the addition of an earthen rampart at their rear and artillery batteries were placed on 'mounts' at intervals around the circuit. Houses were cleared from in front of the walls to create fields of fire, and a series of earthen forts and entrenchments were constructed which enclosed much of the south-eastern suburb but gave no protection to St Sidwell's (which belonged to the royalist-leaning dean and chapter). The city fell to the royalists in September 1643, and over the next two years the fortifications were greatly enhanced, making Exeter one of the most strongly defended cities in the kingdom. A system of 'zig-zag' ramparts and ditches defended the South and East

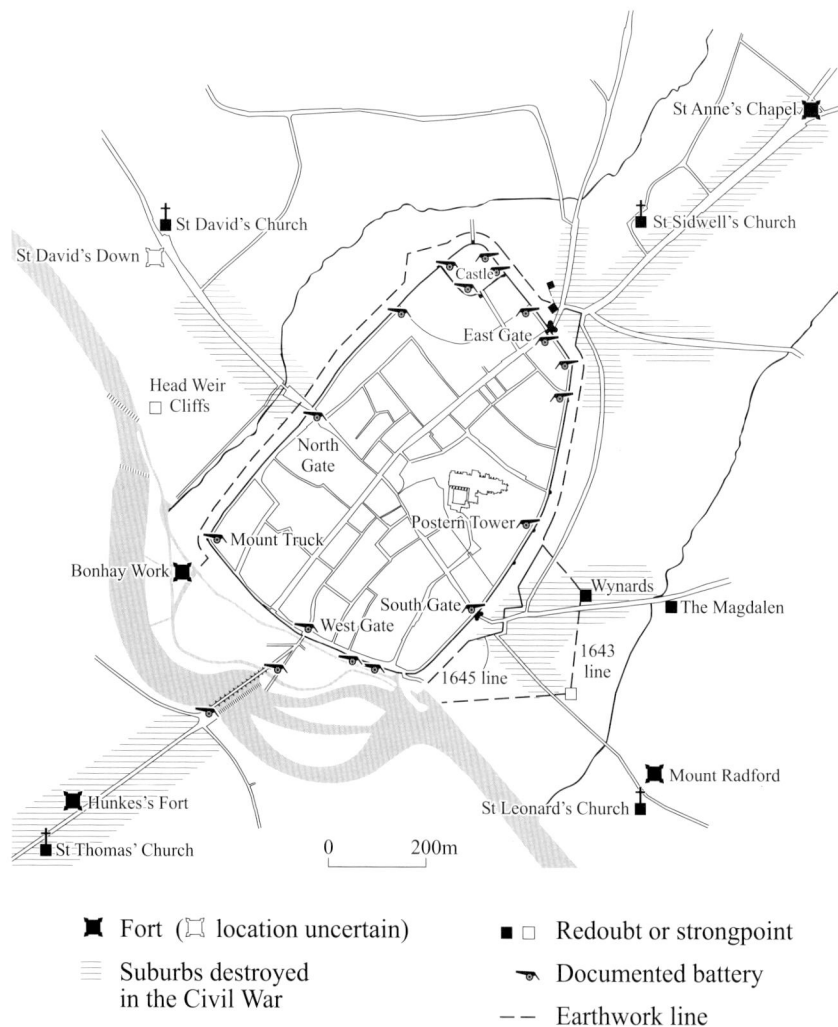

Gates, while four major forts and many lesser strongpoints guarded the main approaches to the city. Exeter fell to the Roundhead army in April 1646 after a long blockade. By this time many Exonians had become homeless refugees, almost every house in the suburbs having been destroyed in the conflict. Whilst quite a number of buildings dating from between the fifteenth and the early seventeenth centuries can still be seen within the walls, very few survive in the suburbs.

MID-EIGHTEENTH-CENTURY EXETER (**Maps 61.14 and 61.15 and Fig. 61.7**)

The rebuilding of houses destroyed in the suburbs during the Civil War proceeded apace during the 1650s and 1660s. By 1670 Exeter's population had returned to around its pre-war level, and the city was entering a period of great prosperity and growth which reached its peak in the second decade of the

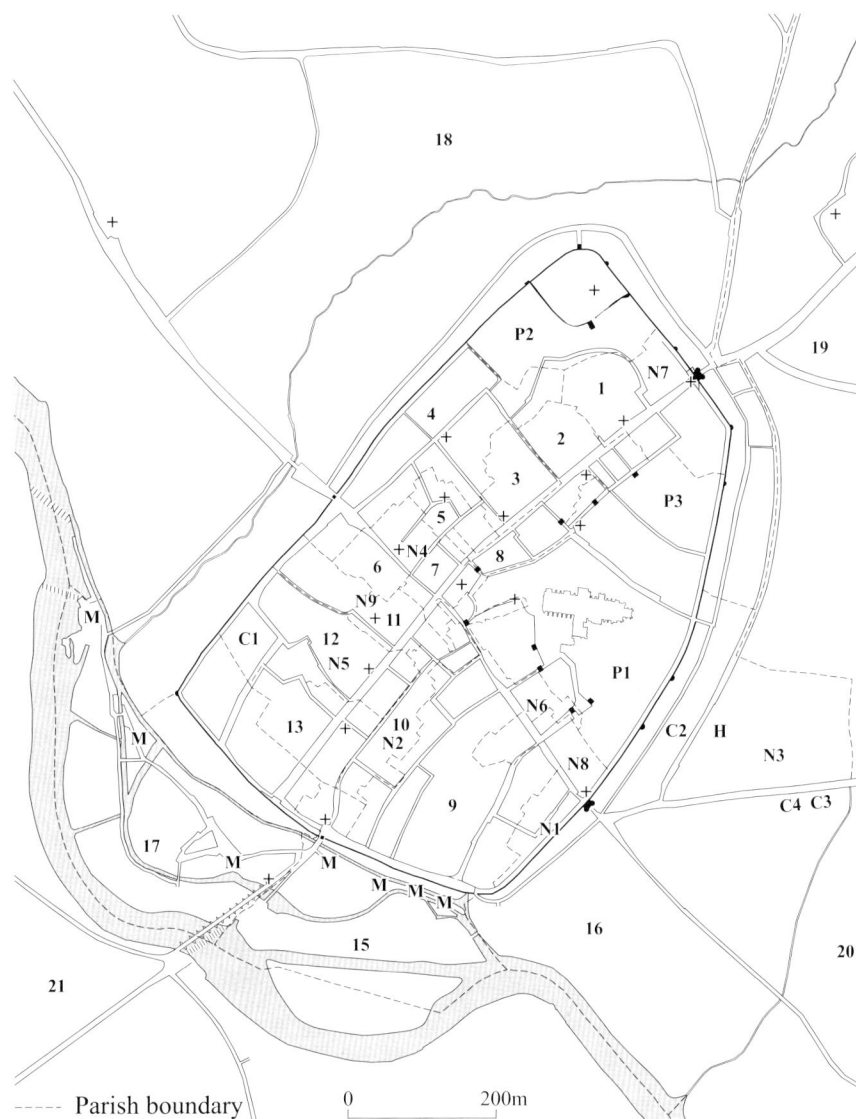

Parishes

1 St Lawrence
2 St Stephen
3 Allhallows Goldsmith Street
4 St Paul
5 St Pancras
6 St Kerrian
7 St Petrock
8 St Martin
9 St Mary Major
10 St George
11 St Mary Arches
12 St Olave
13 St John
14 Allhallows on the wall
15 St Mary Steps
16 Holy Trinity
17 St Edmund
18 St David
19 St Sidwell
20 St Leonard
21 St Thomas

Extra-parochial precincts

P1 The Close
P2 Bradninch
P3 Bedford

Cemeteries

C1 Bartholomew
C2 Southernhay
C3 Jewish cemetery
C4 Dissenters' cemetery

Non-conformist chapels

N1 James Meeting
N2 Bow Meeting
N3 Friends' Meeting House & burial ground
N4 Little Meeting
N5 Mint Meeting
N6 South Street Baptist Church
N7 Independent
N8 George's Meeting
N9 Jewish Synagogue

- - - - Parish boundary

0   200m

eighteenth century. Exeter was now one of the leading English ports, exporting a prodigious volume of serge cloth for sale on the Continent and further afield, notably in the Low Countries but towards the middle of the century increasingly in the markets of southern Europe. It is possible that as much as three-quarters of the working population was employed in some branch of the woollen industry at this time. Mid-eighteenth-century maps show many features connected with the cloth-finishing trades. Most prominent are the extensive rack fields or tenter grounds which occupied every available open space on all sides of the city. Along the leats, in addition to the fulling mills, there were many dye-houses, and in every part of the city were to be seen the small-scale workshops of woolcombers, weavers, sergemakers and pressmen. Cricklepit Mill was rebuilt in the late seventeenth century as a corn mill with an associated fulling mill; only the corn mill remains today, though an early eighteenth-century cloth-drying shed (or 'dryhouse') still survives nearby.

A number of lesser industries flourished in the late seventeenth and early eighteenth centuries. Some of these were dependent on the contacts with foreign markets created by the cloth trade. Large numbers of clay tobacco pipes made in Exeter were sent across the Atlantic, and molasses carried in the return direction was processed in small refineries to make sugar. A glass factory was established at Wear in the 1690s, and lime kilns located at Exeter Quay since the late sixteenth century were transferred at this period to a site opposite the head of the canal. Brick-making was carried out on a small scale at many places in the suburbs. From the 1650s onwards the majority of the buildings erected in Exeter incorporated some brick in their fabric; the finest buildings, such as the Custom House (completed in 1681), were entirely brick-built. Other public buildings erected at this period include the short-lived City Hospital (1741) and the Devon and Exeter Hospital (1743).

Burial in the Cathedral Yard had ceased in 1637, when a new cemetery was opened in Friernhay (Bartholomew Yard) on the site

Map 61.14
Mid-eighteenth-century Exeter.

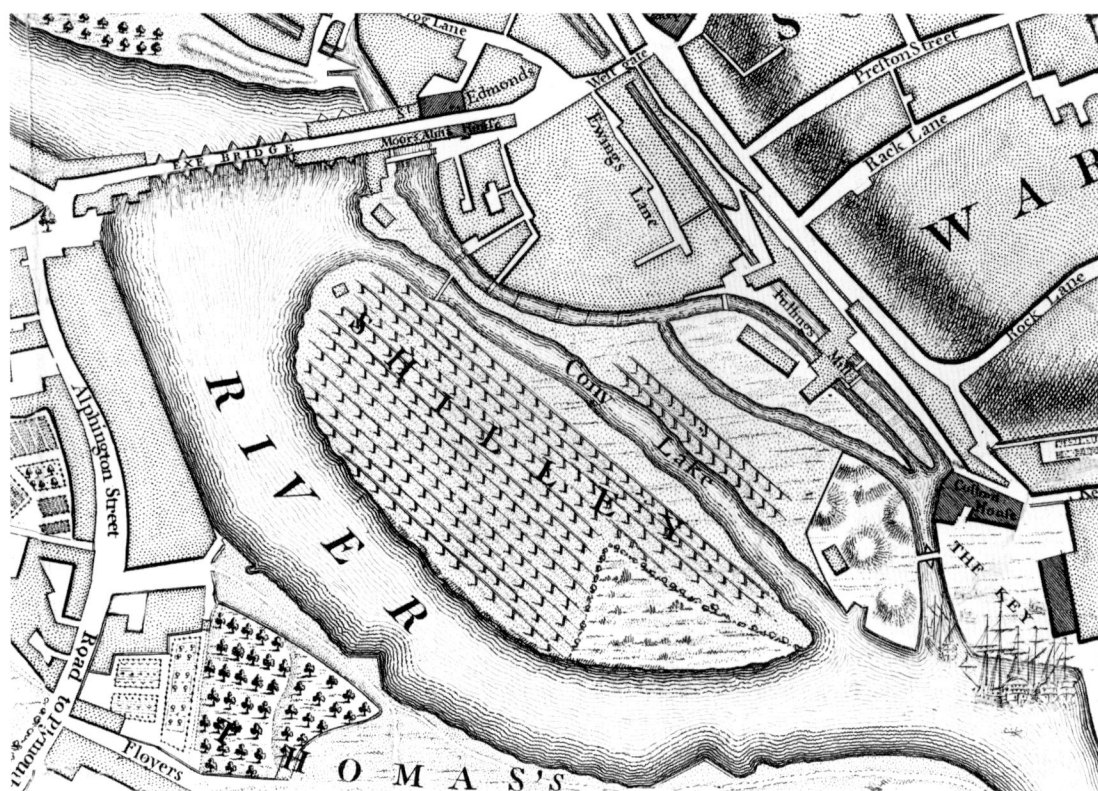

Map 61.15
Detail from John Roque's map of Exeter, 1744, showing the area between the medieval Exe Bridge and Exeter Quay; note the system of leats and the cloth-drying racks on Shilhay.

SOURCE: photograph reproduced by permission of Exeter City Council.

of the first Greyfriars; by 1664 it had become necessary to create a second burial ground at the lower end of Southernhay. The following century saw the advent of places of worship for dissenters: James's Meeting and Bow Meeting House about 1687, the Friend's Meeting House and burial ground in 1692, Little Meeting by 1715, Mint Meeting in 1720, the South Street Baptist Church in 1725, the Independent Chapel by 1744 and George's Meeting in 1760. The Jewish Cemetery, in Magdalen Street, was in use from 1757, and the Synagogue (off Mary Arches Street) was founded in 1763.

Fig. 61.7
View of Exeter Quay and the entrance to the Exeter Canal, from John Roque's map of 1744.

SOURCE: photograph reproduced by permission of Exeter City Council.

EXETER IN 1820 (**Map 61.16**)

Georgian Exeter witnessed a number of major topographical developments. Within the walls, Bedford House was demolished in 1773 and the first two quadrants of Bedford Circus had been erected on its site by 1775. A broad street leading into Southernhay breached the city wall at the south-east end of the circus. Elsewhere around the walls new openings were being created at this period. The New Cut, connecting Southernhay and the Cathedral Close, was made in 1750, and a passage through the wall at Maddocks Row was created between Paul Street and Northernhay Street in 1778. The construction in 1774 of the imposing new County Court within the ancient inner ward of the castle entailed the removal of a considerable section of the wall. The medieval gates were successively removed to ease the flow of traffic through the main streets of the city. The North Gate was the first to be taken down, in 1769, followed by the East Gate in 1784, West Gate and Watergate in 1815, and finally South Gate in 1819 (**Fig. 61.8**). The latter housed the city's prison until the opening of a new House of Correction in Northernhay in 1818. The county prison was erected in St David's parish, to the north of the Longbrook, under an Act passed in 1787. The next decade saw the commencement of the Revolutionary and Napoleonic Wars, and a cavalry barracks was built in the 1790s on a site to the north-east of the prison. Further construction was ordered in the new century, resulting in the artillery barracks adjoining Topsham Road and a westward extension of the cavalry barracks. The old county gaol, which had been located in the outer ward of the castle since the twelfth century, was acquired in 1796 for conversion to the Independence Chapel.

Obstructions removed within the walls include the Great Conduit (1772) and the Close gates, including Broadgate in 1825. Another major development in this period was the construction of the new Exe Bridge, completed in 1778. This connected directly to the lower end of Fore Street via a raised causeway, New Bridge Street. As well as being too narrow for the volume of traffic now needing to use it, the medieval bridge obstructed the flow of the river in times of flood. The western half of the old bridge was therefore removed following the opening of the new bridge; the other nine arches (which survive today) were retained, since they still carried St Edmund's Church and a number of houses.

From 1774 the city council began leasing building plots in Southernhay to local architect-builders, and this area gradually came to be occupied by terraces of large middle-class houses. Barnfield Crescent, a little further out, was built in 1792, and Colleton Crescent, overlooking the river on part of the former Greyfriars site, was built from 1802. Many of these new houses were serviced by a system of brick sewers, much of which is still in use today. Town gas was available to Exeter households from 1817, when the first gas works was established on Bonhay.

This was the great age of the coaching inn. The most prominent of these large hostelries were the New London Inn, built just outside the East Gate in 1794, and the establishment that

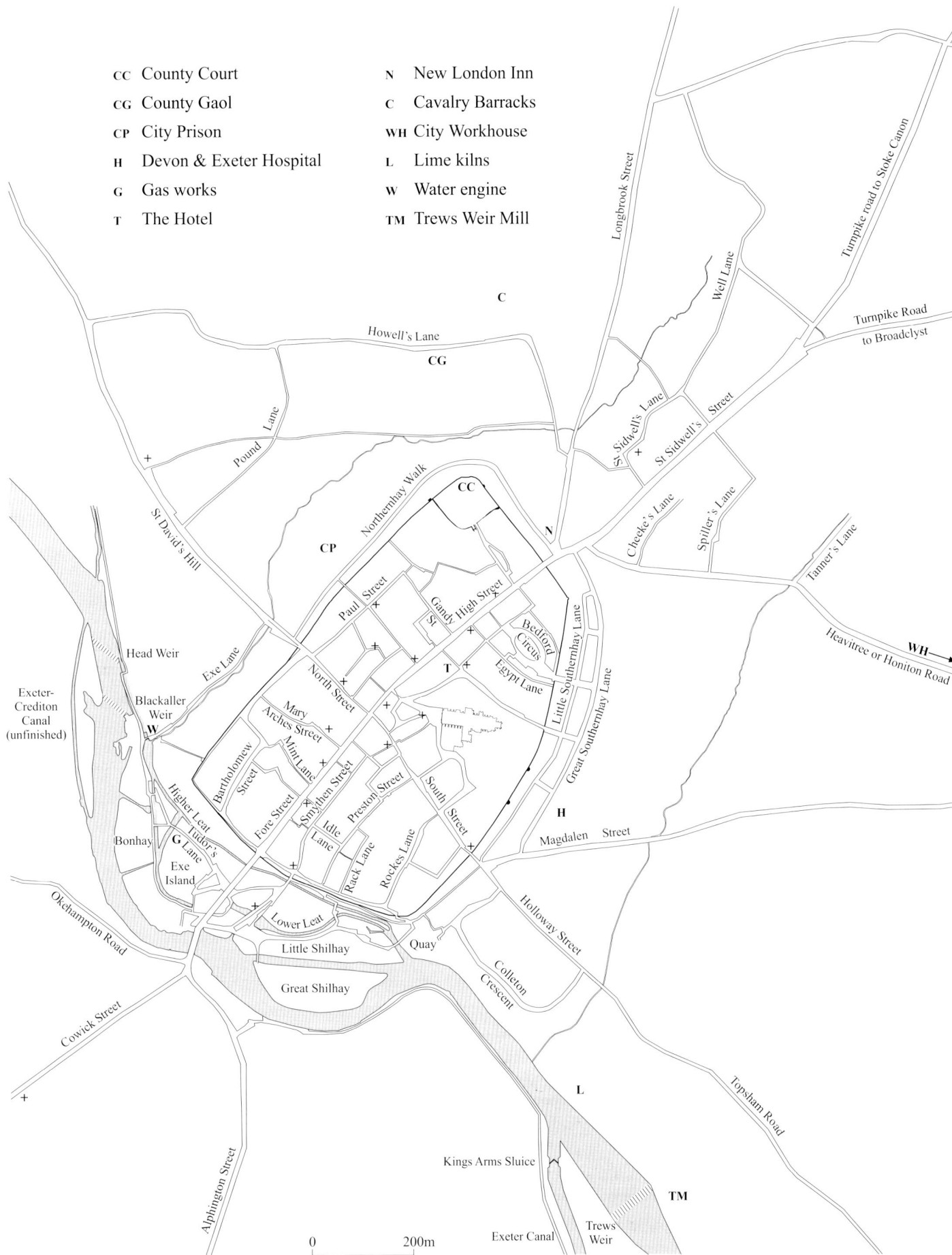

CC County Court
CG County Gaol
CP City Prison
H Devon & Exeter Hospital
G Gas works
T The Hotel

N New London Inn
C Cavalry Barracks
WH City Workhouse
L Lime kilns
W Water engine
TM Trews Weir Mill

C

Longbrook Street

Well Lane

Turnpike road to Stoke Canon

Howell's Lane

CG

Turnpike Road to Broadclyst

Pound Lane

St Sidwell's Lane

St Sidwell's Street

Cheeke's Lane

Spiller's Lane

St David's Hill

Northernhay Walk

CC

N

CP

Tanner's Lane

Paul Street

Giandz High Street

St

Bedford Circus

Little Southernhay Lane

Heavitree or Honiton Road

WH→

Exe Lane

Egypt Lane

Great Southernhay Lane

Head Weir

North Street

T

Exeter-Crediton Canal (unfinished)

Blackaller Weir

W

Mary Arches Street

Bartholomew Street

Mint Lane

Smythen Street

Preston Street

South Street

H

Higher Leat

Tudor's Lane

Bonhay

G

Fore Street

Idle Lane

Rack Lane

Rockes Lane

Magdalen Street

Exe Island

Okehampton Road

Lower Leat

Holloway Street

Little Shilhay

Quay

Colleton Crescent

Great Shilhay

Alphington Street

L

Topsham Road

Kings Arms Sluice

TM

0   200m

Exeter Canal

Trews Weir

Cowick Street

Map 61.16
Exeter, 1820.

Fig. 61.8
'View of the Southgate at Exeter, taken down in 1819', by F. Nash after J. Farington, published T. Cadell, London 1822.

originated in 1769 as Berlon's Hotel and later became the Royal Clarence Hotel. The first banks in the city were founded at this period: the Exeter Bank (1769), Devonshire Bank (1770), City Bank (1786), General Bank (1792) and Western Bank (1793). By the 1790s Exeter could no longer be regarded as an industrial city.

FURTHER READING

For details of excavations in Exeter see Aileen Fox, *Roman Exeter (Isca Dumnoniorum): Excavations in the War-Damaged Areas 1945–7* (Manchester, 1952); P.T. Bidwell, *Roman Exeter: Fortress and Town* (Exeter, 1980) and N. Holbrook and P.T. Bidwell, *Roman Finds from Exeter* (Exeter, 1991). For a more recent account of the early Roman period see C.G. Henderson, 'Exeter (*Isca Dumnoniorum*)', in G. Webster (ed.), *Fortress into City: The Consolidation of Roman Britain, First Century AD* (1988), 91–119.

For a general review of early medieval Exeter see J.P. Allan, C.G. Henderson and R.A. Higham, 'Saxon Exeter', in J. Haslam (ed.), *Anglo-Saxon Towns in Southern England* (Chichester, 1984), 385–414. Finds from this and later periods are described in J.P. Allan, *Medieval and Post-Medieval Finds from Exeter, 1971–80* (Exeter, 1984). For later medieval Exeter see Maryanne Kowaleski, *Local Markets and Regional Trade in Medieval Exeter* (Cambridge, 1995), E.M. Carus-Wilson, *The Expansion of Exeter at the Close of the Middle Ages* (Exeter, 1963) and A.M. Jackson, 'Medieval Exeter, the Exe and the earldom of Devon', in *Transactions of the Devonshire Association*, 104 (1972), 57–79.

For Exeter's water supply see J.P. Allan, *Exeter's Underground Passages* (Exeter, 1994) and Walter Minchington, *Life to the City* (Exeter, 1987). And for Exeter Quay and canal see C.G. Henderson, 'The archaeology of Exeter Quay', *Devon Archaeology*, 4 (1991), 1–15; K.R. Clew, *The Exeter Canal* (Chichester, 1984) and E.A.G. Clark, *The Ports of the Exe Estuary, 1660–1860* (Exeter, 1960).

For Tudor and Stuart Exeter see W.T. MacCaffrey, *Exeter 1540–1640* (Cambridge MA, 1958); Joyce Youings, *Tuckers Hall, Exeter: The History of a Provincial City Company through Five Centuries* (Exeter, 1968). W.G. Hoskins's *Two Thousand Years in Exeter* (Exeter, 1960) remains the most readable general history of the city.

Exeter in the Civil War is covered by Mark Stoyle, 'Exeter in the Civil War', *Devon Archaeology*, 6 (1995) and *From Deliverance to Destruction: Rebellion and Civil War in an English City* (Exeter, 1996). Works covering the seventeenth and eighteenth centuries include W.G. Hoskins, *Industry, Trade and People in Exeter, 1688–1800* (Manchester, 1935, reprinted Exeter, 1968), W.B. Stephens, *Seventeenth-Century Exeter: A Study of Industrial and Commercial Development, 1625–1688* (Exeter, 1958) and Robert Newton, *Eighteenth-Century Exeter* (Exeter, 1984).

62

Exeter and Plymouth

EXETER CATHEDRAL

NICHOLAS ORME AND C.G. HENDERSON

There have been three cathedral buildings at Exeter, each larger and grander than its predecessor—reflecting the growth of an important religious community.

THE ANGLO-SAXON CATHEDRAL (C.G. Henderson; **Map 62.1 and Fig. 62.1**)

A tradition current in the early eleventh century states that the minster at Exeter was refounded by King Æthelstan (925–39). The excavations carried out to the west of the medieval cathedral in the 1970s uncovered the remains of a sizeable late Anglo-Saxon church, aligned south-west/north-east, with an associated burial ground. This was the minster church that became a cathedral in 1050. It is probable that additions had been made to its fabric at the time of the later tenth-century monastic revival, which saw monks installed in the minster in 968; an axial tower, shown on a cathedral seal impression of 1133, and a polygonal eastern apse, uncovered in excavation, perhaps date from this period. Major additions are also likely to have been made to the church between 1046, the date of Bishop Leofric's appointment to the see of Crediton, and 1050, when King Edward and Queen Edith attended his enthronement in what had become his cathedral at Exeter. Attributable to Leofric (1050–72) is a large eastern transept, while a crypt (wherein the first two bishops were interred) in the transept's northern arm must have been built by Leofric or his Norman successor Osbern (1072–1103). Probably also dating from the mid-eleventh century is a broad west front (or 'westwork') flanked by tall stair-turrets, which was depicted on the 1133 seal. The base of the northern turret was found in excavation.

THE SECOND CATHEDRAL (Nicholas Orme; **Map 62.2**)

Bishop Leofric probably intended his cathedral to be staffed by twenty-four canons as well as by lesser clergy (he mentioned boys), and the development of worship and staffing eventually required a larger church. This was begun by Bishop William Warelwast (1107–37) on a site further east in 1114. Built in the

Romanesque style, its eastern half or choir appears to have been finished in 1133 when the clergy moved into it from the older building. The western half or nave was probably completed in the 1170s or 1180s, and the abandoned minster church became the parish church of St Mary Major.

The second cathedral was more spacious than its predecessor, reflecting the need to house a larger number of clergy and the wish to build a house of God on the scale of cathedrals and monasteries elsewhere. Its heart was the choir, where services were performed by the twenty-four canons or rather the twenty-four vicars choral who were recruited to deputize when the canons were absent on business. An ambulatory surrounded most and perhaps all of the choir, enabling processions to be made, and the nave provided room for the laity to watch the services. The whole floor area was available for the burials of bishops and canons. Two towers either side of the choir gave the building a distinctive appearance from outside, their lowest storeys being probably separated from the interior rather than acting as open transepts. In the thirteenth century the building had to accommodate still further liturgical developments. The cult of the Virgin Mary increased in importance and large churches began to include Lady chapels in which daily services could be given in her honour. Exeter had a Lady chapel by 1236, which may have been sited beneath the south tower. Intercessory masses for the dead also became popular and the cathedral staff began to be augmented by chantry priests, privately endowed by wealthy clergy and knights to say masses for their souls. Altars for this purpose were set up in suitable corners of the cathedral—probably seven of them—in addition to the high altar in the choir.

THE THIRD CATHEDRAL (Nicholas Orme; **Maps 62.3 and 62.4**)

In the 1270s the Romanesque cathedral began to be enlarged and rebuilt into the third and final building of today. The work was patronized and probably instituted by Bishop Walter Bronescombe and was carried forward under his successors until it was finished in about 1342. The building was doubtless influenced by that of Salisbury Cathedral (consecrated in 1258), whose Gothic architecture rendered the Romanesque of Exeter old-fashioned and whose

Map 62.1 (right)
The Anglo-Saxon
Cathedral, 1050–1133.

Fig. 62.1 (centre)
A wax impression of the
earliest Exeter Cathedral
chapter seal, possibly
depicting the west front of
the eleventh-century church;
from a document of 1133.

SOURCE: reproduced by
permission of Devon
Record Office.

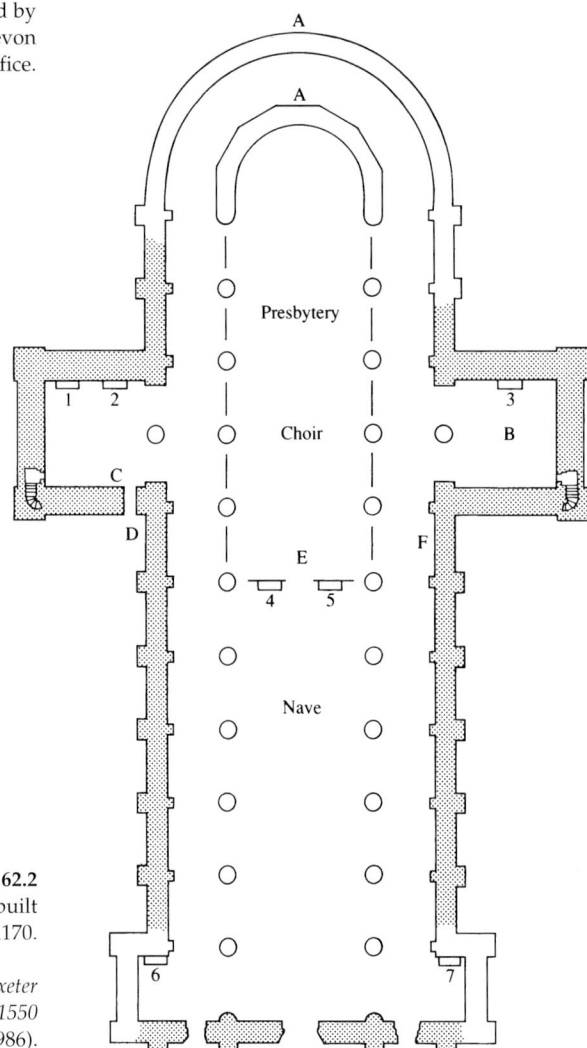

dimensions provided the room that Exeter lacked for a Lady chapel and numerous lesser altars. The third cathedral shared the width of its predecessor, and kept the two side towers, but was half as long again towards the east. The new eastern extension was built first, in the Decorated style, including a Lady chapel; the choir was moved further eastwards, the lower parts of the towers were opened to the interior and made into transepts, and the Romanesque nave was rebuilt in the style of the rest. The new cathedral provided ample room for the clergy and the services.

Presbytery

Choir

Nave

Map 62.2
The second Cathedral built
1114–c.1170.

SOURCE: N.I. Orme, *Exeter
Cathedral As It Was: 1050–1550*
(Exeter, 1986).

A Possible forms of east end
B Possible site of Lady Chapel
C Small door in North tower
D Canons' bread-house
E Approximate site of pulpitum (choir-screen)
F "Briwere door" 13th century

Altars
1 Cross
2 St Paul
3 ? Lady Chapel, later St John Baptist
4 St Mary (nave)
5 St John Baptist, later St Nicholas
6 St Edmund
7 ? St Richard and St Radegund

0   10   20   30m

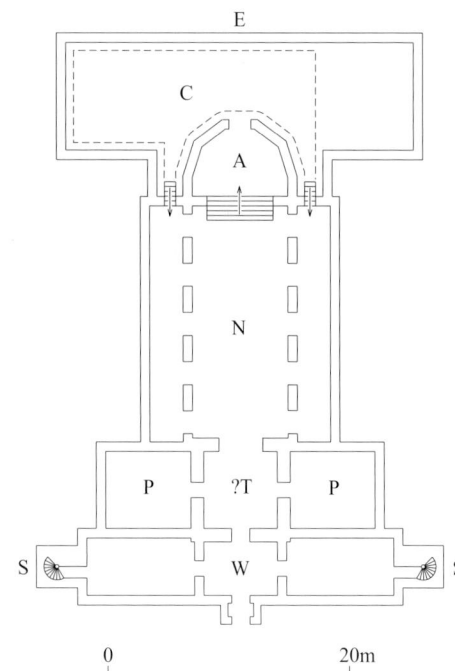

A Apse   S Stair-turret
C Crypt   E Eastern transept
N Nave   T Tower
P Porticus   W Westwork

The choir could accommodate the ninety or so who now attended church each day, the Lady chapel housed a separate cycle of worship, and the number of lesser altars was doubled to fourteen, most of them housed in chapels or enclosures to give them more dignity and privacy. There was a large nave for sermons, more spacious aisles for processions, more ground for burials and plenty of wall space for displaying images.

Paradoxically, the completion of the new cathedral was soon followed by the Black Death of 1348–9, which killed off many clergy and made available parish posts for the survivors which were better paid than the work of cathedral vicars choral and chantry priests. It became more difficult to recruit such men and the number of cathedral clergy declined. The building was now almost too large for those who remained and two of the smaller chapels were eventually converted to other purposes. The erection of ancillary premises continued in the fourteenth and fifteenth centuries, however, including a cloister south of the nave, a library in the north walk of the cloister, and a college in Kalendarhay for the vicars choral who had hitherto lived dispersed (**Map 62.4**). A final period of development followed in the late fifteenth and early sixteenth centuries, in which the recruitment of the lesser clergy improved. More chantries were founded and construction of a house started in 1528 in which the chantry priests could live together. Two new chapels, St George and St Saviour, were built at the east end of the choir aisles in the

Map 62.3
The third Cathedral built
*c.*1270–1342.

SOURCE: as **Map 62.2.**

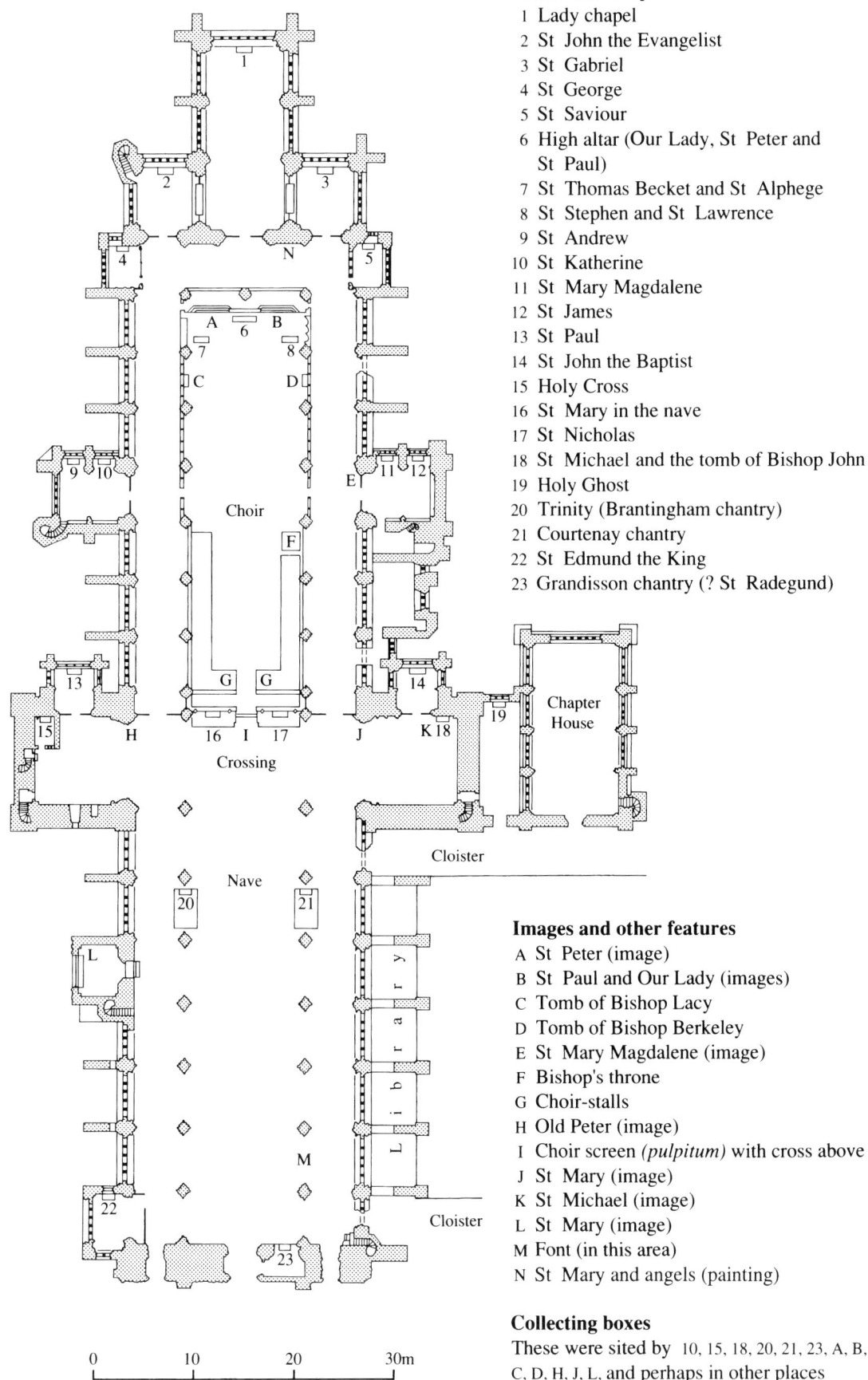

Altars and chapels

1 Lady chapel
2 St John the Evangelist
3 St Gabriel
4 St George
5 St Saviour
6 High altar (Our Lady, St Peter and
  St Paul)
7 St Thomas Becket and St Alphege
8 St Stephen and St Lawrence
9 St Andrew
10 St Katherine
11 St Mary Magdalene
12 St James
13 St Paul
14 St John the Baptist
15 Holy Cross
16 St Mary in the nave
17 St Nicholas
18 St Michael and the tomb of Bishop John
19 Holy Ghost
20 Trinity (Brantingham chantry)
21 Courtenay chantry
22 St Edmund the King
23 Grandisson chantry (? St Radegund)

Images and other features

A St Peter (image)
B St Paul and Our Lady (images)
C Tomb of Bishop Lacy
D Tomb of Bishop Berkeley
E St Mary Magdalene (image)
F Bishop's throne
G Choir-stalls
H Old Peter (image)
I Choir screen (*pulpitum*) with cross above
J St Mary (image)
K St Michael (image)
L St Mary (image)
M Font (in this area)
N St Mary and angels (painting)

Collecting boxes

These were sited by 10, 15, 18, 20, 21, 23, A, B,
C, D, H, J, L, and perhaps in other places

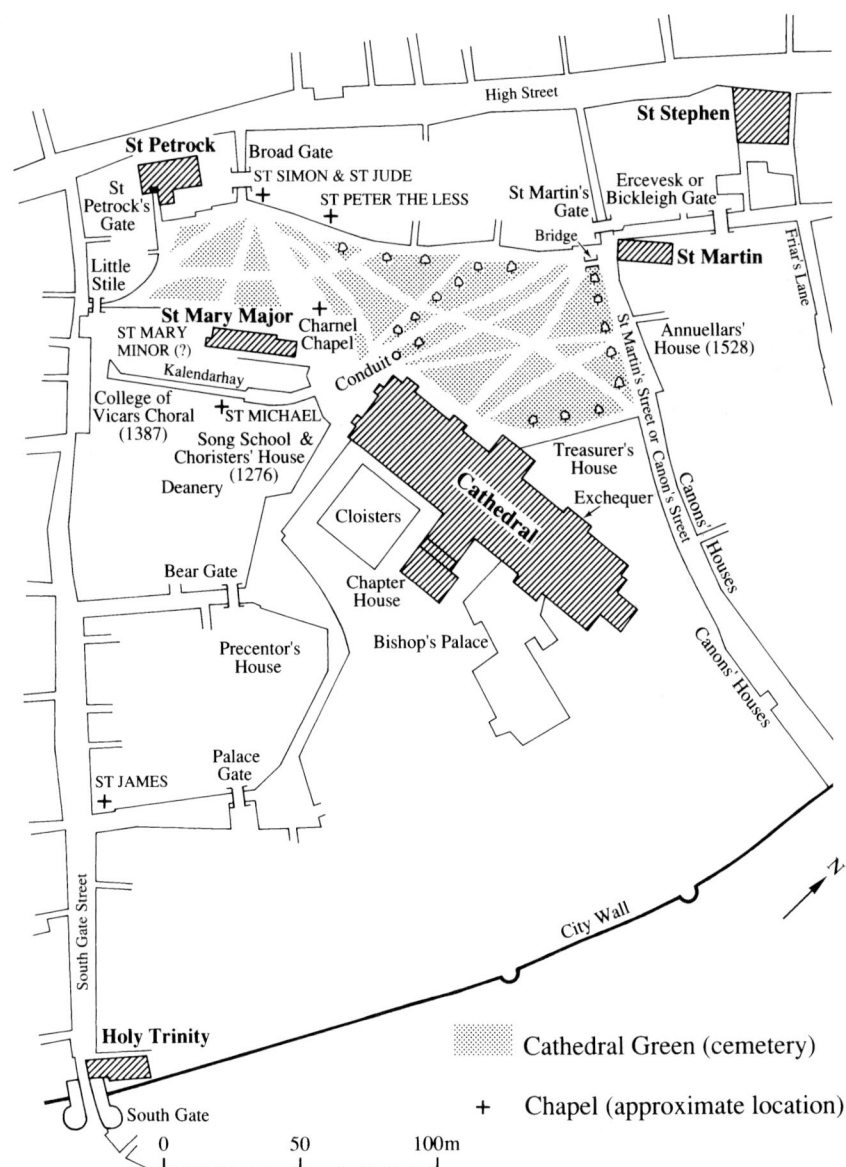

Map 62.4
Exeter Cathedral Close in
the Middle Ages.

SOURCE: as **Map 62.2.**

1510s to provide extra altars for intercessory masses. In the 1520s the cathedral was a prosperous community of clergy and a busy centre of spiritual activities but in 1547–9 it was overtaken by the Reformation. The intercessory masses and the chantry priests were abolished and the vicars choral were reduced in number. Unlike the monasteries, however, the cathedral kept most of its property and stayed wealthy enough to maintain its fabric, though it was too reduced in staffing and activities to think of rebuilding or expansion. This balance of resources ensured the survival of the building in its fourteenth-century form for long enough to reach the modern age of conservation which began in the early nineteenth century.

FURTHER READING

The best general history of the cathedral is Audrey Erskine, V. Hope and L.J. Lloyd, *Exeter Cathedral: A Short History and Description*, 2nd edn (Exeter, 1988). For the Saxon minster, see C.G. Henderson and P.T. Bidwell, 'The Saxon minster at Exeter', in Susan M. Pearce (ed.), *The Early Church in Great Britain and Ireland* (Oxford, 1982), 145–75, and J. Blair and N. Orme, 'The Anglo-Saxon minster and cathedral at Exeter: twin churches?', *Friends of Exeter Cathedral, Annual Report*, 65 (1995), 24–6. The whole pre-Reformation period is discussed in N. Orme, *Exeter Cathedral: As It Was 1050–1550* (Exeter, 1986).

EXETER AND PLYMOUTH

MAP EVIDENCE
OF THE
GROWTH OF EXETER
DURING THE NINETEENTH CENTURY

RICHARD OLIVER

In the nineteenth century the built-over area which outsiders might loosely define as 'Exeter' consisted of three distinct areas, under three different jurisdictions. First, there was the ancient city of Exeter proper, lying to the east of the River Exe on its hilltop. Then to the south-west was St Thomas, which was separated from the city physically by the river and was administratively separate as well: by 1870 it had its own local board of health, and in 1895 it became an urban district. St Thomas was absorbed in the city in 1900. Lastly, to the east lay the village of Heavitree, also latterly an urban district, which only became part of the city of Exeter in 1913. This administrative history is of some importance as most nineteenth-century map-makers concentrated on the city alone, to the exclusion of St Thomas and Heavitree, and so give a somewhat incomplete impression of Exeter's growth.

ORDNANCE SURVEY SMALL-SCALE MAPS OF EXETER (Maps 63.1, 63.2, 63.3 and 63.4)

One exception to this is the Ordnance Survey's one-inch (1:63,360) map. Four editions of this map are reproduced in this chapter to provide a synoptic picture of the city's growth between about 1800 and 1914. The first of these maps (**Map 63.1**), surveyed probably in 1801, shows Heavitree still physically separate from the city.[1] The second (**Map 63.2**), probably revised in the late 1840s, shortly after the arrival of the first railways, shows Heavitree loosely connected to the city by an area of large houses in their own grounds.[2] The third and fourth (**Maps 63.3 and 63.4**) represent the situation in 1888 and 1913 respectively and show the built-up area now fairly continuous for a mile or more from the centre of the historic city, as defined by the cathedral and the castle.

R. BROWN, 'CITY OF EXETER', 1835 (Map 63.5)

Larger-scale maps enable the growth of the city to be shown in greater detail. John Coldridge's map of 1819 shows a state of things roughly comparable with the earliest Ordnance Survey map. Despite the presence of terraces, such as Southernhay, built during the previous half-century, Coldridge's map shows a townscape still markedly medieval in many ways, not least in its road approaches, and in its overall compactness and cohesiveness (see chapter 61).[3] Development beyond the city walls and the quayside is almost entirely confined to ribbon-development along existing roads.

Coldridge's map may have provided a basis for R. Brown's *The City of Exeter* of 1835 (**Map 63.5**).[4] It shows the city at an interesting time, contemporary with the Municipal Corporations Act and shortly after the New North Road had been constructed. It is not quite up to date, as the iron bridge of 1833–4, which considerably eased the approach to North Street, is not shown, but it does show the beginnings of Queen Street as eventually laid out but unnamed, north of the City Prison. Brown's map indicates quite well the move away from ribbon development towards that of laying out streets for residential use rather than for through traffic, for example in the angle of Sidwell Street and Paris Street, and by the streets south of Magdalen Road (the Mount Radford area).

FEATHERSTONE AND CO., 'MAP OF THE CITY OF EXETER', c.1852 (Map 63.6)

In 1840 the city was re-surveyed at a scale of one inch to four chains (1:3168) by John Wood.[5] This map is too large to be illustrated satisfactorily here, but formed the basis of a number of maps produced over the next half-century. One such (**Map 63.6**) was that published by Featherstone and Company in c.1852, at one inch to seven chains (1:5544).[6] It shows much more building in specially laid out residential streets and two of the more notable developments of the early 1840s: Queen Street and the first railways. Up to about the time of Wood's map (1840) Exeter's outward growth had been characterized by the construction of fairly large houses—or 'residences'—suitable for well-to-do tradesmen, retired army and navy officers, public servants and the like. Some of these detached and semi-detached houses are apparent in the south-east part of Featherstone's map, but, like Wood's map, it does not extend far enough north to show the most splendid of all, Pennsylvania Park, which dated from the early 1820s.[7] Two contrasting

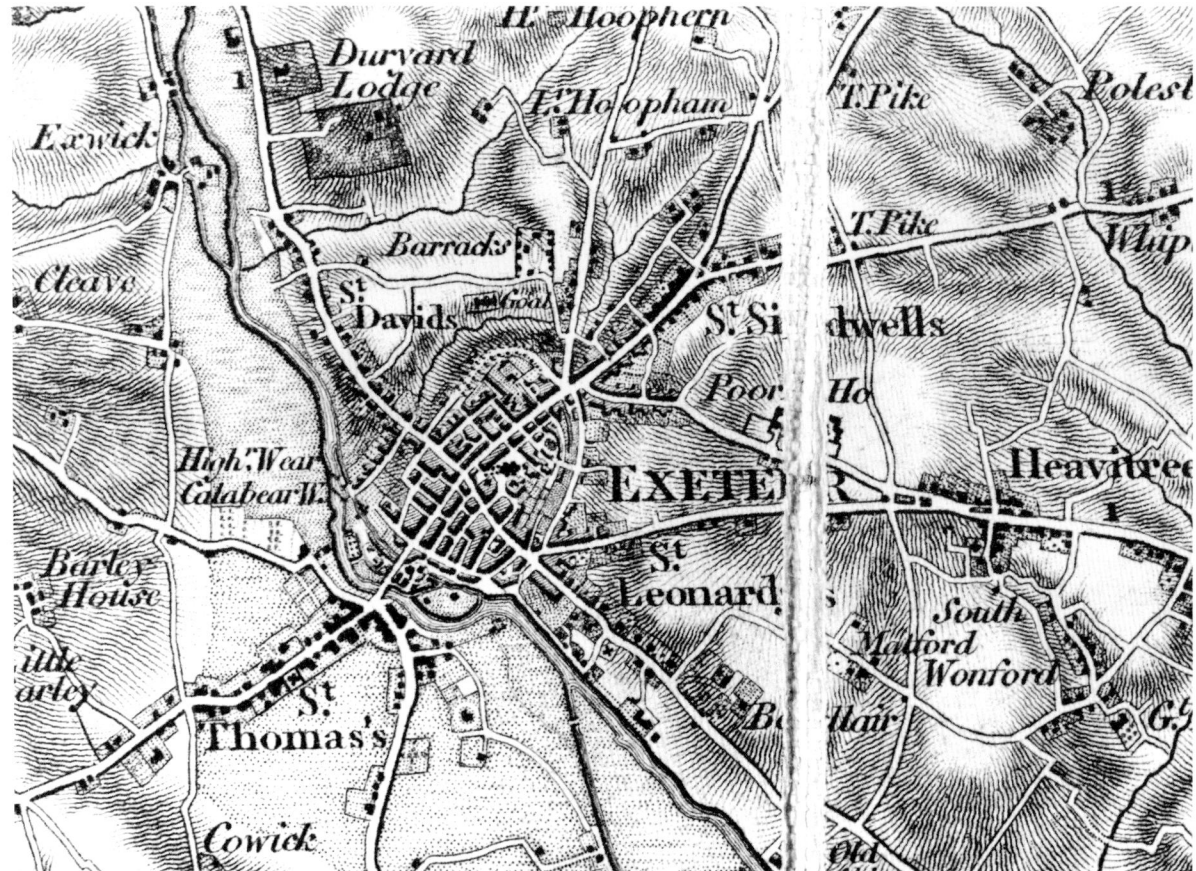

Map 63.1
Ordnance Survey one-inch
Old Series sheet 22
(1809; surveyed 1801).

SOURCE: photograph
reproduced by permission
of University of Exeter,
Department of Geography.

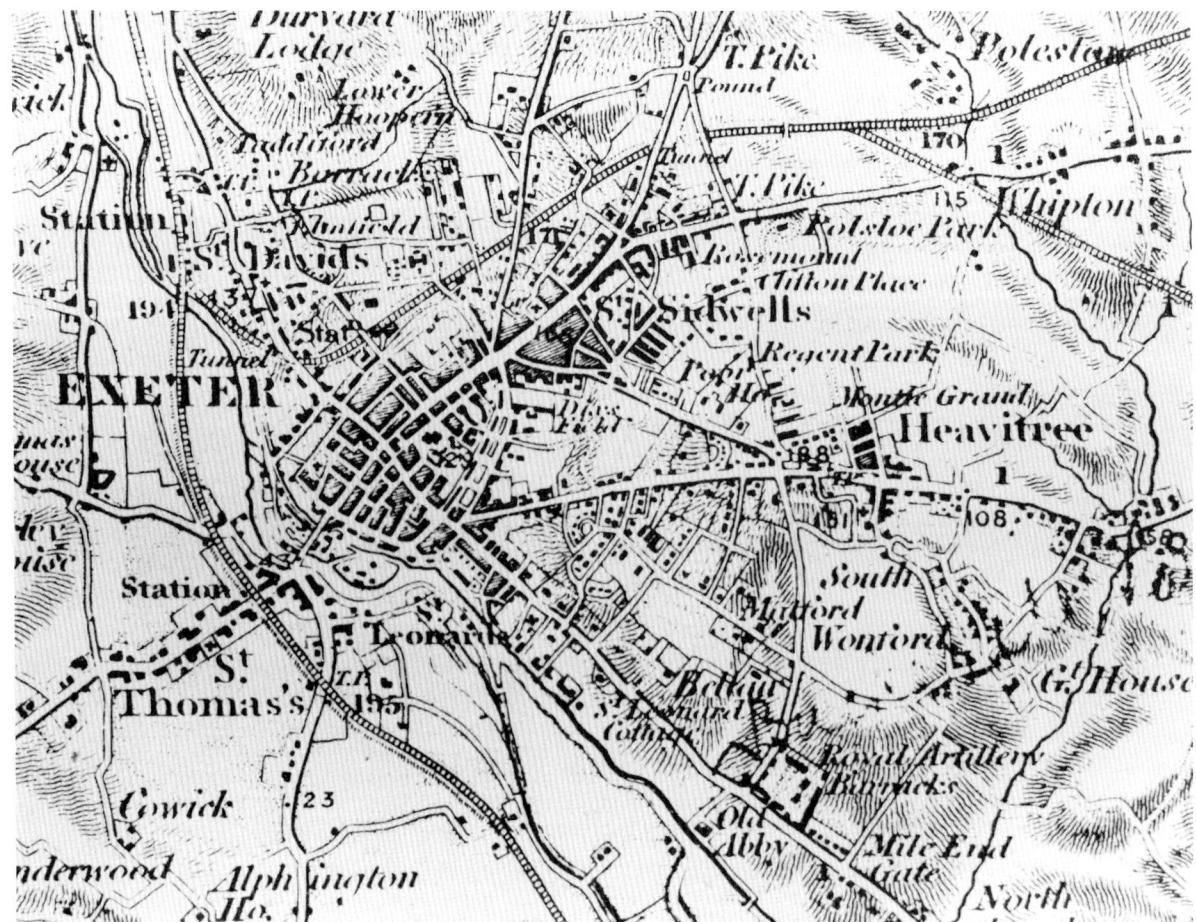

Map 63.2
Ordnance Survey one-inch
Old Series sheet 22
(1809; partly revised c.1848)

SOURCE: photograph
reproduced by permission
of University of Exeter,
Department of Geography.

Map 63.3
Ordnance Survey one-inch
New Series sheet 325
(1894; surveyed 1888).

SOURCE: private collection.

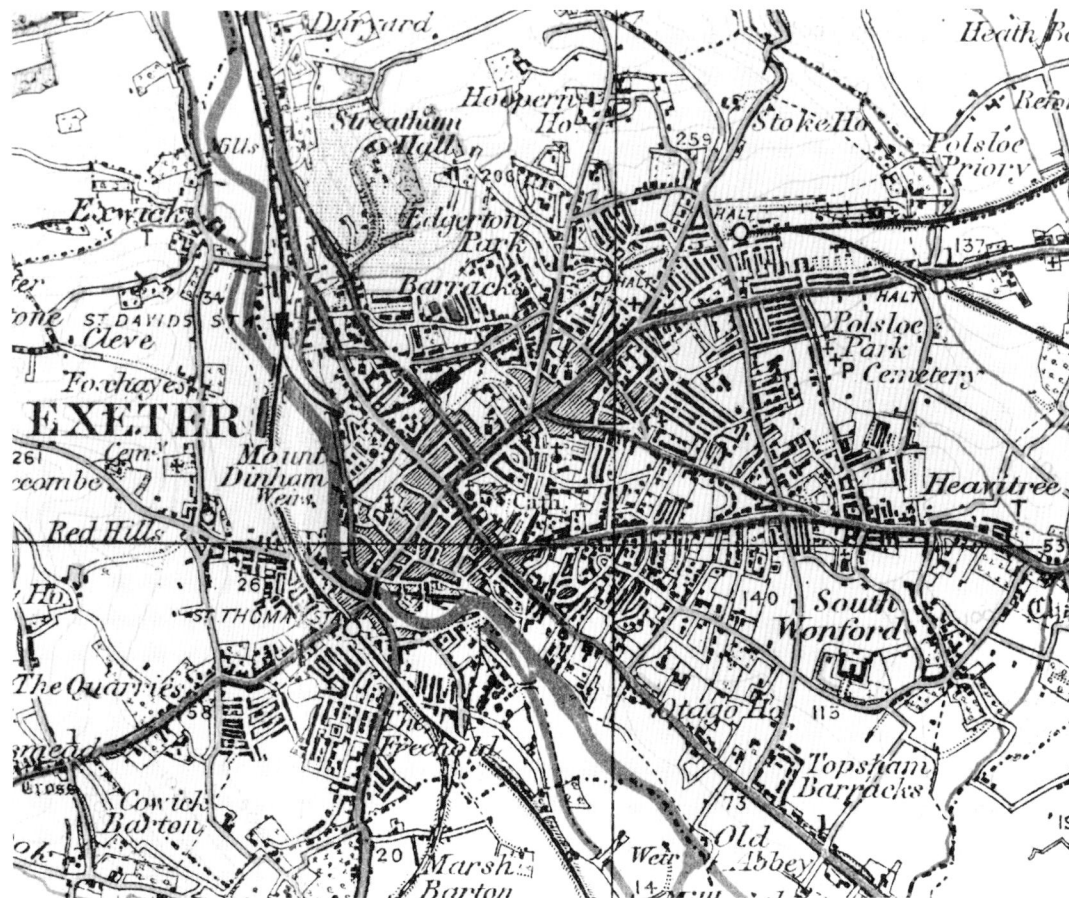

Map 63.4
Ordnance Survey one-inch
Popular Edition
(1919; revised 1912–1913).

SOURCE: private collection.

Map 63.5
The City of Exeter, 1835,
'Drawn by R. Brown',
'Published by R. Colliver,
Holloway Street, Exeter'.

SOURCE: photograph
reproduced by permission
of Westcountry Studies
Library, Exeter.

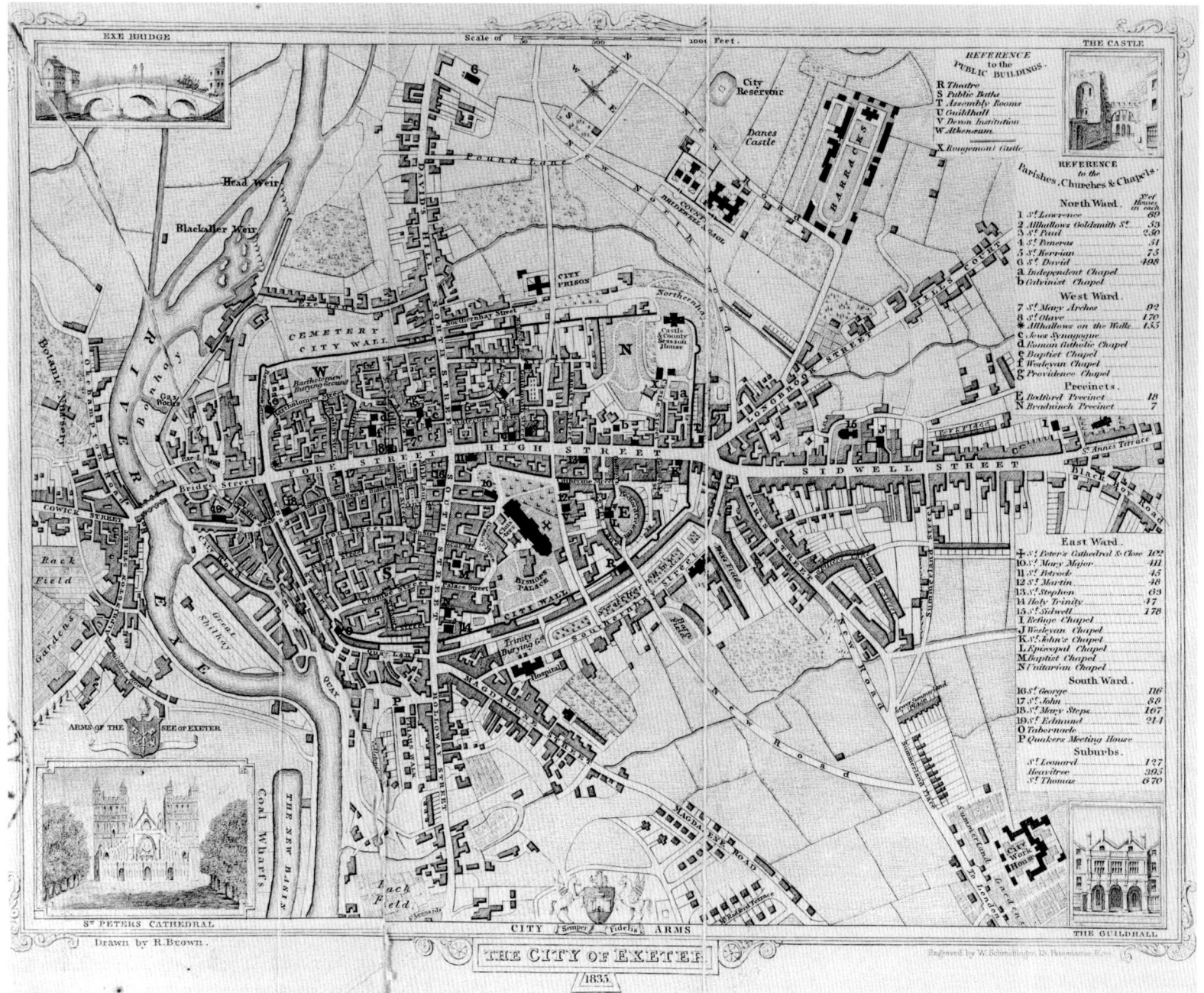

developments are shown on the eastern side of the city: New Town, mostly of fairly modest terrace houses, and the more spacious terraces at Regents Park, which must have appeared even more isolated then than they do today. New Town began in the mid-1830s: at first (as on Wood's map), it was known as Schlater's town, after one of the landowners and developers.[8] The Brick Field to the west of Workhouse Lane (now Polsloe Road) is an indication of how very local building materials were used. Even in this the early railway age, bricks were far too heavy to be transported far, unless for special purposes. The 'sprawl' is offset to an extent by some infill building, for example in the apex of New North Road and Longbrook Street. Beyond this, only scattered building is shown; Union Road towards the north of this map appears still isolated. A constraint not clearly indicated on the maps is the Longbrook valley, between the city wall and the New North Road, which was steep-sided and unsuited for building, and was 'blighted' during the 1840s and 1850s by the prospect of a railway being built through it. A lack of planning controls combined with fragmented landownership could and did result in isolated fields being built over; the only constraint on 'sprawl' was that of walking distance from the city and likely employment.

By the time that Featherstone's map was published in the early 1850s, Exeter's growth was turning away from an emphasis on larger houses for the more prosperous and well-heeled, and towards humbler ones for lower socio-economic groups. By then the city's population growth had very much slowed, and whilst

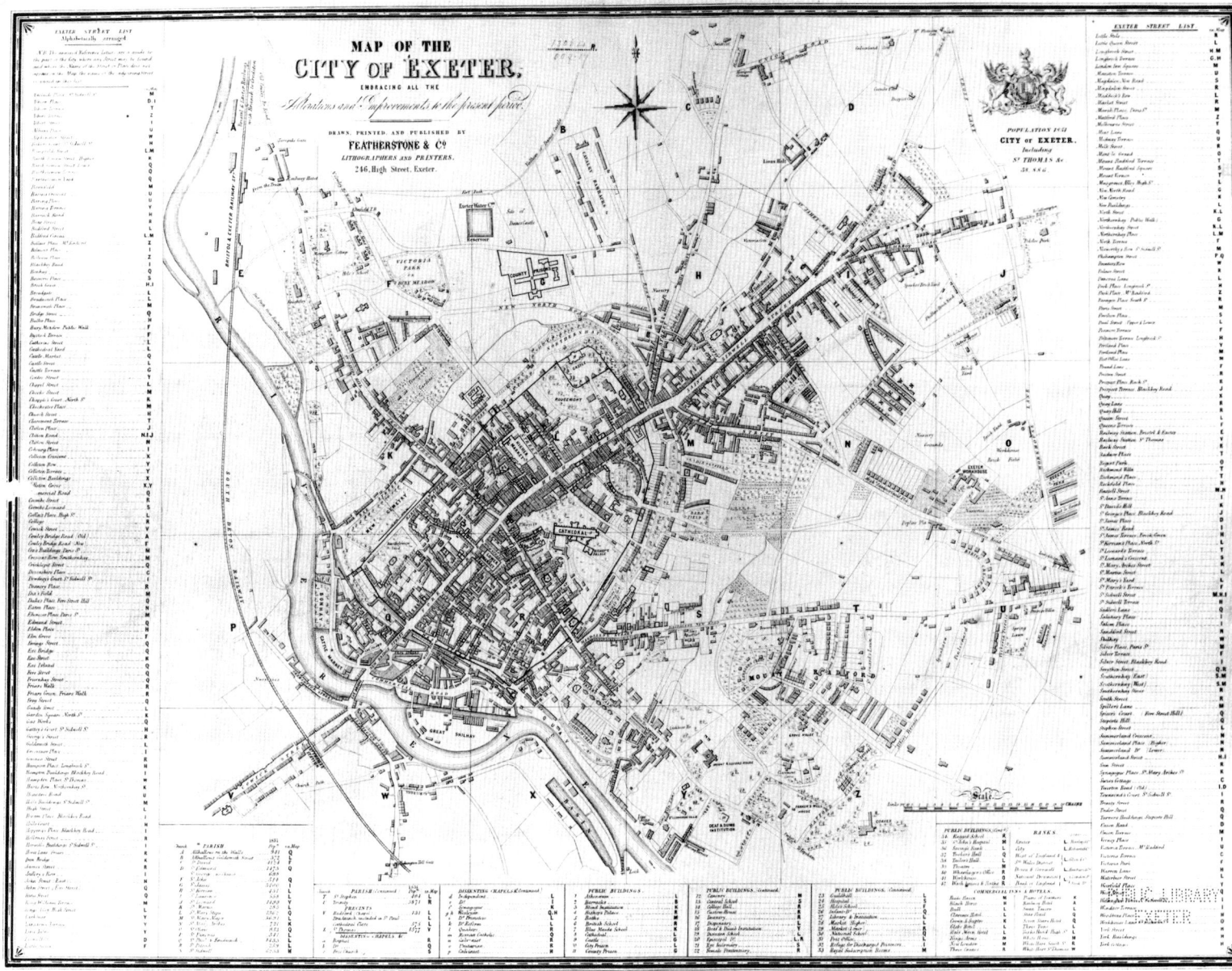

Map 63.6
Map of the City of Exeter ..., n.d.
(*c*.1852), 'published by
Featherstone & Co.,
Lithographers & Printers,
246, High Street, Exeter'.

SOURCE: photograph
reproduced by permission of
Westcountry Studies Library,
Exeter.

Exeter retained its attraction as a centre for retirement, its growth as a service centre only balanced its steep decline as an industrial centre. Although there were still extensive areas of slums, notably in the west part of St Sidwell's and by the river adjoining St Thomas, further developments were usually built to a reasonable standard, thanks to the introduction of local by-laws. These could not control the siting of new buildings, but they could and did control quality, often in quite subtle ways. For example, in 1885 a builder was brought before the St Thomas Local Board of Health for using an incorrect mix of mortar![9]

HENRY BESLEY, 'EXETER', 1880 (Map 63.7)

It is unusual for general purpose maps, such as those used to illustrate this chapter, to show underground services, and these are no exception. Thus they do not show that for much of the nineteenth century Exeter was inadequately drained and inadequately supplied with water.[10] It was to facilitate the sanitary improvement of the city that the Ordnance Survey made a very large scale 1:500 survey of Exeter in 1875–6. Though these maps contain exceptional topographic detail, they are not easy to use to demonstrate overall topographic change. No small-scale version of these maps was produced. The Ordnance Survey was also constrained by the city boundary and so omits most of St Thomas and Heavitree.[11] In its place Besley's map of 1880 is reproduced as **Map 63.7**.[12] This is typical of later Victorian street maps: functional, no doubt, but hardly distinguished as to execution. It appears to derive either from Wood or from Featherstone and, although there has been some revision of developed areas, such as the replacement of the City Prison by the Rougemont Hotel in the late 1870s, it would be

Map 63.7
Exeter, 1880, 'Published by
Henry Besley, Directory Office,
South St, Exeter'.

SOURCE: photograph
reproduced by permission
of Westcountry Studies
Library, Exeter.

REFERENCE
1 Guildhall
2 Theatre
3 Bankruptcy Court &c.
4 Stamp Office
5 Excise Office
6 Custom House
7 Wharfingers Office
8 Baths & Washouses
9 Conduit
10 St Johns Hospital and Blue School
11 West of England Insurance Office
12 Savings Bank
13 Exeter Bank
14 City Bank
15 Devon and Cornwall Bank
16 West of England and South Wales Bank
17 National Provincial Bank
18 Devon and Exeter Institution
19 Athenaeum
20 Royal Subscription Rooms
21 Reformatory
22 Tuckers and Masonic Hall
23 Tailors Hall
24 Episcopal Charity Schools
25 National Schools
26 Diocesan Training College
27 Infant School
28 New London Inn
29 Bude Haven Hotel
30 Clarence Hotel
31 Half Moon
32 White Lion
33 Queens Hotel
34 Globe Inn
35 White Hart Inn
36 Railway Hotel
37 Crown and Sceptre Inn
38 Black Lions Inn

unwise to infer that those details which are common to the two maps had necessarily remained unchanged in the intervening years. Besley's map accompanied a street directory and is at its most useful in showing further building development. Although, in common with maps like Featherstone's, it does not extend north of Union Road or east of Workhouse Lane, it does show development under way at Prospect Park, between Union Road and the London & South Western Railway, which had been opened in 1860 and which occupied the Longbrook valley. Here the map gives an impression of a dynamic landscape: Victoria Street is named and built-up (work had begun *c*.1869); Culverland Road (now Springfield Road) is named, but building has apparently not yet begun, and a road to the east

(now Culverland Road) is in outline with as yet no name.[13] The need to interpret this map with caution is illustrated by the treatment of the road crossing of the railway at Lions Holt (spelled 'Hold' on the map), where both the old route and the diverted route are shown, and by the omission of the two short railway tunnels, under St David's Hill and Old Tiverton Road. There is some evidence that the early Victorian tendency to urban sprawl is being offset by infilling, as on the city side of New Town, though these modest infill developments are offset by more scattered villas around Mount Radford. Not obvious on the reproduction are the indications of public buildings, including the blocks of almshouses which continue to be a feature of the city.

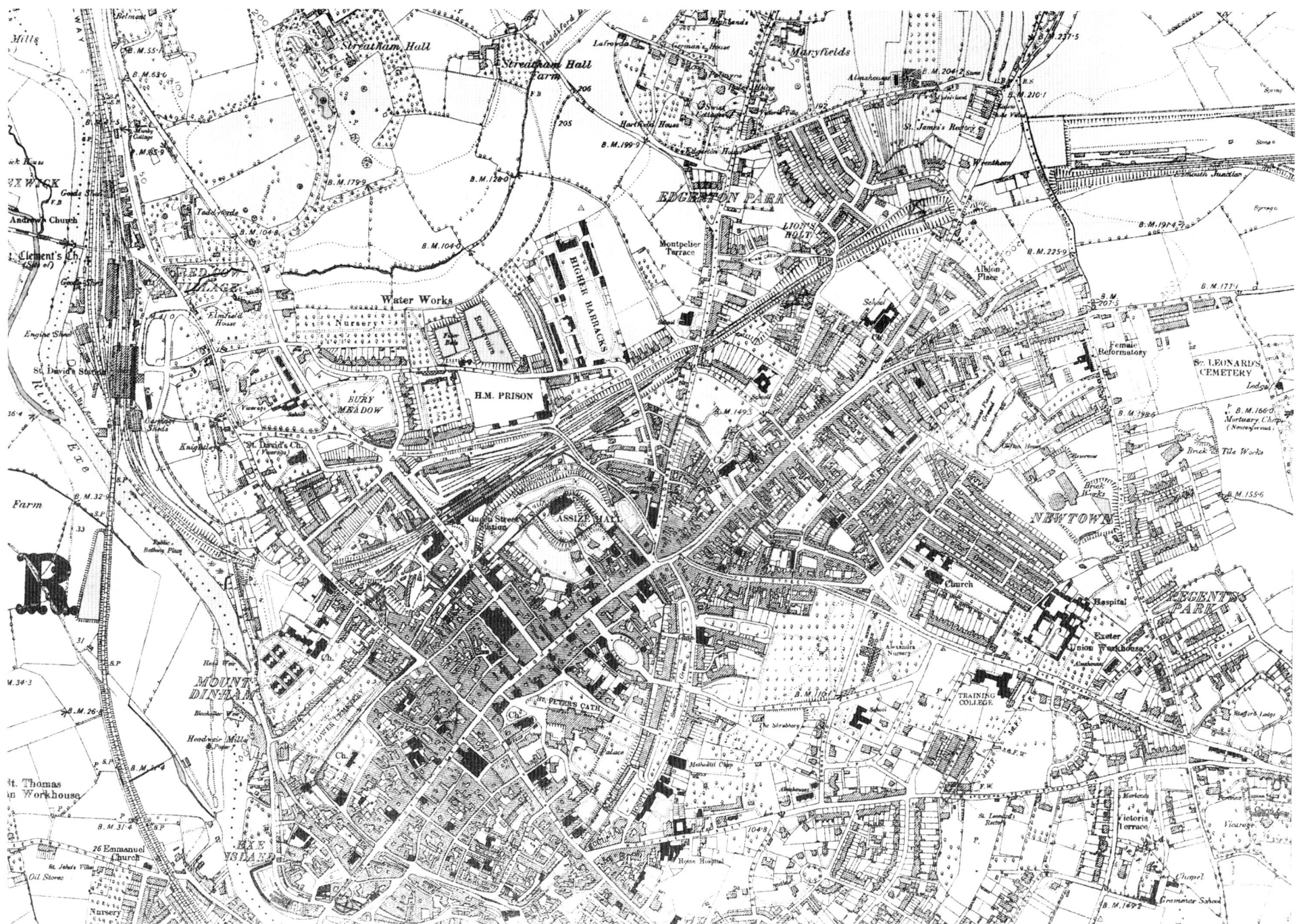

ORDNANCE SURVEY SIX-INCH MAP, SURVEYED 1888 (Map 63.8)

The Ordnance Survey came to Exeter again in 1888 as part of a comprehensive survey of Devon at 1:500 (for larger towns) or 1:2500 (for small towns and rural areas), and this time presented their results in a form amenable to reproduction on the page of a book (**Map 63.8**). By now, the Ordnance Survey's work had the advantage of being continuous across Britain, and being underwritten by the state, and so there are none of the problems of limited cover which affect earlier mapping of the city. Compared with Besley's map of less than a decade earlier, there

would appear to have been considerable aggregate development and, in plan, Exeter is once more assuming some of that compactness which it had shown at the beginning of the century: Pennsylvania Park is still isolated, and there are still orchards around Barnfield Road (to the south-east of the old city), but elsewhere the gaps in building are far fewer. There are two groups of brickworks, on either side of Polsloe Road, which were no doubt busy supplying the start of the Polsloe Park development; a former brick yard to the north-west is now the Belmont Pleasure Ground, one of several open spaces which punctuated building. Development is now dense enough for Heavitree to be, in building if not in administration, an integral

Map 63.8
Ordnance Survey six-inch sheet Devon 80 NW (1890; surveyed 1888).

SOURCE: private collection.

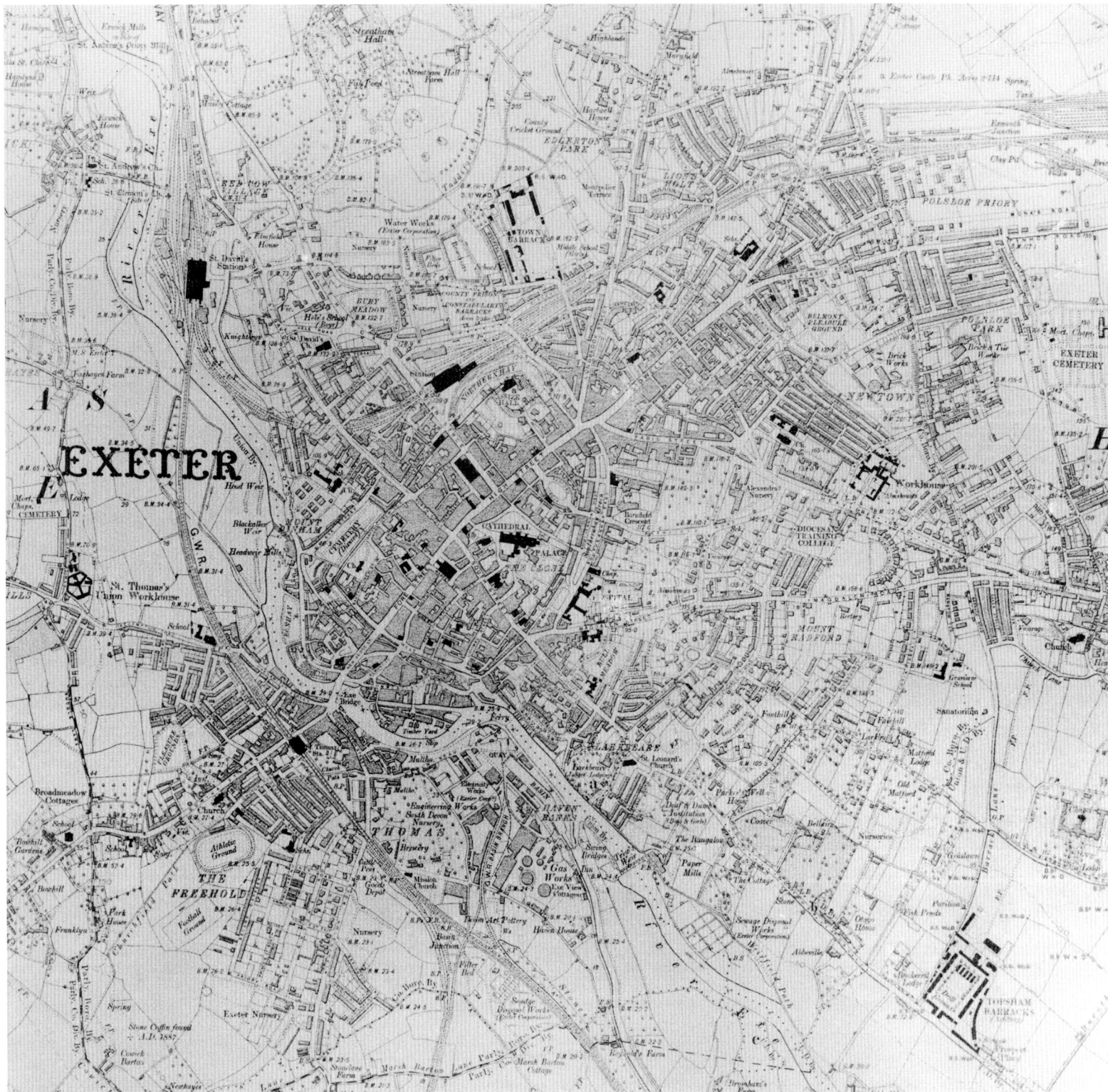

part of Exeter. St Thomas is revealed as having grown since the first Ordnance Survey of 1801, with the old settlement of ribbon development now supplemented by blocks of terrace houses. Being a mix of low-lying land by the river and steep slopes elsewhere, it was unappealing to the builders of detached villas, and the proximity of a gas works did nothing for the *cachet* of the neighbourhood.[14]

ORDNANCE SURVEY SIX-INCH MAP, REVISED 1904 (**Map 63.9**)

Whilst **Map 63.4** shows the ultimate extent of Exeter's growth on the eve of the First World War, the six-inch map revised in 1904 (**Map 63.9**) shows its development at its Edwardian peak of growth. There was now much less scope for infilling, particularly in areas suited to the development of lower-class housing, and so there is a marked trend to expand outwards. Good examples of this are to be seen in the north-east part of the map: the Polsloe Park development of the 1890s is now complete and the Polsloe Priory estate to the north is now being built up. Whatever picturesque aspiration there may be in 'Priory' or 'Park', both consist exclusively of terraced houses, mostly modest in size. The continuing need of many people to live within walking distance of their work is illustrated by the large numbers of railway employees living in these and nearby streets at this time: many of them would have been based at the locomotive depot at Exmouth Junction. At this time the Polsloe developments, though in a structural sense an extension of the city, were still adminstratively part of Heavitree.

FURTHER READING

The development of the city is described in outline in two books by Robert Newton: *Eighteenth-Century Exeter* (Exeter, 1984), especially 120–38, and *Victorian Exeter* (Leicester, 1968), especially 132–48, 299–302. The development of Exeter and its suburbs is investigated in more detail in Joyce C. Miles, 'The rise of suburban Exeter and the naming of its streets and houses' (unpublished Ph.D. thesis, University of Leicester, 1990). This present essay is extensively indebted to these three works.

EXETER
IN THE
TWENTIETH CENTURY

ANDREW GILG

In the twentieth century Exeter experienced two quite distinct periods of development, first a continuation of the period of unplanned growth typical of previous centuries, and then, since 1946, a period of planned growth. It is this last period which is the prime focus of this chapter. Before then Exeter continued to develop as in former times but far more rapidly, so much so that between 1919 and 1940 its areal extent almost doubled (**Map 64.1**), while its population increased by only some 18 per cent. During the inter-war period the city continued to expand away from its historic core, notably to the south-east, while the two barriers to growth, Stoke Hill to the north and the River Exe to the west (**Map 64.1**), continued to exert limiting, albeit reduced, constraints to growth. Growth between 1914 and the mid-1960s was also closely linked to residential status, with the more exclusive areas (detached houses with garage space) being found along Heavitree Road and Pennsylvania Road and in an outer ring of suburbs, while the new, less exclusive areas (semi-detached houses without garage space) were to be found in large bands to the east and west of the city, and to some extent along the Exe, mainly in large local authority housing estates.[1]

PLANS FOR EXETER AND ITS SUB-REGION SINCE 1945 (**Maps 64.2 and 64.3**)

The first plan for Exeter was produced by the celebrated planner Thomas Sharp.[2] It was a non-statutory plan but was very influential, especially for the rebuilding of the bomb-damaged city centre (see below). With regard to the wider city area, Thomas Sharp set down a number of principles that have affected Exeter from that time until today. These included:

(a) major bypass roads to the south-east and south-west of the city and main feeder roads to the city centre, along Pinhoe Road, Heavitree Road, Topsham Road, Hill Barton, Alphington Road and Okehampton Road;

(b) suburban expansion in Pennsylvania, Pinhoe and Countess Wear, but safeguarding Exeter's green hills to the west and north;

(c) industrial development in Marsh Barton, and along the London railway line through Whipton and Pinhoe;

(d) major open space provision around the soon-to-be-born University, along the River Exe, and around Wonford.

The main themes of the Sharp plan of 1946 were followed in Exeter's statutory development plan, the 1963 version of which is shown in **Map 64.2**. For example, the ring road bypass now became the boundary for expansion to the south-east and south. To cater for continued growth, however, the plan had to allocate two further major areas for residential development, the Sylvania and Exwick areas, which have so disfigured the city with their characterless and service-free monotony, albeit just preserving the green hills.

By the 1960s, the creation of new employment outside the city centre, notably in the newly built County Hall, the growing University, and the rapidly expanding Marsh Barton industrial estate, was producing a growing traffic problem, partly because zoning was separating land-use functions leading to daily journeys-to-work of increasing length. Notwithstanding these developments, in 1965 the city centre still dominated the employment picture. In 1969 Exeter City Council and Devon County Council produced a major joint feasibility study which attempted to examine the options available for planning for the estimated growth in the Exeter region from 246,000 people in 1961 to 301,500 in 1991 and 353,500 by 2010.[3] Once again, land within the constraints of the 1946 and 1963 plans was shown to be running out and a discussion document, *A Strategy for Growth,* was produced which proposed major new sites for development not just in Exwick and Sylvania, but on even larger sites to the east and south towards the planned M5 motorway and A30 Pocombe–Peamore link.[4]

In 1975 Devon County Council produced a draft strategy for the Exeter sub-region to the year 2001.[5] The basic argument was the continuing one that Exeter was reaching its limit of size both in terms of a population of around 90,000 and in terms of physical environment, being hemmed in by the green hills to the north and west (which give the city much of its character), to the south by the Exe estuary, and the east by the M5 motorway and prime agricultural land. Therefore, the 1975 document proposed three alternative options: 1) the development of linked communities, basically all the towns in the sub-region; 2) the growth of inland towns; and 3) the growth of selected inland and coastal towns. These options were then further tested as part of the structure planning process (see chapter 55).

Allied to the problems of growth were the problems of traffic congestion, which had been examined in the 1969 feasibility study, but about which little had been done, largely because the option of new roads had to be rejected in the 1970s due to lack of finance and the general political climate which was against new roads. Also, by the late 1970s, the structure plan for Devon was being prepared and Exeter as one of the 'areas of main change' was given a separate section in the plan. The 1977 draft structure plan proposed growth in the south-east and to the west of the M5, a much increased industrial estate at Sowton, and an expanded Marsh Barton industrial estate. In 1984 flesh was put on these bones by the Exeter local plan (**Map 64.3**). The local plan is in many ways an extension of the 1946 and 1963 plans but with a renewed emphasis on infilling as the city continued to strain against its borders.

PLANS FOR EXETER CITY CENTRE (**Map 64.4 and Figs. 64.1, 64.2, 64.3, 64.4 and 64.5**)

Thomas Sharp's aptly named *Exeter Phoenix* of 1946 had a clear vision of not only rebuilding that part of the city damaged by bombing in 1942 (see chapter 23), but also of redevelopment. Instead of advocating the reconstruction of the old city, an approach much favoured on the continent of Europe, not least in West Germany, Sharp proposed the construction of bold new roads, an inner bypass, and a green belt beyond to create in effect a new city centre (**Map 64.4**). Sharp's book is well worth reading, both for its idealistic utopianism and because it provides a basic lesson in planning history, namely that no plan ever achieves all its goals and indeed few plans achieve a majority of their aims. The way in which the city centre was actually rebuilt in subsequent decades is vividly told by Venning, who was closely involved as an employee of the city from 1945 to 1974.[6] In particular, he points to the piecemeal nature of the reconstruction and the length of time involved; the major Debenhams department store building was not constructed until the 1960s (**Fig. 64.1**) and the second Exe bridge not until 1969.

The year 1969 not only marked the effective end of the wartime rebuilding, but also saw the publication of a commissioned report on Exeter city centre by Wilson and Womersley.[7] This examined two key and related issues: first, how to solve Exeter's growing traffic and parking problem, and second, how to improve the rather bleak environment created by the rebuilding. In order to deal with the first problem, Wilson and Womersley proposed an ambitious series of road-building schemes to implement Sharp's inner bypass (**Fig. 64.2**) by the year 2010. This was intended to be achieved by a series of stages, the first of which was the elimination of traffic from the High Street, and the eradication of the bottleneck at the Exe Bridge and the High Street–Paris Street crossroads by the construction of a new north–south route, the building of a major Exe Bridge

roundabout, and a dual carriageway inner bypass to the south of the city. The plan had the merit that the proposed new roads conveniently missed most of the listed buildings and conservation areas that had escaped the blitz. It also had the advantage that the new roads broadly followed the lines of previous city boundaries, notably the Roman walls.

In addition to proposals for a bypass, the 1969 study also proposed the redevelopment of existing buildings along High Street, North Street and Queen Street, and the development of waste land to the south of Paul Street. This area, since known as the Guildhall site, was redeveloped in the 1970s in a rather different way than planned. The fine Higher Market was converted into galleries of small shops, and in the south-east corner the whole block was demolished to make way for an eclectic vernacular style building for a new Marks & Spencer store (**Fig. 64.3**).

At the end of the 1970s, an architectural and urban conservation study recognized that Exeter had not only lost many buildings in the blitz but also just as many in subsequent redevelopments, for example to construct the C&A, Tesco, and Marks & Spencer stores.[8] New emphasis in the 1980s was, therefore, placed on the preservation of the remaining old buildings (**Fig. 64.4**) and, more importantly, on upgrading their environment and finding alternative uses. To this end the 1980s witnessed the upgrading of the Quay, Gandy Street, the Iron Bridge and much sympathetic new development in keeping with the older style, for example a new Regency terrace in Dix's Fields (**Fig. 64.5**).

Partial restraint of the car in the 1980s, the reintroduction of brick and vernacular buildings from the mid-1970s, and the development of tourism in the city give cause for hope in the next millennium, but the continued growth of low-rise, out-of-town developments to the south of the city give cause for concern that Exeter may in the twenty-first century become another formless sprawling city of the type that planning is intended to avoid. If this is to be prevented, some difficult decisions need to be made in the decades to come and alternative sites for development found in south-east Devon. The harsh reality for planners is that the more they succeed in their task of creating both economically viable and attractive environments, the more these places come under assault. In the seeds of success may be the harvest of failure. Exeter in 1999 stands at such a crossroads.

FURTHER READING

B.S. Morgan, 'The residential structure of Exeter', in W.L.D. Ravenhill and K.J. Gregory (eds), *Exeter Essays in Geography* (Exeter, 1971), 21–36; N. Venning, *Exeter: The Blitz and Rebirth of the City: A Pictorial History* (Exeter, 1988).

For **Maps 64.1–64.4** and **Figs 64.1–64.5** see pages 15–21 in the colour section.

EXETER AND PLYMOUTH

PLYMOUTH

MARK BRAYSHAY, CYNTHIA GASKELL BROWN AND JAMES BARBER

THE ORIGINS OF SETTLEMENT (Map 65.1)

When Plymouth became an incorporated borough in the fifteenth century, it had existed as a market settlement for almost two centuries.[1] Indeed its origins may be traced back much further. Excavations at Mount Batten have yielded considerable evidence of Neolithic, Bronze Age and Iron Age settlement.[2] A complete picture of prehistoric settlement in the area now occupied by modern Plymouth has not yet been reconstructed, and **Map 65.1**, therefore, identifies only those locations where evidence of activity has been well established. However, recent research indicates that the extent of settlement was considerably greater than has hitherto been suggested.[3]

Studies at Mount Batten show that the site flourished both as a port and as a centre of industry into the Roman era and it is now believed that a wider network of contemporary sites existed elsewhere in the Plymouth area.[4] Moreover, at the edge of what became medieval Plymouth, excavations in the 1960s yielded more than 100 fragments of Roman tiles buried in the marine silts beneath the reclaimed foreshore of Sutton Harbour.[5] These appear to have come from a late Roman building located near to the site.[6]

The settlements established to the north-west of Sutton Pool became part of a large Saxon royal estate comprising the manors of Sutton, King's Tamerton and Maker.[7] The *Exon Domesday* records that the king's manor of Sutton had land for six ploughs, 2 acres of meadow, 20 acres of pasture, and fifteen sheep. The community consisted of six peasant households and a slave.[8] Although the precise boundaries of the manor are not known, Sutton probably occupied most of the area lying south of the Lipson and Stonehouse creeks: a substantial tract of agricultural land (**Map 65.1**). Soon after 1105, Henry I granted his manors of Maker, King's Tamerton and Sutton to the Valletorts, a family of Norman origin already holding lands in Cornwall who rendered the king support during his campaigns in France.[9]

Beyond Sutton the large area now contained within the boundaries of modern Plymouth was divided by Saxon times into a mosaic of other manors. Indeed, it seems likely that some of the estates, and the settlements within them, are of Celtic origin. The Saxon manor of King's Tamerton has already been noted.

In addition there were the substantial properties of Tamerton (Foliot), Stoke (centred on Stoke Damerel), and Eggbuckland. The remaining smaller manors were Lipson, Leigham, Efford, St Budeaux, Compton, Mutley-Higher, Mutley-Lower, Widey, Manadon, Weston, Burrington, Honicknowle, Whitleigh-East and Whitleigh-West. Stonehouse existed as an extremely small property containing only one peasant farm.[10] During the early medieval period, this basic settlement geography was infilled and extended by the addition of further hamlets and farmsteads (**Map 65.1**).

EARLY PLYMPTON (Map 65.2)

To the east of the city centre, on the other side of the Plym estuary, modern Plymouth extends on to land that in the Saxon period comprised the royal manor of Plympton (with its outlier at Radford), the abbot of Tavistock's manor of Plymstock, and the smaller manors of Hooe, Staddiscombe, Goosewell, Langage and Sparkwell.[11] Of these, Plympton had developed as the most important centre by medieval times.

There were two separate centres in medieval Plympton (**Map 65.2**). First, at the point where the Ridgeway reached the marshy tidal limit of the Plym estuary, there was the priory, originally founded in *c*.900, but closed by the king in 1121 because the 'monks would not give up their concubines'.[12] In the same year, Henry I re-founded the priory for Augustinian canons.[13] Although it was to become one of the richest houses of the order in England, there are now very few visible remains. However, a limited excavation in 1989 revealed the foundations of the south-western range of monastic buildings and further archaeological work has since been completed.[14] St Mary's Church, first built in 1311 as a chapel within the precincts of the priory for worship by parishioners, survives as the most tangible legacy of the presence of the Augustinians in Plympton.

Plympton's second focal point was the castle, erected in the early twelfth century to the south east of the priory by Richard de Redvers, earl of Devon (**Map 65.2**). The town of Plympton 'Erle' or 'Earle' was planted below the castle and its medieval topography is still discernible in the modern townscape.[15] The church,

Main road 1784
Coastline 1784
City of Plymouth boundary 1980
⊙ Prehistoric site

Medieval up to 1500
● Simple settlement
✪ Settlement with church
+ Church
▨ Hamlet or town
▲ Castle
■ Watermill

Post-medieval 1500-1784
○ Simple settlement
▨ Hamlet or town
× Church
□ Watermill
△ Fortification
○ Lime kiln
☾ Limestone quarry

River Tamar

Barrow Site
Barrow Site
Southway
Tamerton Foliot
Looseleigh
Higher Looseleigh
Budshead Mill
Budshead
Whitleigh
Devonport Leat 1797
Little Ernesettle
St Budeaux
Thornbury
Ernesettle
Honicknowle
Crownhill Earthwork
Rock
Kings Tamerton
Knackershole
Coleridge
River Plym
Plym Bridge
Burraton
Manadon
Leigham
Saltash Ferry
Weston Mill
Pennycross Church
Widey Court
Lower Leigham
Kinterbury
Burrington
Ham
Eggbuckland
Boringdon
Plymouth Leat 1591
Prospect
Burleigh
Torr House
Austin
Woodford
Colebrook
Weston Mill Lake
Pound House
Compton
Little Efford
Great Efford
Long Bridge
Marsh Mill
Swilly
1591
Mutley
Earls Mill
Keyham
Higher Swilly
Lipson Mill
Crabtree
Merafield
Plympton St Mary
Ford
The Laira
Priory Mill
Chaddlewood
Mount Pleasant
Devonport Leat
Stoke Damerel
Lipson
Underwood
Plympton St Maurice (Plympton Erle)
River Tamar
Saltram
1593
Stoke Church
Mill
Stonehouse Leat (Drake's) Leat
Stonehouse
Hardwick
Dock
Stonehouse Creek
Plymouth (Drake's) Leat
Plymouth
Tothill
Hamoaze
Mutton Cove
Millbay
Drake's Mills
Sutton Pool
Tothill Mill
Wixenford
Hay
St John's Lake
Chapel
Fisher's Nose
Pomphlett
Devil's Point
Firestone Bay
Mount Batten Port
Cattewater
Ferry
Oreston
Pomphlett Mill
West Stonehouse
Cremyll Ferry
Barn Pool
St Nicholas or Drake's Island
Turnchapel
Plymstock
Elburton
Millbrook Lake
Mount Edgcumbe
The Sound
Hooe
Radford Lake
Stamford Hill Cemetery
Radford
Staddiscombe Round
Staddiscombe

0   1   2km

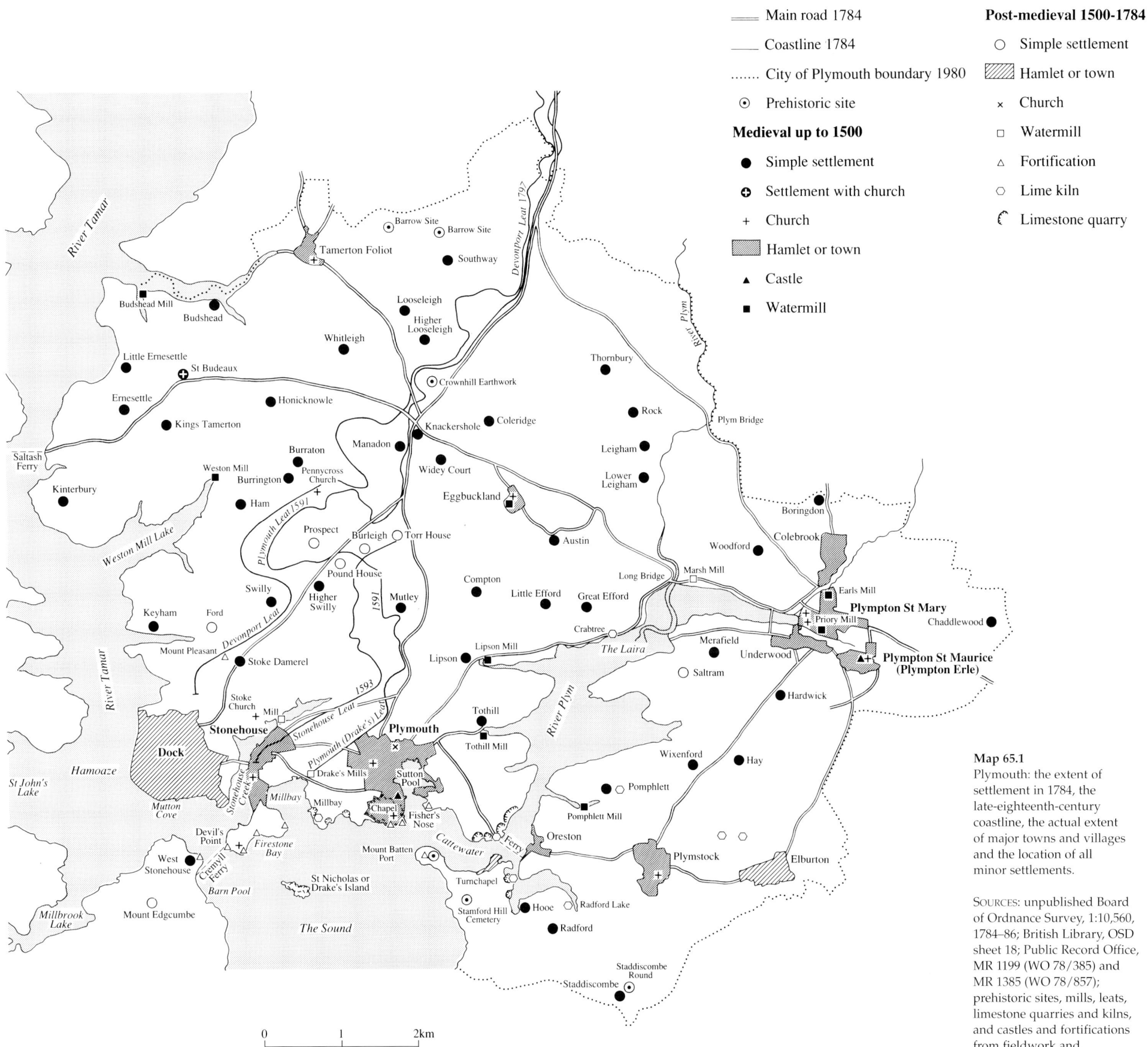

Map 65.1
Plymouth: the extent of settlement in 1784, the late-eighteenth-century coastline, the actual extent of major towns and villages and the location of all minor settlements.

SOURCES: unpublished Board of Ordnance Survey, 1:10,560, 1784–86; British Library, OSD sheet 18; Public Record Office, MR 1199 (WO 78/385) and MR 1385 (WO 78/857); prehistoric sites, mills, leats, limestone quarries and kilns, and castles and fortifications from fieldwork and archaeological records, Plymouth City Museum; and works cited in notes 1 and 13.

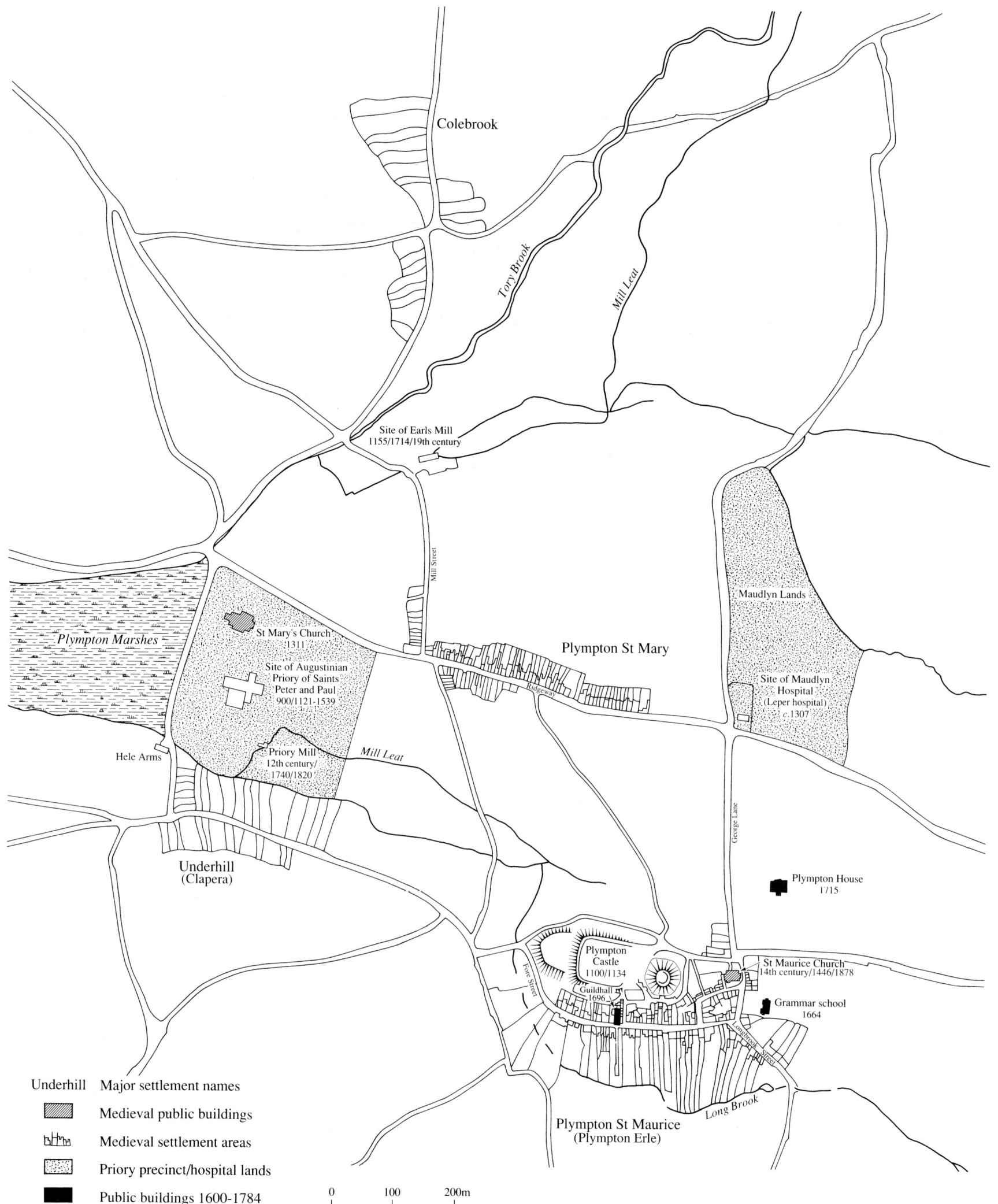

Colebrook

Tory Brook

Mill Leat

Site of Earls Mill
1155/1714/19th century

Mill Street

Maudlyn Lands

Plympton Marshes

St Mary's Church
1311

Plympton St Mary

Site of Augustinian
Priory of Saints
Peter and Paul
900/1121-1539

Site of Maudlyn
Hospital
(Leper hospital)
c.1307

Ridgeway

Hele Arms

Priory Mill
12th century/
1740/1820

Mill Leat

George Lane

Underhill
(Clapera)

Plympton House
1/15

Plympton Castle
1100/1134

St Maurice Church
14th century/1446/1878

Fore Street

Guildhall
1696

Grammar school
1664

Plympton St Maurice
(Plympton Erle)

Long Brook

Underhill   Major settlement names

Medieval public buildings

Medieval settlement areas

Priory precinct/hospital lands

516

Public buildings 1600-1784

0   100   200m

first dedicated to St Thomas of Canterbury, but re-dedicated to St Maurice in the sixteenth century, was also probably built by the Redvers family. In securing a market grant by 1194, Plympton Earle developed its urban role half a century before Plymouth. The waters of the Plym estuary fed the tidal creek which extended to the outskirts of the earl's new town, where a small harbour was built to handle the export of slates and Dartmoor tin. Plympton Earle became one of Devon's four stannary towns in 1328, but by then tin-streaming had already deposited considerable quantities of silt in Dartmoor's rivers and the upper reaches of the Plym estuary gradually became too shallow to allow easy access to the earl's harbour. Cargoes were increasingly landed or loaded at 'Plym-mouth' and the mercantile role of Sutton Harbour emerged to rival that of its older urban neighbour.

To the north of the Ridgeway, at some distance from both of Plympton's centres, land was set aside in the medieval period for a lazar or leper colony. Although the date of its foundation is not known, the Plympton Maudlyn is recorded in the will of Bishop de Bytton of Exeter c.1307, who left provision for the fifty-four lepers accommodated in the hospital.[16] Both the priory and Earl Redvers built a mill in the twelfth century and each was fed by a leat. **Map 65.2** also depicts three of Plympton's most important seventeenth- and early eighteenth-century public buildings.[17]

SUTTON PRIOR AND THE EARLY DEVELOPMENT OF PLYMOUTH (Map 65.3)

Occupying rising ground above the inlet of Sutton Pool and sheltered to the south by the limestone ridge of the Hoe, the site where the two adjacent settlements of Sutton grew up was part of the ancient manor of the Valletort family (**Map 65.3**). Their manor house and park is thought to have been located beyond the site of Frankfort Gate.[18] In line with their activities elsewhere, the Augustinian priors of Plympton controlled the advowson of St Andrew's Church in Sutton and the Valletort family also granted them the right to erect a mill, a mill dam and a fishery in the 'Sour Pool' which lay to the north of Millbay.[19] Building on these modest interests in Sutton, the prior of Plympton petitioned the crown in 1254 for the grant of a weekly market and an annual fair to be held in the settlement lying closest to Sutton Pool which then came to be known as 'Sutton Prior'.[20] Thus although the Valletorts retained their rights as rural landlords at 'Sutton Valletort', the new market town of Sutton Prior was clearly a monastic foundation.

By the 1270s, Sutton Pool had established itself not only as a focus of trade with Europe, but also as an embarkation point for overseas military expeditions.[21] In earlier years it had been possible to draw the light fishing vessels that used Sutton Pool up on to the shelving sandy beaches below the town, but the newly enlarged role of Plymouth as an international port required the construction of the first quays, built out into deeper water. The

natural western shoreline of Sutton Pool in fact ran roughly along the line of what later became Tin Street, Vauxhall Street, Woolster Street and Southside Street (**Map 65.3**). But by the reign of Edward I the long process of reclaiming land and building quays to form 'Sutton harbour' was underway.[22] Precisely when the name Plymouth came regularly into use to replace Sutton is uncertain, but it is first used in fourteenth-century documents. The change may reflect a need by those unfamiliar with the area for a name offering a more recognizable geographical identification. For example, when the Black Prince assembled his expeditionary fleet bound for Gascony in 1355 some vessels certainly sheltered in Sutton Pool, but the majority anchored in 'Plym-mouth'. However, the two names were used interchangeably until the sixteenth century.[23]

In 1411 the inhabitants of Sutton Valletort joined with those residing in the market settlement of Sutton Prior to petition for incorporation as a borough. Together with a desire to secure by legal instrument their independence from Plympton priory, one of the key motives which lay behind the townspeople's request was undoubtedly the right which a charter of incorporation would bring to fortify and defend the town against increasing foreign attacks which had already wrought considerable damage and destruction in the town.[24] It was, however, another twenty-eight years before an act of incorporation was passed which established the borough of Sutton Prior, 'commonly called Plymuthe'.[25] This included not only the existing market town, but also parcels of the hamlet of Sutton Valletort, the tithing of Sutton Raf (the area now known as Coxside and Cattedown) and a parcel of the tithing of Compton.[26]

The charter of incorporation, based on that of Bristol, was granted by Henry VI on 25 July 1440 and gave the town its independent courts, the authority to establish a merchants guild, and the legal right to hold markets on Mondays and Thursdays and two annual fairs, each of four days duration.[27] Mayors were to be elected annually on St Lambert's day (17 September) when the township boundaries were solemnly perambulated.[28] These defined an area of sparsely populated agricultural land lying beyond the built-up area of the town itself and extending northwards to the Lipson and Stonehouse Creeks. On the western side, the borough's lands were marked by the 'great ditch' at Eldad which separated the new township from Stonehouse.[29]

PLYMOUTH'S EARLY TOPOGRAPHY (Map 65.3)

Although the location of the medieval town may be determined with relative ease, understanding its detailed topography is a great deal more difficult. Post-war redevelopment has drastically altered not only major parts of Plymouth's town plan, but through amalgamation and rebuilding, hundreds of historic property boundaries have been obliterated. Several sixteenth- and seventeenth-century maps exist, but the earliest that can help in revealing the details of Plymouth's early topography are those of

Map 65.2 (facing page) Plympton: the medieval settlement pattern and major post-medieval buildings up to 1784.

SOURCES: as **Map 60.1**; Llewellyn Jewitt, *History of Plymouth* (Plymouth, 1873); R. Midmer, *English Medieval Monasteries, 1066–1540: A Summary* (London, 1979).

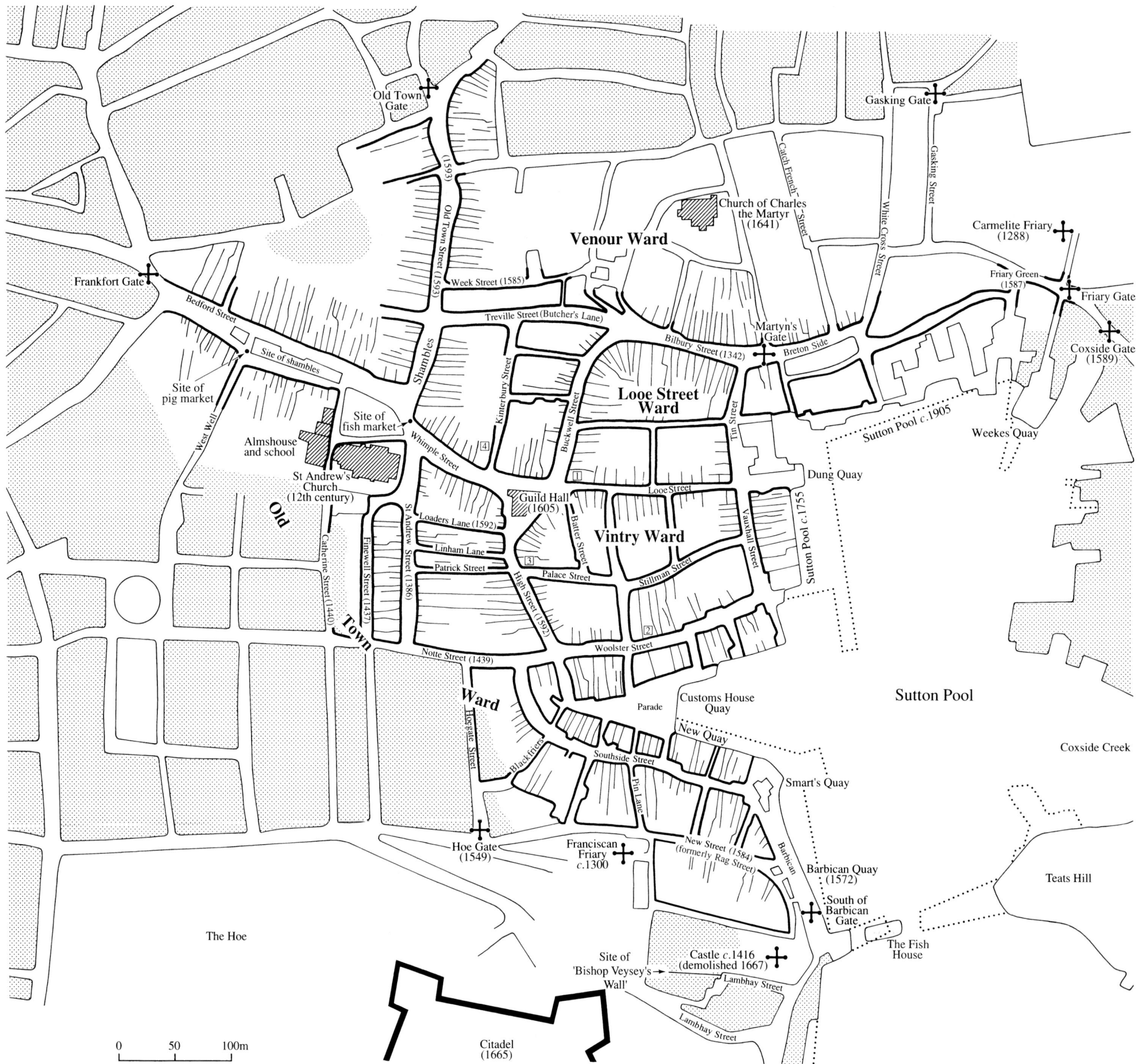

Old Town
Gate

Gasking Gate

Catch French Street

Church of Charles
the Martyr
(1641)

Venour Ward

Carmelite Friary
(1288)

White Cross Street

Gasking Street

Frankfort Gate

Old Town Street (1593)

(1593)

Week Street (1585)

Friary Green
(1587)

Friary Gate

Bedford Street

Treville Street (Butcher's Lane)

Bilbury Street (1342)

Martyn's
Gate

Breton Side

Coxside Gate
(1589)

Shambles

Site of shambles

Kinterbury Street

Buckwell Street

**Looe Street
Ward**

Tin Street

Sutton Pool c.1905

Weekes Quay

Site of
pig market

Site of
fish market

West Well

Whimple Street

4

1

Looe Street

Dung Quay

Almshouse
and school

St Andrew's
Church
(12th century)

Guild Hall
(1605)

Loaders Lane (1592)

Batter Street

Vintry Ward

Stillman Street

Vauxhall Street

Sutton Pool c.1755

Old

Catherine Street (1440)

Finewell Street (1437)

St Andrew Street (1386)

Linham Lane

Patrick Street

3

Palace Street

High Street (1592)

2

Notte Street (1439)

Woolster Street

Town

Hoegate Street

Ward

Customs House
Quay

Sutton Pool

Parade

New Quay

Coxside Creek

Blackfriars

Southside Street

Pin Lane

Smart's Quay

Hoe Gate
(1549)

Franciscan
Friary
c.1300

New Street (1584)
(formerly Rag Street)

Barbican

Barbican Quay
(1572)

Teats Hill

South of
Barbican Gate

The Hoe

Site of
'Bishop Veysey's
Wall'

Castle c.1416
(demolished 1667)

Lambhay Street

The Fish
House

Lambhay Street

0   50   100m

Citadel
(1665)

✛ Site of antiquity

(1288) Date established by or first named in map or document

━━━ Built-up streets *c*.1600

⊥⊥⊥ Conjectural plot boundaries *c*.1600

━━━ Other streets

▒▒▒ Area built up only after *c*.1620

⌐⌐ Quays of Sutton Pool *c*.1755

‥‥‥ Quays of Sutton Pool *c*.1905

① Site of Drake's town house

② Site of Mayor's residence

③ Palace Court

④ Hawkins' family house

Martyn's Gate	(demolished 1789)
Hoe Gate	(demolished 1863)
Old Town Gate	(demolished 1809)
Frankfort Gate	(demolished 1783)
Barbican Gate	(demolished 1803)
East Gate at Coxside	(built 1589, later demolished)
North Gate or Gasking Gate	(demolished 1768)
Friary Gate	(demolished 1763)

Emanuel Bowen (1755), John Manson (1756), Benjamin Donn (1765) and William Gardner (1784).[30] In conjunction with the earliest Ordnance Survey plans at 1:500 (1857) and 1:2,500 (1868), these sources have been employed to reconstruct in a tentative manner the outline of early Plymouth (**Map 65.3**).[31]

Key features of the geography of the site provide certain clues about the early development of Plymouth. Sutton Pool confined the settlement to the east, while the Hoe limited expansion to the south. Indeed, in medieval times, a small stream still drained from the Hoe, roughly along the line of modern Notte Street into Sutton Pool.[32] St Andrew's Church and the line of Old Town Street together occupied a low north–south ridge which originally formed a natural division between Sutton Valletort and Sutton Prior. The maps of Bowen, Manson, Donn and Gardner show that even by the later eighteenth century, cradled naturally in its sheltered location, Plymouth was still largely contained within its medieval and early modern built-up area. It has already been noted that the area defined by the 1440 charter of incorporation comprised the two main centres of Sutton Valletort and Sutton Prior. As late as the sixteenth century, the antiquary John Leland was still able to identify these two distinct settlements within the town. He describes the 'middle and hert of the town cawllid

Sutton Prior' and 'the oldest part of the Toun stoode by north and west sumwhat, … sore decayed'. The latter refers to the vestige of Sutton Valletort located in the 'Old Town' area.[33] The nucleus of Sutton Valletort may be identified to the west of St Andrew's Church along Bedford Street, with an extension northwards along Old Town Street (**Map 65.3**). Although later encroached upon, and possibly used as a shambles in medieval times, Bedford Street was unusually wide, perhaps suggesting an origin as a village pinfold (a pound for stray animals).[34] Scarcely any archaeological work has been conducted in this area which was extensively redeveloped in the Victorian era and, more damagingly, during the post-war reconstruction of Plymouth's city centre.

The original fishing village that became the market town of Sutton Prior lay to the east of St Andrew's Church (itself an eleventh- or early twelfth-century creation), and was focused on a large triangular open space at the junction of Whimple Street, High Street, and the main road to Plympton, namely Buckwell Street. Though later used as the site for a new guildhall, this space was almost certainly the original thirteenth-century market place. At its centre, Tudor maps show a market cross and this is known to have survived at least until 1603 when the accession of James I was proclaimed from its steps.[35] There are frequent references in the borough records to 'shambles', though these do not identify the precise location. By the Tudor period there were, in any case, two or three market places in Plymouth. References in 1606 suggest that the flesh shambles was moved from the main market place to St Andrew's churchyard. However, the butchers soon spread into Old Town Street. There was a pig shambles in Bedford Street, and fish catches were hauled up Looe Street to be sold from stalls along Whimple Street.[36]

Most of the early urban expansion appears to have been focused upon Sutton Prior. Growth occurred both to the south-west where St Andrews Street was formed, and to the north-east in the area of Looe Street.[37] Moreover, archaeological excavations have revealed that reclamation on both the Woolster Street and the Southside Street waterfronts was well underway by the mid to late thirteenth century when substantial stone quays were built out into Sutton Pool and new buildings were erected upon them.[38] The larger buildings were both warehouses and residences built around open courtyards and referred to as palaces.[39] This process of reclamation and expansion continued into the early modern period.[40] In the later sixteenth century, the Parade inlet was reduced in size by the creation of the New Quay and the Customs House Quay, while Smart's Quay and the Barbican Quay were formed to the south. Elsewhere, existing buildings were either replaced or converted, especially those on the waterfront.[41] The large gardens behind houses in the oldest part of the town were infilled with cottages and, on the periphery, especially to the south, new streets were formed and wealthy merchants built new houses.[42] Excavations at North Quay in 1994 have shed further light on the continuing process of waterfront reclamation in the period after the Civil War.[43]

Map 65.3 (facing page)
The topography of Plymouth to *c*.1625. The conjectural alignment of early property boundaries is based primarily upon map evidence; some are confirmed by archaeological excavation.

SOURCES: *Plat of Plymouth*, c.1592, Hatfield CPM I 35; *A True Mapp and Description of the Towne of Plymouth*, 1643, Wenceslas Hollar, Ashmolean Museum Oxford, C, 11, 304; *A Plan of the Town and Citadel of Plymouth*, 1754, Emanuel Bowen, British Library Maps C.10.d.18; *A Plan of the Town and Citadel of Plymouth, 1765*, Benjamin Donn, British Library Maps, C.11.c.7; *Map of Plymouth*, 1776, British Library, Add. MS 60393c; Ordnance Survey 1:2,500 Sheet 123.8 and 123.12 both surveyed 1855–6 and revised 1905 and 1892–3 respectively and published 1907 and 1895.

PLYMOUTH'S FRIARIES (**Map 65.3**)

By the end of the thirteenth century, parcels of land lying well beyond the outer limits of the built-up area had been granted to two separate communities of friars. Their houses were thereafter to become distinctive and important elements in the townscape of Plymouth. The house of 'White Friars' was located on the north side of Sutton Pool to the east of the medieval market town on land granted by John de Valletort to the Bristol Carmelites in 1288.[44] The Valletorts granted further land in 1329.[45] Although it lay outside their jurisdiction, the new friary was, it seems, resented by the Augustinian canons of Plympton, who were perhaps suspicious of the rival order.[46] Certainly, in addition to their spiritual role, there is evidence that the friary was directly involved in business transactions in the town.[47] Moreover, the Carmelites occupied a prime site close not only to the northern beaches of Sutton Pool, but also to the main road out towards Plympton, Ashburton and Exeter. However, it was also somewhat vulnerable and early Tudor maps indicate that the friary precincts, dominated by the lofty spire of its church, were defended by substantial walls on the southern side.[48] The Carmelites appear regularly to have offered hospitality to travellers.[49] Indeed, this use of the buildings was continued even after the Dissolution in 1539 when the friary was the venue for important meetings and for accommodating visiting dignitaries.[50] The buildings were still in use in the late eighteenth century when they served as a hospital for sick soldiers. But all surface trace of the friary was finally removed when a railway goods yard was constructed in the 1870s and Friary Gardens passenger station opened in 1891. However, excavations carried out in the early 1990s have revealed the site of the west gate and a stretch of the precinct walls of the friary. In fact the overall layout of the entire friary site has now been established.[51]

On the other side of the medieval town, a Franciscan friary was established in 1300 in a relatively isolated position behind the Hoe, on the southern side of what is now known as New Street. This thoroughfare was formerly called either Rag Street or, sometimes, Greyfriars Lane.[52] By depending upon alms for their subsistence, by preaching to the townspeople, and by their work for lepers, the Greyfriars also appear to have been resented by Plympton Priory.

By tradition, it was thought that there was also a Dominican friary in medieval Plymouth. Blackfriars Lane appears on the Plymouth inset of Bowen's map of Devonshire published in 1765,[53] and in the nineteenth century the city's local historian, R.N. Worth, drew attention to a rentroll of 1491 which appeared to refer to a Dominican property.[54] More recent work has shown, however, that there was in fact no Dominican friary in Plymouth. The medieval building in Southside Street at the core of the 'Blackfriars Gin Distillery' may instead have been the medieval merchants' guildhall referred to in the borough accounts.[55]

GUILDHALLS (**Map 65.3**)

It follows from the above that Plymouth's earliest guildhall may have been located in Southside Street. This may have been the building destroyed by fire when rebels attacked the town in 1548, thereby destroying the 'towne's evydence'.[56] In 1564 a 'newe gildhall' was built to replace it. This was replaced in turn by a large new guildhall constructed in 1606 in the centre of the town's market place. With a colonnaded market beneath, the building was erected by Thomas Aspey of Nettlecom.[57] Accommodation for felons was provided in the clink below, where there were two dungeons. Though partly rebuilt in 1667, and adorned by the addition of a cupola in 1706, the Jacobean guildhall was not replaced until 1800. But the replacement was never regarded as a success, and Plymouth eventually identified a new site for yet another replacement guildhall, made available by the demolition of a seventeenth-century complex of almshouses known as the Old Hospital of the Poor's Portion. An imposing Victorian gothic guildhall was erected there in 1874. Though bombed in 1942, it was subsequently restored and remains as a distinguished element in Plymouth's twentieth-century townscape.[58]

TUDOR DEFENCES (**Map 65.4**)

Plymouth's historic dual role as both a wealthy commercial port and a base used by fleets of English warships inevitably subjected the town to the periodic threat of foreign invasion. The fortification of the surrounding coasts against hostile attack was, therefore, a key preoccupation. As noted already, damaging foreign raids occurred in the fourteenth century and culminated in 1403 when thirty hostile Breton ships brought an assault force into Plymouth Sound. They launched their attack on the northern shore of Sutton Pool, and were repulsed only after considerable loss of life and property.[59] To deter further enemy incursions, Plymouth's first small castle was erected during the reign of Henry V on the knoll of higher ground overlooking the harbour entrance (**Map 65.4**).[60] It was a simple arrangement of four round towers linked by curtain walling, described by John Leland as a 'castle quadrate'.[61] The structure was demolished in 1667 and the later use of the site as a stone quarry and a rubbish pit appears to have destroyed even the foundations of the walls as well as other kinds of archaeological evidence.[62] However, a street name (Castle Street) and an associated area (the Barbican) mark the perimeter of its site.

Frequent expenditure on repairs to the castle throughout the sixteenth century clearly indicates its continued importance in Plymouth's defences at least until the first Hoe fort was built in the 1590s. The castle was the symbolic key to the town's protection. In times of emergency, the mayor (as commander-in-chief of the borough) took charge of the north-east tower, while the twelve aldermen and twenty-four common councillors were expected to station themselves in the other three.

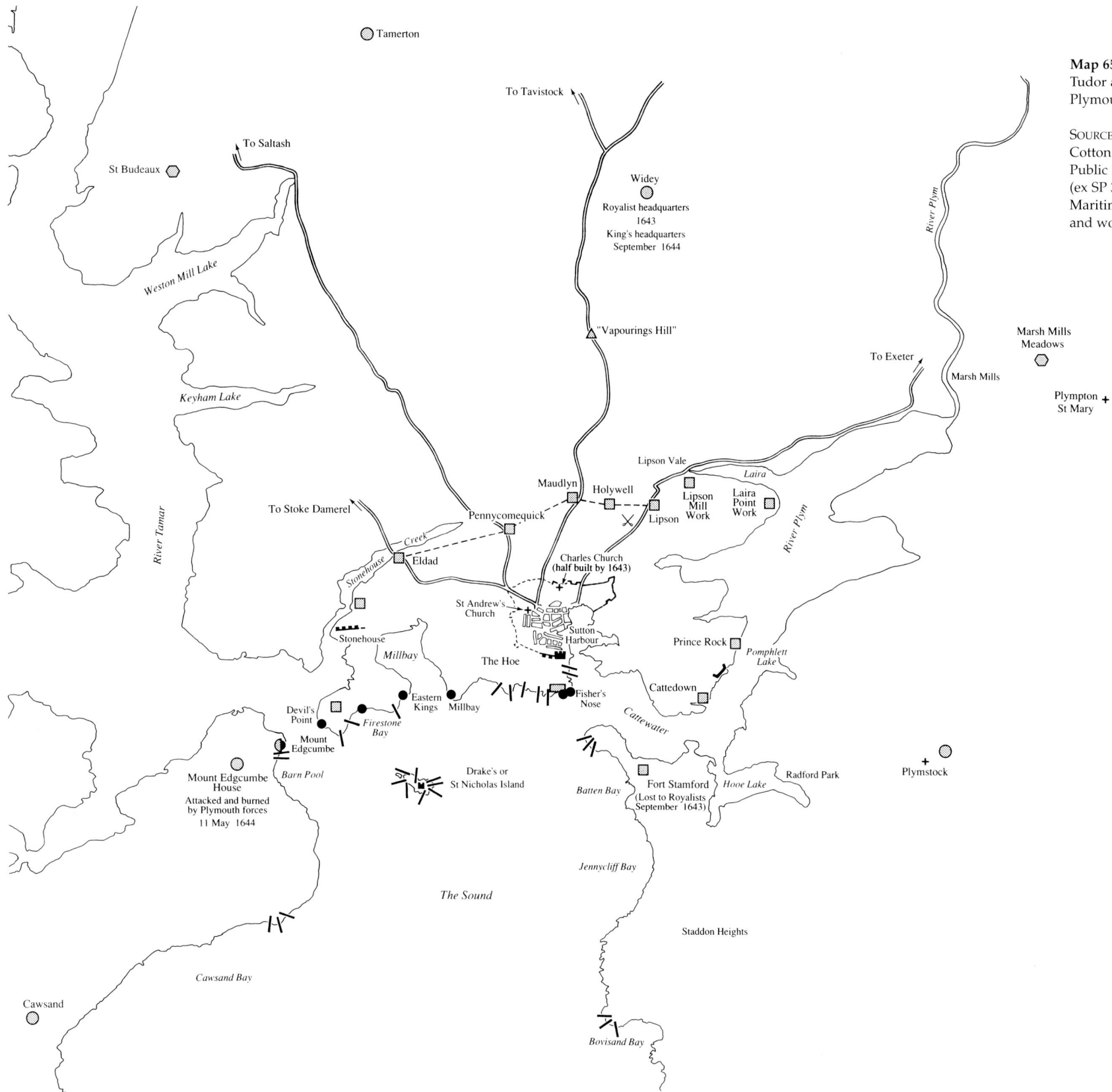

Tamerton

Map 65.4
Tudor and Stuart defences of Plymouth.

SOURCES: British Library, Cotton MS. Aug. I.ii.6; Public Record Office, MPF 6 (ex SP 34/36/4); National Maritime Museum, LAD/11; and works cited in note 60.

To Saltash

To Tavistock

St Budeaux

Widey
Royalist headquarters 1643
King's headquarters September 1644

River Plym

Weston Mill Lake

"Vapourings Hill"

Marsh Mills Meadows

To Exeter

Marsh Mills

Keyham Lake

Plympton St Mary

River Tamar

Lipson Vale

Laira

Maudlyn   Holywell   Lipson Mill Work   Laira Point Work

To Stoke Damerel   Pennycomequick   Lipson

River Plym

Creek

Stonehouse   Eldad   Charles Church (half built by 1643)

St Andrew's Church   Sutton Harbour

Stonehouse   Prince Rock   Pomphlett Lake

Millbay   The Hoe   Cattedown

Devil's Point   Eastern Kings   Millbay   Fisher's Nose

Firestone Bay   Cattewater

Mount Edgcumbe

Barn Pool   Drake's or St Nicholas Island   Fort Stamford (Lost to Royalists September 1643)   Radford Park

Mount Edgcumbe House
Attacked and burned by Plymouth forces 11 May 1644

Batten Bay   Hooe Lake   Plymstock

Jennycliff Bay

The Sound   Staddon Heights

Cawsand Bay

Cawsand

Bovisand Bay

Defences at the time of the Spanish Armada

Medieval castle

Tudor artillery towers

Positions of Field Pieces (of ordnance) depicted on "The Grenville Map" of 1587

Defensive curtain walling

Drake's Island Fort 1548

Plymouth

Defences at the time of the Civil War siege

Fortresses built as a line of outer defences in 1640s

Hoe Fort, built in 1590s

Royalist positions established by Prince Maurice in 1643

Positions established by Sir Richard "Skellum" Grenville April 1644

Vapron (or "Vapourings") Hill where King received salute each morning whilst camped at Widey in 1644

Freedom Fields Battle 3 December 1643

Plym breastwork

Conjectural line of defences linking fortresses

Conjectural inner line of defences

Established inner line of defences

0   1   2km

At the same time as the castle was built, a 'great bulwark' was constructed on the beach at the eastern end of the Hoe. Additional bulwarks were added during the early sixteenth century. Some were enclosed structures with doors, windows and stone gun loops.[63] By 1512 there were five separate bulwarks providing protection for the town. During that year the pier (known as the 'Causey') at the entrance to Sutton Pool was also considerably strengthened and a 'new house' or blockhouse was erected upon it to accommodate guns directed out towards the Cattewater. In 1523 a 'great chain' was purchased for the Causey together with a mast and a smaller chain.[64] These chains were stretched across the entrance to Sutton Harbour in times of danger while the mast perhaps formed a reinforcing boom.

In the early 1540s, when fears of invasion were acute in southern England, the town erected a new blockhouse at the Fishersnose end of the Hoe. This structure is first mentioned in local documents in 1542 but, a year later, following a visit to Plymouth by Sir Richard Cavendish, Master of the King's Works at Dover, more work appears to have been carried out when the blockhouse was probably enlarged.[65] Indeed, by the end of the 1540s, Plymouth was protected by bulwarks and gun emplacements which extended westwards from the Fishersnose tower to another blockhouse overlooking the entrance to Millbay.

Beyond Millbay the foreshore belonged not to Plymouth, but to the Edgcumbe family, which had acquired the manor of Stonehouse by marriage in the 1490s. Plymouth corporation enjoyed good relations with the Edgcumbes in the mid-sixteenth century. In July 1545 Lord Russell reported to the Privy Council that 'Mr Edgcumbe does his utmost to strengthen Plymouth'. From this statement it may be inferred that Edgcumbe was then engaged in building the blockhouse at Wilderness Point on the shores below Mount Edgcumbe deer park. But equally it may be a reference to new work on the foreshore of Stonehouse where more artillery towers are known to have existed. Only two survive (in Firestone Bay and at Devil's Point), while nothing remains of the others such as those at the entrance to Millbay opposite Plymouth's Poltor blockhouse.[66] Though probably modified on subsequent occasions, the Firestone Bay artillery tower probably dates back to the 1490s when it was built to protect the landing beach on the northern side of the Sound and to control the deep-water channel which runs between St Nicholas's (or Drake's) Island and the foreshore (**Fig. 65.3**). Even older is the surviving fragment of the medieval wall which enclosed the southern side of the growing settlement of East Stonehouse.

Permanent fortifications were not erected on St Nicholas's Island until the reign of Edward VI. An indenture between the king and Plymouth's corporation was drawn up on 14 February 1548/9. This appears to have been a direct response to the appeal for royal assistance made by the mayor following criticism in 1547 that the town had made no effort to convert the chapel on the island into a bulwark to protect the deep shipping lane. The crown not only made a grant towards the cost of constructing a fort on the island, but also specified arrangements for a garrison, armaments and supervision. Work was carried out under the direction of Sir Francis Fleming, Lieutenant of the Ordnance, and appears to have been completed in 1549/50, when the first garrison of soldiers was installed.[67] Maintenance costs and the wages of the soldiers were thereafter met directly from customs duties levied on goods traded in the haven. In emergencies, however, the borough was entitled to call upon all the parishes of Devon and Cornwall to make a contribution.

The fortress on St Nicholas's Island was the last major new defence installation to be erected in Plymouth until the reign of Elizabeth I. But from the 1570s, England's political and economic quarrels with Phillip II of Spain thrust the town to the forefront of national affairs. And the weaknesses of local defences against a determined invasion attempt were increasingly recognized. The threat of the Spanish Armada provoked a frenzy of defence preparations in Plymouth. In 1587 Sir Richard Grenville surveyed the Sound and recommended the deployment of men and weapons at strategic points from Cawsand Bay in the west to Bovisand Bay in the east. His proposals were largely adopted and they are shown on **Map 65.4**.[68] Though the borough's defences were not directly tested in 1588, the Armada crisis underlined Plymouth's vulnerability and, within two years, Drake conveyed a petition from Plymouth's burgesses requesting crown assistance in building a fort on the Hoe.

In December 1591 a commission of enquiry chaired by Sir Arthur Champernowne considered plans for both a new fortress on the Hoe and an entire circuit of town walls.[69] The Privy Council called in the military surveyor, Robert Adams, to provide expert guidance. He judged Champernowne's scheme to be too ambitious and calculated that the resources available would permit only the erection of a fort, not town walls. Ultimately, the views of Adams were accepted and an indenture between the borough and Queen Elizabeth followed on 30 May 1592.[70] Thereafter, Robert Adams spent eighteen months in Plymouth as supervisor of the construction work. By February 1595/6, the first garrison was installed, and a month later Sir Ferdinando Gorges was appointed as captain of the new fort, thus ending the role of the mayor as military commander of the borough, and opening a new era for Plymouth as a permanent garrison town.

Several contemporary plans of the Elizabethan fort survive, which clearly show that it not only incorporated much of the earlier alignment of bulwarks and gun emplacements on the foreshore, but also had massive new walls and a dry moat cut into solid rock on its landward side.[71] These latter features were obliterated by the construction of the Royal Citadel in the late seventeenth century and it is hard to gain an impression of the structure in today's much-altered landscape.[72] However, archaeological excavations carried out in 1994 by Exeter Museum located some of Robert Adams's sixteenth-century walls surviving beneath those of the later Citadel.[73]

STUART DEFENCES AND THE CIVIL WAR SIEGE OF PLYMOUTH (**Map 65.4**)

In the seventeenth century, Plymouth's increasing North American trade and her involvement with the Newfoundland fisheries profoundly influenced the town. Men were regularly away for several months each year during the fishing season, returning only in late July and August with their catch.[74] A readiness to travel and a pioneering spirit of independence became a deeply embedded feature of town life. In addition, years of conflict with Catholic Spain rendered Plymouth staunchly Protestant. The calvinist Pilgrim Fathers, who arrived from Southampton in 1620 aboard the *Speedwell* and the *Mayflower*, on their way to find a new life in America, received a sympathetic reception in Plymouth and the town was thereafter a recognized departure point for a steady flow of puritan emigration to New England. Latent anti-government, anti-establishment attitudes in Plymouth were thereby nourished, and when Charles I declared war on Spain in 1625 and ineptly ordered 10,000 ill-prepared and badly equipped troops to embark in the Sound, the outcome only intensified local antipathies. Starving and disease-ridden, the king's army brought plague to Plymouth which decimated the civilian population. Borough accounts record spending on wooden hospitals for 'sicke folke' in 1625. One was erected in 'Mr Robert Trelawnyes grounde' at Lipson, while another three were built at 'Husart' [Hoe Stert, or Mount Batten]. Marginal notes define these as 'pest houses'. Excavations at Mount Batten have now revealed the site of early seventeenth-century burials and it is likely that these were the graves of some of the 1625 plague victims.[75] Recollections of this royal incompetence and insensitivity coloured Plymouth's attitudes towards the king's cause when the Civil War broke out in 1642 (**Map 65.4**). While much of the South West rallied to the king's banner, Plymouth defiantly declared for parliament.[76] The price was a royalist siege which began in November 1642, when Sir Ralph Hopton severed the town's water supply, and lasted until 1645. While there were periods when the Cavaliers encircled the town, at other times access was unimpeded. Nevertheless the Civil War exposed further weaknesses in Plymouth's defences. Until then the major threat had appeared to be a sea-borne invasion, but in the Civil War the danger was greatest on the landward side.[77] In response, early in 1643, the borough authorities hastily ordered the erection of new defences which came to be known as the 'inner line' and the 'outer line'. Since Sir Arthur Champernowne's ambitious 1591 plan to enclose the town with a complete circuit of strong walls had not been realized, an immediate priority was to build defensive gun emplacements on the edge of the built-up area. Recent excavations indicate that these were rather more substantial structures than was once supposed.[78] Their precise alignment to the north-east of Sutton Harbour has now been established and here at least there were substantial stone foundations and almost certainly stone facings.[79]

The intricate topography of Plymouth's township was exploited fully in positioning the outer line of defences. A series of stockaded earthworks was built at strategic locations and linked by a ditch. All the main roads into Plymouth, together with the important crossings of the indented creeks, were thus controlled. There were further fortlets watching the waterfront of Stonehouse, the Cattewater and the Plym. An outlier, Fort Stamford (see chapter 21 and Map 21.3), above Turnchapel, was intended to control Hoe Stert (Mount Batten) and the entrance to Sutton Pool, but in fact this fell to the Cavaliers very early in the siege. But while enemy control of Mount Batten made entry to the Cattewater by shipping more dangerous, the inaccuracy of royalist artillery prevented a complete blockade of the port.

The Civil War defences shown on **Map 65.4** are based on the structures depicted on the well-known map of the town engraved by Wenceslas Hollar which purports to show 'the works and approaches of the enemy at the last siege in 1643'.[80] Though emphasis is placed on the briefly held forward positions of the Cavalier forces, the map also indicates that the royalists established their own military positions in an arc 4 or 5 miles from the centre of the town.[81] But communications between the widely spaced Royalist camps would clearly have been difficult across the hilly terrain which separated them. The town was not, therefore, completely encircled by the enemy. Moreover, ordinary townspeople would not usually have been able to see the enemy positions, though the daily gun-salute fired in honour of King Charles when he came briefly to direct siege operations in person was undoubtedly audible and those on the outskirts of Plymouth probably also saw the smoke. These somewhat theatrical royal ceremonies were derisively dubbed the king's 'vapourings', from which the name Vapron Hill is taken. Some months earlier, however, on 3 December 1643 when the royalists launched a particularly determined assault, the danger in Plymouth was momentarily perilous. The enemy had crossed Lipson Creek during the hours of darkness and took the fortlet at Laira Point by surprise.[82] Reinforcements hurried out from Plymouth, but by then Prince Maurice had advanced with his main force from Compton village and his troops broke through as far as the inner line of defences. A ferocious battle took place in the vicinity of the Lipson fort. After four hours Colonel Gould was able to lead an advance which forced the Cavaliers to retreat. Many drowned as they tried to flee across Lipson Creek against an incoming tide.[83] The scene of the battle is now known as Freedom Fields and Colonel Gould's exploits are recalled in the name of the area above the creek which is known as Mount Gould.

The complex sequence of events in the long Civil War siege of Plymouth cannot be recounted in full here. In any case, apart from place names and street names, little remains in the modern townscape of the structures which played such a prominent part in the defences of the town in this period of danger and turbulence. The attacks in 1644 of Sir Richard 'Skellum' Grenville, and later those led by the king himself, failed to break

Map 65.5
The Royal Citadel, Plymouth.
The map shows the layout of
the Citadel prior to major
developments and extensions
on the eastern flank of
the fortress. The modern
names of the bastions are
used; internal buildings
as in the 1950s.

SOURCES: E. Stuart,
*Lost Landscapes of Plymouth:
Maps, Charts and Plans to 1800*
(Stroud, 1991);
Sir Bernard de Gomme's plans
of the Citadel: British Library,
Add. MS 16371D (1668);
National Maritime Museum,
P/45 (1672).

down the town's defences, and although Grenville tried again in 1645, the tide of the war had turned. On 18 January 1646 the last royalists departed and by 23 March Fairfax and Cromwell were being welcomed into Plymouth.

THE ROYAL CITADEL (**Map 65.5**)

Overseas wars during the Cromwellian period brought some new ship-repairing and victualling business which aided the town's recovery, but notwithstanding the reclamation and new building activity already noted at North Quay, the return to growth and prosperity was relatively slow until the restoration of the monarchy in 1660. Five years later Charles II ordered the construction of Plymouth's Royal Citadel on the Hoe. At first it was planned by Sir Bernard de Gomme to stand as a separate symmetrical structure to the west of the Elizabethan fortress, but

later designs integrated the two, thus forming the peculiarly irregular structure, most of which survives today (**Map 65.5**).[84] Although there were considerable developments in the coastal defences of the Sound during the eighteenth and nineteenth centuries, Plymouth's foreshore continued to be dominated by the Royal Citadel, completed in the 1670s.[85] Today the Citadel is regarded as Britain's principal seventeenth-century fortification, and for a century after its construction it was the most important fortress in the country and commanded by a governor who held the nation's highest-paid military appointment.[86]

THE ROYAL DOCKYARD (**Map 65.6**)

Sir Walter Raleigh first proposed the construction of a naval dockyard on the Hamoaze or lower Tamar in the late 1590s. He argued that the deep, broad estuary offered an excellent

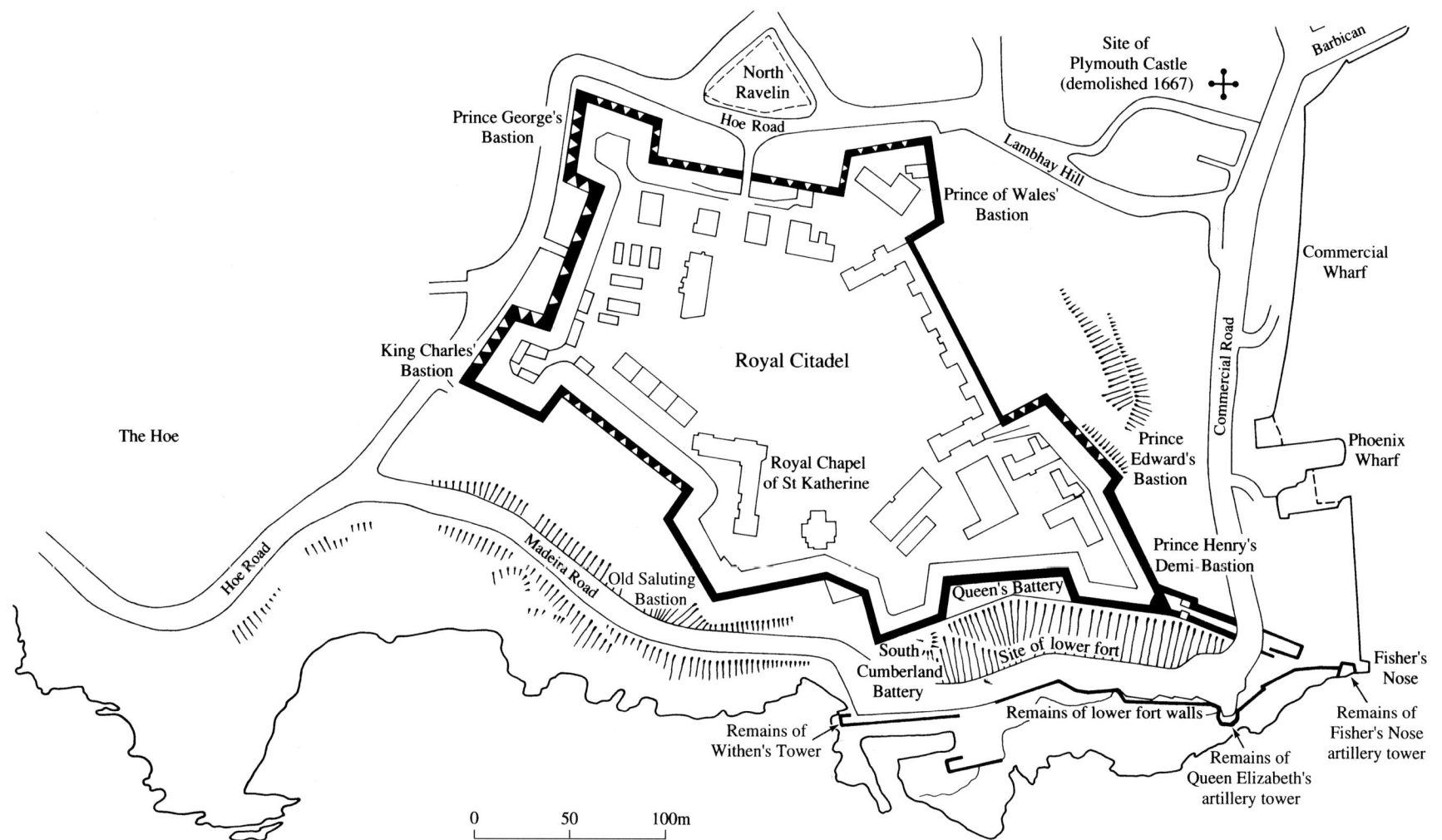

Weston Mill Railway

Viaduct

Weston Mill Lake

Nuclear-powered submarine refit complex

HMS Drake
Royal Naval
Barracks
(1885)

(1-12 = docks)

Dates refer to construction
commencement and
completion

Keyham extension
1895-1907
(Submarine refit complex
fleet maintenance base)

Basin No. 5

Saltash Road

11   10   9   8

12

Goschen Yard

Basin No. 4

Keyham Road

North
Basin
No.3

Keyham Steam Yard
1844-1853
(Frigate complex)

South
Basin
No. 2

7

6

5

Albert Road

Morice Yard
extension   (1950)

Pottery Quay

John St.

Ferry Road

Torpoint
Ferry

Ferry Road

Gun Wharf

Granby Barracks

Morice Yard
1696-1719
(The Ordnance Yard)

North Corner

Raglan
Barracks

North New Dock
(1789)

4

North Dock
(1762)

3

South
Yard
extension

Double Dock
(1727)

2

(1950)

South Yard
1691-1695
(The original dock
and yard)

Basin
No. 1

1

Former
area
of
ropery

South Channel

Mount Wise
Barracks

The 'Ozey'
Ground

Original
Building Slips

Covered Slip
(1814)

Mutton Cove

H a m o a z e

0   250   500m

River Tamar

Hamoaze

Devonport Dockyard
Area shown
on main map

City
centre

R. Plym

Plymouth Sound

0   3km

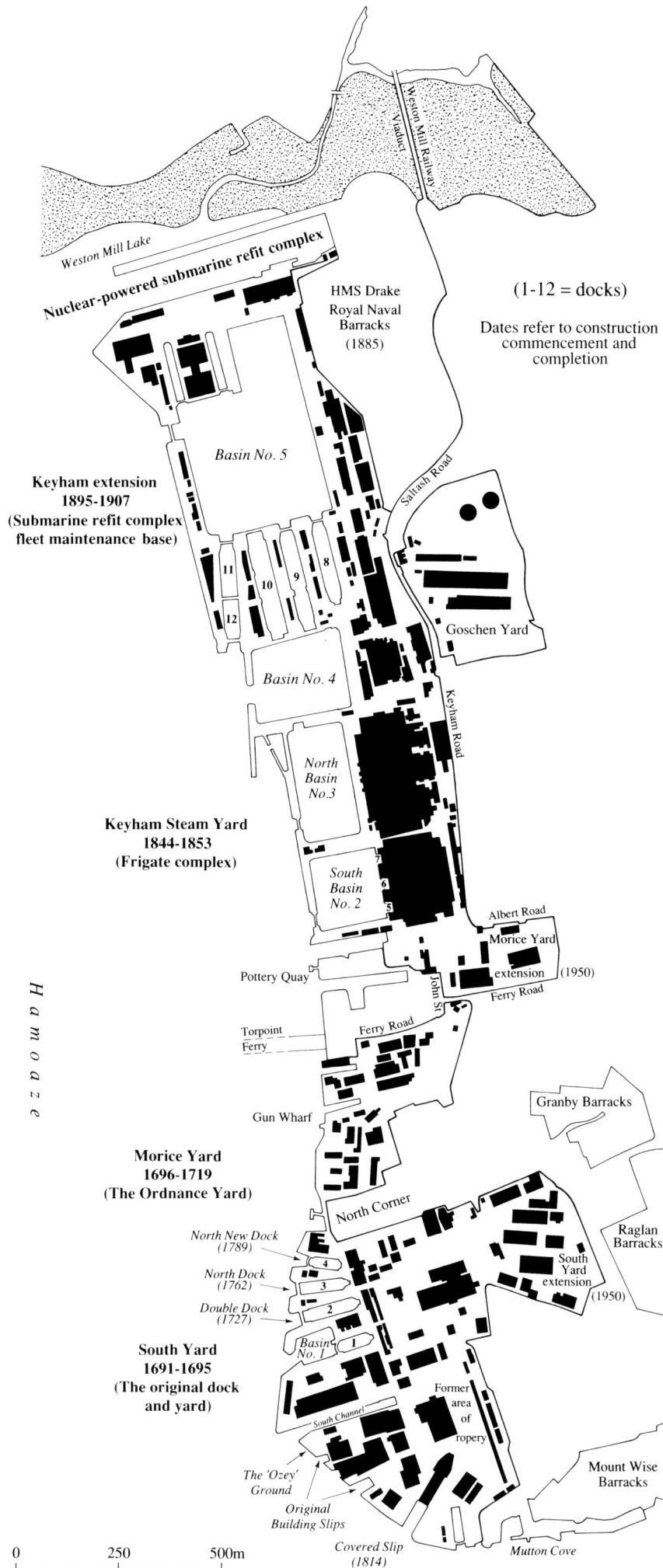

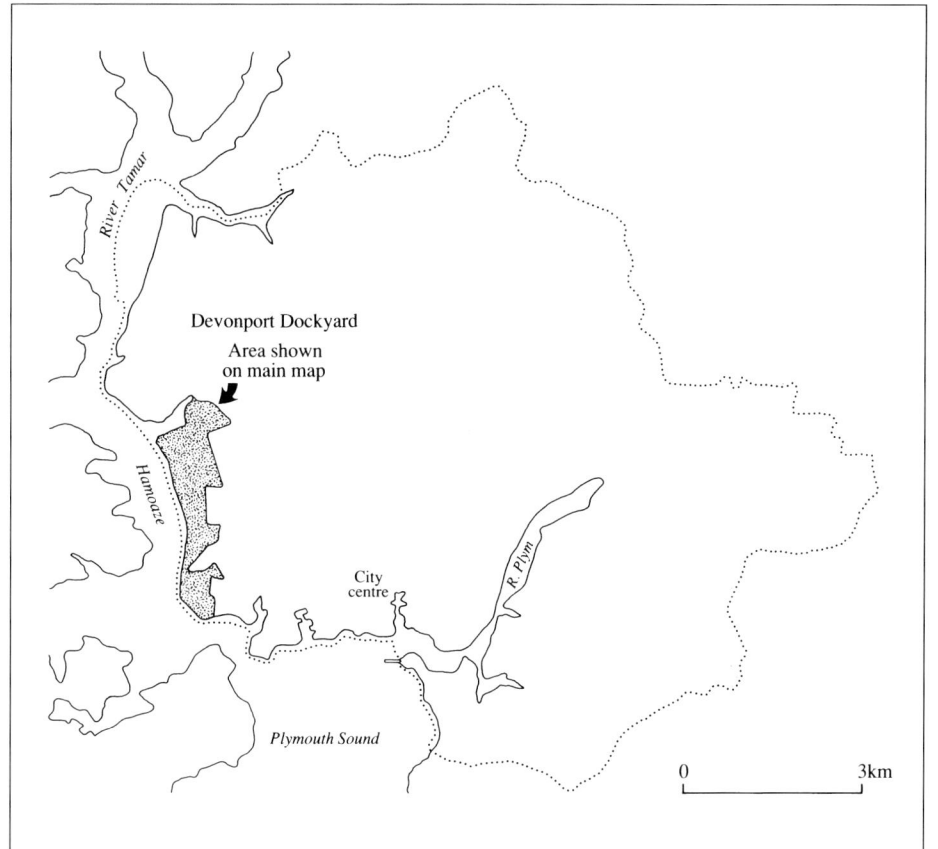

Map 65.6
Devonport Dockyard,
1691–1985. The map shows
the Dockyard just before
privatization and the recent
'scaling-down'.

SOURCES: *HM Dockyard
Plymouth* (1698), British
Library, King's Top., 43,
fos 129–30; *Street Plan of
Plymouth* published c.1939
by Geographia Ltd; maps of
Plymouth, Stonehouse and
Devonport published by
John Tallis (1860), A & C Black
(1870), W.H. Maddock (1881);
and works cited in notes 87
and 88.

525

Map 65.7
Benjamin Donn's *Plan of Stoke Town and Plymouth Dock,* published in 1765.

SOURCE: photograph reproduced by permission of University of Exeter, Department of Geography.

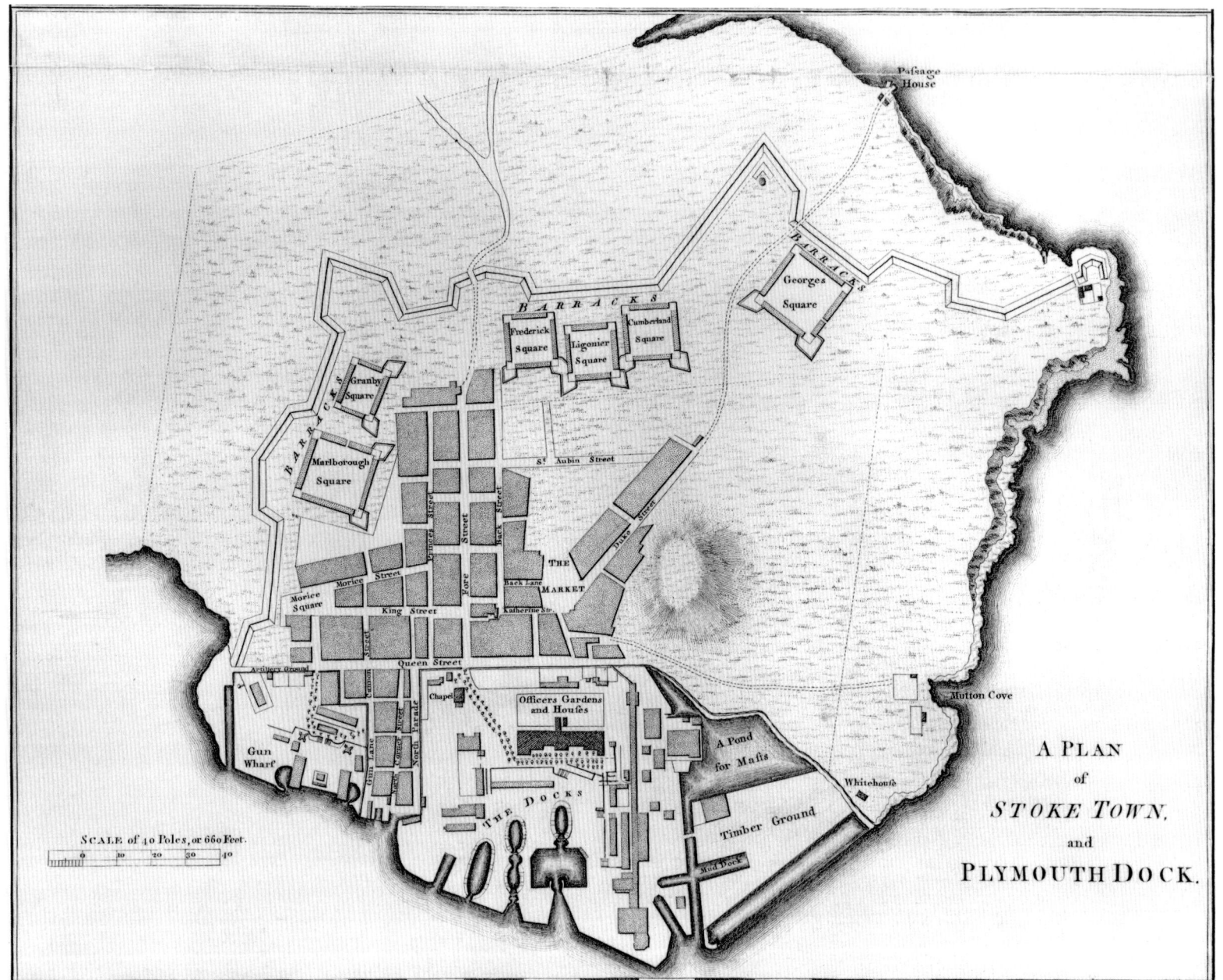

sheltered anchorage which was afforded natural protection by the narrow, awkward entrance from Plymouth Sound.[87] Although nothing came of Raleigh's proposal, it was re-examined in 1662 and again in 1671.[88] William III finally launched the project in 1690 when he sanctioned the appointment of Edmund Dummer, surveyor to the Royal Navy, as supervisor of a stone-built new dock on 40 acres of scrub waste-land leased from Sir Nicholas Morice at Point Froward on the eastern side of the Tamar estuary (**Map 65.6**).[89] By 1695, the first dock and basin were operational and work on related dockyard buildings including a ropehouse, a tarring house, a sail loft and storehouses had commenced. Construction of a terrace of thirteen officers' houses had begun in 1693. In 1696 Sir Nicholas Morice leased another 4.25 acres of land to enable the Ordnance Board, formerly based at Sutton Harbour, to build a

new gun wharf adjacent to the new Dockyard. Expanded in 1719, this facility was thereafter known as Morice Yard.

The Admiralty leased more land from the Morice family which provided space for the construction of building slips to the north of the original basin; three were constructed in 1727, 1762 and 1789. They are now known as number 2, 3 and 4 docks (**Map 65.6**). Reclamation and development of the 'Ozey' Ground marshes to the south commenced in 1773 with the construction of a new slip. Completed a year later, it was roofed over in 1814 and is the oldest covered slip in the country.

By 1749, ownership of the site had passed by marriage to the St Aubyn family, who continued as lessors until the Admiralty finally obtained freehold control a century later.[90] By then all existing space had been used, and there was no room to develop further facilities. Ultimately, the demands of steam-ship technology obliged the Admiralty to purchase more land at Keyham to the north. Approximately 70 acres were acquired and the Steam Yard complex, comprising two basins and associated boiler shops, smithies and stores, was developed between 1844 and 1853.[91] Between 1973 and 1977 the old 'Keyham Steam Yard' was redeveloped as the new covered frigate re-fit complex.

In response to the increasing size and complexity of Royal Navy ships, a further expansion of the Dockyard became necessary at the end of the nineteenth century. The Keyham 'extension' added a further 130 acres to the existing Yard, thereby almost doubling its size. Large-scale reclamation was involved in the creation of a 10.5 acre tidal dock (Basin 4) linked to a series of three parallel graving docks (Nos 8, 9 and 10). Further north, a 35-acre enclosed basin was created behind a new 5,000ft sea wall. Vessels could enter this huge basin (No. 5) through the locks (Nos 11 and 12). Building took more than a decade and the total cost exceeded £4 million.[92]

Submarine refitting work placed new demands on the dockyard and in 1973 work began on remodelling the north-western corner of the Keyham extension in order to provide appropriate facilities. These were completed in 1980 and comprised two separate dry docks and related installations. The latter was dominated by the massive 80-ton cantilever crane designed to lift nuclear core packages between submarines and rail transporters. Keyham extension also houses the Fleet Maintenance Base, which provides facilities and expertise for repairs to be carried out on today's warships whilst in service both at home and abroad.

THE NEW TOWN OF 'PLYMOUTH DOCK' AND THE EMERGENCE OF DEVONPORT (**Map 65.7**)

In the early years, Dockyard artificers were either accommodated in old hulks acquired for the purpose and anchored in the Hamoaze, or were obliged to walk from their lodgings in Stonehouse, Stoke village or even Plymouth. In fact, land for building workers' dwellings near the Royal Dockyard was not released by the Morice family until 1700, but in the following thirty years a community of almost 3,000 was established in the new settlement which was then simply known as Plymouth 'Dock'. The earliest terraces of workers' cottages were erected near North Corner on the narrow strip of available land running up from the water's edge between the north wall of the Dockyard and south wall of the Board of Ordnance (Morice) Yard.[93] This curious narrow 'civilian' corridor survives in today's townscape as a visible legacy of the piecemeal process by which land was released by the Morice family in the early years of Devonport's development. Benjamin Donn's map of 1765 shows North Corner Street (later renamed Cornwall Street), Cannon Street and Gun Lane in this vicinity (**Map 65.7**). Beyond this a grid-iron pattern of streets was developed. The main axis was Fore Street which was aligned on the main Dockyard gates.[94]

Demand for housing outpaced supply, and overcrowding, poor sanitation and inferior dwellings were the hallmarks of the new town.[95] But expansion was rapid and Plymouth Dock quickly developed its own separate identity. Indeed, to underline its new civic aspirations the town commissioned the local architect John Foulston to design a town hall which was erected in 1821.[96] By 1823 its population was greater than that of Plymouth and permission was sought to rename the town Devonport. This was granted by George IV in 1824 and Foulston was again commissioned to design a column to be erected near the town centre in celebration of its new name. Eventually, in 1837, Devonport became a separate municipal corporation.[97] The growth of the Dockyard stimulated the rapid development of Stonehouse. The Edgcumbe and St Aubyn families collaborated in the construction of the toll bridge across Stonehouse Creek providing a direct road link to Dock; it was completed in 1773. Thereafter Lord Edgcumbe leased out land in Stonehouse where the elegant neoclassical terraces in Durnford Street and Emma Place were built. By 1795, in little more than twenty years, Stonehouse had grown from a small village to a town occupied by almost 4,000 people.[98]

EIGHTEENTH- AND NINETEENTH-CENTURY DEFENCES (**Map 65.8**)

Map 65.8 shows the various modifications made in the eighteenth and nineteenth centuries to the defences of Plymouth. Although the map is chiefly concerned with defences built during Georgian and Victorian times, for a century after its completion, the Royal Citadel on the Hoe continued to dominate the approach not only to the anchorage in the Cattewater and to Sutton Pool, but also to the Royal Dockyard on the Hamoaze. The other key element in the defence of the Sound was Drake's Island where the guns and batteries were repeatedly modernized throughout the period.[99] Indeed, the expansion of the Dockyard meant that the protection of the lower Tamar assumed increasing importance. In response, by 1701, two new batteries had been built on Drake's Island, and a third was placed on Western King. Before the start of the Seven Years War, three new batteries facing westwards were built below the Royal Citadel. After the war, in 1756, plans were devised for a defensive line of landward rampart and ditch fortifications to enclose 'Plymouth Dock' entirely, and within a year work had commenced.[100] Known as the 'Dock Lines' (later the Devonport Lines), these in fact provided only limited protection until, in 1853, they were realigned and improved with a deeper ditch and stone revetments (**Map 65.8**).

The imminent threat of a Franco-Spanish invasion in 1779 resulted in the construction of a system of redoubts and earthwork batteries to the designs of Lt Col. Matthew Dixon, RE. While the batteries considerably strengthened the defence of the shipping channel, the redoubts were placed around the Dockyard

Map 65.8
Evolution of the seaward and landward defences of Plymouth from the Royal Citadel, 1665 to the end of the nineteenth century.

SOURCES: Ordnance Survey 1:2,500 maps, 1888; Public Record Office, ZHC 1/2577, MR 1199 (WO 78/385), and MPH 406 (WO 78/1521); British Library, Add. MS 22875 f. 20; and works cited in notes 99–102.

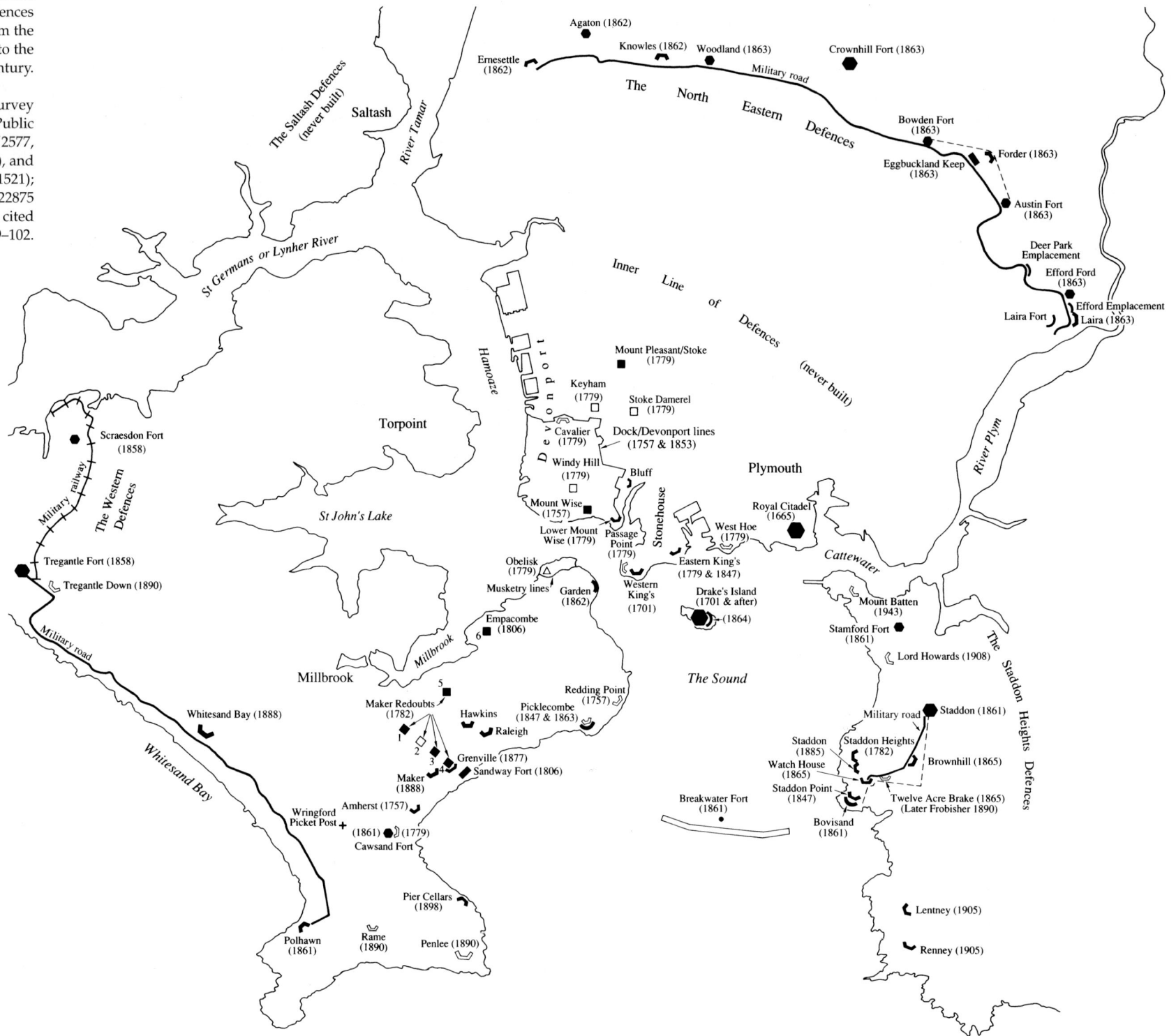

Agaton (1862)
Knowles (1862)
Woodland (1863)
Crownhill Fort (1863)
Ernesettle (1862)
Military road
The North Eastern Defences
The Saltash Defences (never built)
Saltash
River Tamar
St Germans or Lynher River
Bowden Fort (1863)
Forder (1863)
Eggbuckland Keep (1863)
Austin Fort (1863)
Deer Park Emplacement
Efford Ford (1863)
Efford Emplacement
Laira (1863)
Laira Fort
Inner Line of Defences (never built)
Mount Pleasant/Stoke (1779)
Keyham (1779)
Stoke Damerel (1779)
Cavalier (1779)
Dock/Devonport lines (1757 & 1853)
Windy Hill (1779)
Bluff
Plymouth
Royal Citadel (1665)
Hamoaze
Devonport
Torpoint
Mount Wise (1757)
Lower Mount Wise (1779)
Passage Point (1779)
Stonehouse
West Hoe (1779)
River Plym
Cattewater
Scraesdon Fort (1858)
Military railway
The Western Defences
St John's Lake
Tregantle Fort (1858)
Tregantle Down (1890)
Military road
Obelisk (1779)
Musketry lines
Eastern King's (1779 & 1847)
Western King's (1701)
Drake's Island (1701 & after)
(1864)
Mount Batten (1943)
Stamford Fort (1861)
Lord Howards (1908)
The Staddon Heights Defences
Garden (1862)
Empacombe (1806)
6
Millbrook
Millbrook
5
The Sound
Redding Point (1757)
Picklecombe (1847 & 1863)
Maker Redoubts (1782)
Hawkins
Raleigh
1
2
3
4
Grenville (1877)
Maker (1888)
Sandway Fort (1806)
Whitesand Bay (1888)
Military road
Breakwater Fort (1861)
Staddon (1885)
Staddon Heights (1782)
Watch House (1865)
Staddon Point (1847)
Staddon (1861)
Brownhill (1865)
Twelve Acre Brake (1865) (Later Frobisher 1890)
Bovisand (1861)
Wringford Picket Post
Amherst (1757)
(1861)
(1779)
Cawsand Fort
Whitesand Bay
Pier Cellars (1898)
Lentney (1905)
Renney (1905)
Polhawn (1861)
Rame (1890)
Penlee (1890)

⬢ Fortress
⬣ Major fort
● Minor fort
❭ Battery
❭ Casemated battery
■ Redoubt
--- Ditch
— Military road
□ ♫ Battery, redoubt, etc., demolished
(Dates refer to first structure not later re-arming)

0   1   2   3km

and on Maker Heights where they could afford direct protection to ships and installations in the Dockyard itself. Attention was then turned towards the relative vulnerability of the Cornish side of Plymouth Sound. A series of defensive structures was gradually erected to protect Cawsand Bay, the best landing beach in the Sound, and the Heights above it (**Map 65.8**).

Plymouth's defences were only marginally improved during the Napoleonic Wars and further major changes were made only after the Breakwater across the Sound was completed in 1844.[101] Three years later work began on the batteries at Staddon Point, Eastern King and Picklecombe. The threat of French invasion prompted the construction in 1858 of two new forts to the west of Plymouth and, a year later, a Royal Commission was appointed to review the defences of the entire United Kingdom.

Although the recommendations of this Commission were in practice considerably scaled down, during the course of the next ten years a chain of new forts and batteries, disposed in a vast encircling arc stretching from Bovisand in the east to Tregantle in the west, were built to afford better security to the Dockyard and Plymouth. However, by the late 1860s, despite rapid progress in general, no work had commenced on the so-called 'Saltash position'. In fact this gap in the western defences was never closed.[102] Traces of the military roads built to link the various sites still exist and several of the forts survive in good condition. Those forming the north-eastern line were dominated by the massive fort at Crownhill which was defended by a 30ft wide ditch cut into the solid rock. Experience in the Crimean War highlighted the need to modernize coastal defences at home and the result in Plymouth was the construction of four new batteries, the largest of which was at Picklecombe. Thus, by 1870, a network of fifteen principal forts and batteries around Plymouth was virtually complete.[103] There were several smaller batteries, and at Cawsand and Polhawn new batteries had been built for beach defence. But by that time the immediate threat of invasion had been removed and Plymouth's great mid-Victorian military structures are, therefore, sometimes ridiculed as the 'Palmerston follies' (see also chapter 22).[104]

During the final decades of the nineteenth century, a dozen new coast batteries were built further south to protect the entrance to the Sound itself and to counter any attempt by enemy ships to fire their guns directly at the Dockyard from positions in the open channel. In the early 1900s modern breech-loading guns were introduced, and these remained the basis of coastal defence until after the Second World War. But coastal artillery finally became obsolete in 1956; land defences were abandoned after the First World War. Today very few of the installations and structures shown on **Map 65.8** retain their original purpose and many have been obliterated. Several of the magnificent mid-Victorian forts have been sadly neglected in recent decades, though attempts are now at last being made to conserve and interpret Plymouth's extensive military geography and to recognize its historical significance.[105]

EIGHTEENTH- AND NINETEENTH-CENTURY SEA DEFENCES, HARBOUR WORKS AND COMMUNICATIONS (**Map 65.9**)

Major infrastructure developments, often involving remarkable feats of engineering, profoundly affected Plymouth's townscape during the late eighteenth and nineteenth centuries (**Map 65.9**). Their building also created a demand for stone, thereby stimulating the exploitation of local quarries which represented one of Plymouth's largest sources of employment in this period. Local limestone had traditionally been taken by boat to be burnt in lime-kilns located elsewhere on the coast or in the Tamar valley and then sold to dress farmland.[106] Stone was, however, increasingly used not only as a building material, but also for paving the town's streets.[107] By the nineteenth century, although granite was used, local limestone was also in demand for projects such as John Rennie's Royal William Victualling Yard and Brunel's development of Millbay Docks.[108] Other outstanding monuments designed by the leading engineers of the day include John Smeaton's lighthouse erected on the Eddystone reef, 20 miles from Plymouth, and completed in 1759 (**Fig. 65.5**);[109] the great Breakwater in the Sound designed by John Rennie and, as previously noted, completed in 1844; and Brunel's Royal Albert railway bridge linking Cornwall to the broad-gauge railway line to London in 1859 (**Fig. 65.6**).[110]

Notwithstanding short tramways and railways built earlier to move large quantities of stone for specific projects, the development of an elaborate network of public lines began in 1823 with the opening of the Plymouth and Dartmoor Tramway between Princetown and Crabtree.[111] By 1825, this route had been extended to Sutton Harbour. Long-distance passenger links were established in 1848 when the South Devon Railway opened its station at Laira Green; this line was extended to Millbay a year later. Thereafter, a number of railway companies built lines in Plymouth, but in 1876 the South Devon and Cornwall Railways merged to form the Great Western Railway; North Road Station was opened a year later.[112]

URBAN GROWTH IN NINETEENTH-CENTURY PLYMOUTH (**Fig. 65.1 and Maps 65.10 and 65.11**)

Fig. 65.1 depicts aspects of the rapid change and development occurring in Plymouth during the nineteenth century. The total population of the three towns of Plymouth, Stonehouse and Devonport increased more than four times from 43,194 in 1801 to 193,184 a century later. Although not legally united as a single county borough until 1914, and not recognized as a city until 1928, by the mid-Victorian period Plymouth was already physically joined to its neighbours.[113] From late Georgian times a web of new streets had begun to spread across former farm land, thereby trebling the extent of the built-up area by the end of the Victorian era. Indeed, as early as the 1860s, there was a virtually continuous urban area stretching for more than 3 miles from the Plym estuary in the east to the Hamoaze and Tamar in the west (**Fig. 65.1 and**

For **Figs 65.4, 65.5 and 65.9** see pages 23–24 in the colour section.

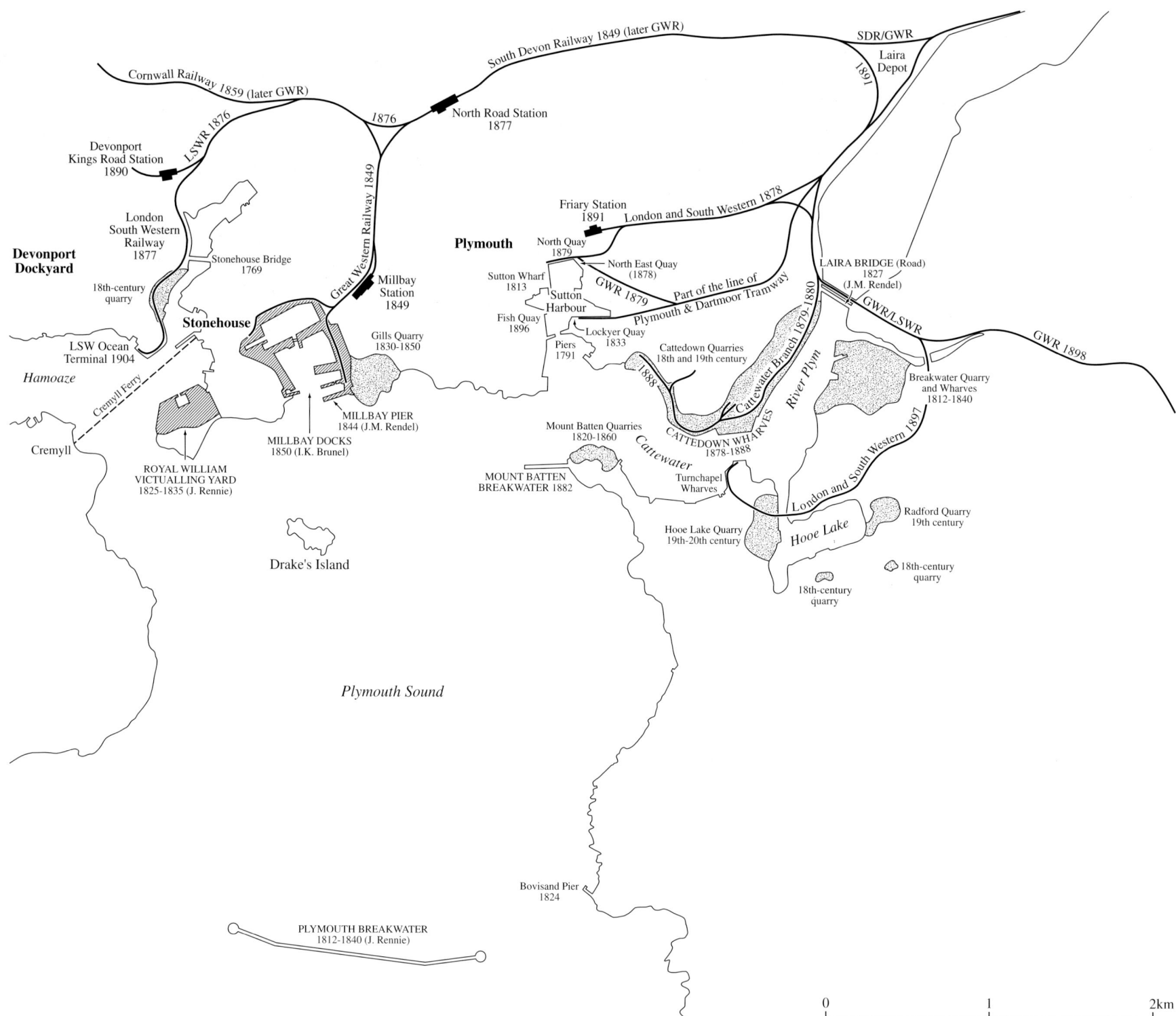

Map 65.9

Eighteenth- and nineteenth-century sea defences, harbour works and communications including the development of the Breakwater, Millbay Docks, Sutton Harbour improvements, the Mount Batten breakwater, the Brunel Railway Bridge, the railway links, Victorian stone quarries, and major piers and wharves. The Plymouth and Dartmoor Tramway (1823) was taken over and rebuilt by South Devon Railway in 1856-7 and in turn taken over by the GWR in 1878. The Cattewater

Branch was originally built by the Plymouth and Dartmoor Railway on behalf of the L&SWR in 1879–80. This line was extended to Cattedown Wharves by 1888.

SOURCES: maps of Plymouth by John Cooke (F. Nile, Stonehouse, 1820); by J. Rapkin, (John Tallis, London, 1860); and by W.H. Maddock (Plymouth, 1892); and works cited in notes 108–12.

Map 65.11). Legal acknowledgement of the urban growth which had occurred came twenty years later when the boundaries of both Plymouth and Devonport were extended.

Population increases outpaced the supply of new housing, and overcrowding in poor quality accommodation became a persistent feature of Victorian Plymouth. Further infilling of rear plots in the older parts of the three towns was common, resulting, for example, in average densities of up to thirty dwellings to the acre in the Sutton Harbour area. In Plymouth, despite the annual addition of an average of 500 new houses, rents were relatively high compared with other cities, and as a result many dwellings were multi-occupied as soon as they were complete.[114] A plethora of speculative builders were eager to erect terraced cottages the moment land was leased for building, the new streets frequently preserving pre-urban field and property boundary patterns (see chapter 54 and **Maps 54.7 and 54.8**). But steep slopes and areas of low-lying marsh meant that there was a shortage of suitable building land, made worse by the slow release for development of some of the encircling private rural estates. Thus, for example, the Barley estate of the Elliots was not made available until the 1860s, while the Bewes family held on to the Beaumont estate until 1890. Houses themselves were overcrowded. The 1851 census records some spectacular examples, such as in New Street where 598 people resided in twenty-three houses, an average of twenty-six persons per dwelling. Other blackspots included Basket Street, Stillman Street and Lower Street.[115] Substantial areas of the 'three towns' rapidly became insanitary and insalubrious. Cholera dealt savage blows in 1832 and 1849. Ultimately, in the 1850s, both Plymouth and Devonport adopted the terms of the 1848 Public Health Act and reforms began, but the scale of the task was formidable. As late as 1871, there were still a number of census enumeration districts where the average number of people in each inhabited dwelling was twelve or more. But by then much work had already been undertaken to remove building encroachments made in the early nineteenth century which had gradually rendered some streets in the poorer quarters impassable by carriage or cart. Indeed, street widening, water and sewer pipe installation, and road surfacing were major preoccupations in the three towns throughout later Victorian times. Yet, nineteenth-century Plymouth, Stonehouse and Devonport had developed a distinctly multi-cellular social geography whereby the

Map 65.10

Map of Plymouth, Stonehouse, Plymouth Dock (Devonport) and Plymouth Sound, 1863. This map, 'constructed and engraved from the Admiralty Surveys of G. H. Swanston, Edinburgh', effectively united the depiction of the 'three towns' with the haven upon which they developed. Information about navigating the Sound and the lower Tamar is shown, as well as details of the townscape. The green fields, with their neat hedges, painted by W. du Buse, c.1690 (**Fig. 65.4**), have been replaced by urban streets and buildings, the marshy inlet to the west of the Hoe has been developed as Millbay Docks, and much of the southern part of the manor of Stoke Damerel has been converted into the Royal Dockyard and its associated settlement.

SOURCE: Mr Timothy Absalom.

531

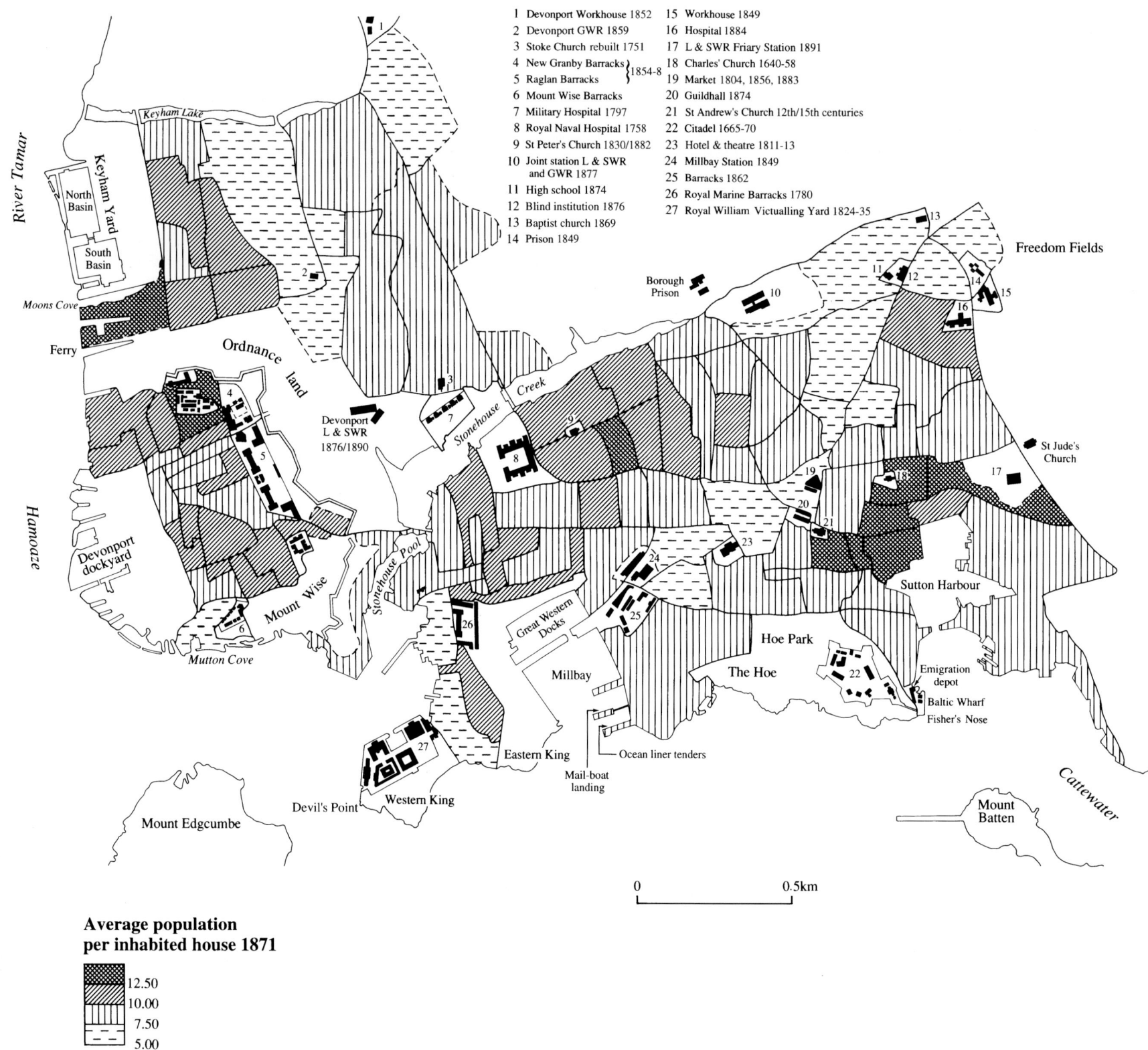

1 Devonport Workhouse 1852
2 Devonport GWR 1859
3 Stoke Church rebuilt 1751
4 New Granby Barracks } 1854-8
5 Raglan Barracks
6 Mount Wise Barracks
7 Military Hospital 1797
8 Royal Naval Hospital 1758
9 St Peter's Church 1830/1882
10 Joint station L & SWR and GWR 1877
11 High school 1874
12 Blind institution 1876
13 Baptist church 1869
14 Prison 1849
15 Workhouse 1849
16 Hospital 1884
17 L & SWR Friary Station 1891
18 Charles' Church 1640-58
19 Market 1804, 1856, 1883
20 Guildhall 1874
21 St Andrew's Church 12th/15th centuries
22 Citadel 1665-70
23 Hotel & theatre 1811-13
24 Millbay Station 1849
25 Barracks 1862
26 Royal Marine Barracks 1780
27 Royal William Victualling Yard 1824-35

**Average population
per inhabited house 1871**

12.50
10.00
7.50
5.00

Map 65.11

Nineteenth-century Plymouth: population and townscape. The map shows the built-up area of Plymouth, Stonehouse and Devonport at the time of the census of 1871. Major public buildings are shown especially where they housed an institution which represented a 'special enumeration district'.

SOURCES: maps as for **Map 65.9**; enumeration district boundaries plotted from written descriptions in manuscript census returns, 1871 (Public Record Office, RG 10/2119–2141).

characteristics and density of population varied considerably from one part to another. The evidence for these local-scale spatial variations can be found in the manuscript census enumerators books—the districts to which such information relates are plotted on **Map 65.11**.

In Stonehouse, eighteenth-century residential development was matched by several major military buildings including the Royal Marine Barracks, the Royal Naval Hospital and the Military Hospital. Elsewhere, as **Maps 65.10 and 65.11** show, major new building projects occurred during the nineteenth century. New workhouses, civilian hospitals, churches serving every denomination, theatres, libraries, and public baths all expressed the energy and confidence of Victorian Plymouth. As noted earlier, a new guildhall was completed in 1874, and municipal offices were erected opposite, thereby forming a narrow civic square bounded at one end by St Andrew's and opening into

Westwell Street at the other. Only the guildhall, in part reconstructed according to the Victorian design, survives; both the Square and the municipal offices were swept away in the blitz and post-war redevelopment.

At the beginning of the nineteenth century, markets were still held as they had been since medieval times under the old guildhall and in Whimple Street, while some of the butchers' stalls were still arranged along the walls of St Andrew's churchyard. However, in 1804 a new red-brick market building was erected on a field to the north of Bedford Street. Opening first in 1807, it was extended twice, in 1856 and 1883. Both Stonehouse and Devonport had independent markets, but neither flourished on a scale comparable with that of Plymouth, whose catchment extended more than 20 miles, deep into Devon and Cornwall.[116]

Taken together, the 'three towns' represented the South West's foremost urban leviathan, with an economy based not

Fig. 65.1
Urban growth, 1801–1981.

SOURCE: published census data.

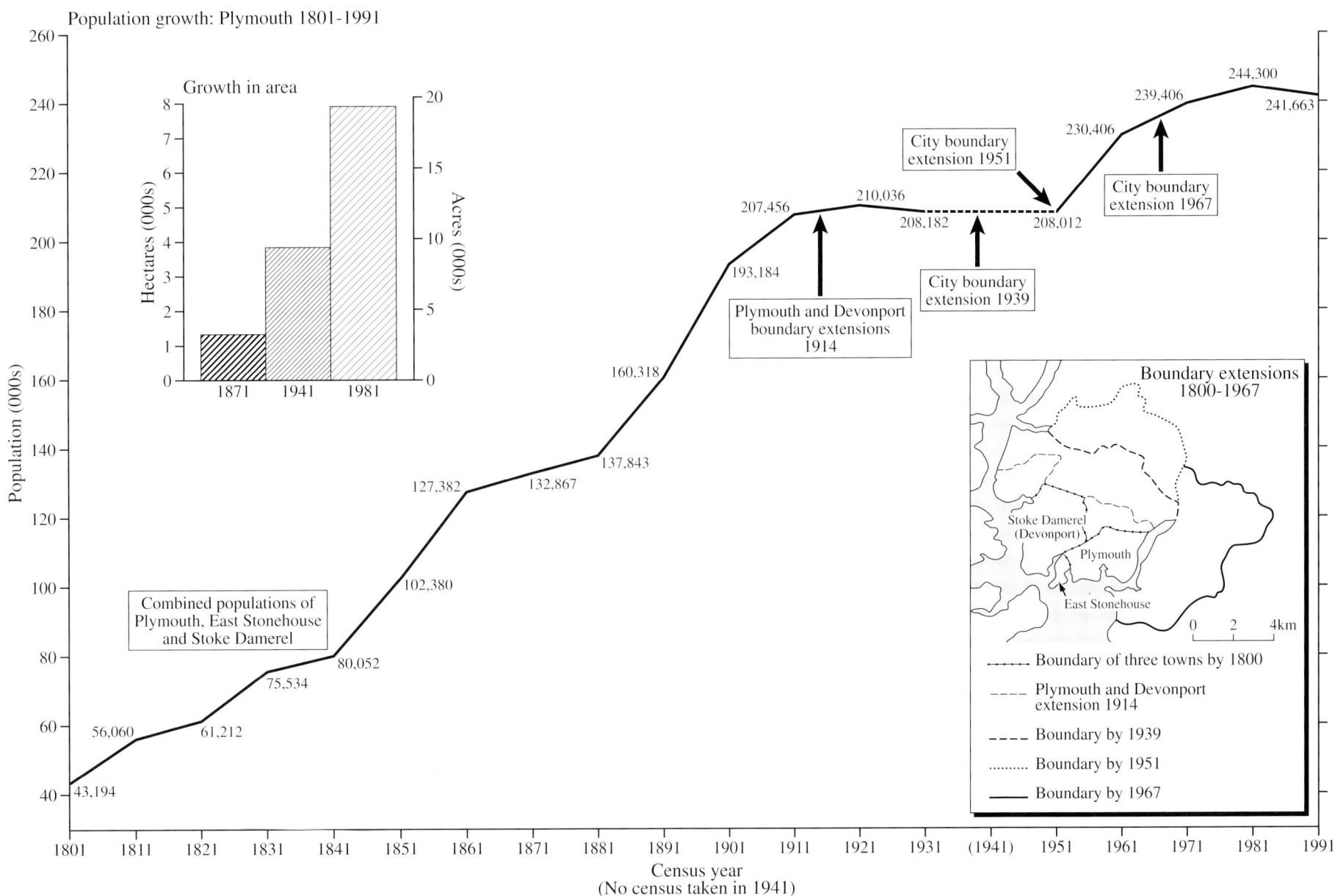

Population growth: Plymouth 1801-1991

Growth in area

Combined populations of Plymouth, East Stonehouse and Stoke Damerel

43,194 · 56,060 · 61,212 · 75,534 · 80,052 · 102,380 · 127,382 · 132,867 · 137,843 · 160,318 · 193,184 · 207,456 · 210,036 · 208,182 · 208,012 · 230,406 · 239,406 · 244,300 · 241,663

Plymouth and Devonport boundary extensions 1914

City boundary extension 1939

City boundary extension 1951

City boundary extension 1967

Boundary extensions 1800-1967

Stoke Damerel (Devonport)
Plymouth
East Stonehouse

0   2   4km

········· Boundary of three towns by 1800
── ── Plymouth and Devonport extension 1914
─ ─ ─ Boundary by 1939
·········· Boundary by 1951
──── Boundary by 1967

Census year
(No census taken in 1941)

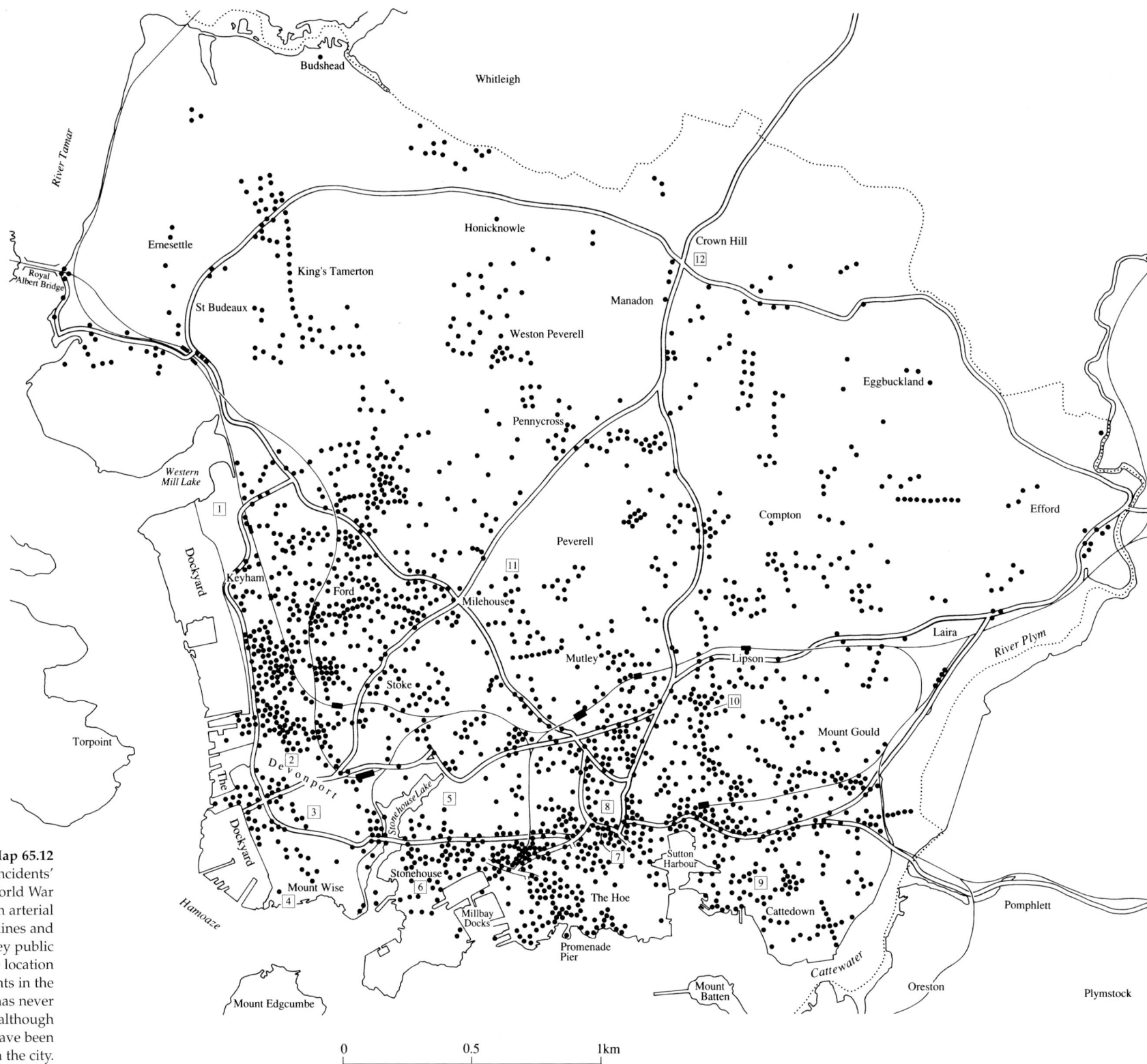

Map 65.12
Location of 'bomb incidents' during the Second World War in relation to the main arterial roads, railway lines and stations, and key public buildings; the precise location of bomb incidents in the Royal Dockyard has never been disclosed, although these are known to have been among the heaviest in the city.

SOURCE: 'Bomb incidents and areas devastated during the heavy raids' plotted on OS 1:10,560 base maps of Plymouth by the staff of the City Engineer and Surveyor, 20 March 1941–29 April 1941 (Plymouth City Museum).

0   0.5   1km

Arterial roads

Railways and stations

Borough boundary 1945

• Bomb

1 Royal Naval Barracks
2 Granby Barracks
3 Raglan Barracks
4 Mount Wise Barracks
5 Royal Naval Hospital
6 Royal Marine Barracks

7 Guildhall, Municipal Offices and St Andrew's Church
8 Market
9 Coxside Gas Works
10 General Hospital
11 Plymouth Argyle Football Stadium
12 Crownhill Barracks

only on the needs of a large permanent garrison of soldiers and naval personnel, but also the activities of Royal Navy and commercial shipbuilding and repairing, mercantile trade and broking, and a wide range of import-based manufacturing industries. Passenger shipping, most notably of emigrants to the colonies, was expanding, and by mid-Victorian times the railways had stimulated the growth of another key source of income, namely tourism.[117]

THE DESTRUCTION OF PLYMOUTH IN THE SECOND WORLD WAR BLITZ (**Map 65.12**)

Until the fall of France in June 1940, Plymouth was beyond the range of enemy aircraft and therefore escaped bombing raids for the first nine months of the war.[118] But on 6 July the first bombs fell. The emergency lasted until just before the allied invasion of Normandy in 1944. In fact the last bombing incident occurred on 30 April 1944. The city was attacked on fifty-nine separate occasions. Only a handful of the buildings of the pre-war town centre escaped destruction while beyond, some 3,754 houses were wrecked beyond repair, and another 18,398 were seriously damaged. Indeed, across the city as a whole, a total of 72,102 homes sustained damage of some kind during the blitz.[119] Official figures released at the end of the war indicate that 1,172 civilians were killed and another 3,269 had been injured. Casualties amongst armed services personnel were not disclosed. The heaviest contributions to the casualty lists came in two periods of concentrated bombing in March and April 1941. During two consecutive nights, 20 and 21 March, sustained attacks caused damage to property estimated at more than £100 million, and killed 336 civilians.[120] Thereafter, during 21, 22, 23, 28 and 29 April, some 1,140 high explosive bombs, seventeen paramines and countless thousands of incendiaries rained down on Plymouth. In fact fire accounted for three-quarters of the damage sustained by the city during the wartime blitz. Central Plymouth had been the key target in the March raids, but in April Devonport was the particular focus of enemy attacks. Another 590 people were killed and 427 were seriously injured. Altogether, during those five nights, the raids lasted for over twenty-three hours and an estimated 2,000 fires had raged across the city. The reddish glow of Plymouth 'in flames' was easily visible from more than 20 miles away.

During the war Plymouth's city council used large scale Ordnance Survey maps to plot accurately the location of every bomb dropped within the urban boundary (except those which fell on the Royal Dockyard, where the extent of the damage was not disclosed). These maps have been used as the basis of **Map 65.12**, which shows clearly that by the end of the war no district in the city remained completely unscathed, though the concentration of bombing on the commercial and retail core of Plymouth and Devonport stands out along with damage to the city's railways and key public utilities.[121]

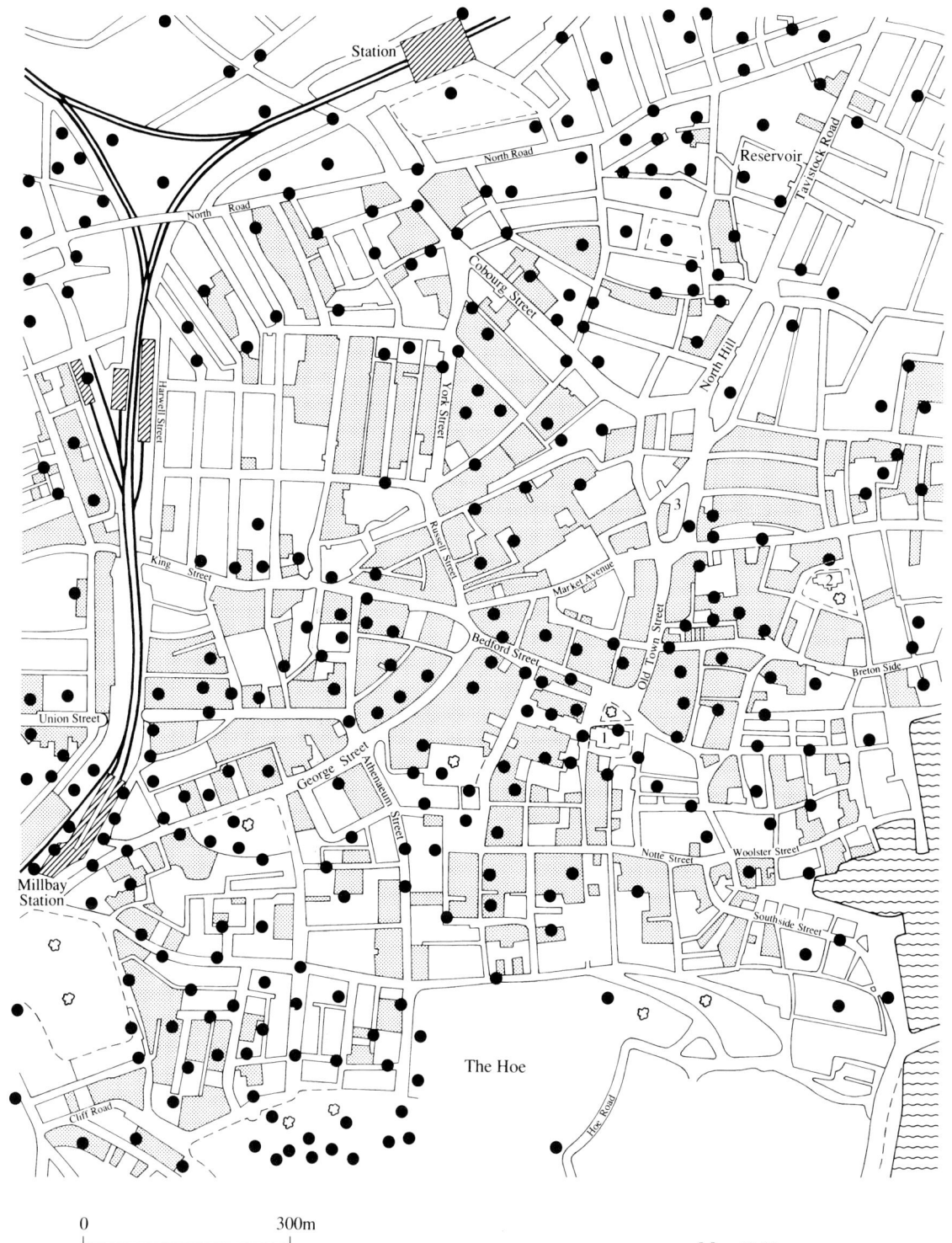

	Location of bombs dropped during Second World War
	Public open spaces
	Derelict/previously cleared sites available for reconstruction
1	St Andrew's Church
2	Charles Church
3	Drake Circus

Map 65.13
The destruction of the city centre in the Second World War.

SOURCES: as **Map 65.12** and 'Properties demolished or beyond repair and slum clearance areas' plotted by the staff of the City Engineer and Surveyor on OS 1:10,560 base maps; 'Map of derelict areas', in J. Paton Watson and P. Abercrombie, *A Plan for Plymouth* (Plymouth, 1943).

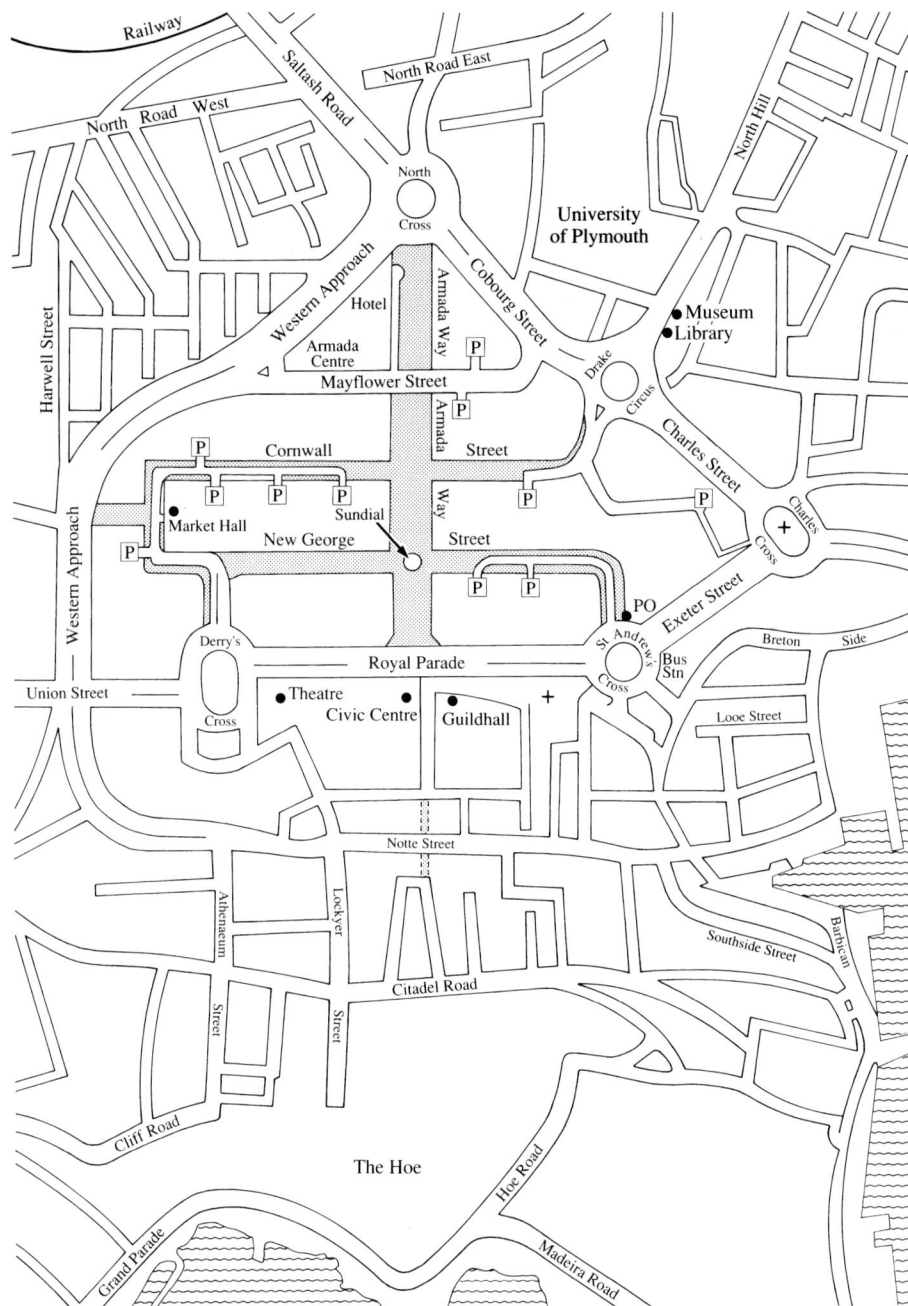

+ Churches (Charles Church & St Andrew's Church)

P Car parks

Pedestrianised shopping streets (since 1988)

0   300m

POST-WAR RECONSTRUCTION (Maps 65.13, 65.14 and 65.15)

Destruction during the blitz presented an unparalleled opportunity to remedy some of the city's pre-war problems, including poor housing, traffic congestion and inadequate open space. Before 1939, largely because of the weakness of the

For **Map 65.14** see page 22 in the colour section.

planning system, there had been only modest progress in correcting some of these defects. It was said, for instance, that Plymouth had the worst traffic congestion problems of any provincial city in southern England. Housing densities were worse than in Liverpool and London's East End.[122] However, although many buildings in the old city centre were devastated by enemy raids, the city council's maps showing exactly where each bomb fell indicate that significant areas were not bombed at all (**Map 65.13**). But as early as July 1941, Lord Reith, then Minister of Buildings and Works, had urged the local authority to plan 'boldly and comprehensively' for the redevelopment of Plymouth.[123] The fate of the old pre-war street pattern was effectively sealed and a decision was taken to replace it with a new design. Viscount Astor, the city's wartime mayor, was determined to seize the chance to redevelop Plymouth along entirely modern lines. At his personal invitation, Professor (later Sir) Patrick Abercrombie was engaged as the consultant on the *Plan for Plymouth*.[124] Abercrombie was professor of town planning at the University of London and he had already been retained to prepare reconstruction plans for the capital. Collaborating closely with the city engineer and surveyor, James Paton Watson, Abercrombie devised the well-known blueprint for Plymouth's reconstruction.[125] Published in 1943, the *Plan* not only contained radical and innovative proposals on housing, retailing, communications, population redistribution, employment and recreation, but also a completely new design for the city centre.[126] Abercrombie and Paton Watson envisaged a formal, neoclassical pattern of buildings and townscape whose 'symmetry and dignity' were supposed to proclaim the advantages of comprehensive planning over piecemeal reconstruction (**Fig. 65.8**).

The new city centre, shown in **Map 65.15**, was intended as a symbol of a 'better tomorrow' and as recompense for the sufferings endured by Plymouth in the war. In sharp contrast to many cities in mainland Europe which were rebuilt to their original design, in Plymouth the complicated tangle of pre-war streets was literally wiped off the map. They were replaced by a grid-iron pattern of lengthy, formal east–west boulevards bisected by the main north–south axis of 'Armada Way' which was intended as a wide processional avenue linking North Cross and the railway station with the Hoe (**Map 65.14**).[127] The plan received local authority approval in 1944 and by 1947 more than 170 acres in the city centre had been compulsorily purchased (at 1939 values), thereby enabling construction work to begin in March.[128] The *Plan for Plymouth* was seen as a landmark in urban planning and reconstruction. Its weaknesses derive in part from the basic design—sometimes described as a 'wilderness of straight lines'—and in part from the way in which it was implemented. The architecture is at best somewhat mediocre. In recent years, however, the pedestrianization of the main shopping streets (completed in 1988) has facilitated the landscaping of the principal thoroughfares, and has done much to soften the harsh rigidity of Abercrombie's geometry.[129]

POST-WAR RESIDENTIAL PLANNING (**Map 65.16**)

Although the proposals for the reconstruction of Plymouth's city centre were the most publicized parts of the 1943 *Plan*, it also contained far-reaching policies for housing and residential development.[130] Abercrombie was an advocate of the ideas and ideals of neighbourhood unit planning so much in vogue in the 1940s and 1950s. As a result, Plymouth today provides some interesting examples of residential areas designed according to neighbourhood unit principles (**Map 65.16**). The theory argued that many of the social and psychological problems which afflicted urban life stemmed from the anonymity and physical scale of cities. By subdividing an urban area into readily identifiable small, village-like communities, a new social cohesion and sense of belonging could be engineered. The notion rested on a nostalgia for pre-industrial rural lifestyles to the extent that in Plymouth the new neighbourhoods were each planned to accommodate only 6,000–10,000 residents in housing arranged in streets focused on a 'village green'.[131] More importantly, each neighbourhood was to have its own community facilities such as shops, schools, libraries, churches, health centres and a meeting hall. **Map 65.16** shows Abercrombie's intended division of Plymouth into neighbourhood units, together with later areas of housing development on the city's northern and eastern edges at Southway, Estover, Plympton and Plymstock. Indeed, the 1943 *Plan for Plymouth* envisaged five completely new purpose-built neighbourhoods and another

Map 65.15 (facing page)
The post-war reconstruction of the city centre: the redesigned street pattern with its rigid geometry and wide shopping boulevards; the pedestrianisation of part of the centre in the late 1980s is also shown.

SOURCE: Ordnance Survey 1:1,250 maps, 1960–82.

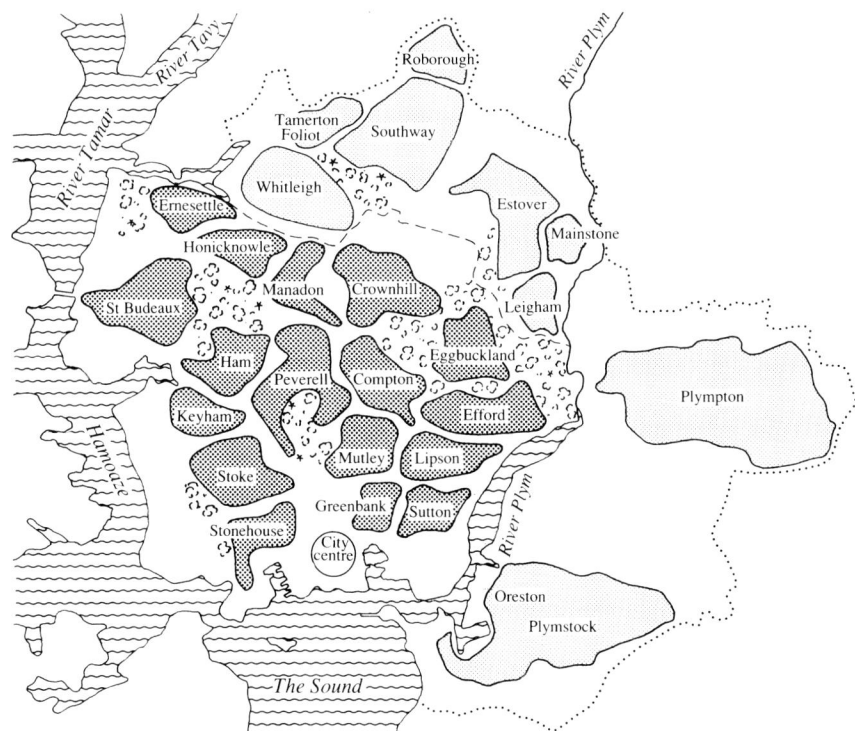

____ 1943 city boundary

.... Present city boundary

Neighbourhoods

▓ 1943 plan for Plymouth

☐ Recent additions

✦ Open space and rural reservation

Map 65.16 (above)
Post-war neighbourhood units.

SOURCES:
J. Paton Watson and P. Abercrombie, *A Plan for Plymouth* (Plymouth, 1943); D.J. Maguire, W.M. Brayshay and B.S. Chalkley, *Plymouth in Maps: A Social and Economic Atlas* (Plymouth, 1987).

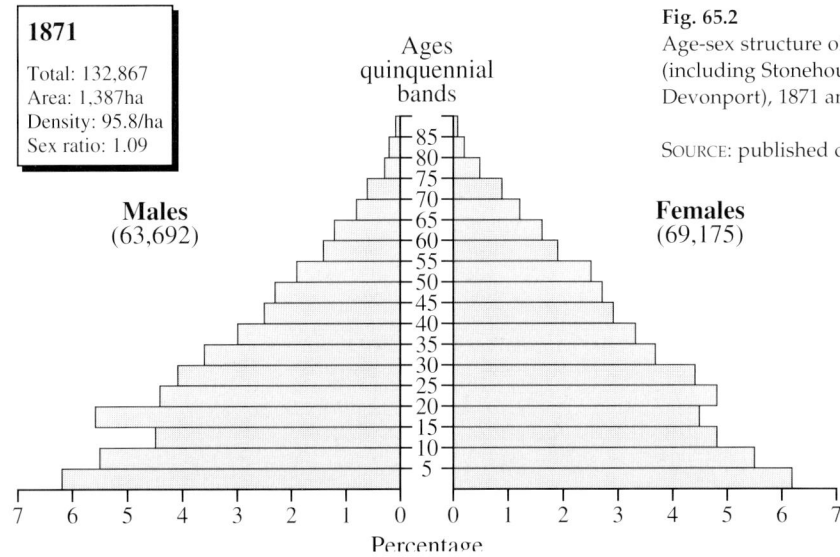

1871
Total: 132,867
Area: 1,387ha
Density: 95.8/ha
Sex ratio: 1.09

Ages quinquennial bands

Males (63,692) **Females** (69,175)

Percentage

1981
Total: 240,652
Area: 7,929ha
Density: 30.4/ha
Sex ratio: 1.04

Ages quinquennial bands

Males (118,003) **Females** (122,649)

Percentage

Fig. 65.2
Age-sex structure of Plymouth (including Stonehouse and Devonport), 1871 and 1981.

SOURCE: published census data.

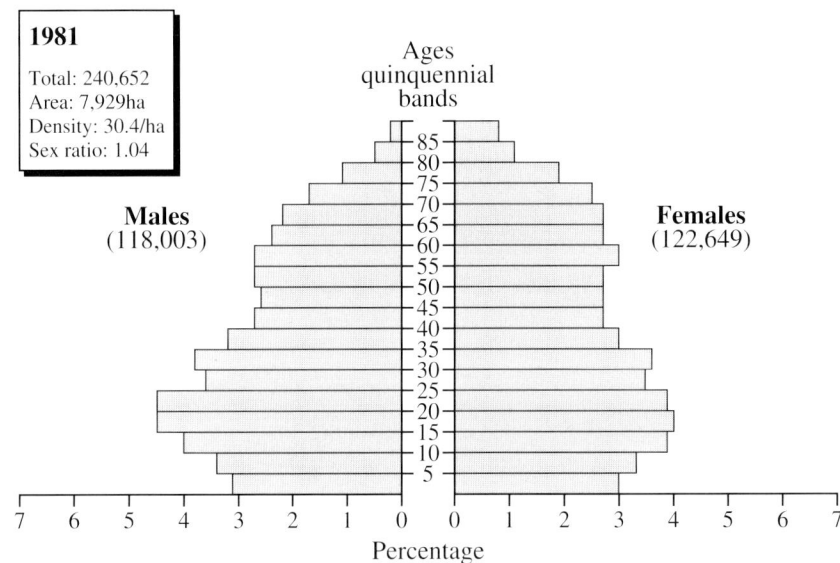

*Area: Plymouth, Stonehouse and Stoke Damerel Registration Districts

Map 65.17
Average number of persons
per household in the
enumeration districts of
Plymouth, 1981.

SOURCE: Census of Population,
1981.

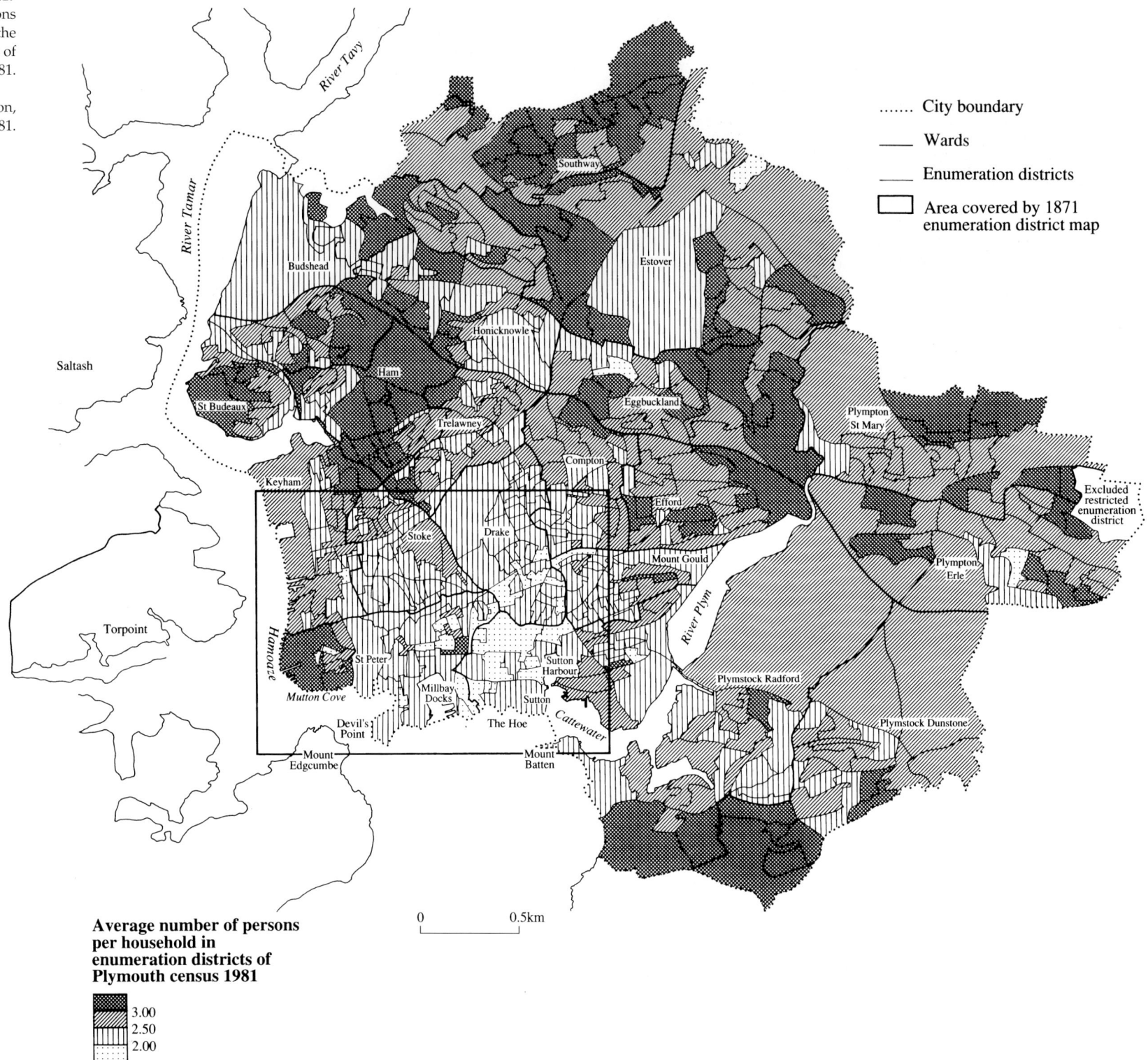

.... City boundary

——— Wards

——— Enumeration districts

☐ Area covered by 1871
enumeration district map

**Average number of persons
per household in
enumeration districts of
Plymouth census 1981**

3.00
2.50
2.00

thirteen were to be based on existing districts within the built-up area. But in fact it proved extremely difficult to reshape the inherited morphology and the practical application of the theory was possible only in the new post-war estates such as Whitleigh, Honicknowle, Efford and Ernesettle. Even here, the constraints of economy meant that only a modest range of neighbourhood facilities were provided and the aesthetic quality of the local authority housing could hardly be described as inspired. Nevertheless, the new 'neighbourhoods' of the city have gradually acquired an individual identity and they have engendered a certain attachment amongst their residents. They represent a rather unusual experiment in urban design aimed at reducing the problems of overcrowding and poor housing conditions in Plymouth's congested pre-war inner city. To that extent the neighbourhood units achieved their objective.

Fig. 65.3 (left)

Firestone Bay artillery tower, Plymouth (fifteenth century). One of the oldest complete military buildings on Plymouth's sea front, this tower protected the deep water passage at the north of Plymouth Sound, between Drake's Island and the main waterfront, which gives access to the Hamoaze and Tamar. The lowest gun-ports (now blocked) were designed to enable the defenders' cannon to inflict maximum damage on enemy ships, namely at the waterline.

SOURCE: University of Plymouth.

For **Figs 65.4, 65.5 and 65.9** see pages 23–24 in the colour section.

Fig. 65.6 (centre)

The Royal Albert Railway Bridge, Plymouth, *c.*1860. Isambard Kingdom Brunel was commissioned to design a bridge to carry the railway from Plymouth, across the Tamar, to Saltash and the rest of Cornwall, and after several early proposals were rejected, the structure he devised in 1848 was accepted. A bridge which not only allowed for the passage of tall ships but was also strong enough to carry steam trains required the engineering genius of Brunel. Its construction took ten years; it was officially opened in 1859.

SOURCE: Plymouth City Museum and Art Gallery.

Fig. 65.7 (below)

Panorama of Plymouth, looking north-east, 1891; pen, wash and watercolour by H.W. Brewer. This remarkably detailed topographical drawing of Plymouth, Stonehouse and Devonport accurately depicts the late-nineteenth-century townscape. The covered slips of South Yard in the Royal Dockyard are clearly visible on the eastern shore of the Hamoaze and the curves of Brunel's Royal Albert railway bridge over the Tamar at Saltash can be seen in the distance further north. The Royal William Victualling Yard appears in the centre foreground of the picture; to the east lies Millbay Dock. The pier is shown on the Hoe waterfront, below Smeaton's Tower. The towers of the Guildhall and St Andrew's Church mark the position of Plymouth's Victorian civic centre located to the north of the Hoe. Sutton Harbour, Mount Batten and the Plym estuary appear on the right of the picture. In the right foreground, Drake's Island is shown with ships passing through the deepest waters of the Sound between the Island and the foreshore. The artist's vantage point appears to have been from above Mount Edgcumbe Park, suggesting perhaps that he may have viewed the prospect from a hot-air balloon.

SOURCE: Plymouth City Museum and Art Gallery.

Fig. 65.8

The Plan for Plymouth, 1943: Traffic Centre at North Road and Coburg Street. This drawing shows the Plan's vision for a 'large traffic circus' providing the main approach to the railway station and the northern end of Armada Way, from which point arrivals were to receive 'a magnificent impression of the expanse of the business district and civic centres and, at the same time, see the Hoe beyond'. The monumental architecture and flag-bedecked towers depicted in the drawing reflected Abercrombie's aim to 'recapture the wonderful continuity of the street scene obtained by Nash and Wood the Younger, as in Old Regent Street and Bath, but in the modern idiom'. To achieve this, Abercrombie argued that Plymouth's new streets should be 'designed as a whole'. Translating these intentions into reality proved difficult, however, and these designs for a grand entrance to the city, to greet arriving railway travellers, have never been fully realized.

SOURCE: author's collection.

POPULATION DISTRIBUTION IN 1981 (**Map 65.17 and Fig. 65.2**)

Occupying 7,924ha (19,580 acres), Plymouth today is the largest urban area in Devon and Cornwall (**Map 65.17**). At 31 persons per hectare, the city's population density is higher (1991) than in Exeter (21), Falmouth (25), Torbay (19), or Truro (16). The general demographic trend in Plymouth during the twentieth century has been one of growth and the only serious downturn came when wartime evacuations lowered the population to about 130,000, a reduction of one-third. Numbers were rapidly restored after 1945 and, following boundary extensions in 1951 (for the neighbourhood units) and in 1967 (to take in Plympton and Plymstock), the population rose to 239,400 in 1971. Growth since then has been modest, but Plymouth was the only one of the country's twenty largest cities which did not experience a fall in numbers between 1971 and 1981. By 1991, however, the impact of counterurbanization, whereby inhabitants of Plymouth had moved to surrounding centres including Saltash

For **Figs 65.4, 65.5** and **65.9** see pages 23–24 in the colour section.

and Ivybridge, meant that the city's population total fell to 241,663.[132] Plymouth's birth and death rates are not far from the national average, but the level of net out-migration is in fact still relatively low.[133]

The spatial distribution of the 1981 population, by average household size, is depicted in **Map 65.17**, which shows Plymouth divided into 522 census enumeration districts (data for the 'restricted' and 'special' districts are excluded).[134] Although the basis of **Map 65.11** is not strictly comparable with **Map 65.17**, by viewing the two maps in conjunction it is possible not only to gauge the substantial areal expansion of the city since 1871, but also to note that larger households (once concentrated in the central districts) are now a feature of Plymouth's peripheral housing estates. In the inner city, by contrast, single-person households are now a key element in the population. Here the age structure tends to be markedly bi-modal, consisting of pensioners on the one hand, and young adults on the other. The latter are attracted by the available supply of rented accommodation and tend to be highly mobile and transient. Pensioner households are more stable and, in the inner city, they often represent a residual population that has been either unable or unwilling to join the general exodus to the suburbs.[135]

Reductions in the birth rate and the gradual process of demographic ageing have produced marked changes in the structure of the city's population since the nineteenth century (**Fig. 65.2**). The potential for rapid and sustained growth which existed in 1871 was manifested in the 'youthful' character of the population pyramid. In 1981, by contrast, a larger proportion of the population was either middle-aged or elderly and the dynamic of growth was considerably weaker. But the impact of demographic ageing has been less marked in Plymouth than elsewhere in south-west England. Thus, between 1971 and 1981 the city was the only district in Devon still experiencing more births than deaths. Large-scale retirement migration which has affected other towns in the region has not occurred in Plymouth. At the census of 1991, therefore, only 18.4 per cent of the city's population was of pensionable age compared with, for example, 22 per cent in Camborne, 26 per cent in Truro, Penzance and Falmouth, 27 per cent in Kingsbridge, and 45 per cent in Sidmouth. In demographic terms, as in many other respects (**Fig. 65.9**), the city of Plymouth contrasts markedly with other parts of the region.

FURTHER READING

Crispin Gill, *Plymouth: A New History, Ice Age to the Elizabethans* (Newton Abbot, 1966) and *Plymouth: A New History, 1603 to the Present Day* (Newton Abbot, 1979); B.S. Chalkley, D. Dunkerley and P. Gripaios (eds), *Plymouth: Maritime City in Transition* (Newton Abbot, 1991); D.J. Maguire, W.M. Brayshay and B.S. Chalkley, *Plymouth in Maps: A Social and Economic Atlas* (Plymouth, 1987).

NOTES

Chapter 1

1. M. Todd, *The South West to AD 1000* (London, 1987), 1.
2. Sir Arthur Conan Doyle, *The Hound of the Baskervilles* (London, 1901).
3. A. Fleming, *The Dartmoor Reaves* (London, 1988).
4. C.J. Caseldine, 'Environmental change in Cornwall during the last 13,000 years', *Cornish Archaeology*, 19 (1980), 3–17; C.J. Caseldine and D.J. Maguire, 'A review of the prehistoric and historic environment on Dartmoor', *Proceedings of the Devon Archaeological Society*, 39 (1981), 1–16; C.J. Caseldine, 'Into the mists? Thoughts on the prehistoric and historic environmental history of Dartmoor', *The Archaeology of Dartmoor: Perspectives from the 1990s, Proceedings of the Devon Archaeological Society*, 52 (1994), 35–47.
5. E.A. Edmonds, M.C. McKeown and M. Williams, *British Regional Geology: South West England* (London, 1985), 52.
6. E.M. Durrance and D.J.C. Laming (eds), *The Geology of Devon* (Exeter, 1985), 12.
7. Edmonds, McKeown and Williams, *British Regional Geology*, 2.

Chapter 2

1. Richard Polwhele, *History of Devonshire*, 3 vols (Exeter, 1793), I, 40.
2. Edwin Welch (ed.), *Plymouth Building Accounts of the Sixteenth and Seventeenth Centuries*, Devon and Cornwall Record Society, new series, 12 (1967), 64.
3. British Geological Survey, *Memoir*, Sheet 324 (1968).
4. Richard Carew of Antony, *The Survey of Cornwall*, ed. F.E. Halliday (London, 1953), 86, 87.
5. Ibid., 87.
6. One of the more unexpected examples is the walling of the Keep at Launceston Castle.
7. Richard Carew, *The Survey of Cornwall*, ed. Lord de Dunstanville (1801), 18, 19, comments on the weatherproof qualities of granite but Tonkin, in a footnote, complains of brittleness in some granite.
8. The term 'killas' is often used for slate-stone in Cornwall; it is subject to a variety of definitions but seems to relate to the softer slate-stones occurring around the granite masses particularly.
9. For south Devon slates see Anne Born, 'Blue slate quarrying in south Devon', *Industrial Archaeology Review*, 11, 1 (1988).
10. For a full discussion of the role of west-country roofing slate well beyond the region, with extensive references to written and archaeological sources, see E.M. Jope and G.C. Dunning, 'The use of blue slate for roofing in medieval England', *Antiquarian Journal*, 34 (July–Oct. 1954), 209–17.
11. L.F. Salzman, *Building in England* (Oxford, 1967), 132.
12. For the full story of the discovery and re-use of this quarry see Deryck Laming, 'The building stone and its quarry', in Michael Swanton (ed.), *Exeter Cathedral: A Celebration* (Exeter, 1991).
13. It is estimated that Devon has at least 20,000 houses built entirely or partly of cob: Devon Historic Buildings Trust, *The Cob Buildings of Devon* (Exeter, 1992).
14. For example, the enthusiastic article by J. Joce, 'Cob cottages for the twentieth century', *Transactions of the Devonshire Association*, 5 (1919), 171. Joce refers to a 'recently built' cob house in Budleigh Salterton, probably Coxen, where the eight men building it took three months to reach wall-plate height.
15. Carew, *Survey of Cornwall*, ed. de Dunstanville, 143; W.G. Hoskins, *Devon* (Newton Abbot, 1954), 266–7.
16. 'Exeter port books', in J.P. Allan, *Medieval and Post-Medieval Finds from Exeter 1971 to 1980*, Exeter Archaeological Reports 3 (Exeter, 1984).

Chapter 3

1. Wessex Archaeology, 'The South West and South of the Thames', Report No. 2, 1992–3, Southern Rivers Palaeolithic Project (Trust for Wessex Archaeology Ltd and English Heritage, 1993); D.A. Roe, 'A gazetteer of British Lower and Middle Palaeolithic sites', *Research Report of the Council for British Archaeology* (1968), 8, 42–8, 256–61; D.A. Roe, *The Lower and Middle Palaeolithic Periods in Britain* (London, 1981); J.B. Campbell, *The Upper Palaeolithic of Britain: A Study of Man and Nature in the Late Ice Age*, 2 vols (Oxford, 1977).
2. R.M. Jacobi, 'Leaf-points and the British Early Upper Palaeolithic', *Archaeological Studies and Research of the University of Liège*, 42 (1990), 271–89; J.J. Wymer, 'Palaeolithic archaeology and the British Quaternary sequence', *Quaternary Science Reviews*, 7 (1988), 79–98.
3. N.J. Shackleton and N.D. Opdyke, 'Oxygen isotope and palaeomagnetic stratigraphy of Equatorial Pacific core V28–238: oxygen isotope temperatures and ice volumes on a 10^5 and 10^6 year scale', *Quaternary Research*, 3 (1973), 39–55.
4. D.Q. Bowen, 'Time and space in the glacial sediment systems of the British Isles', in J. Ehlers, P.L. Gibbard and J. Rose (eds), *Glacial Deposits in Great Britian and Ireland* (Rotterdam, 1991), 3–11.
5. Wymer, 'Palaeolithic archaeology', 93–5.
6. J.C. Campbell and C.G. Sampson, 'A new analysis of Kents Cavern, Devonshire, England', *University of Oregon Anthropological Papers*, 3 (1971); Roe, *Lower and Middle Palaeolithic*.
7. S.N. Collcutt, 'Contextual archaeology: the example of debris flows in caves', in S.N. Colcutt (ed.), *The Palaeolithic of Britain and its Nearest Neighbours* (Sheffield, 1986), 57–8; W. Pengelly, 'The literature of Kents Cavern, Part V: Fifteenth Report to the British Association 1879', *Transactions of the Devonshire Association*, 16 (1884), 412–22; Campbell and Sampson, 'A new analysis', 19; A. Straw, 'Kents Cavern—whence and whither', *Transactions and Proceedings of the Torquay Natural History Society*, 21 (1995), 198–211.
8. Campbell and Sampson, 'A new analysis', 23; Roe, *Lower and Middle Palaeolithic*, 99.
9. C.J. Proctor, 'A British Pleistocene chronology based on Uranium Series and Electron Spin Resonance Dating of Speleothem' (unpublished Ph.D. thesis, University of Bristol, 1994).
10. Roe, *Lower and Middle Palaeolithic*, 103; Wymer, 'Palaeolithic archaeology'; M.B. Roberts, C.B. Stringer and S.A. Parfitt, 'A hominid tibia from Middle Pleistocene sediments at Boxgrove, UK', *Nature*, 369 (1994), 311–13.
11. R.A. Shakesby and N. Stephens, 'The Pleistocene gravels of the Axe valley, Devon', *Transactions of the Devonshire Association*, 116 (1984), 77–88; C.P. Green, 'The Palaeolithic site at Broom, Dorset, 1932–41: from the record of C.E. Bean Esq

FSA', *Proceedings of the Geologists' Association*, 99 (1988), 173–80; Wessex Archaeology, 'The South West', 159–64.

12. Shakesby and Stephens, 'Pleistocene gravels of the Axe valley', 81; and Green, 'The Palaeothic site at Broom', 179.

13. J.J. Wymer, 'A chert handaxe from Chard, Somerset', *Somerset Archaeology and Natural History*, 20 (1976), 101–3.

14. J. Scourse, Appendix, in Shakesby and Stephens, 'Pleistocene gravels of the Axe valley', 86–7.

15. E.M. Durrance, 'The buried channels of the Exe', *Geological Magazine*, 106 (1969), 174–89; E.M. Durrance, 'Gradients of buried channels in Devon', *Proceedings of the Ussher Society*, 3 (1974), 111–19; 1934 Borehole, Wiltshire United Dairies Ltd, Institute of Geological Sciences.

16. Shakesby and Stephens, 'Pleistocene gravels of the Axe valley', 86.

17. C. Stringer and C. Gamble, *In Search of the Neanderthals* (London, 1993), 74–5.

18. Ibid., 147.

19. J.J. Wymer, 'The Palaeolithic', in I. Simmons and M. Tooley (eds), *The Environment in British Prehistory* (London, 1981), 70–1; A. Roberts, personal communication.

20. J.J. Wymer, 'Early man in Britain—time and change', *Modern Geology*, 9 (1985), 261–72; Campbell and Sampson, 'A new analysis', 33–4.

21. A. Rosenfeld, 'The Palaeolithic and Mesolithic', in F. Barlow (ed.), *Exeter and its Region* (Exeter, 1969), 133.

22. Roe, *Lower and Middle Palaeolithic*.

23. Campbell and Sampson, 'A new analysis', 15, 24.

24. Stringer and Gamble, *In Search of the Neanderhals*, 179–94.

25. Campbell, *The Upper Palaeolithic*, II, Table 4.

26. H.G. Dowie and A.H. Ogilvie, 'Kents Cavern, Torquay', *Report of the British Association for the Advancement of Science* (Leeds, 1927), 303–6; radio-carbon dates from R.E.M. Hedges, R.A. Housley, I.A. Law and C.R. Bronk, 'Radio-carbon dates from the Oxford AMS system: Archaeometry datelist 9', *Archaeometry*, 31 (1989), 207–34.

27. P. Berridge and A. Roberts, 'Windmill Hill Cave, Brixham: setting the record straight', *Lithics*, 11 (1990), 24–30; Jacobi, 'Leaf-points', 279.

CHAPTER 4

1. N. Barton, *Stone Age Britain* (London, 1997) contains a thorough discussion of the question of the Lateglacial recolonization of Britain.

2. In the past the term 'Creswellian' was used in error as being synonymous with the entire British Late Upper Palaeolithic. Recent work has redefined the term in a more effective manner: see R.M. Jacobi, 'The Creswellian, Creswell and Cheddar', in R.N.E. Barton, A.J. Roberts and D.A. Roe (eds), *The Late Glacial in North-West Europe: Human Adaptation and Environmental Change at the End of the Pleistocene*, Council for British Archaeology Research Report 77 (London, 1991), 128–40; R. Jacobi, 'The Creswellian" in Britain', in J.-P. Fagnart and A. Thévenin (eds), *Le Tardiglaciaire en Europe du Nord-Ouest*, Actes du 119e Congrés National des Sociétés Historiques et Scientifiques 1994 (Amiens, 1997), 497–505; and R.M. Jacobi and A.J. Roberts, 'A new variant on the Creswellian angle-backed blade', *Lithics*, 13 (1992), 33–9.

3. A preliminary report on the Late Upper Palaeolithic archaeology from this site is contained in R.N.E. Barton, 'The Late Glacial period and Upper Palaeolithic archaeology', in D.J. Charman, R.M. Newnham and D.G. Croot (eds), *The Quaternary of Devon and East Cornwall: Field Guide* (London, 1996), 198–200.

4. The cut-marked human bones are discussed in J. Cook, 'Preliminary report on marked human bones from the 1986–1987 excavations at Gough's Cave, Somerset, England', in Barton, Roberts and Roe (eds), *Late Glacial*, 160–8.

5. For full discussion see R.N.E. Barton and A. Roberts, 'Reviewing the British Late Upper Palaeolithic: new evidence for chronological patterning in the Lateglacial record', *Oxford Journal of Archaeology*, 15, 3 (1996), 245–65.

6. See P. Berridge and A. Roberts, 'The Mesolithic period in Cornwall', *Cornish Archaeology*, 25 (1986), 7–34.

7. See R.N.E. Barton 'Long blade technology in Southern Britain', in C. Bonsall (ed.), *The Mesolithic in Europe: Proceedings of the Third International Symposium, Edinburgh 1985* (Edinburgh, 1989), 264–71; and R.N.E. Barton, 'Technological innovation and continuity at the end of the Pleistocene in Britain', in Barton, Roberts and Roe (eds), *Late Glacial*, 234–45.

8. R.M. Jacobi, 'Aspects of the "Mesolithic Age" in Britain', in S.K. Kozlowski (ed.), *The Mesolithic in Europe: Papers Read at the International Archaeological Symposium on the Mesolithic in Europe* (Warsaw, 1973), 237–66.

9. See papers and references in R.C. Preece (ed.), *Island Britain: A Quaternary Perspective* (London, 1995).

10. P. Berridge, 'A Mesolithic flint adze from the Lizard', *Cornish Archaeology*, 21 (1982), 171.

11. For a description of the sites and discussion of the axes see P.J. Berridge, 'Mesolithic sites in the Yarty valley', *Proceedings of the Devon Archaeological Society*, 34 (1985), 1–21.

12. For example, the microliths hafted as knives from the waterlogged-site of Zamostje 2: V.M. Lozovski, *Zamostje 2: The Last Prehistoric Hunter-Fishers of the Russian Plain* (Belgium, 1996). See also D. Clarke, 'Mesolithic Europe: the economic basis', in G. Sieveking, I.H. Longworth and K.E. Wilson (eds), *Problems in Economic and Social Archaeology* (London, 1976).

13. The three Early Mesolithic typological groups which have been recognized so far are those represented by the assemblages from Star Carr (North Yorkshire), Deepcar (West Yorkshire), and Horsham (Sussex): R.M. Jacobi, 'Northern England in the eighth millennium BC: an essay', in P.A. Mellars (ed.), *The Early Postglacial Settlement of Northern Europe* (London, 1978), 295–332; M. Reynier, 'A stylistic analysis of ten early Mesolithic sites in south east England', in N. Ashton and A. David (eds), *Stories in Stone*, Lithic Studies Society Occasional Paper No. 4 (London, 1994), 199–205; M.J. Reynier 'Radiocarbon dating of Early Mesolithic stone technologies from Great Britain', in Fagnart and Thévenin (eds), *Tardiglaciaire*, 529–42. Most of the Early Mesolithic sites and findspots in both Devon and Cornwall are of Deepcar type. There are, however, a significant number of Star Carr type findspots, especially in the Sidmouth area of east Devon, the Barnstaple area of north Devon, the granitic uplands of Dartmoor and Bodmin Moor, and the Trevose Head area of Cornwall. Although Horsham industries do not seem to extend beyond south-east England, the presence of diagnostic microliths with concave basal retouch ('hollow-based' points) from sites in the interior of east Devon is intriguing. See examples in A. Rosenfeld, 'Palaeolithic and Mesolithic', in F. Barlow (ed.), *Exeter and its Region* (Exeter, 1969), 129–36; and Berridge, 'Mesolithic sites'.

14. The model was first published in R.M. Jacobi, 'Early Flandrian hunters in the South-West', *Proceedings of the Devon Archaeological Society*, 37 (1979), 48–93. It was modified in R.M. Jacobi and C.F. Tebbutt, 'A Late Mesolithic rock-shelter site at High Hurstwood, Sussex', *Sussex Archaeological Collections*, 119 (1981), 1–36.

15. This model is based upon doctoral research by the author. It will be published in the forthcoming excavation report on the Torbryan Caves.

16. See C.J. Caseldine and J. Hatton, 'The development of high moorland on Dartmoor: fire and the influence of Mesolithic activity on vegetation change', in F.M. Chambers (ed.), *Climate Change and Human Impact on the Landscape* (London, 1993), 119–31.

17. A preliminary report on the Mesolithic archaeology from this site is contained in A. Roberts, 'Evidence for late Pleistocene and early Holocene human activity and environmental change from the Torbryan Valley, South Devon', in Charman, Newnham and Croot (eds), *Quaternary of Devon and East Cornwall*, 168–204.

CHAPTER 5

1. The chronology is based on radio-carbon dates which have been calibrated, or adjusted, to calendar years by means of a scale derived from dendrochronology: see, for example, M.J. Aitken, *Science-based Dating in Archaeology* (London, 1990), chapters 3 and 4. S. Piggott, *The Neolithic Cultures of the British Isles* (Cambridge, 1954) remains the classic summary of the traditional cultural-historical approach, and is a valuable source of illustrations.

2. J. Thomas, *Rethinking the Neolithic* (Cambridge, 1991) provides a summary of modern work and theory with comprehensive references.

3. L. Moffett, M.A. Robinson and V. Straker, 'Cereals, fruit and nuts: charred plant remains from Neolithic sites in England and Wales and the Neolithic economy', in A. Milles, D. Williams and N. Gardner (eds), *The Beginnings of Agriculture*, British Archaeological Reports, International Series No. 496 (Oxford, 1989), 243–61.

4. For Dartmoor see C.J. Caseldine and J.M. Hatton, 'Into the mists? Thoughts on the prehistoric and historic environmental history of Dartmoor', *The Archaeology of Dartmoor: Perspectives from the 1990s*, *Proceedings of the Devon Archaeological Society*, 52 (1994), 42–3.

5. Extensive literature in local journals includes H. Miles, 'Flint scatters and prehistoric settlement in Devon', *Proceedings of the Devon Archaeological Society*, 34 (1976), 3–16; G.H. Smith, 'The Lizard project: landscape survey 1978–1983', *Cornish Archaeology*, 26 (1987), 13–68.

6. See, for example, J. Bayley in H. Miles, 'Barrows on the St Austell granite', *Cornish Archaeology*, 14 (1975), 66.

7. A. Fox, *South West England*, 2nd edn (Newton Abbot, 1973), 211–12, gives complete references up to the time of publication; M. Todd, 'Excavations at Hembury Fort, Devon, 1980–3: a summary report', *Antiquaries Journal*, 64 (1984), 251–68; Membury information is from Peter Berridge and Martin Tingle; information on Raddon is from Exeter Archaeology.

8. Thomas, *Rethinking the Neolithic*, 32–8 with references.

9. References to Haldon and Hazard Hill in Fox, *South West England*; recent data on Haldon from Exeter Archaeology.

10. R.J. Mercer, 'Excavations at Carn Brea, Illogan, Cornwall', *Cornish Archaeology*, 20

(1981), 1–204: R.J. Mercer, 'Excavation of a Neolithic enclosure at Helman Tor, Lanlivery', *Cornish Archaeology* (forthcoming).

11. For 'tor forts' see R.J. Silvester, 'The relationship of first millennium settlement to upland areas of the South West', *Proceedings of the Devon Archaeological Society*, 37 (1979), 188–9 and Fig. 5; information about De Lank, Bodmin Moor, and Carn Galver, West Penwith, Cornwall Archaeological Unit.

12. H. Quinnell, *Isles of Scilly Coastal Erosion Project 1989–93: The Pottery and Other Significant Artefacts from Sites with Recorded Stratigraphy*, Archive Report for Cornwall Archaeological Unit (Truro, 1994); and Smith, 'The Lizard project'.

13. See, for example, D.G. Buckley, 'The excavation of two slate cairns at Trevone, Padstow, 1972', *Cornish Archaeology*, 11 (1972), 15.

14. For example, J. Gardiner, 'Lithic distributions and Neolithic settlement patterns in central southern England', in R. Bradley and J. Gardiner (eds), *Neolithic Studies: A Review of Some Recent Research*, British Archaeological Reports, British Series No. 133 (Oxford, 1984), 15–40.

15. D.P.S. Peacock, 'The gabbroic pottery of Cornwall', *Antiquity*, 62 (1988), 302–4, summarizes the arguments, with useful references.

16. For Carn Brea pottery see I.F. Smith in Mercer, 'Excavations at Carn Brea', 178.

17. Re-examination by author (HQ) of collections in Royal Albert Memorial Museum, Exeter.

18. For example, see Mercer, 'Excavations at Carn Brea' and 'Excavation of a Neolithic enclosure at Helman Tor'.

19. T.H.McK. Clough and W.A. Cummins (eds), *Stone Axe Studies, Vol. 2*, Council for British Archaeology Research Report No. 67 (London, 1988) summarizes the conventional analyses, with numerous maps.

20. For a critique of sourcing see P. Berridge, 'Cornish axe factories: fact or fiction', in N. Ashton and A. David (eds), *Stories in Stone* (London, 1994), 45–56.

CHAPTER 6

1. For example, J. Barrett *et al.*, *Landscape, Monuments and Society: The Prehistory of Cranborne Chase* (Cambridge, 1991).

2. C. Tilley, *A Phenomenology of Landscape* (Oxford, 1994).

3. J. Turner, 'Ring cairns, stone circles and related monuments on Dartmoor', *Proceedings of the Devon Archaeological Society*, 48 (1990), 27–80; Corndon Ridge is Turner's A66.

4. For Bodmin see N. Johnson and P. Rose, *Bodmin Moor: An Archaeological Survey, Vol. 1, The Human Landscape to c.1800*

(London, 1994); for Dartmoor see work by the Royal Commission on the Historical Monuments of England in Devon Sites and Monuments Register, and J. Butler, *Dartmoor Atlas of Antiquities*, 5 vols (Exeter, 1991–7).

5. L.V. Grinsell, 'The barrows of North Devon', *Proceedings of the Devon Archaeological Society*, 28 (1970), 95–129; 'Dartmoor barrows', *Proceedings of the Devon Archaeological Society*, 36 (1978), 85–180; 'The barrows of south and east Devon', *Proceedings of the Devon Archaeological Society*, 41 (1983), 5–46.

6. J. Barnatt, *Prehistoric Cornwall: The Ceremonial Monuments* (Wellingborough, 1982) provides much detail for Cornwall; see also R. Mercer, 'The Neolithic in Cornwall', *Cornish Archaeology*, 25 (1986). There are no equivalent summaries for Devon.

7. See note 4 and E. Higginbotham, 'Excavations at Woolley Barrows, Morwenstow', *Cornish Archaeology*, 16 (1977), 10–18; G. Smith, 'A Neolithic long barrow at Uplowman Road, Tiverton', *Proceedings of the Devon Archaeological Society*, 48 (1990), 15–26; P. Ashbee, *The Earthen Long Barrow in Britain*, 2nd edn (London, 1984).

8. J. L'Helgouac'h, 'Les civilisations Néolithiques en Armorique', in J. Guilaine (ed.), *Les Civilisations Néolithiques et Protohistoriques de la France* (Paris, 1976), 365–74.

9. R. Loveday and M. Petchey, 'Oblong ditches: a discussion and some new evidence', *Aerial Archaeology*, 8 (1982), 17–24; F.M. Griffith, 'Some newly discovered ritual monuments in mid Devon', *Proceedings of the Prehistoric Society*, 51 (1985), 310–15.

10. R. Bradley, 'The excavation of an oval barrow beside the Abingdon causewayed enclosure, Oxfordshire', *Proceedings of the Prehistoric Society*, 51 (1992), 127–42.

11. F.M. Griffith, 'Changing perceptions of the context of prehistoric Dartmoor', *The Archaeology of Dartmoor: Perspectives from the 1990s*, *Proceedings of the Devon Archaeological Society*, 52 (1994), Pl. 2.

12. A. Gibson, 'Excavations at the Sarn-y-bryn-caled cursus complex, Welshpool, Powys, and the timber circles of Great Britain and Ireland', *Proceedings of the Prehistoric Society*, 60 (1994), 179–80, provides a summary of current dates.

13. G.E. Daniel, *The Prehistoric Chamber Tombs of England and Wales* (Cambridge, 1950), 237–42, is still a reasonably comprehensive list for Cornwall; for additions in Devon see also Grinsell 'Dartmoor barrows', and J. Turner, 'Chambered cairns, Gidleigh', *Proceedings of the Devon Archaeological Society*, 38 (1980), 117–19.

14. T.G.E. Powell *et al.*, *Megalithic Enquiries in the West of Britain* (Liverpool, 1969) provides a relevant overview.

15. C.A.R. Radford, 'The chambered tomb at Broadsands, Paignton', *Proceedings of the Devon Archaeological and Exploration Society*, 5 (1957–8), 147–67; this megalith produced Neolithic pottery.

16. P. Ashbee, *Ancient Scilly* (Newton Abbot, 1974) provides a useful overview with many plans.

17. A.F. Harding and G.E. Lee, *Henge Monuments and Related Sites of Great Britain*, British Archaeological Reports, British Series No. 175 (Oxford, 1987), 52.

18. Griffith, 'Some newly discovered ritual monuments'.

19. A. Burl, *The Stone Circles of the British Isles* (Yale, 1976), especially 106–28. See also Barnatt, *Prehistoric Cornwall*.

20. D.D. Emmett, 'Stone rows: the traditional view reconsidered', *Prehistoric Dartmoor in its Context*, *Proceedings of the Devon Archaeological Society*, 37 (1979), 94–114, summarizes data both for Dartmoor and the country as a whole.

21. G. Eogan, 'The excavation of a stone alignment and circle at Cholwichtown, Lee Moor', *Proceedings of the Prehistoric Society*, 30 (1964), 25–38.

22. N.V. Quinnell and C.J. Dunn, 'Lithic monuments within the Exmoor National Park: a new survey for management purposes by the Royal Commission on the Historical Monuments of England' (unpublished report, National Monument Record, Swindon, 1992); also L.V. Grinsell, *The Archaeology of Exmoor* (Newton Abbot, 1970), chapter 4.

23. H. Miles, 'Barrows on the St Austell granite', *Cornish Archaeology*, 14 (1975), 5–81; F.M. Griffith, 'Archaeological investigations at Colliford Reservoir, Bodmin Moor, 1977–78', *Cornish Archaeology*, 23 (1984), 47–140.

24. See Miles, 'Barrows on the St Austell granite', 76–7; Johnson and Rose, *Bodmin Moor*; Turner, 'Ring cairns, stone circles and related monuments on Dartmoor'.

25. I. Kinnes, *Round Barrows and Ring Ditches in the British Neolithic* (London, 1979).

26. P.M. Christie, 'A barrow cemetery on Davidstow Moor, Cornwall: wartime excavations by C.K. Croft Andrew', *Cornish Archaeology*, 27 (1988), 164–5.

27. H.S. Green, 'Early Bronze Age burial, territory and population in Milton Keynes, Buckinghamshire, and the Great Ouse Valley', *Archaeological Journal*, 131 (1974), 75–139.

28. Griffith, 'Some newly discovered ritual monuments' and 'Changing perceptions', 92.

29. K. Jarvis, 'The excavation of a ring ditch at Markham Lane, Exminster', *Proceedings of the Devon Archaeological Society*, 34 (1976), 62–7; P. Pearce and P.J. Weddell *Archaeological Evaluation and Excavation at the Tesco Stores Site, Digby Hospital, Rydon Lane, Exeter*, Exeter Museums Archaeological Field Unit Report 94.19 (Exeter, 1994).

30. M. Parker-Pearson, 'The production and distribution of Bronze Age pottery in south-west Britain', *Cornish Archaeology*, 29 (1990), 5–32, contains numerous distribution maps of Bronze Age pottery.

31. For Merrivale see S. Baring-Gould *et al.*, 'Second Report of the Dartmoor Exploration Committee', *Transactions of the Devonshire Association*, 27 (1895), 81–92. *Worth's Dartmoor* (Newton Abbot, 1967) summarizes early work on Dartmoor.

CHAPTER 7

1. F.M. Griffith 'Changing perceptions of the context of prehistoric Dartmoor', *Proceedings of the Devon Archaeological Society*, 52 (1994), 85–100.

2. A.C. Thomas, *Tintagel Castle* (London, 1986); F.M. Griffith, 'Salvage observations at the Dark Age site at Bantham Ham in 1982', *Proceedings of the Devon Archaeological Society*, 44 (1986), 39–58; S.J. Simpson, F.M. Griffith and N. Holbrook, 'The prehistoric, Roman and post-Roman site at Hayes Farm, Clyst Honiton', *Proceedings of the Devon Archaeological Society*, 47 (1989) 1–28.

3. Recent work on Dartmoor is summarized in *The Archaeology of Dartmoor: Perspectives from the 1990s*, *Proceedings of the Devon Archaeological Society*, 52 (1994), while for Bodmin Moor see N. Johnson and P. Rose, *Bodmin Moor: An Archaeological Survey, Vol. 1, The Human Landscape to AD 1800* (London, 1994).

4. Griffith, 'Changing perceptions'. For an example of how the upland/lowland settlement pattern was formerly perceived, see for example I.G. Simmons's plan (Fig. 3) of Bronze Age settlement in *Proceedings of the Prehistoric Society*, 35 (1969), 203–19.

5. A detailed description of types and distribution of hillforts is given in J. Forde-Johnson, *Hillforts of the Iron Age in England and Wales* (Liverpool, 1976). It may be noted, however, that some of the differences in density of 'hillforts' that he observes, particularly in simple univallate enclosures, may be a function of differential survival rather than original distribution.

6. A. Fox, 'Hillslope enclosures and related earthworks in England and Wales', *Archaeological Journal*, 109 (1952), 1–22.

7. H. Miles *et al.*, 'Excavations at Killibury Hillfort, Egloshayle 1975–6', *Cornish Archaeology*, 16 (1977), 89–121; A. Young and K.M. Richardson, 'Report on the excavations at Blackbury Castle', *Proceedings of the Devon Archaeological and Exploration Society*, 5 (1954–5), 43–67.

8. G.J. Wainwright and K. Smith 'The Shaugh Moor project, second report—the enclosure', *Proceedings of the Prehistoric Society*, 46 (1980) 65–122.

9. For a more detailed discussion of this subject in Devon, see Griffith, 'Changing perceptions'. For Cornwall see especially papers by Mercer, Christie and Quinnell in *Cornish Archaeology*, 25 (1986), and N. Johnson and P. Rose, 'Defended settlement in Cornwall—an illustrated discussion', in D. Miles (ed.), *The Romano-British Countryside*, British Archaeological Reports, British Series No. 103 (Oxford, 1982), 151–207.

10. A. Fleming, *The Dartmoor Reaves* (London, 1988); Wainwright, 'The Shaugh Moor project'.

11. A. Gibson, 'The excavation of an Iron Age settlement at Gold Park, Dartmoor', *Proceedings of the Devon Archaeological Society*, 50 (1992), 19–47.

12. J.A. Nowakowski, 'Trethellan Farm, Newquay: the excavation of a lowland Bronze Age settlement and Iron Age cemetery', *Cornish Archaeology*, 30 (1991), 5–242.

13. See, for example, J.A. Nowakowski, 'Archaeology along the hard shoulder', *Cornish Archaeology*, 32 (1993), 146–52.

14. Fleming, *Dartmoor Reaves*; J. Butler, *Dartmoor Atlas of Antiquities*, 5 vols (Exeter, 1991–7). See also *The Archaeology of Dartmoor*, and Johnson and Rose, *Bodmin Moor*.

15. A.C. Thomas, *Gwithian, Ten Years' Work, 1949–58* (Camborne, 1958).

16. F.M. Griffith and P.J. Weddell, 'Ironworking in the Blackdown Hills: results of recent survey', in P. Newman (ed.), *The Archaeology of Mining and Metallurgy in South West Britain* (Matlock, 1996), 27–34.

17. M. Parker-Pearson, 'The production and distribution of Bronze Age pottery in south-west Britain', *Cornish Archaeology*, 29 (1990), 5–32; D.P.S. Peacock, 'A contribution to the study of Glastonbury Ware from south-west Britain', *Antiquaries Journal*, 49 (1969), 41–61.

CHAPTER 8

1. Bronze Age evidence may be analysed using approaches developed by the study of material culture; see, for example, I. Hodder, *Reading the Past* (Cambridge, 1986); C. Tilly (ed.), *Reading Material Culture* (Oxford, 1990); S. Pearce, *Museums, Objects and Collections* (Leicester, 1992).

2. S. Pearce, *The Bronze Age Metalwork of South Western Britain*, British Archaeological Reports, British Series Nos 120(i) and (ii) (Oxford, 1983).

3. The evidence for early metal extraction in Cornwall has been marshalled in R. Penhallurick's *Tin in Antiquity* (London, 1986).

CHAPTER 9

1. B. Cunliffe, *Iron Age Communities in Britain*, 3rd edn (London, 1991) and M. Millett, *The Romanization of Britain* (Cambridge, 1990) provide national overviews. Local studies include A. Fox, *South West England* (London, 1964), chapter 8; A.C. Thomas, 'The character and origins of Roman Dumnonia', in A.C. Thomas (ed.), *Rural Settlement in Roman Britain*, Council for British Archaeology Research Report No. 7 (London, 1966), 74–98; H. Quinnell, 'Cornwall during the Iron Age and Roman period', *Cornish Archaeology*, 25 (1986), 111–34. For different cartographic presentations see Ordnance Survey, *Map of Southern Britain in the Iron Age* (1962) and *Map of Roman Britain*, 4th edn (1978).

2. R. Van Arsdell, *Celtic Coinage of Britain*, (London, 1989), Nos 2, 13, 33, 76.

3. P.M. Christie, 'Cornish souterrains in the light of recent research', *Bulletin of the Institute of Archaeology, University of London*, 16 (1979), 187–213; E.V. Clark, *Cornish Fogous* (London, 1961); I.M. Cooke, *Mother and Sun: The Cornish Fogou* (Penzance, 1993) provides comprehensive referencing.

4. See, for example, N. Edwards, *The Archaeology of Early Medieval Ireland* (London, 1990), 28–33.

5. R. Whimster, *Burial Practices in Iron Age Britain*, British Archaeological Reports, British Series No. 90 (Oxford, 1981), 60–74, 273–85, for a discussion and a gazetteer which describes many doubtful sites; J.A. Nowakowski, 'Trethellan Farm, Newquay: the excavation of a lowland Bronze Age settlement and Iron Age cemetery', *Cornish Archaeology*, 30 (1991), 210–32.

6. S. Pollard, 'A Late Iron Age settlement and a Romano-British villa at Holcombe, near Uplyme, Devon', *Proceedings of the Devon Archaeological Society*, 32 (1974), 59–161; H. Miles, 'The Honeyditches Villa, Seaton', *Britannia*, 8 (1977), 107–48; N. Holbrook, 'Trial excavations at Honeyditches and the nature of the Roman occupation at Seaton', *Proceedings of the Devon Archaeological Society*, 45 (1987), 59–74.

7. K. Jarvis and V. Maxfield, 'The excavation of a first-century Roman farmstead and a Late Neolithic settlement, Topsham, Devon', *Proceedings of the Devon Archaeological Society*, 33 (1975), 209–45, with references to earlier work. S. Brown and N. Holbrook, 'A Roman site at Otterton Point', *Proceedings of the Devon Archaeological Society*, 47 (1989), 29–42.

8. F.M. Griffith, 'A Romano-British villa near Crediton', *Proceedings of the Devon Archaeological Society*, 46 (1988), 137–42.

9. For example, P.T. Bidwell in P.J. Weddell, 'Excavations at 3–5 Lower Fore Street, Exmouth', *Proceedings of the Devon Archaeological Society*, 38 (1980), 113.

10. B. St J. O'Neil, 'The Roman villa at Magor farm, near Camborne, Cornwall', *Journal of the British Archaeological Association*, 39 (1933), 116–75.

11. For example, H.O'N. Hencken, 'The excavation by H.M. Office of Works at Chysauster, Cornwall, 1931', *Archaeologia*, 83 (1933), 237–84. Quinnell, 'Cornwall', 120.

12. Courtyard houses plot based on lists in V. Russell, *West Penwith Survey* (Truro, 1971).

13. Quinnell, 'Cornwall', 126.

14. D.P.S. Peacock, 'A Romano-British salt working site at Trebarveth, St Keverne', *Cornish Archaeology*, 8 (1969), 47–65.

15. G.H. Smith, 'The Lizard project: landscape survey 1978–1983', *Cornish Archaeology*, 26 (1987), 3–68.

16. H. Quinnell, 'Cornish gabbroic pottery: the development of a hypothesis', *Cornish Archaeology*, 26 (1987), 7–12.

17. N. Holbrook and P.T. Bidwell, *Roman Finds from Exeter*, Exeter Archaeological Reports No. 4 (Exeter, 1991), chapter 6.

18. R.D. Penhallurick, *Tin in Antiquity* (London, 1986) for ingots, other tin finds and finds from tinworkings. For finds from Cornwall see *Victoria County Histories: Cornwall*, II, part 5 (1924); parochial checklists in *Proceedings of the West Cornwall Field Club* (1952–61); *Cornish Archaeology* (1962–93); Devon Sites and Monuments Register.

19. B. Cunliffe, *Mount Batten, Plymouth: A Prehistoric and a Roman Port*, Oxford University Committee for Archaeology Monograph 26 (Oxford, 1988).

20. A. Fox, 'Roman objects from Cadbury Castle', *Transactions of the Devonshire Association*, 91 (1959), 105–14; W. Borlase, *Antiquities Historical and Monumental of the County of Cornwall* (London, 1769, republished Wakefield, 1973). A. Ross, *Pagan Celtic Britain* (London, 1967), 24–8.

21. D. Dudley, 'Excavations on Nor'nour in the Isles of Scilly', *Archaeological Journal*, 124 (1968), 1–64; C. Thomas, *Exploration of a Drowned Landscape* (London, 1985), 163–5.

22. Minimum number four in hoard; sources as in note 18; M. Tosdevin provided useful data for the Plymouth area. See Thomas, 'Character and origins', 92.

23. P. Isaac, 'Coin hoards and history in the West', in K. Branigan and P.J. Fowler (eds), *The Roman West Country* (Newton Abbot, 1976), 52–62.

CHAPTER 10

1. J.K. St Joseph, 'Air reconnaissance in Roman Britain 1973–1976', *Journal of Roman Studies*, 67 (1977), 125–61, especially 125–6.

2. M. Todd, 'Excavations at Hembury, Devon, 1980–83: a summary report', *Antiquaries Journal*, 64 (1984), 251–68, especially 261–5.

3. F.M. Griffith, 'Roman military sites in Devon: some recent discoveries', *Proceedings of the Devon Archaeological Society*, 42 (1984), 11–32, especially 20–5; V.A. Maxfield, 'Tiverton Roman Fort: excavations 1982–1986', *Proceedings of the Devon Archaeological Society*, 49 (1991), 25–98.

4. Cullompton: Griffith, 'Roman military sites in Devon', note 1, 13–16; Exeter: see note 7 below.

5. A. Fox and W.L.D. Ravenhill, 'The Roman fort at Nanstallon, Cornwall', *Britannia*, 3 (1972), 56–111.

6. V.A. Maxfield, 'The Roman occupation of south-west England: further light and fresh problems', in W. Hanson and L.J.F. Keppie (eds), *Roman Frontier Studies 1979* (Oxford, 1980), 297–309, especially Appendix 305–7.

7. P. Bidwell, *Roman Exeter: Fortress and Town* (Exeter, 1980), 41; S. Frere, 'Roman Britain in 1990', *Britannia*, 22 (1991), 281–2.

8. N. Holbrook, 'Trial excavations at Honeyditches and the nature of the Roman occupation at Seaton', *Proceedings of the Devon Archaeological Society*, 45 (1987), 59–74, especially 72–3.

9. B. Cunliffe, *Mount Batten, Plymouth: A Prehistoric and Roman Port* (Oxford, 1988).

CHAPTER 11

1. These are fully discussed in A.L.F. Rivet and C. Smith, *The Place-Names of Roman Britain* (London, 1979).

2. I.A. Richmond and O.G.S. Crawford, 'The British section of the Ravenna Cosmography', *Archaeologia*, 93 (1949), 1–50.

3. A.L.F. Rivet, 'Some aspects of Ptolemy's Geography of Britain', in R. Chevallier (ed.), *Littérature gréco-romaine et géographie historique. Mélanges offerts a Roger Dion* (Paris, 1974), 55–82.

4. Rivet and Smith, *Place-Names*, 306–7.

5. A.L.F. Rivet, 'The British section of the Antonine Itinerary', *Britannia*, 1 (1970), 77.

6. M. Todd, *The South West to AD 1000* (London, 1987), 191–2.

7. Rivet and Smith, *Place-Names*, 421–2.

8. Richmond and Crawford, 'British section', 17.

9. L. Dillemann, 'Observations on Chapter V, 31, Britannia in the Ravenna Cosmography', *Archaeologia*, 106 (1979), 65.

10. This is another likely river-name.

11. Todd, *The South West to AD 1000*, 204. The precise location of Nemetostatio is unknown, but the valley of the Taw or its tributaries the Yeo and Mole seem most likely. Rivet and Smith suggest North Tawton (*Place-Names*, 242–5); Todd prefers

Bury Barton ('The Roman fort at Bury Barton, Devonshire', *Britannia*, 16 (1985), 55).

12. Richmond and Crawford, 'British section', 17.

13. Dillemann, 'Observations', 65.

Chapter 12

1. Four stones on Lundy and two near the north Devon–south-west Somerset coast are excluded, as these are best interpreted as a cultural extension of sub-Roman Christianity in the Glamorgan–Gwent region.

2. The numbers that precede the inscriptions refer to entries in R.A.S. Macalister, *Corpus Inscriptionum Insularum Celticarum*, 2 vols (Dublin 1945, 1949) or C. Thomas, *And Shall These Mute Stones Speak? Post-Roman Inscriptions in Western Britain* (Cardiff, 1994).

3. B.L. Olson, *Early Monasteries in Cornwall* (Woodbridge, 1989).

Chapter 13

1. The material for the maps has been taken, in Devon, from J.E.B. Gover, A. Mawer and F.M. Stenton, *The Place-Names of Devon*, 2 vols, English Place-Name Society 8–9 (Cambridge, 1931–2); and for Cornwall, from my own collection of place-name forms for the county, housed at the Institute of Cornish Studies, University of Exeter. The Cornish coverage is rather more detailed than that for Devon but it does not seem that this has affected the distributions greatly.

2. O.J. Padel, *Cornish Place-Name Elements*, English Place-Name Society 56–57 (Nottingham, 1985), 223–6.

3. Gover *et al.*, *The Place-Names of Devon*, I, 80.

4. P.H. Sawyer, *Anglo-Saxon Charters: An Annotated List and Bibliography* (London, 1968), no. 830; text printed by J.B. Davidson, 'On some Anglo-Saxon charters at Exeter', *Journal of the British Archaeological Association*, 39 (1883), 259–303 (at 280–2); discussion by H.P.R. Finberg, 'The Treable Charter', in his *The Early Charters of Devon and Cornwall*, 2nd edn (Leicester, 1963), 21–32, reprinted as 'Hyple's Old Land', in his *Lucerna* (London, 1964), 116–30.

5. It should be noted that one example of *bod*, Bodgate in North Petherwin parish, SX 285 905, just north of the River Ottery, was originally in Devon, though west of the River Tamar; it is in that area, discussed below, which jutted into Cornwall, and which was transferred from Devon in 1974.

6. H.P.R. Finberg, 'The making of a boundary', in W.G. Hoskins and

H.P.R. Finberg, *Devonshire Studies* (London, 1952), 19–39; reprinted in Finberg, *Lucerna*, 161–80.

7. Örjan Svensson, 'The worthy-names of Devon', *Nomina*, 15 (1991–2), 53–9.

8. Ibid., 56–8.

Chapter 14

1. Gildas, *De Excidio Britonum*, in *Gildas, The Ruin of Britain and other Works*, ed. and trans. M. Winterbottom (Chichester, 1978), (Gildas, 28–36), 29.

2. Malcolm Todd, *The South West to AD 1000* (London, 1987), 273.

3. Ann Preston-James and Peter Rose, 'Medieval Cornwall', *Cornish Archaeology*, 25 (1986), 135–85; Charles Thomas, 'The context of Tintagel: a new model for the diffusion of post-Roman Mediterranean imports', *Cornish Archaeology*, 27 (1988), 7–25.

4. Oliver Padel, *A Popular Dictionary of Cornish Place-Names* (Penzance, 1988).

5. Here and later the numbering of charters (as this example, S 1296) is that of Peter H. Sawyer, *Anglo-Saxon Charters: An Annotated List and Bibliography* (London, 1968); Sawyer does not list lost charters and these follow the numbering of H.P.R. Finberg, *The Early Charters of Devon and Cornwall*, Department of English Local History Occasional Paper 2, University of Leicester (Leicester, 1953). S 1296 is translated in D. Whitelock (ed.), *English Historical Documents I, c.500–1042*, 2nd edn (London, 1979), 893 and reproduced in *Anglo-Saxon Charters, Supplementary Vol.1*, ed. S. Keynes (Oxford, 1991), charter no.8.

6. *The Anglo-Saxon Chronicle*, ed. and trans. M.J. Swanton (London, 1996), 'The Peterborough Manuscript (E)', 59. All subsequent quotations are taken from the Swanton edition.

7. Jeremy Haslam, 'The towns of Devon', in J. Haslam (ed.), *Anglo-Saxon Towns in Southern England* (Chichester, 1984), 249–83.

8. M. Chibnall, *Historia Ecclesiastica*, Oxford Medieval Texts Series (Oxford, 1969), 2, 211ff.

9. David J. Hill, *An Atlas of Anglo-Saxon England* (Oxford, 1981), 130.

10. J. Maddicott, 'Trade, industry and the wealth of King Alfred', *Past and Present*, 123 (1989), 3–51.

11. Gildas, *De Excidio Britonum* (28), 29.

12. B.L. Olson, *Early Monasteries in Cornwall* (Woodbridge, 1989), 66–78.

13. Ibid., 8–9, citing *Aldhelmi Opera*, in *Monumenta Germaniae Historica, Auctores Antiquissimi*, ed. Rudolf Ehwald (Berlin, 1913–19), 481.

14. J. Allen, C. Henderson and R. Higham, 'Saxon Exeter', in Haslam (ed.), *Anglo-Saxon Towns in Southern England*, 389–92.

15. H.P.R. Finberg, *West Country Historical Studies* (Newton Abbot, 1969), 44–52. Map

14.2 incorporates minor corrections to Finberg's solution, especially in the west.

16. Haslam, 'The towns of Devon'.

17. B.L. Olson and O.J. Padel, 'A tenth-century list of Cornish parochial saints', *Cambridge Medieval Celtic Studies*, 12 (1986), 68–71.

18. Olson, *Early Monasteries*, 95, 108–9.

19. Charles Thomas, 'Christians, chapels, churches and charters—or, Proto-parochial provisions for the pious in a peninsula (Land's End)', *Landscape History*, 11 (1989), 19–26.

20. P. Chaplais, 'The authenticity of the royal Anglo-Saxon diplomas of Exeter', *Bulletin of the Institute of Historical Research*, 39 (1966), 1–34.

21. Charters granting land to the church of Glastonbury include F 1, 6, 7, 11, 12, S 791; to Sherborne Abbey F 8, 14, 15, 17, 72, 74, S 1296; to the church of Crediton S 255, S 405, S 890.

22. Wendy Davies, 'The Latin charter-tradition in western Britain, Brittany and Ireland in the early medieval period', in D. Whitelock, R. McKitterick and D. Dumville (eds), *Ireland in Early Medieval Europe* (Cambridge, 1982), 59–60.

23. O.J. Padel, 'Two new pre-Conquest charters for Cornwall', *Celtic Studies*, 6 (1978), 20–7.

24. Della Hooke, 'Studies on Devon charter boundaries', *Transactions of the Devonshire Association* (1990), 193–211.

25. F. Rose-Troup, 'The new Edgar charter and the South Hams', *Transactions of the Devonshire Association*, 61 (1929), 249–80; Finberg, *West-Country Historical Studies*, 13–18.

26. Hooke, 'Studies on Devon charter boundaries'.

27. Della Hooke, *Worcestershire Anglo-Saxon Charter-Bounds* (Woodbridge, 1990); *The Pre-Conquest Charter-Bounds of Devon and Cornwall* (Woodbridge, 1994).

28. H.P.R. Finberg, 'Anglo-Saxon England to 1042', in H.P.R. Finberg (ed.), *The Agrarian History of England and Wales, Vol. I, Part 2, AD 43–1042* (Cambridge, 1972), 494.

Chapter 16

1. T.L. Stoate (ed.), *Devon Taxes 1581–1660* (Bristol, 1988).

2. T.L. Stoate (ed.), *Devon Hearth Tax 1674* (Bristol, 1982); T.L. Stoate (ed.), *Cornwall Hearth and Poll Taxes 1660–4* (Bristol, 1981); T. Arkell, 'Cornwall's hearth tax lists (1662–67)', *Cornwall Association of Local Historians Journal* (Autumn 1994), 13–17.

3. A.J. Howard and T.L. Stoate (eds), *The Devon Muster Roll for 1569* (Bristol, 1977); H.L. Douch (ed.), *Cornwall Muster Roll for 1569* (Bristol, 1984); J. Youings, 'Bowmen, billmen and hackbutters: the Elizabethan militia in the South West', in R. Higham (ed.), *Security and Defence in South-West*

England before 1800 (Exeter, 1987), 51–68.

4. N. Pounds, 'Population of Cornwall before the first census', in W. Minchinton (ed.), *Population and Marketing: Two Studies in the History of the South West* (Exeter, 1976), 11–30.

5. *Devon Muster Roll*, xvi, 99, 111, 186–7, 238, 245 and index of missing parishes.

6. In 1560 another Devon muster return had given a county total of 10,528 able and 17,478 'unable but to keep the country', but none of the Cornish hundreds in 1569 had so many 'unable men': the highest known ratio was in Stratton where the unable total was 92 per cent of the able.

7. A.J. Howard (ed.), *Devon Protestation Returns* (n.p., 1973); H.L. Douch (ed.), *Cornwall Protestation Returns 1641* (Bristol, 1974). Some detailed household listings by parish for Pydar hundred in Cornwall and Coleridge and Stanborough hundreds in Devon are given for 1630 and 1623 respectively in Todd Gray (ed.), *Harvest Failure in Cornwall and Devon* (Redruth, 1992), but these only cover grain-producing households.

8. E.A. Wrigley and R.S. Schofield, *Population History of England, 1541–1871: A Reconstruction* (London, 1981).

9. H. Peskett, *Guide to the Parish and Non-Parochial Registers of Devon and Cornwall 1538–1837*, Devon and Cornwall Record Society, extra series, 2 (1979).

10. The 1563 and 1603 censuses are evaluated in *Local Population Studies*, 30 (1983), 55–9; 34 (1985), 66–7; and 49 (1992), 19–31; and J.S. Moore, 'Canterbury visitations and the demography of mid-Tudor Kent', *Southern History*, 15 (1993), 36–85. The chantry certificates of 1547 provide such approximate and contradictory estimates that they have not been used here: see L.S. Snell (ed.), *Chantry Certificates for Cornwall (1547)* (Exeter, 1953).

11. *The Compton Census of 1676: A Critical Edition*, ed. A. Whiteman with Mary Clapinson, British Academy Records of Social and Economic History, new series, 10 (1986), especially 262–93.

12. T. Arkell, 'A method for estimating population totals from the Compton Census returns', in K. Schurer and T. Arkell (eds), *Surveying the People* (Oxford, 1991), 97–116.

13. E. Higgs, *Making Sense of the Census* (London, 1989).

14. Devon Record Office, Exeter, Chanter 225 A–C (1744–6 Visitation Returns of Bishop Claggett); Chanter 228 A–C (1764–6 Visitation Returns of Bishop Keppel); Chanter 232 A–B (1779 Visitation Returns of Bishop Ross); Moger Basket C/4–5 (1768 and 1771 Visitation Returns).

15. N. Pounds, 'Population movement in Cornwall and the rise of mining in the eighteenth century', *Geography*, 28 (1943), 37–46; P. Thomas, 'The population of Cornwall in the eighteenth century', *Journal of the Royal Institution of Cornwall*, 10, 4 (1990), 416–56.

16. Pounds, 'Population of Cornwall' and 'Population movement'.

17. Pounds, 'Population of Cornwall', 16, 21, 23; Arkell, 'Cornwall's hearth tax lists', 16, suggests 'that in the 1660s about 85,000 to 100,000 people lived in Cornwall'.

18. J. Whetter, *Cornwall in the Seventeenth Century: An Economic History of Kernow* (Padstow, 1974), 8–9.

19. Pounds, 'Population of Cornwall', 23; Thomas, 'Population of Cornwall', 428. Rickman's estimates for 1670 Cornwall were 99,031 (from baptisms), 175,007 (from burials) and 106,864 (from marriages); the baptism total at least is compatible with the estimates here: Whiteman (ed.), *Compton Census*, cvii.

20. W.G. Hoskins, *Devon* (London, 1954), 62–5, 169–74, esp. 172.

21. Whiteman (ed.), *Compton Census*, ci, cvii, cix–cx, cxxv, 263, 376–8. A. Whiteman and M. Clapinson, 'The use of the Compton Census for demographic purposes', *Local Population Studies*, 50 (1993), 61–6; they note on p. 64 that more of the Devon returns probably include children than they allowed in their edition.

22. Whiteman (ed.), *Compton Census*, xcviii, ci.

23. Wrigley and Schofield, *Population History*, 532–4.

24. Thomas, 'Population of Cornwall', 428.

25. Wrigley and Schofield, *Population History*, 486 (Chardstock, which they also list under Devon, was in Dorset until 1896 and is not included here).

26. Pounds, 'Population of Cornwall', 16–19. The tax-paying population of Devon in the 1524/5 lay subsidy rolls has now been mapped: see J. Cade and M. Brayshay, 'The taxable wealth and population of Devon parishes in 1524/1525', *History and Computing*, 8 (1996), 105–21.

27. Thomas, 'Population of Cornwall', 429.

28. Pounds, 'Population movement', 39; D. Cullum, 'Society and economy in west Cornwall c.1588–1750' (unpublished Ph.D. thesis, University of Exeter, 1994).

29. D. Souden, '"East, West—Home's Best?": regional patterns in migration in early modern England', in P. Clark and D. Souden (eds) *Migration and Society in Early Modern England* (London, 1987), 292–332.

30. See P. Sharpe, 'Gender-specific demographic adjustment to changing economic circumstances: Colyton 1538–1837' (unpublished Ph.D. thesis, University of Cambridge, 1988); P. Sharpe, 'The total reconstitution method', *Local Population Studies*, 44 (1990), 41–51; P. Sharpe, 'Locating the "missing marryers" in Colyton 1660–1750', *Local Population Studies*, 48 (1992), 49–59; P. Sharpe, 'Literally spinsters', *Economic History Review*, 44 (1991), 46–65; Cullum, 'Society and economy'. See also Minchinton, *Population and Marketing*, and B. Clapp, 'The place of Colyton in English population history', *Devon Historian* (1982), 4–9.

Chapter 17

1. R. Lawton and C.G. Pooley, *Britain 1740–1950: An Historical Geography* (London, 1992); N. Tranter, *Population and Society, 1750–1940* (London, 1985); E. Wrigley and R.S. Schofield, *The Population History of England, 1541–1871: A Reconstruction* (London, 1981).

2. R. Lawton (ed.), *The Census and Social Structure* (London, 1978). See also M. Drake and R. Finnegan (eds), *Studying Family and Community History, Vol. 4, Sources and Methods: A Handbook* (Cambridge, 1984), especially chapter 3.

3. P. Mathias, *Retailing Revolution* (London, 1967).

4. Wrigley and Schofield, *Population History*.

5. Lawton and Pooley, *Britain*, 29.

6. Ibid., 117.

7. C.G. Pooley and I.D. Whyte (eds), *Migrants, Emigrants and Immigrants: A Social History of Migration* (London, 1991).

8. J. Saville, *Rural Depopulation in England and Wales, 1851–1951* (London, 1957).

9. S. Nicholas and P. Shergold, 'Internal migration in England, 1818–1939', *Journal of Historical Geography*, 13 (1987), 155–68.

10. W.G. Hoskins, *Devon* (Exeter, 1992).

11. Ibid.

12. Notes from the *Census of Population*, 1851.

13. M. Brayshay, 'Depopulation and changing household structure in the mining communities of West Cornwall, 1851–1871', *Local Population Studies* 25 (Autumn 1980), 26–41.

14. M. Brayshay, 'Plymouth's past: so worthy and peerless a western port', in B. Chalkley, D. Dunkerley and P. Gripaios (eds), *Plymouth: Maritime City in Transition* (Newton Abbot, 1991), 38–61.

15. M. Brayshay and V. Pointon, 'Migration and the social geography of nineteenth-century Plymouth', *Devon Historian*, 28 (1984), 3–14.

16. Brayshay, 'Depopulation'.

17. *Census of Population*, 1851.

18. Brayshay, 'Depopulation'.

19. Ibid.

20. G. Shaw and A. Williams, 'From bathing hut to theme park: tourism development in south-west England', *Journal of Regional and Local Studies*, 11 (1991), 16–32.

Chapter 18

1. Cornwall County Council, *Development Plan: Report of Survey*, Part 1, *The County* (Truro, 1952).

2. S.W.E. Vince, 'Reflections on the structure and distribution of rural population in England and Wales 1921–31', *Transactions of the Institute of British Geographers*, 18 (1952), 53–76.

3. G.C. Willatts and M.G. Newsom, 'The geographical pattern of population changes in England and Wales 1921–1951', *Geographical Journal*, 119 (1953), 431–54.

4. W. Randolph and S. Robert, 'Beyond decentralisation: the evolution of population distribution in England and Wales 1961–81', *Geoforum*, 14 (1983), 75–102.

5. P.J. Cloke, 'An index of rurality for England and Wales', *Regional Studies*, 11 (1977), 31–46; P.J. Cloke and G. Edwards, *Rurality in England and Wales 1981: A Replication of the 1971 Index and a Contemporary Index of Rurality for England and Wales 1981* (Lampeter, 1985).

Chapter 19

1. Full bibliographical details of historical and archaeological research will be found in A. Preston-Jones and P. Rose, 'Medieval Cornwall', *Cornish Archaeology*, 25 (1986), 135–85, at 169–73 and 179–85; R.A. Higham, 'Devon castles: an annotated list and bibliography', *Proceedings of the Devon Archaeological Society*, 46 (1988), 142–9.

2. R.A. Higham, 'Early castles in Devon (1068–1201)', in *Château Gaillard*, IX–X (Caen, 1982), 101–16.

3. *Victoria County Histories: Devon*, I, ed. W. Page (London, 1906), 573–630; *Victoria County Histories: Cornwall*, I, ed. W. Page (London, 1906), 451–73; D.J.C. King and L. Alcock, 'Ringworks of England and Wales', in *Château Gaillard*, III (Chichester, 1969), 90–127; D.J.C. King, *Castellarium Anglicanum*, 2 vols (London, 1983). Some dubious castle ascriptions are omitted from the maps.

4. R.A. Higham, 'Was there a castle at Axminster?', *Proceedings of the Devon Archaeological Society*, 44 (1986), 182–3.

5. Cornwall Committee for Rescue Archaeology, Tenth Annual Report (1984–85).

6. R.J. Silvester, 'An excavation at Dunkeswell', *Proceedings of the Devon Archaeological Society*, 38 (1980), 53–66.

7. Preston-Jones and Rose, 'Medieval Cornwall', 172. Veryan, listed as a castle ringwork in King, *Castellarium Anglicanum*, I, 77, is now thought to be a 'round' (personal communication from Mr Peter Rose, Cornwall Archaeological Unit).

8. N. Pevsner and B. Cherry (eds), *The Buildings of England: Devon*, 2nd edn (London, 1989), and N. Pevsner and E. Radcliffe (eds), *The Buildings of England: Cornwall*, 2nd edn (London, 1970). See also E.M. Jope (ed.), *Studies in Building History* (London, 1961), 192–222. On moated sites, R.J. Silvester and R.A. Higham, 'Domestic enclosures of probable medieval date', in Silvester, 'An excavation at Dunkeswell',

63–5, and Preston-Jones and Rose, 'Medieval Cornwall', 173.

9. For settlement aspects, see R.A. Higham, 'The castles of medieval Devon' (unpublished Ph.D. thesis, University of Exeter, 1979); Higham 'Early castles'; R.A. Higham, 'Castles in Devon', in S.C. Timms (ed.), *Archaeology of the Devon Landscape* (Exeter, 1980), 70–80; Preston-Jones and Rose, 'Medieval Cornwall'; M. Beresford, *New Towns of the Middle Ages* (London, 1967); M. Beresford and H.P.R. Finberg, *English Medieval Boroughs: A Handlist* (Newton Abbot, 1973); A. Shorter, W. Ravenhill and K. Gregory, *South-West England* (London, 1969), chapters 5 and 6; P. Sheppard, *The Historic Towns of Cornwall* (Truro, 1980).

10. The map is published and discussed in R.A. Higham, 'Excavations at Okehampton Castle, Devon. Part I: the motte and keep', *Proceedings of the Devon Archaeological Society*, 35 (1977), 3–42, at 11–14. For other maps of estates relating to castles in Devon, see Higham 'Castles of medieval Devon'.

11. For what follows on historical aspects, see Higham, 'Castles of medieval Devon' and 'Early castles'; R.A. Higham, 'Public and private defence in the medieval south-west town: town, castle and fort', in R.A. Higham (ed.), *Security and Defence in South-West England before 1800*, Exeter Studies in History 19 (Exeter, 1987), 27–49; R.A. Higham, 'A knight to remember: the building enterprises of Hugh Courtenay (1276–1340)', *Transactions of the Devonshire Association*, 121 (1989), 153–8; Preston-Jones and Rose, 'Medieval Cornwall'; R.A. Brown, H.M. Colvin and A.J. Taylor, *The History of the King's Works: The Middle Ages*, 2 vols (London, 1963); A.D. Saunders, 'The coastal defences of Cornwall', *Archaeological Journal*, 130 (1973), 232–6; O. Padel, 'Tintagel in the twelfth and thirteenth centuries', *Cornish Studies*, 16 (1988), 61–6. For the later earldom and duchy and its castles, see J. Hatcher, *Rural Economy and Society in the Duchy of Cornwall 1300–1500* (Cambridge, 1970), especially 24–5, 45–6, 144–5, 179. On Lostwithiel, see N. Pounds, 'The Duchy Palace at Lostwithiel, Cornwall', *Archaeological Journal*, 136 (1979), 203–17. I am indebted to Dr G. Haslam, former Duchy of Cornwall Office archivist, for help with the history of some of the Cornish sites.

12. Exeter bishop's palace, the episcopal house at Chudleigh, and Buckland Abbey had licences to crenellate. For the walled and towered perimeter of the collegiate church and residences at Glasney, which are shown on a sixteenth-century map of the south coast, see T.C. Peter, *The History of Glasney Collegiate Church* (Camborne, 1903), and Sheppard, *Historic Towns of Cornwall*, 17–20.

13. On towns with murage grants, see H.L. Turner, *Town Defences in England and Wales* (London, 1971); the Ilfracombe grant is in *Calendar of Patent Rolls 1416–1422*, 172; for Plymouth and Dartmouth, Higham, 'Public and private defence'; for Fowey, C. Henderson, 'Fowey', in *Essays in Cornish History* (Oxford, 1935), 26–43, and Sheppard, *Historic Towns of Cornwall*, 35–8; for St Ives, Sheppard, *Historic Towns of Cornwall*, 7–10.

Chapter 22

1. J.R. Kenyon, 'Early artillery fortification in England and Wales', *Archaeological Journal*, 138 (1981), 217–18; A.D. Saunders, *Fortress Britain* (Liphook, 1989), 26–7; A.D. Saunders, *Dartmouth Castle* (London, 1991), 22–4.
2. Kenyon, 'Early artillery fortification', 218–19; F.W. Woodward, *Plymouth's Defences* (Plymouth, 1990), 7; Saunders, *Dartmouth Castle*, 24–5; F. Kitchen, 'The defence of the coast of Cornwall during the reign of Henry VIII', *Cornwall Association of Local Historians Journal*, 23 (April 1992), 3–4; H.M. Colvin (ed.), *The History of the King's Works, Vol. IV, 1485–1660*, Part 2 (London, 1982), 594–5.
3. Colvin (ed.), *King's Works*, 368–83, 594–8.
4. Saunders, *Fortress Britain*, 56.
5. R.A. Erskine, 'The military coast defences of Devon, 1500–1956', in M. Duffy *et al.* (eds), *The New Maritime History of Devon, Vol. I* (London, 1992), 120–1; B. StJ. O'Neil, *Ancient Monuments of the Isles of Scilly* (London, 1961), 24–8; Colvin (ed.), *King's Works*, 384–99, 484–5, 588–95.
6. N. Longmate, *Defending the Island. From Caesar to the Armada* (London, 1990), 583–95; Colvin (ed.), *King's Works*, 412, 486–8, 591–3, 598–602; Saunders, *Fortress Britain*, 64, 66–7; B. Morley and P. Whyte, 'Pendennis Castle', *Archaeological Journal*, 130 (1973), 286–8; Erskine, 'Military coast defences', 121–2.
7. Quotation from B. StJ. O'Neil, *Castles and Cannon* (Oxford, 1960), 114–15; Erskine, 'Military coast defences', 122–3; Saunders, *Fortress Britain*, 79–81; O'Neil, *Ancient Monuments of the Isles of Scilly*, 32–5; F.W. Woodward, *Citadel* (Exeter, 1987), 21–47, 131–4.
8. M.M. Oppenheim, 'The maritime history of Cornwall', *Victoria County Histories: Cornwall* (London, 1906), I, 503; M.M. Oppenheim, *The Maritime History of Devon* (Exeter, 1968), 71, 73–4; N. Longmate, *Island Fortress* (London, 1991), 82–192, 122–3; Saunders, *Dartmouth Castle*, 11–12.
9. K.W. Maurice-Jones, *The History of Coast Artillery in the British Army* (London, 1959), 19, 25, 40, 43; Woodward, *Citadel*, 64–7; F. Kitchen, 'The defence of the

southern coast of Cornwall against the French Revolution', *Cornish Association of Local Historians Journal*, 19 (April 1990), 13; Saunders, *Fortress Britain*, 118; O'Neil, *Ancient Monuments of the Isles of Scilly*, 36–7.
10. Erskine, 'Military coast defences', 123–6; M. Duffy, 'Devon and the naval strategy of the French Wars, 1689–1815', in Duffy *et al.* (eds), *The New Maritime History of Devon, Vol. I*, 182–91.
11. Erskine, 'Military coast defences', 124–7; Saunders, *Fortress Britain*, 124–8; *Berry Head Official Guide Book* (Torquay, n.d.), 14–15; Kitchen, 'The defences of the southern coast of Cornwall', 13–17.
12. Maurice-Jones, *The History of the Coast Artillery*, 97–8, 103–4; Erskine, 'Military coast defences', 136–7; Saunders, *Fortress Britain*, 125, 152; I.V. Stevenson, 'Some West Country defences', *Fort*, 17 (1989), 12–13, 17 and n.14; J. Goodwin, 'Granite towers on St Mary's, Isles of Scilly', *Cornish Archaeology*, 32 (1993), 128–39.
13. Erskine, 'Military coast defences', 126; Woodward, *Plymouth's Defences*, 14; B. Thompson, *The Story of Dartmoor Prison* (London, 1907), 10–12.
14. Longmate, *Island Fortress*, 307–23.
15. Woodward, *Plymouth's Defences*, 15–18; F.W. Woodward, *Forts or Follies? A History of Plymouth's Palmerston Forts* (Tiverton, 1998), 39–44.
16. A.D. Saunders, '"Palmerston's Follies"—a centenary', *Journal of the Royal Artillery*, 87, 3 (1960), 138–44; I.V. Hogg, *Coast Defences of England and Wales 1856–1956* (Newton Abbot, 1974), 13–75.
17. Woodward, *Plymouth's Defences*, 18–29; *Forts or Follies?*, 45–92, 143–77; Saunders, *Fortress Britain*, 133–5, 153–89; Hogg, *Coast Defences*, 22, 74, 168–204; I. Hunt, 'Plymouth sound? The defence of the naval station', *Fort*, 11 (1983), 86–101.
18. Saunders, *Dartmouth Castle*, 27–30; Erskine, 'Military coast defences', 127; A.D. Saunders, 'The coastal defences of Cornwall', *Archaeological Journal*, 130 (1973), 235; Stevenson, 'Some West Country defences', 14, 17–18, 23; C. Gill, *The Isles of Scilly* (Newton Abbot, 1975), 44.
19. Saunders, *Fortress Britain*, 190–2, 202–5; Woodward, *Plymouth's Defences*, 31–6, 46; Hunt, 'Plymouth sound?', 101–5.
20. B. Morley, *The Castles of Pendennis and St Mawes* (London, 1988), 2, 12–13, 23; Stevenson, 'Some West Country defences', 17–19.
21. Stevenson, 'Some West Country defences', 17–19; Saunders, *Fortress Britain*, 192; O'Neil, *Ancient Monuments of the Isles of Scilly*, 37; Gill, *Isles of Scilly*, 44.
22. B. Cherry and N. Pevsner (eds), *The Buildings of England: Devon* (London, 1989), 409, 655; C. Elmsley, 'The military and popular disorder in England, 1790–1801', *Journal of the Society for Army Historical Research*, LXI (1983), 18;

H.L. Douch, *The Book of Truro* (Chesham, 1977), 89; *Berry Head Official Guide Book*, 15–18.
23. Cherry and Pevsner (eds), *Devon*, 325, 655–6, 677; J. Coad, 'Architecture and development of Devonport naval base, 1815–1982' and P. Payton, 'Naval education and training in Devon', in M. Duffy *et al.* (eds), *The New Maritime History of Devon, Vol. 2* (London, 1994).
24. *British Parliamentary Papers*, 1882, XXXVIII, 763, 789; Woodward, *Forts or Follies?*, 93–102.

Chapter 23

1. Gerald Wasley, *Devon at War 1939–45* (Tiverton, 1994).
2. Baroness Sharp, *Dartmoor* (London, 1977); Mark Blacksell and Fiona Reynolds, 'Military training in National Parks: a question of land use conflict and national priorities', in Michael Bateman and Raymond Riley (eds), *The Geography of Defence* (Beckenham, 1987), 215–27.
3. Devon Record Office (DRO), 4318 M/P1, details of the area reserved for manoeuvres on Woodbury, Colaton Raleigh and Aylsebeare Commons
4. Cornwall Record Office (CRO), AD439, details of the Women's Land Army in Cornwall during the Great War, as well as all the quotations, are contained in an anonymous pamphlet.
5. Geoff Worrall, *Exeter Airport: In Peace and War* (Exeter, 1988).
6. DRO, 3935Z/Z27, photograph of Exeter Airport, *c*.1939.
7. Jane Whitcher, 'Evacuation to Torquay in the Second World War', in Ross Whitehead, Carl Brazier and Jane Whitehead (eds), *Reactions in Devon to Invasion* (Exeter, 1992), 51–64.
8. DRO, 3248A/16/1–2/ County of Devon: Government Evacuation Scheme Memorandum, 28 April 1939.
9. Norman Longmate, *How We Lived Then* (London, 1971), 51–2.
10. Ibid., 72.
11. DRO, 44770/Z1–2, map of all the bombs that fell on the city of Exeter in the course of eleven raids.
12. DRO, 4085E/241, photograph of St Lukes College on the morning of 5 May 1942.
13. CRO, CC3/23/2/4, details of all the air raids in Cornwall.
14. Grace Bradbeer, *The Land Changed its Face: The Evacuation of Devon's South Hams 1943–44* (Newton Abbot, 1984), 69.
15. Edwin Hoyt, *The Invasion Before Normandy: The Secret Battle of Slapton Sands* (New York, 1985); Ken Small and Mark Rogerson, *The Forgotten Dead* (London, 1989); Nigel Lewis, *Channel Firing: The Tragedy of Exercise Tiger* (London, 1989).

Chapter 24

1. A.H.A. Hamilton, *Quarter Sessions from Queen Elizabeth to Queen Anne* (London, 1878), 35.
2. *Preliminary Observations*, Census of Population, 1871, lxxi.
3. The main topic of possible dispute in recent years has been the part of the Plymouth 'journey-to-work' area which lies to the west of the Tamar. Strong opposition ('Duchyism') was aroused after 1945 and in the late 1960s to proposals to incorporate Torpoint and Saltash in the City, and these have not been revived.
4. See Local Government Commission, *Recommendations for Devon*, 1994, and Local Government Commission, *Recommendations for Cornwall*, January 1995.
5. At parliamentary level they are 'constituencies'.
6. *The New Local Authorities: Management and Structure* (London, 1972), 33
7. It is, however, important to scrutinize official statistics very carefully because the Isles are sometimes included with Cornwall and sometimes excluded in the printed tables.
8. 1892 order under Local Government Act 1888 (51&52 Vict, c.41), Local Government Act 1933 (23&24 Geo V, c.51), Local Government Act 1972 (c.70), and Isles of Scilly Order 1978.
9. The non-payment of tax on locally generated income and profits was 'a small loophole ... owing to defective machinery of assessment', said R.A. Butler in his budget speech, 1953. The reason was that no land tax commissioners had been appointed in 1798 and this omission had been carried through to income tax administration. See *The Times*, 15, 16, 18, 20 and 21 April, 8 and 15 May 1953. In 1995 a related issue was revived in respect of the payment of road fund tax which had been described in 1953 as a 'voluntary levy'. See *The Times*, 1995.
10. There are about 140 rocky islets uncovered at ebb tide.
11. It was deemed in the 1861 census to be a new parish under the Extra-Parochial Places Act 1857 (20 Vict, c.19), but in practice these legislative provisions were never applied to Lundy and it reverted to being listed as 'extra-parochial'. See, for instance, the census, 1911, II, 1.
12. Parliamentary counties are not considered in this section.
13. The Registrar General's Office was created under Registration of Births, Deaths and Mariages Act 1836 (6&7 Will IV, c.86) and worked through districts based on the poor law union areas created after 1834. As some of these overlapped ancient county boundaries their aggregations did also.

14. Sources for the anomalies: Benjamin Donn's Map (1765); the censuses of 1801, 1811, 1821 and 1831; the first Ordnance Survey (1809); the Parliamentary inquiries of 1825–6; see also the references in the tables.

15. It gained that status by the second charter of Henry VIII in 1537, and the boundaries of the county corporate were defined by Act of Parliament in 1550.

16. Exeter Extension Act 1877.

17. These are: St Leonard (2,588), St Sidwell (16,263), St Martin (115), Close of St Peter's Cathedral (262), St Mary Major (2,480), Holy Trinity (3,410), St David (5,424), St Paul (870), St Lawrence (346), St Stephen (180), Bedford Circus (83), Bradninch (66), All Hallows Goldsmith Street (125), St Pancras (177), St Petrock (176), St Kerrian (269), St George the Martyr (539), St Mary Arches (468), St Olave (533), St John (439), All Hallows on the Walls (805), St Mary Steps (875), St Edmund (982).

18. Ministry of Housing and Local Government, SI 687, 28/3/1963.

19. Eddystone Lighthouse (extra-parochial 1841, population 2), Plymouth Breakwater (Fort) (extra-parochial with Plymouth Lighthouse 1871, population 2).

20. P. Russell, *A History of Torbay* (Torquay, 1960).

21. 'Hundred' was the commonest name for this geographical level but in some other counties these areas were called 'wapentake', 'ward' or 'division'; there were also 'lathes' in Kent and 'rapes' in Sussex, which were intermediate between hundreds and the county.

22. Division of Counties Act 1828 (9 Geo IV, c.43), Petty Sessional Divisions Act 1829 (10 Geo IV, c.46) and Petty Sessional Divisions Act 1836 (6 Will IV, c.12).

23. 1841 was an intermediate year with both traditional and new areas existing side by side. Presumably for comparative purposes over time, some statistics were also presented for hundreds up to 1881.

24. The source for the maps of the hundreds is the census of 1801 and the boundaries have been reconstructed from the lists of parishes.

25. In addition to the traditional divisions of the hundreds that were widely acknowledged, there were others which were not generally accepted as important enough to be census units.

26. Winkleigh, a separate hundred in 1801 and 1811, was listed with North Tawton from 1821 onwards.

27. Irregular boundary lines and exclaves were disliked by reformers because they made continuous administrative action very difficult or impossible.

28. In the following list, names of hundreds are in capitals and parishes forming detached parts of those hundreds are in lower case. AXMINSTER Thorncombe;

BRAUNTON East Buckland, West Buckland, Filleigh; CLISTON Butterleigh; SHERWILL Arlington, Loxhore, Shirwell; HAYTOR Buckland-in-the-Moor, Widecombe-in-the-Moor; LIFTON Sydenham-Damarel, Lamerton; WONFORD Stokeinteignhead, Combinteignhead, Haccombe, St Nicholas, East Ogwell, West Ogwell; HARTLAND Yarnscombe; WEST BUDLEIGH Washfield. The three detached parishes of Braunton hundred were contiguous with each other but separated from the rest by parishes in the Sherwill and South Molton hundreds; the three detached parishes of Sherwill hundred may have been contiguous with the rest at one point but their appearance on the map is one of separation by Braunton parishes; three of the detached parts of Wonford hundred were contiguous, as were the other two, but the groups were separated from each other by part of Haytor hundred and from the nearest part of the main body of Wonford by six 'crow miles'.

29. One reason for treating the divisions as hundreds is that if they are merged they produce aggregate units out of line with the rest of the Cornish hundreds. The rejected divisions likewise are out of line, but in this case they are too small.

30. 1698 Exeter, 1708 Plymouth, 1781 Devonport.

31. Until 1929 in Exeter's case and 1929 in Plymouth's. The various municipal boundary extensions in some cases involved a change in the poor law area, in other cases left the situation unchanged.

32. The following variations were observable in 1871: from Devon to Chard (one), Taunton (one), Wellington (five) and Dulverton (one) in Somerset, St Germans (small parts) and Launceston (six) in Cornwall. The following unions contained a few parishes in other counties: Axminster (five), Holsworthy (one), Tavistock (one) and South Molton (one).

33. These do not include areas that were described only as 'suburban'.

34. From the Royal Commission on Municipal Corporations, Library Microfiche: PP 1835, XXIII.

35. Their existence was acknowledged in the 1851 census, Table 18.

36. In addition to the alleged boroughs mentioned above, the investigation also looked at Bovey Tracey, East Looe, St Cleer and Tavistock.

37. There were also areas which failed either to adopt the relevant acts or to make the general transitions of 1835, 1872 and 1885 to the appropriate status of the time; particularly Bradninch and Plympton Earle of the pre-1835 boroughs, and Topsham of the geographical ' towns' of 1851.

38. *Preliminary Observations*, 1871, xxxi

39. Cornwall: Bossiney, Camelford, Grampound, Kilkhampton, Marazion, Mevagissey, Morwenstow, Newport (St Stephens), Poughill, St German, St Mawes, St Mitchell (Enoder), St Thomas the Apostle, Tregony. Devon: Beer Ferrers, Braunton, Chudleigh, Colyford, Colyton, Dartmoor Forest, Hatherleigh, Hollacombe, Newport (Barnstaple), Newport (Burlescombe).

40. In adddition to the broad directions above, the 'paths' also included changes in parish structure too numerous and, in many cases, too small to be mapped here.

41. 'Town district' is a convenient name for NCBS and UDs.

42. The areas that disappeared into the large urban areas of Devon did not have this has an option.

43. They are for instance of vital importance to the work of Devon County Council in the spheres of planning and public rights of way.

44. *Preliminary Observations*, 1811, ix–xvi, speaks of 'striking irregularities' and 'deformities'; 1861, viii, 'the large amount of correspondence ... to procure correct information as to the boundaries of the various and complex local subdivisions of the country'.

45. Most of the parishes posed no problems, but a small number of 'anomalies' prevented representation as the sort of neat pattern the twentieth century expects.

46. The Ordnance Survey Act 1841 (4&5 Vict, c.30) required the mapping of parish boundaries but it was not until 1848 that mapping in the South West started (at Devonport).

47. *Returns from Clerks of the Peace of Insulated Parcels of Land*, House of Commons, 26 May 1825; *Returns of All Parishes and of All Townships or Similar Divisions of Parishes, which Extend into Two or More Counties*, House of Commons, 17 May 1826.

48. Boundary Act (7&8 Vict, c.61), 1844. This process was described above.

49. Twelve areas appeared in 1841 and 1851, in addition to the lighthouses. The 1841 census did not identify all the geographically distinct places that were recognized by the gazetteers of the nineteenth century. No general description or explanation has been given of the 'ignorance'.

50. The opposite of an extra-parochial place is a condominion, that is, an area shared between two or more other separate areas—not an area in dispute but one where joint jurisdiction is recognized. These included 'lands common to' Philleigh and Ruan Lanihorne, Axminster and Kilmington and Bridestowe and Sourton.

51. 'Appendix' to the *Report* of the 1873 Committee.

52. These were in addition to the changes within the Plymouth and Exeter county borough areas.

53. The Municipal Yearbook, 1994, warns that, as parishes do not have to report a decision to adopt 'town' status, the list may be slightly inaccurate.

CHAPTER 25

1. F.W.S. Craig, *British Parliamentary Election Results, 1950–1970* (Chichester, 1971); *The Times House of Commons, February 1974* (London, 1974); *The Times House of Commons, October 1974* (London, 1974); *The Times House of Commons, May 1979* (London, 1979); *The Times House of Commons, June 1983* (London, 1983); *The Times House of Commons, June 1987* (London, 1987); *The Times House of Commons, April 1992* (London, 1992); *The Times House of Commons, May 1997* (London, 1997).

CHAPTER 26

1. For a general discussion of how and why Domesday was made, see Frank Barlow, 'Domesday Book: an introduction', in Christopher Holdsworth (ed.), *Domesday Essays*, Exeter Studies in History 14 (Exeter, 1986), 17–28; for Domesday scholarship see David Bates, 'Domesday Book, 1086–1986', in *Domesday Essays*, 1–16.

2. W.G. Hoskins, *Devon* (Newton Abbot, 1972), 10.

3. References to Domesday are given first to the Farley, Record Commission edition: all here are to Exchequer Domesday in volume 1. The folio number is followed by a letter 'a' to 'd' for the column on recto or verso. After this number and letter I give in brackets reference to the relevant Phillimore edition, general editor John Morris: i.e. for Devon, volume 9, 2 parts, edited by Caroline and Frank Thorn (Chichester, 1985), and for Cornwall, volume 10, edited by Caroline and Frank Thorn (Chichester, 1979). References for the parishes in Dorset are given in Devon, i after 118c (unnumbered page). For Stockland see also J.P. Traskey, *Milton Abbey: A Dorset Monastery in the Middle Ages* (Tisbury, 1978), 7–8, 203, which includes Stockland in Æthelstan's original endowment of 934; for Axminster see O.J. Reichel, 'The hundreds of Axminster and Axmouth in early times', in F.B. Prideaux (ed.), *Supplements to Transactions of the Devonshire Association*, 4 (1931), 137–85.

4. The bishop's holdings are 101d–102a (Devon, 2.1–24), 120c–d (Cornwall, 2.1–25). Places have been identified from the Phillimore edition and H.C. Darby and G.R. Versey, *Domesday Gazetteer* (Cambridge, 1975).

5. Frank Barlow, *The English Church 1000–1066: A Constitutional History*, 2nd edn (London, 1979), 211–15.

6. See notes to 101d, 102a (Devon 2.2, 2.12, 2.15, 2.13).

7. Frank Barlow *et al.*, *Leofric of Exeter* (Exeter, 1972), 11–12, and Barlow, *The English Church 1000–1066*, 190–1.

8. Hoskins, *Devon*, 223: as he remarks (548 note), the basis upon which Domesday hides, and for that matter ploughlands, are given an equivalent in acres is not precise, but the proportionate size of estates emerges clearly.

9. Cf. W.J. Blair, 'Local churches in Domesday Book and beyond', in J.C. Holt (ed.), *Domesday Studies* (Woodbridge, 1987), 265–78. I have adopted the list for the South West in his 'Secular minster churches in Domesday Book', in Peter Sawyer (ed.), *Domesday Book: A Reassessment* (London, 1985), 104–42, especially the map at 110. See too his 'Minster churches in the landscape', in Della Hooke (ed.), *Anglo-Saxon Settlements* (Oxford, 1988), 35–58. For *Niuuetona* see my comments in 'The Church at Domesday', in Holdsworth (ed.), *Domesday Essays*, 64. Richard Morris, *Churches in the Landscape* (London, 1989), 93–139, discusses the meaning of minster very helpfully. Recently two articles in *Early Medieval Europe* 4 (1995) have reopened the relationship of minsters and local churches: Eric Cambridge and David Rollason, 'Debate: the pastoral organisation of the Anglo-Saxon Church: a review of the "Minster Hypothesis"', 87–104, and John Blair, 'Debate: ecclesiastical organisation and pastoral care in Anglo-Saxon England', 193–212. I am grateful to Dr Julia Crick for drawing these to my attention.

10. Cf. Holdsworth, 'The Church at Domesday', 56–60, and Morris, *Churches in the Landscape*, 140–67.

11. Domesday, 120d–121b, 117d (Cornwall, 4.3–22; Devon, 51.15–16). For its monastic history see David Knowles and R. Neville Hadcock, *Medieval Religious Houses of England and Wales*, 2nd edn (London, 1971), 52, 60.

12. For short accounts see Knowles and Hadcock, *Medieval Religious Houses*, 57, 77: 52, 61. Full studies are H.P.R. Finberg, *Tavistock Abbey: A Study in the Social and Economic History of Devon*, 2nd edn (Newton Abbot, 1969), and John Stephan, *A History of Buckfast Abbey from 1098 to 1968* (Bristol, 1970). See too Christopher Holdsworth, 'Tavistock Abbey in its late 10th century context', in Tom Greeves (ed.), *Vikings '97* (forthcoming).

13. The Domesday Tavistock figures are hard to calculate, and estimates vary. David Knowles, *The Monastic Order in England*, 2nd edn (Cambridge, 1963), 703, makes the total £78.10.0, whereas Finberg, *Tavistock Abbey*, 11, has £58.00. I have

14. Knowles, *Monastic Order*, 702.

15. Finberg, *Tavistock Abbey*, 8–10. His whole discussion of losses after 1066, 6–8, is valuable. See Holdsworth, 'The Church at Domesday', 54, for a comparison with the burden of knight service on other houses.

16. St. Michael's Mount is a probable exception, but it was staffed from Mont-St. Michel; Knowles and Hadcock, *Medieval Religious Houses*, 91. Tavistock's losses are at Domesday, 121c (Cornwall, 3.7); Robert's other depredations are on 121a–b (Cornwall, 4.7, 8–14, 15, 22, 23, 25–9).

17. Uplyme, for example, was given to Glastonbury in the tenth century, see Domesday 103c (Devon, notes to 4.1). For Horton, Cranborne and Battle's holdings in Devon see Domesday 104a–b (Devon, 7–9); for Rouen, Domesday 104b (Devon, 10.1).

18. For details see Holdsworth, 'The Church at Domesday', 55.

19. Details are drawn from Knowles and Hadcock, *Medieval Religious Houses*, unless otherwise stated. For a discussion of the region see Christopher Holdsworth, 'From 1050 to 1307', in Nicholas Orme (ed.), *Unity and Variety: A History of the Church in Devon and Cornwall* (Exeter, 1991), 23–51, especially 35–46.

20. L.E. Elliott-Binns, *Medieval Cornwall* (London, 1955), 160, 163, 148; Frederick Hockey, *Beaulieu King John's Abbey* (Old Woking, 1976), 46–7, 95–6; Christopher Holdsworth, 'Royal Cistercians: Beaulieu, her daughters and Rewley', in P.R. Coss and S.D. Lloyd (eds), *Thirteenth Century England IV*, Proceedings of the Newcastle upon Tyne Conference 1991 (Woodbridge, 1992), 139–50.

21. Figures are rarely given in documents; see Knowles and Hadcock, *Medieval Religious Houses*, under each house.

22. For this class of house see Donald J.A. Matthew, *The Norman Monasteries and their English Possessions* (Oxford, 1962).

23. Edward Miller and John Hatcher, *Medieval England: Rural Society and Economic Change, 1086–1348* (London, 1978), 29.

24. C.H. Lawrence, *Medieval Monasticism: Forms of Religious Life in Western Europe in the Middle Ages*, 2nd edn (London, 1989), 149–273.

25. Earl Baldwin's gift of Tiverton Church was made 'rogatu Ricardi monachi cognati mei': George Oliver, *Monasticon Dioecesis Exoniensis* (Exeter and London, 1846), 193; J.A. Sparks, *In the Shadow of the Blackdowns: Life at the Cistercian Abbey of Dunkeswell and on its Manors and Estates 1201–1539* (Bradford-on-Avon, 1978), 2.

26. Lawrence, *Medieval Monasticism*, 69–71.

27. Oliver, *Monasticon*, 193. Ultimately only part of the church remained in the

priory's control: Robert Bearman (ed.), *Charters of the Redvers Family and the Earldom of Devon 1090–1217*, Devon and Cornwall Record Society, new series, 37 (1944), 95.

28. For Buckfast see my article 'The Cistercians in Devon', in Christopher Harper-Bill, Christopher Holdsworth and Janet Nelson (eds), *Studies in Medieval History presented to R. Allen Brown* (Woodbridge, 1989), 179–81, at 181.

29. The relations of bishops with Cistercians is discussed in ibid., 186–9.

30. For Hartland, besides Knowles and Hadcock, *Medieval Religious Houses*, see the article in Alfred Baudrillart *et al.* (eds), *Dictionnaire d'histoire et de géographie ecclésiastique* (Paris, 1912–), 23, 1989, cols 430–3.

31. For the Cistercian houses see Holdsworth, 'The Cistercians in Devon', and for Torre see Deryck Seymour, *Torre Abbey* (Exeter, 1977), 47–52.

32. Cf. Susan Wood, *English Monasteries and their Patrons in the Thirteenth Century* (Oxford, 1955), and my lecture *The Piper and the Tune: Medieval Patrons and Monks* (Reading, 1991).

33. Knowles and Hadcock, *Medieval Religious Houses*, provide net income, but for its division into categories discussed in the next paragraph I have used Oliver's *Monasticon*.

34. Cf. *The Cartulary of Canonsleigh Abbey* (Harleian MS, no. 3660), ed. Vera C. London, Devon and Cornwall Record Society, new series, 8 (1965).

35. Lawrence, *Monasticism*, 177.

36. A.L. Rowse, *Tudor Cornwall: Portrait of a Society* (London, 1969), 159–60, for some striking figures.

CHAPTER 28

1. H. Miles Brown, *The Church in Cornwall* (Truro, 1964); A. Warne, *Church and Society in Eighteenth-Century Devon* (Newton Abbot, 1969); J. Barry, 'The seventeenth and eighteenth centuries', in N. Orme (ed.), *Unity and Variety: A History of the Church in Devon and Cornwall* (Exeter, 1991), 81–108, 212–16, especially the table of Anglican livings on 101. Several of the arguments sketched here are expanded in my article.

2. Compare A.D. Gilbert, *Religion and Society in Industrial England* (London, 1976) with the much more sophisticated analysis in M. Watts, *The Dissenters: From the Reformation to the French Revolution* (Oxford, 1978). For the sources for Protestant dissenting groups, written and architectural, see *Nonconformist Congregations in Great Britain: A List of Histories and Other Materials in Dr Williams' Library* (London, 1973); C. Stell, *An Inventory of Nonconformist Chapels and Meeting-Houses in South-West England* (London, 1991).

3. G. Oliver, *Collections Illustrating the History of the Catholic Religion in the Counties of Cornwall, Devon, Somerset, Wiltshire and Gloucestershire* (London, 1867); J. Bossy, *The English Catholic Community* (London, 1975); H. Peskett, *Guide to the Parish and Non-Parochial Registers of Devon and Cornwall 1538–1837*, Devon and Cornwall Record Society, extra series, 2 (1979), li–liv, 175–9; *Journal of the South-Western Catholic History Society*.

4. K. Beck, 'Recusancy and nonconformity in Devon and Somerset 1660–1714' (unpublished M.A. thesis, University of Bristol, 1961), 43–75, and especially Appendices I, xv–xvii, and III, xxviii–xxx.

5. *The Compton Census of 1676: A Critical Edition*, ed. A. Whiteman with Mary Clapinson, British Academy Records of Social and Economic History, new series, 10 (1986), 262–93. The Devon evidence is analysed more fully in R. Hole, 'Devonshire Catholics, 1676–1688', *Southern History*, 16 (1994), 85–99.

6. Whiteman (ed.), *Compton Census*, xcviii.

7. K. McGrath, *Catholicism in Devon and Cornwall in 1767* (Buckfast Abbey, 1960), 4. But Warne, *Church and Society*, 88, cites a 1705 return of 221 papists in Devon, and Peskett, *Guide*, 176, gives a Jesuit count of 275 Devon Catholics in 1710.

8. The figures here are taken from McGrath, *Catholicism*. The full national results are now available in E.S. Worrall (ed.), *Returns of Papists 1767*, 2 vols (Catholic Record Society, Occasional Publications 1 and 2, 1989). The national total was 67,916 and Exeter diocese was eighteenth out of twenty-six in the number of papists recorded.

9. J.H. Whyte, 'Vicars apostolic returns of 1773', *Recusant History*, 9 (1968), 205–14. Dorset had 540 and Somerset including Bristol 650.

10. Peskett, *Guide*, lxii–lxvii, 172–4, 229–30, 238; R. Gwynn and A. Grant, 'The Huguenots of Devon', *Transactions of the Devonshire Association*, 117 (1985), 161–94; B. Susser, *The Jews of South-West England* (Exeter, 1993). The attack on La Rochelle in the 1620s also brought a brief Huguenot presence in Plymouth.

11. I. Gowers, 'Puritanism in the county of Devon between 1570 and 1641' (unpublished M.A. thesis, University of Exeter, 1970), 88–99; R.J. Whiting, *The Blind Devotion of the People: Popular Religion and the English Reformation* (Cambridge, 1989); M. Stoyle, *Loyalty and Locality: Popular Allegiance in Devon during the English Civil War* (Exeter, 1994).

12. Whiteman (ed.), *Compton Census*; Devon Record Office, Chanter 225 A–C (1744–6 Visitation Returns of Bishop Claggett); Chanter 228 A–C (1764–6 Visitation Returns of Bishop Keppel); Moger Basket C/4–5 (1768 and 1771 Visitation Returns); Chanter 232 A–B (1779 Visitation Returns).

13. E.A. Windeatt, 'Devonshire Indulgence of 1672', *Transactions of the Devonshire*

Association, 1 (1901–4), 84–91, 159–70, 301; F. Bate, *The Declaration of Indulgence 1672* (London, 1908), Appendix VII, xix–xxiv and lxi–lxiii; A. Brockett, *Nonconformity in Exeter 1650–1875* (Manchester, 1962); Beck, 'Recusancy'; P. Jackson, 'Nonconformity and society in Devon 1660–88' (unpublished Ph.D. thesis, University of Exeter, 1986), especially Appendix 3.

14. Warne, *Church and Society*, 93; Beck, 'Recusancy', 168; Jackson, 'Nonconformity', Appendix 1 (meeting-house licences issued at Devon quarter sessions 1689–99). By 1736 383 licences had been taken out in Devon.

15. The best guide to all these matters is Watts, *Dissenters*. For local studies see J. Murch, *A History of the Presbyterian and General Baptist Churches in the West of England* (London, 1835); D. Jackman, *Baptists in the West Country* (Bridgwater, 1953); R. Ball, *Congregationalism in Cornwall* (London, 1955); A.D. Selleck, 'The history of the Society of Friends in Plymouth and west Devon from 1654 to the early nineteenth century' (unpublished M.A. thesis, University of London, 1959). John Slate's forthcoming Exeter Ph.D. on east Devon Quakers will put our understanding of the geography of the Quakers in that area on a firm footing.

16. Jackson, 'Nonconformity', 356–8, for a list of such 'partially conforming' ministers in Devon.

17. P. Jackson, 'Nonconformity and the Compton Census in late seventeenth-century Devon', in K. Schurer and T. Arkell (eds), *Surveying the People* (Oxford, 1993).

18. In addition to works already cited see M. Towgood, 'An Imperfect Account of the Succession of Ministers in the Several Dissenting Congregations in this County Devonshire' (Dr Williams's Library, MS 24.60, c.1773); G.L. Turner, 'Episcopal returns of 1665–6', *Transactions of the Congregational History Society*, 3 (1907–8), 339–53, and 4 (1909–10), 113–25, 148–58; A. Gordon, *Freedom after Ejection* (Manchester, 1917), 17–20, 30–3; A. Brockett (ed.), *Exeter Assembly Minutes 1691–1717*, Devon and Cornwall Record Society, new series, 6 (1963); M. Goldie, 'James II and the dissenters' revenge: the Commission of Enquiry of 1688', *Historical Research*, 66 (1993), 54–88.

19. Warne, *Church and Society*, 99–100.

20. 'Evans List' (Dr Williams's Library MS 34.4, c.1715–29), fos 16 (Cornwall) and 26–30 (Devon); 'The State of the Dissenting Interest in the Several Counties of England and Wales Collected in the Years 1715 and 1773' (Dr Williams's Library MS 38.5, c.1773), nos 6 (Cornwall) and 9 (Devonshire). The latter is published in 'A view of English nonconformity in 1773', *Transactions of the Congregational History Society*, 5 (1911–12), 210, 212–14, 379–80.

The best discussion of the lists is provided by Watts, *Dissenters*, 500–3. The basic county statistics are reproduced in the table in Barry, 'Seventeenth and eighteenth centuries', 87.

21. See J. Triffitt, 'Believing and belonging: church behaviour in Plymouth and Dartmouth 1710–30', in S. Wright (ed.), *Parish, Church and People* (London, 1988), 179–202. Anglican estimates of dissenting strength in 1669 had included information on 'hearers' for Somerset, but not for Devon: see Beck, 'Recusancy', 110–11 and Appendix IV, xliv–vi.

22. J. Manning, 'Account of Ministers Settled in Devonshire from 1662 to the Present Time with their Number of Hearers in 1715 and 1794' (microfilm copy in Dr Williams's Library, c.1794). See also 'Lists of poor ministers c.1785', *Transactions of the Congregational History Society*, 17 (1952–5), 92–9.

23. Selleck, 'History', 207–9, lists the 181 west Devon members in 1809. Watts, *Dissenters*, 505–6, attempts to supply Quaker figures for 1715 from registration material.

24. F.J. Powicke, 'Arianism and the Exeter Assembly', *Transactions of the Congregational History Society*, 7 (1916–18), 34–43; H.P.R. Finberg, 'A chapter in religious history', in W.G. Hoskins and H.P.R. Finberg, *Devonshire Studies* (London, 1952), 366–95; Brockett, *Nonconformity*; C.E. Welch, 'Dissenters' meeting houses in Plymouth to 1852', *Transactions of the Devonshire Association*, 94 (1962), 579–612; C.E. Welch, 'Andrew Kinsman's churches at Plymouth', *Transactions of the Devonshire Association*, 97 (1965), 212–36; R. Hayden, 'Evangelical Calvinism among eighteenth-century British Baptists' (unpublished Ph.D. thesis, University of Keele, 1991).

25. J.G. Hayman, *A History of the Methodist Revival of the Last Century in its Relations to North Devon*, 2nd edn (London, 1885); G. Davies, *The Early Cornish Evangelicals 1735–60* (London, 1951); H. Miles Brown, 'Methodism and the Church of England in Cornwall 1738–1838' (typescript in Cornwall County Library, 1947); H. Miles Brown, *Episcopal Visitation Returns and Methodism*, Cornish Methodist Historical Association, Occasional Publications 3 (1962); J.H.B. Andrews, 'The rise of the Bible Christians', *Transactions of the Devonshire Association*, 96 (1964), 147–85; T. Shaw, *A History of Cornish Methodism* (Truro, 1967); Warne, *Church and Society*, 126; Peskett, *Guide*, lxvii–lxx; R.F.S. Thorne, *Methodism in Devon: Handlist of Chapels and Records*, Devon Record Office Handlist 2 (1983); R.F.S. Thorne, *Methodism in the South West: Historical Bibliography* (London, 1983); M. Wickes (ed.), *John Wesley in Cornwall* (London, 1985). In 1772–3 there were 1,994 members of the Cornish circuits and 425 of the Devon, in 1789 3,964 and 1,308, and

in 1795 4,470 and 1,006 respectively. Most of the Devon members in 1789 were in the Plymouth area, notably Devonport (805), together with the Tiverton/Cullompton cloth area and among the mineworkers of North Molton, but the next decades saw a growth in north Devon, and a new Ashburton circuit. Even then Methodism remained weak in such centres of old dissent as Barnstaple and Bideford.

CHAPTER 30

1. The very detailed information required for the first pair of maps, for example, was only available for the diocese of Truro and not for its neighbour east of the Tamar.

2. Stating that there has been a marked decline in organized religion in the twentieth century is not meant to imply that such decline necessarily begins with the turn of the century. It is, in fact, very difficult to provide accurate dating for what is necessarily a complex and constantly changing phenomenon.

3. Peter Brierley (ed.), *Prospects for the Nineties: South West. Trends and Tables from the English Church Census* (London, 1991). See also Peter Brierley (ed.), *Christian England* (London, 1991) and Peter Brierley (ed.), *Prospects for the Eighties* (London, 1980 and 1983).

4. The category 'independent' is somewhat problematic, for it includes a wide variety of churches. Amongst these are 'house churches' which initially expanded very rapidly and flourish in the South West as elsewhere. Their growth poses interesting questions for the sociologist of religion: are such churches picking up the disaffected members from the mainline churches or are they making inroads into new sections of the population? It remains very difficult to say.

5. Sadly, the lack of presence does not imply the lack of negative reactions towards racial or religious minorities in the South West, a point nicely caught in the title of the report by Eric Jay on behalf of the Commission for Racial Equality. Eric Jay, *'Keep them in Birmingham': Challenging Racism in the South West* (London, 1992).

6. In Nicholas Orme (ed.), *Unity and Variety: A History of the Church in Devon and Cornwall* (Exeter, 1991), 157–74.

7. The significance of the Anglo-Catholic tendency in the South West for recent ecclesiastical history is obvious; there has been a certain amount of resistance to the ordination of women as priests in the Church of England in both dioceses of the region, but only in Truro did the vote in the Diocesan Synod go against the motion.

8. For a full discussion of the increasing separation of belief from practice in contemporary British religion, together

with its implications for different denominational groups and different parts of the country, see Grace Davie, *Religion in Britain since 1945: Believing without Belonging* (Oxford, 1994).

9. Reflected of course in the closure of churches and chapels. This needs, however, to be understood against a background of rampant and competitive overbuilding, especially between competing groups of nonconformists, or competing groups of Methodists. Following Robin Gill in *The Myth of the Empty Church* (London, 1993), such chapels never were full, nor could they ever have been even if everyone in the population attended simultaneously.

10. These maps were initially prepared as part of the *Rural Church Project*, a four-volume typescript available from the Centre for Rural Studies at the Royal Agricultural College in Cirencester or from the Department of Theology, University of Nottingham. Much of this information, including the maps, has been published in Douglas Davies, Charles Watkins and Michael Winter, *Church and Religion in Rural England* (Edinburgh, 1991). The report of the Archbishops' Commission on Rural Areas was published in 1990 under the title *Faith in the Countryside*. Chapter 8 describes changing patterns of ministry in rural areas. The amount of work involved in producing the detail contained in the maps is considerable. Regrettably, the diocese of Exeter (unlike Truro) was not one of the five included in the original study and it has not proved possible to carry out a similar exercise for this diocese.

11. See Derek Hearl, 'Regional political behaviour: or politics in the English wild west', paper presented to the Political Studies Association Annual Conference, Plymouth, 1988, for a full discussion of the methodological issues involved (paper available from the Department of Politics, University of Exeter).

CHAPTER 31

1. George Oliver, *The Lives of the Bishops of Exeter* (Exeter, 1861), 374–5. The original inventory is now missing. The contents of the medieval Cathedral Library have been analysed in Ian Maxted, *Exeter Cathedral Library: A Concordance of Medieval Catalogues and Surviving Manuscripts* (Exeter, 1987). Nicholas Orme, 'Martin Coeffin, the first Exeter publisher', *Library*, 6th series, 10, 3 (Sept. 1988), 220–30.

2. E.A. Clough, *A Short-Title catalogue, arranged Geographically, of Books printed and distributed … in the English Provincial Towns and in Scotland and Ireland … to … 1700* (London, 1969). This is supplemented to

some extent by Ian Maxted, *Books with Devon Imprints: A Chronological Checklist* (Exeter, 1989).

3. Ian Maxted, 'A common culture? The inventory of Michael Harte, bookseller of Exeter, 1615', in Todd Gray (ed.), *Devon Documents* (Exeter, 1996), 119–28.

4. *The Portledge Papers*, ed. Russell J. Kerr and Ida Coffin Duncan (London, 1928); Michael Treadwell, 'Richard Lapthorne and the London retail book trade, 1683–1697', in *The Book Trade and its Customers 1450–1900* (Winchester, 1997), 205–22.

5. Ian Maxted, 'Four rotten cornbags and some old books', in Robin Myers and Michael Harris (eds), *Sale and Distribution of Books from 1700* (Oxford, 1982), 37–76, especially 38–9.

6. Ian Maxted, 'Exeter book trades: artisan dynasties from Freemen's registers', in *Report of the Seminar on the Provincial Book Trade* (Newcastle upon Tyne, 1983), 5–8.

7. Margaret Cash (ed.), *Devon Inventories of the Sixteenth and Seventeenth Centuries*, Devon and Cornwall Record Society, new series, 11 (1966).

8. Martin Dunsford, *Historical Memoirs of the Town and Parish of Tiverton* (Exeter, 1790), iii–viii.

9. Cornwall County Library, *Catalogue of the Cornwall County Library* (Truro, 1794).

10. Paul Kaufman, 'English book clubs and their social import', in Paul Kaufman (ed.), *Libraries and their Users* (London, 1969), 36–64.

11. Dunsford, *Historical Memoirs*, 59.

12. G.H. Radford, 'Early printing in Devon', *Transactions of the Devonshire Association*, 60 (1928), 51–74.

13. Michael Treadwell and Ian Maxted 'The Exeter printer of 1688', *Devon and Cornwall Notes and Queries*, 36 (1990), 255–9.

14. Maxted, *Books with Devon Imprints*.

15. R.A.J. Potts, 'Early Cornish printers, 1740–1850', *Journal of the Royal Institution of Cornwall*, new series, 4 (1963), 318–19.

16. Winslow Jones, 'The author of "The worthies of Devon" and the Prince family', *Transactions of the Devonshire Association*, 25 (1893), 422; A.A. Brockett, 'The historians of Devon: a bibliographical appreciation' (unpublished essay for Library Association, 1949, typescript held by Westcountry Studies Library, Exeter).

17. M.G. Dickinson, 'Early Exeter printers and booksellers 1669–1741', *Devon and Cornwall Notes and Queries*, 29 (1963), 164–71; Ian Maxted, 'A most barbarous, dreadful and shocking murder …', *Factotum*, Occasional paper 3 (1986), 2–6; Ian Maxted, 'Single sheets from a country town: the example of Exeter', in *Spreading the Word: The Distribution Networks of Print 1550–1850* (Exeter, 1990), 109–29.

18. R.M. Wiles, *Freshest Advices: Early Provincial Newspapers in England* (Columbus, 1965), 129–30.

19. Devon Record Office, 1508M/Devon account books/v.10 fo.122.

20. Wiles, *Freshest Advices*, 290–2.

21. Richard S. Lambert, *The Cobbett of the West: A Study of Thomas Latimer* (London, 1939).

22. *Exeter Gazette*, 28 July 1793, 1.

23. Ian Maxted, *Newspaper Readership in South-West England: An Analysis of the Flindell's Western Luminary Subscription List of 1815* (Exeter, 1996).

24. *British Parliamentary Papers*, 1851, XLII, xvii, 516–29, House of Commons Returns of newspaper stamps 1837–50.

Chapter 33

1. The principal primary sources for this chapter and its maps are: Diocesan Registry papers on licensing of schoolmasters; Charity Commission Reports, 1819, 1 and 3; *British Parliamentary Papers* (*BPP*), 1835, XIIII, Parliamentary Return of Schools, 1833; *BPP*, 1867–8, XXVIII, Endowed Schools Commission Report; *BPP*, 1871, LV, Return of Schools outside municipal boroughs, 1870; Diocesan Board of Education Reports, 1840, 1893; Committee of Council for Education Reports, 1840, 1899; Board of Education Reports, 1899–1901; County directories (Pigot, White, Billing, Kelly); Devon Education Committee staffing and salary records, from 1903; Devon Education Committee minutes, from 1905; school logbooks; school attendance records; school boards' and voluntary managers' minutes.

Chapter 34

1. J.G. Holmes, 'The place of Cornish in east Cornwall, 700–1500 AD' (unpublished paper presented at the 6th International Congress of Celtic Studies, Galway, 1979); K.J. George, 'How many people spoke Cornish traditionally?', *Cornish Studies*, 14 (1986), 67–70.

2. N.J.A. Williams, *Cornish Today: An Examination of the Revived Language* (Sutton Coldfield, 1995), 77.

3. Ibid., 77.

4. James Whetter, *The History of Gorran Haven, Part 1, 0–1800 AD* (St Austell, 1991), 11, 15–16.

5. John Norden, *Description of Cornwall* (London, 1728), 26.

6. Richard Carew of Antony, *The Survey of Cornwall*, ed. F.E. Halliday (London, 1953), 303.

7. Derek R. Williams, *Prying into Every Hole and Corner: Edward Lhuyd in Cornwall in 1700* (Redruth, 1993).

8. R. Morton Nance, 'When was Cornish last spoken traditionally?', *Journal of the Royal Institution of Cornwall*, new series, 7, 1 (1973), 76–82.

Chapter 35

1. Exchequer Domesday Book is actually two manuscripts, both kept in the Public Record Office at Kew: Great Domesday Book which is now bound in two parts and covers thirty-one of the thirty-four counties, and Little Domesday which represents an earlier stage in the Domesday Book Survey and is now in three volumes and contains the returns for the remaining three counties (Essex, Norfolk and Suffolk). The printed edition, transcribed and set in special type by Abraham Farley, is *Domesday Book: seu liber censualis Willelmi primi regis Angliae …*, 2 vols (London, 1783). Photozincographic facsimiles were published by the Ordnance Survey in 1861–3. A colour facsimile of Great Domesday Book was published by Alecto Historical Editions (London, 1986).

2. *The Anglo-Saxon Chronicle: A Revised Translation*, eds D. Whitelock *et al.* (London, 1961), s.a. 1085, 1086.

3. Ibid., s.a. 1084

4. N.E.S.A. Hamilton, *Inquisitio Comitatus Cantabrigiensis … subjicitur Inquisitio Eliensis* (London, 1876), 97; The *Inquisitio Eliensis* states that all the details should be given at three dates: in the time of King Edward, when King William made his grant, and 'now' (1086). In fact only the value is regularly given at three dates in Domesday Book and then only in some counties.

5. The manuscript is in the Cathedral Library, Exeter. The printed edition, by Sir Henry Ellis, is *Libri Censualis, vocati Domesday Book, additamenta ex codic. antiquiss: Exon'. Domesday; Inquisitio Eliensis; Liber Winton'; Boldon Book* (London, 1816).

6. The hundreds of Cornwall and Devon and the problems of their mapping are discussed in articles by F.R. Thorn in the Alecto County Editions and in the editions of C. and F. Thorn, cited under Further Reading, and in F.R. Thorn 'The Identification of Domesday Places in the South-Western Counties of England', *Nomina*, 10 (1986), 41–59.

Chapter 36

1. *De Gestis Regnum Anglorum*, ed. William Stubbs (London, 1887–9), I, 148; *Chronicle of Richard of Devizes of the Time of King Richard the First*, ed. John T. Appleby (London, 1963), 66; Herbert P.R. Finberg, 'The making of a boundary' in his *Lucerna* (London, 1964), 161; George Oliver, *Monasticon Diocesis Exoniensis* (Exeter and London, 1846), 342; *The Itinerary of John Leland*, ed. Thomas Hearne, 3rd edn (London, 1768–9), IX, xxi (for the poem); *The Itinerary of John Leland*, ed. Lucy Toulmin Smith (London, 1907), 178, 315,
172, 224; *The Anglo-Saxon Chronicle*, ed. George N. Garmonsway (London, 1972), 132.

2. Map based upon about 170 demesne accounts, largely from the fourteenth century, listed in Harold S.A. Fox, 'Farming practice and techniques: Devon and Cornwall', in Edward Miller (ed.), *The Agrarian History of England Wales, Vol. III, 1348–1500* (Cambridge, 1991), 305, with some additional information from a few earlier demesne accounts, from inventories and from sources relating to tithe and multure. Gaps on the map simply indicate lack of data.

3. As at Yealmpton, South Pool and Burniere: Public Record Office (PRO) S.C. 6/830/29; Cornwall County Record Office (CRO) AR2/858; British Library (BL) Add. rolls 64,319–2.

4. As at Sampford Courtenay and Bishop's Nympton: Dean and Chapter Archives, Exeter (DCA) V.C. 22,279; Devon Record Office (DRO) W1258M/G/6/10.

5. CRO AR2/901,908, 899; DRO W1258M/D/74/6; Sampford Courtenay and Bishop's Nympton as in note 4 above; Herbert P.R. Finberg, *Tavistock Abbey: A Study in the Social and Economic History of Devon* (Cambridge, 1951), 98; servants' diet at, for example, Sampford Courtenay and Bishop's Nympton as in note 4 above; William J. Blake, 'Hooker's Synopsis Chorographical of Devonshire', *Transactions of the Devonshire Association*, 47 (1915), 345.

6. *Register of Edward the Black Prince*, II (London, 1931), 178. Cornwall was called upon to supply other parts of England with grain during the famine of 1315 onwards: John Hatcher, *Rural Economy and Society in the Duchy of Cornwall 1300–1500* (Cambridge, 1970), 85.

7. The map shows only a sample of the movements, based upon references in manorial accounts. Details of livestock husbandry and animal movements in Fox, 'Farming practice', 315–22.

8. Peter Herring, 'Transhumance in medieval Cornwall', in Harold S.A. Fox (ed.), *Seasonal Settlement* (Leicester, 1996), 35–44; Harold S.A. Fox, 'Occupation of the land: Devon and Cornwall', in Miller (ed.), *Agrarian History*, 168–9; and Nicholas Johnson and Peter Rose (eds), *Bodmin Moor: An Archaeological Survey, Vol. I, The Human Landscape to c.1800* (London, 1994), 80, for survivals of the system on Bodmin Moor; Staffordshire Record Office, D593/A/1/14/4 (for Exmoor); Harold S.A. Fox, 'Medieval Dartmoor as seen through its account rolls', *Proceedings of the Devon Archaeological Society*, 52 (1996), 156–62.

9. Finberg, *Tavistock Abbey*, 89–94; Fox, 'Farming practice', 309–12; Hatcher, *Rural Economy*, 14, 82–4; Harold S.A. Fox, 'Outfield cultivation in Devon and Cornwall: a reinterpretation', in Michael

Havinden (ed.), *Husbandry and Marketing in the South-West, 1500–1800* (Exeter, 1973), 19–38. The Hampshire reference is in C.R. Straton (ed.), *Survey of the Lands of William, First Earl of Pembroke*, 2 vols (1909), I, lxii, and II, 543.

10. Earliest documented references to beat burning and sanding are from 1246 and *c*.1250 respectively but the practices themselves are much older: Finberg, *Tavistock Abbey*, 92; *Calendar of Charter Rolls 1257–1300* (London, 1906), 36. The earliest reference to outfields may be from 958: Herbert P.R. Finberg, 'Ayshford and Boehill', *Transactions of the Devonshire Association*, 103 (1971), 23.

11. Herbert P.R. Finberg, 'The open field in Devon', in W.G. Hoskins and H.P.R. Finberg, *Devonshire Studies* (London, 1952), 279–82; W.G. Hoskins, 'The making of the agrarian landscape', ibid., 314–15; Harold S.A. Fox, 'Tenant farming and tenant farmers', in Miller (ed.), *Agrarian History*, 731–2; Harold S.A. Fox, 'Peasant farmers, patterns of settlement and *pays*: transformations in the landscapes of Devon and Cornwall during the later Middle Ages', in Robert Higham (ed.), *Landscape and Townscape in the South West* (Exeter, 1989), notes 80, 81, 82, 83, 84; Ann Preston-James and Peter Rose, 'Medieval Cornwall', *Cornish Archaeology*, 25 (1986), 152.

12. Harold S.A. Fox, 'The chronology of enclosure and economic development in medieval Devon', *Economic History Review*, 2nd series, 28 (1975), 181–202; Finberg, 'Open field in Devon', 277–8; Fox, 'Tenant farming', 732–3.

13. DRO Z1/10/26; *Register of the Black Prince*, II, 129; DRO 902M/M/13.

14. BL Harl. MS 71f. 63v; Fox, 'Tenant farming', 723–32.

15. Fox, 'Occupation of the land', 161, for decayed mills; and see Fox, 'Peasant farmers', 53, for an illustration of the contrasting fortunes of Cornish coastlands and moors.

16. Even the earliest of demesne accounts (and there are not many of them) reveal the contrasts between regions described in this chapter.

17. Red cattle: PRO S.C. 2/167/34, court near saints Tiburtius and Valerianus, 22 Ric. II; DRO W1258M/D/70, court of 1366. Steer houses and linhays: DRO, Cary MSS, Ashwater court near Invention of the Holy Cross, 4 Hen. VII; DRO C.R. 107. This last document, from Cookbury in *c*.1496, refers to *le pole orium* and *le pole boveriam*, and is the earliest mention known to me of the classic Devonshire linhay with its arcaded open front. Living-in servants: Harold S.A. Fox, 'Servants, cottagers and tied cottages during the later Middle Ages: towards a regional dimension', *Rural History: Economy, Society, Culture*, 6 (1995), 128–33,

and cf. George Vancouver, *General View of the Agriculture of the County of Devon* (London, 1808), 93–4.

18. C. and F. Thorn (eds), *Domesday Book: Cornwall* (Chichester, 1979), 5.2.21. Other examples: William L.D. Ravenhill, 'Cornwall', in H.C. Darby and R. Welldon Finn (eds), *The Domesday Geography of South-West England* (Cambridge, 1967), 344; W.G. Hoskins, 'The highland zone in Domesday Book', in his *Provincial England* (London, 1963), 20–1.

19. Oswald J. Reichel (ed.), *Devon Feet of Fines*, I, Devon and Cornwall Record Society (1912), 14–15.

20. For example, CRO AR2/561 (Hartland); DRO Cary MSS, Ashwater rental of 1346.

21. I have not been able to identify the location of a very small proportion of the free tenements. See also the pioneering study by Maurice W. Beresford, 'Dispersed and grouped settlement in medieval Cornwall', *Agricultural History Review*, 12 (1964), 13–27. For another example see Fox, 'Peasant farmers', 53.

22. For excavated examples of medieval hamlets away from the moorlands: E.M. Jope and R.I. Threlfall, 'Excavation of a medieval settlement at Beere, North Tawton, Devon', *Medieval Archaeology*, I I (1958), 112–39; Guy Beresford, 'Tresmorn St Gennys', *Cornish Archaeology*, 10 (1971), 55–73. West Wortha, at the Roadford Reservoir site, is another example.

23. A very small proportion of sites remain unidentified. The map is more speculative than that of Helstone, for it is based on an early-sixteenth-century rental from which the number of ferlings (farms of about 30 acres) at each site may be calculated. The map is based on the assumption that each ferling represented one farm in about 1300 when the number of farms is likely to have been at its height. Reconstitutions of the landscape of Sampford Courtenay are assisted by other rentals and by accounts, surveys and maps: PRO S.C. 6/1118/6; DCA V.C. 22279; DRO 56/10; King's College Cambridge SAC 67, 90, 122.

24. Fox, 'Medieval Dartmoor', 152–3; Johnson and Rose (eds), *Bodmin Moor*, 107–9. Other examples are Hound Tor on Dartmoor and Garrow on Bodmin Moor: Guy Beresford, 'Three deserted medieval settlements on Dartmoor: a report on the late E. Marie Minter's excavations', *Medieval Archaeology*, 23 (1979), 98–158, including the important appendix on 'The medieval fields' by Peter F. Brandon; Dorothy Dudley and E. Marie Minter, 'The medieval village at Garrow Tor, Bodmin Moor, Cornwall', *Medieval Archaeology*, 6–7 (1962–3), 272–94. See also Andrew Fleming and Nicholas Ralph, 'Medieval settlement and land use on Holne Moor, Dartmoor: the landscape evidence', *Medieval Archaeology*, 26 (1982), 101–37, and Catherine D. Linehan, 'Deserted sites and rabbit-warrens on

Dartmoor, Devon', *Medieval Archaeology*, 10 (1966), 113–44.

25. Peter L. Hull (ed.), *The Caption of Seisin of the Duchy of Cornwall (1377)*, Devon and Cornwall Record Society, 17 (1971), 69.

26. Hoskins, 'Highland zone', 45. Hoskins made this statement after a reconstruction of the settlement of Sampford Courtenay manor in 1086, a reconstruction in which he attempted to place all Domesday villeins in the landscape. His map of the manor (ibid., 28) contains far fewer farmsteads than mine, almost certainly because he failed to plot the bordars of Domesday. The fullness of settlement on my map does nevertheless strengthen his general conclusion.

27. Hoskins, 'Making of the agrarian landscape', 316–24, and, following him, William G.V. Balchin, *Cornwall* (London, 1954), 40, lay great stress on the twelfth century as a formative era in the creation of large numbers of settlement sites, and while this view is true for some regions (especially the moorlands), and in some other cases, it almost certainly underrates the antiquity of many isolated settlements.

28. Sources for Sampford as in note 23 above; DRO Cary MSS, Stoke Fleming rental of 1523. See also Vancouver, *General View*, 97.

29. *Itinerary of John Leland*, ed. Toulmin Smith, 319, 321.

30. For Sidbury DCA 2944 and 2945. Data from the latter rental are mapped in Harold S.A. Fox, 'The field systems of east and south Devon, pt I: East Devon', *Transactions of the Devonshire Association*, 104 (1972), 90, but this map unaccountably failed to show cottagers and smallholders who considerably inflated the size of Sidbury village; for these see Fox, 'Servants', 134–6. Axminster: DRO 123M/TB/159, 191.

31. Charles Henderson, 'A note on the origin of Cornish towns', in his *Essays in Cornish History* (Oxford, 1935), 19. A few settlements in far eastern Cornwall may have been true agricultural villages in the Middle Ages.

32. Frederic W. Maitland, *Domesday Book and Beyond* (Cambridge, 1921), plate following 16.

33. Maps and discussion of Hartland: see Fox, 'Peasant farmers', 50–2, and 'Contraction: desertion and dwindling of dispersed settlement in a Devon parish', *Annual Report of the Medieval Village Research Group*, 31 (1983), 40–2. For a discussion of settlement contraction region by region see Fox, 'Occupation of the land', 164–70.

34. PRO E. 306/2/1, 2/16; Duchy of Cornwall Archives 475; Patricia Christie and Peter Rose, 'Davidstow Moor, Cornwall: the medieval and later sites', *Cornish Archaeology*, 26 (1987), 163–95; Johnson and Rose (eds), *Bodmin Moor*, 83. For

Dartmoor see Linehan, 'Deserted sites', 113–44, and Fox, 'Medieval Dartmoor', 155–6. For Exmoor see Michael Aston, 'Deserted farmsteads on Exmoor and the lay subsidy of 1327 in west Somerset', *Somerset Archaeological and Natural History Society Proceedings*, 127 (1983), 71–104.

35. *Annual Report of the Medieval Settlement Research Group*, 8 (1993), 48, summary by Peter Herring of a Rapid Identification Survey.

36. For example, Charles Henderson, *The 109 Ancient Parishes of the Four Western Hundreds of Cornwall* (Truro, n.d.), 1–2; Gordon R. Dunstan (ed.), *The Register of Edmund Lacy*, IV, Devon and Cornwall Record Society, new series, 16 (1971), 314–18.

37. Patrick Cowley, *The Church Houses: Their Religious and Social Significance* (London, 1970), 16; Joanna Mattingly, 'The medieval parish guilds of Cornwall', *Journal of the Royal Institution of Cornwall*, new series, 10, 3 (1989), 290–329; Richard Carew of Antony, *The Survey of Cornwall*, ed. F.E. Halliday (London, 1953), 141; Alison Hanhan (ed.), *Churchwardens' Accounts of Ashburton, 1479–1580*, Devon and Cornwall Record Society, 15 (1970), 58.

38. A.L. Rowse, *Tudor Cornwall* (London, 1941), 121–2; Mattingly, 'Medieval parish guilds', 313. Julian Cornwall, *Revolt of the Peasantry 1549* (London, 1977), 65–6, describes the events at Sampford Courtenay and their background. A rental of Sampford, more or less contemporary with the revolt (PRO SC 12/22/19), shows that a farmstead occupied by the Underhill family (William Underhill was a leader in the revolt) was one of the remotest sites in the parish.

39. Royal Institution of Cornwall, Henderson transcripts XIX, no. 314 (and see also XXVI, no. 97 for the same terminology); DRO 902M/M/33; Johnson and Rose (eds), *Bodmin Moor*, 93; Peter L. Hull (ed.), *The Cartulary of Launceston Priory*, Devon and Cornwall Record Society, 30 (1987), 208–9; Charles Henderson, 'Twelve Mens' Moor', in his *Essays in Cornish History*, 125–9.

40. J.H. Adams, 'The medieval chapels of Cornwall', *Journal of the Royal Institution of Cornwall*, new series, 3 (1958), 48–65; James Coulter, *The Ancient Chapels of North Devon* (Yarnscombe, 1993); Mattingly, 'Medieval parish guilds', 299.

41. Fox, 'Servants'; comment on Cornwall based upon absence of labourers' cottage holdings in many manors, according to extents and rentals; *Calendar of Inquisitions Miscellaneous, 1377–1388* (London, 1957), 67; Christopher Dyer, 'How urbanized was medieval England?', in Jean-Marie Duvosquel and Erik Thoen (eds), *Peasants and Townsmen in Medieval Europe* (Ghent, 1995), 174, 176 (for the visibility of taxable goods); Peter A.S. Pool, 'The tithings of Cornwall', *Journal of the Royal Institution of*

Cornwall, new series, 8, 4 (1981), 279; Francis Hingeston-Randolph (ed.), *The Register of John de Grandisson* (London and Exeter, 1897), 958.

CHAPTER 37

1. Cited by James Whetter, *Cornwall in the Seventeenth Century: An Economic History of Kernow* (Padstow, 1974), 51.
2. Ibid., 58, note 56.
3. Christopher Morris (ed.), *The Journeys of Celia Fiennes* (London, 1947), 250.
4. Daniel and Samuel Lysons, *Magna Britannia, Vol. III, Cornwall* (London, 1814), 310.
5. *British Parliamentary Papers* (*BPP*), 1801, VI, 813 (140), 41–7; Census of 1801; W.G. Hoskins, *Devon* (London, 1954), 527 and 531.
6. Margaret Cash (ed.), *Devon Inventories of the Sixteenth and Seventeenth Centuries*, Devon and Cornwall Record Society, new series, 11 (1966), 57–9 and 79.
7. Ibid., 47.
8. *BPP*, LXVIII, 363, Agricultural Statistics for 1870, 28; *BPP*, 1908, CXXI, 70, Agricultural Statistics for 1907.
9. William Marshall, *The Rural Economy of the West of England*, 2 vols (London, 1796), II, 225–31.
10. H.S.A. Fox, 'A geographical study of the field systems of Devon and Cornwall' (unpublished Ph.D. thesis, Cambridge University, 1971), passim.
11. Joan Thirsk, 'The common fields', *Past and Present*, 29 (1964).
12. Anthony Powell, *Aubrey and his Friends* (London, 1948), 91, quoting Bodleian Library MS., Aubrey 2, fo. 83.
13. H.P.R. Finberg, *Tavistock Abbey: A Study in the Social and Economic History of Devon* (Cambridge, 1951); Robin Stanes, 'A georgicall account of Devonshire and Cornwalle by Samuel Colepresse (1667)', *Transactions of the Devonshire Association*, 96 (1964), 269–302; Robin Stanes, 'Devon agriculture in the mid-eighteenth century', in M.A. Havinden and Celia M. King (eds), *The South-West and the Land* (Exeter, 1969), 56–8; and William Marshall, *The Rural Economy of the West of England* (London, 1796).
14. Michael Havinden, 'Lime as a means of agricultural improvement: the Devon example', in C.W. Chalklin and M.A. Havinden (eds), *Rural Change and Urban Growth 1502–1800: Essays in English Regional History in Honour of W.G. Hoskins* (London, 1974), 113.
15. Havinden, 'Lime', 116.
16. Whetter, *Cornwall in the Seventeenth Century*, 22.
17. Ibid., 24.
18. Ibid., 25.
19. Against a background of a rising population wheat exports rose from

15,002 quarters in 1697 to 950,002 quarters in 1750: B.R. Mitchell and Phyllis Deane, *Abstract of British Historical Statistics* (Cambridge, 1962), 94.
20. Whetter, *Cornwall in the Seventeenth Century*, 38, 48.
21. Giles V. Harrison, 'The South-West: Dorset, Somerset, Devon and Cornwall', in Joan Thirsk (ed.), *The Agrarian History of England and Wales, Vol. V, Part 1, 1642–1752: Regional Farming Systems* (Cambridge, 1984), 367 and 388–9.
22. Whetter, *Cornwall in the Seventeenth Century*, 31 and 34–6.
23. Harrison, 'The South-West', 389.
24. John Rowe, *Cornwall in the Age of the Industrial Revolution* (Liverpool, 1953), 211 and 231–2.
25. Marshall, *Rural Economy*, II, cited by Whetter, *Cornwall in the Seventeenth Century*, 24.
26. Hoskins, *Devon*, 74–89.
27. F.E. Halliday, *A History of Cornwall* (London, 1959), 172–3.
28. Joseph Foster, *Peerage, Baronetage and Knightage of the British Empire for 1882* (London, 1882), 273.
29. Ibid., 277; Halliday, *A History of Cornwall*, 258.
30. Halliday, *A History of Cornwall*, 213.
31. Ibid., 258.
32. Ibid., 156.

CHAPTER 38

1. William Marshall, *The Rural Economy of the West of England*, 2 vols (London, 1796), II, 226–30.
2. H.S.A. Fox, 'The functioning of bocage landscapes in Devon and Cornwall between 1500 and 1800', in *Les Bocages: Histoire, Ecologie, Economie* (Rennes, 1976), 55–61, 56–7.
3. Charles Vancouver, *General View of the Agriculture of the County of Devon* (London, 1808), 171–2, 177, 179, 183; Henry Tanner, 'The farming of Devonshire', *Journal of the Royal Agricultural Society of England*, 9 (1849), 454–95, 462.
4. R. Morgan, 'The root crop in English agriculture, 1650–1870' (unpublished Ph.D. thesis, University of Reading, 1978), 124, 148–9.
5. A.H. Shorter, W.L.D. Ravenhill and K.J. Gregory, *South-West England* (London, 1969), 136–7, Figure 35.
6. Vancouver, *General View*, 142.
7. S.A.H. Wilmot, 'Landownership, farm structure and agrarian change in south-west England, 1800–1900: regional experience and national ideals' (unpublished Ph.D. thesis, University of Exeter, 1988), 185–7.
8. Marshall, *Rural Economy*, II, 119; R. Trow-Smith, *A History of British Livestock Husbandry, 1700–1900* (London,

1959), 98; Vancouver, *General View*, 325; William Youatt, *The Complete Grazier and Farmer's and Cattle-breeder's Assistant*, 11th edn (London, 1864), 7–8.
9. W.F. Karkeek, 'On the farming of Cornwall', *Journal of the Royal Agricultural Society of England*, 6 (1845), 400–62, 448–51.
10. *British Parliamentary Papers* (*BPP*), 1882, XV, Royal Commission on the Depressed Condition of the Agricultural Interest, 1881, Report by Assistant Commissioner W.C. Little, 11–15.
11. A.H.D. Acland, *Memoirs and Letters of the Right Honourable Sir Thomas Dyke Acland* (London, 1902), 311.
12. *BPP*, 1852–3, Population Census of Great Britain, 1851.
13. M.A. Reed (ed.), *Discovering Past Landscapes* (London, 1984) provides a useful review of some of the problems involved in using these statistical sources for the study of nineteenth-century agriculture. On the defects of the tithe files regarding livestock data see Roger J.P. Kain and Hugh C. Prince, *The Tithe Surveys of England and Wales* (Cambridge, 1985), 143–5.
14. Crop combinations are calculated by comparing the actual percentage of a parish's cropland cultivated in particular crops with a theoretical percentage based on the assumption, in turn, of a monoculture, a two-crop combination, a three-crop combination and so on. Crop combinations give an indication of the dominance of particular crops in each parish. The method of calculation is described in Wilmot, 'Landownership, farm structure and agrarian change', 163 and 208 n.143.
15. M. Overton, 'The 1801 crop returns for Cornwall', in M.A. Havinden (ed.), *Husbandry and Marketing in the Southwest, 1500–1815* (Exeter, 1973), 39–62, 47–8.
16. Tanner, 'The farming of Devonshire', 463.
17. Karkeek, 'Farming of Cornwall', 422–3, 432.
18. F. Punchard, 'Farming in Devon and Cornwall', *Journal of the Royal Agricultural Society of England*, 51 (1890), 511–36, 522.
19. R.J.P. Kain, *An Atlas and Index of the Tithe Files of Mid-Nineteenth-Century England and Wales* (Cambridge, 1986), 215, 225.
20. Vancouver, *General View*, 19–20, 23–4, 276–8.
21. J.G. Cornish, *Reminiscences of Country Life* (London, 1939), 131.
22. *BPP*, 1882, XV, Royal Commission on the Depressed Condition of the Agricultural Interest, 1881, Report by Assistant Commissioner W.C. Little, 19, 23.
23. W. Harwood-Long, 'Factors affecting some types of farming in Devon and Cornwall', *Journal of the Royal Agricultural Society of England*, 94 (1933), 42–61, 60.
24. Punchard, 'Farming in Devon and Cornwall', 526–8; Shorter, Ravenhill and Gregory, *South-West England*, 138.
25. Public Record Office, Kew, IR18/1163, Tithe file for East Buckland; Tanner, 'The

farming of Devonshire', 495; James Caird, *English Agriculture, 1850–1* (London, 1852), 48.
26. Karkeek, 'Farming of Cornwall', 402–3, 426; Tanner, 'The farming of Devonshire', 462; W.H. Gamlen, 'Agriculture in north east Devon, fifty or sixty years ago', *Transactions of the Devonshire Association*, 12 (1880), 380–6; A.D.M. Phillips, *The Underdraining of Farmland in England during the Nineteenth Century* (Cambridge, 1989), 72, Figure 3.8; 74, Table 3.1; 224–28.
27. Punchard, 'Farming in Devon and Cornwall', 536.

CHAPTER 39

1. Ministry of Agriculture Fisheries and Food, *A Century of Agricultural Statistics 1866–1966* (London, 1968).
2. J.T. Coppock, *An Agricultural Atlas of England and Wales* (London, 1964 and 1974); G. Clark, D.J. Knowles and H.L. Phillips, 'The accuracy of the agricultural census', *Geography*, 68 (1983), 115–20.
3. G.H. Peters, *Agriculture: Reviews of United Kingdom Statistical Sources* (London, 1988); J.T. Coppock and L.F. Gebbett, *Land Use and Town and Country Planning: Reviews of United Kingdom Statistical Sources* (Oxford, 1978).
4. P.J. Perry, *British Farming in the Great Depression* (Newton Abbot, 1973).
5. G.M. Robinson, *Agricultural Change* (Edinburgh, 1988).
6. L.D. Stamp, *Land of Britain, Part 92, Devonshire* (London, 1951); B.S. Robertson, *Land of Britain, Part 91, Cornwall* (London, 1951).
7. A.H. Shorter, W.L.D. Ravenhill and K.J. Gregory, *South-West England* (London, 1969).
8. Coppock, *Agricultual Atlas*.
9. Ibid.
10. C. Watkins, 'Sources for the assessment of British woodland change in the twentieth century', *Applied Geography*, 5 (1985), 151–66.
11. M.L. Parry, *Surveys of Moorland and Roughland Change* (Birmingham, 1981).
12. M. Blacksell and A. Gilg, *The Countryside: Planning and Change* (London, 1981).
13. Ibid.

CHAPTER 40

1. Daniel Defoe, *A Tour through the Whole Island of Great Britain*, revised Everyman edn, ed. G.D.H. Cole and D.C. Browning (London, 1962), I, 221.
2. Louis F. Salzman, 'Mines and stannaries', in James F. Willard, William A. Morris and William H. Dunham (eds), *The English Government at Work, 1327–1336, Vol. III* (Cambridge MA, 1950), 77–8;

Calendar of Patent Rolls, 1461–7 (London, 1897), 482; *Rotuli Parliamentorum*, I (London, 1783), 312.

3. John R. Maddicott, 'Trade, industry and the wealth of King Alfred', *Past and Present*, 123 (1989), 3–51.

4. John Hatcher, *English Tin Production and Trade Before 1550* (Oxford, 1973), 18–26.

5. Christopher Dyer, *Standards of Living in the Later Middle Ages: Social Change in England c.1200–1520* (Cambridge, 1989), 174, 207.

6. Two examples from among very many are tolls from the 'moors of Bre and Carn' taken by lords of Tihidy and the expenses paid to the 'collector of the lord's tin' on the Arundell estate: Cornwall Record Office (CRO), AR2/839, 927.

7. Maps in John Hatcher, *Rural Economy and Society in the Duchy of Cornwall 1300–1500* (Cambridge, 1970), xiv, and Thomas Greeves, 'Four Devon stannaries: a comparative study of tinworking in the sixteenth century', in Todd Gray, Margery Rowe and Audrey Erskine (eds), *Tudor and Stuart Devon: The Common Estate and Government: Essays Presented to Joyce Youings* (Exeter, 1992), 41, 48.

8. For coinage see George R. Lewis, *The Stannaries: A Study of the English Tin Miner* (Cambridge MA, 1924), 149–52; Herbert P.R. Finberg, 'The stannary of Tavistock', *Transactions of the Devonshire Association*, 81 (1949), 160; Greeves, 'Four Devon stannaries', 52–8. Coinage rolls in print: Finberg, 'Stannary of Tavistock', 173–82, and Sir John Maclean, 'Stannary roll, 34th Edward I (1305–6), with introductory remarks thereon and on similar rolls', *Journal of the Royal Institution of Cornwall*, 3 (1868–70), 238–59 (with a facsimile).

9. The population estimate for Bodmin is very rough and is based upon William of Worcester's abstract from the register of Bodmin Priory: *William Worcestre: Itineraries*, ed. John H. Harvey (Oxford, 1969), 95. The abstract states that 1,500 people died there in the plague of 1348, although this is probably an exaggeration. I have assumed a mortality of 50 per cent. For Lostwithiel, see Adolphus Ballard, *British Borough Charters 1042–1216* (Cambridge, 1913), 217.

10. Peter C. Herring, 'An exercise in landscape history: pre-Norman and medieval Brown Willy and Bodmin Moor, Cornwall' (unpublished M.Phil. thesis, University of Sheffield, 1986), II, 146; David Austin, G.A.M. Gerrard and Thomas A.P. Greeves, 'Tin and agriculture in the Middle Ages and beyond: landscape archaeology in St Neot Parish, Cornwall', *Cornish Archaeology*, 28 (1989), 27; G.A.M. Gerrard, 'The early Cornish tin industry: an archaeological and historical survey' (unpublished Ph.D. thesis, University of Wales, 1986), 126, 161.

11. John Norden, *Speculi Britanniae Pars: A Topographical and Historical Description of*

Cornwall (Newcastle, 1966, based upon the first printed edition of 1728), 41, 50; *The Itinerary of John Leland*, ed. Lucy Toulmin Smith (London, 1907), parts 1–3, 193.

12. Lewis, *Stannaries*, 40.

13. Herbert P.R. Finberg, *Tavistock Abbey: A Study in the Social and Economic History of Devon* (Cambridge, 1969), 189.

14. For the wealth of material in the stannary court rolls see Hatcher, *English Tin*, 63–4. There is yet no adequate guide to listings of exempt tinners. Examples are Public Record Office (PRO) E.179/95/28-32 (for c.1373) and E.179/95/12 (for c.1337). A small portion of the listing for 1337 is printed in Ethel Lega-Weekes, 'Neighbours of North Wyke, pt.1', *Transactions of the Devonshire Association*, 33 (1901), 423–5.

15. Ian Blanchard, 'The miner and agricultural community in late medieval England', *Agricultural History Review*, 20 (1972), 93–106; John Hatcher, 'Myths, miners and agricultural communities', *Agricultural History Review*, 22 (1974), 54–61; Ian Blanchard, 'Rejoinder: *stannator fabulosus*', *Agricultural History Review*, 22 (1974), 62–74.

16. Hatcher, *English Tin*, 59–82; Hatcher, 'Myths', 54–61.

17. Hatcher, *English Tin*, 59–66; CRO, AR2/927; PRO S.C. 6/823/17.

18. Blanchard, 'Rejoinder', 71; West Devon Record Office, 70/53.

19. Comments on seasonality before 1348 are based upon early coinage rolls: PRO E.101/260/24 (Devon) and 261/1 (Cornwall). Late summer and early autumn coinage is especially restricted in the first of these sources. Comments on later seasonality based upon: Finberg, *Tavistock Abbey*, 189 (statistics between 1456 and 1487); Hatcher, *English Tin*, 78; Hatcher, 'Myths', 59. See also Greeves, 'Four Devon stannaries', 53.

20. PRO E.179/95/28-32 and Josiah C. Russell, *British Medieval Population* (Albuquerque, 1948), 132, on the assumption, almost certainly unrealistic, that as many women as men paid the poll tax of 1377.

21. Hatcher, *Rural Economy*, 32. For the contribution of women see Hatcher, *English Tin*, 47, and for the later situation Alfred K. Hamilton Jenkin, *The Cornish Miner* (London, 1948), 236–9.

22. Lords often valued and accounted for turbaries along with toll of tin, for example at Hamatethy and Enniscaven: PRO C.139/93; CRO, AR2/899. For Cornish and Devonian *carbonarii* see Harold S.A. Fox, 'Medieval Dartmoor as seen through its account rolls', *Proceedings of the Devon Archaeological Society*, 52 (1994), 162–4.

23. William J. Blake, 'Hooker's Synopsis Chorographical of Devonshire', *Transactions of the Devonshire Association*, 47 (1915), 346; Eleanora M. Carus-Wilson and Olive Coleman, *England's Export Trade 1275–1547* (Oxford, 1963), 144–5.

All my references to exports are from this source.

24. For example, in the late 1480s and in the 1510s. For the same conclusion, reached from a different direction, see Maryanne Kowaleski, *Local Markets and Regional Trade in Medieval Exeter* (Cambridge, 1995), 233.

25. Kowaleski, *Local Markets*, 273–4; Eleanora M. Carus-Wilson, *The Expansion of Exeter at the Close of the Middle Ages*, Harte Memorial Lecture (Exeter, 1963), 26–8; Longleat House MS 10665.

26. Howard L. Gray, 'The production and exportation of English woollens in the fourteenth century', *English Historical Review*, 39 (1924), 34. For inaccuracies in Gray's figures for Devon (which are too low) see Kowaleski, *Local Markets*, 23.

27. For a craftsman's wage see Dyer, *Standards of Living*, 226. The average price of a small Devonshire cloth at the end of the fourteenth century was about 4s 6d, while the cost of fulling commonly came to about one-eighth of the price of the finished article: Richard P. Chope, 'The aulnager in Devon', *Transactions of the Devonshire Association*, 44 (1912), 577; Richard H. Britnell, *Growth and Decline in Colchester, 1300–1525* (Cambridge, 1986), 58–62. With a charge of, say, 6d for the fulling of one cloth, each fulling mill would have to process 180 cloths in order to raise £4 plus a rent of 10s; 80 mills would have to process 14,400 Devonshire cloths, the equivalent of 3,600 cloths of assize.

28. Kowaleski, *Local Markets*, 21; *Calendar of Patent Rolls 1313–1317* (London, 1898), 344; Louis F. Salzman, *English Industries of the Middle Ages* (Oxford, 1923), 197.

29. Anthony R. Bridbury, *Medieval English Clothmaking: An Economic Survey* (London, 1982), 59. For fulling mills in Devon and Cornwall before 1300 see Chope, 'The aulnager', 568–9; Eleanora M. Carus-Wilson, 'An industrial revolution of the thirteenth century', *Economic History Review*, 11 (1941), 48–9; Charles Henderson, 'Cornish tucking mills and windmills', in his *Essays in Cornish History* (Oxford, 1935), 204–10. The Tiverton reference is in Kenneth Ugawa, 'The economic development of some Devon manors in the thirteenth century', *Transactions of the Devonshire Association*, 94 (1962), 657.

30. Harold S.A. Fox, 'Farming practice and techniques: Devon and Cornwall', in Edward Miller (ed.), *The Agrarian History of England and Wales, Vol. IV, 1348–1500* (Cambridge, 1991), 320.

31. *Rotuli Parliamentorum*, V (London, 1783), 621; Eileen Power, 'The wool trade in the fifteenth century', in Eileen Power and Michael M. Postan (eds), *Studies in English Trade in the Fifteenth Century* (London, 1933), 367; J.P. Bischoff, '"I cannot do't without counters": fleece weights and sheep breeds in late thirteenth- and

early fourteenth-century England', *Agricultural History*, 57 (1983), 143–60.

32. These maps do not purpose to show *all* medieval mills; the complete picture will never be known because of the patchy nature of the sources. But there is no reason to suppose that the maps give an unreliable impression of regional variation in the distribution of mills.

33. PRO S.C.6/823/28; CRO, AR2/446, 450; Finberg, *Tavistock Abbey*, 154, and see also *Calendar of Inquisitions Post Mortem, Henry VII*, I (London, 1898), 224.

34. All of the east Devon vales had medieval mills. The Otter may be taken as an example. At Ottery St Mary, according to an account of 1452–3, there were no less than five fulling mills; also racks (used for stretching the cloth), a 'washing place' and a 'place for putting fleeces': British Library (BL) Add. roll 13975. Other mills in the vale of the Otter were at Harpford, Feniton, Otterton and Honiton: PRO C.133/89/3; BL Add. MS 28,838; PRO S.C.2/167/46; PRO C.132/3/10.

35. Typical was the Dartmoor-border manor of Shaugh Prior where farm holdings were on the small side (15 to 20 acres), where a fulling mill is recorded by 1347 and a 'teysell mill' by 1578: BL Add. MS 21,605; West Devon Record Office 70/82.

36. PRO C.132/31; E.179/95/12; E.101/338/11; C.139/94; C.139/123; *Calendar of Inquisitions Miscellaneous, 1392–99* (London, 1963), 125; T.L. Stoate (ed.), *Devon Lay Subsidy Rolls, 1524–7* (Bristol, 1979), 63–4; Frederic M. Eden, *The State of the Poor* (London, 1928), 174.

37. PRO E.179/95/15.

38. The very large literature, and its criticisms, is summed up in Leslie A. Clarkson, *Proto-industrialization: The First Phase of Industrialization?* (London, 1985) and in a special issue of *Continuity and Change*, 8, 2 (1993).

39. Hatcher, *English Tin*, 141; Kowaleski, *Local Markets*, 233.

40. Finberg, *Tavistock Abbey*, 154, 189; Bridget Cherry and Nikolaus Pevsner (eds), *The Buildings of England: Devon* (London, 1989), 781.

41. Ian Blanchard, 'Industrial employment and the rural land market 1380–1520', in Richard M. Smith (ed.), *Land, Kinship and Life-cycle* (Cambridge, 1984), 231–2; G.F.R. Spenceley, 'The origins of the English pillow lace industry', *Agricultural History Review*, 21 (1973), 90; Thomas Westcote, *A View of Devonshire in MDCXXX, with a Pedigree of Most of its Gentry* (Exeter, 1845), 253, where it is stated of Combe Martin that the 'greatest trade and profit is the making of shoemakers' thread, … furnishing therewith the most part of the shire'; Combe Martin Local History Group, *Out of the World and into Combe Martin* (Combe Martin, 1989), 33–7 for the fitful history of the silver mines, and 62–7 for hemp.

42. The classic exposition of this model of the origins of rural industry is Joan Thirsk, 'Industries in the countryside', in F. Jack Fisher (ed.), *Essays in the Economic History of Tudor and Stuart England in Honour of R. H. Tawney* (Cambridge, 1961), 70–88. For the pastoral economy of east Devon see Fox, 'Farming practice', 318–19.

43. Bridbury, *Medieval English Clothmaking*, 29.

44. Case study of Sidbury contained in Harold S.A. Fox, 'Servants, cottagers and tied cottages during the later Middle Ages', *Rural History: Economy, Society, Culture*, 6 (1995), 134–6, 145–7.

45. For these trends see Fox, 'Farming practice', 317–20, and 'Servants, cottagers and tied cottages', 128–36.

46. Blanchard, 'Industrial employment', 229 n.8; Bridbury, *Medieval English Clothmaking*, 48. But see Kowaleski, *Local Markets*, 23, for fraud even at this time.

47. The map should be read with some caution. It does not show places missed by the aulnager, e.g. Ottery St Mary and Chagford, for which see Kowaleski, *Local Markets*, 23.

48. Great Torrington and Pilton: *Itinerary of John Leland*, parts 1–3, 172, 300. South Molton: PRO C.134/99 and Kowaleski, *Local Markets*, 25 n.68, using the evidence of surnames in court rolls and lay subsidies. Barnstaple: John R. Chanter and Thomas Wainwright, *Reprint of the Barnstaple Records* (Barnstaple, 1900), I, 110, 115, and J. F. Chanter, 'The Barnstaple goldsmiths' guild, with some notes on the early history of the town', *Transactions of the Devonshire Association*, 49 (1917), 166–75. Chulmleigh: PRO C.133/62/7 and S.C. 6/1118/6 (fulling mill); *Calendar of Patent Rolls, 1361–64* (London, 1912), 480 (a cloth merchant there).

49. Blake, 'Hooker's Synopsis', 346.

50. PRO E.101/338/11; Dean and Chapter Archives, 4927–45, 5108–10; information on will from M. Kowaleski.

51. *Calendar of Inquisitions Post Mortem*, XI (London, 1935), 302.

52. For Lane see Eleanora Carus-Wilson, 'The significance of the secular sculptures in the Lane Chapel, Cullompton', *Medieval Archaeology*, 1 (1957), 113–17. For similar conclusions for clothing industries elsewhere in England see Eleanora M. Carus-Wilson, 'The woollen industry before 1550', *Victoria County Histories: Wiltshire*, IV (London, 1959), 127–8, 134–7; Larry R. Poos, *A Rural Society after the Black Death: Essex 1350–1525* (Cambridge, 1991), 68–9.

53. Dean and Chapter Archives, 4835; CRO, AR2/494.

54. PRO C.138/52; BL, Add. MS 40,730.

Chapter 41

1. J. Hatcher, *English Tin Production and Trade Before 1550* (Oxford, 1973).

2. R.D. Penhallurick, *Tin in Antiquity* (London, 1986); G.R. Lewis, *The Stannaries: A Study of the Medieval Tin Miners of Cornwall and Devon* (Truro, 1908; new edn 1965); Hatcher, *English Tin Production*; J. Whetter, *Cornwall in the Seventeenth Century: An Economic History of Kernow* (Padstow, 1974).

3. Royal Institution of Cornwall, Charles Henderson's Calenders.

4. Cornwall Record Office, DD.EN 1031.

5. G.A.M. Gerrard, 'The early Cornish tin industry: an archaeological and historical survey' (unpublished Ph.D. thesis, University of Wales, 1986).

6. S. Gerrard, 'Streamworking in medieval Cornwall', *Journal of the Trevithick Society*, 14 (1987), 7–29.

7. Richard Carew of Antony, *The Survey of Cornwall*, ed. F.E. Halliday (New York, 1969); G. Agricola, *De Re Metallica*, ed. H.C. and L.H. Hoover (New York, 1950); Anon, 'An Accompt of some Mineral Observations touching the Mines of Cornwall and Devon; etc', *Philosophical Transactions of the Royal Society of London*, 5 (1670), 2096–113.

8. T.A.P. Greeves, 'The Devon tin industry 1450–1750: an archaeological and historical survey' (unpublished Ph.D. thesis, University of Exeter, 1981); S. Gerrard, 'Retallack: a late medieval tin milling complex in the parish of Constantine, and its Cornish context', *Cornish Archaeology*, 24 (1985), 175–82.

9. Greeves, 'The Devon tin industry'.

10. Ibid., 194.

11. British Library, Harleian 6380; D.B. Barton, *Essays in Cornish Mining History*, II (Truro, 1971).

Chapter 42

1. Joyce Youings, *Tuckers Hall Exeter: The History of a Provincial City Company through Five Centuries* (Exeter, 1968), 1–5.

2. A.H. Shorter, W.L.D. Ravenhill and K.J. Gregory, *South-West England* (London, 1969), 160.

3. P.J. Bowden, *The Wool Trade in Tudor and Stuart England* (London, 1962), 28 and 29.

4. Youings, *Tuckers Hall*, 2.

5. Ibid., 2.

6. Eric Kerridge, *Textiles Manufactures in Early Modern England* (Manchester, 1985), 18.

7. Bowden, *Wool Trade*, 33.

8. Youings, *Tuckers Hall*, 3; W.G. Hoskins, *Industry, Trade and People in Exeter, 1688–1800* (Manchester, 1935), 36–41.

9. Hoskins, *Industry, Trade and People*, 41–4.

10. Youings, *Tuckers Hall*, 86.

11. James Whetter, *Cornwall in the Seventeenth Century: An Economic History of Kernow* (Padstow, 1974), 115 for exports from Cornish ports.

12. Youings, *Tuckers Hall*, 87.

13. M.A. Havinden, J. Quéniart and J. Stanyer (eds), *Centre and Periphery: Brittany and Cornwall and Devon Compared* (Exeter, 1991), 78–9.

14. Hoskins, *Industry, Trade and People*, 86 and 156.

15. W.G. Hoskins, *Devon* (London, 1954), 141; John Yallop, *The History of The Honiton Lace Industry* (Exeter, 1992).

16. Shorter, Ravenhill and Gregory, *South-West England*, 137, and M.A. Havinden, 'Lime as a means of agricultural improvement', in C.W. Chalklin and M.A. Havinden (eds), *Rural Change and Urban Growth: Essays in English Regional History in Honour of W.G. Hoskins* (London, 1974), 108.

17. Havinden, 'Lime', 111–13.

18. Ibid., 106–15.

19. Sir Anthony (or his elder brother, John) Fitzherbert, *The Boke of Surveying and Improvements* (1523), fo. 42 (dorse).

20. Cited by Eric Kerridge in *The Agricultural Revolution* (London, 1967), 249.

21. F.W. Emery, 'West Glamorgan farming, c.1580–1620', *National Library of Wales Journal*, 9 (1955), 399–400.

22. Tristram Risdon, *The Chorographical Description or Survey of the County of Devon* (1811, but written in 1630), 11; and R.G.F. Stanes (ed.), 'A georgical account of Devonshire and Cornwalle, Samuel Colepresse (1667)', *Transactions of the Devonshire Association*, 96 (1964), 279–81.

23. Charles Hadfield, *The Canals of South West England* (Newton Abbot, 1967), 142–74.

24. Havinden, 'Lime', 118.

25. William Marshall, *The Rural Economy of the West of England*, 2 vols (London, 1796), I, 56–7.

26. Havinden, 'Lime', 110.

27. Shorter, Ravenhill and Gregory, *South-West England*, 159.

28. W.G. Hoskins, *Two Thousand Years in Exeter* (Exeter, 1960), 53.

29. Leslie A. Clarkson, 'The leather crafts in Tudor and Stuart England', *Agricultural History Review*, 14, 1 (1966), 25–39.

30. Leslie A. Clarkson, 'The organisation of the English leather industry in the late sixteenth and seventeenth centuries', *Economic History Review*, 2nd series, 13, 2 (1960), 245–56.

31. Whetter, *Cornwall in the Seventeenth Century*, 123.

32. Ibid., 124.

33. Daniel Defoe, *A Tour through Great Britain* (1724), cited in R. Pearse Chope (ed.), *Early Tours in Devon and Cornwall* (Newton Abbot, 1967), 157. Although not published until 1724, much of the work was written earlier. For Defoe's tour in Cornwall in 1705–6, see James Sutherland, *Defoe* (London, 1937), 152–3.

34. D.M. Trethowan, 'The leather trade of Tavistock', *Devon and Cornwall Notes and Queries*, 30 (1965–7), 220–1.

35. Giles V. Harrison, 'The South West', in Joan Thirsk (ed.), *The Agrarian History of*

England and Wales, Vol. V, Part 1, 1640–1750: Regional Farming Systems (Cambridge, 1984), 387.

36. Robert Newton, *Eighteenth Century Exeter* (Exeter, 1984), 4.

37. Shorter, Ravenhill and Gregory, *South-West England*, 159.

38. Alfred H. Shorter, *Paper Mills and Paper Makers in England, 1495–1800* (Hilversum, Holland, 1957) for the maps and his *Paper Making in the British Isles* (Newton Abbot, 1971), 13 and 98.

39. Shorter, *Paper Making in the British Isles*, 13–15.

40. D.C. Coleman, *The British Paper Industry, 1495–1860* (Oxford, 1958), 4, 23 and Figure 2 opposite 90.

41. Shorter, *Paper Mills*, 156.

42. Ibid., 152–9.

43. Shorter, Ravenhill and Gregory, *South-West England*, 162–4.

44. Coleman, *The British Paper Industry*, 219; Shorter, *Paper Mills*, 149–59.

Chapter 43

1. See W. Borlase, *The Natural History of Cornwall* (1758) and T.A. Morrison, *Cornwall's Central Mines: The Southern District 1810–1895* (Penzance, 1983), 3. See also D.B. Barton, *A Guide to the Mines of West Cornwall* (Truro, 1963), and *A Historical Survey of The Mines and Mineral Railways of East Cornwall and West Devon* (Truro, 1971); A.K. Hamilton Jenkin, *Mines and Miners of Cornwall* (Truro, 1962); B. Atkinson, *Mining Sites in Cornwall and South West Devon*, 2 vols (Redruth, 1988 and 1994).

2. D. and S. Lysons, *Magna Britannia, Vol. VI, Devonshire* (1822), cclxxxi–cclxxxix.

3. Ibid., cclxxi.

4. Ibid., cclxxxix; R. Burt and I. Wilkie, 'Manganese mining in South-west England', *Journal of the Trevithick Society*, 11 (1984), 18–40.

5. M. Atkinson, R. Burt and P. Waite, *Dartmoor Mines: The Mines of the Granite Mass* (Exeter, 1978); T.A.P. Greaves, *Tin Mines and Tin Miners of Dartmoor* (Exeter, 1986); A.K. Hamilton Jenkins, *Mines of Devon I: The Southern Area* (Newton Abbot, 1974), *II: North and East of Dartmoor* (Exeter, 1981); P.H.G. Richardson, *Mines of Dartmoor and Tamar Valley* (Sheffield, 1992).

6. B.R. Mitchell and P. Deane, *An Abstract of British Historical Statistics* (Cambridge, 1962), 155, 159.

7. C.J. Schmitz, *World Non-Ferrous Metal Production and Prices 1700–1976* (London, 1979), 269–70.

8. D.B. Barton, *Copper Mining in Cornwall and Devon* (Truro, 1978), 45–67.

9. R. Burt, P. Waite and R. Burnley, *Devon and Somerset Mines* (Exeter, 1984), xii and xxiv.

10. J.M. Slader, *Days of Renown* (Bracknell, 1965) and T.C. Cantrill, R.L. Sherlock and H. Dewey, *Iron Ores: Sundry Unbedded Ores of Durham, East Cumberland, North Wales, Derbyshire, the Isle of Man, Bristol District and Somerset, Devon and Cornwall,* Special Reports on the Mineral Resources of Great Britain, IX, Memoirs of the Geological Survey (London, 1919).

11. See R. Burt, P. Waite and R. Burnley, *Cornish Mines* (Exeter, 1987) and *Devon and Somerset Mines.*

12. See C.J. Schmitz 'The early growth of the Devon barytes industry, 1835–1875' and 'The development and decline of the Devon barytes industry, 1875–1958', *Transactions of the Devonshire Association,* 106 (1974), 59–76, and 109 (1977), 117–33.

CHAPTER 44

1. Cornwall County Council, *Development Plan 1952: Report of Survey Part One—The County* (Truro, 1952), 77.

2. Exeter University College of the South West, *Devon and Cornwall: A Preliminary Survey* (Exeter, 1947).

3. Devon County Council, *The County Development Plan (The County Area): Second Review: Analysis of the Survey* (Exeter, 1970).

4. Cornwall County Council, *Structure Plan* (Truro, 1979).

5. A.W. Gilg, 'Rural employment', in G.E. Cherry (ed.), *Rural Planning Problems* (Glasgow, 1976), 110–25.

CHAPTER 45

1. Cyril Noall, *A History of the Cornish Mail- and Stage-Coaches* (Truro, 1963), 14–16; W.T. Jackman, *The Development of Transportation in Modern England* (London, 1962), 304–5, 686; M.C. Lowe, 'The turnpike trusts in Devon and their roads: 1753–1889', *Transactions of the Devonshire Association,* 122 (1990), 47–69; William Buckingham, *A Turnpike Key* (Exeter, 1885), 8–9.

2. Jackman, *Development of Transportation,* 134–8, 683–6; Buckingham, *Turnpike Key,* 38.

3. Quoted in Noall, *History,* 18.

4. Noall, *History,* 16, 22–3; Philip Bagwell, *The Transport Revolution* (London, 1974), 42; Buckingham, *Turnpike Key,* 38; Jackman, *Development of Transportation,* 136.

5. Daphne du Maurier, *Frenchman's Creek* (London, 1941), 12; Noall, *History,* 23.

6. Celia Fiennes, *Through England on a Side Saddle* (1698), quoted by Lowe, 'Turnpike trusts', 66.

7. Thomas Martyn, *A New Improved Map of Cornwall* (c.1748); Thomas Kitchen *A New Improved Map of Cornwall* (1750) (copied on one sheet from Martyn); Benjamin Donn, *A Map of the County of Devon 1765* (Exeter, 1965).

8. William Albert, *The Turnpike Road System in England 1663–1840* (Cambridge, 1974), chapters 2 and 3; John Kanefsky, *Devon Tollhouses* (Exeter, 1976), 7–8; Lowe, 'Turnpike trusts', 47–8.

9. Buckingham, *Turnpike Key,* 11–12; Kanefsky, *Devon Tollhouses,* 33; Lowe, 'Turnpike trusts', 48–9.

10. Noall, *History,* 18–20, 91.

11. Ibid., 18; Lowe, 'Turnpike trusts', 50; *British Parliamentary Papers* (*BPP*), 1840, XVII, Appendix to the Report of the Commissioners on Roads etc in England and Wales, 51–8, 87–101, 106, 307–9. Ordnance Survey, 1st Edition (1809–20); Greenwood's maps of Devon (1826) and Cornwall (1827).

12. Kanefsky, *Devon Tollhouses,* 13; Lowe, 'Turnpike trusts', 51–2; Jackman, *Development of Transportation,* 138–40, 705–7.

13. Lowe, 'Turnpike trusts', 54–7; Kanefsky, *Devon Tollhouses,* 8.

14. Buckingham, *Turnpike Key,* 22–4; Lowe, 'Turnpike trusts', 58–9.

15. Bagwell, *Transport Revolution,* 48–9.

16. Noall, *History,* 18, 91; Lowe, 'Turnpike trusts', 54–8; *BPP,* 1840, XVII, Appendix to the Report of the Commissioners on Roads etc in England and Wales, 51–8, 87–101, 106, 307–9; Annual returns of Turnpike Trust income etc. in *BPP.*

17. Kanefsky, *Devon Tollhouses,* 8 and map; Lowe, 'Turnpike trusts', 54–8; Noall, *History,* 27, 38–40.

18. E.A. MacDermot and C.R. Clinker, *History of the Great Western Railway* (London, 1964), Appendix I; Noall, *History,* 81–5.

19. John Kanefsky, 'Railway competition and turnpike roads in east Devon, 1843–48', *Transactions of the Devonshire Association,* 109 (1977), 59–72; Lowe, 'Turnpike trusts', 63–4; Annual Returns of Turnpike Trust income etc. in *BPP.*

20. Noall, *History,* 82–5; Annual Returns of Turnpike Trust income etc. in *BPP.*

21. Buckingham, *Turnpike Key,* 36–7; Kanefsky, *Devon Tollhouses,* 15–16; Lowe, 'Turnpike trusts', 65–6; Annual Returns of Turnpike Trust income etc. in *BPP.*

CHAPTER 46

1. Jack Simmons, 'The railway in Cornwall, 1835–1914', *Journal of the Royal Institution of Cornwall,* new series, 9 (1982), 11–29.

2. Most of the line to Lyme Regis ran through the county of Devon.

3. The tramways serving Dartmoor were built to 4ft 6in (1.372m) gauge.

CHAPTER 47

1. Frank Booker, *The Great Western Railway* (Newton Abbot, 1977), 103.

2. Hamilton Ellis, *British Railway History* (London, 1959), 264.

3. Ibid., 311–13.

4. Ibid., 231; Booker, *The Great Western Railway,* 108.

5. Booker, *The Great Western Railway,* 150.

6. Ellis, *British Railway History,* 335.

7. David Norman Smith, *The Railway and its Passengers: A Social History* (Newton Abbot, 1988), 69, 168.

8. For a discussion of the railway companies between the wars see Michael Bonavia, *Railway Policy between the Wars* (Manchester, 1982) and Booker, *The Great Western Railway,* 146–7, 162.

9. Smith, *Railway and its Passengers,* 69, 168.

10. Department of Transport, *Transport Statistics Great Britain 1965–1975* (London, 1977), Table 66.

11. Patrick Whitehouse and David St John Thomas, *The Great Western Railway* (Newton Abbot, 1984), 112; British Railways Board, *The Reshaping of British Railways,* 2 vols (London, 1963), 15.

12. Whitehouse and Thomas, *Great Western,* 174.

13. Transport Users Consultative Committee for the South-Western Area, *Annual Reports* (Bristol, 1978–83).

14. Alan Powell, *Feniton and the Railway* (Feniton, 1993).

15. Cornwall County Council, *Development Plan 1952: Report of Survey* (Truro, 1952); Michael Hawkins, *Roads in Devon* (Exeter, 1987), 35.

16. *Transport Statistics Great Britain 1969–1979* (1980), 72; Devon County Council, *Devon to 2011 County Structure Plan First Review: The Issues* (Exeter, 1993), 48.

17. Cornwall County Council, *Development Plan 1952: Report of Survey,* 37; Cornwall County Council, *Traffic in Cornwall 1991* (Truro, 1992), 5; Devon County Council, *Devon in Figures 1979* (Exeter, 1980).

18. Hawkins, *Roads in Devon,* 43, 46.

19. *Transport Statistics Great Britain 1969–1979,* 9.

20. Cornwall County Council, *Traffic in Cornwall 1991,* 18; *Transport Statistics Great Britain 1969–1979,* 28.

21. *Transport Policy: A Consultation Document,* 2 vols (London, 1976), I, 31–2.

22. Illustrated in V.R. Belsey, *British Roads: Cornwall Past and Present* (Peterborough, 1994), 16.

23. North Devon District Council, *North Devon District Local Plan: Transport* (Barnstaple, 1993), para 2.4.

24. Devon County Council, *Towards a Sustainable Transport Policy for Devon* (Exeter, 1993), 10.

25. *Transport Statistics Great Britain 1969–1979,* 34.

26. Cornwall County Council, *Development Plan: Report of the Survey,* 39, Fig. 11.

27. Cornwall County Council, *Traffic in Cornwall 1991,* 3.

28. Cornwall County Council, *Structure Plan: Report of the Survey,* Table 19.

29. Cornwall County Council, *Traffic in Cornwall 1991,* 16.

30. Devon County Council, *Devon County Structure Plan Monitoring Report 1988,* 3.

31. Devon County Council, *Devon in 2001* (Exeter, 1988), 17; Cornwall County Council, *Traffic in Cornwall 1991,* 5.

32. Cornwall County Council, *Traffic in Cornwall 1991,* 16.

33. Devon County Council, *Devon in Figures 1985* (Exeter, 1986), 83.

34. Devon County Council, *Devon in Figures 1979,* 57.

35. Cornwall County Council, *Traffic in Cornwall 1991,* 13, and *Structure Plan: Report of the Survey,* Table 40.

36. *Keeping Plymouth Mobile* (Exeter, 1992), Appendix A, para 3.2.2.

37. Devon County Council, *Towards a Sustainable Transport Policy for Devon,* 5; repeated in *Devon to 2011 County Structure Plan: First Review: The Issues,* 48.

CHAPTER 48

1. Richard Carew, *The Survey of Cornwall,* ed. Lord de Dunstanville (1811), 104, 97.

2. Todd Gray (ed.), *Harvest Failure in Cornwall and Devon: The Book of Orders and the Corn Surveys of 1623 and 1630–1* (Camborne, 1992), xxxiii.

3. Todd Gray (ed.), *Devon Household Accounts, 1627–59,* Devon and Cornwall Record Society, 38 (1995–6), I–II.

4. William J. Blake, 'Hooker's Synopsis Chorographical of Devonshire', *Transactions of the Devonshire Association,* 47 (1915), 348.

5. Cyril Noall, *Cornish Seines and Seiners: A History of the Pilchard Fishing Industry* (Truro, 1972), 24.

6. Plymouth and West Devon Record Office, 975, 978, 1010, 1020, 1082 and 1836, photocopies of Corsini papers.

7. Carew, *Survey of Cornwall.*

8. John Norden, *A Topographical and Historical Description of Cornwall* (Newcastle upon Tyne, 1966), 19.

9. Noall, *Cornish Seines and Seiners,* 37–59.

10. Thomas Westcote, *A View of Devonshire in 1630* (London, 1811), 67–8.

11. Devon Record Office (DRO), ECA Book 57, 158–71.

12. James Whetter, *Cornwall in the Seventeenth Century: An Economic History of Kernow* (Padstow, 1974), 87.

13. Tristram Risdon, *Chorographical Description or Survey of the County of Devon* (1811), 351–2.

14. Norden, *Description of Cornwall,* 51.

15. Todd Gray, 'Devon fisheries and early Stuart northern New England', 140–2, in Michael Duffy *et al.* (eds), *The New Maritime History of Devon, Vol. I* (London, 1992).

16. R.H. Tawney and Eileen Power, *Tudor Economic Documents* (London, 1951), III, 245–6; British Library, Harleian 6238, fo. 211.

17. Wendy Childs, 'Diverse lands far beyond the seas: the overseas trade of the South West in the Later Middle Ages', *Mariner's Mirror* (forthcoming).

18. Joyce Youings with Peter W. Cornford, 'Seafaring and maritime trade in sixteenth-century Devon', in Duffy *et al.* (eds), *The New Maritime History of Devon, Vol. I*, 102.

19. Public Record Office (PRO), SP14/157/24.

20. Michael MacCarthy-Morrogh, *The Munster Plantation: English Migration to Southern Ireland* (Oxford, 1986), 155–8, 223–4.

21. DRO, Chanter 854b, fo. 288, and Chanter 855, fo. 494.

22. Gillian Cell, *English Enterprise in Newfoundland* (Toronto, 1969); DRO, Chanter 855b, fos 13–14 and, Chanter 855a, fos 478–9.

23. J. Scantlebury, 'John Rashleigh and the Newfoundland cod fishery, 1608–20', *Journal of the Royal Institution of Cornwall*, 8, 1 (1978), 61–71.

24. David J. Starkey, 'The West Country–Newfoundland fishery and the manning of the Royal Navy', in Robert Higham (ed.), *Security and Defence in South West England before 1800* (Exeter, 1987), 93–101.

25. James Phinney Baxter, *The Trelawny Papers* (Portland ME, 1884).

26. Plymouth and West Devon Record Office, parish records of Newton Ferrers.

27. Richard Carew, *The Survey of Cornwall* (1602), 27.

28. Ibid., 136.

29. PRO, SP16/282/120.

30. PRO, 16/101/50, 16/100/40, 16/100/47, 16/60/15.

31. PRO, SP16/282/135.

Chapter 49

1. This and the following paragraph are mainly drawn from Stephen Fisher, 'Devon's maritime trade and shipping, 1680–1780', in Michael Duffy et al. (eds), *The New Maritime History of Devon, Vol. I* (London, 1992), 232.

2. British Library, Add. MS 9293.

3. On this section see David J. Starkey, 'Devonians and the Newfoundland trade', in Duffy et al. (eds), *The New Maritime History of Devon, Vol. I*, 163–71.

4. Ibid., 168.

5. On these naval affairs see Michael Duffy, 'Devon and the naval strategy of the French wars, 1689–1815', in Duffy et al. (eds), *The New Maritime History of Devon, Vol. I*, 182–91.

6. Ibid., 187.

7. Ibid., 185.

8. For a discussion of this subject, see N.A.M. Rodger, 'Devon men and the Navy, 1689–1815', in Duffy et al. (eds), *The New Maritime History of Devon, Vol. I*, 209–15.

9. David J. Starkey, *British Privateering Enterprise in the Eighteenth Century* (Exeter, 1990).

Chapter 50

1. H.J. Dyos and D.H. Aldcroft, *British Transport: An Economic Survey from the Seventeenth Century to the Twentieth* (Leicester, 1969), 234.

2. Sidney Pollard and Paul Robertson, *The British Shipbuilding Industry, 1870–1914* (Cambridge MA, 1979), 45.

3. G. Jackson, 'The ports' and 'The shipping industry', in Michael J. Freeman and Derek H. Aldcroft (eds), *Transport in the Victorian Age* (Manchester, 1988), 218–83.

4. Pollard and Robertson, *British Shipbuilding*, 49–69; Simon Ville (ed.), *Shipbuilding in the United Kingdom in the Nineteenth Century: A Regional Approach* (St John's, 1993).

5. Rupert C. Jarvis, 'Sources for the history of ports', *Journal of Transport History*, 1st series, 3 (1957–8), 76–93.

6. Grahame Farr, *The Ship Registers of the Port of Hayle, 1864–1882* (London, 1975), vi–vii; Grahame Farr, *Shipbuilding in North Devon* (London, 1976), 4–5.

7. Public Record Office, CUS 17/11.

8. *British Parliamentary Papers* (*BPP*), 1843, LII, Accounts Relating to Shipping, 382; *Annual Statements of Navigation and Shipping*, 1881, 1910.

9. *Annual Statement of Navigation and Shipping*, 1895.

10. Jackson, 'Ports', 238–9; Sarah Palmer, 'The British coal export trade, 1850–1913', in David Alexander and Rosemary Ommer (eds), *Volumes not Values: Canadian Sailing Ships and World Trades* (St John's, 1979), 331–54.

11. P.S. Bagwell and J.A. Armstrong, 'Coastal shipping', in Freeman and Aldcroft (eds), *Transport in the Victorian Age*, 183–90.

12. *BPP*, 1824, XVIII, An Account of the Number of Vessels, 319–63; *BPP*, 1830, XXVII, An Account of the Several Counties in England and Wales into which Coals have been Brought Coastwise in 1829, 131; Devon Record Office, 3258A Add/HA/A1, Teignmouth Harbour Commission, Import Book, 1853–63; for example see *BPP*, 1911, CI, Coal received Coastwise in 1910, 758.

13. *BPP*, 1882, LXII, A Return of the Names of the Port and Harbour Authorities in the United Kingdom in August 1881, 409.

14. E. Delderfield, *The Exmouth Dock Company: One Hundred Years of Progress, 1865–1965* (Exmouth, 1965).

15. See C.H. Ward-Jackson, *Ships and Shipbuilders of a Westcountry Seaport: Fowey, 1786–1939* (Truro, 1986) and H.J. Trump, *Teignmouth: A Maritime History* (Chichester, 1986).

16. Ward-Jackson, *Westcountry Seaport*, 25–7; Edmund Vale, *The Harveys of Hayle: Engine*

Builders, Shipwrights and Merchants of Cornwall (Truro, 1966); Crispin Gill, *Sutton Harbour* (Plymouth, n.d.); Ray Freeman, *Dartmouth: A New History of the Port and its People* (Dartmouth, 1973), 108–117.

17. Gill, *Sutton Harbour*; Martin Langley and Edwina Small, *Millbay Docks* (Exeter, 1987); C.B.M. Sillick, 'The city-port of Plymouth: an essay in geographical interpretation' (unpublished Ph.D. thesis, University of London, 1938); David J. Starkey, 'The ports, sea-borne trade and shipping industry of south Devon, 1786–1914' and Crispin Gill, 'Ocean liners at Plymouth', in Michael Duffy et al. (eds), *The New Maritime History of Devon, Vol. II* (London, 1994), 32–47, 226–35.

18. See E.A.G. Clark, *The Ports of the Exe Estuary, 1660–1860: A Study in Historical Geography* (Exeter, 1968), 12–14.

19. For instance, Ward-Jackson, *Westcountry Seaport*, 63–4; Starkey, 'Ports, sea-borne trade'.

20. Jackson, 'Shipping industry'.

21. P.M. Heaton, *Reardon Smith Line: The History of a South Wales Shipping Venture* (Newport, 1984); Plymouth Customs House, Plymouth Custom House Ship Registers, 1890–1914. I am grateful to Robin Craig for this information.

22. K.J. O'Donoghue and H.S. Appleyard, *Hain of St Ives* (Kendal, 1986); Robin Craig, 'Steamship enterprise in Devon, 1852–1920', in Duffy et al. (eds), *The New Maritime History of Devon, Vol. II*, 98, n.36.

23. Craig, 'Steamship enterprise', 91–8.

24. O'Donoghue and Appleyard, *Hain*, 8–13.

25. *BPP*, 1805, VIII, An Account of the Number of Shipwrights … Employed in the Merchant Yards, 467–86. See David J. Starkey, 'Shipbuilding in the South West during the Napoleonic War', *Maritime South West*, 6 (1993), 5–15.

26. Basil Greenhill, *The Evolution of the Wooden Ship* (London, 1988), 87–8.

27. David J. Starkey, 'Devon's shipbuilding industry, 1786–1970', in Duffy et al. (eds), *The New Maritime History of Devon, Vol. II*, 78–90; Ward-Jackson, *Westcountry Seaport*, 105–6.

28. *BPP*, 1813–14, VIII, An Account of His Majesty's Ships Launched from Private Yards …, 498. I am grateful to George Hogg for information provided on Cornish shipbuilding.

29. Farr, *Shipbuilding*, 6–18; Starkey, 'Devon's shipbuilding industry'.

30. See the description of Holman's Yard in the *Western Times*, 21 December 1865, cited in Clive N. Ponsford, *Topsham and the Exe Estuary* (Exeter, 1979), 35–6; and that of David Banks's Yard at Plymouth, *Exeter Flying Post*, 26 April 1865.

31. *Port Statistics 1990* (1991), 171–5.

32. *BPP*, 1882, LXII, A Return of the Names of the Port and Harbour Authorities in the United Kingdom in August 1881, 409; *Port Statistics 1990*.

33. D.H. Aldcroft, 'The eclipse of British coastal shipping 1913–1921', *Journal of Transport History*, 1st series, 6 (1963), 24–38.

34. P. Ford and J.A. Bound, *Coastwise Shipping and the Small Ports* (Oxford, 1951).

35. Mark Porter, 'Devon's port industry since 1914', in Duffy et al. (eds), *The New Maritime History of Devon, Vol. II*, 235–42.

36. Gill, 'Ocean liners at Plymouth', 226–34.

37. *Annual Statement of Navigation and Shipping 1911*.

38. Robb Robinson, *Trawling: the Rise and Fall of the British Trawl Fishery* (Exeter, 1996); Porter, 'Devon's fishing industry'.

39. *Port Statistics 1990*, 134–5.

40. Information kindly supplied by Janet Cusack.

41. Information kindly supplied by Janet Cusack.

42. David J Starkey, 'The shipbuilding industry of southwest England, 1790–1913', in Ville (ed.), *Shipbuilding in the United Kingdom*, 75–110.

43. Crispin Gill, 'Yachts, marinas, and the maritime leisure industry', in Duffy et al. (eds), *The New Maritime History of Devon, Vol. II*, 262–4.

Chapter 51

1. Herbert P.R. Finberg, 'The genesis of the Gloucestershire towns' in his edited collection *Gloucestershire Studies* (Leicester, 1957), 64; Rodney H. Hilton, 'Medieval market towns and simple commodity production', *Past and Present*, 109 (1985), 3–23; Somerset Record Office DDWO/46/1; Public Record Office (PRO) SC6/823/38; Maryanne Kowaleski, *Local Markets and Regional Trade in Medieval Exeter* (Cambridge, 1995), 142; see note 30 for goldsmiths.

2. J.S. Brewer and Charles Trice Martin (eds), *Registrum Malmesburiense*, 2 vols (London, 1879–80), II, 34; *Calendar of Patent Rolls, 1232–1247* (London, 1906), 495; Hugh Peskett, 'The borough charter of Bovey Tracey', *Devon and Cornwall Notes and Queries*, 32 (1971–3), 176; Dean and Chapter Archives, Dawlish account rolls recording repairs to the market house at East Teignmouth; PRO S.C. 6/830/4 and other accounts, repair of market house at Sidmouth; Devon Record Office (DRO) W1258M/D/52/2 for tolsels, booths for collection of tolls, at Tavistock and Denbury.

3. Adolphus Ballard and James Tait, *British Borough Charters, 1216–1307* (Cambridge, 1923), 16, 56, 73, 96, 111, 116, 150, 190, 196, 216, 250, 289, 293, 352; Adolphus Ballard, *British Borough Charters, 1042–1216* (Cambridge, 1913), 67, 193.

4. Henry Spencer Toy, *The History of Helston* (Oxford, 1936), 438–9; Richard and Otho

B. Peter, *The Histories of Launceston and Dunheved* (Plymouth, 1885), 73.

5. Toy, *Helston*, 439–41; Ballard and Tait, *British Borough Charters*, 312.

6. Dom. John Stéphan, 'Some notes on Ashburton Grammar School', *Transactions of the Devonshire Association*, 91 (1959), 120–1; Toy, *Helston*, 484–7. For portreeves see Herbert P.R. Finberg, 'The borough of Tavistock: its origins and early history', in W.G. Hoskins and H.P.R. Finberg (eds), *Devonshire Studies* (London, 1952), 182–3.

7. For an archaic portreeve and a 'mayor' at North Tawton and Week St Mary see Daniel and Samuel Lysons, *Magna Britannia: Devonshire* (London, 1822), 481, and Ann Preston-Jones and Peter Rose, 'Week St Mary, town and castle', *Cornish Archaeology*, 31 (1992), 143.

8. Hugh R. Watkin, *Dartmouth, Vol. I, Pre-Reformation* (1935), 4–9; *The Great Roll of the Pipe, 1176–1177* (London, 1905), 10; Toy, *Helston*, 18, 439; Charles Henderson, 'Helston' in his *Essays in Cornish History* (Oxford, 1935), 67–79, especially 70. Liskeard had a market recorded in Domesday Book.

9. Maurice W. Beresford and Herbert P.R. Finberg, *English Medieval Boroughs: A Handlist* (Newton Abbot, 1973); Maurice Beresford, 'English medieval boroughs: a hand-list: revisions', *Urban History Yearbook* (1981), 62. I shall publish the list of modifications in another place.

10. Beresford and Finberg, *English Medieval Boroughs*, 49.

11. Maps in Joseph Bettey, 'From the Norman Conquest to the Reformation', in Michael Aston (ed.), *Aspects of the Medieval Landscape of Somerset* (Taunton, 1988), 54; Michael Aston, 'Post-Roman central places in Somerset', in Eric Grant (ed.), *Central Places, Archaeology and History* (Sheffield, 1986), 67; Norman Scarfe, 'Medieval and later markets', in David Dymond and Edward Martin (eds), *An Historical Atlas of Suffolk* (Ipswich, 1988), 61.

12. Maps in Herbert P.R. Finberg, *Tavistock Abbey* (Cambridge, 1951), facing 41; Christopher Holdsworth, 'From 1050 to 1307', in N. Orme (ed.), *Unity and Variety: A History of the Church in Devon and Cornwall* (Exeter, 1991), 52; Chris Given-Wilson, *The English Nobility in the Late Middle Ages* (London, 1987), xiv–xv; John Hatcher, *Rural Economy and Society in the Duchy of Cornwall, 1300–1500* (Cambridge, 1970), xiv.

13. Beresford and Finberg, *English Medieval Boroughs*, 54. See also Maurice Beresford, *New Towns of the Middle Ages* (London, 1967), 418: 'If communications were slowed down by steep and rugged terrain, the number of nightly resting places on major routes was necessarily increased'.

14. For the concept see John Hatcher, 'A diversified economy: later medieval Cornwall', *Economic History Review*, 22 (1969), 208–27.

15. *Curia Regis Rolls, 1219–20* (London, 1938), 267–80; ancient petition cited in Finberg, *Tavistock Abbey*, 178; *Calendar of Patent Rolls, 1327–30* (London, 1891), 367, 379; John Hatcher, *English Tin Production and Trade before 1550* (Oxford, 1973), 140–1 for the maritime trade in tin, albeit in the fifteenth century. Chagford belongs to the class of unchartered towns mentioned later in this chapter.

16. Details in Harold S.A. Fox, *Fishers and Fishing along the South Devon Coast: A Study in Social and Settlement History* (forthcoming). A full reconstruction of the fish trade in the later fourteenth century is in Kowaleski, *Local Markets*, 307–21.

17. Edward M. Jope and G.C. Dunning, 'The use of blue slate for roofing in medieval England', *Antiquaries Journal*, 34 (1954), 214; Louis F. Salzman, *English Industries of the Middle Ages* (Oxford, 1923), 197; Crispin Gill, *Plymouth: A New History, Ice Age to the Elizabethans* (Newton Abbot, 1979), 68.

18. Concentrations of industry in the countryside were often associated with large numbers of landless cottagers who, by definition, needed markets in order to provide food. Examples are cottages occupied by cloth-makers at Uffculme and Menheniot and slate quarriers at Charleton: PRO C.138/52; British Library (BL) Add. MS. 40, 730; PRO C.139/94.

19. John R. Maddicott, 'Trade, industry and the wealth of King Alfred', *Past and Present*, 123 (1989), 25–6; Hatcher, *English Tin*, 23–5; Richard Pearse, *The Ports and Harbours of Cornwall* (St Austell, 1963), 28–30; Austin Lane Poole, *From Domesday Book to Magna Carta* (Oxford, 1951), 96, where read 'River Fowey' for 'Fowey'.

20. W.G. Hoskins, *Two Thousand Years in Exeter* (Exeter, 1960), 28; Kowaleski, *Local Markets*, 223; *The Great Roll of the Pipe for the First Year of the Reign of King John* (London, 1933), 196; John P. Allan, *Medieval and Post-Medieval Finds from Exeter, 1971–1980* (Exeter, 1984), 17.

21. Eleanora Carus-Wilson, 'The effects of the acquisition and of the loss of Gascony on the English wine trade', in her *Medieval Merchant Venturers* (London, 1967), 269–70.

22. Hatcher, *English Tin*, 23.

23. Victualling in thoroughfare boroughs gave rise, for example, to the occupation of cook in Camelford and inn-keeper (several) in Colyford: PRO E. 142/41; BL Add. roll 13771.

24. Week St Mary: Preston-Jones and Rose, 'Week St Mary'. Camelford: survey of burgages in PRO E. 142/141. Wadebridge: Beresford, *New Towns*, 413–4, and Peter Sheppard, *The Historic Towns of Cornwall: An Archaeological Survey* (Truro, 1980), 61. St Columb Major: the urban character of the place is evident from rentals which show a concentration there of a relatively large number of tenements without much agricultural land attached to them, e.g. Cornwall Record Office (CRO) AR2/1339.

25. Richard Carew of Antony, *The Survey of Cornwall*, ed. F.E. Halliday (London, 1953), 226; Henderson and Coates, *Old Cornish Bridges*, 14–15.

26. Granite: suggestive is the presence of several people bearing the name Stonehewer in the borough of Penknight which adjoined Lostwithiel: PRO E.142/41. For gorse there are no medieval references to trade in this humble but useful commodity, but William Marshall in the eighteenth century described Tavistock as 'long … a market for furze feed': *The Rural Economy of the West of England*, 2 vols (London, 1796), II, 9.

27. Tintagel: Oliver J. Padel, 'Tintagel in the twelfth and thirteenth centuries', *Cornish Studies*, 16 (1988), 61–6, and for the borough's charter and burgesses see Ballard and Tait, *British Borough Charters*, 5, 22, 13, 247, 250, 265; Peter L. Hull (ed.), *The Caption of Seisin of the Duchy of Cornwall*, Devon and Cornwall Record Society, new series, 17 (1971), 33–4; PRO E. 142/41. Trematon: there is some uncertainty about the location of this minute borough as in Beresford, *New Towns*, 411–13, and Hull (ed.), *Caption of Seisin*, lii, but it was almost certainly adjacent to the castle of the same name. Wiscombe: DRO 123M/TB/436, 442–5.

28. Maryanne Kowaleski, 'The grain trade in fourteenth-century Exeter', in Edwin B. DeWindt (ed.), *The Salt of Common Life* (Kalamazoo, 1995), 2; Edward Miller and John Hatcher, *Medieval England: Towns, Commerce and Crafts* (London, 1995), 263, 274.

29. Helston: Sir John Maclean, 'Poll tax account for the County of Cornwall', *Journal of the Royal Institution of Cornwall*, 4 (1871–3), 40. Tavistock: DRO W1258M/D/84/2. Saltash: PRO E.152/8. Penryn: Ballard and Tait, *British Borough Charters*, 55, and Francis C. Hingeston-Randolph, *The Register of Walter de Stapledon* (London and Exeter, 1892), 27. Camelford: PRO E. 142/41. Colyford: PRO. C. 133/6.

30. Rodney H. Hilton, 'The small town as part of peasant society', in his *English Peasantry in the Later Middle Ages* (Oxford, 1975), 79. Bradninch and Tavistock occupations: PRO S.C. 11/802; DRO W1258/D/84/2; PRO E.179/95/12, 32. Goldsmiths were to be found at Exeter, Barnstaple, Totnes, Bodmin, Truro and Lostwithiel: Kowaleski, *Local Markets*, 163; J.F. Chanter, 'The Barnstaple goldsmiths' guild, with some notes on the early history of the town', *Transactions of the Devonshire Association*, 49 (1917), 173–4;

31. Bodleian Library Top. Devon d.5 f. 15v.; DRO 123M/TB/160; BL Arundel MS 17 f. 49–49v.; DRO 123M/TB/167. For the medieval village and borough see Harold S.A. Fox, 'Field systems of east and south Devon. Pt. I: east Devon', *Transactions of the Devonshire Association*, 104 (1972), 112–13. For small town medieval architecture see John R.L. Thorp, 'Two hall houses in a late medieval terrace: 8–12 Fore Street, Silverton', *Proceedings of the Devonshire Archaeological Society*, 40 (1982), 171–80.

32. By expressing the urban component (Devon only) in Table 51.1 as a percentage of the total population in *c*.1300, derived by backwards extrapolation from the poll tax figure for 1377 for the whole country. The note to Table 51.1 explains the treatment of the poll tax figures; unrealistically I have assumed no urban rural differentiation in mortality, 1300–77.

33. Christopher Dyer, 'How urbanized was medieval England?', in Jean-Marie Duvosquel and Erik Thoen (eds), *Peasants and Townsmen in Medieval Europe* (Ghent, 1995), 20. For an even lower figure see Richard H. Britnell, *The Commercialisation of English Society 1000–1500* (Cambridge, 1993), 115.

34. Devon markets meticulously listed in Kowaleski, *Local Markets*, 353–70; Cornish ones incompletely listed in John Richardson, *The Local Historian's Encyclopedia* (New Barnet, 1974), 243–51.

35. For these unchartered towns in general see Christopher Dyer, 'The hidden trade of the Middle Ages: evidence from the West Midlands of England', *Journal of Historical Geography*, 18 (1992), 141–57. Exmouth: BL Cott. Faust. A II, fos 69–72, revealing a large concentration of landless houses and cottages here in 1388; Peter J. Weddell, 'The excavation of medieval and later houses and St Margaret's chapel at Exmouth, 1982–1984', *Proceedings of the Devon Archaeological Society*, 44 (1986), 107–41. Kingswear: as early as the twelfth century there were landless properties here described as houses or messuages; the minute size of the parish means that their occupiers must have had non-agricultural occupations: Watkin, *Totnes Priory and Medieval Town*, I, 94–6, and Deryck Seymour, *Torre Abbey* (Exeter, 1977), 214. St Ives: Maclean, 'Poll tax', 39. Landulph: J.H. Adams, 'The port of Landhelp', *Devon and Cornwall Notes and Queries*, 32 (1973), 210–14. For Wadebridge and Camelford see note 24 above.

36. Very usefully summarized by Alan Dyer, *Decline and Growth in English Towns*

(Basingstoke and London, 1991).

37. Chiefly Leland, Camden, Norden, Carew, Hals, Tonkin, Hooker in MS, Westcote and Risdon; also very useful, and said to be reliably informed by local contacts, is R. Blome, *Britannia* (1673). For the approach, and maps of the South West's markets in the early modern period, see Alan Everitt, 'The marketing of agricultural produce', in Joan Thirsk (ed.), *The Agrarian History of England and Wales, Vol. IV, 1500–1640* (Cambridge, 1967), 466–592, and Alan D. Dyer, 'The market towns of southern England', *Southern History*, 1 (1979), 123–34.

38. Davies Gilbert, *The Parochial History of Cornwall* (London, 1838), 340.

39. For South Zeal see T.L. Stoate (ed.), *Devon Lay Subsidy Rolls, 1524–7* (Bristol, 1979).

40. *William Worcestre: Itineraries*, ed. John H. Harvey (Oxford, 1969), 95; PRO SC6/1138/2.

41. DRO W1258M/D84/2 and Stoate (ed.), *Devon Lay Subsidy Rolls 1524–7*, 152–3.

42. This would be true of Exeter, said to have had a population of around 3,000 in 1377 and of between 7,000 and 9,000 in the 1520s; it is likely that a growth rate of this magnitude would have been spread over many decades: Walter T. MacCaffrey, *Exeter, 1540–1640*, 2nd edn (Harvard, 1975), 289–90; Kowaleski, *Local Markets*, 371–5. Most towns associated with clothing or tinning in their hinterlands, grew in the latter half of the fifteenth century and the first decades of the sixteenth, but the phenomenal expansion attributed to Crediton in Dyer, *Decline and Growth*, 72, is too high because it is based upon the 1524–5 lay subsidy figure for the large parish, not the borough.

Chapter 52

1. S.M. Pearce, 'Church and society in south Devon A.D. 350–700', *Proceedings of the Devon Archaeological Society*, 40 (1982), 1–18; A. Preston-Jones and P. Rose, 'Medieval Cornwall', *Cornish Archaeology*, 25 (1986), 135–85.

2. M.W. Beresford, *New Towns of the Middle Ages* (London, 1967).

3. M.R.G. Conzen, 'Alnwick, Northumberland: a study in town-plan analysis', *Papers of the Institute of British Geographers*, 27 (1960).

4. D. Portman, *Exeter Houses, 1400–1700* (Exeter, 1966), 3–22; M. Laithwaite, 'Totnes houses, 1500–1800', in P. Clark (ed.), *The Transformation of English Provincial Towns* (London, 1984), 62–98; V. O'Neill and P. Russell, 'The old house known as No. 5 Higher Street, Dartmouth', *Transactions of the Devonshire Association*, 83 (1951), 267–71: P. Russell and A.W. Everett, 'The old house known as No. 13 Higher Street, Dartmouth', *Transactions of the Devonshire*

Association, 91 (1959), 107–11; J.R.L. Thorp, 'Two hall houses in a late medieval terrace, 8–12 Fore Street, Silverton', *Proceedings of the Devon Archaeological Society*, 40 (1982), 171–80.

5. C. Henderson and E. Jervoise, *Old Devon Bridges* (Exeter, 1938); C. Henderson and H. Coates, *Old Cornish Bridges and Streams* (London, 1928); I. Burrow, 'The town defences of Exeter', *Transactions of the Devonshire Association*, 109 (1977), 13–40; C.F. Rea, 'The bastewalls of Totnes', *Transactions of the Devonshire Association*, 56 (1924), 202–13.

6. P. Russell, *Dartmouth: A History of the Port and Town* (London, 1950), 21–59.

7. P.J. Weddell, 'The excavation of medieval and later houses and St Margaret's Chapel at Exmouth, 1982–1984', *Proceedings of the Devon Archaeological Society*, 44 (1986), 107–40.

8. E.L. Jones, 'Fire disasters, the special case of east Devon', *Devon Historian*, 20 (1980), 11–17; E.L. Jones, S. Porter and M. Turner, 'A gazetteer of English urban fire disasters, 1500–1900', *Historical Geography Research Series*, 13 (1984).

9. J. Haslam, 'The towns of Devon', in J. Haslam (ed.), *Anglo-Saxon Towns in Southern England* (Chichester, 1984), 249–83.

10. Beresford, *New Towns*, 405–6.

11. St Martin's was a very early church as it has a documented 'stow' place-name; see Preston-Jones and Rose, 'Medieval Cornwall'.

12. Such beach trading sites can have very early medieval origins as the site at Bantham, Devon shows; see F.M. Griffith, 'Salvage observations at the Dark Age site at Bantham Ham, Thurlestone, in 1982', *Proceedings of the Devon Archaeological Society*, 44 (1986), 39–58.

13. Beresford, *New Towns*, 411.

14. Ibid., 424; Brewer was trying to build up a substantial feudal estate in east Devon.

15. R.J. Silvester and P.T. Bidwell, 'A Roman site at Woodbury, Axminster', *Proceedings of the Devon Archaeological Society*, 42 (1984), 33–58, provides plans and the air photographic evidence for a fort located to the south-east of the town.

16. Pearce, 'Church and society'.

17. W.G. Hoskins, *Devon* (Newton Abbot, 1972), 324–5.

18. R.A. Higham, 'Was there a castle at Axminster?', *Proceedings of the Devon Archaeological Society*, 44 (1986), 182–3; see also Devon Record Office (DRO), 123M/TB 159.

19. DRO, 123M/TB 160.

Chapter 53

1. In Cornwall and parts of Devon without nucleated settlement the hamlet where the parish church was situated was called the 'church town', but this usage, though

an important contemporary sense of town, has been ignored here. For this usage in Devon as well as the better-known Cornwall examples, see R. Polwhele, *History of Devonshire*, 3 vols (1793–1806, reprinted Dorking, 1977), II, 60.

2. T. Arkell, 'Establishing population totals for small towns from the 1851 census' (unpublished paper, 1992); Dr Arkell has been of enormous assistance in the production of these figures, sharing data and advice. For the general issues discussed here see also P. Clark and J. Hosking (eds), *Population Estimates of Small Towns 1550–1851*, revised edn (Leicester, 1993). Although this analysis extends and corrects the material presented in the latter volume, it (and the earlier edition) formed an invaluable basis for this work. For parish boundaries see H. Peskett, *Guide to the Parish and Non-Parochial Registers of Devon and Cornwall 1538–1837*, Devon and Cornwall Record Society, extra series, 2 (1979) and for the census returns E. Higgs, *Making Sense of the Census* (London, 1989). Earlier estimates for Devon and Cornwall can be found in W.G. Hoskins, *Devon* (Newton Abbot, 1954), 104–22, and J. Whetter, *Cornwall in the Seventeenth Century: An Economic History of Kernow* (Padstow, 1974), 8–13.

3. See the invaluable guide in J. Gibson and M. Medlycott, *Local Census Listings 1522–1930: Holdings in the British Isles* (Birmingham, 1992), although this only covers listings with names: many more summaries of census findings survive, often referred to in printed accounts.

4. See R. Pickard, *Population and Epidemics of Exeter in Pre-Census Times* (Exeter, 1947) and M. Dunsford, *Historical Memoirs of Town and Parish of Tiverton* (Exeter, 1790), 35–57 and 462–4, respectively. Exeter's population is also discussed in W.J. Harte and W.G. Hoskins, 'Estimates of Exeter's population', *Devon and Cornwall Notes and Queries*, 20 (1938–9), 210–14, 242–7; W.B. Stephens, 'Population of Exeter in 1641–2', *Devon and Cornwall Notes and Queries*, 33 (1975), 141–3. Eighteenth-century register material for Falmouth, Launceston, Exeter and Plymouth is given with the 1801 census. For Tavistock see C. Barham, 'Remarks on the abstracts of the parish registers of Tavistock, Devon', *Journal of the Statistical Society of London*, 4 (1841), 34–49, which also discusses a 1781 census on page 35. The survival and whereabouts of registers are shown in Peskett, *Guide*.

5. A.J. Howard (ed.), *Devon Protestation Returns* (n.p., 1973); H.L. Douch (ed.), *Cornwall Protestation Returns 1641* (Bristol, 1974).

6. *The Compton Census of 1676: A Critical Edition*, ed. A. Whiteman, British Academy Records of Social and Economic

History, new series, 10 (1986). The use of this and other late-seventeenth-century sources is discussed in K. Schurer and T. Arkell (eds), *Surveying the People* (Oxford, 1992).

7. Devon Record Office (DRO), Chanter 225 A–C (1744–6 Visitation Returns of Bishop Claggett); Chanter 228 A–C (1764–6 Visitation Returns of Bishop Keppel); Chanter 232 A–B (1779 Visitation Returns of Bishop Ross); Moger Basket C/4–5 (1768 and 1771 Visitation Returns). Systematic use of this material for the region has been largely confined to Cornwall: see N. Pounds, 'Population movement in Cornwall and the rise of mining in the eighteenth century', *Geography*, 28 (1943), 37–46; N. Pounds, 'Population of Cornwall before the first census', in W. Minchinton (ed.), *Population and Marketing: Two Studies in the History of the South West* (Exeter, 1976), 11–30; P. Thomas, 'The population of Cornwall in the eighteenth century', *Journal of the Royal Institution of Cornwall*, 10, 4 (1990), 416–56. The 1821 returns are printed in M. Cook (ed.), *The Diocese of Exeter in 1821*, Devon and Cornwall Record Society, new series, 3–4 (1958–60).

8. T.L. Stoate (ed.), *Cornwall Hearth and Poll Taxes 1660–4* (Bristol, 1981); T.L. Stoate (ed.), *Devon Hearth Tax 1674* (Bristol, 1982); T.L. Stoate (ed.), *Devon Taxes 1581–1660* (Bristol, 1988).

9. See: (1) R.P. Chope, *Early Tours in Devon and Cornwall* (1918, reprinted Newton Abbot, 1967) and many other traveller's accounts, e.g. those reprinted in *Devon and Cornwall Notes and Queries* and the *Journal of the Royal Institution of Cornwall*; (2) county histories such as: R. Carew, *The Survey of Cornwall*, ed. F.E. Halliday (London, 1953); J. Norden, *Description of Cornwall* (1728, reprinted Newcastle, 1967); T. Risdon, *Chorographical Survey of County of Devon*, ed. J.T. (1811, reprinted Barnstaple, 1970); T. Westcote, *View of Devonshire in 1630*, ed. G. Oliver and P. Jones (Exeter, 1845); the parochial histories of Hans and Tonkin, included in *A Complete Parochial History of the County of Cornwall*, 4 vols (Truro, 1867–72); W. Borlase, *Natural History of Cornwall* (Oxford, 1758); W. Chapple, *Review of Part of Risdon's Survey of Devon* (Exeter, 1785, reprinted Barnstaple, 1970); R. Polwhele, *History of Cornwall*, II (Falmouth, 1803), 205–23, IV (1806), 121, 138–44, and VII (Truro, 1806); Polwhele, *History of Devonshire*; D. and S. Lysons, *Magna Britannia, Vol. III, Cornwall* (London, 1814); D. and S. Lysons, *Magna Britannia, Vol. VI, Devonshire* (London, 1822); (3) topographical dictionaries such as A. Brice, *The Grand Gazeteer* (Exeter, 1759); S. Lewis, *Topographical Dictionary of England*, 4 vols (1831); W. White, *History, Gazeteer and Directory of Devonshire* (1850, reprinted Newton Abbot, 1968);

(4) surveys such as T.L. Stoate (ed.), *Cornwall Manorial Rentals and Surveys* (Bristol, 1988), and N. Pounds (ed.), *Parliamentary Survey of the Duchy of Cornwall*, Devon and Cornwall Record Society, new series, 25 and 27 (1982–4); (5) agricultural surveys: W. Marshall, *Rural Economy of the West of England*, 2 vols (1796, reprinted Newton Abbot, 1970); C. Vancouver, *A General View of Agriculture of County of Devon* (1808, reprinted Newton Abbot, 1969); G.B Worgan, *General View of Agriculture of County of Cornwall* (1811).

10. In 1630 Westcote noted of Plymouth that it was 'in every way so esteemed by Cornishmen they would claim it for their own' (*View*, 382).

11. For the context of these changes see M. Duffy *et al.* (eds), *New Maritime History of Devon, Vol. I* (London, 1992) and J. Travis, *Rise of Devon's Seaside Resorts 1750–1900* (Exeter, 1993).

12. See D. Souden, 'Migrants and population structure of later seventeenth-century provincial cities and market towns', in P. Clark (ed.), *Transformation of English Provincial Towns 1600–1800* (London, 1984), 133–68, and D. Souden, '"East, West—Home's Best?": regional patterns in migration in early modern England', in P. Clark and D. Souden (eds), *Migration and Society in Early Modern England* (London, 1987), 292–332. Calculations provided for me by Dr Arkell show sex ratios for Cornish towns in the censuses between 1801 and 1851 reaching as low as 663 (part of Falmouth), often falling in the 770–870 range and in the case of the great majority of towns falling well below the county ratio of just over 900 men per 1,000 women. Only Bodmin, St Stephens by Launceston and St Martin by Looe, all only partly urban, recorded more men than women.

13. See A. Everitt, 'The marketing of agricultural produce', in J. Thirsk (ed.), *Agrarian History of England and Wales, Vol. IV, 1500–1640* (Cambridge, 1967), 466–592, at 470–1ff.; J. Chartres, 'Marketing of agricultural produce in metropolitan western England in the late seventeenth and eighteenth centuries', in M. Havinden (ed.), *Husbandry and Marketing in the South West 1500–1800* (Exeter, 1973), 63–74; A. Dyer, 'Market towns of southern England 1500–1700', *Southern History*, 1 (1979), 123–34; J. Chartres, 'Marketing of agricultural produce', in J. Thirsk (ed.) *Agrarian History of England and Wales, Vol. V, Part 2* (Cambridge, 1985), 406–502, at 410–11ff. Primary sources include the works cited in note 9, especially Risden, Westcote, and Brice; J. Adams, *Index Villaris* (1680); *Universal British Directory*, 5 vols (1793–8, reprinted 1993); Vancouver, *General View*, 2–3; Worgan, *General View*, table facing 2 and 163–4; Lysons, *Cornwall*, xxxviii–xxxix;

Lysons, *Devonshire*, xxxiv–xxxv, together with the various maps that show market towns, such as those in Norden, *Description*; Joel Gascoyne, *Map of Cornwall 1699*, ed. W. Ravenhill and O. Padel, Devon and Cornwall Record Society, new series, 34 (1991); Borlase, *Natural History*, new map of Cornwall opposite 1; Benjamin Donn, *A Map of the County of Devon 1765*, ed. W. Ravenhill (Exeter, 1975).

14. Polwhele, *History of Devonshire*, I, 312; Lysons, *Devonshire*, xxxiv.

15. Stoate, *Cornwall Manorial Rentals*, 47; Lysons, *Cornwall*, 69.

16. Historical Manuscripts Commission, *Twelfth Report Appendix 6: House of Lords 1689–90*, 298–302 (charters granted 1680–8); Lysons, *Cornwall*, xxxvi; Lewis, *Topographical Dictionary*; *British Parliamentary Papers*, 1835, XXIII, Appendix to Report from Commissioners on Municipal Corporations in England and Wales, especially 108–9; R.N. Worth, 'Common seals of Devon', *Transactions of the Devonshire Association*, 6 (1873), 79–100; R.N. Worth, 'Styles and titles of municipal corporations of Devon', *Transactions of the Devonshire Association*, 9 (1877), 407–8; M. Weinbaum, *British Borough Charters 1307–1660* (Cambridge, 1943); H. Finberg, 'Boroughs of Devon', *Devon and Cornwall Notes and Queries*, 24 (1950–1), 203–9; W.G. Hoskins, *Devon* (London, 1954), 111.

17. Carew, *Survey of Cornwall*, 137, 157–9.

18. What follows is based largely on the constituency entries in *House of Commons 1509–1558*, ed. S.T. Bindoff (History of Parliament Trust, 1982); *House of Commons 1558–1603*, ed. P. Hasler (History of Parliament Trust, 1981); *House of Commons 1660–1690*, ed. B. Henning (History of Parliament Trust, 1983); *House of Commons 1715–1754*, ed. R. Sedgwick (History of Parliament Trust, 1970); *House of Commons 1754–1790*, ed. L. Namier and J. Brooke (History of Parliament Trust, 1964); J.H. Philbin, *Parliamentary Representation 1832 England and Wales* (New Haven, 1965). There are many useful articles on Devon's parliamentary boroughs in *Transactions of the Devonshire Association*, notably those by J.J. Alexander between 1909 and 1930. For Cornwall see the studies by W.P. Courtney, W.T. Lawrance, A. de C. Glubb and, most recently, H. Spencer Toy, *The Cornish Pocket Borough* (Penzance, 1968).

19. See G. Haslam, 'The duchy and parliamentary representation in Cornwall 1547–1640', *Journal of the Royal Institution of Cornwall*, 8, 3 (1980), 224–42; D. Dean, 'Locality and parliament: the legislative activities of Devon's MPs during the reign of Elizabeth', in T. Gray (ed.), *Tudor and Stuart Devon* (Exeter, 1992), 75–95.

20. These maps are based on references in note 18, together with B. Willis, *Notitia*

Parliamentaria (1716); D. Hirst, *Representative of the People?* (Cambridge, 1975), 213–26; J. Cannon, *Parliamentary Reform 1640–1832* (Cambridge, 1972), 278–89; W. Speck, *Tory and Whig* (1970), Appendix E, 126–31; MS report on boroughs in Cornwall (*c*.1747) in Royal Institute of Cornwall BRA/A437/29; Public Record Office, Chatham Papers 30/8/83 T. Pitt on the Present State of the Boroughs in the County of Cornwall (Oct. 1740). I owe the last two references to the kindness of John More.

21. *Universal British Directory*, IV, 368–9 (St Germans). See also R. Eliot, 'Pocket boroughs of the Eliot family', *Journal of the Royal Institution of Cornwall*, 9 (1982–5), 321–49.

22. An excellent example of this is provided by Defoe in Chope, *Early Tours*, 145–6, 160, 169–70.

23. See P. Borsay, *The English Urban Renaissance* (Oxford, 1989); M. Girouard, *The English Town* (New Haven and London, 1990).

24. Leland in Chope, *Early Tours*, 1–82, e.g. 13 (Bossiney), 26 (Newlyn), 38 (St Mawes), 48 (Polperro), 79 (Otterton).

CHAPTER 54

1. Office of Population Censuses and Surveys, *Census 1981: Key Statistics for Urban Areas, The South West and Wales* (London, 1984). This deals with the concept and definition of urban areas.

2. Peter Clark and Paul Slack, *English Towns in Transition, 1500–1700* (Oxford, 1979), 1.

3. Office of Population Censuses and Surveys, *Census 1991: Cornwall and the Isles of Scilly Ward and Civil Parish Monitor* (London, 1993) and *Census 1991, Devon Ward and Civil Parish Monitor* (London, 1994).

4. See, for example, Walter Ison, *The Georgian Buildings of Bath* (London, 1948), 46; John Summerson, *Georgian London* (London, 1978), 19 and 172; Ewart Johns, *British Townscapes* (London, 1965), 27 and 130. For a discussion of the ideas and methods of morphological analysis see Peter J. Larkham and Andrew N. Jones, *A Glossary of Urban Form*, Historical Geography Research Series No. 26 (1991) and J.W.R. Whitehand, *The Making of the Urban Landscape* (Oxford, 1992), 1–13. See also T.R. Slater (ed.), *The Built Form of Western Cities* (Leicester, 1990).

5. A.E. Smailes, *The Geography of Towns* (London, 1953), 92. See also A.E. Smailes, 'Some reflections on the geographical description and analysis of townscape', *Transactions and Papers of the Institute of British Geographers*, 21 (1955), 99–115.

6. H.P.R. Finberg, *Tavistock Abbey: A Study in the Social and Economic History of Devon* (Cambridge, 1951), ix–x.

7. H.P.R. Finberg, 'The borough of Tavistock', in W.G. Hoskins and H.P.R. Finberg (eds), *Devonshire Studies* (London, 1952); for a recent in-depth study of population change in Tavistock see M.C. Phillips, 'Demographic and social structural change in Tavistock, 1741–1871, and its relation to the mining industry' (unpublished Ph.D. thesis, University of Exeter, 1987).

8. W.G. Hoskins, *Devon* (Newton Abbot, 1972), 486.

9. Mark Brayshay, 'The duke of Bedford's model cottages in Tavistock, 1840–1870', *Transactions of the Devonshire Association*, 114 (1982), 115–31.

10. These figures are based on measurements from the Ordnance Survey 1:10,000 map of Tavistock (SX 47 SE).

11. A.E. Smailes, 'Some reflections'; M.R.G. Conzen, 'Alnwick, Northumberland: a study in town-plan analysis', *Transactions of the Institute of British Geographers*, 27 (1960); M.R.G. Conzen, 'Historical townscapes in Britain: a problem in applied geography', in J.W. House (ed.), *Northern Geographical Essays in Honour of G.H.J. Daysh* (Newcastle upon Tyne, 1966), 56–78; M.R.G. Conzen, 'The plan analysis of an English city centre', in K. Norberg (ed.), *Proceedings of the IGU Symposium on Urban Geography* (Lund, 1962), 383–414.

12. See, for example, N.P. Brooks and G. Whittington, 'Planning and growth in the medieval Scottish burgh: the example of St. Andrews', *Transactions of the Institute of British Geographers*, new series, 2 (1977), 278–95; P.E. Jones, 'The development of the urban morphology of central Bangor, Gwynedd, in the early nineteenth century', *National Library of Wales Journal*, 26 (1989), 133–64. Larkham and Jones, *A Glossary of Urban Form*; Slater (ed.), *The Built Form of Western Cities*.

13. Ewart Johns, 'Urban design in Dawlish and Chelston', in K.J. Gregory and W.L.D. Ravenhill (eds), *Exeter Essays in Geography in Honour of Arthur Davies* (Exeter, 1971), 201–8. See also J.W.R. Whitehand, 'Background to the urban morphogenetic tradition', in J.W.R. Whitehand (ed.), *The Urban Landscape: Historical Development and Management. Papers by M. R. G. Conzen* (London, 1981), 1–24.

14. W.I. Leeson-Day, *Holsworthy*, Devonshire Association Parochial Histories No. 2. (Torquay, 1934).

15. Ibid., 7–9.

16. T.R. Slater, *The Analysis of Burgages in Medieval Towns*, Department of Geography, University of Birmingham, Working Paper No. 4 (Birmingham, 1980), 1.

17. T.R. Slater, 'The analysis of burgage patterns in medieval towns', *Area*, 13 (1981), 211–16.

18. Conzen, 'Alnwick', 56–65. See also Harold Carter, *An Introduction to Urban Historical Geography* (London, 1983), 146–8.

19. J.W.R. Whitehand, 'Fringe belts: a neglected aspect of urban geography',

Transactions of the Institute of British Geographers, 41 (1967), 230–2; J.W.R. Whitehand, 'The changing nature of the urban fringe: a time perspective', in J.H. Johnson (ed.), *Suburban Growth: Geographical Processes at the Edge of the Western City* (London, 1974), 31–52; J.W.R. Whitehand, 'Conzenian ideas: extension and development', in Whitehand (ed.), *The Urban Landscape*, 132–9.

20. Jones, 'The urban morphology of central Bangor', 135.

21. Harold Carter, 'A decision-making approach to town plan analysis: a case study of Llandudno', in H. Carter and W.K.D. Davies (eds), *Urban Essays: Studies in the Geography of Wales* (London, 1970), 67.

22. J.D. Chambers, *A Century of Nottingham History* (Nottingham, 1952).

23. David Ward, 'The pre-urban cadaster and the urban pattern of Leeds', *Annals of the Association of American Geographers*, 52 (1962), 150–66.

24. M.J. Mortimer, 'Landownership and urban growth in Bradford and environs in the West Riding conurbation, 1850–1950', *Transactions of the Institute of British Geographers*, 46 (1969), 105–19; J.M. Rawcliffe, 'Bromley: Kentish market town to London suburb, 1841–81', in F.M.L. Thompson (ed.), *The Rise of Suburbia* (Leicester, 1982), 93–156; J. Springett, 'Landowners and urban development: the Ramsden estate and nineteenth-century Huddersfield', *Journal of Historical Geography*, 8 (1982), 129–44.

25. Johns, 'Urban design', 204.

26. For background information on the unpublished Ordnance Survey six-inch drawings of Plymouth see J.B. Harley and Yolande O'Donoghue, *The Old Series Ordnance Survey Maps of England and Wales* (Lympne, 1977), xiv–xvi; see also J.B. Harley, 'The birth of the topographical survey', in W.A. Seymour (ed.), *A History of the Ordnance Survey* (Folkestone, 1980), 45.

27. Crispin Gill, *Plymouth: A New History, 1603 to the Present Day* (Newton Abbot, 1979), 146–8.

28. The Ordnance Survey 1:2,500 sheet CXXIII.8 was surveyed 1855–6. The version used (in Map 54.7) was published in 1888, but amended by the local authority in the 1890s. The second illustration (Map 54.8) shows the same area depicted in the revised Ordnance Survey map published in 1907.

29. Devon Record Office, Tithe Survey and Apportionment: The Tything of Compton Gifford, 1840.

30. Geoffrey Holmes, *Sidmouth: A History* (Sidmouth, 1987), 48.

31. Hoskins, *Devon*, 477.

32. *Census 1991, Devon Ward and Civil Parish Monitor*, 12.

33. Lois Lamplugh, *A History of Ilfracombe* (Chichester, 1984), 9.

34. Richard Rodger, *Housing in Urban Britain, 1780–1914* (London, 1989), 32–4.

35. C.A. Forster, *Court Housing in Kingston-Upon-Hull*, University of Hull Occasional Papers in Geography No. 19. (Hull, 1972).

36. W.M. Brayshay, 'The demography of three west Cornwall mining communities, 1851–71: a society in decline' (unpublished Ph.D. thesis, University of Exeter, 1977), 123.

37. Ibid., 326–9; see also M. Brayshay, 'Depopulation and changing household structure in the mining communities of west Cornwall', *Local Population Studies*, 25 (1980), 26–41.

38. Public Record Office (PRO), RG 10/2315/4, Census Enumerators Books, Redruth, 1871.

39. PRO, RG 10/2343/6, Census Enumerators Books, St Just in Penwith, 1871.

40. In 1871 in the hamlet of Boscaswell as a whole there were only nine farmers and two (landless) farm labourers. Three of the farmers combined agriculture with tin-mining. Only two had holdings of 20 acres or more and the average 'farm' was less than 8.5 acres.

CHAPTER 55

1. W.H. Thompson, *Devon: A Survey for the Council for the Preservation of Rural England* (London, 1932).

2. Devon County Council, *The Challenge: The Motorway into Devon: The need for Action Now* (Exeter, 1971).

3. Devon County Council, *Report of Survey for the Structure Plan* (Exeter, 1977).

4. Devon County Council, *Structure Plan: Policies and Proposals 1981* (Exeter, 1981).

5. Devon County Council, *Structure Plan Incorporating the First Alteration: Explanatory Memorandum* (Exeter, 1987).

6. W.H. Thompson, *Cornwall—Coast, Moors and Valleys: A Survey* (London, 1930).

7. Cornwall County Council, *Development Plan 1952: Report of Survey. Part One—The County; Part Two—The Towns* (Truro, 1952).

8. Cornwall County Council, *Cornwall County Structure Plan: Explanatory Memorandum* (Truro, 1981).

9. A.W. Gilg, 'Regional planning in the South West', *Social and Economic Administration*, 9 (1975), 220–5; G. Shaw and A. Williams, 'The regional structure of structure plans', *Planning Outlook*, 23 (1981), 2–7.

10. South West Economic Planning Council, *A Strategic Settlement Pattern for the South West* (London, 1974).

11. Gilg, 'Regional planning in the South West'.

CHAPTER 57

1. J.K. Walton, 'The seaside resorts of England and Wales, 1900–1950', in G. Shaw and A. Williams (eds), *The Rise and Fall of British Coastal Resorts* (London, 1997); John Travis, *The Rise of Devon's Seaside Resorts, 1750–1900* (Exeter, 1993).

2. A.B. Granville, *The Spas of England and Principal Sea Bathing Places*, 2 vols (1841, reprinted 1971).

3. Granville's visit to Torquay is related by R. Manning-Sanders, *Seaside England* (London, 1951), 102–3.

4. A. and P. Burton, *The Green Bag Travellers: Britain's First Tourists* (London, 1978), 1.

5. J. Walvin, *Leisure and Society, 1830–1950* (London, 1978), chapter 6.

6. J.K. Walton, *The English Seaside Resort: A Social History, 1750–1914* (Leicester, 1983), chapter 3.

7. W. White, *History, Gazetteer and Directory of Devonshire* (Sheffield, 1850), 445–6.

8. J. Pike, *Torquay, Torbay: A Bibliographical Guide* (Torquay, 1973).

9. R. Pearse, *The Land Beside the Celtic Sea: Aspects of Cornwall's Past* (Truro, n.d.).

10. H. Fox (ed.), *Mariana's Diary: A Record of a Holiday at Falmouth* (Falmouth, n.d.); for a discussion of early tourism see J. Mattingly and J. Palmer (eds), *From Pilgrimage to Package Tour* (Falmouth, 1992)

11. Fox (ed.), *Mariana's Diary*.

12. Walton, *The English Seaside Resort*, chapter 3.

13. Jack Simmons, 'The railway in Cornwall, 1835–1914', *Journal of the Royal Institution of Cornwall*, new series, 9, 1 (1982), 11–30.

14. This is described in Simmons, 'The railway in Cornwall'; see also P. Brendon, *Thomas Cook: 150 Years of Popular Tourism* (London, 1991).

15. Simmons, 'The railway in Cornwall'.

16. This information is discussed in D.H. Matthews, *Tourist Attractions in Cornwall: An Overview* (West Country Tourist Board, 1988), chapter 2.

17. White, *History, Gazetteer and Directory*, 230.

18. Ibid., 232.

19. Ibid., 231.

20. Pike, *Torquay, Torbay*.

21. Ibid.

22. Walton, *The English Seaside Resort*, 68.

23. Ibid., 69.

24. Ibid., 81.

25. J.K. Walton, *The Blackpool Landlady: A Social History* (Manchester, 1984), 184.

26. W. Crossing, *Guide to Dartmoor* (Exeter, 1914), preface.

27. H.W.J. Heck, *Survey of the Holiday Industry* (Truro, 1966), 17.

28. M. Fraser, *Holiday Haunts 1935: GWR Official Guide to Holiday Resorts* (London, 1935).

29. Ibid., 367.

30. Ibid., 193.

31. Ibid., 356.

32. Shaw and Williams, *The Rise and Fall of British Coastal Resorts*.

33. Cornwall County Council, *County Structure Plan, Topic Report: The Holiday Industry* (Truro, 1976).

34. F.M.M. Lewes, *The Holiday Industry of Devon and Cornwall* (London, 1970).

35. Cornwall County Council, *County Structure Plan: Report of Survey* (Truro, 1979).

36. Devon County Council, *County Structure Plan: First Alteration, 1981 Data Base* (Exeter, 1983).

37. Devon County Council, *County Structure Plan: Report of Survey* (Exeter, 1977).

38. Heck, *Survey of the Holiday Industry*.

39. Lewes, *The Holiday Industry of Devon and Cornwall*.

40. Heck, *Survey of the Holiday Industry*.

41. Devon County Council, *Devon County Development Plan* (Exeter, 1961).

42. Lewes, *The Holiday Industry of Devon and Cornwall*.

43. G. Shaw, A. Williams and J. Greenwood, *Tourism and the Economy of Cornwall* (Exeter, 1987); D. Matthews, 'Recent trends in tourism in the South West', in G. Shaw and A. Williams (eds), *Tourism and Development* (Exeter, 1987); Devon County Council, *Devon Tourism Review, 1989* (Exeter, 1990).

44. J. Greenwood, A. Williams and G. Shaw, *Cornwall Visitor Surveys 1987 and 1988* (Exeter, 1988, 1989).

45. Shaw, Williams and Greenwood, *Tourism and the Economy of Cornwall*.

46. Devon County Council, *Devon Tourism Review, 1989*.

47. Matthews, 'Recent trends in tourism'.

48. Devon County Council, *Devon Tourism Review, 1989*.

49. Shaw, Williams and Greenwood, *Tourism and the Economy of Cornwall*.

50. Greenwood, Williams and Shaw, *Cornwall Visitor Surveys*.

51. R. Perry, 'Cultural tourism in Cornwall', in Shaw and Williams (eds), *Tourism and Development*.

CHAPTER 58

1. Defined here as a firm possessing five or more retail outlets.

2. Letter from Montague Burton to Mr Embden, Montague Burton Archive, West Yorkshire Archive Service, Leeds.

3. G. Shaw 'The evolution and impact of large scale retailing in Britain', in J. Benson and G. Shaw (eds), *The Evolution of Retail Systems, c.1800–1914* (Leicester, 1992).

4. Ibid.

5. M. Winstanley, *The Shopkeeper's World 1830–1914* (Manchester, 1983).

6. J.B. Jefferys, *Retail Trading in Great Britain 1850–1950* (Cambridge, 1954).

7. A. Alexander, 'The evolution of multiple retailing in Britain 1870–1950: a geographical analysis' (unpublished Ph.D. thesis, University of Exeter, 1994).

8. M. Phillips, 'Market exchange and social relations: the practices of food circulation in and to the three towns of Plymouth, Devonport and Stonehouse 1800–c.1870' (unpublished Ph.D. thesis, University of Exeter, 1991).

9. The West of England as defined by Shaw incorporates the counties of Cornwall, Devon, Dorset, Somerset, Gloucestershire and Wiltshire. See G. Shaw, 'Retail patterns', in J. Langton and R.J. Morris (eds), *Atlas of Industrialising Britain 1780–1914* (London, 1986).

10. Jefferys's definition of the South West of England incorporates the counties of Cornwall, Devon, Dorset, Somerset and Gloucestershire (*Retail Trading*).

11. Alexander 'Evolution of multiple retailing'.

12. Ibid.

13. For the purposes of consistency, data for Plymouth, Devonport and Stonehouse have been combined in Maps 58.1, 58.2 and 58.3.

14. Jefferys, *Retail Trading*.

15. P. Mathias, *Retailing Revolution* (London, 1967).

16. Ibid.

17. Ibid.

18. Ibid.

19. A. Alexander, 'Retail revolution: the spread of multiple retailers in South West England', *Journal of Regional and Local Studies*, 13 (1993), 39–54.

20. Data taken from alternative directory publications suggest some slight variation in terms of number and location of grocery and provision retailers. The overall trend that they present is, nonetheless, very similar. The authors acknowledge the assistance of Richard Oliver in identifying the precise location of retailers in each of these streets.

21. Jefferys, *Retail Trading*.

CHAPTER 59

1. S. Yeo, 'Socialism, the state and some oppositional Englishness', in R. Colls and P. Dodd (eds), *Englishness, Politics and Culture 1880–1920* (London, 1986), 338.

2. M. Purvis, 'The development of co-operative retailing in England and Wales 1851–1901: a geographical study', *Journal of Historical Geography*, 16 (1990), 314–31.

3. J. Tann, 'Co-operative corn milling: self-help during the grain crises of the Napoleonic Wars', *Agricultural History Review*, 28 (1980), 48.

4. B. Jones, *Co-operative Production* 1 (Oxford, 1894), 38.

5. *British Parliamentary Papers* (BPP), 1880, LXVIII, 458, Reports of the Chief Registrar of Friendly Societies for the year ending 31 December 1879, Part 1(B) Industrial and Provident Societies; *BPP*, 1893–4, LXXXIV, 248, Reports of the Chief Registrar of Friendly Societies for the year ending 31 December 1892, Part B Industrial and Provident Societies; *BPP*, 1899, XCI, 310, Reports of the Chief Registrar of Friendly Societies for the year ending 31 December 1898, Part B Industrial and Provident Societies.

6. *Howitt's Journal*, 22 May 1847; *Christian Socialist*, 25 October 1851.

7. *Trades Free Press*, 8 August 1829, 20 February 1830; *Co-operative Magazine*, April 1826, May 1826, June 1826, January 1827; D. Hardy, *Alternative Communities in Nineteenth-Century England* (London, 1979), 46–8, 242, suggests that the projected community might have been located in the vicinity of Rochbeare.

8. *Christian Socialist*, 15 March 1851, 29 March 1851, 26 April 1851, 3 January 1852.

9. *Christian Socialist*, 13 September 1851.

10. See the report from St Ive Cross: *Co-operator*, 15 August 1865.

11. *Co-operator*, 29 April 1871.

12. For example, men associated with the Amalgamated Society of Carpenters and Joiners were involved in initiating the revival of co-operation in Exeter in 1883–4: Exeter Co-operative and Industrial Society Ltd, *History of the Society 1885–1935* (Plymouth, 1935).

13. Co-operative Union, *Co-operative Annual Congress Reports* 1890 and 1892.

14. *BPP*, 1903, LXVI, 160, Reports of the Chief Registrar of Friendly Societies for the year ending 31 December 1902, Part B Industrial and Provident Societies; *BPP*, 1906, LXII, 464, Reports of the Chief Registrar of Friendly Societies for the year ending 31 December 1905, Part B Industrial and Provident Societies; *BPP*, 1905, LXXV, 168, Reports of the Chief Registrar of Friendly Societies for the year ending 31 December 1904, Part B Industrial and Provident Societies; *BPP*, 1914, LXXVI, 422, Reports of the Chief Registrar of Friendly Societies for the year ending 31 December 1913, Part B Industrial and Provident Societies.

15. Co-operative Union, *Co-operative Directory for 1900* (Manchester, 1900).

CHAPTER 60

1. J.A. Dawson, 'Market services in the UK', in R.J. Johnston and V. Gardiner (eds), *The Changing Geography of the UK*, 2nd edn (London, 1992), chapter 9.

2. G. Rowley, *Let's Talk Shop: Relocational Trends in British Retailing* (Hatfield, 1986).

3. C. Guy, 'Accessibility to multiple-owned grocery stores in Cardiff: a description and evaluation of recent changes', *Planning, Practice and Research*, 1 (1987), 9–15.

4. Dawson, 'Market services in the UK'.

5. C. Guy, *The Retail Development Process* (London, 1994).

6. Dawson, 'Market services in the UK'.

7. G. Shaw, 'Retail development strategies and the British shopping hierarchy', in G. Johnson (ed.), *Business Strategy and Retailing* (Chichester, 1987), chapter 15.

8. R. Schiller, 'Out of town exodus', in E. McFadyen (ed.), *The Changing Face of British Retailing* (London, 1987), 64–73.

9. Exeter City Council, *Exeter District Plan* (Exeter, 1981).

10. Shaw, 'Retail development strategies'.

11. Department of the Environment, *Vital to Viable Town Centres: Meeting the Challenge* (London, 1994).

12. Devon County Council, *Shopping* (Exeter, 1987).

13. R.L. Davies and D.J. Bennison, 'The impact of town shopping schemes in Britain', *Progress in Planning*, 14 (1981).

14. Plymouth City Council, *Tomorrow's Plymouth: City Centre Shopping Strategy* (Plymouth, 1987).

15. Cornwall County Council, *Retail Database* (Truro, 1993).

CHAPTER 63

1. Yolande Hodson, *Ordnance Surveyors' Drawings 1789–c.1840* (Reading, 1989), 98.

2. The writer's inference, on the basis of cartobibliographical experience of the Ordnance Survey 1:63,360 Old Series.

3. *A Plan of the City of Exeter … Made from Actual Measurements in the years 1818 and 1819 By J. Coldridge Surveyor*, Devon Record Office, ECA. The map is somewhat damaged, and the panel which covered the built-up part of St Thomas is now missing, and elsewhere the map does not extend much beyond the built-up area.

4. *The City of Exeter, 1835, Drawn by R. Brown, Published by R. Colliver, Holloway Street, Exeter*, West Country Studies Library, Exeter. It is noticeable that the Coldridge and Brown maps are virtually co-extensive.

5. John Wood, *Map of Exeter, from Actual Survey* (1840). Although it was printed, the map is quite rare: there is a rather worn copy in Exeter City Library. Having compared it with Coldridge's map, the author is satisfied as to Wood's claim to be the surveyor. Furthermore, it covers a larger area than that which Coldridge mapped.

6. Dated 1852 in West Country Studies Library, Exeter catalogue; PM B/EXE/1852/FEA.

7. Joyce C. Miles, 'The rise of suburban Exeter and the naming of its streets and houses' (unpublished Ph.D. thesis, University of Leicester, 1990), 59.

8. Robert Newton, *Victorian Exeter* (Leicester, 1968), 141.

9. Miles, 'Rise of suburban Exeter', 219.

10. Newton, *Victorian Exeter*, 136.

11. See J.B. Harley and J.B. Manterfield, 'The Ordnance Survey 1:500 plans of Exeter, 1874–1877', *Devon and Cornwall Notes and Queries*, 24 (1978), 63–75.

12. *Exeter*, 1880, 'published by Henry Besley, Directory Office, South St, Exeter'. There is a copy of this map in the West Country Studies Library, Exeter.

13. Miles, 'Rise of suburban Exeter', 163; Newton, *Victorian Exeter*, 142.

14. Miles, 'Rise of suburban Exeter', 219.

CHAPTER 64

1. B.S. Morgan, 'The residential structure of Exeter', in W.L.D. Ravenhill and K.J. Gregory (eds), *Exeter Essays in Geography* (Exeter, 1971), 21–36.

2. Thomas Sharp, *Exeter Phoenix* (London, 1946).

3. Exeter City Council and Devon County Council, *Exeter and District Joint Feasibility Study* (Exeter, 1969).

4. Exeter City Council, *A Strategy for Growth* (Exeter, 1974).

5. Devon County Council, *Towards 2001: The Future of the Exeter Sub-Region* (Exeter, 1975).

6. N. Venning, *Exeter: The Blitz and Rebirth of the City: A Pictorial History* (Exeter, 1988).

7. Exeter City Council, *Exeter Central Area: A Report by Wilson and Womersley Partners* (Exeter, 1969).

8. Exeter City Council, *Central Conservation Study* (Exeter, 1977).

CHAPTER 65

1. R.N. Worth, 'The early commerce of Plymouth', *Journal of the Plymouth Institution*, 6 (1876–78), 290–334; R.N. Worth, 'On the Plymouth municipal records', *Journal of the British Archaeological Society*, 39 (1883), 110–18; R.N. Worth, 'The earlier municipal history of Plymouth', *Transactions of the Devonshire Association*, 16 (1884), 725–41.

2. Barry Cunliffe (ed.), *Mount Batten, Plymouth: A Prehistoric and Roman Port*, Oxford University Committee for Archaeology, Monograph No. 26 (Oxford, 1988).

3. Some authors have always argued that Plymouth was settled in prehistoric times. See Crispin Gill, *Plymouth: A New History, Ice Age to the Elizabethans* (Newton Abbot, 1966), 13–21.

4. For example, Plymouth archaeologists have recently found evidence of Roman occupation at Elburton. See also E.R. Harris, 'The village of St Budeaux', *Proceedings of the Plymouth Athenaeum*, 5 (1978–82), 35–7.

5. Paul Bidwell, 'Roman pottery and tiles', in Cynthia Gaskell Brown (ed.), *The Medieval Waterfront: Woolster Street* (Plymouth, 1986), 13.

6. Bidwell, 'Roman pottery', 13.

7. Worth, 'The earlier municipal history of Plymouth', 724.

8. Caroline and Frank Thorn (eds), *Domesday Book: Devon* (Chichester, 1985), 1, 20.

9. Gill, *Plymouth: A New History*, 46.

10. Thorn, *Domesday*, 29, 7.
11. Worth, 'Earlier municipal history of Plymouth', 725–8.
12. Gill, *Plymouth: A New History*, 52.
13. Bridget Cherry and Nikolaus Pevsner (eds), *The Buildings of England: Devon*, 2nd edn (London, 1989), 683; Roy Midmer, *English Mediaeval Monasteries, 1066–1540* (London, 1979), 256. The priory was dissolved in 1539.
14. Limited investigations were completed inside the Priory precincts in 1994. Personal communication, Plymouth City Archaeological Officer.
15. Maurice Beresford, *New Towns of the Middle Ages*, 2nd edn (Stroud, 1988), 425; A. Mills, 'Plympton St Maurice: a town within a city', *Proceedings of the Plymouth Athenaeum*, 5 (1978–82), 62–8; R.A. Higham, S. Goddard and M. Rouillard, 'Plympton Castle, Devon', *Proceedings of the Devon Archaeological Society*, 43 (1985), 59–75; J. Brooking Rowe, 'Plympton Castle', *Journal of the Plymouth Institution*, 6 (1876–8), 246–74.
16. D. Vellacott, 'Plymouth's hospitals: The living past', *Proceedings of the Plymouth Athenaeum*, 6 (1982–7), 53–4. Other lazar houses existed in medieval Plymouth at Mutley and possibly near the Franciscan friary.
17. Cherry and Pevsner, *Devon*, 684.
18. Gill, *Plymouth: A New History*, 47; Worth, 'Plymouth municipal records', 111.
19. Worth, 'The earlier municipal history of Plymouth', 729.
20. This was granted by Henry III: Worth, 'Earlier municipal history of Plymouth', 729.
21. Henry Whitfield, *Plymouth and Devonport in Times of Peace and War* (Plymouth, 1900), 12–15; Gill, *Plymouth: A New History*, 64–94; Worth, 'Early commerce of Plymouth', 290–4.
22. James Barber, 'Buildings and quays', in Gaskell Brown (ed.), *Medieval Waterfront*, 11–12; Keith Ray, 'Sutton-Super-Plymouth: a medieval port', in K. Ray (ed.), *Archaeological Investigations and Research in Plymouth, Vol. 1, 1992–93*, Plymouth Archaeology Occasional Publication No. 2 (Plymouth, 1995), 63.
23. Whitfield, *Plymouth and Devonport*, 13–14; see also Graham Fairclough, *St Andrew's Street: Plymouth Excavations 1976*, Plymouth Museum Archaeological Series No. 2 (Plymouth, 1979), 7.
24. Worth, 'The earlier municipal history of Plymouth', 743; Whitfield, *Plymouth and Devonport*, 16.
25. Worth, 'The earlier municipal history of Plymouth', 744, and 'Plymouth municipal records', 110–11.
26. Gill, *Plymouth: A New History*, 94–6.
27. Ibid., 97; C.E. Welch, *Plymouth City Charters, 1439–1835: A Catalogue* (Plymouth, 1962).
28. Mark Brayshay, 'Plymouth's past: so worthy and peerless a western port', in

B.S. Chalkley, D. Dunkerley and P. Gripaios (eds), *Plymouth: Maritime City in Transition* (Newton Abbot, 1991), 40, 234; Nicholas J. Casley, *The Medieval Incorporation of Plymouth and a Survey of the Borough Bounds*, Old Plymouth Society, new series, 5 (Plymouth, 1997), 45–78.
29. Gill, *Plymouth: A New History*, 98.
30. Elisabeth Stuart, *Lost Landscapes of Plymouth: Maps, Charts and Plans to 1800* (Stroud, 1991), 114, 116–17, 122–3, 144–7.
31. For a discussion of the problems involved in establishing early property boundaries in Plymouth see James Barber, 'Architectural and documentary evidence', in Fairclough (ed.), *St Andrew's Street Excavations*, 9–22.
32. James Barber, 'Excavations at Woolster Street, Plymouth, 1963–1969', *Proceedings of the Plymouth Athenaeum*, 2 (1971), 77.
33. R. Pearse Chope, *Early Tours in Devon and Cornwall*, originally published in *Devon and Cornwall Notes and Queries*, 9, 2 (1918) (reprinted Newton Abbot, 1968), 53–4. *The Itinerary of John Leland*, ed. Lucy Toulmin-Smith (London, 1964), 212–13.
34. Although by then encroached upon, eighteenth-century maps show this wide street very clearly. See, for example, 'A Plan of the Royal Citadel and Town of Plymouth', British Library (BL) Add. MS 60393 C.
35. BL Cotton MS Augustus I i 35–6, 38–9; Hatfield CPM I 41; see also Plymouth and West Devon Record Office (PWDRO) W46 fo. 9^V. James I was proclaimed king from 'the Market Crosse here in Plymouth'.
36. Worth, 'The early commerce of Plymouth', 328–9.
37. Fairclough (ed.), *St Andrew's Street Excavations*, 8–9; Gill, *Plymouth: A New History*, 98; James Barber, 'No. 33 St Andrew's Street, Plymouth: a record during restoration', *Transactions of the Devonshire Association*, 105 (1973), 37–58. St Andrew's Street is first recorded in 1386 in an unspecified deed referred to by R.N. Worth, *History of Plymouth* (Plymouth, 1890), 365.
38. Barber, 'Buildings and quays', 11; Barber, 'Excavations at Woolster Street', 76–82; James Barber, 'No. 4 Vauxhall Street: a seventeenth-century merchant's house of "gallery and back block" type', *Transactions of the Devonshire Association*, 105 (1973), 17–36.
39. Barber, 'No. 4 Vauxhall Street', describes the two seventeenth-century tenements built on the site of a medieval palace known as 'Foxhole' mentioned in the Borough rentals: PWDRO, W170, 1608/9.
40. Gill, *Plymouth: A New History*, 199.
41. James Barber's work at Woolster Street showed the succession of buildings erected on this site; Gill, *Plymouth: A New History*, 146, 199.
42. One which survives is the so-called 'Prysten House' behind St Andrew's Church. Once thought to have

accommodated Augustinian chantry priests and preaching canons from Plympton Priory, recent work has indicated that this was the dwelling of a wealthy merchant, Thomas Yogge, who built it in 1498. See Jennifer Barber, 'Yogge's house or Prysten house?', *Transactions of the Devonshire Association*, 105 (1973), 75–86.
43. Archaeological work at North Quay in 1995 by Exeter Museums Archaeological Field Unit revealed evidence of seventeenth-century reclamation.
44. Jennifer Barber, 'New light on the Plymouth friaries', *Transactions of the Devonshire Association*, 105 (1973), 59–74. The 1288 grant of land is identified in Public Record Office (PRO), C143/12/3. See also J. Brooking Rowe, 'The ecclesiastical history of old Plymouth', *Journal of the Plymouth Institution*, 4 (1869–73), 320–59. The Plymouth Carmelite friary was the only one established by this Order in Devon and Cornwall.
45. The additional grant in 1329 is specified in PRO, C143/207/19.
46. Barber, 'Plymouth friaries', 60.
47. For example in 1511 the 'White Frers' sold timber planks to the borough to be used in repairs to the castle (PWDRO, W130, f. 95).
48. These are depicted very clearly on Tudor maps. See, for example, BL Cotton MS Augustus I i 35–6, 38–9; Hatfield CPM I 60.
49. Gill, *Plymouth: A New History*, 72; Barber, 'Plymouth friaries', 61; Brooking Rowe, 'Ecclesiastical history of Plymouth', 335.
50. Gill, *Plymouth: A New History*, 99, 129–30.
51. The 'Friary Court' excavations were directed for the Exeter Museums Archaeological Field Unit by Chris Henderson; C.G. Henderson, 'Excavations at Plymouth Whitefriars, 1989–94', in Ray (ed.), *Archaeological Investigations and Research in Plymouth*, 47–58.
52. Barber, 'Plymouth friaries', 66–7.
53. Emanuel Bowen and John Kitchin, *Atlas of England and Wales* (London, 1755).
54. Worth, 'Early commerce of Plymouth', 303.
55. Barber, 'Plymouth friaries', 69–70; S.R. Blaylock, 'An architectural survey of the late-medieval hall at Plymouth Gin Distillery', *Proceedings of the Devon Archaeological Society*, 43 (1985), 121–5.
56. Gill, *Plymouth: A New History*, 134; PWDRO, W46 fo. 5^V; Worth, 'Plymouth municipal records', 114. Worth notes that a Totnes man, Nicholas Goodridge, set fire to a record chest in the Council Chamber in 1601/2, and the town thereby lost further documents through this act of arson; see PWDRO W132/49 f.139.
57. Edwin Welch, *Plymouth Building Accounts*, Devon and Cornwall Record Society, 12 (1967). E. Arnold, 'Guildhalls: past and present', *Proceedings of the Plymouth Athenaeum*, 6 (1982–87), 65–9; see also PWDRO, W132 fo. 155^V and W137.

58. Arnold, 'Guildhalls', 68–9.
59. Worth, 'Plymouth municipal records', 113; Gill, *Plymouth: A New History*, 88–9; Whitfield, *Plymouth and Devonport*, 13–14.
60. F.W. Woodward, *Plymouth's Defences: A Short History* (Ivybridge, 1990), 7; Gill, *Plymouth: A New History*, 92–3; Mark Brayshay, 'Plymouth's coastal defences in the year of the Spanish Armada', *Transactions of the Devonshire Association*, 119 (1987), 166–96.
61. *Itinerary of John Leland*, ed. Toulmin Smith, 212.
62. Plymouth's Municipal Records in 1666/7 refer to the castle 'lately demolished' (PWDRO, W133 fo. 37). See also Cynthia Gaskell Brown, *Castle Street: The Pottery*, Plymouth Museum Archaeological Series No. 1 (Plymouth, 1979), 1–2. A full report on the excavations carried out in 1959 by Miss V.B. Leger at the site of Plymouth Castle has never been produced.
63. PWDRO, W130 Old Audit Book, fo. 58.
64. PWDRO, W130 Old Audit Book, fos 158–9.
65. H.M. Colvin, *The History of the King's Works, Vol. IV, 1485–1660*, Part II (London, 1982), 369–98; WDRO, W130, Old Audit Book, fo. 219.
66. Mark Brayshay, 'Tudor artillery towers and their role in the defence of Plymouth in 1588', *Devon Historian* (Autumn 1987), 3–14.
67. PRO, SP10, 6/24; Mark Brayshay and Keith Ray, 'The Tudor and Stuart fortification of the Island of St Nicholas in Plymouth Sound', in Mark Blacksell, Judith Matthews and Peter Sims (eds), *Environmental Management and Change in Plymouth* (Plymouth, 1998), 159–82.
68. Sir Richard Grenville's survey yielded a fine coloured manuscript map: PRO, MPF 6 (ex SP46, 36/4).
69. See Hatfield CPM I 35.
70. PRO, SP12 242/31 f. 65–66. Edwin Welch, 'Pilchards and Plymouth Fort', *Devon and Cornwall Notes and Queries*, 30 (1967), 260–2.
71. Woodward, *Plymouth's Defences*, 7–8; Stuart, *Lost Landscapes of Plymouth*, 81–5. See also R.A. Preston, *Gorges of Plymouth Fort* (Toronto, 1953) and James Barber, 'New light on old Plymouth', *Proceedings of the Plymouth Athenaeum*, 4 (1973–79), 55–66.
72. F.W. Woodward, *Citadel: A History of the Royal Citadel of Plymouth* (Exeter, 1987), 14–39.
73. Personal communication from Dr Keith Ray, Plymouth City Archaeological Officer; see also J. Sykes, 'Citadel reveals secrets of Drake's old fort', *Western Evening Herald*, 10 September 1994.
74. W.B. Stephens, 'The west-country ports and the struggle for the Newfoundland fisheries', *Transactions of the Devonshire Association*, 88 (1956), 125–6; W.B. Stephens, 'The foreign trade of Plymouth and the Cornish ports in the early seventeenth century', *Transactions of the Devonshire Association*, 101 (1969), 25–37; Todd Gray,

'Devon's fisheries and early Stuart northern New England', in M. Duffy *et al.* (eds), *The New Maritime History of Devon, Vol. I* (London, 1992), 142.

75. G.E. Larks, 'The early history of medicine in Plymouth', *Proceedings of the Plymouth Athenaeum*, 2 (1965–9), 48–60. See also PWDRO, W138 (358/6/108) Overseers' Rates and Accounts, 1611–42: 'An account of all monies collected and all donations for the relief of the sick during the prevalence of the pestilence, 1625–1627'; references to Plymouth's 1625 'pest houses' may be found in PWDRO, W132/49 Plymouth Receivers Accounts, fo. 208^V; see also John Hawkes, 'The Mount Batten Development and Archaeology Project', in Ray (ed.), *Archaeological Investigations and Research in Plymouth*, 23.

76. Eugene A. Andriette, *Devon and Exeter in the Civil War* (Newton Abbot, 1971), 123–34; Winfrid Scutt, *The Siege of Plymouth* (Plymouth, 1980), 3–4.

77. R.N. Worth, 'The siege accounts of Plymouth', *Transactions of the Devonshire Association*, 17 (1885), 216–39; Woodward, *Citadel*, 4–6.

78. The 'Friary Court' excavations directed for the Exeter Museums Archaeological Field Unit by Chris Henderson included work on 'Resolution Fort' on the north east corner of the 'inner line' of siege defences; Henderson, 'Excavations at Plymouth Whitefriars, 1989–94', 52–8.

79. Parts of the structure revealed during the excavations were preserved as part of the Tay Homes 'Friary Court' housing development.

80. National Maritime Museum LAD/11.

81. Crispin Gill, *Plymouth: A New History, 1603 to the Present Day* (Newton Abbot, 1979), 23–31; Harris, 'St Budeaux', 36–7; Arnold, 'Plymstock', 65–6.

82. Gill, *Plymouth: 1603 to the Present Day*, 27–8.

83. Whitfield, *Plymouth and Devonport*, 104.

84. Woodward, *Citadel*, 24–35.

85. Woodward, *Citadel*, 59–70. See also R.A. Erskine, 'The military coast defences of Devon, 1500–1956', in Duffy *et al.* (eds), *New Maritime History of Devon, Vol. I*, 123–4.

86. Woodward, *Citadel*, 64.

87. Walter Ralegh, *Observations touching the Royal Navy and Sea Service* (London, 1650); Jonathan Coad, 'The development and organisation of Plymouth Dockyard', in Duffy *et al.* (eds), *New Maritime History of Devon, Vol. I*, 192–200; Whitfield, *Plymouth and Devonport*, 148.

88. Gill, *Plymouth: 1603 to the Present Day*, 50–1; Leona Waddell, *A History and Bibliography of Devonport Dockyard* (Plymouth, 1982), 4–6.

89. Coad, 'Development of Plymouth Dockyard', 192; Whitfield, *Plymouth and Devonport*, 148. Dummer's plans and progress reports are in the British Library: BL, Kings MS 40/8–10, Lansdowne MS 847/79–83, Add. MSS 9329.

90. Gill, *Plymouth: 1603 to the Present Day*, 53; Coad, 'Development of Plymouth Dockyard', 194–5.

91. Jonathan Coad, 'Architecture and development of Devonport naval base', in M. Duffy *et al.* (eds), *The New Maritime History of Devon, Vol. II* (London, 1994), 169–70.

92. Coad, 'Architecture and development of Devonport', 172–3.

93. A.J. Marsh, 'The local community and the operation of Plymouth Dockyard, 1689–1763', in Duffy *et al.* (eds), *New Maritime History of Devon, Vol. I*, 205–6; Whitfield, *Plymouth and Devonport*, 150. See also Crispin Gill, 'Plymouth in the eighteenth century', *Proceedings of the Plymouth Athenaeum*, 6 (1982–87), 76–80.

94. Gill, *Plymouth: 1603 to the Present Day*, 58. Severely damaged during the air raids of the Second World War, most of this original core area of Devonport was absorbed by the 'South Yard' in an expansion of the Dockyard in 1950.

95. Whitfield, *Plymouth and Devonport*, 189–90.

96. John Dawe, 'John Foulston', *Proceedings of the Plymouth Athenaeum*, 2 (1965–69), 20–7.

97. Gill, *Plymouth: 1603 to the Present Day*, 107. C.E. Welch, 'Municipal reform in Plymouth', *Transactions of the Devonshire Association*, 96 (1964).

98. Gill, 'Plymouth in the eighteenth century', 77–9. R.N. Worth, *History of Plymouth from the Earliest Period to the Present Time* (Plymouth, 1890), 336.

99. Woodward, *Plymouth's Defences*, 10–11; F.W. Woodward, *Drake's Island*, Devon Archaeology No. 5 (Exeter, 1991), 12–24; a full inventory and survey of the city's defence installations may be found in Andrew Pye and Freddy Woodward, *The Historic Defences of Plymouth*, Exeter Archaeology and Fortress Study Group South West (Truro, 1996).

100. Erskine, 'The military coast defences of Devon', 123–4; R.N. Worth, *History of the Town and Borough of Devonport* (Plymouth, 1870), 14–16, 70.

101. L.H. Merrett, 'A most important undertaking: the building of the Plymouth breakwater', *Maritime History*, 5 (1977), 136–47; J. Naish, 'Joseph Whidbey and the building of the Plymouth Breakwater', *Mariner's Mirror*, 78 (1992), 37–56.

102. Ian V. Hogg, *Coast Defences of England and Wales, 1856–1956* (Newton Abbot, 1974), 168–204; Woodward, *Plymouth's Defences*, 18–31.

103. Woodward, *Plymouth's Defences*, 25, 46.

104. Woodward, *Plymouth's Defences*, 39.

105. Pye and Woodward, *The Historic Defences of Plymouth*, Foreword by Andrew Saunders, ix–x; also see 25–7. The Landmark Trust has restored and opened Crownhill fort to the public.

106. Brenda Hammerton, 'Plymstock, Oreston and Hooe', in Cynthia Gaskell Brown (ed.), *Industrial Archaeology of Plymouth* (Plymouth, 1980), 21–4.

107. Mark Brayshay and Vivien Pointon, 'Local politics and public health in mid-nineteenth-century Plymouth', *Medical History*, 27 (1983), 176–7; Mark Brayshay, 'The reform of urban management and the shaping of Plymouth's mid-Victorian landscape', in Robert Higham (ed.), *Landscape and Townscape in the South West* (Exeter, 1989), 112; A.W. Gale, *The Building Stones of Devon* (Exeter, 1992), 8–10.

108. Gaskell Brown (ed.), *Industrial Archaeology of Plymouth*, 7–8, 27–32, 35–6; Coad, 'Architecture and development of Devonport', 167–8; Michael Nix, *The Royal William Victualling Yard: An Illustrated History of Naval Victualling in Plymouth* (Plymouth, 1997); Crispin Gill, 'Ocean liners at Plymouth', in Duffy *et al.* (eds), *New Maritime History of Devon, Vol. II*, 226–9.

109. L.H. Merrett, *Smeaton's Tower and the Plymouth Breakwater*, reprinted *from Maritime History*, 5, 2 (1978), 1–11. Smeaton's lighthouse was ultimately replaced at the Eddystone and was brought back to Plymouth to be re-erected on the Hoe.

110. Gaskell Brown (ed.), *Industrial Archaeology*, 7–8.

111. See, for example, Helen Coxon, 'Train ferry at the breakwater', *Plymouth Council of Social Service* (1959), 9–13.

112. R.J. Sellick, 'Plymouth and Dartmoor Railway', *Plymouth Council of Social Service* (1957), 17–21; H. Liddle, 'Railways of Plymouth', in Gaskell Brown (ed.), *Industrial Archaeology*, 51–2; Clive Charlton and Richard Gibb, 'Transport in and around Plymouth', in Chalkley *et al.* (eds), *Plymouth: Maritime City in Transition*, 144–7.

113. Vivien F.T. Pointon, 'Mid-Victorian Plymouth: a social geography' (unpublished Ph.D. thesis, Polytechnic South West, 1989), 212–15.

114. Also see Ann Chiswell, 'The nature of urban overcrowding', *Local Historian*, 16 (1984), 156–60. Relative rent levels are discussed by M.J. Daunton, *House and Home in the Victorian City: Working-Class Housing, 1850–1914* (London, 1983), 82–3.

115. Brayshay and Pointon, 'Local politics and public health', 165; Mark Brayshay and Vivien Pointon, 'Migration and the social geography of mid-nineteenth-century Plymouth', *Devon Historian*, 28 (1984), 6.

116. Pointon, 'Mid-Victorian Plymouth', 304–6.

117. Mark Brayshay, 'The emigration trade in nineteenth century Devon', in Duffy *et al.* (eds), *New Maritime History of Devon, Vol. II*, 108–18; Stephen Essex, 'Tourism in Plymouth', in Chalkley *et al.* (eds), *Plymouth: Maritime City in Transition*, 118–26.

118. John Goodridge, 'No half-and-half affair: the plan for Plymouth, 1943', in Mark Brayshay (ed.), *Post-War Plymouth: Planning and Reconstruction*, South West Papers in Geography, Occasional Series No. 8 (Plymouth, 1983), 7–21.

119. Western Morning News Company, *Plymouth Blitz: The Story of the Raids* (Plymouth, n.d.), 16–28.

120. *Western Independent*, 23 March 1941.

121. D.J. Maguire, W.M. Brayshay and B.S. Chalkley, *Plymouth in Maps: A Social and Economic Atlas* (Plymouth, 1987), 14–15; Brian Chalkley and John Goodridge, 'The 1943 plan for Plymouth: war-time vision and post-war realities', in Chalkley *et al.* (eds), *Plymouth: Maritime City in Transition*, 63.

122. Chalkley and Goodridge, '1943 plan for Plymouth', 66.

123. *Western Evening Herald*, 5 July 1941.

124. Goodridge, 'No half-and-half affair', 12–13.

125. Western Morning News Company, *Plymouth Blitz*, 60–1.

126. James Paton Watson and Patrick Abercrombie, *Plan for Plymouth* (Plymouth, 1943), 1–11.

127. Brian Chalkley, 'The plan for the city centre', in Brayshay (ed.), *Post-War Plymouth*, 23–7.

128. Gill, *Plymouth: 1603 to the Present Day*, 202–4.

129. A study of the implementation and impact of Plymouth's pedestrianization has been carried out by J. Meaton, 'Pedestrianisation in Plymouth: the effect on car users' accessibility to and within the traffic-free zone' (unpublished Ph.D. thesis, Polytechnic South West, Plymouth, 1990).

130. Paton Watson and Abercrombie, *Plan*, 85.

131. Brian Chalkley, 'Plymouth's housing and employment: the 1943 plan and its effects', in Brayshay (ed.), *Post-War Plymouth*, 36–7.

132. Office of Population Censuses and Surveys, *1991 Census County Report: Devon (part 1)* (London, 1992), 17–18.

133. Office of Population Censuses and Surveys, Census 1981, SASPAC (Small Area Statistics) Plymouth; see also Chris Shepley, 'Planning Plymouth's future', in Chalkley *et al.* (eds), *Plymouth: Maritime City in Transition*, 212–14.

134. These are districts with either very few residents or where the Ministry of Defence presence requires secrecy.

135. Maguire, Brayshay and Chalkley, *Plymouth in Maps*, 26–30.

Penwith

1 St Just-in-Penwith
2 Sennen
3 St Levan
4 St Buryan
5 Sancreed
6 Morvah
7 Zennor
8 Madron
9 Paul
10 Penzance
11 Ludgvan
12 Towednack
13 St Ives
14 Hayle
15 Gwinear-Gwithian
16 St Erth
17 St Hilary
18 Marazion
19 Perranuthnoe
20 St Michael's Mount

Kerrier

21 Camborne
22 Illogan
23 Portreath
24 Carn Brea
25 Redruth
26 St Day
27 Carharrack
28 Lanner
29 Crowan
30 Breage
31 Germoe
32 Sithney
33 Wendron
34 Stithians
35 St Gluvias
36 Mabe
37 Constantine
38 Gweek
39 Helston
40 Porthleven
41 Gunwalloe
42 Mawgan-in-Meneage
43 Cury
44 St Martin-in-Meneage
45 Manaccan
46 St Anthony-in-Meneage
47 Mullion
48 Landewednack
49 Grade-Ruan
50 St Keverne

51 Budock
52 Mawnan

Carrick

53 Gwennap
54 Chacewater
55 St Agnes
56 Perranzabuloe
57 Cubert
58 St Newlyn East
59 Ladock
60 St Erme
61 St Allen
62 Kenwyn
63 Truro
64 St Clement
65 Probus
66 Cuby
67 Veryan
68 Tregoney
69 Ruanlanihorne
70 St Michael Penkevil
71 Kea
72 Feock
73 Perranarworthal
74 Mylor
75 Penryn
76 Falmouth
77 St Just-in-Roseland
78 Philleigh
79 Gerrans

Restormel

80 Crantock
81 Newquay
82 Colan
83 Mawgan-in-Pydar
84 St Columb Major
85 St Wenn
86 Roche
87 Luxulyan
88 Lanlivery
89 Lostwithiel
90 St Sampson
91 Fowey
92 Tywardreath
93 St Blaise
94 Treverbyn
95 St Dennis
96 St Enoder
97 St Stephen-in-Brannel
98 St Mewan
99 Grampound with Creed
100 St Ewe
101 St Michael Caerhays
102 St Goran
103 Mevagissey

Caradon

104 Lanteglos
105 St Veep
106 St Winnow
107 Broadoak
108 Boconnoc
109 Lanreath
110 Lansallos
111 Pelynt
112 Duloe
113 South Hill
114 Dobwalls & Trewidland
115 Warleggan
116 St Neot
117 St Cleer
118 Liskeard
119 St Keyne
120 Menheniot
121 Morval
122 Looe
123 St Martin-by-Looe
124 St Germans
125 Sheviock
126 Antony
127 St John
128 Millbrook
129 Torpoint
130 Maker-with-Rame
131 Landulph
132 Pillaton
133 Botusfleming
134 Saltash
135 Landrake with St Erney
136 Quethiock
137 St Ive
138 Linkinhorne
139 South Hill
140 Callington
141 St Mellion
142 St Dominick
143 Calstock

North Cornwall

144 Morwenstow
145 Kilkhampton
146 Bude-Stratton
147 Launcells
148 Marhamchurch
149 Poundstock
150 Week St Mary
151 Whitstone
152 St Gennys
153 Jacobstow
154 North Tamerton
155 St Juliot
156 Otterham
157 Warbstow
158 North Petherwin
159 Boyton
160 Trevalga
161 Forrabury & Minster
162 Lesnewth
163 Treneglos
164 Tremaine
165 Werrington
166 Tintagel
167 Davidstow
168 St Clether
169 Tresmeer
170 Egloskerry
171 St Stephens by Launceston Rural

172 Launceston
173 St Teath
174 Camelford
175 Advent
176 Altarnun
177 Laneast
178 Trewen
179 St Thomas the Apostle Rural
180 South Petherwin
181 Lawhitton
182 St Endellion
183 St Kew
184 Michaelstow
185 St Breward
186 St Tudy
187 Lewannick
188 St Merryn
189 Padstow
190 St Minver Lowlands
191 St Minver Highlands
192 St Eval
193 St Ervan
194 St Issey
195 St Breock
196 Wadebridge

197 Egloshayle
198 St Mabyn
199 Blisland
200 Withiel
201 Lanivet
202 Helland
203 Bodmin
204 Lanhydrock
205 Cardinham
206 North Hill
207 Lezant
208 Stokeclimsland

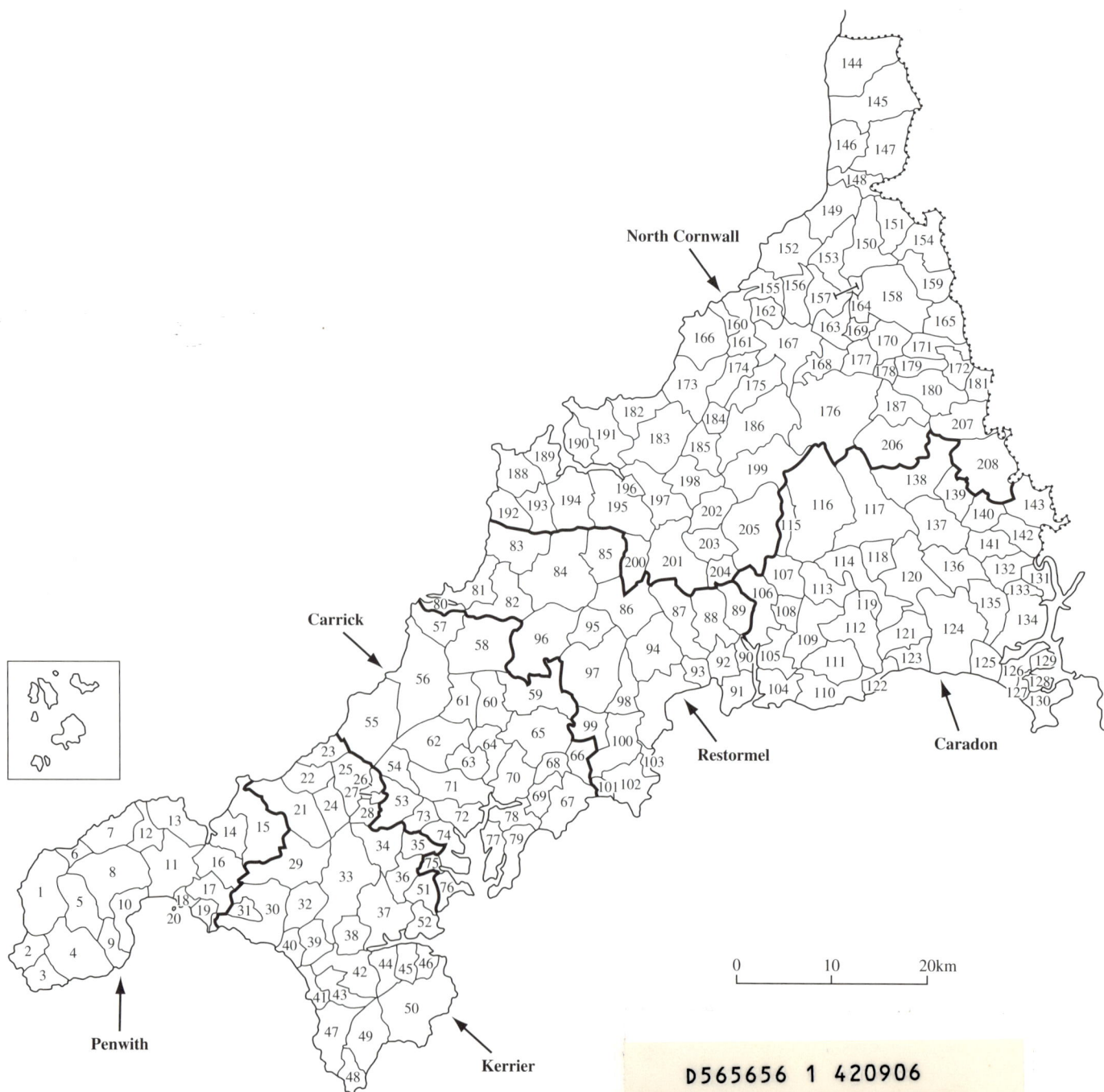

The parishes and districts of Devon (facing page) and Cornwall, 1999.

0   10   20km